a LANGE medical book

Review of

Medical Physiology

twenty-first edition

William F. Ganong, MD
Jack and DeLoris Lange Professor of Physiology Emeritus
University of California
San Francisco

Lange Medical Books/McGraw-Hill
Medical Publishing Division

New York   Chicago   San Francisco   Lisbon   London   Madrid   Mexico City
Milan   New Deli   San Juan   Seoul   Singapore   Sydney   Toronto

Review of Medical Physiology, Twenty-First Edition

234567890 DOC/DOC 09876543

ISBN 0-07-140236-5
ISSN 0892-1253

Notice

Medicine is an ever-changing science. As new research and clinical experience broaden our knowledge, changes in treatment and drug therapy are required. The author and the publisher of this work have checked with sources believed to be reliable in their efforts to provide information that is complete and generally in accord with the standards accepted at the time of publication. However, in view of the possibility of human error or changes in medical sciences, neither the author nor the publisher nor any other party who has been involved in the preparation or publication of this work warrants that the information contained herein is in every respect accurate or complete, and they disclaim all responsibility for any errors or omissions or for the results obtained from use of the information contained in this work. Readers are encouraged to confirm the information contained herein with other sources. For example and in particular, readers are advised to check the product information sheet included in the package of each drug they plan to administer to be certain that the information contained in this work is accurate and that changes have not been made in the recommended dose or in the contraindications for administration. This recommendation is of particular importance in connection with new or infrequently used drugs.

The book was set in Adobe Garamond by Rainbow Graphics.
The editors were Janet Foltin, Janene Matragrano, Jim Ransom, and Karen Davis.
The production supervisor was Phil Galea.
The cover designer was Mary McKeon.
The art manager was Charissa Baker.
The art coordinator was Becky Hainz-Baxter.
The illustrators were Linda F. Harris, Shirley Bortoli, and Teshin Associates.
The index was prepared by Katherine Pitcoff.
RR Donnelley was printer and binder.

This book is printed on acid-free paper.

INTERNATIONAL EDITION ISBN 0-07-121765-7
Exclusive rights by The McGraw-Hill Companies, Inc., for manufacture and export. This book cannot be re-exported from the country to which it is consigned by McGraw-Hill. The International Edition is not available in North America.

Contents

Preface

This book is designed to provide a concise summary of mammalian and, particularly, of human physiology that medical students and others can use by itself or can supplement with readings in other texts, monographs, and reviews. Pertinent aspects of general and comparative physiology are also included. Summaries of relevant anatomic considerations will be found in each section, but this book is written primarily for those who have some knowledge of anatomy, chemistry, and biochemistry. Examples from clinical medicine are given where pertinent to illustrate physiologic points. In many of the chapters, physicians desiring to use this book as a review will find short discussions of important symptoms produced by disordered function.

Review of Medical Physiology also includes a self-study section to help students review for Board and other examinations and an appendix that contains general references, a discussion of statistical methods, a glossary of abbreviations, acronyms, and symbols commonly used in physiology, and several useful tables. The index is comprehensive and specifically designed for ease in locating important terms, topics, and concepts.

In writing this book, the author has not been able to be complete and concise without also being dogmatic. I believe, however, that the conclusions presented without detailed discussion of the experimental data on which they are based are supported by the bulk of the current evidence. Much of this evidence can be found in the papers cited in the credit lines accompanying the illustrations. Further discussions of particular subjects and information on subjects not considered in detail can be found in the references listed at the end of each section. Information about serial review publications that provide up-to-date discussion of various physiologic subjects is included in the note on general references in the appendix. In the interest of brevity and clarity, I have in most instances omitted the names of the many investigators whose work made possible the view of physiology presented here. This omission is in no way intended to slight their contributions, but including their names and specific references to original papers would greatly increase the length of the book.

In this twenty-first edition, as in previous editions, the entire book has been thoroughly revised, with a view to eliminating errors, incorporating suggestions of readers, updating concepts, and discarding material that is no longer relevant. In this way, the book has been kept concise while remaining as up-to-date and accurate as possible. The coverage of physiology related to obesity has been reorganized and expanded, with consideration of different aspects of this important subject in Chapters 14, 17, and 19. The availability of the complete structure of the human genome has permitted rapid expansion of knowledge about the genetic causes of disease, and physiologic aspects of this topic have been expanded. There is also more detailed consideration of circadian rhythms, intracellular regulation of insulin secretion, vascular smooth muscle, the Na^+-K^+-$2Cl^-$ cotransporter, and the control of exercise tolerance. New information has been added on dyslexia, PPARs, ghrelin, and adipokines.

The self-study section has been updated, and more emphasis has been placed on physiology in relation to disease, in keeping with the current trend in the United States Medical Licensing Examinations (USMLE).

I am greatly indebted to the many individuals who helped with the preparation of this book. Those who were especially helpful in the preparation of the twenty-first edition include Drs. Stephen McPhee, Roger Nicoll, and Narayan Rao. Andrea Chase provided invaluable secretarial assistance, and, as always, my wife made important contributions. Jim Ransom, who edited the first edition of this book 40 years ago, came back again and did an excellent job of editing this edition. Many associates and friends provided unpublished illustrative materials, and numerous authors and publishers generously granted permission to reproduce illustrations from other books and journals. I also thank all the students and others who took the time to write to me offering helpful criticisms and suggestions. Such comments are always welcome, and I solicit additional corrections and criticisms, which may be addressed to me at

Department of Physiology
University of California
San Francisco, CA 94143-0444 USA

Since this book was first published in 1963, the following translations have been published: Bulgarian, Chinese (two independent translations), Czech (two editions), French (two independent translations), German (four editions), Greek (two editions), Hungarian, Indonesian (three editions), Italian (nine editions), Japanese (seventeen editions), Korean, Malaysian, Polish (two editions), Portuguese (seven editions), Serbo-Croatian, Spanish (eighteen editions), Turkish (two editions), and Ukranian. Various foreign English language editions have been published, and the book has been recorded in English on tape for use by the blind. The tape recording is available from Recording for the Blind, Inc., 20 Rozsel Road, Princeton, NJ 08540 USA. For computer users, the book is now available, along with several other titles in the Lange Medical Books series, in STAT!-Ref, a searchable Electronic Medical Library (*http://www.statref.com*), from Teton Data Systems, P.O. Box 4798, Jackson, WY 83001 USA. More information about this and other Lange and McGraw-Hill books, including addresses of the publisher's international offices, is available on McGraw-Hill's web site, *www.AccessMedBooks.com*.

William F. Ganong, MD

San Francisco
March 2003

SECTION I

Introduction

The General & Cellular Basis of Medical Physiology

1

INTRODUCTION

In unicellular organisms, all vital processes occur in a single cell. As the evolution of multicellular organisms has progressed, various cell groups have taken over particular functions. In humans and other vertebrate animals, the specialized cell groups include a gastrointestinal system to digest and absorb food; a respiratory system to take up O_2 and eliminate CO_2; a urinary system to remove wastes; a cardiovascular system to distribute food, O_2, and the products of metabolism; a reproductive system to perpetuate the species; and nervous and endocrine systems to coordinate and integrate the functions of the other systems. This book is concerned with the way these systems function and the way each contributes to the functions of the body as a whole.

This chapter presents general concepts and principles that are basic to the function of all the systems. It also includes a short review of fundamental aspects of cell physiology. Additional aspects of cellular and molecular biology are considered in the relevant chapters on the various organs.

GENERAL PRINCIPLES

Organization of the Body

The cells that make up the bodies of all but the simplest multicellular animals, both aquatic and terrestrial, exist in an "internal sea" of **extracellular fluid (ECF)** enclosed within the integument of the animal. From this fluid, the cells take up O_2 and nutrients; into it, they discharge metabolic waste products. The ECF is more dilute than present-day seawater, but its composition closely resembles that of the primordial oceans in which, presumably, all life originated.

In animals with a closed vascular system, the ECF is divided into two components: the **interstitial fluid** and the circulating **blood plasma.** The plasma and the cellular elements of the blood, principally red blood cells, fill the vascular system, and together they constitute the **total blood volume.** The interstitial fluid is that part of the ECF that is outside the vascular system, bathing the cells. The special fluids lumped together as transcellular fluids are discussed below. About a third of the **total body water (TBW)** is extracellular; the remaining two-thirds are intracellular **(intracellular fluid).**

Body Composition

In the average young adult male, 18% of the body weight is protein and related substances, 7% is mineral, and 15% is fat. The remaining 60% is water. The distribution of this water is shown in Figure 1–1.

The intracellular component of the body water accounts for about 40% of body weight and the extracellular component for about 20%. Approximately 25% of the extracellular component is in the vascular system (plasma = 5% of body weight) and 75% outside the blood vessels (interstitial fluid = 15% of body weight). The total blood volume is about 8% of body weight.

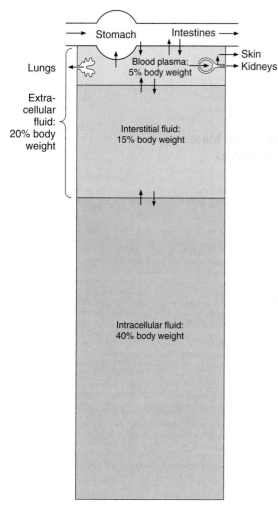

Figure 1–1. Body fluid compartments. Arrows represent fluid movement. Transcellular fluids, which constitute a very small percentage of total body fluids, are not shown.

Measurement of Body Fluid Volumes

It is theoretically possible to measure the size of each of the body fluid compartments by injecting substances that will stay in only one compartment and then calculating the volume of fluid in which the test substance is distributed (the **volume of distribution** of the injected material). The volume of distribution is equal to the amount injected (minus any that has been removed from the body by metabolism or excretion during the time allowed for mixing) divided by the concentration of the substance in the sample. *Example:* 150 mg of sucrose is injected into a 70 kg man. The plasma sucrose

level after mixing is 0.01 mg/mL, and 10 mg has been excreted or metabolized during the mixing period. The volume of distribution of the sucrose is

$$\frac{150\ mg\ -\ 10\ mg}{0.01\ mg/mL} = 14{,}000\ mL$$

Since 14,000 mL is the space in which the sucrose was distributed, it is also called the **sucrose space.**

Volumes of distribution can be calculated for any substance that can be injected into the body provided the concentration in the body fluids and the amount removed by excretion and metabolism can be accurately measured.

Although the principle involved in such measurements is simple, a number of complicating factors must be considered. The material injected must be nontoxic, must mix evenly throughout the compartment being measured, and must have no effect of its own on the distribution of water or other substances in the body. In addition, either it must be unchanged by the body during the mixing period, or the amount changed must be known. The material also should be relatively easy to measure.

Plasma Volume, Total Blood Volume, & Red Cell Volume

Plasma volume has been measured by using dyes that become bound to plasma protein—particularly Evans blue (T-1824). Plasma volume can also be measured by injecting serum albumin labeled with radioactive iodine. Suitable aliquots of the injected solution and plasma samples obtained after injection are counted in a scintillation counter. An average value is 3500 mL (5% of the body weight of a 70 kg man, assuming unit density).

If one knows the plasma volume and the hematocrit (ie, the percentage of the blood volume that is made up of cells), the **total blood volume** can be calculated by multiplying the plasma volume by

$$\frac{100}{100 - hematocrit}$$

Example: The hematocrit is 38 and the plasma volume 3500 mL. The total blood volume is

$$3500 \times \frac{100}{100 - 38} = 5645\ mL$$

The **red cell volume** (volume occupied by all the circulating red cells in the body) can be determined by subtracting the plasma volume from the total blood volume. It may also be measured independently by injecting tagged red blood cells and, after mixing has occurred, measuring the fraction of the red cells that is tagged. A commonly used tag is ^{51}Cr, a radioactive isotope of chromium that is attached to the cells by incubating them in a suitable chromium solution. Isotopes of iron and phosphorus (^{59}Fe and ^{32}P) and antigenic tagging have also been employed.

Extracellular Fluid Volume

The ECF volume is difficult to measure because the limits of this space are ill defined and because few substances mix rapidly in all parts of the space while remaining exclusively extracellular. The lymph cannot be separated from the ECF and is measured with it. Many substances enter the cerebrospinal fluid (CSF) slowly because of the blood-brain barrier (see Chapter 32). Equilibration is slow with joint fluid and aqueous humor and with the ECF in relatively avascular tissues such as dense connective tissue, cartilage, and some parts of bone. Substances that distribute in ECF appear in glandular secretions and in the contents of the gastrointestinal tract. Because they are separated from the rest of the ECF, these fluids—as well as CSF, the fluids in the eye, and a few other special fluids—are called **transcellular fluids.** Their volume is relatively small.

Perhaps the most accurate measurement of ECF volume is that obtained by using inulin, a polysaccharide with a molecular weight of 5200. Mannitol and sucrose have also been used to measure ECF volume. A generally accepted value for ECF volume is 20% of the body weight, or about 14 L in a 70 kg man (3.5 L = plasma; 10.5 L = interstitial fluid).

Interstitial Fluid Volume

The interstitial fluid space cannot be measured directly, since it is difficult to sample interstitial fluid and since substances that equilibrate in interstitial fluid also equilibrate in plasma. The volume of the interstitial fluid can be calculated by subtracting the plasma volume from the ECF volume. The ECF volume/intracellular fluid volume ratio is larger in infants and children than it is in adults, but the absolute volume of ECF in children is, of course, smaller than in adults. Therefore, dehydration develops more rapidly and is frequently more severe in children than in adults.

Intracellular Fluid Volume

The intracellular fluid volume cannot be measured directly, but it can be calculated by subtracting the ECF volume from the TBW. TBW can be measured by the same dilution principle used to measure the other body spaces. Deuterium oxide (D_2O, heavy water) is most frequently used. D_2O has slightly different properties from those of H_2O, but in equilibration experiments for measuring body water it gives accurate results. Tritium oxide and aminopyrine have also been used for this purpose.

The water content of lean body tissue is constant at 71–72 mL/100 g of tissue, but since fat is relatively free of water, the ratio of TBW to body weight varies with the amount of fat present. TBW is somewhat lower in women than men, and in both sexes, the values tend to decrease with age (Table 1–1).

Units for Measuring Concentration of Solutes

In considering the effects of various physiologically important substances and the interactions between them, the number of molecules, electrical charges, or particles of a substance per unit volume of a particular body fluid are often more meaningful than simply the weight of the substance per unit volume. For this reason, concentrations are frequently expressed in moles, equivalents, or osmoles.

Moles

A mole is the gram-molecular weight of a substance, ie, the molecular weight of the substance in grams. Each mole (mol) consists of approximately 6×10^{23} molecules. The millimole (mmol) is 1/1000 of a mole, and the micromole (mmol) is 1/1,000,000 of a mole. Thus, 1 mol of NaCl = 23 + 35.5 g = 58.5 g, and 1 mmol = 58.5 mg. The mole is the standard unit for expressing the amount of substances in the SI unit system (see Appendix).

The molecular weight of a substance is the ratio of the mass of one molecule of the substance to the mass of one-twelfth the mass of an atom of carbon-12. Since molecular weight is a ratio, it is dimensionless. The dalton (Da) is a unit of mass equal to one-twelfth the mass of an atom of carbon-12, and 1000 Da = 1 kilodalton (kDa). The kilodalton, which is sometimes expressed simply as K, is a useful unit for expressing the molecular mass of proteins. Thus, for example, one can speak of a 64 K protein or state that the molecular mass of

Table 1–1. Total body water (as percentage of body weight) in relation to age and sex.

Age	Male	Female
10–18	59%	57%
18–40	61%	51%
40–60	55%	47%
Over 60	52%	46%

the protein is 64,000 Da. However, since molecular weight is a dimensionless ratio, it is incorrect to say that the molecular weight of the protein is 64 kDa.

Equivalents

The concept of electrical equivalence is important in physiology because many of the important solutes in the body are in the form of charged particles. One equivalent (eq) is 1 mol of an ionized substance divided by its valence. One mole of NaCl dissociates into 1 eq of Na^+ and 1 eq of Cl^-. One equivalent of Na^+ = 23 g; but 1 eq of Ca^{2+} = 40 g/2 = 20 g. The milliequivalent (meq) is 1/1000 of 1 eq.

Electrical equivalence is not necessarily the same as chemical equivalence. A gram equivalent is the weight of a substance that is chemically equivalent to 8.000 g of oxygen. The normality (N) of a solution is the number of gram equivalents in 1 liter. A 1 N solution of hydrochloric acid contains 1 + 35.5 g/L = 36.5 g/L.

pH

The maintenance of a stable hydrogen ion concentration in the body fluids is essential to life. The pH of a solution is the logarithm to the base 10 of the reciprocal of the H^+ concentration ($[H^+]$), ie, the negative logarithm of the $[H^+]$. The pH of water at 25 °C, in which H^+ and OH^- ions are present in equal numbers, is 7.0 (Figure 1–2). For each pH unit less than 7.0, the $[H^+]$ is increased tenfold; for each pH unit above 7.0, it is decreased tenfold.

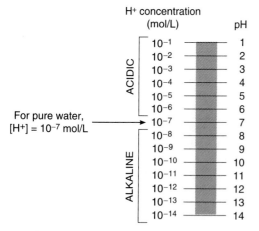

H+ concentration (mol/L) — **pH**

	H+ concentration (mol/L)	pH
ACIDIC	10^{-1}	1
	10^{-2}	2
	10^{-3}	3
	10^{-4}	4
	10^{-5}	5
	10^{-6}	6
For pure water, $[H^+] = 10^{-7}$ mol/L	10^{-7}	7
	10^{-8}	8
	10^{-9}	9
ALKALINE	10^{-10}	10
	10^{-11}	11
	10^{-12}	12
	10^{-13}	13
	10^{-14}	14

Figure 1–2. **pH.** (Reproduced, with permission, from Alberts B et al: *Molecular Biology of the Cell*, 4th ed. Garland Science, 2002.)

Buffers

Intracellular and extracellular pH are generally maintained at very constant levels. For example, the pH of the ECF is 7.40, and in health, this value usually varies less than ±0.05 pH unit. Body pH is stabilized by the **buffering capacity** of the body fluids. A buffer is a substance that has the ability to bind or release H^+ in solution, thus keeping the pH of the solution relatively constant despite the addition of considerable quantities of acid or base. One buffer in the body is carbonic acid. This acid is only partly dissociated into H^+ and bicarbonate: $H_2CO_3 \rightleftarrows H^+ + HCO_3^-$. If H^+ is added to a solution of carbonic acid, the equilibrium shifts to the left and most of the added H^+ is removed from solution. If OH^- is added, H^+ and OH^- combine, taking H^+ out of solution. However, the decrease is countered by more dissociation of H_2CO_3, and the decline in H^+ concentration is minimized. Other buffers include the blood proteins and the proteins in cells. The quantitative aspects of buffering and the respiratory and renal adjustments that operate with buffers to maintain a stable ECF pH of 7.40 are discussed in Chapter 39.

Diffusion

Diffusion is the process by which a gas or a substance in solution expands, because of the motion of its particles, to fill all of the available volume. The particles (molecules or atoms) of a substance dissolved in a solvent are in continuous random movement. A given particle is equally likely to move into or out of an area in which it is present in high concentration. However, since there are more particles in the area of high concentration, the total number of particles moving to areas of lower concentration is greater; ie, there is a **net flux** of solute particles from areas of high to areas of low concentration. The time required for equilibrium by diffusion is proportionate to the square of the diffusion distance. The magnitude of the diffusing tendency from one region to another is directly proportionate to the cross-sectional area across which diffusion is taking place and the **concentration gradient,** or **chemical gradient,** which is the difference in concentration of the diffusing substance divided by the thickness of the boundary (**Fick's law of diffusion**). Thus,

$$J = -DA \frac{\Delta c}{\Delta x}$$

where J is the net rate of diffusion, D is the diffusion coefficient, A is the area, and $\Delta c/\Delta x$ is the concentration gradient. The minus sign indicates the direction of diffusion. When considering movement of molecules from a higher to a lower concentration, $\Delta c/\Delta x$ is negative, so multiplying by −DA gives a positive value. The permeabilities of the boundaries across which diffusion occurs in the body vary, but diffusion is still a major force affecting the distribution of water and solutes.

Osmosis

When a substance is dissolved in water, the concentration of water molecules in the solution is less than that in pure water, since the addition of solute to water results in a solution that occupies a greater volume than does the water alone. If the solution is placed on one side of a membrane that is permeable to water but not to the solute and an equal volume of water is placed on the other, water molecules diffuse down their concentration gradient into the solution (Figure 1–3). This process—the diffusion of **solvent** molecules into a region in which there is a higher concentration of a **solute** to which the membrane is impermeable—is called **osmosis.** It is an important factor in physiologic processes. The tendency for movement of solvent molecules to a region of greater solute concentration can be prevented by applying pressure to the more concentrated solution. The pressure necessary to prevent solvent migration is the **osmotic pressure** of the solution.

Osmotic pressure, like vapor pressure lowering, freezing-point depression, and boiling-point elevation, depends upon the number rather than the type of particles in a solution; ie, it is a fundamental colligative property of solutions. In an **ideal solution,** osmotic pressure (P) is related to temperature and volume in the same way as the pressure of a gas:

$$P = \frac{nRT}{V}$$

where n is the number of particles, R is the gas constant, T is the absolute temperature, and V is the volume. If T is held constant, it is clear that the osmotic pressure is proportionate to the number of particles in solution per unit volume of solution. For this reason, the concentration of osmotically active particles is usually expressed in **osmoles.** One osmole (osm) equals the gram-molecular weight of a substance divided by the number of freely moving particles that each molecule liberates in solution. The milliosmole (mosm) is 1/1000 of 1 osm.

If a solute is a nonionizing compound such as glucose, the osmotic pressure is a function of the number of glucose molecules present. If the solute ionizes and forms an ideal solution, each ion is an osmotically active particle. For example, NaCl would dissociate into Na^+ and Cl^- ions, so that each mole in solution would supply 2 osm. One mole of Na_2SO_4 would dissociate into Na^+, Na^+, and SO_4^{2-}, supplying 3 osm. However, the body fluids are not ideal solutions, and although the dissociation of strong electrolytes is complete, the number of particles free to exert an osmotic effect is reduced owing to interactions between the ions. Thus, it is actually the effective concentration (**activity**) in the body fluids rather than the number of equivalents of an electrolyte in solution that determines its osmotic effect. This is why, for example, 1 mmol of NaCl per liter in the body fluids contributes somewhat less than 2 mosm of osmotically active particles per liter. The more concentrated the solution, the greater the deviation from an ideal solution.

The osmolal concentration of a substance in a fluid is measured by the degree to which it depresses the freezing point, with 1 mol of an ideal solution depressing the freezing point 1.86 Celsius degrees. The number of milliosmoles per liter in a solution equals the freezing point depression divided by 0.00186. The **osmolarity** is the number of osmoles per liter of solution—eg, plasma—whereas the **osmolality** is the number of osmoles per kilogram of solvent. Therefore, osmolarity is affected by the volume of the various solutes in the solution and the temperature, while the osmolality is not. Osmotically active substances in the body are dissolved in water, and the density of water is 1, so osmolal concentrations can be expressed as osmoles per liter (osm/L) of water. In this book, osmolal (rather than osmolar) concentrations are considered, and osmolality is expressed in milliosmoles per liter (of water).

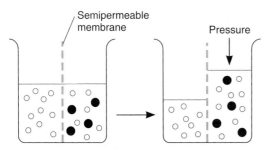

Figure 1–3. Diagrammatic representation of osmosis. Water molecules are represented by small open circles, solute molecules by large solid circles. In the diagram on the left, water is placed on one side of a membrane permeable to water but not to solute, and an equal volume of a solution of the solute is placed on the other. Water molecules move down their concentration gradient into the solution, and, as shown in the diagram on the right, the volume of the solution increases. As indicated by the arrow on the right, the osmotic pressure is the pressure that would have to be applied to prevent the movement of the water molecules.

Note that although a homogeneous solution contains osmotically active particles and can be said to have an osmotic pressure, it can exert an osmotic pressure only when it is in contact with another solution across a membrane permeable to the solvent but not to the solute.

Osmolal Concentration of Plasma: Tonicity

The freezing point of normal human plasma averages −0.54 °C, which corresponds to an osmolal concentration in plasma of 290 mosm/L. This is equivalent to an osmotic pressure against pure water of 7.3 atmospheres. The osmolality might be expected to be higher than this, because the sum of all the cation and anion equivalents in plasma is over 300. It is not this high because plasma is not an ideal solution and ionic interactions reduce the number of particles free to exert an osmotic effect. Except when there has been insufficient time after a sudden change in composition for equilibrium to occur, all fluid compartments of the body are in or nearly in osmotic equilibrium. The term **tonicity** is used to describe the osmolality of a solution relative to plasma. Solutions that have the same osmolality as plasma are said to be **isotonic;** those with greater osmolality are **hypertonic;** and those with lesser osmolality are **hypotonic.** All solutions that are initially isosmotic with plasma—ie, that have the same actual osmotic pressure or freezing-point depression as plasma—would remain isotonic if it were not for the fact that some solutes diffuse into cells and others are metabolized. Thus, a 0.9% saline solution remains isotonic because there is no net movement of the osmotically active particles in the solution into cells and the particles are not metabolized. On the other hand, a 5% glucose solution is isotonic when initially infused intravenously, but glucose is metabolized, so the net effect is that of infusing a hypotonic solution.

It is important to note the relative contributions of the various plasma components to the total osmolal concentration of plasma. All but about 20 of the 290 mosm in each liter of normal plasma are contributed by Na^+ and its accompanying anions, principally Cl^- and HCO_3^-. Other cations and anions make a relatively small contribution. Although the concentration of the plasma proteins is large when expressed in grams per liter, they normally contribute less than 2 mosm/L because of their very high molecular weights. The major nonelectrolytes of plasma are glucose and urea, which in the steady state are in equilibrium with cells. Their contributions to osmolality are normally about 5 mosm/L each but can become quite large in hyperglycemia or uremia. The total plasma osmolality is important in assessing dehydration, overhydration, and other fluid and electrolyte abnormalities. Hyper-

osmolality can cause coma (hyperosmolar coma; see Chapter 19). Because of the predominant role of the major solutes and the deviation of plasma from an ideal solution, one can ordinarily approximate the plasma osmolality within a few milliosmoles per liter by using the following formula, in which the constants convert the clinical units to millimoles of solute per liter:

$$\begin{array}{cccc} \text{Osmolality} = & 2[Na^+] & + 0.055[\text{Glucose}] + 0.36[\text{BUN}] \\ \text{(mosm/L)} & \text{(meq/L)} & \text{(mg/dL)} & \text{(mg/dL)} \end{array}$$

BUN is the blood urea nitrogen. The formula is also useful in calling attention to abnormally high concentrations of other solutes. An observed plasma osmolality (measured by freezing-point depression) that greatly exceeds the value predicted by this formula probably indicates the presence of a foreign substance such as ethanol, mannitol (sometimes injected to shrink swollen cells osmotically), or poisons such as ethylene glycol or methanol (components of antifreeze).

Regulation of Cell Volume

Unlike plant cells, which have rigid walls, animal cell membranes are flexible. Therefore, animal cells swell when exposed to extracellular hypotonicity and shrink when exposed to extracellular hypertonicity. However, cell swelling activates channels in the cell membrane that permit increased efflux of K^+, Cl^-, and small organic solutes referred to collectively as **organic osmolytes.** Water follows these osmotically active particles out of the cell, and the cell volume returns to normal. Ion channels and other membrane transport proteins are discussed in detail in a later section of this chapter.

Nonionic Diffusion

Some weak acids and bases are quite soluble in cell membranes in the undissociated form, whereas they cross membranes with difficulty in the ionic form. Consequently, if molecules of the undissociated substance diffuse from one side of the membrane to the other and then dissociate, there is appreciable net movement of the undissociated substance from one side of the membrane to another. This phenomenon, which occurs in the gastrointestinal tract (see Chapter 25) and kidneys (see Chapter 38), is called **nonionic diffusion.**

Donnan Effect

When there is an ion on one side of a membrane that cannot diffuse through the membrane, the distribution of other ions to which the membrane is permeable is affected in a predictable way. For example, the negative charge of a nondiffusible anion hinders diffusion of the diffusible cations and favors diffusion of the diffusible anions. Consider the following situation,

$$
\begin{array}{c|c}
\underline{X} & \underline{Y} \\
\quad m & \\
K^+ & K^+ \\
Cl^- & Cl^- \\
Prot^- &
\end{array}
$$

in which the membrane (m) between compartments X and Y is impermeable to $Prot^-$ but freely permeable to K^+ and Cl^-. Assume that the concentrations of the anions and of the cations on the two sides are initially equal. Cl^- diffuses down its concentration gradient from Y to X, and K^+ moves with the negatively charged Cl^-, maintaining electroneutrality on side Y. Therefore, at equilibrium,

$$[K^+_X] > [K^+_Y]$$

Furthermore,

$$[K^+_X] + [Cl^-_X] + [Prot^-_X] > [K^+_Y] + [Cl^-_Y]$$

ie, there are more osmotically active particles on side X than on side Y.

Donnan and Gibbs showed that in the presence of a nondiffusible ion, the diffusible ions distribute themselves so that at equilibrium, their concentration ratios are equal:

$$\frac{[K^+_X]}{[K^+_Y]} = \frac{[Cl^-_Y]}{[Cl^-_X]}$$

Cross-multiplying,

$$[K^+_X][Cl^-_X] = [K^+_Y][Cl^-_Y]$$

This is the **Gibbs-Donnan equation.** It holds for any pair of cations and anions of the same valence.

The Donnan effect on the distribution of ions has three effects in the body. First, because of proteins ($Prot^-$) in cells, there are more osmotically active particles in cells than in interstitial fluid, and since animal cells have flexible walls, osmosis would make them swell and eventually rupture if it were not for Na^+-K^+ ATPase pumping ions back out of cells (see below). Thus, normal cell volume and pressure depend on Na^+-K^+ ATPase. Second, because at equilibrium there is an asymmetric distribution of permeant ions across the membrane (m in the example used here), there will be an electrical difference across the membrane whose magnitude can be determined by the Nernst equation (see below). In the example used here, side X will be negative relative to side Y. The charges line up along the membrane with the concentration gradient for Cl^-

exactly balanced by the oppositely directed electrical gradient, and the same holds true for K^+. Third, since there are more proteins in plasma than in interstitial fluid, there is a Donnan effect on ion movement across the capillary wall (see below).

Forces Acting on Ions

The forces acting across the cell membrane on each ion can be analyzed mathematically. Chloride ions are present in higher concentration in the ECF than in the cell interior, and they tend to diffuse along this **concentration gradient** into the cell. The interior of the cell is negative relative to the exterior, and chloride ions are pushed out of the cell along this **electrical gradient.** An equilibrium is reached at which Cl^- influx and Cl^- efflux are equal. The membrane potential at which this equilibrium exists is the **equilibrium potential.** Its magnitude can be calculated from the **Nernst equation,** as follows:

$$E_{Cl} = \frac{RT}{FZ_{Cl}} \ln \frac{[Cl_o^-]}{[Cl_i^-]}$$

where

E_{Cl} = equilibrium potential for Cl^-
R = gas constant
T = absolute temperature
F = the faraday (number of coulombs per mole of charge)
Z_{Cl} = valence of Cl^- (−1)
$[Cl_o^-]$ = Cl^- concentration outside the cell
$[Cl_i^-]$ = Cl^- concentration inside the cell

Converting from the natural log to the base 10 log and replacing some of the constants with numerical values, the equation becomes

$$E_{Cl} = 61.5 \log \frac{[Cl_i^-]}{[Cl_o^-]} \text{ at } 37\ ^\circ C$$

Note that in converting to the simplified expression the concentration ratio is reversed because the −1 valence of Cl^- has been removed from the expression.

E_{Cl}, calculated from the values in Table 1–2, is −70 mV, a value identical to the measured resting membrane potential of −70 mV. Therefore, no forces other than those represented by the chemical and electrical gradients need be invoked to explain the distribution of Cl^- across the membrane.

A similar equilibrium potential can be calculated for K^+:

$$E_K = \frac{RT}{FZ_K} \ln \frac{[K_o^+]}{[K_i^+]} = 61.5 \log \frac{[K_o^+]}{[K_i^+]} \text{ at } 37\ ^\circ C$$

where

> E_K = equilibrium potential for K^+
> Z_K = valence of K^+ (+1)
> $[K_o^+]$ = K^+ concentration outside the cell
> $[K_i^+]$ = K^+ concentration inside the cell
> R, T, and F as above

In this case, the concentration gradient is outward and the electrical gradient inward. In mammalian spinal motor neurons, E_K is −90 mV (Table 1–2). Since the resting membrane potential is −70 mV, there is somewhat more K^+ in the neurons than can be accounted for by the electrical and chemical gradients.

The situation for Na^+ is quite different from that for K^+ and Cl^-. The direction of the chemical gradient for Na^+ is inward, to the area where it is in lesser concentration, and the electrical gradient is in the same direction. E_{Na} is +60 mV (Table 1–2). Since neither E_K nor E_{Na} is at the membrane potential, one would expect the cell to gradually gain Na^+ and lose K^+ if only passive electrical and chemical forces were acting across the membrane. However, the intracellular concentration of Na^+ and K^+ remain constant because there is active transport of Na^+ out of the cell against its electrical and concentration gradients, and this transport is coupled to active transport of K^+ into the cell (see below).

The magnitude of the membrane potential at any given time depends, of course, upon the distribution of Na^+, K^+, and Cl^- and the permeability of the membrane to each of these ions. An equation that describes this relationship with considerable accuracy is the **Goldman constant-field equation:**

$$V = \frac{RT}{F} \ln \frac{P_{K^+}[K_o^+] + P_{Na^+}[Na_o^+] + P_{Cl^-}[Cl_i^-]}{P_{K^+}[K_i^+] + P_{Na^+}[Na_i^+] + P_{Cl^-}[Cl_o^-]}$$

where V is the membrane potential, R is the gas constant, T is the absolute temperature, F is the faraday,

and P_{K^+}, P_{Na^+}, and P_{Cl^-} are the permeabilities of the membrane to K^+, Na^+, and Cl^-. The brackets signify concentration, and i and o refer to the inside and outside of the cell, respectively. Since P_{Na^+} is low relative to P_{K^+} in the resting cells, Na^+ contributes little to the value of V. As would be predicted from the Goldman equation, changes in external Na^+ concentration produce little change in the resting membrane potential whereas increases in external K^+ concentration decrease it.

The situation in skeletal muscle and various other types of cells is similar. In muscle, for example, the resting membrane potential is about −90 mV. E_{Cl} is −86 mV, E_K is −100 mV, and E_{Na} is +55 mV.

Genesis of the Membrane Potential

The distribution of ions across the cell membrane and the nature of this membrane provide the explanation for the membrane potential. The concentration gradient for K^+ facilitates its movement out of the cell via K^+ channels, but its electrical gradient is in the opposite (inward) direction. Consequently, an equilibrium is reached in which the tendency of K^+ to move out of the cell is balanced by its tendency to move into the cell, and at that equilibrium there is a slight excess of cations on the outside and anions on the inside. This condition is maintained by Na^+-K^+ ATPase, which pumps K^+ back into the cell and keeps the intracellular concentration of Na^+ low. The Na^+-K^+ pump is also electrogenic, because it pumps three Na^+ out of the cell for every two K^+ it pumps in; thus, it also contributes a small amount to the membrane potential by itself. It should be emphasized that the number of ions responsible for the membrane potential is a minute fraction of the total number present and that the total concentrations of positive and negative ions are equal everywhere except along the membrane. Na^+ influx does not compensate for the K^+ efflux because the K^+ channels (see below) make the membrane more permeable to K^+ than to Na^+.

FUNCTIONAL MORPHOLOGY OF THE CELL

Revolutionary advances in the understanding of cell structure and function have been made through use of the techniques of modern cellular and molecular biology. There have also been major advances in the study of embryology and development at the cellular level. Developmental biology and the details of cell biology are beyond the scope of this book. However, a basic knowledge of cell biology is essential to an understanding of the organ systems in the body and the way they function.

Table 1–2. Concentration of some ions inside and outside mammalian spinal motor neurons.

Ion	Concentration (mmol/L of H_2O)		Equilibrium Potential (mV)
	Inside Cell	Outside Cell	
Na^+	15.0	150.0	+60
K^+	150.0	5.5	−90
Cl^-	9.0	125.0	−70

Resting membrane potential = −70 mV

The specialization of the cells in the various organs is very great, and no cell can be called "typical" of all cells in the body. However, a number of structures (**organelles**) are common to most cells. These structures are shown in Figure 1–4. Many of them can be isolated by ultracentrifugation combined with other techniques. When cells are homogenized and the resulting suspension is centrifuged, the nuclei sediment first, followed by the mitochondria. High-speed centrifugation that generates forces of 100,000 times gravity or more causes a fraction made up of granules called the **microsomes** to sediment. This fraction includes organelles such as the ribosomes and peroxisomes.

Cell Membrane

The membrane that surrounds the cell is a remarkable structure. It is made up of lipids and proteins and is semipermeable, allowing some substances to pass through it and excluding others. However, its permeability can also be varied because it contains numerous regulated ion channels and other transport proteins that can change the amounts of substances moving across it. It is generally referred to as the **plasma membrane.** The nucleus is also surrounded by a membrane of this type, and the organelles are surrounded by or made up of a membrane.

Although the chemical structure of membranes and their properties vary considerably from one location to another, they have certain common features. They are generally about 7.5 nm (75 Angstrom units) thick. They are made up primarily of protein and lipids. The chemistry of proteins and lipids is discussed in Chapter 17. The major lipids are phospholipids such as phosphatidylcholine and phosphatidylethanolamine. The shape of the phospholipid molecule is roughly that of a clothespin (Figure 1–5). The head end of the molecule contains the phosphate portion and is relatively soluble in water (polar, **hydrophilic**). The tails are relatively insoluble (nonpolar, **hydrophobic**). In the membrane,

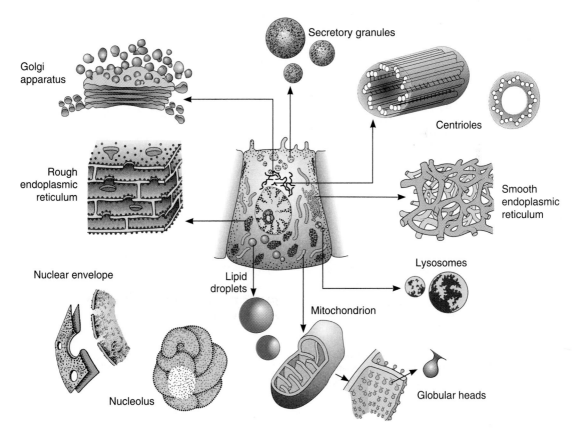

Figure 1–4. Diagram showing a hypothetical cell in the center as seen with the light microscope. It is surrounded by various organelles. (After Bloom and Fawcett. Reproduced, with permission, from Junqueira LC, Carneiro J, Kelley RO: *Basic Histology,* 9th ed. McGraw-Hill, 1998.)

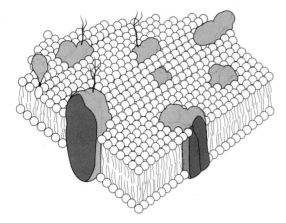

Figure 1–5. Biologic membrane. The phospholipid molecules each have two fatty acid chains **(wavy lines)** attached to a phosphate head **(open circle).** Proteins are shown as irregular colored globules. Many are integral proteins, which extend through the membrane, but peripheral proteins are attached to the inside (not shown) and outside of the membrane, sometimes by glycosylphosphatidylinositol (GPI) anchors.

the hydrophilic ends of the molecules are exposed to the aqueous environment that bathes the exterior of the cells and the aqueous cytoplasm; the hydrophobic ends meet in the water-poor interior of the membrane. In **prokaryotes** (cells such as bacteria in which there is no nucleus), the membranes are relatively simple, but in **eukaryotes** (cells containing nuclei), cell membranes contain various glycosphingolipids, sphingomyelin, and cholesterol.

There are many different proteins embedded in the membrane. They exist as separate globular units and many pass through the membrane **(integral proteins),** whereas others **(peripheral proteins)** stud the inside and outside of the membrane (Figure 1–5). The amount of protein varies with the function of the membrane but makes up on average 50% of the mass of the membrane; ie, there is about one protein molecule per 50 of the much smaller phospholipid molecules. The proteins in the membranes carry out many functions. Some are **cell adhesion molecules** that anchor cells to their neighbors or to basal laminas. There are proteins that function as **pumps,** actively transporting ions across the membrane. Other proteins function as **carriers,** transporting substances down electrochemical gradients by facilitated diffusion. Still others are **ion channels,** which, when activated, permit the passage of ions into or out of the cell. The role of the pumps, carriers, and ion channels in transport across the cell membrane is discussed below. Proteins in another

group function as **receptors** that bind neurotransmitters and hormones, initiating physiologic changes inside the cell. Proteins also function as **enzymes,** catalyzing reactions at the surfaces of the membrane. In addition, some glycoproteins function in antibody processing and distinguishing self from nonself (see Chapter 27).

The uncharged, hydrophobic portions of the proteins are usually located in the interior of the membrane, whereas the charged, hydrophilic portions are located on the surfaces. Peripheral proteins are attached to the surfaces of the membrane in various ways. One common way is attachment to glycosylated forms of phosphatidylinositol. Proteins held by these **glycosyl-phosphatidylinositol (GPI) anchors** (Figure 1–5) include enzymes such as alkaline phosphatase, various antigens, a number of cell adhesion molecules, and three proteins that combat cell lysis by complement (see Chapter 27). Over 40 GPI-linked cell surface proteins have now been described. Other proteins are **lipidated,** ie, they have specific lipids attached to them (Figure 1–6). They may be **myristolated, palmitoylated,** or **prenylated,** ie, attached to geranylgeranyl or farnesyl groups.

The protein structure—and particularly the enzyme content—of biologic membranes varies not only from cell to cell but also within the same cell. For example, there are different enzymes embedded in cell membranes than in mitochondrial membranes. In epithelial cells, the enzymes in the cell membrane on the mucosal surface differ from those in the cell membrane on the basal and lateral margins of the cells; ie, the cells are **polarized.** This is what makes transport across epithelia possible (see below). The membranes are dynamic structures, and their constituents are being constantly renewed at different rates. Some proteins are anchored to the cytoskeleton, but others move laterally in the membrane. For example, receptors move in the membrane and aggregate at sites of endocytosis (see below).

Underlying most cells is a thin, fuzzy layer plus some fibrils that collectively make up the **basement membrane** or, more properly, the **basal lamina.** The basal lamina and, more generally, the extracellular matrix are made up of many proteins that hold cells together, regulate their development, and determine their growth. These include collagens, laminins (see below), fibronectin, tenascin, and proteoglycans.

Mitochondria

Although the morphology of mitochondria varies somewhat from cell to cell, each mitochondrion (Figure 1–4) is in essence a sausage-shaped structure. It is made up of an outer membrane and an inner membrane. The latter is folded to form shelves **(cristae).** The space between

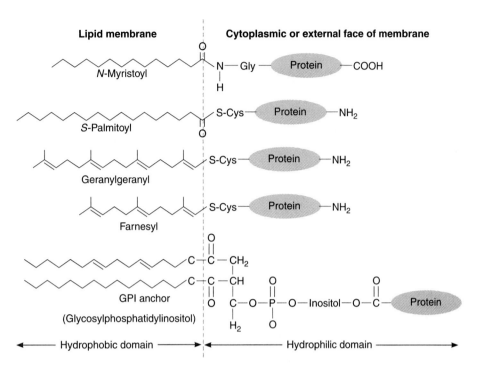

Figure 1–6. Protein linkages to membrane lipids. Some are linked by their amino terminals, others by their carboxyl terminals. Many are attached via glycosylated forms of phosphatidylinositol (GPI anchors). (Reproduced, with permission, from Fuller GM, Shields D: *Molecular Basis of Medical Cell Biology*. McGraw-Hill, 1998.)

the two membranes is called the intracristal space, and the space inside the inner membrane is called the matrix space. The mitochondria are the power-generating units of the cell and are most plentiful and best developed in parts of cells where energy-requiring processes take place. The chemical reactions occurring in them are discussed in detail in Chapter 17. The outer membrane of each mitochondrion is studded with the enzymes concerned with biologic oxidations, providing raw materials for the reactions occurring inside the mitochondrion. In the interior, the enzymes that convert the products of carbohydrate, protein, and fat metabolism to CO_2 and water are located in the internal mitochondrial membrane. Four complex enzymes are involved (Figure 1–7): **NADH dehydrogenase, succinic dehydrogenase, cytochrome c,** and **cytochrome oxidase.** During these reactions, protons (H^+) are pumped from the matrix to the intracristal space, establishing a proton gradient. Protons diffusing back along this gradient drive the synthesis of **adenosine triphosphate (ATP)** by the enzyme **ATP synthase.** ATP is the energy-rich triphosphate which provides the energy for many key metabolic processes, not only in animals (see Chapter 17) but in bacteria and plants as well, which synthesize it in differ-

ent ways. The coupling of oxidation with formation of ATP in the mitochondria of animals is called **oxidative phosphorylation.**

ATP synthase is a unique enzyme made up of many subunits with a base embedded in the inner mitochon-

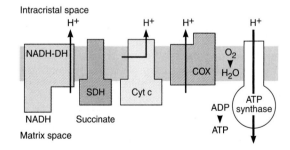

Figure 1–7. Principal enzymes in the inner mitochondrial membrane that are involved in oxidative phosphorylation. NADH-DH, NADH dehydrogenase; SDH, succinic dehydrogenase; c, cytochrome c; COX, cytochrome oxidase. (Modified from Saraste M: Oxidative phosphorylation at the *fin de siècle*. Science 1999;283:1488).

dria membrane, a neck, and a globular head in the mitochondrial matrix. Most of the neck and the base actually rotate as ATP is formed.

It is almost certain that mitochondria were once autonomous microorganisms that developed a symbiotic relation with and became incorporated into ancestral eukaryotic cells. Consistent with this origin is the fact that mitochondria have their own genome. There is much less DNA in the mitochondrial genome than in the nuclear genome (see below), and 99% of the proteins in the mitochondria are the products of nuclear genes, but mitochondrial DNA is responsible for certain key components of the pathway for oxidative phosphorylation. Specifically, human mitochondrial DNA is a double-stranded circular molecule containing 16,569 base pairs (compared with over a billion in nuclear DNA). It codes for 13 protein subunits that are associated with proteins encoded by nuclear genes to form four enzyme complexes plus two ribosomal RNAs and 22 transfer RNAs (see below) that are needed for protein production by the intramitochondrial ribosomes.

Sperm contributes no mitochondria to the zygote; therefore, mitochondria come from the ovum and inheritance is strictly maternal. There is no effective DNA repair system in the mitochondria, and the mutation rate for mitochondrial DNA is over ten times the rate for nuclear DNA. A large number of relatively rare diseases have now been traced to mutations in mitochondrial DNA. These include for the most part disorders of tissues with high metabolic rates in which energy production is defective as a result of abnormalities in the production of ATP.

Evidence is accumulating that mammalian cell death (apoptosis) is initiated in the mitochondria—or at least involves them at an early stage (see below).

Lysosomes

In the cytoplasm of the cell, there are large, somewhat irregular structures surrounded by membrane. The interior of these structures, which are called **lysosomes,** is more acidic than the rest of the cytoplasm, and external material such as endocytosed bacteria as well as worn-out cell components are digested in them. Some of the enzymes involved are listed in Table 1–3.

When a lysosomal enzyme is congenitally absent, the lysosomes become engorged with the material the enzyme normally degrades. This eventually leads to one of the **lysosomal storage diseases.** For example, α-galactosidase A deficiency causes Fabry's disease, and β-galactocerebrosidase deficiency causes Gaucher's disease. These diseases are rare, but they are serious and can be fatal. Another example is the lysosomal storage disease called Tay-Sachs disease, which causes mental retardation and blindness.

Table 1–3. Some of the enzymes found in lysosomes and the cell components that are their substrates.

Enzyme	Substrate
Ribonuclease	RNA
Deoxyribonuclease	DNA
Phosphatase	Phosphate esters
Glycosidases	Complex carbohydrates; glycosides and polysaccharides
Arylsulfatases	Sulfate esters
Collagenase	Proteins
Cathepsins	Proteins

Peroxisomes

Peroxisomes are found in the microsomal fraction of cells. They are 0.5 mm in diameter and are surrounded by a membrane. This membrane contains a number of peroxisome-specific proteins that are concerned with transport of substances into and out of the matrix of the peroxisome. The matrix contains more than 40 enzymes, which operate in concert with enzymes outside the peroxisome to catalyze a variety of anabolic and catabolic reactions. Several years ago, a number of synthetic compounds were found to cause proliferation of peroxisomes by acting on receptors in the nuclei of cells. These receptors (**PPARs**) are members of the nuclear receptor superfamily, which includes receptors for steroid hormones, thyroid hormones, certain vitamins, and a number of other substances (see below). When activated, they bind to DNA, producing changes in the production of mRNAs. Three PPAR receptors—α, β, and γ—have been characterized. PPAR-α and PPAR-γ have received the most attention because PPAR-γ's are activated by feeding and initiate increases in enzymes involved in energy storage, whereas PPAR-α's are activated by fasting and increase energy-producing enzyme activity. Thiazolidinediones are synthetic ligands for PPAR-γ's and they increase sensitivity to insulin, though their use in diabetes has been limited by their toxic side effects. Fibrates, which lower circulating triglycerides, are ligands for PPAR-α's.

Cytoskeleton

All cells have a **cytoskeleton,** a system of fibers that not only maintains the structure of the cell but also permits it to change shape and move. The cytoskeleton is made up primarily of microtubules, intermediate filaments, and microfilaments, along with proteins that anchor them and tie them together. In addition, proteins and organelles move along microtubules and microfilaments from one part of the cell to another propelled by molecular motors.

Microtubules (Figures 1–8 and 1–9) are long, hollow structures with 5-nm walls surrounding a cavity 15 nm in diameter. They are made up of two globular protein subunits, α- and β-tubulin. A third subunit, γ-tubulin, is associated with the production of microtubules by the centrosomes (see below). The α and β subunits form heterodimers (Figure 1–9), which aggregate to form long tubes made up of stacked rings, with each ring usually containing 13 subunits. The tubules also contain other proteins that facilitate their formation. The assembly of microtubules is facilitated by warmth and various other factors, and disassembly is facilitated by cold and other factors. The end where assembly predominates is called the + end, and the end where disassembly predominates is the − end. Both processes occur simultaneously in vitro.

Because of their constant assembly and disassembly, microtubules are a dynamic portion of the cell skeleton. They provide the tracks for transport of vesicles, organelles such as secretory granules, and mitochondria from one part of the cell to another. They also form the spindle, which moves the chromosomes in mitosis. Microtubules can transport in both directions.

Microtubule assembly is prevented by colchicine and vinblastine. The anticancer drug **paclitaxel (Taxol)** binds to microtubules and makes them so stable that organelles cannot move. Mitotic spindles cannot form, and the cells die.

Intermediate filaments are 8–14 nm in diameter and are made up of various subunits. Some of these filaments connect the nuclear membrane to the cell membrane. They form a flexible scaffolding for the cell and help it resist external pressure. In their absence, cells rupture more easily; and when they are abnormal in humans, blistering of the skin is common.

Microfilaments (Figure 1–8) are long solid fibers 4–6 nm in diameter. They are made up of actin, the protein that by its interaction with myosin brings about contraction of muscle (see Chapter 3). Actin and its mRNA (see below) are present in all types of cells. It is the most abundant protein in mammalian cells, sometimes accounting for as much as 15% of the total protein in the cell. Its structure is highly conserved; for example, 88% of the amino acid sequences in yeast and rabbit actin are identical. The actin molecules (**G-actin**) polymerize in vivo to form **F-actin,** the long filamentous chains that are the microfilaments. They also depolymerize in vivo, with polymerization often occurring at one end of a microfilament (the + end, as in microtubules) and depolymerization at the other (the −

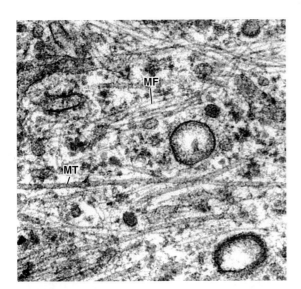

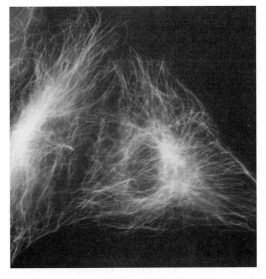

Figure 1–8. **Left:** Electron micrograph of the cytoplasm of a fibroblast, showing microfilaments (MF) and microtubules (MT). (Reproduced, with permission, from Junqueira LC, Carneiro J, Kelley RO: *Basic Histology,* 9th ed. McGraw-Hill, 1998.) **Right:** Distribution of microtubules in fibroblasts. The cells are treated with a fluorescently labeled antibody to tubulin, making microtubules visible as the light-colored structures. (Reproduced, with permission, from Connolly J et al: Immunofluorescent staining of cytoplasmic and spindle microtubules in mouse fibroblasts with antibody to τ protein. Proc Natl Acad Sci U S A 1977;74:2437.)

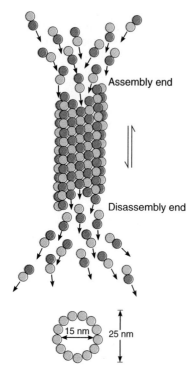

Figure 1–9. Assembly and disassembly of a microtubule by aggregation and disaggregation of dimers made up of α- and β-tubulin. (From Sloboda RD: The role of microtubules in cell structure and cell division. Am Sci 1980;68:290. Reprinted by permission of *American Scientist,* journal of Sigma Xi, the Scientific Research Society.)

end). They attach to various parts of the cytoskeleton (Figure 1–10). They reach to the tips of the microvilli on the epithelial cells of the intestinal mucosa. They are also abundant in the lamellipodia that cells put out when they crawl along surfaces. The actin filaments interact with integrin receptors and form **focal adhesion complexes** (focal adhesions; see below) which serve as points of traction with the surface over which the cell pulls itself.

Molecular Motors

The molecular motors that move proteins, organelles, and other cell parts (their **cargo**) to all parts of the cell are 100–500 kDa ATPases that attach to the cargo. Their heads form cross-bridges to the microtubules or other actin polymers, and they convert ATP into energy that produces bending of the cross-bridges, causing the ATPase molecules to move. There are two types of molecular motors: those producing motion along microtubules and those producing motion along actin (Table 1–4). Examples are shown in Figure 1–11, but each type is a superfamily, with many forms throughout the animal kingdom.

The conventional form of **kinesin** is a double-headed molecule that moves its cargo toward the + ends of microtubules. One head binds to the microtubule and then bends its neck while the other head swings forward and binds, producing almost continuous movement. Some kinesins are associated with mitosis and meiosis. Other kinesins perform other functions, including, in some instances, moving cargo to the − end of microtubules.

Dyneins have two heads, with their neck pieces embedded in a complex of proteins (Figure 1–11). **Cyto-**

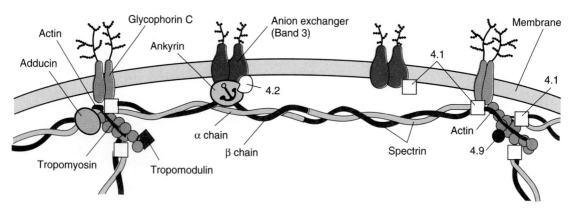

Figure 1–10. Membrane-cytoskeleton attachments in the red blood cell, showing the various proteins that anchor actin microfilaments to the membrane. Some are identified by numbers (4.1, 4.2, 4.9), whereas others have received names. (Reproduced, with permission, from Luna EJ, Hitt AL: Cytoskeleton-plasma membrane interactions. Science 1992; 258:955.)

Table 1–4. Examples of molecular motors.

Microtubule-based
Conventional Kinesin
Dyneins
Actin-based
Myosins I–V

plasmic dynein has a function like that of conventional kinesin, except that it moves particles and membranes to the − end of the microtubules. **Axonemal dynein** oscillates and is responsible for the beating of flagella and cilia (see below).

Myosins form cross-bridges to actin filaments and the myosin heads move, generating force. This produces movement that literally ranges from contraction of intestinal villi to cell migration and contraction of all the skeletal and other muscles in the human body. The myosin superfamily is divided into 15 classes. Myosin-I and myosin-II are shown in Figure 1–11. Myosin-I is associated with the actin in many cells, whereas myosin-II (discussed in detail in Chapter 3) is the form present in skeletal muscle.

Myosin molecules have globular heads, which contain the ATPase activity, and tails of various lengths (Figure 1–11). **Myosin-I** molecules have a single head.

Along with actin, myosin-I is often associated with cell membranes. **Myosin-II** molecules have two heads, but only one head is active in a given molecule.

Centrosomes

Near the nucleus in the cytoplasm of eukaryotic animal cells is a **centrosome**. The centrosome is made up of two **centrioles** and surrounding amorphous **pericentriolar material.** The centrioles are short cylinders arranged so that they are at right angles to each other. Microtubules in groups of three run longitudinally in the walls of each centriole (Figure 1–4). There are nine of these triplets spaced at regular intervals around the circumference.

The centrosomes are **microtubule-organizing centers (MTOCs)** that contain γ-tubulin. The microtubules grow out of this γ-tubulin in the pericentriolar material. When a cell divides, the centrosomes duplicate themselves, and the pairs move apart to the poles of the mitotic spindle, where they monitor the steps in cell division. In multinucleate cells, there is a centrosome near each nucleus.

Cilia

There are various types of projections from cells. True **cilia** are dynein-driven motile processes that are used by unicellular organisms to propel themselves through the water and by multicellular organisms to propel mu-

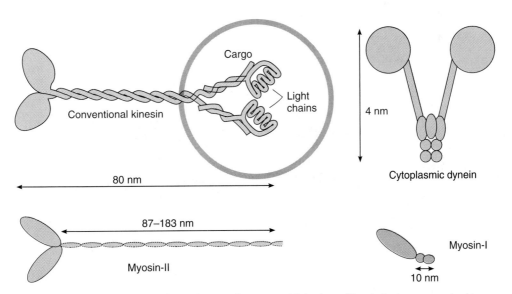

Figure 1–11. Examples of the three types of molecular motors. Ubiquitous kinesin is shown attached to cargo, in this case a membrane-bound organelle. All the motors have motor heads that hydrolyze ATP and use the energy to flex the heads on their necks. Note that the dimensions vary, and the motors are not drawn to the same scale.

cus and other substances over the surface of various epithelia. They resemble centrioles in having an array of nine tubular structures in their walls, but they have in addition a pair of microtubules in the center, and there are two rather than three microtubules in each of the nine circumferential structures. The **basal granule,** on the other hand, is the structure to which each cilium is anchored. It has nine circumferential triplets, like a centriole, and there is evidence that basal granules and centrioles are interconvertible.

Cell Adhesion Molecules

Cells are attached to the basal lamina and to each other by **cell adhesion molecules (CAMs)** that are prominent parts of the intercellular connections described below. These adhesion proteins have attracted great attention in recent years because they are important in embryonic development and formation of the nervous system and other tissues; in holding tissues together in adults; in inflammation and wound healing; and in the metastasis of tumors. Many pass through the cell membrane and are anchored to the cytoskeleton inside the cell. Some bind to like molecules on other cells (homophilic binding), whereas others bind to other molecules (heterophilic binding). Many bind to **laminins,** a family of large cross-shaped molecules with multiple receptor domains in the extracellular matrix.

Nomenclature in the CAM field is somewhat chaotic, partly because the field is growing so rapidly and partly because of the extensive use of acronyms, as in other areas of modern biology. However, the CAMs can be divided into four broad families: (1) **integrins,** heterodimers that bind to various receptors; (2) adhesion molecules of the **IgG superfamily** of immunoglobulins; (3) **cadherins,** Ca^{2+}-dependent molecules that mediate cell-to-cell adhesion by homophilic reactions; and (4) **selectins,** which have lectin-like domains that bind carbohydrates. The functions of CAMs in granulocytes and platelets are described in Chapter 27, and their roles in inflammation and wound healing are discussed in Chapter 33.

The CAMs not only fasten cells to their neighbors—they also transmit signals into and out of the cell. Cells that lose their contact with the extracellular matrix via integrins have a higher rate of apoptosis (see below) than anchored cells, and interactions between integrins and the cytoskeleton are involved in cell movement.

Intercellular Connections

Two types of junctions form between the cells that make up tissues: junctions that fasten the cells to one another and to surrounding tissues, and junctions that permit transfer of ions and other molecules from one cell to another. The types of junctions that tie cells together and endow tissues with strength and stability include the **tight junction,** which is also known as the **zonula occludens.** The **desmosome** and **zonula adherens** (Figure 1–12) hold cells together, and the **hemidesmosome** and **focal adhesion** attach cells to their basal laminas. Tight junctions between epithelial cells are also essential for transport of ions across epithelia. The junction by which molecules are transferred is the **gap junction.**

Tight junctions characteristically surround the apical margins of the cells in epithelia such as the intestinal mucosa, the walls of the renal tubules, and the choroid plexus. They are made up of ridges—half from one cell and half from the other—which adhere so strongly at cell junctions that they almost obliterate the space between the cells. They permit the passage of some ions and solute, and the degree of this "leakiness" varies. Extracellular fluxes of ions and solute across epithelia at these junctions are a significant part of overall ion and solute flux. In addition, tight junctions prevent the movement of proteins in the plane of the membrane, helping to maintain the different distribution of transporters and channels in the apical and basolateral cell membranes that make transport across epithelia possible (see above and Chapters 25 and 38).

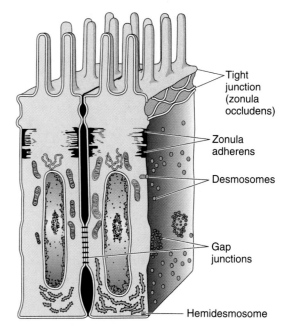

Figure 1–12. Intercellular junctions in the mucosa of the small intestine. Focal adhesions are not shown in detail.

In epithelial cells, each zonula adherens is usually a continuous structure on the basal side of the zonula occludens, and it is a major site of attachment for intracellular microfilaments. It contains cadherins.

Desmosomes are patches characterized by apposed thickenings of the membranes of two adjacent cells. Attached to the thickened area in each cell are intermediate filaments, some running parallel to the membrane and others radiating away from it. Between the two membrane thickenings, there is filamentous material in the intercellular space that includes cadherins and the extracellular portions of several other transmembrane proteins.

Hemidesmosomes look like half-desmosomes that attach cells to the underlying basal lamina and are connected intracellularly to intermediate filaments. However, they contain integrins rather than cadherins. Focal adhesions also attach cells to their basal laminas. As noted above, they are labile structures associated with actin filaments inside the cell, and they play an important role in cell movement.

Gap Junctions

At gap junctions, the intercellular space narrows from 25 nm to 3 nm, and hexagonal arrays of protein units called **connexons** in the membrane of each cell are lined up with one another (Figure 1–13). Each connexon is made up of six subunits surrounding a channel that, when lined up with the channel in the corresponding connexon in the adjacent cell, permits

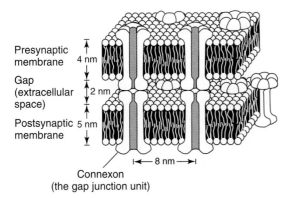

Presynaptic membrane 4 nm
Gap (extracellular space) 2 nm
Postsynaptic membrane 5 nm
|← 8 nm →|
Connexon (the gap junction unit)

Figure 1–13. Gap junction. Note that each connexon is made up of six subunits and that each connexon in the membrane of one cell lines up with a connexon in the membrane of the neighboring cell, forming a channel through which substances can pass from one cell to another without entering the ECF. (Reproduced, with permission, from Kandel ER, Schwartz JH, Jessell TM [editors]: *Principles of Neural Science,* 4th ed. McGraw-Hill, 2000.)

substances to pass between the cells without entering the ECF. The diameter of the channel is normally about 2 nm, which permits the passage of ions, sugars, amino acids, and other solutes with molecular weights up to about 1000. Gap junctions thus permit the rapid propagation of electrical activity from cell to cell (see Chapter 4) and the exchange of various chemical messengers. The diameter of each channel is regulated by intracellular Ca^{2+}, and an increase in Ca^{2+} concentration causes the subunits to slide together, reducing the diameter of the channel. The diameter may also be regulated by pH and voltage.

Connexon isoforms are encoded by at least 13 different genes in rodents, and it seems likely that there are at least this many different genes that are involved in humans. There is considerable variation in distribution of the isoforms from one tissue to another. In mice, the ovum is connected to granulosa cells by tight junctions, and mice deficient in one type of connexon fail to ovulate and complete meiosis. A number of diseases in humans have been linked to mutant connexons. One is the X-linked form of **Charcot-Marie-Tooth disease,** a peripheral neuropathy, and another is **heterotaxia,** in which there are multiple abnormalities with failure to establish normal left–right asymmetry. It is not known how connexon abnormalities lead to these diseases, and at least for heterotaxia, other genes are probably involved.

Nucleus & Related Structures

A nucleus is present in all eukaryotic cells that divide. If a cell is cut in half, the anucleate portion eventually dies without dividing. The nucleus is made up in large part of the **chromosomes,** the structures in the nucleus that carry a complete blueprint for all the heritable species and individual characteristics of the animal. Except in germ cells, the chromosomes occur in pairs, one originally from each parent (see Figure 23–2). Each chromosome is made up of a giant molecule of **deoxyribonucleic acid (DNA).** The DNA strand is about 2 m long, but it can fit in the nucleus because at intervals it is wrapped around a core of histone proteins to form a **nucleosome.** There are about 25 million nucleosomes in each nucleus. Thus, the structure of the chromosomes has been likened to a string of beads. The beads are the nucleosomes, and the linker DNA between them is the string. The whole complex of DNA and proteins is called **chromatin.** During cell division, the coiling around histones is loosened, probably by acetylation of the histones, and pairs of chromosomes become visible, but between cell divisions only clumps of chromatin can be discerned in the nucleus. The ultimate units of heredity are the **genes** on the chromosomes (see below), and each gene is a portion of the DNA molecule.

During normal cell division by **mitosis,** the chromosomes duplicate themselves and then divide in such a way that each daughter cell receives a full complement **(diploid number)** of chromosomes. During their final maturation, germ cells undergo a division in which half the chromosomes go to each daughter cell (see Chapter 23). This reduction division **(meiosis)** is actually a two-stage process, but the important consideration is that as a result of it, mature sperms and ova contain half the normal number (the **haploid number**) of chromosomes. When a sperm and ovum unite, the resultant cell **(zygote)** has a full diploid complement of chromosomes, one-half from the female parent and one-half from the male. The chromosomes undergo recombination, which mixes maternal and paternal genes.

The nucleus of most cells contains a **nucleolus** (Figure 1–4), a patchwork of granules rich in **ribonucleic acid (RNA).** In some cells, the nucleus contains several of these structures. Nucleoli are most prominent and numerous in growing cells. They are the site of synthesis of ribosomes, the structures in the cytoplasm in which proteins are synthesized (see below).

The interior of the nucleus has a skeleton of fine filaments that are attached to the **nuclear membrane,** or **envelope** (Figure 1–4), which surrounds the nucleus. This membrane is a double membrane, and spaces between the two folds are called **perinuclear cisterns.** The membrane is permeable only to ions and small molecules. However, it contains **nuclear pore complexes.** Each complex has eightfold symmetry and is made up of about 100 proteins organized to form a tunnel through which transport of proteins and mRNA occurs. There are many transport pathways, and proteins called **importins** and **exportins** have been isolated and characterized. A protein named **Ran** appears to play an organizing role. Much current research is focused on transport into and out of the nucleus, and a more detailed understanding of these processes should emerge in the near future.

Endoplasmic Reticulum

The **endoplasmic reticulum** is a complex series of tubules in the cytoplasm of the cell (Figure 1–4). The outer limb of its membrane is continuous with a segment of the nuclear membrane, so in effect this part of the nuclear membrane is a cistern of the endoplasmic reticulum. The tubule walls are made up of membrane. In **rough** or **granular endoplasmic reticulum,** granules called **ribosomes** are attached to the cytoplasmic side of the membrane, whereas in **smooth** or **agranular endoplasmic reticulum,** the granules are absent. Free ribosomes are also found in the cytoplasm. The granular endoplasmic reticulum is concerned with protein synthesis and the initial folding of polypeptide chains with the formation of disulfide bonds. The agranular endoplasmic reticulum is the site of steroid synthesis in steroid-secreting cells and the site of detoxification processes in other cells. As the sarcoplasmic reticulum (see Chapter 3), it plays an important role in skeletal and cardiac muscle.

Ribosomes

The ribosomes in eukaryotes measure approximately 22 by 32 nm. Each is made up of a large and a small subunit called, on the basis of their rates of sedimentation in the ultracentrifuge, the 60S and 40S subunits. The ribosomes are complex structures, containing many different proteins and at least three ribosomal RNAs (see below). They are the sites of protein synthesis. The ribosomes that become attached to the endoplasmic reticulum synthesize all transmembrane proteins, most secreted proteins, and most proteins that are stored in the Golgi apparatus, lysosomes, and endosomes. All these proteins have a hydrophobic signal peptide at one end (see below). The polypeptide chains that form these proteins are extruded into the endoplasmic reticulum. The free ribosomes synthesize cytoplasmic proteins such as hemoglobin (see Chapter 27) and the proteins found in peroxisomes and mitochondria.

The **Golgi apparatus,** which is involved in processing proteins formed in the ribosomes, and secretory granules, vesicles, and endosomes are discussed below in the context of protein synthesis and secretion.

STRUCTURE & FUNCTION OF DNA & RNA

The Genome

DNA is found in bacteria, in the nuclei of eukaryotic cells, and in mitochondria. It is made up of two extremely long nucleotide chains containing the bases adenine (A), guanine (G), thymine (T), and cytosine (C) (Figure 1–14). The chemistry of these purine and pyrimidine bases and of nucleotides is discussed in Chapter 17. The chains are bound together by hydrogen bonding between the bases, with adenine bonding to thymine and guanine to cytosine. The resultant double-helical structure of the molecule is shown in Figure 1–15. An indication of the complexity of the molecule is the fact that the DNA in the human haploid genome (the total genetic message) is made up of 3×10^9 base pairs.

DNA is the component of the chromosomes that carry the "genetic message," the blueprint for all the heritable characteristics of the cell and its descendants. Each chromosome contains a segment of the DNA

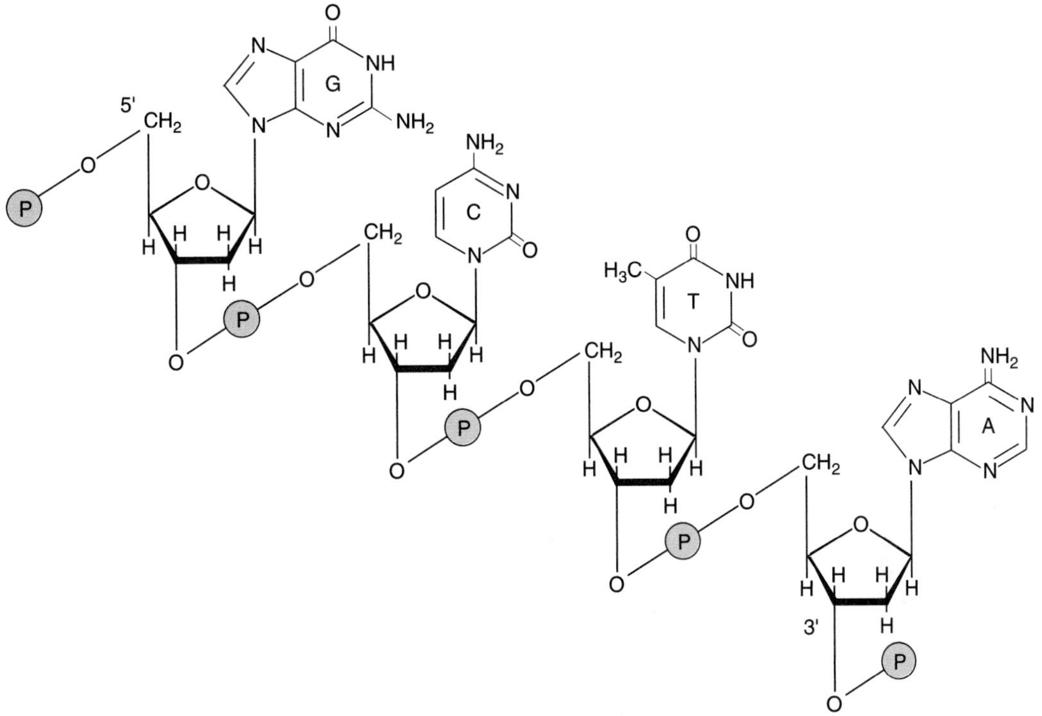

Figure 1–14. Segment of the structure of the DNA molecule in which the purine and pyrimidine bases adenine (A), thymine (T), cytosine (C), and guanine (G) are held together by a phosphodiester backbone between 2′-deoxyribosyl moieties attached to the nucleobases by an N-glycosidic bond. Note that the backbone has a polarity (ie, a 5′ and a 3′ direction). (Reproduced, with permission, from Murray RK et al: *Harper's Biochemistry,* 26th ed. McGraw-Hill, 2003.)

double helix. The genetic message is encoded by the sequence of purine and pyrimidine bases in the nucleotide chains. The text of the message is the order in which the amino acids are lined up in the proteins manufactured by the cell. The message is transferred to ribosomes, the sites of protein synthesis in the cytoplasm, by RNA. RNA differs from DNA in that it is single-stranded, has uracil in place of thymine, and its sugar moiety is ribose rather than 2′-deoxyribose (see Chapter 17). The proteins formed from the DNA blueprint include all the enzymes, and these in turn control the metabolism of the cell. A gene used to be defined as the amount of information necessary to specify a single protein molecule. However, the protein encoded by a single gene may be subsequently divided into several different physiologically active proteins. In addition, different mRNAs can be formed from a gene, with each mRNA dictating formation of a different protein. Genes also contain promoters, DNA sequences that facilitate the formation of RNA. **Mutations** occur when the base sequence in the DNA is altered by x-rays, cosmic rays, or other mutagenic agents.

The Human Genome

When the human genome was finally mapped several years ago, there was considerable surprise that it contained only about 30,000 genes and not the 50,000 or more that had been expected. Yet humans differ quite markedly from their nearest simian relatives. The explanation appears to be that rather than a greater number of genes in humans, there is a greater number of mRNAs—perhaps as many as 85,000. The implications of this increase are discussed below.

DNA Polymorphism

The protein-coding genes (exons) make up only 3% of the human genome; the remaining 97% is made up of introns (see below) and other DNA of unsettled or unknown function. This 97% is sometimes called **junk DNA.** A characteristic of human DNA is its structural variability from one individual to another. Most of the variations occur in noncoding regions, but they can also occur in coding regions, where they can be silent

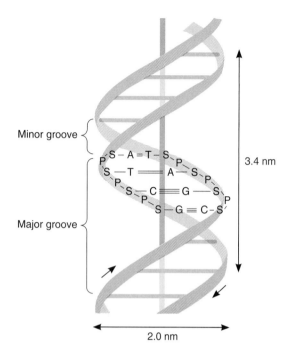

Minor groove

Major groove

3.4 nm

2.0 nm

Figure 1–15. Double-helical structure of DNA, with adenine (A) bonding to thymine (T) and cytosine (C) to guanine (G). (Reproduced, with permission, from Murray RK et al: *Harper's Biochemistry,* 26th ed. McGraw-Hill, 2003.)

or expressed as a detectable alteration in a protein. A common form of these variations is variable repetition of base pairs (tandem repeats) from one to hundreds of times. This variation alters the length of the DNA chain between points where it is cut by various restriction enzymes, so that there is **restriction fragment length polymorphism (RFLP)** in the DNA fragments from different individuals. Consequently, analysis of RFLP in a population gives a pattern that is in effect a **DNA fingerprint.** The value of DNA fingerprinting has been improved by additional specialized techniques. The chance of obtaining identical DNA patterns by using these techniques in individuals who are not identical twins varies with the number of enzymes used, the relatedness of the individuals, and other factors, and there has been debate about the appropriate statistics to use for analysis. However, the possibility that an RFLP match is due to chance has been estimated at 1 in 1,000,000 to 1 in 100,000. Furthermore, RFLP analysis can be carried out on small specimens of semen, blood, or other tissue, and multiple copies of pieces of DNA can be made by using the **polymerase chain reaction (PCR),** an ingenious technique for making DNA copy itself. Therefore, DNA fingerprint-

ing is of obvious value in investigating crimes and determining paternity, although reliable techniques must be used and the results interpreted with care. RFLP analysis is also of value in studying animal and human evolution and in identifying the chromosomal location of genes causing inherited diseases.

Mitosis

At the time of each somatic cell division **(mitosis),** the two DNA chains separate, each serving as a template for the synthesis of a new complementary chain. **DNA polymerase** catalyzes this reaction. One of the double helices thus formed goes to one daughter cell and one goes to the other, so the amount of DNA in each daughter cell is the same as that in the parent cell.

Telomeres

Cell replication involves not only DNA polymerase but a special reverse transcriptase that synthesizes the short repeats of DNA that characterize the ends **(telomeres)** of chromosomes. Without this transcriptase and related enzymes known collectively as **telomerase,** somatic cells lose DNA as they divide 40–60 times and then become senescent and undergo apoptosis (see below). On the other hand, cells with high telomerase activity, which includes most cancer cells, can in theory keep multiplying indefinitely. Not surprisingly, there has been considerable interest in the telomerase mechanism, both in terms of aging and in terms of cancer. However, it now seems clear that the mechanism for replicating chromosome ends is complex, and much additional research will be needed before a complete understanding is achieved and therapeutic applications emerge.

Meiosis

In germ cells, reduction division **(meiosis)** takes place during maturation. The net result is that one of each pair of chromosomes ends up in each mature germ cell; consequently, each mature germ cell contains half the amount of chromosomal material found in somatic cells. Therefore, when a sperm unites with an ovum, the resulting zygote has the full complement of DNA, half of which came from the father and half from the mother. The chromosomal events that occur at the time of fertilization are discussed in detail in Chapter 23. The term "ploidy" is sometimes used to refer to the number of chromosomes in cells. Normal resting diploid cells are **euploid** and become **tetraploid** just before division. **Aneuploidy** is the condition in which a cell contains other than the haploid number of chromosomes or an exact multiple of it, and this condition is common in cancerous cells.

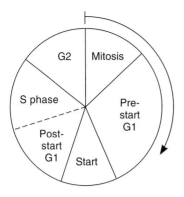

Figure 1–16. Sequence of events during the cell cycle.

Cell Cycle

Obviously, the initiation of mitosis and normal cell division depends on the orderly occurrence of events during what has come to be called the **cell cycle.** A diagram of these events is shown in Figure 1–16. There is

intense interest in the biochemical machinery that produces mitosis, in part because of the obvious possibility of its relation to cancer. When DNA is damaged, entry into mitosis is inhibited, giving the cell time to repair the DNA; failure to repair damaged DNA leads to cancer. The cell cycle is regulated by proteins called **cyclins** and **cyclin-dependent protein kinases,** which phosphorylate other proteins. However, the regulation is complex, and a detailed analysis of it is beyond the scope of this book.

Transcription & Translation

The strands of the DNA double helix not only replicate themselves, but also serve as templates by lining up complementary bases for the formation in the nucleus of **messenger RNA (mRNA), transfer RNA (tRNA),** the RNA in the ribosomes (rRNA), and various other RNAs. The formation of mRNA is called **transcription** (Figure 1–17) and is catalyzed by various forms of **RNA polymerase.** Usually after some **posttranscriptional processing** (see below), mRNA moves to the cy-

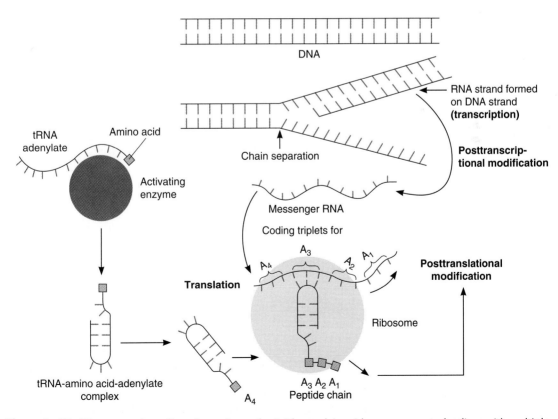

Figure 1–17. Diagrammatic outline of protein synthesis. The nucleic acids are represented as lines with multiple short projections representing the individual bases.

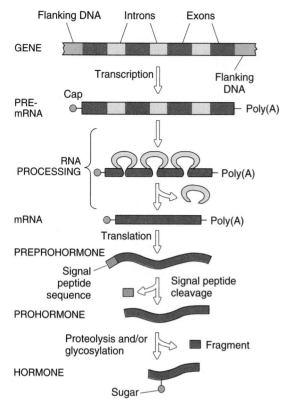

Figure 1–19. Transcription, posttranscriptional modification of mRNA, translation in the ribosomes, and posttranslational processing in the formation of hormones and other proteins. Cap, cap site. (Modified and reproduced, with permission, from Baxter JD: Principles of endocrinology. In: *Cecil Textbook of Medicine*, 16th ed. Wyngaarden JB, Smith LH Jr [editors]. Saunders, 1982.)

toplasm and dictates the formation of the polypeptide chain of a protein (**translation**). This process occurs in the ribosomes. tRNA attaches the amino acids to mRNA. The mRNA molecules are smaller than the DNA molecules, and each represents a transcript of a small segment of the DNA chain. The molecules of tRNA contain only 70–80 nitrogenous bases, compared with hundreds in mRNA and 3 billion in DNA.

It is worth noting that DNA is responsible for the maintenance of the species; it passes from one generation to the next in germ cells. RNA, on the other hand, is responsible for the production of the individual; it transcribes the information coded in the DNA and forms a mortal individual, a process that has been called "budding off from the germ line."

Genes

Information is accumulating at an accelerating rate about the structure of genes and their regulation. The structure of a typical eukaryotic gene is shown in diagrammatic form in Figure 1–18. It is made up of a strand of DNA that includes coding and noncoding regions. In eukaryotes, unlike prokaryotes, the portions of the genes that dictate the formation of proteins are usually broken into several segments (**exons**) separated by segments that are not translated (**introns**). A pre-mRNA is formed from the DNA, and then the introns and sometimes some of the exons are eliminated in the nucleus by posttranscriptional processing, so that the final mRNA which enters the cytoplasm is made up of exons (Figure 1–19). Introns are eliminated and exons are joined by several different processes. The introns of some genes are eliminated by **spliceosomes,** complex units that are made up of small RNAs and proteins. Other introns are eliminated by **self-splicing** by the RNA they contain. Two different mechanisms produce self-splicing. RNA can catalyze other reactions as well and there is great interest today in the catalytic activity of RNA.

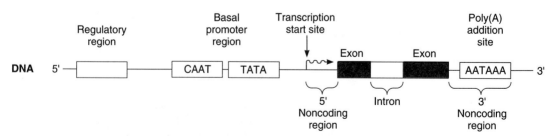

Figure 1–18. Diagram of the components of a typical eukaryotic gene. The coding region (consisting of introns and exons) is flanked by noncoding regions. The 5'-flanking region contains stretches of DNA that interact with proteins to facilitate or inhibit transcription. The 3'-flanking region contains the poly(A) addition site. (Reproduced, with permission, from Murray RK et al: *Harper's Biochemistry*, 26th ed. McGraw-Hill, 2003.)

Because of introns and splicing, more than one mRNA is formed from the same gene. As noted above, the formation of multiple proteins from one gene is perhaps one of the explanations of the surprisingly small number of genes in the human genome. Other physiologic functions of the introns are still unsettled, though they may foster changes in the genetic message and thus aid evolution.

Near the transcription start site of the gene is a **promoter,** which is the site at which RNA polymerase and its cofactors bind. It often includes a TATA sequence **(TATA box),** which ensures that transcription starts at the proper point. Farther out in the 5′ region are **regulatory elements,** which include enhancer and silencer sequences. It has been estimated that there are an average of five regulatory sites per gene. Regulatory sequences are sometimes found in the 3′-flanking region as well, and there is evidence that sequences in this region can also affect the function of other genes.

Regulation of Gene Expression

Each nucleated somatic cell in the body contains the full genetic message, yet there is great differentiation and specialization in the functions of the various types of adult cells. Only small parts of the message are normally transcribed. Thus, the genetic message is normally maintained in a repressed state. However, genes are controlled both spatially and temporally. What turns on genes in one cell and not in other cells? What turns on genes in a cell at one stage of development and not at other, inappropriate stages? What maintains orderly growth in cells and prevents the uncontrolled growth that we call cancer? Obviously, DNA sequences such as the TATA box promote orderly transcription of the gene of which they are a part (cis regulation). However, the major key to selective gene expression is the proteins that bind to the regulatory regions of the gene and increase or shut off its activity. These **transcription factors** are products of other genes and hence mediate trans regulation. They are extremely numerous and include activated steroid hormone receptors and many other factors.

It is common for stimuli such as neurotransmitters that bind to the cell membrane to initiate chemical events which activate **immediate-early genes.** These in turn produce transcription factors that act on other genes. The best-characterized immediate-early genes are *c-fos,* and *c-jun.* The proteins produced by these genes, **c-Fos, c-Jun,** and several related proteins, form homodimer or heterodimer transcription factors which bind to a specific DNA regulatory sequence called an **AP-1 site** (Figure 1–20). Some of the dimers enhance transcription, and others inhibit it. The appearance of c-Fos, c-Jun, and related proteins is such a common

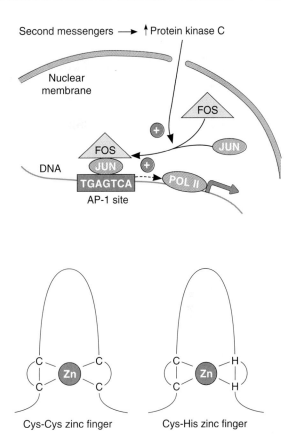

Figure 1–20. **Top:** Activation of genes by second messengers. Increased protein kinase C causes production of c-Fos and c-Jun by immediate-early genes. The c-Fos–c-Jun heterodimer binds to an AP-1 site, in this case activating RNA polymerase II (Pol II) and increasing transcription of other genes. (Courtesy of DG Gardner.) **Bottom:** Zinc fingers. The curved lines represent polypeptide chains of proteins that bind to DNA, and the straight lines indicate coordinate binding of zinc to cysteines (C) or cysteines and histidines (H). (Reproduced, with permission, from Murray RK et al: *Harper's Biochemistry,* 26th ed. McGraw-Hill, 2003.)

sign of cell activation that immunocytochemistry for them or measurement of their mRNAs is used to determine which cells in the nervous system and elsewhere are activated by particular stimuli.

Over 80% of the known transcription factors have one of four DNA-binding motifs. The most common is the **zinc finger** motif, in which characteristically shaped complexes are formed by coordinate binding of Zn^{2+} between two cysteine and two histidine residues or between four cysteine residues. Various transcription factors contain 2–37 of these zinc fingers, which mediate the binding to DNA. Another motif is the **leucine zipper,** in which α-helical regions of dimers have regularly spaced leucine residues that interact with one another to form a coiled coil. Extensions of the dimer beyond the zippered region are rich in arginine and lysine, and these bind to DNA. The other common DNA-binding motifs are helix-turn-helix and helix-loop-helix structures.

It is now possible by using molecular biologic techniques to augment the function of particular genes, to transfer human genes into animals, and to disrupt the function of single genes **(gene knockout).** The gene knockout technique is currently being used in numerous experiments.

Protein Synthesis

The process of protein synthesis is a complex but fascinating one that, as noted above, involves four steps: transcription, posttranscriptional modification, translation, and posttranslational modification. The various steps are summarized in simplified form in Figure 1–19.

When suitably activated, transcription of the gene starts at the **cap site** (Figure 1–19) and ends about 20 bases beyond the signal sequence AATAAA. The RNA transcript is capped in the nucleus by addition of 7-methylguanosine triphosphate to the 5′ end; this cap is necessary for proper binding to the ribosome (see below). A **poly(A) tail** of about 100 bases is added to the untranslated segment at the 3′ end. The function of the poly(A) tail is unsettled, but it may help maintain the stability of the mRNA. The pre-mRNA formed by capping and addition of the poly(A) tail is then processed by elimination of the introns (Figure 1–19), and once this posttranscriptional modification is complete, the mature mRNA moves to the cytoplasm. Posttranscriptional modification of the pre-mRNA is a regulated process, and, as noted above, differential splicing can occur, with the formation of more than one mRNA from a single pre-mRNA.

When a definitive mRNA reaches a ribosome in the cytoplasm, it dictates the formation of a polypeptide chain. Amino acids in the cytoplasm are activated by combination with an enzyme and adenosine monophosphate (adenylate), and each **activated amino acid** then combines with a specific molecule of tRNA. There is at least one tRNA for each of the 20 unmodified amino acids found in large quantities in the body proteins of animals (see Chapter 17), but there is more than one tRNA for some amino acids. The tRNA-amino acid-adenylate complex is next attached to the mRNA template, a process that occurs in the ribosomes. This process is shown diagrammatically in Figure 1–17. The tRNA "recognizes" the proper spot to attach on the mRNA template because it has on its active end a set of three bases that are complementary to a set of three bases in a particular spot on the mRNA chain. The genetic code is made up of such **triplets,** sequences of three purine or pyrimidine bases (or both); each triplet stands for a particular amino acid.

Translation starts in the ribosomes with an AUG (transcribed from ATG in the gene) which codes for methionine. The amino terminal amino acid is then added, and the chain is lengthened one amino acid at a time. The mRNA attaches to the 40S subunit of the ribosome during protein synthesis; the polypeptide chain being formed attaches to the 60S subunit; and the tRNA attaches to both. As the amino acids are added in the order dictated by the triplet code, the ribosome moves along the mRNA molecule like a bead on a string. Translation stops at one of three stop, or nonsense, codons—UGA, UAA, or UAG—and the polypeptide chain is released. The tRNA molecules are used again. The mRNA molecules are also reused approximately 10 times before being replaced.

Typically, there is more than one ribosome on a given mRNA chain at one time. The mRNA chain plus its collection of ribosomes is visible under the electron microscope as an aggregation of ribosomes called a **polyribosome (polysome).**

At least in theory, synthesis of particular proteins can be stopped by administering **antisense oligonucleotides,** short synthetic stretches of bases complementary to segments of the bases on the mRNA for the protein. These bind to the mRNA, blocking translation. Early results with this technology were disappointing because of nonspecific binding and immune responses, but research continues and there is hope for products that will be useful in the treatment of a variety of diseases, including cancer.

Posttranslational Modification

After the polypeptide chain is formed, it is modified to the final protein by one or more of a combination of reactions that include hydroxylation, carboxylation, glycosylation, or phosphorylation of amino acid residues; cleavage of peptide bonds that converts a larger polypeptide to a smaller form; and the folding and packaging of the protein into its ultimate, often complex configuration.

It has been claimed that a typical eukaryotic cell synthesizes about 10,000 different proteins during its lifetime. How do these proteins get to the right locations in the cell? Synthesis starts in the free ribosomes. As noted above, most proteins that are going to be secreted or stored in organelles and most transmembrane proteins have at their amino terminal a **signal peptide (leader** sequence) that guides them into the endoplasmic reticulum. It is made up of 15–30 predominantly hydrophobic amino acid residues. The signal peptide, once synthesized, binds to a **signal recognition particle (SRP),** a complex molecule made up of six polypeptides and 7S RNA, one of the small RNAs. The SRP stops translation until it binds to a **translocon,** a pore in the endoplasmic reticulum that is a heterotrimeric structure made up of Sec 61 proteins. The ribosome also binds, and the signal peptide leads the growing peptide chain into the cavity of the endoplasmic reticulum (Figure 1–21). The signal peptide is next cleaved from the rest of the peptide by a signal peptidase while the rest of the peptide chain is still being synthesized.

The signals that direct nascent proteins to some of the other parts of the cell are fashioned in the Golgi apparatus (see below) and involve specific modifications of the carbohydrate residues on glycoproteins.

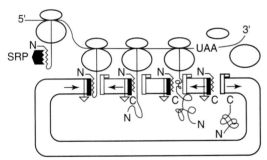

Figure 1–21. Translation of protein into endoplasmic reticulum according to the signal hypothesis. The ribosomes synthesizing a protein move along the mRNA from the 5′ to the 3′ end. When the signal peptide of a protein destined for secretion, the cell membrane, or lysosomes emerges from the large unit of the ribosome, it binds to a signal recognition particle (SRP), and this arrests further translation until it binds to the translocon on the endoplasmic reticulum. This frees the SRP, which is recycled in the cytoplasm. Binding to the ribosome receptor also occurs, and a tunnel opens, permitting the growing protein chain to enter the endoplasmic reticulum. The signal peptide is removed by signal peptidase. At the termination of protein synthesis, the two subunits of the ribosome dissociate and the carboxyl terminal enters the endoplasmic reticulum. N, amino end of protein; C, carboxyl end of protein. (Reproduced, with permission, from Perara E, Lingappa VR: Transport of proteins into and across the endoplasmic reticulum membrane. In: *Protein Transfer and Organelle Biogenesis.* Das RC, Robbins PW [editors]. Academic Press, 1988.)

Secreted Proteins

Many and perhaps all proteins that are secreted by cells are synthesized as larger proteins, and polypeptide sequences are cleaved off from them during maturation. In the case of the hormones, these larger forms are called **preprohormones** and **prohormones** (Figures 1–19 and 1–22). Parathyroid hormone (see Chapter 21) is a good example. It is synthesized as a molecule containing 115 amino acid residues (preproparathyroid hormone). The leader sequence, 25 amino acid residues at the amino terminal, is rapidly removed to form proparathyroid hormone. Before secretion, an additional 6 amino acids are removed from the amino terminal to form the secreted molecule. The function of the 6-amino-acid fragment is unknown.

Although most secreted polypeptides and proteins have a leader sequence that targets them to the endoplasmic reticulum and are secreted by exocytosis (see below), a growing list of proteins that are secreted lack a signal sequence. These include in humans the cytokines interleukin-1α (IL-1α) and IL-1β, three growth factors, and various factors involved in hemostasis. Secretion probably occurs via ATP-dependent membrane transporters. There is a large family of these ATP-binding-cassette (ABC) transport proteins, and they transport ions and other substances as well as proteins between organelles and across cell membranes. In general, they are made up of two cytoplasmic ATP-binding domains and two membrane domains, each of which probably spans the membrane and in general contains six long α-helical sequences (Figure 1–23). The **cystic fibrosis transmembrane conductance regulator (CFTR)** is one of those ABC transport proteins that also has a region for regulation by cAMP. It transports Cl⁻ and is abnormal in individuals with cystic fibrosis (see Chapter 37).

Protein Folding

Protein folding is an additional posttranslational modification. It is a complex process that is dictated primarily by the sequence of the amino acids in the polypeptide chain. In some instances, however, nascent proteins associate with other proteins called **chaperones,** which prevent inappropriate contacts with other proteins and ensure that the final "proper" conformation of the nascent protein is reached. Misfolded proteins and other proteins targeted for degradation are conjugated to **ubiquitin** and broken down in the organelles called 26S proteasomes (see Chapter 17).

Apoptosis

In addition to dividing and growing under genetic control, cells can die and be absorbed under genetic con-

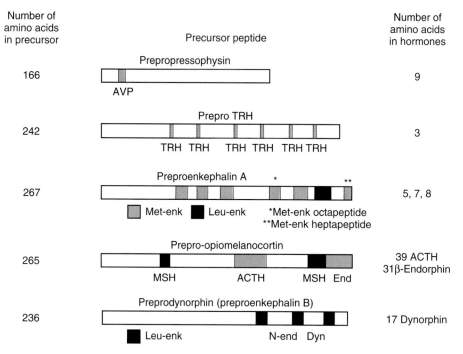

Number of amino acids in precursor	Precursor peptide	Number of amino acids in hormones
	Prepropressophysin	
166	AVP	9
	Prepro TRH	
242	TRH TRH TRH TRH TRH TRH	3
	Preproenkephalin A	
267	Met-enk Leu-enk *Met-enk octapeptide **Met-enk heptapeptide	5, 7, 8
	Prepro-opiomelanocortin	
265	MSH ACTH MSH End	39 ACTH 31β-Endorphin
	Preprodynorphin (preproenkephalin B)	
236	Leu-enk N-end Dyn	17 Dynorphin

Figure 1–22. Examples of large precursors (preprohormones) for small peptide hormones. See also Figure 14–11. TRH, thyrotropin-releasing hormone; AVP, arginine vasopressin; Met-enk, met-enkephalin; Leu-enk, leu-enkephalin; MSH, melanocyte-stimulating hormone; ACTH, adrenocorticotropic hormone; End, β-endorphin; Dyn, dynorphin; N-end, neoendorphin.

trol. This process is called **programmed cell death,** or **apoptosis** (Gr *apo* "away" + *ptosis* "fall"). It can be called "cell suicide" in the sense that the cell's own

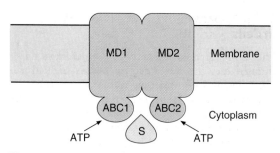

Figure 1–23. General structure of eukaryotic ABC transporter proteins that move ions, other substances, and proteins across membranes. ABC1 and ABC2, ATP-binding domains; MD1 and MD2, membrane domains; S, substrate. (Modified from Kuchler K, Thorner J: Secretion of peptides and proteins lacking hydrophobic signal sequences: The role of adenosine triphosphate-driven membrane translocators. Endocr Rev 1992;13:499.)

genes play an active role in its demise. It should be distinguished from necrosis ("cell murder"), in which healthy cells are destroyed by external processes such as inflammation.

Apoptosis is a very common process during development and in adulthood. In the central nervous system, large numbers of neurons are produced and then die during the remodeling that occurs during development and synapse formation (see Chapter 4). In the immune system, apoptosis gets rid of inappropriate clones of immunocytes (see Chapter 27) and is responsible for the lytic effects of glucocorticoids on lymphocytes (see Chapter 20). Apoptosis is also an important factor in processes such as removal of the webs between the fingers in fetal life and regression of duct systems in the course of sexual development in the fetus (see Chapter 23). In adults, it participates in the cyclic breakdown of the endometrium that leads to menstruation (see Chapter 23). In epithelia, cells that lose their connections to the basal lamina and neighboring cells undergo apoptosis. This is responsible for the death of the enterocytes sloughed off the tips of intestinal villi (see Chapter 26). Abnormal apoptosis probably occurs in autoimmune disease, neurodegenerative diseases, and cancer. It is interesting that apoptosis occurs in invertebrates, including nematodes and insects. However, its molecular mechanism is much more complex than that in vertebrates.

The final common pathway bringing about apoptosis is activation of **caspases,** a group of cysteine proteases. Thirteen of these have been characterized to date in mammals. They exist in cells as inactive proenzymes until activated by the cellular machinery. The net result is DNA fragmentation, cytoplasmic and chromatin condensation, and eventually membrane bleb formation, with cell breakup and removal of the debris by phagocytes.

Apoptosis can be triggered by external and internal stimuli. One ligand that activates receptors triggering apoptosis is **Fas,** a transmembrane protein that projects from natural killer cells and T lymphocytes (see Chapter 27) but also exists in a circulating from. Another is tumor necrosis factor.

Between initiating stimuli and caspase activation is a complex network of excitatory and inhibitory intracellular proteins. One of the important pathways goes through the mitochondria, which release **cytochrome c** and a protein called **smac/DIABLO.** Cytochrome c acts with several cytoplasmic proteins to facilitate caspase activation. Smac/DIABLO binds to several inhibiting proteins, lifting the inhibition of caspase-9 and thus increasing apoptotic activity.

Molecular Medicine

Fundamental research in molecular aspects of genetics, regulation of gene expression, and protein synthesis has been paying off in clinical medicine at a rapidly accelerating rate.

One early dividend was an understanding of the mechanisms by which antibiotics exert their effects. Almost all act by inhibiting protein synthesis at one or another of the steps described above. Antiviral drugs act in a similar way; for example, acyclovir and ganciclovir act by inhibiting DNA polymerase. Some of these drugs have this effect primarily in bacteria, but others inhibit protein synthesis in the cells of other animals, including mammals. This fact makes antibiotics of great value for research as well as for treatment of infections.

Single genetic abnormalities that cause over 600 human diseases have now been identified. Many of the diseases are rare, but others are more common and some cause conditions that are severe and eventually fatal. Examples include the defectively regulated Cl⁻ channel in cystic fibrosis (see above and Chapter 34) and the unstable **trinucleotide repeats** in various parts of the genome that cause Huntington's disease, the fragile X syndrome, and several other neurologic diseases (see Chapter 12).

Abnormalities in mitochondrial DNA can also cause human diseases such as Leber's hereditary optic neuropathy and some forms of cardiomyopathy. Not surprisingly, genetic aspects of cancer are probably receiving the greatest current attention. Some cancers are caused by **oncogenes,** genes which are carried in the genomes of cancer cells and are responsible for producing their malignant properties. These genes are derived by somatic mutation from closely related **proto-oncogenes,** which are normal genes that control growth. Over 100 oncogenes have been described. Another group of genes produce proteins that suppress tumors, and more than 10 of these **tumor suppressor genes** have been described. The most studied of these is the *P53* gene (*p53* in some publications) on human chromosome 17. The P53 protein produced by this gene triggers apoptosis. It is also a nuclear transcription factor that appears to increase production of a 21-kDa protein which blocks two cell cycle enzymes, slowing the cycle and permitting repair of mutations and other defects in DNA. The *P53* gene is mutated in up to 50% of human cancer patients, with the production of P53 proteins that fail to slow the cell cycle and permit other mutations in DNA to persist. The accumulated mutations eventually cause cancer.

Gene therapy is still in its infancy, but various ingenious approaches to the problem of getting genes into cells are now being developed. One that is already in clinical trials for some diseases involves removal of cells from the patient with the disease, transfection of the cells with normal genes in vitro, and reinjection of the cells into the patient as an autotransplant. Another is insertion of appropriate genes into relatively benign viruses that are then administered to patients to carry the genes to the cells they invade.

Golgi Apparatus & Vesicular Transport in Cells

The Golgi apparatus is a collection of membrane-enclosed sacs (cisterns) that are stacked like dinner plates (Figure 1–4). There are usually about six sacs in each apparatus, but there may be more. One or more Golgi apparatuses are present in all eukaryotic cells, usually near the nucleus. The Golgi apparatus is a polarized structure, with cis and trans sides (Figure 1–24). Membranous vesicles containing newly synthesized proteins bud off from the granular endoplasmic reticulum and fuse with the cistern on the cis side of the apparatus. The proteins are then passed via other vesicles to the middle cisterns and finally to the cistern on the trans side, from which vesicles branch off into the cytoplasm. From the trans Golgi, vesicles shuttle to the lysosomes and to the cell exterior via a constitutive and a nonconstitutive pathway, both involving exocytosis

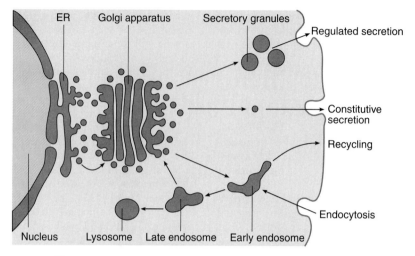

Figure 1–24. Pathways involved in protein processing in cells.

(see below). Mannose-6-phosphate receptors (MPRs) on the lysosomes capture hydrolytic enzymes destined for these organelles. Conversely, vesicles are pinched off from the cell membrane by endocytosis (see below) and pass to endosomes and eventually to the lysosomes. From the lysosomes, some of the proteins are shuttled back to the trans Golgi. The initial glycosylation of proteins occurs with the attachment of preformed oligosaccharides in the endoplasmic reticulum, but these oligosaccharides are altered to a variety of different carbohydrate moieties in the Golgi apparatus.

Quality Control

The processes involved in protein synthesis, folding, and migration to the various parts of the cell are so complex that it is remarkable that more errors and abnormalities do not occur. This is due to mechanisms at each level that are responsible for "quality control." Damaged DNA is detected and repaired or bypassed. The various RNAs are also checked during the translation process. Finally, when the protein chains are in the endoplasmic reticulum and Golgi apparatus, defective structure is detected and the abnormal proteins are degraded in lysosomes and proteasomes. The net result is a remarkable accuracy in the production of the proteins needed for normal body function.

TRANSPORT ACROSS CELL MEMBRANES

Transport across cell membranes is accomplished primarily by exocytosis, endocytosis, movement through ion channels, and primary and secondary active transport.

Exocytosis

Proteins that are secreted by cells move from the endoplasmic reticulum to the Golgi apparatus, and from the trans Golgi, they are extruded into secretory granules or vesicles (Figure 1–24). The granules and vesicles move to the cell membrane. Their membrane then fuses to the cell membrane (Figure 1–25), and the area of fusion breaks down. This leaves the contents of the granules or vesicles outside the cell and the cell membrane intact. The extrusion process is called **exocytosis.** It requires Ca^{2+} and energy, along with docking proteins (see below and Chapter 4).

Note that there are two pathways by which secretion from the cell occurs (Figure 1–24). In the **nonconstitutive pathway,** proteins from the Golgi apparatus initially enter secretory granules, where processing of prohormones to the mature hormones occurs before exocytosis. The other pathway, the **constitutive pathway,** involves the prompt transport of proteins to the cell membrane in vesicles, with little or no processing or storage. The nonconstitutive pathway is sometimes called the **regulated pathway,** but this term is misleading because the output of proteins by the constitutive pathway is also regulated.

Endocytosis

Endocytosis is the reverse of exocytosis. There are various types. **Phagocytosis** ("cell eating") is the process by which bacteria, dead tissue, or other bits of material visible under the microscope are engulfed by cells such as the polymorphonuclear leukocytes of the blood. The material makes contact with the cell membrane, which

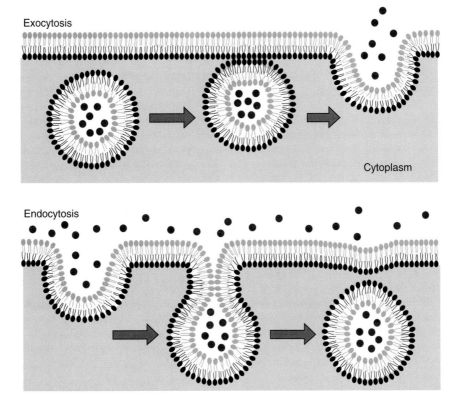

Exocytosis

Cytoplasm

Endocytosis

Figure 1–25. Exocytosis and endocytosis. Note that in exocytosis the cytoplasmic sides of two membranes fuse, whereas in endocytosis two noncytoplasmic sides fuse. (Reproduced, with permission, from Alberts B et al: *Molecular Biology of the Cell*, 4th ed. Garland Science, 2002.)

then invaginates. The invagination is pinched off, leaving the engulfed material in the membrane-enclosed vacuole and the cell membrane intact. **Pinocytosis** ("cell drinking") is essentially the same process, the difference being that the substances ingested are in solution and not visible under the microscope.

Endocytosis can be constitutive or clathrin-mediated. **Constitutive endocytosis** is not a specialized process, whereas **clathrin-mediated endocytosis** occurs at membrane indentations where the protein **clathrin** accumulates. Clathrin molecules have the shape of a triskelion, with three legs radiating from a central hub (Figure 1–26). As endocytosis progresses, the clathrin molecules form a geometric array that surrounds the endocytotic vesicle. At the neck of the vesicle, a guanosine triphosphatase protein called **dynamin** is involved, either directly or indirectly, in pinching off the vesicle, and this protein has been called a "pinchase." Once the complete vesicle is formed, the clathrin falls off and the three-legged proteins recycle to form another vesicle. The vesicle fuses with and dumps its contents into an

early endosome (Figure 1–24). From the early endosome, a new vesicle can bud off and return to the cell membrane (see Figure 4–4). Alternatively, the early endosome can become a **late endosome** and fuse with a

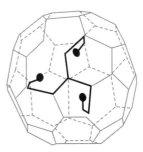

Figure 1–26. Clathrin molecule on the surface of an endocytotic vesicle. Note the characteristic triskelion shape and the fact that with other clathrin molecules it forms a net supporting the vesicle.

lysosome (Figure 1–24) in which the contents are digested by the lysosomal proteases.

Clathrin-mediated endocytosis is responsible for the internalization of many receptors and the ligands bound to them—including, for example, nerve growth factor and low-density lipoproteins (LDL; see Chapter 17). It also plays a major role in synaptic function (see Chapter 4).

It is apparent that exocytosis adds to the total amount of membrane surrounding the cell, and if membrane were not removed elsewhere at an equivalent rate, the cell would enlarge. However, removal of cell membrane occurs by endocytosis, and such exocytosis-endocytosis coupling maintains the surface area of the cell at its normal size.

Rafts & Caveolae

Some areas of the cell membrane are especially rich in cholesterol and sphingolipids and have been called **rafts.** These rafts are probably the precursors of flask-shaped membrane depressions called **caveolae** (little caves) when their walls become infiltrated with a protein called **caveolin** that resembles clathrin. Three isoforms of caveolin—caveolins-1, -2, and -3—have been identified. There is considerable debate about the functions of rafts and caveolae, with evidence that they are involved in cholesterol regulation and transcytosis (see below). However, mice in which the gene for caveolin-1 is knocked out are relatively healthy, with only some poorly understood blood vessel and pulmonary abnormalities.

Mechanisms Involved in Vesicle Transport

Considerable progress has been made in analyzing the biochemical basis of vesicle formation, transport, and docking in cells. It is convenient to consider transport within the cell and exocytosis and endocytosis together in this regard, since very similar processes are involved.

It now appears that all vesicles involved in transport have protein coats. In addition to the caveolin, there are at least four types of coat proteins: **AP-1 clathrin, AP-2 clathrin, COPI, and COPII.** In humans, 53 coat complex subunits have been identified. Vesicles that transport proteins from the trans Golgi to lysosomes have AP-1 clathrin coats, and endocytotic vesicles that transport to endosomes have AP-2 clathrin coats. Vesicles that transport between the endoplasmic reticulum and the Golgi have COPI and COPII coats. Certain amino acid sequences or attached groups on the transported proteins ticket the proteins for particular locations. For example, the amino acid sequence Asn-Pro-any amino acid-Tyr tickets transport from the cell surface to the endosomes, and, as noted above, mannose-6-phosphate

groups ticket transfer from the Golgi to MPR receptors on the lysosomes. Dynamin appears to be involved in the formation of vesicles in the Golgi as well as in endocytosis from the cell surface. Much of the work on proteins involved in vesicle docking and fusion has been done on synaptic vesicles, and these are discussed in Chapter 4. In general, vesicles move along microtubules, and when a vesicle reaches its target, it docks when **V-snare proteins** on the vesicle latch with **T-snare proteins** on the target. Each target has a unique set of T-snare proteins, thus ensuring that only the vesicle with the corresponding set of V-snare proteins will dock. Various small GTP-binding proteins of the Rab family (see below) are associated with the various types of vesicles. They appear to guide and facilitate orderly attachments of these vesicles. In humans, there are 60 Rab proteins and 35 snare proteins.

Distribution of Ions & Other Substances Across Cell Membranes

The unique properties of the cell membranes are responsible for the differences in the composition of intracellular and interstitial fluid. Specific values for one mammalian tissue are shown in Table 1–2. Average values for humans are shown in Figure 1–27.

Membrane Permeability & Membrane Transport Proteins

Small, nonpolar molecules—including O_2 and N_2—and small uncharged polar molecules such as CO_2 diffuse across the lipid membranes of cells. However, the membranes have very limited permeability to other substances. Instead, they cross the membranes by endocytosis and exocytosis and by passage through highly specific **transport proteins,** transmembrane proteins that form channels for ions or transport substances such as glucose, urea, and amino acids. The limited permeability applies even to water, with simple diffusion being supplemented throughout the body with various water channels (**aquaporins;** see Chapters 14 and 38). For reference, the sizes of ions and other biologically important substances are summarized in Table 1–5.

An important technique that has permitted major advances in our knowledge about transport proteins is **patch clamping.** A micropipette is placed on the membrane of a cell and forms a tight seal to the membrane. The patch of membrane under the pipette tip usually contains only a few transport proteins, and they can be studied in detail (Figure 1–28). The cell can be left intact (**cell-attached patch clamp).** Alternatively, the patch can be pulled loose from the cell, forming an **inside-out patch.** A third alternative is to suck out the

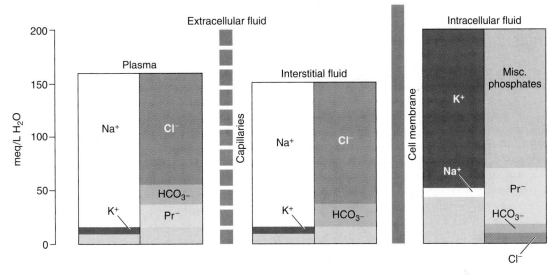

Figure 1–27. Electrolyte composition of human body fluids. Note that the values are in meq/L of water, not of body fluid. (Reproduced, with permission, from Johnson LR [editor]: *Essential Medical Physiology.* Raven Press, 1992.)

patch with the micropipette still attached to the rest of the cell membrane, providing direct access to the interior of the cell (**whole cell recording**).

Table 1–5. Size of hydrated ions and other substances of biologic interest.[1]

Substance	Atomic or Molecular Weight	Radius (nm)
Cl⁻	35	0.12
K⁺	39	0.12
H₂O	18	0.12
Ca²⁺	40	0.15
Na⁺	23	0.18
Urea	60	0.23
Li⁺	7	0.24
Glucose	180	0.38
Sucrose	342	0.48
Inulin	5000	0.75
Albumin	69,000	7.50

[1]Data from Moore EW: *Physiology of Intestinal Water and Electrolyte Absorption.* American Gastroenterological Association, 1976.

Some transport proteins are simple aqueous **ion channels,** though many of these have special features that make them effective for a given substance such as Ca^{2+} or, in the case of aquaporins, for water. Some of these transport proteins are continuously open, whereas others are **gated;** ie, they have gates that open or close. Some are gated by alterations in membrane potential (**voltage-gated**), whereas others are opened or closed when they bind a ligand (**ligand-gated**). The ligand is often external, eg, a neurotransmitter or a hormone. However, it can also be internal; intracellular Ca^{2+}, cAMP, lipids, or one of the G proteins produced in cells (see below) can bind directly to channels and activate them. Some channels are also opened by mechanical stretch, and these mechanosensitive channels play an important role in cell movement. A typical voltage-gated channel is the Na^+ channel (see below), and a typical ligand-gated channel is the acetylcholine receptor (see Chapter 4).

Other transport proteins are **carriers** that bind ions and other molecules and then change their configuration, moving the bound molecule from one side of the cell membrane to the other. Molecules move from areas of high concentration to areas of low concentration (down their **chemical gradient**), and cations move to negatively charged areas whereas anions move to positively charged areas (down their **electrical gradient**). When carrier proteins move substances in the direction of their chemical or electrical gradients, no energy input is required and the process is called **facilitated diffusion.** A typical example is glucose transport by the

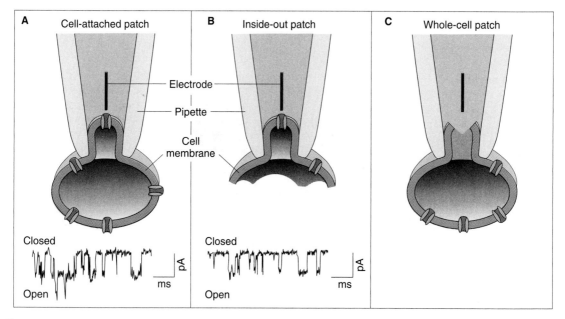

Figure 1–28. Types of patch clamps used to study activity of ion channels across a cell membrane. In A and B, the changes in membrane current with time are also shown. (Modified from Ackerman MJ, Clapham DE: Ion channels: Basic science and clinical disease. N Engl J Med 1997;336:1575.)

glucose transporter, which moves glucose down its concentration gradient from the ECF to the cytoplasm of the cell (see Chapter 19). Other carriers transport substances against their electrical and chemical gradients. This form of transport requires energy and is called **active transport.** In animal cells, the energy is provided almost exclusively by hydrolysis of ATP (see above and Chapter 17). Not surprisingly, therefore, the carrier molecules are ATPases, enzymes that catalyze the hydrolysis of ATP. One of these ATPases is **sodium-potassium-activated adenosine triphosphatase** (Na^+-K^+ ATPase), which is also known as the Na^+-K^+ pump. There are also H^+-K^+ ATPases in the gastric mucosa (see Chapter 26) and the renal tubules (see Chapter 38). Ca^{2+} ATPase pumps Ca^{2+} out of cells. Proton ATPases acidify many intracellular organelles, including parts of the Golgi complex and lysosomes. F-ATPases are present in mitochondria and synthesize ATP. Some cell membranes contain ATPases that transport Ca^{2+}.

Some of the transport proteins are called **uniports,** because they transport only one substance. Others are called **symports,** because transport requires the binding of more than one substance to the transport protein and the substances are transported across the membrane together. An example is the symport in the intestinal mucosa that is responsible for the cotransport by facilitated diffusion of Na^+ and glucose from the intestinal lumen into mucosal cells (see Chapter 25). Other transporters are called **antiports** because they exchange one substance for another. The Na^+-K^+ ATPase mentioned above is a typical antiport; it moves three Na^+ out of the cell in exchange for each two K^+ that it moves into the cell.

Ion Channels

There are ion channels for K^+, Na^+, Ca^{2+}, and Cl^-, and each exists in multiple forms with diverse properties. Most are made up of identical or very similar subunits. A Ca^{2+} channel and a Na^+ channel are shown in extended diagrammatic form in Figure 1–29, and an acetylcholine-gated ion channel is shown in Figure 4–18. Figure 1–30 shows the multiunit structure of various channels in diagrammatic cross-section. Most K^+ channels are tetramers, with each of the four subunits forming part of the pore through which K^+ ions pass. Fast-inactivating voltage-gated K^+ channels have a unique feature. Each of the tetramers has a polypeptide ball structure on the end of a polypeptide chain, and the ball moves into the channel, producing inactivation (Figure 1–31). There are ball-and-chain structures on all four tetramers even though there is only one K^+ pore. In the acetylcholine ion channel and other ligand-gated cation or anion channels, five subunits

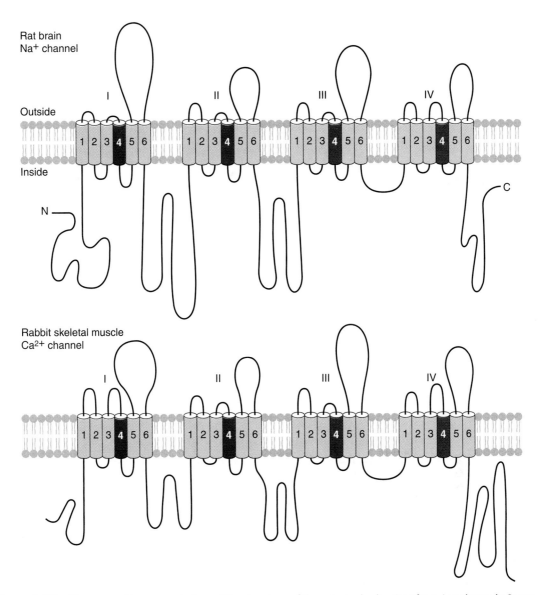

Figure 1–29. Diagrammatic representation of the structure of the principal subunits of two ion channels. Some Na+ and Ca2+ channels have additional subunits. The Arabic numbers identify the α-helical domains that cross the cell membrane. H5 domain not shown. (After Catterall WK. Modified and reproduced from Hall ZW: *An Introduction to Molecular Neurobiology.* Sinauer, 1992.)

make up the pore. Members of the CLC family of Cl⁻ channels are dimers, but they have two pores, one in each subunit. Finally, aquaporins are tetramers with a water pore in each of the subunits. Recently, a number of ion channels with intrinsic enzyme activity have been cloned.

More than 30 different voltage-gated or cyclic nucleotide-gated Na+ and Ca2+ channels of this type have been described. The toxins tetrodotoxin (TTX) and saxitoxin (STX) bind to the Na+ channels and block them. The number and distribution of the Na+ channels can be determined by tagging them with labeled TTX or STX carrying a suitable label and analyzing the distribution of the label.

Another family of Na+ channels with a different structure has been found in the apical membranes of

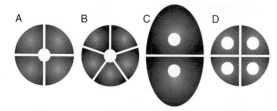

Figure 1–30. Different ways in which ion channels form pores. Many K+ channels are tetramers (A), with each protein subunit forming part of the channel. In ligand-gated cation and anion channels (B) such as the acetylcholine receptor, five identical or very similar subunits form the channel. Cl− channels from the CLC family are dimers (C), with an intracellular pore in each subunit. Aquaporin water channels (D) are tetramers with an intracellular channel in each subunit. (Reproduced, with permission, from Jentsch TJ: Chloride channels are different. Nature 2002;415:276.)

epithelial cells in the kidneys, colon, lungs, and brain. Those **epithelial sodium channels (ENaCs)** are made up of three subunits encoded by three different genes. Each of the subunits probably spans the membrane twice, and the amino terminal and carboxyl terminal are located inside the cell. The α subunit transports Na+, whereas the β and γ subunits do not. However, the addition of the β and γ subunits increases Na+ transport through the α subunit. ENaCs are inhibited by the diuretic amiloride, which binds to the α subunit, and they used to be called **amiloride-inhibitable Na+** channels. The ENaCs in the kidney play an im-

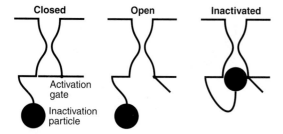

Figure 1–31. Fast inactivation of voltage-gated K+ channels. Depolarization opens an activation gate, and then the amino terminal of the subunit swings into the channel, stopping conductance despite continuing depolarization. After repolarization of the membrane, the resting conformation of the protein is restored. (Reproduced, with permission, from Antz C, Fakler B: Fast inactivation of voltage-gated K+ channels: from cartoon to structure. News Physiol Sci 1998;13:177.)

portant role in the regulation of ECF volume by aldosterone (see Chapter 38). ENaC knockout mice are born alive but promptly die because they cannot pump Na+ and hence water out of their lungs.

There are several types of Cl− channels in humans. The CLC dimeric channels described above (Figure 1–29) are found in plants, bacteria, and animals, and there are nine different CLC genes in humans. Other Cl− channels have the same pentameric form as the acetylcholine receptor; examples include the $GABA_A$ and glycine receptors in the CNS (see Chapter 4). The CFTR receptor that is mutated in cystic fibrosis is also a Cl− channel. Ion channel mutations cause a variety of **channelopathies**—diseases that mostly affect muscle and brain tissue and produce episodic paralyses or convulsions.

Na+-K+ ATPase

As noted above, Na+-K+ ATPase catalyzes the hydrolysis of ATP to ADP and uses the energy to extrude three Na+ from the cell and take two K+ into the cell for each mole of ATP hydrolyzed. It is an **electrogenic pump** in that it moves three positive charges out of the cell for each two that it moves in, and it is therefore said to have a **coupling ratio** of 3:2. It is found in all parts of the body. Its activity is inhibited by ouabain and related digitalis glycosides used in the treatment of heart failure. It is a heterodimer made up of an α subunit with a molecular weight of approximately 100,000 and a β subunit with a molecular weight of approximately 55,000. Both extend through the cell membrane (Figure 1–32). Separation of the subunits eliminates activity. However, the β subunit is a glycoprotein, whereas Na+ and K+ transport occur through the α subunit. The β subunit has a single membrane-spanning domain and three extracellular glycosylation sites, all of which appear to have attached carbohydrate residues. These residues account for one-third of its molecular weight. The α subunit probably spans the cell membrane ten times, with the amino and carboxyl terminals both located intracellularly. This subunit has intracellular Na+- and ATP-binding sites and a phosphorylation site; it also has extracellular binding sites for K+ and ouabain. When Na+ binds to the α subunit, ATP also binds and is converted to ADP, with a phosphate being transferred to Asp 376, the phosphorylation site. This causes a change in the configuration of the protein, extruding Na+ into the ECF. K+ then binds extracellularly, dephosphorylating the α subunit, which returns to its previous conformation, releasing K+ into the cytoplasm.

The α and β subunits are heterogeneous, with $α_1$, $α_2$, and $α_3$ subunits and $β_1$, $β_2$, and $β_3$ subunits described so far. The $α_1$ isoform is found in the membranes of most cells, whereas $α_2$ is present in muscle,

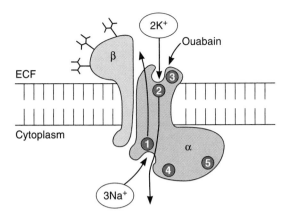

Figure 1–32. Na^+-K^+ ATPase. The intracellular portion of the α subunit has a Na^+-binding site (1), a phosphorylation site (4), and an ATP-binding site (5). The extracellular portion has a K^+-binding site (2) and a ouabain-binding site (3). (From Horisberger J-D et al: Structure-function relationship of Na-K-ATPase. Annu Rev Physiol 1991;53:565. Reproduced, with permission, from the *Annual Review of Physiology*, vol. 53. Copyright © 1991 by Annual Reviews Inc.)

heart, adipose tissue, and brain, and $α_3$ is present in heart and brain. The $β_1$ subunit is widely distributed but is absent in certain astrocytes, vestibular cells of the inner ear, and glycolytic fast-twitch muscles. The fast-twitch muscles contain only $β_2$ subunits. The different α and β subunit structures of Na^+-K^+ ATPase in various tissues probably represents specialization for specific tissue functions.

Regulation of Na^+-K^+ ATPase Activity

The amount of Na^+ normally found in cells is not saturating, so if it increases, the pump activity and hence the amount of Na^+ extruded from the cell increase. Pump activity is affected by second messengers produced in cells, including cAMP and diacylglycerol (DAG; see below); the magnitude and direction of the observed effects vary with the experimental conditions. Thyroid hormones increase pump activity by a genomic action to increase the formation of Na^+-K^+ ATPase molecules. Aldosterone also increases the number of pumps, although this effect is probably secondary (see Chapters 20 and 38). Dopamine in the kidney inhibits the pump by phosphorylating it, causing a natriuresis. Insulin increases pump activity, probably by a variety of different mechanisms. Finally, Na^+-K^+ ATPase is anchored in the membrane by the cytoskeleton, and it is interesting that G-actin (see above) increases pump activity.

Secondary Active Transport

In many situations, the active transport of Na^+ is coupled to the transport of other substances (**secondary active transport**). For example, the luminal membranes of mucosal cells in the small intestine contain a symport that transports glucose into the cell only if Na^+ binds to the protein and is transported into the cell at the same time. From the cells, the glucose enters the blood. The electrochemical gradient for Na^+ is maintained by the active transport of Na^+ out of the mucosal cell into ECF (see Chapter 25). Other examples are shown in Figure 1–33. In the heart, Na^+-K^+ ATPase indirectly affects Ca^{2+} transport. An antiport in the membranes of cardiac muscle cells normally exchanges intracellular Ca^{2+} for extracellular Na^+. The rate of this exchange is proportionate to the concentration gradient for Na^+ across the cell membrane. If the operation of the Na^+-K^+ ATPase is inhibited (eg, by ouabain), the intracellular Na^+ concentration increases, the Na^+ gradient across the cell membrane decreases, and Ca^{2+} extrusion decreases. The resulting increase in intracellular Ca^{2+} concentra-

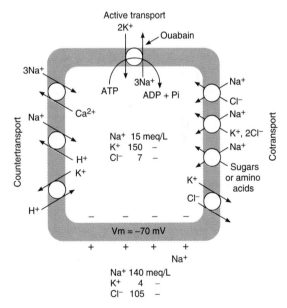

Figure 1–33. Composite diagram of main secondary effects of active transport of Na^+ and K^+. Na^+-K^+ ATPase converts the chemical energy of ATP hydrolysis into maintenance of an inward gradient for Na^+ and an outward gradient for K^+. The energy of the gradients is used for countertransport, cotransport, and maintenance of the membrane potential. (Reproduced, with permission, from Skou JC: The Na-K pump. News Physiol Sci 1992;7:95.)

tion facilitates the contraction of cardiac muscle (positively inotropic effect; see Chapter 3).

Active transport of Na^+ and K^+ is one of the major energy-using processes in the body. On the average, it accounts for about 24% of the energy utilized by cells, and in neurons it accounts for 70%. Thus, it accounts for a large part of the basal metabolism.

Transport Across Epithelia

In the gastrointestinal tract, the pulmonary airways, the renal tubules, and other structures, substances enter one side of a cell and exit another, producing movement of the substance from one side of the epithelium to the other. For transepithelial transport to occur, the cells need to be bound by tight junctions and, obviously, have different ion channels and transport proteins in different parts of their membranes. Most of the instances of secondary active transport cited in the preceding paragraph involve transepithelial movement of ions and other molecules.

THE CAPILLARY WALL

Filtration

The capillary wall separating plasma from interstitial fluid is different from the cell membranes separating interstitial fluid from intracellular fluid because the pressure difference across it makes **filtration** a significant factor in producing movement of water and solute. By definition, filtration is the process by which fluid is forced through a membrane or other barrier because of a difference in pressure on the two sides.

Oncotic Pressure

The structure of the capillary wall varies from one vascular bed to another (see Chapter 30). However, in skeletal muscle and many other organs, water and relatively small solutes are the only substances that cross the wall with ease. The apertures in the junctions between the endothelial cells are too small to permit plasma proteins and other colloids to pass through in significant quantities. The colloids have a high molecular weight but are present in large amounts. Small amounts cross the capillary wall by vesicular transport (see below), but their effect is slight. Therefore, the capillary wall behaves like a membrane impermeable to colloids, and these exert an osmotic pressure of about 25 mm Hg. The colloid osmotic pressure due to the plasma colloids is called the **oncotic pressure.** Filtration across the capillary membrane as a result of the hydrostatic pressure head in the vascular system is opposed by the oncotic pressure. The way the balance between the hydrostatic and oncotic pressures controls exchanges across the capillary wall is considered in detail in Chapter 30.

Transcytosis

There are vesicles in the cytoplasm of endothelial cells, and tagged protein molecules injected into the bloodstream have been found in the vesicles and in the interstitium. This indicates that small amounts of protein are transported out of capillaries across endothelial cells by endocytosis on the capillary side followed by exocytosis on the interstitial side of the cells. The transport mechanism makes use of coated vesicles that appear to be coated with caveolin and is called **transcytosis, vesicular transport,** or **cytopempsis.**

INTERCELLULAR COMMUNICATION

Cells communicate with each other via chemical messengers. Within a given tissue, some messengers move from cell to cell via gap junctions (see above) without entering the ECF. In addition, cells are affected by chemical messengers secreted into the ECF. These chemical messengers bind to protein receptors on the surface of the cell or, in some instances, in the cytoplasm or the nucleus, triggering sequences of intracellular changes that produce their physiologic effects. There are three general types of intercellular communication mediated by messengers in the ECF: (1) **neural communication,** in which neurotransmitters are released at synaptic junctions from nerve cells and act across a narrow synaptic cleft on a postsynaptic cell (see Chapter 4); (2) **endocrine communication,** in which hormones and growth factors reach cells via the circulating blood (see Chapters 18–24); and (3) **paracrine communication,** in which the products of cells diffuse in the ECF to affect neighboring cells that may be some distance away (Figure 1–34). In addition, cells secrete chemical messengers that in some situations bind to receptors on the same cell, ie, the cell that secreted the messenger (**autocrine communication**). The chemical messengers include amines, amino acids, steroids, polypeptides, and in some instances lipids, purine nucleotides, and pyrimidine nucleotides. It is worth noting that in various parts of the body, the same chemical messenger can function as a neurotransmitter, a paracrine mediator, a hormone secreted by neurons into the blood (neural hormone), and a hormone secreted by gland cells into the blood.

An additional form of intercellular communication is called **juxtacrine communication.** Some cells express multiple repeats of growth factors such as **transforming growth factor alpha (TGFα)** extracellularly on transmembrane proteins that provide an anchor to the cell. Other cells have TGFα receptors. Conse-

	GAP JUNCTIONS	SYNAPTIC	PARACRINE AND AUTOCRINE	ENDOCRINE
	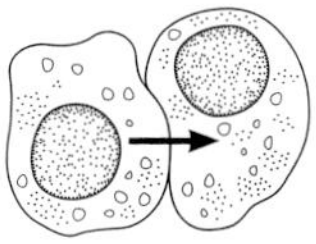	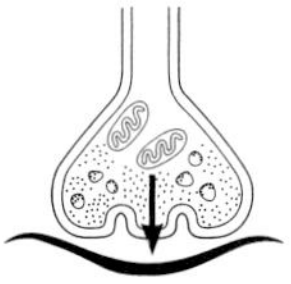	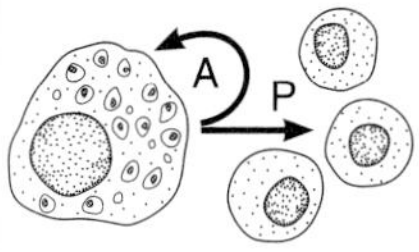	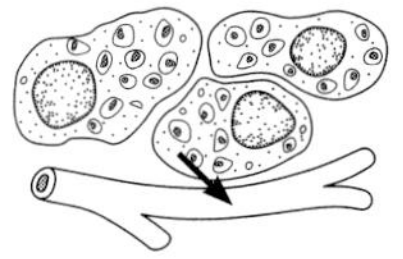
Message transmission	Directly from cell to cell	Across synaptic cleft	By diffusion in interstitial fluid	By circulating body fluids
Local or general	Local	Local	Locally diffuse	General
Specificity depends on	Anatomic location	Anatomic location and receptors	Receptors	Receptors

Figure 1–34. Intercellular communication by chemical mediators. A, autocrine; P, paracrine.

quently, TGFα anchored to a cell can bind to a TGFα receptor on another cell, linking the two. This could be important in producing local foci of growth in tissues.

Radioimmunoassay

Antibodies to the polypeptides and proteins are readily produced, and, by using special techniques, it is possible to make antibodies to the other chemical messengers as well. The antibodies can be used to measure the messengers in body fluids and in tissue extracts by **radioimmunoassay.** This technique depends on the fact that the naturally occurring, unlabeled ligand and added radioactive ligand compete to bind to an antibody to the ligand. The greater the amount of unlabeled ligand in the specimen being analyzed, the more it competes and the smaller the amount of radioactive ligand that binds to the antibody. Radioimmunoassays are extensively used in research and in clinical medicine.

Receptors for Hormones, Neurotransmitters, & Other Ligands

Many of the receptors for chemical messengers have now been isolated and characterized. These proteins are not static components of the cell, but their numbers increase and decrease in response to various stimuli, and their properties change with changes in physiologic conditions. When a hormone or neurotransmitter is present in excess, the number of active receptors generally decreases **(down-regulation),** whereas in the presence of a deficiency of the chemical messenger, there is an increase in the number of active receptors **(up-regulation).** Angiotensin II in its actions on the adrenal cortex is an exception; it increases rather than decreases the number of its receptors in the adrenal. In

the case of receptors in the membrane, receptor-mediated endocytosis is responsible for down-regulation in some instances; ligands bind to their receptors, and the ligand-receptor complexes move laterally in the membrane to coated pits, where they are taken into the cell by endocytosis **(internalization).** This decreases the number of receptors in the membrane. Some receptors are recycled after internalization, whereas others are replaced by de novo synthesis in the cell. Another type of down-regulation is desensitization, in which receptors are chemically modified in ways that make them less responsive (see Chapter 4).

Mechanisms by Which Chemical Messengers Act

The principal mechanisms by which chemical messengers exert their intracellular effects are summarized in Table 1–6. Ligands such as acetylcholine bind directly to ion channels in the cell membrane, changing their conductance. Thyroid and steroid hormones, 1,25-dihydroxycholecalciferol, and retinoids enter cells and act on one or another member of a family of structurally related cytoplasmic or nuclear receptors. The activated receptor binds to DNA and increases transcription of selected mRNAs. Many other ligands in the ECF bind to receptors on the surface of cells, and many of them trigger the release of intracellular mediators such as cAMP, IP_3, and DAG (see below) that initiate changes in cell function. Consequently, the extracellular ligands are called **"first messengers"** and the intracellular mediators are called **"second messengers."**

Second messengers bring about many short-term changes in cell function by altering enzyme function, triggering exocytosis, etc, but they also alter transcription of various genes. They do this in part by activating

Table 1–6. Principal mechanisms by which chemical messengers in the ECF bring about changes in cell function.

Mechanism	Examples
Open or close ion channels in cell membrane	Acetylcholine on nicotinic cholinergic receptor; norepinephrine on K^+ channel in the heart
Act via cytoplasmic or nuclear receptors to increase transcription of selected mRNAs	Thyroid hormones, retinoic acid, steroid hormones
Activate phospholipase C with intracellular production of DAG, IP_3, and other inositol phosphates	Angiotensin II, norepinephrine via α_1-adrenergic receptor, vasopressin via V_1 receptor
Activate or inhibit adenylyl cyclase, causing increased or decreased intracellular production of cAMP	Norepinephrine via β_1-adrenergic receptor (increased cAMP); norepinephrine via α_2-adrenergic receptor (decreased cAMP)
Increase cGMP in cell	ANP; NO (EDRF)
Increase tyrosine kinase activity of cytoplasmic portions of transmembrane receptors	Insulin, EGF, PDGF, M-CSF
Increase serine or threonine kinase activity	TGFβ, MAPKs

transcription factors already present in the cell, and these activated factors induce the transcription of immediate-early genes (Figure 1–20). The transcription factors that are the products of the immediate-early genes then activate other genes which produce more long-term effects.

When activated, many of the membrane receptors initiate release of second messengers or other intracellular events via GTP-binding proteins (G proteins; see below). The second messengers generally activate **protein kinases,** enzymes that catalyze the phosphorylation of tyrosine or serine and threonine residues in proteins. More than 300 protein kinases have been described. Some of the principal ones that are important in mammals are summarized in Table 1–7. Addition of phosphate groups changes the configuration of the proteins, altering their functions and consequently the functions of the cell. In some instances, eg, the insulin receptor, the intracellular portions of the receptors themselves are protein kinases, and in some instances, they phosphorylate themselves (autophosphorylation). Other receptors, eg, cytokine receptors, are not protein kinases themselves but readily initiate phosphorylation of many intracellular proteins. Obviously, **phosphatases** are also important, since removal of a phosphate group inactivates some transport proteins or enzymes whereas it activates others.

Stimulation of Transcription

When thyroid and steroid hormones, 1,25-dihydroxycholecalciferol, and retinoids bind to their receptors inside cells, the conformation of the receptor protein is changed and a DNA-binding domain is exposed (Figure 1–35). The receptor-hormone complex moves to

Table 1–7. Principal protein kinases.

Phosphorylate serine and/or threonine residues
Calmodulin-dependent
Myosin light-chain kinase
Phosphorylase kinase
Ca^{2+}/calmodulin kinase I
Ca^{2+}/calmodulin kinase II
Ca^{2+}/calmodulin kinase III
Calcium-phospholipid-dependent
Protein kinase C (seven subspecies)
Cyclic nucleotide-dependent
cAMP-dependent kinase (protein kinase A; two subspecies)
cGMP-dependent kinase
Phosphorylate tyrosine residues
Insulin receptor, EGF receptor, PDGF receptor, and M-CSF receptor have tyrosine kinase activity

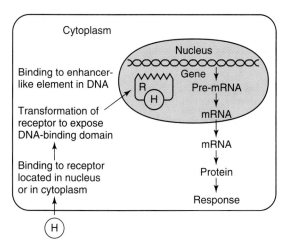

Figure 1–35. Mechanism of action of steroid and thyroid hormones. H, hormone; R, receptor.

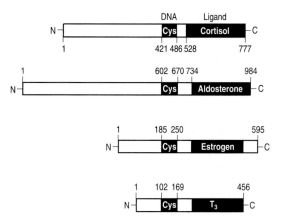

Figure 1–36. Structure of human glucocorticoid, mineralocorticoid, α-estrogen, and β-T₃ receptors. Note that each receptor has a cysteine-rich DNA-binding domain and a ligand-binding domain at or near the carboxyl terminal, with considerable variability in the amino terminal part of the protein. The numbers identify amino acid residues.

DNA, where it binds to enhancer elements in the untranslated 5′-flanking portions of certain genes. The estrogen and the triiodothyronine (T_3) receptors bind hormones in the nucleus. The T_3 receptors also bind thyroxine (T_4), but with less affinity. The glucocorticoid receptor is located mainly in the cytoplasm but migrates promptly to the nucleus as soon as it binds its ligand. The initial location of the other receptors that act in this fashion is unsettled. In any case, binding of the receptor-hormone complex to DNA increases the transcription of mRNAs encoded by the gene to which it binds. The mRNAs are translated in the ribosomes, with the production of increased quantities of proteins that alter cell function.

At least for the glucocorticoid, estrogen, and progesterone receptors, the receptor is bound to the **heat shock protein** Hsp90 and other proteins in the absence of the steroid, and it appears that the heat shock protein covers the DNA-binding domain. When the steroid binds to the receptor, the conformation change releases the heat shock protein, exposing the DNA-binding domain.

Heat shock proteins are a group of intracellular proteins whose amounts increase when cells are exposed to heat and other stresses, and they help the cells survive a variety of stresses. Consequently, it is probably more appropriate to call them **stress proteins.**

Structure of Receptors

The structures of the human glucocorticoid and mineralocorticoid receptors are shown in Figure 1–36. Two estrogen receptors (α and β) and two T_3 receptors (α and β) have been identified; the α estrogen receptor

and the β T_3 receptor are shown in the figure. All these receptors are part of a superfamily of receptors that have in common a highly conserved cysteine-rich DNA-binding domain; a ligand-binding domain at or near the carboxyl terminal of the receptor; and a relatively variable, poorly conserved amino terminal region. When a ligand binds to one of them, it becomes a transcription factor and binds to DNA via zinc fingers (see above). Other receptors in the family include the receptors for progesterone, androgen, and 1,25-dihydroxycholecalciferol. Many other factors that regulate genes act via receptors of this type in species ranging from fruit flies to humans, and over 70 members of this receptor superfamily have been described. Ligands are now known for about half of these, but the remaining half are **orphan receptors,** for which the ligands are still unidentified. Retinoic acid, which is a derivative of retinol (vitamin A), has an extensive role in fetal development, and there are three retinoic acid receptors, α, β, and γ, encoded by two families of retinoic acid receptors, **RAR** and **RXR**, each with α, β, and γ forms. T_3 receptors form homodimers before binding to DNA, but heterodimers with retinoic receptors also form and bind, and their actions are complex (see Chapter 18).

Rapid Actions of Steroids

Some of the actions of steroids are much more rapid than those known to be mediated via binding to DNA. Examples include the rapid increase in the Ca^{2+} concentration in sperm heads that is produced by proges-terone and prompt steroid-induced alteration in the functions of various neurons. This has led to the hypothesis that there are **nongenomic actions** of steroids which are mediated by putative membrane receptors and second messengers inside the cells (see below). There is molecular biologic evidence for the existence of these receptors, though detailed information about them is still lacking. Steroids also bind to GABA$_A$ receptors, facilitating their action (see Chapter 4).

Intracellular Ca²⁺

Ca^{2+} regulates a very large number of physiologic processes that are as diverse as proliferation, neural signaling, learning, contraction, secretion, and fertilization, so regulation of intracellular Ca^{2+} is of great importance. The free Ca^{2+} concentration in the cytoplasm at rest is maintained at about 100 nmol/L. The Ca^{2+} concentration in the interstitial fluid is about 12,000 times the cytoplasmic concentration, ie, 1,200,000 nmol/L, so there is a marked inwardly directed concentration gradient as well as an inwardly directed electrical gradient. Much of the intracellular Ca^{2+} is bound by the endoplasmic reticulum and other organelles (Figure 1–37), and these organelles provide a store from which Ca^{2+} can be mobilized via ligand-gated channels to increase the concentration of free Ca^{2+} in the cytoplasm. Increased cytoplasmic Ca^{2+} binds to and activates calcium-binding proteins, and these in turn activate a number of protein kinases.

Ca^{2+} enters cells through many different Ca^{2+} channels. Some of these are ligand-gated and others are voltage-gated. There appear to be stretch-activated channels as well. The voltage-gated Ca^{2+} channels are often divided into T (transient) or L (long-lasting) types depending on whether they do or do not inactivate during maintained depolarization.

Ca^{2+} is pumped out of cells in exchange for two H^+ by a Ca^{2+}-H^+ ATPase, and it is transported out of cells by an antiport driven by the Na^+ gradient that exchanges three Na^+ for each Ca^{2+}.

Many second messengers act by increasing the cytoplasmic Ca^{2+} concentration. The increase is produced by releasing Ca^{2+} from intracellular stores—primarily the endoplasmic reticulum—or by increasing the entry of Ca^{2+} into cells, or by both mechanisms. IP_3 is the major second messenger that causes Ca^{2+} release from the endoplasmic reticulum. In many tissues, transient release of Ca^{2+} from internal stores into the cytoplasm triggers opening of a population of Ca^{2+} channels in the cell membrane (**store-operated Ca²⁺ channels; SOCCs**). The resulting Ca^{2+} influx replenishes the intracellular Ca^{2+} supply and refills the endoplasmic reticulum. The exact identity of the SOCCs is still unknown, and there is debate about the signal from the endoplasmic reticulum that opens them. However, evidence is accumulating that IP_3 is responsible for both the internal release from the endoplasmic reticulum and the activation of the SOCCs.

Calcium-Binding Proteins

Many different Ca^{2+}-binding proteins have been described, including **troponin, calmodulin,** and **calbindin.** Troponin is the Ca^{2+}-binding protein involved in contraction of skeletal muscle (see Chapter 3). Calmodulin contains 148 amino acid residues (Figure 1–38) and has four Ca^{2+}-binding domains. It is unique in that residue 115 is trimethylated, and it is extensively conserved, being found in plants as well as animals. When calmodulin binds Ca^{2+}, it is capable of activating five different calmodulin-dependent kinases (Table 1–7). One of these is **myosin light-chain kinase,** which phosphorylates myosin. This brings about contraction in smooth muscle. Another is **phosphorylase kinase,** which activates phosphorylase (see Chapter 17). Ca^{2+}/calmodulin kinases I and II are concerned with synaptic function, and Ca^{2+}/calmodulin kinase III is concerned with protein synthesis. Another calmodulin-activated protein is **calcineurin,** a phosphatase that inactivates Ca^{2+} channels by dephosphorylating them. It also plays a role in activating T cells

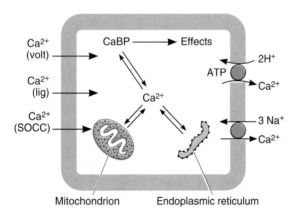

Figure 1–37. Ca^{2+} metabolism in mammalian cells. Ca^{2+} is stored in the endoplasmic reticulum and mitochondria and can be released from them to replenish cytoplasmic Ca^{2+}. Calcium-binding proteins (CaBP) bind cytoplasmic Ca^{2+} and, when activated in this fashion, bring about a variety of physiologic effects. Ca^{2+} enters the cells via voltage-gated (volt) and ligand-gated (lig) Ca^{2+} channels and SOCCs. It is transported out of the cell by a Ca^{2+}-H^+ ATPase and an Na^+-Ca^{2+} antiport.

Figure 1–38. Structure of calmodulin from bovine brain. Single-letter abbreviations are used for the amino acid residues (Table 17–2). Note the four calcium domains (dark residues) flanked on either side by stretches of α helix. (Reproduced, with permission, from Cheung WY: Calmodulin: An overview. Fed Proc 1982;41:2253.)

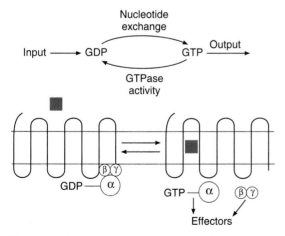

Figure 1–39. Heterotrimeric G proteins. **Top:** Summary of overall reaction. **Bottom:** When the ligand (square) binds to the serpentine receptor in the cell membrane, GTP replaces GDP on the α subunit. GTP-α separates from the βγ subunit and GTP-α and βγ both activate various effectors, producing physiologic effects. The intrinsic GTPase activity of GTP-α then converts GTP to GDP, and the α, β, and γ subunits reassociate.

and is inhibited by some immunosuppressants (see Chapter 27).

Mechanisms of Diversity of Ca²⁺ Actions

It may seem difficult to understand how intracellular Ca^{2+} can have so many varied effects as a second messenger. Part of the explanation is that Ca^{2+} may have different effects at low and at high concentrations. The ion may be in high concentration at the site of its release from an organelle or a channel (**Ca²⁺ sparks**) and at a subsequent lower concentration after it diffuses throughout the cell. Some of the changes it produces can outlast the rise in intracellular Ca^{2+} concentration because of the way it binds to some of the Ca^{2+}-binding proteins. In addition, once released, intracellular Ca^{2+} concentrations frequently oscillate at regular intervals, and there is evidence that the frequency and, to a lesser extent, the amplitude of those oscillations codes information for effector mechanisms. Finally, increases in intracellular Ca^{2+} concentration can spread from cell to cell in waves, producing coordinated events such as the rhythmic beating of cilia in epithelial tissue.

G Proteins

A common way to translate a signal to a biologic effect inside cells is by way of nucleotide regulatory proteins (**G proteins**) that bind GTP. GTP is the guanosine analog of ATP (see Chapter 17). When the signal reaches a G protein, the protein exchanges GDP for GTP. The GTP-protein complex brings about the effect. The inherent GTPase activity of the protein then converts GTP to GDP, restoring the resting state. The GTPase activity is accelerated by a family of RGS (regulators of G protein signaling) proteins that accelerate the formation of GDP.

Small G proteins are involved in many cellular functions. Members of the Rab family of these proteins regulate the rate of vesicle traffic between the endoplasmic reticulum, the Golgi apparatus, lysosomes, endosomes, and the cell membrane (see above). Another family of small GTP-binding proteins, the Rho/Rac family, mediates interactions between the cytoskeleton and cell membrane, and a third family, the Ras family, regulates growth by transmitting signals from the cell membrane to the nucleus. The members of these three families are related to the product of the *ras* protooncogene.

Another family of G proteins, the larger **heterotrimeric G proteins,** couple cell surface receptors to catalytic units that catalyze the intracellular formation of second messengers or couple the receptors directly to ion channels. These G proteins are made up of three subunits designated α, β, and γ (Figure 1–39). The α subunit is bound to GDP. When a ligand binds to a G-coupled receptor, this GDP is exchanged for GTP and the α subunit separates from the combined β and γ subunits. The separated α subunit brings about many biologic effects. The β and γ subunits do not separate from each other, and βγ also activates a variety of effectors. The intrinsic GTPase activity of the α subunit then converts GTP to GDP, and this leads to reassociation of the α with the βγ subunit and termination of effector activation.

Heterotrimeric G proteins relay signals from over 1000 receptors, and their effectors in the cells include ion channels and enzymes. Examples are listed in Table 1–8. There are 16 α, 6 β, and 12 γ genes, so a large number of subunits are produced, and they can combine in various ways. They can be divided into five families, each with a relatively characteristic set of effectors. The families are G_s, G_i, G_t, G_q, and G_{13}.

Many G proteins are modified by having specific lipids attached to them, ie, they are **lipidated** (Figure 1–6). Trimeric G proteins may be myristolated, palmitoylated, or prenylated. Small G proteins may be prenylated.

Table 1–8. Some of the ligands for receptors coupled to G proteins.

Class	Ligand
Neurotransmitters	Epinephrine Norepinephrine Dopamine 5-Hydroxytryptamine Histamine Acetylcholine Adenosine Opioids
Tachykinins	Substance P Neurokinin A Neuropeptide K
Other peptides	Angiotensin II Arginine vasopressin Oxytocin VIP, GRP, TRH, PTH
Glycoprotein hormones	TSH, FSH, LH, hCG
Arachidonic acid derivatives	Thromboxane A_2
Other	Odorants Tastants Endothelins Platelet-activating factor Cannabinoids Light

Serpentine Receptors

All the heterotrimeric G protein-coupled receptors that have been characterized to date are proteins that span the cell membrane seven times **(serpentine receptors).** These receptors may be palmitoylated. A very large number have been cloned, and their functions are multiple and diverse. The structures of two of them are shown in Figure 1–40. In general, small ligands bind to the amino acid residues in the membrane, whereas large polypeptide and protein ligands bind to the extracellular domains, which are bigger and better developed in the receptors for polypeptides and proteins. It is generally amino acid residues in the third cytoplasmic loop, the loop nearest the carboxyl terminal, that interact with the G proteins.

Inositol Triphosphate & Diacylglycerol as Second Messengers

The link between membrane binding of a ligand that acts via Ca^{2+} and the prompt increase in the cytoplasmic Ca^{2+} concentration is often **inositol triphosphate (inositol 1,4,5-triphosphate; IP_3).** When one of these ligands binds to its receptor, activation of the receptor produces activation of phospholipase C on the inner surface of the membrane via G_q. There are at least eight isoforms of phospholipase C (PLC), and the $PLC\beta_1$ and $PLC\beta_2$ forms are activated by G proteins. They catalyze the hydrolysis of phosphatidylinositol 4,5-diphosphate (PIP_2) to form IP_3 and **diacylglycerol (DAG)** (Figure 1–41). Tyrosine kinase-linked receptors (see below) can also produce IP_3 and DAG by activating $PLC\gamma_1$. The IP_3 diffuses to the endoplasmic reticulum, where it triggers the release of Ca^{2+} into the cytoplasm (Figure 1–42). The IP_3 receptor resembles the ryanodine receptor which is the Ca^{2+} channel in the sarcoplasmic reticulum of skeletal muscle (see Chapter 3), except that the IP_3 receptor is half as large. DAG is also a second messenger; it stays in the cell membrane, where it activates one of the seven subspecies of **protein kinase C** (Table 1–7). Examples of ligands that act via these second messengers are listed in Table 1–8.

The precursor of PIP_2 is phosphatidylinositol (Figure 1–41). This phospholipid is found in relatively small amounts in the inner lamella of the cell membrane. It is first converted to phosphatidyl 4-phosphate (PIP) and then to PIP_2, the derivative that is hydrolyzed to form IP_3 and the DAG. Other inositol phosphates are also formed in cells, but their function is uncertain. The IP_3 is metabolized by stepwise dephosphorylation to inositol. DAG is converted to phosphatidic acid and then to cytosine diphosphate (CDP) diacylglycerol, which combines with inositol to form phosphatidylinositol, completing the cycle.

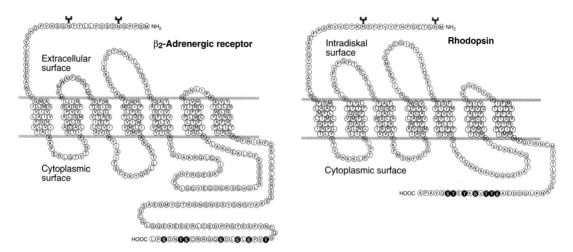

Figure 1–40. Structure of the β_2-adrenergic receptor and rhodopsin. The individual amino acid residues are identified by their single-letter codes, and the dark-colored residues are sites of phosphorylation. The y-shaped symbols on N residues identify glycosylation sites. Note the extracellular amino terminal, the intracellular carboxyl terminal, and, in light color, the seven membrane-spanning portions of each protein. (Reproduced, with permission, from Benovic JL et al: Light-dependent phosphorylation of rhodopsin by β-adrenergic receptor kinase. Reprinted by permission from Nature 1986;321:869. Copyright © 1986 by Macmillan Magazines Ltd.)

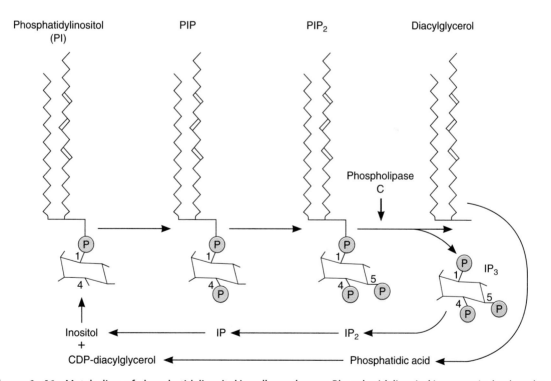

Figure 1–41. Metabolism of phosphatidylinositol in cell membranes. Phosphatidylinositol is successively phosphorylated to form phosphatidylinositol 4-phosphate (PIP), then phosphatidylinositol 4,5-diphosphate (PIP$_2$). Phospholipase Cβ_1 and β_2 catalyze the breakdown of PIP$_2$ to inositol 1,4,5-triphosphate (IP$_3$) and diacylglycerol. Other inositol phosphates and phosphatidylinositol derivatives can also be formed. IP$_3$ is dephosphorylated to inositol, and diacylglycerol is metabolized to cytosine diphosphate (CDP)-diacylglycerol. CDP-diacylglycerol and inositol then combine to form **phosphatidylinositol, completing the cycle.** (Modified from Berridge MJ: Inositol triphosphate and diacylglycerol as second messengers. Biochem J 1984;220:345.)

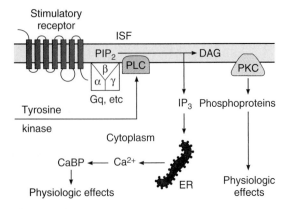

Figure 1–42. Diagrammatic representation of release of inositol triphosphate (IP$_3$) and diacylglycerol (DAG) as second messengers. Binding of ligand to G protein-coupled receptor activates phospholipase C (PLC) β$_1$ or β$_2$. Alternatively, activation of receptors with intracellular tyrosine kinase domains can activate PLCγ$_1$. The resulting hydrolysis of PIP$_2$ produces IP$_3$, which releases Ca^{2+} from the endoplasmic reticulum (ER), and DAG, which activates protein kinase C (PKC). CaBP, Ca^{2+}-binding proteins. ISF, interstitial fluid.

Cyclic AMP

Another important second messenger is **cyclic AMP (cAMP)** (Figure 1–43). Some of the many ligands that act via this compound are listed in Table 1–6. Cyclic AMP is cyclic adenosine 3′,5′-monophosphate. It is formed from ATP by the action of the enzyme **adenylyl cyclase** and converted to physiologically inactive 5′-AMP by the action of the enzyme phosphodiesterase.

Cyclic AMP activates one of the cyclic nucleotide-dependent protein kinases (**protein kinase A**) that, like protein kinase C, catalyzes the phosphorylation of proteins, changing their conformation and altering their activity. A typical example is the activation of phosphorylase kinase in the liver by epinephrine via cAMP and protein kinase A (see Figure 17–13). In addition, the active catalytic subunit of PKA moves to the nucleus and phosphorylates the **cAMP-responsive element-binding protein (CREB)** This transcription factor then binds to DNA and alters transcription of a number of genes.

Cyclic AMP is metabolized by a phosphodiesterase. This phosphodiesterase is inhibited by methylxanthines such as caffeine and theophylline; consequently, these compounds augment hormonal and transmitter effects mediated via cAMP.

Activation of Adenylyl Cyclase

Five components are involved in the mechanism by which ligands bring about changes in the intracellular concentration of cAMP: a catalytic unit, adenylyl cyclase, which catalyzes the conversion of ATP to cAMP; stimulatory and inhibitory receptors; and stimulatory and inhibitory G proteins that link the receptor to the catalytic unit (Figure 1–44). Like the receptors, adenylyl cyclase is a transmembrane protein, and it crosses the membrane 12 times. Eight isoforms of this enzyme have been described, and, combined with the many different forms of G proteins, this permits the cAMP pathway to be customized to specific tissue needs. When the appropriate ligand binds to a stimulatory receptor, a G$_s$ α subunit activates one of the adenylyl cyclases. Conversely, when the appropriate ligand binds

Adenosine triphosphate (ATP) → AC → **cAMP** → PD → **5′-Adenosine monophosphate (5′-AMP)**

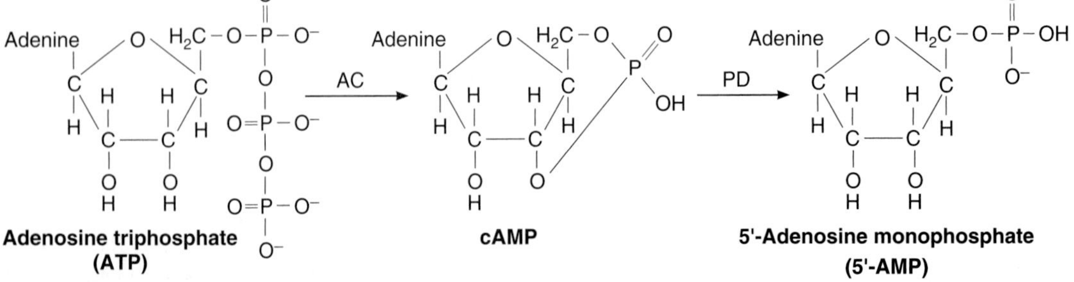

Figure 1–43. Formation and metabolism of cAMP. AC, adenylyl cyclase; PD, phosphodiesterase.

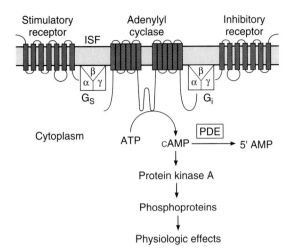

Figure 1–44. The cAMP system. Activation of adenylyl cyclase catalyzes the conversion of ATP to cAMP. Cyclic AMP activates protein kinase A, which phosphorylates proteins, producing physiologic effects. Stimulatory ligands bind to stimulatory receptors and activate adenylyl cyclase via G_s. Inhibitory ligands inhibit adenylyl cyclase via inhibitory receptors and G_i. ISF, interstitial fluid.

to the inhibitory receptor, a G_i α subunit inhibits adenylyl cyclase. The receptors are specific, responding at low threshold to only one or a select group of related ligands. However, heterotrimeric G proteins mediate the stimulatory and inhibitory effects produced by many different ligands. In addition, there is cross-talk between the phospholipase C system and the adenylyl cyclase system, and several of the isoforms of adenylyl cyclase are stimulated by calmodulin. Finally, the effects of protein kinase A and protein kinase C are very widespread. Given this complexity, how are specific responses to specific stimuli obtained? The answer lies in part in tethering of the G proteins, adenylyl cyclase, and the protein kinases to the cytoskeleton so that local microdomains are created. Some of this tethering is carried out by lipid products (Figure 1–6).

Some cAMP escapes from cells upon stimulation by certain hormones, but the amounts are small compared with the intracellular concentration, and only small amounts of extracellular cAMP enter cells.

Two bacterial toxins have important effects on adenylyl cyclase that are mediated by G proteins. The A subunit of **cholera toxin** catalyzes the transfer of ADP-ribose to an arginine residue in the middle of the α subunit of G_s. This inhibits its GTPase activity, producing prolonged stimulation of adenylyl cyclase (see Chapter 25). **Pertussis toxin** catalyzes ADP-ribosylation of a cysteine residue near the carboxyl terminal of the α sub-

unit of G_i. This inhibits the function of G_i. In addition to the implications of these alterations in disease, both toxins are used for fundamental research on G protein function. The drug forskolin stimulates adenylyl cyclase activity by a direct action on the enzyme.

Guanylyl Cyclase

Another cyclic nucleotide of physiologic importance is **cyclic guanosine monophosphate** (cyclic GMP; cGMP). Cyclic GMP is important in vision. Light acts on rhodopsin in rods; rhodopsin is linked to phosphodiesterase by G_{t1}; and activation of phosphodiesterase accelerates conversion of cGMP to 5′-GMP (see Chapter 8). A similar process occurs in the cones. In addition, there are cGMP-regulated ion channels, and cGMP activates cGMP-dependent kinase (Table 1–7), producing a number of physiologic effects.

Guanylyl cyclases are a family of enzymes that catalyze the formation of cGMP. They exist in two forms (Figure 1–45). In one form, there is an extracellular amino terminal domain that is a receptor, single trans-

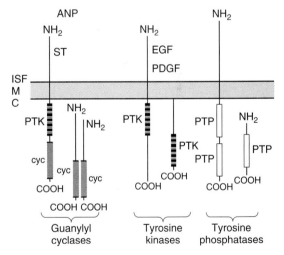

Figure 1–45. Diagrammatic representation of guanylyl cyclases, tyrosine kinases, and tyrosine phosphatases. ANP, atrial natriuretic peptide; C, cytoplasm; cyc, cyclase domain; EGF, epidermal growth factor; ISF, interstitial fluid; M, cell membrane; PDGF, platelet-derived growth factor; PTK, tyrosine kinase domain; PTP, tyrosine phosphatase domain; ST, *E coli* enterotoxin. (Modified and reproduced, with permission, from Koesling D, Böhme E, Schultz G: Guanylyl cyclases, a growing family of signal transducing enzymes. FASEB J 1991;5:2785.)

membrane domain, and a cytoplasmic carboxyl terminal that has a tyrosine kinase-like and a guanylyl cyclase catalytic domain. Three such guanylyl cyclases have been characterized. Two are receptors for ANP (ANPR-A and ANPR-B; see Chapter 24), and a third binds an *Escherichia coli* enterotoxin and the gastrointestinal polypeptide guanylin (see Chapter 26). The other form of guanylyl cyclase is soluble, contains heme, and is totally intracellular. There appear to be several isoforms of the intracellular enzyme. They are activated by nitric oxide (NO) and NO-containing compounds. NO is the endothelium-derived relaxing factor (EDRF), an intercellular messenger formed from arginine, that plays an important role in regulating the diameter of blood vessels (see Chapter 31). NO is also involved in many other functions, including penile erection (see Chapter 23) and synaptic transmission in the brain (see Chapter 4). Recent evidence indicates that in addition to its effects mediated by cGMP, NO can activate Ca^{2+}-dependent K^+ channels by a direct action on the channels.

Phosphatases

Numerous phosphatases that remove phosphate groups from proteins are found in cells. Frequently these are closely associated with or coupled to tyrosine kinases and serine-threonine kinases. Two examples are shown in Figure 1–45.

Growth Factors

Growth factors have become increasingly important in many different aspects of physiology. They are polypeptides and proteins that are conveniently divided into three groups. One group is made up of agents that foster the multiplication or development of various types of cells; nerve growth factor (see Chapter 2), insulin-like growth factor I (IGF-I; see Chapter 22), activins and inhibins (see Chapter 23), and epidermal growth factor (EGF) are examples, and more than 20 have been described. The cytokines are a second group. These factors are produced by macrophages and lymphocytes and are important in regulation of the immune system (see Chapter 27). Again, more than 20 have been described. The third group is made up of the colony-stimulating factors that regulate proliferation and maturation of red and white blood cells. They are also discussed in Chapter 27.

Receptors for EGF, platelet-derived growth factor (PDGF), and many of the other factors that foster cell multiplication and growth have a single membrane-spanning domain with an intracellular tyrosine kinase domain (Figure 1–46). When ligand binds to the receptor, the tyrosine kinase domain autophosphorylates itself. Some of the receptors dimerize when they bind their ligands, and the intracellular tyrosine kinase domains cross-phosphorylate each other. One of the path-

ways activated by phosphorylation leads, through the product of the *ras* proto-oncogene and several MAP kinases, directly to the production of transcription factors in the nucleus that alter gene expression. This important direct path from the cell surface to the nucleus is shown diagrammatically in Figure 1–46. Note that Ras is one of the small G proteins that requires binding to GTP for activation.

Receptors for the cytokines and the colony-stimulating factors differ from the other growth factors in that most of them do not have tyrosine kinase domains in their cytoplasmic portions and some have lit-

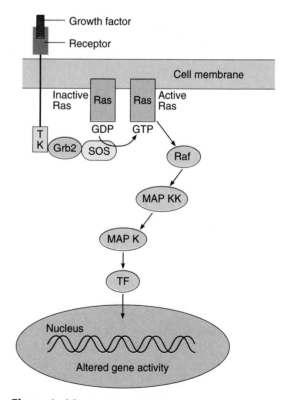

Figure 1–46. One of the direct pathways by which growth factors alter gene activity. TK, tyrosine kinase domain; Grb2, Ras activator controller; Sos, Ras activator; Ras, product of the *ras* gene; MAP K, mitogen-activated protein kinase; MAP KK, MAP kinase kinase; TF, transcription factors. There is cross talk between this pathway and the cAMP pathway, as well as cross talk with the IP_3-DAG pathway.

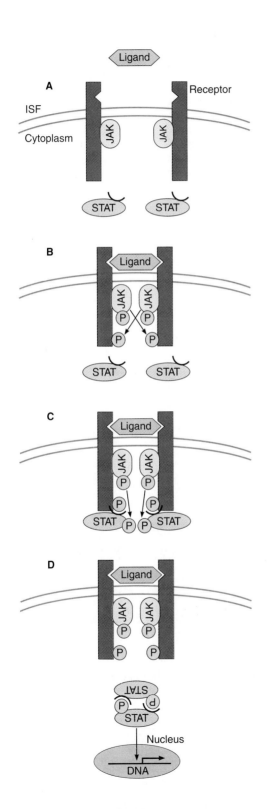

tle or no cytoplasmic tail. However, they initiate tyrosine kinase activity in the cytoplasm. In some instances, this involves binding to the associated transmembrane protein gp130 (see Chapter 27). In particular, they activate the so-called Janus tyrosine kinases (**JAKs**) in the cytoplasm (Figure 1–47). These in turn phosphorylate signal transducer and activator of transcription (**STAT**) proteins. The phosphorylated STATs form homo- and heterodimers and move to the nucleus, where they act as transcription factors. There are four known mammalian JAKs and seven known STATs. The JAK-STAT pathway is also activated by growth hormone (see Figure 22–4) and is another important direct path from the cell surface to the nucleus. However, it should be emphasized that both the Ras and the JAK-STAT pathways are complex and there is cross talk between them and the phospholipase C and cAMP pathways.

Another family of receptors binds transforming growth factor β (TGFβ) and related polypeptides. These receptors have serine-threonine kinase activity, and their effects are mediated by SMADs, intracellular proteins that when phosphorylated move to the nucleus, bind to DNA, and, with other factors, initiate transcription of various genes.

As noted above, integrins also initiate phosphorylation of proteins that enter the nucleus and alter gene transcription.

Note that a common theme is activation of transcription factors that without activation are "locked" in the cytoplasm. Once activated, the transcription factor moves to the nucleus and alters gene transcription. Additional examples include NF-AT (see Figure 27–13) and NF-$_\kappa$B (see Chapters 20 and 33).

Receptor & G Protein Diseases

Many diseases are being traced to mutations of the genes for receptors. For example, loss-of-function receptor mutations that cause disease have been reported

Figure 1–47. Signal transaction via the JAK-STAT pathway. **A:** Ligand binding leads to dimerization of receptor. **B:** Activation and tyrosine phosphorylation of JAKs. **C:** JAKs phosphorylate STATs. **D:** STATs dimerize and move to nucleus, where they bind to response elements on DNA. (Modified from Takeda K, Kishimoto T, Akira S: STAT6: Its role in interleukin 4-mediated biological functions. J Mol Med 1997;75:317.)

for the 1,25-dihydroxycholecalciferol receptor (see Chapter 21) and the insulin receptor (see Chapter 19). Certain other diseases are caused by production of antibodies against receptors. Thus, antibodies against TSH receptors cause Graves' disease (see Chapter 18), and antibodies against nicotinic acetylcholine receptors cause myasthenia gravis (see Chapter 4).

An example of loss of function of a receptor is the type of **nephrogenic diabetes insipidus** that is due to loss of the ability of mutated V_2 vasopressin receptors to mediate concentration of the urine (see Chapters 14 and 38). Mutant receptors can gain as well as lose function. A gain-in-function mutation of the Ca^{2+} receptor (see Chapter 21) causes excess inhibition of parathyroid hormone secretion and **familial hypercalciuric hypocalcemia.** G proteins can also undergo loss-of-function or gain-of-function mutations that cause disease (Table 1–9). In one form of pseudohypoparathyroidism, a mutated G_s α fails to respond to parathyroid hormone, producing the symptoms of hypoparathyroidism without any decline in circulating parathyroid hormone. **Testotoxicosis** is an interesting disease that combines

gain and loss of function. In this condition, an activating mutation of G_s α causes excess testosterone secretion and prepubertal sexual maturation. However, this mutation is temperature-sensitive and is active only at the relatively low temperature of the testes (33 °C; see Chapter 23). At 37 °C, the normal temperature of the rest of the body, it is replaced by loss of function, with the production of hypoparathyroidism and decreased responsiveness to TSH. A different activating mutation in G_s α is associated with the rough-bordered areas of skin pigmentation and hypercortisolism in the McCune-Albright syndrome. This mutation occurs during fetal development, creating a mosaic of normal and abnormal cells. A third mutation in G_s α reduces its intrinsic GTPase activity. As a result, it is much more active than normal, and excess cAMP is produced. This causes hyperplasia and eventually neoplasia in somatotrope cells of the anterior pituitary. Forty percent of somatotrope tumors causing acromegaly (see Chapter 22) have cells containing a somatic mutation of this type.

Table 1–9. Diseases caused by loss- or gain-of-function mutations of heterotrimeric G-protein-coupled receptors and G proteins.[1]

Site	Type of Mutation	Disease
Receptor		
Cone opsins	Loss	Color blindness
Rhodopsin	Loss	Congenital night blindness Two forms of retinitis pigmentosa
V_2 vasopressin	Loss	X-linked nephrogenic diabetes insipidus
ACTH	Loss	Familial glucocorticoid deficiency
LH	Gain	Familial male precocious puberty
TSH	Gain	Familial nonautoimmune hyperthyroidism
TSH	Loss	Familial hypothyroidism
Ca^{2+}	Gain	Familial hypercalciuric hypocalcemia
Thromboxane A_2	Loss	Congenital bleeding
Endothelin B	Loss	Hirschsprung's disease
G protein		
G_s α	Loss	Pseudohypothyroidism type 1a
G_s α	Gain/loss	Testotoxicosis
G_s α	Gain (mosaic)	McCune-Albright syndrome
G_s α	Gain	Somatotroph adenomas with acromegaly
G_i α	Gain	Ovarian and adrenocortical tumors

[1]Modified from Lem J: Diseases of G-protein-coupled signal transduction pathways: the mammalian visual system as a model. Semin Neurosci 1998;9:232.

HOMEOSTASIS

The actual environment of the cells of the body is the interstitial component of the ECF. Since normal cell function depends upon the constancy of this fluid, it is not surprising that in multicellular animals, an immense number of regulatory mechanisms have evolved to maintain it. To describe "the various physiologic arrangements which serve to restore the normal state, once it has been disturbed," W.B. Cannon coined the term **homeostasis.** The buffering properties of the body fluids and the renal and respiratory adjustments to the presence of excess acid or alkali are examples of homeostatic mechanisms. There are countless other examples, and a large part of physiology is concerned with regulatory mechanisms that act to maintain the constancy of the internal environment. Many of these regulatory mechanisms operate on the principle of negative feedback; deviations from a given normal set point are detected by a sensor, and signals from the sensor trigger compensatory changes that continue until the set point is again reached.

AGING

Aging is a general physiologic process that is as yet poorly understood. In the United States, life expectancy has increased from 47 years in 1900 to about 75 years today. However, this increase is due for the most part to improved treatment and prevention of infections and other causes of early death, so that more people survive into their 70s. In the meantime, the maximum human life span of 100–110 years has increased little if at all. Aging affects cells and the systems made up of them, as well as tissue components such as collagen, and numerous theories have been advanced to explain the phenomenon.

One theory of aging holds that tissues age as a result of random mutations in the DNA of somatic cells, with consequent introduction of cumulative abnormalities. Others hold that cumulative abnormalities are produced by increased cross-linkage of collagen and other proteins, possibly as the end result of the nonenzymatic combination of glucose with amino groups on these molecules. A third theory envisions aging as the cumulative result of damage to tissues by free radicals formed in them. It is interesting in this regard that species with longer life spans produce more **superoxide dismutase,** an enzyme that inactivates oxygen-free radicals (see Chapter 27).

Evidence in favor of cumulative DNA abnormalities is the recent demonstration that in **Werner's syndrome,** a condition in which humans age at a markedly accelerated rate, the genetic abnormality is mutation of a gene coding for a **DNA helicase,** one of the enzymes that helps split the DNA strands before replication. This abnormality would be expected to produce unusually rapid accumulation of chromosomal damage. Mice that lack one of the components of telomerase (see above) age rapidly and have many of the defects that are characteristic of Werner's syndrome in humans. In addition, as human cells age, there is a large accumulation of point mutations in the portion of their mitochondrial DNA that controls its reproduction. This could lead to defective energy production or, possibly, increases in the free radicals in cells.

It is now established that in experimental animals, a chronically decreased caloric intake prolongs life, and this could be true in humans as well. One possible explanation for this effect of **caloric restriction** is decreased metabolism, with decreased formation of protein cross-links and decreased production of free radicals. It may be relevant in this regard that in yeasts, worms, and flies, mutations in the homologs of one of the mammalian insulin pathways causes a dramatic prolongation of their lifespan. However, the exact cause of the lengthened life span produced by caloric restriction remains to be determined.

In aging humans, there are declines in the circulating levels of some sex hormones, the adrenal androgen dehydroepiandrosterone and its sulfate, and growth hormone. Replacement therapy with estrogens and progesterone in women (see Chapter 23) decreases the incidence of osteoporosis and heart disease. Replacement therapy with testosterone (see Chapter 23), dehydroepiandrosterone (see Chapter 20), and growth hormone (see Chapter 22) each has some salutary effects, but each also has undesirable side effects, and there is little if any evidence that they prolong life.

REFERENCES FOR SECTION I: INTRODUCTION

Albert B et al: *Molecular Biology of the Cell*, 4th ed. Garland Science, 2002.

Berridge MJ, Bootman MD, Lipp P: Calcium—a life and death signal. Nature 1998;395:645.

Blackhorn EH: Telomere states and cell fates. Nature 2000;408:53.

Cannon WB: *The Wisdom of the Body.* Norton, 1932.

Chang L, Karin M: Mammalian signaling cascades. Nature 2001;410:37.

Downward J: The ins and outs of signaling. Nature 2001;411:759.

Dynlacht BD: Regulation of transcription by proteins that control the cell cycle. Nature 1997;389:149.

Farfel Z, Bourne HR, Iiri T: The expanding spectrum of G protein diseases. N Engl J Med 1999;340:1012.

Giancotti FG, Ruoslahti E: Integrin signaling. Science 1999;285:1028.

Göhrlich D, Kutay V: Transport between the cell nucleus and the cytoplasm. Annu Rev Cell Dev Biology 1999; 15:607.

Hoffman JE, Jamieson JD (editors): *Handbook of Physiology:* Section 14: *Cell Physiology.* Oxford Univ Press, 1997.

Housman DE: DNA on trial: The molecular basis of DNA fingerprinting. N Engl J Med 1995;332:534.

Howard J: *Mechanisms of Motor Proteins and the Cytoskeleton.* Sinauer/Palgrave, 2001.

Huntley SM (editor): Frontiers in cell biology: quality control. (Special Section.) Science 1999;286:1881.

Jentsch TJ et al: Molecular structure and physiological function of chloride channels. Physiol Rev 2002;82:503.

Johns DR: The other human genome: Mitochondrial DNA and disease. Nat Med 1996;2:1065.

Kaznetsov G, Nigam SJ: Folding of secretory and membrane proteins. N Engl J Med 1998;339:1688.

Kersten S, Desvergne B, Wahil W: Roles of PPAPs in health and disease. Nature 2000;405:421.

Klug A, Schwabe JWR: Zinc fingers. FASEB J 1995;9:597.

Kmiec EB: Gene therapy. Am Scientist 1999;87:240.

Kuchler K, Thorner J: Secretion of peptides and proteins lacking hydrophobic signal sequences: The role of adenosine triphosphate-driven membrane translocators. Endocr Rev 1992;13:497.

Lamberts SWJ, Van den Beld AW, Van der Lely A-J: The endocrinology of aging. Science 1997;278:419.

Making sense of channel diversity. Nat Neurosci 1998;1:169.

Mukherjee S, Ghosh RN, Maxfield FR: Endocytosis. Physiol Rev 1997;77:759.

Ray LB, Gough NR: Orienteering strategies for a signaling maze. Science 2002;296:1632.

Ray LB: The science of signal transduction. Science 1999;284:755.

Rebbechi MJ, Pentyala SN: Structure, function, and control of phosphoinositol-specific phospholipase C. Physiol Rev 2000;80:1291.

Rothman JE, Wieland FT: Protein sorting by transport vesicles. Science 1996;272:227.

Russell JM: Sodium-potassium-chloride cotransport. Physiol Rev 2000;20:211.

Scriver CR et al (editors): *The Metabolic and Molecular Bases of Inherited Disease,* 8th ed. McGraw-Hill, 2001.

Strehler E, Zacharias DA: Role of alternative splicing on generation of diversity among plasma membrane calcium pumps. Physiol Rev 2001;81:21.

Wallace DG: Mitochondrial diseases in man and mouse. Science 1999;283:1482.

Weindruch R, Schal RS: Caloric intake and aging. N Engl J Med 1997;337:986.

Yellen G: The voltage-gated potassium channel and their relatives. Nature 2002;419:35.

Zhu X, Birnbaumer L: Calcium channels formed by mammalian Trp homology. News Physiol Sci 1998;11:211.

SECTION II
Physiology of Nerve & Muscle Cells

Excitable Tissue: Nerve

2

INTRODUCTION

The human central nervous system (CNS) contains about 10^{11} (100 billion) neurons. It also contains 10–50 times this number of glial cells. It is a complex organ; it has been calculated that 40% of the human genes participate, at least to a degree, in its formation. The neurons, the basic building blocks of the nervous system, have evolved from primitive neuroeffector cells that respond to various stimuli by contracting. In more complex animals, contraction has become the specialized function of muscle cells, whereas integration and transmission of nerve impulses have become the specialized functions of neurons. This chapter is concerned with the ways these neurons are excited and the way they integrate and transmit impulses.

NERVE CELLS

Morphology

Neurons in the mammalian central nervous system come in many different shapes and sizes (Figure 2–1). However, most have the same parts as the typical spinal motor neuron illustrated in Figure 2–2. This cell has five to seven processes called **dendrites** that extend outward from the cell body and arborize extensively. Particularly in the cerebral and cerebellar cortex, the dendrites have small knobby projections called **dendritic spines.** A typical neuron also has a long fibrous **axon** that originates from a somewhat thickened area of the cell body, the **axon hillock.** The first portion of the axon is called the **initial segment.** The axon divides into terminal branches, each ending in a number of

synaptic knobs. The knobs are also called **terminal buttons** or **axon telodendria.** They contain granules or vesicles in which the synaptic transmitters secreted by the nerves are stored (see Chapter 4).

The axons of many neurons are myelinated, ie, they acquire a sheath of **myelin,** a protein-lipid complex that is wrapped around the axon (Figure 2–3). Outside the CNS, the myelin is produced by Schwann cells, glia-like cells found along the axon. Myelin forms when a Schwann cell wraps its membrane around an axon up to 100 times. The myelin is then compacted when the extracellular portions of a membrane protein called protein zero (P_0) lock to the extracellular portions of P_0 in the apposing membrane. Various mutations in the gene for P_0 cause peripheral neuropathies; 29 different mutations have been described that cause symptoms ranging from mild to severe. The myelin sheath envelops the axon except at its ending and at the **nodes of Ranvier,** periodic 1-μm constrictions that are about 1 mm apart. The insulating function of myelin is discussed below. Not all mammalian neurons are myelinated; some are **unmyelinated,** ie, are simply surrounded by Schwann cells without the wrapping of the Schwann cell membrane around the axon that produces myelin. Most neurons in invertebrates are unmyelinated.

In the CNS of mammals, most neurons are myelinated, but the cells that form the myelin are oligodendrogliocytes rather than Schwann cells (Figure 2–3). Unlike the Schwann cell, which forms the myelin between two nodes of Ranvier on a single neuron, oligodendrogliocytes send off multiple processes that form myelin on many neighboring axons. In multiple sclerosis, a crippling autoimmune disease, there is patchy de-

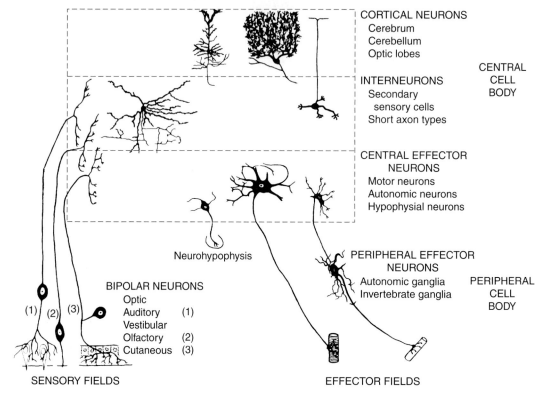

CORTICAL NEURONS
 Cerebrum
 Cerebellum
 Optic lobes

INTERNEURONS
 Secondary
  sensory cells
 Short axon types

CENTRAL
CELL
BODY

CENTRAL EFFECTOR
 NEURONS
 Motor neurons
 Autonomic neurons
 Hypophysial neurons

Neurohypophysis

PERIPHERAL EFFECTOR
 NEURONS
 Autonomic ganglia
 Invertebrate ganglia

PERIPHERAL
CELL
BODY

BIPOLAR NEURONS
 Optic
 Auditory   (1)
 Vestibular
 Olfactory   (2)
 Cutaneous   (3)

(1) (2) (3)

SENSORY FIELDS

EFFECTOR FIELDS

Figure 2–1. Some of the types of neurons in the mammalian nervous system. (Reproduced, with permission, from Bodian D: Introductory survey of neurons. Cold Spring Harbor Symp Quant Biol 1952;17:1.)

struction of myelin in the CNS. The loss of myelin is associated with delayed or blocked conduction in the demyelinated axons.

The dimensions of some neurons are truly remarkable. For spinal neurons supplying the muscles of the foot, for example, it has been calculated that if the cell body were the size of a tennis ball, the dendrites of the cell would fill a large room and the axon would be up to 1.6 km (almost a mile) long although only 13 mm (half an inch) in diameter.

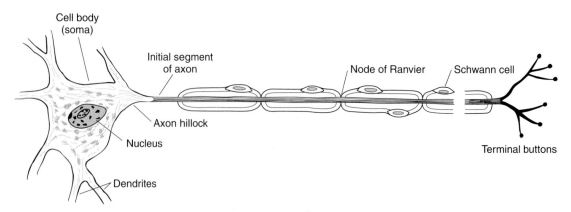

Cell body
(soma)

Initial segment
of axon

Node of Ranvier

Schwann cell

Axon hillock

Nucleus

Terminal buttons

Dendrites

Figure 2–2. Motor neuron with myelinated axon.

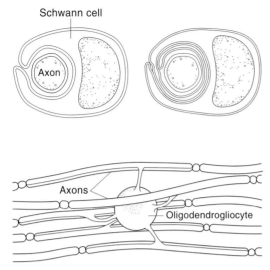

Figure 2–3. **Top:** Relation of Schwann cells to axons in peripheral nerves. On the left is an unmyelinated axon, and on the right is a myelinated axon. Note that the cell membrane of the Schwann cell has wrapped itself around and around the axon.
Bottom: Myelination of axons in the central nervous system by oligodendrogliocytes. One oligodendrogliocyte sends processes to up to 40 axons.

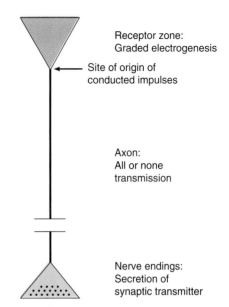

Figure 2–4. Functional organization of neurons. Nonconducted local potentials are integrated in the receptor zone, and action potentials are initiated at a site close to the receptor zone (arrow). The action potentials are conducted along the axon to the nerve endings, where they cause release of synaptic transmitters.

The conventional terminology used above for the parts of a neuron works well enough for spinal motor neurons and interneurons, but there are problems in terms of "dendrites" and "axons" when it is applied to other types of neurons found in the nervous system. From a functional point of view (see below and Chapters 4 and 5), neurons generally have four important zones. There is a receptor, or dendritic zone, where multiple local potential changes generated by synaptic connections are integrated (Figure 2–4); a site where propagated action potentials are generated (the initial segment in spinal motor neurons, the initial node of Ranvier in cutaneous sensory neurons); an axonal process that transmits propagated impulses to the nerve endings; and the nerve endings, where action potentials cause the release of synaptic transmitters. The cell body is often located at the dendritic zone end of the axon, but it can be within the axon (eg, auditory neurons) or attached to the side of the axon (eg, cutaneous neurons; see Figure 2–1). Its location makes no difference as far as the receptor function of the dendritic zone and the transmission function of the axon are concerned.

It should be noted that the size and complexity of the dendritic trees on neurons varies markedly (Figure 2–1; see also Figures 11–1 and 12–17). In addition to integrated passive electrical activity, propagated action potentials appear to be generated in dendrites in some special situations. Furthermore, new research suggests that dendrites have more complex functions. This topic is discussed in greater detail in Chapter 4.

Protein Synthesis & Axoplasmic Transport

Nerve cells are secretory cells, but they differ from other secretory cells in that the secretory zone is generally at the end of the axon, far removed from the cell body. There are few if any ribosomes in axons and nerve terminals, and all necessary proteins are synthesized in the endoplasmic reticulum and Golgi apparatus of the cell body and then transported along the axon to the synaptic knobs by the process of **axoplasmic flow.** Thus, the cell body maintains the functional and anatomic integrity of the axon; if the axon is cut, the part distal to the cut degenerates (**wallerian degeneration**). **Anterograde transport** occurs along microtubules. The molecular motors involved are discussed in Chapter 1. Fast transport occurs at about 400 mm/d, and slow anterograde transport occurs at 0.5–10 mm/d. **Retrograde transport** in the opposite direction also occurs along microtubules at about 200 mm/d. Synaptic vesicles recycle in the membrane, but some used vesicles are carried back to the cell body and deposited in lysosomes. Some of the material taken up at the ending by endocytosis, including nerve growth factor (see below) and various viruses, is also transported back to the cell body.

EXCITATION & CONDUCTION

Nerve cells have a low threshold for excitation. The stimulus may be electrical, chemical, or mechanical. Two types of physicochemical disturbances are produced: local, nonpropagated potentials called, depending on their location, **synaptic, generator,** or **electrotonic potentials;** and propagated disturbances, the **action potentials** (or **nerve impulses**). These are the only electrical responses of neurons and other excitable tissues, and they are the main language of the nervous system. They are due to changes in the conduction of ions across the cell membrane that are produced by alterations in ion channels.

The impulse is normally transmitted (**conducted**) along the axon to its termination. Nerves are not "telephone wires" that transmit impulses passively; conduction of nerve impulses, although rapid, is much slower than that of electricity. Nerve tissue is in fact a relatively poor passive conductor, and it would take a potential of many volts to produce a signal of a fraction of 1 V at the other end of a 1-m axon in the absence of active processes in the nerve. Conduction is an active, self-propagating process, and the impulse moves along the nerve at a constant amplitude and velocity. The process is often compared to what happens when a match is applied to one end of a train of gunpowder; by igniting the powder particles immediately in front of it, the flame moves steadily down the train to its end.

The electrical events in neurons are rapid, being measured in **milliseconds (ms);** and the potential changes are small, being measured in **millivolts (mV).** In addition to development of microelectrodes with a tip diameter of less than 1 μm, the principal advances that made detailed study of the electrical activity in nerves possible were the development of electronic amplifiers and the cathode ray oscilloscope. Modern amplifiers magnify potential changes 1000 times or more, and the cathode ray oscilloscope provides an almost inertia-less and almost instantaneously responding "lever" for recording electrical events.

The Cathode Ray Oscilloscope

The **cathode ray oscilloscope (CRO)** is used to measure the electrical events in living tissue. In the CRO, electrons emitted from a cathode are directed into a focused beam that strikes the face of the glass tube in which the cathode is located (Figure 2–5). The face is coated with one of a number of substances (phosphors) that emit light when struck by electrons. A vertical metal plate is placed on either side of the electron beam. When a voltage is applied across these plates, the negatively charged electrons are drawn toward the positively charged plate and repelled by the negatively charged plate. If the voltage applied to the vertical plates (X plates) is increased slowly and then reduced suddenly and increased again, the beam moves steadily toward the positive plate, snaps back to its former position, and moves toward the positive plate again. Application of a "saw-tooth voltage" of this type thus causes the beam to sweep across the face of the tube, and the speed of the sweep is proportionate to the rate of rise of the applied voltage.

Another set of plates (Y plates) is arranged horizontally, with one plate above and one below the beam. Voltages applied to these plates deflect the beam up and down as it sweeps across the face of the tube, and the magnitude of the vertical deflection is proportionate to the potential difference between the horizontal plates. When these plates are connected to electrodes on a nerve, any changes in potential occurring in the nerve are recorded as vertical deflections of the beam as it moves across the tube.

Recording From Single Axons

Mammalian axons are relatively small (20 μm or less in diameter) and are difficult to separate from other axons, but giant unmyelinated nerve cells exist in a num-

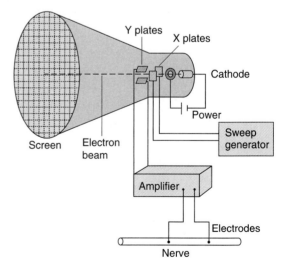

Figure 2–5. Cathode ray oscilloscope. Simplified diagram of the principal connections when arranged to record potential changes in a nerve.

ber of invertebrate species. Such giant cells are found, for example, in crabs *(Carcinus)* and cuttlefish *(Sepia),* but the largest known axons are found in the squid *(Loligo).* The neck region of the muscular mantle of the squid contains single axons up to 1 mm in diameter. The fundamental properties of these long axons are similar to those of mammalian axons.

Resting Membrane Potential

When two electrodes are connected through a suitable amplifier to a CRO and placed on the surface of a single axon, no potential difference is observed. However, if one electrode is inserted into the interior of the cell, a constant potential difference is observed, with the inside negative relative to the outside of the cell at rest. This **resting membrane potential** is found in almost all cells. Its genesis is discussed below and in Chapter 1. In neurons, it is usually about −70 mV.

Latent Period

If the axon is stimulated and a conducted impulse occurs, a characteristic series of potential changes known as the **action potential** is observed as the impulse passes the external electrode (Figure 2–6).

When the stimulus is applied, there is a brief irregular deflection of the baseline, the **stimulus artifact.**

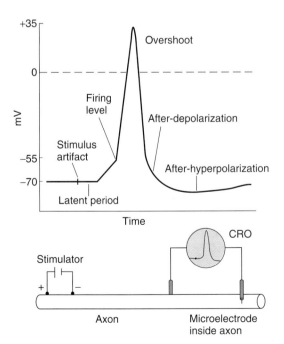

Figure 2–6. Action potential in a neuron recorded with one electrode inside the cell.

This artifact is due to current leakage from the stimulating electrodes to the recording electrodes. It usually occurs despite careful shielding, but it is of value because it marks on the cathode ray screen the point at which the stimulus was applied.

The stimulus artifact is followed by an isopotential interval (**latent period**) that ends with the start of the action potential and corresponds to the time it takes the impulse to travel along the axon from the site of stimulation to the recording electrodes. Its duration is proportionate to the distance between the stimulating and recording electrodes and inversely proportionate to the speed of conduction. If the duration of the latent period and the distance between the electrodes are known, the speed of conduction in the axon can be calculated. For example, assume that the distance between the cathodal stimulating electrode and the exterior electrode in Figure 2–6 is 4 cm. The cathode is normally the stimulating electrode (see below). If the latent period is 2 ms, the speed of conduction is 4 cm/2 ms, or 20 m/s.

Action Potential

The first manifestation of the approaching action potential is a beginning depolarization of the membrane. After an initial 15 mV of depolarization, the rate of depolarization increases. The point at which this change in rate occurs is called the **firing level** or sometimes the **threshold.** Thereafter, the tracing on the oscilloscope rapidly reaches and **overshoots** the isopotential (zero potential) line to approximately +35 mV. It then reverses and falls rapidly toward the resting level. When repolarization is about 70% completed, the rate of repolarization decreases and the tracing approaches the resting level more slowly. The sharp rise and rapid fall are the **spike potential** of the axon, and the slower fall at the end of the process is the **after-depolarization.** After reaching the previous resting level, the tracing overshoots slightly in the hyperpolarizing direction to form the small but prolonged **after-hyperpolarization.** When recorded with one electrode in the cell, the action potential is called **monophasic,** because it is primarily in one direction.

The proportions of the tracing in Figure 2–6 are intentionally distorted to illustrate the various components of the action potential. A tracing with the components plotted on exact temporal and magnitude scales for a mammalian neuron is shown in Figure 2–7. Note that the rise of the action is so rapid that it fails to show clearly the change in depolarization rate at the firing level, and also that the after-hyperpolarization is only about 1–2 mV in amplitude although it lasts about 40 ms. The duration of the after-depolarization is about 4 ms in this instance. It is shorter and less

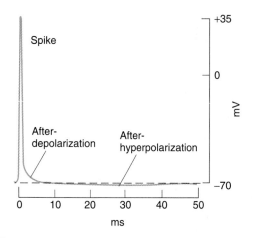

Figure 2–7. Diagram of the complete action potential of a large mammalian myelinated fiber, drawn without time or voltage distortion to show the proportions of the components.

prominent in many other neurons. Changes may occur in the after-polarizations without changes in the rest of the action potential. For example, if the nerve has been conducting repetitively for a long time, the after-hyperpolarization is usually quite large.

"All-or-None" Law

If an axon is arranged for recording as shown in Figure 2–6, with the recording electrodes at an appreciable distance from the stimulating electrodes, it is possible to determine the minimal intensity of stimulating current (**threshold intensity**) that, acting for a given duration, will just produce an action potential. The threshold intensity varies with the duration; with weak stimuli it is long, and with strong stimuli it is short. The relation between the strength and the duration of a threshold stimulus is called the **strength-duration curve.** Slowly rising currents fail to fire the nerve because the nerve adapts to the applied stimulus, a process called **accommodation.**

Once threshold intensity is reached, a full-fledged action potential is produced. Further increases in the intensity of a stimulus produce no increment or other change in the action potential as long as the other experimental conditions remain constant. The action potential fails to occur if the stimulus is subthreshold in magnitude, and it occurs with a constant amplitude and form regardless of the strength of the stimulus if the stimulus is at or above threshold intensity. The action potential is therefore "all or none" in character and is said to obey the **all-or-none law.**

Electrotonic Potentials, Local Response, & Firing Level

Although subthreshold stimuli do not produce an action potential, they do have an effect on the membrane potential. This can be demonstrated by placing recording electrodes within a few millimeters of a stimulating electrode and applying subthreshold stimuli of fixed duration. Application of such currents with a cathode leads to a localized depolarizing potential change that rises sharply and decays exponentially with time. The magnitude of this response drops off rapidly as the distance between the stimulating and recording electrodes is increased. Conversely, an anodal current produces a hyperpolarizing potential change of similar duration. These potential changes are called **electrotonic potentials,** those produced at a cathode being **catelectrotonic** and those at an anode **anelectrotonic.**

The anelectronic potential is proportionate to the applied anodal current. The catelectronic potential is roughly proportionate at low applied cathodal current, but as the strength of the current is increased, the response is greater due to the increasing addition of a **local response** of the membrane (Figure 2–8). Finally, at 7–15 mV of depolarization, the **firing level,** runaway depolarization, and a spike potential result.

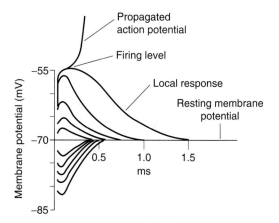

Figure 2–8. Electrotonic potentials and local response. The changes in the membrane potential of a neuron following application of stimuli of 0.2, 0.4, 0.6, 0.8, and 1.0 times threshold intensity are shown superimposed on the same time scale. The responses below the horizontal line are those recorded near the anode, and the responses above the line are those recorded near the cathode. The stimulus of threshold intensity was repeated twice. Once it caused a propagated action potential (top line), and once it did not.

Changes in Excitability During Electrotonic Potentials & the Action Potential

During the action potential as well as during catelectrotonic and anelectrotonic potentials and the local response, there are changes in the threshold of the neuron to stimulation. Hyperpolarizing anelectrotonic responses elevate the threshold and depolarizing catelectrotonic potentials lower it as they move the membrane potential closer to the firing level. During the local response, the threshold is lowered, but during the rising and much of the falling phases of the spike potential, the neuron is refractory to stimulation. This **refractory period** is divided into an **absolute refractory period,** corresponding to the period from the time the firing level is reached until repolarization is about one-third complete, and a **relative refractory period,** lasting from this point to the start of after-depolarization. During the absolute refractory period, no stimulus, no matter how strong, will excite the nerve, but during the relative refractory period, stronger than normal stimuli can cause excitation. During after-depolarization, the threshold is again decreased, and during after-hyperpolarization, it is increased. These changes in threshold are correlated with the phases of the action potential in Figure 2–9.

Electrogenesis of the Action Potential

The nerve cell membrane is polarized at rest, with positive charges lined up along the outside of the membrane and negative charges along the inside. During the action potential, this polarity is abolished and for a brief period is actually reversed (Figure 2–10). Positive charges from the membrane ahead of and behind the action potential flow into the area of negativity represented by the action potential ("current sink"). By drawing off positive charges, this flow decreases the polarity of the membrane ahead of the action potential. Such electrotonic depolarization initiates a local response, and when the firing level is reached, a propagated response occurs that in turn electrotonically depolarizes the membrane in front of it.

Saltatory Conduction

Conduction in myelinated axons depends upon a similar pattern of circular current flow. However, myelin is an effective insulator, and current flow through it is negligible. Instead, depolarization in myelinated axons jumps from one node of Ranvier to the next, with the current sink at the active node serving to electrotonically depolarize to the firing level the node ahead of the action potential (Figure 2–10). This jumping of depolarization from node to node is called **saltatory conduction.** It is a rapid process, and myelinated axons

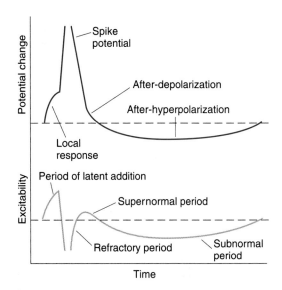

Figure 2–9. Relative changes in excitability of a nerve cell membrane during the passage of an impulse. Note that excitability is the reciprocal of threshold. (Modified and reproduced, with permission, from Morgan CT: *Physiological Psychology.* McGraw-Hill, 1943.)

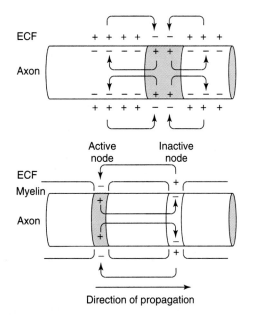

Figure 2–10. Local current flow (movement of positive charges) around an impulse in an axon. **Top:** Unmyelinated axon. **Bottom:** Myelinated axon.

conduct up to 50 times faster than the fastest unmyelinated fibers.

Orthodromic & Antidromic Conduction

An axon can conduct in either direction. When an action potential is initiated in the middle of it, two impulses traveling in opposite directions are set up by electrotonic depolarization on either side of the initial current sink.

In a living animal, impulses normally pass in one direction only, ie, from synaptic junctions or receptors along axons to their termination. Such conduction is called **orthodromic.** Conduction in the opposite direction is called **antidromic.** Since synapses, unlike axons, permit conduction in one direction only, any antidromic impulses that are set up fail to pass the first synapse they encounter (see Chapter 4) and die out at that point.

Biphasic Action Potentials

The descriptions of the resting membrane potential and action potential outlined above are based on recording with two electrodes, one on the surface of the axon and the other inside the axon. If both recording electrodes are placed on the surface of the axon, there is no potential difference between them at rest. When the nerve is stimulated and an impulse is conducted past the two electrodes, a characteristic sequence of potential changes results. As the wave of depolarization reaches the electrode nearest the stimulator, this electrode becomes negative relative to the other electrode (Figure 2–11). When the impulse passes to the portion of the nerve between the two electrodes, the potential returns to zero, and then, as it passes the second electrode, the first electrode becomes positive relative to the second. It is conventional to connect the leads in such a way that when the first electrode becomes negative relative to the second, an upward deflection is recorded. Therefore, the record shows an upward deflection followed by an isoelectric interval and then a downward deflection. This sequence is called a **biphasic action potential** (Figure 2–11). The duration of the isoelectric interval is proportionate to the speed of conduction of the nerve and the distance between the two recording electrodes.

Conduction in a Volume Conductor

Because the body fluids contain large quantities of electrolytes, the nerves in the body function in a conducting medium that is often called a **volume conductor.** The monophasic and biphasic action potentials described above are those seen when an axon is stimulated in a nonconducting medium outside the body. The po-

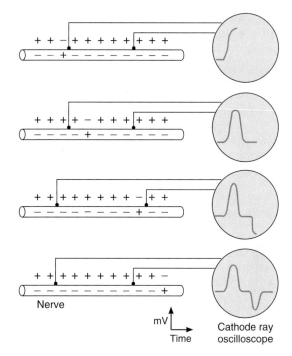

Figure 2–11. Biphasic action potential. Both recording electrodes are on the outside of the nerve membrane.

tential changes observed during extracellular recording in a volume conductor are basically similar to these action potentials, but they are complicated by the effects of current flow in the volume conductor. These effects are complex and are influenced by such factors as the orientation of the electrodes relative to the direction the action potential is moving and the distance between the recording electrode over active tissue and the indifferent electrode. In general, when an action potential is recorded in a volume conductor, there are positive deflections before and after the negative spike.

IONIC BASIS OF EXCITATION & CONDUCTION

The cell membranes of nerves, like those of other cells, contain many different types of ion channels. Some of these are voltage-gated and others are ligand-gated. It is the behavior of these channels, and particularly Na^+ and K^+ channels, that explains the electrical events in nerves.

Ionic Basis of Resting Membrane Potential

As pointed out in Chapter 1, Na^+ is actively transported out of neurons and other cells and K^+ is actively transported into cells. K$^+$ moves out of cells and Na$^+$ moves in, but because of K$^+$ channels, K$^+$ permeability at rest is greater than Na$^+$ permeability. Therefore, K$^+$ channels maintain the resting membrane potential. With catelectronic currents, some of the voltage-activated Na$^+$ channels become active, and when the firing level is reached, the voltage-activated Na$^+$ channels overwhelm the K$^+$ and other channels and a spike potential results.

Ionic Fluxes During the Action Potential

The changes in membrane conductance of Na$^+$ and K$^+$ that occur during the action potentials are shown in Figure 2–12. The conductance of an ion is the reciprocal of its electrical resistance in the membrane and is a measure of the membrane permeability to that ion.

The equilibrium potential for Na$^+$ in mammalian neurons, calculated by using the Nernst equation, is about +60 mV. The membrane potential moves toward this value but does not reach it during the action potential, primarily because the increase in Na$^+$ conductance is short-lived. The Na$^+$ channels rapidly enter a closed state called the **inactivated state** and remain in this state for a few milliseconds before returning to the resting state. In addition, the direction of the electrical gradient for Na$^+$ is reversed during the overshoot because the membrane potential is reversed, and this limits its Na$^+$ influx. A third factor producing repolarization

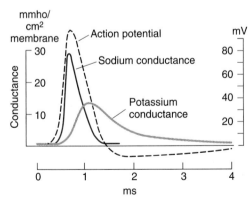

Figure 2–12. Changes in Na$^+$ and K$^+$ conductance during the action potential in giant squid axon. The dotted line represents the action potential superimposed on the same time coordinate. Note that the initial electrotonic depolarization initiates the change in Na$^+$ conductance, which in turn adds to the depolarization. (Redrawn and reproduced, with permission, from Hodgkin AL: Ionic movements and electrical activity in giant nerve fibers. Proc R Soc Lond Ser B 1958;143:1.)

is the opening of voltage-gated K$^+$ channels. This opening is slower and more prolonged than the opening of the Na$^+$ channels, and consequently, much of the increase in K$^+$ conductance comes after the increase in Na$^+$ conductance. The net movement of positive charge out of the cell due to K$^+$ efflux at this time helps complete the process of repolarization. The slow return of the K$^+$ channels to the closed state also explains the after-hyperpolarization.

Decreasing the external Na$^+$ concentration decreases the size of the action potential but has little effect on the resting membrane potential. The lack of much effect on the resting membrane potential would be predicted from the Goldman equation (see Chapter 1), since the permeability of the membrane to Na$^+$ at rest is relatively low. Conversely, increasing the external K$^+$ concentration decreases the resting membrane potential.

Although Na$^+$ enters the nerve cell and K$^+$ leaves it during the action potential, the number of ions involved is minute relative to the total numbers present. The fact that the nerve gains Na$^+$ and loses K$^+$ during activity has been demonstrated experimentally, but significant differences in ion concentrations can be measured only after prolonged, repeated stimulation.

The slower opening and delayed closing of the voltage-gated K$^+$ channels also explain accommodation. If depolarization occurs rapidly, the opening of the Na$^+$ channels overwhelms the repolarizing forces, but if the induced depolarization is produced slowly, the opening of K$^+$ channels balances the gradual opening of Na$^+$ channels, and an action potential does not occur.

A decrease in extracellular Ca^{2+} concentration increases the excitability of nerve and muscle cells by decreasing the amount of depolarization necessary to initiate the changes in the Na$^+$ and K$^+$ conductance that produce the action potential. Conversely, an increase in extracellular Ca^{2+} concentration "stabilizes the membrane" by decreasing excitability.

Distribution of Ion Channels in Myelinated Neurons

As noted in Chapter 1, various substances that bind to Na$^+$ and K$^+$ channels can be labeled and used to identify the locations of the channels in the cell membrane. Voltage-gated Na$^+$ channels are highly concentrated in the nodes of Ranvier and the initial segment in myelinated neurons. The initial segment and, in sensory neurons, the first node of Ranvier are the sites where impulses are normally generated, and the other nodes of Ranvier are the sites to which the impulses jump during saltatory conduction. The number of Na$^+$ channels per square micrometer of membrane in myelinated mammalian neurons has been estimated to be 50–75 in the cell body, 350–500 in the initial segment, less than 25 on the surface of the myelin, 2000–12,000 at the nodes of Ranvier, and 20–75 at the axon terminals. Along the axons of unmyelinated neurons, the number is about 110. In many myelinated neurons, the Na^+ channels are flanked by K^+ channels that are involved in repolarization.

Energy Sources & Metabolism of Nerve

The major part of the energy requirement of nerve— about 70%—is the portion used to maintain polarization of the membrane by the action of Na^+-K^+ ATPase. During maximal activity, the metabolic rate of nerve doubles; by comparison, the metabolic rate of skeletal muscle increases as much as 100-fold. Inhibition of lactic acid production does not influence nerve function.

Like muscle, nerve has a resting heat while inactive, an initial heat during the action potential, and a recovery heat that follows activity. However, in nerve, the recovery heat after a single impulse is about 30 times the initial heat. There is some evidence that the initial heat is produced during the after-depolarization rather than the spike. The metabolism of muscle is discussed in detail in Chapter 3.

PROPERTIES OF MIXED NERVES

Peripheral nerves in mammals are made up of many axons bound together in a fibrous envelope called the **epineurium.** Potential changes recorded extracellularly from such nerves therefore represent an algebraic summation of the all-or-none action potentials of many axons. The thresholds of the individual axons in the nerve and their distance from the stimulating electrodes vary. With subthreshold stimuli, none of the axons are stimulated and no response occurs. When the stimuli are of threshold intensity, axons with low thresholds fire and a small potential change is observed. As the intensity of the stimulating current is increased, the axons with higher thresholds are also discharged. The electrical response increases proportionately until the stimulus is strong enough to excite all of the axons in the nerve. The stimulus that produces excitation of all the axons is the **maximal stimulus,** and application of greater, **supramaximal** stimuli produces no further increase in the size of the observed potential.

Compound Action Potentials

Another property of mixed nerves, as opposed to single axons, is the appearance of multiple peaks in the action potential. The multipeaked action potential is called a **compound action potential** (Figure 2–13). It has a

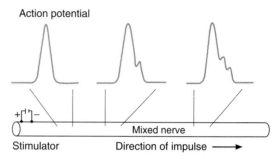

Figure 2–13. Compound action potential. The drawing shows the record obtained with recording electrodes at various distances from the stimulating electrodes along a mixed nerve.

unique shape because a mixed nerve is made up of families of fibers with various speeds of conduction. Therefore, when all the fibers are stimulated, the activity in fast-conducting fibers arrives at the recording electrodes sooner than the activity in slower fibers; and the farther away from the stimulating electrodes the action potential is recorded, the greater is the separation between the fast and slow fiber peaks. The number and size of the peaks vary with the types of fibers in the particular nerve being studied. If less than maximal stimuli are used, the shape of the compound action potential also depends upon the number and type of fibers stimulated.

NERVE FIBER TYPES & FUNCTION

Erlanger and Gasser divided mammalian nerve fibers into A, B, and C groups, further subdividing the A group into α, β, γ, and δ fibers. By comparing the neurologic deficits produced by careful dorsal root section and other nerve-cutting experiments with the histologic changes in the nerves, the functions and histologic characteristics of each of the families of axons responsible for the various peaks of the compound action potential have been established. In general, the greater the diameter of a given nerve fiber, the greater its speed of conduction. The large axons are concerned primarily with proprioceptive sensation, somatic motor function, conscious touch, and pressure, while the smaller axons subserve pain and temperature sensations and autonomic function. In Table 2–1, the various fiber types are listed with their diameters, electrical characteristics, and functions. The dorsal root C fibers conduct some impulses generated by touch and other cutaneous receptors in addition to impulses generated by pain and temperature receptors.

Further research has shown that not all the classically described lettered components are homogeneous,

Table 2–1. Nerve fiber types in mammalian nerve.[1]

Fiber Type	Function	Fiber Diameter (mm)	Conduction Velocity (m/s)	Spike Duration (ms)	Absolute Refractory Period (ms)
A					
α	Proprioception; somatic motor	12–20	70–120		
β	Touch, pressure, motor	5–12	30–70	0.4–0.5	0.4–1
γ	Motor to muscle spindles	3–6	15–30		
δ	Pain, cold, touch	2–5	12–30		
B	Preganglionic autonomic	<3	3–15	1.2	1.2
C					
Dorsal root	Pain, temperature, some mechano-reception, reflex responses	0.4–1.2	0.5–2	2	2
Sympathetic	Postganglionic sympathetics	0.3–1.3	0.7–2.3	2	2

[1]A and B fibers are myelinated; C fibers are unmyelinated.

and a numerical system (Ia, Ib, II, III, IV) has been used by some physiologists to classify sensory fibers. Unfortunately, this has led to confusion. A comparison of the number system and the letter system is shown in Table 2–2.

In addition to variations in speed of conduction and fiber diameter, the various classes of fibers in peripheral nerves differ in their sensitivity to hypoxia and anesthetics (Table 2–3). This fact has clinical as well as physiologic significance. Local anesthetics depress transmission in the group C fibers before they affect the touch fibers in the A group. Conversely, pressure on a nerve can cause loss of conduction in large-diameter motor, touch, and pressure fibers while pain sensation remains relatively intact. Patterns of this type are sometimes seen in individuals who sleep with their arms under their heads for long periods, causing compression of the nerves in the arms. Because of the association of deep sleep with alcoholic intoxication, the syndrome is commonest on weekends and has acquired the interesting name Saturday night or Sunday morning paralysis.

Table 2–2. Numerical classification sometimes used for sensory neurons.

Number	Origin	Fiber Type
Ia	Muscle spindle, annulo-spiral ending.	A α
Ib	Golgi tendon organ.	A α
II	Muscle spindle, flower-spray ending; touch, pressure.	A β
III	Pain and cold receptors; some touch receptors.	A δ
IV	Pain, temperature, and other receptors.	Dorsal root C

Table 2–3. Relative susceptibility of mammalian A, B, and C nerve fibers to conduction block produced by various agents.

Susceptibility to:	Most Susceptible	Intermediate	Least Susceptible
Hypoxia	B	A	C
Pressure	A	B	C
Local anesthetics	C	B	A

NEUROTROPHINS

Trophic Support of Neurons

A number of proteins that are necessary for survival and growth of neurons have been isolated and studied. Some of these **neurotrophins** are products of the muscles or other structures that the neurons innervate, but others are produced by astrocytes. These proteins bind to receptors at the endings of a neuron. They are internalized and then transported by retrograde transport to the neuronal cell body, where they foster the production of proteins associated with neuronal development, growth, and survival. Other neurotrophins are produced in neurons and transported anterogradely to the nerve ending, where they maintain the integrity of the postsynaptic neuron.

Receptors

Four established neurotrophins and their high-affinity receptors are listed in Table 2–4. Each of these **Trk receptors** dimerizes, and this initiates autophosphorylation in the cytoplasmic tyrosine kinase domains of the receptors. There is an additional low-affinity NGF receptor that is a 75-kDa protein and is called p75NTR. This receptor binds all four of the listed neurotrophins with equal affinity. There is some evidence that it can form a heterodimer with Trk A monomer and that the dimer has increased affinity and specificity for NGF. However, it now appears that p75NTR homodimers that bind neurotrophins can mediate responses by themselves and that those responses include production of apoptosis, an effect opposite to the usual growth-promoting and nurturing effects of neurotrophins.

Nerve Growth Factor

The first neurotrophin to be characterized was **nerve growth factor (NGF),** a protein growth factor that is necessary for the growth and maintenance of sympathetic neurons and some sensory neurons. It is present in a broad spectrum of animal species, including humans, and is found in many different tissues. In male mice, there is a particularly high concentration in the submandibular salivary glands, and the level is reduced by castration to that seen in females. The factor is made up of two α, two β, and two γ subunits. The β subunits, each of which has a molecular weight of 13,200, have all the nerve growth-promoting activity, the α subunits have trypsin-like activity, and the γ subunits are serine proteases. The function of the proteases is unknown. The structure of the β unit of NGF resembles that of insulin.

NGF is picked up by neurons in the extracerebral organs they innervate and is transported in retrograde

Table 2–4. Neurotrophins.

Neurotrophin	Receptor
Nerve growth factor (NGF)	Trk A
Brain-derived neuro-trophic factor (BDNF)	Trk B
Neurotrophin 3 (NT-3)	Trk C, less on Trk A and Trk B
Neurotrophin 4/5 (NT-4/5)	Trk B

fashion from the endings of the neurons to their cell bodies. It is also present in the brain and appears to be responsible for the growth and maintenance of cholinergic neurons in the basal forebrain and striatum. Injection of antiserum against NGF in newborn animals leads to near total destruction of the sympathetic ganglia; it thus produces an **immunosympathectomy.** There is evidence that the maintenance of neurons by NGF is due to a reduction in apoptosis.

Other Neurotrophins

Brain-derived neurotrophic factor (BDNF), neurotrophin 3 (NT-3), NT-4/5, and NGF each maintain a different pattern of neurons, preventing apoptosis, although there is some overlap. Disruption of NT-3 by gene knockout causes a marked loss of cutaneous mechanoreceptors (see Chapter 7), even in heterozygotes. BDNF acts rapidly and can actually depolarize neurons. BDNF-deficient mice lose peripheral sensory neurons and have severe degenerative changes in their vestibular ganglia and blunted LTP (see Chapter 4). An NT-6 has also been described.

Other Factors Affecting Neuronal Growth

Schwann cells and astrocytes produce **ciliary neurotrophic factor (CNTF).** This factor promotes the survival of damaged and embryonic spinal cord neurons and may prove to be of value in treating human diseases in which motor neurons degenerate. **Glial cell line-derived neurotrophic factor (GDNF)** maintains midbrain dopaminergic neurons in vitro. However, GDNF knockouts have dopaminergic neurons that appear normal. Instead, they have no kidneys and fail to develop an enteric nervous system (see Chapter 26). Another factor that enhances the growth of neurons is **leukemia inhibitory factor (LIF).** In addition, neurons as well as other cells respond to **insulin-like** growth factor I (IGF-I) and the various forms of **transforming growth factor (TGF), fibroblast growth factor (FGF),** and **platelet-derived growth factor (PDGF).** Thus, the regulation of neuronal growth is a complex process.

NEUROGLIA

In addition to neurons, the nervous system contains glial cells (neuroglia). Glial cells are very numerous; as noted above, there are 10–50 times as many glial cells as neurons. The Schwann cells that invest axons in peripheral nerves are classified as glia. In the CNS, there are three main types of neuroglia. **Microglia** (Figure 2–14) consists of scavenger cells that resemble tissue macrophages. They probably come from the bone marrow and enter the nervous system from the circulating blood vessels. **Oligodendrogliocytes** are involved in myelin formation (Figure 2–3). **Astrocytes,** which are found throughout the brain, are of two subtypes. **Fi-** brous astrocytes, which contain many intermediate filaments, are found primarily in white matter. **Protoplasmic astrocytes** are found in gray matter and have granular cytoplasm. Both types send processes to blood vessels, where they induce capillaries to form the tight junctions that form the blood-brain barrier (see Chapter 32). They also send processes that envelop synapses and the surface of nerve cells. They have a membrane potential that varies with the external K⁺ concentration but do not generate propagated potentials. They produce substances that are trophic to neurons, and they help maintain the appropriate concentration of ions and neurotransmitters by taking up K^+ and the neurotransmitters glutamate and γ-aminobutyrate (GABA; see Chapter 4).

The interaction between astrocytes and glutaminergic neurons is shown in Figure 2–15. Released glutamate is taken up by astrocytes and converted to glutamine, which passes back to the neurons and is converted back to glutamate, which is released as the synaptic transmitter.

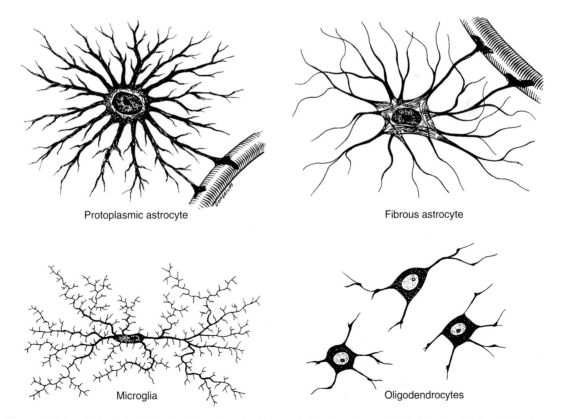

Protoplasmic astrocyte

Fibrous astrocyte

Microglia

Oligodendrocytes

Figure 2–14. Glial cells in the brain. (Reproduced, with permission, from Junqueira LC, Carneiro J, Kelley RO: *Basic Histology,* 9th ed. McGraw-Hill, 1998.)

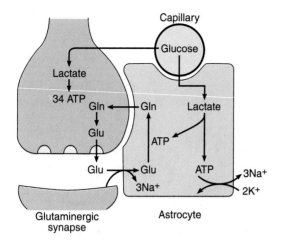

Figure 2–15. The glutamate-glutamine cycle through glutaminergic neurons and astrocytes. Glutamate released into the synaptic cleft is taken up by a Na⁺-dependent glutamate transporter, and in the astrocyte it is converted to glutamine. The glutamine enters the neuron and is converted to glutamate. Glucose is transported out of capillaries and enters astrocytes and neurons. In astrocytes, it is metabolized to lactate, producing two ATPs. One of these powers the conversion of glutamate to glutamine, and the other is used by Na⁺-K⁺ ATPase to transport three Na⁺ out of the cell in exchange for two K⁺. In neurons, the glucose is metabolized further through the citric acid cycle, producing 34 ATPs.

Astroglia has receptors for many neurotransmitters, and there is evidence that glial receptors are needed for the proper interaction between Bergman glia and glutaminergic neurons in the cerebellum. Beyond this, ideas about the importance of glia are increasing in number. However, to date, the ideas are long on speculation and short on facts.

Excitable Tissue: Muscle

3

INTRODUCTION

Muscle cells, like neurons, can be excited chemically, electrically, and mechanically to produce an action potential that is transmitted along their cell membrane. Unlike neurons, they have a contractile mechanism that is activated by the action potential. The contractile proteins actin and myosin are abundant in muscle, where they bring about contraction. They are found in many different types of cells, and as described in Chapter 1, the actin-binding protein myosin and actin make up one of the molecular motors that converts the energy of ATP hydrolysis into movement of one cellular component along another.

Muscle is generally divided into three types, **skeletal, cardiac,** and **smooth,** though smooth muscle is not a homogeneous single category. Skeletal muscle makes up the great mass of the somatic musculature. It has well-developed cross-striations, does not normally contract in the absence of nervous stimulation, lacks anatomic and functional connections between individual muscle fibers, and is generally under voluntary control. Cardiac muscle also has cross-striations, but it is functionally syncytial and contracts rhythmically in the absence of external innervation owing to the presence in the myocardium of pacemaker cells that discharge spontaneously. Smooth muscle lacks cross-striations. The type found in most hollow viscera is functionally syncytial and contains pacemakers that discharge irregularly. The type found in the eye and in some other locations is not spontaneously active and resembles skeletal muscle.

SKELETAL MUSCLE

MORPHOLOGY

Organization

Skeletal muscle is made up of individual muscle fibers that are the "building blocks" of the muscular system in the same sense that the neurons are the building blocks of the nervous system. Most skeletal muscles begin and end in tendons, and the muscle fibers are arranged in parallel between the tendinous ends, so that the force of contraction of the units is additive. Each muscle fiber is a single cell that is multinucleated, long, cylindric, and surrounded by a cell membrane, the **sarcolemma** (Figure 3–1). There are no syncytial bridges between cells. The muscle fibers are made up of myofibrils, which are divisible into individual filaments. The filaments are made up of the contractile proteins.

The contractile mechanism in skeletal muscle depends on the proteins **myosin-II** (molecular weight 460,000), **actin** (molecular weight 43,000), **tropomyosin** (molecular weight 70,000), and **troponin.** Troponin is made up of three subunits, **troponin I, troponin T,** and **troponin C.** The three subunits have molecular weights ranging from 18,000 to 35,000. Other important proteins in muscle are involved in maintaining the contractile proteins in appropriate relation to each other and connected to the extracellular matrix.

Striations

Differences in the refractive indexes of the various parts of the muscle fiber are responsible for the characteristic cross-striations seen in skeletal muscle. The parts of the cross-striations are identified by letters (Figure 3–2). The light I band is divided by the dark Z line, and the dark A band has the lighter H band in its center. A transverse M line is seen in the middle of the H band, and this line plus the narrow light areas on either side of it are sometimes called the pseudo-H zone. The area between two adjacent Z lines is called a **sarcomere.** The orderly arrangement of actin, myosin, and related proteins that produce this pattern is shown in Figure 3–3. The thick filaments, which are about twice the diameter of the thin filaments, are made up of myosin; the thin filaments are made up of actin, tropomyosin, and troponin. The thick filaments are lined up to form the A bands, whereas the array of thin filaments forms the less dense I bands. The lighter H bands in the center of the A bands are the regions where, when the muscle is relaxed, the thin filaments do not overlap the thick filaments. The Z lines transect the fibrils and connect to the thin filaments. If a transverse section through the A band is examined under the electron mi-

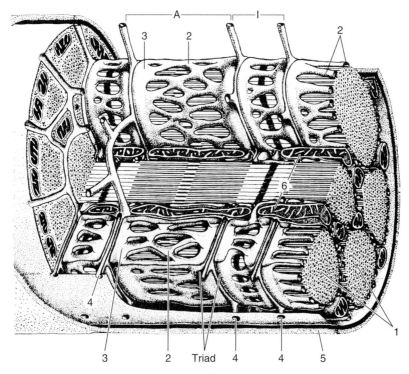

Figure 3–1. Mammalian skeletal muscle. A single muscle fiber surrounded by its sarcolemma has been cut away to show individual myofibrils (1). The cut surface of the myofibrils shows the arrays of thick and thin filaments. The sarcoplasmic reticulum (2) with its terminal cisterns (3) surrounds each myofibril. The T system of tubules (4), which invaginates from the sarcolemma, contacts the myofibrils between the A and I bands twice in every sarcomere. The T system and the adjacent cisterns of the sarcoplasmic reticulum constitute a triad. A basal lamina (5) surrounds the sarcolemma. (6) Mitochondria. (Modified and reproduced, with permission, from Krstic RV: *Ultrastructure of the Mammalian Cell.* Springer, 1979.)

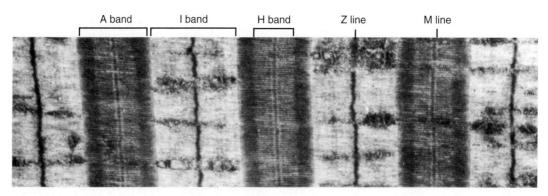

Figure 3–2. Electron micrograph of human gastrocnemius muscle. The various bands and lines are identified at the top. (× 13,500.) (Courtesy of SM Walker and GR Schrodt.)

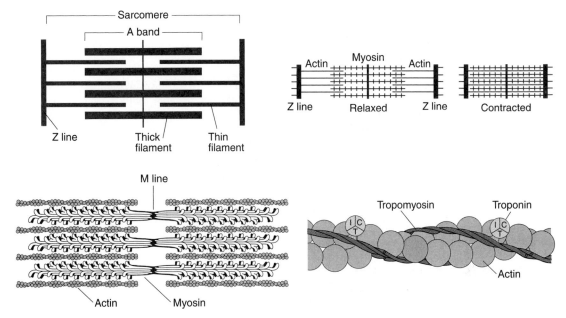

Figure 3–3. Top left: Arrangement of thin (actin) and thick (myosin) filaments in skeletal muscle. **Top right:** Sliding of actin on myosin during contraction so that Z lines move closer together. **Bottom left:** Detail of relation of myosin to actin. Note that myosin thick filaments reverse polarity at the M line in the middle of the sarcomere. (Modified from Alberts B et al: *Molecular Biology of the Cell,* 2nd ed. Garland, 1989.) **Bottom right:** Diagrammatic representation of the arrangement of actin, tropomyosin, and the three subunits of troponin (I, C, and T). The structure of an individual myosin-II molecule is shown in Figure 1–11.

croscope, each thick filament is found to be surrounded by six thin filaments in a regular hexagonal pattern.

The form of myosin found in muscle is myosin-II, with two globular heads and a long tail (see Figure 1–11). The heads and necks of the myosin molecules form cross-links to actin. Myosin contains heavy chains and light chains, and its heads are made up of the light chains and the amino terminal portions of the heavy chains. These heads contain an actin-binding site and a catalytic site that hydrolyzes ATP (see below). The myosin molecules are arranged symmetrically on either side of the center of the sarcomere, and it is this arrangement that creates the light areas in the pseudo-H zone. The M line is the site of the reversal of polarity of the myosin molecules in each of the thick filaments. At these points, there are slender cross-connections that hold the thick filaments in proper array. There are several hundred myosin molecules in each thick segment.

The thin filaments are polymers made up of two chains of actin that form a long double helix. Tropomyosin molecules are long filaments located in the groove between the two chains in the actin (Figure 3–3). Each thin filament contains 300–400 actin molecules and 40–60 tropomyosin molecules. Troponin

molecules are small globular units located at intervals along the tropomyosin molecules. Troponin T binds the other troponin components to tropomyosin, troponin I inhibits the interaction of myosin with actin (see below), and troponin C contains the binding sites for the Ca^{2+} that initiates contraction.

Actinin, which has a molecular weight of 190,000, binds actin to the Z lines. **Titin,** a large protein, connects the Z lines to the M lines and provides the scaffolding for the sarcomere. It contains two kinds of folded domains, and these provide muscle with its elasticity. At first when the muscle is stretched there is relatively little resistance as the domains unfold, but with further stretch there is a rapid increase in resistance that protects the structure of the sarcomere. **Desmin** binds the Z lines to the plasma membrane.

Sarcotubular System

The muscle fibrils are surrounded by structures made up of membrane that appear in electron photomicrographs as vesicles and tubules. These structures form the **sarcotubular system,** which is made up of a **T system** and a **sarcoplasmic reticulum.** The T system of transverse tubules, which is continuous with the membrane of the muscle fiber, forms a grid perforated by the individual muscle fibrils (Figure 3–1). The space between the two layers of the T system is an extension of the extracellular space. The sarcoplasmic reticulum, which forms an irregular curtain around each of the fibrils, has enlarged **terminal cisterns** in close contact with the T system at the junctions between the A and I bands. At these points of contact, the arrangement of the central T system with a cistern of the sarcoplasmic reticulum on either side has led to the use of the term **triads** to describe the system. The function of the T system, which is continuous with the sarcolemma, is the rapid transmission of the action potential from the cell membrane to all the fibrils in the muscle. The sarcoplasmic reticulum is concerned with Ca^{2+} movement and muscle metabolism (see below).

Dystrophin-Glycoprotein Complex

A large protein called **dystrophin** (molecular weight 427,000) forms a rod that connects the thin actin filaments to the transmembrane protein β-**dystroglycan** in the sarcolemma, and β-dystroglycan is connected to **laminin** in the extracellular matrix by α-**dystroglycan** (Figure 3–4). The dystroglycans are in turn associated with a complex of four transmembrane glycoproteins: α-, β-, γ-, and δ-sarcoglycan. It appears that this **dys-**

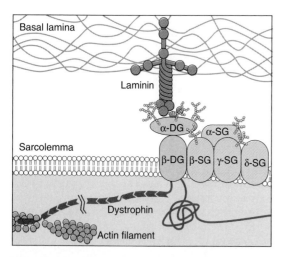

Figure 3–4. The dystrophin-glycoprotein complex. Dystrophin connects actin to the two members of the dystroglycan (DG) complex, α- and β-dystroglycan, and these in turn connect to the merosin subunit of laminin in the extracellular matrix. The sarcoglycan (SG) complex of four glycoproteins, α-, β-, γ-, and δ-sarcoglycan, are associated with the dystroglycan complex.

trophin-glycoprotein complex provides structural support and strength to the muscle fibril and transmits the force generated by contraction to the cytoskeleton. It may well have additional functions. The complex is of special interest because congenital defects in it cause many of the different forms of muscular dystrophy (see below). The dystrophin gene, which is one of the largest known genes, is also expressed in cardiac and smooth muscle and, with an alternative promoter and first exon, in the brain.

ELECTRICAL PHENOMENA & IONIC FLUXES

Electrical Characteristics of Skeletal Muscle

The electrical events in skeletal muscle and the ionic fluxes underlying them are similar to those in nerve, although there are quantitative differences in timing and magnitude. The resting membrane potential of skeletal muscle is about −90 mV. The action potential lasts 2–4 ms and is conducted along the muscle fiber at about 5 m/s. The absolute refractory period is 1–3 ms long, and the after-polarizations, with their related changes in threshold to electrical stimulation, are relatively prolonged. The initiation of impulses at the myoneural junction is discussed in Chapter 4.

Although the electrical properties of the individual fibers in a muscle do not differ sufficiently to produce anything resembling a compound action potential, there are slight differences in the thresholds of the various fibers. Furthermore, in any stimulation experiment, some fibers are farther from the stimulating electrodes than others. Therefore, the size of the action potential recorded from a whole-muscle preparation is proportionate to the intensity of the stimulating current between threshold and maximal current intensities.

Ion Distribution & Fluxes

The distribution of ions across the muscle fiber membrane is similar to that across the nerve cell membrane. The values for the various ions and their equilibrium potentials are shown in Table 3–1. As in nerves, depolarization is a manifestation of Na^+ influx, and repolarization is a manifestation of K^+ efflux (as described in Chapter 2 for nerves).

CONTRACTILE RESPONSES

It is important to distinguish between the electrical and mechanical events in muscle. Although one response does not normally occur without the other, their physiologic basis and characteristics are different. Muscle

Table 3–1. Steady-state distribution of ions in the intracellular and extracellular compartments of mammalian skeletal muscle, and the equilibrium potentials for these ions.[1]

| Ion[2] | Concentration (mmol/L) | | Equilibrium Potential (mV) |
	Intracellular Fluid	Extracellular Fluid	
Na$^+$	12	145	+65
K$^+$	155	4	−95
H$^+$	13×10^{-5}	3.8×10^{-5}	−32
Cl$^-$	3.8	120	−90
HCO$_3^-$	8	27	−32
A$^-$	155	0	....
Membrane potential = −90 mV			

[1]Data from Ruch TC, Patton HD (editors): *Physiology and Biophysics,* 19th ed. Saunders, 1965.

[2]A$^-$ represents organic anions. The value for intracellular Cl$^-$ is calculated from the membrane potential, using the Nernst equation.

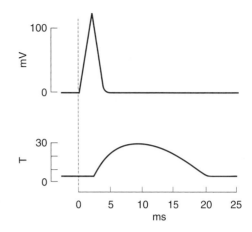

Figure 3–5. The electrical and mechanical responses of a mammalian skeletal muscle fiber to a single maximal stimulus. The electrical response (mV potential change) and the mechanical response (T, tension in arbitrary units) are plotted on the same abscissa (time).

fiber membrane depolarization normally starts at the motor end-plate, the specialized structure under the motor nerve ending (see Chapter 4); the action potential is transmitted along the muscle fiber and initiates the contractile response.

The Muscle Twitch

A single action potential causes a brief contraction followed by relaxation. This response is called a **muscle twitch.** In Figure 3–5, the action potential and the twitch are plotted on the same time scale. The twitch starts about 2 ms after the start of depolarization of the membrane, before repolarization is complete. The duration of the twitch varies with the type of muscle being tested (see below). "Fast" muscle fibers, primarily those concerned with fine, rapid, precise movement, have twitch durations as short as 7.5 ms. "Slow" muscle fibers, principally those involved in strong, gross, sustained movements, have twitch durations up to 100 ms.

Molecular Basis of Contraction

The process by which the shortening of the contractile elements in muscle is brought about is a sliding of the thin filaments over the thick filaments. The width of the A bands is constant, whereas the Z lines move closer together when the muscle contracts and farther apart when it is stretched (Figure 3–3).

The sliding during muscle contraction occurs when the myosin heads bind firmly to actin, bend at the junction of the head with the neck, and then detach. Myosin-II molecules are dimers with two heads, but the heads are independent and only one attaches to actin at a time. This "power stroke" depends on the simultaneous hydrolysis of ATP. There is still debate about the exact details, and there may be several steps involved, but the overall cycle is as shown in Figure 3–6. Many heads cycle at or near the same time, and they cycle repeatedly, producing gross muscle contraction. Each power stroke shortens the sarcomere about 10 nm. Each thick filament has about 500 myosin heads, and each head cycles about five times per second during a rapid contraction.

The process by which depolarization of the muscle fiber initiates contraction is called **excitation-contraction coupling.** The action potential is transmitted to all the fibrils in the fiber via the T system (Table 3–2). It triggers the release of Ca^{2+} from the terminal cisterns, the lateral sacs of the sarcoplasmic reticulum next to the T system. Ca^{2+} initiates contraction by binding to troponin C. In resting muscle, troponin I is tightly bound to actin and tropomyosin covers the sites where myosin heads bind to actin. Thus, the troponin-tropomyosin complex constitutes a "relaxing protein" that inhibits the interaction between actin and myosin. When the Ca^{2+} released by the action potential binds to troponin C, the binding of troponin I to actin is presumably weakened,

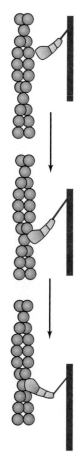

Figure 3–6. Power stroke of myosin in skeletal muscle. The myosin head detaches from actin **(top),** moves several nm along the actin strand, and reattaches **(middle).** The head then flexes on the neck of the myosin molecule **(bottom),** moving the myosin along the actin strand. There is an ATP-binding site 3.5 nm behind the actin-binding site on the head. ATP hydrolysis occurs during the power stroke, though the details of the relation of ATP to the stroke remain unsettled. (Modified from Irving M, Goldman YE: Another step ahead for myosin. Nature 1999;398:463.)

Table 3–2. Sequence of events in contraction and relaxation of skeletal muscle.

Steps in contraction[1]
(1) Discharge of motor neuron.
(2) Release of transmitter (acetylcholine) at motor end-plate.
(3) Binding of acetylcholine to nicotinic acetylcholine receptors.
(4) Increased Na^+ and K^+ conductance in end-plate membrane.
(5) Generation of end-plate potential.
(6) Generation of action potential in muscle fibers.
(7) Inward spread of depolarization along T tubules.
(8) Release of Ca^{2+} from terminal cisterns of sarcoplasmic reticulum and diffusion to thick and thin filaments.
(9) Binding of Ca^{2+} to troponin C, uncovering myosin-binding sites on actin.
(10) Formation of cross-linkages between actin and myosin and sliding of thin on thick filaments, producing shortening.
Steps in relaxation
(1) Ca^{2+} pumped back into sarcoplasmic reticulum.
(2) Release of Ca^{2+} from troponin.
(3) Cessation of interaction between actin and myosin.

[1]Steps 1–6 in contraction are discussed in Chapter 4.

and this permits the tropomyosin to move laterally (Figure 3–7). This movement uncovers binding sites for the myosin heads. ATP is then split and contraction occurs. Seven myosin-binding sites are uncovered for each molecule of troponin that binds a calcium ion.

Shortly after releasing Ca^{2+}, the sarcoplasmic reticulum begins to reaccumulate it by actively transporting it into the longitudinal portions of the reticulum. The Ca^{2+} then diffuses into the terminal cisterns, where it is stored until released by the next action potential. Once the Ca^{2+} concentration outside the reticulum has been lowered sufficiently, chemical interaction between myosin and actin ceases and the muscle relaxes. Note that ATP provides the energy for both contraction and relaxation. If transport of Ca^{2+} into the reticulum is inhibited, relaxation does not occur even though there are no more action potentials; the resulting sustained contraction is called a **contracture.**

Depolarization of the T tubule membrane activates the sarcoplasmic reticulum via **dihydropyridine receptors,** named for the drug dihydropyridine, which blocks them (Figure 3–8). They are voltage-gated Ca^{2+} channels in the T tubule membrane. In cardiac muscle, influx of Ca^{2+} via these channels triggers the release of Ca^{2+} stored in the sarcoplasmic reticulum, but in skeletal muscle, Ca^{2+} entry from the ECF by this route is not required for Ca^{2+} release. Instead, the dihydropyri-

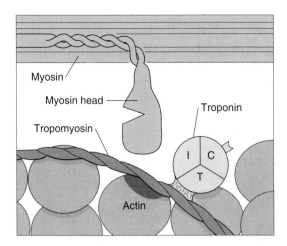 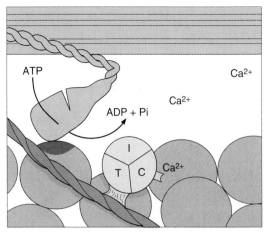

Figure 3–7. Initiation of muscle contraction by Ca^{2+}. When Ca^{2+} binds to troponin C, tropomyosin is displaced laterally, exposing the binding site for myosin on actin (dark area). The myosin head then binds, ATP is hydrolyzed, and the configuration of the head and neck region of myosin changes. For simplicity, only one of the two heads of the myosin-II molecule is shown.

dine receptor serves as the voltage sensor and trigger that unlocks release of Ca^{2+} from the nearby sarcoplasmic reticulum. The Ca^{2+} channel in the sarcoplasmic reticulum that opens to permit the outpouring of Ca^{2+}

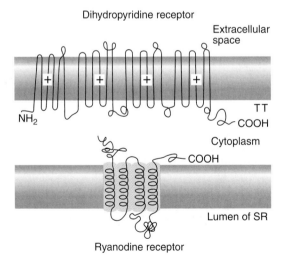

Figure 3–8. Relation of the T tubule (TT) to the sarcoplasmic reticulum in Ca^{2+} transport. In skeletal muscle, the voltage-gated dihydropyridine receptor in the T tubule triggers Ca^{2+} release from the sarcoplasmic reticulum (SR) via the ryanodine receptor. (After Takeshima. Modified from Zigmond MJ et al [editors]: *Fundamental Neuroscience*. Academic Press, 1999.)

is not voltage-gated and is called the **ryanodine receptor** (Figure 3–8) because it is locked in the open position by the plant alkaloid ryanodine. It is closely related to the IP_3 receptor, a ligand-gated Ca^{2+} channel that, when it binds IP_3, permits Ca^{2+} to enter the cytoplasm from the endoplasmic reticulum (see Chapter 1). The pump that moves Ca^{2+} back into the reticulum, producing relaxation, is a Ca^{2+}-Mg^{2+} ATPase.

The events involved in muscle contraction and relaxation are summarized in Table 3–2.

Types of Contraction

Muscular contraction involves shortening of the contractile elements, but because muscles have elastic and viscous elements in series with the contractile mechanism, it is possible for contraction to occur without an appreciable decrease in the length of the whole muscle (Figure 3–9). Such a contraction is called **isometric** ("same measure" or length). Contraction against a constant load, with approximation of the ends of the muscle, is **isotonic** ("same tension"). Note that since work is the product of force times distance, isotonic contractions do work whereas isometric contractions do not. In other situations, muscle can do negative work while lengthening against a constant weight.

Summation of Contractions

The electrical response of a muscle fiber to repeated stimulation is like that of nerve. The fiber is electrically refractory only during the rising and part of the falling

A

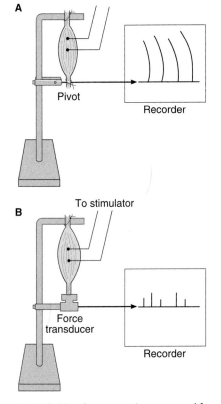

Pivot

Recorder

To stimulator

B

Force
transducer

Recorder

Figure 3–9. **A:** Muscle preparation arranged for recording isotonic contractions. **B:** Preparation arranged for recording isometric contractions. In **A,** the muscle is fastened to a writing lever that swings on a pivot. In **B,** it is attached to an electronic transducer that measures the force generated without permitting the muscle to shorten.

phase of the spike potential. At this time, the contraction initiated by the first stimulus is just beginning. However, because the contractile mechanism does not have a refractory period, repeated stimulation before relaxation has occurred produces additional activation of the contractile elements and a response that is added to the contraction already present. This phenomenon is

known as **summation of contractions.** The tension developed during summation is considerably greater than that during the single muscle twitch. With rapidly repeated stimulation, activation of the contractile mechanism occurs repeatedly before any relaxation has occurred, and the individual responses fuse into one continuous contraction. Such a response is called a **tetanus (tetanic contraction).** It is a **complete tetanus** when there is no relaxation between stimuli and an **incomplete tetanus** when there are periods of incomplete relaxation between the summated stimuli. During a complete tetanus, the tension developed is about four times that developed by the individual twitch contractions. The development of an incomplete and a complete tetanus in response to stimuli of increasing frequency is shown in Figure 3–10.

The stimulation frequency at which summation of contractions occurs is determined by the twitch duration of the particular muscle being studied. For example, if the twitch duration is 10 ms, frequencies less than 1/10 ms (100/s) cause discrete responses interrupted by complete relaxation, and frequencies greater than 100/s cause summation.

Treppe

When a series of maximal stimuli is delivered to skeletal muscle at a frequency just below the tetanizing frequency, there is an increase in the tension developed during each twitch until, after several contractions, a uniform tension per contraction is reached. This phenomenon is known as **treppe,** or the "staircase" phenomenon (G *Treppe* "staircase"). It also occurs in cardiac muscle. Treppe is believed to be due to increased availability of Ca^{2+} for binding to troponin C. It should not be confused with summation of contractions and tetanus.

Relation Between Muscle Length, Tension, & Velocity of Contraction

Both the tension that a muscle develops when stimulated to contract isometrically (the **total tension**) and the **passive tension** exerted by the unstimulated mus-

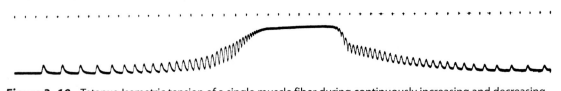

Figure 3–10. Tetanus. Isometric tension of a single muscle fiber during continuously increasing and decreasing stimulation frequency. Dots at the top are at intervals of 0.2 s.

cle vary with the length of the muscle fiber. This relationship can be studied in a whole skeletal muscle preparation such as that shown in Figure 3–9. The length of the muscle can be varied by changing the distance between its two attachments. At each length, the passive tension is measured, the muscle is then stimulated electrically, and the total tension is measured. The difference between the two values at any length is the amount of tension actually generated by the contractile process, the **active tension.** The records obtained by plotting passive tension and total tension against muscle length are shown in Figure 3–11. Similar curves are obtained when single muscle fibers are studied. The length of the muscle at which the active tension is maximal is usually called its **resting length.** The term comes originally from experiments demonstrating that the length of many of the muscles in the body at rest is the length at which they develop maximal tension.

The observed length-tension relation in skeletal muscle is explained by the sliding filament mechanism of muscle contraction. When the muscle fiber contracts isometrically, the tension developed is proportionate to the number of cross-linkages between the actin and the myosin molecules. When muscle is stretched, the overlap between actin and myosin is reduced and the number of cross-linkages is therefore reduced. Conversely, when the

muscle is appreciably shorter than resting length, the distance the thin filaments can move is reduced.

The velocity of muscle contraction varies inversely with the load on the muscle. At a given load, the velocity is maximal at the resting length and declines if the muscle is shorter or longer than this length.

Fiber Types

Although skeletal muscle fibers resemble one another in a general way, skeletal muscle is a very heterogeneous tissue made up of fibers that vary in myosin ATPase activity, contractile speed, and other properties. The fibers fall roughly into two types, type I and type II, although each of these types is itself a spectrum. The properties of type I and type II fibers are summarized in Table 3–3.

Muscles containing many type I fibers are called **red muscles** because they are darker than other muscles. The red muscles, which respond slowly and have a long latency, are adapted for long, slow, posture-maintaining contractions. The long muscles of the back are red muscles. **White muscles,** which contain mostly type II fibers, have short twitch durations and are specialized for fine, skilled movement. The extraocular muscles and some of the hand muscles contain many type II fibers and are generally classified as white muscles.

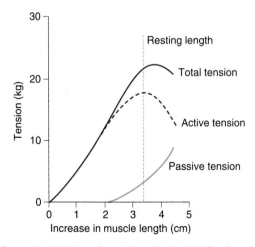

Figure 3–11. Length-tension relationship for the human triceps muscle. The passive tension curve measures the tension exerted by this skeletal muscle at each length when it is not stimulated. The total tension curve represents the tension developed when the muscle contracts isometrically in response to a maximal stimulus. The active tension is the difference between the two.

Table 3–3. Classification of fiber types in skeletal muscles.[1]

	Type I	Type II
Other names	Slow; oxidative; red	Fast; glycolytic; white
Myosin isoenzyme ATPase rate	Slow	Fast
Ca^{2+}-pumping capacity of sarcoplasmic reticulum	Moderate	High
Diameter	Moderate	Large
Glycolytic capacity	Moderate	High
Oxidative capacity (correlates with content of mitochondria, capillary density, myoglobin content)	High	Low

[1]Modified from Murphy RA: Muscle. In: *Physiology,* 2nd ed. Berne RM, Levy MN (editors). Mosby, 1988.

Protein Isoforms in Muscle & Their Genetic Control

The differences in the fibers that make up muscles stem from differences in the proteins in them. Most of these are encoded by multigene families. Ten different **isoforms** of the myosin heavy chains (MHCs) have been characterized. Isoforms have similar biologic activity but different amino acid compositions. There are also fast and slow isoforms of each of the two types of light chains. It appears that there is only one form of actin, but there are multiple isoforms of tropomyosin and all three components of troponin. The general composition of slow and fast fibers is summarized in Table 3–4. Multiple isoforms are also found in cardiac muscle.

The expression of the MHCs is precisely regulated during development, and alterations in expression play a major role in the assembly of muscles. In adults, the proportions of the various isoforms determine the functional characteristics of each muscle. In addition, changes in muscle function can be produced by alterations in activity, innervation, and hormonal milieu, and these changes are generally produced by alterations in the transcription of MHC genes. Examples include conversion of fast to slow skeletal muscle when muscle innervation is changed (see below) and marked alter-

Table 3–4. Myofibrillar protein isoforms in adult extrafusal fibers in the rat.[1,2]

	Slow	Fast
Myosin heavy chain	HCI(=HCβ$_{card}$), HCI$_{ton}$, HCα$_{card}$	HCIIb, HCIId, HCIIa, HC$_{eom}$, HC$_{sf}$, HC$_{emb}$, HC$_{neo}$
Myosin light chains		
Alkali	1s$_a$, 1s$_b$	1f, 3f
Phosphorylatable	2s	2f
Tropomyosin	α$_s$ < β	α$_f$ < β
Troponin		
TN-T	1s, 2s	1f, 2f, 3f, 4f
TN-I	s	f
TN-C	s	f

[1]Modified from Pette D, Staron RS: The molecular diversity of mammalian muscle fibers. *News Physiol Sci* 1993;8:153.

[2]Abbreviations are as follows: heavy-chain (HC) isoforms: HCI, type I; HCII, type II; HCα$_{card}$, HCβ$_{card}$, cardiac; HCI$_{ton}$, tonic; HC$_{eom}$, extraocular; HC$_{sf}$, superfast; HC$_{emb}$, embryonic; HC$_{neo}$, neonatal. Troponin (TN) subunits: s, slow; f, fast; TN-T, tropomyosin-binding; TN-I, inhibitory; TN-C, Ca^{2+}-binding.

ations in the isoforms in skeletal and cardiac muscle in hypothyroidism.

ENERGY SOURCES & METABOLISM

Muscle contraction requires energy, and muscle has been called "a machine for converting chemical energy into mechanical work." The immediate source of this energy is the energy-rich organic phosphate derivatives in muscle; the ultimate source is the intermediary metabolism of carbohydrate and lipids. The hydrolysis of ATP to provide the energy for contraction has been discussed above.

Phosphorylcreatine

ATP is resynthesized from ADP by the addition of a phosphate group. Some of the energy for this endothermic reaction is supplied by the breakdown of glucose to CO_2 and H_2O, but there also exists in muscle another energy-rich phosphate compound that can supply this energy for short periods. This compound is **phosphorylcreatine** (see Figure 17–21), which is hydrolyzed to creatine and phosphate groups with the release of considerable energy. At rest, some ATP in the mitochondria transfers its phosphate to creatine, so that a phosphorylcreatine store is built up. During exercise, the phosphorylcreatine is hydrolyzed at the junction between the myosin heads and the actin, forming ATP from ADP and thus permitting contraction to continue.

Carbohydrate & Lipid Breakdown

At rest and during light exercise, muscles utilize lipids in the form of free fatty acids (FFA; see Chapter 17) as their energy source. As the intensity of exercise increases, lipids alone cannot supply energy fast enough and so utilization of carbohydrate becomes the predominant component in the muscle fuel mixture. Thus, during exercise, much of the energy for phosphorylcreatine and ATP resynthesis comes from the breakdown of glucose to CO_2 and H_2O. An outline of the major metabolic pathways involved is presented in Chapter 17. For the purposes of the present discussion, it is sufficient to point out that glucose in the bloodstream enters cells, where it is degraded through a series of chemical reactions to pyruvate. Another source of intracellular glucose, and consequently of pyruvate, is glycogen, the carbohydrate polymer that is especially abundant in liver and skeletal muscle. When adequate O_2 is present, pyruvate enters the citric acid cycle and is metabolized—through this cycle and the so-called respiratory enzyme pathway—to CO_2 and H_2O. This process is called **aerobic glycolysis.** The metabolism of glucose or glycogen to CO_2 and H_2O liberates sufficient energy to form large quantities of ATP from ADP. If O_2 supplies are insufficient, the pyruvate formed from glucose does not enter the tricarboxylic acid cycle but is reduced to lactate. This process of **anaerobic glycolysis** is associated with the net production of much smaller quantities of energy-rich phosphate bonds, but it does not require the presence of O_2. The various reactions involved in supplying energy to skeletal muscle are summarized in Figure 3–12.

The Oxygen Debt Mechanism

During muscular exercise, the muscle blood vessels dilate and blood flow is increased so that the available O_2 supply is increased. Up to a point, the increase in O_2 consumption is proportionate to the energy expended, and all the energy needs are met by aerobic processes. However, when muscular exertion is very great, aerobic resynthesis of energy stores cannot keep pace with their utilization. Under these conditions, phosphorylcreatine is still used to resynthesize ATP. Some ATP synthesis is accomplished by using the energy released by the anaerobic breakdown of glucose to lactate. Use of the anaerobic pathway is self-limiting because in spite of rapid diffusion of lactate into the bloodstream, enough accumulates in the muscles to eventually exceed the capacity of the tissue buffers and produce an enzyme-inhibiting decline in pH. However, for short periods, the presence of an anaerobic pathway for glucose breakdown permits muscular exertion of a far greater magnitude than would be possible without it. For example, in a 100-meter dash that takes 10 seconds, 85% of the energy consumed is derived anaerobically; in a 2-mile race that takes 10 minutes, 20% of the energy is derived anaerobically; and in a long-distance race that takes 60 minutes, only 5% of the energy comes from anaerobic metabolism.

After a period of exertion is over, extra O_2 is consumed to remove the excess lactate, replenish the ATP and phosphorylcreatine stores, and replace the small amounts of O_2 that have come from myoglobin. The amount of extra O_2 consumed is proportionate to the extent to which the energy demands during exertion exceeded the capacity for the aerobic synthesis of energy stores, ie, the extent to which an **oxygen debt** was incurred. The O_2 debt is measured experimentally by determining O_2 consumption after exercise until a constant, basal consumption is reached and subtracting the basal consumption from the total. The amount of this debt may be six times the basal O_2 consumption, which indicates that the subject is capable of six times the exertion that would have been possible without it. Obviously, the maximal debt can be incurred rapidly or slowly; violent exertion is possible for only short periods, whereas less strenuous exercise can be carried on for longer periods.

Trained athletes are able to increase the O_2 consumption of their muscles to a greater degree than untrained individuals and are able to utilize FFA more effectively. Consequently, they are capable of greater exertion without depleting their glycogen stores and increasing their lactate production. Because of this, they contract smaller oxygen debts for a given amount of exertion. They have also learned to gorge on carbohydrates for several days before a competitive event, increasing their muscle stores of glycogen. This alone can greatly increase their endurance.

Rigor

When muscle fibers are completely depleted of ATP and phosphorylcreatine, they develop a state of rigidity called **rigor**. When this occurs after death, the condition is called rigor mortis. In rigor, almost all of the myosin heads attach to actin but in an abnormal, fixed, and resistant way.

Heat Production in Muscle

Thermodynamically, the energy supplied to a muscle must equal its energy output. The energy output appears in work done by the muscle, in energy-rich phosphate bonds formed for later use, and in heat. The overall mechanical efficiency of skeletal muscle (work done/total energy expenditure) ranges up to 50% while lifting a weight during isotonic contraction and is essentially 0% during isometric contraction. Energy storage in phos-

$$ATP + H_2O \rightarrow ADP + H_3PO_4 + 7.3 \text{ kcal}$$

$$\text{Phosphorylcreatine} + ADP \rightleftharpoons \text{Creatine} + ATP$$

$$\text{Glucose} + 2 \text{ ATP (or glycogen} + 1 \text{ ATP)}$$

$$\xrightarrow[\text{Anaerobic}]{} 2 \text{ Lactic acid} + 4 \text{ ATP}$$

$$\text{Glucose} + 2 \text{ ATP (or glycogen} + 1 \text{ ATP)}$$

$$\xrightarrow[\text{Oxygen}]{} 6 \text{ CO}_2 + 6 \text{ H}_2\text{O} + 40 \text{ ATP}$$

$$\text{FFA} \xrightarrow[\text{Oxygen}]{} CO_2 + H_2O + ATP$$

Figure 3–12. Energy released by hydrolysis of 1 mol of ATP and reactions responsible for resynthesis of ATP. The amount of ATP formed per mole of free fatty acid (FFA) oxidized is large but varies with the size of the FFA. For example, complete oxidation of 1 mol of palmitic acid generates 140 mol of ATP.

phate bonds is a small factor. Consequently, heat production is considerable. The heat produced in muscle can be measured accurately with suitable thermocouples.

Resting heat, the heat given off at rest, is the external manifestation of basal metabolic processes. The heat produced in excess of resting heat during contraction is called the **initial heat.** This is made up of **activation heat,** the heat that muscle produces whenever it is contracting, and **shortening heat,** which is proportionate in amount to the distance the muscle shortens. Shortening heat is apparently due to some change in the structure of the muscle during shortening.

Following contraction, heat production in excess of resting heat continues for as long as 30 minutes. This **recovery heat** is the heat liberated by the metabolic processes that restore the muscle to its precontraction state. The recovery heat of muscle is approximately equal to the initial heat; ie, the heat produced during recovery is equal to the heat produced during contraction.

If a muscle that has contracted isotonically is restored to its previous length, extra heat in addition to recovery heat is produced **(relaxation heat).** External work must be done on the muscle to return it to its previous length, and relaxation heat is mainly a manifestation of this work.

PROPERTIES OF SKELETAL MUSCLES IN THE INTACT ORGANISM

Effects of Denervation

In the intact animal or human, healthy skeletal muscle does not contract except in response to stimulation of its motor nerve supply. Destruction of this nerve supply causes muscle atrophy. It also leads to abnormal excitability of the muscle and increases its sensitivity to circulating acetylcholine (denervation hypersensitivity; see Chapter 4). Fine, irregular contractions of individual fibers **(fibrillations)** appear. This is the classic picture of a **lower motor neuron lesion.** If the motor nerve regenerates, the fibrillations disappear. Usually, the contractions are not visible grossly, and they should not be confused with **fasciculations,** which are jerky, visible contractions of groups of muscle fibers that occur as a result of pathologic discharge of spinal motor neurons.

The Motor Unit

Since the axons of the spinal motor neurons supplying skeletal muscle each branch to innervate several muscle fibers, the smallest possible amount of muscle that can contract in response to the excitation of a single motor neuron is not one muscle fiber but all the fibers supplied by the neuron. Each single motor neuron and the muscle fibers it innervates constitute a **motor unit.** The number of muscle fibers in a motor unit varies. In muscles such as those of the hand and those concerned with motion of the eye—ie, muscles concerned with fine, graded, precise movement—there are three to six muscle fibers per motor unit. On the other hand, values of 120–165 fibers per unit have been reported in cat leg muscles, and some of the large muscles of the back in humans probably contain even more.

Each spinal motor neuron innervates only one kind of muscle fiber, so that all of the muscle fibers in a motor unit are of the same type. On the basis of the type of muscle fiber they innervate (Table 3–3), and thus on the basis of the duration of their twitch contraction, motor units are divided into fast and slow units. In general, slow muscle units are innervated by small, slowly conducting motor neurons and fast units by large, rapidly conducting motor neurons (**size principle**). In large limb muscles, the small, slow units are first recruited in most movements, are resistant to fatigue, and are the most frequently used units. The fast units, which are more easily fatigued, are generally recruited with more forceful movements.

The differences between types of muscle units are not inherent but are determined by, among other things, their activity. When the nerve to a slow muscle is cut and replaced with the nerve to a fast muscle, the nerve regenerates and innervates the slow muscle. However, the muscle becomes fast and there are corresponding changes in its muscle protein isoforms and myosin ATPase activity. This change is due to changes in the pattern of activity of the muscle; in stimulation experiments, changes in the expression of MHC genes and consequently of MHC isoforms can be produced by changes in the pattern of electrical activity used to stimulate the muscle.

Electromyography

Activation of motor units can be studied by **electromyography,** the process of recording the electrical activity of muscle on a cathode ray oscilloscope. This may be done in unanesthetized humans by using small metal disks on the skin overlying the muscle as the pick-up electrodes or by using hypodermic needle electrodes. The record obtained with such electrodes is the **electromyogram (EMG).** With needle electrodes, it is usually possible to pick up the activity of single muscle fibers. A typical EMG is shown in Figure 3–13.

Factors Responsible for Grading of Muscular Activity

It has been shown by electromyography that there is little if any spontaneous activity in the skeletal muscles of normal individuals at rest. With minimal voluntary activity a few motor units discharge, and with increasing

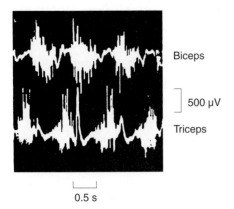

Figure 3–13. Electromyographic tracings from human biceps and triceps muscles during alternate flexion and extension of the elbow. (Courtesy of BC Garoutte.)

voluntary effort more and more are brought into play. This process is sometimes called **recruitment of motor units.** Gradation of muscle response is therefore in part a function of the number of motor units activated. In addition, the frequency of discharge in the individual nerve fibers plays a role, the tension developed during a tetanic contraction being greater than that during individual twitches. The length of the muscle is also a factor. Finally, the motor units fire asynchronously, ie, out of phase with each other. This asynchronous firing causes the individual muscle fiber responses to merge into a smooth contraction of the whole muscle.

The Strength of Skeletal Muscles

Human skeletal muscle can exert 3–4 kg of tension per cm^2 of cross-sectional area. This figure is about the same as that obtained in a variety of experimental animals and seems to be constant for all mammalian species. Since many of the muscles in humans have a relatively large cross-sectional area, the tension they can develop is quite large. The gastrocnemius, for example, not only supports the weight of the whole body during climbing but resists a force several times this great when the foot hits the ground during running or jumping. An even more striking example is the gluteus maximus, which can exert a tension of 1200 kg. The total tension that could be developed if all muscles in the body of an adult man pulled together is approximately 22,000 kg (nearly 25 tons).

Body Mechanics

Body movements are generally organized in such a way that they take maximal advantage of the physiologic

principles outlined above. For example, the attachments of the muscles in the body are such that many of them are normally at or near their resting length when they start to contract. In muscles that extend over more than one joint, movement at one joint may compensate for movement at another in such a way that relatively little shortening of the muscle occurs during contraction. Nearly isometric contractions of this type permit development of maximal tension per contraction. The hamstring muscles extend from the pelvis over the hip joint and the knee joint to the tibia and fibula. Hamstring contraction produces flexion of the leg on the thigh. If the thigh is flexed on the pelvis at the same time, the lengthening of the hamstrings across the hip joint tends to compensate for the shortening across the knee joint. In the course of various activities, the body moves in a way that takes advantage of this. Such factors as momentum and balance are integrated into body movement in ways that make possible maximal motion with minimal muscular exertion. One net effect is that the stress put on tendons and bones is rarely over 50% of their failure strength, protecting them from damage.

In walking, each limb passes rhythmically through a support or stance phase when the foot is on the ground and a swing phase when the foot is off the ground. The support phases of the two legs overlap, so that there are two periods of double support during each cycle. There is a brief burst of activity in the leg flexors at the start of each step, and then the leg is swung forward with little more active muscular contraction. Therefore, the muscles are active for only a fraction of each step, and walking for long periods causes relatively little fatigue.

A young adult walking at a comfortable pace moves at a velocity of about 80 m/min and generates a power output of 150–175 W per step. A group of young adults asked to walk at their most comfortable rate selected a velocity close to 80 m/min, and it was found that they had selected the velocity at which their energy output was minimal. Walking more rapidly or more slowly took more energy.

Even though walking is a complex activity, it is common knowledge that it is carried out more or less automatically. Experiments in animals indicate that it is organized in preprogrammed nerve pathways within the spinal cord and is activated by some sort of command signal in a fashion which is analogous to the initiation of patterns of activity by the discharge of **command neurons** in invertebrates.

Disease of Muscle

As noted above, mutations in the genes coding for the various components of the dystrophin-glycoprotein complex cause **muscular dystrophy,** a syndrome characterized by progressive muscle weakness. Some of the many forms of this disease cause great disability and are eventually fatal. **Duchenne's muscular dystrophy** is X-linked, occurs in about 1 of every 3000 male infants, and is usually fatal by the age of 30. It is caused by mutations of the dystrophin gene that cause dystrophin to be absent from the muscles. In a milder form of the disease, **Becker's muscular dystrophy,** dystrophin is present but altered or reduced in amount. Limb-girdle muscular dystrophy of various types is associated with mutations of the genes coding for the sarcoglycans.

At neuromuscular junctions, dystrophin is replaced by **utrophin,** a similar protein that is coded by a different gene. In dystrophin-deficient mice genetically programmed to overproduce utrophin, there is marked improvement in muscle function, indicating that utrophin has largely taken over for dystrophin. Not surprisingly, this has led to an active search for drugs or other means of turning on the utrophin gene in humans with Duchenne's muscular dystrophy.

Mutations in the gene for desmin, another structural protein (see above), also cause skeletal and cardial myopathies.

Mutations in genes that code for enzymes involved in the metabolism of carbohydrates, fats, and proteins to CO_2 and H_2O in muscle and the production of ATP cause **metabolic myopathies.** One of these, McArdle's syndrome, is considered in Chapter 17, but there are many others. They have many different manifestations, depending on the particular genetic defect, but all have in common exercise intolerance and the possibility of muscle breakdown due to accumulation of toxic metabolites.

In the various forms of clinical **myotonia,** muscle relaxation is prolonged after voluntary contraction. The myotonias are due to abnormal genes on chromosome 7, 17, or 19, which produce abnormalities of Na^+ or Cl^- channels. The possibility of implanting normal muscle genes in patients with muscular dystrophy is currently the subject of intensive research, but many problems are as yet unsolved.

Muscle Development

Although consideration of embryonic development is beyond the scope of this book, it is worth noting that remarkable advances in the understanding of genetic control of muscle development have been made in recent years. **Myogenin** is a transcription factor that is central to this process. It induces fibroblasts to become muscle cells, and when mice made homozygous for a mutant myogenin gene are born, they die be-

cause they lack muscles, including the muscles necessary for breathing.

CARDIAC MUSCLE

MORPHOLOGY

The striations in cardiac muscle are similar to those in skeletal muscle, and Z lines are present. There are large numbers of elongated mitochondria in close contact with the muscle fibrils. The muscle fibers branch and interdigitate, but each is a complete unit surrounded by a cell membrane. Where the end of one muscle fiber abuts on another, the membranes of both fibers parallel each other through an extensive series of folds. These areas, which always occur at Z lines, are called **intercalated disks** (Figure 3–14). They provide a strong union between fibers, maintaining cell-to-cell cohesion, so that the pull of one contractile unit can be transmitted along its axis to the next. Along the sides of the muscle fibers next to the disks, the cell membranes of adjacent fibers fuse for considerable distances, forming gap junctions. These junctions provide low-resistance bridges for the spread of excitation from one fiber to another (see Chapter 1). They permit cardiac muscle to function as if it were a syncytium, even though there are no protoplasmic bridges between cells. The T system in cardiac muscle is located at the Z lines rather than at the A-I junction, where it is located in mammalian skeletal muscle.

ELECTRICAL PROPERTIES

Resting Membrane & Action Potentials

The resting membrane potential of individual mammalian cardiac muscle cells is about −90 mV (interior negative to exterior). Stimulation produces a propagated action potential that is responsible for initiating contraction. Depolarization proceeds rapidly, and an overshoot is present, as in skeletal muscle and nerve, but this is followed by a plateau before the membrane potential returns to the baseline (Figure 3–15). In mammalian hearts, depolarization lasts about 2 ms, but the plateau phase and repolarization last 200 ms or more. Repolarization is therefore not complete until the contraction is half over. With extracellular recording, the electrical events include a spike and a later wave that resemble the QRS complex and T wave of the ECG.

As in other excitable tissues, changes in the external K^+ concentration affect the resting membrane potential of cardiac muscle, whereas changes in the external Na^+ concentration affect the magnitude of the action poten-

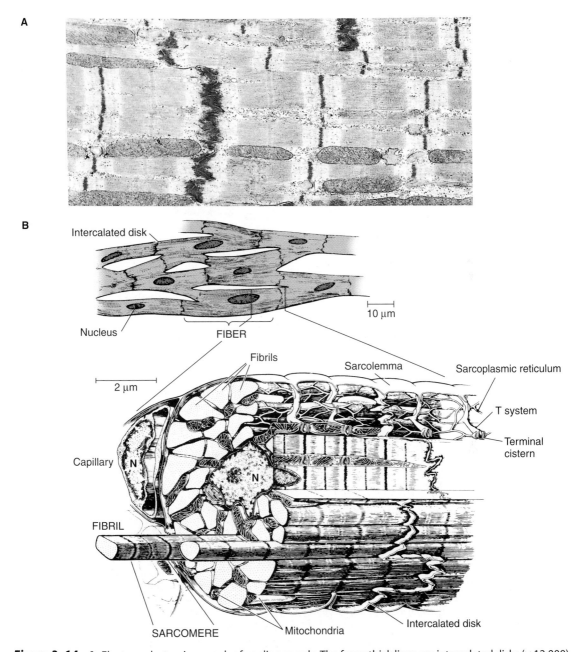

A

B

Intercalated disk

Nucleus

FIBER

10 μm

Fibrils

Sarcolemma

Sarcoplasmic reticulum

T system

Terminal cistern

2 μm

Capillary

N

N

FIBRIL

SARCOMERE

Mitochondria

Intercalated disk

Figure 3–14. **A:** Electron photomicrograph of cardiac muscle. The fuzzy thick lines are intercalated disks (× 12,000). (Reproduced, with permission, from Bloom W, Fawcett DW: A Textbook of Histology, 10th ed. Saunders, 1975.) **B:** Diagrams of cardiac muscle as seen under the light microscope and the electron microscope. N, nucleus. (Reproduced, with permission, from Braunwald E, Ross J, Sonnenblick EH: Mechanisms of contraction of the normal and failing heart. N Engl J Med 1967;277:794. Courtesy of Little, Brown, Inc.)

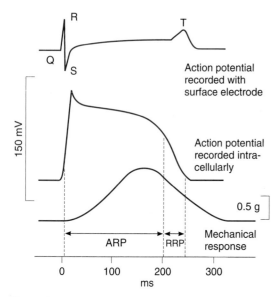

Figure 3–15. Action potentials and contractile response of mammalian cardiac muscle fiber plotted on the same time axis. ARP, absolute refractory period; RRP, relative refractory period.

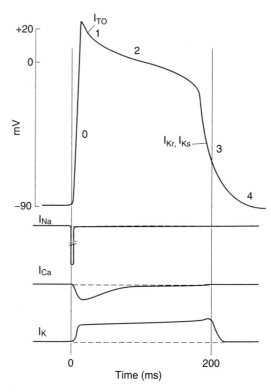

Figure 3–16. Top: Phases of the action potential of a cardiac muscle fiber. 0, depolarization; 1, initial rapid repolarization; 2, plateau phase; 3, late rapid repolarization; 4, baseline. **Bottom:** Diagrammatic summary of Na$^+$, Ca^{2+}, and cumulative K$^+$ currents during the action potential. Inward current down, outward current up. Individual K$^+$ currents that make up the cumulative K$^+$ current in humans and the approximate points at which they contribute are indicated on the diagram of the action potential. See text for details.

tial. The initial rapid depolarization and the overshoot (phase 0) are due to opening of voltage-gated Na$^+$ channels similar to that occurring in nerve and skeletal muscle (Figure 3–16). The initial rapid repolarization (phase 1) is due to closure of Na$^+$ channels. The subsequent prolonged plateau (phase 2) is due to a slower but prolonged opening of voltage-gated Ca^{2+} channels. Final repolarization (phase 3) to the resting membrane potential (phase 4) is due to closure of the Ca^{2+} channels and K$^+$ efflux through various types of K$^+$ channels. The types of ion channels in cardiac muscle are summarized in Table 3–5.

The voltage-gated Na$^+$ channel in cardiac muscle has two gates—an outer gate that opens at the start of depolarization, at a membrane potential of −70 to −80 mV; and an inner gate that then closes and precludes further influx until the action potential ends (Na$^+$ channel inactivation). The slow Ca^{2+} channel is activated at a membrane potential of −30 to −40 mV. There are three types of K$^+$ channels that produce repolarization. The first produces a transient, early outward current (I_{TO}) that produces an early incomplete repolarization. The second is inwardly rectifying, ie, at plateau potentials it allows K$^+$ influx but resists K$^+$ efflux, and only at lower membrane potentials does it permit K$^+$ efflux. The current it produces is called I_{Kr}. The third type is a slowly activating (delayed rectifying) type that produces a current called I_{Ks}. The sum of I_{Kr}

and I_{Ks} is a small net outward current that increases with time and produces repolarization.

The subunits that make up the K$^+$ channels responsible for I_{Kr} cross the membrane six times and are the product of *HERG* (for **human ether-a-go-go-related gene**). This gene is similar to a gene in the fruit fly *Drosophila* which, when mutated, made the flies shake their legs when they were anesthetized with ether. The channel responsible for I_{Ks} is made up of a protein that crosses the membrane six times combined with a small protein called **minK** (because of its size) that has only a single membrane-spanning domain. The relation of the various K$^+$ channels and the Na$^+$ channel to the long QT syndrome and cardiac arrhythmias is discussed in Chapter 28.

Table 3–5. Cardiac ion channels.

Voltage-gated channels
 Na^+
 $T\ Ca^{2+}$
 $L\ Ca^{2+}$
 K^+
  Inward rectifying
  Delayed rectifying
  Transient outward

Ligand-gated K^+ channels
 Ca^{2+}-activated
 Na^+-activated
 ATP-sensitive
 Acetylcholine-activated
 Arachidonic acid-activated

MECHANICAL PROPERTIES

Contractile Response

The contractile response of cardiac muscle begins just after the start of depolarization and lasts about 1.5 times as long as the action potential (Figure 3–15). The role of Ca^{2+} in excitation-contraction coupling is similar to its role in skeletal muscle (see above). However, as noted above, it is the influx of extracellular Ca^{2+} that is triggered by activation of the dihydropyridine channels in the T system, rather than depolarization per se, that triggers release of stored Ca^{2+} from the sarcoplasmic reticulum.

During phases 0–2 and about half of phase 3 (until the membrane potential reaches approximately −50 mV during repolarization), cardiac muscle cannot be excited again; ie, it is in its **absolute refractory period** (Figure 3–15). It remains relatively refractory until phase 4. Therefore, tetanus of the type seen in skeletal muscle cannot occur. Of course, tetanization of cardiac muscle for any length of time would have lethal consequences, and in this sense, the fact that cardiac muscle cannot be tetanized is a safety feature.

Isoforms

Cardiac muscle is generally slow and has relatively low ATPase activity. Its fibers are dependent on oxidative metabolism and hence on a continuous supply of O_2. The human heart contains both the α and the β isoforms of the myosin heavy chain (α MHC and β MHC). β MHC has lower myosin ATPase activity than α MHC. Both are present in the atria, with the α isoform predominating, whereas only the β isoform is found in the ventricle. The effects of thyroid hormones on the isoforms are discussed in Chapter 18.

Thyroid hormones also produce changes in MHC isoforms in skeletal muscle, but the effects are more complex.

Correlation Between Muscle Fiber Length & Tension

The relation between initial fiber length and total tension in cardiac muscle is similar to that in skeletal muscle; there is a resting length at which the tension developed upon stimulation is maximal. In the body, the initial length of the fibers is determined by the degree of diastolic filling of the heart, and the pressure developed in the ventricle is proportionate to the total tension developed (**Starling's law of the heart;** see Chapter 29). Thus, the developed tension (Figure 3–17) increases as the diastolic volume increases until it reaches a maximum (ascending limb of Starling curve), then tends to decrease (descending limb of Starling curve). However, unlike skeletal muscle, the decrease in developed tension at high degrees of stretch is not due to a decrease in the number of cross-bridges between actin and myosin, because even severely dilated hearts are not stretched to this degree. The descending limb is due instead to beginning disruption of the myocardial fibers. The homeostatic value of Starling's law is discussed in Chapter 29.

The force of contraction of cardiac muscle is also increased by catecholamines (see Chapters 13 and 20), and this increase occurs without a change in muscle length. The increase, which is called the positively inotropic effect of catecholamines, is mediated via inner-

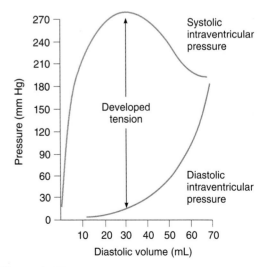

Figure 3–17. Length-tension relationship for cardiac muscle. The values are for canine heart.

vated β_1-adrenergic receptors and cyclic AMP (see Chapter 1). The heart also contains noninnervated β_2-adrenergic receptors, which also act via cyclic AMP, but their inotropic effect is smaller and is maximum in the atria. Cyclic AMP activates protein kinase A, and this leads to phosphorylation of the voltage-dependent Ca^{2+} channels, causing them to spend more time in the open state. Cyclic AMP also increases the active transport of Ca^{2+} to the sarcoplasmic reticulum, thus accelerating relaxation and consequently shortening systole. This is important when the cardiac rate is increased, because it permits adequate diastolic filling (see Chapter 29).

Digitalis glycosides increase cardiac contractions by inhibiting the Na^+-K^+ ATPase in cell membranes of the muscle fibers. The resultant increase in the level of intracellular Na^+ decreases the Na^+ gradient across the cell membrane. This decreases Na^+ influx and hence Ca^{2+} efflux via the Na^+-Ca^{2+} exchange antiport in the cell membranes (see Chapter 1). The intracellular Ca^{2+} concentration increases, and this in turn increases the strength of contraction of the cardiac muscle. A second action of digitalis glycosides is to increase the slow inward Ca^{2+} current during the action potential, and this also increases the intracellular Ca^{2+} concentration.

Cardiac Hypertrophy

Cardiac muscle, like skeletal muscle, undergoes hypertrophy when the load on it is chronically increased—as it is, for example, in hypertension. It can also hypertrophy when there are mutations in the genes coding for various proteins that make up the contractile apparatus, weakening the strength of contraction and producing **hypertrophic cardiomyopathy.** Over 100 different mutations in seven different proteins (myosin heavy chain, two myosin light chains, tropomyosin, troponin T, myosin-binding protein C, and troponin I) have been reported in patients with this disease. On the other hand, the mutations in the dystrophin gene that are seen in Duchenne's or Becker's muscular dystrophy (see above) prevent proper force generation by the heart, which dilates without hypertrophy (**dilated cardiomyopathy**) and eventually fails. Mutations in the gene for actin have also been reported to cause dilated cardiomyopathy. Cardiomyopathies due to mutations generally develop early in life, but some do not become manifest until middle age or later.

The calcineurin-NF-AT pathway originally described in T cells of the immune system also mediates hypertrophy of cardiac muscle. This system, which is blocked by cyclosporine, is described in Chapter 27. There is also some evidence that the pathway is involved in skeletal muscle hypertrophy.

METABOLISM

Mammalian hearts have an abundant blood supply, numerous mitochondria, and a high content of myoglobin, a muscle pigment that may function as an O_2 storage mechanism (see Chapter 35). Normally, less than 1% of the total energy liberated is provided by anaerobic metabolism. During hypoxia, this figure may increase to nearly 10%; but under totally anaerobic conditions, the energy liberated is inadequate to sustain ventricular contractions. Under basal conditions, 35% of the caloric needs of the human heart are provided by carbohydrate, 5% by ketones and amino acids, and 60% by fat. However, the proportions of substrates utilized vary greatly with the nutritional state. After ingestion of large amounts of glucose, more lactate and pyruvate are used; during prolonged starvation, more fat is used. Circulating free fatty acids normally account for almost 50% of the lipid utilized. In untreated diabetics, the carbohydrate utilization of cardiac muscle is reduced and that of fat increased. The factors affecting the O_2 consumption of the human heart are discussed in Chapter 29.

PACEMAKER TISSUE

The heart continues to beat after all nerves to it are sectioned; indeed, if the heart is cut into pieces, the pieces continue to beat. This is because of the presence in the heart of specialized pacemaker tissue that can initiate repetitive action potentials. The pacemaker tissue makes up the conduction system that normally spreads impulses throughout the heart. It is characterized by an unstable membrane potential that slowly decreases after each impulse until the firing level is reached and another impulse is generated (see Chapter 28).

SMOOTH MUSCLE

MORPHOLOGY

Smooth muscle is distinguished anatomically from skeletal and cardiac muscle because it lacks visible cross-striations. Actin and myosin-II are present, and they slide on each other to produce contraction. However, they are not arranged in regular arrays, as in skeletal and cardiac muscle, and so the striations are absent. Instead of Z lines, there are **dense bodies** in the cytoplasm and attached to the cell membrane, and these are bound by α-actinin to actin filaments. Smooth muscle also contains tropomyosin, but troponin appears to be absent. The isoforms of actin and myosin differ from those in skeletal muscle. There is a sarcoplasmic reticulum, but it is poorly developed. In general, smooth muscles contain few mitochondria and depend to a large extent on glycolysis for their metabolic needs.

Types

There is considerable variation in the structure and function of smooth muscle in different parts of the body. In general, smooth muscle can be divided into **visceral smooth muscle** and **multi-unit smooth muscle.** Visceral smooth muscle occurs in large sheets, has low-resistance bridges between individual muscle cells, and functions in a syncytial fashion. The bridges, like those in cardiac muscle, are junctions where the membranes of the two adjacent cells fuse to form gap junctions. Visceral smooth muscle is found primarily in the walls of hollow viscera. The musculature of the intestine, the uterus, and the ureters are examples. Multiunit smooth muscle is made up of individual units without interconnecting bridges. It is found in structures such as the iris of the eye, in which fine, graded contractions occur. It is not under voluntary control, but it has many functional similarities to skeletal muscle.

Growth

Because proliferation of vascular smooth muscle cells contributes to atherosclerosis, hypertension, and thickening of blood vessel walls when the endothelium is damaged, there has been considerable interest in the regulation of smooth muscle growth. The regulation is complex and involves a variety of different growth factors. Catecholamines and angiotensin II stimulate growth and proliferation, whereas glucocorticoids inhibit growth. Other possible regulatory factors include arachidonic acid derivatives, adenosine, heparinoids, and serotonin.

VISCERAL SMOOTH MUSCLE

Electrical & Mechanical Activity

Visceral smooth muscle is characterized by the instability of its membrane potential and by the fact that it shows continuous, irregular contractions that are independent of its nerve supply. This maintained state of partial contraction is called **tonus** or **tone.** The membrane potential has no true "resting" value, being relatively low when the tissue is active and higher when it is inhibited, but in periods of relative quiescence it averages about −50 mV. Superimposed on the membrane potential are waves of various types (Figure 3–18). There are slow sine wave-like fluctuations a few millivolts in magnitude and spikes that sometimes overshoot the zero potential line and sometimes do not. In

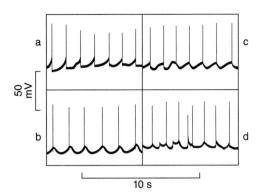

Figure 3–18. Spontaneous electrical activity in individual smooth muscle cells of teniae coli of guinea pig colon. **a:** Pacemaker type; **b:** sinusoidal waves with action potentials on the rising phases; **c:** sinusoidal waves with action potentials on the falling phases; **d:** mixture of pacemaker and sinusoidal waves and action potentials. (Reproduced, with permission, from Bulbring E: Physiology and pharmacology of intestinal smooth muscle. Lectures on the Scientific Basis of Medicine 1957;7:374.)

many tissues, the spikes have a duration of about 50 ms. However, in some tissues the action potentials have a prolonged plateau during repolarization, like the action potentials in cardiac muscle. The spikes may occur on the rising or falling phases of the sine wave oscillations. There are, in addition, pacemaker potentials similar to those found in the cardiac pacemakers. However, in visceral smooth muscle, these potentials are generated in multiple foci that shift from place to place. Spikes generated in the pacemaker foci are conducted for some distance in the muscle. Because of the continuous activity, it is difficult to study the relation between the electrical and mechanical events in visceral smooth muscle, but in some relatively inactive preparations, a single spike can be generated. The muscle starts to contract about 200 ms after the start of the spike and 150 ms after the spike is over. The peak contraction is reached as long as 500 ms after the spike. Thus, the excitation-contraction coupling in visceral smooth muscle is a very slow process compared with that in skeletal and cardiac muscle, in which the time from initial depolarization to initiation of contraction is less than 10 ms.

Molecular Basis of Contraction

Ca^{2+} is involved in the initiation of contraction of smooth muscle, as it is in skeletal muscle. However, visceral smooth muscle generally has a poorly developed sarcoplasmic reticulum, and the increase in intracellular Ca^{2+} concentration that initiates contraction is due primarily to Ca^{2+} influx from the ECF via voltage-gated and ligand-gated Ca^{2+} channels. In addition, the myosin in smooth muscle must be phosphorylated for activation of the myosin ATPase. Phosphorylation and dephosphorylation of myosin also occur in skeletal muscle, but phosphorylation is not necessary for activation of the ATPase. In smooth muscle, Ca^{2+} binds to calmodulin (see Chapter 1), and the resulting complex activates **calmodulin-dependent myosin light chain kinase.** This enzyme catalyzes the phosphorylation of the myosin light chain on serine at position 19. The phosphorylation allows the myosin ATPase to be activated, and actin slides on myosin, producing contraction. This is in contrast to skeletal and cardiac muscle, where contraction is triggered by the binding of Ca^{2+} to troponin C.

Myosin is dephosphorylated by **myosin light chain phosphatase** in the cell. This enzyme is inhibited when it is phosphorylated and activated by dephosphorylation. It is dephosphorylated by a rho-associated kinase that is activated by ligands which produce inhibition of smooth muscle activity. However, dephosphorylation of myosin light chain kinase does not necessarily lead to relaxation of the smooth muscle. Various mechanisms are involved. One appears to be a **latch bridge** mechanism by which myosin cross-bridges remain attached to actin for some time after the cytoplasmic Ca^{2+} concentration falls. This produces sustained contraction with little expenditure of energy, which is especially important in vascular smooth muscle. Relaxation of the muscle presumably occurs when there is final dissociation of the Ca^{2+}-calmodulin complex or when some other mechanism comes into play. The events leading to contraction and relaxation of visceral smooth muscle are summarized in Table 3–6. The events in multi-unit smooth muscle are generally similar.

It is worth highlighting the differences between cardiac and vascular smooth muscle, since both are involved in cardiovascular control. In the heart, the responses are phasic, with contraction alternating with relaxation, whereas in smooth muscle, contraction is often tonic because of the latch bridge mechanism. Furthermore, increases in intracellular cyclic AMP levels increase the force of contraction of cardiac muscle, whereas cyclic AMP relaxes vascular smooth muscle because it inhibits the phosphorylation of myosin light chain kinase.

Stimulation

Visceral smooth muscle is unique in that, unlike other types of muscle, it contracts when stretched in the absence of any extrinsic innervation. Stretch is followed

Table 3–6. Sequence of events in contraction and relaxation of visceral smooth muscle.[1]

(1) Binding of acetylcholine to muscarinic receptors
(2) Increased influx of Ca^{2+} into the cell
(3) Activation of calmodulin-dependent myosin light chain kinase
(4) Phosphorylation of myosin
(5) Increased myosin ATPase activity and binding of myosin to actin
(6) Contraction
(7) Dephosphorylation of myosin by myosin light chain phosphatase
(8) Relaxation, or sustained contraction due to the latch bridge and other mechanisms

[1]Compare with Table 3–2.

by a decline in membrane potential, an increase in the frequency of spikes, and a general increase in tone.

If epinephrine or norepinephrine is added to a preparation of intestinal smooth muscle arranged for recording of intracellular potentials in vitro, the membrane potential usually becomes larger, the spikes decrease in frequency, and the muscle relaxes (Figure 3–19). Norepinephrine is the chemical mediator released at noradrenergic nerve endings (see Chapter 4), and stimulation of the noradrenergic nerves to the preparation produces inhibitory potentials (see Chapter 4). Stimulation of the noradrenergic nerves to the intestine inhibits contractions in vivo. Norepinephrine exerts both α and β actions (see Chapter 4) on the

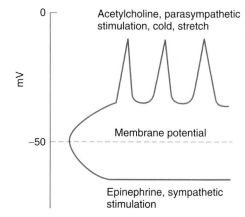

Figure 3–19. Effects of various agents on the membrane potential of intestinal smooth muscle.

muscle. The β action, reduced muscle tension in response to excitation, is mediated via cyclic AMP (see Chapter 1) and is probably due to increased intracellular binding of Ca^{2+}. The α action, which is also inhibition of contraction, is associated with increased Ca^{2+} efflux from the muscle cells.

Acetylcholine has an effect opposite to that of norepinephrine on the membrane potential and contractile activity of intestinal smooth muscle. If acetylcholine is added to the fluid bathing a smooth muscle preparation in vitro, the membrane potential decreases and the spikes become more frequent (Figure 3–18). The muscle becomes more active, with an increase in tonic tension and the number of rhythmic contractions. The effect is mediated by phospholipase C and IP_3, which increases the intracellular Ca^{2+} concentration. In the intact animal, stimulation of cholinergic nerves causes release of acetylcholine, excitatory potentials (see Chapter 4), and increased intestinal contractions. In vitro, similar effects are produced by cold and stretch.

Function of the Nerve Supply to Smooth Muscle

The effects of acetylcholine and norepinephrine on visceral smooth muscle serve to emphasize two of its important properties: (1) its spontaneous activity in the absence of nervous stimulation, and (2) its sensitivity to chemical agents released from nerves locally or brought to it in the circulation. In mammals, visceral muscle usually has a dual nerve supply from the two divisions of the autonomic nervous system. The structure and function of the contacts between these nerves and smooth muscle are discussed in Chapter 4. The function of the nerve supply is not to initiate activity in the muscle but rather to modify it. Stimulation of one division of the autonomic nervous system usually increases smooth muscle activity, whereas stimulation of the other decreases it. However, in some organs, noradrenergic stimulation increases and cholinergic stimulation decreases smooth muscle activity; in others, the reverse is true.

Relation of Length to Tension; Plasticity

Another special characteristic of smooth muscle is the variability of the tension it exerts at any given length. If a piece of visceral smooth muscle is stretched, it first exerts increased tension (see above). However, if the muscle is held at the greater length after stretching, the tension gradually decreases. Sometimes the tension falls to or below the level exerted before the muscle was stretched. It is consequently impossible to correlate length and developed tension accurately, and no resting length can be assigned. In some ways, therefore, smooth muscle behaves more like a viscous mass than a rigidly structured tissue, and it is this property that is referred to as the **plasticity** of smooth muscle.

The consequences of plasticity can be demonstrated in intact humans. For example, the tension exerted by the smooth muscle walls of the bladder can be measured at different degrees of distention as fluid is infused into the bladder via a catheter, as shown in Figure 38–26. Initially there is relatively little increase in tension as volume is increased, because of the plasticity of the bladder wall. However, a point is eventually reached at which the bladder contracts forcefully.

MULTI-UNIT SMOOTH MUSCLE

Unlike visceral smooth muscle, multi-unit smooth muscle is nonsyncytial and contractions do not spread widely through it. Because of this, the contractions of multi-unit smooth muscle are more discrete, fine, and localized than those of visceral smooth muscle. Like visceral smooth muscle, multi-unit smooth muscle is very sensitive to circulating chemical substances and is normally activated by chemical mediators (acetylcholine and norepinephrine) released at the endings of its motor nerves. Norepinephrine in particular tends to persist in the muscle and to cause repeated firing of the muscle after a single stimulus rather than a single action potential. Therefore, the contractile response produced is usually an irregular tetanus rather than a single twitch. When a single twitch response is obtained, it resembles the twitch contraction of skeletal muscle except that its duration is ten times as long.

Synaptic & Junctional Transmission | 4

INTRODUCTION

The all-or-none type of conduction seen in axons and skeletal muscle has been discussed in Chapters 2 and 3. Impulses are transmitted from one nerve cell to another cell at **synapses** (Figure 4–1). These are the junctions where the axon or some other portion of one cell (the **presynaptic cell**) terminates on the dendrites, soma, or axon of another neuron (Figure 4–2) or in some cases a muscle or gland cell (the **postsynaptic cell**). Transmission at most synaptic junctions is chemical; the impulse in the presynaptic axon causes secretion of a **neurotransmitter** such as acetylcholine or serotonin. This chemical mediator binds to receptors on the surface of the postsynaptic cell, and this triggers events that open or close channels in the membrane of the postsynaptic cell. At some of the junctions, however, transmission is electrical, and at a few conjoint synapses it is both electrical and chemical. In any case, transmission is not a simple jumping of one action potential from the presynaptic to the postsynaptic cell. The effects of discharge at individual synaptic endings can be excitatory or inhibitory, and when the postsynaptic cell is a neuron, the summation of all the excitatory and inhibitory effects determines whether an action potential is generated. Thus, synaptic transmission is a complex process that permits the grading and adjustment of neural activity necessary for normal function.

In electrical synapses, the membranes of the presynaptic and postsynaptic neurons come close together, and gap junctions form between the cells (see Chapter 1). Like the intercellular junctions in other tissues, these junctions form low-resistance bridges through which ions pass with relative ease. Electrical and conjoint synapses occur in mammals, and there is electrical coupling, for example, between some of the neurons in the lateral vestibular nucleus. However, since most synaptic transmission is chemical, consideration in this chapter is limited to chemical transmission unless otherwise specified.

Transmission from nerve to muscle resembles chemical synaptic transmission from one neuron to another. The **neuromuscular junction,** the specialized area where a motor nerve terminates on a skeletal muscle fiber, is the site of a stereotyped transmission process. The contacts between autonomic neurons and smooth and cardiac muscle are less specialized, and transmission in these locations is a more diffuse process.

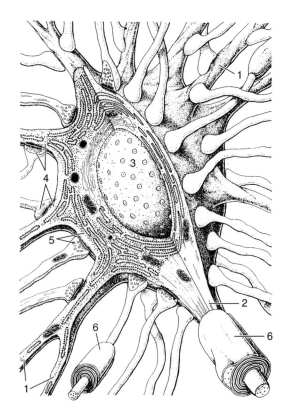

Figure 4–1. Synapses on a typical motor neuron. The neuron has dendrites (1), an axon (2), and a prominent nucleus (3). Note that rough endoplasmic reticulum extends into the dendrites but not into the axon. Many different axons converge on the neuron, and their terminal buttons form axodendritic (4) and axosomatic (5) synapses. (6) Myelin sheath. (Reproduced, with permission, from Krstic RV: *Ultrastructure of the Mammalian Cell.* Springer, 1979.)

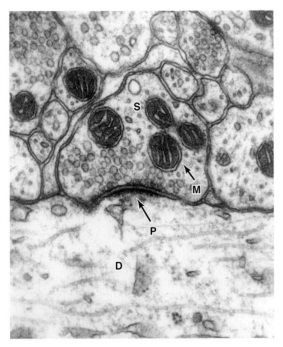

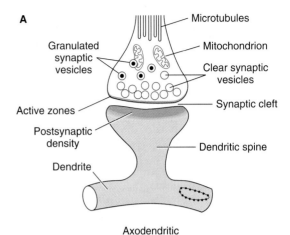

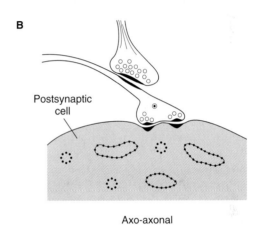

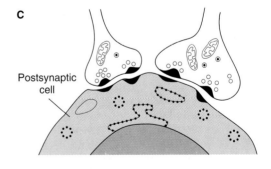

Figure 4–2. Electron photomicrograph of synaptic knob (S) ending on the shaft of a dendrite (D) in the central nervous system. P, postsynaptic density; M, mitochondrion. (× 56,000; courtesy of DM McDonald.)

SYNAPTIC TRANSMISSION

FUNCTIONAL ANATOMY

Types of Synapses

There is considerable variation in the anatomic structure of synapses in various parts of the mammalian nervous system. The ends of the presynaptic fibers are generally enlarged to form **terminal buttons (synaptic knobs)** (Figure 4–1). In the CNS, some are wrapped in astrocyte processes, which seem to confer regularity and efficiency (see Chapter 2). In the cerebral and cerebellar cortex, endings are commonly located on dendrites (Figure 4–2) and frequently on **dendritic spines,** which are small knobs projecting from dendrites (Figure 4–3). In some instances, the terminal branches of the axon of the presynaptic neuron form a basket or net around the soma of the postsynaptic cell ("basket cells" of the cerebellum and autonomic ganglia). In other locations, they intertwine with the dendrites of the postsynaptic cell (climbing fibers of the cerebellum) or end on the dendrites directly (apical dendrites of cortical

Figure 4–3. Axodendritic, axo-axonal, and axosomatic synapses. Many presynaptic neurons terminate on dendritic spines, as shown at the top, but some also end directly on the shafts of dendrites. Note the presence of clear and granulated synaptic vesicles in endings and clustering of clear vesicles at active zones, shown longitudinally in A and in cross section in B and C.

pyramids) or on the axons (axo-axonal endings). On average, each neuron divides to form over 2000 synaptic endings, and since there are 10^{11} neurons in the human central nervous system (CNS), it follows that there are about 2×10^{14} synapses. Obviously, the number of pathways that an impulse can take in such a complex network is extremely large. In the spinal cord, the presynaptic endings are closely applied to the soma and the proximal portions of the dendrites of the postsynaptic neuron. The number of synaptic knobs applied to a single spinal motor neuron has been calculated to be about 10,000, with 2000 on the cell body and 8000 on the dendrites. Indeed, there are so many knobs that the neuron appears to be encrusted with them. The portion of the membrane covered by any single synaptic knob is small, but the synaptic knobs are so numerous that, in aggregate, the area covered by them all is often 40% of the soma membrane area (Figure 4–1) and 75% of the dendritic membrane area. It has been calculated that in the cerebral cortex, 98% of the synapses are on dendrites and only 2% are on cell bodies. Many of the endings that are excitatory in the cerebral and cerebellar cortex end on dendritic spines. Many of the spines have narrower necks than heads, and their electrical properties are affected by the ratio of the head to neck area. The ratio of synapses to neurons in the human forebrain has been calculated to be 40,000:1.

Pre- & Postsynaptic Structure & Function

Each presynaptic terminal of a chemical synapse is separated from the postsynaptic structure by a synaptic cleft that is 20–40 nm wide. Across the synaptic cleft, there are many neurotransmitter receptors in the postsynaptic membrane, and usually a postsynaptic thickening called the **postsynaptic density** (Figures 4–2 and 4–3). The postsynaptic density is an ordered complex of specific receptors, binding proteins, and enzymes induced by postsynaptic effects.

Inside the presynaptic terminal there are many mitochondria; there are also many membrane-enclosed vesicles, which contain the neurotransmitters. There are three kinds of **synaptic vesicles:** small, clear synaptic vesicles that contain acetylcholine, glycine, GABA, or glutamate (see below); small vesicles with a dense core that contain catecholamines; and large vesicles with a dense core that contain neuropeptides. The vesicles and the proteins contained in their walls are synthesized in the Golgi apparatus in the neuronal cell body and migrate down the axon to the endings by fast axoplasmic transport. The neuropeptides in the large dense-core vesicles must also be produced by the protein-synthesizing machinery in the cell body. However, the small clear vesicles and the small dense-core vesicles recycle in

the ending. They are loaded with transmitter in the ending, fuse with the membrane, and discharge the transmitter by exocytosis, then are retrieved by endocytosis. They enter endosomes and are budded off the endosome and refilled, starting the cycle over again. The steps involved are shown in Figure 4–4.

The large dense-core vesicles are located throughout the presynaptic terminals that contain them and release their neuropeptide contents by exocytosis from all parts of the terminal. On the other hand, the small vesicles are located near the synaptic cleft and fuse to the membrane and discharge their contents very rapidly into the cleft at areas of membrane thickening called **active zones** (Figure 4–3). The active zones contain many proteins and rows of calcium channels.

Ca^{2+} is the key to synaptic vesicle fusion and discharge. An action potential reaching the presynaptic terminal opens voltage-gated Ca^{2+} channels, and the resulting Ca^{2+} influx triggers release. The Ca^{2+} content is then restored to the resting level by rapid sequestration and removal from the cell, primarily by a Ca^{2+}-Na^+ antiport.

As noted in Chapter 1, vesicle budding, fusion, and discharge of contents with subsequent retrieval of vesicle membrane are fundamental processes occurring in all cells. They occur inside cells, eg, in the Golgi apparatus, and at the cell membrane in the form of exocytosis and endocytosis. Similar mechanisms are involved, with specializations for particular situations. The carefully regulated processes involved in the case of synaptic vesicle discharge and membrane recovery are summarized in Figure 4–4. The details of the processes by which synaptic vesicles fuse with the cell membrane are still being worked out, but they involve the **v-snare** protein **synaptobrevin** in the vesicle membrane locking with the **t-snare** protein **syntaxin** in the cell membrane (Figure 4–5).

It is interesting and clinically relevant that several deadly toxins which block neurotransmitter release are zinc endopeptidases that cleave and hence inactivate proteins in the fusion-exocytosis complex. Tetanus toxin and botulinum toxins B, D, F, and G act on synaptobrevin, and botulinum toxin C acts on syntaxin. Botulinum toxins A and B act on SNAP-25 (Figure 4–5). Clinically, tetanus toxin causes spastic paralysis by blocking presynaptic transmitter release in the CNS and botulism causes flaccid paralysis by blocking the release of acetylcholine at the neuromuscular junction. On the positive side, however, local injection of small doses of botulinum toxin has proved effective in the treatment of a wide variety of conditions characterized by muscle hyperactivity. Examples include injection into the lower esophageal sphincter to relieve achalasia and injection into facial muscles to remove wrinkles.

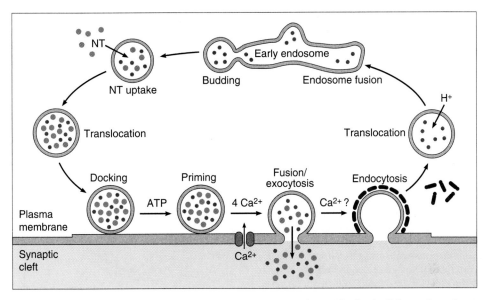

Figure 4–4. Small synaptic vesicle cycle in presynaptic nerve terminals. Vesicles bud off the early endosome and then fill with neurotransmitter (NT; top left). They then move to the plasma membrane, dock, and become primed. Upon arrival of an action potential at the ending, Ca^{2+} influx triggers fusion and exocytosis of the granule contents to the synaptic cleft. The vesicle wall is then coated with clathrin and taken up by endocytosis. In the cytoplasm, it fuses with the early endosome, and the cycle is ready to repeat. (Reproduced, with permission, from Südhof TC: The synaptic vesicle cycle: a cascade of protein-protein interactions. Nature 1995;375:645. Copyright © by Macmillan Magazines Ltd.)

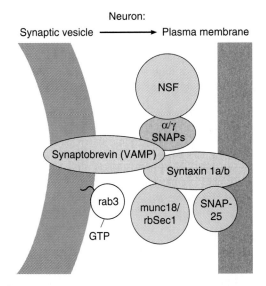

Figure 4–5. Main proteins that interact to produce synaptic vesicle docking and fusion in nerve endings. (Reproduced, with permission, from Ferro-Novick S, John R: Vesicle fusion from yeast to man. Nature 1994;370:191. Copyright © by Macmillan Magazines Ltd.)

Convergence & Divergence

Only a few of the synaptic knobs on a postsynaptic neuron are endings of any single presynaptic neuron. The inputs to the cell are multiple. In spinal motor neurons, for example, some inputs come directly from the dorsal root, some from the long descending spinal tracts, and many from **interneurons,** the short interconnecting neurons of the spinal cord. Thus, many presynaptic neurons converge on any single postsynaptic neuron. Conversely, the axons of most presynaptic neurons divide into many branches that diverge to end on many postsynaptic neurons. **Convergence** and **divergence** are the anatomic substrates for facilitation, occlusion, and reverberation (see below).

One-Way Conduction

Synapses generally permit conduction of impulses in one direction only, from the presynaptic to the postsynaptic neurons. An impulse conducted antidromically up the axons of the ventral root dies out after depolarizing the cell bodies of the spinal motor neurons. Since axons conduct in either direction with equal facility, the one-way gate at the synapses is necessary for orderly neural function. Chemical mediation at synaptic junctions explains one-way conduction. The mediator is located in the synaptic knobs of the presynaptic fibers but not in the postsynaptic membrane. Therefore, an impulse arriving at the postsynaptic membrane cannot release synaptic mediator. Progression of impulse traffic occurs only when the action potential arrives in the presynaptic terminals and causes secretion of stored chemical transmitter.

Synaptic Development

A fascinating question that has attracted a great deal of attention is how, during development, neurons find the "right" targets and make the "right" synaptic connections. Growing axons have **growth cones** at their tips which migrate through tissues. These cones are guided along the way by attractants and repellents and, in the CNS, by certain types of glia. Details of development are beyond the scope of this book. In general, however, a superfamily of proteins called **semaphorins** is involved. More synapses form than are needed, and the "inappropriate" ones disappear as many neurons undergo apoptosis (see Chapter 1) during development.

ELECTRICAL EVENTS IN POSTSYNAPTIC NEURONS

Penetration of an anterior horn cell is a good example of the techniques used to study postsynaptic electrical activity. It is achieved by advancing a microelectrode through the ventral portion of the spinal cord. Puncture of a cell membrane is signaled by the appearance of a steady 70-mV potential difference between the microelectrode and an electrode outside the cell. The cell can be identified as a spinal motor neuron by stimulating the appropriate ventral root and observing the electrical activity of the cell. Such stimulation initiates an antidromic impulse (see Chapter 2) that is conducted to the soma and stops at this point. Therefore, the presence of an action potential in the cell after antidromic stimulation indicates that the cell that has been penetrated is a motor neuron rather than an interneuron. Activity in some of the presynaptic terminals impinging on the impaled spinal motor neuron (Figure 4–6) can be initiated by stimulating the dorsal roots.

Excitatory Postsynaptic Potentials

Single stimuli applied to the sensory nerves in the experimental situation described above characteristically do not lead to the formation of a propagated action potential in the postsynaptic neuron. Instead, the stimulation produces either a transient partial depolarization or a transient hyperpolarization.

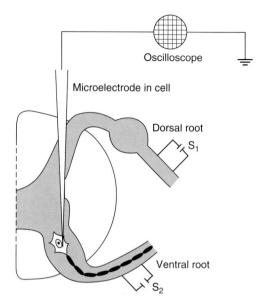

Figure 4–6. Arrangement of recording electrodes and stimulators for studying synaptic activity in spinal motor neurons in mammals. One stimulator (S2) is used to produce antidromic impulses for identifying the cell; the other (S1) is used to produce orthodromic stimulation via reflex pathways.

The initial depolarizing response produced by a single stimulus to the proper input begins about 0.5 ms after the afferent impulse enters the spinal cord. It reaches its peak 1–1.5 ms later and then declines exponentially. During this potential, the excitability of the neuron to other stimuli is increased, and consequently the potential is called an **excitatory postsynaptic potential (EPSP).**

The EPSP is produced by depolarization of the postsynaptic cell membrane immediately under the presynaptic ending. The excitatory transmitter opens Na^+ or Ca^{2+} ion channels in the postsynaptic membrane, producing an inward current. The area of current flow thus created is so small that it does not drain off enough positive charges to depolarize the whole membrane. Instead, an EPSP is inscribed. The EPSP due to activity in one synaptic knob is small, but the depolarizations produced by each of the active knobs summate.

Summation may be **spatial** or **temporal.** When activity is present in more than one synaptic knob at the same time, spatial summation occurs and activity in one synaptic knob is said to **facilitate** activity in another to approach the firing level. Temporal summation occurs if repeated afferent stimuli cause new EPSPs before previous EPSPs have decayed. Obviously, the longer the time constant for the EPSP, the greater the opportunity for summation. Spatial and temporal facilitation are illustrated in Figure 4–7. The EPSP is therefore not an all-or-none response but is proportionate in size to the strength of the afferent stimulus. If the EPSP is large enough to reach the firing level of the cell, a full-fledged action potential is produced.

Synaptic Delay

When an impulse reaches the presynaptic terminals, there is an interval of at least 0.5 ms, the **synaptic delay,** before a response is obtained in the postsynaptic neuron. The delay following maximal stimulation of the presynaptic neuron corresponds to the latency of the EPSP and is due to the time it takes for the synaptic mediator to be released and to act on the membrane of the postsynaptic cell. Because of it, conduction along a chain of neurons is slower if there are many synapses in the chain than if there are only a few. Since the minimum time for transmission across one synapse is 0.5 ms, it is also possible to determine whether a given reflex pathway is monosynaptic or polysynaptic (contains more than one synapse) by measuring the delay in transmission from the dorsal to the ventral root across the spinal cord.

Inhibitory Postsynaptic Potentials

EPSPs are produced by stimulation of some inputs, but stimulation of other inputs produces hyperpolarizing responses. Like the EPSPs, they peak 1–1.5 ms after the stimulus and decrease exponentially with a **time constant** (time to decay to 1/e, or 1/2.718 of maximum) of about 3 ms (Figure 4–8). During this potential, the excitability of the neuron to other stimuli is decreased; consequently, it is called an **inhibitory postsynaptic potential (IPSP).** Spatial summation of IPSPs occurs, as shown by the increasing size of the response as the strength of an inhibitory afferent volley is increased. Temporal summation also occurs. This type of inhibition is called **postsynaptic** or **direct inhibition.**

An IPSP can be produced by a localized increase in Cl^- transport. When an inhibitory synaptic knob becomes active, the released transmitter triggers the opening of Cl^- channels in the area of the postsynaptic cell membrane under the knob. Cl^- moves down its concentration gradient. The net effect is the transfer of negative charge into the cell, so that the membrane potential increases.

The decreased excitability of the nerve cell during the IPSP is due to movement of the membrane potential away from the firing level. Consequently, more excitatory (depolarizing) activity is necessary to reach the firing level. The fact that an IPSP is mediated by Cl^- can be demonstrated by repeating the stimulus while varying the resting membrane potential of the postsynaptic cell and holding it with a voltage clamp. When the membrane potential is set at E_{Cl}, the potential dis-

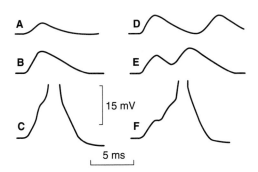

Figure 4–7. Spatial summation **(A–C)** and temporal summation **(D–F)** of EPSPs. Records are potential changes recorded with one electrode inside the post-synaptic cell. **A–C:** Afferent volleys of increasing strength were delivered. More and more synaptic knobs were activated, and in C, the firing level was reached and an action potential generated. **D–F:** Two different volleys of the same strength were delivered, but the time interval between them was shortened. In F, the firing level was reached and an action potential generated.

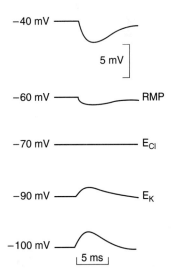

Figure 4–8. IPSP due to increased Cl^- influx produced by stimulation when the membrane potential is set at various values with a voltage clamp. RMP, resting membrane potential of this neuron. Note that when the voltage is set at E_{Cl} the IPSP disappears and that at higher membrane voltages it becomes positive.

appears (Figure 4–8), and at more negative membrane potentials, it becomes positive.

IPSPs can also be produced by opening of K^+ channels, with movement of K^+ out of the postsynaptic cell. In addition, they can be produced by closure of Na^+ or Ca^{2+} channels.

Slow Postsynaptic Potentials

In addition to the EPSPs and IPSPs described above, slow EPSPs and IPSPs have been described in autonomic ganglia, cardiac and smooth muscle, and cortical neurons. These postsynaptic potentials have a latency of 100–500 ms and last several seconds. The slow EPSPs are generally due to decreases in K^+ conductance, and the slow IPSPs are due to increases in K^+ conductance. In sympathetic ganglia, there is also a late slow EPSP that has a latency of 1–5 seconds and lasts 10–30 minutes. This potential is also due, at least in part, to decreased K^+ conductance, and the transmitter responsible for the potential is a peptide very closely related to GnRH, the hormone secreted by neurons in the hypothalamus that stimulates LH secretion (see Chapter 14).

Generation of the Action Potential in the Postsynaptic Neuron

The constant interplay of excitatory and inhibitory activity on the postsynaptic neuron produces a fluctuating membrane potential that is the algebraic sum of the hyperpolarizing and depolarizing activity. The soma of the neuron thus acts as a sort of integrator. When the 10–15 mV of depolarization sufficient to reach the firing level is attained, a propagated spike results. However, the discharge of the neuron is slightly more complicated than this. In motor neurons, the portion with the lowest threshold for the production of the cell of a full-fledged action potential is the **initial segment,** the portion of the axon at and just beyond the axon hillock. This unmyelinated segment is depolarized or hyperpolarized electrotonically by the current sinks and sources under the excitatory and inhibitory synaptic knobs. It is the first part of the neuron to fire, and its discharge is propagated in two directions: down the axon and back into the soma. Retrograde firing of the soma in this fashion probably has value in "wiping the slate clean" for subsequent renewal of the interplay of excitatory and inhibitory activity on the cell.

Function of the Dendrites

For many years, the standard view has been that dendrites are simply the sites of current sources or sinks that electrotonically change the membrane potential at the initial segment; ie, they are merely extensions of the soma that expand the area available for integration. When the dendritic tree of a neuron is extensive and has multiple presynaptic knobs ending on it, there is room for a great interplay of inhibitory and excitatory activity.

Recent data indicate that dendrites contribute to neural function in more complex ways. Action potentials can be recorded in dendrites. In many instances, these are initiated in the initial segment and conduction in a retrograde fashion, but propagated action potentials are initiated in some dendrites. Further research has demonstrated the malleability of dendritic spines. Not only do they increase during development (Figure 4–9), but the dendritic spines appear, change, and even disappear over a time scale of minutes and hours, not days and months. Also, although protein synthesis occurs mainly in the soma with its nucleus, strands of mRNA migrate into the dendrites. There each can become associated with a single ribosome in a dendritic spine and produce proteins, which alters the effects of input from individual glutaminergic synapses on the spine. The receptors involved are NMDA and AMPA receptors.

Here and elsewhere, proteins called **neurolignins** may moderate actual synapse formation, although the process needs further study. However, the selective changes in the dendritic spine mediate one form of learning and long-term potentiation (LTP).

Electrical Transmission

At synaptic junctions where transmission is electrical, the impulse reaching the presynaptic terminal generates

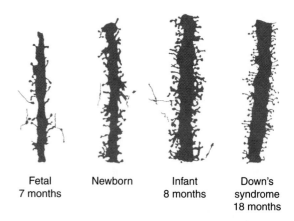

| Fetal 7 months | Newborn | Infant 8 months | Down's syndrome 18 months |

Figure 4–9. Spines on apical dendrites of large pyramidal neurons in the human cerebral cortex. Note that the numbers of spines increase rapidly from birth to 8 months of age, and that in Down's syndrome, the spines are thin and small. (Modified from Shepherd GM: *Neurobiology,* 2nd ed. Oxford Univ Press, 1988.)

an EPSP in the postsynaptic cell that, because of the low-resistance bridge between the two, has a much shorter latency than the EPSP at a synapse where transmission is chemical. In conjoint synapses, there is both a short-latency response and a longer-latency, chemically mediated postsynaptic response.

INHIBITION & FACILITATION AT SYNAPSES

Direct & Indirect Inhibition

Inhibition in the CNS can be postsynaptic or presynaptic. **Postsynaptic inhibition** during the course of an IPSP is called **direct inhibition** because it is not a consequence of previous discharges of the postsynaptic neuron. Various forms of **indirect inhibition,** inhibition due to the effects of previous postsynaptic neuron discharge, also occur. For example, the postsynaptic cell can be refractory to excitation because it has just fired and is in its refractory period. During after-hyperpolarization it is also less excitable. In spinal neurons, especially after repeated firing, this after-hyperpolarization may be large and prolonged.

Postsynaptic Inhibition in the Spinal Cord

The various pathways in the nervous system that are known to mediate postsynaptic inhibition are discussed in Chapter 6, but one illustrative example is presented here. Afferent fibers from the muscle spindles (stretch receptors) in skeletal muscle are known to pass directly to the spinal motor neurons of the motor units supplying the same muscle. Impulses in this afferent supply cause EPSPs and, with summation, propagated responses in the postsynaptic motor neurons. At the same time, IPSPs are produced in motor neurons supplying the antagonistic muscles. This latter response is mediated by branches of the afferent fibers that end on Golgi bottle neurons. These interneurons, in turn, secrete the inhibitory transmitter glycine at synapses on the proximal dendrites or cell bodies of the motor neurons that supply the antagonist (Figure 4–10). Therefore, activity in the afferent fibers from the muscle spindles excites the motor neurons supplying the muscle from which the impulses come and inhibits those supplying its antagonists (**reciprocal innervation**).

Presynaptic Inhibition & Facilitation

Another type of inhibition occurring in the CNS is **presynaptic inhibition,** a process mediated by neurons that end on excitatory endings, forming **axo-axonal synapses** (Figure 4–3). The neurons responsible for postsynaptic and presynaptic inhibition are compared in Figure 4–11. Three mechanisms of presynaptic inhi-

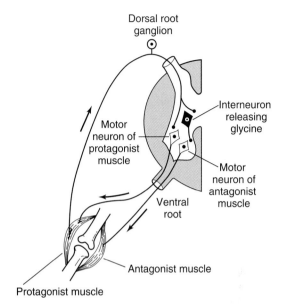

Figure 4–10. Diagram illustrating the anatomic connections responsible for inhibiting the antagonists to a muscle contracting in response to stretch. Activity is initiated in the spindle in the protagonist muscle. Impulses pass directly to the motor neurons supplying the same muscle and, via branches, to inhibitory interneurons that end on the motor neurons of the antagonist muscle.

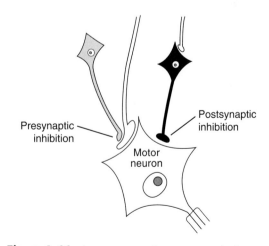

Figure 4–11. Arrangement of neurons producing presynaptic and postsynaptic inhibition. The neuron producing presynaptic inhibition is shown ending on an excitatory synaptic knob. Many of these neurons actually end higher up along the axon of the excitatory cell.

bition have been described. First, activation of the presynaptic receptors increases Cl⁻ conductance, and this has been shown to decrease the size of the action potentials reaching the excitatory ending (Figure 4–12). This in turn reduces Ca^{2+} entry and consequently the amount of excitatory transmitter released. Voltage-gated K^+ channels are also opened, and the resulting K^+ efflux also decreases the Ca^{2+} influx. Finally, there is evidence for direct inhibition of transmitter release independent of Ca^{2+} influx into the excitatory ending.

The first transmitter to be shown to produce presynaptic inhibition was GABA. Acting via $GABA_A$ receptors (see below), GABA increases Cl⁻ conductance. $GABA_B$ receptors are also present in the spinal cord and appear to mediate presynaptic inhibition via a G protein that produces an increase in K^+ conductance. Baclofen, a $GABA_B$ agonist, is effective in the treatment of the spasticity of spinal cord injury and multi-

ple sclerosis, particularly when administered intrathecally via an implanted pump. Other transmitters also mediate presynaptic inhibition by G protein-mediated effects on Ca^{2+} channels and K^+ channels.

Conversely, **presynaptic facilitation** is produced when the action potential is prolonged (Figure 4–12) and the Ca^{2+} channels are open for a longer period. The molecular events responsible for the production of presynaptic facilitation mediated by serotonin in the sea snail *Aplysia* have been worked out in detail. Serotonin released at an axo-axonal ending increases intraneuronal cAMP levels, and the resulting phosphorylation of one group of K^+ channels closes the channels, slowing repolarization and prolonging the action potential.

Organization of Inhibitory Systems

Presynaptic and postsynaptic inhibition are usually produced by stimulation of certain systems converging on a given postsynaptic neuron ("afferent inhibition"). Neurons may also inhibit themselves in a negative feedback fashion ("negative feedback inhibition"). For instance, each spinal motor neuron regularly gives off a recurrent collateral that synapses with an inhibitory interneuron which terminates on the cell body of the spinal neuron and other spinal motor neurons (Figure 4–13). This particular inhibitory neuron is sometimes called a Renshaw cell after its discoverer. Impulses generated in the motor neuron activate the inhibitory interneuron to secrete inhibitory mediator, and this slows

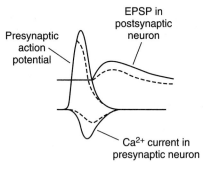

Presynaptic inhibition

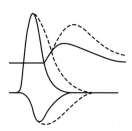

Presynaptic facilitation

Figure 4–12. Effects of presynaptic inhibition and facilitation on the action potential and the Ca^{2+} current in the presynaptic neuron and the EPSP in the postsynaptic neuron. In each case, the solid lines are the controls and the dashed lines the records obtained during inhibition or facilitation. (Modified from Kandel ER, Schwartz JH, Jessell TM [editors]. *Principles of Neural Science,* 4th ed. McGraw-Hill, 2000.)

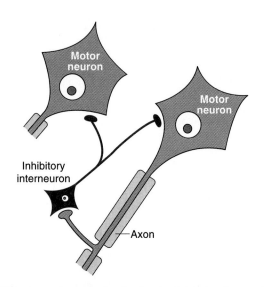

Figure 4–13. Negative feedback inhibition of a spinal motor neuron via an inhibitory interneuron (Renshaw cell).

or stops the discharge of the motor neuron. Similar inhibition via recurrent collaterals is seen in the cerebral cortex and limbic system. Presynaptic inhibition due to descending pathways that terminate on afferent pathways in the dorsal horn may be involved in the "gating" of pain transmission (see Chapter 7).

Another type of inhibition is seen in the cerebellum. In this part of the brain, stimulation of basket cells produces IPSPs in the Purkinje cells (see Chapter 12). However, the basket cells and the Purkinje cells are excited by the same parallel-fiber excitatory input. This arrangement, which has been called "feed-forward inhibition," presumably limits the duration of the excitation produced by any given afferent volley.

Summation & Occlusion

The interplay between excitatory and inhibitory influences at synaptic junctions in a nerve net illustrates the integrating and modulating activity of the nervous system.

In the hypothetical nerve net shown in Figure 4–14, neurons A and B converge on X, and neuron B diverges on X and Y. A stimulus applied to A or to B will set up an EPSP in X. If A and B are stimulated at the same time and action potentials are produced, two areas of depolarization will be produced in X and their actions will sum. The resultant EPSP in X will be twice as large as that produced by stimulation of A or B alone, and the membrane potential may well reach the firing level of X. The effect of the depolarization caused by the impulse in A is facilitated by that due to activity in B, and vice versa; spatial facilitation has taken place. In this case, Y has not fired, but its excitability has been increased, and it is easier for activity in neuron C to fire Y during the EPSP. Y is therefore said to be in the **sub-liminal fringe** of X. More generally stated, neurons are in the subliminal fringe if they are not discharged by an afferent volley (not in the **discharge zone**) but do have their excitability increased. The neurons that have few active knobs ending on them are in the subliminal fringe, and those with many are in the discharge zone. Inhibitory impulses show similar temporal and spatial facilitation and subliminal fringe effects.

If action potentials are produced repeatedly in neuron B, X and Y will discharge as a result of temporal summation of the EPSPs that are produced. If C is stimulated repeatedly, Y and Z will discharge. If B and C are fired repeatedly at the same time, X, Y, and Z will discharge. Thus, the response to stimulation of B and C together is not as great as the sum of responses to stimulation of B and C separately, because B and C both end on neuron Y. This decrease in expected response, due to presynaptic fibers sharing postsynaptic neurons, is called **occlusion.**

Excitatory and inhibitory subliminal effects and occlusive phenomena can have pronounced effects on transmission in any given pathway. Because of these effects, temporal patterns in peripheral nerves are usually altered as they pass through synapses on the way to the brain. These effects may also explain such important phenomena as referred pain (see Chapter 7).

Neuromodulation

The term **modulation** is often used in physiology in such a loose sense that it adds little to knowledge of function. However, the term **neuromodulation** has a place in neurobiology when it is strictly defined as a nonsynaptic action of a substance on neurons that alters their sensitivity to synaptic stimulation or inhibition. Neuromodulation is frequently produced by neuropeptides and by circulating steroids and steroids produced in the nervous system (neurosteroids; see below).

CHEMICAL TRANSMISSION OF SYNAPTIC ACTIVITY

Implications

The fact that transmission at most synapses is chemical is of great physiologic and pharmacologic importance. Nerve endings have been called biological transducers that convert electrical energy into chemical energy. In broad terms, this conversion process involves the synthesis of the transmitter agents, their storage in synaptic vesicles, and their release by the nerve impulses into the synaptic cleft. The secreted transmitters then act on appropriate receptors on the membrane of the postsynaptic cell and are rapidly removed from the synaptic cleft

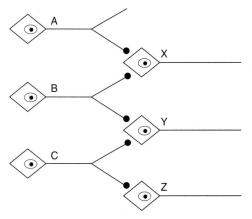

Figure 4–14. Simple nerve net. Neurons A, B, and C have excitatory endings on neurons X, Y, and Z.

by diffusion, metabolism, and, in many instances, re-uptake into the presynaptic neuron. All these processes, plus the postreceptor events in the postsynaptic neuron, are regulated by many physiologic factors and at least in theory can be altered by drugs. Therefore, pharmacologists should be able to develop drugs that regulate not only somatic and visceral motor activity but also emotions, behavior, and all the other complex functions of the brain.

Chemistry of Transmitters

One suspects that a substance is a transmitter if it is unevenly distributed in the nervous system and its distribution parallels that of its receptors and synthesizing and catabolizing enzymes. Additional evidence includes demonstration that it is released from appropriate brain regions in vitro and that it produces effects on single target neurons when applied to their membranes by means of a micropipette (microiontophoresis). Many transmitters and enzymes involved in their synthesis and catabolism have been localized in nerve endings by **immunocytochemistry,** a technique in which antibodies to a given substance are labeled and applied to brain and other tissues. The antibodies bind to the substance, and the location of the substance is then determined by locating the label with the light microscope or electron microscope. **In situ hybridization histochemistry,** which permits localization of the mRNAs for particular synthesizing enzymes or receptors, has also been a valuable tool.

Identified neurotransmitters can be divided into broad categories or families based on their chemical structure; some are amines, some are amino acids, and many are polypeptides. Some are purines, and NO and CO (see below) are gases. In addition, some derivatives of arachidonic acid may be transmitters. It is worth noting that most of these substances are not only released into synaptic clefts, where they produce highly localized effects. In other situations, they diffuse into the ECF around the synapse and exert effects at some distance from their site of release (paracrine communication; see Chapter 1). In some cases, they are also released by neurons into the bloodstream as hormones. A somewhat arbitrary compilation of most of the substances currently known or suspected to be synaptic mediators or neuromodulators is presented in Table 4–1.

Receptors

Cloning and other molecular biology techniques have permitted spectacular recent advances in knowledge about the structure and function of receptors for neurotransmitters and other chemical messengers. The in-

dividual receptors, along with their ligands, are discussed in the following parts of this chapter. However, five themes have emerged that should be mentioned in this introductory discussion.

First, in every instance studied in detail to date, it has become clear that for each ligand there are many subtypes of receptors. Thus, for example, norepinephrine acts on α_1 and α_2 receptors, and three of each subtype have been cloned. In addition, there are β_1, β_2, and β_3 receptors. Obviously, this multiplies the possible effects of a given ligand and makes its effects in a given cell more selective.

Second, there are receptors on the presynaptic as well as the postsynaptic elements for many secreted transmitters. These **presynaptic receptors,** or **autoreceptors,** often inhibit further secretion of the ligand, providing feedback control. For example, norepinephrine acts on α_2 presynaptic receptors to inhibit norepinephrine secretion. However, autoreceptors can also facilitate the release of neurotransmitters.

Third, although there are many ligands and many subtypes of receptors for each ligand, the receptors tend to group in large families as far as structure and function are concerned. Many are serpentine receptors that act via G proteins and protein kinases (see Chapter 1) to produce their effects. Others are ion channels. The receptors for a group of selected, established neurotransmitters are listed in Table 4–2, along with their principal second messengers and, where established, their net effect on channels. It should be noted that this table is an oversimplification. For example, activation of α_2 receptors decreases intracellular cAMP concentrations, but there is evidence that the G protein activated by α_2 presynaptic receptors acts directly on Ca^{2+} channels to inhibit norepinephrine release by decreasing the Ca^{2+} increase.

Fourth, receptors are concentrated in clusters in postsynaptic structures close to the endings of neurons that secrete the neurotransmitters specific for them. This is generally due to the presence of specific binding proteins for them. In the case of nicotinic acetylcholine receptors at the neuromuscular junction, the protein is **rapsyn,** and in the case of excitatory glutaminergic receptors, a family of **PB2-binding proteins** are involved. $GABA_A$ receptors are associated with the protein **gephyrin,** which also binds glycine receptors, and $GABA_C$ receptors are bound to the cytoskeleton in the retina by the protein **MAP-1B.** At least in the case of $GABA_A$ receptors, the binding protein **gephyrin** is located in clumps in the postsynaptic membrane. With activity, the free receptors move rapidly to the gephyrin and bind to it, creating membrane clusters. Gephyrin binding slows and restricts their further movement. Presumably, during neural inactivity, the receptors are unbound and move again.

Table 4–1. Neurotransmitters and neuromodulators in the nervous system of mammals.[1]

Substance	Location
Acetylcholine	Myoneural junction; preganglionic autonomic endings, postganglionic sympathetic sweat gland, and muscle vasodilator endings; many parts of brain; endings of some amacrine cells in retina.
Amines	
Dopamine	SIF cells in sympathetic ganglia; striatum, median eminence, and other parts of hypothalamus; limbic system; parts of neocortex; endings of some interneurons in retina.
Norepinephrine	Most postganglionic sympathetic endings; cerebral cortex, hypothalamus, brain stem, cerebellum, spinal cord.
Epinephrine	Hypothalamus, thalamus, periaqueductal gray, spinal cord.
Serotonin	Hypothalamus, limbic system, cerebellum, spinal cord; retina.
Histamine	Hypothalamus, other parts of brain.
Excitatory amino acids	
Glutamate	Cerebral cortex, brain stem.
Aspartate	Visual cortex.
Inhibitory amino acids	
Glycine	Neurons mediating direct inhibition in spinal cord, brain stem, forebrain; retina.
Gamma-aminobutyrate (GABA)	Cerebellum; cerebral cortex; neurons mediating presynaptic inhibition; retina.
Polypeptides	
Substance P, other tachykinins	Endings of primary afferent neurons mediating nociception; many parts of brain; retina.
Vasopressin	Posterior pituitary; medulla; spinal cord.
Oxytocin	Posterior pituitary; medulla; spinal cord.
CRH	Median eminence of hypothalamus; other parts of brain.
TRH	Median eminence of hypothalamus; other parts of brain; retina.
GRH	Median eminence of hypothalamus.
Somatostatin	Median eminence of hypothalamus; other parts of brain; substantia gelatinosa; retina.
GnRH	Median eminence of hypothalamus; circumventricular organs; preganglionic autonomic endings; retina.
Endothelins	Posterior pituitary, brain stem.
Enkephalins	Substantia gelatinosa, many other parts of CNS; retina.
β-Endorphin, other derivatives of pro-opiomelanocortin	Hypothalamus, thalamus, brain stem; retina.
Endomorphins	Thalamus, hypothalamus, striatum.
Dynorphins	Periaqueductal gray, rostroventral medulla, substantia gelatinosa.
Cholecystokinin (CCK-4 and CCK-8)	Cerebral cortex; hypothalamus; retina.

(continues)

Table 4–1. Neurotransmitters and neuromodulators in the nervous system of mammals.[1] *(continued)*

Substance	Location
Vasoactive intestinal peptide	Postganglionic cholinergic neurons; some sensory neurons; hypothalamus; cerebral cortex; retina.
Neurotensin	Hypothalamus; retina.
Gastrin-releasing peptide	Hypothalamus.
Gastrin	Hypothalamus; medulla oblongata.
Motilin	Neurohypophysis; cerebral cortex, cerebellum.
Secretin	Hypothalamus, thalamus, olfactory bulb, brain stem, cerebral cortex, septum, hippocampus, striatum.
Glucagon derivatives	Hypothalamus; retina.
Calcitonin gene-related peptide-α)	Endings of primary afferent neurons; taste pathways; sensory nerves; medial forebrain bundle.
Neuropeptide Y	Noradrenergic, adrenergic, and other neurons in medulla, periaqueductal gray, hypothalamus, autonomic nervous system.
Activins	Brain stem.
Inhibins	Brain stem.
Angiotensin II	Hypothalamus, amygdala, brain stem, spinal cord.
FMRF amide	Hypothalamus, brain stem.
Galanin	Hypothalamus, hippocampus, midbrain, spinal cord.
Atrial natriuretic peptide	Hypothalamus, brain stem.
Brain natriuretic peptide	Hypothalamus, brain stem.
Other polypeptides	Especially hypothalamus.
Pyrimidine	
UTP	Autonomic nervous system.
Purines	
Adenosine	Neocortex, olfactory cortex, hippocampus, cerebellum.
ATP	Autonomic ganglia, habenula.
Gases	
NO, CO	CNS.
Lipids	
Anandamide	Hippocampus, basal ganglia, cerebellum.

[1]Transmitter functions have not been proved for some of the polypeptides.

Table 4–2. Mechanism of action of selected nonpeptide neurotransmitters.

Transmitter	Receptor	Second Messenger	Net Channel Effects
Acetylcholine	Nicotinic	. . .	↑Na^+, other small ions
	M_1	↑IP_3, DAG	↑Ca^{2+}
	M_2 (cardiac)	↓Cyclic AMP	↑K^+
	M_3	↓Cyclic AMP	
	M_4 (glandular)	↑IP_3, DAG	
	M_5	↑IP_3, DAG	
Dopamine	D_1, D_5	↑Cyclic AMP	
	D_2	↓Cyclic AMP	↑K^+, ↓Ca^{2+}
	D_3, D_4	↓Cyclic AMP	
Norepinephrine	$\alpha_{1A}, \alpha_{1B}, \alpha_{1D}$	↑IP_3, DAG	↓K^+
	$\alpha_{2A}, \alpha_{2B}, \alpha_{2C}$	↓Cyclic AMP	↑K^+, ↓Ca^{2+}
	β_1	↑Cyclic AMP	
	β_2	↑Cyclic AMP	
	β_3	↑Cyclic AMP	
5HT[1]	$5HT_{1A}$	↓Cyclic AMP	↑K^+
	$5HT_{1B}$	↓Cyclic AMP	
	$5HT_{1D}$	↓Cyclic AMP	↓K^+
	$5HT_{2A}$	↑IP_3, DAG	↓K^+
	$5HT_{2C}$	↑IP_3, DAG	
	$5HT_3$	. . .	↑Na^+
	$5HT_4$	↑Cyclic AMP	
Adenosine	A_1	↓Cyclic AMP	
	A_2	↑Cyclic AMP	
Glutamate	Metabotropic[2]		
	Ionotropic		
	AMPA, Kainate	. . .	↑Na^+
	NMDA	. . .	↑Na^+, Ca^{2+}
GABA	$GABA_A$	. . .	↑Cl^-
	$GABA_B$	↑IP_3, DAG	↑K^+, ↓Ca^{2+}

[1]$5HT_{1E}, 5HT_{1F}, 5HT_{2B}, 5HT_{5A}, 5HT_{5B}, 5HT_6,$ and $5HT_7$ receptors also cloned.
[2]Eleven subtypes identified; all decrease cAMP or increase IP_3 and DAG, except one, which increases cAMP.

Fifth, prolonged exposure to their ligands causes most receptors to become unresponsive, ie, to undergo **desensitization.** This can be of two types: **homologous desensitization,** with loss of responsiveness only to the particular ligand and maintained responsiveness of the cell to other ligands; and **heterologous desensitization,** in which the cell becomes unresponsive to other ligands as well. Desensitization in β-adrenergic receptors has been studied in considerable detail. One form involves phosphorylation of the carboxyl terminal region of the receptor by a specific β-adrenergic receptor kinase (**β-ARK**), assisted by another protein, **β-arrestin.** β-Arrestin also binds to clathrin, promoting receptor endocytosis. In addition, the β-adrenergic receptor can be desensitized by phosphorylation in the carboxyl terminal region by PKA (see Chapter 1). Internalization of β$_2$-adrenergic receptors triggers an increase in receptor mRNA degradation so that the number of receptors is decreased (**down-regulation).**

Reuptake

In recent years, it has become clear that there is **reuptake** from the synaptic cleft into the cytoplasm of the presynaptic neuron (Figure 4–15) of most and possibly all amine and amino acid neurotransmitters by the neurons that secrete them. The high-affinity uptake systems employ two families of transporter proteins. One

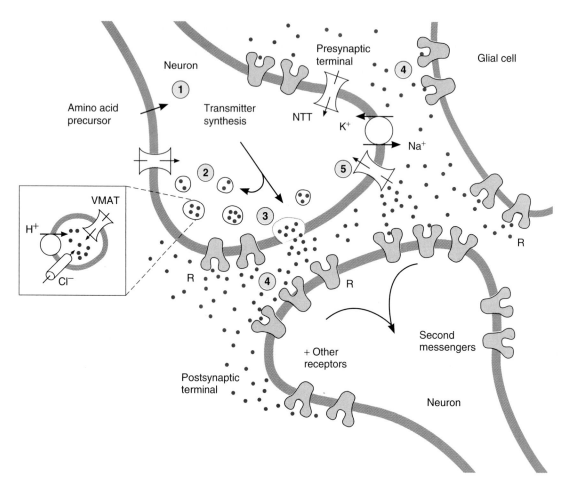

Figure 4–15. Fate of monoamines secreted at synaptic junctions. In each monoamine-secreting neuron, the monoamine is synthesized in the cytoplasm and the secretory granules (1) and its concentration in secretory granules is maintained (2) by the two vesicular monoamine transporters (VMAT). The monoamine is secreted by exocytosis of the granules (3), and it acts (4) on receptors (Y-shaped structures labeled R). Many of these receptors are postsynaptic, but some are presynaptic and some are located on glia. In addition, there is extensive reuptake into the cytoplasm of the presynaptic terminal (5) via the monoamine neurotransmitter transporter (NTT) for the monoamine that is synthesized in the neuron. (Reproduced, with permission, from Hoffman BJ et al: Distribution of monoamine neurotransmitter transporters in the rat brain. Front Neuroendocrinol 1998;19:187.)

family has 12 transmembrane domains and cotransports the transmitter with Na^+ and Cl^-. Members of this family include transporters for norepinephrine, dopamine, serotonin, GABA, and glycine, as well as transporters for proline, taurine, and the acetylcholine precursor choline. In addition, there may be an epinephrine transporter. The other family is made up of at least three transporters that mediate glutamate uptake by neurons and two others. These latter transporters are coupled to the cotransport of Na^+ and the countertransport of K^+, and they are not dependent on Cl^-

transport. There is a debate about their structure, and they may have 6, 8, or 10 transmembrane domains. One of them transports glutamate into glia rather than neurons (see Chapter 2).

There are in addition two vesicular monoamine transporters, VMAT1 and VMAT2. They are coded by different genes but have extensive homology. Both have a broad specificity, moving dopamine, norepinephrine, epinephrine, serotonin, and histamine from the cytoplasm into secretory granules. Both are inhibited by reserpine, which accounts for the marked monoamine depletion produced by this drug. Like the neurotransmitter membrane transporter family, they have 12 transmembrane domains, but they have little homology to the membrane transporters. There is also uptake by a vesicular GABA transporter (VGAT) that moves GABA and glycine into vesicles and a vesicular acetylcholine transporter (see below).

Reuptake is a major factor in terminating the action of transmitters, and when it is inhibited, the effects of transmitter release are increased and prolonged. This has clinical consequences. For example, several effective antidepressant drugs are inhibitors of the reuptake of amine transmitters, and cocaine is believed to inhibit dopamine reuptake. Glutamate uptake into neurons and glia is important because glutamate is an excitotoxin that can kill cells by overstimulating them (see below). There is evidence that during ischemia and anoxia, loss of neurons is increased because glutamate reuptake is inhibited.

PRINCIPAL NEUROTRANSMITTER SYSTEMS

Synaptic physiology is a rapidly expanding, complex field that cannot be covered in detail in this book. In addition, emphasis is shifting from transmitters per se to interactions of their receptors. However, it is appropriate to summarize information about the principal neurotransmitters and their receptors.

Acetylcholine

The relatively simple structure of acetylcholine, which is the acetyl ester of choline, is shown in Figure 4–16. It exists, largely enclosed in small, clear synaptic vesicles, in high concentration in the terminal buttons of neurons that release acetylcholine (**cholinergic** neurons).

Acetylcholine Synthesis

Synthesis of acetylcholine involves the reaction of choline with acetate. Choline is an important amine that is also the precursor of the membrane phospholipids phosphatidylcholine and sphingomyelin and the signaling phospholipids platelet-activating factor and sphingosylphosphorylcholine. There is an active uptake of choline via a transporter into cholinergic neurons (Figure 4–17). Choline is also synthesized in neurons. The acetate is activated by the combination of acetate groups with reduced coenzyme A. The reaction between active acetate (acetyl-coenzyme A, acetyl-CoA) and choline is catalyzed by the enzyme **choline acetyltransferase.** This enzyme is found in high concentration in the cytoplasm of cholinergic nerve endings. Acetylcholine is then taken up into synaptic vesicles by a vesicular transporter, VAChT.

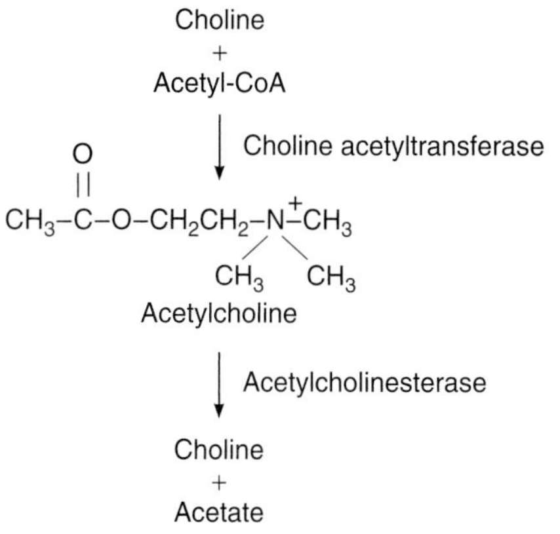

Choline
+
Acetyl-CoA

$\downarrow$ Choline acetyltransferase

$$CH_3-\overset{\overset{\displaystyle O}{\|}}{C}-O-CH_2CH_2-\overset{+}{N}\overset{\diagup}{\diagdown}\overset{CH_3}{CH_3}$$
Acetylcholine

$\downarrow$ Acetylcholinesterase

Choline
+
Acetate

Figure 4–16. Biosynthesis and catabolism of acetylcholine.

Cholinesterases

Acetylcholine must be rapidly removed from the synapse if repolarization is to occur. The removal occurs by way of hydrolysis of acetylcholine to choline and acetate, a reaction catalyzed by the enzyme **acetylcholinesterase.** This enzyme is also called **true** or **specific cholinesterase.** Its greatest affinity is for acetylcholine, but it also hydrolyzes other choline esters. There are a variety of es-

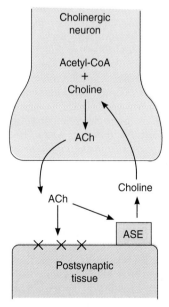

Figure 4–17. Biochemical events at cholinergic endings. ACh, acetylcholine; ASE, acetylcholinesterase; X, receptor. Compare with Figures 4–21 and 4–25.

terases in the body. One found in plasma is capable of hydrolyzing acetylcholine but has different properties from acetylcholinesterase. It is therefore called **pseudocholinesterase** or **nonspecific cholinesterase.** The plasma moiety is partly under endocrine control and is affected by variations in liver function. On the other hand, the specific cholinesterase molecules are clustered in the postsynaptic membrane of cholinergic synapses. Hydrolysis of acetylcholine by this enzyme is rapid enough to explain the observed changes in Na^+ conductance and electrical activity during synaptic transmission.

Acetylcholine Receptors

Historically, acetylcholine receptors have been divided into two main types on the basis of their pharmacologic properties. Muscarine, the alkaloid responsible for the toxicity of toadstools, has little effect on the receptors in autonomic ganglia but mimics the stimulatory action of acetylcholine on smooth muscle and glands. These actions of acetylcholine are therefore called **muscarinic actions,** and the receptors involved are **muscarinic cholinergic receptors.** They are blocked by the drug atropine. In sympathetic ganglia, small amounts of acetylcholine stimulate postganglionic neurons and large amounts block transmission of impulses from pre- to postganglionic neurons. These actions are unaffected by atropine but mimicked by nicotine. Consequently, these actions of acetylcholine are **nicotinic actions** and the receptors are **nicotinic cholinergic receptors.** Nicotinic receptors are subdivided into those found in muscle at neuromuscular junctions and those found in autonomic ganglia and the central nervous system. Both muscarinic and nicotinic acetylcholine receptors are found in large numbers in the brain.

The nicotinic acetylcholine receptors are members of a superfamily of ligand-gated ion channels that also includes the $GABA_A$ and glycine receptors and some of the glutamate receptors. They are made up of multiple subunits coded by different genes. Each nicotinic cholinergic receptor is made up of five subunits that form a central channel which, when the receptor is activated, permits the passage of Na^+ and other cations. The five subunits came from a menu of 16 known subunits, $\alpha_1-\alpha_9$, $\beta_2-\beta_5$, γ, δ, and ε, coded by 16 different genes. Some of the receptors are homomeric, eg, those that contain five α_7 subunits, but most are heteromeric. The muscle type nicotinic receptor found in the fetus is made up of two α_1 subunits, a β_1 subunit, an α subunit, and a δ subunit (Figure 4–18). In adult mammals, the δ subunit is replaced by an ε subunit, which decreases the channel open time but increases its conductance. The nicotinic cholinergic receptors in autonomic ganglia are heteromers that usually contain α_3 subunits in combination with others, and the nicotinic

receptors in the brain are made up of many other subunits. Many of the nicotinic cholinergic receptors in the brain are located presynaptically on glutamine-secreting axon terminals (see below), and they facilitate the release of this transmitter. However, others are postsynaptic. Some are located on structures other than neurons, and some seem to be free in the interstitial fluid, ie, they are perisynaptic in location.

There is a binding site for acetylcholine on each α subunit, and when an acetylcholine molecule binds to each of them, they induce a configurational change in the protein so that the channel opens. This increases the conductance of Na^+ and other cations, and the resulting influx of Na^+ produces a depolarizing potential. A prominent feature of neuronal nicotinic cholinergic receptors is their high permeability to Ca^{2+}, suggesting

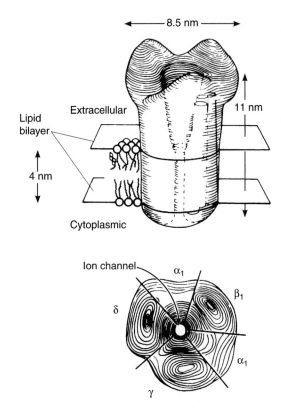

Figure 4–18. Diagram of fetal nicotinic acetylcholine receptor as viewed from the side **(above)** and from the top **(below).** $\alpha_1, \beta_1, \gamma, \delta$: receptor subunits. (From McCarthy MP et al: Molecular biology of the acetylcholine receptor. Annu Rev Neurosci 1986;9:383. Reproduced, with permission, from the Annual Review of Neuroscience, vol 9. Copyright © 1986 by Annual Reviews Inc.)

their involvement in synaptic facilitation and learning (see below).

Muscarinic cholinergic receptors are very different from nicotinic cholinergic receptors. Five types, encoded by five separate genes, have been cloned. The exact status of M5 is uncertain, but the remaining four all are serpentine receptors coupled via G proteins to adenylyl cyclase, K^+ channels, or phospholipase C (Table 4–2). The nomenclature of these receptors has not been standardized, but the receptor designated M_1 in Table 4–2 is abundant in the brain. The M_2 receptor is found in the heart (see Chapter 28). The M_4 receptor is found in pancreatic acinar and islet tissue, where it mediates increased secretion of pancreatic enzymes and insulin. The M_3 and M_4 receptors are both found in smooth muscle.

Norepinephrine & Epinephrine

The chemical transmitter present at most sympathetic postganglionic endings is norepinephrine (levarterenol). It is stored in the synaptic knobs of the neurons that secrete it in characteristic small vesicles which have a dense core (granulated vesicles; see above). Norepi-

nephrine and its methyl derivative, epinephrine, are secreted by the adrenal medulla (see Chapter 20), but epinephrine is not a mediator at postganglionic sympathetic endings. The endings of sympathetic postganglionic neurons in smooth muscle are discussed below; each neuron has multiple varicosities along its course, and each of these varicosities appears to be a site at which norepinephrine is secreted. There are also norepinephrine-secreting, dopamine-secreting, and epinephrine-secreting neurons in the brain (see Chapter 15). Norepinephrine-secreting neurons are properly called **noradrenergic neurons,** although the term **adrenergic neurons** is also applied. However, it seems appropriate to reserve the latter term for epinephrine-secreting neurons. Dopamine-secreting neurons are called **dopaminergic neurons.**

Biosynthesis & Release of Catecholamines

The principal **catecholamines** found in the body—norepinephrine, epinephrine, and dopamine—are formed by hydroxylation and decarboxylation of the amino acid tyrosine (Figure 4–19). Some of the tyrosine is formed from phenylalanine, but most is of di-

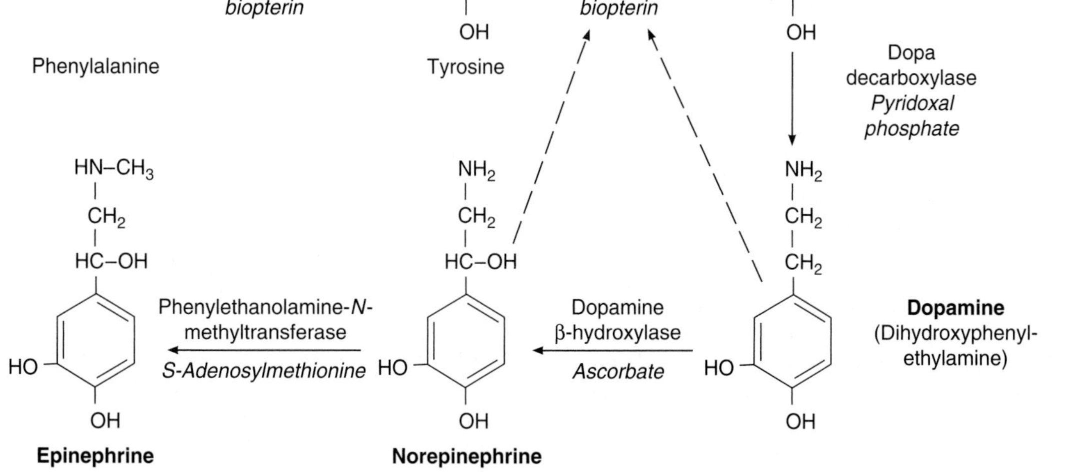

Figure 4–19. Biosynthesis of catecholamines. The dashed lines indicate inhibition of tyrosine hydroxylase by norepinephrine and dopamine. Essential cofactors are shown in italics.

etary origin. **Phenylalanine hydroxylase** is found primarily in the liver. Tyrosine is transported into catecholamine-secreting neurons and adrenal medullary cells by a concentrating mechanism. It is converted to dopa and then to dopamine in the cytoplasm of the cells by **tyrosine hydroxylase** and **dopa decarboxylase.** The decarboxylase, which is also called aromatic L-amino acid decarboxylase, is very similar but probably not identical to 5-hydroxytryptophan decarboxylase. The dopamine then enters the granulated vesicles, within which it is converted to norepinephrine by **dopamine β-hydroxylase (DBH).** L-Dopa is the isomer involved, but the norepinephrine that is formed is in the D configuration. The rate-limiting step in synthesis is the conversion of tyrosine to dopa. Tyrosine hydroxylase, which catalyzes this step, is subject to feedback inhibition by dopamine and norepinephrine, thus providing internal control of the synthetic process. The cofactor for tyrosine hydroxylase is **tetrahydrobiopterin,** which is converted to dihydrobiopterin when tyrosine is converted to dopa.

Some neurons and adrenal medullary cells also contain the cytoplasmic enzyme **phenylethanolamine-N-methyltransferase (PNMT),** which catalyzes the conversion of norepinephrine to epinephrine. In these cells, norepinephrine apparently leaves the vesicles, is converted to epinephrine, and then enters other storage vesicles.

In granulated vesicles, norepinephrine and epinephrine are bound to ATP and associated with a protein called **chromogranin A.** In some but not all noradrenergic neurons, the large granulated vesicles also contain neuropeptide Y (see below). Chromogranin A is a 49-kDa protein that is also found in many other endocrine and neuroendocrine cells, and it may play some sort of general role in hormone storage or secretion. A related protein, chromogranin B, is formed in some tissues. Plasma levels of chromogranin A are increased in patients with a variety of different endocrine tumors.

The catecholamines are transported into the granulated vesicles by two vesicular transports (see above), and these transporters are inhibited by the drug reserpine.

Catecholamines are released from autonomic neurons and adrenal medullary cells by exocytosis (see Chapter 1). Since they are present in the granulated vesicles, ATP, chromogranin A, and the dopamine β-hydroxylase that is not membrane-bound are released with norepinephrine and epinephrine. The half-life of circulating dopamine β-hydroxylase is much longer than that of the catecholamines, and circulating levels of this substance are affected by genetic and other factors in addition to the rate of sympathetic activity. Circulating levels of chromogranin A appear to be a better index of sympathetic activity.

Phenylpyruvic Oligophrenia

Phenylpyruvic oligophrenia, or phenylketonuria, is a disorder characterized by severe mental deficiency and the accumulation in the blood, tissues, and urine of large amounts of phenylalanine and its keto acid derivatives. It is usually due to decreased function resulting from mutation of the gene for phenylalanine hydroxylase (Figure 4–19). This gene is located on the long arm of chromosome 12. Catecholamines are still formed from tyrosine, and the mental retardation is largely due to accumulation of phenylalanine and its derivatives in the blood. Therefore, it can be treated with considerable success by markedly reducing the amount of phenylalanine in the diet.

The condition can also be caused by tetrahydrobiopterin deficiency. Since tetrahydrobiopterin is a cofactor for tyrosine hydroxylase and tryptophan hydroxylase (see below) as well as phenylalanine hydroxylase, cases due to tetrahydrobiopterin deficiency have catecholamine and serotonin deficiencies in addition to hyperphenylalaninemia. These cause hypotonia, inactivity, and developmental problems. They are treated with tetrahydrobiopterin, levodopa, and 5-hydroxytryptophan in addition to a low-phenylalanine diet.

Catabolism of Catecholamines

Norepinephrine, like other amine and amino acid transmitters, is removed from the synaptic cleft by binding to postsynaptic receptors, binding to presynaptic receptors (Figure 4–15), reuptake into the presynaptic neurons, or catabolism (Figure 4–20). Reuptake is a major mechanism in the case of norepinephrine, and the hypersensitivity of sympathetically denervated structures is explained in part on this basis. After the noradrenergic neurons are cut, their endings degenerate; consequently, there is no reuptake, and more norepinephrine from other sources is available to stimulate the receptors on the autonomic effectors.

Epinephrine and norepinephrine are metabolized to biologically inactive products by oxidation and methylation. The former reaction is catalyzed by **monoamine oxidase (MAO)** and the latter by **catechol-O-methyltransferase (COMT)** (Figure 4–20). MAO is located on the outer surface of the mitochondria. There are two isoforms of MAO, MAO-A and MAO-B, which differ in substrate specificity and sensitivity to drugs. Both are found in neurons. MAO is widely distributed, being particularly plentiful in the nerve endings at which catecholamines are secreted. COMT is also widely distributed, particularly in the liver, kidneys, and smooth muscles. In the brain, it is present in glial cells, and small amounts are found in postsynaptic neurons, but none is found in presynaptic noradrenergic

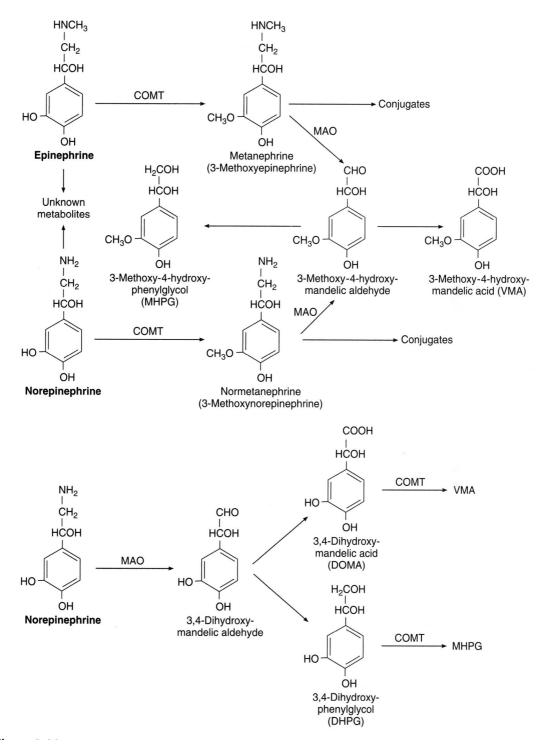

Figure 4–20. **Top:** Catabolism of extracellular epinephrine and norepinephrine. The main site of catabolism is the liver. The conjugates are mostly glucuronides and sulfates. MHPG is also conjugated. **Bottom:** Catabolism of norepinephrine in noradrenergic nerve endings. The acid and the glycol formed by MAO enter the extracellular fluid and are subsequently O-methylated to VMA and MHPG. Epinephrine in nerve endings is presumably catabolized in the same way.

neurons. Consequently, there are two different patterns of catecholamine metabolism.

Extracellular epinephrine and norepinephrine are for the most part O-methylated, and measurement of the concentrations of the O-methylated derivatives normetanephrine and metanephrine in the urine is a good index of the rate of secretion of norepinephrine and epinephrine. The O-methylated derivatives that are not excreted are largely oxidized, and 3-methoxy-4-hydroxymandelic acid (vanillylmandelic acid, VMA) (Figure 4–20) is the most plentiful catecholamine metabolite in the urine. Small amounts of the O-methylated derivatives are also conjugated to sulfates and glucuronides.

In the noradrenergic nerve terminals, on the other hand, some of the norepinephrine is being constantly converted by intracellular MAO (Figure 4–21) to the physiologically inactive deaminated derivatives, 3,4-dihydroxymandelic acid (DOMA) and its corresponding glycol (DHPG). These are subsequently converted to their corresponding *O*-methyl derivatives, VMA and MHPG (Figure 4–20).

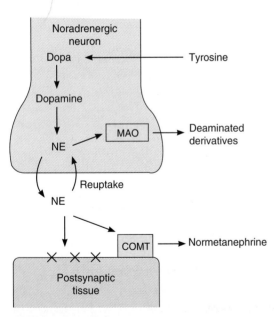

Figure 4–21. Biochemical events at noradrenergic endings. NE, norepinephrine; COMT, catechol-*O*-methyltransferase; MAO, monoamine oxidase; X, receptor. For clarity, the presynaptic receptors have been omitted. Note that MAO is intracellular, so that norepinephrine is being constantly deaminated in noradrenergic endings. COMT acts primarily on secreted norepinephrine. Compare with Figures 4–17 and 4–25.

Alpha & Beta Receptors

Epinephrine and norepinephrine both act on α and β receptors, with norepinephrine having a greater affinity for α-adrenergic receptors and epinephrine for β-adrenergic receptors. As noted above, the α and β receptors are typical serpentine receptors linked to G proteins, and there are multiple forms of each. They are closely related to the cloned receptors for dopamine and serotonin and to muscarinic acetylcholine receptors.

Imidazoline Receptors

The imidazoline **clonidine** (Figure 4–22) lowers blood pressure when administered centrally. It is an α_2 agonist and was initially thought to act on presynaptic α_2 receptors, reducing central norepinephrine discharge. However, its structure resembles that of **imidazoline,** and it binds to imidazoline receptors with higher affinity than to α_2 adrenergic receptors. A subsequent search led to the discovery that there were imidazoline receptors in the nucleus tractus solitarius and the ventrolateral medulla (VLM; see Chapter 31). Administration of imidazolines lowers blood pressure and has a depressive effect. However, the full significance of these observations remains to be explored.

Dopamine

In the small intensely fluorescent (SIF) cells in autonomic ganglia (see Chapter 13) and in certain parts of the brain (see Chapter 15), catecholamine synthesis stops at dopamine (Figure 4–19), and this catecholamine is secreted as a synaptic transmitter. There is an active reuptake of dopamine via an Na$^+$- and Cl$^-$-dependent transporter (see above). Dopamine is metabolized to inactive compounds by MAO and COMT (Figure 4–23) in a manner analogous to the inactivation of norepinephrine. DOPAC and HVA are also conjugated, primarily to sulfates.

Five different dopamine receptors have been cloned, and several of these exist in multiple forms. This provides for variety in the type of responses produced by dopamine. Most but perhaps not all of the responses to these receptors are mediated by heterotrimeric G proteins. Recent research indicates that one of the two

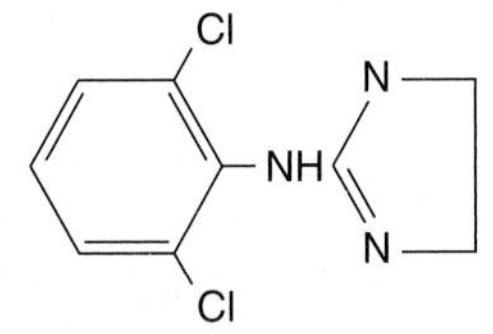

Figure 4–22. Clonidine.

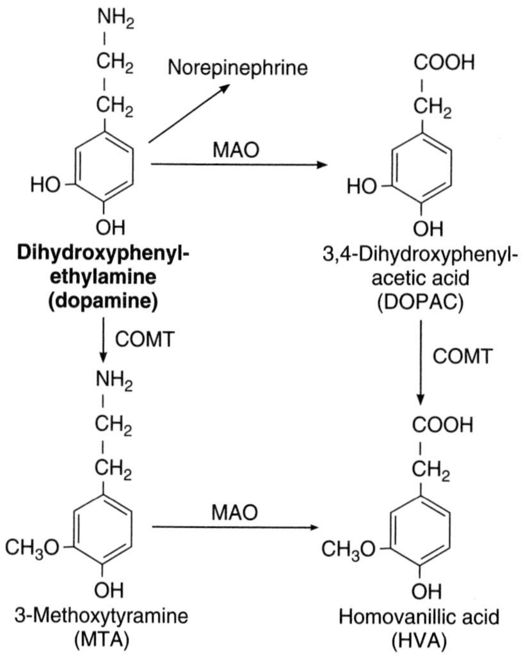

Figure 4–23. Catabolism of dopamine. As in other oxidative deaminations catalyzed by MAO, aldehydes are formed first and then oxidized in the presence of aldehyde dehydrogenase to the corresponding acids (DOPAC and HVA). The aldehydes are also reduced to 3,4-dihydroxyphenylethanol (DOPET) and 3-methoxy-4-hydroxyphenylethanol. DOPAC and HVA form sulfate conjugates.

forms of D_2 receptors can form a heterodimer with the somatostatin SST5 receptor (see below), further increasing the dopamine response menu. D_3 receptors are highly localized, especially to the nucleus accumbens (see Chapter 15). D_4 receptors have a greater affinity than the other dopamine receptors for the "atypical" antipsychotic drug clozapine, which is effective in schizophrenia but produces fewer extrapyramidal side effects than the other major tranquilizers do.

Serotonin

Serotonin (5-hydroxytryptamine; 5-HT) is present in highest concentration in blood platelets and in the gastrointestinal tract, where it is found in the enterochromaffin cells and the myenteric plexus (see Chapter 26). Lesser amounts are found in the brain and in the retina.

Serotonin is formed in the body by hydroxylation and decarboxylation of the essential amino acid tryptophan (Figures 4–24 and 4–25). Normally, the hydroxylase is not saturated; consequently, increased intake of

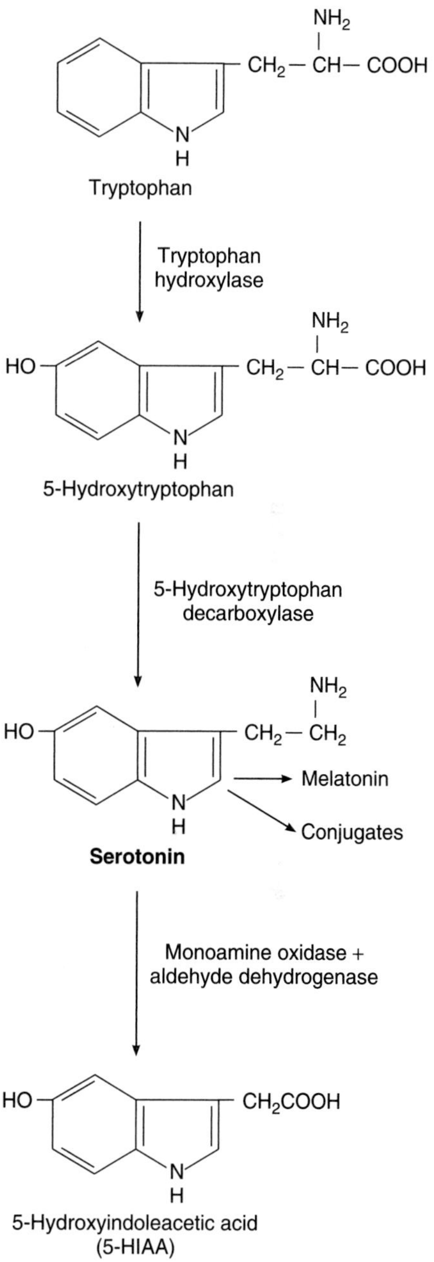

Figure 4–24. Biosynthesis and catabolism of serotonin (5-hydroxytryptamine). The enzyme that catalyzes the decarboxylation of 5-hydroxytryptophan is very similar but probably not identical to the enzyme that catalyzes the decarboxylation of dopa. Tetrahydrobiopterin is a cofactor for the action of tryptophan hydroxylase. The details of the formation of melatonin are shown in Figure 24–11.

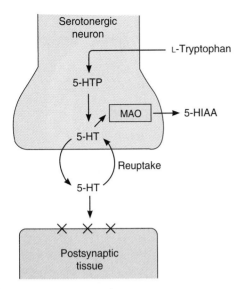

Figure 4–25. Biochemical events at serotonergic synapses. Compare with Figures 4–17 and 4–21. 5-HTP, 5-hydroxytryptophan; 5-HT, 5-hydroxytryptamine (serotonin); 5-HIAA, 5-hydroxyindoleacetic acid; X, serotonin receptor. For clarity, the presynaptic receptors have been omitted.

tryptophan in the diet can increase brain serotonin content. After release from serotonergic neurons, much of the released serotonin is recaptured by an active reuptake mechanism (Figure 4–25) and inactivated by MAO (Figure 4–24) to form 5-hydroxyindoleacetic acid (5-HIAA). This substance is the principal urinary metabolite of serotonin, and urinary output of 5-HIAA is used as an index of the rate of serotonin metabolism in the body. In the pineal gland, serotonin is converted to melatonin (see Chapter 24).

The number of cloned and characterized serotonin receptors continues to increase rapidly. Currently, there are 5-HT$_1$, 5-HT$_2$, 5-HT$_3$, 5-HT$_4$, 5-HT$_5$, 5-HT$_6$, and 5-HT$_7$ receptors. Within the 5-HT$_1$ group, there are 5-HT$_{1A}$, 5-HT$_{1B}$, 5-HT$_{1D}$, 5-HT$_{1E}$, and 5-HT$_{1F}$ subtypes. Within the 5-HT$_2$ group, there are 5-HT$_{2A}$, 5-HT$_{2B}$, and 5-HT$_{2C}$ (formerly called 5-HT$_{1C}$) subtypes. There are two 5-HT$_5$ subtypes, 5-HT$_{5A}$ and 5-HT$_{5B}$. Most of these receptors are coupled to G proteins and affect adenylyl cyclase or phospholipase C (Table 4–2). However, the 5-HT$_3$ receptors, like nicotinic cholinergic receptors, are ion channels. Some of the serotonin receptors are presynaptic, and others are postsynaptic.

5-HT$_{2A}$ receptors mediate platelet aggregation and smooth muscle contraction. Mice in which the gene for 5-HT$_{2C}$ receptors has been knocked out are obese as a result of increased food intake despite normal responses to leptin (see Chapter 14), and they are prone to fatal seizures. 5-HT$_3$ receptors are present in the gastrointestinal tract and the area postrema and are related to vomiting (see Chapter 14). 5-HT$_4$ receptors are also present in the gastrointestinal tract, where they facilitate secretion and peristalsis, and in the brain. 5-HT$_6$ and 5-HT$_7$ receptors in the brain are distributed throughout the limbic system, and the 5-HT$_6$ receptors have a high affinity for antidepressant drugs (see Chapter 15).

Histamine

Histaminergic neurons have their cell bodies in the tuberomammillary nucleus of the posterior hypothalamus (see Figure 15–6), and their axons project to all parts of the brain, including the cerebral cortex and the spinal cord. Thus, the histaminergic system resembles the noradrenergic, adrenergic, dopaminergic, and serotonergic systems, with projections from relatively few cells to all parts of the CNS.

Histamine is also found in cells in the gastric mucosa (see Chapter 6) and in heparin-containing cells called **mast cells** that are plentiful in the anterior and posterior lobes of the pituitary gland.

Histamine is formed by decarboxylation of the amino acid histidine (Figure 4–26). The enzyme that catalyzes this step differs from the L-aromatic amino acid decarboxylases that decarboxylate 5-hydroxytryptophan and L-dopa. Histamine is converted to methylhistamine or, alternatively, to imidazoleacetic acid. The latter reaction is quantitatively less important in humans. It requires the enzyme **diamine oxidase (histaminase)** rather than MAO, even though MAO catalyzes the oxidation of methylhistamine to methylimidazoleacetic acid.

There are three known types of histamine receptors, H$_1$, H$_2$, and H$_3$, and all three are found in peripheral tissues and the brain. Most, if not all, of the H$_3$ receptors are presynaptic, and they mediate inhibition of the release of histamine and other transmitters via a G protein. H$_1$ receptors activate phospholipase C, and H$_2$ receptors increase the intracellular cAMP concentration. The function of the histaminergic systems in the brain is uncertain, but histamine has been related to arousal, sexual behavior, regulation of the secretion of some anterior pituitary hormones, blood pressure, drinking, and pain thresholds. They are also involved in the sensation of itch (see Chapter 7).

Excitatory Amino Acids: Glutamate & Aspartate

The amino acid **glutamate** is the main excitatory transmitter in the brain and spinal cord, and it has been cal-

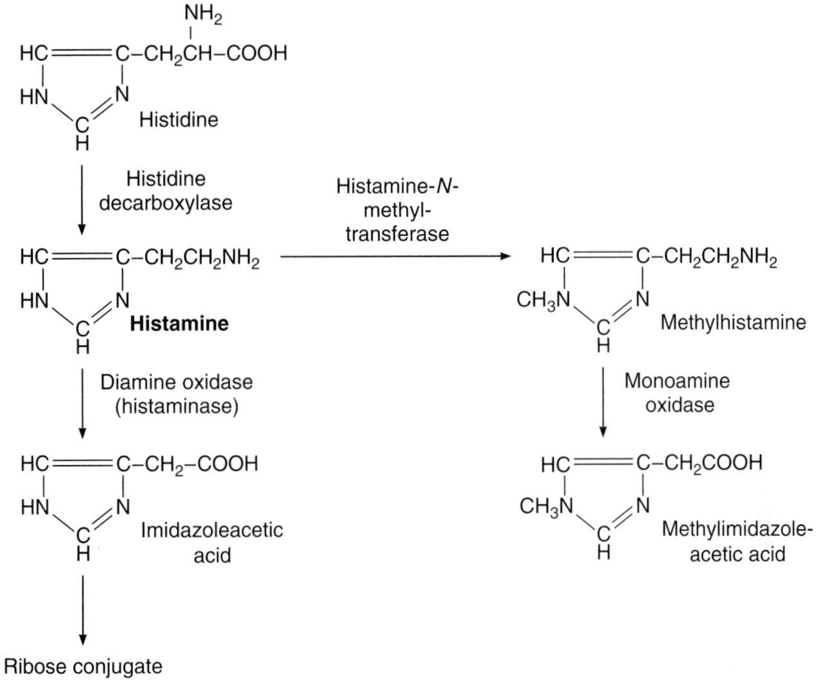

Figure 4–26. Synthesis and catabolism of histamine.

culated that it is the transmitter responsible for 75% of the excitatory transmission in the brain. Aspartate is apparently a transmitter in pyramidal cells and spiny stellate cells in the visual cortex, but it has not been studied in as great detail. Glutamate is formed by reductive amination of the Krebs cycle intermediate α-ketoglutarate (Figure 4–27) in the cytoplasm. The reaction is reversible, but in glutaminergic neurons glutamate is concentrated in synaptic vesicles by the vesicle-bound transporter **BPN1.** The cytoplasmic store of glutamine is enriched by three transporters that import glutamate from the interstitial fluid, and two additional transporters carry glutamate into astrocytes, where it is converted to glutamines and passed on to glutaminergic neurons (see Chapter 2). Uptake into neurons and astrocytes is the main mechanism for removal of glutamate from synapses.

Glutamate receptors are of two types: **metabotropic receptors** and **ionotropic receptors.** The metabotropic receptors are serpentine G protein-coupled receptors that increase intracellular IP_3 and DAG levels or decrease intracellular cAMP levels. Eleven different subtypes have been identified (Table 4–2). They are both presynaptic and postsynaptic and are widely distributed in the brain. They appear to be involved in the production of synaptic plasticity, particularly in the hippocampus and the cerebellum. Knockout of the gene for one

of these receptors, one of the forms of mGluR1, causes severe motor incoordination and deficits in spatial learning.

The ionotropic receptors are ligand-gated ion channels that resemble the nicotinic cholinergic receptors (see above) and the GABA and glycine receptors (see below). There are three general types, each named for the congeners of glutamate to which they respond in maximum fashion. These are the **kainate receptors** (kainate is an acid isolated from seaweed), the **AMPA receptors** (for α-amino-3-hydroxy-5-methylisoxazole-4-propionate), and the **NMDA receptors** (for *N*-methyl-D-aspartate). Like the nicotinic, GABA, and glycine ionotropic receptors, they are made up of multiple subunits. Four AMPA, five kainate, and six NMDA subunits have been identified, each coded by a different gene. The receptors were thought to be pentamers, but some may be tetramers, and their exact stoichiometry is unsettled.

The kainate receptors are simple ion channels that, when open, permit Na^+ influx and K^+ efflux. There are two populations of AMPA receptors, one a simple Na^+ channel and one which also passes Ca^{2+}. The balance between the two in a given synapse can be shifted by activity.

The NMDA receptor is also a cation channel, but it permits passage of relatively large amounts of Ca^{2+}, and

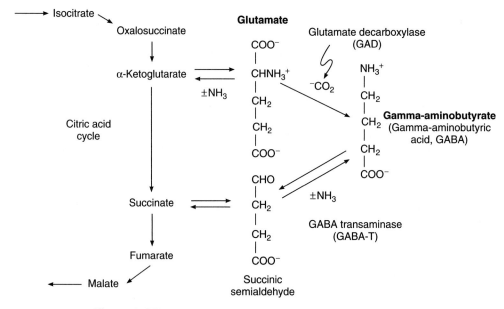

Figure 4–27. Formation and metabolism of glutamate and GABA.

it is unique in several ways. First, glycine facilitates its function by binding to it, and glycine appears to be essential for its normal response to glutamate (Figure 4–28). Second, when glutamate binds to it, it opens, but at normal membrane potentials, its channel is blocked by an Mg^{2+} ion. This block is removed only when the neuron containing the receptor is partially depolarized by activation of AMPA or other channels that produce rapid depolarization via other synaptic circuits. Third, phencyclidine and ketamine, which produce amnesia and a feeling of dissociation from the environment, bind to another site inside the channel. Most target neurons for glutamate have both AMPA and NMDA receptors. Kainate receptors are located presynaptically on GABA-secreting nerve endings and postsynaptically at various localized sites in the brain. Kainate and AMPA receptors are found in glia as well as neurons, but it appears that NMDA receptors occur only in neurons.

There is a high concentration of NMDA receptors in the hippocampus, and blockade of these receptors prevents **long-term potentiation,** a long-lasting facilitation of transmission in neural pathways following a brief period of high-frequency stimulation (see below). Thus, these receptors may well be involved in memory and learning.

Glutamate and some of its synthetic congeners are unique in that when they act on neuronal cell bodies, they can produce so much Ca^{2+} influx that neurons die. This is the reason why microinjections of these ex-

citotoxins are used in research to produce discrete lesions that destroy neuronal cell bodies without affecting neighboring axons.

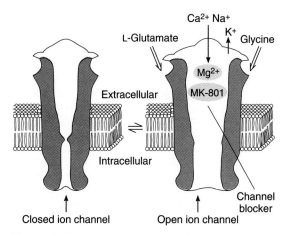

Figure 4–28. Diagrammatic representation of the NMDA receptor. When glycine and glutamate bind to the receptor, the closed ion channel **(left)** opens, but at the resting membrane potential, the channel is blocked by Mg^{2+} **(right).** This block is removed if partial depolarization is produced by other inputs to the neuron containing the receptor, and Ca^{2+} and Na^+ enter the neuron. Blockade can also be produced by the drug dizocilpine maleate (MK-801).

Evidence is accumulating that excitotoxins play a significant role in the damage done to the brain by a stroke (see Chapter 32). Glutamate is usually cleared from the brain ECF by Na^+-dependent uptake systems in neurons and glia. When a cerebral artery is occluded, the cells in the severely ischemic area die. Surrounding partially ischemic cells may survive but lose their ability to maintain the transmembrane Na^+ gradient that drives the glutamate uptake. Therefore, ECF glutamate accumulates to the point that there is excitotoxic damage and cell death around the completely infarcted area (the **penumbra**). The implications of these changes in terms of the treatment of stroke are discussed in Chapter 33.

Inhibitory Amino Acids: Gamma-Aminobutyrate

Gamma-aminobutyric acid (GABA) is the major inhibitory mediator in the brain, where it is the transmitter at 20% of CNS synapses. It is also present in the retina and is the mediator responsible for presynaptic inhibition (see above).

GABA, which exists as γ-aminobutyrate in the body fluids, is formed by decarboxylation of glutamate (Figure 4–27). The enzyme that catalyzes this reaction is **glutamate decarboxylase (GAD),** which is present in nerve endings in many parts of the brain. GABA is metabolized primarily by transamination to succinic semialdehyde and thence to succinate in the citric acid cycle (see Chapter 17). **GABA transaminase (GABA-T)** is the en-zyme that catalyzes the transamination. Pyridoxal phosphate, a derivative of the B complex vitamin pyridoxine, is a cofactor for GAD and GABA-T. There is in addition an active reuptake of GABA via the GABA transporter (see above). A vesicular GABA transporter (VGAT) transports GABA and glycine into secretory vesicles.

Autoimmunity to GAD appears to cause the **stiff-man syndrome (SMS),** a disease characterized by fluctuating but progressive muscle rigidity and painful muscle spasms, presumably due to GABA deficiency. It is interesting that GAD is also present in structures resembling synaptic vesicles in the insulin-secreting B cells of the pancreas, and GABA may be a paracrine mediator in the islets (see Chapter 19). The autoimmune disease type 1 diabetes is characterized by destruction of B cells, and the most abundant autoantibodies in this condition are against GAD. However, SMS is rare whereas type 1 diabetes is common, and not all patients with SMS have type 1 diabetes. Thus, the relation between the two diseases remains unsettled.

Three types of GABA receptors have been described: $GABA_A$, $GABA_B$, and $GABA_C$. The $GABA_A$ and $GABA_B$ receptors are widely distributed in the CNS, whereas in adult vertebrates the $GABA_C$ receptors are found almost exclusively in the retina. The $GABA_A$ and $GABA_C$ receptors are ion channels made up of five subunits surrounding a pore, like the nicotinic acetylcholine receptors and many of the glutamate receptors. In this case, the ion is Cl^- (Figure 4–29). The $GABA_B$ receptors are metabotropic and are cou-

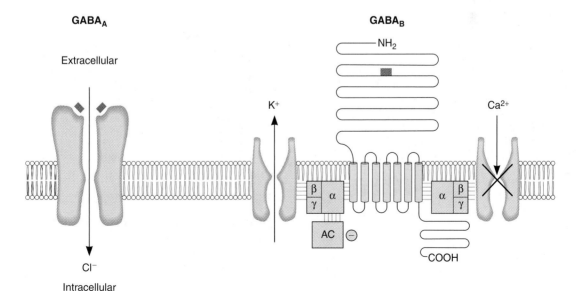

Figure 4–29. Diagram of $GABA_A$ and $GABA_B$ receptors, showing their principal actions. Note that the G protein which mediates the effects of $GABA_B$ receptors is a heterodimer. (Reproduced, with permission, from Bowery NG, Brown DA: The cloning of $GABA_B$ receptors. Nature 1997;386:223. Copyright © 1997 by Macmillan Magazines Ltd.)

pled to heterotrimeric G proteins that increase conductance in K^+ channels, inhibit adenylyl cyclase, and inhibit Ca^{2+} influx. Increases in Cl^- influx and K^+ efflux and decreases in Ca^{2+} influx all hyperpolarize neurons, producing an IPSP. The G protein mediation of $GABA_B$ receptor effects is unique in that a G protein heterodimer, rather than a single protein, is involved.

The $GABA_C$ receptors are relatively simple in that they are pentamers of three ρ subunits in various combinations. On the other hand, the $GABA_A$ receptors are pentamers made up of various combinations of six α subunits, four β, four γ, one δ, and one ε. This endows them with considerably different properties from one location to another.

A recent observation of considerable interest is that there is a chronic low-level stimulation of $GABA_A$ receptors in the CNS that is aided by GABA in the interstitial fluid. This background stimulation cuts down on the "noise" caused by incidental discharge of the billions of neural units and greatly improves the signal-to-noise ratio in the brain.

The increase in Cl^- conductance produced by $GABA_A$ receptors is potentiated by the benzodiazepines, drugs that have marked antianxiety activity and are also effective muscle relaxants, anticonvulsants, and sedatives. Benzodiazepines bind to the α subunits. Diazepam and other benzodiazepines are used throughout the world. At least in part, barbiturates and alcohol also act by facilitating Cl^- conductance through the Cl^- channel. Metabolites of the steroid hormones progesterone and deoxycorticosterone bind to $GABA_A$ receptors and increase Cl^- conductance. It has been known for many years that progesterone and deoxycorticosterone are sleep-inducing and anesthetic in large doses, and these effects are due to their action on $GABA_A$ receptors.

A second class of benzodiazepine receptors is found in steroid-secreting endocrine glands and other peripheral tissues, and hence these receptors are called **peripheral benzodiazepine receptors.** They may be involved in steroid biosynthesis, possibly performing a function like that of the StAR protein (see Chapter 20) in moving steroids into the mitochondria. Another possibility is a role in the regulation of cell proliferation. Peripheral type benzodiazepine receptors are also present in astrocytes in the brain, and they are found in brain tumors.

Glycine

By its action on the NMDA receptors, glycine has an excitatory effect in the brain. However, glycine is also responsible in part for direct inhibition, primarily in the brain stem and spinal cord. Like GABA, it acts by increasing Cl^- conductance. Its action is antagonized

by strychnine. The clinical picture of convulsions and muscular hyperactivity produced by strychnine emphasizes the importance of postsynaptic inhibition in normal neural function. The glycine receptor responsible for inhibition is a Cl^- channel. It is a pentamer made up of two subunits, the ligand-binding α subunit and the structural β subunit. Recently, solid evidence has been presented that there are three kinds of neurons responsible for direct inhibition in the spinal cord: neurons that secrete glycine, neurons that secrete GABA, and neurons that secrete both. Presumably, neurons that secrete only glycine have the glycine transporter GLYT2, those that secrete only GABA have GAD, and those that secrete glycine and GABA have both. This third type of neuron is of special interest because the neurons seem to have glycine and GABA in the same vesicles.

Anesthesia

The mechanism of action of general anesthetics has been a mystery. However, it now appears that alcohols, barbiturates, and volatile inhaled anesthetics as well act on ion channel receptors and specifically on $GABA_A$ and glycine receptors to increase Cl^- conductance.

Substance P & Other Tachykinins

Substance P is a polypeptide containing 11 amino acid residues that is found in the intestine, various peripheral nerves, and many parts of the CNS. Its structure is shown in Table 26–2. It is one of a family of six mammalian polypeptides called tachykinins that differ at the amino terminal end but have in common the carboxyl terminal sequence of Phe-X-Gly-Leu-Met-NH_2, where X is Val, His, Lys, or Phe. The members of the family are listed in Table 4–3. There are many related tachykinins in other vertebrates and in invertebrates.

The mammalian tachykinins are encoded by two genes. The **neurokinin B gene** encodes only one known polypeptide, neurokinin B. The **substance**

Table 4–3. Mammalian tachykinins.

Gene	Polypeptide Products	Receptors
SP/NKA	Substance P Neurokinin A Neuropeptide K Neuropeptide α Neurokinin A (3–10)	Substance P (NK-1) Neuropeptide K (NK-2)
NKB	Neurokinin B	Neurokinin B (NK-3)

P/neurokinin A gene encodes the remaining five polypeptides. Three are formed by alternative processing of the primary RNA and two by posttranslational processing.

There are three neurokinin receptors. Two of these, the substance P and the neuropeptide K receptors, are serpentine receptors that act via G proteins. Activation of the substance P receptor causes activation of phospholipase C and increased formation of IP_3 and DAG.

Substance P is found in high concentration in the endings of primary afferent neurons in the spinal cord, and it is probably the mediator at the first synapse in the pathways for slow pain (see Chapter 7). It is also found in high concentration in the nigrostriatal system, where its concentration is proportionate to that of dopamine, and in the hypothalamus, where it may play a role in neuroendocrine regulation. Upon injection into the skin, it causes redness and swelling, and it is probably the mediator released by nerve fibers that is responsible for the axon reflex (see Chapter 32). In the intestine, it is involved in peristalsis (see Chapter 26). It has recently been reported that a centrally active NK-1 receptor antagonist has antidepressant activity in humans. This antidepressant effect takes time to develop, like the effect of the antidepressants that affect brain monoamine metabolism (see Chapter 15), but the NK-1 inhibitor does not alter brain monoamines in experimental animals. The functions of the other tachykinins are unsettled.

Opioid Peptides

The brain and the gastrointestinal tract contain receptors that bind morphine. The search for endogenous ligands for these receptors led to the discovery of two closely related pentapeptides, called **enkephalins** (Table 4–4), that bind to these opioid receptors. One contains methionine **(met-enkephalin),** and one contains leucine **(leu-enkephalin).** These and other peptides that bind to opioid receptors are called **opioid peptides.** The enkephalins are found in nerve endings in the gastrointestinal tract and many different parts of the brain, and they appear to function as synaptic transmitters. They are found in the substantia gelatinosa and have analgesic activity when injected into the brain stem. They also decrease intestinal motility (see Chapter 26).

Like other small peptides, the opioid peptides are synthesized as part of larger precursor molecules (see Chapter 1). More than 20 active opioid peptides have been identified. Unlike other peptides, however, the opioid peptides have a number of different precursors. Each has a prepro form and a pro form from which the signal peptide has been cleaved. The three precursors that have been characterized, and the opioid peptides they produce are shown in Table 4–4. **Proenkephalin** was first identified in the adrenal medulla (see Chapter 20), but it is also the precursor for met-enkephalin and leu-enkephalin in the brain. Each proenkephalin molecule contains four met-enkephalins, one leu-enkephalin, one octapeptide, and one heptapeptide. **Pro-opiomelanocortin,** a large precursor molecule found in the anterior and intermediate lobes of the pituitary gland and the brain, contains β-endorphin, a polypeptide of 31 amino acid residues that has met-enkephalin at its amino terminal (see Chapter 22). Other shorter endorphins may also be produced, and the precursor molecule also produces ACTH and MSHs. There are separate enkephalin-secreting and β-endorphin-secreting systems of neurons in the brain (see Chapter 15). β-Endorphin is also secreted into the bloodstream by the pituitary gland. A third precursor molecule is **prodynorphin,** a protein that contains three leu-enkephalin residues associated with dynorphin and neoendorphin. Dynorphin 1-17 is found in the duodenum and dynorphin 1-8 in the posterior pi-

Table 4–4. Opioid peptides and their precursors.

Precursor	Opioid Peptides	Structures
Proenkephalin (see Chapter 20)	Met-enkephalin Leu-enkephalin Octapeptide Heptapeptide	Tyr-Gly-Gly-Phe-Met_5 Tyr-Gly-Gly-Phe-Leu_5 Tyr-Gly-Gly-Phe-Met-Arg-Gly-Leu_8 Tyr-Gly-Gly-Phe-Met-Arg-Phe_7
Pro-opiomelanocortin (see Chapter 22)	β-Endorphin Other endorphins	See Chapter 22 See Chapter 22
Prodynorphin	Dynorphin 1–8 Dynorphin 1–17 α-Neoendorphin β-Neoendorphin	Tyr-Gly-Gly-Phe-Leu-Arg-Arg-Ile_8 Tyr-Gly-Gly-Phe-Leu-Arg-Arg-Ile-Arg-Pro-Lys-Leu-Lys-Trp-Asp-Asn-Gln_{17} Tyr-Gly-Gly-Phe-Leu-Arg-Lys-Tyr-Pro-Lys_{10} Tyr-Gly-Gly-Phe-Leu-Arg-Lys-Tyr-Pro_9

tuitary and hypothalamus. Alpha- and β-neoendorphins are also found in the hypothalamus. The reasons for the existence of multiple opioid peptide precursors and for the presence of the peptides in the circulation as well as in the brain and the gastrointestinal tract are presently unknown.

Enkephalins are metabolized primarily by two peptidases: enkephalinase A, which splits the Gly-Phe bond, and enkephalinase B, which splits the Gly-Gly bond. Aminopeptidase, which splits the Tyr-Gly bond, also contributes to their metabolism.

Opioid receptors have been studied in detail, and three are now established: μ, κ, and δ. They differ in physiologic effects (Table 4–5), distribution in the brain and elsewhere, and affinity for various opioid peptides. All three are serpentine receptors coupled to G_q, and all inhibit adenylyl cyclase. Subtypes of μ and κ receptors probably exist. Activation of μ receptors increases K^+ conductance, hyperpolarizing central neurons and primary afferents. Activation of κ receptors and δ receptors closes Ca^{2+} channels.

The affinities of individual ligands for the three types of receptors are summarized in Figure 4–30. Endorphins bind only to μ receptors, the main receptors that mediate analgesia. Other opioid peptides bind to multiple opioid receptors. The pharmacology of morphine is discussed in Chapter 7.

Other Polypeptides

Numerous other polypeptides are found in the brain. Among these are the hypophysiotropic hormones (see

Table 4–5. Physiologic effects produced by stimulation of opiate receptors.

Receptor	Effect
μ	Analgesia
	Site of action of morphine
	Respiratory depression
	Constipation
	Euphoria
	Sedation
	Increased secretion of growth hormone and prolactin
	Miosis
κ	Analgesia
	Diuresis
	Sedation
	Miosis
	Dysphoria
δ	Analgesia

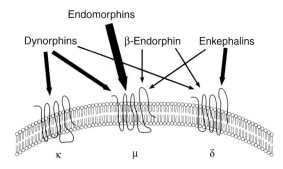

Figure 4–30. Opioid receptors. The ligands for the κ, μ, and δ receptors are shown with the width of the arrows proportionate to the affinity of the receptor for each ligand. (Reproduced, with permission, from Julius DJ: Another spark for the masses? Nature 1997;386:442. Copyright © 1997 by Macmillan Magazines Ltd.)

Chapter 14), which are found in different parts of the nervous system, and many (perhaps all) of them function as neurotransmitters as well as hormones. Preprosomatostatin is processed to two polypeptides, somatostatin 14 (see Figure 14–18) and somatostatin 28 (Figure 4–31). They occur together in tissues. Somatostatin is found in various parts of the brain, where it apparently functions as a neurotransmitter with effects on sensory input, locomotor activity, and cognitive

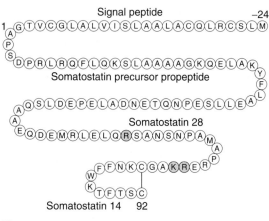

Figure 4–31. Human preprosomatostatin. The basic residues at which cleavage occurs to form somatostatin 14 and somatostatin 28 at the amino terminal are shown in color. Single-letter codes for amino acid residues. (Reproduced, with permission, from Reisine T, Bell GI: Molecular biology of somatostatin receptors. Endocr Rev 1995;16:427. Copyright © 1995 by The Endocrine Society.)

function. In the hypothalamus, it is the growth hormone–inhibiting hormone secreted into the portal hypophysial vessels (see Chapter 14); in the endocrine pancreas, it inhibits insulin secretion and the secretion of other pancreatic hormones (see Chapter 19); and in the gastrointestinal tract, it is an important inhibitory gastrointestinal hormone (see Chapter 26). Both somatostatin 28 and somatostatin 14 are biologically active, but somatostatin 28 is more active than somatostatin 14 in inhibiting insulin secretion. A family of five different somatostatin receptors have been identified (SSTR1 through SSTR5). All are G protein-coupled. They inhibit adenylyl cyclase and exert various other effects on intracellular messenger systems. It appears that SSTR2 mediates cognitive effects and inhibition of growth hormone secretion whereas SSTR5 mediates the inhibition of insulin secretion.

Vasopressin and oxytocin are not only secreted as hormones but also are present in neurons that project to the brain stem and spinal cord. The brain contains bradykinin, angiotensin II, and endothelin (see Chapters 24 and 31). The gastrointestinal hormones VIP, CCK-4, and CCK-8 (see Chapter 26) are also found in the brain. There are two kinds of CCK receptors in the brain, CCK-A and CCK-B. CCK-8 acts at both binding sites, whereas CCK-4 acts at the CCK-B sites (see Chapters 14 and 26). Gastrin, neurotensin, galanin, and gastrin-releasing peptide are also found in the gastrointestinal tract and brain. The neurotensin and the VIP receptors have been cloned and shown to be serpentine receptors. The hypothalamus contains both gastrin 17 and gastrin 34 (see Chapter 26). VIP produces vasodilation and is found in vasomotor nerve fibers. The functions of these peptides in the nervous system are unknown.

Calcitonin gene-related peptide (CGRP) is a polypeptide that in rats and humans exists in two forms, CGRPα and CGRPβ. In humans, these two forms differ by only three amino acid residues, yet they are encoded by different genes. In rats, and presumably in humans, CGRPβ is present in the gastrointestinal tract, whereas CGRPα is found in primary afferent neurons, neurons by which taste impulses project to the thalamus, and neurons in the medial forebrain bundle. It is also present along with substance P in the branches of primary afferent neurons that end near blood vessels. CGRP-like immunoreactivity is present in the circulation, and injection of CGRP causes vasodilation. CGRPα and the calcium-lowering hormone calcitonin (see Chapter 21) are both products of the calcitonin gene. However, in the thyroid gland, splicing produces the mRNA that codes for calcitonin, whereas in the brain, alternative splicing produces the mRNA that codes for CGRPα. CGRP has little effect on Ca^{2+} metabolism, and calcitonin is only a weak vasodilator.

Neuropeptide Y is a polypeptide containing 36 amino acid residues that is closely related to pancreatic polypeptide (see Chapter 19). It is present in many parts of the brain and the autonomic nervous system. In the autonomic nervous system, although not in the brain, much of it is located in noradrenergic neurons, from which it is released by high-frequency stimulation. It augments the vasoconstrictor effects of norepinephrine. Circulating neuropeptide Y from sympathetic nerves increases with severe exercise in humans. In the hypothalamus, it mediates increased appetite and increases in food intake (see Chapter 14). Y_1, Y_2, Y_4, Y_5, and Y_6 receptors for this polypeptide have been cloned.

Purine & Pyrimidine Transmitters

After extended debate, it now seems clear that ATP, uridine, adenosine, and adenosine metabolites are neurotransmitters. ATP in the ECF is the ATP released with norepinephrine, dopamine, GABA, glutamate, acetylcholine, and histamine when they are secreted by neurons. Adenosine is a neuromodulator that acts as a general CNS depressant. Adenosine is also a vasodilator in the heart (see Chapter 32) and has additional widespread effects throughout the body. It acts on four receptors, A_1, A_{2A}, A_{2B}, and A_3. All are serpentine receptors that are G protein-coupled and increase (A_{2A} and A_{2B}) or decrease (A_1 and A_3) cAMP concentrations. The stimulatory effects of coffee and tea are due to blockade of adenosine receptors by caffeine and theophylline. Currently, there is considerable interest in the potential use of A_1 antagonists to decrease excessive glutamate release and thus to minimize the effects of strokes.

ATP is also becoming established as a transmitter, and it has widespread receptor-mediated effects in the body. It appears that soluble nucleotidases are released with ATP, and these accelerate its removal after it has produced its effects. Four purinergic receptors that bind ATP have been characterized: P2Y and P2U, which activate PLC via G proteins; and P2X and P2Z, which are ligand-gated ion channels. Three subtypes of P2X have been identified: $P2X_1$, $P2X_2$, and $P2X_3$. $P2X_1$ and $P2X_2$ receptors are present in the dorsal horn, whereas PTX receptors are found only in dorsal root and trigeminal ganglia, indicating a role for ATP in sensory transmission. In addition, there is a P2T receptor, which appears to be an ion channel activated by ADP. ATP has now been shown to mediate rapid synaptic responses in the autonomic nervous system and a fast response in the habenula. There are also purinergic receptors on glial cells.

Cannabinoids

Two receptors with a high affinity for Δ^9-tetrahydrocannabinol (THC), the psychoactive ingredient in marijuana, have been cloned. The CB_1 receptor triggers a G protein-mediated decrease in intracellular cAMP levels and is common in central pain pathways as well as in parts of the cerebellum, hippocampus, and cerebral cortex. The endogenous ligand for the receptor is **anandamide,** a derivative of arachidonic acid (Figure 4–32). This compound mimics the euphoria, calmness, dream states, drowsiness, and analgesia produced by marijuana. There are also CB_1 receptors in peripheral tissues, and blockade of these receptors reduces the vasodilator effect of anandamide. However, it appears that the vasodilator effect is indirect. A CB_2 receptor has also been cloned, and its endogenous ligand may be **palmitoylethanolamide (PEA).** However, the physiologic role of this compound is unsettled.

Gases

Nitric oxide (NO), a compound released by the endothelium of blood vessels as endothelium-derived relaxing factor (EDRF), is also produced in the brain. Its synthesis from arginine, a reaction catalyzed in the brain by one of the 3 forms of NO synthase, is discussed in Chapter 31 (see Figure 31–1). It activates guanylyl cyclase (see Chapter 1), and, unlike other transmitters, it is a gas, which crosses cell membranes with ease and binds directly to guanylyl cyclase. It may be the signal by which postsynaptic neurons communicate with presynaptic endings in LTP and LTD (see below). NO synthase requires NADPH, and it is now known that NADPH-diaphorase (NDP), for which a histochemical stain has been available for many years, is NO synthase. Thus, it is easy to stain for NO synthase in the brain and other tissues.

Carbon monoxide (CO) is another gas that is probably a transmitter in the brain. It is formed in the course of the metabolism of heme (see Chapter 27) by a subtype of heme oxygenase (HO) designated HO2 (Figure 4–33), and, like NO, it activates guanylyl cyclase.

Other Substances

Prostaglandins are derivatives of arachidonic acid (see Chapter 17) that are found in the nervous system. They are present in nerve-ending fractions of brain ho-

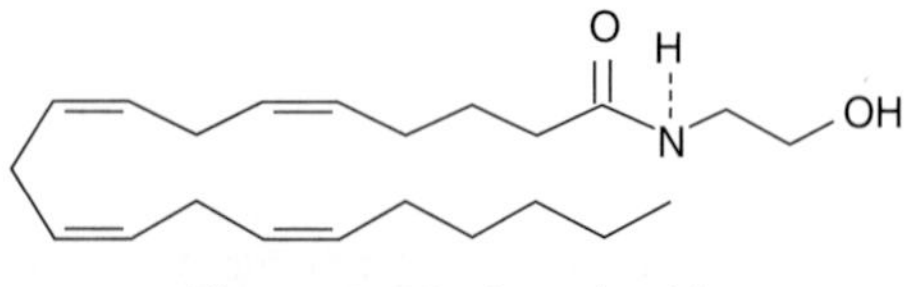

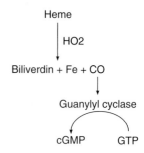

Figure 4–33. Formation and action of CO in vivo.

mogenates and are released from neural tissue in vitro. A putative prostaglandin transporter with 12 membrane-spanning domains has been described. However, prostaglandins appear to exert their effects by modulating reactions mediated by cAMP rather than by functioning as synaptic transmitters.

Many steroids are **neuroactive steroids;** ie, they affect brain function, although they are not neurotransmitters in the usual sense. Circulating steroids enter the brain with ease, and there are numerous sex steroid and glucocorticoid receptors in neurons. In addition to acting in the established fashion by binding to DNA (genomic effects), some steroids seem to act rapidly by a direct effect on cell membranes (nongenomic effects). The role of steroids in neuroendocrine control is discussed in Chapter 14. The effects of steroids on GABA receptors is discussed above. Evidence has now accumulated that the brain can produce some hormonally active steroids from simpler steroid precursors, and the term **neurosteroids** has been coined to refer to these products. Progesterone facilitates the formation of myelin (see Chapter 2), but the exact role of most steroids in the regulation of brain function remains to be determined.

Cotransmitters

Numerous examples have now been described in which neurons contain and secrete two and even three transmitters. The **cotransmitters** in these situations are often a catecholamine or serotonin plus a polypeptide, and examples of coexistence of a polypeptide with GABA or acetylcholine have been described. Coexistence of two polypeptides and coexistence of GABA with various catecholamines or acetylcholine also occur. Some neurons in the brain stem contain serotonin, substance P, and TRH. Many cholinergic neurons contain VIP, and many noradrenergic and adrenergic neurons contain ATP and neuropeptide Y. Neurons containing multiple trans-

Figure 4–32. Anandamide.

mitters often exist side by side with neurons containing a single transmitter. It has been suggested that low-frequency discharge of the neurons releases the small transmitter molecules, whereas high-frequency stimulation releases the polypeptide cotransmitters as well; however, this has not been proved. The physiologic significance of cotransmitters is still obscure. However, the VIP secreted with acetylcholine potentiates the postsynaptic actions of acetylcholine, and neuropeptide Y potentiates some of the actions of norepinephrine.

SYNAPTIC PLASTICITY & LEARNING

Short- and long-term changes in synaptic function can occur as a result of the history of discharge at a synapse; ie, synaptic conduction can be strengthened or weakened on the basis of past experience. These changes are of great interest because they obviously represent forms of learning and memory (see Chapter 16). They can be presynaptic or postsynaptic in location.

Posttetanic Potentiation

One form of plastic change is **posttetanic potentiation,** the production of enhanced postsynaptic potentials in response to stimulation. This enhancement lasts up to 60 seconds and occurs after a brief (tetanizing) train of stimuli in the presynaptic neuron. The tetanizing stimulation causes Ca^{2+} to accumulate in the presynaptic neuron to such a degree that the intracellular binding sites that keep cytoplasmic Ca^{2+} low are overwhelmed.

Habituation

When a stimulus is benign and is repeated over and over, the response to the stimulus gradually disappears **(habituation).** This is associated with decreased release of neurotransmitter from the presynaptic terminal because of decreased intracellular Ca^{2+}. The decrease in intracellular Ca^{2+} is due to a gradual inactivation of Ca^{2+} channels. It can be short-term, or it can be prolonged if exposure to the benign stimulus is repeated many times.

Sensitization

Sensitization is the prolonged occurrence of augmented postsynaptic responses after a stimulus to which an animal has become habituated is paired once or several times with a noxious stimulus. At least in the sea snail *Aplysia,* the noxious stimulus causes discharge of serotonergic neurons that end on the presynaptic endings

of sensory neurons. Thus, sensitization is due to presynaptic facilitation (see above).

Sensitization may occur as a transient response, or if it is reinforced by additional pairings of the noxious stimulus and the initial stimulus, it can exhibit features of short-term or long-term memory. The short-term prolongation of sensitization is due to a Ca^{2+}-mediated change in adenylyl cyclase that leads to a greater production of cAMP. The long-term potentiation also involves protein synthesis and growth of the presynaptic and postsynaptic neurons and their connections.

Long-Term Potentiation

Long-term potentiation (LTP) is a rapidly developing persistent enhancement of the postsynaptic potential response to presynaptic stimulation after a brief period of rapidly repeated stimulation of the presynaptic neuron. It resembles posttetanic potentiation but is much more prolonged and can last for days. Unlike posttetanic potentiation, it is initiated by an increase in intracellular Ca^{2+} in the postsynaptic rather than the presynaptic neuron. It occurs in many parts of the nervous system but has been studied in greatest detail in the hippocampus. There are two forms in the hippocampus: mossy fiber LTP, which is presynaptic and independent of NMDA receptors; and Schaffer collateral LTP, which is postsynaptic and NMDA receptor-dependent. The hypothetical basis of the latter form is summarized in Figure 4–34. The basis of mossy fiber LTP is unsettled, though it appears to include cAMP and I_h, a hyperpolarization-activated cation channel.

Other parts of the nervous system have not been as well studied, but it is interesting that NMDA-independent LTP can be produced in GABAergic neurons in the amygdala.

Long-Term Depression

Long-term depression (LTD) was first noted in the hippocampus but was subsequently shown to be present throughout the brain in the same fibers as LTP. In the cerebellum, LTD of climbing fibers causes decreased firing of parallel fibers (see Chapter 12) and may be a mechanism by which the cerebellum learns.

LTD is the opposite of LTP. It resembles LTP in many ways, but it is characterized by a decrease in synaptic strength. It is produced by slower stimulation of presynaptic neurons and is associated with a smaller rise in intracellular Ca^{2+} than occurs in LTP. It is believed to be due to dephosphorylation of AMPA receptors, decreasing their conductance and facilitating their movement away from the synaptic plasma membrane

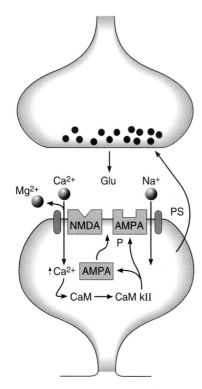

Figure 4–34. Production of LTP in Schaffer collaterals in the hippocampus. Glutamate (Glu) released from the presynaptic neuron binds to AMPA and NMDA receptors in the membrane of the dendrite. The depolarization triggered by activation of the AMPA receptors relieves the Mg^{2+} block in the NMDA receptor channel, and Ca^{2+} enters the neuron with Na^+. The increase in cytoplasmic Ca^{2+} activates calmodulin (CaM), which in turn activates Ca^{2+}/calmodulin kinase II (CaM kII). The kinase phosphorylates the AMPA receptor (P), increasing its conductance, and probably moves more AMPA receptors into the synaptic cell membrane. In addition, a chemical signal (PS) may pass to the presynaptic neuron, producing a long-term increase in the quantal release of glutamate. (Courtesy of R Nicoll.)

NEUROMUSCULAR TRANSMISSION

NEUROMUSCULAR JUNCTION

Anatomy

As the axon supplying a skeletal muscle fiber approaches its termination, it loses its myelin sheath and divides into a number of terminal buttons or end-feet (Figure 4–35). The endfeet contain many small, clear vesicles that contain acetylcholine, the transmitter at these junctions. The endings fit into depressions in the **motor end plate,** the thickened portion of the muscle membrane of the junction. Underneath the nerve ending, the muscle membrane of the end plate is thrown into **junctional folds.** The space between the nerve and the thickened muscle membrane is comparable to the synaptic cleft at synapses. The whole structure is known as the **neuromuscular** or **myoneural junction.** Only one nerve fiber ends on each end plate, with no convergence of multiple inputs.

Sequence of Events During Transmission

The events occurring during transmission of impulses from the motor nerve to the muscle (see Table 3–2) are somewhat similar to those occurring at other synapses. The impulse arriving in the end of the motor neuron increases the permeability of its endings to Ca^{2+}. Ca^{2+} enters the endings and triggers a marked increase in exocytosis of the acetylcholine-containing vesicles. The acetylcholine diffuses to the muscle type nicotinic acetylcholine receptors (Figure 4–18), which are concentrated at the tops of the junctional folds of the membrane of the motor end plate. Binding of acetylcholine to these receptors increases the Na^+ and K^+ conductance of the membrane, and the resultant influx of Na^+ produces a depolarizing potential, the **end plate potential.** The current sink created by this local potential depolarizes the adjacent muscle membrane to its firing level. Acetylcholine is then removed from the synaptic cleft by acetylcholinesterase, which is present in high concentration at the neuromuscular junction. Action potentials are generated on either side of the end plate and are conducted away from the end plate in both directions along the muscle fiber. The muscle action potential, in turn, initiates muscle contraction, as described in Chapter 3.

End Plate Potential

An average human end plate contains about 15–40 million acetylcholine receptors. Each nerve impulse releases about 60 acetylcholine vesicles, and each vesicle contains about 10,000 molecules of the neurotransmitter. This amount is enough to activate about ten times the number of acetylcholine receptors needed to produce a full end plate potential. Therefore, a propagated response in the muscle is regularly produced, and this large response obscures the end plate potential. However, the end plate potential can be seen if the tenfold safety factor is overcome and the potential is reduced to a size that is insufficient to fire the adjacent muscle membrane. This can be accomplished by administra-

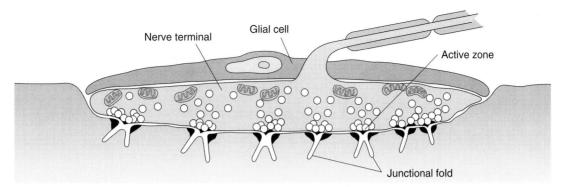

Figure 4–35. Neuromuscular junction. Note that the clear vesicles containing acetylcholine are most numerous at active zones in the nerve terminal. The zones are located over junctional folds in the motor end plate. (Reprinted by permission of the publishers from Dowling JE: *Neurons and Networks: An Introduction to Neuroscience.* The Belknap Press of Harvard University Press. Copyright © 1992 by the President and Fellows of Harvard College.)

tion of small doses of curare, a drug that competes with acetylcholine for binding to muscle type nicotinic acetylcholine receptors. The response is then recorded only at the end plate region and decreases exponentially away from it. Under these conditions, end plate potentials can be shown to undergo temporal summation.

Quantal Release of Transmitter

Small quanta ("packets") of acetylcholine are released randomly from the nerve cell membrane at rest, each producing a minute depolarizing spike called a **miniature end plate potential,** which is about 0.5 mV in amplitude. The size of the quanta of acetylcholine released in this way varies directly with the Ca^{2+} concentration and inversely with the Mg^{2+} concentration at the end plate. When a nerve impulse reaches the ending, the number of quanta released increases by several orders of magnitude, and the result is the large end plate potential that exceeds the firing level of the muscle fiber.

Quantal release of acetylcholine similar to that seen at the myoneural junction has been observed at other cholinergic synapses, and quantal release of other transmitters probably occurs at noradrenergic, glutaminergic, and other synaptic junctions.

Myasthenia Gravis & Lambert-Eaton Syndrome

Myasthenia gravis is a serious and sometimes fatal disease in which skeletal muscles are weak and tire easily. It is caused by the formation of circulating antibodies to the muscle type of nicotinic acetylcholine receptors. These antibodies destroy some of the receptors and bind others to neighboring receptors, triggering their

removal by endocytosis (see Chapter 1). The reason for the development of autoimmunity to acetylcholine receptors in this disease is still unknown.

Another condition that resembles myasthenia gravis is **Lambert-Eaton syndrome.** In this condition, muscle weakness is caused by antibodies against one of the Ca^{2+} channels in the nerve endings at the neuromuscular junction. This decreases the normal Ca^{2+} influx that causes acetylcholine release. However, muscle strength increases with prolonged contractions as more Ca^{2+} is released.

NERVE ENDINGS IN SMOOTH & CARDIAC MUSCLE

Anatomy

The postganglionic neurons in the various smooth muscles that have been studied in detail branch extensively and come in close contact with the muscle cells (Figure 4–36). Some of these nerve fibers contain clear vesicles and are cholinergic, whereas others contain the characteristic dense-core vesicles that are known to contain norepinephrine. There are no recognizable end plates or other postsynaptic specializations. The nerve fibers run along the membranes of the muscle cells and sometimes groove their surfaces. The multiple branches of the noradrenergic and, presumably, the cholinergic neurons are beaded with enlargements (**varicosities**) that are not covered by Schwann cells and contain synaptic vesicles (Figure 4–36). In noradrenergic neurons, the varicosities are about 5 μm apart, with up to 20,000 varicosities per neuron. Transmitter is apparently liberated at each varicosity, ie, at many locations along each axon. This arrangement permits one neuron

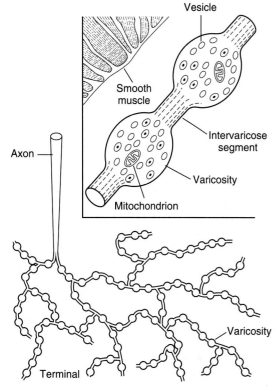

Vesicle

Smooth
muscle

Axon

Intervaricose
segment

Varicosity

Mitochondrion

Varicosity

Terminal

Figure 4–36. Endings of postganglionic autonomic neurons on smooth muscle. (Reproduced, with permission, from Kandel ER, Schwartz JH [editors]: *Principles of Neural Science,* 2nd ed. Elsevier, 1985.)

to innervate many effector cells. The type of contact in which a neuron forms a synapse on the surface of another neuron or a smooth muscle cell and then passes on to make similar contacts with other cells is called a **synapse en passant.**

In the heart, cholinergic and noradrenergic nerve fibers end on the sinoatrial node, the atrioventricular node, and the bundle of His. Noradrenergic fibers also innervate the ventricular muscle. The exact nature of the endings on nodal tissue is not known. In the ventricle, the contacts between the noradrenergic fibers and the cardiac muscle fibers resemble those found in smooth muscle.

Electrical Responses

Individual boutons of intact preganglionic cholinergic neurons have been studied in autonomic ganglia. In this location, there appear to be presynaptic receptors for cotransmitters released with the acetylcholine. In smooth muscles in which noradrenergic discharge is ex-

citatory, stimulation of the noradrenergic nerves produces discrete partial depolarizations that look like small end plate potentials and are called **excitatory junction potentials (EJPs).** These potentials summate with repeated stimuli. Similar EJPs are seen in tissues excited by cholinergic discharges. In tissues inhibited by noradrenergic stimuli, hyperpolarizing **inhibitory junction potentials (IJPs)** are produced by stimulation of the noradrenergic nerves.

These electrical responses are observed in many smooth muscle cells when a single nerve is stimulated, but their latency varies. This finding is consistent with the synapse en passant arrangement described above, but it could also be explained by transmission of the junction responses from cell to cell across low-resistance junctions or by diffusion of transmitter from its site of release to many smooth muscle cells.

DENERVATION HYPERSENSITIVITY

When the motor nerve to skeletal muscle is cut and allowed to degenerate, the muscle gradually becomes extremely sensitive to acetylcholine. This **denervation hypersensitivity** or **supersensitivity** is also seen in smooth muscle. Smooth muscle, unlike skeletal muscle, does not atrophy when denervated, but it becomes hyperresponsive to the chemical mediator that normally activates it. Denervated exocrine glands, except for sweat glands, also become hypersensitive. A good example of denervation hypersensitivity is the response of the denervated iris. If the postganglionic sympathetic nerves to one iris are cut in an experimental animal and, after several weeks, norepinephrine is injected intravenously, the denervated pupil dilates widely. A much smaller, less prolonged response is observed on the intact side.

The reactions triggered by section of an axon are summarized in Figure 4–37. Hypersensitivity of the postsynaptic structure to the transmitter previously secreted by the axon endings is a general phenomenon, largely due to the synthesis or activation of more receptors. There is in addition orthograde degeneration (**wallerian degeneration;** see Chapter 2) and retrograde degeneration of the axon stump to the nearest collateral (**sustaining collateral**). A series of changes occur in the cell body that include a decrease in Nissl substance (chromatolysis). The nerve then starts to regrow, with multiple small branches projecting along the path the axon previously followed (regenerative sprouting). Axons sometimes grow back to their original targets, especially in locations like the neuromuscular junction. However, nerve regeneration is generally limited because axons often become entangled in the area of tissue damage at the site where they were disrupted. In interesting recent experiments, this difficulty has

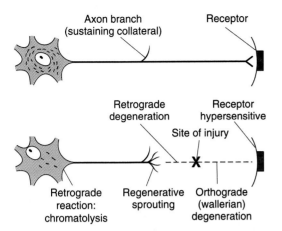

Figure 4–37. Summary of changes occurring in a neuron and the structure it innervates when its axon is crushed or cut at the point marked X.

been reduced by administration of neurotrophins. For example, sensory neurons torn when dorsal nerve roots are avulsed from the spinal cord regrow and form functional connections in the spinal cord if the experimental animals are treated with NGF, neurotrophin 3, or GDNF (see Chapter 2).

When higher centers in the nervous system are destroyed, the activity of the lower centers they control is generally increased ("release phenomenon"). The increased activity may be due in part to denervation hypersensitivity of the lower centers.

Hypersensitivity is limited to the structures immediately innervated by the destroyed neurons and fails to develop in neurons and muscle farther "downstream." Suprasegmental spinal cord lesions do not lead to hypersensitivity of the paralyzed skeletal muscles to acetylcholine, and destruction of the preganglionic autonomic nerves to visceral structures does not cause hypersensitivity of the denervated viscera. This fact has practical implications in the treatment of diseases due to spasm of the blood vessels in the extremities. For example, if the upper extremity is sympathectomized by removing the upper part of the ganglion chain and the stellate ganglion, the hypersensitive smooth muscle in the vessel walls is stimulated by circulating norepinephrine, and episodic vasospasm continues to occur. However, if preganglionic sympathectomy of the arm is performed by cutting the ganglion chain below the third ganglion (to interrupt ascending preganglionic fibers) and the white rami of the first three thoracic nerves, no hypersensitivity results.

Denervation hypersensitivity has multiple causes. As noted in Chapter 1, a deficiency of a given chemical messenger generally produces an up-regulation of its receptors. Another factor is lack of reuptake of secreted neurotransmitters.

In denervated skeletal muscle, there is an increase in the area of the muscle membrane sensitive to acetylcholine. Normally, only the end plate region contains nicotinic acetylcholine receptors, and they are of the adult ε subunit-containing type (see above). After denervation, the sensitivity of the end plate is no greater, but acetylcholine receptors of the fetal γ subunit-containing type appear over large portions of the muscle membrane. These disappear and the sensitivity returns to normal if the nerve regrows. During development there is a similar dispersion of acetylcholine receptors. These aggregate at the neuromuscular junction when a motor nerve reaches the muscle and secretes a protein called **agrin,** which binds to a muscle protein called muscle-specific kinase (**MuSK**). Agrin presumably plays the same role during reinnervation after nerve injury.

Initiation of Impulses in Sense Organs

INTRODUCTION

Information about the internal and external environment reaches the CNS via a variety of **sensory receptors.** These receptors are transducers that convert various forms of energy in the environment into action potentials in neurons. The characteristics of these receptors, the way they generate impulses in afferent neurons, and the general principles or "laws" that apply to sensation are considered in this chapter.

SENSE ORGANS & RECEPTORS

It is worth noting that the term "receptor" is used in physiology to refer not only to sensory receptors but also, in a very different sense, to proteins that bind neurotransmitters, hormones, and other substances with great affinity and specificity as a first step in initiating specific physiologic responses (see Chapter 1). The sensory receptor may be part of a neuron or a specialized cell that generates action potentials in neurons. The receptor is often associated with nonneural cells that surround it, forming a **sense organ.** The forms of energy converted by the receptors include, for example, mechanical (touch-pressure), thermal (degrees of warmth), electromagnetic (light), and chemical energy (odor, taste, and O_2 content of blood). The receptors in each of the sense organs are adapted to respond to one particular form of energy at a much lower threshold than other receptors respond to this form of energy. The particular form of energy to which a receptor is most sensitive is called its **adequate stimulus.** The adequate stimulus for the rods and cones in the eye, for example, is light. Receptors do respond to forms of energy other than their adequate stimulus, but the threshold for these nonspecific responses is much higher. Pressure on the eyeball will stimulate the rods and cones, for example, but the threshold of these receptors to pressure is much higher than the threshold of the pressure receptors in the skin.

THE SENSES

Sensory Modalities

Because the sensory receptors are specialized to respond to one particular form of energy and because many variables in the environment are perceived, it follows that there must be many different types of receptors. We learn in elementary school that there are "five senses," but the inadequacy of this dictum is apparent if we list the major sensory modalities and their receptors in humans. The first 11 modalities in Table 5–1 are conscious senses. There are, in addition, a large number of sensory receptors which relay information that does not reach consciousness. For example, the muscle spindles provide information about muscle length, and other receptors provide information about such variables as the arterial blood pressure, the temperature of the blood in the head, and the pH of the cerebrospinal fluid. The existence of other receptors of this type is suspected, and future research will undoubtedly add to the list of "unconscious senses." Furthermore, any listing of the senses is bound to be arbitrary. The rods and cones, for example, respond maximally to light of different wavelengths, and there are different cones for each of the three primary colors. There are five different modalities of taste—sweet, salt, sour, bitter, and umami. Sounds of different pitches are heard primarily because different groups of hair cells in the organ of Corti are activated maximally by sound waves of different frequencies. Whether these various responses to light, taste, and sound should be considered separate senses is a semantic question that in the present context is largely academic.

Classifications of Sense Organs

Numerous attempts have been made to classify the senses into groups, but none has been entirely successful. Traditionally, the special senses are smell, vision,

Table 5–1. Principal sensory modalities.

Sensory Modality[1]	Receptor	Sense Organ
Vision	Rods and cones	Eye
Hearing	Hair cells	Ear (organ of Corti)
Smell	Olfactory neurons	Olfactory mucous membrane
Taste	Taste receptor cells	Taste bud
Rotational acceleration	Hair cells	Ear (semicircular canals)
Linear acceleration	Hair cells	Ear (utricle and saccule)
Touch-pressure	Nerve endings	Probably nerve endings
Warmth	Nerve endings	Probably nerve endings
Cold	Nerve endings	Probably nerve endings
Pain	Nerve endings	Probably nerve endings
Joint position and movement	Nerve endings	Various[2]
Muscle length	Nerve endings	Muscle spindle
Muscle tension	Nerve endings	Golgi tendon organ
Arterial blood pressure	Nerve endings	Stretch receptors in carotid sinus and aortic arch
Central venous pressure	Nerve endings	Stretch receptors in walls of great veins, atria
Inflation of lung	Nerve endings	Stretch receptors in lung parenchyma
Temperature of blood in head	Neurons in hypothalamus	
Arterial P_{O_2}	Glomus cells	Carotid and aortic bodies
pH of CSF	Receptors on ventral surface of medulla oblongata	
Osmotic pressure of plasma	Cells in OVLT and possibly other circumventricular organs in anterior hypothalamus	
Arteriovenous blood glucose difference	Cells in hypothalamus (glucostats)	

[1]The first 11 are conscious sensations.
[2]See text.

hearing, rotational and linear acceleration, and taste; the cutaneous senses are those with receptors in the skin; and the visceral senses are those concerned with perception of the internal environment. Pain from visceral structures is usually classified as a visceral sensation. Another classification of the various receptors divides them into (1) teleceptors ("distance receivers"), which are concerned with events at a distance; (2) exteroceptors, which are concerned with the external environment near at hand; (3) interoceptors, which are concerned with the internal environment; and (4) proprioceptors, which provide information about the position of the body in space at any given instant. However, the conscious component of proprioception ("body image") is actually synthesized from information coming not only from receptors in and around joints but also from cutaneous touch and pressure receptors. Certain other special terms are sometimes used. The cutaneous receptors for touch and pressure are **mechanoreceptors.** Potentially harmful stimuli such as pain, extreme heat, and extreme cold are said to be mediated by **nociceptors.** The term **chemoreceptor** is used to refer to those receptors that are stimulated by a change in the chemical composition of the environment in which they are located. These include receptors for taste and smell as well as visceral receptors such as those sensitive to changes in the plasma level of O_2, pH, and osmolality.

Cutaneous Sense Organs

There are four cutaneous senses: touch-pressure (pressure is sustained touch), cold, warmth, and pain. Considerable effort has been expended in describing putative receptors for these sensors in the skin and subcutaneous tissues. However, it has been established that large areas of skin have all four sensations with no visually recognizable specialized receptors. This indicates that the receptors are probably on naked nerve endings. It may be that when they are present, some of the histologic endings mediate subtypes of the four basic sensations. For example, the pacinian corpuscle is a touch receptor, but it responds only to transient touch, not to sustained pressure, because the corpuscle rapidly adapts and dissipates the pressure; this makes it an excellent mechanism for detecting benign tactile stimuli.

A major advance in this field has been the cloning of three receptors for cutaneous sensations—one for moderate cold (CMR-1) and two for extreme heat (VR1 and VRL-1). VR1 is clearly a nociceptor receptor, and VRL-1 is probably a nociceptor receptor as well. These receptors are discussed in more detail in Chapter 7. All three are ion channels and are members of the **transient receptor potential (TRP)** subfamily. Aside from the fact that activation of the cool receptor causes an influx of Ca^{2+}, little is known about the ionic basis of the initial depolarization they produce. In the cutaneous receptors in general, depolarization could be due to inhibition of K^+ channels, activation of degenerin Na^+ channels, or inhibition of Na^+-K^+ ATPase, but the choice between these possibilities has not been made.

GENERATION OF IMPULSES IN DIFFERENT NERVES

The problem of how receptors convert energy into action potentials in the sensory nerves that innervate them has been the subject of intensive study. In the complex sense organs such as those concerned with vision, hearing, equilibrium, and taste, there are separate receptor cells and synaptic junctions between receptors

and afferent nerves. However, in the cutaneous senses, only the pacinian corpuscle has been studied in detail.

Pacinian Corpuscles

As noted above, the pacinian corpuscles are touch receptors. Because of their relatively large size and accessibility, they can be isolated, studied with microelectrodes, and subjected to microdissection. Each capsule consists of the straight, unmyelinated ending of a sensory nerve fiber, 2 μm in diameter, surrounded by concentric lamellas of connective tissue that give the organ the appearance of a minute cocktail onion. The myelin sheath of the sensory nerve begins inside the corpuscle. The first node of Ranvier is also located inside, whereas the second is usually near the point at which the nerve fiber leaves the corpuscle (Figure 5–1).

Generator Potentials

Recording electrodes can be placed on the sensory nerve as it leaves a pacinian corpuscle and graded pressure applied to the corpuscle. When a small amount of pressure is applied, a nonpropagated depolarizing potential resembling an EPSP is recorded. This is called the **generator potential** or **receptor potential.** As the pressure is increased, the magnitude of the receptor potential increases. When the magnitude of the generator potential is about 10 mV, an action potential is generated in the sensory nerve. As the pressure is further increased, the generator potential becomes even larger and the sensory nerve fires repetitively.

Source of the Generator Potential

By microdissection techniques, it has been shown that removal of the connective tissue lamellas from the unmyelinated nerve ending in a pacinian corpuscle does not abolish the generator potential. When the first node of Ranvier is blocked by pressure or narcotics, the generator potential is unaffected but conducted impulses are abolished (Figure 5–1). When the sensory nerve is sectioned and the nonmyelinated terminal is allowed to degenerate, no generator potential is formed. These and other experiments have established that the generator potential is produced in the unmyelinated nerve terminal. The receptor therefore converts mechanical energy into an electrical response, the magnitude of which is proportionate to the intensity of the stimulus. The generator potential in turn depolarizes the sensory nerve at the first node of Ranvier. Once the firing level is reached, an action potential is produced and the membrane repolarizes. If the generator potential is great enough, the neuron fires again as soon as it repolarizes, and it continues to fire as long as the generator potential is large enough to bring the mem-

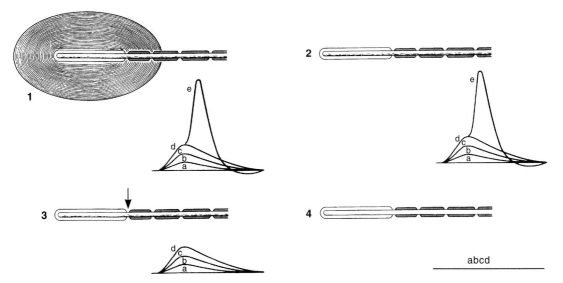

Figure 5–1. Demonstration that the generator potential in a pacinian corpuscle originates in the nonmyelinated nerve terminal. **1:** The electrical responses to a pressure of 1× (record a), 2× (b), 3× (c), and 4× (d) were recorded. The strongest stimulus produced an action potential in the sensory nerve (e). **2:** Similar responses persisted after removal of the connective tissue capsule, except that the responses were more prolonged because of partial loss of adaptation. **3:** The generator responses persisted but the action potential was absent when the first node of Ranvier was blocked by pressure or with narcotics (arrow). **4:** All responses disappeared when the sensory nerve was cut and allowed to degenerate before the experiment.

brane potential of the node to the firing level. Thus, the node converts the graded response of the receptor into action potentials, the frequency of which is proportionate to the magnitude of the applied stimuli.

Similar generator potentials in the muscle spindle have been studied. The relation between muscle length, which determines stimulus intensity in the spindle, and the size of the generator potential is shown in Figure 5–2, along with the relation between muscle length and frequency of action potentials in the afferent nerve fiber from the spindle. The frequency of the action potentials is generally related to the intensity of the stimulus by a power function (see below). Generator potentials occur in other sense organs, and in all, they appear to initiate depolarization in the sensory nerve fibers from the organ.

Adaptation

When a maintained stimulus of constant strength is applied to a receptor, the frequency of the action potentials in its sensory nerve declines over time. This phenomenon is known as **adaptation** or **desensitization.** The degree to which adaptation occurs varies from one sense to another. Adaptation has been known to occur for many years (Figure 5–3). It is pre-

sumably a receptor phenomenon, although with the possible exception of the pacinian corpuscle this has not been proved. Nevertheless, the literature speaks of **rapidly adapting** (phasic) **receptors** and **slowly adapting** (tonic) **receptors.** Light touch appears to have rapidly adapting receptors, for example, whereas spindle and nociceptor input is slowly adapting. This appears to have some value to the individual. Thus, light touch would be distracting if it were persistent; and, conversely, slow adaptation of spindle input is needed to maintain posture. Similarly, input from nociceptors provides a warning that would lose its value if it adapted and disappeared.

"CODING" OF SENSORY INFORMATION

There are variations in the speed of conduction and other characteristics of sensory nerve fibers (see Chapter 2), but action potentials are similar in all nerves. The action potentials in the nerve from a touch receptor, for example, are essentially identical to those in the nerve from a warmth receptor. This raises the question of why stimulation of a touch receptor causes a sensation of touch and not of warmth. It also raises the question of how it is possible to tell whether the touch is light or heavy.

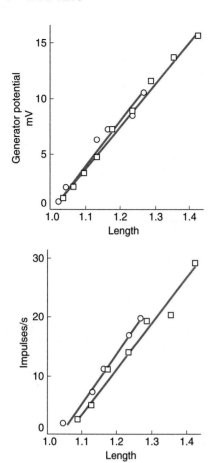

Figure 5–2. Relation between muscle length and size of generator potential **(top)** and impulse frequency **(bottom)** in crayfish stretch receptor. Squares and circles indicate values in two different preparations. (Reproduced, with permission, from Terzuolo CA, Washizu Y: Relation between stimulus strength, generator potential, and impulse frequency in stretch receptor of crustacea. J Neurophysiol 1962;25:56.)

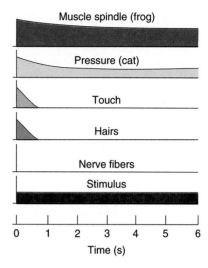

Figure 5–3. Adaptation. The height of the curve in each case indicates the frequency of the discharge in afferent nerve fibers at various times after beginning sustained stimulation. (Reproduced, with permission, from Adrian ED: *Basis of Sensation.* Christophers, 1928.)

Doctrine of Specific Nerve Energies

The sensation evoked by impulses generated in a receptor depends in part upon the specific part of the brain they ultimately activate. The specific sensory pathways are discrete from sense organ to cortex. Therefore, when the nerve pathways from a particular sense organ are stimulated, the sensation evoked is that for which the receptor is specialized no matter how or where along the pathway the activity is initiated. This principle, first enunciated by Müller in 1835, has been given the rather cumbersome name of the **doctrine of specific nerve energies.** For example, if the sensory nerve from a pacinian corpuscle in the hand is stimulated by pressure at the elbow or by irritation from a tumor in the brachial plexus, the sensation evoked is touch. Similarly, if a fine enough electrode could be inserted into the appropriate fibers of the dorsal columns of the spinal cord, the thalamus, or the postcentral gyrus of the cerebral cortex, the sensation produced by stimulation would be touch. This doctrine has been questioned from time to time; furthermore, it is not certain where the nociceptor pathway that signals both heat and pain (see Chapter 7) fits in. Nevertheless, the general principle of specific nerve energies remains one of the cornerstones of sensory physiology.

Projection

No matter where a particular sensory pathway is stimulated along its course to the cortex, the conscious sensation produced is referred to the location of the receptor. This principle is called the **law of projection.** Cortical stimulation experiments during neurosurgical procedures on conscious patients illustrate this phenomenon. For example, when the cortical receiving area for impulses from the left hand is stimulated, the patient reports sensation in the left hand, not in the head. Another dramatic example is seen in amputees. Some of these patients may complain, often bitterly, of pain and proprioceptive sensations in the absent limb **(phantom limb).** The ends of the nerves cut at the time of amputation often form nerve tangles called neuromas. These may discharge spontaneously or when pressure is put on them. The impulses that are generated are in nerve fibers that previously came from sense organs in the amputated limb, and the sensations evoked are projected to where the receptors used to be. However, there is evidence that plasticity in sensory systems within the CNS (see Chapter 7) is also involved in the phantom limb phenomenon.

Intensity Discrimination

There are two ways in which information about intensity of stimuli is transmitted to the brain: by variation in the frequency of the action potentials generated by the activity in a given receptor, and by variation in the number of receptors activated. It has long been taught that the magnitude of the sensation felt is proportionate to the log of the intensity of the stimulus (**Weber-Fechner law**). It now appears, however, that a power function more accurately describes this relation. In other words,

$$R = KS^A$$

where R is the sensation felt, S is the intensity of the stimulus, and, for any specific sensory modality, K and A are constants. The frequency of the action potentials generated by a stimulus to a sensory nerve fiber is also related to the intensity of the initiating stimulus by a power function. An example of this relation is shown in Figure 5–2, in which the exponent is approximately 1.0. Another example is shown in Figure 5–4, in which the calculated exponent is 0.52. Current evidence indicates that in the CNS the relation between stimulus and sensation is linear; consequently, it appears that for any given sensory modality, the relation between sensation and stimulus intensity is determined primarily by the properties of the peripheral receptors themselves.

Sensory Units

The term "sensory unit" is applied to a single sensory axon and all its peripheral branches. These branches vary in number but may be numerous, especially in the cutaneous senses. The **receptive field** of a sensory unit is the area from which a stimulus produces a response in that unit. In the cornea and adjacent sclera of the eye, the surface area supplied by a single sensory unit is 50–200 mm². Generally, the areas supplied by one unit overlap and interdigitate with the areas supplied by others.

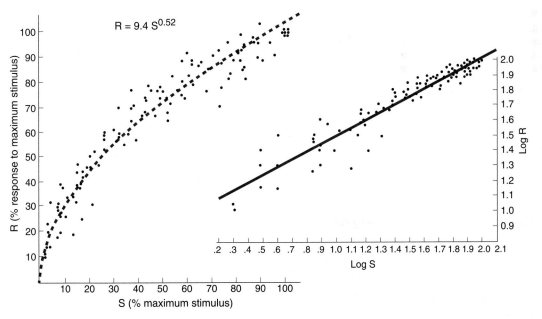

Figure 5–4. Relation between magnitude of touch stimulus (S) and frequency of action potentials in sensory nerve fibers (R). Dots are individual values from cats and are plotted on linear coordinates (**left**) and log-log coordinates (**right**). The equation shows the calculated power function relationship between R and S. (Reproduced, with permission, from Werner G, Mountcastle VB: Neural activity in mechanoreceptive cutaneous afferents. Stimulus-response relations, Weber functions, and information transmission. J Neurophysiol 1965;28:359.)

Recruitment of Sensory Units

As the strength of a stimulus is increased, it tends to spread over a large area and generally not only activates the sense organs immediately in contact with it but also "recruits" those in the surrounding area. Furthermore, weak stimuli activate the receptors with the lowest thresholds, whereas stronger stimuli also activate those with higher thresholds. Some of the receptors activated are part of the same sensory unit, and impulse frequency in the unit therefore increases. Because of overlap and interdigitation of one unit with another, however, receptors of other units are also stimulated, and consequently more units fire. In this way, more afferent pathways are activated, and this is interpreted in the brain as an increase in intensity of the sensation.

REFERENCES FOR SECTION II: PHYSIOLOGY OF NERVE & MUSCLE CELLS

Akil H et al: Endogenous opioids: Biology and function. Annu Rev Neurosci 1984;7:223.

Baulieu E-E: Neurosteroids: Of the nervous system, by the nervous system, for the nervous system. Recent Prog Horm Res 1997;22:1.

Bennett MR: Synaptic transmission at single boutons in sympathetic ganglia. News Physiol Sci 2000;15:98.

Carlsson A: A paradigm shift in brain research. Science 2001;294:1021.

Chen H, He Z, Tessier-Lavigne M: Axon guidance mechanisms: semiphorins as simultaneous repellents and anti-repellents. Nat Neurosci 1998;1:436.

Cooper JR, Bloom FE, Roth RH: *The Biochemical Basis of Neuropharmacology,* 6th ed. Oxford Univ Press, 1991.

Dunwiddie TV: The role and regulation of adenosine in the central nervous system. Annu Rev Neurosci 2001;24:32.

Fung YC: *Biomechanics,* 2nd ed. Springer, 1993.

Gordon AM, Regnier M, Homsher E: Skeletal and cardiac muscle contractile activation: tropomyosin "rocks and rolls." News Physiol Sci 2001;16:49.

Greengard P: The neurobiology of slow synaptic transmission. Science 2001;294:1024.

Griggs RC, Mendell JR, Miller RG: *Evaluation and Treatment of Myopathies.* Davis, 1994.

Hallett M: One man's poison—clinical applications of botulinum toxin. N Engl J Med 1999;341:188.

Hannah MJ, Schmidt AA, Huttner WB: Synaptic vesicle biogenesis. Annu Rev Cell Dev Biol 1999;15:723.

Hille B: *Ionic Channels of Excitable Membranes,* 2nd ed. Sinauer, 1992.

Hollmann M, Heinemann S: Cloned glutamate receptors. Annu Rev Neurosci 1994;17:31.

Horowitz A et al: Mechanisms of smooth muscle contraction. Physiol Rev 1996;76:967.

Iversen LL: *The Science of Marijuana.* Oxford Univ Press, 2000.

Jones DA, Round JM: *Skeletal Muscle in Health and Disease: A Textbook of Muscle Physiology.* Manchester Univ Press, 1990.

Malenka RC, Nicoll RA: Long-term potentiation—a decade of progress? Science 1999;285:1870.

Männistö P, Kaakkola S: Catechol-O-methyltransferase (COMT): Biochemistry, molecular biology, pharmacology, and clinical efficiency of the new selective COMT inhibitors. Pharmacol Rev 1999;51:593.

Mountcastle VB: *Perceptual Neuroscience.* Harvard Univ Press, 1999.

Noseworthy JH et al: Multiple sclerosis. N Engl J Med 2000;343:938.

Pette D, Staron RS: The molecular diversity of mammalian muscle fibers. News Physiol Sci 1993;8:153.

Sanguinetti MC, Keating MT: Role of delayed rectifier potassium channels in cardiac repolarization and arrhythmias. News Physiol Sci 1997;12:152.

Schiavo G, Matteoli M, Montecucco C: Neurotoxins affecting neuroexocytosis. Physiol Rev 2000;80:717.

Schuman EM, Madison DV: Nitric oxide and synaptic function. Annu Rev Neurosci 1994;17:153.

Somlyo AP, Somlyo AV: Signal transduction and regulation in smooth muscle. Nature 1994;372:231.

Stern P, Marx J (editors): Dendrites: Beautiful, complex, and diverse specialists. Science 2000;290:735.

Südhof TC: The synaptic vesicle cycle: A cascade of protein-protein interactions. Nature 1995;375:645.

Thoenen H: Neurotrophins and neuronal plasticity. Science 1995;270:593.

Vidal C, Changeux J-P: Neuronal nicotinic acetylcholine receptors in the brain. News Physiol Sci 1996;11:202.

Wimalawansa SJ: Calcitonin gene-related peptide and its receptors: molecular genetics, physiology, pathophysiology, and therapeutic potentials. Endocr Rev 1996;17:533.

SECTION III
Functions of the Nervous System

Reflexes

6

INTRODUCTION

The basic unit of integrated reflex activity is the **reflex arc.** This arc consists of a sense organ, an afferent neuron, one or more synapses in a central integrating station or sympathetic ganglion, an efferent neuron, and an effector. In mammals, the connection between afferent and efferent somatic neurons is generally in the brain or spinal cord. The afferent neurons enter via the dorsal roots or cranial nerves and have their cell bodies in the dorsal root ganglia or in the homologous ganglia on the cranial nerves. The efferent fibers leave via the ventral roots or corresponding motor cranial nerves. The principle that in the spinal cord the dorsal roots are sensory and the ventral roots are motor is known as the **Bell-Magendie law.**

Activity in the reflex arc starts in a sensory receptor with a receptor potential whose magnitude is proportionate to the strength of the stimulus (Figure 6–1). This generates all-or-none action potentials in the afferent nerve, the number of action potentials being proportionate to the size of the generator potential. In the CNS, the responses are again graded in terms of EPSPs and IPSPs at the synaptic junctions (see Chapter 4).

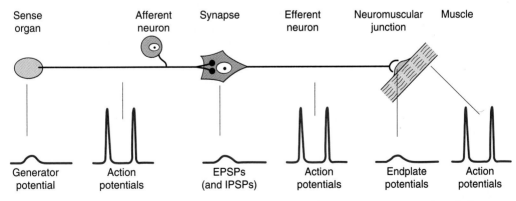

Figure 6–1. The reflex arc. Note that at the receptor and in the CNS there is a nonpropagated graded response that is proportionate to the magnitude of the stimulus. The response at the neuromuscular junction is also graded, though under normal conditions it is always large enough to produce a response in skeletal muscle. On the other hand, in the portions of the arc specialized for transmission (afferent and efferent axons, muscle membrane), the responses are all-or-none action potentials.

All-or-none responses are generated in the efferent nerve. When these reach the effector, they again set up a graded response. When the effector is smooth muscle, responses summate to produce action potentials in the smooth muscle, but when the effector is skeletal muscle, the graded response is always adequate to produce action potentials that bring about muscle contraction (see Chapter 4). It should be emphasized that the connection between the afferent and efferent neurons is usually in the CNS, and activity in the reflex arc is modified by the multiple inputs converging on the efferent neurons.

The simplest reflex arc is one with a single synapse between the afferent and efferent neurons. Such arcs are **monosynaptic,** and reflexes occurring in them are **monosynaptic reflexes.** Reflex arcs in which one or more interneurons are interposed between the afferent and efferent neurons are **polysynaptic,** the number of synapses in the arcs varying from two to many hundreds. In both types, but especially in polysynaptic reflex arcs, activity is modified by spatial and temporal facilitation, occlusion, subliminal fringe effects, and other effects.

MONOSYNAPTIC REFLEXES: THE STRETCH REFLEX

When a skeletal muscle with an intact nerve supply is stretched, it contracts. This response is called the **stretch reflex.** The stimulus that initiates the reflex is stretch of the muscle, and the response is contraction of the muscle being stretched. The sense organ is the muscle spindle. The impulses originating in the spindle are conducted in the CNS by fast sensory fibers that pass directly to the motor neurons which supply the same muscle. The neurotransmitter at the central synapse is glutamate. Stretch reflexes are the best known and studied monosynaptic reflexes in the body.

Clinical Examples

Tapping the patellar tendon elicits the **knee jerk,** a stretch reflex of the quadriceps femoris muscle, because the tap on the tendon stretches the muscle. A similar contraction is observed if the quadriceps is stretched manually. Stretch reflexes can also be elicited from most of the large muscles of the body. Tapping on the tendon of the triceps brachii, for example, causes an extensor response at the elbow as a result of reflex contraction of the triceps; tapping on the Achilles tendon causes an ankle jerk due to reflex contraction of the gastrocnemius; and tapping on the side of the face causes a stretch reflex in the masseter. Other examples of stretch reflexes are listed in neurology textbooks.

Structure of Muscle Spindles

Each muscle spindle consists of up to about 10 muscle fibers enclosed in a connective tissue capsule. These fibers are more embryonal in character and have less distinct striations than the rest of the fibers in the muscle. They are called **intrafusal fibers** to distinguish them from the **extrafusal fibers,** the regular contractile units of the muscle. The intrafusal fibers are in parallel with the rest of the muscle fibers because the ends of the capsule of the spindle are attached to the tendons at either end of the muscle or to the sides of the extrafusal fibers.

There are two types of intrafusal fibers in mammalian muscle spindles. The first type contains many nuclei in a dilated central area and is therefore called a **nuclear bag fiber** (Figure 6–2). Typically there are two nuclear bag fibers per spindle, nuclear bag fiber 1 with a low level of myosin ATPase activity and nuclear bag fiber 2 with a high level of myosin ATPase activity. The second fiber type, the **nuclear chain fiber,** is thinner and shorter and lacks a definite bag. There are four or more of these fibers per spindle. Their ends connect to the sides of the nuclear bag fibers. The ends of the intrafusal fibers are contractile, whereas the central portions probably are not.

There are two kinds of sensory endings in each spindle. The **primary (annulospiral) endings** are the terminations of rapidly conducting group Ia afferent fibers (see Table 2–2). One branch of the Ia fiber innervates nuclear bag fiber 1, whereas another branch innervates nuclear bag fiber 2 and nuclear chain fibers. These sensory fibers wrap around the center of the nuclear bag and nuclear chain fibers. The **secondary (flower-spray) endings** are terminations of group II sensory fibers and are located nearer the ends of the intrafusal fibers but only on nuclear chain fibers.

The spindles have a motor nerve supply of their own. These nerves are 3–6 μm in diameter, constitute about 30% of the fibers in the ventral roots, and belong in Erlanger and Gasser's A γ group. Because of their characteristic size, they are called the γ efferents of Leksell or the **small motor nerve system.** They go exclusively to the spindles. In addition, larger β motor neurons innervate both intrafusal and extrafusal fibers. The endings of the γ efferent fibers are of two histologic types. There are motor end plates (**plate endings**) on the nuclear bag fibers, and there are endings that form extensive networks (**trail endings**) primarily on the nuclear chain fibers.

The spindles produce two kinds of sensory nerve patterns, dynamic and static (see below), and both γ and β motor axons produce two functional types of re-

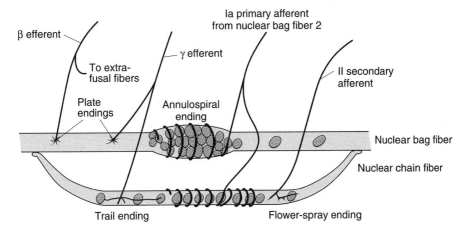

Figure 6–2. Diagrammatic representation of the main components of mammalian muscle spindle. Each spindle has a capsule and usually contains two nuclear bag fibers and four or more nuclear chain fibers.

sponses. Stimulation of one type increases dynamic responses (**dynamic fusimotor axons**), and stimulation of the other increases static discharge at constant length (**static fusiform axons**).

Central Connections of Afferent Fibers

It can be proved experimentally that the Ia fibers from the primary endings end directly on motor neurons supplying the extrafusal fibers of the same muscle. The time between the application of the stimulus and the response is the **reaction time.** In humans, the reaction time for a stretch reflex such as the knee jerk is 19–24 ms. Weak stimulation of the sensory nerve from the muscle, known to stimulate only Ia fibers, causes a contractile response with a similar latency. Since the conduction velocities of the afferent and efferent fiber types are known and the distance from the muscle to the spinal cord can be measured, it is possible to calculate how much of the reaction time was taken up by conduction to and from the spinal cord. When this value is subtracted from the reaction time, the remainder, called the **central delay,** is the time taken for the reflex activity to traverse the spinal cord. In humans, the central delay for the knee jerk is 0.6–0.9 ms, and figures of similar magnitude have been found in experimental animals. Since the minimal synaptic delay is 0.5 ms (see Chapter 4), only one synapse could have been traversed.

Muscle spindles also make connections that cause muscle contraction via polysynaptic pathways, and the afferents involved are probably those from the secondary endings. However, group II fibers also make monosynaptic connections to the motor neurons and make a small contribution to the stretch reflex.

Function of Muscle Spindles

When the muscle spindle is stretched, its sensory endings are distorted and receptor potentials are generated. These in turn set up action potentials in the sensory fibers at a frequency proportionate to the degree of stretching. The spindle is in parallel with the extrafusal fibers, and when the muscle is passively stretched, the spindles are also stretched. This initiates reflex contraction of the extrafusal fibers in the muscle. On the other hand, the spindle afferents characteristically stop firing when the muscle is made to contract by electrical stimulation of the nerve fibers to the extrafusal fibers because the muscle shortens while the spindle does not (Figure 6–3).

Thus, the spindle and its reflex connections constitute a feedback device that operates to maintain muscle length; if the muscle is stretched, spindle discharge increases and reflex shortening is produced, whereas if the muscle is shortened without a change in γ efferent discharge, spindle discharge decreases and the muscle relaxes.

Primary endings on the nuclear bag fibers and nuclear chain fibers are both stimulated when the spindle is stretched, but the pattern of response differs. The nerves from the endings in the nuclear bag region show a **dynamic response;** ie, they discharge most rapidly while the muscle is being stretched and less rapidly during sustained stretch (Figure 6–4). The nerves from the primary endings on the nuclear chain fibers show a **static response;** ie, they discharge at an increased rate throughout the period when a muscle is stretched. Thus, the primary endings respond to both changes in length and changes in the rate of stretch. The response of the primary ending to the phasic as well as the static

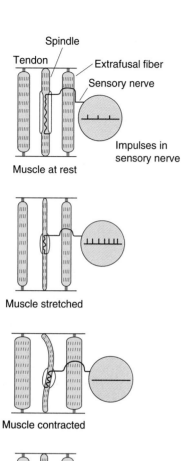

Muscle at rest

Muscle stretched

Muscle contracted

Increased γ efferent discharge

Increased γ efferent
discharge—muscle stretched

Figure 6–3. Effect of various conditions on muscle spindle discharge.

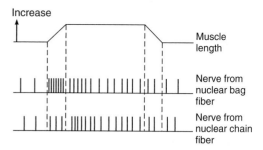

Figure 6–4. Response of spindle afferents to muscle stretch. The bottom two lines represent the number of discharges in afferent nerves from the primary endings on nuclear bag and nuclear chain fibers as the muscle is stretched and then permitted to return to its original length.

events in the muscle is important because the prompt, marked phasic response helps to dampen oscillations caused by conduction delays in the feedback loop regulating muscle length. There is normally a small oscillation in this feedback loop. This **physiologic tremor** has a frequency of approximately 10 Hz. However, the tremor would be worse if it were not for the sensitivity of the spindle to velocity of stretch.

Effects of Gamma Efferent Discharge

Stimulation of the γ efferent system produces a very different picture from that produced by stimulation of the extrafusal fibers. Such stimulation does not lead directly to detectable contraction of the muscles because the intrafusal fibers are not strong enough or plentiful enough to cause shortening. However, stimulation does cause the contractile ends of the intrafusal fibers to shorten and therefore stretches the nuclear bag portion of the spindles, deforming the annulospiral endings and initiating impulses in the Ia fibers. This in turn can lead to reflex contraction of the muscle. Thus, muscle can be made to contract via stimulation of the α motor neurons that innervate the extrafusal fibers or the γ efferent neurons that initiate contraction indirectly via the stretch reflex.

When the rate of γ efferent discharge is increased, the intrafusal fibers are shorter than the extrafusal ones. If the whole muscle is stretched during stimulation of the γ efferents, additional action potentials are generated by the additional stretch of the nuclear bag region, and the rate of discharge in the Ia fibers is further increased (Figure 6–3). Increased γ efferent discharge thus increases spindle sensitivity, and the sensitivity of the spindles to stretch varies with the rate of γ efferent discharge.

There is considerable evidence of increased γ efferent discharge along with the increased discharge of the α motor neurons that initiates α movements. Because of this "α-γ linkage," the spindle shortens with the muscle, and spindle discharge may continue throughout the contraction. In this way, the spindle remains capable of responding to stretch and reflexly adjusting motor neuron discharge throughout the contraction.

The existence of dynamic and static γ and β efferents is mentioned above. Stimulation of the dynamic efferents increases spindle sensitivity to the rate of change of stretch. Stimulation of the static efferents increases spindle sensitivity to steady, maintained stretch. It is thus possible to adjust separately the spindle responses to phasic and static events.

Control of Gamma Efferent Discharge

The motor neurons of the γ efferent system are regulated to a large degree by descending tracts from a number of areas in the brain. Via these pathways, the sensitivity of the muscle spindles and hence the threshold of the stretch reflexes in various parts of the body can be adjusted and shifted to meet the needs of postural control (see Chapter 12).

Other factors also influence γ efferent discharge. Anxiety causes an increased discharge, a fact that probably explains the hyperactive tendon reflexes sometimes seen in anxious patients. In addition, unexpected movement is associated with a greater efferent discharge. Stimulation of the skin, especially by noxious agents, increases γ efferent discharge to ipsilateral flexor muscle spindles while decreasing that to extensors and produces the opposite pattern in the opposite limb. It is well known that trying to pull the hands apart when the flexed fingers are hooked together facilitates the knee jerk reflex (Jendrassik's maneuver), and this may

also be due to increased γ efferent discharge initiated by afferent impulses from the hands.

Reciprocal Innervation

When a stretch reflex occurs, the muscles that antagonize the action of the muscle involved (antagonists) relax. This phenomenon is said to be due to **reciprocal innervation.** Impulses in the Ia fibers from the muscle spindles of the protagonist muscle cause postsynaptic inhibition of the motor neurons to the antagonists. The pathway mediating this effect is bisynaptic. A collateral from each Ia fiber passes in the spinal cord to an inhibitory interneuron (Golgi bottle neuron) that synapses directly on one of the motor neurons supplying the antagonist muscles. This example of postsynaptic inhibition is discussed in Chapter 4, and the pathway is illustrated in Figure 4–9.

Inverse Stretch Reflex

Up to a point, the harder a muscle is stretched, the stronger is the reflex contraction. However, when the tension becomes great enough, contraction suddenly ceases and the muscle relaxes. This relaxation in response to strong stretch is called the **inverse stretch reflex** or **autogenic inhibition.**

The receptor for the inverse stretch reflex is in the **Golgi tendon organ** (Figure 6–5). This organ consists of a net-like collection of knobby nerve endings among the fascicles of a tendon. There are 3–25 muscle fibers per tendon organ. The fibers from the Golgi tendon organs make up the Ib group of myelinated, rapidly conducting sensory nerve fibers (see Table 2–2). Stimulation of these Ib fibers leads to the production of IPSPs on the motor neurons that supply the muscle from which the fibers arise. The Ib fibers end in the spinal cord on inhibitory interneurons that, in turn, terminate

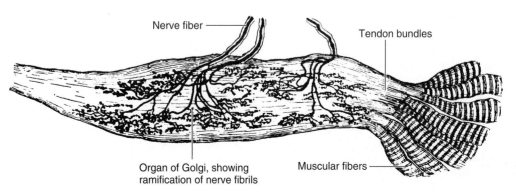

Nerve fiber

Tendon bundles

Organ of Golgi, showing ramification of nerve fibrils

Muscular fibers

Figure 6–5. **Golgi tendon organ.** (Reproduced, with permission, from Goss CM [editor]: *Gray's Anatomy of the Human Body,* 29th ed. Lea & Febiger, 1973.)

directly on the motor neurons (Figure 6–6). They also make excitatory connections with motor neurons supplying antagonists to the muscle.

Since the Golgi tendon organs, unlike the spindles, are in series with the muscle fibers, they are stimulated by both passive stretch and active contraction of the muscle. The threshold of the Golgi tendon organs is low. The degree of stimulation by passive stretch is not great, because the more elastic muscle fibers take up much of the stretch, and this is why it takes a strong stretch to produce relaxation. However, discharge is regularly produced by contraction of the muscle, and the Golgi tendon organ thus functions as a transducer in a feedback circuit that regulates muscle force in a fashion analogous to the spindle feedback circuit that regulates muscle length.

The importance of the primary endings in the spindles and the Golgi tendon organs in regulating the velocity of the muscle contraction, muscle length, and muscle force is illustrated by the fact that section of the afferent nerves to an arm causes the limb to hang loosely in a semiparalyzed state. The organization of the

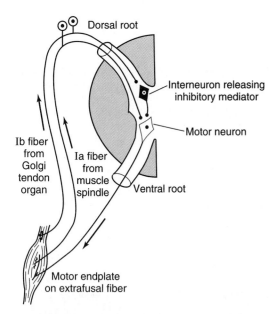

Figure 6–6. Diagram illustrating the pathways responsible for the stretch reflex and the inverse stretch reflex. Stretch stimulates the spindle, and impulses pass up the Ia fiber to excite the motor neuron. It also stimulates the Golgi tendon organ, and impulses passing up the Ib fiber activate the interneuron to release the inhibitory mediator glycine. With strong stretch, the resulting hyperpolarization of the motor neuron is so great that it stops discharging.

Labels in figure: Dorsal root · Interneuron releasing inhibitory mediator · Motor neuron · Ib fiber from Golgi tendon organ · Ia fiber from muscle spindle · Ventral root · Motor endplate on extrafusal fiber

system is shown in Figure 6–7, and the interaction of spindle discharge, tendon organ discharge, and reciprocal innervation in determining the rate of discharge of a motor neuron is shown in Figure 6–8.

Muscle Tone

The resistance of a muscle to stretch is often referred to as its **tone** or **tonus.** If the motor nerve to a muscle is cut, the muscle offers very little resistance and is said to be **flaccid.** A **hypertonic (spastic)** muscle is one in which the resistance to stretch is high because of hyperactive stretch reflexes. Somewhere between the states of flaccidity and spasticity is the ill-defined area of normal tone. The muscles are generally **hypotonic** when the rate of γ efferent discharge is low and hypertonic when it is high.

Lengthening Reaction

When the muscles are hypertonic, the sequence of moderate stretch → muscle contraction, strong stretch → muscle relaxation is clearly seen. Passive flexion of the elbow, for example, meets immediate resistance as a result of the stretch reflex in the triceps muscle. Further stretch activates the inverse stretch reflex. The resistance to flexion suddenly collapses, and the arm flexes. Continued passive flexion stretches the muscle again, and the sequence may be repeated. This sequence of resistance followed by give when a limb is moved passively is known clinically as the **clasp-knife effect** because of its resemblance to the closing of a pocket knife. The physiologic name for it is the **lengthening reaction** because it is the response of a spastic muscle (in the example cited, the triceps) to lengthening.

Clonus

Another finding characteristic of states in which increased γ efferent discharge is present is **clonus.** This neurologic sign is the occurrence of regular, rhythmic contractions of a muscle subjected to sudden, maintained stretch. Ankle clonus is a typical example. This is initiated by brisk, maintained dorsiflexion of the foot, and the response is rhythmic plantar flexion at the ankle. The stretch reflex–inverse stretch reflex sequence described above may contribute to this response. However, it can occur on the basis of synchronized motor neuron discharge without Golgi tendon organ discharge. The spindles of the tested muscle are hyperactive, and the burst of impulses from them discharges all the motor neurons supplying the muscle at once. The consequent muscle contraction stops spindle discharge. However, the stretch has been maintained, and as soon as the muscle relaxes it is again stretched and the spindles stimulated.

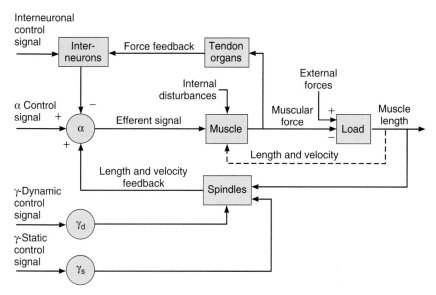

Figure 6–7. Block diagram of peripheral motor control system. The dashed line indicates the nonneural feedback from muscle that limits length and velocity via the inherent mechanical properties of muscle. γ_d, dynamic γ motor neurons; γ_s, static γ motor neurons. (Reproduced, with permission, from Houk J in: *Medical Physiology*, 13th ed. Mountcastle VB [editor]. Mosby, 1974.)

POLYSYNAPTIC REFLEXES: THE WITHDRAWAL REFLEX

Polysynaptic reflex paths branch in a complex fashion (Figure 6–9). The number of synapses in each of their branches is variable. Because of the synaptic delay incurred at each synapse, activity in the branches with fewer synapses reaches the motor neurons first, followed by activity in the longer pathways. This causes prolonged bombardment of the motor neurons from a single stimulus and consequently prolonged responses. Furthermore, as shown in Figure 6–9, some of the branch pathways turn back on themselves, permitting activity to reverberate until it becomes unable to cause a propagated transsynaptic response and dies out. Such **reverberating circuits** are common in the brain and spinal cord.

Withdrawal Reflex

The withdrawal reflex is a typical polysynaptic reflex that occurs in response to a noxious and usually painful stimulation of the skin or subcutaneous tissues and muscle. The response is flexor muscle contraction and inhibition of extensor muscles, so that the part stimulated is flexed and withdrawn from the stimulus. When a strong stimulus is applied to a limb, the response includes not only flexion and withdrawal of that limb but also extension of the opposite limb. This **crossed ex-**

tensor response is properly part of the withdrawal reflex. Strong stimuli in experimental animals generate activity in the interneuron pool which spreads to all four extremities. This is difficult to demonstrate in normal animals but is easily demonstrated in an animal in which the modulating effects of impulses from the brain have been abolished by prior section of the spinal cord (**spinal animal**). For example, when the hind limb of a spinal cat is pinched, the stimulated limb is withdrawn, the opposite hind limb extended, the ipsilateral forelimb extended, and the contralateral forelimb flexed. This spread of excitatory impulses up and down the spinal cord to more and more motor neurons is called **irradiation of the stimulus,** and the increase in the number of active motor units is called **recruitment of motor units.**

Importance of the Withdrawal Reflex

Flexor responses can be produced by innocuous stimulation of the skin or by stretch of the muscle, but strong flexor responses with withdrawal are initiated only by stimuli that are noxious or at least potentially harmful to the animal. These stimuli are therefore called **nociceptive stimuli.** Sherrington pointed out the survival value of the withdrawal response. Flexion of the stimulated limb gets it away from the source of irritation, and extension of the other limb supports the body. The pattern assumed by all four extremities puts

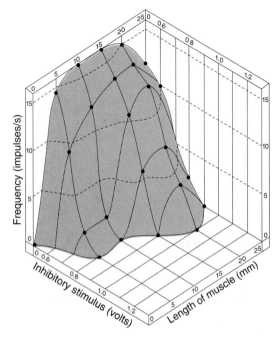

Figure 6–8. Discharge frequency of a motor neuron supplying a leg muscle in a cat. Discharge frequency is plotted against muscle length at various magnitudes of inhibition produced by stimulation of the nerve from the antagonist to the muscle. (Reproduced, with permission, from Henneman E et al: Excitability and inhibitability of motor neurons of different sizes. J Neurophysiol 1965;28:599.)

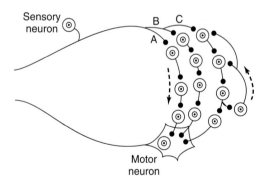

Figure 6–9. Diagram of polysynaptic connections between afferent and efferent neurons in the spinal cord. The dorsal root fiber activates pathway A with three interneurons, pathway B with four interneurons, and pathway C with four interneurons. Note that one of the interneurons in pathway C connects to a neuron that doubles back to other interneurons, forming reverberating circuits.

the animal in position to run away from the offending stimulus. Withdrawal reflexes are **prepotent;** ie, they preempt the spinal pathways from any other reflex activity taking place at the moment.

Many of the characteristics of polysynaptic reflexes can be demonstrated by studying the withdrawal reflex in the laboratory. A weak noxious stimulus to one foot evokes a minimal flexion response; stronger stimuli produce greater and greater flexion as the stimulus irradiates to more and more of the motor neuron pool supplying the muscles of the limb. Stronger stimuli also cause a more prolonged response. A weak stimulus causes one quick flexion movement; a strong stimulus causes prolonged flexion and sometimes a series of flexion movements. This prolonged response is due to prolonged, repeated firing of the motor neurons. The repeated firing is called **after-discharge** and is due to continued bombardment of motor neurons by impulses arriving by complicated and circuitous polysynaptic paths.

As the strength of a noxious stimulus is increased, the reaction time is shortened. Spatial and temporal facilitation occurs at synapses in the polysynaptic pathway. Stronger stimuli produce more action potentials per second in the active branches and cause more branches to become active; summation of the EPSPs to the firing level therefore occurs more rapidly.

Local Sign

The exact flexor pattern of the withdrawal reflex in a limb varies with the part of the limb that is stimulated. If the medial surface of the limb is stimulated, for example, the response will include some abduction, whereas stimulation of the lateral surface will produce some adduction with flexion. The reflex response in each case generally serves to effectively remove the limb from the irritating stimulus. This dependence of the exact response on the location of the stimulus is called **local sign.** The degree to which local sign determines the particular pattern is illustrated in Figure 6–10.

Fractionation & Occlusion

Another characteristic of the withdrawal response is the fact that supramaximal stimulation of any of the sensory nerves from a limb never produces as strong a contraction of the flexor muscles as that elicited by direct electrical stimulation of the muscles themselves. This indicates that the afferent inputs **fractionate** the motor neuron pool; ie, each input goes to only part of the motor neuron pool for the flexors of that particular extremity. On the other hand, if all the sensory inputs are dissected out and stimulated one after the other, the

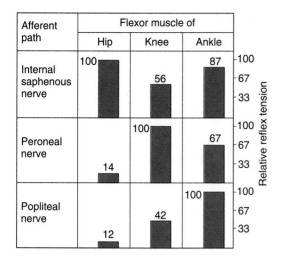

Afferent path	Flexor muscle of		
	Hip	Knee	Ankle
Internal saphenous nerve	100	56	87
Peroneal nerve	14	100	67
Popliteal nerve	12	42	100

Figure 6–10. The importance of local sign in determining the character of the withdrawal response in a leg. When afferent fibers in each of the three nerves on the left were stimulated, hip, knee, and ankle flexors contracted but the relative tension that developed in each case (shaded bars) varied.

sum of the tension developed by stimulation of each is greater than that produced by direct electrical stimulation of the muscle or stimulation of all inputs at once. This indicates that the various afferent inputs share some of the motor neurons and that occlusion (see Chapter 4) occurs when all inputs are stimulated at once.

Other Polysynaptic Reflexes

There are many polysynaptic reflexes in addition to the withdrawal reflex, all with similar properties. Numerous polysynaptic reflexes that relate to specific regulatory functions are described in other sections of this book, and comprehensive lists can be found in neurology textbooks.

GENERAL PROPERTIES OF REFLEXES

It is apparent from the preceding description of the properties of monosynaptic and polysynaptic reflexes that reflex activity is stereotyped and specific in terms of both the stimulus and the response; a particular stimulus elicits a particular response.

Adequate Stimulus

The stimulus that triggers a reflex is generally very precise. This stimulus is called the **adequate stimulus** for

the particular reflex. A dramatic example is the scratch reflex in the dog. This spinal reflex is adequately stimulated by multiple linear touch stimuli such as those produced by an insect crawling across the skin. The response is vigorous scratching of the area stimulated. (Incidentally, the precision with which the scratching foot goes to the site of the irritant is a good example of local sign.) If the multiple touch stimuli are widely separated or not in a line, the adequate stimulus is not produced and no scratching occurs. Fleas crawl, but they also jump from place to place. This jumping separates the touch stimuli so that an adequate stimulus for the scratch reflex is not produced. It is doubtful if the flea population would long survive without the ability to jump.

Final Common Path

The motor neurons that supply the extrafusal fibers in skeletal muscles are the efferent side of many reflex arcs. All neural influences affecting muscular contraction ultimately funnel through them to the muscles, and they are therefore called the **final common paths.** Numerous inputs converge on them. Indeed, the surface of the average motor neuron and its dendrites accommodates about 10,000 synaptic knobs. There are at least five inputs from the same spinal segment to a typical spinal motor neuron. In addition to these, there are excitatory and inhibitory inputs, generally relayed via interneurons, from other levels of the spinal cord and multiple long descending tracts from the brain. All of these pathways converge on and determine the activity in the final common paths.

Central Excitatory & Inhibitory States

The spread up and down the spinal cord of subliminal fringe effects from excitatory stimulation has already been mentioned. Direct and presynaptic inhibitory effects can also be widespread. These effects are generally transient. However, the spinal cord also shows prolonged changes in excitability, possibly because of activity in reverberating circuits or prolonged effects of synaptic mediators. The terms **central excitatory state** and **central inhibitory state** have been used to describe prolonged states in which excitatory influences overbalance inhibitory influences and vice versa. When the central excitatory state is marked, excitatory impulses irradiate not only to many somatic areas of the spinal cord but also to autonomic areas. In chronically paraplegic humans, for example, a mild noxious stimulus may cause, in addition to prolonged withdrawal-extension patterns in all four limbs, urination, defecation, sweating, and blood pressure fluctuations (**mass reflex**).

Habituation & Sensitization of Reflex Responses

The fact that reflex responses are stereotyped does not exclude the possibility of their being modified by experience. Examples include habituation and sensitization, which are discussed in terms of synaptic function in Chapter 4 and in terms of their relation to learning and memory in Chapter 16.

Cutaneous, Deep, & Visceral Sensation

INTRODUCTION

The sense organs for mechanical stimulation (touch and pressure), warmth, cold, and pain have been discussed in Chapter 5, and the types of neurons that carry impulses generated in them to the CNS are listed in Table 2–1. These primary afferent neurons have their cell bodies in the dorsal root ganglia or equivalent ganglia in cranial nerves. They enter the spinal cord or brain stem and make polysynaptic reflex connections to motor neurons at many levels as well as connections that relay impulses to the cerebral cortex. Each of the sensations they mediate is considered in this chapter.

PATHWAYS

The dorsal horns are divided on the basis of histologic characteristics into laminas I–VII, with I being the most superficial and VI the deepest. Lamina II and part of lamina III make up the **substantia gelatinosa,** a lightly stained area near the top of each dorsal horn. There are three types of primary afferent fibers that mediate cutaneous sensation: (1) large myelinated Aα and Aβ fibers that transmit impulses generated by mechanical stimuli; (2) small myelinated Aδ fibers, some of which transmit impulses from cold receptors and nociceptors that mediate fast pain (see below) and some of which transmit impulses from mechanoreceptors; and (3) small unmyelinated C fibers that are concerned primarily with pain and temperature. However, there are also a few C fibers that transmit impulses from mechanoreceptors. The orderly distribution of these fibers in the dorsal columns and the various layers of the dorsal horn is shown in Figure 7–1.

The principal direct pathways to the cerebral cortex for the cutaneous senses are shown in Figure 7–2. Fibers mediating fine touch and proprioception ascend in the dorsal columns to the medulla, where they synapse in the gracile and cuneate nuclei. The second-order neurons from the gracile and cuneate nuclei cross the midline and ascend in the medial lemniscus to end in the ventral posterior nucleus and related specific sensory relay nuclei of the thalamus (see Chapter 11). This

ascending system is frequently called the **dorsal column** or **lemniscal system.**

Other touch fibers, along with those mediating temperature and pain, synapse on neurons in the dorsal horn. The axons from these neurons cross the midline and ascend in the anterolateral quadrant of the spinal cord, where they form the **anterolateral system** of ascending fibers. Others ascend more dorsally. In general, touch is associated with the ventral spinothalamic tract whereas pain and temperature are associated with the lateral spinothalamic tract, but there is no rigid localization of function. Some of the fibers of the anterolateral system end in the specific relay nuclei of the thalamus; others project to the midline and intralaminar nonspecific projection nuclei. There is a major input from the anterolateral systems into the mesencephalic reticular formation. Thus, sensory input activates the reticular activating system, which in turn maintains the cortex in the alert state (see Chapter 11).

Collaterals from the fibers that enter the dorsal columns pass to the dorsal horn. These collaterals may modify the input into other cutaneous sensory systems,

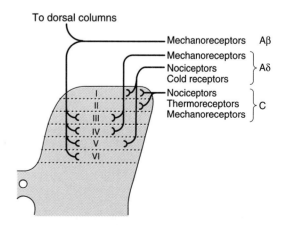

Figure 7–1. Schematic representation of the terminations of the three types of primary afferent neurons in the various layers of the dorsal horn of the spinal cord.

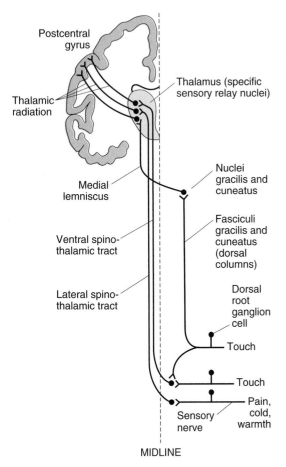

Figure 7–2. Touch, pain, and temperature pathways from the trunk and limbs. Recent evidence indicates that thermal sensations also project to the ipsilateral insular cortex, and this area may even be the primary thermal cortex. The anterolateral system (ventral and lateral spinothalamic and related ascending tracts) also projects to the mesencephalic reticular formation and the nonspecific thalamic nuclei.

including the pain system. The dorsal horn represents a "gate" in which impulses in the sensory nerve fibers are translated into impulses in ascending tracts, and it appears that passage through this gate is dependent on the nature and pattern of impulses reaching the substantia gelatinosa and its environs. This gate is also affected by impulses in descending tracts from the brain. The relation of the gate to pain is discussed below.

Axons of the spinothalamic tracts from sacral and lumbar segments of the body are pushed laterally by axons crossing the midline at successively higher levels. On the other hand, sacral and lumbar dorsal column

fibers are pushed medially by fibers from higher segments (Figure 7–3). Consequently, both of these ascending systems are laminated, with cervical, thoracic, lumbar, and sacral segments represented from medial to lateral in the anterolateral pathways and sacral to cervical segments from medial to lateral in the dorsal columns. Because of this lamination, tumors arising outside the spinal cord first compress the spinothalamic fibers from sacral and lumbar areas, causing the early symptom of loss of pain and temperature sensation in the sacral region. Intraspinal tumors cause loss of sensation first in higher segments.

The fibers within the lemniscal and anterolateral systems are joined in the brain stem by fibers mediating sensation from the head. Pain and temperature impulses are relayed via the spinal nucleus of the trigeminal nerve, and touch and proprioception mostly via the main sensory and mesencephalic nuclei of this nerve.

Cortical Representation

Mapping of cortical areas involved in sensation has been carried out in experimental animals and during neurosurgical procedures in humans, but it has also been carried out more recently in intact humans by techniques such as positron emission tomography (PET) and functional magnetic resonance imaging (fMRI). These techniques, which are described in Chapter 32 and referenced in the Appendix, have led to major advances not only in sensory physiology but also in all aspects of cortical function in normal humans.

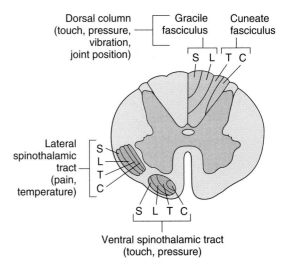

Figure 7–3. Cross section of spinal cord, showing location of ascending sensory pathways. Note that each is laminated. S, sacral; L, lumbar; T, thoracic; C, cervical.

From the specific sensory nuclei of the thalamus, neurons carrying sensory information project in a highly specific way to the two somatic sensory areas of the cortex: somatic sensory area I (SI) in the postcentral gyrus and somatic sensory area II (SII) in the wall of the sylvian fissure. In addition, SI projects to SII. SI corresponds to Brodmann's areas 1, 2, and 3. Brodmann was a histologist who painstakingly divided the cerebral cortex into numbered areas based on their histologic characteristics.

The arrangement of the thalamic fibers in SI is such that the parts of the body are represented in order along the postcentral gyrus, with the legs on top and the head at the foot of the gyrus (Figure 7–4). Not only is there detailed localization of the fibers from the various parts of the body in the postcentral gyrus, but also the size of the cortical receiving area for impulses from a particular part of the body is proportionate to the number of receptors in the part. The relative sizes of the cortical receiving areas are shown dramatically in Figure 7–5, in which the proportions of the homunculus have been distorted to correspond to the size of the cortical receiving areas for each. Note that the cortical areas for sensation from the trunk and back are small, whereas very large areas are concerned with impulses from the hand and the parts of the mouth concerned with speech.

Studies of the sensory receiving area emphasize the very discrete nature of the point-for-point localization of peripheral areas in the cortex and provide further evidence for the general validity of the doctrine of specific

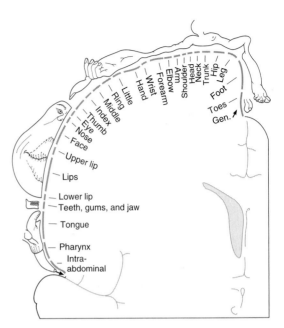

Figure 7–5. Sensory homunculus, drawn overlying a coronal section through the postcentral gyrus. Gen., genitalia. (Reproduced, with permission, from Penfield W, Rasmussen G: *The Cerebral Cortex of Man.* Macmillan, 1950.)

nerve energies (see Chapter 5). Stimulation of the various parts of the postcentral gyrus gives rise to sensations projected to appropriate parts of the body. The sensations produced are usually numbness, tingling, or a sense of movement, but with fine enough electrodes it has been possible to produce relatively pure sensations of touch, warmth, and cold. The cells in the postcentral gyrus are organized in vertical columns, like cells in the visual cortex (see Chapter 8). The cells in a given column are all activated by afferents from a given part of the body, and all respond to the same sensory modality.

SII is located in the superior wall of the sylvian fissure, the fissure that separates the temporal from the frontal and parietal lobes. The head is represented at the inferior end of the postcentral gyrus, and the feet at the bottom of the sylvian fissure. The representation of the body parts is not as complete or detailed as it is in the postcentral gyrus.

Cortical Plasticity

It is now clear that the extensive neuronal connections described in the previous paragraphs are not innate and immutable but can be changed relatively rapidly by experience to reflect the use of the represented area. For example, if a digit is amputated in a monkey, the corti-

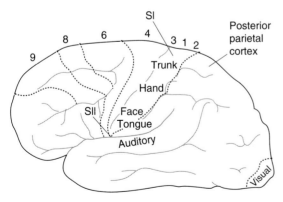

Figure 7–4. Brain areas concerned with somatic sensation, and some of the cortical receiving areas for other sensory modalities in the human brain. The numbers are those of Brodmann's cortical areas. The primary auditory area is actually located in the sylvian fissure on the top of the superior temporal gyrus and is not normally visible.

cal representation of the neighboring digits spreads into the cortical area that was formerly occupied by the representation of the amputated digit. Conversely, if the cortical area representing a digit is removed, the somatosensory map of the digit moves to the surrounding cortex. Extensive, long-term deafferentation of limbs leads to even more dramatic shifts in somatosensory representation in the cortex, with, for example, the limb cortical area responding to touching the face. The explanation of these shifts appears to be that cortical connections of sensory units to the cortex have extensive convergence and divergence, with connections that can become weak with disuse and strong with use. In rats, the basal forebrain appears to be involved; when discharge of neurons from this area to the cortex is increased at the same time an auditory stimulus is applied, the auditory sensory area ends up bigger than when the stimuli are paired at a slow basal forebrain discharge rate.

Plasticity of this type occurs not only with input from cutaneous receptors but also with input in other sensory systems. For example, in cats with small lesions of the retina, the cortical area for the blinded spot begins to respond to light striking other areas of the retina. Development of the adult pattern of retinal projections to the visual cortex is another example of this plasticity. At a more extreme level, experimentally routing visual input to the auditory cortex during development creates visual receptive fields in the auditory system.

Plastic changes of the type described above in experimental animals also occur in humans. For example, in some individuals who have had an arm amputated, touching the face causes sensations projected to the missing arm. PET scanning also documents plastic changes, sometimes from one sensory modality to another. Thus, for example, tactile and auditory stimuli increase metabolic activity in the visual cortex in blind individuals. Conversely, deaf individuals respond faster and more accurately than normal individuals to moving stimuli in the visual periphery. Plasticity also occurs in the motor cortex (see Chapter 12). These findings illustrate the malleability of the brain and its ability to adapt.

Effects of Cortical Lesions

Ablation of SI in animals causes deficits in position sense and in the ability to discriminate size and shape. Ablation of SII causes deficits in learning based on tactile discrimination. Ablation of SI causes deficits in sensory processing in SII, whereas ablation of SII has no gross effect on processing in SI. Thus, it seems clear that SI and SII process sensory information in series rather than in parallel and that SII is concerned with

further elaboration of sensory data. SI also projects to the posterior parietal cortex (Figure 7–4), and lesions of this association area produce complex abnormalities of spatial orientation on the contralateral side of the body (see Chapter 16).

It is worth emphasizing that in experimental animals and humans, cortical lesions do not abolish somatic sensation. Proprioception and fine touch are most affected by cortical lesions. Temperature sensibility is less affected, and pain sensibility is only slightly affected. Thus, perception is possible in the absence of the cortex.

Principles of Sensory Physiology

The important general principles that relate to the physiology of sensory systems have been discussed in detail in Chapter 5. Another principle that applies to cutaneous sensation is that of punctate representation. If the skin is carefully mapped, millimeter by millimeter, with a fine hair, a sensation of touch is evoked from spots overlying touch receptors. None is evoked from the intervening areas. Similarly, temperature sensations are produced by stimulation of the skin only over the spots where the sense organs for these modalities are located.

TOUCH

As noted in Chapter 5, pressure is maintained touch. Touch is present in areas that have no visible specialized receptors. However, pacinian corpuscles and possibly other putative receptors may subsume special functions related to touch. Touch receptors are most numerous in the skin of the fingers and lips and relatively scarce in the skin of the trunk. There are many receptors around hair follicles in addition to those in the subcutaneous tissues of hairless areas. When a hair is moved, it acts as a lever with its fulcrum at the edge of the follicle, so that slight movements of the hairs are magnified into relatively potent stimuli to the nerve endings around the follicles. The stiff vibrissae on the snouts of some animals are highly developed examples of hairs that act as levers to magnify tactile stimuli.

The Na^+ channel BNC1 is closely associated with touch receptors. This channel is one of the **degenerins,** so called because when they are hyperexpressed they cause the neurons they are in to degenerate. However, it is not known if BNC1 is part of the receptor complex or the neural fiber at the point of initiation of the spike potential. The receptor may be opened mechanically by pressure on the skin.

The Aβ sensory fibers that transmit impulses from touch receptors to the central nervous system are 5–12 μm in diameter and have conduction velocities of

30–70 m/s. Some touch impulses are also conducted via C fibers.

Touch information is transmitted in both the lemniscal and anterolateral pathways, so that only very extensive lesions completely interrupt touch sensation. However, there are differences in the type of touch information transmitted in the two systems. When the dorsal columns are destroyed, vibratory sensation and proprioception are reduced, the touch threshold is elevated, and the number of touch-sensitive areas in the skin is decreased. In addition, localization of touch sensation is impaired. An increase in touch threshold and a decrease in the number of touch spots in the skin are also observed after interrupting the spinothalamic tracts, but the touch deficit is slight and touch localization remains normal. The information carried in the lemniscal system is concerned with the detailed localization, spatial form, and temporal pattern of tactile stimuli. The information carried in the spinothalamic tracts, on the other hand, is concerned with poorly localized, gross tactile sensations.

PROPRIOCEPTION

Proprioceptive information is transmitted up the spinal cord in the dorsal columns. A good deal of the proprioceptive input goes to the cerebellum, but some passes via the medial lemnisci and thalamic radiations to the cortex. Diseases of the dorsal columns produce ataxia because of the interruption of proprioceptive input to the cerebellum.

There is some evidence that proprioceptive information passes to consciousness in the anterolateral columns of the spinal cord. Conscious awareness of the positions of the various parts of the body in space depends in part upon impulses from sense organs in and around the joints. The organs involved are slowly adapting "spray" endings, structures that resemble Golgi tendon organs, and probably pacinian corpuscles in the synovia and ligaments. Impulses from these organs, touch receptors in the skin and other tissues, and muscle spindles are synthesized in the cortex into a conscious picture of the position of the body in space. Microelectrode studies indicate that many of the neurons in the sensory cortex respond to particular movements, not just to touch or static position. In this regard, the sensory cortex is organized like the visual cortex (see Chapter 8).

TEMPERATURE

Mapping experiments show that there are discrete cold-sensitive and heat-sensitive spots in the skin. There are four to ten times as many cold-sensitive as heat-sensitive spots. Cold receptors respond from 10 °C to 38 °C

and heat receptors from 30 °C to over 45 °C. The afferents for cold are Aδ and C fibers, whereas the afferents for heat are C fibers. Temperature has generally been regarded as closely related to touch, but new evidence indicates that in addition to ending in the postcentral gyrus, thermal fibers from the thalamus end in the ipsilateral insular cortex. It has even been suggested that this is the true primary thermal receiving area.

Three receptors involved in temperature perception have been cloned. The receptor for moderate cold is the **cold-** and **menthol-sensitive receptor 1 (CMR 1).** Two receptors respond to high, potentially noxious heat: VR1, which also responds to the pain-producing chemical **capsaicin** and is clearly a nociceptor; and VRL-1, a closely related receptor that does not respond to capsaicin but is probably a nociceptor as well. All three are members of the TRP family of cation channels (see Chapter 5). The receptor that responds to moderate heat (warmth receptor) could be the ATP P2X receptor because injection of ATP causes a feeling of warmth, and mice in which the P2X receptor gene has been knocked out do not show the activity in the spinal cord normally produced by mild skin warming.

Because the sense organs are located subepithelially, it is the temperature of the subcutaneous tissues that determines the responses. Cool metal objects feel colder than wooden objects of the same temperature because the metal conducts heat away from the skin more rapidly, cooling the subcutaneous tissues to a greater degree.

PAIN

Pain differs from other sensations in that it sounds a warning that something is wrong, and it preempts other signals. It turns out to be immensely complex because when pain is prolonged and tissue is damaged, central nociceptor pathways are facilitated and reorganized. There is much still to be learned, but in a general way it is convenient to talk about **physiologic** or **acute pain** and two pathologic states, **inflammatory pain** and **neuropathic pain.**

Receptors & Pathways

The sense organs for pain are the naked nerve endings found in almost every tissue of the body. Pain impulses are transmitted to the CNS by two fiber systems. One nociceptor system is made up of small myelinated Aδ fibers 2–5 μm in diameter, which conduct at rates of 12–30 m/s. The other consists of unmyelinated C fibers 0.4–1.2 μm in diameter. These latter fibers are found in the lateral division of the dorsal roots and are often called dorsal root C fibers. They conduct at the low rate of 0.5–2 m/s. Both fiber groups end in the dorsal horn; Aδ fibers terminate primarily on neurons in laminas I and V, whereas the dorsal root C fibers terminate on neurons in laminas I and II. The synaptic transmitter secreted by primary afferent fibers subserving fast mild pain (see below) is glutamate, and the transmitter subserving slow severe pain is substance P.

The synaptic junctions between the peripheral nociceptor fibers and the dorsal horn cells in the spinal cord are the sites of considerable plasticity. For this reason, the dorsal horn has been called a **gate,** where pain impulses can be "gated," ie, modified.

Some of the axons of the dorsal horn neurons end in the spinal cord and brain stem. Others enter the anterolateral system, including the lateral spinothalamic tract. A few ascend in the posterolateral portion of the cord. Some of the ascending fibers project to the ventral posterior nuclei, which are the specific sensory relay nuclei of the thalamus, and from there to the cerebral cortex. PET and fMRI studies in normal humans indicate that pain activates cortical areas SI, SII, and the cingulate gyrus on the side opposite the stimulus. In addition, the mediofrontal cortex, the insular cortex, and the cerebellum are activated.

Pain was called by Sherrington "the physical adjunct of an imperative protective reflex." Painful stimuli generally initiate potent withdrawal and avoidance responses. Furthermore, pain is unique among the sensations in that it has a "built-in" unpleasant affect.

Fast & Slow Pain

The presence of two pain pathways, one slow and one fast, explains the physiologic observation that there are two kinds of pain. A painful stimulus causes a "bright," sharp, localized sensation followed by a dull, intense, diffuse, and unpleasant feeling. These two sensations are variously called fast and slow pain or first and second pain. The farther from the brain the stimulus is applied, the greater the temporal separation of the two components. This and other evidence make it clear that fast pain is due to activity in the Aδ pain fibers whereas slow pain is due to activity in the C pain fibers.

Receptors & Stimuli

An important recent event was the isolation of **vanilloid receptor-1 (VR1).** Vanillins are a group of compounds, including capsaicin, that cause pain. This necessitated revision of the concept that a single pathway carries pain and only pain to the cerebral cortex. The VR1 receptors respond not only to the pain-causing agents such as capsaicin but also to protons and to potentially harmful temperatures above 43 °C. Another receptor, **VRL-1,** which responds to temperatures above 50 °C but not to capsaicin, has been isolated

from C fibers. There may be many types of receptors on single peripheral C fiber endings, so single fibers can respond to many different noxious stimuli. However, the different properties of the VR1 and the VRL-1 receptors make it likely that there are many different nociceptor C fibers systems as well.

Subcortical Perception & Affect

There is considerable evidence that sensory stimuli are perceived in the absence of the cerebral cortex, and this is especially true of pain. The cortical receiving areas are apparently concerned with the discriminative, exact, and meaningful interpretation of pain and some of its emotional components, but perception alone does not require the cortex.

Deep Pain

The main difference between superficial and deep sensibility is the different nature of the pain evoked by noxious stimuli. This is probably due to a relative deficiency of Aδ nerve fibers in deep structures, so there is little rapid, bright pain. In addition, deep pain and visceral pain are poorly localized, nauseating, and frequently associated with sweating and changes in blood pressure. Pain can be elicited experimentally from the periosteum and ligaments by injecting hypertonic saline into them. The pain produced in this fashion initiates reflex contraction of nearby skeletal muscles. This reflex contraction is similar to the muscle spasm associated with injuries to bones, tendons, and joints. The steadily contracting muscles become ischemic, and ischemia stimulates the pain receptors in the muscles (see below). The pain in turn initiates more spasm, setting up a vicious circle.

Muscle Pain

If a muscle contracts rhythmically in the presence of an adequate blood supply, pain does not usually result. However, if the blood supply to a muscle is occluded, contraction soon causes pain. The pain persists after the contraction until blood flow is reestablished.

These observations are difficult to interpret except in terms of the release during contraction of a chemical agent (Lewis's **"P factor"**) that causes pain when its local concentration is high enough. When the blood supply is restored, the material is washed out or metabolized. The identity of the P factor is not settled, but it could be K^+.

Clinically, the substernal pain that develops when the myocardium becomes ischemic during exertion (angina pectoris) is a classic example of the accumulation of P factor in a muscle. Angina is relieved by rest because this decreases the myocardial O_2 requirement and permits the blood supply to remove the factor. Intermittent claudication, the pain produced in the leg muscles of persons with occlusive vascular disease, is another example. It characteristically comes on while the patient is walking and disappears upon resting.

Visceral Pain

In addition to being poorly localized, unpleasant, and associated with nausea and autonomic symptoms, visceral pain often radiates or is referred to other areas.

The autonomic nervous system, like the somatic, has afferent components, central integrating stations, and effector pathways. The receptors for pain and the other sensory modalities present in the viscera are simi-

lar to those in skin, but there are marked differences in their distribution. There are no proprioceptors in the viscera, and few temperature and touch sense organs. Pain receptors are present, although they are more sparsely distributed than in somatic structures.

Afferent fibers from visceral structures reach the CNS via sympathetic and parasympathetic pathways. Their cell bodies are located in the dorsal roots and the homologous cranial nerve ganglia. Specifically, there are visceral afferents in the facial, glossopharyngeal, and vagus nerves; in the thoracic and upper lumbar dorsal roots; and in the sacral roots (Figure 7–6). There may also be visceral afferent fibers from the eye in the trigeminal nerve. At least some substance P-containing afferents make connections via collaterals to postgan-

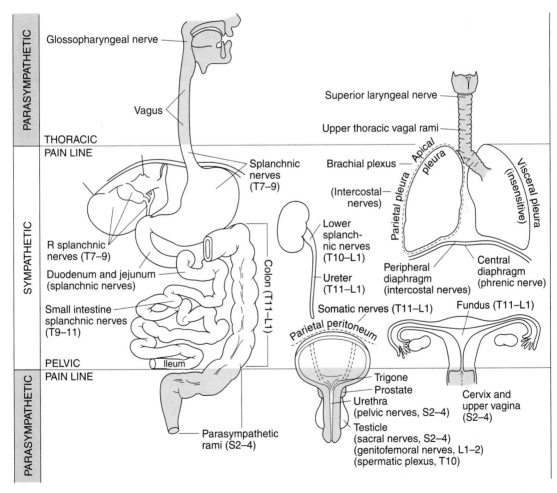

Figure 7–6. Pain innervation of the viscera. Pain afferents from structures above the thoracic pain line and below the pelvic pain line traverse parasympathetic pathways. (After White JC. Reproduced, with permission, from Ruch TC in: *Physiology and Biophysics,* 19th ed. Ruch TC, Patton HD [editors]. Saunders, 1965.)

glionic sympathetic neurons in collateral sympathetic ganglia such as the inferior mesenteric ganglion. These connections may play a part in reflex control of the viscera independent of the CNS.

In the CNS, visceral sensation travels along the same pathways as somatic sensation in the spinothalamic tracts and thalamic radiations, and the cortical receiving areas for visceral sensation are intermixed with the somatic receiving areas.

Stimulation of Pain Fibers

As almost everyone knows from personal experience, visceral pain can be very severe. The receptors in the walls of the hollow viscera are especially sensitive to distention of these organs. Such distention can be produced experimentally in the gastrointestinal tract by inflation of a swallowed balloon attached to a tube. This produces pain that waxes and wanes (intestinal colic) as the intestine contracts and relaxes on the balloon. Similar colic is produced in intestinal obstruction by the contractions of the dilated intestine above the obstruction. When a viscus is inflamed or hyperemic, relatively minor stimuli cause severe pain. This is probably a form of primary hyperalgesia (see below). Traction on the mesentery is also claimed to be painful, but the significance of this observation in the production of visceral pain is not clear.

Muscle Spasm & Rigidity

Visceral pain, like deep somatic pain, initiates reflex contraction of nearby skeletal muscle. This reflex spasm is usually in the abdominal wall and makes the abdominal wall rigid. It is most marked when visceral inflammatory processes involve the peritoneum. However, it can occur without such involvement. The spasm protects the underlying inflamed structures from inadvertent trauma. Indeed, this reflex spasm is sometimes called "guarding."

Referred Pain

Irritation of a viscus frequently produces pain which is felt not in the viscus but in some somatic structure that may be a considerable distance away. Such pain is said to be **referred** to the somatic structure. Deep somatic pain may also be referred, but superficial pain is not. When visceral pain is both local and referred, it sometimes seems to spread (**radiate**) from the local to the distant site.

Obviously, a knowledge of referred pain and the common sites of pain referral from each of the viscera is of great importance to the physician. Perhaps the best known example is referral of cardiac pain to the inner aspect of the left arm. Other dramatic examples include

pain in the tip of the shoulder caused by irritation of the central portion of the diaphragm and pain in the testicle due to distention of the ureter. Additional instances abound in the practice of medicine, surgery, and dentistry. However, sites of reference are not stereotyped, and unusual reference sites occur with considerable frequency. Heart pain, for instance, may be purely abdominal, may be referred to the right arm, and may even be referred to the neck. Referred pain can be produced experimentally by stimulation of the cut end of a splanchnic nerve.

Dermatomal Rule

When pain is referred, it is usually to a structure that developed from the same embryonic segment or dermatome as the structure in which the pain originates. This principle is called the **dermatomal rule.** For example, during embryonic development, the diaphragm migrates from the neck region to its adult location between the chest and the abdomen and takes its nerve supply, the phrenic nerve, with it. One-third of the fibers in the phrenic nerve are afferent, and they enter the spinal cord at the level of the second to fourth cervical segments, the same location at which afferents from the tip of the shoulder enter. Similarly, the heart and the arm have the same segmental origin, and the testicle has migrated with its nerve supply from the primitive urogenital ridge from which the kidney and ureter has developed.

Cause

The main cause of referred pain appears to be plasticity in the CNS coupled with convergence of peripheral and visceral pain fibers on the same second-order neuron that projects to the brain. Peripheral and visceral neurons do not converge in laminas I–VI of the dorsal horn but do converge in lamina VII (Figure 7–7). In addition, lamina VII neurons receive afferents from both sides of the body—a requirement if convergence is to explain referral to the side opposite that of the source of pain. The peripheral pain fibers normally do not fire the second-order neurons, but when the visceral stimulus is prolonged there is facilitation of the peripheral fibers. They now stimulate the second-order neurons, and of course the brain cannot determine whether the stimulus came from the viscera or from the area of referral.

Central Inhibition & Counterirritants

It is well known that soldiers wounded in the heat of battle may feel no pain until the battle is over (**stress analgesia).** Many people have learned from practical experience that touching or shaking an injured area de-

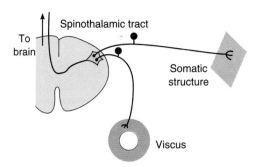

Figure 7–7. Diagram of the way in which convergence in lamina VII of the dorsal horn may cause referred pain.

creases the pain of the injury. Stimulation with an electric vibrator at the site of pain also gives some relief. The relief is due primarily to inhibition of pain pathways in the dorsal horn gate by stimulation of large-diameter touch-pressure afferents. The same mechanism is probably responsible for the efficacy of counterirritants. Stimulation of the skin over an area of visceral inflammation produces some relief of the pain due to the visceral disease. The old-fashioned mustard plaster works on this principle.

Inflammatory Pain

After anything more than a minor injury, **inflammatory pain** sets in and persists until the injury heals. Characteristically, stimuli in the injured area that would normally cause only minor pain produce an exaggerated response **(hyperalgesia)** and normally innocuous stimuli such as touch cause pain **(allodynia).** Inflammation of any type causes the release of many different cytokines and growth factors (the "inflammatory soup") in the inflamed area. Many of these facilitate perception and transmission in cutaneous areas as well as in the dorsal horn. This is what causes the hyperalgesia and allodynia.

Neuropathic Pain

Neuropathic pain may occur when nerve fibers are injured. Commonly, it is excruciating and a difficult condition to treat. It occurs in various forms in humans. One is pain in addition to other sensations in a limb that has been amputated (phantom limb; see Chapter 5). In **causalgia,** there is spontaneous burning pain long after seemingly trivial injuries. The pain is often accompanied by hyperalgesia and allodynia. **Reflex sympathetic dystrophy** is often present as well. In this condition, the skin in the affected area is thin and shiny, and there is increased hair growth. Research in

animals indicates that nerve injury leads to sprouting and eventual overgrowth of noradrenergic sympathetic nerve fibers into the dorsal root ganglia of the sensory nerves from the injured area. Sympathetic discharge then brings on pain. Thus, it appears that the periphery has been short-circuited and that the relevant altered fibers are being stimulated by norepinephrine at the dorsal root ganglion level. Alpha-adrenergic blockade produces relief of causalgia-type pain in humans, though for unknown reasons α_1-adrenergic blockers are more effective than α_2-adrenergic blocking agents.

Surgical procedures undertaken to relieve severe pain include cutting the nerve from the site of injury or **anterolateral cordotomy,** in which the spinothalamic tracts are carefully cut. However, the effects of these procedures are transient at best if the periphery has been short-circuited by sympathetic or other reorganization of the central pathways.

Pain can often be handled by administration of analgesic drugs in adequate doses, though this is not always the case. The most effective of these agents is morphine.

Action of Morphine & Enkephalins

Morphine is particularly effective when given intrathecally. The receptors that bind morphine and the "body's own morphines," the opioid peptides, are discussed in Chapter 4.

There are at least three nonmutually exclusive sites at which opioids could act to produce analgesia: peripherally, at the site of an injury; in the dorsal horn "gate," where nociceptive fibers synapse on dorsal root ganglion cells; and at more rostral sites in the brain stem. Opioid receptors are produced in dorsal root ganglion cells and migrate both peripherally and centrally along their nerve fibers. In the periphery, inflammation causes the production of opioid peptides by immune cells, and these presumably act on the receptors in the afferent nerve fibers to reduce the pain that would otherwise be felt. The opioid receptors in the dorsal horn region could act presynaptically to decrease release of substance P, although presynaptic nerve endings have not been identified. Finally, injections of morphine into the periaqueductal gray of the midbrain relieve pain by activating descending pathways that produce inhibition of primary afferent transmission in the dorsal horn. There is evidence that this activation occurs via projections from the periaqueductal gray to the nearby raphe magnus nucleus and that descending serotonergic fibers from this nucleus mediate the inhibition. However, the mechanism by which serotonin inhibits transmission in the dorsal horn is unsettled.

Morphine is, of course, an addicting drug in that it causes **tolerance,** defined as the need for an increasing dose to cause a given analgesic or other effect; and **dependence,** defined as a compulsive need to keep taking the drug at almost any cost. Despite intensive study, relatively little is known about the brain mechanisms that cause tolerance and dependence. However, the two can be separated. Absence of β-**arrestin-2** blocks tolerance but has no effect on dependence. β-Arrestin-2 is a member of a family of proteins that phosphorylate and thus alter heterotrimeric G proteins.

Placebos appear to be capable of producing the release of endogenous opioids, and this helps to relieve pain. Their effects are inhibited in part by morphine antagonists such as naloxone. Acupuncture at a location distant from the site of a pain also acts by releasing endorphins. Acupuncture at the site of the pain appears to act primarily in the same way as touching or shaking (see above). There appears to be a component of stress analgesia that is mediated by endogenous opioids, because in experimental animals, some forms of stress analgesia are prevented by naloxone. However, other forms are unaffected, and so other components are also involved.

Acetylcholine

Epibatidine, a cholinergic agonist first isolated from the skin of a frog, is a potent nonopioid analgesic agent, and even more potent synthetic congeners of this compound have been developed. Their effects are blocked by cholinergic blocking drugs, and as yet there is no evidence that they are addictive. Conversely, the analgesic effect of nicotine is reduced in mice lacking the α_4 and β_2 nicotine cholinergic receptor subunits. These observations make it clear that a nicotinic cholinergic mechanism is involved in the regulation of pain, though its exact role remains to be determined.

Cannabinoids

As noted in Chapter 4, the cannabinoids anandamide and PEA are produced endogenously and bind to CB_1 and CB_2 receptors, respectively. Anandamide has now been shown to have definite analgesic effects, and there are anandamide-containing neurons in the periaqueductal gray and other areas concerned with pain. When PEA is administered, it acts peripherally to augment the analgesic effects of anandamide.

OTHER SENSATIONS

Itching (pruritus) is not much of a problem for normal individuals, but severe itching that is difficult to treat occurs in diseases such as chronic renal failure, some forms of liver disease, atopic dermatitis, and HIV infection. For many years, convincing evidence for an itch-

specific neural system was not obtained, so hypotheses were advanced that itch was due to a specific pattern of discharge in other systems. There are indeed many interactions, particularly with pain, but itch-specific fibers have been demonstrated in the spinothalamic tract. This and other evidence has caused the pendulum to swing back toward the idea of an itch-specific path.

Itch & Tickle

Relatively mild stimulation, especially if produced by something that moves across the skin, produces itch and tickle. Itch spots can be identified on the skin by careful mapping; they are especially common in regions where there are many naked endings of unmyelinated fibers. Scratching relieves itching because it activates large, fast-conducting afferents that gate transmission in the dorsal horn in a manner analogous to the inhibition of pain by stimulation of similar afferents (see above). It is interesting that a tickling sensation is usually regarded as pleasurable, whereas itching is annoying and pain is unpleasant.

Itching can be produced not only by repeated local mechanical stimulation of the skin but also by a variety of chemical agents. Histamine produces intense itching, and injuries cause its liberation in the skin. However, in most instances of itching, endogenous histamine does not appear to be the responsible agent; doses of histamine that are too small to produce itching still produce redness and swelling on injection into the skin, and severe itching frequently occurs without any visible change in the skin. The kinins cause severe itching.

"Synthetic Senses"

The cutaneous senses for which separate neural pathways exist are touch, warmth, cold, pain, and probably itching. Combinations of these sensations, patterns of stimulation, and, in some cases, cortical components are synthesized into the sensations of vibratory sensation, two-point discrimination, and stereognosis.

Vibratory Sensibility

When a vibrating tuning fork is applied to the skin, a buzzing or thrill is felt. The sensation is most marked over bones, but it can be felt when the tuning fork is placed in other locations. The receptors involved are the receptors for touch, especially pacinian corpuscles, but a time factor is also necessary. A pattern of rhythmic pressure stimuli is interpreted as vibration. The impulses responsible for the vibrating sensation are carried in the dorsal columns. Degeneration of this part of the spinal cord occurs in poorly controlled diabetes, pernicious anemia, some vitamin deficiencies, and occasionally other conditions; elevation of the threshold for vibratory stimuli is an early symptom of this degeneration. Vibratory sensation and proprioception are closely related; when one is depressed, so is the other.

Two-Point Discrimination

The minimal distance by which two touch stimuli must be separated to be perceived as separate is called the **two-point threshold.** It depends upon touch plus the cortical component of identifying one or two stimuli. Its magnitude varies from place to place on the body and is smallest where the touch receptors are most abundant. Points on the back, for instance, must be separated by 65 mm or more before they can be distinguished as separate points, whereas on the fingers two stimuli can be resolved if they are separated by as little as 3 mm. On the hands, the magnitude of the two-point threshold is also small. However, the peripheral neural basis of discriminating two points is not completely understood, and in view of the extensive interdigitation and overlapping of the sensory units, it is probably complex.

Stereognosis

The ability to identify objects by handling them without looking at them is called **stereognosis.** Normal persons can readily identify objects such as keys and coins of various denominations. This ability obviously depends upon relatively intact touch and pressure sensation and is compromised when the dorsal columns are damaged. It also has a large cortical component; impaired stereognosis is an early sign of damage to the cerebral cortex and sometimes occurs in the absence of any detectable defect in touch and pressure sensation when there is a lesion in the parietal lobe posterior to the postcentral gyrus.

Vision

8

INTRODUCTION

The eyes are complex sense organs that have evolved from primitive light-sensitive spots on the surface of invertebrates. Within its protective casing, each eye has a layer of receptors, a lens system that focuses light on these receptors, and a system of nerves that conducts impulses from the receptors to the brain. The way these components operate to set up conscious visual images is the subject of this chapter.

ANATOMIC CONSIDERATIONS

The principal structures of the eye are shown in Figure 8–1. The outer protective layer of the eyeball, the **sclera,** is modified anteriorly to form the transparent **cornea,** through which light rays enter the eye. Inside the sclera is the **choroid,** a layer that contains many of the blood vessels which nourish the structures in the eyeball. Lining the posterior two-thirds of the choroid is the **retina,** the neural tissue containing the receptor cells.

The **crystalline lens** is a transparent structure held in place by a circular **lens ligament (zonule).** The zonule is attached to the thickened anterior part of the choroid, the **ciliary body.** The ciliary body contains circular muscle fibers and longitudinal muscle fibers that attach near the corneoscleral junction. In front of the lens is the pigmented and opaque **iris,** the colored portion of the eye. The iris contains circular muscle fibers that constrict and radial fibers that dilate the **pupil.** Variations in the diameter of the pupil can produce up to fivefold changes in the amount of light reaching the retina.

The space between the lens and the retina is filled primarily with a clear gelatinous material called the **vitreous (vitreous humor). Aqueous humor,** a clear liquid which nourishes the cornea and lens, is produced in the ciliary body by diffusion and active transport from plasma. It flows through the pupil and fills the anterior chamber of the eye. It is normally reabsorbed through a network of trabeculae into the **canal of Schlemm,** a venous channel at the junction between the iris and the cornea (anterior chamber angle). Obstruction of this outlet leads to increased intraocular pressure. Increased intraocular pressure does not cause **glaucoma,** a degenerative disease in which there is loss of retinal ganglia cells, and a substantial minority of the patients with this disease have normal intraocular pressure (10–20 mm Hg). However, increased pressure makes glaucoma worse, and treatment is aimed at lowering the pressure. One cause of increased pressure is decreased permeability through the trabeculae **(open-angle glaucoma),** and another is forward movement of the iris, obliterating the angle **(angle-closure glaucoma).** Glaucoma can be treated with β-adrenergic blocking drugs or carbonic anhydrase inhibitors, both of which decrease the

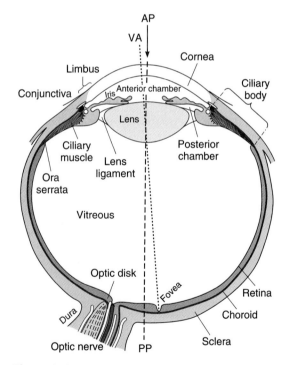

Figure 8–1. Horizontal section of the right eye. AP, anterior pole; PP, posterior pole; VA, visual axis.
(Reproduced, with permission, from Warwick R: *Eugene Wolff's Anatomy of the Eye and Orbit,* 7th ed. Saunders, 1977.)

production of aqueous humor, or with cholinergic agonists, which increase aqueous outflow.

Retina

The retina extends anteriorly almost to the ciliary body. It is organized in ten layers and contains the **rods** and **cones,** which are the visual receptors, plus four types of neurons: **bipolar cells, ganglion cells, horizontal cells,** and **amacrine cells** (Figure 8–2). The rods and cones, which are next to the choroid, synapse with bipolar cells, and the bipolar cells synapse with ganglion cells. The axons of the ganglion cells converge

and leave the eye as the optic nerve. Horizontal cells connect receptor cells to the other receptor cells in the outer plexiform layer. Amacrine cells connect ganglion cells to one another in the inner plexiform layer. They have no axons, and their processes make both pre- and postsynaptic connections with neighboring neural elements. There is considerable overall convergence of receptors on bipolar cells and of bipolar cells on ganglion cells (see below). Gap junctions also connect retinal neurons to one another, and the permeability of these gap junctions is regulated.

Since the receptor layer of the retina rests on the **pigment epithelium** next to the choroid, light rays

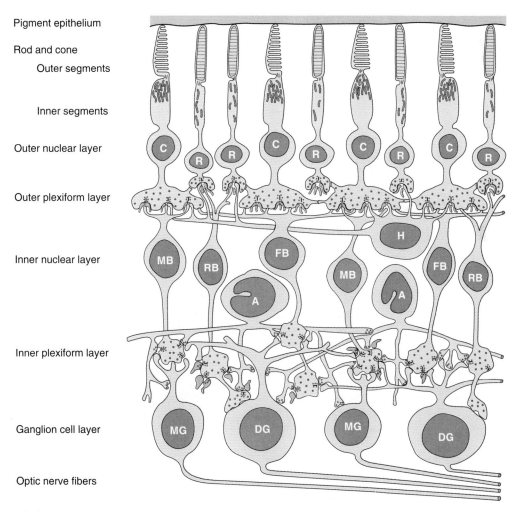

Pigment epithelium

Rod and cone
   Outer segments

Inner segments

Outer nuclear layer

Outer plexiform layer

Inner nuclear layer

Inner plexiform layer

Ganglion cell layer

Optic nerve fibers

Figure 8–2. Neural components of the extrafoveal portion of the retina. C, cone; R, rod; MB, RB, and FB, midget, rod, and flat bipolar cells; DG and MG, diffuse and midget ganglion cells; H, horizontal cells; A, amacrine cells. (Reproduced, with permission, from Dowling JE, Boycott BB: Organization of the primate retina: Electron microscopy. Proc R Soc Lond Ser B [Biol] 1966;166:80.)

must pass through the ganglion cell and bipolar cell layers to reach the rods and cones. The pigment epithelium absorbs light rays, preventing the reflection of rays back through the retina. Such reflection would produce blurring of the visual images.

The neural elements of the retina are bound together by glial cells called Müller cells. The processes of these cells form an internal limiting membrane on the inner surface of the retina and an external limiting membrane in the receptor layer.

The optic nerve leaves the eye and the retinal blood vessels enter it at a point 3 mm medial to and slightly above the posterior pole of the globe. This region is visible through the ophthalmoscope as the **optic disk** (Figure 8–3). There are no visual receptors overlying the disk, and consequently this spot is blind (the **blind spot**).

Near the posterior pole of the eye, there is a yellowish pigmented spot, the **macula lutea.** This marks the location of the **fovea centralis,** a thinned-out, rod-free portion of the retina which is present in humans and other primates. In it, the cones are densely packed, and each synapses to a single bipolar cell which in turn synapses on a single ganglion cell, providing a direct pathway to the brain. There are very few overlying cells and no blood vessels. Consequently, the fovea is the point where visual acuity is greatest. When attention is attracted to or fixed on an object, the eyes are normally moved so that light rays coming from the object fall on the fovea.

The arteries, arterioles, and veins in the superficial layers of the retina near its vitreous surface can be seen through the ophthalmoscope. Since this is the one place in the body where arterioles are readily visible, ophthalmoscopic examination is of great value in the diagnosis and evaluation of diabetes mellitus, hypertension, and other diseases that affect blood vessels. The retinal vessels supply the bipolar and ganglion cells, but the receptors are nourished for the most part by the capillary plexus in the choroid. This is why retinal detachment is so damaging to the receptor cells.

Neural Pathways

The axons of the ganglion cells pass caudally in the **optic nerve** and **optic tract** to end in the **lateral geniculate body,** a part of the thalamus (Figure 8–4). The fibers from each nasal hemiretina decussate in the **optic chiasm.** In the geniculate body, the fibers from the nasal half of one retina and the temporal half of the other synapse on the cells whose axons form the **geniculocalcarine tract.** This tract passes to the occipital lobe of the cerebral cortex.

The primary visual receiving area (**primary visual cortex,** Brodmann's area 17; also known as V1), is located principally on the sides of the calcarine fissure (Figure 8–5). The organization of the primary visual cortex is discussed below.

Some ganglion cell axons pass from the optic tract to the pretectal region of the midbrain and the superior colliculus, where they form connections that mediate pupillary reflexes and eye movements. The frontal cortex is also concerned with eye movement, and especially its refinement. The bilateral **frontal eye fields** in this part of the cortex are concerned with control of saccades (see below), and an area just anterior to these

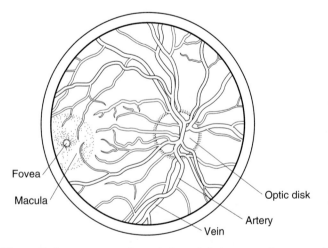
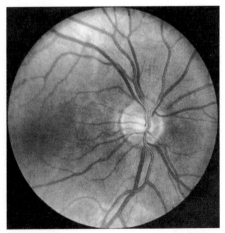

Figure 8–3. Retina seen through the ophthalmoscope in a normal human. The diagram on the left identifies the landmarks in the photograph on the right. (Reproduced, with permission, from Vaughan D, Asbury T, Riordan-Eva P: *General Ophthalmology,* 15th ed. McGraw-Hill, 1998.)

Fovea

Macula

Optic disk

Artery

Vein

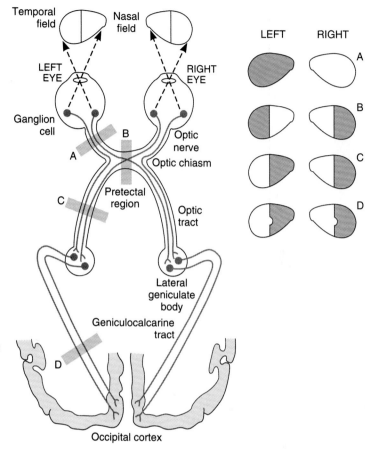

Figure 8–4. Visual pathways. Transection of the pathways at the locations indicated by the letters causes the visual field defects shown in the diagrams on the right (see text). Occipital lesions may spare the fibers from the macula (as in D) because of the separation in the brain of these fibers from the others subserving vision (see Figure 8–5).

fields is concerned with vergence and the near response. The frontal areas concerned with vision probably project to the nucleus reticularis tegmentalis pontinus, and from there to the other brain stem nuclei mentioned above.

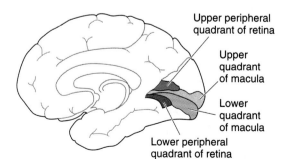

Figure 8–5. Medial view of the human right cerebral hemisphere showing projection of the retina on the occipital cortex around the calcarine fissure.

Other axons pass directly from the optic chiasm to the suprachiasmatic nuclei in the hypothalamus, where they form connections that synchronize a variety of endocrine and other circadian rhythms with the light-dark cycle (see Chapter 14).

The brain areas activated by visual stimuli have been investigated in monkeys and humans by PET and other imaging techniques (see Chapter 32). Activation occurs not only in the occipital lobe but also in parts of the inferior temporal cortex, the posteroinferior parietal cortex, portions of the frontal lobe, and the amygdala. The subcortical structures activated in addition to the lateral geniculate body include the superior colliculus, pulvinar, caudate nucleus, putamen, and claustrum.

Receptors

Each rod and cone is divided into an outer segment, an inner segment that includes a nuclear region, and a synaptic zone (Figure 8–6). The outer segments are modified cilia and are made up of regular stacks of flattened saccules or disks composed of membrane. These

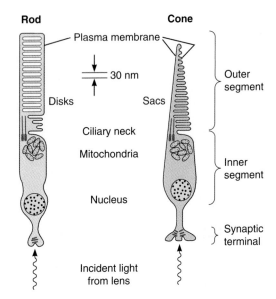

Figure 8–6. Schematic diagram of a rod and a cone. (Reproduced, with permission, from Lamb TD: Electrical responses of photoreceptors. In: *Recent Advances in Physiology*. No. 10. Baker PF [editor]. Churchill Livingstone, 1984.)

saccules and disks contain the photosensitive compounds that react to light, initiating action potentials in the visual pathways (see below). The inner segments are rich in mitochondria. The rods are named for the thin, rod-like

appearance of their outer segments. Cones generally have thick inner segments and conical outer segments, although their morphology varies from place to place in the retina. In cones, the saccules are formed in the outer segments by infoldings of the cell membrane, but in rods, the disks are separated from the cell membrane.

Rod outer segments are being constantly renewed by formation of new disks at the inner edge of the segment and phagocytosis of old disks from the outer tip by cells of the pigment epithelium. Cone renewal is a more diffuse process and appears to occur at multiple sites in the outer segments.

In the extrafoveal portions of the retina, rods predominate (Figure 8–7), and there is a good deal of convergence. Flat bipolar cells (Figure 8–2) make synaptic contact with several cones, and rod bipolar cells make synaptic contact with several rods. Since there are approximately 6 million cones and 120 million rods in each human eye but only 1.2 million nerve fibers in each optic nerve, the overall convergence of receptors through bipolar cells on ganglion cells is about 105:1. However, it is worth noting that there is divergence from this point on; there are twice as many fibers in the geniculocalcarine tracts as in the optic nerves, and in the visual cortex the number of neurons concerned with vision is 1000 times the number of fibers in the optic nerves.

Eye Muscles

The eye is moved within the orbit by six ocular muscles (Figure 8–8). These are innervated by the oculomotor,

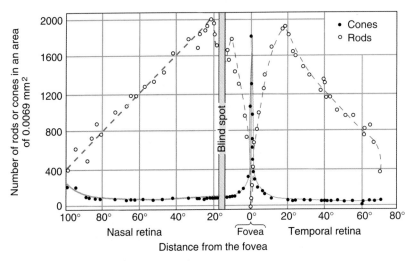

Figure 8–7. Rod and cone density along the horizontal meridian through the human retina. A plot of the relative acuity of vision in the various parts of the light-adapted eye would parallel the cone density curve; a similar plot of relative acuity of the dark-adapted eye would parallel the rod density curve.

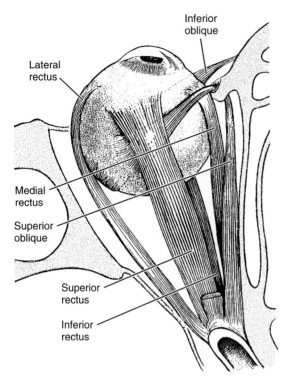

Lateral rectus

Inferior oblique

Medial rectus

Superior oblique

Superior rectus

Inferior rectus

Figure 8–8. The six extraocular muscles, viewed from the top. (Modified from Dox I, Melloni BJ, Eisner GM: *Melloni's Illustrated Medical Dictionary.* Williams & Williams, 1979.)

trochlear, and abducens nerves. The muscles and the directions in which they move the eyeball are discussed at the end of this chapter.

Protection

The eye is well protected from injury by the bony walls of the orbit. The cornea is moistened and kept clear by tears that course from the **lacrimal gland** in the upper portion of each orbit across the surface of the eye to empty via the **lacrimal duct** into the nose. Blinking helps keep the cornea moist.

One of the most important characteristics of the visual system is its ability to function over a wide range of light intensity. When one goes from near darkness to bright sunlight, light intensity increases by 10 log units, ie, by a factor of 10 billion. One factor reducing the fluctuation in intensity is the diameter of the pupil; when this is reduced from 8 mm to 2 mm, its area decreases by a factor of 16 and light intensity at the retina is reduced by more than 1 log unit.

Another factor in reacting to fluctuations in intensity is the presence of two types of receptors. The rods are extremely sensitive to light and are the receptors for night vision (**scotopic vision**). The scotopic visual apparatus is not capable of resolving the details and boundaries of objects or determining their color. The cones have a much higher threshold, but the cone system has a much greater acuity and is the system responsible for vision in bright light (**photopic vision**) and for color vision. There are thus two kinds of inputs to the CNS from the eye: input from the rods and input from the cones. The existence of these two kinds of input, each working maximally under different conditions of illumination, is called the **duplicity theory.** In addition, both the rods and the cones undergo adaptation (see below).

THE IMAGE-FORMING MECHANISM

The eyes convert energy in the visible spectrum into action potentials in the optic nerve. The wavelengths of visible light range from approximately 397 nm to 723 nm. The images of objects in the environment are focused on the retina. The light rays striking the retina generate potentials in the rods and cones. Impulses initiated in the retina are conducted to the cerebral cortex, where they produce the sensation of vision.

Principles of Optics

Light rays are bent (refracted) when they pass from one medium into a medium of a different density, except when they strike perpendicular to the interface. Parallel light rays striking a biconvex lens (Figure 8–9) are refracted to a point (**principal focus**) behind the lens. The principal focus is on a line passing through the centers of curvature of the lens, the **principal axis.** The distance between the lens and the principal focus is the **principal focal distance.** For practical purposes, light rays from an object that strike a lens more than 6 m (20 ft) away are considered to be parallel. The rays from an object closer than 6 m are diverging and are therefore brought to a focus farther back on the principal axis than the principal focus (Figure 8–9). Biconcave lenses cause light rays to diverge.

The greater the curvature of a lens, the greater its refractive power. The refractive power of a lens is conveniently measured in **diopters,** the number of diopters being the reciprocal of the principal focal distance in meters. For example, a lens with a principal focal distance of 0.25 m has a refractive power of 1/0.25, or 4 diopters. The human eye has a refractive power of approximately 60 diopters at rest.

Accommodation

When the ciliary muscle is relaxed, parallel light rays striking the optically normal (**emmetropic**) eye are

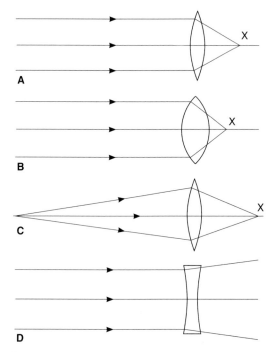

Figure 8–9. Refraction of light rays by lenses: **A:** Biconvex lens. **B:** Biconvex lens of greater strength than **A**. **C:** Same lens as **A**, showing effect on light rays from a near point. **D:** Biconcave lens. The center line in each case is the principal axis. X is the principal focus.

brought to a focus on the retina. As long as this relaxation is maintained, rays from objects closer than 6 m from the observer are brought to a focus behind the retina, and consequently the objects appear blurred. The problem of bringing diverging rays from close objects to a focus on the retina can be solved by increasing the distance between the lens and the retina or by increasing the curvature or refractive power of the lens. In bony fish, the problem is solved by increasing the length of the eyeball, a solution analogous to the manner in which the images of objects closer than 6 m are focused on the film of a camera by moving the lens away from the film. In mammals, the problem is solved by increasing the curvature of the lens.

The process by which the curvature of the lens is increased is called **accommodation.** At rest, the lens is held under tension by the lens ligaments. Because the lens substance is malleable and the lens capsule has considerable elasticity, the lens is pulled into a flattened shape. When the gaze is directed at a near object, the ciliary muscle contracts. This decreases the distance between the edges of the ciliary body and relaxes the lens ligaments, so that the lens springs into a more convex

shape. In young individuals, the change in shape may add as many as 12 diopters to the refractive power of the eye. The relaxation of the lens ligaments produced by contraction of the ciliary muscle is due partly to the sphincter-like action of the circular muscle fibers in the ciliary body and partly to the contraction of longitudinal muscle fibers that attach anteriorly, near the corneoscleral junction. When these fibers contract, they pull the whole ciliary body forward and inward. This motion brings the edges of the ciliary body closer together.

The change in lens curvature during accommodation affects principally the anterior surface of the lens (Figure 8–10). This can be demonstrated by a simple experiment first described many years ago. If an observer holds an object in front of the eyes of an individual who is looking into the distance, three reflections of the object are visible in the subject's eye. A clear, small upright image is reflected from the cornea; a larger, fainter upright image is reflected from the anterior surface of the lens; and a small inverted image is reflected from the posterior surface of the lens. If the subject then focuses on an object nearby, the large, faint upright image becomes smaller and moves toward the other upright image, whereas the third image changes very little. The change in size of the second image is due to the increase in curvature of the reflecting surface, the anterior surface of the lens (Figure 8–10). The fact that the small upright image does not change and the inverted image changes very little shows that the corneal curvature is unchanged and that the curvature of the posterior lens surface is changed very little by accommodation.

Near Point

Accommodation is an active process, requiring muscular effort, and can therefore be tiring. Indeed, the ciliary muscle is one of the most used muscles in the

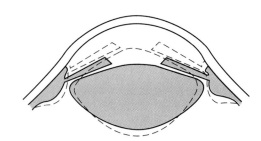

Figure 8–10. Accommodation. The solid lines represent the shape of the lens, iris, and ciliary body at rest, and the dashed lines represent the shape during accommodation.

body. The degree to which the lens curvature can be increased is, of course, limited, and light rays from an object very near the individual cannot be brought to a focus on the retina even with the greatest of effort. The nearest point to the eye at which an object can be brought into clear focus by accommodation is called the **near point of vision.** The near point recedes throughout life, slowly at first and then rapidly with advancing age, from approximately 9 cm at age 10 to approximately 83 cm at age 60. This recession is due principally to increasing hardness of the lens, with a resulting loss of accommodation (Figure 8–11) due to the steady decrease in the degree to which the curvature of the lens can be increased. By the time a normal individual reaches age 40–45, the loss of accommodation is usually sufficient to make reading and close work difficult. This condition, which is known as **presbyopia,** can be corrected by wearing glasses with convex lenses.

The Near Response

In addition to accommodation, the visual axes converge and the pupil constricts when an individual looks at a near object. This three-part response—accommodation, convergence of the visual axes, and pupillary constriction—is called the **near response.**

Other Pupillary Reflexes

When light is directed into one eye, the pupil constricts **(pupillary light reflex).** The pupil of the other eye also constricts **(consensual light reflex).** The optic nerve fibers that carry the impulses initiating these pupillary responses leave the optic nerves near the lateral geniculate bodies. On each side, they enter the midbrain via the brachium of the superior colliculus and terminate in the pretectal nucleus. From this nucleus, the second-order neurons project to the ipsilateral Edinger-Westphal nucleus and the contralateral Edinger-Westphal nucleus. The third-order neurons pass from this nucleus to the ciliary ganglion in the oculomotor nerve, and the fourth-order neurons pass from this ganglion to the ciliary body. This pathway is dorsal to the pathway for the near response. Consequently, the light response is sometimes lost while the response to accommodation remains intact **(Argyll Robertson pupil).** One cause of this abnormality is CNS syphilis, but the Argyll Robertson pupil is also seen in other diseases producing selective lesions in the midbrain.

Retinal Image

In the eye, light is actually refracted at the anterior surface of the cornea and at the anterior and posterior surfaces of the lens. The process of refraction can be represented diagrammatically, however, without introducing any appreciable error, by drawing the rays of light as if all refraction occurs at the anterior surface of the cornea. Figure 8–12 is a diagram of such a "reduced" or "schematic" eye. In this diagram, the **nodal**

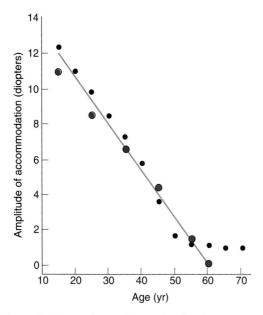

Figure 8–11. Decline in the amplitude of accommodation in humans with advancing age. The different symbols identify data from different studies.
(Reproduced, with permission, from Fisher RF: Presbyopia and the changes with age in the human crystalline lens. J Physiol 1973;228:765.)

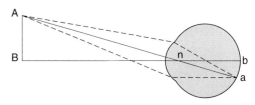

Figure 8–12. Reduced eye. n, nodal point. AnB and anb are similar triangles. In this reduced eye, the nodal point is 15 mm from the retina. All refraction is assumed to take place at the surface of the cornea, 5 mm from the nodal point, between a medium of density 1.000 (air) and a medium of density 1.333 (water). The dotted lines represent rays of light diverging from A and refracted at the cornea so that they are focused on the retina at a.

point (optical center of the eye) coincides with the junction of the middle and posterior third of the lens, 15 mm from the retina. This is the point through which the light rays from an object pass without refraction. All other rays entering the pupil from each point on the object are refracted and brought to a focus on the retina. If the height of the object (AB) and its distance from the observer (Bn) are known, the size of its retinal image can be calculated, because AnB and anb in Figure 8–12 are similar triangles. The angle AnB is the **visual angle** subtended by object AB. It should be noted that the retinal image is inverted. The connections of the retinal receptors are such that from birth any inverted image on the retina is viewed right side up and projected to the visual field on the side opposite to the retinal area stimulated. This perception is present in infants and is innate. If retinal images are turned right side up by means of special lenses, the objects viewed look as if they are upside down.

Common Defects of the Image-Forming Mechanism

In some individuals, the eyeball is shorter than normal and the parallel rays of light are brought to a focus behind the retina. This abnormality is called **hyperopia** or farsightedness (Figure 8–13). Sustained accommodation, even when viewing distant objects, can partially compensate for the defect, but the prolonged muscular effort is tiring and may cause headaches and blurring of vision. The prolonged convergence of the visual axes associated with the accommodation may lead eventually to squint **(strabismus)** (see below). The defect can be corrected by using glasses with convex lenses, which aid the refractive power of the eye in shortening the focal distance.

In **myopia** (nearsightedness), the anteroposterior diameter of the eyeball is too long. Myopia is said to be genetic in origin. However, in experimental animals it can be produced by changing refraction during development. In humans, there is a positive correlation between sleeping in a lighted room before the age of two and the subsequent development of myopia. Thus, the shape of the eye appears to be determined in part by the refraction presented to it. In young adult humans the extensive close work involved in activities such as studying accelerates the development of myopia. This defect can be corrected by glasses with biconcave lenses, which make parallel light rays diverge slightly before they strike the eye.

Astigmatism is a common condition in which the curvature of the cornea is not uniform. When the curvature in one meridian is different from that in others, light rays in that meridian are refracted to a different focus, so that part of the retinal image is blurred. A similar defect may be produced if the lens is pushed out of alignment or the curvature of lens is not uniform, but these conditions are rare. Astigmatism can usually be corrected with cylindric lenses placed in such a way that they equalize the refraction in all meridians. **Presbyopia** has been mentioned above.

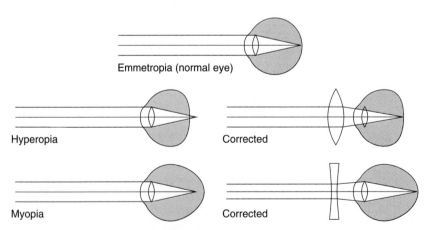

Figure 8–13. Common defects of the optical system of the eye. In hyperopia, the eyeball is too short and light rays come to a focus behind the retina. A biconvex lens corrects this by adding to the refractive power of the lens of the eye. In myopia, the eyeball is too long and light rays focus in front of the retina. Placing a biconcave lens in front of the eye causes the light rays to diverge slightly before striking the eye, so that they are brought to a focus on the retina.

THE PHOTORECEPTOR MECHANISM

Genesis of Electrical Responses

The potential changes that initiate action potentials in the retina are generated by the action of light on photosensitive compounds in the rods and cones. When light is absorbed by these substances, their structure changes, and this change triggers a sequence of events that initiates neural activity.

The eye is unique in that the receptor potentials of the photoreceptors and the electrical responses of most of the other neural elements in the retina are local, graded potentials, and it is only in the ganglion cells that all-or-none action potentials transmitted over appreciable distances are generated. The responses of the rods, cones, and horizontal cells are hyperpolarizing (Figure 8–14), and the responses of the bipolar cells are either hyperpolarizing or depolarizing, whereas amacrine cells produce depolarizing potentials and spikes that may act as generator potentials for the propagated spikes produced in the ganglion cells.

The cone receptor potential has a sharp onset and offset, whereas the rod receptor potential has a sharp onset and slow offset. The curves relating the amplitude of receptor potentials to stimulus intensity have similar shapes in rods and cones, but the rods are much more sensitive. Therefore, rod responses are proportionate to stimulus intensity at levels of illumination that are below the threshold for cones. On the other hand, cone responses are proportionate to stimulus intensity at high levels of illumination when the rod responses are maximal and cannot change. This is why cones generate good responses to changes in light intensity above background but do not represent absolute illumination well, whereas rods detect absolute illumination.

Ionic Basis of Photoreceptor Potentials

Na$^+$ channels in the outer segments of the rods and cones are open in the dark, so current flows from the inner to the outer segment (Figure 8–15). Current also flows to the synaptic ending of the photoreceptor. Na$^+$-K$^+$ ATPase in the inner segment maintains ionic equilibrium. Release of synaptic transmitter is steady in the dark. When light strikes the outer segment, the reactions that are initiated close some of the Na^2 channels, and the result is a hyperpolarizing receptor potential. The hyperpolarization reduces the release of synaptic transmitter, and this generates a signal that ultimately leads to action potentials in ganglion cells. The action potentials are transmitted to the brain.

Photosensitive Compounds

The photosensitive compounds in the eyes of humans and most other mammals are made up of a protein

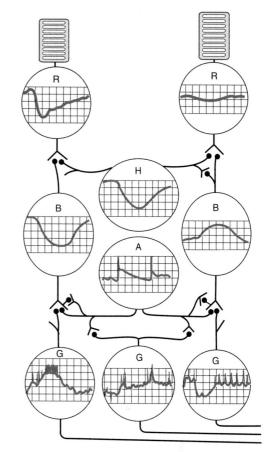

Figure 8–14. Intracellularly recorded responses of cells in the retina to light. The synaptic connections of the cells are also indicated. The rod (R) on the left is receiving a light flash, whereas the rod on the right is receiving steady, low-intensity illumination. H, horizontal cell; B, bipolar cells; A, amacrine cell; G, ganglion cell. (Reproduced, with permission, from Dowling JE: Organization of vertebrate retinas. Invest Ophthalmol 1970;9:655.)

called an **opsin**, and **retinene$_1$**, the aldehyde of vitamin A$_1$. The term retinene$_1$ is used to distinguish this compound from retinene$_2$, which is found in the eyes of some animal species. Since the retinenes are aldehydes, they are also called **retinals**. The A vitamins themselves are alcohols and are therefore called **retinols**.

Rhodopsin

The photosensitive pigment in the rods is called **rhodopsin** or **visual purple**. Its opsin is called **scotopsin**. Rhodopsin has a peak sensitivity to light at a wavelength of 505 nm.

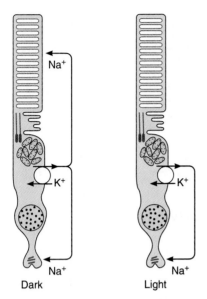

Figure 8–15. Effect of light on current flow in visual receptors. In the dark, Na$^+$ channels in the outer segment are held open by cGMP. Light leads to increased conversion of cGMP to 5'-GMP, and some of the channels close. This produces hyperpolarization of the synaptic terminal of the photoreceptor.

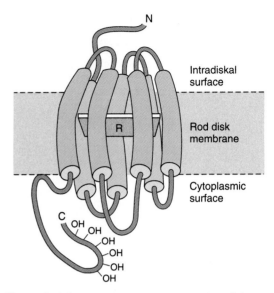

Figure 8–16. Diagrammatic representation of the structure of rhodopsin, showing the position of retinene$_1$ (R) in the rod disk membrane.

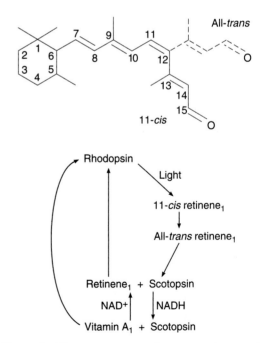

Figure 8–17. **Top:** Structure of retinene$_1$, showing the 11-cis configuration (unbroken lines) and the all-trans configuration produced by light (dashed lines). **Bottom:** Effects of light on rhodopsin.

Human rhodopsin has a molecular weight of 41,000. It is found in the membranes of the rod disks and makes up 90% of the total protein in these membranes. It is one of the many serpentine receptors coupled to G proteins (see Figure 1–40). Retinene$_1$ is parallel to the surface of the membrane (Figure 8–16) and is attached to a lysine residue at position 296 in the seventh transmembrane domain.

In the dark, the retinene$_1$ in rhodopsin is in the 11-cis configuration. The only action of light is to change the shape of the retinene, converting it to the all-trans isomer (Figure 8–17). This in turn alters the configuration of the opsin, and the opsin change activates the associated heterotrimeric G protein, which in this case is called **transducin** or **G$_{t1}$**. The G protein exchanges GDP for GTP, and the α subunit separates. This subunit remains active until its intrinsic GTPase activity hydrolyzes the GTP. Termination of the activity of transducin is also accelerated by its binding of β-arrestin (see Chapter 4). The α subunit activates cGMP phosphodiesterase, which converts cGMP to 5'-GMP (Figure 8–18). cGMP normally acts directly on Na$^+$ channels to maintain them in the open position, so the decline in the cytoplasmic cGMP concentration causes some Na$^+$ channels to close. This produces the hyperpolarizing potential.

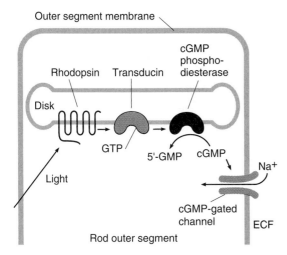

Figure 8–18. Initial steps in phototransduction in rods. Light activates rhodopsin, which activates transducin to bind GTP. This activates phosphodiesterase, which catalyzes the conversion of cGMP to 5′-GMP. The resulting decrease in the cytoplasmic cGMP concentration causes cGMP-gated ion channels to close.

The cascade of reactions described in the preceding paragraph occurs very rapidly and amplifies the light signal. The amplification helps explain the remarkable sensitivity of rod photoreceptors; these receptors are capable of producing a detectable response to as little as one photon of light.

After $retinene_1$ is converted to the all-*trans* configuration, it separates from the opsin (bleaching). Some of the rhodopsin is regenerated directly, while some of the $retinene_1$ is reduced by the enzyme alcohol dehydrogenase in the presence of NADH to vitamin A_1, and this in turn reacts with scotopsin to form rhodopsin (Figure 8–17). All of these reactions except the formation of the all-*trans* isomer of $retinene_1$ are independent of the light intensity, proceeding equally well in light or darkness. The amount of rhodopsin in the receptors therefore varies inversely with the incident light level.

Cone Pigments

There are three different kinds of cones in primates. These receptors subserve color vision and respond maximally to light at wavelengths of 440, 535, and 565 nm (see below). Each contains $retinene_1$ and an opsin. The opsin resembles rhodopsin and spans the cone membrane seven times but has a characteristic structure in each type of cone. As noted above, the cell membrane of cones is invaginated to form the saccules, but there are no separate intracellular disks like those in rods.

The details of the responses of cones to light are probably similar to those in rods. Light activates $retinene_1$, and this activates G_{t2}, a G protein that differs somewhat from rod transducin. G_{t2} in turn activates phosphodiesterase, catalyzing the conversion of cGMP to 5′-GMP. This results in closure of Na^+ channels between the extracellular fluid and the cone cytoplasm, a decrease in intracellular Na^+ concentration, and hyperpolarization of the cone synaptic terminals.

The sequence of events in photoreceptors by which incident light leads to production of a signal in the next succeeding neural unit in the retina is summarized in Figure 8–19.

Resynthesis of Cyclic GMP

Light reduces the concentration of Ca^{2+} as well as that of Na^+ in photoreceptors. The resulting decrease in Ca^{2+} concentration activates guanylyl cyclase, which generates more cGMP. It also inhibits the light-activated phosphodiesterase. Both actions speed recovery, restoring the Na^+ channels to their open position.

Synaptic Mediators in the Retina

Many different synaptic transmitters are found in the retina. These include acetylcholine, glutamate, dopamine, serotonin, GABA, glycine, substance P, somatostatin,

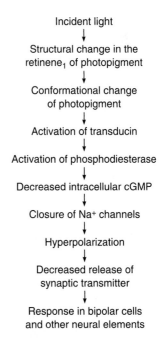

Figure 8–19. Sequence of events involved in phototransduction in rods and cones.

TRH, GnRH, enkephalins, β-endorphin, CCK, VIP, neurotensin, and glucagon (see Chapter 4). Kainate receptors mediate the synaptic responses between cones and one type of bipolar cells. Amacrine cells are the only cells that secrete acetylcholine in the retina. At least in some vertebrate species, dopamine is secreted by cells along the border between the inner nuclear and inner plexiform layers (Figure 8–2) and spreads through the retina by diffusion. Among other things, it affects the structure of gap junctions. These junctions allow current to pass freely through horizontal cells in the dark, enlarging the receptive fields of the photoreceptors. Light reduces the current flow, decoupling the horizontal cells, and this decoupling appears to be due to increased release of dopamine in daylight.

Image Formation

In a sense, the processing of visual information in the retina involves the formation of three images. The first image, formed by the action of light on the photoreceptors, is changed to a second image in the bipolar cells, and this in turn is converted to a third image in the ganglion cells. In the formation of the second image, the signal is altered by the horizontal cells, and in the formation of the third, it is altered by the amacrine cells. There is little change in the impulse pattern in the lateral geniculate bodies, so the third image reaches the occipital cortex.

A characteristic of the bipolar and ganglion cells (as well as the lateral geniculate cells and the cells in layer 4 of the visual cortex) is that they respond best to a small, circular stimulus and that, within their receptive field, an annulus of light around the center (surround illumination) inhibits the response to the central spot (Figure 8–20). The center can be excitatory with an inhibitory surround (an "on-center" cell) or inhibitory with an excitatory surround (an "off-center" cell). The inhibition of the center response by the surround is probably due to inhibitory feedback from one photoreceptor to another mediated via horizontal cells. Thus, activation of nearby photoreceptors by addition of the annulus triggers horizontal cell hyperpolarization, which in turn inhibits the response of the centrally activated photoreceptors. The inhibition of the response to central illumination by an increase in surrounding illumination is an example of **lateral** or **afferent inhibition**—that form of inhibition in which activation of a particular neural unit is associated with inhibition of the activity of nearby units. It is a general phenomenon in mammalian sensory systems and helps to sharpen the edges of a stimulus and improve discrimination.

Electroretinogram

The electrical activity of the eye has been studied by recording fluctuations in the potential difference between an electrode in the eye and another on the back of the eye or, in humans, an electrode on the cornea and another on the skin of the head. A flash of light produces a characteristic sequence of waves: rapid a and b waves due to the electrical activity in the retina and a slower c wave, which is generated in the pigment epithelium. This **electroretinogram (ERG)** is helpful in the diagnosis of diseases in which visualization of the retina is difficult because the ocular fluids are cloudy.

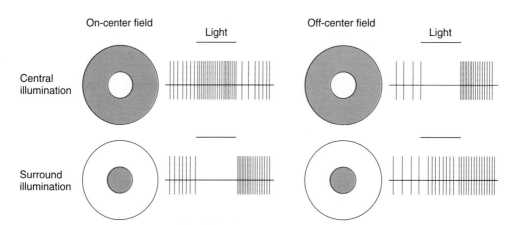

Figure 8–20. Responses of retinal ganglion cells to light on the portions of their receptive fields indicated in white. Beside each receptive-field diagram is a diagram of the ganglion cell response, indicated by extracellularly recorded action potentials. Note that in three of the four situations, there is increased discharge when the light is turned off. (Modified from Kandel E, Schwartz JH, Jessell TM [editors]: *Principles of Neural Science,* 4th ed. McGraw-Hill, 2000.)

The ERG is also useful in congenital retinal dystrophies in which the retina appears normal by ophthalmoscopy.

RESPONSES IN THE VISUAL PATHWAYS & CORTEX

Pathways to the Cortex

The axons of retinal ganglion cells project a detailed spatial representation of the retina on the lateral geniculate body. Each geniculate body contains six well-defined layers (Figure 8–21). Layers 3–6 have small cells and are called parvocellular, whereas layers 1 and 2 have large cells and are called magnocellular. On each side, layers 1, 4, and 6 receive input from the contralateral eye whereas layers 2, 3, and 5 receive input from the ipsilateral eye. In each layer, there is a precise point-for-point representation of the retina, and all six layers are in register so that along a line perpendicular to the layers, the receptive fields of the cells in each layer are almost identical. It is worth noting that only 10–20% of the input to the lateral geniculate nucleus comes from the retina. There are in addition major inputs from the visual cortex and other brain regions. The feedback pathway from the visual cortex has been shown to be involved in visual processing related to the perception of orientation and motion.

Two kinds of ganglion cells can be distinguished in the retina: large ganglion cells (magno, or M cells), which add responses from different kinds of cones and are concerned with movement and stereopsis; and small ganglion cells (parvo, or P cells), which subtract input from one type of cone from input from another and are concerned with color, texture, and shape. The M ganglion cells project to the magnocellular portion of the lateral geniculate, whereas the P ganglion cells project to the parvocellular portion (Figure 8–22).

From the lateral geniculate nucleus, a magnocellular pathway and a parvocellular pathway project to the visual cortex. The magnocellular pathway, from layers 1 and 2 (Figure 8–21), carries signals for detection of movement, depth and flicker. The parvocellular pathway, from layers 3–6, carries signals for color vision, texture, shape, and fine detail.

Cells in the interlaminar region of the lateral geniculate nucleus also receive input from P ganglion cells,

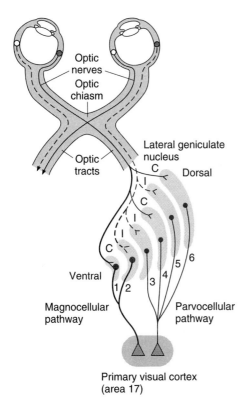

Figure 8–21. Ganglion cell projections from the right hemiretina of each eye to the right lateral geniculate body and from this nucleus to the right primary visual cortex. Note the six layers of the geniculate. P ganglion cells project to layers 3–6, and M ganglion cells project to layers 1 and 2. The ipsilateral (I) and contralateral (C) eyes project to alternate layers. Not shown are the interlaminar area cells, which project via a separate component of the P pathway to blobs in the visual cortex. (Modified from Kandel ER, Schwartz JH, Jessell TM [editors]: *Principles of Neural Science,* 4th ed. McGraw-Hill, 2000.)

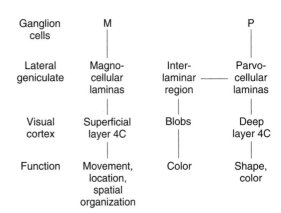

Ganglion cells	M		P
Lateral geniculate	Magno-cellular laminas	Inter-laminar region	Parvo-cellular laminas
Visual cortex	Superficial layer 4C	Blobs	Deep layer 4C
Function	Movement, location, spatial organization	Color	Shape, color

Figure 8–22. Organization of the visual pathways. M ganglion cells project to magnocellular laminas of the lateral geniculate nuclei, and P ganglion cells project to parvocellular laminas. P cells contact interlaminar cells as well, probably via dendrites.

probably via dendrites of interlaminar cells that penetrate the parvocellular layers. They project via a separate component of the P pathway to the blobs in the visual cortex (see below).

Primary Visual Cortex

Just as the ganglion cell axons project a detailed spatial representation of the retina on the lateral geniculate body, the lateral geniculate body projects a similar point-for-point representation on the primary visual cortex (Figure 8–5). In the visual cortex, there are many nerve cells associated with each fiber. Like the rest of the neocortex, the visual cortex has six layers (see Figure 11–1). The axons from the lateral geniculate nucleus that form the magnocellular pathway end in layer 4, specifically in its deepest part, layer 4C. Many of the axons that form the parvocellular pathway also end in layer 4C. However, the axons from the interlaminar region end in layers 2 and 3.

Layers 2 and 3 of the cortex contain clusters of cells about 0.2 mm in diameter that, unlike the neighboring cells, contain a high concentration of the mitochondrial enzyme cytochrome oxidase. The clusters have been named **blobs** (Figure 8–22). They are arranged in a mosaic in the visual cortex and are concerned with color vision. However, the parvocellular pathway also carries color opponent data to the deep part of layer 4 (see below).

Like the ganglion cells, the lateral geniculate neurons and the neurons in layer 4 of the visual cortex respond to stimuli in their receptive fields with on centers and inhibitory surrounds or off centers and excitatory surrounds. A bar of light covering the center is an effective stimulus for them because it stimulates all the center and relatively little of the surround. However, the bar has no preferred orientation and, as a stimulus, is equally effective at any angle.

The responses of the neurons in other layers of the visual cortex are strikingly different. So-called **simple cells** in these locations respond to bars of light, lines, or edges, but only when they have a particular orientation. When, for example, a bar of light is rotated as little as 10 degrees from the preferred orientation, the firing rate of the simple cell is usually decreased, and if the stimulus is rotated much more, the response disappears. There are also **complex cells,** which resemble simple cells in requiring a preferred orientation of a linear stimulus but are less dependent upon the location of a stimulus in the visual field than the simple cells and the cells in layer 4. They often respond maximally when a linear stimulus is moved laterally without a change in its orientation. They probably receive input from the simple cells.

If a microelectrode is inserted perpendicularly into the visual cortex and passed through the various layers, the orientation preference of the neurons is the same. Thus, the visual cortex, like the somatosensory cortex (see Chapter 7), is arranged in vertical columns that are concerned with orientation **(orientation columns).** Each is about 1 mm in diameter. However, the orientation preferences of neighboring columns differ in a systematic way; as one moves from column to column across the cortex, there are sequential changes in orientation preference of 5–10 degrees (Figure 8–23). Thus, it seems likely that for each ganglion cell receptive field in the visual field, there is a collection of columns in a small area of visual cortex representing the possible preferred orientations at small intervals throughout the full 360 degrees. The simple and complex cells have been called **feature detectors** because they respond to and analyze certain features of the stimulus. Feature detectors are also found in the cortical areas for other sensory modalities.

The orientation columns can be mapped with the aid of radioactive 2-deoxyglucose. The uptake of this glucose derivative is proportionate to the activity of neurons (see Chapter 32). When this technique is employed in animals exposed to uniformly oriented visual stimuli such as vertical lines, the brain shows a remarkable array of intricately curved but evenly spaced orientation columns over a large area of the visual cortex.

Another feature of the visual cortex is the presence of **ocular dominance columns.** The geniculate cells and the cells in layer 4 receive input from only one eye, and the layer 4 cells alternate with cells receiving input from the other eye. If a large amount of a radioactive amino acid is injected into one eye, the amino acid is incorporated into protein and transported by axoplasmic flow to the ganglion cell terminals, across the geniculate synapses, and along the geniculocalcarine fibers to the visual cortex. In layer 4, labeled endings from the injected eye alternate with unlabeled endings from the uninjected eye. The result, when viewed from above, is a vivid pattern of stripes that covers much of the visual cortex (Figure 8–24) and is separate from and independent of the grid of orientation columns.

About half the simple and complex cells receive an input from both eyes. The inputs are identical or nearly so in terms of the portion of the visual field involved and the preferred orientation. However, they differ in strength, so that between the cells to which the input is totally from the ipsilateral or the contralateral eye, there is a spectrum of cells influenced to different degrees by both eyes.

Thus, the primary visual cortex segregates information about color from that concerned with form and movement, combines the input from the two eyes, and converts the visual world into short line segments of various orientations.

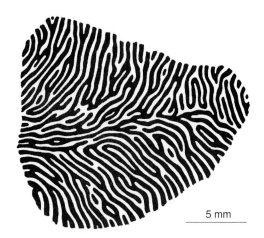

Figure 8–24. Reconstruction of ocular dominance columns in a subdivision of layer 4 of a portion of the right visual cortex of a rhesus monkey. Dark stripes represent one eye, light stripes the other. (Reproduced, with permission, from LeVay S, Hubel DH, Wiesel TN: The pattern of ocular dominance columns in macaque visual cortex revealed by a reduced silver stain. J Comp Neurol 1975;159:559.)

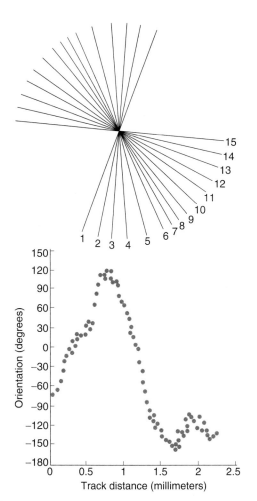

Figure 8–23. **Top:** Orientation preferences of 15 neurons encountered as a microelectrode penetrates the visual cortex obliquely. The preferred orientation changes steadily in a counterclockwise direction. **Bottom:** Results of a similar experiment plotted against distance traveled by the electrode. In this case, there are a number of reversals in the direction of rotation. (Modified and reproduced, with permission, from Hubel DH, Wiesel TN: Sequence regularity of orientation columns in the monkey striate cortex. J Comp Neurol 1974;158:267.)

Other Cortical Areas Concerned With Vision

As mentioned above, the primary visual cortex (V1) projects to many other parts of the occipital lobes and other parts of the brain. These are often identi-fied by number (V2, V3, etc) or by letters (LO, MT, etc). The distribution of some of these in the human brain is shown in Figure 8–25, and their putative functions are listed in Table 8–1. Studies of these areas have been carried out in monkeys trained to do various tasks and then fitted with implanted microelectrodes. In addition, the availability of PET and fMRI scanning (see Chapter 16 and Appendix) has made it possible to conduct sophisticated experiments on visual cognition and other cortical visual functions in normal, conscious humans. The visual projections from V1 can be divided roughly into a **dorsal** or **parietal pathway,** concerned primarily with motion, and a **ventral** or **temporal pathway,** concerned with shape and recognition of forms and faces. In addition, connections to the sensory areas are important. For example visual responses in the occipital cortex to an object are better if the object is felt at the same time. There are many other relevant connections to other systems.

Area V8 appears to be uniquely concerned with color vision in humans.

It is apparent from the preceding paragraphs that there is parallel processing of visual information along multiple paths. In some as yet unknown way, all the information is eventually pulled together into what we experience as a conscious visual image.

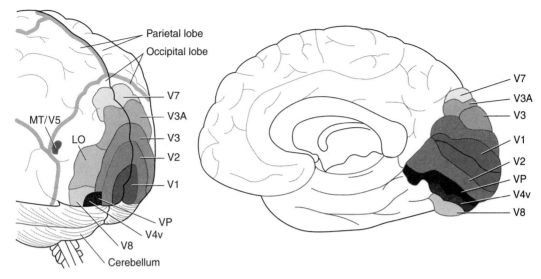

Figure 8–25. Some of the main areas to which the primary visual cortex (V1) projects in the human brain. Lateral and medial views. See also Table 8–1. (Modified from Logothetis N: Vision: a window on consciousness. Sci Am [Nov] 1999;281:99.)

COLOR VISION

Characteristics of Color

Colors have three attributes: **hue, intensity,** and **saturation** (degree of freedom from dilution with white). For any color there is a **complementary color** that, when properly mixed with it, produces a sensation of white. Black is the sensation produced by the absence of light, but it is probably a positive sensation, because

Table 8–1. Functions of visual projection areas in the human brain.[1]

V1	Primary visual cortex; receives input from lateral geniculate nucleus, begins processing in terms of orientation, edges, etc
V2, V3, VP	Continued processing, larger visual fields
V3A	Motion
V4v	Unknown
MT/V5	Motion; put to control of movement
LO	Recognition of large objects
V7	Unknown
V8	Color vision

[1]Modified from Logothetis N: Vision: a window on consciousness. Sci Am (Nov) 1999;281:99.

the blind eye does not "see black"; it "sees nothing." Such phenomena as successive and simultaneous contrasts, optical tricks that produce a sensation of color in the absence of color, negative and positive after-images, and various psychologic aspects of color vision are also pertinent. Detailed discussion of these phenomena is beyond the scope of this book.

Another observation of basic importance is the demonstration that the sensation of white, any spectral color, and even the extraspectral color, purple, can be produced by mixing various proportions of red light (wavelength 723–647 nm), green light (575–492 nm), and blue light (492–450 nm). Red, green, and blue are therefore called the **primary colors.** A third important point is that the color perceived depends in part on the color of other objects in the visual field. Thus, for example, a red object is seen as red if the field is illuminated with green or blue light but as pale pink or white if the field is illuminated with red light.

Retinal Mechanisms

The **Young-Helmholtz theory** of color vision in humans postulates the existence of three kinds of cones, each containing a different photopigment and maximally sensitive to one of the three primary colors, with the sensation of any given color being determined by the relative frequency of the impulses from each of these cone systems. The correctness of this theory has been demonstrated by the identification and chemical characterization of each of the three pigments. One pigment (the blue-sensitive or short-wave pigment) absorbs light maximally in the blue-violet portion of the spectrum (Figure 8–26). Another (the green-sensitive or middle-wave pigment) absorbs maximally in the green portion. The third (the red-sensitive or long-wave pigment) absorbs maximally in the yellow portion. Blue, green, and red are the primary colors, but the cones with their maximal sensitivity in the yellow portion of the spectrum are sensitive enough in the red portion to respond to red light at a lower threshold than green. This is all the Young-Helmholtz theory requires.

The gene for human rhodopsin is on chromosome 3, and the gene for the blue-sensitive S cone pigment is on chromosome 7. The other two cone pigments are encoded by genes arranged in tandem on the q arm of the X chromosome. The green-sensitive M and red-sensitive L pigments are very similar in structure; their opsins show 96% homology of amino acid sequences, whereas each of these pigments has only about 43% homology with the opsin of blue-sensitive pigment, and all three have about 41% homology with rhodopsin. Many mammals are **dichromats;** ie, they have only two cone pigments, a short-wave and a long-wave pigment. Old World monkeys, apes, and humans are trichromats, with separate middle- and long-wave pigments—in all probability because there was duplication of the ancestral long-wave gene followed by divergence.

It now appears that, in addition, there is variation in the human population in the red, long-wave pigment. It has been known for some time that responses to the **Rayleigh match,** the amounts of red and green light that a subject mixes to match a monochromatic orange, are bimodal. This correlates with new evidence that

62% of otherwise color-normal individuals have serine at site 180 of their long-wave cone opsin, whereas 38% have alanine. The absorption curve of the subjects with serine at position 180 peaks at 556.7 nm, and they are more sensitive to red light, whereas the absorption curve of the subjects with alanine at position 180 peaks at 552.4 nm.

Neural Mechanisms

Color is mediated by ganglion cells that subtract or add input from one type of cone to input from another type. Processing in the ganglion cells and the lateral geniculate nucleus produces impulses that pass along three types of neural pathways that project to V1: a red-green pathway that signals differences between L- and M-cone responses; a blue-yellow pathway that signals differences between S-cone and the sum of L- and M-cone responses; and a luminance pathway that signals the sum of L- and M-cone responses. These pathways project to the blobs and the deep portion of layer 4C of V1. From the blobs and layer 4, color information is projected to V8. However, it is not known how V8 converts color input into the sensation of color.

Color Blindness

There are numerous tests for detecting color blindness. The most commonly used routine tests are the Ishihara charts. These charts and similar polychromatic plates are plates on which are printed figures made up of colored spots on a background of similarly shaped colored spots. The figures are intentionally made up of colors that are liable to look the same as the background to an individual who is color-blind.

Some color-blind individuals are unable to distinguish certain colors, whereas others have only a color weakness. The suffix "-anomaly" denotes color weakness and the suffix "-anopia" color blindness. The prefixes "prot-," "deuter-," and "tri-" refer to defects of the red, green, and blue cone systems, respectively. Individuals with normal color vision and those with protanomaly, deuteranomaly, and tritanomaly are called **trichromats;** they have all three cone systems, but one may be weak. **Dichromats** are individuals with only two cone systems; they may have protanopia, deuteranopia, or tritanopia. **Monochromats** have only one cone system. Dichromats can match their color spectrum by mixing only two primary colors, and monochromats match theirs by varying the intensity of only one.

Color blindness is most frequently inherited. However, it also occurs in individuals with lesions of V8 (see

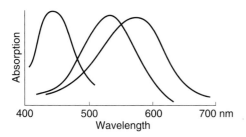

Figure 8–26. Absorption spectra of the three cone pigments in the human retina. The S pigment that peaks at 440 nm senses blue, and the M pigment that peaks at 535 nm senses green. The remaining L pigment peaks in the yellow portion of the spectrum, at 565 nm, but its spectrum extends far enough into the long wavelengths to sense red. (Reproduced, with permission, from Michael CR: Color vision. N Engl J Med 1973;288:724.)

above) who have **achromatopsia.** In addition, transient blue-green color weakness occurs as a side effect in individuals taking sildenafil (Viagra) for the treatment of erectile dysfunction because this drug inhibits the retinal form of phosphodiesterase (see Chapter 23).

Inheritance of Color Blindness

Abnormal color vision is present as an inherited abnormality in Caucasian populations in about 8% of the males and 0.4% of the females. Tritanomaly and tritanopia are rare and show no sexual selectivity. However, about 2% of the color-blind males are dichromats who have protanopia or deuteranopia, and about 6% are anomalous trichromats in whom the red-sensitive or the green-sensitive pigment is shifted in its spectral sensitivity. These abnormalities are inherited as recessive and X-linked characteristics; ie, they are due to an abnormal gene on the X chromosome. Since all of the male's cells except germ cells contain one X and one Y chromosome in addition to the 44 somatic chromosomes (see Chapter 23), color blindness is present in males if the X chromosome has the abnormal gene. On the other hand, the normal female's cells have two X chromosomes, one from each parent, and since these abnormalities are recessive, females show a defect only when both X chromosomes contain the abnormal gene. However, female children of a man with X-linked color blindness are carriers of the color blindness and pass the defect on to half of their sons. Therefore, X-linked color blindness skips generations and appears in males of every second generation. Hemophilia, Duchenne's muscular dystrophy, and many other inherited disorders are caused by mutant genes on the X chromosome.

The common occurrence of deuteranomaly and protanomaly is probably due to the arrangement of the genes for the green-sensitive and red-sensitive cone pigments. They are located near each other in a head-to-tail tandem array on the q arm of the X chromosome and are prone to unequal homologous recombination (unequal crossing over) during development of the germ cells. This produces hybrid pigments with shifted spectral sensitivities, and a number of such hybrids have been characterized.

OTHER ASPECTS OF VISUAL FUNCTION

Dark Adaptation

The truly remarkable range of luminance to which the human eye responds has been mentioned above, and is summarized in Figure 8–27. If a person spends a considerable length of time in brightly lighted surroundings and then moves to a dimly lighted environment, the retinas slowly become more sensitive to light as the

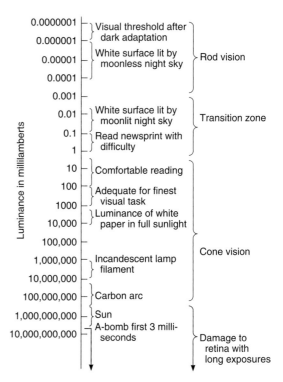

Figure 8–27. Range of luminance to which the human eye responds, with the receptive mechanisms involved. (Reproduced, with permission, by courtesy of Campbell FW, from Bell GH, Emslie-Smith D, Paterson CR: *Textbook of Physiology and Biochemistry,* 9th ed. Churchill Livingstone, 1976.)

individual becomes "accustomed to the dark." This decline in visual threshold is known as **dark adaptation.** It is nearly maximal in about 20 minutes, although there is some further decline over longer periods. On the other hand, when one passes suddenly from a dim to a brightly lighted environment, the light seems intensely and even uncomfortably bright until the eyes adapt to the increased illumination and the visual threshold rises. This adaptation occurs over a period of about 5 minutes and is called **light adaptation,** although, strictly speaking, it is merely the disappearance of dark adaptation.

There are actually two components to the dark adaptation response (Figure 8–28). The first drop in visual threshold, rapid but small in magnitude, is known to be due to dark adaptation of the cones because when only the foveal, rod-free portion of the retina is tested, the decline proceeds no further. In the peripheral portions of the retina, a further drop occurs as a result of adaptation of the rods. The total change in threshold

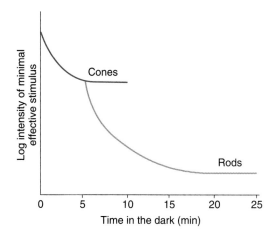

Figure 8–28. Dark adaptation. The curve shows the change in the intensity of a stimulus necessary to just excite the retina in dim light as a function of the time the observer has been in the dark.

between the light-adapted and the fully dark-adapted eye is very great.

Radiologists, aircraft pilots, and others who need maximal visual sensitivity in dim light can avoid having to wait 20 minutes in the dark to become dark-adapted if they wear red goggles when in bright light. Light wavelengths in the red end of the spectrum stimulate the rods to only a slight degree while permitting the cones to function reasonably well. Therefore, a person wearing red glasses can see in bright light during the time it takes for the rods to become dark-adapted.

The time required for dark adaptation is determined in part by the time required to build up the rhodopsin stores. In bright light, much of the pigment is continuously being broken down, and some time is required in dim light for accumulation of the amounts necessary for optimal rod function. However, dark adaptation also occurs in the cones, and additional factors are undoubtedly involved.

Effect of Vitamin Deficiencies on the Eye

In view of the importance of vitamin A in the synthesis of retinene$_1$, it is not surprising that avitaminosis A produces visual abnormalities. Among these, one of the earliest to appear is night blindness **(nyctalopia).** This fact first called attention to the role of vitamin A in rod function, but concomitant cone degeneration also occurs as vitamin A deficiency develops. Prolonged deficiency is associated with anatomic changes in the rods and cones followed by degeneration of the neural layers of the retina. Treatment with vitamin A

can restore retinal function if given before the receptors are destroyed.

Other vitamins, especially those of the B complex, are necessary for the normal functioning of the retina and other neural tissues.

Visual Acuity

Saccadic eye movement (see below) is one of the many factors that determine **visual acuity.** This parameter of vision should not be confused with **visual threshold.** Visual threshold is the minimal amount of light that elicits a sensation of light; visual acuity is the degree to which the details and contours of objects are perceived. Although there is evidence that other measures are more accurate, visual acuity is usually defined in terms of the **minimum separable**—ie, the shortest distance by which two lines can be separated and still be perceived as two lines. Clinically, visual acuity is often determined by the use of the familiar Snellen letter charts viewed at a distance of 20 ft (6 m). The individual being tested reads aloud the smallest line distinguishable. The results are expressed as a fraction. The numerator of the fraction is 20, the distance at which the subject reads the chart. The denominator is the greatest distance from the chart at which a normal individual can read the smallest line the subject can read. Normal visual acuity is 20/20; a subject with 20/15 visual acuity has better than normal vision (not farsightedness); and one with 20/100 visual acuity has subnormal vision. The Snellen charts are designed so that the height of the letters in the smallest line a normal individual can read at 20 ft subtends a visual angle of 5 minutes. Each of the lines in the letters are separated by 1 minute of arc. Thus, the minimum separable in a normal individual corresponds to a visual angle of about 1 minute.

Visual acuity is a complex phenomenon and is influenced by a large variety of factors. These include optical factors such as the state of the image-forming mechanisms of the eye, retinal factors such as the state of the cones, and stimulus factors including illumination, brightness of the stimulus, contrast between the stimulus and the background, and length of time the subject is exposed to the stimulus.

Critical Fusion Frequency

The time-resolving ability of the eye is determined by measuring the **critical fusion frequency (CFF),** the rate at which stimuli can be presented and still be perceived as separate stimuli. Stimuli presented at a higher rate than the CFF are perceived as continuous stimuli. Motion pictures move because the frames are presented at a rate above the CFF, and movies begin to flicker when the projector slows down.

Visual Fields & Binocular Vision

The visual field of each eye is the portion of the external world visible out of that eye. Theoretically, it should be circular, but actually it is cut off medially by the nose and superiorly by the roof of the orbit (Figure 8–29). Mapping the visual fields is important in neurologic diagnosis. The peripheral portions of the visual fields are mapped with an instrument called a **perimeter,** and the process is referred to as **perimetry.** One eye is covered while the other is fixed on a central point. A small target is moved toward this central point along selected meridians, and, along each, the location where the target first becomes visible is plotted in degrees of arc away from the central point (Figure 8–29). The central visual fields are mapped with a **tangent screen,** a black felt screen across which a white target is moved. By noting the locations where the target disappears and reappears, the blind spot and any **objective scotomas** (blind spots due to disease) can be outlined.

The central parts of the visual fields of the two eyes coincide; therefore, anything in this portion of the field is viewed with **binocular vision.** The impulses set up in the two retinas by light rays from an object are fused at the cortical level into a single image (**fusion**). The points on the retina on which the image of an object must fall if it is to be seen binocularly as a single object are called **corresponding points.** If one eye is gently pushed out of the line while staring fixedly

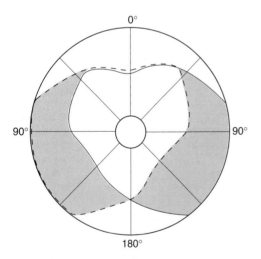

Figure 8–29. Monocular and binocular visual fields. The dashed line encloses the visual field of the left eye; the solid line, that of the right eye. The common area (heart-shaped clear zone in the center) is viewed with binocular vision. The colored areas are viewed with monocular vision.

at an object in the center of the visual field, double vision (**diplopia**) results; the image on the retina of the eye that is displaced no longer falls on the corresponding point.

Binocular vision has an important role in the perception of depth. However, depth perception also has numerous monocular components, such as the relative sizes of objects, the degree one looks down at them, their shadows, and, for moving objects, their movement relative to one another (movement parallax).

Effect of Lesions in the Optic Pathways

The anatomy of the pathways from the eyes to the brain is shown in Figure 8–4. Lesions along these pathways can be localized with a high degree of accuracy by the effects they produce in the visual fields.

The fibers from the nasal half of each retina decussate in the optic chiasm, so that the fibers in the optic tracts are those from the temporal half of one retina and the nasal half of the other. In other words, each optic tract subserves half of the field of vision. Therefore, a lesion that interrupts one optic nerve causes blindness in that eye, but a lesion in one optic tract causes blindness in half of the visual field (Figure 8–4). This defect is classified as a **homonymous** (same side of both visual fields) **hemianopia** (half-blindness). Lesions affecting the optic chiasm, such as pituitary tumors expanding out of the sella turcica, cause destruction of the fibers from both nasal hemiretinas and produce a **heteronymous** (opposite sides of the visual fields) **hemianopia.** Since the fibers from the maculas are located posteriorly in the optic chiasm, hemianopic scotomas develop before there is complete loss of vision in the two hemiretinas. Selective visual field defects are further classified as bitemporal, binasal, and right or left.

The optic nerve fibers from the upper retinal quadrants subserving vision in the lower half of the visual field terminate in the medial half of the lateral geniculate body, while the fibers from the lower retinal quadrants terminate in the lateral half. The geniculocalcarine fibers from the medial half of the lateral geniculate terminate on the superior lip of the calcarine fissure, while those from the lateral half terminate on the inferior lip. Furthermore, the fibers from the lateral geniculate body that subserve macular vision separate from those that subserve peripheral vision and end more posteriorly on the lips of the calcarine fissure (Figure 8–5). Because of this anatomic arrangement, occipital lobe lesions may produce discrete quadrantic visual field defects (upper and lower quadrants of each half visual field). **Macular sparing,** ie, loss of peripheral vision with intact macular vision, is also common with occipital lesions (Figure 8–4), because the macular representation is separate from that of the peripheral fields and very large relative to that of the peripheral fields. Therefore, occipital lesions must extend considerable distances to destroy macular as well as peripheral vision. Bilateral destruction of the occipital cortex in humans causes subjective blindness. However, there is appreciable **blindsight,** ie, residual responses to visual stimuli even though they do not reach consciousness. For example, when these individuals are asked to guess where a stimulus is located during perimetry, they respond with much more accuracy than can be explained by chance. There is considerable discrimination of movement, flicker, orientation, and even color. Similar biasing of responses can be produced by stimuli in the blind areas in patients with hemianopia due to lesions in the visual cortex.

The fibers to the pretectal region that subserve the reflex pupillary constriction produced by shining a light into the eye leave the optic tracts near the geniculate bodies. Therefore, blindness with preservation of the pupillary light reflex is usually due to bilateral lesions behind the optic tract.

EYE MOVEMENTS

The directions in which each of the eye muscles move the eye are summarized in Figure 8–30. Since the oblique muscles pull medially, their actions vary with the position of the eye. When the eye is turned nasally,

the obliques elevate and depress it, whereas the superior and inferior recti rotate it; when the eye is turned temporally, the superior and inferior recti elevate and depress it and the obliques rotate it.

Since much of the visual field is binocular, it is clear that a very high order of coordination of the movements of the two eyes is necessary if visual images are to fall at all times on corresponding points in the two retinas and diplopia is to be avoided.

There are four types of eye movements, each controlled by a different neural system but sharing the same final common path, the motor neurons that supply the external ocular muscles (Figure 8–31). **Saccades,** sudden jerky movements, occur as the gaze shifts from one object to another. They bring new objects of interest onto the fovea and reduce adaptation in the visual pathway that would occur if gaze were fixed on a single object for long periods. **Smooth pursuit movements** are tracking movements of the eyes as they follow moving objects. **Vestibular movements,** adjustments that occur in response to stimuli initiated in the semicircular canals, maintain visual fixation as the head moves. **Convergence movements** bring the visual axes toward each other as attention is focused on objects near the observer. The similarity to a man-made track-

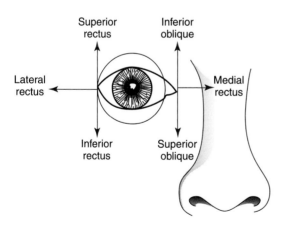

Figure 8–30. Extraocular muscles subserving the six cardinal positions of gaze. The eye is adducted by the medial rectus and abducted by the lateral rectus. The adducted eye is elevated by the inferior oblique and depressed by the superior oblique; the abducted eye is elevated by the superior rectus and depressed by the inferior rectus. (Reproduced, with permission, from Greenberg DA, Aminoff MJ, Simon RP: *Clinical Neurology,* 5th ed. McGraw-Hill, 2002.)

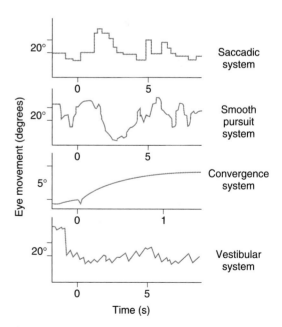

Figure 8–31. Types of eye movements. (Modified and reproduced, with permission, from Robinson DA: Eye movement control in primates. Science 1968;161:1219. Copyright © 1968 by the American Association for the Advancement of Science.)

ing system on an unstable platform such as a ship is apparent: saccadic movements seek out visual targets; pursuit movements follow them as they move about; and vestibular movements stabilize the tracking device as the platform on which the device is mounted (ie, the head) moves about. In primates, these eye movements depend on an intact visual cortex. Saccades are programmed in the frontal cortex and the superior colliculi and pursuit movements in the cerebellum.

Superior Colliculi

The superior colliculi, which regulate saccades, are innervated by M fibers from the retina. They also receive extensive innervation from the cerebral cortex. In each superior colliculus, there is a map of visual space plus a map of the body surface and a map for sound in space. There is a motor map that projects to the regions of the brain stem that control eye movements. There are also projections via the tectopontine tract to the cerebellum and via the tectospinal tract to areas concerned with reflex movements of the head and neck. The superior colliculi are constantly active positioning the eyes, and they have one of the highest rates of blood flow and metabolism of any region in the brain (see Chapter 32).

Strabismus

Abnormalities of the coordinating mechanisms can be due to a variety of causes. When visual images no longer fall on corresponding retinal points, **strabismus** (squint) is said to be present. Successful treatment of some types of strabismus is possible by careful surgical shortening of some of the eye muscles, by eye muscle training exercises, and by the use of glasses with prisms that bend the light rays sufficiently to compensate for the abnormal position of the eyeball. However, subtle defects in depth perception persist, and it has been suggested that congenital abnormalities of the visual tracking mechanisms may cause both the strabismus and the defective depth perception.

When visual images chronically fall on noncorresponding points in the two retinas in children under age 6, one is eventually suppressed (**suppression scotoma**) and diplopia disappears. This suppression is a cortical phenomenon, and it usually does not develop in adults. It is important to institute treatment before age 6 in children with one visual image suppressed, because if the suppression persists, there is permanent loss of visual acuity in the eye generating the suppressed image. A similar suppression with subsequent permanent loss of visual acuity can occur in children in whom vision in one eye is blurred or distorted owing to a refractive error. The loss of vision in these cases is called **amblyopia ex anopsia,** a term that refers to uncorrectable loss of visual acuity which is not directly due to organic disease of the eye. In infant monkeys, covering one eye with a patch for 3 months causes a loss of ocular dominance columns; input from the remaining eye spreads to take over all the cortical cells, and the patched eye becomes functionally blind. Comparable changes presumably occur in children with strabismus.

Hearing & Equilibrium

9

INTRODUCTION

Receptors for two sensory modalities, hearing and equilibrium, are housed in the ear. The external ear, the middle ear, and the cochlea of the inner ear are concerned with hearing. The semicircular canals, the utricle, and the saccule of the inner ear are concerned with equilibrium. Receptors in the semicircular canals detect rotational acceleration, receptors in the utricle detect linear acceleration in the horizontal direction, and receptors in the saccule detect linear acceleration in the vertical direction. The receptors for hearing and equilibrium are hair cells, and there are six groups of hair cells in each inner ear: one in each of the three semicircular canals, one in the utricle, one in the saccule, and one in the cochlea.

ANATOMIC CONSIDERATIONS

External & Middle Ear

The external ear funnels sound waves to the **external auditory meatus.** In some animals, the ears can be moved like radar antennas to seek out sound. From the meatus, the **external auditory canal** passes inward to the **tympanic membrane** (eardrum) (Figure 9–1).

The middle ear is an air-filled cavity in the temporal bone that opens via the **auditory (eustachian) tube** into the nasopharynx and through the nasopharynx to the exterior. The tube is usually closed, but during swallowing, chewing, and yawning it opens, keeping the air pressure on the two sides of the eardrum equalized. The three **auditory ossicles,** the

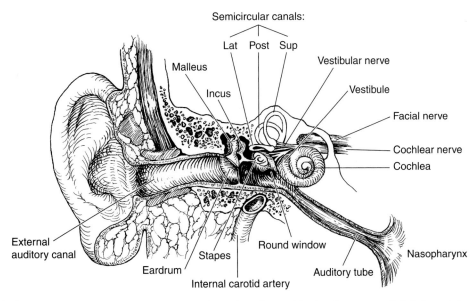

Figure 9–1. The human ear. To make the relationships clear, the cochlea has been turned slightly and the middle ear muscles have been omitted. Sup, superior; Post, posterior; Lat, lateral.

malleus, incus, and **stapes,** are located in the middle ear. The **manubrium** (handle of the malleus) is attached to the back of the tympanic membrane. Its head is attached to the wall of the middle ear, and its short process is attached to the incus, which in turn articulates with the head of the stapes. The stapes is named for its resemblance to a stirrup. Its **foot plate** is attached by an annular ligament to the walls of the **oval window** (Figure 9–2). Two small skeletal muscles, the **tensor tympani** and the **stapedius,** are also located in the middle ear. Contraction of the former pulls the manubrium of the malleus medially and decreases the vibrations of the tympanic membrane; contraction of the latter pulls the footplate of the stapes out of the oval window.

Inner Ear

The inner ear **(labyrinth)** is made up of two parts, one within the other. The **bony labyrinth** is a series of channels in the petrous portion of the temporal bone. Inside these channels, surrounded by a fluid called **perilymph,** is the **membranous labyrinth** (Figure 9–3). This membranous structure more or less duplicates the shape of the bony channels. It is filled with a fluid called **endolymph,** and there is no communication between the spaces filled with endolymph and those filled with perilymph.

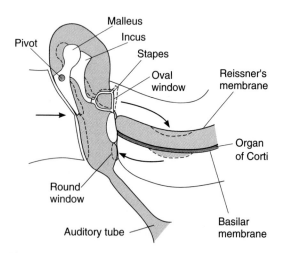

Figure 9–2. Schematic representation of the auditory ossicles and the way their movement translates movements of the tympanic membrane into a wave in the fluid of the inner ear. The wave is dissipated at the round window. The movements of the ossicles, the membranous labyrinth, and the round window are indicated by dashed lines.

Cochlea

The cochlear portion of the labyrinth is a coiled tube which in humans is 35 mm long and makes 23/4 turns. Throughout its length, the basilar membrane and Reissner's membrane divide it into three chambers **(scalae)** (Figure 9–4). The upper **scala vestibuli** and the lower **scala tympani** contain perilymph and communicate with each other at the apex of the cochlea through a small opening called the **helicotrema.** At the base of the cochlea, the scala vestibuli ends at the oval window, which is closed by the footplate of the stapes. The scala tympani ends at the **round window,** a foramen on the medial wall of the middle ear that is closed by the flexible **secondary tympanic membrane.** The **scala media,** the middle cochlear chamber, is continuous with the membranous labyrinth and does not communicate with the other two scalae. It contains endolymph (Figures 9–3 and 9–4).

Organ of Corti

Located on the basilar membrane is the **organ of Corti,** the structure that contains the hair cells which are the auditory receptors. This organ extends from the apex to the base of the cochlea and consequently has a spiral shape. The processes of the hair cells pierce the tough, membrane-like **reticular lamina** that is supported by the **rods of Corti** (Figure 9–4). The hair cells are arranged in four rows: three rows of **outer hair cells** lateral to the tunnel formed by the rods of Corti, and one row of **inner hair cells** medial to the tunnel. There are 20,000 outer hair cells and 3500 inner hair cells in each human cochlea. Covering the rows of hair cells is a thin, viscous, but elastic **tectorial membrane** in which the tips of the hairs of the outer but not the inner hair cells are embedded. The cell bodies of the afferent neurons that arborize around the bases of the hair cells are located in the **spiral ganglion** within the **modiolus,** the bony core around which the cochlea is wound. Ninety to 95% of these afferent neurons innervate the inner hair cells; only 5–10% innervate the more numerous outer hair cells, and each neuron innervates several of these outer cells. By contrast, most of the efferent fibers in the auditory nerve (see below) terminate on the outer hair cells rather than on the inner hair cells. The axons of the afferent neurons that innervate the hair cells form the auditory (cochlear) division of the vestibulocochlear acoustic nerve and terminate in the **dorsal** and **ventral cochlear nuclei** of the medulla oblongata. The total number of afferent and efferent fibers in each auditory nerve is approximately 28,000.

In the cochlea, there are tight junctions between the hair cells and the adjacent phalangeal cells; these prevent endolymph from reaching the bases of the

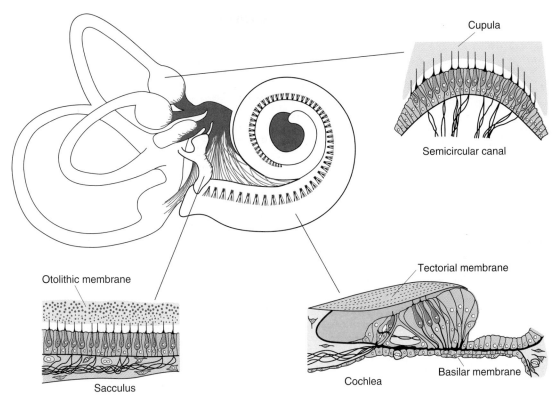

Figure 9–3. Human membranous labyrinth, with enlargements of the structures in which hair cells are embedded. (Reproduced, with permission, from Hudspeth AJ: How the ear's works work. Nature 1989;341:397. Copyright © 1989 by Macmillan Magazines Ltd.)

cells. However, the basilar membrane is relatively permeable to perilymph in the scala tympani, and consequently, the tunnel of the organ of Corti and the bases of the hair cells are bathed in perilymph. Because of similar tight junctions, the arrangement is similar for the hair cells in other parts of the inner ear; ie, the processes of the hair cells are bathed in endolymph, whereas their bases are bathed in perilymph.

Central Auditory Pathways

From the cochlear nuclei, auditory impulses pass via a variety of pathways to the **inferior colliculi,** the centers for auditory reflexes, and via the **medial geniculate body** in the thalamus to the **auditory cortex.** Others enter the reticular formation (Figure 9–5). Information from both ears converges on each superior olive, and at all higher levels most of the neurons respond to inputs from both sides. The primary auditory cortex, Brodmann's area 41, is in the superior portion of the temporal lobe. In humans, it is located

in the sylvian fissure (see Figure 7–4) and is not normally visible on the surface of the brain. In the primary auditory cortex, most neurons respond to inputs from both ears, but there are also strips of cells that are stimulated by input from the contralateral ear and inhibited by input from the ipsilateral ear. There are several additional auditory receiving areas, just as there are several receiving areas for cutaneous sensation (see Chapter 7). The auditory association areas adjacent to the primary auditory receiving area are widespread (see below). The **olivocochlear bundle** is a prominent bundle of efferent fibers in each auditory nerve that arises from both the ipsilateral and the contralateral superior olivary complex and ends primarily around the bases of the outer hair cells of the organ of Corti.

Semicircular Canals

On each side of the head, the semicircular canals are perpendicular to each other, so that they are oriented in the three planes of space. Inside the bony canals, the

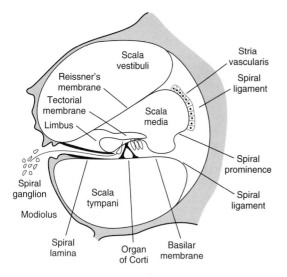

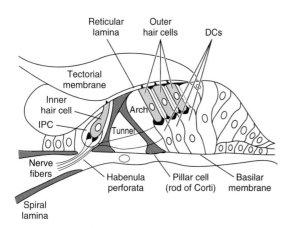

Figure 9–4. Top: Cross section of the cochlea, showing the organ of Corti and the three scalas of the cochlea. **Bottom:** Structure of the organ of Corti, as it appears in the basal turn of the cochlea. DC, outer phalangeal cells (Deiters' cells) supporting outer hair cells; IPC, inner phalangeal cell supporting inner hair cell. (Reproduced, with permission, from Pickels JO: *An Introduction to the Physiology of Hearing,* 2nd ed. Academic Press, 1988.)

membranous canals are suspended in perilymph. A receptor structure, the **crista ampullaris,** is located in the expanded end **(ampulla)** of each of the membranous canals. Each crista consists of hair cells and sustentacular cells surmounted by a gelatinous partition **(cupula)** that closes off the ampulla (Figure 9–3). The processes of the hair cells are embedded in the cupula, and the bases of the hair cells are in close contact with the affer-

ent fibers of the vestibular division of the vestibulocochlear nerve.

Utricle & Saccule

Within each membranous labyrinth, on the floor of the utricle, there is an **otolithic organ (macula).** Another macula is located on the wall of the saccule in a semivertical position. The maculas contain sustentacular cells and hair cells, surmounted by an otolithic membrane in which are embedded crystals of calcium carbonate, the **otoliths** (Figure 9–3). The otoliths, which are also called **otoconia** or **ear dust,** range from 3 to 19 μm in length in humans and are more dense than the endolymph. The processes of the hair cells are embedded in the membrane. The nerve fibers from the hair cells join those from the cristae in the vestibular division of the vestibulocochlear nerve.

Neural Pathways

The cell bodies of the 19,000 neurons supplying the cristae and maculas on each side are located in the vestibular ganglion. Each vestibular nerve terminates in the ipsilateral four-part vestibular nucleus and in the flocculonodular lobe of the cerebellum (Figure 9–6). Fibers from the semicircular canals end primarily in the superior and medial divisions of the vestibular nucleus and project mainly to nuclei controlling eye movement. Fibers from the utricle and saccule end predominantly in the lateral division (Deiters' nucleus), which projects to the spinal cord. They also end in the descending nucleus, which projects to the cerebellum and the reticular formation. The vestibular nuclei also project to the thalamus and from there to two parts of the primary somatosensory cortex.

HAIR CELLS

Structure

The hair cells in the inner ear have a common structure (Figure 9–7). Each is embedded in an epithelium made up of supporting or sustentacular cells, with the basal end in close contact with afferent neurons. Projecting from the apical end are 30–150 rod-shaped processes, or hairs. Except in the cochlea, one of these, the **kinocilium,** is a true but nonmotile cilium with nine pairs of microtubules around its circumference and a central pair of microtubules (see Chapter 1). It is one of the largest processes and has a clubbed end. The kinocilium is lost in the hair cells of the cochlea in adult mammals. However, the other processes, which are called **stereocilia,** are present in all hair cells. They have cores composed of parallel filaments of actin. The actin is coated with various isoforms of myosin. Within

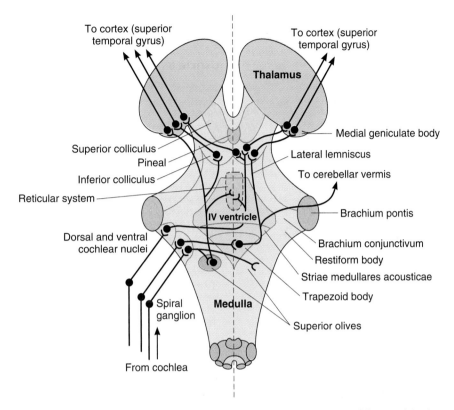

Figure 9–5. Simplified diagram of main auditory pathways superimposed on a dorsal view of the brain stem. Cerebellum and cerebral cortex removed.

the clump of processes on each cell, there is an orderly structure. Along an axis toward the kinocilium, the stereocilia increase progressively in height; along the perpendicular axis, all the stereocilia are the same height.

Electrical Responses

The membrane potential of the hair cells is about −60 mV. When the stereocilia are pushed toward the kinocilium, the membrane potential is decreased to about −50 mV. When the bundle of processes is pushed in the opposite direction, the cell is hyperpolarized. Displacing the processes in a direction perpendicular to this axis provides no change in membrane potential, and displacing the processes in directions that are intermediate between these two directions produces depolarization or hyperpolarization that is proportionate to the degree to which the direction is toward or away from the kinocilium. Thus, the hair processes provide a mechanism for generating changes in membrane potential that are proportionate to the direction and distance the hair moves.

Genesis of Action Potentials in Afferent Nerve Fibers

As noted above, the processes of the hair cells project into the endolymph whereas the bases are bathed in perilymph. This arrangement is necessary for the normal production of generator potentials. The perilymph is formed mainly from plasma. On the other hand, endolymph is formed in the scala media by the stria vascularis and has a high concentration of K^+ and a low concentration of Na^+ (Figure 9–8). Cells in the stria vascularis have a high concentration of Na^+-K^+ ATPase. In addition, it appears that there is a unique electrogenic K^+ pump in the stria vascularis, which accounts for the fact that the scala media is electrically positive by 85 mV relative to the scala vestibuli and scala tympani.

Very fine processes called **tip links** (Figure 9–9) tie the tip of each stereocilium to the side of its higher neighbor, and at the junction there appear to be mechanically sensitive cation channels in the higher process. When the shorter stereocilia are pushed toward the higher, the open time of these channels increases.

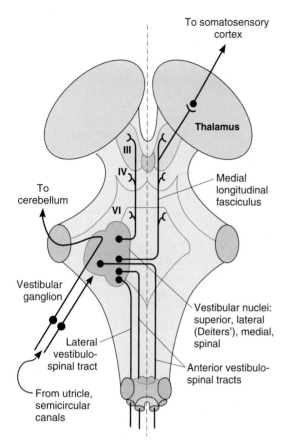

Figure 9–6. Principal vestibular pathways superimposed on a dorsal view of the brain stem. Cerebellum and cerebral cortex removed.

K⁺—the most abundant cation in endolymph—and Ca^{2+} enter via the channel and produce depolarization. There is still considerable uncertainty about subsequent events. However, one hypothesis is that a molecular motor in the higher neighbor next moves the channel toward the base, releasing tension in the tip link (Figure 9–9). This causes the channel to close and permits restoration of the resting state. The motor apparently is myosin-based (see Chapter 1).

Depolarization of hair cells causes them to release a neurotransmitter, probably glutamine, which initiates depolarization of neighboring afferent neurons.

The K⁺ that enters hair cells via the mechanically sensitive cation channels is recycled (Figure 9–8). It enters sustentacular cells and then passes on to other sustentacular cells by way of tight junctions. In the cochlea, it eventually reaches the stria vascularis and is secreted back into the endolymph, completing the cycle.

HEARING

Sound Waves

Sound is the sensation produced when longitudinal vibrations of the molecules in the external environment, ie, alternate phases of condensation and rarefaction of the molecules, strike the tympanic membrane. A plot of these movements as changes in pressure on the tympanic membrane per unit of time is a series of waves (Figure 9–10), and such movements in the environment are generally called sound waves. The waves travel through air at a speed of approximately 344 m/s (770 miles/h) at 20 °C at sea level. The speed of sound increases with temperature and with altitude. Other media in which humans occasionally find themselves also conduct sound waves but at different speeds. For example, the speed of sound is 1450 m/s at 20 °C in fresh water and is even greater in salt water. It is said that the whistle of the blue whale is as loud as 188 decibels (see below) and is audible for 500 miles.

Generally speaking, the **loudness** of a sound is correlated with the **amplitude** of a sound wave and its **pitch** with the **frequency** (number of waves per unit of time). The greater the amplitude, the louder the sound; and the greater the frequency, the higher the pitch. However, pitch is determined by other poorly understood factors in addition to frequency, and frequency affects loudness, since the auditory threshold is lower at some frequencies than others (see below). Sound waves that have repeating patterns, even though the individual waves are complex, are perceived as musical sounds; aperiodic nonrepeating vibrations cause a sensation of noise. Most musical sounds are made up of a wave with a primary frequency that determines the pitch of the sound plus a number of harmonic vibrations (**overtones**) that give the sound its characteristic **timbre** (quality). Variations in timbre permit us to identify the sounds of the various musical instruments even though they are playing notes of the same pitch.

The amplitude of a sound wave can be expressed in terms of the maximum pressure change at the eardrum, but a relative scale is more convenient. The **decibel scale** is such a scale. The intensity of a sound in **bels** is the logarithm of the ratio of the intensity of that sound and a standard sound. A decibel (dB) is 0.1 bel. Therefore,

$$\text{Number of dB} = 10 \log \frac{\text{intensity of sound}}{\text{intensity of standard sound}}$$

Sound intensity is proportionate to the square of sound pressure. Therefore,

$$\text{Number of dB} = 20 \log \frac{\text{pressure of sound}}{\text{pressure of standard sound}}$$

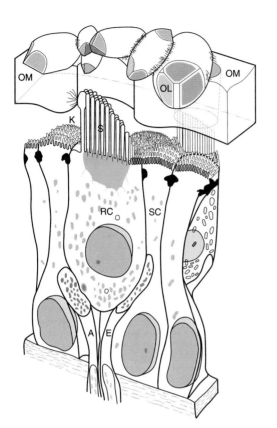

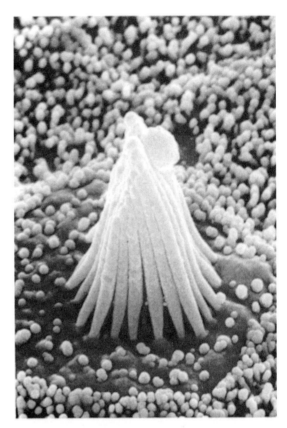

Figure 9–7. **Left:** Structure of a hair cell in the saccule of a frog, showing its relation to the otolithic membrane (OM). K, kinocilium; S, stereocilia; RC, hair cell with afferent (A) and efferent (E) nerve fibers; OL, otolith; SC, supporting cell. (Reproduced, with permission, from Hillman DE: Morphology of peripheral and central vestibular systems. In: Llinas R, Precht W [editors]: *Frog Neurobiology.* Springer, 1976.) **Right:** Scanning electron photomicrograph of processes on a hair cell in the saccule of a frog. The otolithic membrane has been removed. The small projections around the hair cell are microvilli on supporting cells. (Courtesy of AJ Hudspeth.)

The standard sound reference level adopted by the Acoustical Society of America corresponds to 0 decibels at a pressure level of $0.000204 \times$ dyne/cm^2, a value that is just at the auditory threshold for the average human. In Figure 9–11, the decibel levels of various common sounds are compared. It is important to remember that the decibel scale is a log scale. Therefore, a value of 0 decibels does not mean the absence of sound but a sound level of an intensity equal to that of the standard. Furthermore, the 0- to 140-decibel range from threshold pressure to a pressure that is potentially damaging to the organ of Corti actually represents a 10^7- (10 million)-fold variation in sound pressure. Put another way, atmospheric pressure at sea level is 15 lb/in^2 or 1 bar, and the range from the threshold of hearing to potential damage to the cochlea is 0.0002–2000 µbar.

The sound frequencies audible to humans range from about 20 to a maximum of 20,000 cycles per second (cps, Hz). In other animals, notably bats and dogs, much higher frequencies are audible. The threshold of the human ear varies with the pitch of the sound (Figure 9–12), the greatest sensitivity being in the 1000- to 4000-Hz range. The pitch of the average male voice in conversation is about 120 Hz and that of the average female voice about 250 Hz. The number of pitches that can be distinguished by an average individual is about 2000, but trained musicians can improve on this figure considerably. Pitch discrimination is best in the 1000- to 3000-Hz range and is poor at high and low pitches.

Masking

It is common knowledge that the presence of one sound decreases an individual's ability to hear other

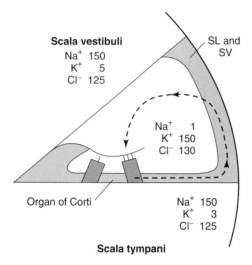

Figure 9-8. Ionic composition of perilymph in the scala vestibuli, endolymph in the scala media, and peri-lymph in the scala tympani. SL, spiral ligament. SV, stria vascularis. The dashed arrow indicates the path by which K⁺ recycles from the hair cells to the supporting cells to the spiral ligament and is then secreted back into the endolymph by cells in the stria vascularis.

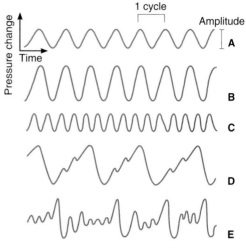

Figure 9-10. Characteristics of sound waves. **A** is the record of a pure tone. **B** has a greater amplitude and is louder than **A. C** has the same amplitude as **A** but a greater frequency, and its pitch is higher. **D** is a complex wave form that is regularly repeated. Such patterns are perceived as musical sounds, whereas waves like that shown in **E,** which have no regular pattern, are perceived as noise.

sounds. This phenomenon is known as **masking.** It is believed to be due to the relative or absolute refractoriness of previously stimulated auditory receptors and nerve fibers to other stimuli. The degree to which a given tone masks other tones is related to its pitch. The masking effect of the background noise in all but

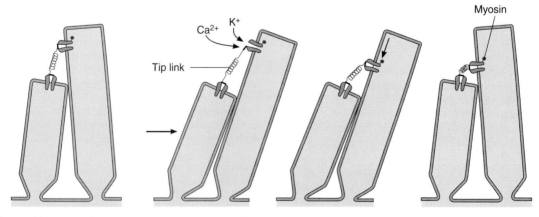

Figure 9-9. Schematic representation of the role of tip links in the responses of hair cells. When the hair cells are pushed toward the tallest stereocilium, the tip line is stretched and opens an ion channel in its taller neighbor. The channel next is moved down the taller stereocilium by a molecular motor, so the tension on the tip link is released. When the hairs return to the resting position, the motor moves back up the stereocilium. (Modified and reproduced, with permission, from Kandel ER, Schwartz JH, Jessel TM [editors]: *Principles of Neuroscience,* 4th ed. McGraw-Hill, 2000.)

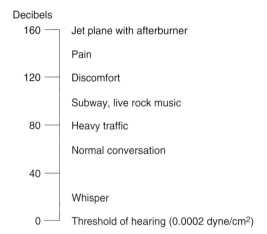

Figure 9–11. Decibel scale for common sounds.

the most carefully soundproofed environments raises the auditory threshold by a definite and measurable amount.

Sound Transmission

The ear converts sound waves in the external environment into action potentials in the auditory nerves. The waves are transformed by the eardrum and auditory ossicles into movements of the footplate of the stapes. These movements set up waves in the fluid of the inner ear. The action of the waves on the

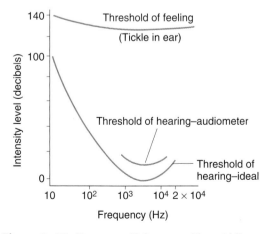

Figure 9–12. Human audibility curve. The middle curve is that obtained by audiometry under the usual conditions. The lower curve is that obtained under ideal conditions. At about 140 decibels (top curve), sounds are felt as well as heard.

organ of Corti generates action potentials in the nerve fibers.

Functions of the Tympanic Membrane & Ossicles

In response to the pressure changes produced by sound waves on its external surface, the tympanic membrane moves in and out. The membrane therefore functions as a **resonator** that reproduces the vibrations of the sound source. It stops vibrating almost immediately when the sound wave stops; ie, it is very nearly **critically damped.** The motions of the tympanic membrane are imparted to the manubrium of the malleus. The malleus rocks on an axis through the junction of its long and short processes, so that the short process transmits the vibrations of the manubrium to the incus. The incus moves in such a way that the vibrations are transmitted to the head of the stapes. Movements of the head of the stapes swing its footplate to and fro like a door hinged at the posterior edge of the oval window. The auditory ossicles thus function as a lever system that converts the resonant vibrations of the tympanic membrane into movements of the stapes against the perilymph-filled scala vestibuli of the cochlea (Figure 9–2). This system increases the sound pressure that arrives at the oval window, because the lever action of the malleus and incus multiplies the force 1.3 times and the area of the tympanic membrane is much greater than the area of the footplate of the stapes. There are losses of sound energy as a result of resistance, but it has been calculated that at frequencies below 3000 Hz, 60% of the sound energy incident on the tympanic membrane is transmitted to the fluid in the cochlea.

Tympanic Reflex

When the middle ear muscles—the tensor tympani and the stapedius—contract, they pull the manubrium of the malleus inward and the footplate of the stapes outward. This decreases sound transmission. Loud sounds initiate a reflex contraction of these muscles generally called the **tympanic reflex.** Its function is protective, preventing strong sound waves from causing excessive stimulation of the auditory receptors. However, the reaction time for the reflex is 40–160 ms, so it does not protect against brief intense stimulation such as that produced by gunshots.

Bone & Air Conduction

Conduction of sound waves to the fluid of the inner ear via the tympanic membrane and the auditory ossicles, the main pathway for normal hearing, is called **ossicular conduction.** Sound waves also initiate vibrations of the secondary tympanic membrane that closes the round window. This process, unimportant in normal hearing, is **air conduction.** A third type of conduction, **bone conduction,** is the transmission of vibrations of the bones of the skull to the fluid of the inner ear. Considerable bone conduction occurs when tuning forks or other vibrating bodies are applied directly to the skull. This route also plays a role in transmission of extremely loud sounds.

Traveling Waves

The movements of the footplate of the stapes set up a series of traveling waves in the perilymph of the scala vestibuli. A diagram of such a wave is shown in Figure 9–13. As the wave moves up the cochlea, its height increases to a maximum and then drops off rapidly. The distance from the stapes to this point of maximum height varies with the frequency of the vibrations initiating the wave. High-pitched sounds generate waves that reach maximum height near the base of the cochlea; low-pitched sounds generate waves that peak near the apex. The bony walls of the scala vestibuli are rigid, but Reissner's membrane is flexible. The basilar membrane is not under tension, and it also is readily depressed into the scala tympani by the peaks of waves in the scala vestibuli. Displacements of the fluid in the scala tympani are dissipated into air at the round window. Therefore, sound produces distortion

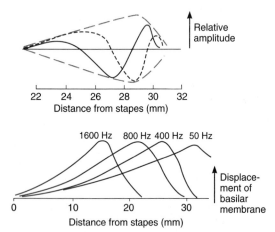

Figure 9–13. Traveling waves. **Top:** The solid and the short-dashed lines represent the wave at two instants of time. The long-dashed line shows the "envelope" of the wave formed by connecting the wave peaks at successive instants. **Bottom:** Displacement of the basilar membrane by the waves generated by stapes vibration of the frequencies shown at the top of each curve.

of the basilar membrane, and the site at which this distortion is maximal is determined by the frequency of the sound wave. The tops of the hair cells in the organ of Corti are held rigid by the reticular lamina, and the hairs of the outer hair cells are embedded in the tectorial membrane (Figure 9–4). When the stapes moves, both membranes move in the same direction, but they are hinged on different axes, so there is a shearing motion that bends the hairs. The hairs of the inner hair cells are not attached to the tectorial membrane, but they are apparently bent by fluid moving between the tectorial membrane and the underlying hair cells.

Functions of the Inner & Outer Hair Cells

The inner hair cells are the primary sensory cells that generate action potentials in the auditory nerves, and presumably they are stimulated by the fluid movements noted above.

The outer hair cells, on the other hand, have a different function. These respond to sound, like the inner hair cells, but depolarization makes them shorten and hyperpolarization makes them lengthen. They do this over a very flexible part of the basal membrane, and this action somehow increases the amplitude and clarity of sounds. These changes in outer hair cells occur in parallel with changes in **prestin,** a membrane protein, and this protein may well be the motor protein of outer hair cells.

The outer hair cells receive cholinergic innervation via an efferent component of the auditory nerve, and acetylcholine hyperpolarizes the cells. However, the physiologic function of this innervation is unknown.

Action Potentials in Auditory Nerve Fibers

The frequency of the action potentials in single auditory nerve fibers is proportionate to the loudness of the sound stimuli. At low sound intensities, each axon discharges to sounds of only one frequency, and this frequency varies from axon to axon depending upon the part of the cochlea from which the fiber originates. At higher sound intensities, the individual axons discharge to a wider spectrum of sound frequencies (Figure 9–14)—particularly to frequencies lower than that at which threshold simulation occurs.

The major determinant of the pitch perceived when a sound wave strikes the ear is the place in the organ of Corti that is maximally stimulated. The traveling wave set up by a tone produces peak depression of the basilar membrane, and consequently maximal receptor stimulation, at one point. As noted above, the distance between this point and the stapes is inversely related to the pitch of the sound, low tones

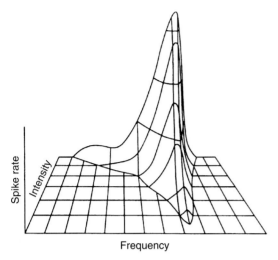

Figure 9–14. Relation of discharge rate (spike rate) in auditory nerve fiber to frequency and intensity of sound. Because the diagram represents the type of response seen in many different individual fibers, no numbers are given on the scales. (Modified from Kiang NYS: Peripheral neural processing of auditory information. In: *Handbook of Physiology.* Section 1, *The Nervous System,* vol 3, part 2. Brookhart JM, Mountcastle VB [editors]. American Physiological Society, 1984.)

producing maximal stimulation at the apex of the cochlea and high tones producing maximal stimulation at the base. The pathways from the various parts of the cochlea to the brain are distinct. An additional factor involved in pitch perception at sound frequencies of less than 2000 Hz may be the pattern of the action potentials in the auditory nerve. When the frequency is low enough, the nerve fibers begin to respond with an impulse to each cycle of a sound wave. The importance of this **volley effect,** however, is limited; the frequency of the action potentials in a given auditory nerve fiber determines principally the loudness, rather than the pitch, of a sound.

Although the pitch of a sound depends primarily on the frequency of the sound wave, loudness also plays a part; low tones (below 500 Hz) seem lower and high tones (above 4000 Hz) seem higher as their loudness increases. Duration also affects pitch to a minor degree. The pitch of a tone cannot be perceived unless it lasts for more than 0.01 s, and with durations between 0.01 and 0.1 s, pitch rises as duration increases. Finally, the pitch of complex sounds that include harmonics of a given frequency is still perceived even when the primary frequency (missing fundamental) is absent.

Auditory Responses of Neurons in the Medulla Oblongata

The response of individual second-order neurons in the cochlear nuclei to sound stimuli are like those of the individual auditory nerve fibers. The frequency at which sounds of the lowest intensity evoke a response varies from unit to unit; with increased sound intensities, the band of frequencies to which a response occurs becomes wider. The major difference between the responses of the first- and second-order neurons is the presence of a sharper "cutoff" on the low-frequency side in the medullary neurons. This greater specificity of the second-order neurons is probably due to some sort of inhibitory process in the brain stem, but how it is achieved is not known.

Primary Auditory Cortex

The pathways from the cochlea to the auditory cortex are described in the first section of this chapter. Impulses ascend from the dorsal and ventral cochlear nuclei through complex paths that are both crossed and uncrossed. In animals, there is an organized pattern of tonal localization in the primary auditory cortex (area 41), as if the cochlea had been unrolled upon it. In humans, low tones are represented anterolaterally and high tones posteromedially in the auditory cortex. However, it is pitch and not frequency per se that is coded in the auditory cortex, because when a complex sound with a missing fundamental is presented (see above), the part of the cortex that is stimulated is the part corresponding to the perceived pitch. Thus, processing of pure frequencies into pitch must occur at a subcortical level.

Other Cortical Areas Concerned With Audition

The increasing availability of PET scanning and fMRI (see Chapter 16) has led to rapid increases in knowledge about auditory association areas in humans. The auditory pathways in the cortex resemble the visual pathways in that there is increasingly complex processing of auditory information along them. An interesting observation is that although the auditory areas look very much the same on the two sides of the brain, there is marked hemispheric specialization. For example, Brodmann's area 22 is concerned with the processing of auditory signals related to speech. During language processing, it is much more active on the left side than on the right side. Area 22 on the right side is more concerned with melody, pitch, and sound intensity. There is also great plasticity in the auditory pathways, and, like the visual and somasthetic pathways, they are modified by experience. Examples of auditory plasticity in humans include the observation that in individuals who become deaf before language skills are fully developed, viewing sign language activates auditory association areas. Conversely, individuals who become blind early in life are demonstrably better at localizing sound than individuals with normal eyesight.

Musicians provide additional examples of cortical plasticity. In these individuals, there is an increase in the size of the auditory areas activated by musical tones. In addition, violinists have altered somatosensory representation of the areas to which the fingers they use in playing their instruments project. Musicians also have larger cerebellums than nonmusicians, presumably because of learned precise finger movements.

A portion of the posterior superior temporal gyrus known as the **planum temporale** (Figure 9–15) is regularly larger in the left than in the right cerebral hemisphere, particularly in right-handed individuals. This area appears to be involved in language-related auditory processing. A curious observation which is presently unexplained is that the planum temporale is even larger than normal on the left side, ie, the asymmetry is greater, in musicians and others who have perfect pitch. The general subject of cortical asymmetry is discussed in Chapter 16.

Sound Localization

Determination of the direction from which a sound emanates in the horizontal plane depends upon detecting the difference in time between the arrival of the stimulus in the two ears and the consequent difference in phase of the sound waves on the two sides; it also depends upon the fact that the sound is louder on the side closest to the source. The detectable time difference, which can be as little as 20 μs, is said to be the most important factor at frequencies below 3000 Hz and the loudness difference the most important at frequencies above 3000 Hz. Neurons in the auditory cortex that receive input from both ears respond maximally or minimally when the time of arrival of a stimulus at one ear is delayed by a fixed period relative to the time of arrival at the other ear. This fixed period varies from neuron to neuron.

Sounds coming from directly in front of the individual differ in quality from those coming from behind because each pinna (the visible portion of the exterior ear) is turned slightly forward. In addition, reflections of the sound waves from the pinnal surface change as sounds move up or down, and the change in the sound waves is the primary factor in locating sounds in the vertical plane. Sound localization is markedly disrupted by lesions of the auditory cortex.

Audiometry

Auditory acuity is commonly measured with an **audiometer.** This device presents the subject with pure tones of various frequencies through earphones. At each frequency, the threshold intensity is determined and plotted on a graph as a percentage of normal hearing. This provides an objective measurement of the degree of deafness and a picture of the tonal range most affected.

Deafness

Clinical deafness may be due to impaired sound transmission in the external or middle ear (**conduction deafness**) or to damage to the hair cells or neural pathways (**nerve deafness**). The two can be differentiated by a number of simple tests with a tuning fork. Three of these tests, named for the individuals who developed

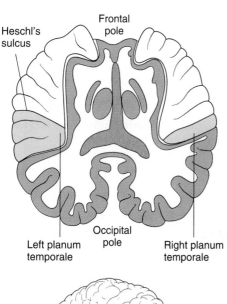

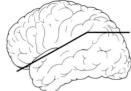

Figure 9–15. Left and right planum temporale in a brain sectioned horizontally along the plane of the sylvian fissure. Plane of section shown in the insert at the bottom. (Reproduced, with permission, from Kandel ER, Schwartz JH, Jessel TM [editors]: *Principles of Neural Science,* 3rd ed. McGraw-Hill, 1991.)

them, are outlined in Table 9–1. The Weber and Schwabach tests demonstrate the important masking effect of environmental noise on the auditory threshold.

Among the causes of conduction deafness are plugging of the external auditory canals with wax or foreign bodies, destruction of the auditory ossicles, thickening of the eardrum following repeated middle ear infections, and abnormal rigidity of the attachments of the stapes to the oval window. Aminoglycoside antibiotics such as streptomycin and gentamicin obstruct the mechanosensitive channels in the stereocilia of hair cells and can cause the cells to degenerate, producing nerve deafness and abnormal vestibular function. Damage to the outer hair cells by prolonged exposure to noise is associated with hearing loss. Other causes include tumors of the vestibulocochlear nerve and cerebellopontine angle, and vascular damage in the medulla. **Presbycusis,** the gradual hearing loss associated with aging, affects more than one-third of those over 75 and is probably due to gradual cumulative loss of hair cells and neurons.

Deafness due to genetic mutations occurs in about 0.1% of newborns. In 30% of the cases, it is associated with abnormalities in other systems **(syndromic deafness),** but in the remaining 70% it is the only apparent abnormality **(nonsyndromic deafness).** There is evidence that nonsyndromic deafness due to some mutations can first appear in adults rather than children, so the incidence is higher than 0.1% and may account for many of the 16% of all adults who have significant hearing impairment. In the past few years, a remarkably large number of mutations that cause deafness have been described. This not only has added to knowledge about the pathophysiology of deafness, but characterization of the normal products of the genes has provided valuable information about the physiology of hearing. It is now estimated that the products of 100 or more genes are essential for normal hearing, and deafness loci have been described in all but five of the 24 human chromosomes.

Interesting examples of proteins which when mutated cause deafness include connexon 26. The defect this produces in the function of connexons (see Chapter 1) presumably prevents the normal recycling of K^+ through the sustenacular cells (Figure 9–8). Mutations in three nonmuscle myosins (see Chapter 1) cause deafness. These are myosin-VIIa, associated with the actin in the hair cell processes; myosin-Ib, which is probably part of the "adaptation motor" that adjusts tension on the tip links (see above); and myosin-VI, which is essential in some way for the formation of normal cilia. Deafness is also associated with mutant forms of α-tectin, one of the major proteins in the tectorial membrane.

An example of syndromic deafness is **Pendred's syndrome,** in which a mutant sulfate transport protein causes deafness and goiter. Another example is one form of the **long QT syndrome** in which there is a mutation of one of the K^+ channel proteins, **KVLQT1.** In the stria vascularis, the normal form of this protein is essential for maintaining the high K^+ concentration in endolymph, and in the heart it helps maintain a normal QT interval. Individuals who are homozygous for mutant KVLQT1 are deaf and predisposed to the ventricular arrhythmias and sudden death that characterize the long QT syndrome (see Chapter 28). The newly

Table 9–1. Common tests with a tuning fork to distinguish between nerve and conduction deafness.

	Weber	Rinne	Schwabach
Method	Base of vibrating tuning fork placed on vertex of skull.	Base of vibrating tuning fork placed on mastoid process until subject no longer hears it, then held in air next to ear.	Bone conduction of patient compared with that of normal subject.
Normal	Hears equally on both sides.	Hears vibration in air after bone conduction is over.	
Conduction deafness (one ear)	Sound louder in diseased ear because masking effect of environmental noise is absent on diseased side.	Vibrations in air not heard after bone conduction is over.	Bone conduction better than normal (conduction defect excludes masking noise).
Nerve deafness (one ear)	Sound louder in normal ear.	Vibration heard in air after bone conduction is over, as long as nerve deafness is partial.	Bone conduction worse than normal.

discovered membrane protein **barttin,** mutations of which can cause deafness and the renal manifestations of Bartter's syndrome, is discussed in Chapter 39.

VESTIBULAR FUNCTION

Responses to Rotational Acceleration

Rotational acceleration in the plane of a given semicircular canal stimulates its crista. The endolymph, because of its inertia, is displaced in a direction opposite to the direction of rotation. The fluid pushes on the cupula, deforming it. This bends the processes of the hair cells (Figure 9–6). When a constant speed of rotation is reached, the fluid spins at the same rate as the body and the cupula swings back into the upright position. When rotation is stopped, deceleration produces displacement of the endolymph in the direction of the rotation, and the cupula is deformed in a direction opposite to that during acceleration. It returns to mid position in 25–30 seconds. Movement of the cupula in one direction commonly causes increased impulse traffic in single nerve fibers from its crista, whereas movement in the opposite direction commonly inhibits neural activity (Figure 9–16).

Rotation causes maximal stimulation of the semicircular canals most nearly in the plane of rotation. Since the canals on one side of the head are a mirror image of those on the other side, the endolymph is displaced toward the ampulla on one side and away from it on the other. The pattern of stimulation reaching the brain therefore varies with the direction as well as the plane of rotation. Linear acceleration probably fails to dis-

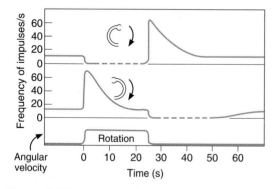

Figure 9–16. Ampullary responses to rotation. Average time course of impulse discharge from the ampulla of two semicircular canals during rotational acceleration, steady rotation, and deceleration. (Reproduced, with permission, from Adrian ED: Discharges from vestibular receptors in the cat. J Physiol [Lond] 1943;101:389.)

place the cupula and therefore does not stimulate the cristae. However, there is considerable evidence that when one part of the labyrinth is destroyed, other parts take over its functions. Experimental localization of labyrinthine functions is therefore difficult.

The vestibular nuclei are primarily concerned with maintaining the position of the head in space. The tracts that descend from these nuclei mediate head-on-neck and head-on-body adjustments (see Chapter 12). The ascending connections to cranial nerve nuclei are largely concerned with eye movements.

Nystagmus

The characteristic jerky movement of the eye observed at the start and end of a period of rotation is called **nystagmus.** It is actually a reflex that maintains visual fixation on stationary points while the body rotates, although it is not initiated by visual impulses and is present in blind individuals. When rotation starts, the eyes move slowly in a direction opposite to the direction of rotation, maintaining visual fixation (**vestibulo-ocular reflex, VOR**). When the limit of this movement is reached, the eyes quickly snap back to a new fixation point and then again move slowly in the other direction. The slow component is initiated by impulses from the labyrinths; the quick component is triggered by a center in the brain stem. Nystagmus is frequently horizontal (ie, the eyes move in the horizontal plane), but it can also be vertical, when the head is tipped sidewise during rotation, or rotatory, when the head is tipped forward. By convention, the direction of eye movement in nystagmus is identified by the direction of the quick component. The direction of the quick component during rotation is the same as that of the rotation, but the **postrotatory nystagmus** that occurs owing to displacement of the cupula when rotation is stopped is in the opposite direction. Clinically, nystagmus is seen at rest in patients with lesions of the brain stem.

Responses to Linear Acceleration

In mammals, the utricular and saccular maculas respond to linear acceleration. In general, the utricle responds to horizontal acceleration and the saccule to vertical acceleration. The otoliths are more dense than the endolymph, and acceleration in any direction causes them to be displaced in the opposite direction, distorting the hair cell processes and generating activity in the nerve fibers. The maculas also discharge tonically in the absence of head movement, because of the pull of gravity on the otoliths. The impulses generated from these receptors are partly responsible for reflex righting of the head and other important postural adjustments discussed in Chapter 12.

Although most of the responses to stimulation of the maculas are reflex in nature, vestibular impulses also reach the cerebral cortex. These impulses are presumably responsible for conscious perception of motion and supply part of the information necessary for orientation in space. **Vertigo** is the sensation of rotation in the absence of actual rotation and is a prominent symptom when one labyrinth is inflamed.

Caloric Stimulation

The semicircular canals can be stimulated by instilling water that is hotter or colder than body temperature into the external auditory meatus. The temperature difference sets up convection currents in the endolymph, with consequent motion of the cupula. This technique of **caloric stimulation,** which is sometimes used diagnostically, causes nystagmus, vertigo, and nausea. To avoid these symptoms when irrigating the ear canals in the treatment of ear infections, it is important to be sure that the fluid used is at body temperature.

Spatial Orientation

Orientation in space depends in part upon input from the vestibular receptors, but visual cues are also important. Pertinent information is also supplied by impulses from proprioceptors in joint capsules, which supply data about the relative position of the various parts of the body, and impulses from cutaneous exteroceptors, especially touch and pressure receptors. These four inputs are synthesized at a cortical level into a continuous picture of the individual's orientation in space.

Motion Sickness

The nausea, blood pressure changes, sweating, pallor, and vomiting that are the well-known symptoms of **motion sickness** are produced by excessive vestibular stimulation. They are probably due to reflexes mediated via vestibular connections in the brain stem and the flocculonodular lobe of the cerebellum (see Chapter 12).

Space motion sickness, the nausea, vomiting, and vertigo that occur in astronauts, develops when they are first exposed to microgravity and often wears off after a few days of space flight. It can then recur with reentry, as the force of gravity increases again. It is believed to be due to mismatches in neural input created by changes in the input from some parts of the vestibular apparatus and other gravity sensors without corresponding changes in the other spatial orientation inputs.

Smell & Taste

10

INTRODUCTION

Smell and taste are generally classified as visceral senses because of their close association with gastrointestinal function. Physiologically, they are related to each other. The flavors of various foods are in large part a combination of their taste and smell. Consequently, food may taste "different" if one has a cold that depresses the sense of smell. Both taste and smell receptors are chemoreceptors that are stimulated by molecules in solution in mucus in the nose and saliva in the mouth. However, these two senses are anatomically quite different. The smell receptors are distance receptors (teleceptors), and the smell pathways have no relay in the thalamus. The taste pathways pass up the brain stem to the thalamus and project to the postcentral gyrus along with those for touch and pressure sensibility from the mouth.

SMELL

Olfactory Mucous Membrane

The olfactory receptor cells are located in a specialized portion of the nasal mucosa, the yellowish-pigmented **olfactory mucous membrane.** In dogs and other animals in which the sense of smell is highly developed (macrosmatic animals), the area covered by this membrane is large; in microsmatic animals such as humans, it is small. In humans, it covers an area of 5 cm² in the roof of the nasal cavity near the septum (Figure 10–1). It contains supporting cells and progenitor cells for the olfactory receptors. Interspersed between these cells are 10–20 million receptor cells. Each olfactory receptor is a neuron, and the olfactory mucous membrane is said to be the place in the body where the nervous system is closest to the external world. Each neuron has a short, thick dendrite with an expanded end called an olfactory rod (Figure 10–2). From these rods, cilia project to the surface of the mucus. The cilia are unmyelinated processes about 2 μm long and 0.1 μm in diameter. There are 10–20 cilia per receptor neuron. The axons of the olfactory receptor neurons pierce the cribriform plate of the ethmoid bone and enter the olfactory bulbs.

The olfactory neurons, like the taste receptor cells (see below) but unlike most other neurons, are constantly being replaced with a half-time of a few weeks. The olfactory renewal process is carefully regulated, and there is evidence that in this situation, a bone morphogenic protein (BMP) exerts an inhibitory effect. BMPs are a large family of growth factors originally described as promoters of bone growth but now known to act on most tissues in the body during development, including many types of nerve cells.

The olfactory mucous membrane is constantly covered by mucus. This mucus is produced by Bowman's glands, which are just under the basal lamina of the membrane.

Olfactory Bulbs

In the olfactory bulbs, the axons of the receptors contact the primary dendrites of the **mitral cells** and **tufted cells** (Figure 10–3) to form the complex globu-

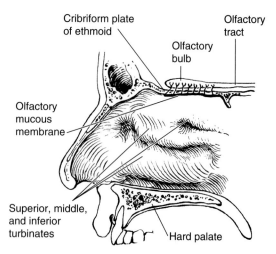

Figure 10–1. Olfactory mucous membrane. (Reproduced, with permission, from Waxman SG: *Neuroanatomy with Clinical Correlations,* 25th ed. McGraw-Hill, 2003.)

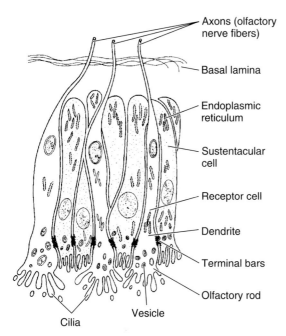

Figure 10–2. Structure of the olfactory mucous membrane.

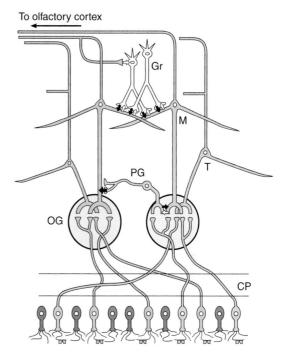

Figure 10–3. Basic neural circuits in the olfactory bulb. Note that olfactory receptor cells with one type of odorant receptor project to one olfactory glomerulus (OG) and olfactory receptor cells with another type of receptor project to a different olfactory glomerulus. CP, cribriform plate; PG, periglomerular cell; M, mitral cell; T, tufted cell; Gr, granule cell. (Modified from Mori K, Nagao H, Yoshihara Y: The olfactory bulb: coding and processing of odor molecular information. Science 1999;286:711.)

lar synapses called **olfactory glomeruli.** The tufted cells are smaller than the mitral cells and have thinner axons, but both types send axons into the olfactory cortex, and they appear to be similar from a functional point of view. In addition to mitral and tufted cells, the olfactory bulbs contain **periglomerular cells,** which are inhibitory neurons connecting one glomerulus to another, and **granule cells,** which have no axons and make reciprocal synapses with the lateral dendrites of the mitral and tufted cells (Figure 10–3). At these synapses, the mitral or tufted cell excites the granule cell by releasing glutamate, and the granule cell side of the synapse in turn inhibits the mitral or tufted cell by releasing GABA. The physiologic significance of this arrangement is discussed below.

Olfactory Cortex

The axons of the mitral and tufted cells pass posteriorly through the **intermediate olfactory stria** and the **lateral olfactory stria** to the **olfactory cortex.** The axons terminate on the apical dendrites of pyramidal cells in the olfactory cortex. In humans, sniffing activates the piriform cortex, but smells with or without sniffing activate the lateral and anterior orbitofrontal gyri of the frontal lobe. The orbitofrontal activation is generally greater on the right side than the left. Thus, the cortical representation of olfaction is asymmetric. Other fibers

project to the amygdala, which is probably involved with the emotional responses to olfactory stimuli, and to the entorhinal cortex, which is concerned with olfactory memories.

Olfactory Thresholds & Discrimination

Olfactory receptors respond only to substances that are in contact with the olfactory epithelium and are dissolved in the thin layer of mucus that covers it. The olfactory thresholds for the representative substances shown in Table 10–1 illustrate the remarkable sensitivity of the olfactory receptors to some substances. For example, methyl mercaptan, one of the substances in garlic, can be smelled at a concentration of less than 500 pg/L of air. In addition, olfactory discrimination is remarkable; for example, humans can recognize more than 10,000 different odors. On the other hand, determination of differences in the intensity of any given

Table 10–1. Some olfactory thresholds.

Substance	mg/L of Air
Ethyl ether	5.83
Chloroform	3.30
Pyridine	0.03
Oil of peppermint	0.02
Iodoform	0.02
Butyric acid	0.009
Propyl mercaptan	0.006
Artificial musk	0.00004
Methyl mercaptan	0.0000004

odor is poor. The concentration of an odor-producing substance must be changed by about 30% before a difference can be detected. The comparable visual discrimination threshold is a 1% change in light intensity. The direction from which a smell comes may be indicated by the slight difference in the time of arrival of odoriferous molecules in the two nostrils.

Odor-producing molecules are generally small, containing from 3 or 4 to 18–20 carbon atoms, and molecules with the same number of carbon atoms but different structural configurations have different odors. Relatively high water and lipid solubility are characteristic of substances with strong odors.

Signal Transduction

The olfactory system has received considerable attention in recent years because of the intriguing biological question of how a simple sense organ such as the olfactory mucosa and its brain representation that apparently lacks a high degree of complexity can mediate discrimination of more than 10,000 different odors. One part of the answer to this question is that there are many different odorant receptors.

In mice, there are about 1000 different odorant receptors, and the number in humans is comparable. This means that 1% or more of the genome is devoted to making odorant receptors. The genes for the receptor make up the largest gene family so far described in mammals—larger than the immunoglobulin and T cell receptor gene families combined. All the odorant receptors are coupled to heterotrimeric G proteins. Some act via adenylyl cyclase and cAMP, and others act via phospholipase C and the products of phosphatidylinositol

hydrolysis. Most of them open cation channels, causing an inward-directed Ca^{2+} current.

A second part of the answer to the question how 10,000 different odors can be detected lies in the neural organization of the olfactory pathway. In mice, there are 2 million olfactory sensory neurons, and each expresses only one of the thousand different odorant receptors. Each neuron expressing a given receptor projects to two of the 1800 glomeruli (Figure 10–3). This provides a distinct two-dimensional map in the olfactory bulb that is unique to the odorant. The mitral cells with their glomeruli project to different parts of the olfactory cortex.

In the olfactory glomeruli, there is lateral inhibition mediated by periglomerular cells and granule cells. This sharpens and focuses olfactory signals. In addition, the extracellular field potential in each glomerulus oscillates, and the granule cells appear to regulate the frequency of the oscillation. The exact function of the oscillation is unknown, but it probably also helps to focus the olfactory signals reaching the cortex.

Odorant-Binding Proteins

In contrast to the low threshold for olfactory stimulation when the olfactory mucous membrane is intact, single olfactory receptors that have been patch-clamped have a relatively high threshold and a long latency. In addition, lipophilic odor-producing molecules must traverse the hydrophilic mucus in the nose to reach the receptors. These facts led to the suggestion that the olfactory mucus might contain one or more odorant-binding proteins (OBP) that concentrate the odorants and transfer them to the receptors. An 18-kDa OBP that is unique to the nasal cavity has been isolated, and other related proteins probably exist. The protein has considerable homology to other proteins in the body that are known to be carriers for small lipophilic molecules. A similar binding protein appears to be associated with taste (see below).

Vomeronasal Organ

In rodents and various other mammals, the nasal cavity contains another patch of olfactory mucous membrane located along the nasal septum in a well-developed **vomeronasal organ.** This structure is concerned with the perception of odors that act as **pheromones** (see Chapter 15). Its receptors project to the **accessory olfactory bulb** and from there primarily to areas in the amygdala and hypothalamus that are concerned with reproduction and ingestive behavior. Vomeronasal input has major effects on these functions. An example is pregnancy block in mice; the pheromones of a male from a different strain prevent pregnancy as a result of mating with that male, but mating with a mouse of the same strain does not produce blockade. The vomeronasal organ has about 30 serpentine odorant receptors that differ quite markedly in structure from those in the rest of the olfactory epithelium.

The organ is not well developed in humans, but there is an anatomically separate and biochemically unique area of olfactory mucous membrane in a pit in the anterior third of the nasal septum which appears to be a homologous structure. There is evidence for the existence of pheromones in humans, and there is a close relationship between smell and sexual function. The perfume ads bear witness to this. The sense of smell is said to be more acute in women than in men, and in women it is most acute at the time of ovulation. Smell and, to a lesser extent, taste have a unique ability to trigger long-term memories, a fact noted by novelists and documented by experimental psychologists.

Sniffing

The portion of the nasal cavity containing the olfactory receptors is poorly ventilated in humans. Most of the air normally moves smoothly over the turbinates with each respiratory cycle, although eddy currents pass some air over the olfactory mucous membrane. These eddy currents are probably set up by convection as cool air strikes the warm mucosal surfaces. The amount of air reaching this region is greatly increased by sniffing, an action that includes contraction of the lower part of the nares on the septum, deflecting the airstream upward. Sniffing is a semi-reflex response that usually occurs when a new odor attracts attention.

Role of Pain Fibers in the Nose

Naked endings of many trigeminal pain fibers are found in the olfactory mucous membrane. They are stimulated by irritating substances, and an irritative, trigeminally mediated component is part of the characteristic "odor" of such substances as peppermint, menthol, and chlorine. These endings are also responsible for initiating sneezing, lacrimation, respiratory inhibition, and other reflex responses to nasal irritants.

Adaptation

It is common knowledge that when one is continuously exposed to even the most disagreeable odor, perception of the odor decreases and eventually ceases. This sometimes beneficent phenomenon is due to the fairly rapid adaptation, or desensitization, that occurs in the olfactory system. It is mediated by Ca^{2+} acting via calmodulin on **cyclic nucleotide-gated (CNG)** ion channels. When CNG A4 is knocked out, adaptation is slowed.

Abnormalities

Abnormalities of olfaction include **anosmia** (absence of the sense of smell), **hyposmia** (diminished olfactory sensitivity), and **dysosmia** (distorted sense of smell). Several dozen different anosmias have been detected in humans. They are presumably due in each case to absence or disrupted function of one of the many members of the odorant receptor family. Olfactory thresholds increase with advancing age, and more than 75% of humans over the age of 80 have an impaired ability to identify smells. Anosmia associated with hypogonadism (Kallmann's syndrome) is discussed in Chapters 14 and 23.

TASTE

RECEPTOR ORGANS & PATHWAYS

Taste Buds

The taste buds, the sense organs for taste, are ovoid bodies measuring 50–70 μm. Each taste bud is made up of four types of cells (Figure 10–4): basal cells; type 1 and 2 cells, which are sustentacular cells; and type 3 cells, which are the gustatory receptor cells that make synaptic connections to sensory nerve fibers. The type 3 cells have a microvillus which projects into the taste pore, an opening to the oral cavity. The necks of the sustentacular and taste cells are connected to each other and to the surrounding epithelial cells by tight junc-

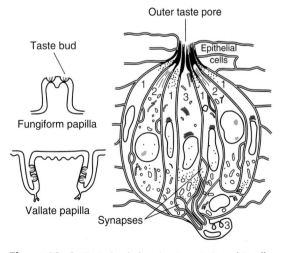

Figure 10–4. Taste bud, showing type 1, 2, and 3 cells. The locations of taste buds on fungiform and vallate papillae are shown on the left. (Modified from Shepherd GM: *Neurobiology*, 2nd ed. Oxford Univ Press, 1988.)

tions, so that the only part of the gustatory receptor cells exposed to the fluids in the oral cavity is their microvillus. Each taste bud is innervated by about 50 nerve fibers, and conversely, each nerve fiber receives input from an average of five taste buds. The basal cells arise from the epithelial cells surrounding the taste bud. They differentiate into new receptor cells, and the old receptor cells are continuously replaced with a half-time of about 10 days. If the sensory nerve is cut, the taste buds it innervates degenerate and eventually disappear. However, if the nerve regenerates, the cells in the neighborhood become organized into new taste buds, presumably as a result of some sort of chemical inductive effect from the regenerating fiber.

In humans, the taste buds are located in the mucosa of the epiglottis, palate, and pharynx and in the walls of the **fungiform** and **vallate papillae** of the tongue. The fungiform papillae are rounded structures most numerous near the tip of the tongue; the vallate papillae are prominent structures arranged in a V on the back of the tongue. There are up to five taste buds per fungiform papilla, and they are usually located at the top of the papilla (Figure 10–4). The larger vallate papillae each contain up to 100 taste buds, usually located along the sides of the papillae. The small conical **filiform papillae** that cover the dorsum of the tongue do not usually contain taste buds. There are a total of about 10,000 taste buds.

Taste Pathways

The sensory nerve fibers from the taste buds on the anterior two-thirds of the tongue travel in the chorda tympani branch of the facial nerve, and those from the posterior third of the tongue reach the brain stem via the glossopharyngeal nerve (Figure 10–5). The fibers from areas other than the tongue reach the brain stem via the vagus nerve. On each side, the myelinated but relatively slowly conducting taste fibers in these three nerves unite in the gustatory portion of the **nucleus tractus solitarius** in the medulla oblongata (Figure 10–5). From there, axons of second-order neurons ascend in the ipsilateral medial lemniscus and, in primates, pass directly to the ventral posteromedial nucleus of the thalamus. From the thalamus, the axons of the third-order neurons pass in the thalamic radiation to the face area of the somatosensory cortex in the ipsilateral postcentral gyrus. They also pass to the anterior part of the insula. The relevant insular cortex is anterior to the face area of the postcentral gyrus and is probably the area that mediates conscious perception of taste and taste discrimination.

Basic Taste Modalities

In humans there are five established basic tastes: sweet, sour, bitter, salt, and umami. It used to be thought that

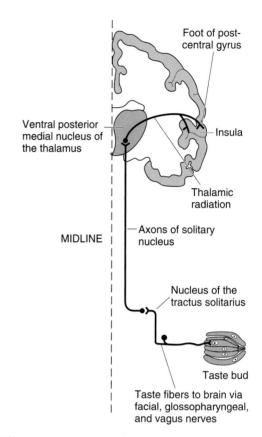

Figure 10–5. Diagram of taste pathways.

there were special areas on the surface of the tongue for each of the first four of these sensations, but it is now clear that all are sensed from all parts of the tongue and adjacent structures. It has also become clear that the afferent nerves to the nucleus tractus solitarius contain fibers from all types of taste buds, without any clear localization of types.

The fifth taste sense, **umami,** was recently added to the four classic tastes. This taste has actually been known for almost 100 years, and it became established once its receptor was identified (see below). It is triggered by glutamate and particularly by the monosodium glutamate (MSG) used so extensively in Asian cooking. The taste is pleasant and sweet but differs from the standard sweet taste.

Taste Receptors & Transduction

The putative receptors for taste are shown diagrammatically in Figure 10–6. The salty taste is triggered by NaCl. The main receptor is the ENaC. Like ENaC receptors elsewhere in the body, the receptors in the oral

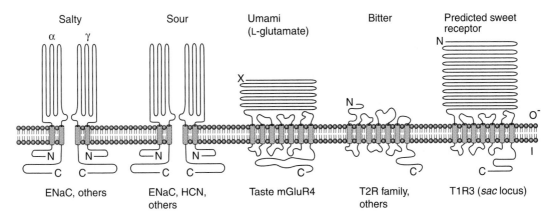

Figure 10–6. Taste receptors. (Modified from Lindemann B: Receptors and transduction in taste. Nature 2001;413:219.)

cavity are inhibited by amiloride (see Chapters 1 and 38). However, the inhibition on the tongue is incomplete, leading to the suspicion that there are additional salt receptors in the taste buds. The entry of Na$^+$ into the salt receptors depolarizes them and triggers release of glutamate, which depolarizes the surrounding afferent neurons.

The sour taste is triggered by protons. The ENaCs permit the entry of protons and may be responsible, at least in part, for the sour taste. However, **HCN,** a hyperpolarization-activated cyclic nucleotide-gated cation channel and other receptors may also be involved.

Umami taste is due to activation of a truncated metabotopic glutamate receptor, **mGluR4,** in the taste buds, and the agonists are purine 5-ribonucleotides such as IMP and GMP in the food. The way this produces depolarization is unsettled.

Bitter taste is produced by a variety of unrelated compounds. Many of these are poisons, and bitter taste serves as a warning to avoid them. It was originally thought that there was a single bitter receptor linked to the heterotrimeric G protein **gustducin.** However, there are many G protein-linked receptors in the human genome that are found in taste receptor cells (T$_2$R family) and are stimulated by bitter substances such as strychnine. The exact number of G protein-linked receptors in taste buds that respond to bitter agonists is unsettled, but there may be as many as 24 in humans. It appears that at least some of these receptors and perhaps all of them act independently, ie, that there are multiple different bitter pathways.

Some compounds that taste bitter bypass the receptor and act directly on the G protein; quinine is an example. Others inhibit the phospholipase that metabolizes cGMP. Gustducin lowers cyclic nucleotides and increases the formation of DAG and inositol phosphates. Either action could lead to depolarization.

Substances that taste sweet also act via the G protein gustducin. Sugars taste sweet, but so do compounds such as **saccharin** that have an entirely different structure. It appears at present that natural sugars such as sucrose and synthetic sweeteners act via different receptors on gustducin. Like the bitter-responsive receptors, sweet-responsive receptors act via cyclic nucleotides and inositol phosphate metabolism.

Taste Thresholds & Intensity Discriminations

The ability of humans to discriminate differences in the intensity of tastes, like intensity discrimination in olfaction, is relatively crude. A 30% change in the concentration of the substance being tasted is necessary before an intensity difference can be detected. The threshold concentrations of substances to which the taste buds respond vary with the particular substance (Table 10–2).

Table 10–2. Some taste thresholds.

Substance	Taste	Threshold Concentration (µmol/L)
Hydrochloric acid	Sour	100
Sodium chloride	Salt	2000
Strychnine hydrochloride	Bitter	1.6
Glucose	Sweet	80,000
Sucrose	Sweet	10,000
Saccharin	Sweet	23

A protein that binds taste-producing molecules has been cloned. It is produced by Ebner's glands—glands that secrete mucus into the cleft around vallate papillae (Figure 10–4)—and probably has a concentrating and transport function similar to that of the OBP described in the section on olfaction.

Flavor

The almost infinite variety of tastes so dear to the gourmet are mostly synthesized from the five basic taste components. In some cases, a desirable taste includes an element of pain stimulation (eg, "hot" sauces). In addition, smell plays an important role in the overall sensation produced by food, and the consistency (or texture) and temperature of foods also contribute to their "flavor."

Variation & After-Effects

Taste exhibits after-reactions and contrast phenomena that are similar in some ways to visual after-images and contrasts. Some of these are chemical "tricks," but others may be true central phenomena. A taste modifier protein, **miraculin,** has been discovered in a plant. When applied to the tongue, this protein makes acids taste sweet.

Animals, including humans, form particularly strong aversions to novel foods if eating the food is followed by illness. The survival value of such aversions is apparent in terms of avoiding poisons.

Abnormalities

Abnormalities of taste include **ageusia** (absence of the sense of taste), **hypogeusia** (diminished taste sensitivity), and **dysgeusia** (disturbed sense of taste). Many different diseases can produce hypogeusia. In addition, drugs such as captopril and penicillamine, which contain sulfhydryl groups, cause temporary loss of taste sensation. The reason for this effect of sulfhydryl compounds is not known.

Alert Behavior, Sleep, & the Electrical Activity of the Brain

11

INTRODUCTION

Most of the various sensory pathways described in Chapters 7–10 relay impulses from sense organs via three- and four-neuron chains to particular loci in the cerebral cortex. The impulses are responsible for perception and localization of individual sensations. However, they must be processed in the awake brain to be perceived. At least in mammals, there is a spectrum of behavioral states ranging from deep sleep through light sleep, REM sleep, and the two awake states: relaxed awareness and awareness with concentrated attention. There are patterns of brain electrical activity that correlate with each of these states, including electroencephalographic (EEG) patterns. In recent years, feedback oscillations within the cerebral cortex and between the thalamus and the cortex have received attention as producers of the EEG and possible determinants of the behavioral state. Arousal and the awake patterns of the EEG and thalamic discharges can be produced by sensory stimulation and by impulses ascending in the reticular core of the midbrain. Sleep and sleep patterns can be produced by stimulating the basal forebrain and other "sleep zones." This chapter is concerned with the various awake and sleep states and the electrical activity that underlies them.

THE THALAMUS & THE CEREBRAL CORTEX

Thalamic Nuclei

On developmental and topographic grounds, the thalamus can be divided into three parts: the epithalamus, the dorsal thalamus, and the ventral thalamus. The **epithalamus** has connections to the olfactory system, and the projections and functions of the **ventral thalamus** are undetermined. The **dorsal thalamus** can be divided into nuclei that project diffusely to the whole neocortex and nuclei that project to specific discrete portions of the neocortex and limbic system. The nuclei that project to all parts of the neocortex are the midline and intralaminar nuclei. The nuclei of the dorsal thalamus

that project to specific areas include the specific sensory relay nuclei and the nuclei concerned with efferent control mechanisms. The **specific sensory relay nuclei** include the medial and lateral geniculate bodies, which relay auditory and visual impulses to the auditory and visual cortices; and the ventrobasal group of nuclei, which relay somatesthetic information to the postcentral gyrus. The **nuclei concerned with efferent control mechanisms** include several nuclei that are concerned with motor function. They receive input from the basal ganglia and the cerebellum and project to the motor cortex. Also included in this group are the anterior nuclei, which receive afferents from the mamillary bodies and project to the limbic cortex.

Cortical Organization

The neocortex is generally arranged in six layers (Figure 11–1). The neurons are mostly pyramidal cells with extensive vertical dendritic trees (Figures 11–1 and 11–2) that may reach to the cortical surface. The axons of these cells usually give off recurrent collaterals that turn back and synapse on the superficial portions of the dendritic trees. Afferents from the specific nuclei of the thalamus terminate primarily in cortical layer 4, whereas the nonspecific afferents are distributed to layers 1–4.

THE RETICULAR FORMATION & THE RETICULAR ACTIVATING SYSTEM

The **reticular formation,** the phylogenetically old reticular core of the brain, occupies the midventral portion of the medulla and midbrain. It is primarily an anatomic area made up of various neural clusters and fibers with discrete functions. For example, it contains the cell bodies and fibers of many of the serotonergic, noradrenergic, and adrenergic systems that are discussed in Chapter 15. It also contains many of the areas concerned with regulation of heart rate, blood pressure, and respiration that are discussed in Chapters 31 and 36. Some of the descending fibers in it inhibit trans-

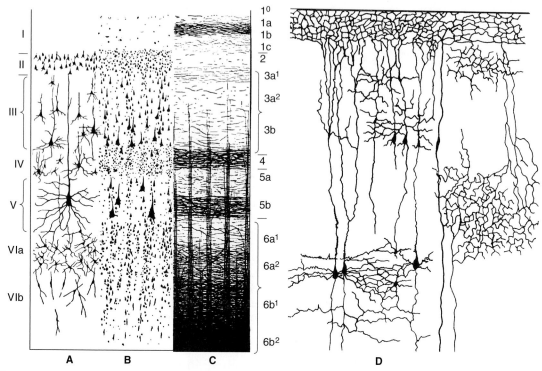

Figure 11–1. Structure of the cerebral cortex. The cortical layers are indicated by the numbers. **A:** Golgi stain showing neuronal cell bodies and dendrites. **B:** Nissl stain, showing cell bodies. **C:** Weigert stain, showing myelinated nerve fibers. **D:** Drawing of neural connections. (A, B, and C reproduced, with permission, from Ranson SW, Clark SL: *The Anatomy of the Nervous System,* 10th ed. Saunders, 1959.)

mission in sensory pathways in the spinal cord (see Chapter 7). Various reticular areas and the pathways from them are concerned with spasticity and adjustment of stretch reflexes (see Chapter 12). The reticular activating system (RAS) and related components of the brain concerned with consciousness and sleep are considered in this chapter.

The RAS is a complex polysynaptic pathway. Collaterals funnel into it not only from the long ascending sensory tracts but also from the trigeminal, auditory, and visual systems and the olfactory system. The complexity of the neuron net and the degree of convergence in it abolish modality specificity, and most reticular neurons are activated with equal facility by different sensory stimuli. The system is therefore **nonspecific,** whereas the classic sensory pathways are **specific** in that the fibers in them are activated by only one type of sensory stimulation. Part of the RAS bypasses the thalamus to project diffusely to the cortex. Another part ends in the intralaminar and related thalamic nuclei and, from them, is projected diffusely and nonspecifically to the whole neocortex (Figure 11–3).

EVOKED CORTICAL POTENTIALS

The electrical events that occur in the cortex after stimulation of a sense organ can be monitored with an exploring electrode connected to another electrode at an indifferent point some distance away. A characteristic response is seen in animals under barbiturate anesthesia, which eliminates much of the background electrical activity. If the exploring electrode is over the primary receiving area for the particular sense, a surface-positive wave appears with a latency of 5–12 ms. This is followed by a small negative wave, and then there is frequently a larger, more prolonged positive deflection with a latency of 20–80 ms. The first positive-negative wave sequence is the **primary evoked potential;** the second is the **diffuse secondary response.**

The primary evoked potential is highly specific in its location and can be observed only where the pathways from a particular sense organ end. An electrode on the pial surface of the cortex samples activity to a depth of only 0.3–0.6 mm. The primary response is negative rather than positive when it is recorded with a

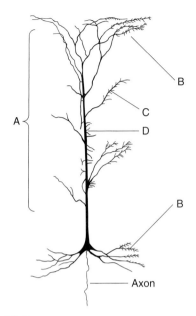

Figure 11–2. Neocortical pyramidal cell, showing the distribution of neurons that terminate on it. **A** denotes nonspecific afferents from the reticular formation and the thalamus; **B** denotes recurrent collaterals of pyramidal cell axons; **C** denotes commissural fibers from mirror image sites in the contralateral hemisphere; **D** denotes specific afferents from thalamic sensory relay nuclei. (Reproduced, with permission, from Chow KL, Leiman AL: The structural and functional organization of the neocortex. Neurosci Res Program Bull 1970;8:157.)

microelectrode inserted in layers 2–6 of the underlying cortex, and the negative wave within the cortex is followed by a positive wave. The negative-positive sequence indicates depolarization on the dendrites and somas of the cells in the cortex, followed by hyperpolarization. The positive-negative wave sequence recorded from the surface of the cortex occurs because the superficial cortical layers are positive relative to the initial negativity, then negative relative to the deep hyperpolarization. In unanesthetized animals or humans, the primary evoked potential is largely obscured by the spontaneous activity of the brain, but it can be demonstrated by superimposing multiple traces so that the background activity is averaged out. It is somewhat more diffuse in unanesthetized animals but still well localized compared with the diffuse secondary response.

The surface-positive diffuse secondary response, unlike the primary, is not highly localized. It appears at the same time over most of the cortex and is due to activity in projections from the midline and related thalamic nuclei.

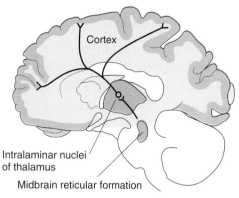

Figure 11–3. Diagram showing the ascending reticular system in the human midbrain, its projections to the intralaminar nuclei of the thalamus, and the output from the intralaminar nuclei to many parts of the cerebral cortex. Activation of these areas is shown by PET scanning when subjects shift from a relaxed awake state to an attention-demanding task. (Data from Kinomura S et al: Activation by attention of the human reticular formation and thalamic intralaminar nuclei. Science 1996;271:512.)

THE ELECTROENCEPHALOGRAM

The background electrical activity of the brain in unanesthetized animals was first described in the 19th century. Subsequently it was analyzed in systematic fashion by the German psychiatrist Hans Berger, who introduced the term **electroencephalogram (EEG)** to denote the record of the variations in brain potential. The EEG can be recorded with scalp electrodes through the unopened skull or with electrodes on or in the brain. The term **electrocorticogram (ECoG)** is sometimes used for the record obtained with electrodes on the pial surface of the cortex.

EEG records may be **bipolar** or **unipolar.** Bipolar records show fluctuations in potential between two cortical electrodes; unipolar records show potential differences between a cortical electrode and a theoretically indifferent electrode on some part of the body distant from the cortex.

Alpha Rhythm

In adult humans who are awake but at rest with the mind wandering and the eyes closed, the most prominent component of the EEG is fairly regular pattern of waves at a frequency of 8–12 Hz and an amplitude of 50–100 μV when recorded from the scalp. This pattern is the **alpha rhythm.** It is most marked in the parieto-occipital area, though it is sometimes observed in other locations. A similar rhythm has been observed in a wide variety of mammalian species (Figure 11–4). In the cat it is slightly more rapid than in the human, and there are other minor variations from species to species, but in all mammals the pattern is remarkably similar.

Other Rhythms

In addition to the dominant rhythm, 18–30 Hz patterns of lower amplitude are sometimes seen over the frontal regions. This **beta rhythm** may be a harmonic of the alpha. **Gamma oscillations** at 30–80 Hz are often seen when an individual is aroused and focuses attention on something. This is often replaced by irregular fast activity as the individual initiates motor activity in response to the stimulus. A pattern of large-amplitude, regular 4–7 Hz waves called the **theta rhythm** occurs in children and is generated in the hippocampus in experimental animals (see below). Large, slow waves with a frequency of less than 4 Hz are sometimes called **delta waves.**

Variations in the EEG

In humans, the frequency of the dominant EEG rhythm at rest varies with age. In infants, there is fast, beta-like activity, but the occipital rhythm is a slow 0.5–2 Hz pattern. During childhood this latter rhythm speeds up, and the adult alpha pattern gradually appears during adolescence. The frequency of the alpha rhythm is decreased by a low blood glucose level, a low body temperature, a low level of adrenal glucocorticoid hormones, and a high arterial partial pressure of CO_2 (Pa_{CO_2}). It is increased by the reverse conditions. Forced overbreathing to lower the Pa_{CO_2} is sometimes used clinically to bring out latent EEG abnormalities.

Alpha Block

When attention is focused on something, the alpha rhythm is replaced by fast, somewhat irregular low-voltage activity. This phenomenon is called **alpha block.** A breakup of the alpha pattern is also produced by any form of sensory stimulation (Figure 11–5) or mental concentration such as solving arithmetic problems. A common term for this replacement of the regular alpha rhythm with irregular low-voltage activity is

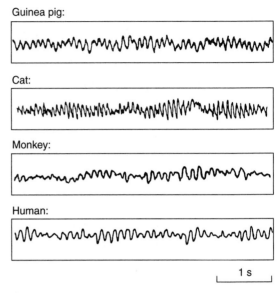

Guinea pig:

Cat:

Monkey:

Human:

1 s

Figure 11–4. EEG records showing the alpha rhythm from four different species.

the **arousal** or **alerting response,** because it is correlated with the aroused, alert state. It has also been called **desynchronization,** because it represents breaking up of the obviously synchronized neural activity necessary to produce regular waves. However, the rapid EEG activity seen in the alert state is also synchronized, but at a higher rate. Therefore, the term "desynchronization" is misleading.

Sleep Patterns

There are two different kinds of sleep: **rapid eye movement (REM) sleep** and **non-REM (NREM),** or **slow-wave sleep.** NREM sleep is divided into four stages. A person falling asleep first enters stage 1, which is characterized by low-amplitude, high-frequency EEG activity (Figure 11–6). Stage 2 is marked by the appearance of **sleep spindles.** These are bursts of alpha-like, 10–14 Hz, 50 μV waves. In stage 3, the pattern is one of lower frequency and increased amplitude of the EEG waves. Maximum slowing with large waves is seen in stage 4. Thus, the characteristic of deep sleep is a pattern of rhythmic slow waves, indicating marked **synchronization.**

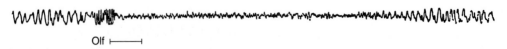

Olf ├────┤

Figure 11–5. Cortical EEG of a rabbit, showing alerting response produced by an olfactory stimulus (Olf).

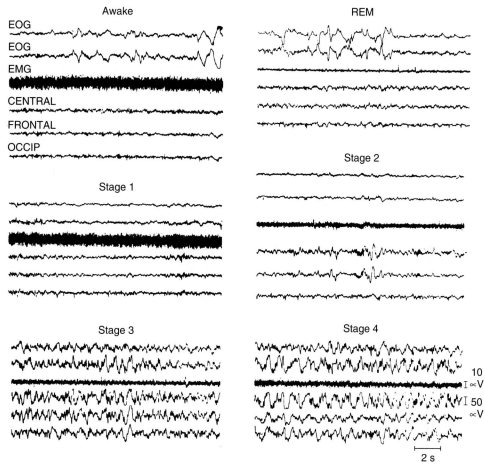

Figure 11–6. EEG and muscle activity during various stages of sleep. EOG, electro-oculogram registering eye movements; EMG, electromyogram registering skeletal muscle activity; CENTRAL, FRONTAL, OCCIP, three EEG leads. Note the low muscle tone with extensive eye movements in REM sleep. (Reproduced, with permission, from Kales A et al: Sleep and dreams: Recent research on clinical aspects. Ann Intern Med 1968;68:1078.)

REM Sleep

The high-amplitude slow waves seen in the EEG during sleep are sometimes replaced by rapid, low-voltage EEG activity, which in primates, including humans, resembles that seen in stage 1 sleep. However, sleep is not interrupted; indeed, the threshold for arousal by sensory stimuli and by stimulation of the reticular formation is elevated. This condition is sometimes called **paradoxical sleep,** since the EEG activity is rapid. There are rapid, roving movements of the eyes during paradoxical sleep, and it is for this reason that it is also called REM sleep. There are no such movements in slow-wave sleep, which consequently is often called NREM sleep. Another characteristic of REM sleep is the occurrence of large phasic potentials, in groups of

three to five, that originate in the pons and pass rapidly to the lateral geniculate body and from there to the occipital cortex. For this reason, they are called **ponto-geniculo-occipital (PGO) spikes.** There is a marked reduction in the tone of the skeletal muscles in the neck (Figure 11–6) during REM sleep. Other muscles keep their tone, but there is a locus ceruleus-dependent relative paralysis of voluntary activity. In cats with locus ceruleus lesions, REM sleep is associated with thrashing about, as if they were acting out their dreams.

PHYSIOLOGIC BASIS OF THE EEG, CONSCIOUSNESS, & SLEEP

The EEG is the record of electrical activity of cortical neural units in a volume conductor (see Chapter 2). It is usually recorded through the skull and scalp and is therefore of much lower voltage than it would be if recorded directly from the cortex. As noted above in reference to evoked potentials, recording from the cortical surface or scalp registers a positive wave when net current flow is toward the electrode and a negative wave when net current flow is away from the surface.

Cortical Dipoles

The dendrites of the cortical cells are a forest of similarly oriented, densely packed units in the superficial layers of the cerebral cortex (Figure 11–1). Propagated potentials can be generated in dendrites (see Chapter 4). In addition, recurrent axon collaterals end on dendrites in the superficial layers. As excitatory and inhibitory endings on the dendrites of each cell become active, current flows into and out of these current sinks and sources from the rest of the dendritic processes and the cell body. The cell-dendrite relationship is therefore that of a constantly shifting dipole. Current flow in this dipole produces wave-like potential fluctuations in a volume conductor (Figure 11–7). When the sum of the dendritic activity is negative relative to the cell, the cell is hypopolarized and hyperexcitable; when it is positive, the cell is hyperpolarized and less excitable. The cere-

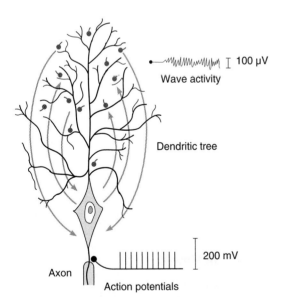

Figure 11–7. Diagrammatic comparison of the electrical responses of the axon and the dendrites of a large cortical neuron. Current flow to and from active synaptic knobs on the dendrites produces wave activity, while all-or-none action potentials are transmitted along the axon.

bellar cortex and the hippocampus are two other parts of the CNS where many complex, parallel dendritic processes are located subpially over a layer of cells. In both areas, there is characteristic rhythmic fluctuation in surface potential similar to that observed in the cortical EEG.

Coordination of Neural Activity

If activity were random in individual cortical dipoles, discharges would cancel out and no waves would be produced. Thus, a coordination mechanism is necessary to produce the waves of the EEG as recorded from the scalp. There is solid new evidence for inhibitory networks of neurons in the brain that secrete GABA and are able to make neurons coordinate their activity. Some of these networks are intracortical and may be involved in learning. Others contain excitatory as well as inhibiting neurons and produce reciprocal current flow between the thalamus and the cerebral cortex.

Information about the effects of these **thalamocortical oscillations** is presented in Figure 11–8. The midline thalamic neurons are hyperpolarized and discharge only in sleep spindle-like phasic bursts during slow wave sleep. During the aware state, they are partially depolarized and fire tonically at rapid rates. This is associated with a more rapid firing rate of cortical neurons.

The state of the thalamic neurons can be shifted from hyperpolarized phasic firing to depolarized tonic firing by sensory stimulation that produces arousal. Conversely, the neurons can be shifted from depolarized to hyperpolarized by stimulation of sleep zones (see below). There is reason to believe that when the neurons are hyperpolarized and firing only on phasic bursts, activity in the thalamocortical oscillations prevents cortical neurons from receiving or processing specific inputs.

Mechanisms Producing EEG Arousal

Replacement of a rhythmic EEG pattern with low-voltage, rapid activity is produced by stimulation of the specific sensory systems up to the level of the midbrain, but stimulation of these systems above the midbrain, stimulation of the specific sensory relay nuclei of the thalamus, or stimulation of the cortical receiving areas themselves does not produce the alerting response. On the other hand, high-frequency stimulation of the midbrain reticular formation produces the EEG alerting response (Figure 11–9) and arouses a sleeping animal. Large bilateral lesions of the lateral and superior portions of the midbrain that interrupt the medial lemnisci and other ascending specific sensory systems fail to prevent EEG alerting produced by sensory stimulation,

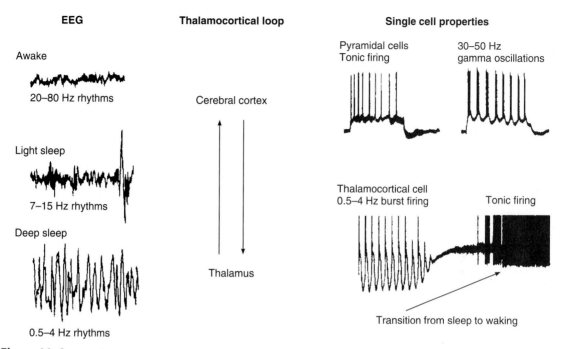

Figure 11–8. Correlation between behavioral states, EEG, and single cell responses in the cerebral cortex and thalamus. (Modified from McCormick DA: Are thalamocortical rhythms the Rosetta stone of a subset of neurological disorders? Nat Med 1999;12:1349.)

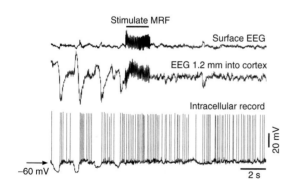

Figure 11–9. Simultaneously recorded surface EEG, EEG 1.2 mm into the cortex, and responses of a cortical neuron recorded with an intracellular electrode in a lightly anesthetized cat. Stimulation of the midbrain reticular formation (MRF) at 300 Hz at the bar produced an alerting response. Note that the rate of discharge of the intracortical neuron increased during and after the stimulation. (Reproduced, with permission, from Steriade M, Amzica F, Contreras D: Synchronization of fast (30–40 Hz) spontaneous cortical rhythms during brain activation. J Neurosci 1996;16:392.)

but lesions in the midbrain tegmentum that disrupt the RAS without damaging the specific systems are associated with a slow-wave pattern that is unaffected by sensory stimulation. Animals with the former type of lesion are awake; those with the latter type are comatose for long periods. Patients with lesions that interrupt the RAS are also somnolent or comatose. Thus, the ascending activity responsible for the EEG alerting response following sensory stimulation passes up the specific sensory systems to the midbrain, enters the RAS via collaterals, and continues through the interlaminar nuclei of thalamus and the nonspecific thalamic projection system to the cortex.

Genesis of Slow-Wave Sleep

Slow waves in the EEG and their behavioral correlate, slow-wave sleep, can be produced by stimulation of at least three subcortical regions. The **diencephalic sleep zone** is in the posterior hypothalamus and the nearby intralaminar and anterior thalamic nuclei. The stimulus frequency must be about 8 Hz; faster stimuli produce arousal. This finding need not be confusing; the important point is that low-frequency stimulation produces one response, whereas high-frequency stimulation produces another. The second zone is the **medullary synchronizing zone** in the reticular formation of the medulla oblongata at the level of the nucleus of the tractus solitarius. Stimulation of this zone, like stimulation of the diencephalic sleep zone, produces sleep if the frequency is low but arousal if the frequency is high. The mechanism by which these effects are produced is not known, but it presumably involves pathways that ascend to the thalamus. The third synchronizing region is the **basal forebrain sleep zone.** This zone includes the preoptic area and the diagonal band of Broca. It differs from the other two zones in that stimulation of the basal forebrain zone produces slow waves and sleep whether the stimulating frequency is high or low.

It is worth noting that stimulation of afferents from mechanoreceptors in the skin at rates of 10 Hz or less also produces sleep in animals, apparently via the brain stem, and it is of course common knowledge that regularly repeated monotonous stimuli put humans to sleep.

On the other hand, slow-wave sleep is under marked circadian control. The roles of the suprachiasmatic nuclei of the hypothalamus in the regulation of sleep and other circadian rhythms are discussed in Chapter 14

There has been considerable debate about the relation of serotonergic neurons in the brain (see Chapter 15) to sleep, but it now appears that serotonin agonists suppress sleep and that the serotonin antagonist ritanserin increases slow-wave sleep in humans. The concentration of adenosine changes in some sleep areas during sleep, so it has been argued that it is a sleep-producing factor. This is consistent with the well-known alerting effects of caffeine, which is an adenosine antagonist. Another hypothesis holds that release of prostaglandin D_2 (PGD_2; see Chapter 17) in the medial preoptic area of the hypothalamus causes increased slow-wave sleep and REM sleep whereas release of PGE_2 causes wakefulness. A lipid produced by the brain is said to have sleep-inducing properties. Several investigators have argued that a peptide produced in the brain is responsible for sleep. However, there is disagreement about which peptide is the putative **sleep peptide,** and its physiologic role, if any, is uncertain.

Concomitants of REM Sleep

Humans aroused at a time when they show the EEG characteristics of REM sleep generally report that they were dreaming, whereas individuals awakened from slow-wave sleep do not. This observation and other evidence indicate that REM sleep and dreaming are closely associated. The tooth-grinding **(bruxism)** that

occurs in some individuals is also associated with dreaming. REM sleep is found in all species of mammals and birds that have been studied, but it probably does not occur in other classes of animals.

If humans are awakened every time they show REM sleep, then permitted to sleep without interruption, they show a great deal more than the normal amount of REM sleep for a few nights. Relatively prolonged REM deprivation does not seem to have adverse psychologic effects. However, experimental animals completely deprived of REM sleep for long periods lose weight in spite of increased caloric intake and eventually die, indicating that REM sleep has some not yet understood but important homeostatic role. On the other hand, deprivation of slow-wave sleep produces similar changes, so they may not be specifically related to REM sleep.

Genesis of REM Sleep

The low-voltage rapid rhythm of the cerebral cortex during REM sleep resembles that during the EEG alerting response and is presumably generated in the same way. The main difference between REM sleep and wakefulness is that dream consciousness is characterized by bizarre imagery and illogical thoughts, and dreams are generally not stored in memory. The reason for this difference is unknown. However, PET scanning of humans in REM sleep shows increased activity in the pontine area, the amygdalas, and the anterior cingulate gyrus but decreased activity in the prefrontal and parietal cortex. Activity in visual association areas is increased, but there is a decrease in the primary visual cortex. This is consistent with increased emotion and operation of a closed neural system cut off from the areas that relate brain activity to the external world.

The mechanism that triggers REM sleep is located in the pontine reticular formation. PGO spikes originate in the lateral pontine tegmentum. The spikes are due to discharge of cholinergic neurons. It now appears that discharge of noradrenergic neurons in the locus ceruleus and serotonergic neurons in the midbrain raphe contributes to wakefulness and that these neurons are silent when cholinergic PGO spike discharge initiates REM sleep. Reserpine, which depletes serotonin and catecholamines, blocks slow-wave sleep and some aspects of REM sleep but increases PGO spike activity. Barbiturates decrease the amount of REM sleep.

Distribution of Sleep Stages

In a typical night of sleep, a young adult first enters NREM sleep, passes through stages 1 and 2, and spends 70–100 minutes in stages 3 and 4. Sleep then lightens, and an REM period follows. This cycle is repeated at intervals of about 90 minutes throughout the night (Figure 11–10). The cycles are similar, though there is less stage 3 and 4 sleep and more REM sleep toward morning. Thus, there are four to six REM periods per night. REM sleep occupies 80% of total sleep time in premature infants (Figure 11–11) and 50% in full-term neonates. Thereafter, the proportion of REM sleep falls rapidly and plateaus at about 25% until it falls further in old age. Children have more total sleep time and stage 4 sleep than adults.

Sleep Disorders

In experimental animals, sleep is necessary for certain forms of learning. Learning sessions do not improve performance in these forms until a period of slow-wave or slow-wave plus REM sleep has occurred. However, it is not known why sleep is necessary, and there is as yet no clinical correlate to this experimental observation.

Insomnia, which may be defined as the subjective problem of insufficient or nonrestorative sleep despite

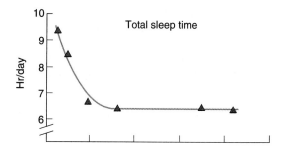
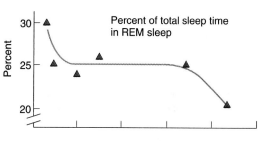
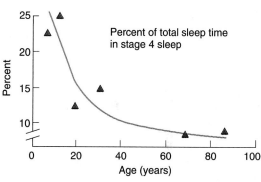

Figure 11–11. Changes in human sleep pattern with age. Each plot shows data points for the ages of 6, 10, 21, 30, 69, and 84 years. (Data from Kandel ER, Schwartz JH, Jessel TM [editors]: *Principles of Neural Science,* 4th ed. McGraw-Hill, 2000.)

an adequate opportunity for sleep, occurs at one time or another in almost all adults. Persistent insomnia can be due to many different mental and medical conditions. It can be relieved temporarily by "sleeping pills," especially benzodiazepines, but prolonged use of any of these pills is unwise because they compromise daytime performance and can be habit-forming.

Fatal familial insomnia is a progressive prion disease that occurs in inherited and sporadic forms. It is characterized by worsening insomnia, impaired autonomic and motor functions, dementia, and death. Patients with the disease have severe neuronal loss and gliosis in the ventral and mediodorsal nuclei of the

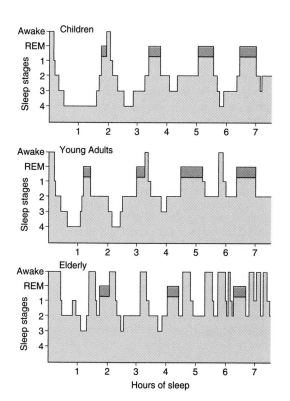

Figure 11–10. Normal sleep cycles at various ages. REM sleep is indicated by the darker colored areas. (Reproduced, with permission, from Kales AM, Kales JD: Sleep disorders. N Engl J Med 1974;290:487.)

thalamus and the olives in the medulla oblongata. Prion diseases are encephalopathies that are transmissible to animals and now include in humans several forms of Creutzfeldt-Jakob disease, Gerstmann-Sträussler-Scheinker disease, and kuru, which is associated with ritual cannibalism.

Sleepwalking (**somnambulism),** bed-wetting (**nocturnal enuresis),** and night terrors occur during slow-wave sleep or, more specifically, during arousal from slow-wave sleep. They are not associated with REM sleep. Episodes of sleepwalking are more common in children than in adults and occur predominantly in males. They may last several minutes. Somnambulists walk with their eyes open and avoid obstacles, but when awakened they cannot recall the episodes.

Narcolepsy is a disease in which there is episodic sudden loss of muscle tone and an eventually irresistible urge to sleep during daytime activities. In some cases it has been shown to start with the sudden onset of REM sleep. REM sleep almost never occurs without previous slow-wave sleep in normal individuals. Familial transmission of this condition in humans is rare, but some Doberman and Labrador dogs have narcolepsy and were found to have a defect in one of the receptors for **orexins (hypocretins)** in the hypothalamus. These peptides also increase appetite (see Chapter 14). Knockout of the orexin pathway in mice causes narcolepsy, and although there does not seem to be a genetic defect of this pathway in humans with narcolepsy, they have very low CSF levels of the orexins.

Sleep apnea is discussed in Chapter 37.

REM behavior disorder is a newly recognized condition in which hypotonia fails to occur during REM sleep. Consequently, patients with this condition, like cats with locus ceruleus lesions, "act out their dreams." They thrash about and may even jump out of bed, ready to do battle with imagined aggressors. This disorder usually responds to treatment with benzodiazepines. Other specific sleep disorders have also been delineated by studies in sleep laboratories.

Clinical Uses of the EEG

The EEG is sometimes of value in localizing pathologic processes. When a collection of fluid overlies a portion of the cortex, activity over this area may be damped. This fact may aid in diagnosing and localizing conditions such as subdural hematomas. Lesions in the cor-

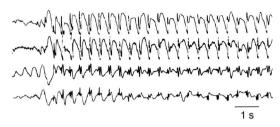

Figure 11–12. Petit mal epilepsy. Record of four cortical EEG leads from a 6-year-old boy who, during the recording, had one of his "blank spells" in which he was transiently unaware of his surroundings and blinked his eyelids. Time is indicated by the horizontal calibration line. (Reproduced, with permission, from Waxman SG: *Neuroanatomy with Clinical Correlations,* 25th ed. McGraw-Hill, 2003.)

tex cause local formation of irregular or slow waves that can be picked up in the EEG leads. Epileptogenic foci sometimes generate high-voltage waves that can be localized.

Epilepsy is a syndrome with multiple causes. In some forms, there are characteristic EEG patterns during seizures; between attacks, however, abnormalities are often difficult to demonstrate. Seizures are now divided into those that arise from one cerebral hemisphere (**partial** or **local seizures**) and those that involve both hemispheres simultaneously (**general-onset seizures).** Each category is further subdivided. Generalized seizures of the tonic-clonic (**grand mal**) type are characterized by loss of consciousness, which usually occurs without warning. This is followed by a tonic phase, with sustained contraction of limb muscles, followed by a clonic phase with symmetric jerking of the limbs as a result of alternating contraction and relaxation. There is fast EEG activity during the tonic phase. Slow waves, each preceded by a spike, occur at the time of each clonic jerk. For a while after the attack, slow waves are present. Similar changes are seen in experimental animals during convulsions produced by electric shocks. Absence (**petit mal**) seizures are one of the forms of generalized seizures characterized by a momentary loss of responsiveness. They are associated with 3/s doublets, each consisting of a typical spike and rounded wave (Figure 11–12).

Control of Posture & Movement

INTRODUCTION

Somatic motor activity depends ultimately upon the pattern and rate of discharge of the spinal motor neurons and homologous neurons in the motor nuclei of the cranial nerves. These neurons, the final common paths to skeletal muscle, are bombarded by impulses from an immense array of pathways. There are many inputs to each spinal motor neuron from the same spinal segment (see Chapter 6). Numerous suprasegmental inputs also converge on these cells from other spinal segments, the brain stem, and the cerebral cortex. Some of these inputs end directly on the motor neurons, but many exert their effects via interneurons or via the γ efferent system to the muscle spindles and back through the Ia afferent fibers to the spinal cord. It is the integrated activity of these multiple inputs from spinal, medullary, midbrain, and cortical levels that regulates the posture of the body and makes coordinated movement possible.

The inputs converging on the motor neurons subserve three semidistinct functions: they bring about voluntary activity; they adjust body posture to provide a stable background for movement; and they coordinate the action of the various muscles to make movements smooth and precise. The patterns of voluntary activity are planned within the brain, and the commands are sent to the muscles primarily via the **corticospinal** and **corticobulbar systems.** Posture is continually adjusted not only before but also during movement by **posture-regulating systems.** Movement is smoothed and coordinated by the medial and intermediate portions of the cerebellum (**spinocerebellum**) and its connections. The **basal ganglia** and the lateral portions of the cerebellum (**neocerebellum**) are part of a feedback circuit to the premotor and motor cortex that is concerned with planning and organizing voluntary movement.

GENERAL PRINCIPLES

Organization

Motor output is of two types: reflexive, or involuntary, and voluntary. Some would add as a subdivision of reflex responses rhythmic responses such as swallowing, chewing, scratching, and walking, which are largely involuntary but subject to voluntary adjustment and control.

Much is still unknown about the control of voluntary movement. To move a limb, for example, the brain must plan a movement, arrange appropriate motion at many different joints at the same time, and adjust the motion by comparing plan with performance. The motor system "learns by doing," and performance improves with repetition. This involves synaptic plasticity.

Nevertheless, there is considerable evidence for the general motor control scheme shown in Figure 12–1. Commands for voluntary movement originate in corti-

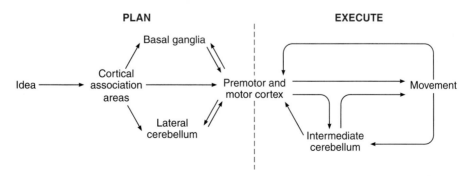

Figure 12–1. Control of voluntary movement.

cal association areas. The movements are planned in the cortex as well as in the basal ganglia and the lateral portions of the cerebellar hemispheres, as indicated by increased electrical activity before the movement. The basal ganglia and cerebellum both funnel information to the premotor and motor cortex by way of the thalamus. Motor commands from the motor cortex are relayed in large part via the corticospinal tracts to the spinal cord and the corresponding corticobulbar tracts to motor neurons in the brain stem. However, collaterals from these pathways and a few direct connections from the motor cortex end on brain stem nuclei, which also project to motor neurons in the brain stem and spinal cord. These pathways can also mediate voluntary movement. Movement sets up alterations in sensory input from the special senses and from muscles, tendons, joints, and the skin. This feedback information, which adjusts and smoothes movement, is relayed directly to the motor cortex and to the spinocerebellum. The spinocerebellum projects in turn to the brain stem. The main brain stem pathways that are concerned with posture and coordination are the rubrospinal, reticulospinal, tectospinal, and vestibulospinal tracts and corresponding projections to motor neurons in the brain stem.

Control of Axial & Distal Muscles

Another theme that is important in motor control is that in the brain stem and spinal cord, medial or ventral pathways and neurons are concerned with the control of muscles of the trunk and proximal portions of the limbs, whereas lateral pathways are concerned with the control of muscles in the distal portions of the limbs. The axial muscles are concerned with postural adjustments and gross movements, whereas the distal limb muscles are those that mediate fine, skilled movements. Thus, for example, the neurons in the medial portion of the ventral horn innervate the proximal limb muscles, particularly the flexors, whereas the lateral ventral horn neurons innervate the distal limb muscles. Similarly, the ventral corticospinal tract and the medial descending paths from the brain stem (the tectospinal, reticulospinal, and vestibulospinal tracts) are concerned with adjustments of proximal muscles and posture, whereas the lateral corticospinal tract and the rubrospinal tract are concerned with distal limb muscles and, particularly in the case of the lateral corticospinal tract, with skilled voluntary movements. Phylogenetically, the medial pathways are old, whereas the lateral pathways are new.

Other Terms

Because the fibers of the lateral corticospinal tract form the pyramids in the medulla, the corticospinal pathways have often been referred to as the **pyramidal system.**

The rest of the descending brain stem and spinal pathways that do not pass through the pyramids and are concerned with postural control have been called the **extrapyramidal system.** However, the ventral corticospinal pathway does not go through the pyramids, many pyramidal fibers are concerned with other functions, and the system that used to be called extrapyramidal is made up of many different pathways with multiple functions. Consequently, the terms pyramidal and extrapyramidal are misleading, and it seems wise to drop them.

In addition, the motor system has often been divided into **upper** and **lower motor neurons.** Lesions of the lower motor neurons—the spinal and cranial motor neurons that directly innervate the muscles—are associated with flaccid paralysis, muscular atrophy, and absence of reflex responses. The syndrome of spastic paralysis and hyperactive stretch reflexes in the absence of muscle atrophy is said to be due to destruction of the "upper motor neurons," the neurons in the brain and spinal cord that activate the motor neurons. However, there are three types of "upper motor neurons" to consider. Lesions in many of the posture-regulating pathways cause spastic paralysis, but lesions limited to the corticospinal and corticobulbar tracts produce weakness **(paresis)** rather than paralysis, and the affected musculature is generally hypotonic. Cerebellar lesions produce incoordination. The unmodified term "upper motor neuron" is therefore confusing.

CORTICOSPINAL & CORTICOBULBAR SYSTEM

ANATOMY & FUNCTION
Tracts

The nerve fibers that pass from the motor cortex to the cranial nerve nuclei form the **corticobulbar tract.** The nerve fibers that cross the midline in the medullary pyramids and form the **lateral corticospinal tract** make up about 80% of the fibers in the corticospinal pathway. The remaining 20% make up the **anterior** or **ventral corticospinal tract** (Figure 12–2), which does not cross the midline until it reaches the level of the muscles it controls. At this point, its fibers end on interneurons that make contact with motor nerves on both sides of the body. The lateral corticospinal tract is concerned with skilled movements, and in humans its fibers end directly on the motor neurons.

Cortical Motor Areas

The cortical areas from which the corticospinal and corticobulbar system originates are generally held to be

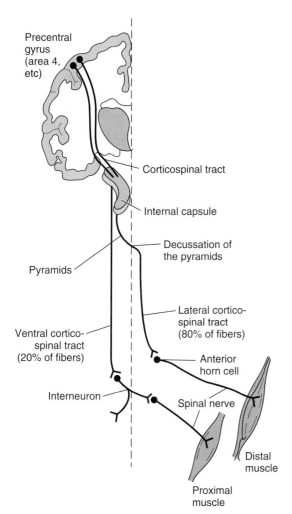

Figure 12–2. The corticospinal tracts.

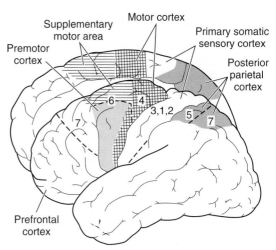

Figure 12–3. Medial **(above)** and lateral **(below)** views of the human cerebral cortex, showing the motor cortex (Brodmann's area 4) and other areas concerned with control of voluntary movement, along with the numbers assigned to the regions by Brodmann. (Reproduced, with permission, from Kandel ER, Schwartz JH, Jessell TM [editors]: *Principles of Neural Science,* 4th ed. McGraw-Hill, 2000.)

those where stimulation produces prompt discrete movement. The best known is the **motor cortex (M1)** in the precentral gyrus (Figure 12–3). However, there is a **supplementary motor area** on and above the superior bank of the cingulate sulcus on the medial side of the hemisphere that reaches to the **premotor cortex** on the lateral surface of the brain (Figure 12–3). Motor responses are also produced by stimulation of somatic sensory area I in the postcentral gyrus and by stimulation of somatic sensory area II in the wall of the sylvian fissure (see Chapter 7). These observations fit with the fact that 30% of the fibers making up the corticospinal and corticobulbar tracts come from the motor cortex but 30% come from the premotor cortex and 40% from the parietal lobe, especially the somatic sensory area.

By means of stimulation experiments in patients undergoing craniotomy under local anesthesia, it has been possible to outline most of the motor projections from the motor cortex. These have been confirmed in unanesthetized unoperated humans by PET scanning and fMRI (Figure 12–4). The various parts of the body are represented in the precentral gyrus, with the feet at the top of the gyrus and the face at the bottom (Figure 12–5). The facial area is represented bilaterally, but the rest of the representation is unilateral, the cortical motor area controlling the musculature on the opposite side of the body. The cortical representation of each body part is proportionate in size to the skill with which the part is used in fine, voluntary movement. The areas involved in speech and hand movements are especially large in the cortex; use of the pharynx, lips, and tongue to form words and of the fingers and apposable thumbs to manipulate the environment are activities in which humans are especially skilled.

The conditions under which the human stimulation studies were performed precluded stimulation of the banks of the sulci and other inaccessible areas. Meticulous study has shown that in monkeys, there is a regular representation of the body, with the axial musculature and the proximal portions of the limbs represented along the anterior edge of the precentral gyrus and the distal part of the limbs along the posterior edge. An-

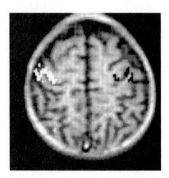

Figure 12–4. Hand area of motor cortex demonstrated by fMRI in a 7-year-old boy. Changes in activity associated with squeezing a rubber ball with the right hand are shown in white; with the left hand in black. In this particular instance, activation of the left motor cortex by activity of the left hand (see text) is not visible. (Reproduced, with permission, from Waxman SG: *Neuroanatomy with Clinical Correlations,* 25th ed. McGraw-Hill, 2003.)

other feature of M1 is the presence of considerable overlap in the muscles that are innervated, as well as innervation of synergic muscles separated by considerable distances. There has been debate about whether individual muscles or movements are represented in M1, and the most recent evidence indicates that both are represented. The cells in the cortical motor areas are arranged in columns. The cells in each column receive fairly extensive sensory input from the peripheral area in which they produce movement, providing the basis for feedback control of movement. Some of this input may be direct, and some is relayed from somatic sensory area I in the postcentral gyrus.

Cerebral dominance, which is discussed in detail in Chapter 16, also affects the motor cortex in humans. Moving the fingers of the left hand is associated mainly with activation of the right motor cortex and vice versa, as measured by imaging techniques (see Chapter 32). However, moving the fingers of the left hand also activates the left motor cortex, particularly in individuals who are right-handed. This correlates with the fact that lesions of the left motor cortex cause motor dysfunction in the left hand as well as the right hand, whereas lesions of the right motor cortex have little effect on the right hand.

Plasticity

A striking discovery made possible by PET and functional MRI (fMRI) is that in intact experimental animals and humans, the motor cortex shows the same kind of plasticity as the sensory cortex (see Chapter 7).

Thus, for example, the finger areas of the contralateral motor cortex enlarge as a pattern of rapid finger movement is learned with the fingers of one hand; this change is detectable at 1 week and maximal at 4 weeks. Cortical areas of output to other muscles also increase in size when motor learning involves these muscles. When a small focal ischemic lesion is produced in the hand area of the motor cortex of monkeys, the hand area may reappear, with return of motor function, in an adjacent undamaged part of the cortex. Thus, the maps of the motor cortex are not immutable, and they change with experience.

Supplementary Motor Area

For the most part, the supplementary motor area projects to the motor cortex. It appears to be involved primarily in programming motor sequences. Lesions of this area in monkeys produce awkwardness in performing complex activities and difficulty with bimanual coordination.

When human subjects count to themselves without speaking, the motor cortex is quiescent, but when they

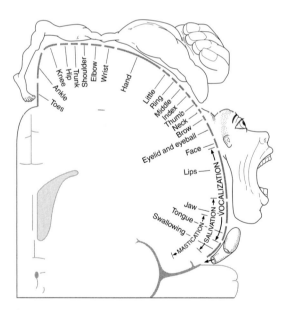

Figure 12–5. Motor homunculus. The figure represents, on a coronal section of the precentral gyrus, the location of the cortical representation of the various parts. The size of the various parts is proportionate to the cortical area devoted to them. Compare with Figure 7–5. (Reproduced, with permission, from Penfield W, Rasmussen G: *The Cerebral Cortex of Man.* Macmillan, 1950.)

speak the numbers aloud as they count, blood flow increases in the motor cortex and the supplementary motor area. Thus, the supplementary motor area as well as the motor cortex is involved in voluntary movement when the movements being performed are complex and involve planning. Blood flow increases whether or not a planned movement is carried out. The increase occurs whether the movement is performed by the contralateral or the ipsilateral hand.

Premotor Cortex

The premotor cortex projects to the brain stem areas concerned with postural control and to the motor cortex as well as providing part of the corticospinal and corticobulbar output. Its function is still incompletely understood, but it may be concerned with setting posture at the start of a planned movement and with getting the individual ready to perform.

Posterior Parietal Cortex

In addition to providing fibers that run in the corticospinal and corticobulbar tracts, the somatic sensory area and related portions of the posterior parietal lobe project to the premotor area. Lesions of the somatic sensory area cause defects in motor performance that are characterized by inability to execute learned sequences of movements such as eating with a knife and fork. Some of the neurons in area 5 (Figure 12–3) are concerned with aiming the hands toward an object and manipulating it, whereas some of the neurons in area 7 are concerned with hand-eye coordination.

Role in Movement

The corticospinal and corticobulbar system is the primary pathway for the initiation of skilled voluntary movement. This does not mean that movement—even skilled movement—is impossible without it. Nonmammalian vertebrates have essentially no corticospinal and corticobulbar system, but they move with great agility. Cats and dogs stand, walk, run, and even eat if food is presented to them after complete destruction of this system. Only in primates are relatively marked deficits produced.

Careful section of the pyramids producing highly selective destruction of the lateral corticospinal tract in laboratory primates produces prompt and sustained loss of the ability to grasp small objects between two fingers and to make isolated movements of the wrists. However, the animal can still use the hand in a gross fashion and can stand and walk. These deficits are consistent with loss of control of the distal musculature of the limbs, which is concerned with fine skilled movements. On the other hand, lesions of the ventral corticospinal tract produce axial muscle deficits that cause difficulty with balance, walking, and climbing.

Effects on Stretch Reflexes

Section of the pyramids in monkeys produces prolonged hypotonia and flaccidity rather than spasticity. The anatomic arrangements in humans are such that disease processes rarely, if ever, damage the corticospinal and corticobulbar tracts without also destroying posture-regulating pathways. When spasticity is present, it is probably caused by damage to these latter pathways rather than to the corticospinal and corticobulbar tracts.

Damage to the lateral corticospinal tract in humans produces the **Babinski sign:** dorsiflexion of the great toe and fanning of the other toes when the lateral aspect of the sole of the foot is scratched. Except in infancy, the normal response to this stimulation is plantar flexion in all the toes. The Babinski sign is believed to be a flexor withdrawal reflex that is normally held in check by the lateral corticospinal system. It is of value in the localization of disease processes, but its physiologic significance is unknown.

POSTURE-REGULATING SYSTEMS

The posture-regulating mechanisms are multiple. They involve a whole series of nuclei and many structures, including the spinal cord, the brain stem, and the cerebral cortex. They are concerned not only with static posture but also, in concert with the corticospinal and corticobulbar systems, with the initiation and control of movement.

Integration

At the spinal cord level, afferent impulses produce simple reflex responses. At higher levels in the nervous system, neural connections of increasing complexity mediate increasingly complicated motor responses. This principle of levels of motor integration is illustrated in Table 12–1. In the intact animal, the individual motor responses are fitted into or "submerged" in the total pattern of motor activity. When the neural axis is transected, the activities integrated below the section are cut off or **released** from the "control of higher brain centers" and often appear to be accentuated. Release of this type, long a cardinal principle in neurology, may be due in some situations to removal of an inhibitory control by higher neural centers. A more important cause of the apparent hyperactivity is loss of differentiation of the reaction, so that it no longer fits into the broader pattern of motor activity. An additional factor may be denervation hypersensitivity of the centers below the

Table 12–1. Summary of levels involved in various neural functions.

Functions	Normal	Decorticate[2]	Midbrain	Hindbrain (Decerebrate)[3]	Spinal	Decerebellate	Level of Integration
Initiative, memory, etc.	+	0	0	0	0	+	Cerebral cortex required
Conditioned reflexes	+	+[4]	0	0	0	+	Cerebral cortex facilitates
Emotional responses	+	++	0	0	0	+	Hypothalamus, limbic system
Locomotor reflexes	+	++	+	0	0	Incoordinate	Midbrain, thalamus
Righting reflexes	+	+	++	0	0	Incoordinate	Midbrain
Antigravity reflexes	+	+	+	++	0	Incoordinate	Medulla
Respiration	+	+	+	+	0	+	Lower medulla
Spinal reflexes[5]	+	+	+	+	++	+	Spinal cord

[1]0 = absent; + = present; ++ = accentuated.
[2]Cerebral cortex removed.
[3]Brain stem transected at the top of the pons.
[4]Conditioned reflexes are more difficult to establish in decorticate than in normal animals.
[5]Other than stretch reflexes.

transection, but the role of this component remains to be determined.

Postural Control

It is impossible to separate postural adjustments from voluntary movement in any rigid way, but it is possible to differentiate a series of postural reflexes (Table 12–2) that not only maintain the body in an upright, balanced position but also provide the constant adjustments necessary to maintain a stable postural background for voluntary activity. These adjustments include maintained **static** reflexes and dynamic, short-term **phasic** reflexes. The former involve sustained contraction of the musculature, whereas the latter involve transient movements. Both are integrated at various levels in the CNS from the spinal cord to the cerebral cortex and are effected largely through various motor pathways. A major factor in postural control is variation in the threshold of the spinal stretch reflexes, which is caused in turn by changes in the excitability of motor neurons and, indirectly, by changes in the

rate of discharge in the γ efferent neurons to muscle spindles.

SPINAL INTEGRATION

The responses of animals and humans after spinal cord transection of the cervical region illustrate the integration of reflexes at the spinal level. The individual spinal reflexes are discussed in Chapter 6.

Spinal Shock

In all vertebrates, transection of the spinal cord is followed by a period of **spinal shock** during which all spinal reflex responses are profoundly depressed. During this period the resting membrane potential of the spinal motor neurons is 2–6 mV greater than normal. Subsequently, reflex responses return and become relatively hyperactive. The duration of spinal shock is proportionate to the degree of encephalization of motor function in the various species. In frogs and rats it lasts

Table 12–2. Principal postural reflexes.

Reflex	Stimulus	Response	Receptor	Integrated In
Stretch reflexes	Stretch	Contraction of muscle	Muscle spindles	Spinal cord, medulla
Positive supporting (magnet) reaction	Contact with sole or palm	Foot extended to support body	Proprioceptors in distal flexors	Spinal cord
Negative supporting reaction	Stretch	Release of positive supporting reaction	Proprioceptors in extensors	Spinal cord
Tonic labyrinthine reflexes	Gravity	Contraction of limb extensor muscles	Otolithic organs	Medulla
Tonic neck reflexes	Head turned: (1) To side (2) Up (3) Down	Change in pattern of extensor contraction: (1) Extension of limbs on side to which head is turned (2) Hind legs flex (3) Forelegs flex	Neck proprioceptors	Medulla
Labyrinthine righting reflexes	Gravity	Head kept level	Otolithic organs	Midbrain
Neck righting reflexes	Stretch of neck muscles	Righting of thorax and shoulders, then pelvis	Muscle spindles	Midbrain
Body on head righting reflexes	Pressure on side of body	Righting of head	Exteroceptors	Midbrain
Body on body righting reflexes	Pressure on side of body	Righting of body even when head held sideways	Exteroceptors	Midbrain
Optical righting reflexes	Visual cues	Righting of head	Eyes	Cerebral cortex
Placing reactions	Various visual, exteroceptive, and proprioceptive cues	Foot placed on supporting surface in position to support body	Various	Cerebral cortex
Hopping reactions	Lateral displacement while standing	Hops, maintaining limbs in position to support body	Muscle spindles	Cerebral cortex

for minutes; in dogs and cats it lasts for 1–2 hours; in monkeys it lasts for days; and in humans it usually lasts for a minimum of 2 weeks.

The cause of spinal shock is uncertain. Cessation of tonic bombardment of spinal neurons by excitatory impulses in descending pathways undoubtedly plays a role, but the subsequent return of reflexes and their eventual hyperactivity also have to be explained. The recovery of reflex excitability may be due to the development of denervation hypersensitivity to the mediators released by the remaining spinal excitatory endings. Another possibility for which there is some evidence is the sprouting of collaterals from existing neurons, with the formation of additional excitatory endings on interneurons and motor neurons.

The first reflex response to appear as spinal shock wears off in humans is frequently a slight contraction of the leg flexors and adductors in response to a noxious stimulus. In some patients, the knee jerks come back first. The interval between cord transection and the beginning return of reflex activity is about 2 weeks in the absence of any complications, but if complications are present it is much longer. It is not known why infection, malnutrition, and other complications of cord transection inhibit spinal reflex activity.

Complications of Cord Transection

Management of paraplegic and quadriplegic humans presents complex problems. Like all immobilized patients, they develop a negative nitrogen balance and catabolize large amounts of body protein. The weight of the body compresses the circulation to the skin over bony prominences, so that unless the patient is moved frequently the skin breaks down at these points and **decubitus ulcers** form. The ulcers heal poorly and are prone to infection because of body protein depletion. The tissues that are broken down include the protein matrix of bone, and this plus the immobilization cause Ca^{2+} to be released in large amounts. This leads to hypercalcemia and hypercalciuria, and calcium stones often form in the urinary tract. The stones and the paralysis of bladder function both cause urinary stasis, which predisposes to urinary tract infection, the most common complication of spinal cord injury. Therefore, the prognosis in patients with transected spinal cords used to be very poor, and death from septicemia, uremia, or inanition occurred in up to 80% of cases. Since World War II, however, the use of antibiotics and meticulous attention to nutrition, fluid balance, skin care, bladder function, and general nursing care have reduced mortality to 6% in major treatment centers and made it possible for many of these patients to survive and lead meaningful lives.

A treatment that fosters recovery and minimizes loss of function after spinal cord injury is acute administration of large doses of glucocorticoids. They should be given as early as possible after the spinal cord injury, then discontinued because of the well-established deleterious effects of long-term treatment with large doses of glucocorticoids (see Chapter 20). Their immediate value in spinal cord and injury cases is probably due to reduction of the inflammatory response in the damaged tissue.

In the meantime, the search continues for ways to get axons of neurons in the spinal cord to regrow across the site of transection. Administration of neurotrophins shows some promise in experimental animals (see Chapter 4), and so does implantation of embryonic stem cells at the site of injury. Another possibility being explored is bypassing the site of the cord injury with brain-computer interface devices. However, these approaches are still a long way from routine clinical use.

Responses in Chronic Spinal Animals & Humans

Once the spinal reflexes begin to reappear after spinal shock, their threshold steadily drops. In chronically quadriplegic humans, the threshold of the withdrawal reflex is especially low. Even minor noxious stimuli may cause not only prolonged withdrawal of one extremity but marked flexion-extension patterns in the other three limbs. Repeated flexion movements may occur

for prolonged periods, and contractures of the flexor muscles develop. Stretch reflexes are also hyperactive, as are more complex reactions based on this reflex. For example, if a finger is placed on the sole of the foot of an animal after the spinal cord has been transected (**spinal animal**), the limb usually extends, following the finger as it is withdrawn. This **magnet reaction (positive supporting reaction)** involves proprioceptive as well as tactile afferents and transforms the limb into a rigid pillar to resist gravity and support the animal. On the basis of the positive supporting reaction, spinal cats and dogs can be made to stand, albeit awkwardly, for as long as 2–3 minutes.

If the cord section is incomplete, the flexor spasms initiated by noxious stimuli can be associated with bursts of pain that are particularly bothersome. They can be treated with considerable success with baclofen, a $GABA_B$ receptor agonist that crosses the blood-brain barrier and facilitates inhibition (see Chapter 4). This treatment is also of benefit in patients with spasticity due to lesions of the brain stem or internal capsule (see below).

Locomotion Generator

Not only can spinal cats and dogs be made to stand, but circuits intrinsic to the spinal cord produce walking movements when stimulated in a suitable fashion. There are two **pattern generators** for locomotion in the spinal cord, one in the cervical and one in the lumbar region. However, this does not mean that spinal animals or humans can walk without stimulation; the pattern generator has to be turned on by tonic discharge of a discrete area in the midbrain, the mesencephalic locomotor region, and of course this is only possible in patients with incomplete spinal cord transection. Interestingly, the generators can also be turned on in experimental animals by administration of the norepinephrine precursor L-dopa (levodopa) after complete section of the spinal cord. Progress is being made in teaching spinal humans to take a few steps by placing them, with support, on a treadmill.

Autonomic Reflexes

Reflex contractions of the full bladder and rectum occur in spinal animals and humans, although the bladder is rarely emptied completely. Hyperactive bladder reflexes can keep the bladder in a shrunken state long enough for hypertrophy and fibrosis of its wall to occur. Blood pressure is generally normal at rest, but the precise feedback regulation normally supplied by the baroreceptor reflexes is absent and wide swings in pressure are common. Bouts of sweating and blanching of the skin also occur.

Sexual Reflexes

Other reflex responses are present in the spinal animal, but in general they are only fragments of patterns that are integrated in the normal animal into purposeful sequences. The sexual reflexes are an example. Coordinated sexual activity depends upon a series of reflexes integrated at many neural levels and is absent after cord transection. However, genital manipulation in male spinal animals and humans produces erection and even ejaculation. In female spinal dogs, vaginal stimulation causes tail deviation and movement of the pelvis into the copulatory position.

Mass Reflex

In chronic spinal animals, afferent stimuli irradiate from one reflex center to another. When even a relatively minor noxious stimulus is applied to the skin, it may irradiate to autonomic centers and produce evacuation of the bladder and rectum, sweating, pallor, and blood pressure swings in addition to the withdrawal response. This distressing **mass reflex** can sometimes be used to give paraplegic patients a degree of bladder and bowel control. They can be trained to initiate urination and defecation by stroking or pinching their thighs, thus producing an intentional mass reflex.

MEDULLARY COMPONENTS

In experimental animals in which the hindbrain and spinal cord are isolated from the rest of the brain by transection of the brain stem at the superior border of the pons, the most prominent finding is marked spasticity of the body musculature. The operative procedure is called **decerebration,** and the resulting pattern of spasticity is called **decerebrate rigidity.** Decerebration produces no phenomenon akin to spinal shock, and the rigidity develops as soon as the brain stem is transected.

Mechanism of Decerebrate Rigidity

On analysis, decerebrate rigidity is found to be spasticity due to diffuse facilitation of stretch reflexes (see Chapter 6). The facilitation is due to two factors: increased general excitability of the motor neuron pool and an increase in the rate of discharge in the γ efferent neurons.

Supraspinal Regulation of Stretch Reflexes

The brain areas that facilitate and inhibit stretch reflexes are shown in Figure 12–6. These areas generally act by increasing or decreasing spindle sensitivity (Figure 12–7). The large facilitatory area in the brain stem reticular formation discharges spontaneously, or possibly in response to afferent input like the RAS. However, the

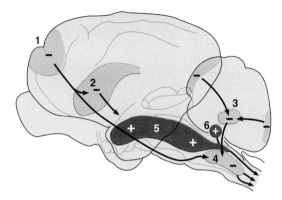

Figure 12–6. Areas in the cat brain where stimulation produces facilitation (plus signs) or inhibition (minus signs) of stretch reflexes. 1, motor cortex; 2, basal ganglia; 3, cerebellum; 4, reticular inhibitory area; 5, reticular facilitatory area; 6, vestibular nuclei.

smaller brain stem area that inhibits γ efferent discharge is driven instead by fibers from the cerebral cortex and the cerebellum. The inhibitory area in the basal ganglia may act through descending connections, as shown in Figure 12–6, or by stimulating the cortical inhibitory center. From the reticular inhibitory and facilitatory areas, impulses descend in the lateral funiculus of the spinal cord. When the brain stem is transected at the level of the top of the pons, the effects of two of the three in-

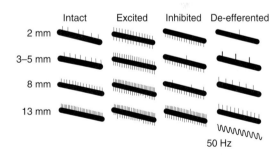

Figure 12–7. Response of a single afferent fiber from a muscle spindle to various degrees of muscle stretch. Numbers at left indicate the amount of stretch. The upward deflections are action potentials; the downward deflections are stimulus artifacts. Records were obtained before brain stimulation (first column), during stimulation of brain areas facilitating (second column) and inhibiting (third column) stretch reflexes, and after section of the motor nerve (fourth column).
(Reproduced, with permission, from Eldred E, Granit R, Merton PA: Supraspinal control of muscle spindles. J Physiol [Lond] 1953;122:498.)

hibitory areas that drive the reticular inhibitory center are removed. Discharge of the facilitatory area continues, but that of the inhibitory area is decreased. Consequently, the balance of facilitatory and inhibitory impulses converging on the γ efferent neurons shifts toward facilitation. Gamma efferent discharge is increased, and stretch reflexes become hyperactive. The cerebellar inhibitory area is still present, and in decerebrate animals, removal of the cerebellum increases the rigidity. The influence of the cerebellum is complex, however, and the net effect of destruction of the cerebellum in humans is hypotonia rather than spasticity.

The vestibulospinal and some related descending pathways are also facilitatory to stretch reflexes and promote rigidity. Unlike the reticular pathways, they pass primarily in the anterior funiculus of the spinal cord, and the rigidity due to increased discharge in them is not abolished by deafferentation of the muscles. This indicates that this rigidity is due to a direct action on the α motor neurons to increase their excitability, rather than an effect mediated through the small motor nerve system, which would, of course, be blocked by deafferentation.

Significance of Decerebrate Rigidity

In cats and dogs, the spasticity produced by decerebration is most marked in the extensor muscles. Sherrington pointed out that these are the muscles with which the cat and dog resist gravity; the decerebrate posture in these animals is, as he put it, "a caricature of the normal standing position." What has been uncovered by decerebration, then, are the tonic, static postural reflex mechanisms that support the animal against gravity. Additional evidence that this is the correct interpretation of the phenomenon comes from the observation that decerebration in the sloth, an arboreal animal that hangs upside down from branches most of the time, causes rigidity in flexion. In humans, the pattern in true decerebrate rigidity is extensor in all four limbs, like that in cats and dogs. Apparently, human beings are not far enough removed from their quadruped ancestors to have changed the pattern in their upper extremities even though the main antigravity muscles of the arms in the upright position are flexors. However, decerebrate rigidity is rare in humans, and the defects that produce it are usually incompatible with life. The more common pattern of extensor rigidity in the legs and moderate flexion in the arms is actually **decorticate rigidity** due to lesions of the cerebral cortex, with most of the brain stem intact (Figure 12–8).

Tonic Labyrinthine Reflexes

In the decerebrate animal, the pattern of rigidity in the limbs varies with the position. No righting responses

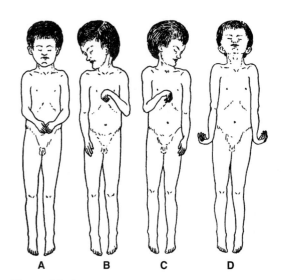

Figure 12–8. Human decorticate rigidity (**A–C**) and true decerebrate rigidity (**D**). In **A** the patient is lying supine with the head unturned. In **B** and **C**, the tonic neck reflex patterns produced by turning of the head to the right or left are shown. (Reproduced, with permission, from Fulton JF [editor]: *Textbook of Physiology*, 17th ed. Saunders, 1955.)

are present, and the animal stays in the position in which it is put. If the animal is placed on its back, the extension of all four limbs is maximal. As the animal is turned to either side, the rigidity decreases, and when it is prone, the rigidity is minimal though still present. These changes in rigidity, the **tonic labyrinthine reflexes,** are initiated by the action of gravity on the otolithic organs and are effected via the vestibulospinal tracts. They are rather surprising in view of the role of rigidity in standing, and their exact physiologic significance remains obscure.

An interesting indication of the overall importance of the vestibular apparatus in postural control is the effects of labyrinthectomy in cats. A normal cat can learn to walk along a rotating beam with little difficulty. After unilateral labyrinthectomy, this ability is lost, but it returns in about 6 weeks as a result of adaptive changes in the remaining postural pathways. However, if both labyrinths are destroyed, the ability to walk the beam never returns.

Tonic Neck Reflexes

If the head of a decerebrate animal is moved relative to the body, changes in the pattern of rigidity occur. If the head is turned to one side, the limbs on that side ("jaw limbs") become more rigidly extended while the contralateral limbs become less so. This is the position often assumed by a normal animal looking to one side. Flexion of the head causes flexion of the forelimbs and continued extension of the hind limbs, the posture of an animal looking into a hole in the ground. Extension of the head causes flexion of the hind limbs and extension of the forelimbs, the posture of an animal looking over an obstacle. These responses are the **tonic neck reflexes.** They are initiated by stretch of the proprioceptors in the upper part of the neck, and they can be sustained for long periods.

MIDBRAIN COMPONENTS

After section of the neural axis at the superior border of the midbrain **(midbrain animal),** extensor rigidity like that seen in the decerebrate animal is present only when the animal lies quietly on its back. In the decerebrate animal, the rigidity, which is a static postural reflex, is prominent because there are no modifying phasic postural reflexes. Chronic midbrain animals can rise to the standing position, walk, and right themselves. While the animals are engaged in these phasic activities, the static phenomenon of rigidity is not seen.

Righting Reflexes

Righting reflexes operate to maintain the normal standing position and keep an animal's head upright. These reflexes are a series of responses integrated for the most part in the nuclei of the midbrain.

When the midbrain animal is held by its body and tipped from side to side, the head stays level in response to the **labyrinthine righting reflexes.** The stimulus is tilting of the head, which stimulates the otolithic organs; the response is compensatory contraction of the neck muscles to keep the head level. If the animal is laid on its side, the pressure on that side of the body initiates reflex righting of the head even if the labyrinths have been destroyed. This is the **body on head righting reflex.** If the head is righted by either of these mechanisms and the body remains tilted, the neck muscles are stretched. Their contraction rights the thorax and initiates a wave of similar stretch reflexes that pass down the body, righting the abdomen and the hindquarters **(neck righting reflexes).** Pressure on the side of the body may cause body righting even if the head is prevented from righting **(body on body righting reflex).**

In cats, dogs, and primates, visual cues can initiate **optical righting reflexes** that right the animal in the absence of labyrinthine or body stimulation. Unlike the other righting reflexes, these responses depend upon an intact cerebral cortex.

In intact humans, the operation of these reflexes maintains the head in a stable position and the eyes fixed on visual targets despite movements of the body and the jerks and jolts of everyday life. The responses are initiated by vestibular stimulation, stretching of neck muscles, and movement of visual images on the retina, and the responses are the vestibulo-ocular reflex (VOR; see Chapter 8) and other remarkably precise reflex contractions of the neck and extraocular muscles.

Grasp Reflex

When a primate in which the brain tissue above the thalamus has been removed lies on its side, the limbs next to the supporting surface are extended. The upper limbs are flexed, and the hand on the upper side grasps firmly any object brought in contact with it (grasp reflex). This whole response is probably a supporting reaction that steadies the animal and aids in pulling it upright.

Other Midbrain Responses

Animals with intact midbrains show pupillary light reflexes if the optic nerves are also intact. Nystagmus, the reflex response to rotational acceleration described in Chapter 9, is also present. If a blindfolded animal is lowered rapidly, its forelegs extend and its toes spread. This response to linear acceleration is a **vestibular placing reaction** that prepares the animal to land on the floor.

CORTICAL COMPONENTS

Effects of Decortication

Removal of the cerebral cortex **(decortication)** produces little motor deficit in many species of mammals. In primates, the deficit is more severe but movement is still possible. Decorticate animals have all the reflex patterns of midbrain animals. In addition, decorticate animals are easier to maintain than midbrain animals because temperature regulation and other visceral homeostatic mechanisms integrated in the hypothalamus (see Chapter 14) are present. The most striking defect is inability to react in terms of past experience. With certain special types of training, conditioned reflexes can be established in the absence of the cerebral cortex. However, under normal laboratory conditions, there is no evidence that learning or conditioning occurs.

Decorticate Rigidity

Moderate rigidity is present in the decorticate animal as a result of the loss of the cortical area that inhibits γ efferent discharge via the reticular formation. Like the rigidity present after transection of the neural axis anywhere above the top of the midbrain, this **decorticate rigidity** is obscured by phasic postural reflexes and is seen only when the animal is at rest. Decorticate rigidity is seen on the hemiplegic side in humans after hemorrhages or thromboses in the internal capsule. Probably because of their anatomy, the small arteries in the internal capsule are especially prone to rupture or thrombotic obstruction, so this type of decorticate rigidity is common. Sixty percent of intracerebral hemorrhages occur in the internal capsule, as opposed to 10% in the cerebral cortex, 10% in the pons, 10% in the thalamus, and 10% in the cerebellum.

The exact site of origin in the cerebral cortex of the fibers that inhibit stretch reflexes is a subject of debate. Under certain experimental conditions, stimulation of the anterior edge of the precentral gyrus is said to cause inhibition of stretch reflexes and cortically evoked movements. This region, which also projects to the basal ganglia, has been named area 4s or the **suppressor strip.**

Hopping & Placing Reactions

Two types of postural reactions, the **hopping** and **placing reactions,** are seriously disrupted by decortication. The former are the hopping movements that keep the limbs in position to support the body when a standing animal is pushed laterally. The latter are the reactions that place the foot firmly on a supporting surface. They can be initiated in a blindfolded animal held suspended in the air by touching the supporting surface with any part of the foot. Similarly, when the snout or vibrissae of a suspended animal touches a table, the animal immediately places both forepaws on the table; and if one limb of a standing animal is pulled out from under it, the limb is promptly replaced on the supporting surface. The vestibular placing reaction has already been mentioned. In cats, dogs, and primates, the limbs are extended to support the body when the animal is lowered toward a surface it can see.

BASAL GANGLIA

Anatomic Considerations

The term **basal ganglia** is generally applied to five structures on each side of the brain: the **caudate nucleus, putamen,** and **globus pallidus,** three large nuclear masses underlying the cortical mantle (Figure 12–9), and the functionally related **subthalamic nucleus** (body of Luys) and **substantia nigra.** The globus pallidus is divided into an external and an internal segment. The substantia nigra is divided into a **pars compacta** and a **pars reticulata.** Parts of the thalamus are

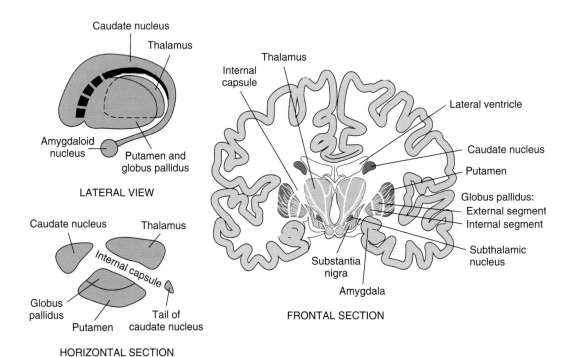

Figure 12–9. The basal ganglia.

intimately related to the basal ganglia. The caudate nucleus and the putamen are frequently called the **striatum;** the putamen and the globus pallidus are sometimes called the **lenticular nucleus** (Table 12–3).

The main afferent connections to the basal ganglia terminate in the striatum (Figure 12–10). They include the **corticostriate projection** from all parts of the cerebral cortex. There is also a projection from the centromedian nucleus of the thalamus to the striatum.

The connections between the parts of the basal ganglia include a dopaminergic nigrostriatal projection from the pars compacta of the substantia nigra to the striatum and a corresponding GABAergic projection from the striatum to the pars reticulata of the substantia nigra. The caudate nucleus and the putamen project to both segments of the globus pallidus. The external segment of the globus pallidus projects to the subthalamic nucleus, which in turn projects to both segments of the globus pallidus and the substantia nigra.

The principal output from the basal ganglia is from the internal segment of the globus pallidus via the **thalamic fasciculus** to the ventral lateral, ventral anterior, and centromedian nuclei of the thalamus. From the thalamic nuclei, fibers project to the prefrontal and premotor cortex. The substantia nigra also projects to the thalamus. These connections, along with the probable synaptic transmitters involved, are summarized in Figure 12–10. There are a few additional projections to the habenula and the superior colliculus. However, the main feature of the connections of the basal ganglia is that the cerebral cortex projects to the striatum, the striatum to the internal segment of the globus pallidus, the internal segment of the globus pallidus to the thalamus, and the thalamus back to the cortex, completing a loop. The output from the internal segment of the globus pallidus to the thalamus is inhibitory, whereas the output from the thalamus to the cerebral cortex is excitatory.

The striatum is made up of a unique mosaic of **patches** or **striosomes** composed of nerve endings in a **matrix** that receives other endings. The neurons of the corticostriate projection that originate in the deep por-

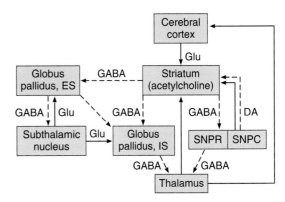

Figure 12–10. Diagrammatic representation of the principal connections of the basal ganglia. Solid lines indicate excitatory pathways, dashed lines inhibitory pathways. The transmitters are indicated in the pathways, where they are known. Glu, glutamate; DA, dopamine. Acetylcholine is the transmitter produced by interneurons in the striatum. SNPR, substantia nigra, pars reticulata; SNPC, substantia nigra, pars compacta; ES, external segment; IS, internal segment. The subthalamic nucleus also projects to the pars compacta of the substantia nigra; this pathway has been omitted for clarity.

tion of layer 5 of the cortex terminate in the patches, whereas the neurons that originate in layers 2 and 3 and the superficial part of layer 5 end primarily in the matrix. Neurons with their cell bodies in patches project in large part to dopaminergic neurons in the pars compacta of the substantia nigra, whereas many of the neurons with their cell bodies in the matrix project to GABAergic neurons in the pars reticulata of the substantia nigra. However, the physiologic significance of these connections is uncertain.

Metabolic Considerations

The metabolism of the basal ganglia is unique in a number of ways. These structures have a high O_2 consumption. The copper content of the substantia nigra and the nearby locus ceruleus is particularly high. In Wilson's disease, a genetic autosomal recessive disorder of copper metabolism in which the plasma level of the copper-binding protein **ceruloplasmin** is usually low, there is chronic copper intoxication and severe degeneration of the lenticular nucleus.

Function

Our knowledge of the precise functions of the basal ganglia is still rudimentary. Lesions in the basal gan-

Table 12–3. The basal ganglia.

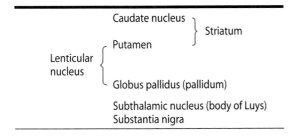

glia of animals have relatively little effect. However, recording studies have made it clear that neurons in the basal ganglia, like those in the lateral portions of the cerebellar hemispheres, discharge before movements begin. These observations, plus careful analysis of the effects of diseases of the basal ganglion in humans and the effects of drugs that destroy dopaminergic neurons in animals (see below), have led to the concept that the basal ganglia are involved in the planning and programming of movement or, more broadly, in the processes by which an abstract thought is converted into voluntary action (Figure 12–1). They discharge via the thalamus to areas related to the motor cortex, and the corticospinal pathways provide the final common pathway to the motor neurons. In addition, the field potentials in the basal ganglia oscillate, and it has been suggested that the oscillations may have functions like the putative functions of the oscillations of the thalamocortical circuits (see Chapter 11).

The basal ganglia also play a role in some cognitive processes, and these are particularly the province of the caudate nucleus. Possibly because of the interconnections of this nucleus with the frontal portions of the neocortex, lesions of the caudate disrupt performance on tests involving object reversal and delayed alternation. In addition, lesions of the head of the left but not the right caudate nucleus and nearby white matter in humans are associated with a dysarthric form of aphasia that resembles but is different from Wernicke's aphasia (see Chapter 16).

Diseases of the Basal Ganglia in Humans

It is interesting that even though lesions in the basal ganglia in experimental animals have little apparent effect, disease processes affecting these ganglia in humans produce marked and characteristic abnormalities of motor function. Disorders of movement associated with diseases of the basal ganglia in humans are of two general types: **hyperkinetic** and **hypokinetic.** The hyperkinetic conditions, those in which there is excessive and abnormal movement, include chorea, athetosis, and ballism. Hypokinetic abnormalities include akinesia and bradykinesia.

Chorea is characterized by rapid, involuntary "dancing" movements. **Athetosis** is characterized by continuous, slow writhing movements. Choreiform and athetotic movements have been likened to the start of voluntary movements occurring in an involuntary, disorganized way. In **ballism,** there are involuntary movements that are flailing, intense, and violent.

Akinesia is difficulty in initiating movement and decreased spontaneous movement. **Bradykinesia** is slowness of movement.

Huntington's Disease

The initial detectable damage in Huntington's disease is to medium spiny neurons in the caudate and putamen. An early sign is a jerky trajectory of the hand when reaching to touch a spot, especially toward the end of the reach. Later, hyperkinetic choreiform movements appear and gradually increase until they incapacitate the patient. Speech becomes slurred and then incomprehensible, and there is progressive dementia followed by death, usually within 10–15 years after the onset of symptoms. In the basal ganglion, three biochemically distinct pathways normally operate in a balanced fashion: (1) the nigrostriatal dopaminergic system, (2) the intrastriatal cholinergic system, and (3) the GABAergic system, which projects from the striatum to the globus pallidus and substantia nigra. In Huntington's disease, there is loss of the intrastriatal GABAergic and cholinergic neurons. The loss of the GABAergic pathway to the external pallidum releases inhibition, permitting the hyperkinetic features of the disease to develop. Degeneration of the nigrostriatal dopaminergic system causes Parkinson's disease (see below).

Huntington's disease is inherited as an autosomal dominant disorder, and its onset is usually between the ages of 30 and 50. The abnormal gene responsible for the disease is located near the end of the short arm of chromosome 4. It normally contains 11–34 cytosine-adenine-guanine (CAG) repeats, each coding for glutamine. In patients with Huntington's disease, this number is increased to 42–86 or more copies, and the greater the number of repeats, the earlier the age of onset and the more rapid the progression of the disease. The gene codes for **huntingtin,** a protein of unknown function. Poorly soluble protein aggregates form in cell nuclei and elsewhere, but it is uncertain whether these cause symptoms. However, it appears there is loss of the function of huntingtin that is proportionate to the size of the CAG insert. At present, no effective treatment is clinically available, and the disease is uniformly fatal. However, there are a few glimmers of hope. In animal models of the disease, intrastriatal grafting of fetal striatal tissue improves cognitive performance. In addition, tissue capsase-1 activity is increased in the brains of humans and animals with the disease, and in mice in which the gene for this apoptosis-regulating enzyme has been knocked out, progression of the disease is slowed.

Huntington's disease is one of an increasing number of human genetic diseases affecting the nervous system that are characterized by **trinucleotide repeat** expansion. Most of these involve CAG repeats (Table 12–4), but one involves CGG repeats and another involves CTG repeats. All these are in exons. However, a GAA repeat in an intron has been shown to be associated

Table 12–4. Examples of trinucleotide repeat diseases.

Disease	Expanded Trinucleotide Repeat	Affected Protein
Huntington's disease	CAG	Huntingtin
Spinocerebellar ataxia, type 1	CAG	Ataxin 1
Spinocerebellar ataxia, type 2	CAG	Ataxin 2
Spinocerebellar ataxia, type 3	CAG	Ataxin 3
Spinocerebellar ataxia, type 6	CAG	α_{1A} subunit of Ca^{2+} channel
Spinocerebellar ataxia, type 7	CAG	Ataxin 7
Dentatorubral-pallidoluysian atrophy	CAG	Atrophin
Spinobulbar muscular atrophy	CAG	Androgen receptor
Fragile X syndrome	CGG	FMR-1
Myotonic dystrophy	CTG	DM protein kinase
Friedreich's ataxia	GAA	Frataxin

with Friedreich's ataxia. There is also preliminary evidence that increased numbers of a 12-nucleotide repeat are associated with a rare form of epilepsy.

Parkinson's Disease (Paralysis Agitans)

Parkinson's disease has both hypokinetic and hyperkinetic features. In this condition, which was originally described by James Parkinson and is named for him, the nigrostriatal dopaminergic neurons degenerate. The fibers to the putamen are most severely affected. Parkinsonism now occurs in sporadic idiopathic form in many middle-aged and elderly individuals and is one of the most common neurodegenerative diseases; it is estimated to occur in 1–2% of individuals over age 65. Familial cases also occur, but these are uncommon. There is a steady loss of dopaminergic neurons and dopamine receptors with age in the basal ganglia in normal individuals, and it is apparently an acceleration of these losses that precipitates parkinsonism. Symp-

toms appear when 60–80% of the nigrostriatal dopaminergic neurons are lost. Parkinsonism is also seen as a complication of treatment with the phenothiazine group of tranquilizer drugs and other drugs that block D_2 dopamine receptors. It can be produced in rapid and dramatic form by injection of MPTP (Figure 12–11). This effect was discovered by chance when a drug dealer in northern California supplied some of his clients with a homemade preparation of "synthetic heroin" that contained MPTP. MPTP is a prodrug that is metabolized in astrocytes by the enzyme monoamine oxidase B to produce a potent oxidant, MPP^+. In rodents, MPP^+ is rapidly removed from the brain, but in primates, it is removed more slowly and is taken up by the dopamine transporter into dopaminergic neurons in the substantia nigra, which it destroys without affecting other dopaminergic neurons to any appreciable degree. Consequently, MPTP can be used to produce parkinsonism in monkeys, and its availability has accelerated research on the function of the basal ganglia.

The hypokinetic features of Parkinson's disease are **akinesia** and **bradykinesia,** and the hyperkinetic features are **rigidity** and **tremor.** The absence of motor activity and the difficulty in initiating voluntary movements are striking. There is a decrease in **associated movements,** the normal, unconscious movements such as swinging of the arms during walking, the panorama of facial expressions related to the emotional content of thought and speech, and the multiple "fidgety" actions and gestures that occur in all of us. The rigidity is different from spasticity because there is increased motor neuron discharge to both the agonist and antagonist muscles. Passive motion of an extremity meets with a plastic, dead-feeling resistance that has been likened to bending a lead pipe and is therefore called **lead pipe rigidity.** Sometimes there is a series of "catches" during passive motion (**cogwheel rigidity**), but the sudden loss of resistance seen in a spastic extremity is absent. The tremor, which is present at rest and disappears

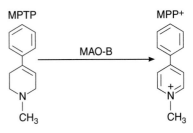

Figure 12–11. Conversion of 1-methyl-4-phenyl-1,2,5,6-tetrahydropyridine (MPTP) to 1-methyl-4-phenylpyridinium (MPP^+) by monoamine oxidase B (MAO-B).

with activity, is due to regular, alternating, 8 Hz contractions of antagonistic muscles.

A current view of the pathogenesis of Parkinson's disease is that there is an imbalance between excitation and inhibition in the basal ganglia, created by the loss of dopaminergic inhibition of the putamen (Figure 12–12). The resulting increase in inhibitory output to the external segment of the globus pallidus decreases inhibitory output from the subthalamic nucleus, and this increases the excitatory output from this nucleus to the internal segment of the globus pallidus. This in turn increases the inhibitory output from this segment to the thalamus, causing a reduction in excitatory drive to the cerebral cortex.

Treatment

An important consideration in Parkinson's disease is the balance between the excitatory discharge of cholinergic interneurons and the inhibitory dopaminergic input in the striatum. Some improvement is produced by decreasing the cholinergic influence with anticholinergic drugs. More dramatic improvement is produced by ad-

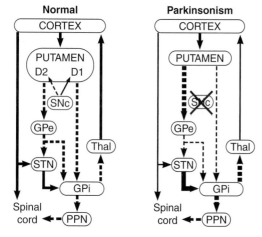

Figure 12–12. Basal ganglia-thalamocortical circuitry in Parkinson's disease. Solid arrows indicate excitatory outputs and dashed arrows inhibitory outputs. The strength of each output is indicated by the width of the arrow. The striatum contains D_1 dopamine receptors that increase direct output to the internal segment of the globus pallidus (GPi) and D_2 receptors that inhibit indirect output to the GPi via the putamen, external segment of the globus pallidus (GPe), and subthalamic nuclei (STN). SNc, substantia nigra; thal, thalamus; PPN, pedunculopontine nuclei. (Reproduced, with permission, from Grafton SC, DeLong M: Tracing the brain's circuitry with functional imaging. Nat Med 1997;3:602.)

ministration of L-dopa (levodopa). Unlike dopamine, this dopamine precursor crosses the blood-brain barrier (see Chapter 15) and helps repair the dopamine deficiency. However, the degeneration of these neurons continues, and in 5–7 years the beneficial effects of L-dopa disappear.

Surgical treatment by making lesions in the internal segment of the globus pallidus (pallidotomy) or in the subthalamic nucleus helps to restore the output balance toward normal (Figure 12–12). Surgical outcomes have been further improved by implanting electrodes attached to subcutaneous stimulators and administering high-frequency current, which produces temporary disruption of circuits at the electrode tip on demand.

Another surgical approach is to implant dopamine-secreting tissue in or near the basal ganglia. Transplants of the patient's own adrenal medullary tissue or carotid body works for a while, apparently by functioning as a sort of dopamine minipump, but long-term results have been disappointing. Results with transplantation of fetal striatal tissue have been better, and there is evidence that the transplanted cells not only survive but make appropriate connections in the host's basal ganglia. However, some patients with transplants develop severe involuntary movements (dyskinesias).

Recent research has indicated that neurotrophic factors (see Chapter 2) benefit the nigrostriatal neurons, and local injection of glial cell line-derived neurotrophic factor (GDNF) attached to a lentivirus vector so that it penetrates cells has produced promising results in monkeys.

In familial Parkinson's disease, the genes for three proteins are mutated. Two of the three proteins, **α-synuclein** and **barkin,** interact and are found in Lewy bodies—inclusion bodies in neurons that are found in all forms of Parkinson's disease. However, the significance of these findings is still unsettled.

CEREBELLUM

Anatomic Divisions

The cerebellum sits astride the main sensory and motor systems in the brain stem (Figure 12–13). It is connected to the brain stem on each side by a **superior peduncle** (brachium conjunctivum), **middle peduncle** (brachium pontis), and **inferior peduncle** (restiform body). The medial **vermis** and lateral **cerebellar hemispheres** are more extensively folded and fissured than the cerebral cortex; the cerebellum weighs only 10% as much as the cerebral cortex, but its surface area is about 75% of that of the cerebral cortex. Anatomically, the cerebellum is divided into three parts by two transverse

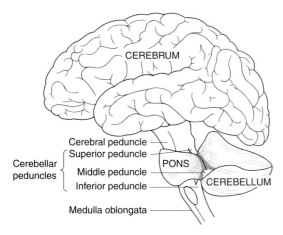

Figure 12–13. Diagrammatic representation of the principal parts of the brain. The parts are distorted to show the cerebellar peduncles and the way the cerebellum, pons, and middle peduncle form a "napkin ring" around the brain stem. (Reproduced, with permission, from Goss CM [editor]: *Gray's Anatomy of the Human Body,* 27th ed. Lea & Febiger, 1959.)

fissures. The posterolateral fissure separates the medial nodulus and the lateral flocculus on either side from the rest of the cerebellum, and the primary fissure divides the remainder into an anterior and a posterior lobe. Lesser fissures divide the vermis into smaller sections, so that it contains ten primary lobules numbered I–X from superior to inferior. These lobules are identified by name and number in Figure 12–14.

Functional Divisions

From a functional point of view, the cerebellum is also divided into three parts, but in a different way (Figure 12–15). The nodulus in the vermis and the flanking flocculus in the hemisphere on each side form the **flocculonodular lobe** or **vestibulocerebellum.** This lobe, which is phylogenetically the oldest part of the cerebellum, has vestibular connections and is concerned with equilibrium and learning-induced changes in the VOR (see Chapter 9). The rest of the vermis and the adjacent medial portions of the hemispheres form the **spinocerebellum,** the region that receives proprioceptive input from the body as well as a copy of the "motor plan" from the motor cortex. By comparing plan with performance, it smoothes and coordinates movements that are ongoing. The vermis projects to the brain stem area concerned with control of axial and proximal limb muscles, whereas the hemispheres project the brain stem areas concerned with control of distal limb muscles. The lateral portions of the cerebellar hemispheres

are called the **neocerebellum.** They are the newest from a phylogenetic point of view, reaching their greatest development in humans. They interact with the motor cortex in planning and programming movements.

Organization

The cerebellum has an external **cerebellar cortex** separated by white matter from the **deep cerebellar nuclei.** Its primary afferent inputs, the mossy and climbing fibers (see below), send collaterals to the deep nuclei and pass to the cortex (Figure 12–16). There are four deep nuclei: the **dentate,** the **globose,** the **emboliform,** and the **fastigial** nuclei. The globose and the emboliform nuclei are sometimes lumped together as the **interpositus nucleus.** Most of the vestibulocerebellar output passes directly to the brain stem, but the rest of the cerebellar cortex projects to the deep nuclei, which in turn project to the brain stem. Thus, the deep nuclei provide the only output for the spinocerebellum and the neocerebellum. The medial portion of the spinocerebellum projects to the fastigial nuclei and from there to the brain stem. The adjacent hemispheric portions of the spinocerebellum project to the emboliform and globose nuclei and from there to the brain stem. The neocerebellum projects to the dentate nucleus and from there either directly or indirectly to the ventrolateral nucleus of the thalamus.

The cerebellar cortex contains only five types of neurons: Purkinje, granule, basket, stellate, and Golgi cells. It has three layers (Figure 12–17): an external molecular layer, a Purkinje cell layer that is only one cell thick, and an internal granular layer. The **Purkinje cells** are among the biggest neurons in the body. They have very extensive dendritic arbors that extend throughout the molecular layer. Their axons, which are the only output from the cerebellar cortex, generally pass to the deep nuclei. The cerebellar cortex also contains **granule cells,** which receive input from the mossy fibers and innervate the Purkinje cells. The granule cells have their cell bodies in the granular layer. Each sends an axon to the molecular layer, where the axon bifurcates to form a T. The branches of the T are straight and run long distances. Consequently, they are called **parallel fibers.** The dendritic trees of the Purkinje cells are markedly flattened (Figure 12–17) and oriented at right angles to the parallel fibers. The parallel fibers thus make synaptic contact with the dendrites of many Purkinje cells, and the parallel fibers and Purkinje dendritic trees form a grid of remarkably regular proportions.

The other three types of neurons in the cerebellar cortex are in effect inhibitory interneurons. The **basket cells** (Figure 12–17) are located in the molecular layer.

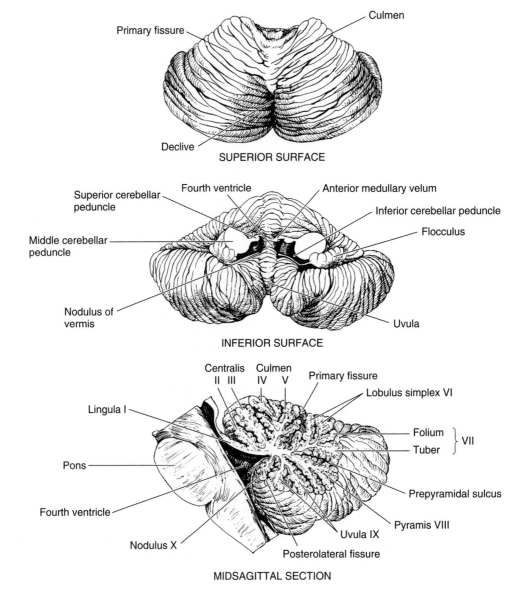

Figure 12–14. Superior and inferior views and sagittal section of the human cerebellum. The ten principal lobules are identified by name and by number (I–X).

They receive input from the parallel fibers, and each projects to many Purkinje cells. Their axons form a basket around the cell body and axon hillock of each Purkinje cell they innervate. The **stellate cells** are similar to the basket cells but more superficial in location. The **Golgi cells** are located in the granular layer. Their dendrites, which project into the molecular layer, receive input from the parallel fibers. Their cell bodies receive input via collaterals from the incoming mossy fibers

and the Purkinje cells. Their axons project to the dendrites of the granule cells.

As noted above, the two main inputs to the cerebellar cortex are **climbing fibers** and **mossy fibers.** Both are excitatory (Figure 12–16). The climbing fibers come from a single source, the inferior olivary nuclei. Each projects to the primary dendrites of a Purkinje cell, around which it entwines like a climbing plant. Proprioceptive input to the inferior olivary nuclei

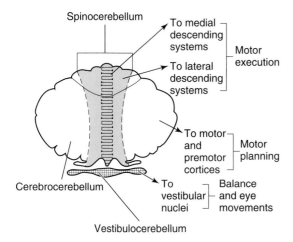

Figure 12–15. Functional divisions of the cerebellum. (Modified from Kandel ER, Schwartz JH, Jessell TM [editors]: *Principles of Neural Science,* 4th ed. McGraw-Hill, 2000.)

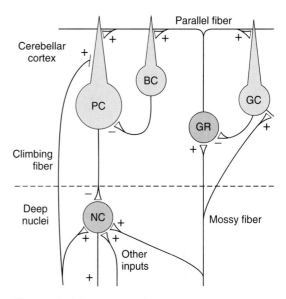

Figure 12–16. Diagram of neural connections in the cerebellum. + and − signs indicate whether endings are excitatory or inhibitory. BC, basket cell; GC, Golgi cell; GR, granule cell; NC, cell in deep nucleus; PC, Purkinje cell. Note that PCs and BCs are inhibitory. The connections of the stellate cells, which are not shown, are similar to those of the basket cells, except that they end for the most part on Purkinje cell dendrites.

comes from all over the body. On the other hand, the mossy fibers provide direct proprioceptive input from all parts of the body plus input from the cerebral cortex via the pontine nuclei to the cerebellar cortex. They end on the dendrites of granule cells in complex synaptic groupings called **glomeruli.** The glomeruli also contain the inhibitory endings of the Golgi cells mentioned above.

The fundamental circuits of the cerebellar cortex are thus relatively simple (Figure 12–16). Climbing fiber inputs exert a strong excitatory effect on single Purkinje cells, whereas mossy fiber inputs exert a weak excitatory effect on many Purkinje cells via the granule cells. The basket and stellate cells are also excited by granule cells via the parallel fibers, and their output inhibits Purkinje cell discharge (feed-forward inhibition). Golgi cells are excited by the mossy fiber collaterals, Purkinje cell collaterals, and parallel fibers, and they inhibit transmission from mossy fibers to granule cells. The transmitter secreted by the stellate, basket, Golgi, and Purkinje cells appears to be GABA, whereas the granule cells probably secrete glutamate. GABA acts via $GABA_A$ receptors, but the combinations of subunits in these receptors (see Chapter 4) vary from one cell type to the next. The granule cell is unique in that it appears to be the only type of neuron in the CNS that has a $GABA_A$ receptor containing the $\alpha6$ subunit.

The output of the Purkinje cells is in turn inhibitory to the deep cerebellar nuclei. As noted above, these nuclei also receive excitatory inputs via collaterals from the mossy and climbing fibers. It is interesting, in view of their inhibitory Purkinje cell input, that the output of the deep cerebellar nuclei to the brain stem and thalamus is always excitatory. Thus, almost all the cerebellar circuitry seems to be concerned solely with modulating or timing the excitatory output of the deep cerebellar nuclei to the brain stem and thalamus.

The primary afferent systems that converge to form the mossy fiber or climbing fiber input to the cerebellum are summarized in Table 12–5.

Flocculonodular Lobe

Animals in which the **flocculonodular lobe** has been destroyed walk in a staggering fashion on a broad base. They tend to fall and are reluctant to move without support. Similar defects are seen in children as the earliest signs of a midline cerebellar tumor that arises from cell rests in the nodule. Early in its course, it produces damage that is generally localized to the flocculonodular lobe.

Selective ablation of the flocculonodular lobe in dogs abolishes the syndrome of **motion sickness** (see Chapter 9), whereas extensive lesions in other parts of the cerebellum and the rest of the brain fail to affect it.

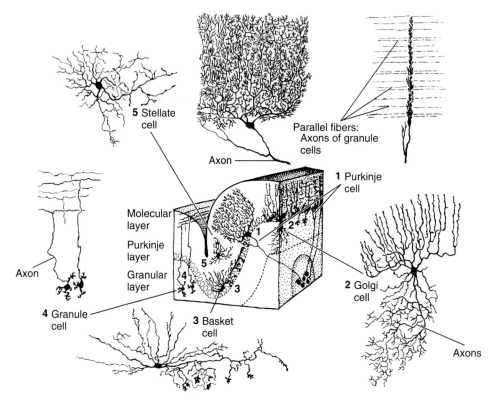

Figure 12–17. Location and structure of neurons in the cerebellar cortex. (Reproduced, with permission, from Kuffler SW, Nicholls JG, Martin AR: *From Neuron to Brain,* 2nd ed. Sinauer, 1984.)

Effects on Stretch Reflexes

Stimulation of the cerebellar areas that receive proprioceptive input sometimes inhibits and sometimes facilitates movements evoked by stimulation of the cerebral cortex. Lesions in folia I–VI and the paramedian areas in experimental animals cause spasticity localized to the part of the body that is represented in the part of the cerebellum destroyed. However, hypotonia is characteristic of cerebellar destruction in humans.

Effects on Movement

Except for the changes in stretch reflexes, experimental animals and humans with lesions of the cerebellar hemispheres show no abnormalities as long as they are at rest. However, pronounced abnormalities are apparent when they move. There is no paralysis and no sensory deficit, but all movements are characterized by a marked **ataxia,** a defect defined as incoordination due to errors in the rate, range, force, and direction of movement. With circumscribed lesions, the ataxia may be localized to one part of the body. If only the cortex of the cerebellum is involved, the movement abnormal-

ities gradually disappear as **compensation** occurs. Lesions of the cerebellar nuclei produce more generalized defects, and the abnormalities are permanent. For this reason, care should be taken to avoid damaging the nuclei when surgical removal of the parts of the cerebellum is necessary.

Other signs of cerebellar deficit in humans provide additional illustrations of the importance of the cerebellum in the control of movement. Ataxia is manifest not only in the wide-based, unsteady, "drunken" gait of patients but also in defects of the skilled movements involved in the production of speech, so that slurred or **scanning speech** results. Other voluntary movements are also highly abnormal. For example, attempting to touch an object with a finger results in overshooting to one side or the other. This **dysmetria,** which is also called **past-pointing,** promptly initiates a gross corrective action, but the correction overshoots to the other side. Consequently, the finger oscillates back and forth. This oscillation is the **intention tremor** of cerebellar disease. Unlike the resting tremor of parkinsonism, it is absent at rest; however, it appears whenever the patient attempts to perform some voluntary action. Another

Table 12–5. Function of principal afferent systems to the cerebellum.[1]

Afferent Tracts	Transmits
Vestibulocerebellar	Vestibular impulses from labyrinths, direct and via vestibular nuclei
Dorsal spinocerebellar	Proprioceptive and exteroceptive impulses from body
Ventral spinocerebellar	Proprioceptive and exteroceptive impulses from body
Cuneocerebellar	Proprioceptive impulses, especially from head and neck
Tectocerebellar	Auditory and visual impulses via inferior and superior colliculi
Pontocerebellar	Impulses from motor and other parts of cerebral cortex via pontine nuclei
Olivocerebellar	Proprioceptive input from whole body via relay in inferior olive

[1]The olivocerebellar pathway projects to the cerebellar cortex via climbing fibers. The rest of the listed paths project via mossy fibers. Several other pathways transmit impulses from nuclei in the brain stem to the cerebellar cortex and to the deep nuclei, including a serotonergic input from the raphe nuclei to the granular and molecular layers and a noradrenergic input from the locus ceruleus to all three layers.

characteristic of cerebellar disease is inability to "put on the brakes," ie, to stop movement promptly. Normally, for example, flexion of the forearm against resistance is quickly checked when the resistance force is suddenly broken off. The patient with cerebellar disease cannot brake the movement of the limb, and the forearm flies backward in a wide arc. This abnormal response is known as the **rebound phenomenon,** and similar impairment is detectable in other motor activities. This is one of the important reasons these patients show **adiadochokinesia,** the inability to perform rapidly alternating opposite movements such as repeated pronation and supination of the hands. Finally, patients with cerebellar disease have difficulty performing actions that involve simultaneous motion at more than one joint. They dissect such movements and carry them out one joint at a time, a phenomenon known as **decomposition of movement.**

The Cerebellum & Learning

The cerebellum is concerned with learned adjustments that make coordination easier when a given task is performed over and over. As a motor task is learned, activity in the brain shifts from the prefrontal areas to the parietal and motor cortex and the cerebellum. The basis of the learning in the cerebellum is probably the input via the olivary nuclei. It is worth noting in this regard that each Purkinje cell receives inputs from 250,000 to 1 million mossy fibers, but each has only a single climbing fiber from the inferior olive, and this fiber makes 2000–3000 synapses on the Purkinje cell. Climbing fiber activation produces a large, complex spike in the Purkinje cell; and this spike in some way produces long-term modification of the pattern of mossy fiber input to that particular Purkinje cell. Climbing fiber activity is increased when a new movement is being learned, and selective lesions of the olivary complex abolish the ability to produce long-term adjustments in certain motor responses. The role of the cerebellum in adjusting the VOR and other forms of reflexive memory is discussed in Chapter 16.

Mechanisms

Although the functions of the flocculonodular lobe, spinocerebellum, and neocerebellum are relatively clear and the cerebellar circuits are simple, the exact ways their different parts carry out their functions are still unknown. The relation of the electrical events in the cerebellum to its function in motor control is another interesting problem. The cerebellar cortex has a basic, 150–300/s, 200-µV electrical rhythm and, superimposed on this, a 1000–2000/s component of smaller amplitude. The frequency of the basic rhythm is thus more than 10 times as great as that of the similarly recorded cerebral cortical alpha rhythm. Incoming stimuli generally alter the amplitude of the cerebellar rhythm, like a broadcast signal modulating a carrier frequency in radio transmission. However, the significance of these electrical phenomena in terms of cerebellar function is not known.

The Autonomic Nervous System

INTRODUCTION

The autonomic nervous system, like the somatic nervous system, is organized on the basis of the reflex arc. Impulses initiated in visceral receptors are relayed via afferent autonomic pathways to the CNS, integrated within it at various levels, and transmitted via efferent pathways to visceral effectors. This organization deserves emphasis because the functionally important afferent components have often been ignored. The visceral receptors and afferent pathways have been considered in Chapters 5 and 7 and the major autonomic effector, smooth muscle, in Chapter 3. The efferent pathways to the viscera are the subject of this chapter. Autonomic integration in the CNS is considered in Chapter 14.

ANATOMIC ORGANIZATION OF AUTONOMIC OUTFLOW

The peripheral motor portions of the autonomic nervous system are made up of **preganglionic** and **postganglionic neurons** (Figures 13–1 and 13–2). The cell bodies of the preganglionic neurons are located in the visceral efferent intermediolateral gray column (IML) of the spinal cord or the homologous motor nuclei of the cranial nerves. Their axons are mostly myelinated, relatively slowly conducting B fibers. The axons synapse on the cell bodies of postganglionic neurons that are located in all cases outside the CNS. Each preganglionic axon diverges to an average of eight or nine postganglionic neurons. In this way, autonomic output is diffused. The axons of the postganglionic neurons,

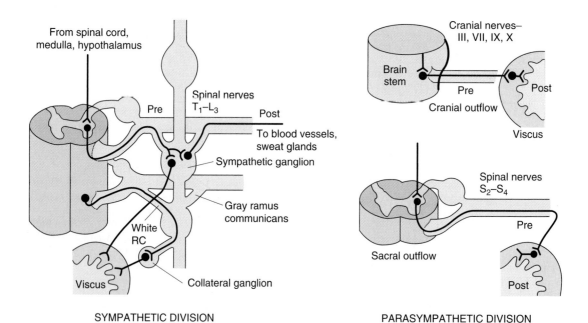

Figure 13–1. Autonomic nervous system. Pre, preganglionic neuron; Post, postganglionic neuron; RC, ramus communicans.

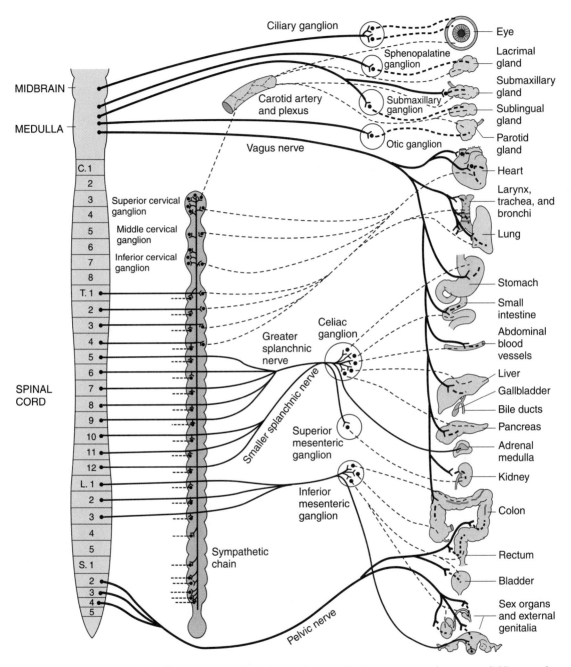

Figure 13–2. Diagram of the efferent autonomic pathways. Preganglionic neurons are shown as solid lines, and postganglionic neurons as dashed lines. The heavy lines are parasympathetic fibers; the light lines are sympathetic. (Modified and reproduced, with permission, from Youmans W: *Fundamentals of Human Physiology*, 2nd ed. Year Book, 1962.)

mostly unmyelinated C fibers, end on the visceral effectors.

Anatomically, the autonomic outflow is divided into two components: the **sympathetic** and **parasympathetic divisions** of the autonomic nervous system. In the gastrointestinal tract, these both communicate with the **enteric nervous system,** and this is sometimes called a third division of the autonomic nervous system.

Sympathetic Division

The axons of the sympathetic preganglionic neurons leave the spinal cord with the ventral roots of the first thoracic to the third or fourth lumbar spinal nerves. They pass via the **white rami communicantes** to the **paravertebral sympathetic ganglion chain,** where most of them end on the cell bodies of the postganglionic neurons. The axons of some of the postganglionic neurons pass to the viscera in the various sympathetic nerves. Others reenter the spinal nerves via the **gray rami communicantes** from the chain ganglia and are distributed to autonomic effectors in the areas supplied by these spinal nerves. The postganglionic sympathetic nerves to the head originate in the **superior, middle,** and **stellate ganglia** in the cranial extension of the sympathetic ganglion chain and travel to the effectors with the blood vessels. Some preganglionic neurons pass through the paravertebral ganglion chain and end on postganglionic neurons located in **collateral ganglia** close to the viscera. Parts of the uterus and the male genital tract are innervated by a special system of **short noradrenergic neurons** with cell bodies in ganglia in or near these organs, and the preganglionic fibers to these postganglionic neurons presumably go all the way to the organs (Figure 13–2). In addition, at least in rats, there are **intrinsic cardiac adrenergic cells (ICA cells).** These cells contain epinephrine and norepinephrine and account for about 15% of the total catecholamine content of the heart. Their exact function is unsettled, but gene knockout experiments indicate that catecholamines are essential for normal development of the heart.

Parasympathetic Division

The **cranial outflow** of the parasympathetic division supplies the visceral structures in the head via the oculomotor, facial, and glossopharyngeal nerves, and those in the thorax and upper abdomen via the vagus nerves. The **sacral outflow** supplies the pelvic viscera via the pelvic branches of the second to fourth sacral spinal nerves. The preganglionic fibers in both outflows end on short postganglionic neurons located on or near the visceral structures (Figure 13–2).

CHEMICAL TRANSMISSION AT AUTONOMIC JUNCTIONS

Transmission at the synaptic junctions between pre- and postganglionic neurons and between the postganglionic neurons and the autonomic effectors is chemically mediated. The principal transmitter agents involved are **acetylcholine** and **norepinephrine,** although **dopamine** is also secreted by interneurons in the sympathetic ganglia and **GnRH** is secreted by some of the preganglionic neurons (Table 13–1). GnRH mediates a slow excitatory response (see below). In addition, there are cotransmitters in autonomic neurons; for example, VIP is released with acetylcholine, and ATP and neuropeptide Y with norepinephrine. The chemistry of all these transmitters and the receptors on which they act are discussed in Chapter 4. VIP causes bronchodilation, and there may be a separate VIP-secreting **nonadrenergic noncholinergic nervous system** innervating bronchial smooth muscle (see Chapter 34).

Chemical Divisions of the Autonomic Nervous System

On the basis of the chemical mediator released, the autonomic nervous system can be divided into **cholinergic** and **noradrenergic divisions** (Table 13–2). The neurons that are cholinergic are (1) all preganglionic neurons, (2) the anatomically parasympathetic postganglionic neurons, (3) the anatomically sympathetic postganglionic neurons which innervate sweat glands, and (4) the anatomically sympathetic neurons which end on blood vessels in skeletal muscles and produce vasodilation when stimulated (sympathetic vasodilator nerves; see Chapter 31). The remaining postganglionic sympathetic neurons are noradrenergic or, apparently, adrenergic in the case of the ICA cells. The adrenal medulla is essentially a sympathetic ganglion in which the postganglionic cells have lost their axons and se-

Table 13–1. Fast and slow responses of postganglionic neurons in sympathetic ganglia.

Potential	Duration	Mediator	Receptor
Fast EPSP	30 ms	Acetylcholine	Nicotinic cholinergic
Slow IPSP	2 s	Dopamine	D_2
Slow EPSP	30 s	Acetylcholine	M_2 cholinergic
Late slow EPSP	4 min	GnRH	GnRH

Table 13-2. Responses of effector organs to autonomic nerve impulses and circulating catecholamines.[1]

Effector Organs	Cholinergic Impulse Response	Noradrenergic Impulses	
		Receptor Type[2]	Response
Eyes			
Radial muscle of iris	...	α_1	Contraction (mydriasis)
Sphincter muscle of iris	Contraction (miosis)		...
Ciliary muscle	Contraction for near vision	β_2	Relaxation for far vision
Heart			
S-A node	Decrease in heart rate, vagal arrest	β_1, β_2	Increase in heart rate
Atria	Decrease in contractility and (usually) increase in conduction velocity	β_1, β_2	Increase in contractility and conduction velocity
A-V node	Decrease in conduction velocity	β_1, β_2	Increase in conduction velocity
His-Purkinje system	Decrease in conduction velocity	β_1, β_2	Increase in conduction velocity
Ventricles	Decrease in contractility	β_1, β_2	Increase in contractility
Arterioles			
Coronary	Dilation	α_1, α_2	Constriction
		β_2	Dilation
Skin and mucosa	Dilation	α_1, α_2	Constriction
Skeletal muscle	Dilation	α_1	Constriction
		β_2	Dilation
Cerebral	Dilation	α_1	Constriction
Pulmonary	Dilation	α_1	Constriction
		β_2	Dilation
Abdominal viscera	...	α_1	Constriction
		β_2	Dilation
Salivary glands	Dilation	α_1, α_2	Constriction
Renal	...	α_1, α_2	Constriction
		β_1, β_2	Dilation
Systemic veins	...	α_1, α_2	Constriction
		β_2	Dilation
Lungs			
Bronchial muscle	Contraction	β_2	Relaxation
Bronchial glands	Stimulation	α_1	Inhibition
		β_2	Stimulation

(continued)

Table 13–2. Responses of effector organs to autonomic nerve impulses and circulating catecholamines.[1] *(continued)*

Effector Organs	Cholinergic Impulse Response	Noradrenergic Impulses	
		Receptor Type[2]	**Response**
Stomach			
Motility and tone	Increase	$\alpha_1, \alpha_2, \beta_2$	Decrease (usually)
Sphincters	Relaxation (usually)	α_1	Contraction (usually)
Secretion	Stimulation	α_2	Inhibition
Intestine			
Motility and tone	Increase	$\alpha_1, \alpha_2, \beta_1, \beta_2$	Decrease (usually)
Sphincters	Relaxation (usually)	α_1	Contraction (usually)
Secretion	Stimulation	α_2	Inhibition
Gallbladder and ducts	Contraction	β_2	Relaxation
Urinary bladder			
Detrusor	Contraction	β_2	Relaxation (usually)
Trigone and sphincter	Relaxation	α_1	Contraction
Ureters			
Motility and tone	Increase (?)	α_1	Increase (usually)
Uterus	Variable[3]	α_1	Contraction (pregnant)
		β_2	Relaxation (pregnant and nonpregnant)
Male sex organs	Erection	α_1	Ejaculation
Skin			
Pilomotor muscles	...	α_1	Contraction
Sweat glands	Generalized secretion	α_1	Slight, localized secretion[4]
Spleen capsule	...	α_1	Contraction
		β_2	Relaxation
Adrenal medulla	Secretion of epinephrine and norepinephrine		...
Liver	...	α_1, β_2	Glycogenolysis
Pancreas			
Acini	Increased secretion	α	Decreased secretion
Islets	Increased insulin and glucagon secretion	α_2	Decreased insulin and glucagon secretion
		β_2	Increased insulin and glucagon secretion

Table 13–2. Responses of effector organs to autonomic nerve impulses and circulating catecholamines.[1] *(continued)*

Effector Organs	Cholinergic Impulse Response	Noradrenergic Impulses	
		Receptor Type[2]	Response
Salivary glands	Profuse, watery secretion	α_1	Thick, viscous secretion
		β	Amylase secretion
Lacrimal glands	Secretion	α	Secretion
Nasopharyngeal glands	Secretion	…	…
Adipose tissue	…	$\alpha_1, \beta_1, \beta_3$	Lipolysis
Juxtaglomerular cells	…	β_1	Increased renin secretion
Pineal gland	…	β	Increased melatonin synthesis and secretion

[1]Modified from Hardman JG et al (editors): *Goodman and Gilman's The Pharmacological Basis of Therapeutics,* 10th ed. McGraw-Hill, 2001.

[2]Where a receptor subtype is not specified, data are as yet inadequate for characterization.

[3]Depends on stage of menstrual cycle, amount of circulating estrogen and progesterone, pregnancy, and other factors.

[4]On palms of hands and in some other locations ("adrenergic sweating").

crete norepinephrine, epinephrine, and some dopamine directly into the bloodstream. The cholinergic preganglionic neurons to these cells have consequently become the secretomotor nerve supply of this gland.

Transmission in Sympathetic Ganglia

At least in experimental animals, the responses produced in postganglionic neurons by stimulation of their preganglionic innervation include not only a rapid depolarization (**fast EPSP**) that generates action potentials (Table 13–1) but also a prolonged inhibitory postsynaptic potential (**slow IPSP**), a prolonged excitatory postsynaptic potential (**slow EPSP**), and a **late slow EPSP** (see Chapter 4). The late slow EPSP is very prolonged, lasting minutes rather than milliseconds. These slow responses apparently modulate and regulate transmission through the sympathetic ganglia. The initial depolarization is produced by acetylcholine via a nicotinic receptor. The slow IPSP is probably produced by dopamine, which is secreted by an interneuron within the ganglion. The interneuron is excited by activation of an M_2 muscarinic receptor. The interneurons that secrete dopamine are the small, intensely fluorescent cells (**SIF cells**) in the ganglia. The production of the slow IPSP does not appear to be mediated via cAMP, suggesting that a D_2 receptor is involved (see Chapter 4). The slow EPSP is produced by acetylcholine acting on a muscarinic receptor on the membrane of the postganglionic neuron. The late

slow EPSP is produced by GnRH or a peptide closely resembling it.

RESPONSES OF EFFECTOR ORGANS TO AUTONOMIC NERVE IMPULSES

General Principles

The effects of stimulation of the noradrenergic and cholinergic postganglionic nerve fibers to the viscera are listed in Table 13–2. The smooth muscle in the walls of the hollow viscera is generally innervated by both noradrenergic and cholinergic fibers, and activity in one of these systems increases the intrinsic activity of the smooth muscle whereas activity in the other decreases it. However, there is no uniform rule about which system stimulates and which inhibits. In the case of sphincter muscles, both noradrenergic and cholinergic innervations are excitatory, but one supplies the constrictor component of the sphincter and the other the dilator.

There is usually no acetylcholine in the circulating blood, and the effects of localized cholinergic discharge are generally discrete and of short duration because of the high concentration of acetylcholinesterase at cholinergic nerve endings. Norepinephrine spreads farther and has a more prolonged action than acetylcholine. Norepinephrine, epinephrine, and dopamine are all found in plasma (see Chapter 20). The epinephrine and some of the dopamine come from the adrenal medulla, but most of the norepinephrine diffuses into the bloodstream from noradrenergic nerve endings. Metabolites of norepinephrine and dopamine also enter the circulation, some from the sympathetic nerve endings and some from smooth muscle cells (Figure 13–3). The metabolism of catecholamines is discussed in Chapter 4. It is worth noting that even when MAO and COMT are both inhibited, the metabolism of norepinephrine is still rapid. However, inhibition of reuptake prolongs its half-life.

Cholinergic Discharge

In a general way, the functions promoted by activity in the cholinergic division of the autonomic nervous system are those concerned with the vegetative aspects of day-to-day living. For example, cholinergic action favors digestion and absorption of food by increasing the activity of the intestinal musculature, increasing gastric secretion, and relaxing the pyloric sphincter. For this reason, and to contrast it with the "catabolic" noradrenergic division, the cholinergic division is sometimes called the **anabolic nervous system.**

The function of the VIP released from postganglionic cholinergic neurons is unsettled, but there is evidence that it facilitates the postsynaptic actions of acetylcholine. Since VIP is a vasodilator, it may also increase blood flow in target organs.

Noradrenergic Discharge

The noradrenergic division discharges as a unit in emergency situations. The effects of this discharge are of considerable value in preparing the individual to cope with the emergency, although it is important to avoid the teleologic fallacy involved in the statement that the system discharges in order to do this. For example, noradrenergic discharge relaxes accommodation and dilates the pupils (letting more light into the eyes), accelerates the heartbeat and raises the blood pressure (providing better perfusion of the vital organs and muscles), and constricts the blood vessels of the skin (which limits bleeding from wounds). Noradrenergic discharge also leads to lower thresholds in the reticular formation (reinforcing the alert, aroused state) and to elevated plasma glucose and free fatty acid levels (supplying more en-

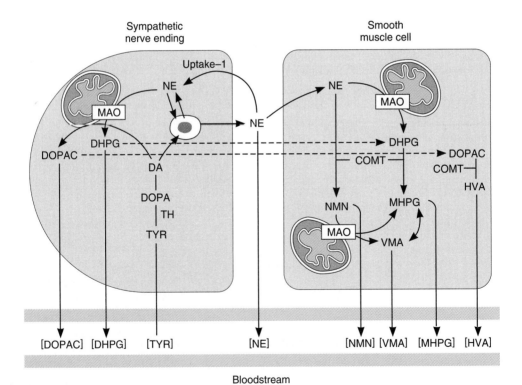

Figure 13–3. Catecholamine metabolism in the sympathetic nervous system. TYR, tyrosine; TH, tyrosine hydroxylase; DOPA, dihydroxyphenylalanine; DA, dopamine; NE, norepinephrine; NMN, normetanephrine. For other abbreviations, see Figures 4–19 and 4–20. (Courtesy of DS Goldstein.)

ergy). On the basis of effects like these, Cannon called the emergency-induced discharge of the noradrenergic nervous system the "preparation for flight or fight."

The emphasis on mass discharge in stressful situations should not obscure the fact that the noradrenergic autonomic fibers also subserve other functions. For example, tonic noradrenergic discharge to the arterioles maintains arterial pressure, and variations in this tonic discharge are the mechanism by which carotid sinus feedback regulation of blood pressure is effected. In addition, sympathetic discharge is decreased in fasting animals and increased when fasted animals are refed. These changes may explain the decrease in blood pressure and metabolic rate produced by fasting and the opposite changes produced by feeding.

The small granulated vesicles in postganglionic noradrenergic neurons contain ATP and norepinephrine, and the large granulated vesicles contain neuropeptide Y. There is evidence that low-frequency stimulation promotes release of ATP whereas high-frequency stimulation causes release of neuropeptide Y. There are purinergic receptors in the viscera, and evidence is accumulating that ATP is a mediator in the autonomic nervous system along with norepinephrine. However, its exact role is unsettled.

Autonomic Pharmacology

The junctions in the peripheral autonomic motor pathways are a logical site for pharmacologic manipulation

Table 13–3. Some drugs and toxins that affect autonomic acitvity.[1]

Site of Action	Compounds That Augment Autonomic Activity	Compounds That Depress Autonomic Activity

Sympathetic and parasympathetic ganglia	**Stimulate postganglionic neurons**
	  Nicotine
	  Dimethyphenylpiperazinium
	Inhibit acetylcholinesterase
	  DFP (diisopropyl fluorophosphate)
	  Physostigmine (Eserine)
	  Neostigmine (Prostigmin)
	  Parathion
		Block conduction
		  Hexamethonium (C-6)
		  Mecamylamine (Inversine)
		  Pentolinium
		  Trimethaphan (Arfonad)
		  High concentrations of acetylcholine

Endings of postganglionic noradrenergic neurons	**Release norepinephrine**
	  Tyramine
	  Ephedrine
	  Amphetamine
		Block norepinephrine synthesis
		  Metyrosine (Demser)
		Interfere with norepinephrine storage
		  Reserpine
		  Guanethidine[2] (Ismelin)
		Prevent norepinephrine release
		  Bretylium (Bretylol)
		  Guanethidine[2] (Ismelin)
		Form false transmitters
		  Methyldopa (Aldomet)

Muscarinic receptors		Atropine, scopolamine

α receptors	**Stimulate α_1 receptors**
	  Methoxamine (Vasoxyl)
	  Phenylephrine (Neo-Synephrine)
		Block α receptors
		  Phenoxybenzamine (Dibenzyline)
		  Phentolamine (Regitine)
		  Prazosin (Minipress) (blocks α_1)
		  Yohimbine (blocks α_2)

β receptors	**Stimulate β receptors**
	  Isoproterenol (Isuprel)
		Block β receptors
		  Propranolol (Inderal) and others (blocks β_1 and β_2)
		  Atenolol (Tenormin) and others (blocks β_1)
		  Butoxamine (blocks β_2)

[1]Only the principal actions are listed.
[2]Guanethidine is believed to have two principal actions.

of visceral function because transmission across them is chemical. The transmitter agents are synthesized, stored in the nerve endings, and released near the neurons, muscle cells, or gland cells on which they act. They bind to receptors on these cells, thus initiating their characteristic actions, and they are then removed from the area by reuptake or metabolism. Each of these steps can be stimulated or inhibited, with predictable consequences. In noradrenergic endings, certain drugs also cause the formation of compounds that replace norepinephrine in the granules, and these weak or inactive "false transmitters" are released instead of norepinephrine by the action potentials reaching the endings.

Some of the drugs and toxins that affect the activity of the autonomic nervous system and the mechanisms by which they produce their effects are listed in Table 13–3. Compounds with muscarinic actions include congeners of acetylcholine and drugs that inhibit acetylcholinesterase. Among the latter are the insecticide parathion and diisopropyl fluorophosphate (DFP), a component of the so-called nerve gases, which kill by producing massive inhibition of acetylcholinesterase.

Central Regulation of Visceral Function

INTRODUCTION

The levels of autonomic integration within the CNS are arranged, like their somatic counterparts, in a hierarchy. Simple reflexes such as contraction of the full bladder are integrated in the spinal cord (see Chapter 12). More complex reflexes are the subject of this chapter. Those that regulate respiration and blood pressure are integrated in the medulla oblongata. Those that control pupillary responses to light and accommodation are integrated in the midbrain. Many of the complex autonomic mechanisms that maintain the chemical constancy and temperature of the internal environment are integrated in the hypothalamus. The hypothalamus also functions with the limbic system as a unit that regulates emotional and instinctual behavior, and these aspects of hypothalamic function are discussed in the next chapter.

MEDULLA OBLONGATA

Control of Respiration, Heart Rate, & Blood Pressure

The medullary areas for the autonomic reflex control of the circulation, heart, and lungs are called the **vital centers** because damage to them is usually fatal. The afferent fibers to these centers originate in a number of instances in specialized visceral receptors. The specialized receptors include not only those of the carotid and aortic sinuses and bodies but also receptor cells that are located in the medulla itself. The motor responses are graded and delicately adjusted and include somatic as well as visceral components. The details of the reflexes themselves are discussed in the chapters on the regulation of the circulation and respiration.

Other Medullary Autonomic Reflexes

Swallowing, coughing, sneezing, gagging, and vomiting are also reflex responses integrated in the medulla ob-

longata. Swallowing is controlled by a **central program generator** in the medulla. It is initiated by the voluntary act of propelling what is in the mouth toward the back of the pharynx (see Chapter 26) and involves carefully timed responses of the respiratory as well as the gastrointestinal system. Coughing is initiated by irritation of the lining of the trachea and extrapulmonary bronchi. The glottis closes, and strong contraction of the respiratory muscles builds up intrapulmonary pressure, whereupon the glottis suddenly opens, causing an explosive discharge of air (see Chapter 36). Sneezing is a somewhat similar response to irritation of the nasal epithelium. It is initiated by stimulation of pain fibers in the trigeminal nerves.

Vomiting

Vomiting is another example of the way visceral reflexes integrated in the medulla include coordinated and carefully timed somatic as well as visceral components. Vomiting starts with salivation and the sensation of nausea. Reverse peristalsis empties material from the upper part of the small intestine into the stomach. The glottis closes, preventing aspiration of vomitus into the trachea. The breath is held in mid inspiration. The muscles of the abdominal wall contract, and because the chest is held in a fixed position, the contraction increases intra-abdominal pressure. The lower esophageal sphincter and the esophagus relax, and the gastric contents are ejected.

The "vomiting center" in the reticular formation of the medulla (Figure 14–1) really consists of various scattered groups of neurons in this region that control the different components of the vomiting act.

Afferents

Irritation of the mucosa of the upper gastrointestinal tract causes vomiting. Impulses are relayed from the mucosa to the medulla over visceral afferent pathways in the sympathetic nerves and vagi. Afferents from the vestibular nuclei mediate the nausea and vomiting of motion sickness. Other afferents presumably reach the

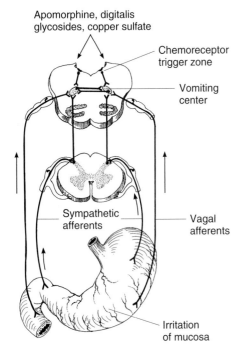

Figure 14–1. Afferent pathways for the vomiting reflex, showing the chemoreceptor trigger zone in the medulla.

vomiting control areas from the diencephalon and limbic system, because emetic responses to emotionally charged stimuli also occur. Thus, we speak of "nauseating smells" and "sickening sights."

Chemoreceptor cells in the medulla initiate vomiting when they are stimulated by certain circulating chemical agents. The **chemoreceptor trigger zone** in which these cells are located (Figure 14–1) is in the **area postrema,** a V-shaped band of tissue on the lateral walls of the fourth ventricle near the obex. This structure is one of the circumventricular organs (see Chapter 32) and is more permeable to many substances than the underlying medulla. Lesions of the area postrema have little effect on the vomiting response to gastrointestinal irritation or motion sickness but abolish the vomiting that follows injection of apomorphine and a number of other emetic drugs. Such lesions also decrease vomiting in uremia and radiation sickness, both of which may be associated with endogenous production of circulating emetic substances.

There are 5-HT_3 receptors in the small intestine, and serotonin (5-HT) released from enterochromaffin cells appears to initiate impulses in afferents that trigger vomiting. In addition, there are dopamine D_2 receptors and 5-HT_3 receptors in the area postrema and adjacent nucleus of the solitary tract. 5-HT_3 antagonists such as ondansetron and D_2 antagonists such as chlorpromazine and haloperidol are effective antiemetic agents. Corticosteroids, cannabinoids, and benzodiazepines, alone or in combination with 5-HT_3 and D_2 antagonists, are also useful in treatment of the vomiting produced by chemotherapy. The mechanisms of action of corticosteroids and cannabinoids are unknown, whereas the benzodiazepines probably reduce the anxiety associated with chemotherapy.

HYPOTHALAMUS

ANATOMIC CONSIDERATIONS

The hypothalamus (Figure 14–2) is the portion of the anterior end of the diencephalon that lies below the hypothalamic sulcus and in front of the interpeduncular nuclei. It is divided into a variety of nuclei and nuclear areas.

Relation to the Pituitary Gland

There are neural connections between the hypothalamus and the posterior lobe of the pituitary gland and vascular connections between the hypothalamus and the anterior lobe. Embryologically, the posterior pituitary arises as an evagination of the floor of the third ventricle. It is made up in large part of the endings of axons that arise from cell bodies in the supraoptic and paraventricular nuclei and pass to the posterior pituitary (Figure 14–3) via the **hypothalamohypophysial tract.** Most of the supraoptic fibers end in the posterior lobe itself, whereas some of the paraventricular fibers end in the median eminence. The anterior and intermediate lobes of the pituitary arise in the embryo from Rathke's pouch, an evagination from the roof of the pharynx (see Figure 22–1). Sympathetic nerve fibers reach the anterior lobe from its capsule, and parasympathetic fibers reach it from the petrosal nerves, but few if any nerve fibers pass to it from the hypothalamus. However, the **portal hypophysial vessels** form a direct vascular link between the hypothalamus and the anterior pituitary. Arterial twigs from the carotid arteries and circle of Willis form a network of fenestrated capillaries called the **primary plexus** on the ventral surface of the hypothalamus (Figure 14–2). Capillary loops also penetrate the median eminence. The capillaries drain into the sinusoidal portal hypophysial vessels that carry blood down the pituitary stalk to the capillaries of the anterior pituitary. This system begins and ends in capillaries without going through the heart and is therefore a true portal system. In birds and some mam-

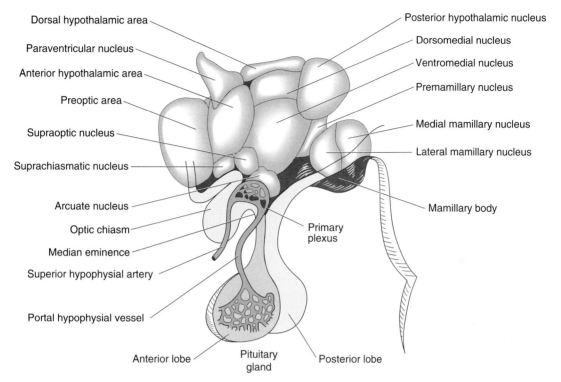

Dorsal hypothalamic area

Paraventricular nucleus

Anterior hypothalamic area

Preoptic area

Supraoptic nucleus

Suprachiasmatic nucleus

Arcuate nucleus

Optic chiasm

Median eminence

Superior hypophysial artery

Portal hypophysial vessel

Posterior hypothalamic nucleus

Dorsomedial nucleus

Ventromedial nucleus

Premamillary nucleus

Medial mamillary nucleus

Lateral mamillary nucleus

Mamillary body

Primary plexus

Anterior lobe

Pituitary gland

Posterior lobe

Figure 14–2. Human hypothalamus, with a superimposed diagrammatic representation of the portal hypophysial vessels.

mals, including humans, there is no other anterior hypophysial arterial supply except capsular vessels and anastomotic connections from the capillaries of the posterior pituitary. The **median eminence** is generally defined as the portion of the ventral hypothalamus from which the portal vessels arise. This region is "outside the blood-brain barrier" (see Chapter 32).

Afferent & Efferent Connections of the Hypothalamus

The principal afferent and efferent neural pathways to and from the hypothalamus are mostly unmyelinated. Many connect the hypothalamus to the limbic system. There are also important connections between the hypothalamus and nuclei in the midbrain tegmentum, pons, and hindbrain.

Norepinephrine-secreting neurons with their cell bodies in the hindbrain end in many different parts of the hypothalamus (see Figure 15–6). Paraventricular neurons that probably secrete oxytocin and vasopressin project in turn to the hindbrain and the spinal cord. Neurons that secrete epinephrine have their cell bodies in the hindbrain and end in the ventral hypothalamus.

There is an intrahypothalamic system of dopamine-secreting neurons which have their cell bodies in the arcuate nucleus and end on or near the capillaries that form the portal vessels in the median eminence. Serotonin-secreting neurons project to the hypothalamus from the raphe nuclei.

HYPOTHALAMIC FUNCTION

The major functions of the hypothalamus are summarized in Table 14–1. Some are fairly clear-cut visceral reflexes, and others include complex behavioral and emotional reactions; however, all involve a particular response to a particular stimulus. It is important to keep this in mind in considering hypothalamic function.

RELATION TO AUTONOMIC FUNCTION

Many years ago, Sherrington called the hypothalamus "the head ganglion of the autonomic system." Stimulation of the hypothalamus produces autonomic responses, but the hypothalamus does not seem to be concerned with the regulation of visceral function per

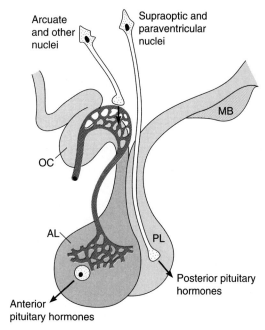

Figure 14–3. Secretion of hypothalamic hormones. The hormones of the posterior lobe (PL) are released into the general circulation from the endings of supraoptic and paraventricular neurons, whereas hypophysiotropic hormones are secreted into the portal hypophysial circulation from the endings of arcuate and other hypothalamic neurons. AL, anterior lobe; MB, mamillary bodies; OC, optic chiasm.

se. Rather, the autonomic responses triggered in the hypothalamus are part of more complex phenomena such as eating as well as rage and other emotions. For example, stimulation of various parts of the hypothalamus, especially the lateral areas, produces diffuse sympathetic discharge and increased adrenal medullary secretion, the mass sympathetic discharge seen in animals exposed to stress (the flight or fight reaction; see Chapter 13).

It has been claimed that there are separate hypothalamic areas for the control of epinephrine and norepinephrine secretion. Differential secretion of one or the other of these adrenal medullary catecholamines does occur in certain situations (see Chapter 20), but the selective increases are small.

RELATION TO SLEEP

The basal forebrain sleep zone includes parts of the hypothalamus. These areas and the overall physiology of sleep and wakefulness are discussed in Chapter 11.

RELATION TO CYCLIC PHENOMENA

Most if not all living cells in plants and animals have rhythmic fluctuations in their function that are about 24 hours in length—ie, they are **circadian** (L *circa* "about" + *dia* "day"). Normally they become entrained, ie, synchronized to the day-night light cycle in the environment. If they are not entrained, they become progressively more out of phase with the light-dark cycle because they are longer or shorter than 24 hours.

In mammals, including humans, there are circadian rhythms in most of the cells in the body. In the liver these are influenced by the pattern of food intake, but in almost all other cells the rhythms are entrained by the paired **suprachiasmatic nuclei (SCN),** one on either side, above the optic chiasm (Figure 14–2). These nuclei receive information about the light-dark cycle via a special neural pathway, the **retinohypothalamic fibers** that pass from the optic chiasm to the SCN. Efferents from the SCN initiate neural and humoral signals that entrain a wide variety of well-known circadian rhythms. These include the rhythms in the secretion of ACTH (see Figure 20–19) and other pituitary hormones, the sleep-wake cycle, activity patterns, and the secretion of the pineal hormone melatonin (see Figure 24–13). The nocturnal peaks in the secretion of melatonin appear to be an important hormonal signal entraining other cells in the body (see Chapter 24).

Neurons in the SCN discharge rhythmically when removed from the body and cultured in vitro. Genes in these neurons are activated diurnally, and their protein products enter the cytoplasm, where they modify cell function and thus neuronal discharge. The proteins are then modified and return to the nucleus, where they act in a negative feedback fashion to inhibit the activity of the genes that produced them. Considerable progress has been made in identifying the genes in mammals. There are at least four of them, and there appear to be two interacting negative feedback loops regulating them.

There is evidence that there are two different peaks of circadian activity in the SCN. This may correlate with the observation that exposure to bright light can either advance, delay, or have no effect on the sleep-wake cycle in humans depending on the time of day when it is experienced. During the usual daytime it has no effect, but just after dark it delays the onset of the sleep period, and just before dawn it accelerates the onset of the next sleep period. Injections of melatonin have similar effects. In experimental animals, exposure to light turns on immediate-early genes (see Chapter 1) in the suprachiasmatic neurons, but only at times during the circadian cycle when light is capable of influencing entrainment. Stimulation during the day is ineffective.

Table 14–1. Summary of principal hypothalamic regulatory mechanisms.

Function	Afferents From	Integrating Areas
Temperature regulation	Temperature receptors in the skin, deep tissues, spinal cord, hypothalamus, and other parts of the brain	Anterior hypothalamus, response to heat; posterior hypothalamus, response to cold
Neuroendocrine control of:		
Catecholamines	Limbic areas concerned with emotion	Dorsal and posterior hypothalamus
Vasopressin	Osmoreceptors, "volume receptors," others	Supraoptic and paraventricular nuclei
Oxytocin	Touch receptors in breast, uterus, genitalia	Supraoptic and paraventricular nuclei
Thyroid-stimulating hormone (thyrotropin, TSH) via TRH	Temperature receptors in infants, perhaps others	Paraventricular nuclei and neighboring areas
Adrenocorticotropic hormone (ACTH) and β-lipotropin (β-LPH) via CRH	Limbic system (emotional stimuli); reticular formation ("systemic" stimuli); hypothalamic and anterior pituitary cells sensitive to circulating blood cortisol level; suprachiasmatic nuclei (diurnal rhythm)	Paraventricular nuclei
Follicle-stimulating hormone (FSH) and luteinizing hormone (LH) via GnRH	Hypothalamic cells sensitive to estrogens, eyes, touch receptors in skin and genitalia of reflex ovulating species	Preoptic area; other areas
Prolactin via PIH and PRH	Touch receptors in breasts, other unknown receptors	Arcuate nucleus; other areas (hypothalamus inhibits secretion)
Growth hormone via somatostatin and GRH	Unknown receptors	Periventricular nucleus, arcuate nucleus
"Appetitive" behavior		
Thirst	Osmoreceptors, probably located in the organum vasculosum of the lamina terminalis; angiotensin II uptake in the subfornical organ	Lateral superior hypothalamus
Hunger	Glucostat cells sensitive to rate of glucose utilization; leptin receptors; receptors for other polypeptides	Ventromedial, arcuate, and paraventricular nuclei; lateral hypothalamus
Sexual behavior	Cells sensitive to circulating estrogen and androgen, others	Anterior ventral hypothalamus plus, in the male, piriform cortex
Defensive reactions (fear, rage)	Sense organs and neocortex, paths unknown	Diffuse, in limbic system and hypothalamus
Control of body rhythms	Retina via retinohypothalamic fibers	Suprachiasmatic nuclei

HUNGER

Feeding & Satiety

Body weight depends on the balance between caloric intake and utilization of calories. Obesity results when the former exceeds the latter. Regulation of food intake is considered in this chapter since the hypothalamus and related parts of the brain play key roles in this process. Obesity is considered in detail in Chapter 17, and the relation of obesity to diabetes mellitus is discussed in Chapter 19.

Food intake is regulated not only on a meal-to-meal basis but also in a way that generally maintains weight at a given set point. If animals are made obese by force-feeding and then permitted to eat as they wish, their spontaneous food intake decreases until their weight falls to the control (Figure 14–4). Conversely, if animals are starved and then permitted to eat freely, their spontaneous food intake increases until they regain the lost weight. It is common knowledge that the same thing happens in humans. Dieters can lose weight when caloric intake is reduced but when they discontinue their diets, 95% of them regain the weight they lost. Similarly, during recovery from illness, food intake is increased in a catch-up fashion until lost weight is regained. Catch-up growth in children is discussed in Chapter 22, and the relation of food intake to longevity is discussed in Chapter 1.

Role of the Hypothalamus

Hypothalamic regulation of the appetite for food depends primarily upon the interaction of two areas: a lateral **"feeding center"** in the bed nucleus of the medial forebrain bundle at its junction with the pallidohypothalamic fibers, and a medial **"satiety center"** in the ventromedial nucleus (Figure 14–2). Stimulation of the feeding center evokes eating behavior in conscious animals, and its destruction causes severe, fatal anorexia in otherwise healthy animals. Stimulation of the ventromedial nucleus causes cessation of eating, whereas lesions in this region cause hyperphagia and, if the food supply is abundant, the syndrome of **hypothalamic obesity** (Figure 14–5). Destruction of the feeding center in rats with lesions of the satiety center causes anorexia, which indicates that the satiety center functions by inhibiting the feeding center. It appears that the feeding center is chronically active and that its activity is transiently inhibited by activity in the satiety center after the ingestion of food. However, it is not certain that the feeding center and the satiety center simply control the desire for food. For example, rats with ventromedial lesions gain weight for a while, but their food intake then levels off. After their intake reaches a plateau, their appetite mechanism operates to maintain their new, higher weight.

Since the discovery of leptin and its receptors (see below), there has been a rapid accumulation of infor-

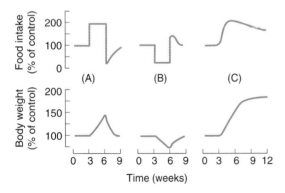

Figure 14–4. Effects of changes in food intake and ventromedial hypothalamic lesions on spontaneous food intake and body weight. **A:** Rats were force-fed for weeks 3–6, then permitted free access to food. **B:** Rats were partially starved for weeks 3–6, then permitted free access to food. **C:** Bilateral ventromedial hypothalamic lesions were produced at 3 weeks and the rats allowed free access to food throughout. (Reproduced, with permission, from Stricker EM: Hyperphagia. N Engl J Med 1978;298:1010.)

Figure 14–5. Hypothalamic obesity. The animal on the right, in which bilateral lesions were placed in the ventromedial nuclei 4 months previously, weighs 1080 g. The control animal on the left weighs 520 g. (Reproduced, with permission, from Stevenson JAF in: *The Hypothalamus*. Haymaker W, Anderson E, Nauta WJH [editors], Thomas, 1969.)

mation about hypothalamic peptides that appear to be involved in the regulation of appetite (Table 14–2). Although more than a dozen have been implicated, it is not possible as yet to relate these peptides to the operation of the feeding and satiety areas described above. However, some patterns are beginning to emerge.

One important polypeptide is **neuropeptide Y** (see Chapter 4). When injected into the hypothalamus, this 36-amino-acid polypeptide increases food intake, and inhibitors of neuropeptide Y synthesis decrease food intake. Neuropeptide Y-containing neurons have their cell bodies in the arcuate nuclei and project to the paraventricular nuclei. Neuropeptide Y mRNA in the hypothalamus increases during feeding and decreases during satiety. Neuropeptide Y exerts its effect through three known receptors—Y1, Y2, and Y5—all coupled to G proteins. Activation of the Y5 receptor increases food intake, but the situation is complex because activation of the Y2 receptor has an apparent inhibitory effect. Knockout of the neuropeptide Y gene does not produce marked effects on feeding, indicating that other pathways are also involved, but knocking out the neuropeptide Y gene in leptin-deficient ob/ob mice (see below) causes them to eat less and expend more energy than ob/ob controls that have intact neuropeptide Y genes.

A potentially important recent observation is that accumulation of malonyl-CoA in the tissues inhibits food intake. This substance is produced from acetyl-CoA and is converted to fatty acids by fatty acid synthase. In experimental animals, accumulation of malonyl-CoA causes a decrease in appetite, weight loss, and a rapid decrease in fat stores. The inhibitor also causes a marked drop in hypothalamic neuropeptide Y mRNA, although why the malonyl-CoA acts on hypothalamic neurons is as yet unknown. It will be interesting to see if nontoxic inhibitors of fatty acid synthase can be developed that are suitable for use in humans.

Table 14–2. Principal hypothalamic polypeptides that appear to play a role in the regulation of the appetite for food.

Increase food intake
Neuropeptide Y
Orexin-A and orexin-B
MCH
Ghrelin
Decrease food intake
α-MSH
CART
CRH

Polypeptides that increase food intake include **orexin-A** and **orexin-B,** derived from the same gene by alternate splicing. They act on two receptors. Orexins are synthesized in neurons located in the lateral hypothalamus. They are also of interest because a mutation in one of the orexin receptor genes causes narcolepsy in Doberman and Labrador dogs.

Another polypeptide that increases food intake in mammals is **melanin-concentrating hormone (MCH),** a polypeptide which is secreted by the pituitary in fish and is involved in the control of their skin color (see Chapter 22). In mammals, its mRNA is found only in the lateral hypothalamus and the zona incerta.

On the other hand, pro-opiomelanocortin (POMC) derivatives, particularly α-MSH (see Chapter 22), decrease food intake. There are four established receptors for these derivatives: MC1-R, which is involved in skin pigmentation; MC2-R, which is involved in adrenal glucocorticoid production; MC3-R, which is associated with the control of sebaceous gland secretion; and MC4-R, which mediates the effects on appetite. A mutant strain of obese mice called **agouti** overproduces the agouti protein that inhibits the action of α-MSH on the MC4 receptor.

Another neuropeptide in the hypothalamus that inhibits food intake is CART (cocaine- and amphetamine-regulated transcript). CRH, the brain hormone that stimulates ACTH secretion (see below), also inhibits food intake.

The polypeptide ghrelin is secreted by the stomach and binds to the growth hormone secretagogue (GHS) receptor in the anterior pituitary (see Chapter 22). However, it is also found in the hypothalamus, and GHS receptors are found in various parts of the brain stem. Systemically administered and intraventricular ghrelin both increase body weight. Circulating ghrelin is decreased by eating and increased during fasting.

Catecholamines are also involved in the regulation of body weight. Mice in which norepinephrine and epinephrine synthesis is prevented by knocking out the gene for dopamine β-hydroxylase have increased food intake. Interestingly, they do not become obese because they also have an unexplained simultaneous increase in metabolic rate. Amphetamine and related drugs used clinically to suppress appetite presumably act by releasing norepinephrine in the CNS. Mice in which the $5HT_{2C}$ receptor is knocked out become obese, indicating that serotonin is also involved in the regulation of food intake.

Afferent Mechanisms

Four main hypotheses about afferent mechanisms that are involved in the control of food intake have been advanced, and they are not mutually exclusive. The **lipostatic hypothesis** holds that adipose tissue produces a humoral signal that is proportionate to the amount of fat and acts on the hypothalamus to decrease food intake and increase energy output. The **gut peptide hypothesis** postulates that food in the gastrointestinal tract causes the release of one or more polypeptides which act on the hypothalamus to inhibit food intake. The **glucostatic hypothesis** holds that increased glucose utilization in the hypothalamus produces a sensation of satiety. The **thermostatic hypothesis** holds that a fall in body temperature below a given set point stimulates appetite and a rise above the set point inhibits appetite.

Leptin

Cloning of the *ob* gene in mice, rats, and humans has focused attention on the lipostatic hypothesis. Mice that are homozygous for a defective *ob* gene (ob/ob mice) do not become sated after eating and become obese and diabetic. The product of this gene, which is produced primarily in fat cells, is a circulating protein that contains 167 amino acids and has been named **leptin**, from the Greek word for thin. This hormone acts on the hypothalamus to decrease food intake and increase energy consumption. It appears to decrease the activity of neuropeptide Y neurons that increase appetite and to increase the activity of POMC-secreting neurons.

Another gene, *db*, produces the leptin receptor, and mice of the db/db strain, in which this gene is defective, are also obese but have high circulating leptin levels because they lack leptin receptors. The leptin receptor gene produces several alternatively spliced forms of the receptor. The long form, which is found in the hypothalamus, has a single transmembrane domain and an intracellular domain which resembles that of several cytokines, including IL-6 (see Chapters 1 and 27). This form mediates the central appetite- and energy-regulating effects of leptin, and it is especially plentiful in the arcuate nuclei. Gold thioglucose, which has been known for a long time to cause obesity in mice, destroys the arcuate nuclei.

To reach its central site of action, circulating leptin must cross the blood-brain barrier. A short form of the leptin receptor is abundant in brain microvessels and probably is involved in transport of leptin into the brain.

The physiologically active components of marijuana (cannabinoids) that are found in the body increase appetite by an action on their CB_1 receptors (see Chapter 4). The anorexiant action of leptin is antagonized by CB_1 receptor blockade. Leptin activates the enzyme phosphatidylinositol-3-hydroxykinase in hypothalamic

cells, and inhibition of this enzyme blocks the anorexiant effect of leptin. Leptin also increases the activity of SOCS3 (suppressor of cytokine signaling-3) in neuropeptide Y neurons, and SOCS3 suppresses further leptin receptor signaling, suggesting a turn-off mechanism. However, the physiologic significance of all these observations is not yet clear.

Another interesting observation is that when infused into the cerebral ventricles, leptin causes bone loss—an action that seems to be independent of its action on appetite. However, its role in bone metabolism is as yet unknown.

Leptin receptors are found in various peripheral tissues as well as the brain. In rodents, the decrease in plasma leptin produced by fasting is associated with inhibition of the onset of puberty, depressed thyroid function, and increased glucocorticoid secretion, and it has been suggested that these are adaptive responses to the shortage of calories signaled by the decrease in leptin. Another interesting observation is that there are leptin receptors in brown adipose tissue, and there is evidence that leptin increases the activity of uncoupling proteins (see Chapter 17), thus producing a direct peripheral increase in energy expenditure.

In summary, then, leptin operates as part of a feedback loop by which the size of the body's fat depots can operate through a humoral link to regulate food intake (Figure 14–6).

In humans, inactivating mutations of the leptin gene are rare but have been reported. They cause obesity that starts early in life. However, plasma leptin levels are increased in obese humans with normal leptin genes in direct proportion to the percentage of body fat, and there is a similar positive correlation between the leptin mRNA concentration in adipose tissue and percentage of body fat. Thus, it appears that at least in many cases, human obesity is like that occurring in

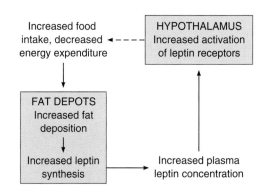

Figure 14–6. Feedback control of fat depots by leptin. Solid arrows indicate stimulation; dashed arrow indicates inhibition.

db/db mice rather than ob/ob mice; ie, there is a defect in the leptin receptor gene. Of course, there are other possibilities, including defective transport of leptin into the brain or defects in the mechanisms activated by the human gene. Further research is necessary to settle this point.

Gut Peptides

Gastrointestinal hormones that have been claimed to inhibit food intake include GRP, glucagon, somatostatin, and CCK. The idea that food entering the gastrointestinal tract triggers the release from the mucosa of substances which act on the brain to produce satiety is intriguing. The effects of leptin appear to be relatively prolonged, and it may be that gut peptides provide short-term, meal-to-meal control of food.

Much of the anorectic effect of circulating CCK is apparently due to an action on visceral receptors, since it is markedly reduced by subdiaphragmatic vagotomy. There are CCK receptors in the area postrema, one of the circumventricular organs (see Chapter 32), but the role of the area postrema in the regulation of food intake is unsettled. Injection of CCK into the hypothalamus also decreases food intake. It is possible that peripheral and central sites are both involved. The peripheral CCK receptors are mostly **CCK-A receptors,** whereas the brain contains both CCK-A and **CCK-B receptors** (see Chapter 26). Selective antagonists of both types of receptors inhibit satiety. However, CCK-B antagonists are 100 times more potent in inhibiting satiety than are CCK-A antagonists. Thus the central CCK receptors seem to be more important.

Glucose

The activity of the satiety center in the ventromedial nuclei is probably governed in part by glucose utilization of the neurons in it. It has been postulated that when their glucose utilization is low—and consequently when the arteriovenous blood glucose difference across them is low—their activity is decreased. Under these conditions, the activity of the feeding center is unchecked and the individual is hungry. When utilization is high, the activity of these glucostats is increased, the feeding center is inhibited, and the individual feels sated. This glucostatic hypothesis of appetite regulation is supported by an appreciable body of experimental data. For example, food intake is rapidly increased by intraventricular administration of compounds such as 2-deoxyglucose that decrease glucose utilization in cells. Hypoglycemia is an appetite stimulant, and it decreases glucose utilization by reducing the amount of glucose reaching the cells. Polyphagia (increased food intake) is seen in diabetes mellitus, in

which blood glucose is high but cellular utilization is low because of the insulin deficiency (see Chapter 19).

Other Factors Affecting Food Intake

Food intake is increased in cold weather and decreased in warm weather. However, there is little evidence that body temperature is a major regulator of food intake.

Distention of the gastrointestinal tract inhibits appetite, and contractions of an empty stomach (**hunger contractions**) stimulate appetite, but denervation of the stomach and intestines does not affect the amount of food eaten. Especially in humans, cultural factors, environment, and past experiences related to the sight, smell, and taste of food also affect food intake.

Brown fat, a special form of body fat that has an extensive sympathetic innervation, may also contribute to the regulation of body weight. It is discussed above and in Chapter 17.

Long-Term Regulation of Appetite

The net effect of all the appetite-regulating mechanisms in normal adult animals and humans is an adjustment of food intake to the point where caloric intake balances energy expenditures, with the result that body weight is maintained. Children are notorious for their uneven food intake, their appetite for certain foods, and their unwillingness to eat others. However, over time they balance food intake with energy expenditure for immediate needs and growth and they grow and develop at a normal pace. Humans gain weight with advancing age, but this is normally a slow, carefully regulated process. One investigator calculated that the average woman gains 11 kg between the ages of 25 and 65. Considering that the total food intake of a woman over the 40-year period is more than 18 metric tons, the error in food intake over energy expenditure that produces the weight gain is less than 0.03%.

THIRST

Another appetitive mechanism under hypothalamic control is thirst. Drinking is regulated by plasma osmolality and ECF volume in much the same fashion as vasopressin secretion (see below). Water intake is increased by increased effective osmotic pressure of the plasma (Figure 14–7), by decreases in ECF volume, and by psychologic and other factors. Osmolality acts via **osmoreceptors,** receptors that sense the osmolality of the body fluids. These osmoreceptors are located in the anterior hypothalamus.

Decreases in ECF volume also stimulate thirst by a pathway independent of that mediating thirst in response to increased plasma osmolality (Figure 14–8). Thus, hemorrhage causes increased drinking even

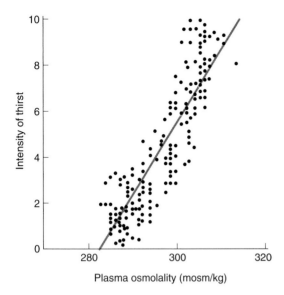

Figure 14–7. Relation of plasma osmolality to thirst in healthy adult humans during infusion of hypertonic saline. The intensity of thirst is measured on a special analog scale. (Reproduced, with permission, from Thompson CJ et al: The osmotic thresholds for thirst and vasopressin release are similar in healthy humans. Clin Sci Lond 1986;71:651.)

though there is no change in the osmolality of the plasma. The effect of ECF volume depletion on thirst is mediated in part via the renin-angiotensin system (see Chapter 24). Renin secretion is increased by hypovolemia and results in an increase in circulating angiotensin II. The angiotensin II acts on the **subfornical organ,** a specialized receptor area in the diencephalon (see Figure 32–7), to stimulate the neural areas concerned with thirst. There is some evidence that it acts on the **organum vasculosum of the lamina terminalis (OVLT)** as well. These areas are highly permeable and are two of the circumventricular organs located "outside the blood-brain barrier" (see Chapter 32). However, drugs which block the action of angiotensin II do not completely block the thirst response to hypovolemia, and it appears that the baroreceptors in the heart and blood vessels are also involved.

The intake of liquids is increased during eating **(prandial drinking).** The increase has been called a learned or habit response, but it has not been investigated in detail. One factor is any increase in plasma osmolality that occurs as the food is absorbed. Another may be an action of one or more gastrointestinal hormones directly on the subfornical organ. There is now some evidence that gastrointestinal hormones stimu-

late the same subfornical neurons that respond to angiotensin II.

Whenever the sensation of thirst is obtunded, either by direct damage to the diencephalon or by depressed or altered states of consciousness, patients stop drinking adequate amounts of fluid. Dehydration results if appropriate measures are not instituted to maintain water balance. If the protein intake is high, the products of protein metabolism cause an osmotic diuresis (see Chapter 38), and the amounts of water required to maintain hydration are large. Most cases of **hypernatremia** are actually due to simple dehydration in patients with psychoses or cerebral disease who do not or cannot increase their water intake when their thirst mechanism is stimulated.

Other Factors Regulating Water Intake

A number of other well-established factors contribute to the regulation of water intake. Psychologic and social factors are important. Dryness of the pharyngeal mucous membrane causes a sensation of thirst. Patients in whom fluid intake must be restricted sometimes get appreciable relief of thirst by sucking ice chips or a wet cloth.

Dehydrated dogs, cats, camels, and some other animals rapidly drink just enough water to make up their water deficit. They stop drinking before the water is absorbed (while their plasma is still hypertonic), so some kind of pharyngeal gastrointestinal "metering" must be involved. There is some evidence that humans have a similar metering ability, though it is not well developed.

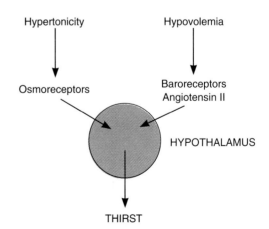

Figure 14–8. Diagrammatic representation of the way in which changes in plasma osmolality and changes in ECF volume affect thirst by separate pathways.

CONTROL OF POSTERIOR PITUITARY SECRETION

Vasopressin & Oxytocin

In most mammals, the hormones secreted by the posterior pituitary gland are **arginine vasopressin (AVP)** and **oxytocin.** In hippopotamuses and most pigs, arginine in the vasopressin molecule is replaced by lysine to form **lysine vasopressin.** The posterior pituitaries of some species of pigs and marsupials contain a mixture of arginine and lysine vasopressin. The posterior lobe hormones are nonapeptides with a disulfide ring at one end (Figure 14–9).

Biosynthesis, Intraneuronal Transport, & Secretion

The hormones of the posterior pituitary gland are synthesized in the cell bodies of the magnocellular neurons in the supraoptic and paraventricular nuclei and transported down the axons of these neurons to their endings in the posterior lobe, where they are secreted in response to electrical activity in the endings. Some of the neurons make oxytocin and others make vasopressin, and oxytocin-containing and vasopressin-containing cells are found in both nuclei.

Oxytocin and vasopressin are typical **neural hormones,** ie, hormones secreted into the circulation by nerve cells. This type of neural regulation is compared with other types in Figure 14–10. The term **neurosecretion** was originally coined to describe the secretion of hormones by neurons, but the term is somewhat misleading, because it appears that all neurons secrete chemical messengers (see Chapter 1).

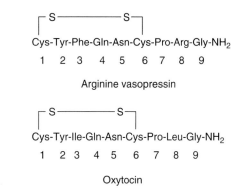

Figure 14–9. Arginine vasopressin and oxytocin.

Like other peptide hormones, the posterior lobe hormones are synthesized as part of larger precursor molecules. Vasopressin and oxytocin each have a characteristic **neurophysin** associated with them in the granules in the neurons that secrete them, neurophysin I in the case of oxytocin and neurophysin II in the case of vasopressin. The neurophysins were originally thought to be binding polypeptides, but it now appears that they are simply parts of the precursor molecules. The precursor for arginine vasopressin, **prepropressophysin,** contains a 19-amino acid residue leader sequence followed by arginine vasopressin, neurophysin II, and a glycopeptide (Figure 14–11). **Preprooxyphysin,** the precursor for oxytocin, is a similar but smaller molecule that lacks the glycopeptide.

The precursor molecules are synthesized in the ribosomes of the cell bodies of the neurons. They have their leader sequences removed in the endoplasmic reticu-

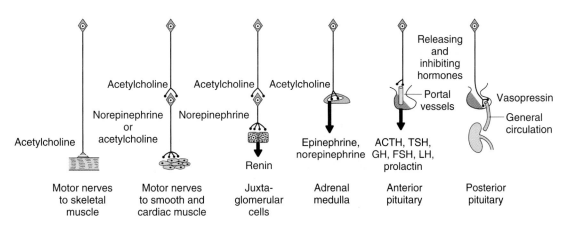

Figure 14–10. Neural control mechanisms. In the two situations on the left, neurotransmitters act at nerve endings on muscle; in the two in the middle, neurotransmitters regulate the secretion of endocrine glands; and in the two on the right, neurons secrete hormones into the hypophysial portal or general circulation.

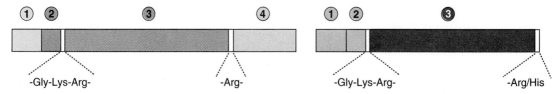

① Signal peptide	19 aa	① Signal peptide	19 aa
② Vasopressin	9 aa	② Oxytocin	9 aa
③ Neurophysin II	95 aa	③ Neurophysin I	93 aa
④ Glycopeptide	39 aa		

-Gly-Lys-Arg-   -Arg-   -Gly-Lys-Arg-   -Arg/His

Figure 14–11. Structure of bovine prepropressophysin **(left)** and prepro-oxyphysin **(right).** Gly in the 10 position of both peptides is necessary for amidation of the Gly residue in position 9. aa, amino acid residues. (Reproduced, with permission, from Richter D: Molecular events in expression of vasopressin and oxytocin and their cognate receptors. Am J Physiol 1988;255:F207.)

lum, are packaged into secretory granules in the Golgi apparatus, and are transported down the axons by axoplasmic flow to the endings in the posterior pituitary. The secretory granules, called **Herring bodies,** are easy to stain in tissue sections, and they have been extensively studied. Cleavage of the precursor molecules occurs as they are being transported, and the storage granules in the endings contain free vasopressin or oxytocin and the corresponding neurophysin. In the case of vasopressin, the glycopeptide is also present. All these products are secreted, but the functions of the components other than the established posterior pituitary hormones are unknown.

Electrical Activity of Magnocellular Neurons

The oxytocin-secreting and vasopressin-secreting neurons also generate and conduct action potentials, and action potentials reaching their endings trigger release of hormone from them by Ca^{2+}-dependent exocytosis. At least in anesthetized rats, these neurons are silent at rest or discharge at low, irregular rates (0.1–3 spikes per second). However, their response to stimulation varies (Figure 14–12). Stimulation of the nipples causes a synchronous, high-frequency discharge of the oxytocin neurons after an appreciable latency. This discharge causes release of a pulse of oxytocin and consequent milk ejection (see below). On the other hand, stimulation of the vasopressin-secreting neurons by a stimulus such as hemorrhage causes an initial steady increase in firing rate followed by a prolonged pattern of phasic discharge in which periods of high-frequency discharge

alternate with periods of electrical quiescence **(phasic bursting).** These phasic bursts are generally not synchronous in different vasopressin-secreting neurons. They are well suited to maintain a prolonged increase in the output of vasopressin, as opposed to the synchronous, relatively short, high-frequency discharge of oxytocin-secreting neurons in response to stimulation of the nipples.

Vasopressin & Oxytocin in Other Locations

Vasopressin-secreting neurons are found in the suprachiasmatic nuclei, and vasopressin and oxytocin are also found in the endings of neurons that project from the paraventricular nuclei to the brain stem and spinal cord. These neurons appear to be involved in cardiovascular control. In addition, vasopressin and oxytocin are synthesized in the gonads and the adrenal cortex and there is oxytocin in the thymus. The functions of the peptides in these organs are unsettled.

Vasopressin Receptors

There are at least three kinds of vasopressin receptors: V_{1A}, V_{1B}, and V_2. All are G protein-coupled. The V_{1A} and V_{1B} receptors act through phosphatidylinositol hydrolysis to increase the intracellular Ca^{2+} concentration. The V_2 receptors act through G_s to increase cAMP levels.

Effects of Vasopressin

Because one of its principal physiologic effects is the retention of water by the kidney, vasopressin is often

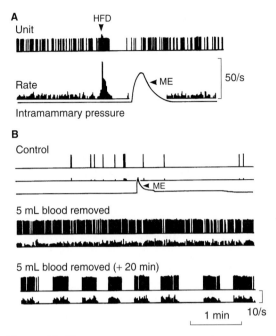

A

Unit

HFD

Rate

50/s

ME

Intramammary pressure

B

Control

ME

5 mL blood removed

5 mL blood removed (+ 20 min)

1 min 10/s

Figure 14–12. Responses of magnocellular neurons to stimulation. The tracings show individual extracellularly recorded action potentials, discharge rates, and intramammary duct pressure. **A:** Response of an oxytocin-secreting neuron. HFD, high-frequency discharge; ME, milk ejection. Stimulation of nipples started before the onset of recording. **B:** Responses of a vasopressin-secreting neuron, showing no change in the slow firing rate in response to stimulation of nipples and a prompt increase in the firing rate when 5 mL of blood was drawn, followed by typical phasic discharge. (Tracings from Wakerly JB: Hypothalamic neurosecretory function: Insights from electrophysiological studies of the magnocellular nuclei. IBRO News 1985;4:15.)

called the **antidiuretic hormone (ADH).** It increases the permeability of the collecting ducts of the kidney, so that water enters the hypertonic interstitium of the renal pyramids (see Chapter 38). The urine becomes concentrated, and its volume decreases. The overall effect is therefore retention of water in excess of solute; consequently, the effective osmotic pressure of the body fluids is decreased. In the absence of vasopressin, the urine is hypotonic to plasma, urine volume is increased, and there is a net water loss. Consequently, the osmolality of the body fluid rises.

The mechanism by which vasopressin exerts its antidiuretic effect is activated by V_2 receptors and involves insertion of protein water channels in the luminal membranes of the principal cells of the collecting

ducts. In many situations in the body, movement of water by simple diffusion is augmented by movement through water channels. They are now called **aquaporins,** and five have been identified. Aquaporin-1, -2, and -3 are found in the kidneys (see Chapter 38); aquaporin-4 is found in the brain; and aquaporin-5 is found in salivary and lacrimal glands and in the respiratory tract. The vasopressin-responsive water channel in the collecting ducts is aquaporin-2. These channels are stored in endosomes inside the cells, and vasopressin causes their rapid translocation to the luminal membranes.

V_{1A} receptors mediate the vasoconstrictor effect of vasopressin, and vasopressin is a potent stimulator of vascular smooth muscle in vitro. However, relatively large amounts of vasopressin are needed to raise blood pressure in vivo, because vasopressin also acts on the brain to cause a decrease in cardiac output. The site of this action is the **area postrema,** one of the circumventricular organs (see Chapter 32). Hemorrhage is a potent stimulus to vasopressin secretion and the blood pressure fall after hemorrhage is more marked in animals that have been treated with synthetic peptides which block the pressor action of vasopressin. Consequently, it appears that vasopressin does play a role in blood pressure homeostasis.

V_{1A} receptors are also found in the liver and the brain. Vasopressin causes glycogenolysis in the liver, and, as noted above, it is a neurotransmitter in the brain and spinal cord.

The V_{1B} receptors (also called V_3 receptors) appear to be unique to the anterior pituitary, where they mediate increased ACTH secretion from the corticotropes.

Synthetic Agonists & Antagonists

Synthetic peptides that have selective actions and are more active than naturally occurring vasopressin and oxytocin have been produced by altering the amino acid residues. For example, 1-deamino-8-D-arginine vasopressin (desmopressin; DDAVP) has very high antidiuretic activity with little pressor activity, making it valuable in the treatment of vasopressin deficiency (see below). Antagonists that selectively block the pressor or antidiuretic activity of vasopressin have also been synthesized.

Metabolism

Circulating vasopressin is rapidly inactivated, principally in the liver and kidneys. It has a **biologic half-life** (time required for inactivation of half a given amount) of approximately 18 minutes in humans. Its effects on the kidney develop rapidly but are of short duration.

Control of Vasopressin Secretion: Osmotic Stimuli

Vasopressin is stored in the posterior pituitary and released into the bloodstream by impulses in the nerve fibers that contain the hormone. The factors affecting its secretion are summarized in Table 14–3. When the effective osmotic pressure of the plasma is increased above the normal 285 mosm/kg, the rate of discharge of these neurons increases and vasopressin secretion is increased (Figure 14–13). At 285 mosm/kg, plasma vasopressin is at or near the limits of detection by available assays, but there is probably a further decrease when plasma osmolality is below this level. Vasopressin secretion is regulated by osmoreceptors located in the anterior hypothalamus. They are outside the blood-brain barrier and appear to be located in the circumventricular organs, primarily the organum vasculosum of the lamina terminalis (see Chapter 32). The osmotic threshold for thirst (Figure 14–8) is the same as or slightly greater than the threshold for increased vasopressin secretion (Figure 14–13), and it is still uncertain whether the same osmoreceptors mediate both effects.

Vasopressin secretion is thus controlled by a delicate feedback mechanism that operates continuously to defend the osmolality of the plasma. Significant changes in secretion occur when osmolality is changed as little as 1%. In this way, the osmolality of the plasma in normal individuals is maintained very close to 285 mosm/L.

Table 14–3. Summary of stimuli affecting vasopressin secretion.

Vasopressin Secretion Increased	Vasopressin Secretion Decreased
Increased effective osmotic pressure of plasma	Decreased effective osmotic pressure of plasma
Decreased extracellular fluid volume	Increased extracellular fluid volume
Pain, emotion, "stress," exercise	Alcohol
Nausea and vomiting	
Standing	
Clofibrate, carbamazepine	
Angiotensin II	

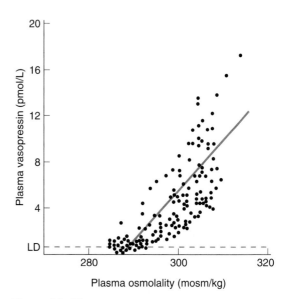

Figure 14–13. Relation between plasma osmolality and plasma vasopressin in healthy adult humans during infusion of hypertonic saline. LD, limit of detection. (Reproduced, with permission, from Thompson CJ et al: The osmotic thresholds for thirst and vasopressin are similar in healthy humans. Clin Sci [Colch] 1986;71:651.)

Volume Effects

ECF volume also affects vasopressin secretion. Vasopressin secretion is increased when ECF volume is low and decreased when ECF volume is high (Table 14–3). There is an inverse relationship between the rate of vasopressin secretion and the rate of discharge in afferents from stretch receptors in the low- and high-pressure portions of the vascular system. The low-pressure receptors are those in the great veins, right and left atria, and pulmonary vessels; the high-pressure receptors are those in the carotid sinuses and aortic arch (see Chapter 31). The exponential increases in plasma vasopressin produced by decreases in blood pressure are documented in Figure 14–14. However, the low-pressure receptors monitor the fullness of the vascular system, and moderate decreases in blood volume that decrease central venous pressure without lowering arterial pressure can also increase plasma vasopressin.

Thus, the low-pressure receptors are the primary mediators of volume effects on vasopressin secretion. Impulses pass from them via the vagi to the nucleus of the tractus solitarius (NTS). An inhibitory pathway projects from the NTS to the caudal ventrolateral medulla (CVLM), and there is a direct excitatory pathway from the CVLM to the hypothalamus. Angiotensin II reinforces the response to hypovolemia and

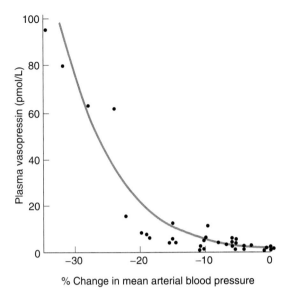

Figure 14–14. Relation of mean arterial blood pressure to plasma vasopressin in healthy adult humans in whom a progressive decline in blood pressure was induced by infusion of graded doses of the ganglionic blocking drug trimethaphan. The relation is exponential rather than linear. (Drawn from data in Baylis PH: Osmoregulation and control of vasopressin secretion in healthy humans. Am J Physiol 1987;253:R671.)

hypotension by acting on the circumventricular organs to increase vasopressin secretion (see Chapter 32).

Hypovolemia and hypotension produced by conditions such as hemorrhage release large amounts of vasopressin, and in the presence of hypovolemia, the osmotic response curve is shifted to the left (Figure 14–15). Its slope is also increased. The result is water retention and reduced plasma osmolality. This includes hyponatremia, since Na^+ is the most abundant osmotically active component of the plasma.

Other Stimuli Affecting Vasopressin Secretion

A variety of stimuli in addition to osmotic pressure changes and ECF volume aberrations increase vasopressin secretion. These include pain, nausea, surgical stress, and some emotions (Table 14–3). Nausea is associated with particularly large increases in vasopressin secretion. Alcohol decreases vasopressin secretion.

Clinical Implications

In various clinical conditions, volume and other nonosmotic stimuli bias the osmotic control of vasopressin

secretion. For example, patients who have had surgery may have elevated levels of plasma vasopressin because of pain and hypovolemia, and this may cause them to develop a low plasma osmolality and dilutional hyponatremia.

In the **syndrome of "inappropriate" hypersecretion of antidiuretic hormone (SIADH),** vasopressin is responsible not only for dilutional hyponatremia but also for loss of salt in the urine when water retention is sufficient to expand the ECF volume, reducing aldosterone secretion (see Chapter 20). This occurs in patients with cerebral disease ("cerebral salt wasting") and pulmonary disease ("pulmonary salt wasting"). Hypersecretion of vasopressin in patients with pulmonary diseases such as lung cancer may be due in part to the interruption of inhibitory impulses in vagal afferents from the stretch receptors in the atria and great veins. However, a significant number of lung tumors and some other cancers secrete vasopressin. Patients with

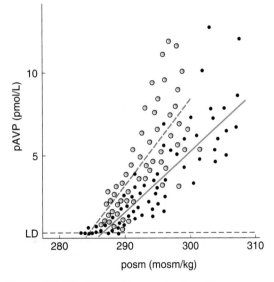

Figure 14–15. Effect of hypovolemia and hypervolemia on the relation between plasma vasopressin (pAVP) and plasma osmolality (posm). Seven blood samples were drawn at various times from ten normal men when hypovolemia was induced by water deprivation (colored circles, dashed line) and again when hypervolemia was induced by infusion of hypertonic saline (black circles, solid line). Linear regression analysis defined the relationship pAVP = 0.52 (posm − 283.5) for water deprivation and pAVP = 0.38 (posm − 285.6) for hypertonic saline. LD, limit of detection. Note the steeper curve as well as the shift of the intercept to the left during hypovolemia. (Courtesy of CJ Thompson.)

inappropriate hypersecretion of vasopressin have been successfully treated with demeclocycline, an antibiotic that reduces the renal response to vasopressin.

Diabetes insipidus is the syndrome that results when there is a vasopressin deficiency or when the kidneys fail to respond to the hormone.

Causes of vasopressin deficiency include disease processes in the supraoptic and paraventricular nuclei, the hypothalamohypophysial tract, or the posterior pituitary gland. It has been estimated that 30% of the clinical cases are due to neoplastic lesions of the hypothalamus, either primary or metastatic; 30% are post-traumatic; 30% are idiopathic; and the remainder are due to vascular lesions, infections, systemic diseases such as sarcoidosis that affect the hypothalamus, or mutations in the gene for prepropressophysin. The disease that develops after surgical removal of the posterior lobe of the pituitary may be temporary if only the distal ends of the supraoptic and paraventricular fibers are damaged, because the fibers recover, make new vascular connections, and begin to secrete vasopressin again. The symptoms of diabetes insipidus are passage of large amounts of dilute urine (**polyuria**) and the drinking of large amounts of fluid (**polydipsia**), provided the thirst mechanism is intact. It is the polydipsia that keeps these patients healthy. If their sense of thirst is depressed for any reason and their intake of dilute fluid decreases, they develop dehydration that can be fatal. In patients who have an incomplete rather than total vasopressin deficiency, treatment with drugs that increase vasopressin secretion, such as clofibrate, has proved to be of value. Chlorpropamide is also of value because it increases the renal response to vasopressin.

Another cause of diabetes insipidus is inability of the kidneys to respond to vasopressin (**nephrogenic diabetes insipidus**). In one form of this condition, congenital defects in the V_2 receptor as a result of various mutations in its gene prevent vasopressin from generating cAMP. This condition is X-linked, because the V_2 gene is on the X chromosome. In the other form of the condition, mutations in the autosomal gene for aquaporin-2 produce nonfunctional versions of this water channel. It is interesting that aquaporin-2 normally occurs in the urine, and in diabetes insipidus due to vasopressin deficiency there is a prompt rise in urinary aquaporin-2 when a vasopressin agonist is injected. However, there is no rise in either form of nephrogenic diabetes.

The amelioration of diabetes insipidus produced by the development of concomitant anterior pituitary insufficiency is discussed in Chapter 22.

Effects of Oxytocin

Oxytocin acts primarily on the breasts and uterus, though it appears to be involved in luteolysis as well (see Chapter 23). A G protein-coupled serpentine oxytocin receptor has been identified in human myometrium, and a similar or identical receptor is found in mammary tissue and the ovary. It triggers increases in intracellular Ca^{2+} levels.

In mammals, oxytocin causes contraction of the **myoepithelial cells,** smooth-muscle-like cells that line the ducts of the breast. This squeezes the milk out of the alveoli of the lactating breast into the large ducts (sinuses) and thence out of the nipple (**milk ejection**). Many hormones acting in concert are responsible for breast growth and the secretion of milk into the ducts (see Chapter 23), but milk ejection in most species requires oxytocin.

The Milk Ejection Reflex

Milk ejection is normally initiated by a neuroendocrine reflex. The receptors involved are the touch receptors, which are plentiful in the breast—especially around the nipple. Impulses generated in these receptors are relayed from the somatic touch pathways to the supraoptic and paraventricular nuclei. Discharge of the oxytocin-containing neurons causes secretion of oxytocin from the posterior pituitary (Figure 14–12). The infant suckling at the breast stimulates the touch receptors, the nuclei are stimulated, oxytocin is released, and the milk is expressed into the sinuses, ready to flow into the mouth of the waiting infant. In lactating women, genital stimulation and emotional stimuli also produce oxytocin secretion, sometimes causing milk to spurt from the breasts.

Other Actions of Oxytocin

Oxytocin causes contraction of the smooth muscle of the uterus. The sensitivity of the uterine musculature to oxytocin is enhanced by estrogen and inhibited by progesterone. The inhibitory effect of progesterone is due to direct action of the steroid on uterine oxytocin receptors. In late pregnancy, the uterus becomes very sensitive to oxytocin coincident with a marked increase in the number of oxytocin receptors and oxytocin receptor mRNA (see Chapter 23). Oxytocin secretion is increased during labor. After dilation of the cervix, descent of the fetus down the birth canal initiates impulses in the afferent nerves that are relayed to the supraoptic and paraventricular nuclei, causing secretion of sufficient oxytocin to enhance labor (see Figure 23–39). The amount of oxytocin in plasma is normal at the onset of labor. It is possible that the marked increase in oxytocin receptors at this time causes normal oxytocin levels to initiate contractions, setting up a positive feedback. However, the amount of oxytocin in the uterus is also increased, and locally produced oxytocin may also play a role.

Oxytocin may also act on the nonpregnant uterus to facilitate sperm transport. The passage of sperm up the female genital tract to the uterine tubes, where fertilization normally takes place, depends not only on the motile powers of the sperm but also, at least in some species, on uterine contractions. The genital stimulation involved in coitus releases oxytocin, but it has not been proved that it is oxytocin which initiates the rather specialized uterine contractions that transport the sperm. The secretion of oxytocin is increased by stressful stimuli and, like that of vasopressin, is inhibited by alcohol.

Circulating oxytocin increases at the time of ejaculation in males, and it is possible that this increase causes increased contraction of the smooth muscle of the vas deferens, propelling sperm toward the urethra.

CONTROL OF ANTERIOR PITUITARY SECRETION

Anterior Pituitary Hormones

The anterior pituitary secretes six hormones: **adrenocorticotropic hormone (corticotropin, ACTH), thyroid-stimulating hormone (thyrotropin, TSH), growth hormone, follicle-stimulating hormone (FSH), luteinizing hormone (LH),** and **prolactin (PRL).** An additional polypeptide, β-lipotropin (β-LPH), is secreted with ACTH, but its physiologic role is unknown. The actions of the anterior pituitary hormones are summarized in Figure 14–16. The hor-

mones are discussed in detail in the chapters on the endocrine system. The hypothalamus plays an important stimulatory role in regulating the secretion of ACTH, β-LPH, TSH, growth hormone, FSH, and LH. It also regulates prolactin secretion, but its effect is predominantly inhibitory rather than stimulatory.

Nature of Hypothalamic Control

Anterior pituitary secretion is controlled by chemical agents carried in the portal hypophysial vessels from the hypothalamus to the pituitary. These substances used to be called releasing and inhibiting factors, but now they are commonly called **hypophysiotropic hormones.** The latter term seems appropriate since they are secreted into the bloodstream and act at a distance from their site of origin. They do not escape into the general circulation to any degree, but they are in high concentration in portal hypophysial blood.

Hypophysiotropic Hormones

There are six established hypothalamic releasing and inhibiting hormones (Figure 14–17): **corticotropin-releasing hormone (CRH); thyrotropin-releasing hormone (TRH); growth hormone-releasing hormone (GRH); growth hormone-inhibiting hormone (GIH;** now generally called **somatostatin); luteinizing hormone-releasing hormone (LHRH),** now generally known as **gonadotropin-releasing hormone (GnRH);** and **prolactin-inhibiting hormone (PIH).** In addi-

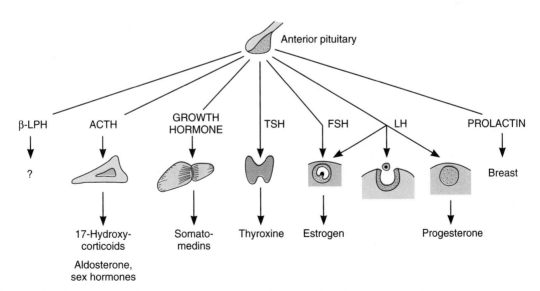

Figure 14–16. Anterior pituitary hormones. In women, FSH and LH act in sequence on the ovary to produce growth of the ovarian follicle, ovulation, and formation and maintenance of the corpus luteum. In men, FSH and LH control the functions of the testes. Prolactin stimulates lactation.

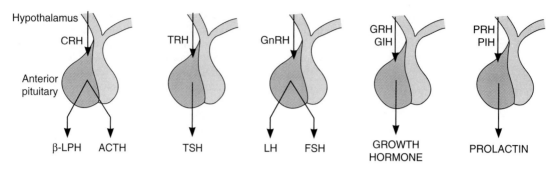

Figure 14–17. Effects of hypophysiotropic hormones on the secretion of anterior pituitary hormones.

tion, hypothalamic extracts contain prolactin-releasing activity, and a **prolactin-releasing hormone (PRH)** has been postulated to exist. TRH, VIP, and several other polypeptides found in the hypothalamus stimulate prolactin secretion, but it is uncertain whether one or more of these peptides is the physiologic PRH. Recently, an orphan receptor was isolated from the anterior pituitary, and the search for its ligand led to the isolation of a 31-amino-acid polypeptide from the human hypothalamus. This polypeptide stimulated prolactin secretion by an action on the anterior pituitary receptor, but additional research is needed to determine if it is the physiologic PRH. GnRH stimulates the secretion of FSH as well as that of LH, and it seems unlikely that there is a separate follicle-stimulating hormone-releasing hormone.

The structures of the six established hypophysiotropic hormones are shown in Figure 14–18. The structures of the genes and preprohormones for TRH, GnRH, somatostatin, CRH, and GRH are known. PreproTRH contains six copies of TRH (see Figure 1–22). Several other preprohormones may contain

other hormonally active peptides in addition to the hypophysiotropic hormones.

The area from which the hypothalamic releasing and inhibiting hormones are secreted is the median eminence of the hypothalamus. This region contains few nerve cell bodies, but there are many nerve endings in close proximity to the capillary loops from which the portal vessels originate.

The locations of the cell bodies of the neurons that project to the external layer of the median eminence and secrete the hypophysiotropic hormones are shown in Figure 14–19, which also shows the location of the neurons secreting oxytocin and vasopressin. The GnRH-secreting neurons are primarily in the medial preoptic area, the somatostatin-secreting neurons are in the periventricular nuclei, the TRH-secreting and CRH-secreting neurons are in the medial parts of the paraventricular nuclei, and the GRH-secreting and dopamine-secreting neurons are in the arcuate nuclei.

Most, if not all, of the hypophysiotropic hormones affect the secretion of more than one anterior pituitary hormone (Figure 14–17). The FSH-stimulating activity

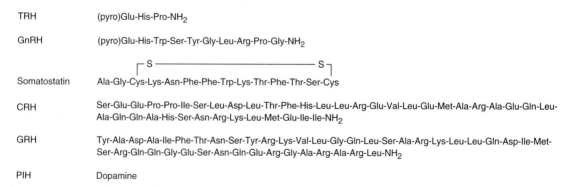

TRH	(pyro)Glu-His-Pro-NH$_2$
GnRH	(pyro)Glu-His-Trp-Ser-Tyr-Gly-Leu-Arg-Pro-Gly-NH$_2$
Somatostatin	Ala-Gly-Cys-Lys-Asn-Phe-Phe-Trp-Lys-Thr-Phe-Thr-Ser-Cys
CRH	Ser-Glu-Glu-Pro-Pro-Ile-Ser-Leu-Asp-Leu-Thr-Phe-His-Leu-Leu-Arg-Glu-Val-Leu-Glu-Met-Ala-Arg-Ala-Glu-Gln-Leu-Ala-Gln-Gln-Ala-His-Ser-Asn-Arg-Lys-Leu-Met-Glu-Ile-Ile-NH$_2$
GRH	Tyr-Ala-Asp-Ala-Ile-Phe-Thr-Asn-Ser-Tyr-Arg-Lys-Val-Leu-Gly-Gln-Leu-Ser-Ala-Arg-Lys-Leu-Leu-Gln-Asp-Ile-Met-Ser-Arg-Gln-Gln-Gly-Glu-Ser-Asn-Gln-Glu-Arg-Gly-Ala-Arg-Ala-Arg-Leu-NH$_2$
PIH	Dopamine

Figure 14–18. Structure of hypophysiotropic hormones in humans. Preprosomatostatin is processed to a tetradecapeptide (somatostatin 14, [SS14], shown above) and also to a polypeptide containing 28 amino acid residues (SS28).

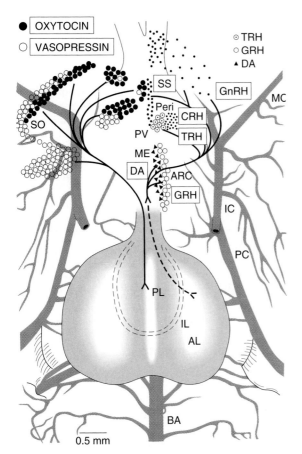

Figure 14–19. Location of cell bodies of hypophysiotropic hormone-secreting neurons projected on a ventral view of the hypothalamus and pituitary of the rat. AL, anterior lobe; ARC, arcuate nucleus; BA, basilar artery; DA, dopamine; IC, internal carotid artery; IL, intermediate lobe; MC, middle cerebral artery; ME, median eminence; PC, posterior cerebral artery; Peri, periventricular nucleus; PL, posterior lobe; PV, paraventricular nucleus; SO, supraoptic nucleus. The names of the hormones are enclosed in boxes. (Courtesy of LW Swanson and ET Cunningham Jr.)

the autonomic nervous system (see Chapter 4). In addition, somatostatin is found in the pancreatic islets (see Chapter 19), GRH is secreted by pancreatic tumors, and somatostatin and TRH are found in the gastrointestinal tract (see Chapter 26).

Receptors for most of the hypophysiotropic hormones are serpentine and coupled to G proteins. There are two human CRH receptors: hCRH-RI, and hCRH-RII. The latter differs from the former in having a 29-amino-acid insert in its first cytoplasmic loop. The physiologic role of hCRH-RII is unsettled, though it is found in many parts of the brain. In addition, there is a **CRH-binding protein** in the peripheral circulation that inactivates CRH. It is also found in the cytoplasm of corticotropes in the anterior pituitary, and in this location it might play a role in receptor internalization. However, the exact physiologic role of this protein is unknown. Other hypophysiotropic hormones do not have known binding proteins.

Significance & Clinical Implications

Research delineating the multiple neuroendocrine regulatory functions of the hypothalamus is important because it explains how endocrine secretion is made appropriate to the demands of a changing environment. The nervous system receives information about changes in the internal and external environment from the sense organs. It brings about adjustments to these changes through effector mechanisms that include not only somatic movement but also changes in the rate at which hormones are secreted.

The manifestations of hypothalamic disease are neurologic defects, endocrine changes, and metabolic abnormalities such as hyperphagia and hyperthermia. The relative frequencies of the signs and symptoms of hypothalamic disease in one large series of cases are shown in Table 14–4. The possibility of hypothalamic pathology should be kept in mind in evaluating all patients with pituitary dysfunction, especially those with isolated deficiencies of single pituitary tropic hormones.

A condition of considerable interest in this context is **Kallmann's syndrome,** the combination of hypogonadism due to low levels of circulating gonadotropins (**hypogonadotropic hypogonadism**) with partial or complete loss of the sense of smell (**hyposmia** or **anosmia**). Embryologically, GnRH neurons develop in the nose and migrate up the olfactory nerves and then through the brain to the hypothalamus. If this migration is prevented by congenital abnormalities in the olfactory pathways, the GnRH neurons do not reach the hypothalamus and pubertal maturation of the gonads fails to occur. The syndrome is most common in men, and the cause in many cases is mutation of the *KALIG1* gene, a gene on the X chromosome that codes

of GnRH has been mentioned above. TRH stimulates the secretion of prolactin as well as TSH. Somatostatin inhibits the secretion of TSH as well as growth hormone. It does not normally inhibit the secretion of the other anterior pituitary hormones, but it inhibits the abnormally elevated secretion of ACTH in patients with Nelson's syndrome. CRH stimulates the secretion of ACTH and β-LPH.

Hypophysiotropic hormones function as neurotransmitters in other parts of the brain, the retina, and

Table 14–4. Symptoms and signs in 60 autopsied patients with hypothalamic disease.[1]

Symptoms and Signs	Percentage of Cases
Endocrine and metabolic findings	
Precocious puberty	40
Hypogonadism	32
Diabetes insipidus	35
Obesity	25
Abnormalities of temperature regulation	22
Emaciation	18
Bulimia	8
Anorexia	7
Neurologic findings	
Eye signs	78
Pyramidal and sensory deficits	75
Headache	65
Extrapyramidal signs	62
Vomiting	40
Psychic disturbances, rage attacks, etc.	35
Somnolence	30
Convulsions	15

[1]Data from Bauer HG: Endocrine and other clinical manifestations of hypothalamic disease. J Clin Endocrinol 1954; 14:13. See also Kahana L et al: Endocrine manifestations of intracranial extrasellar lesions. J Clin Endocrinol 1962;22:304.

for what is apparently an adhesion molecule necessary for normal development of the olfactory nerve on which the GnRH neurons migrate into the brain. However, the condition also occurs in women and can be due to other genetic abnormalities.

TEMPERATURE REGULATION

In the body, heat is produced by muscular exercise, assimilation of food, and all the vital processes that contribute to the basal metabolic rate (see Chapter 17). It is lost from the body by radiation, conduction, and vaporization of water in the respiratory passages and on the skin. Small amounts of heat are also removed in the urine and feces. The balance between heat production and heat loss determines the body temperature. Be-

cause the speed of chemical reactions varies with the temperature and because the enzyme systems of the body have narrow temperature ranges in which their function is optimal, normal body function depends upon a relatively constant body temperature.

Invertebrates generally cannot adjust their body temperatures and so are at the mercy of the environment. In vertebrates, mechanisms for maintaining body temperature by adjusting heat production and heat loss have evolved. In reptiles, amphibia, and fish, the adjusting mechanisms are relatively rudimentary, and these species are called "cold-blooded" (**poikilothermic**) because their body temperature fluctuates over a considerable range. In birds and mammals, the "warm-blooded" (**homeothermic**) animals, a group of reflex responses that are primarily integrated in the hypothalamus operate to maintain body temperature within a narrow range in spite of wide fluctuations in environmental temperature. The hibernating mammals are a partial exception. While awake, they are homeothermic, but during hibernation, their body temperature falls.

Normal Body Temperature

In homeothermic animals, the actual temperature at which the body is maintained varies from species to species and, to a lesser degree, from individual to individual. In humans, the traditional normal value for the oral temperature is 37 °C (98.6 °F), but in one large series of normal young adults, the morning oral temperature averaged 36.7 °C, with a standard deviation of 0.2 °C. Therefore, 95% of all young adults would be expected to have a morning oral temperature of 36.3–37.1 °C (97.3–98.8 °F; mean ± 1.96 standard deviations; see Appendix). Various parts of the body are at different temperatures, and the magnitude of the temperature difference between the parts varies with the environmental temperature (Figure 14–20). The extremities are generally cooler than the rest of the body. The temperature of the scrotum is carefully regulated at 32 °C. The rectal temperature is representative of the temperature at the core of the body and varies least with changes in environmental temperature. The oral temperature is normally 0.5 °C lower than the rectal temperature, but it is affected by many factors, including ingestion of hot or cold fluids, gum-chewing, smoking, and mouth breathing.

The normal human core temperature undergoes a regular circadian fluctuation of 0.5–0.7 °C. In individuals who sleep at night and are awake during the day (even when hospitalized at bed rest), it is lowest at about 6 AM and highest in the evenings (Figure 14–21). It is lowest during sleep, is slightly higher in the awake but relaxed state, and rises with activity. In

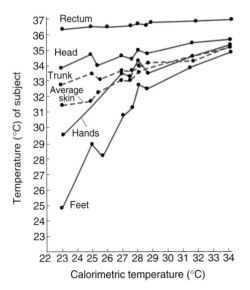

Figure 14–20. Temperatures of various parts of the body of a naked subject at various ambient temperatures in a calorimeter. (Redrawn and reproduced, with permission, from Hardy JD, DuBois EF: Basal metabolism, radiation, convection and vaporization at temperatures of 22–35 °C. J Nutr 1938;15:477.)

women, there is an additional monthly cycle of temperature variation characterized by a rise in basal temperature at the time of ovulation (see Chapter 23 and Figure 23–28). Temperature regulation is less precise in young children, and they may normally have a temperature that is 0.5 °C or so above the established norm for adults.

During exercise, the heat produced by muscular contraction accumulates in the body, and the rectal temperature normally rises as high as 40 °C (104 °F). This rise is due in part to the inability of the heat-dissipating mechanisms to handle the greatly increased amount of heat produced, but there is evidence that in addition there is an elevation of the body temperature at which the heat-dissipating mechanisms are activated during exercise (Figure 14–22). Body temperature also rises slightly during emotional excitement, probably owing to unconscious tensing of the muscles. It is chronically elevated by as much as 0.5 °C when the metabolic rate is high, as in hyperthyroidism, and lowered when the metabolic rate is low, as in hypothyroidism (Figure 14–21). Some apparently normal adults chronically have a temperature above the normal range (constitutional hyperthermia).

Heat Production

Heat production and energy balance are discussed in Chapter 17. A variety of basic chemical reactions contribute to body heat production at all times. Ingestion of food increases heat production because of the specific dynamic action of the food (see Chapter 17), but the major source of heat is the contraction of skeletal muscle (Table 14–5). Heat production can be varied by endocrine mechanisms in the absence of food intake or muscular exertion. Epinephrine and norepinephrine produce a rapid but short-lived increase in heat production; thyroid hormones produce a slowly developing but prolonged increase. Furthermore, sympathetic discharge is decreased during fasting and increased by feeding.

A source of considerable heat, particularly in infants, is **brown fat** (see Chapter 17). This fat has a high rate of metabolism, and its thermogenic function has been likened to that of an electric blanket.

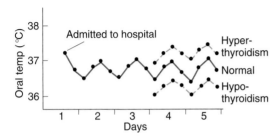

Figure 14–21. Typical temperature chart of a hospitalized patient who does not have a febrile disease. Note the slight rise in temperature, due to excitement and apprehension, at the time of admission to the hospital, and the regular circadian temperature cycle.

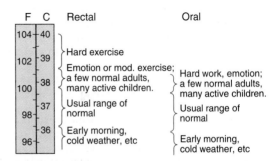

Figure 14–22. Ranges in rectal and oral temperatures seen in normal humans. (Reproduced, with permission, from DuBois EF: *Fever and the Regulation of Body Temperature.* Thomas, 1948.)

Table 14–5. Body heat production and heat loss.

Body heat is produced by:	
Basic metabolic processes	
Food intake (specific dynamic action)	
Muscular activity	

Body heat is lost by:	Percentage of heat lost at 21 °C
Radiation and conduction	70
Vaporization of sweat	27
Respiration	2
Urination and defecation	1

Heat Loss

The processes by which heat is lost from the body when the environmental temperature is below body temperature are listed in Table 14–5. **Conduction** is heat exchange between objects or substances at different temperatures that are in contact with one another. A basic characteristic of matter is that its molecules are in motion, with the amount of motion proportionate to the temperature. These molecules collide with the molecules in cooler objects, transferring thermal energy to them. The amount of heat transferred is proportionate to the temperature difference between the objects in contact **(thermal gradient).** Conduction is aided by **convection,** the movement of molecules away from the area of contact. Thus, for example, an object in contact with air at a different temperature changes the specific gravity of the air, and since warm air rises and cool air falls, a new supply of air is brought into contact with the object. Of course, convection is greatly aided if the object moves about in the medium or the medium moves past the object, eg, if a subject swims through water or a fan blows air through a room. **Radiation** is transfer of heat by infrared electromagnetic radiation from one object to another at a different temperature with which it is not in contact. When an individual is in a cold environment, heat is lost by conduction to the surrounding air and by radiation to cool objects in the vicinity. Conversely, of course, heat is transferred to an individual and the heat load is increased by these processes when the environmental temperature is above body temperature. Note that because of radiation, an individual can feel chilly in a room with cold walls even though the room is relatively warm. On a cold but sunny day, the heat of the sun reflected off bright objects exerts an appreciable warming effect. It is the heat reflected from the snow, for example, that makes it possible to ski in fairly light clothes even though the air temperature is below freezing.

Since conduction occurs from the surface of one object to the surface of another, the temperature of the skin determines to a large extent the degree to which the body heat is lost or gained. The amount of heat reaching the skin from the deep tissues can be varied by changing the blood flow to the skin. When the cutaneous vessels are dilated, warm blood wells into the skin, whereas in the maximally vasoconstricted state, heat is held centrally in the body. The rate at which heat is transferred from the deep tissues to the skin is called the **tissue conductance.** Birds have a layer of feathers next to the skin, and most mammals have a significant layer of hair or fur. Heat is conducted from the skin to the air trapped in this layer and from the trapped air to the exterior. When the thickness of the trapped layer is increased by fluffing the feathers or erection of the hairs **(horripilation),** heat transfer across the layer is reduced and heat losses (or, in a hot environment, heat gains) are decreased. "Goose pimples" are the result of horripilation in humans; they are the visible manifestation of cold-induced contraction of the piloerector muscles attached to the rather meager hair supply. Humans usually supplement this layer of hair with one or more layers of clothes. Heat is conducted from the skin to the layer of air trapped by the clothes, from the inside of the clothes to the outside, and from the outside of the clothes to the exterior. The magnitude of the heat transfer across the clothing, a function of its texture and thickness, is the most important determinant of how warm or cool the clothes feel, but other factors, especially the size of the trapped layer of warm air, are important also. Dark clothes absorb radiated heat, and light-colored clothes reflect it back to the exterior.

The other major process transferring heat from the body in humans and other animals that sweat is vaporization of water on the skin and mucous membranes of the mouth and respiratory passages. Vaporization of 1 g of water removes about 0.6 kcal of heat. A certain amount of water is vaporized at all times. This **insensible water loss** amounts to 50 mL/h in humans. When sweat secretion is increased, the degree to which the sweat vaporizes depends upon the humidity of the environment. It is common knowledge that one feels hotter on a humid day. This is due in part to the decreased vaporization of sweat, but even under conditions in which vaporization of sweat is complete, an individual in a humid environment feels warmer than an individual in a dry environment. The reason for this difference is not known, but it seems related to the fact that in the humid environment sweat spreads over a greater area of skin before it evaporates. During muscular exertion in a hot environment, sweat secretion reaches values as high as 1600 mL/h, and in a dry atmosphere, most of this sweat is vaporized. Heat loss by vaporization of water therefore varies from 30 to over 900 kcal/h.

Some mammals lose heat by **panting.** This rapid, shallow breathing greatly increases the amount of water vaporization in the mouth and respiratory passages and therefore the amount of heat lost. Because the breathing is shallow, it produces relatively little change in the composition of alveolar air (see Chapter 34).

The relative contribution of each of the processes that transfer heat away from the body (Table 14–5) varies with the environmental temperature. At 21 °C, vaporization is a minor component in humans at rest. As the environmental temperature approaches body temperature, radiation losses decline and vaporization losses increase.

Temperature-Regulating Mechanisms

The reflex and semireflex thermoregulatory responses in humans are listed in Table 14–6. They include autonomic, somatic, endocrine, and behavioral changes. One group of responses increases heat loss and decreases heat production; the other decreases heat loss and increases heat production. In general, exposure to heat stimulates the former group of responses and inhibits the latter, whereas exposure to cold does the opposite.

Curling up "in a ball" is a common reaction to cold in animals and has a counterpart in the position some people assume on climbing into a cold bed. Curling up decreases the body surface exposed to the environment. Shivering is an involuntary response of the skeletal

Table 14–6. Temperature-regulating mechanisms.

Mechanisms activated by cold
Increase heat production
Shivering
Hunger
Increased voluntary activity
Increased secretion of norepinephrine and epinephrine
Decrease heat loss
Cutaneous vasoconstriction
Curling up
Horripilation
Mechanisms activated by heat
Increase heat loss
Cutaneous vasodilation
Sweating
Increased respiration
Decrease heat production
Anorexia
Apathy and inertia

muscles, but cold also causes a semiconscious general increase in motor activity. Examples include foot stamping and dancing up and down on a cold day. Increased catecholamine secretion is an important endocrine response to cold. Mice unable to make norepinephrine and epinephrine because their dopamine β-hydroxylase gene is knocked out do not tolerate cold; they have deficient vasoconstriction and are unable to increase thermogenesis in brown adipose tissue through UCP 1 (see Chapter 17). TSH secretion is increased by cold and decreased by heat in laboratory animals, but the change in TSH secretion produced by cold in adult humans is small and of questionable significance. It is common knowledge that activity is decreased in hot weather—the "it's too hot to move" reaction.

Thermoregulatory adjustments involve local responses as well as more general reflex responses. When cutaneous blood vessels are cooled, they become more sensitive to catecholamines and the arterioles and venules constrict. This local effect of cold directs blood away from the skin. Another heat-conserving mechanism that is important in animals living in cold water is heat transfer from arterial to venous blood in the limbs. The deep veins (**venae comitantes**) run alongside the arteries supplying the limbs, and heat is transferred from the warm arterial blood going to the limbs to the cold venous blood coming from the extremities (**countercurrent exchange;** see Chapter 38). This keeps the tips of the extremities cold but conserves body heat.

The reflex responses activated by cold are controlled from the posterior hypothalamus. Those activated by warmth are controlled primarily from the anterior hypothalamus, although some thermoregulation against heat still occurs after decerebration at the level of the rostral midbrain. Stimulation of the anterior hypothalamus causes cutaneous vasodilation and sweating, and lesions in this region cause hyperthermia, with rectal temperatures sometimes reaching 43 °C (109.4 °F). Posterior hypothalamic stimulation causes shivering, and the body temperature of animals with posterior hypothalamic lesions falls toward that of the environment.

There is some evidence that serotonin in primates and humans is a synaptic mediator in the areas activated by cold and that norepinephrine plays a similar role in those activated by heat. However, there are marked species variations in the temperature responses to these amines. Peptides may also be involved, but the details of the central synaptic connections concerned with thermoregulation are still unknown.

Afferents

The hypothalamus is said to integrate body temperature information from sensory receptors (primarily cold receptors) in the skin, deep tissues, spinal cord, extrahypothalamic portions of the brain, and the hypothalamus itself. Each of these five inputs contributes about 20% of the information that is integrated. There are threshold temperatures for each of the main temperature-regulating responses, and when the threshold is reached, the response begins. The threshold is 37 °C for sweating and vasodilation, 36.8 °C for vasoconstriction, 36 °C for nonshivering thermogenesis, and 35.5 °C for shivering. The threshold for vasodilation and sweating is demonstrated in Figure 14–23, in which these responses are plotted against the temperature of the interior of the head.

Fever

Fever is perhaps the oldest and most universally known hallmark of disease. It occurs not only in mammals but also in birds, reptiles, amphibia, and fish. When it occurs in homeothermic animals, the thermoregulatory mechanisms behave as if they were adjusted to main-tain body temperature at a higher than normal level, ie, "as if the thermostat had been reset" to a new point above 37 °C. The temperature receptors then signal that the actual temperature is below the new set point, and the temperature-raising mechanisms are activated. This usually produces chilly sensations due to cutaneous vasoconstriction and occasionally enough shivering to produce a shaking chill. However, the nature of the response depends on the ambient temperature. The temperature rise in experimental animals injected with a pyrogen is due mostly to increased heat production if they are in a cold environment and mostly to decreased heat loss if they are in a warm environment.

The pathogenesis of fever is summarized in Figure 14–24. Toxins from bacteria such as endotoxin act on monocytes, macrophages, and Kupffer cells to produce cytokines that act as **endogenous pyrogens (EPs).** There is good evidence that IL-1β, IL-6, β-IFN, γ-IFN, and TNF-α (see Chapter 27) can act independently to produce fever. These cytokines are polypeptides, and it is unlikely that circulating cytokines penetrate the brain. Instead, there is evidence that they act on the OVLT, one of the circumventricular organs (see Chapter 32). This in turn activates the preoptic area of the hypothalamus. Cytokines are also produced by cells in the CNS when these are stimulated by infection, and these may act directly on the thermoregulatory centers.

The fever produced by cytokines is probably due to local release of prostaglandins in the hypothalamus. Intrahypothalamic injection of prostaglandins produces fever. In addition, the antipyretic effect of aspirin is exerted directly on the hypothalamus, and aspirin inhibits

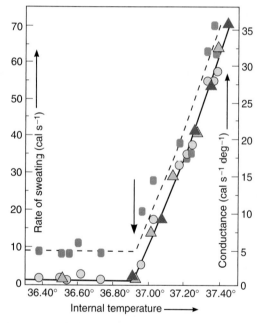

Figure 14–23. Quantitative relations in humans between the temperature of the interior of the head (internal temperature) and cutaneous blood flow (squares, scale at the right) and sweating (circles and triangles, scale at the left). The arrow points to the sharp threshold at which these parameters start to rise. In this subject, the threshold was at 36.9 °C. (Reproduced, with permission, from Benzinger TH: Receptor organs and quantitative mechanisms of human temperature control in a warm environment. Fed Proc 1960;19:32.)

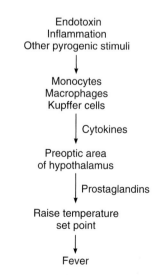

Figure 14–24. Pathogenesis of fever.

prostaglandin synthesis. PGE_2 is one of the prostaglandins that causes fever. It acts on four subtypes of prostaglandin receptors—EP_1, EP_2, EP_3, and EP_4—and knockout of the EP3 receptor impairs the febrile response to PGE_2, IL-1β, and bacterial lipopolysaccharide (LPS).

The benefit of fever to the organism is uncertain. It is presumably beneficial, because it has evolved and persisted as a response to infections and other diseases. Many microorganisms grow best within a relatively narrow temperature range, and a rise in temperature inhibits their growth. In addition, antibody production is increased when body temperature is elevated. Before the advent of antibiotics, fevers were artificially induced for the treatment of neurosyphilis and proved to be beneficial. Hyperthermia benefits individuals infected with anthrax, pneumococcal pneumonia, leprosy, and various fungal, rickettsial, and viral diseases. Hyperthermia also slows the growth of some tumors. However, very high temperatures are harmful. When the rectal temperature is over 41 °C (106 °F) for prolonged periods, some permanent brain damage results. When it is over 43 °C, heat stroke develops and death is common.

In **malignant hyperthermia,** various mutations of the gene coding for the ryanodine receptor (see Chapter 3) lead to excess Ca^{2+} release during muscle contraction triggered by stress. This in turn leads to contractures of the muscles, increased muscle metabolism, and a great increase in heat production in muscle. The increased heat production causes a marked rise in body temperature that is fatal if not treated.

Periodic fevers also occur in humans with mutations in the gene for **pyrin,** a protein found in neutrophils; the gene for mevalonate kinase, an enzyme involved in cholesterol synthesis; and the gene for the type 1 TNF receptor, which is involved in inflammatory responses. However, it is not known how any of these three mutant gene products cause fever.

Hypothermia

In hibernating mammals, body temperature drops to low levels without causing any ill effects that are demonstrable upon subsequent arousal. This observation led to experiments on induced hypothermia. When the skin or the blood is cooled enough to lower the body temperature in nonhibernating animals and in humans, metabolic and physiologic processes slow down. Respiration and heart rate are very slow, blood pressure is low, and consciousness is lost. At rectal temperatures of about 28 °C, the ability to spontaneously return the temperature to normal is lost, but the individual continues to survive and, if rewarmed with external heat, returns to a normal state. If care is taken to prevent the formation of ice crystals in the tissues, the body temperature of experimental animals can be lowered to subfreezing levels without producing any damage that is detectable after subsequent rewarming.

Humans tolerate body temperatures of 21–24 °C (70–75 °F) without permanent ill effects, and induced hypothermia has been used in surgery. On the other hand, accidental hypothermia due to prolonged exposure to cold air or cold water is a serious condition and requires careful monitoring and prompt rewarming.

Neural Basis of Instinctual Behavior & Emotions

INTRODUCTION

Emotions have both mental and physical components. They involve **cognition,** an awareness of the sensation and usually its cause; **affect,** the feeling itself; **conation,** the urge to take action; and **physical changes** such as hypertension, tachycardia, and sweating. The hypothalamus and limbic systems are intimately concerned with emotional expression and with the genesis of emotions.

This chapter reviews the physiologic basis of emotion, sexual behavior, fear, rage, and motivation. It also considers the relation of major neurotransmitter systems in the brain to these processes.

ANATOMIC CONSIDERATIONS

The term **limbic lobe** or **limbic system** is applied to the part of the brain that consists of a rim of cortical tissue around the hilum of the cerebral hemisphere and a group of associated deep structures—the amygdala, the hippocampus, and the septal nuclei (Figures 15–1 and 15–2). The region was formerly called the rhinencephalon because of its relation to olfaction, but only a small part of it is actually concerned with smell.

Histology

The limbic cortex is phylogenetically the oldest part of the cerebral cortex. Histologically, it is made up of a primitive type of cortical tissue called **allocortex,** which in most regions has only three layers and surrounds the hilum of the hemisphere. There is a second ring of transitional cortex called **juxtallocortex** between the allocortex and the neocortex. It has three to six layers and is found in regions such as the cingulate gyrus and the insula. The cortical tissue of the remaining nonlimbic portions of the hemisphere is called **neocortex.** It generally has six layers (see Chapter 11) and is the most highly developed type. The actual extent of the allocortical and juxtallocortical areas has changed little as mammals have evolved, but these regions have been overshadowed by the immense growth of the neocortex, which reaches its greatest development in humans (Figure 15–1).

Afferent & Efferent Connections

The major connections of the limbic system are shown in Figure 15–2. The fornix connects the hippocampus to the mamillary bodies, which are in turn connected to the anterior nuclei of the thalamus by the mamillothalamic tract. The anterior nuclei of the thalamus project to the cingulate cortex, and from the cingulate cortex there are connections to the hippocampus, completing a complex closed circuit. This circuit was originally described by Papez and has been called the Papez circuit.

Correlations Between Structure & Function

One characteristic of the limbic system is the paucity of the connections between it and the neocortex. However, from a functional point of view, neocortical activity does modify emotional behavior and vice versa. On the other hand, one of the characteristics of emotion is that it cannot be turned on and off at will.

Another characteristic of limbic circuits is their prolonged after-discharge following stimulation. This may explain in part the fact that emotional responses are generally prolonged rather than evanescent and outlast the stimuli that initiate them.

LIMBIC FUNCTIONS

Stimulation and ablation experiments indicate that in addition to its role in olfaction (see Chapter 10), the limbic system is concerned with autonomic responses. Along with the hypothalamus, it is also concerned with sexual behavior, the emotions of rage and fear, and motivation.

Limbic stimulation produces autonomic effects, particularly changes in blood pressure and respiration.

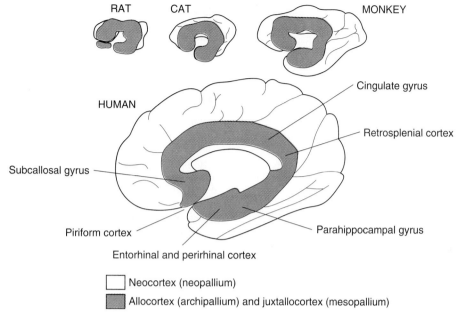

Figure 15–1. Relation of the limbic cortex (allocortex and juxtallocortex) to the neocortex in rats, cats, monkeys, and humans.

These responses are elicited from many limbic structures, and there is little evidence of localization of autonomic responses. This suggests that the autonomic effects are part of more complex phenomena, particularly emotional and behavioral responses.

SEXUAL BEHAVIOR

Mating is a basic but complex phenomenon in which many parts of the nervous system are involved. Copulation itself is made up of a series of reflexes integrated in spinal and lower brain stem centers, but the behavioral components that accompany it, the urge to copulate, and the coordinated sequence of events in the male and female that lead to pregnancy are regulated to a large degree in the limbic system and hypothalamus. Learning plays a part in the development of mating behavior, particularly in primates, but in nonprimate mammals, courtship and successful mating can occur with no previous sexual experience. The basic responses are therefore innate and are undoubtedly present in all mammals. However, in humans, the sexual functions have become extensively encephalized and conditioned by social and psychic factors. The basic physiologic mechanisms of sexual behavior in experimental animals are therefore considered first and then compared with the responses in humans.

Relation to Endocrine Function

In nonprimate mammals, removal of the gonads leads eventually to decreased or absent sexual activity in both the male and the female—although the loss is slow to develop in the males of some species. Injections of gonadal hormones in castrated animals revive sexual activity. Testosterone in the male and estrogen in the female have the most marked effect. Large doses of testosterone and other androgens in castrated females initiate female behavior, and large doses of estrogens in castrated males trigger male mating responses. It is unsettled why responses appropriate to the sex of the animal occur when the hormones of the opposite sex are injected.

In women, ovariectomy does not necessarily reduce libido (defined in this context as sexual interest and drive) or sexual ability. Postmenopausal women continue to have sexual relations, often without much change in frequency from their premenopausal pattern. However, adrenal androgens are still present in these women (see Chapter 26). Testosterone, for example, increases libido in males, and so does estrogen used to treat diseases such as carcinoma of the prostate. The behavioral pattern that was present before treatment is stimulated but not redirected. Thus, administration of testosterone to homosexuals intensifies their homosexual drive but does not convert it to a heterosexual drive.

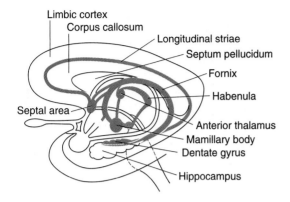

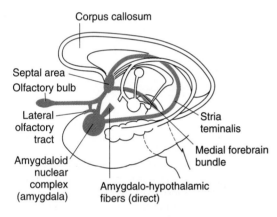

Figure 15–2. Principal connections of the limbic system. **Top:** Hippocampal system. **Bottom:** Olfactory and amygdaloid connections. (Reproduced, with permission, from Waxman SG: *Neuroanatomy with Clinical Correlations*, 25th ed. McGraw-Hill, 2003.)

Neural Control in the Male

In male animals, removal of the neocortex generally inhibits sexual behavior. However, cats and monkeys with bilateral limbic lesions localized to the piriform cortex overlying the amygdala develop a marked intensification of sexual activity. They not only mount adult females; they also mount immature females and other males and attempt to copulate with animals of other species and with inanimate objects. The extent to which these animal studies are applicable to men are uncertain, though there have been a few reports of hypersexuality in men with bilateral amygdaloid lesions.

The hypothalamus is also involved in the control of sexual activity in males. Stimulation along the medial forebrain bundle and in neighboring hypothalamic areas causes penile erection with considerable emotional display in monkeys. In castrated rats, intrahypothalamic implants of testosterone restore the complete pattern of sexual behavior.

Neural Control in the Female

In mammals, the sexual activity of the female is cyclic. Most of the time, the female avoids the male and repulses his sexual advances. Periodically, however, there is an abrupt change in behavior and the female seeks out the male, attempting to mate. These short episodes of **heat** or **estrus** are so characteristic that the sexual cycle in mammalian species that do not menstruate is named the **estrous cycle.**

In captivity, monkeys and apes mate at any time; but in the wild state, the females accept the male more frequently at the time of ovulation. In women, sexual activity occurs throughout the menstrual cycle, but careful studies indicate that, as in other primates, there is more spontaneous female-initiated sexual activity at about the time of ovulation.

In female sheep, discrete anterior hypothalamic lesions abolish behavioral heat (Figure 15–3) without affecting the regular pituitary-ovarian cycle (see Chapter 23).

Implantation of minute amounts of estrogen in the anterior hypothalamus causes heat in ovariectomized rats (see Figure 23–31). Implantation in other parts of

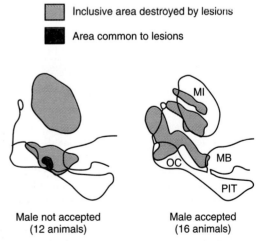

Figure 15–3. Sites of hypothalamic lesions that blocked behavioral heat without affecting ovarian cycles in ewes. MB, mamillary body; MI, massa intermedia; OC, optic chiasm; PIT, pituitary. (From data of MT Clegg and WF Ganong.)

the brain and outside the brain does not have this effect.

Effects of Sex Hormones in Infancy on Adult Behavior

In female experimental animals, exposure to sex steroids in utero or during early postnatal development causes marked abnormalities of sexual behavior when the animals reach adulthood. Female rats treated with a single relatively small dose of androgen before the fifth day of life do not have normal heat periods when they mature. They do not show the cyclic release of pituitary gonadotropins characteristic of the adult female but rather the tonic, steady secretion characteristic of the adult male; their brains have been "masculinized" by the single brief exposure to androgens. They also show increased male sexual behavior. Conversely, male rats castrated at birth develop the female pattern of cyclic gonadotropin secretion and show considerable female sexual behavior when given doses of ovarian hormones that do not have this effect in intact males. Thus, the development of a "female hypothalamus" depends simply on the absence of androgens in early life rather than on exposure to female hormones.

Rats are particularly immature at birth, and animals of other species in which the young are more fully developed at birth do not show these changes when exposed to androgens during the postnatal period. However, these animals develop genital abnormalities when exposed to androgens in utero (see Chapter 23). Female monkeys exposed to androgens in utero do not lose the female pattern of gonadotropin secretion but do develop abnormalities of sexual behavior in adulthood. Exposure of human females to androgens in utero does not change the cyclic pattern of gonadotropin secretion in adulthood (see Chapter 23). However, masculinizing effects on behavior do occur.

Pheromones

Substances produced by an animal that act at a distance to produce hormonal, behavioral, or other physiologic changes in another animal of the same species have been called **pheromones.** The sex attractants of certain insects are particularly well-known examples. The odorant pheromones that act via the vomeronasal organ play a prominent role in the sexual and dominance behavior of rodents. In primates, including humans, pheromones also have effects. For example, women who are good friends or roommates tend to synchronize their menstrual cycles, and armpit odor of women has been shown to be capable of modifying the menstrual cycle. Also, infants prefer pads wiped on breast or axillary areas of their own mothers over pads from unfamiliar women.

Maternal Behavior

Maternal behavior is depressed by lesions of the cingulate and retrosplenial portions of the limbic cortex in animals. Hormones do not appear to be necessary for its occurrence, but prolactin, which is secreted in large amounts during pregnancy and lactation, facilitates it. In addition, there is evidence that prolactin facilitates paternal behavior.

In female mice, knockout of the *fos-B* gene, one of four immediate early *fos* genes (see Chapter 1), is associated with failure to retrieve and care for pups after delivery. The neglected pups die, though if they are nourished by a normal foster mother they develop and flourish. Thus, genetic factors also appear to be involved in the control of maternal behavior.

OTHER EMOTIONS

Fear

The fear reaction can be produced in conscious animals by stimulation of the hypothalamus and the amygdaloid nuclei. Conversely, after destruction of the amygdalas, the fear reaction and its autonomic and endocrine manifestations are absent in situations in which they would normally be evoked. A dramatic example is the reaction of monkeys to snakes. Monkeys are normally terrified by snakes. After bilateral temporal lobectomy, monkeys approach snakes without fear, pick them up, and even eat them.

There is considerable evidence that the amygdaloid nuclei are concerned with the encoding of memories that evoke fear. Fear learning is blocked when LTP (see Chapter 4) is disrupted in pathways to the amygdalas. In humans with amygdala damage, there are deficient fear responses to auditory and visual stimuli. In normal humans, viewing faces that have fearful expressions activates the left amygdala. The degree of activation is proportionate to the intensity of fear in the facial expression, and happy faces fail to produce a response.

Anxiety

Anxiety is a normal emotion in appropriate situations, but excessive anxiety and anxiety in inappropriate situations can be disabling. Anxiety is associated with a bilateral increase in blood flow in a discrete portion of the anterior end of each temporal lobe. It is relieved by benzodiazepines, which bind to $GABA_A$ receptors and increase the Cl^- conductance of these ion channels. Unfortunately, benzodiazepines also cause sedation, and other effects as well. However, it has now been possible to separate the antianxiety effect from the others by pharmacologic means. The data indicate that it is the α_2 GABA$_A$ receptor (see Chapter 4) that mediates anxiety.

Rage & Placidity

Most animals, including humans, maintain a balance between rage and its opposite, the emotional state that for lack of a better name is referred to here as placidity. Major irritations make normal individuals "lose their temper," but minor stimuli are ignored. In animals with certain brain lesions, this balance is altered. Some lesions produce a state in which the most minor stimuli evoke violent episodes of rage; others produce a state in which the most traumatic and anger-provoking stimuli fail to ruffle the animal's abnormal calm.

Rage responses to minor stimuli are observed after removal of the neocortex and after destruction of the ventromedial hypothalamic nuclei and septal nuclei in animals with intact cerebral cortices. These were once thought to be only the physical motor manifestations of rage without the emotion and came to be called "sham rage." However, this appears to be incorrect. The rage reactions appear to be unpleasant to animals because the animals become conditioned against the place where the experiment was conducted. These also occur in humans who have sustained damage to the hypothalamus. On the other hand, bilateral destruction of the amygdaloid nuclei in monkeys causes a state of abnormal placidity. The placidity produced by amygdaloid lesions in animals is converted into rage by subsequent destruction of the ventromedial nuclei of the hypothalamus. Rage can also be produced by stimulation of an area extending back through the lateral hypothalamus to the central gray area of the midbrain.

Gonadal hormones appear to affect aggressive behavior. In male animals, aggression is decreased by castration and increased by androgens. It is also conditioned by social factors; it is more prominent in males that live with females and increases when a stranger is introduced into an animal's territory.

Rage may be related to violence. There has been considerable interest in the search for brain abnormalities in murderers and others who commit violent acts. Studies of this sort are difficult to conduct because of the multiple differences between the groups studied. For instance, people who have committed violent acts are usually confined whereas the normal controls are free. In addition, the studies are subject to inappropriate or premature use in current cases. However, it does appear that violent criminals generally have lower activity in the prefrontal cortex than normals.

Disgust

A recent report highlights the way the brain recognizes emotions. This was a report that in a patient with selective lesions of the left insula and putamen there was difficulty recognizing disgust on the faces or in the voices of people the patient met. Other emotions were recognized without difficulty. Patients with Huntington's disease have similar symptoms and damage in the same area. The insula is known to be activated by unpleasant tastes, so perhaps this response evolved into activation of the area by faces showing disgust.

MOTIVATION & ADDICTION

Self-Stimulation

If an animal is placed in a box with a pedal or bar that can be pressed, the animal sooner or later accidentally presses it. If the bar is connected in such a way that each press delivers a stimulus to an electrode implanted in certain parts of the brain, the animal returns to the bar and presses it again and again. Pressing the bar soon comes to occupy most of the animal's time. Some animals go without food and water and others overcome major obstacles to press the bar for brain stimulation. Rats press the bar 5000–12,000 times per hour, and monkeys have been clocked at 17,000 bar presses per hour. On the other hand, when the electrode is in certain other areas, the animals avoid pressing the bar, and stimulation of these areas is a potent unconditioned stimulus for the development of conditioned avoidance responses.

The points where stimulation leads to repeated bar pressing are located in a medial band of tissue extending from the frontal cortex through the hypothalamus to the midbrain tegmentum (Figure 15–4). The most responsive area is the dopaminergic pathway from the ventral tegmental area to the nucleus accumbens (see below). The points where stimulation is avoided are in the lateral portion of the posterior hypothalamus, the dorsal midbrain, and the entorhinal cortex. The latter points are sometimes close to points where bar pressing is repeated, but they are part of a separate system. The areas where bar pressing is repeated are much more extensive than those where it is avoided. It has been calculated that in rats repeated pressing is obtained from 35% of the brain, avoidance from 5%, and indifferent responses (neither repetition nor avoidance) from 60%.

It is obvious that some effect of the stimulation causes the animals to stimulate themselves again and again, but what the animals feel is, of course, unknown. There are a number of reports of bar-pressing experiments in humans with chronically implanted electrodes. Most of the subjects were schizophrenics or epileptics, but a few were patients with visceral neoplasms and intractable pain. Like animals, humans

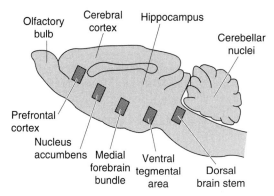

Figure 15–4. Areas where stimulation leads to repeated bar pressing projected on a parasagittal view of the rat brain. The regions where high rates of self-stimulation are produced are indicated by the rectangles. (Modified from Routtenberg A: The reward system of the brain. Sci Am [Nov] 1978; 239:154.)

press the bar repeatedly. They generally report that the sensations evoked are pleasurable, using phrases like "relief of tension" and "a quiet, relaxed feeling" to describe the experience. However, they rarely report "joy" or "ecstasy," and some persons with the highest self-stimulation rates cannot tell why they keep pushing the bar. When the electrodes are in the areas where stimulation is avoided, patients report sensations ranging from vague fear to terror. It is probably wise, therefore, to avoid vivid terms and call the brain systems involved the **reward** or **approach system** and the **punishment** or **avoidance system.**

Drugs that block postsynaptic D_3 dopaminergic receptors reduce the rate of self-stimulation, and dopamine agonists increase it. The main site of the relevant receptors is the **nucleus accumbens.**

Studies of the kind described above provide physiologic evidence that behavior is motivated not only by reduction or prevention of an unpleasant affect but also by primary rewards such as those produced by stimulation of the reward system. Stimulation of this system provides a potent motivation for learning mazes or performing other tasks.

Addiction

Addiction, defined as the repeated compulsive use of a substance despite negative health consequences, can be produced by a variety of different drugs. Not surprisingly, addiction is associated with the reward system, and particularly with the **nucleus accumbens,** located at the base of the striatum (Figure 15–5). The mesocor-

tical dopaminergic neurons that project from the midbrain to this nucleus and the frontal cortex (see below) are also involved. The best-studied addictive drugs are opiates such as morphine and heroin, cocaine, amphetamine, ethyl alcohol, and nicotine. All these affect the brain in different ways, but all have in common the fact that they increase the amount of dopamine available to act on D_3 receptors in the nucleus accumbens. Thus, acutely they stimulate the reward system of the brain. On the other hand, long-term addiction involves the development of tolerance, ie, the need for increasing amounts of a drug to produce a "high." In addition, withdrawal produces psychologic and physical symptoms. Injections of β-noradrenergic antagonists or α_2-noradrenergic agonists in the bed nucleus of the stria terminalis reduce the symptoms of opioid withdrawal, and so do bilateral lesions of the lateral tegmental noradrenergic fibers (see below). However, little else is known about the basis of withdrawal symptoms and tolerance.

One of the characteristics of addiction is the tendency of addicts to relapse after treatment. For opiate addicts, for example, the relapse rate in the first year is about 80%. Relapse often occurs upon exposure to sights, sounds, and situations that were previously associated with drug use. An interesting recent observation that may be relevant in this regard is that as little as a single dose of an addictive drug facilitates release of excitatory neurotransmitters in brain areas concerned with memory. The medial frontal cortex, the hippocampus, and the amygdala are concerned with memory, and they all project to the nucleus accumbens.

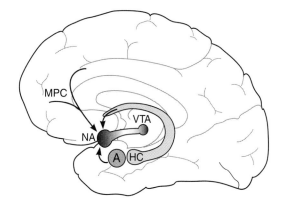

Figure 15–5. Key brain areas involved in addiction. The ventral tegmental area (VTA) projects via the mesocortical dopaminergic system to the nucleus accumbens (NA). The medial prefrontal cortex (MPC), the hippocampus (HC), and the amygdala (A) send excitatory glutaminergic projections to the nucleus accumbens.

BRAIN CHEMISTRY & BEHAVIOR

Drugs that modify human behavior include **hallucinogenic agents,** drugs that produce hallucinations and other manifestations of the psychoses; **tranquilizers,** drugs that allay anxiety and various psychiatric symptoms; and **antidepressants,** drugs that elevate mood and increase interest and drive. These and many other drugs act by modifying transmission at synaptic junctions in the brain. The chemistry of the known and suspected synaptic transmitters in the peripheral as well as the CNS is discussed in Chapter 4. This chapter is concerned primarily with their relation to instinctual behavior and emotions.

Aminergic Systems in the Brain

There are four large aminergic systems in the brain that have in common the presence of their cell bodies in rel-atively few locations with multiple branched axons projecting to almost all parts of the nervous system (Figure 15–6). These are the serotonergic, noradrenergic, adrenergic, and histaminergic systems. Dopaminergic neurons have their cell bodies in more locations, but their axons also project to many different areas.

Serotonin

Serotonin-containing neurons have their cell bodies in the midline raphe nuclei of the brain stem and project to portions of the hypothalamus, the limbic system, the neocortex, the cerebellum, and the spinal cord (Figure 15–6).

The hallucinogenic agent lysergic acid diethylamide (LSD) is a serotonin agonist that produces its effects by activating 5-HT$_2$ receptors (see Chapter 4) in the brain. The transient hallucinations and other mental aberrations produced by this drug were discovered

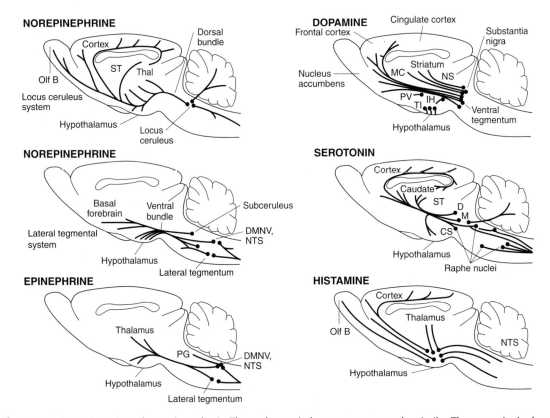

Figure 15–6. Aminergic pathways in rat brain. The pathways in humans appear to be similar. The two principal noradrenergic systems (locus ceruleus and lateral tegmental) are shown separately. Olf B, olfactory bulb; Thal, thalamus; ST, stria terminalis; DMNV, dorsal motor nucleus of vagus; NTS, nucleus of tractus solitarius; PG, periaqueductal gray; NS, nigrostriatal system; MC, mesocortical system; PV, periventricular system; IH, incertohypothalamic system; TI, tuberoinfundibular system; D, M, and CS, dorsal, medial, and central superior raphe nuclei.

when the chemist who synthesized it inhaled some by accident. Its discovery called attention to the correlation between behavior and variations in brain serotonin content. Psilocin, a substance found in certain mushrooms, and N,N-dimethyltryptamine (DMT) are also hallucinogenic and, like serotonin, are derivatives of tryptamine. 2,5-Dimethoxy-4-methyl-amphetamine (DOM) and mescaline and its congeners, the other true hallucinogens, are phenylethyl-amines rather than indolamines. However, all these hallucinogens appear to exert their effects by binding to $5\text{-}HT_2$ receptors. 3,4-Methylenedioxymethamphetamine, a drug known as MDMA or **ecstasy,** is a popular drug of abuse. It produces euphoria, but this is followed by difficulty in concentrating, depression, and, in monkeys, insomnia. The drug causes release of serotonin followed by serotonin depletion; the euphoria may be due to the release and the later symptoms to the depletion.

As noted in Chapter 11, it now appears that serotonin agonists suppress sleep. Serotonergic neurons discharge rapidly in the awake state, slowly during drowsiness, more slowly with bursts during sleep, and not at all during REM sleep (see Chapter 11).

Many other functions for brain serotonin have been proposed. Serotonin may play an excitatory role in the regulation of prolactin secretion (see Chapter 23). There is evidence that descending serotonergic fiber systems inhibit transmission in pain pathways in the dorsal horns. In addition, there is a prominent serotonergic innervation of the suprachiasmatic nuclei of the hypothalamus, and serotonin may be involved in the regulation of circadian rhythms (see Chapter 14). Pathophysiologically, there is evidence that discharge in serotonergic neurons in the dorsal raphe nucleus causes migraine, and antimigraine drugs inhibit the firing of dorsal raphe neurons.

In patients who are depressed, the primary serotonin metabolite 5-HIAA (see Figure 4–23) is low in CSF. It used to be argued that depression is caused by low extracellular norepinephrine in the brain (see below), and drugs which inhibited norepinephrine reuptake were of considerable value in the treatment of depression. However, these drugs also inhibit serotonin reuptake, and drugs such as fluoxetine (Prozac), which inhibit serotonin reuptake without affecting norepinephrine reuptake, are equally effective as antidepressants. Thus, the focus in clinical depression has shifted from norepinephrine to serotonin. It is interesting that all those drugs must be administered for 4–6 weeks before their antidepressive activity becomes manifest, indicating that it is some secondary effect of reuptake inhibition rather than the inhibition itself which produces the improved mood.

For unknown reasons, a number of other compounds unrelated to serotonin seem to cause depression. Blockade of the NK-1 receptors that mediate effects of substance P also relieves depression by an as yet unknown mechanism.

In mice in which the gene for MAO type A (see Chapter 4) has been knocked out and in humans with mutant MAO A genes, aggressive behavior is increased. In the knockout mice, there is a marked increase in brain serotonin. Animals in which the $5\text{-}HT_{1B}$ autoreceptor is knocked out also show increased aggressive behavior.

Norepinephrine

The cell bodies of the norepinephrine-containing neurons in the brain are located in the locus ceruleus and other nuclei in the pons and medulla. From the locus ceruleus, the axons of the noradrenergic neurons form the **locus ceruleus system.** They descend into the spinal cord, enter the cerebellum, and ascend to innervate the paraventricular, supraoptic, and periventricular nuclei of the hypothalamus, the thalamus, the basal telencephalon, and the entire neocortex (Figure 15–6). From cell bodies in the dorsal motor nucleus of the vagus, the nucleus of the tractus solitarius, and areas in the dorsal and lateral tegmentum, the axons of the noradrenergic neurons form a **lateral tegmental system** that projects to the spinal cord, the brain stem, all of the hypothalamus, and the basal telencephalon. The ascending fibers from the locus ceruleus form the **dorsal noradrenergic bundle,** whereas the ascending fibers of the lateral tegmental system form the **ventral noradrenergic bundle** (Figure 15–6).

Drugs that increase extracellular norepinephrine levels in the brain elevate mood, and drugs that decrease extracellular norepinephrine levels cause depression. However, as noted above, emphasis has now shifted from norepinephrine to serotonin in the pathogenesis of depression. In addition, individuals with congenital dopamine β-hydroxylase (DBH) deficiency are normal as far as mood is concerned. Of course the situation relative to any monoamine and brain function is complicated because high levels of extracellular neurotransmitters can have secondary effects, particularly on receptors.

The normal function of the locus ceruleus system remains a mystery, although its electrical activity is increased by unexpected sensory stimuli and it may be related to behavioral vigilance. The ventral tegmental noradrenergic system is involved in regulation of the secretion of vasopressin and oxytocin, and it adjusts the secretion of the hypophysiotropic hormones that regulate the secretion of anterior pituitary hormones (see Chapter 14). Norepinephrine and serotonin both appear to be involved in the control of body temperature.

Epinephrine

There is a system of phenylethanolamine-*N*-methyl-transferase (PNMT)-containing neurons with cell bodies in the medulla that project to the hypothalamus. Those neurons secrete epinephrine, but their function is uncertain. Epinephrine-secreting neurons also project to the thalamus, periaqueductal gray, and spinal cord. There are appreciable quantities of tyramine in the CNS, but no function has been assigned to this agent.

Dopamine

There are many dopaminergic systems in the brain. It is convenient to divide them into ultrashort, intermediate, and long systems on the basis of the length of their axons. The ultrashort dopaminergic neurons include the cells between the inner nuclear and the inner plexiform layers in the retina (see Chapter 8) and the periglomerular cells in the olfactory bulb (see Chapter 10). Intermediate-length dopamine cells include the **tuberoinfundibular system** (Figures 15–6 and 15–7), which secretes the dopamine into the portal hypophysial vessels that inhibits prolactin secretion, the **incertohypothalamic system,** which links the hypothalamus and the lateral septal nuclei, and the **medullary periventricular** group of neurons scattered along the walls of the third and fourth ventricles. The long dopamine systems (Figure 15–6) are the **nigrostriatal system,** which projects from the substantia nigra to the striatum and is involved in motor control (see Chapter 12), and the **mesocortical system,** which projects from the midbrain tegmentum to the limbic and frontal cortex and the olfactory tubercle, the nucleus accumbens, and related limbic subcortical areas. Recent studies by PET scanning (see Chapter 32) in normal humans show that there is a steady loss of dopamine receptors in the basal ganglia with age. The loss is greater in men than in women.

The relation of the mesocortical system, and particularly its ventral tegmental-nucleus accumbens portion, to addiction has been discussed above. There is a large amount of evidence that a defect in the mesocortical system is responsible for the development of at least some of the symptoms of schizophrenia. Attention was initially focused on overstimulation of limbic D_2 dopamine receptors. Amphetamine, which causes release of dopamine as well as norepinephrine in the brain, causes a schizophrenia-like psychosis; brain levels of D_2 receptors are said to be elevated schizophrenics; and there is a clear positive correlation between the antischizophrenic activity of many drugs and their ability to block D_2 receptors. However, several recently developed drugs are effective antipsychotic agents but bind D_2 receptors to a limited degree. Instead, they bind to

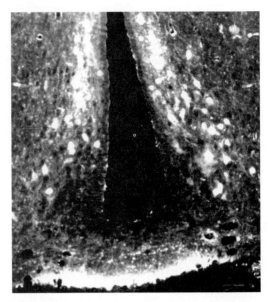

Figure 15–7. Tuberoinfundibular dopaminergic system. Transverse section of rat hypothalamus (fluorescent stain). Note the dopaminergic cell bodies in the arcuate nuclei on either side of the third ventricle and, at the bottom, the dopaminergic endings on the portal hypophysial vessels in the external layer of the median eminence. (Reproduced, with permission, from Hökfelt T, Fuxe K: On the morphology and the neuroendocrine role of the hypothalamic catecholamine neurons. In: *Brain-Endocrine Interaction.* Knigge K, Scott D, Weindl A [editors]. Karger, 1972.)

D_4 receptors, and there is active ongoing research into the possibility that these receptors are abnormal in individuals with schizophrenia.

Histamine

As noted in Chapter 4, histaminergic neurons have their cell bodies in the tuberomamillary nucleus in the ventral portion of the posterior hypothalamus. The axons of these neurons project to all parts of the brain (Figure 15–6). The function of this diffuse histaminergic system is unknown, but there is evidence linking brain histamine to arousal, sexual behavior, blood pressure, drinking, pain thresholds, and regulation of the secretion of several anterior pituitary hormones.

Acetylcholine

Acetylcholine is distributed throughout the CNS, with high concentrations in the cerebral cortex, thalamus, and various nuclei in the basal forebrain. The distribution of choline acetyltransferase and acetylcholinesterase parallels that of acetylcholine. Most of the acetylcholinesterase is in neurons, but some is found in glia. Pseudocholinesterase is found in many parts of the CNS. As noted in Chapter 4, there are nicotinic and muscarinic cholinergic receptors of various types in the brain. There are multiple nicotinic cholinergic subunit genes, and the proteins they code make up pentameric heterodimers of varying composition. In the brain, there are both postsynaptic and presynaptic nicotinic cholinergic receptors, and their composition varies from place to place.

The availability of antibodies specific for choline acetyltransferase has permitted mapping the cholinergic pathways in the brain by immunocytochemical techniques. The distribution of cholinergic neurons resembles that of the monoaminergic systems in that some of the cholinergic neurons project diffusely to much of the brain but differs in that there are also cholinergic interneurons and short cholinergic systems throughout the CNS. There is a large projection from the nucleus basalis of Meynert and adjacent nuclei to the amygdala and the entire neocortex, and these projections are involved in motivation, perception, and cognition. There is extensive cell loss in this projection in Alzheimer's disease (see Chapter 16). The PGO spike system responsible for REM sleep is cholinergic. The relation of nicotine to addictive behavior has been mentioned above. In large doses, muscarinic blocking agents such as atropine can cause hallucinations, and scopolamine is a sedative.

Somatic motor neurons, preganglionic autonomic neurons, and some postganglionic autonomic neurons are cholinergic. Cortical levels of acetylcholinesterase are greater in rats raised in a complex environment than in rats raised in isolation, but the significance of this type of correlation is uncertain. As noted in Chapter 12, acetylcholine is an excitatory transmitter in the basal ganglia, whereas dopamine is an inhibitory transmitter in these structures.

Opioid Peptides

There are three types of opioid peptide-secreting neurons in the brain that produce one of the three opioid peptide precursor molecules (see Table 4–4). There are

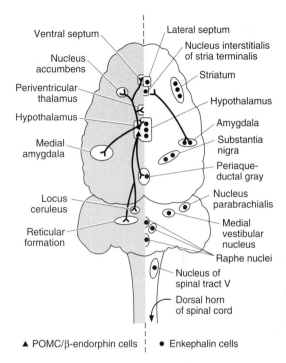

▲ POMC/β-endorphin cells ┊ ● Enkephalin cells

Figure 15–8. Distribution of β-endorphin neurons **(left)** and enkephalin neurons **(right)** in the brain. (Reproduced, with permission, from Barchas JD et al: Behavioral neurochemistry: Neuroregulatory and behavioral states. Science 1978;200:964. Copyright © 1978 by the American Association for the Advancement of Science.)

in addition two endomorphins whose precursors are as yet unknown. Proenkephalin-producing neurons are scattered throughout the brain, whereas pro-opiomelanocortin-producing neurons have their cell bodies in the arcuate nuclei and project to the thalamus and parts of the brain stem (Figure 15–8). Prodynorphin-producing neurons are located primarily in the hypothalamus, limbic system, and brain stem. The peptides they secrete are involved in various functions including, presumably, the phenomenon of tolerance and addiction produced by morphine, but the details are uncertain.

"Higher Functions of the Nervous System": Conditioned Reflexes, Learning, & Related Phenomena

16

INTRODUCTION

Somatic and visceral inputs to the brain and output from it have been described in previous chapters. Topics that have been discussed include specific inputs and outputs; alert, awake states, and sleep; and the functions of the limbic-hypothalamic circuit in maintenance of homeostatic equilibriums and regulation of instinctual and emotional behavior. There remain the phenomena called, for lack of a better or more precise term, the "higher functions of the nervous system": learning, memory, judgment, language, and the other functions of the mind. These phenomena are the subject of this chapter.

METHODS

A revolution in our understanding of brain function in humans has been brought about by the development and now the widespread availability of PET scanning, functional MRI (fMRI), and related techniques. PET is often used to measure local glucose metabolism and fMRI to measure local amounts of oxygenated blood. These techniques make it possible to determine the activity of the various parts of the brain in completely intact normal humans and in humans with many different diseases. They have been used to study not only simple responses but complex aspects of learning, memory, and perception. The physiologic basis of these techniques is discussed in Chapter 32. An example of their use to study the functions of the cerebral cortex is shown in Figure 16–1. Other examples of their use to study normal physiology and the pathophysiology of disease are cited in Chapters 7–15.

Other techniques that have provided information on cortical function include stimulation of the exposed cerebral cortex in conscious humans undergoing neurosurgical procedures and, in a few instances, studies with chronically implanted electrodes. Valuable information has also been obtained from investigations in laboratory primates, but it is worth remembering that in addition to the difficulties in communicating with them, the brain of the rhesus monkey is only one-fourth the size of the brain of the chimpanzee, our nearest primate relative, and the chimpanzee brain is in turn one-fourth the size of the human brain.

LEARNING & MEMORY

A characteristic of animals and particularly of humans is the ability to alter behavior on the basis of experience. **Learning** is acquisition of the information that makes this possible, and **memory** is the retention and storage of that information. The two are obviously closely related and should be considered together.

Forms

From a physiologic point of view, memory is appropriately divided into explicit and implicit forms (Table 16–1). **Explicit memory,** which is also called declarative or recognition memory, is associated with consciousness—or at least awareness—and is dependent for its retention on the hippocampus and other parts of the medial temporal lobes of the brain. It is divided into the memory for events (episodic memory) and the memory for words, rules, and language, etc (semantic memory). **Implicit memory** does not involve awareness and is also called nondeclarative or reflexive memory. Its retention does not involve processing in the hippocampus, at least in most instances, and it includes, among other things, skills, habits, and conditioned reflexes (Table 16–1). However, explicit memories initially required for activities such as riding a bicycle can become implicit once the task is thoroughly learned.

Explicit memory and many forms of implicit memory involve (1) **short-term memory**, which lasts sec-

A. Looking at words

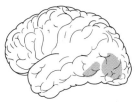

B. Listening to words

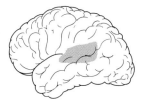

C. Speaking words

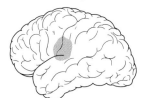

D. Thinking of words

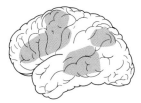

Figure 16–1. Drawings of PET scans of the left cerebral hemisphere showing areas of greatest neuronal activation when subjects performed various activities. **A:** Looking at words activated the primary visual cortex and part of the visual association cortex. **B:** Listening to words activated an area at the junction of the temporal and the parietal cortex. **C:** Speaking words activated Broca's area and the adjacent frontal lobe. **D:** Thinking about words activated large areas, including much of the frontal lobe. (Based on PET scans in Kandel ER, Schwartz JH, Jessell TM: *Essentials of Neural Science and Behavior.* McGraw-Hill, 1995)

onds to hours, during which processing in the hippocampus and elsewhere lays down long-term changes in synaptic strength; and (2) **long-term memory,** which stores memories for years and sometimes for life. During short-term memory, the memory traces are subject to disruption by trauma and various drugs, whereas long-term memory traces are remarkably resistant to disruption. **Working memory** is a form of short-term memory that keeps information available, usually for very short periods, while the individual plans action based on it.

Table 16–1. Types of memory.

Explicit
Episodic
Semantic
Implicit
Nonassociative
Habituation
Sensitization
Associative
Classic conditioning
Operant conditioning
Skills and habits
Priming

Implicit Memory

As noted in Table 16–1, implicit memory includes skills and habits which, once acquired, become unconscious and automatic. It also includes **priming,** which is facilitation of recognition of words or objects by prior exposure to them. An example is improved recall of a word when presented with the first few letters of it.

The other forms of implicit memory can be divided into nonassociative and associative forms. In **nonassociative learning,** the organism learns about a single stimulus. In **associative learning,** the organism learns about the relation of one stimulus to another.

Habituation & Sensitization

Habituation is a simple form of learning in which a neutral stimulus is repeated many times. The first time it is applied, it is novel and evokes a reaction (the orienting reflex or "what is it?" response). However, it evokes less and less electrical response as it is repeated. Eventually, the subject becomes habituated to the stimulus and ignores it. **Sensitization** is in a sense the opposite reaction. A repeated stimulus produces a greater response if it is coupled one or more times with an unpleasant or a pleasant stimulus. It is common knowledge that intensification of the **arousal value** of stimuli occurs in humans. The mother who sleeps through many kinds of noise but wakes promptly when her baby cries is an example.

Habituation is a classic example of nonassociative learning. A classic example of associative learning is a conditioned reflex.

Conditioned Reflexes

A conditioned reflex is a reflex response to a stimulus that previously elicited little or no response, acquired by repeatedly pairing the stimulus with another stimulus that normally does produce the response. In Pavlov's classic experiments, the salivation normally induced by placing meat in the mouth of a dog was studied. A bell was rung just before the meat was placed in the dog's mouth, and this was repeated a number of times until the animal would salivate when the bell was rung even though no meat was placed in its mouth. In this experiment, the meat placed in the mouth was the **unconditioned stimulus (US),** the stimulus that normally produces a particular innate response. The **conditioned stimulus (CS)** was the bell-ringing. After the CS and US had been paired a sufficient number of times, the CS produced the response originally evoked only by the US. The CS had to precede the US. Another example is shown in Figure 16–2. This is so-called classic conditioning. An immense number of somatic, visceral, and other neural changes can be made to occur as con-

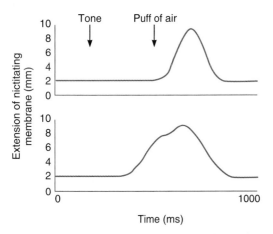

Figure 16–2. Conditioned reflex in a rabbit. At the first arrow a tone is sounded (CS), and at second arrow a gentle puff of air (US) is blown on the rabbit's eye. At first the nictitating membrane extends only in response to the puff of air **(top trace).** After repeated pairing, however, the nictitating membrane starts to contract in response to the tone, before the puff of air **(bottom trace).** (Modified from Alkon DL: Memory storage and neural systems. Sci Am [July] 1989;261:42.)

ditioned reflex responses. Conditioning of visceral responses is often called **biofeedback.** The changes that can be produced include alterations in heart rate and blood pressure, and conditioned decreases in blood pressure have been advocated for the treatment of hypertension. However, the depressor responses that are produced in this fashion are small.

If the CS is presented repeatedly without the US, the conditioned reflex eventually dies out. This process is called **extinction** or **internal inhibition.** If the animal is disturbed by an external stimulus immediately after the CS is applied, the conditioned response may not occur **(external inhibition).** However, if the conditioned reflex is **reinforced** from time to time by again pairing the CS and US, the conditioned reflex persists indefinitely.

As noted in Chapter 15, conditioned reflexes are difficult to form unless the US is associated with a pleasant or unpleasant affect. Stimulation of the brain reward system is a powerful US (pleasant or **positive reinforcement**), and so is stimulation of the avoidance system or a painful shock to the skin (unpleasant or **negative reinforcement**).

Operant conditioning is a form of conditioning in which the animal is taught to perform some task ("operate on the environment") in order to obtain a reward or avoid punishment. The US is the pleasant or unpleasant event, and the CS is a light or some other signal that alerts the animal to perform the task. Conditioned motor responses that permit an animal to avoid an unpleasant event are called **conditioned avoidance reflexes.** For example, an animal is taught that by pressing a bar it can prevent an electric shock to the feet. Another example is **food aversion conditioning.** An animal exposed to the taste of a food develops a strong aversion to the food if the tasting is coupled with injection of a drug that produces nausea or illness. Similar aversion responses occur in humans. These conditioned responses are very strong, can sometimes be learned with a single pairing of the CS and the US, and, unlike other conditioned responses, will develop when the CS and US are separated by an hour or more. The survival value of food aversion conditioning is obvious in terms of avoiding poisons, and it is not surprising that the brain is probably genetically "programmed" to facilitate the development of food aversion responses.

Intercortical Transfer of Memory

If a cat or monkey is conditioned to respond to a visual stimulus with one eye covered and then tested with the blindfold transferred to the other eye, it performs the conditioned response. This is true even if the optic chiasm has been cut, making the visual input from each eye go only to the ipsilateral cortex. If, in addition to the optic chiasm, the anterior and posterior commissures and the corpus callosum are sectioned ("split-brain animal"), no memory transfer occurs. Partial callosal section experiments indicate that the memory transfer occurs in the anterior portion of the corpus callosum. Similar results have been obtained in humans in whom the corpus callosum is congenitally absent or in whom it has been sectioned surgically in an effort to control epileptic seizures. This demonstrates that the neural coding necessary for "remembering with one eye what has been learned with the other" has been transferred to the opposite cortex via the commissures. There is evidence for similar transfer of information acquired through other sensory pathways.

Molecular Basis of Memory

The key to memory is alteration in the strength of selected synaptic connections. In all but the simplest of cases, the alteration involves protein synthesis and activation of genes. This occurs during the change from short-term working memory to long-term memory. In animals, acquisition of long-term learned responses is prevented if, within 5 minutes after each training session, the animals are anesthetized, given electroshock, subjected to hypothermia, or given drugs, antibodies, or oligonucleotides that block the synthesis of proteins. If these interventions are performed 4 hours after the training sessions, there is no effect on acquisition.

The human counterpart of this phenomenon is the loss of memory for the events immediately preceding brain concussion or electroshock therapy (**retrograde amnesia**). This amnesia encompasses longer periods than it does in experimental animals—sometimes many days—but remote memories remain intact.

The biochemical events involved in habituation and sensitization in *Aplysia* and other invertebrates have been worked out in considerable detail, and these events, along with those underlying several forms of **long-term potentiation (LTP)** and **long-term depression (LTD)**, are discussed in Chapter 4.

The biochemical events involved in habituation and sensitization in *Aplysia* and other invertebrates have been worked out in considerable detail. As described in Chapter 4, habituation is due to a decrease in Ca^{2+} in the sensory endings that mediate the response to a particular stimulus, and sensitization is due to prolongation of the action potential in these endings with a resultant increase in intracellular Ca^{2+} that facilitates release of neurotransmitter by exocytosis.

Classic conditioning also occurs in *Aplysia,* and in mammals, in the isolated spinal cord. In *Aplysia,* the US acts presynaptically on the endings of neurons activated by the CS. This leaves free Ca^{2+} in the cell, leading to a long-term change in the adenylyl cyclase molecule, so that when this enzyme is activated by the CS, more cAMP is produced. This in turn closes K^+ channels and prolongs action potentials by the mechanism described in Chapter 4. The key point in this case is the temporal association, with the US coming soon after the CS.

In *Aplysia,* there are morphologic correlates to learning and memory. For example, 40% of the relevant sensory terminals normally contain active zones, whereas in habituated animals, 10% have active zones, and in sensitized animals, 65% have active zones. Long-term memory leads to activation of genes that produce increases in synaptic contacts.

Encoding Implicit Memory in Mammals

Without doubt, molecular events similar to those occurring in *Aplysia* underlie some aspects of implicit memory in mammals. However, events involving various parts of the CNS also contribute. Some investigators argue that the striatum is involved, and it is known that learning of some habit tasks is disrupted by lesions of the basal ganglia. There is other evidence that the cerebellum is involved. For example, the vestibulo-ocular reflex (VOR), the reflex that maintains visual fixation while the head is moving (see Chapter 9), can be adjusted to new eye positions, and this plasticity is abolished by lesions of the flocculus. In addition, conditioning of an eye blink reflex by using a puff of air on the eye as the US and a tone as the CS (Figure 16–2) is prevented by lesions of the interpositus nucleus. In this case, it appears that impulses set up by the US act via the inferior olive and climbing fibers to the cerebellar cortex to alter the Purkinje cell response to the tone arriving via the pontine nuclei and mossy fibers. Climbing-fiber-mediated modification of mossy-fiber-driven Purkinje cell discharge is also responsible for plastic changes in the VOR and learned muscle movements.

Encoding Explicit Memory

Encoding explicit memories involves working memory in the frontal lobes and unique processing in the hippocampus.

Working Memory

As noted above, working memory keeps incoming information available for a short time while deciding what to do with it. It is that form of memory which permits us, for example, to look up a telephone number, then remember the number while we pick up the telephone and dial the number. It consists of what has been called a **central executive** located in the prefrontal cortex, and two "rehearsal systems," a **verbal system** for retaining verbal memories, and a parallel **visuospatial system** for retaining visual and spatial aspects of objects. The executive steers information into these rehearsal systems.

Hippocampus & Medial Temporal Lobe

Working memory areas are connected to the hippocampus and the adjacent parahippocampal portions of the medial temporal cortex (Figure 16–3). In humans, bilateral destruction of the ventral hippocampus or Alzheimer's disease and similar disease processes that destroy its CA1 neurons cause striking defects in short-term memory. So do bilateral lesions of the same area in monkeys. Humans with such destruction have intact working memory and remote memory. Their implicit memory processes are generally intact. They perform adequately in terms of conscious memory as long as they concentrate on what they are doing. However, if they are distracted for even a very short period, all memory of what they were doing and proposed to do is lost. They are thus capable of new learning and retain old prelesion memories, but they cannot form new long-term memories.

The hippocampus is closely associated with the overlying parahippocampal cortex in the medial frontal lobe (Figure 12–3). Memory processes have now been studied not only with fMRI but with measurement of evoked potentials (event-related potentials; ERPs) in epileptic patients with implanted electrodes. When subjects recall words, there is increased activity in their left frontal lobe and their left parahippocampal cortex, but when they recall pictures or scenes, there is activity in their right frontal lobe and the parahippocampal cortex on both sides.

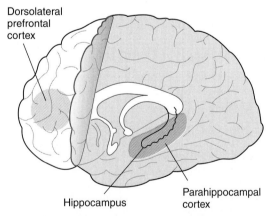

Figure 16–3. Areas concerned with encoding explicit memories. (Modified from Russ MD: Memories are made of this. Science 1998;281:1151.)

The connections of the hippocampus to the diencephalon are also involved in memory. Some alcoholics with brain damage develop impairment of recent memory, and the memory loss correlates well with the presence of pathologic changes in the mamillary bodies, which have extensive efferent connections to the hippocampus via the fornix. The mamillary bodies project to the anterior thalamus via the mamillothalamic tract, and in monkeys, lesions of the thalamus cause loss of recent memory. From the thalamus, the fibers concerned with memory project to the prefrontal cortex and from there to the basal forebrain. From the basal forebrain, there is a diffuse cholinergic projection to all the neocortex, the amygdala, and the hippocampus from the **nucleus basalis of Meynert.** There is a severe loss of these fibers in Alzheimer's disease (see below).

The amygdala is closely associated with the hippocampus and is concerned with encoding emotions related to memories. As described in Chapter 15, amygdaloid lesions make animals less fearful. In normal humans, events associated with strong emotions are remembered better than events without an emotional charge, but in patients with bilateral lesions of the amygdala, this difference is absent.

Confabulation is an interesting though poorly understood condition that sometimes occurs in individuals with lesions of the ventromedial portions of the frontal lobes. These individuals perform poorly on memory tests, but they spontaneously describe events that never occurred. This has been called "honest lying."

New Brain Cells?

It is now established that the traditional view that brain cells are not added after birth is wrong; new neurons form from stem cells throughout life in two areas: the olfactory bulb and the hippocampus. Since the hippocampus is concerned with new memories, they could be related to new brain cells. There is now evidence that reduction in the number of new neurons formed reduces at least one form of hippocampal memory production. However, there is a great deal more to be done before the relation of new cells to memory processing can be considered established.

Long-Term Memory

While the encoding process for short-term explicit memory involves the hippocampus, long-term memories are stored in various parts of the neocortex. Apparently, the various parts of the memories—visual, olfactory, auditory, etc—are located in the cortical regions concerned with these functions, and the pieces are tied together by long-term changes in the strength of transmission at relevant synaptic junctions so that all the components are brought to consciousness when the memory is recalled.

Once long-term memories have been established, they can be recalled or accessed by a large number of different associations. For example, the memory of a vivid scene can be evoked not only by a similar scene but also by a sound or smell associated with the scene and by words such as "scene," "vivid," and "view." Thus, there must be multiple routes or keys to each stored memory. Furthermore, many memories have an emotional component or "color"—ie, in simplest terms, memories can be pleasant or unpleasant.

Strangeness & Familiarity

It is interesting that stimulation of some parts of the temporal lobes in humans causes a change in interpretation of one's surroundings. For example, when the stimulus is applied, the subject may feel strange in a familiar place or may feel that what is happening now has happened before. The occurrence of a sense of familiarity or a sense of strangeness in appropriate situations probably helps the normal individual adjust to the environment. In strange surroundings, one is alert and on guard, whereas in familiar surroundings, vigilance is relaxed. An inappropriate feeling of familiarity with new events or in new surroundings is known clinically as the **déjà vu phenomenon,** from the French words meaning "already seen." The phenomenon occurs from time to time in normal individuals, but it also may occur as an aura (a sensation immediately preceding a seizure) in patients with temporal lobe epilepsy.

Summary

In summary, there is still much to be learned about the encoding of explicit memory. However, according to current views, information from the senses is temporarily stored in various areas of the prefrontal cortex as working memory. It is also passed to the medial temporal lobe, and specifically to the parahippocampal gyrus. From there, it enters the hippocampus and is processed in a way that is not yet understood. At this time, the activity is vulnerable, as described above. Output from the hippocampus leaves via the subiculum and the entorhinal cortex and somehow binds together and strengthens circuits in many different neocortical areas, forming over time the stable remote memories that can now be triggered by many different cues.

Alzheimer's Disease & Senile Dementia

Alzheimer's disease is characterized by progressive loss of short-term memory followed by general loss of cognitive and other brain functions and, eventually, death.

It was originally characterized in middle-aged people, and similar deterioration in elderly individuals is technically senile dementia of the Alzheimer type though it is frequently just called Alzheimer's disease as well. Most cases are sporadic, but some are familial. The disease accounts for 50–60% of cases of senile dementia. Patients with Alzheimer's disease eventually require around-the-clock care. Since 10–15% of the population over age 65 and almost 50% of the population over 85 have some degree of dementia, the condition is not only a serious medical problem but an economic load of increasing magnitude as the number of old people in populations of developed countries increases.

Early changes in Alzheimer's disease include atrophy of the hippocampus and entorhinal cortex, demonstrable by MRI up to 2 years before a definitive diagnosis can be made.

The cytopathologic hallmarks of the disease are intracellular **neurofibrillary tangles,** made up in part of hyperphosphorylated forms of the tau protein that normally binds to microtubules (see Chapter 1), and extracellular **senile plaques,** which have a core of β-**amyloid** peptides ($A\beta$) surrounded by altered nerve fibers and reactive glial cells. The ultimate cause of Alzheimer's disease is unsettled. However, the $A\beta$ peptides are products of a normal protein, **amyloid precursor protein (APP),** which projects from nerve cells. When this protein is hydrolyzed abnormally by the enzyme γ-**secretase,** which acts on the portion of the APP that crosses the cell membrane, two $A\beta$ peptides are formed, one containing 40 amino acid residues and the other 42 residues. These form plaques, and a case can be made for the subsequent scenario shown in Table 16–2. It is interesting in this regard that nonsteroidal anti-inflammatory drugs have been reputed to have beneficial effects in Alzheimer's disease, though opinion on this point is not unanimous.

It is worth noting that selective degeneration with aging can occur in three different types of cells in the CNS, causing three different progressive, crippling, and eventually fatal diseases. Degeneration of the hip-

Table 16–2. Possible sequence of events in the pathogenesis of Alzheimer's disease.

Aggregation of $A\beta_{40}$ and $A\beta_{42}$ peptides
Formation of plaques
Inflammatory reaction
Oxidative damage
Formation of tangles
Loss of synapses and neurons and therefore of neuro transmitters such as acetylcholine
Dementia

pocampal neurons is associated with Alzheimer's disease, as noted above; degeneration of the dopaminergic neurons in the substantia nigra is associated with Parkinson's disease (see Chapter 12); and degeneration of the cholinergic motor neurons in the brain stem and spinal cord is associated with one form of **amyotrophic lateral sclerosis (ALS).** This last disease is often called Lou Gehrig's disease because Gehrig, a famous American baseball player, died of it. All three diseases are mainly sporadic but have familial forms as well. Five percent of the cases of ALS are familial, and in 40% of these, there is a mutation in the gene for Cu/Zn superoxide dismutase (SOD-1) on chromosome 21. A defective *SOD-1* gene could permit free radicals to accumulate and kill neurons. Degenerative diseases of the nervous system due to expanded trinucleotide repeats are discussed in Chapter 12.

FUNCTIONS OF THE NEOCORTEX

Memory and learning are functions of large parts of the brain, but the centers controlling some of the other "higher functions of the nervous system," particularly the mechanisms related to language, are more or less localized to the neocortex. Speech and other intellectual functions are especially well developed in humans—the animal species in which the neocortical mantle is most highly developed.

Anatomic Considerations

There are three living species with brains larger than a human's (the porpoise, the elephant, and the whale), but in humans, the ratio between brain weight and body weight far exceeds that of any of the other three species. From the comparative point of view, the most prominent gross feature of the human brain is the immense growth of the three major **association areas:** the **frontal,** in front of the premotor area; the **parietal-temporal-occipital,** between the somatesthetic and visual cortices, extending into the posterior portion of the temporal lobe; and the **temporal,** extending from the lower portion of the temporal lobe to the limbic system (Figure 16–4). The proportions of the various parts of the brain are similar in the brains of apes and humans, but the human brain is larger, so the absolute size of the association areas is greater. The association areas are part of the six-layered neocortical mantle of gray matter that spreads over the lateral surfaces of the cerebral hemispheres from the concentric allocortical and juxtallocortical rings around the hilum (see Chapter 15).

The neuronal connections within the neocortex form a complicated network (see Figure 11–2). The descending axons of the larger cells in the pyramidal cell

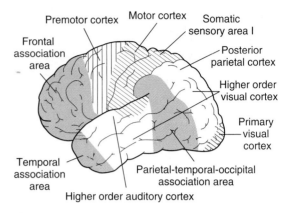

Figure 16–4. Lateral view of the human cerebral cortex, showing the primary sensory and motor areas and the association areas.

layer give off collaterals that feed back via association neurons to the dendrites of the cells from which they originate, laying the foundation for complex feedback control. The recurrent collaterals also connect to neighboring cells. The large, complex dendrites of the deep cells receive specific and nonspecific thalamic afferents, reticular afferents, and association fibers from other cortical areas. Specific thalamic afferents end in layer IV of the cortex. The plasticity of cortical connections and their ability to adapt are discussed in Chapter 7.

Complementary Specialization of the Hemispheres Versus "Cerebral Dominance"

One group of functions more or less localized to the neocortex in humans consists of those related to language, ie, to understanding the spoken and printed word and to expressing ideas in speech and writing. It is a well-established fact that human language functions depend more on one cerebral hemisphere than on the other. This hemisphere is concerned with categorization and symbolization and has often been called the **dominant hemisphere.** However, it is clear that the other hemisphere is not simply less developed or "nondominant"; instead, it is specialized in the area of spatiotemporal relations. It is this hemisphere that is concerned, for example, with the identification of objects by their form and the recognition of musical themes. It also plays a primary role in the recognition of faces. Consequently, the concept of "cerebral dominance" and a dominant and nondominant hemisphere has been replaced by a concept of complementary specialization of the hemispheres, one for sequential-analytic processes (the **categorical hemisphere**) and one for visuospatial relations (the **representational hemisphere**). The categorical hemisphere is concerned with language functions, but hemispheric specialization is also present in monkeys, so it antedates the evolution of language.

Lesions in the categorical hemisphere produce language disorders, whereas extensive lesions in the representational hemisphere do not. Instead, lesions in the representational hemisphere produce **astereognosis**—inability to identify objects by feeling them—and other agnosias. **Agnosia** is the general term used for the inability to recognize objects by a particular sensory modality even though the sensory modality itself is intact. Lesions producing these defects are generally in the parietal lobe. Especially when they are in the representational hemisphere, lesions of the inferior parietal lobule, a region in the posterior part of the parietal lobe that is close to the occipital lobe, cause **unilateral inattention** and **neglect.** Individuals with such lesions do not have any apparent primary visual, auditory, or somatesthetic defects, but they ignore stimuli from the contralateral portion of their bodies or the space around these portions. This leads to failure to care for half their bodies and, in extreme cases, to situations in which individuals shave half their faces, dress half their bodies, or read half of each page. This inability to put together a picture of visual space on one side is due to a shift in visual attention to the side of the brain lesion and can be improved if not totally corrected by wearing eyeglasses that contain prisms.

Hemispheric specialization extends to other parts of the cortex as well. Patients with lesions in the categorical hemisphere are disturbed about their disability and often depressed, whereas patients with lesions in the representational hemisphere are sometimes unconcerned and even euphoric. Other examples of specialization are mentioned elsewhere in this book.

Hemispheric specialization is related to handedness. Handedness appears to be genetically determined. In 96% of right-handed individuals, who constitute 91% of the human population, the left hemisphere is the dominant or categorical hemisphere, and in the remaining 4%, the right hemisphere is dominant. In approximately 15% of left-handed individuals, the right hemisphere is the categorical hemisphere and in 15%, there is no clear lateralization. However, in the remaining 70% of left-handers, the left hemisphere is the categorical hemisphere. It is interesting that learning disabilities such as **dyslexia,** an impaired ability to learn to read, are 12 times as common in left-handers as they are in right-handers, possibly because some fundamental abnormality in the left hemisphere led to a switch in handedness early in development. However, the spatial talents of left-handers may be well above average; a disproportionately large number of artists, musicians, and mathematicians are left-handed. For unknown reasons, left-handers have slightly but significantly shorter life spans than right-handers.

There are anatomic differences between the two hemispheres that may correlate with the functional differences. As noted in Chapter 9, the **planum temporale**, an area of the superior temporal gyrus that is involved in language-related auditory processing, is regularly larger on the left side than the right. It is also larger on the left in the brain of chimpanzees, even though language is almost exclusively a human trait. Imaging studies show that other portions of the upper surface of the left temporal lobe are larger in right-handed individuals, and the right frontal lobe is normally thicker than the left and that the left occipital lobe is wider and protrudes across the midline. Portions of the upper surface of the left temporal lobe are regularly larger in right-handed individuals. There are also chemical differences between the two sides of the brain. For example, there is a higher concentration of dopamine in the nigrostriatal pathway on the left side in right-handed humans and a higher concentration on the right in left-handers. The physiologic significance of these differences is not known.

In patients with schizophrenia, MRI studies have demonstrated reduced volumes of gray matter on the left side in the anterior hippocampus, amygdala, parahippocampal gyrus, and posterior superior temporal gyrus. The degree of reduction in the left superior temporal gyrus correlates with the degree of disordered thinking in the disease. There are also apparent abnormalities of dopaminergic systems (see Chapter 15) and cerebral blood flow (see Chapter 32) in this disease.

Physiology of Language

Language is one of the fundamental bases of human intelligence and a key part of human culture. The primary brain areas concerned with language are arrayed along and near the sylvian fissure (lateral cerebral sulcus) of the categorical hemisphere. A region at the posterior end of the superior temporal gyrus called **Wernicke's area** (Figure 16–5) is concerned with comprehension of auditory and visual information. It projects via the **arcuate fasciculus** to **Broca's area** (area 44) in the frontal lobe immediately in front of the inferior end of the motor cortex. Broca's area processes the information received from Wernicke's area into a detailed and coordinated pattern for vocalization and then projects the pattern via a speech articulation area in the insula to the motor cortex, which initiates the appropriate movements of the lips, tongue, and larynx to produce speech. The probable sequence of events that occurs when a subject names a visual object is shown in Figure 16–6. The angular gyrus behind Wernicke's area appears to process information from words

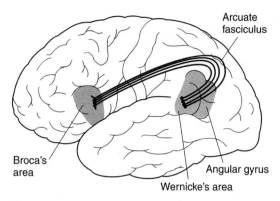

Figure 16–5. Location of some of the areas that in the categorical hemisphere are concerned with language functions.

that are read in such a way that they can be converted into the auditory forms of the words in Wernicke's area.

It is interesting that in individuals who learn a second language in adulthood, fMRI reveals that the portion of Broca's area concerned with it is adjacent to but separate from the area concerned with the native language. However, in children who learn two languages early in life, there is only a single area involved

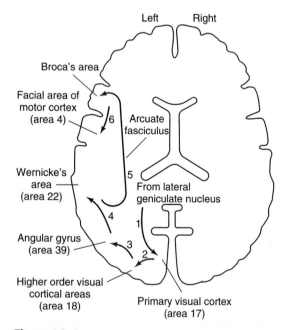

Figure 16–6. Path taken by impulses when a subject names a visual object, projected on a horizontal section of the human brain.

with both. It is well known, of course, that children acquire fluency in a second language more easily than adults.

Language Disorders

Aphasias are abnormalities of language functions that are not due to defects of vision or hearing or to motor paralysis. They are caused by lesions in the categorical hemisphere. The most common cause is embolism or thrombosis of a cerebral blood vessel. Many different classifications of the aphasias have been published, but a convenient classification divides them into **fluent, nonfluent,** and **anomic aphasias.** In nonfluent aphasia, the lesion is in Broca's area (Table 16–3). Speech is slow, and words are hard to come by. Patients with severe damage to this area are limited to two or three words with which to express the whole range of meaning and emotion. Sometimes the words retained are those which were being spoken at the time of the injury or vascular accident that caused the aphasia.

In one form of fluent aphasia, the lesion is in Wernicke's area. In this condition, speech itself is normal and sometimes the patients talk excessively. However, what they say is full of jargon and neologisms that make little sense. The patient also fails to comprehend the meaning of spoken or written words, so other aspects of the use of language are compromised.

Another form of fluent aphasia is a condition in which patients can speak relatively well and have good auditory comprehension but cannot put parts of words together or conjure up words. This is called **conduction aphasia** because it was thought to be due to lesions of the arcuate fasciculus connecting Wernicke's and Broca's areas. However, it now appears that it is

Table 16–3. Aphasias. Characteristic responses of patients with lesions in various areas when shown a picture of a chair.[1]

Type of Aphasia and Site of Lesion	Characteristic Naming Errors
Nonfluent (Broca's area)	"Tssair"
Fluent (Wernicke's area)	"Stool" or "choss" (neologism)
Fluent (areas 40, 41, and 42; conduction aphasia)	"Flair . . . no, swair . . . tair."
Anomic (angular gyrus)	"I know what it is . . . I have a lot of them."

[1]Modified from Goodglass H: Disorders of naming following brain injury. Am Sci 1980;68:647.

due to lesions in and around the auditory cortex (areas 40, 41, and 42).

When there is a lesion damaging the angular gyrus in the categorical hemisphere without affecting Wernicke's or Broca's areas, there is no difficulty with speech or the understanding of auditory information, but there is trouble understanding written language or pictures, because visual information is not processed and transmitted to Wernicke's area. The result is a condition called **anomic aphasia.**

Dyslexia, which is a broad term applied to impaired ability to read, is frequently due to an inherited abnormality that affects 5% of the population. Its cause is unknown, though two pathogenic theories have been advanced. One is that there is reduced ability to recall speech sounds, so there is trouble translating them mentally into sound units **(phonemes).** Another is that there is a defect in the magnocellular portion of the visual system (see Chapter 8) that slows processing and also leads to phonemic deficit. In any case, decreased blood flow in the angular gyrus in the categorical hemisphere is commonly seen.

More selective speech defects have now been described. For example, lesions limited to the left temporal pole (area 38) cause inability to retrieve names of places and persons but preserves the ability to retrieve common nouns, ie, the names of nonunique objects. The ability to retrieve verbs and adjectives is also intact.

The isolated lesions that cause the selective defects described above occur in some patients, but brain destruction is often more general. Consequently, more than one form of aphasia is often present. Frequently, the aphasia is general **(global),** involving both receptive and expressive functions. In this situation, speech is scant as well as nonfluent. Writing is abnormal in all aphasias in which speech is abnormal, but the neural circuits involved are not known. In addition, deaf subjects lose their ability to communicate in sign language if they develop a lesion in the categorical hemisphere.

Although aphasias are produced by lesions of the categorical hemisphere, lesions in the representational hemisphere also have effects. For example, they may impair the ability to tell a story or make a joke. They may also impair a subject's ability to get the point of a joke and, more broadly, to comprehend the meaning of differences in inflection and the "color" of speech. This is one more example of the way the hemispheres are specialized rather than simply being dominant and nondominant.

Stuttering has been found to be associated with right cerebral dominance and widespread overactivity in the cerebral cortex and cerebellum. This includes increased activity of the supplementary motor area. Stimulation of part of this area has been reported to produce **laughter,** with the duration and intensity of the laughter proportionate to the intensity of the stimulus.

Recognition of Faces

An important part of the visual input (see Chapter 8) goes to the inferior temporal lobe, where representations of objects, particularly faces, are stored (Figure 16–7). Faces are particularly important in distinguishing friends from foes and the emotional state of those seen. In humans, storage and recognition of faces is more strongly represented in the right inferior temporal lobe in right-handed individuals, though the left lobe is also active. Lesions in this area cause **prosopagnosia,** the inability to recognize faces. Patients with this abnormality can recognize forms and reproduce them. They can recognize people by their voices, and many of them show autonomic responses when they see familiar as opposed to unfamiliar faces. However, they cannot identify the familiar faces they see. The left hemisphere is also involved, but the role of the right hemisphere is primary. The presence of an autonomic response to a familiar face in the absence of recognition has been explained by postulating the existence of a separate dorsal pathway for processing information about faces that leads to recognition at only a subconscious level.

Localization of Other Functions

Use of fMRI and PET scanning combined with study of patients with strokes and head injuries has provided further insights—or at least glimpses—into the ways serial processing of sensory information produces cognition, reasoning, comprehension, and language. Analysis of the brain regions involved in arithmetic calculations has highlighted two areas. In the inferior por-

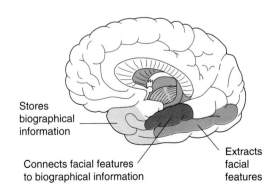

Stores biographical information

Connects facial features to biographical information

Extracts facial features

Figure 16–7. Areas in the right cerebral hemisphere, in right-handed individuals, that are concerned with recognition of faces. (Modified from Szpir M: Accustomed to your face. Am Sci 1992;80:539.)

tion of the left frontal lobe there is an area concerned with number facts and exact calculations. Frontal lobe lesions can cause **acalculia,** a selective impairment of mathematical ability. In the areas around the intraparietal sulci of the parietal lobes bilaterally, there are areas concerned with visuospatial representations of numbers and, presumably, finger counting.

Two right-sided subcortical structures play a role in accurate navigation in humans. One is the right hippocampus, which is concerned with learning where places are located, and the other is the right caudate nucleus, which facilitates movement to the places. Men have larger brains than women and are said to have superior spatial skills and ability to navigate. It has been suggested, partly in jest, that the greater brain weight of men is due to more neural components involved in getting from place to place and that this is why men resist asking directions when lost, whereas women do not hesitate to seek help.

Other defects seen in patients with localized cortical lesions include, for example, the inability to name animals, though the ability to name other living things and objects is intact. One patient with a left parietal lesion had difficulty with the second half but not the first half of words. Some patients with parieto-occipital lesions write only with consonants and omit vowels. The pattern that emerges from studies of this type is one of precise sequential processing of information in localized brain areas. Additional research of this type should greatly expand our understanding of the functions of the neocortex.

Experimental Neurosis

Animals can be conditioned to respond to one stimulus and not to another even when the two stimuli are very much alike. However, when the stimuli are so nearly identical that they cannot be distinguished, the animal becomes upset, whines, fails to cooperate, and tries to escape. Pavlov called these symptoms the **experimental neurosis.** One may quarrel about whether this reaction is a true neurosis in the psychiatric sense, but the term is convenient. If connections between the frontal lobes and the rest of the brain are cut, animals still fail to discriminate but their failure does not upset them.

Because of those results in animals, **prefrontal lobotomy** and various other procedures aimed at cutting the connections between the frontal lobes and deeper portions of the brain were at one time used in humans. In some mental patients, tensions resulting from real or imagined failures of performance and the tensions caused by delusions, compulsions, and phobias are so great as to be incapacitating. Lobotomy may reduce the tension. The delusions and other symptoms are still there, but they no longer bother the patient. A similar

lack of concern over severe pain led to the use of lobotomy in treating patients with intractable pain (see Chapter 7). Unfortunately, this lack of concern often extends to other aspects of the environment, including relations with associates, social amenities, and even toilet habits. It is damage to the orbitofrontal cortex that appears to cause this lack of concern.

REFERENCES FOR SECTION III: FUNCTIONS OF THE NERVOUS SYSTEM

Ahima RS et al: Leptin regulation of neuroendocrine systems. Front Neuroendocrinol 2000;21:263.

Bouchard C, Bray GA (editors): *Regulation of Body Weight: Biological and Behavioral Mechanisms.* Wiley, 1996.

Bronisch FW: *The Clinically Important Reflexes.* Grune & Stratton, 1952.

Caterina MJ et al: Impaired nociception and pain sensation in mice lacking the capsaicin receptor. Science 2000;288:306.

Cordo P, Harnaud S (editors): *Movement Control.* Cambridge Univ Press, 1994.

Ditunno JF Jr, Formal CF: Chronic spinal cord injury. N Engl J Med 1994;330:550.

Dunnett SB, Björklund A: Prospects for new restorative and neuroprotective treatment in Parkinson's disease. Nature 1999;399(Suppl):A32.

Flier JS, Maratos-Flier E: The stomach speaks—ghrelin and weight regulation. N Engl J Med 2002;346:1662.

Gilman S: Imaging the brain. N Engl J Med 1998;338:812.

Greenfield SA: *Journey to the Centers of the Mind: Toward a Science of Consciousness.* Freeman, London, 1995.

Harrington A (editor): *The Placebo Effect: An Interdisciplinary Exploration.* Harvard Univ Press, 1999.

Hof PR, Mobbs CV: *Functional Neurobiology of Aging.* Academic Press, 2001.

Ivry RB, Robertson LC: *The Two Sides of Perception.* MIT Press, 1997.

Jouvet M: *The Paradox of Sleep: The Story of Dreaming.* MIT Press, 1999.

Kandel ER, Schwartz JH, Jessell TM (editors): *Principles of Neural Science,* 4th ed. McGraw-Hill, 2000.

Kandel ER: The molecular biology of memory: a dialogue between genes and synapses. Science 2001;294:1028.

Karmiloff K, Karmiloff-Smith A: *Pathways to Language: From Fetus to Adolescent.* Harvard Univ Press, 2001.

Klockgether T, Dichgans J: Trinucleotide repeats and hereditary ataxias. Nat Med 1997;3:149.

Kupfer DJ, Reynolds CF III: Management of insomnia. N Engl J Med 1997;336:341.

Lamberts SWJ, Hofland LJ, Nobels FRE: Neuroendocrine tumor markers. Front Neuroendocrinol 2001;22:309.

LeVay S: *The Sexual Brain.* MIT Press, 1993.

Li YM: γ-Secretase. Molecular Interventions 2001;1:198.

Lindemann B: Receptors and transduction in taste. Nature 2001;413:219

Locke JL: *The Child's Path to Spoken Language.* Harvard Univ Press, 1993.

Manji HK, Drevets WK, Charney DS: The cellular neurobiology of depression. Nat Med 2000;7:541.

Menon ST, Han M, Sakmar TP: Rhodopsin: structural basis of molecular physiology. Physiol Rev 2001;81:1659.

Minke B, Cook B: TRP channel proteins and signal transduction. Physiol Rev 2002;82:429.

Oyster CW: *The Human Eye: Structure and Function.* Sinauer, 1999.

Plomin R, Owen MJ, McGuffin P: The genetic basis of complex human behaviors. Science 1994;264:1733.

Prusiner SB: Shattuck lecture—Neurodegenerative diseases and prions. N Engl J Med 2001;344:1516.

Raymond JL, Lisberger SG, Mank MD: The cerebellum: A neuronal learning machine. Science 1996;272:225.

Robbins TW, Everett BJ: Drug addiction: bad habits add up. Nature 1999;398:567.

Schradin C, Anzenberger G: Prolactin, the hormone of paternity. News Physiol Sci 1999;14:223.

Selkoe DJ: Translating cell biology into therapeutic advances in Alzheimer's disease. Nature 1999;399(Suppl):A23.

Shaywitz S: Dyslexia. N Engl J Med 1998;338:307.

Siegel JM: Narcolepsy. Sci Am [Jan] 2000;282:76

Smith EE, Jonides J: Storage and executive processes in the frontal lobe. Science 1999;283:1657.

Stern P, Marks J (editors): Making sense of scents. (Special Section.) Science 1999;286:703.

Stux G, Pomeranz B: *Basics of Acupuncture,* 2nd ed. Springer, 1991.

Thaker GK, Carpenter WT Jr: Advances in schizophrenia. Nat Med 2001;7:667.

Tramo MJ: Mask of the hemispheres. Science 2001;291:54.

Waldman SD: *Interventional Pain Management,* 2nd ed. Saunders, 2001

Weinberger DR: Anxiety at the frontier of molecular medicine. N Engl J Med 2001;344:1247.

Weller A: Communication through body odour. Nature 1998;392:120.

Willems PJ: Genetic causes of hearing loss. N Engl J Med 2000;342:1101.

Wilson RA, Keil FC: *The MIT Encyclopedia of the Cognitive Sciences.* MIT Press, 1999.

Woolf CJ, Salter MW: Neuronal plasticity: increasing the gain in pain. Science 2000;288:1765.

SECTION IV
Endocrinology, Metabolism, & Reproductive Function

Energy Balance, Metabolism, & Nutrition

17

INTRODUCTION

The endocrine system, like the nervous system, adjusts and correlates the activities of the various body systems, making them appropriate to the changing demands of the external and internal environment. Endocrine integration is brought about by ductless glands and transported in the circulation to target cells. Other types of chemical messengers are discussed in Chapter 1. Some of the hormones are amines, and others are amino acids, polypeptides, proteins, or steroids.

The hormones regulate metabolic processes. The term **metabolism,** meaning literally "change," is used to refer to all the chemical and energy transformations that occur in the body.

The animal organism oxidizes carbohydrates, proteins, and fats, producing principally CO_2, H_2O, and the energy necessary for life processes. CO_2, H_2O, and energy are also produced when food is burned outside the body. However, in the body, oxidation is not a one-step, semiexplosive reaction but a complex, slow, step-wise process called **catabolism,** which liberates energy in small, usable amounts. Energy can be stored in the body in the form of special energy-rich phosphate compounds and in the form of proteins, fats, and complex carbohydrates synthesized from simpler molecules. Formation of these substances by processes that take up rather than liberate energy is called **anabolism.** This chapter sets the stage for consideration of endocrine function by providing a brief summary of the production and utilization of energy and the metabolism of carbohydrates, proteins, and fats.

ENERGY METABOLISM

Metabolic Rate

The amount of energy liberated by the catabolism of food in the body is the same as the amount liberated when food is burned outside the body. The energy liberated by catabolic processes in the body is used for maintaining body functions, digesting and metabolizing food, thermoregulation, and physical activity. It appears as external work, heat, and energy storage:

$$\frac{\text{Energy}}{\text{output}} = \frac{\text{External}}{\text{work}} + \frac{\text{Energy}}{\text{storage}} + \text{Heat}$$

The amount of energy liberated per unit of time is the **metabolic rate.** Isotonic muscle contractions perform work at a peak efficiency approximating 50%:

$$\text{Efficiency} = \frac{\text{Work done}}{\text{Total energy expended}}$$

Essentially all of the energy of isometric contractions appears as heat, because little or no external work (force multiplied by the distance that the force moves a mass) is done (see Chapter 3). Energy is stored by forming energy-rich compounds. The amount of energy storage varies, but in fasting individuals it is zero or negative. Therefore, in an adult individual who has not eaten recently and who is not moving (or growing, reproducing, or lactating), all of the energy output appears as heat.

Calories

The standard unit of heat energy is the **calorie (cal),** defined as the amount of heat energy necessary to raise the temperature of 1 g of water 1 degree, from 15 °C to 16 °C. This unit is also called the gram calorie, small calorie, or standard calorie. The unit commonly used in physiology and medicine is the **Calorie (kilocalorie; kcal),** which equals 1000 cal.

Calorimetry

The energy released by combustion of foodstuffs outside the body can be measured directly **(direct calorimetry)** by oxidizing the compounds in an apparatus such as a **bomb calorimeter,** a metal vessel surrounded by water inside an insulated container. The food is ignited by an electric spark. The change in the temperature of the water is a measure of the calories produced. Similar measurements of the energy released by combustion of compounds in living animals and humans are much more complex, but calorimeters have been constructed that can physically accommodate human beings. The heat produced by their bodies is measured by the change in temperature of the water in the walls of the calorimeter.

The caloric values of the common foodstuffs, as measured in a bomb calorimeter, are found to be 4.1 kcal/g of carbohydrate, 9.3 kcal/g of fat, and 5.3 kcal/g of protein. In the body, similar values are obtained for carbohydrate and fat, but the oxidation of protein is incomplete, the end products of protein catabolism being urea and related nitrogenous compounds in addition to CO_2 and H_2O (see below). Therefore, the caloric value of protein in the body is only 4.1 kcal/g.

Indirect Calorimetry

Energy production can also be calculated by measuring the products of the energy-producing biologic oxidations—ie, CO_2, H_2O, and the end products of protein catabolism produced—but this is difficult. However, O_2 is not stored, and except when an O_2 debt is being incurred, the amount of O_2 consumption per unit of time is proportionate to the energy liberated by metabolism. Consequently, measurement of O_2 consumption **(indirect calorimetry)** is used to determine the metabolic rate.

Respiratory Quotient (RQ)

The **respiratory quotient (RQ)** is the ratio in the steady state of the volume of CO_2 produced to the volume of O_2 consumed per unit of time. It should be distinguished from the **respiratory exchange ratio (R),** which is the ratio of CO_2 to O_2 at any given time whether or not equilibrium has been reached. R is affected by factors other than metabolism. RQ and R can be calculated for reactions outside the body, for individual organs and tissues, and for the whole body. The RQ of carbohydrate is 1.00, and that of fat is about 0.70. This is because H and O are present in carbohydrate in the same proportions as in water, whereas in the various fats, extra O_2 is necessary for the formation of H_2O.

Carbohydrate:

$$C_6H_{12}O_6 + 6O_2 \rightarrow 6CO_2 + 6H_2O$$
$$\text{(glucose)}$$

$$RQ = 6/6 = 1.00$$

Fat:

$$2C_{51}H_{98}O_6 + 145O_2 \rightarrow 102CO_2 + 98H_2O$$
$$\text{(tripalmitin)}$$

$$RQ = 102/145 = 0.703$$

Determining the RQ of protein in the body is a complex process, but an average value of 0.82 has been calculated. The approximate amounts of carbohydrate, protein, and fat being oxidized in the body at any given time can be calculated from the RQ and the urinary nitrogen excretion. RQ and R for the whole body differ in various conditions. For example, during hyperventilation, R rises because CO_2 is being blown off. During severe exercise, R may reach 2.00 because CO_2 is being blown off and lactic acid from anaerobic glycolysis is being converted to CO_2 (see below). After exercise, R may fall for a while to 0.50 or less. In metabolic acidosis, R rises because respiratory compensation for the acidosis causes the amount of CO_2 expired to rise (see Chapter 39). In severe acidosis, R may be greater than 1.00. In metabolic alkalosis, R falls.

The O_2 consumption and CO_2 production of an organ can be calculated at equilibrium by multiplying its blood flow per unit of time by the arteriovenous differences for O_2 and CO_2 across the organ, and the RQ can then be calculated. Data on the RQ of individual organs are of considerable interest in drawing inferences about the metabolic processes occurring in them. For example, the RQ of the brain is regularly 0.97–0.99, indicating that its principal but not its only fuel is carbohydrate. During secretion of gastric juice, the stomach has a negative R because it takes up more CO_2 from the arterial blood than it puts into the venous blood (see Chapter 26).

Measuring the Metabolic Rate

In determining the metabolic rate, O_2 consumption is usually measured with some form of oxygen-filled spirometer and a CO_2-absorbing system. Such a device is illustrated in Figure 17–1. The spirometer bell is connected to a pen that writes on a rotating drum as the bell moves up and down. The slope of a line joining the ends of each of the spirometer excursions is proportionate to the O_2 consumption. The amount of O_2 (in milliliters) consumed per unit of time is corrected to standard temperature and pressure (see Chapter 34) and then converted to energy production by multiplying by 4.82 kcal/L of O_2 consumed.

Factors Affecting the Metabolic Rate

The metabolic rate is affected by many factors (Table 17–1). The most important is muscular exertion. O_2 consumption is elevated not only during exertion but also for as long afterward as is necessary to repay the O_2 debt (see Chapter 3). Recently ingested foods also increase the metabolic rate because of their **specific dynamic action (SDA).** The SDA of a food is the oblig-atory energy expenditure that occurs during its assimilation into the body. An amount of protein sufficient to provide 100 kcal increases the metabolic rate a total of 30 kcal; a similar amount of carbohydrate increases it 6 kcal; and a similar amount of fat, 4 kcal.

Table 17–1. Factors affecting the metabolic rate.

Muscular exertion during or just before measurement
Recent ingestion of food
High or low environmental temperature
Height, weight, and surface area
Sex
Age
Growth
Reproduction
Lactation
Emotional state
Body temperature
Circulating levels of thyroid hormones
Circulating epinephrine and norepinephrine levels

This means, of course, that the calories available from the three foods are in effect reduced by this amount; the energy used in their assimilation must come from the food itself or from the body energy stores. The cause of the SDA, which may last up to 6 hours, is uncertain.

Another factor that stimulates metabolism is the environmental temperature. The curve relating the metabolic rate to the environmental temperature is U-shaped. When the environmental temperature is lower than body temperature, heat-producing mechanisms such as shivering are activated and the metabolic rate rises. When the temperature is high enough to raise the body temperature, there is a general acceleration of metabolic processes, and the metabolic rate rises about 14% for each Celsius degree of elevation.

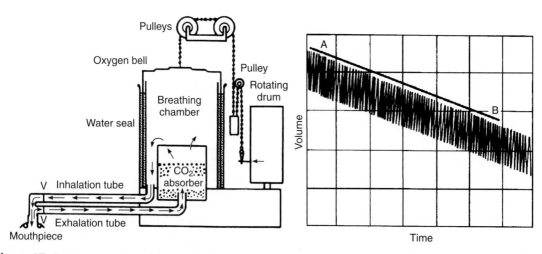

Figure 17–1. Diagram of a modified Benedict apparatus, a recording spirometer used for measuring human O_2 consumption, and the record obtained with it. The slope of the line AB is proportionate to the O_2 consumption. V: one-way check valve.

The metabolic rate determined at rest in a room at a comfortable temperature in the thermoneutral zone 12–14 hours after the last meal is called the **basal metabolic rate (BMR).** This value falls about 10% during sleep and up to 40% during prolonged starvation. The rate during normal daytime activities is, of course, higher than the BMR because of muscular activity and food intake. The **maximum metabolic rate** reached during exercise is often said to be ten times the BMR, but trained athletes can increase their metabolic rate as much as 20-fold.

The BMR of a man of average size is about 2000 kcal/d. Large animals have higher absolute BMRs, but the ratio of BMR to body weight in small animals is much greater. One variable that correlates well with the metabolic rate in different species is the body surface area. This would be expected, since heat exchange occurs at the body surface. The actual relation to body weight (W) would be

$$BMR = 3.52W^{0.67}$$

However, repeated measurements by numerous investigators have come up with a higher exponent, averaging 0.75.

$$BMR = 3.52W^{0.75}$$

Thus, the slope of the line relating metabolic rate to body weight is steeper than it would be if the relation were due solely to body area (Figure 17–2). The cause of the greater slope has been much debated but remains unsettled.

For clinical use, the BMR is usually expressed as a percentage increase or decrease above or below a set of generally used standard normal values. Thus, a value of +65 means that the individual's BMR is 65% above the standard for that age and sex. Factors affecting the BMR are listed in Table 17–1.

The decrease in metabolic rate is part of the explanation of why, when an individual reduces, weight loss is initially rapid and then slows down.

Energy Balance

The first law of thermodynamics, the principle which states that energy is neither created nor destroyed when it is converted from one form to another, applies to living organisms as well as inanimate systems. One may therefore speak of an **energy balance** between caloric intake and energy output. If the caloric content of the food ingested is less than the energy output—ie, if the balance is negative—endogenous stores are utilized. Glycogen, body protein, and fat are catabolized, and the individual loses weight. If the caloric value of the

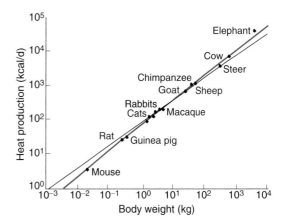

Figure 17–2. Correlation between metabolic rate and body weight, plotted on logarithmic scales. The slope of the colored line is 0.75. The black line represents the way surface area increases with weight for geometrically similar shapes and has a slope of 0.67. (Modified from Kleiber M and reproduced, with permission, from McMahon TA: Size and shape in biology. Science 1973;179: 1201. Copyright © 1973 by the American Association for the Advancement of Science.)

food intake exceeds energy loss due to heat and work and the food is properly digested and absorbed—ie, if the balance is positive—energy is stored, and the individual gains weight.

To balance basal output so that the energy-consuming tasks essential for life can be performed, the average adult must take in about 2000 kcal/d. Caloric requirements above the basal level depend upon the individual's activity. The average sedentary student (or professor) needs another 500 kcal, whereas a lumberjack needs up to 3000 additional kcal per day.

INTERMEDIARY METABOLISM

General Considerations

The end products of the digestive processes discussed in Chapters 25 and 26 are for the most part amino acids, fat derivatives, and the hexoses fructose, galactose, and glucose. These compounds are absorbed and metabolized in the body by various routes. The details of their metabolism are the concern of biochemistry and are not considered here. However, an outline of carbohydrate, protein, and fat metabolism is included for completeness and because some knowledge of the pathways involved is essential to an understanding of the action of thyroid, pancreatic, and adrenal hormones.

General Plan of Metabolism

The short-chain fragments produced by hexose, amino acid, and fat catabolism are very similar. From this **common metabolic pool** of intermediates, carbohydrates, proteins, and fats can be synthesized, although the conversion of fats to carbohydrates is limited (see below). Alternatively, the fragments can enter the citric acid cycle, a sort of final common pathway of catabolism, in which they are broken down to hydrogen atoms and CO_2. The hydrogen atoms are oxidized to form water by a chain of flavoprotein and cytochrome enzymes.

Energy Transfer

Much of the energy liberated by catabolism is not used directly by cells but is applied instead to the formation of bonds between phosphoric acid residues and certain organic compounds. Because the energy of bond formation in some of these phosphates is particularly high, relatively large amounts of energy (10–12 kcal/mol) are released when the bond is hydrolyzed. Compounds containing such bonds are called **high-energy phosphate compounds.** Not all organic phosphates are of the high-energy type. Many, like glucose 6-phosphate, are low-energy phosphates that on hydrolysis liberate 2–3 kcal/mol. Some of the intermediates formed in carbohydrate metabolism are high-energy phosphates, but the most important high-energy phosphate compound is **adenosine triphosphate (ATP).** This ubiquitous molecule (Figure 17–3) is the energy storehouse of the body. Upon hydrolysis to adenosine diphosphate (ADP), it liberates energy directly to such processes as muscle contraction, active transport, and the synthesis of many chemical compounds. Loss of another phosphate to form adenosine monophosphate (AMP) releases more energy. Another energy-rich phosphate compound found in muscle is **creatine phosphate (phosphorylcreatine; CrP)** (see below). Another group of high-energy compounds are the thioesters, the acyl derivatives of mercaptans. **Coenzyme A (CoA)** is a widely distributed mercaptan containing adenine, ribose, pantothenic acid, and thioethanolamine (Figure 17–4). Reduced CoA (usually abbreviated HS-CoA) reacts with acyl groups (R–CO–) to form R–CO–S–CoA derivatives. A prime example is the reaction of HS-CoA with acetic acid to form acetylcoenzyme A (acetyl-CoA), a compound of pivotal importance in intermediary metabolism. Because acetyl-CoA has a much higher energy content than acetic acid, it combines readily with substances in reactions that would otherwise require outside energy. Acetyl-CoA is therefore often called "active acetate." From the point of view of energetics, formation of 1 mol of any acyl-CoA compound is equivalent to the formation of 1 mol of ATP.

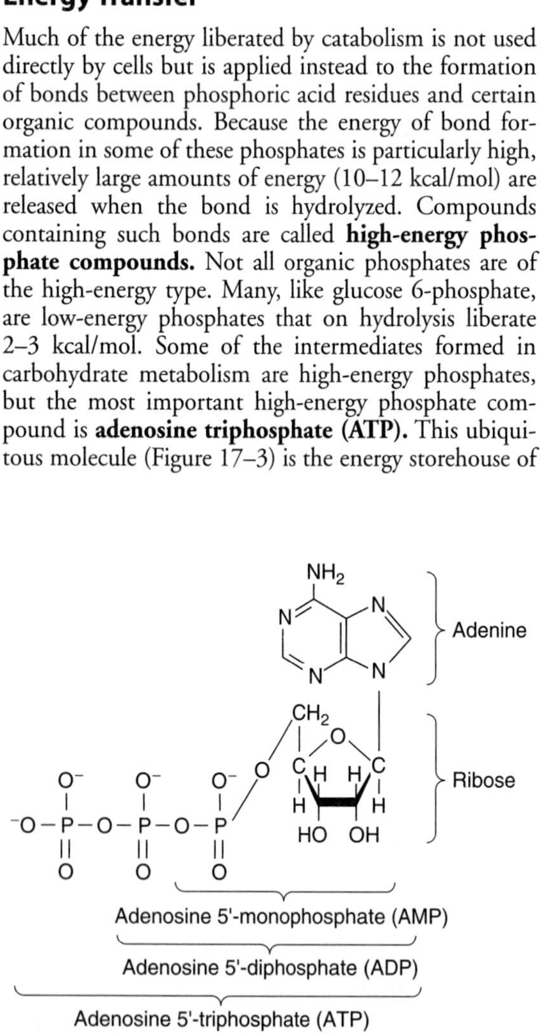

Biologic Oxidations

Oxidation is the combination of a substance with O_2, or loss of hydrogen, or loss of electrons. The corresponding reverse processes are called **reduction.** Biologic oxidations are catalyzed by specific enzymes. Cofactors (simple ions) or coenzymes (organic, nonprotein substances) are accessory substances that usually act as carriers for products of the reaction. Unlike the enzymes, the coenzymes may catalyze a variety of reactions.

A number of coenzymes serve as hydrogen acceptors. One common form of biologic oxidation is removal of hydrogen from an R–OH group, forming R=O. In such dehydrogenation reactions, nicotinamide adenine dinucleotide (NAD^+) and dihydronicotinamide adenine dinucleotide phosphate ($NADP^+$) pick up hydrogen, forming dihydronicotinamide adenine dinucleotide (NADH) and dihydronicotinamide adenine dinucleotide phosphate (NADPH) (Figure 17–5). The hydrogen is then transferred to the flavoprotein-cytochrome system, reoxidizing the NAD^+ and $NADP^+$. Flavin adenine dinucleotide (FAD) is formed when riboflavin is phosphorylated, forming flavin mononucleotide (FMN). FMN then combines with AMP, forming the dinucleotide. FAD can accept hydrogens in a similar fashion, forming its hydro (FADH) and dihydro ($FADH^2$) derivatives.

Figure 17–3. Energy-rich adenosine derivatives. (Reproduced, with permission, from Murray RK et al: *Harper's Biochemistry,* 26th ed. McGraw-Hill, 2003.)

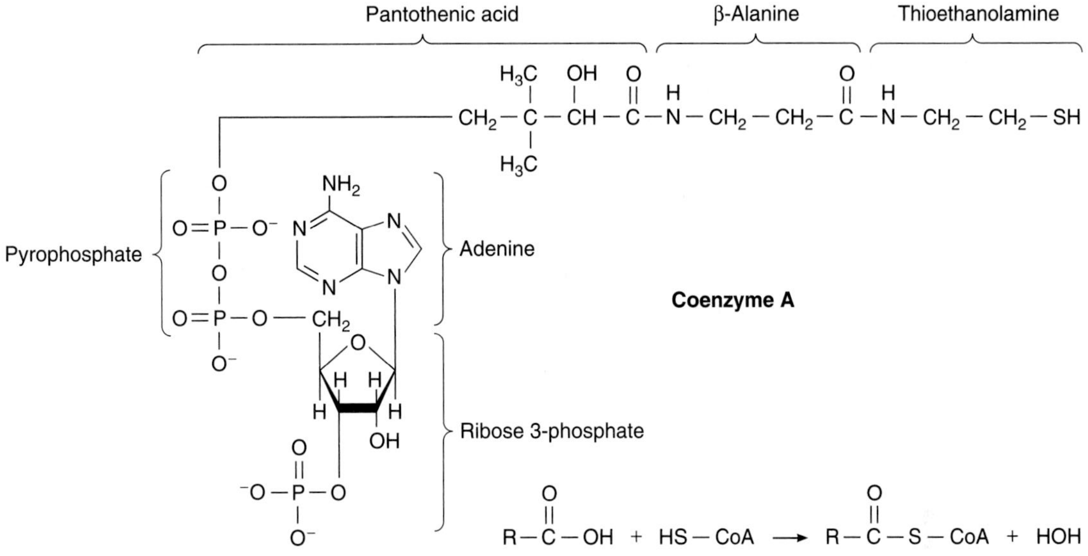

Figure 17–4. **Left:** Formula of reduced CoA (HS-CoA). **Right:** Formula for reaction of CoA with biologically important compounds to form thioesters. R, rest of molecule.

The flavoprotein-cytochrome system is a chain of enzymes that transfers hydrogen to oxygen, forming water. This process occurs in the mitochondria. Each enzyme in the chain is reduced and then reoxidized as the hydrogen is passed down the line. Each of the enzymes is a protein with an attached nonprotein prosthetic group. The final enzyme in the chain is cytochrome c oxidase, which transfers hydrogens to O_2, forming H_2O. It contains two atoms of Fe and three of Cu and has 13 subunits.

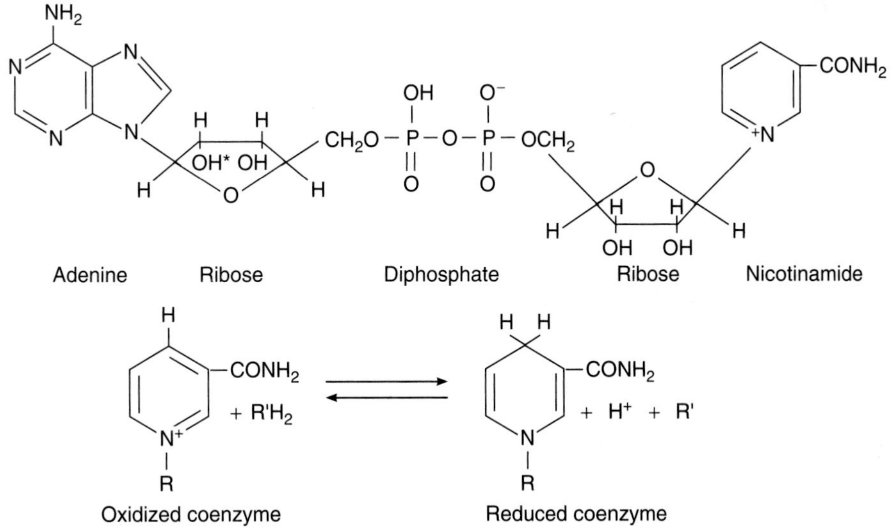

Figure 17–5. **Top:** Formula of the oxidized form of nicotinamide adenine dinucleotide (NAD+). Nicotinamide adenine dinucleotide phosphate (NADP+) has an additional phosphate group at the location marked by the asterisk. **Bottom:** Reaction by which NAD+ and NADP+ become reduced to form NADH and NADPH. R, remainder of molecule; R′, hydrogen donor.

Oxidative Phosphorylation

Production of ATP associated with oxidation by the flavoprotein-cytochrome system is called **oxidative phosphorylation.** ATP can also be formed in other situations when the release of free energy by a chemical reaction is great. This is called **oxidation at the substrate level.** Oxidative phosphorylation involves the transfer of protons across an insulating membrane (the inner membrane that forms the cristae of the mitochondria), the transfer being driven by oxidation in the respiratory chain (Figure 17–6). This creates an electrochemical potential difference across the membrane, and the transport of protons from the intracristal space back into the matrix space (see Chapter 1) drives an ATPase in the membrane (**ATP synthase,** F-ATPase) that converts ADP and inorganic phosphate (Pi) to ATP (Figure 17–7). This complex enzyme or a closely related enzyme is found in the membranes of bacteria, where it is driven by oxidation, and in plants, where it is driven by light (photosynthesis). It has an F_1 and an F_0 portion and is made up of more than a dozen subunits.

Oxidative phosphorylation depends on an adequate supply of ADP and consequently is in part under a form of feedback control; the more rapid the utilization of ATP in the tissues, the greater the accumulation of ADP and, consequently, the higher the rate of oxidative

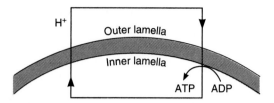

Figure 17–7. Simplified diagram of transport of protons across the inner and outer lamellas of the inner mitochondrial membrane by the electron transport system (flavoprotein-cytochrome system), with return movement of protons down the proton gradient, generating ATP.

phosphorylation. Other factors regulating the rate of generation of ATP include the rate of delivery of fats, lactate, and glucose derivatives to the interior of the mitochondria and the availability of O_2.

Ninety percent of the O_2 consumption in the basal state is mitochondrial, and 80% of this is coupled to ATP synthesis. About 27% of the ATP is used for protein synthesis, and about 24% is used by Na^+-K^+ ATPase, 9% by gluconeogenesis, 6% by Ca^{2+} ATPase, 5% by myosin ATPase, and 3% by ureagenesis. In addition to its function in energy transfer, ATP is the precursor of cAMP (see Chapter 1).

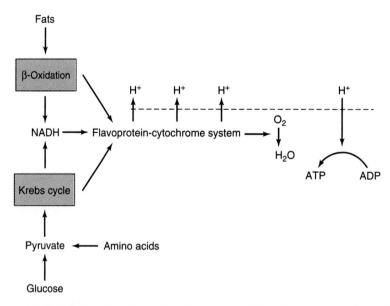

Figure 17–6. Summary of oxidative phosphorylation. Metabolism of fats, glucose, and amino acids provides protons (H^+) that are translocated across the inner mitochondrial membrane (dashed line). Diffusion of these protons down the concentration gradient created in this fashion drives ATP synthase to convert ADP to ATP. The key enzymes involved are shown in Figure 1–7.

CARBOHYDRATE METABOLISM

Dietary carbohydrates are for the most part polymers of hexoses, of which the most important are glucose, galactose, and fructose (Figure 17–8). Most of the monosaccharides occurring in the body are the D isomers. The principal product of carbohydrate digestion and the principal circulating sugar is glucose. The normal fasting level of plasma glucose in peripheral venous blood is 70–110 mg/dL (3.9–6.1 mmol/L). In arterial blood, the plasma glucose level is 15–30 mg/dL higher than in venous blood.

Once it enters the cells, glucose is normally phosphorylated to form glucose 6-phosphate. The enzyme that catalyzes this reaction is **hexokinase.** In the liver, there is in addition an enzyme called **glucokinase,** which has greater specificity for glucose and which, unlike hexokinase, is increased by insulin and decreased in starvation and diabetes. The glucose 6-phosphate is either polymerized into glycogen or catabolized. The steps involved are outlined in Figure 17–9. The process of glycogen formation is called **glycogenesis,** and glycogen breakdown is called **glycogenolysis.** Glycogen, the storage form of glucose, is present in most body tissues, but the major supplies are in the liver and skeletal muscle. The breakdown of glucose to pyruvate or lactate (or both) is called **glycolysis.** Glucose catabolism proceeds via cleavage through fructose to trioses or via oxidation and decarboxylation to pentoses. The pathway to pyruvate through the trioses is the **Embden-Meyerhof pathway,** and that through 6-phosphogluconate and the pentoses is the **direct oxidative pathway (hexose monophosphate shunt)** (Figure 17–9). Pyruvate is converted to acetyl-CoA. Interconversions between carbohydrate, fat, and protein include conversion of the glycerol from fats to dihydroxyacetone phosphate and conversion of a number of amino acids with carbon skeletons resembling intermediates in the Embden-Meyerhof pathway and citric acid cycle to

these intermediates by deamination. In this way, and by conversion of lactate to glucose, nonglucose molecules can be converted to glucose **(gluconeogenesis).** Glucose can be converted to fats through acetyl-CoA, but since the conversion of pyruvate to acetyl-CoA, unlike most reactions in glycolysis, is irreversible (Figure 17–10), fats are not converted to glucose via this pathway. There is therefore very little net conversion of fats to carbohydrate in the body because, except for the quantitatively unimportant production from glycerol, there is no pathway for conversion.

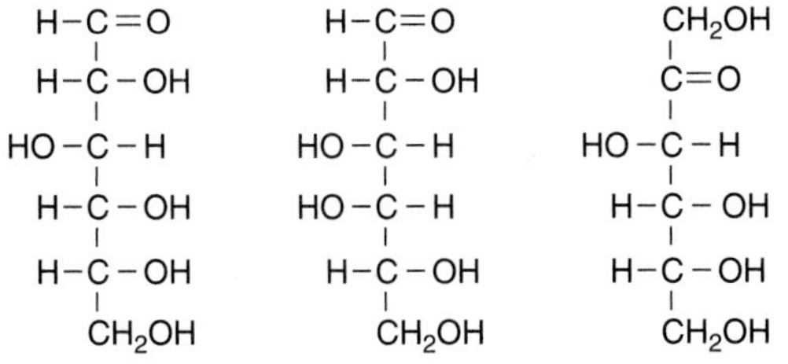

Citric Acid Cycle

The **citric acid cycle** (Krebs cycle, tricarboxylic acid cycle) is a sequence of reactions in which acetyl-CoA is metabolized to CO_2 and H atoms. Acetyl-CoA is first condensed with the anion of a four-carbon acid, oxaloacetate, to form citrate and HS-CoA. In a series of seven subsequent reactions, $2CO_2$ molecules are split off, regenerating oxaloacetate (Figure 17–10). Four pairs of H atoms are transferred to the flavoprotein-cytochrome chain, producing 12ATP and $4H_2O$, of which $2H_2O$ is used in the cycle. The citric acid cycle is the common pathway for oxidation to CO_2 and H_2O of carbohydrate, fat, and some amino acids. The major entry into it is through acetyl-CoA, but a number of amino acids can be converted to citric acid cycle intermediates by deamination. The citric acid cycle requires O_2 and does not function under anaerobic conditions.

Energy Production

The net production of energy-rich phosphate compounds during the metabolism of glucose and glycogen to pyruvate depends on whether metabolism occurs via the Embden-Meyerhof pathway or the hexose monophosphate shunt. By oxidation at the substrate level, the conversion of 1 mol of phosphoglyceraldehyde to phosphoglycerate generates 1 mol of ATP, and the conversion of 1 mol of phosphoenolpyruvate to pyruvate generates another. Since 1 mol of glucose 6-phosphate produces, via the Embden-Meyerhof pathway, 2 mol of phosphoglyceraldehyde, 4 mol of ATP are generated per mole of glucose metabolized to pyruvate. All these reactions occur in the absence of O_2 and consequently represent anaerobic production of energy. However, 1 mol of ATP is used in forming fructose 1,6-diphosphate from fructose 6-phosphate and 1 mol in phosphorylating glucose when it enters the cell. Consequently, when pyruvate is formed anaerobically from glycogen, there is a *net* production of 3 mol of ATP per mole of glucose 6-phosphate; however, when

H–C=O
|
H–C–OH
|
HO–C–H
|
H–C–OH
|
H–C–OH
|
CH₂OH

D-Glucose

H–C=O
|
H–C–OH
|
HO–C–H
|
HO–C–H
|
H–C–OH
|
CH₂OH

D-Galactose

CH₂OH
|
C=O
|
HO–C–H
|
H–C–OH
|
H–C–OH
|
CH₂OH

D-Fructose

Figure 17–8. Structure of principal dietary hexoses. The naturally occurring D isomers are shown.

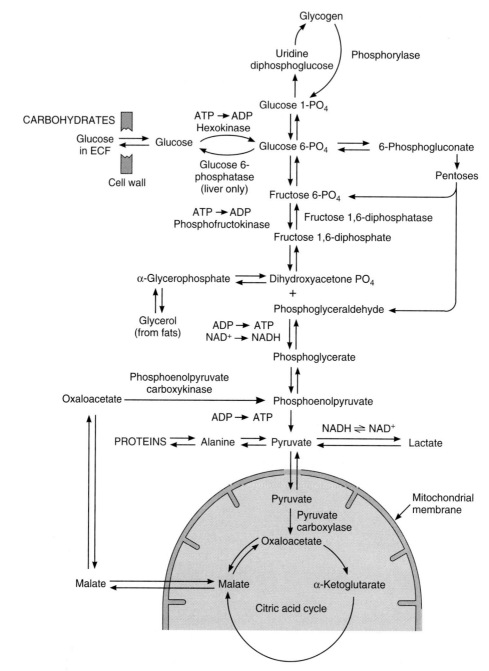

Figure 17–9. Outline of the metabolism of carbohydrate in cells, showing some of the principal enzymes involved.

pyruvate is formed from 1 mol of blood glucose, the net gain is only 2 mol of ATP.

A supply of NAD^+ is necessary for the conversion of phosphoglyceraldehyde to phosphoglycerate. Under anaerobic conditions (anaerobic glycolysis), a block of glycolysis at the phosphoglyceraldehyde conversion step might be expected to develop as soon as the available NAD^+ is converted to NADH. However, pyruvate can accept hydrogen from NADH, forming NAD^+ and lactate.

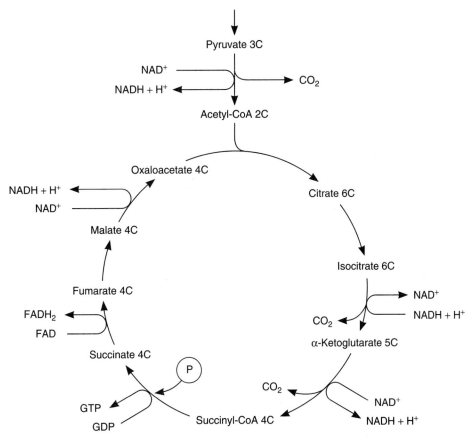

Figure 17–10. Citric acid cycle. The numbers (6C, 5C, etc) indicate the number of carbon atoms in each of the intermediates. The conversion of pyruvate to acetyl-CoA and each turn of the cycle provide four NADH and one FADH$_2$ for
oxidation via the flavoprotein-cytochrome chain plus formation of one GTP that is readily converted to ATP.
(Modified and reproduced, with permission, from Alberts BM et al: *Molecular Biology of the Cell*, 2nd ed. Garland, 1989.)

$$\text{Pyruvate} + \text{NADH} \leftrightarrows \text{Lactate} + \text{NAD+}$$

In this way, glucose metabolism and energy production
can continue for a while without O$_2$. The lactate that accumulates is converted back to pyruvate when the O$_2$
supply is restored, NADH transferring its hydrogen to
the flavoprotein-cytochrome chain.

During aerobic glycolysis, the net production of ATP
is 19 times as great as the two ATPs formed under anaerobic conditions. Six ATPs are formed by oxidation via
the flavoprotein-cytochrome chain of the two NADHs
produced when 2 mol of phosphoglyceraldehyde are
converted to phosphoglycerate (Figure 17–9); six ATPs
are formed from the two NADHs produced when 2 mol
of pyruvate are converted to acetyl-CoA; and 24 ATPs
are formed during the subsequent two turns of the citric
acid cycle. Of these, 18 are formed by oxidation of six
NADHs, four by oxidation of two FADH$_2$s, and two by

oxidation at the substrate level when succinyl-CoA is
converted to succinate. This reaction actually produces
GTP, but the GTP is converted to ATP. Thus, the net
production of ATP per mol of blood glucose metabolized aerobically via the Embden-Meyerhof pathway and
citric acid cycle is $2 + [2 \times 3] + [2 \times 3] + [2 \times 12] = 38$.

Glucose oxidation via the hexosemonophosphate
shunt generates large amounts of NADPH. A supply of
this reduced coenzyme is essential for many metabolic
processes. The pentoses formed in the process are building blocks for nucleotides (see below). The amount of
ATP generated depends upon the amount of NADPH
converted to NADH and then oxidized.

"Directional-Flow Valves"

Metabolism is regulated by a variety of hormones and
other factors. To bring about any net change in a par-

ticular metabolic process, regulatory factors obviously must drive a chemical reaction in one direction. Most of the reactions in intermediary metabolism are freely reversible, but there are a number of "directional-flow valves," ie, reactions that proceed in one direction under the influence of one enzyme or transport mechanism and in the opposite direction under the influence of another. Five examples in the intermediary metabolism of carbohydrate are shown in Figure 17–11. The different pathways for fatty acid synthesis and catabolism (see below) are another example. Regulatory factors exert their influence on metabolism by acting directly or indirectly at these "directional-flow valves."

Glycogen Synthesis & Breakdown

Glycogen is a branched glucose polymer with two types of glycoside linkages: 1:4α and 1:6α (Figure 17–12). It

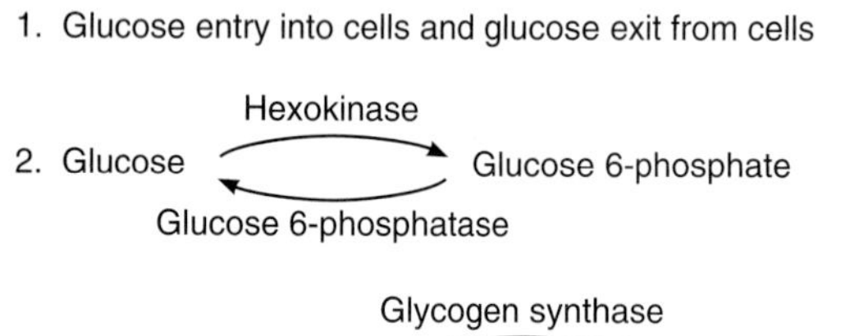

Figure 17–11. Five examples of "directional-flow valves" in carbohydrate metabolism; reactions that proceed in one direction by one mechanism and in the other direction by a different mechanism. The double line in example 5 represents the mitochondrial membrane. Pyruvate is converted to malate in mitochondria, and the malate diffuses out of the mitochondria to the cytosol, where it is converted to phosphoenolpyruvate.

is synthesized on **glycogenin,** a protein primer, from glucose 1-phosphate via uridine diphosphoglucose (UDPG). The enzyme **glycogen synthase** catalyses the final synthetic step. The availability of glycogenin is one of the factors determining the amount of glycogen synthesized. The breakdown of glycogen in 1:4α linkage is catalyzed by phosphorylase, whereas another enzyme catalyzes the breakdown of glycogen in 1:6α linkage.

Phosphorylase is activated in part by the action of epinephrine on β_2-adrenergic receptors in the liver. This in turn initiates a sequence of reactions that provides a classic example of hormonal action via cAMP (Figure 17–13). Protein kinase A is activated by cAMP and catalyzes the transfer of a phosphate group to phosphorylase kinase, converting it to its active form. The phosphorylase kinase in turn catalyzes the phosphorylation and consequent activation of phosphorylase. Inactive phosphorylase is known as phosphorylase b (dephosphophosphorylase), and activated phosphorylase as phosphorylase a (phosphophosphorylase).

Activation of protein kinase A by cAMP not only increases glycogen breakdown but also inhibits glycogen synthesis. Glycogen synthase (Figure 17–12) is active in its dephosphorylated form and inactive when phosphorylated, and it is phosphorylated along with phosphorylase kinase when protein kinase A is activated.

Glycogen is also broken down by the action of catecholamines on α_1-adrenergic receptors in the liver. This breakdown is mediated by intracellular Ca^{2+} and involves an activation of phosphorylase kinase that is independent of cAMP.

Because the liver contains the enzyme **glucose 6-phosphatase,** much of the glucose 6-phosphate that is formed in this organ can be converted to glucose and enter the bloodstream, raising the plasma glucose level. The kidneys can also contribute to the elevation. Other tissues do not contain this enzyme, so in them a large proportion of the glucose 6-phosphate is catabolized via the Embden-Meyerhof pathway and hexose monophosphate shunt pathway. Increased glucose catabolism in skeletal muscle causes a rise in the blood lactate level (see Chapter 3).

By stimulating adenylyl cyclase, epinephrine causes activation of the phosphorylase in liver and skeletal muscle. The consequences of this activation are a rise in the plasma glucose and blood lactate levels. Glucagon has a similar action, but it exerts its effect only on the phosphorylase in the liver. Consequently, glucagon causes a rise in plasma glucose without any change in blood lactate.

McArdle's Syndrome

In the clinical condition known as **McArdle's syndrome** or **myophosphorylase deficiency glycogenosis,** glycogen accumulates in skeletal muscles because of

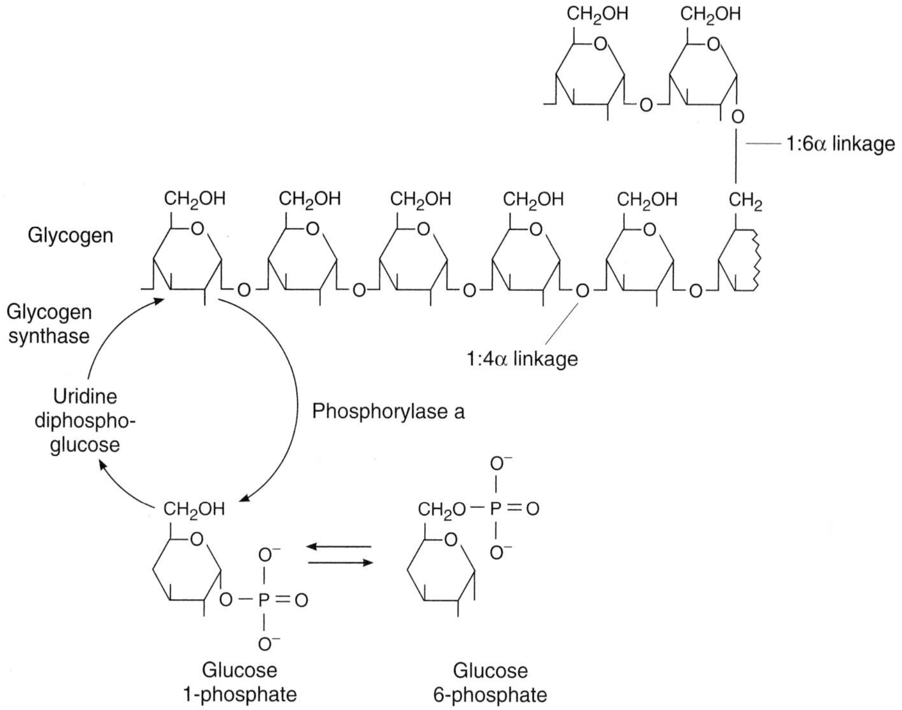

Figure 17–12. Glycogen formation and breakdown. The activation of phosphorylase a is summarized in Figure 17–13.

a deficiency of muscle phosphorylase. Patients with this disease develop muscle pain and stiffness on exertion, and they have a greatly reduced exercise tolerance; they cannot break down their muscle glycogen to provide the energy for muscle contraction (see Chapter 3), and the glucose reaching their muscles from the bloodstream is sufficient only for the demands of very mild exercise. They respond with a normal rise in plasma glucose when given glucagon or epinephrine, which indicates that their hepatic phosphorylase is normal.

The "Hepatic Glucostat"

There is a net uptake of glucose by the liver when the plasma glucose is high and a net discharge when it is low. The liver thus functions as a sort of "glucostat," maintaining a constant circulating glucose level. This function is not automatic; glucose uptake and glucose discharge are affected by the actions of numerous hormones (see below and Chapter 19).

Renal Handling of Glucose

In the kidneys, glucose is freely filtered; but at normal plasma glucose levels, all but a very small amount is reabsorbed in the proximal tubules (see Chapter 38).

When the amount filtered increases, reabsorption increases, but there is a limit to the amount of glucose the proximal tubules can reabsorb. When the tubular maximum for glucose (Tm_G) is exceeded, appreciable amounts of glucose appear in the urine (**glycosuria**). The **renal threshold** for glucose, the arterial blood level at which glycosuria appears, is reached when the glucose concentration in venous plasma is usually about 180 mg/dL, but it may be higher if the glomerular filtration rate is low.

Glycosuria

Glycosuria occurs when the plasma glucose level is elevated because of relative insulin deficiency (diabetes mellitus) or because of excessive glycogenolysis after physical or emotional trauma. In some individuals, the glucose transport mechanism in the renal tubules is congenitally defective, so that glycosuria is present at normal plasma glucose levels.

Factors Determining the Plasma Glucose Level

The plasma glucose level at any given time is determined by the balance between the amount of glucose

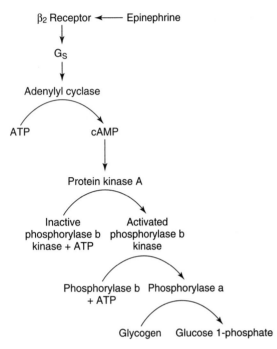

Figure 17–13. Cascade of reactions by which epinephrine activates phosphorylase. Glucagon has a similar action in liver but not in skeletal muscle.

entering the bloodstream and the amount leaving it. The principal determinants are therefore the dietary intake; the rate of entry into the cells of muscle, adipose tissue, and other organs; and the glucostatic activity of the liver (Figure 17–14). Five percent of ingested glucose is promptly converted into glycogen in the liver,

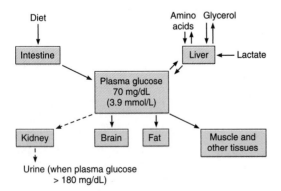

Figure 17–14. Plasma glucose homeostasis. Notice the glucostatic function of the liver, as well as the loss of glucose in the urine when the renal threshold is exceeded (dashed arrows).

and 30–40% is converted into fat. The remainder is metabolized in muscle and other tissues. During fasting, liver glycogen is broken down and the liver adds glucose to the bloodstream. With more prolonged fasting, glycogen is depleted and there is increased gluconeogenesis from amino acids and glycerol in the liver. There is a modest decline in plasma glucose to about 60 mg/dL during prolonged starvation in normal individuals, but symptoms of hypoglycemia do not occur because gluconeogenesis prevents any further fall.

Carbohydrate Homeostasis in Exercise

In a 70-kg man, carbohydrate reserves total about 2500 kcal, stored in 400 g of muscle glycogen, 100 g of liver glycogen, and 20 g of glucose in extracellular fluid. In contrast, 112,000 kcal (about 80% of body fuel supplies) are stored in fat and the remainder in protein. Resting muscle utilizes fatty acids for its metabolism, and so does muscle after exercise. In the fasting human at rest, the brain accounts for 70–80% of the glucose utilized, and red blood cells account for most of the rest.

During exercise, the caloric needs of muscle are initially met by glycogenolysis in muscle and increased uptake of glucose. Plasma glucose initially rises with increased hepatic glycogenolysis but may fall with strenuous, prolonged exercise. There is an increase in gluconeogenesis (Figure 17–15). Plasma insulin falls, and plasma glucagon and epinephrine rise. After exercise, liver glycogen is replenished by additional gluconeogenesis and a decrease in hepatic glucose output.

Regulation of Gluconeogenesis

Recent evidence indicates that **PGC-1,** a transcriptional coactivator, has a central role in regulation of hepatic gluconeogenesis; a transcriptional coactivator is a molecule that does not bind DNA by itself but provides a functional link between gene regulators and mRNA synthesis. PGC-1 is strongly induced by fasting and in streptozocin-induced diabetes, ob/ob mice, and liver insulin-receptor knockouts; gluconeogenesis is increased in all these conditions. In vitro, PGC-1 induces gluconeogenic enzymes, and its full activity requires activation of glucocorticoid receptors, which also promote gluconeogenesis.

Metabolism of Hexoses Other Than Glucose

Other hexoses that are absorbed from the intestine include galactose, which is liberated by the digestion of lactose and converted to glucose in the body; and fructose, part of which is ingested and part produced by

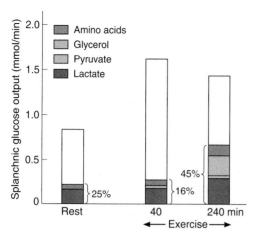

Figure 17–15. Splanchnic (hepatic) glucose output, showing output due to glycogenolysis (open bars) and output presumably due to gluconeogenesis (in brackets). The values for gluconeogenesis are measured values for splanchnic uptake of the various gluconeogenetic precursors. (Reproduced, with permission, from Felig P, Wahren J: Fuel homeostasis in exercise. N Engl J Med 1975;293:1078.)

hydrolysis of sucrose. After phosphorylation, galactose reacts with uridine diphosphoglucose (UDPG) to form uridine diphosphogalactose. The uridine diphosphogalactose is converted back to UDPG, and the UDPG functions in glycogen synthesis (Figure 17–12). This reaction is reversible, and conversion of uridine diphosphoglucose to uridine diphosphogalactose provides the galactose necessary for formation of glycolipids and mucoproteins when dietary galactose intake is inadequate. The utilization of galactose, like that of glucose, is dependent upon insulin (see Chapter 19). In the inborn error of metabolism known as **galactosemia,** there is a congenital deficiency of galactose 1-phosphate uridyl transferase, the enzyme responsible for the reaction between galactose 1-phosphate and UDPG, so that ingested galactose accumulates in the circulation. Serious disturbances of growth and development result. Treatment with galactose-free diets improves this condition without leading to galactose deficiency, because the enzyme necessary for the formation of uridine diphosphogalactose from UDPG is present.

Fructose is converted in part to fructose 6-phosphate and then metabolized via fructose 1,6-diphosphate (Figure 17–9). The enzyme catalyzing the formation of fructose 6-phosphate is hexokinase, the same enzyme that catalyzes the conversion of glucose to glucose 6-phosphate. However, much more fructose is converted to fructose 1-phosphate in a reaction cat-

alyzed by fructokinase. Most of the fructose 1-phosphate is then split into dihydroxyacetone phosphate and glyceraldehyde. The glyceraldehyde is phosphorylated, and it and the dihydroxyacetone phosphate enter the pathways for glucose metabolism. Since the reactions proceeding through phosphorylation of fructose in the 1 position can occur at a normal rate in the absence of insulin, it has been recommended that fructose be given to diabetics to replenish their carbohydrate stores. However, most of the fructose is metabolized in the intestines and liver, so its value in replenishing carbohydrate elsewhere in the body is limited.

Fructose 6-phosphate can also be phosphorylated in the 2 position, forming fructose 2,6-diphosphate. This compound is an important regulator of hepatic gluconeogenesis. When the fructose 2,6-diphosphate level is high, conversion of fructose 6-phosphate to fructose 1,6-diphosphate is facilitated, and thus breakdown of glucose to pyruvate is increased. A decreased level of fructose 2,6-diphosphate facilitates the reverse reaction and consequently aids gluconeogenesis. One of the actions of the protein kinase produced by the action of glucagon is to decrease hepatic fructose 2,6-diphosphate (see Chapter 19).

PROTEIN METABOLISM

Proteins

Proteins are made up of large numbers of amino acids (Figure 17–16) linked into chains by **peptide bonds** joining the amino group of one amino acid to the carboxyl group of the next. In addition, some proteins contain carbohydrates (glycoproteins) and lipids (lipoproteins). Smaller chains of amino acids are called **peptides** or **polypeptides.** The boundaries between peptides, polypeptides, and proteins are not well defined, but in this book, chains containing 2–10 amino acid residues are called peptides, chains containing more than 10 but fewer than 100 amino acid residues are called polypeptides, and chains containing 100 or more amino acid residues are called proteins. The term "oligopeptide," which is employed by others to refer to small peptides, is not used.

The order of the amino acids in the peptide chains is called the **primary structure** of a protein. The chains are twisted and folded in complex ways, and the term **secondary structure** of a protein refers to the spatial arrangement produced by the twisting and folding. A common secondary structure is a regular coil with 3.7 amino acid residues per turn (α-helix). Another common secondary structure is a β-sheet. An antiparallel β-sheet is formed when extended polypeptide chains fold back and forth on each other and hydrogen bonding occurs between the peptide bonds on neighboring

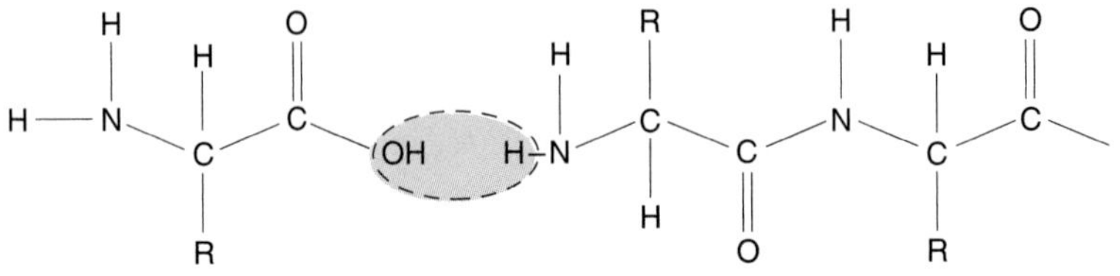

Figure 17–16. Amino acid structure and formation of peptide bonds. The dashed line shows how the peptide bonds are formed, with the production of H_2O. R, remainder of the amino acid. For example, in glycine, R = H; in gluta-mate, R = $-(CH_2)_2-COO^-$.

chains. Parallel β-sheets between polypeptide chains also occur. In the so-called ribbon models of proteins (see Figure 27–8), the α-helices are shown as coils and the β-sheets as parallel arrows.

The **tertiary structure** of a protein is the arrangement of the twisted chains into layers, crystals, or fibers. Many protein molecules are made of subunits (eg, hemoglobin; see Chapter 27), and the term **quaternary structure** is used to refer to the arrangement of the subunits.

Amino Acids

The amino acids that are found in proteins are shown in Table 17–2. These amino acids are identified by the three-letter abbreviations or the single-letter abbreviations shown in the table. Various other important amino acids such as ornithine, 5-hydroxytryptophan, L-dopa, taurine, and thyroxine (T_4) occur in the body but are not found in proteins. In higher animals, the L isomers of the amino acids are the only naturally occurring forms. The L isomers of hormones such as thyroxine are much more active than the D isomers. The amino acids are acidic, neutral, or basic in reaction, depending upon the relative proportions of free acidic (–COOH) or basic (–NH$_2$) groups in the molecule.

Some of the amino acids are **nutritionally essential amino acids,** ie, they must be obtained in the diet, whereas others can be synthesized in vivo in amounts sufficient to meet metabolic needs (see below).

The Amino Acid Pool

Although small amounts of proteins are absorbed from the gastrointestinal tract and some peptides are also absorbed, most ingested proteins are digested and their constituent amino acids absorbed. The body's own proteins are being continuously hydrolyzed to amino acids and resynthesized. The turnover rate of endogenous proteins averages 80–100 g/d, being highest in the intestinal mucosa and practically nil in collagen. The amino acids formed by endogenous protein breakdown are identical to those derived from ingested protein. With the latter, they form a common **amino acid pool** that supplies the needs of the body (Figure 17–17). In the kidney, most of the filtered amino acids are reabsorbed. During growth, the equilibrium between amino acids and body proteins shifts toward the latter, so that synthesis exceeds breakdown. At all ages, a small amount of protein is lost as hair. In women, small amounts are lost in the menstrual flow. Some small proteins are lost in the urine, and there are unreabsorbed protein digestive secretions in the stools. These losses are made up by synthesis from the amino acid pool.

Specific Metabolic Functions of Amino Acids

Thyroid hormones, catecholamines, histamine, serotonin, melatonin, and intermediates in the urea cycle are formed from specific amino acids. Methionine and cysteine provide the sulfur contained in proteins, CoA, taurine, and other biologically important compounds. Methionine is converted into S-adenosylmethionine, which is the active methylating agent in the synthesis of compounds such as epinephrine. It is a major donor of biologically labile methyl groups, but methyl groups can also be synthesized from a derivative of formic acid bound to folic acid derivatives if the diet contains adequate amounts of folic acid and cyanocobalamin.

Urinary Sulfates

Oxidation of cysteine is the ultimate source of most of the sulfates in the urine. Most of the urinary excretion is in the form of **sulfate** (SO_4^{2-}) accompanied by corresponding amounts of cation (Na^+, K^+, NH_4, or H^+). The **ethereal sulfates** in the urine are organic sulfate esters ($R-O-SO_3H$) formed in the liver from endogenous and exogenous phenols, including estrogens and other steroids, indoles, and drugs.

Table 17–2. Amino acids found in proteins. Those in bold type are the nutritionally essential amino acids. The generally accepted three-letter and one-letter abbreviations for the amino acids are shown in parentheses.

Amino acids with aliphatic side chains
  Glycine (Gly, G)
  Alanine (Ala, A)
  Valine (Val, V)
  Leucine (Leu, L)
  Isoleucine (Ile, I)
Hydroxyl-substituted amino acids
  Serine (Ser, S)
  Threonine (Thr, T)
Sulfur-containing amino acids
  Cysteine (Cys, C)
  Methionine (Met, M)
Selenocysteine[1]
Amino acids with aromatic ring side chains
  Phenylalanine (Phe, F)
  Tyrosine (Tyr, Y)
  Tryptophan (Trp, W)
Amino acids with acidic side chains, or their amides
  Aspartic acid (Asp, D)
  Asparagine (Asn, N)
  Glutamine (Gln, Q)
  Glutamic acid (Glu, E)
  γ-Carboxyglutamic acid[2] (Gla)
Amino acids with side chains containing basic groups
  Arginine[3] (Arg, R)
  Lysine (Lys, K)
  Hydroxylysine[2] (Hyl)
  Histidine[3] (His, H)
Imino acids (contain imino group but no amino group)
  Proline (Pro, P)
  4-Hydroxyproline[2] (Hyp)
  3-Hydroxyproline[2]

[1]Selenocysteine is a rare amino acid in which the sulfur of cysteine is replaced by selenium. The codon TGA is usually a stop codon, but in certain situations it codes for selenocysteine.

[2]There are no tRNAs for these four amino acids; they are formed by posttranslational modification of the corresponding unmodified amino acid in peptide linkage. There are tRNAs for selenocysteine and the remaining 20 amino acids, and they are incorporated into peptides and proteins under direct genetic control.

[3]Arginine and histidine are sometimes called "semi-essential"—they are not necessary for maintenance of nitrogen balance but are needed for normal growth.

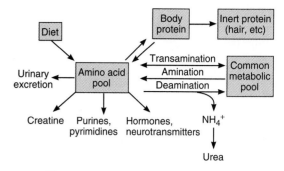

Figure 17–17. Amino acid metabolism.

Deamination, Amination, & Transamination

Interconversions between amino acids and the products of carbohydrate and fat catabolism at the level of the common metabolic pool and the citric acid cycle involve transfer, removal, or formation of amino groups. **Transamination** reactions, conversion of one amino acid to the corresponding keto acid with simultaneous conversion of another keto acid to an amino acid, occur in many tissues:

$$\text{Alanine} + \alpha\text{-Ketoglutarate} \rightleftarrows \text{Pyruvate} + \text{Glutamate}$$

The **transaminases** involved are also present in the circulation. When damage to many active cells occurs as a result of a pathologic process, serum transaminase levels rise. An example is the rise in **plasma aspartate aminotransferase (AST)** following myocardial infarction.

Oxidative deamination of amino acids occurs in the liver. An imino acid is formed by dehydrogenation, and this compound is hydrolyzed to the corresponding keto acid, with production of NH_4^+:

$$\text{Amino acid} + NAD^+ \rightarrow \text{Imino acid} + NADH + H^+$$

$$\text{Imino acid} + H_2O \rightarrow \text{Keto acid} + NH_4^+$$

NH_4^+ is in equilibrium with NH_3. Amino acids can also take up NH_4^+, forming the corresponding amide. An example is the binding of NH_4^+ in the brain by glutamate (Figure 17–18). The reverse reaction occurs in the kidney, with conversion of NH_4^+ to NH_3 and secretion of NH_3 into the urine. The NH_3 reacts with H^+ in the urine to form NH_4^+, thus permitting more H^+ to be secreted into the urine (see Chapter 38). Interconversions between the amino acid pool and the common metabolic pool are summarized in Figure 17–19. Leucine, isoleucine, phenylalanine, and tyrosine are said to be **ketogenic** because they are converted to the ketone body acetoacetate (see below). Alanine and many other amino acids are **glucogenic** or **gluconeogenic;** ie, they give rise to compounds that can readily be converted to glucose.

Urea Formation

Most of the NH_4^+ formed by deamination of amino acids in the liver is converted to urea, and the urea is excreted in the urine. The NH_4^+ forms carbamoyl phosphate, and in the mitochondria it is transferred to ornithine, forming citrulline. The enzyme involved is ornithine carbamoyltransferase. Citrulline is converted to arginine, after which urea is split off and ornithine is regenerated (urea cycle; Figure 17–20). Most of the urea is formed in the liver, and in severe liver disease the blood urea nitrogen (BUN) falls and blood NH_3 rises. Congenital deficiency of ornithine carbamoyltransferase can also lead to NH_3 intoxication, even in individuals who are heterozygous for this deficiency.

Creatine & Creatinine

Creatine is synthesized in the liver from methionine, glycine, and arginine. In skeletal muscle, it is phosphorylated to form **phosphorylcreatine** (Figure 17–21), which is an important energy store for ATP synthesis (see Chapter 3). The ATP formed by glycolysis and oxidative phosphorylation reacts with creatine to form ADP and large amounts of phosphorylcreatine. During exercise, the reaction is reversed, maintaining the supply of ATP, which is the immediate source of the energy for muscle contraction. Some athletes ingest creatine as a dietary supplement and claim that it enhances their performance in sprints and other forms of vigorous short-term exertion.

The creatinine in the urine is formed from phosphorylcreatine. Creatine is not converted directly to creatinine. The rate of creatinine excretion is relatively constant from day to day. Indeed, creatinine output is sometimes measured as a check on the accuracy of the urine collections in metabolic studies; an average daily creatinine output is calculated, and the values for the daily output of other substances are corrected to what they would have been at this creatinine output.

Creatinuria occurs normally in children, in women during and after pregnancy, and occasionally in nonpregnant women. There is very little, if any, creatine in the urine of normal men, but appreciable quantities are excreted in any condition associated with extensive muscle breakdown. Thus, creatinuria occurs in starvation, thyrotoxicosis, poorly controlled diabetes mellitus, and the various primary and secondary diseases of muscle **(myopathies).**

Purines & Pyrimidines

The physiologically important **purines** and **pyrimidines** are shown in Figure 17–22. **Nucleosides**—purines or pyrimidines combined with ribose—are components not only of a variety of coenzymes and related substances (NAD^+, $NADP^+$, ATP, UDPG, etc) but of RNA and DNA as well (Table 17–3). The structure and function of DNA and RNA and their roles in protein synthesis are discussed in Chapter 1.

Nucleic acids in the diet are digested and their constituent purines and pyrimidines absorbed, but most of the purines and pyrimidines are synthesized from amino acids, principally in the liver. The nucleotides and RNA and DNA are then synthesized. RNA is in dynamic equilibrium with the amino acid pool, but DNA, once formed, is metabolically stable throughout life.

The purines and pyrimidines released by the breakdown of nucleotides may be reused or catabolized. Minor amounts are excreted unchanged in the urine. The pyrimidines are catabolized to CO_2 and NH_3, and the purines are converted to uric acid.

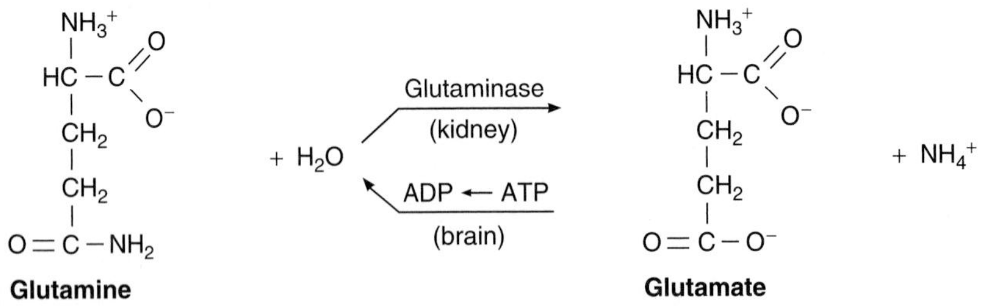

Figure 17–18. Release and uptake of NH_4^+ by interconversion of glutamine and glutamate. NH_4^+ is in equilibrium with NH_3. The reaction goes predominantly to the right in the kidney, and NH_3 is secreted into the urine. The reaction goes predominantly to the left in the brain, removing NH_3, which is toxic to nerve cells.

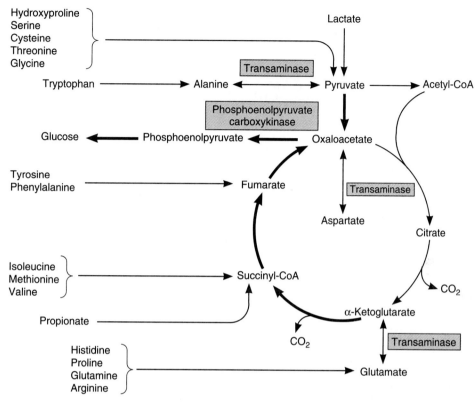

Figure 17–19. Involvement of the citric acid cycle in transamination and gluconeogenesis. The bold arrows indicate the main pathway of gluconeogenesis. (Reproduced, with permission, from Murray RK et al: *Harper's Biochemistry,* 26th ed. McGraw-Hill, 2003.)

Protein Degradation

Like protein synthesis, protein degradation is a carefully regulated, complex process. It has been estimated that overall, up to 30% of newly produced proteins are abnormal. Aged normal proteins also need to be removed as they are replaced. Conjugation of proteins to the 74-amino-acid polypeptide **ubiquitin** tickets them for degradation. This polypeptide is highly conserved and is present in species ranging from bacteria to humans. Ubiquitination of cytoplasmic proteins, including integral proteins of the endoplasmic reticulum, tickets the proteins for degradation in multisubunit proteolytic particles, the **26S proteasomes.** Ubiquitination of membrane proteins such as the growth hormone receptor marks them for degradation in lysosomes.

There is an obvious balance between the rate of production of a protein and its destruction, so ubiquitin conjugation is of major importance in cell biology and, for example, regulation of the cell cycle (see Chapter 1). The rates at which individual proteins are metabolized vary, and the body has mechanisms by which abnormal proteins are recognized and degraded more rapidly than normal body constituents. For example, abnormal hemoglobins are metabolized rapidly in individuals with congenital hemoglobinopathies (see Chapter 27).

The rate of protein degradation is decreased during hypertrophy in exercised skeletal muscle and increased during atrophy in denervated or unused skeletal muscle. In addition, the rate of protein degradation is a factor in the determination of organ size (eg, the rate of degradation of liver protein is markedly reduced during the compensatory hypertrophy that follows partial hepatectomy).

Ubiquitin tags proteins for degradation, but it can also ticket proteins to various destinations within the cell. In some of these instances, the ubiquitin is in the middle of the protein and not at its end.

Uric Acid

Uric acid is formed by the breakdown of purines and by direct synthesis from 5-phosphoribosyl pyrophos-

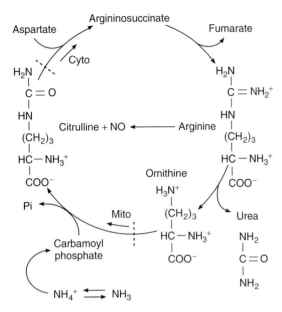

Figure 17–20. Urea cycle. Cyto, cytoplasm; Mito, mitochondrion. Note that production of carbamoyl phosphate and its conversion to citrulline occurs in the mitochondria. Note also that arginine can be converted to NO and citrulline in a reaction catalyzed by the various forms of nitric oxide synthase (NOS; see Chapter 31).

Purine nucleus

Adenine: 6-Aminopurine

Guanine: 1-Amino-6-oxypurine

Hypoxanthine: 6-Oxypurine

Xanthine: 2,6-Dioxypurine

Pyrimidine nucleus

Cytosine: 4-Amino-2-oxypyrimidine

Uracil: 2,4-Dioxypyrimidine

Thymine: 5-Methyl-2,4-dioxypyrimidine

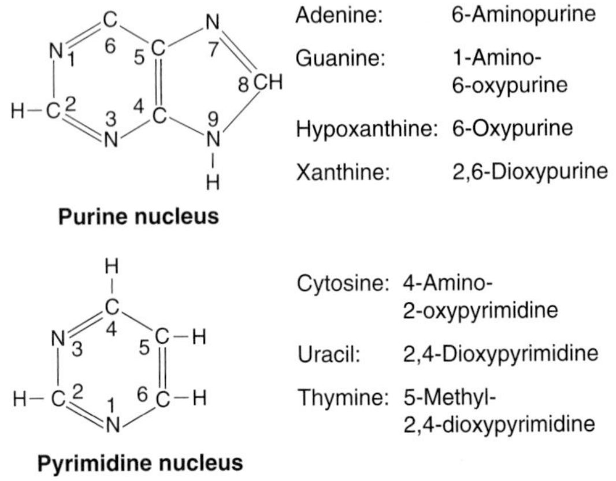

Figure 17–22. Principal physiologically important purines and pyrimidines. Oxypurines and oxypyrimidines may form enol derivatives (hydroxypurines and hydroxypyrimidines) by migration of hydrogen to the oxygen substituents.

phate (5-PRPP) and glutamine (Figure 17–23). In humans, uric acid is excreted in the urine, but in other mammals, uric acid is further oxidized to allantoin before excretion. The normal blood uric acid level in humans is approximately 4 mg/dL (0.24 mmol/L). In the kidney, uric acid is filtered, reabsorbed, and secreted.

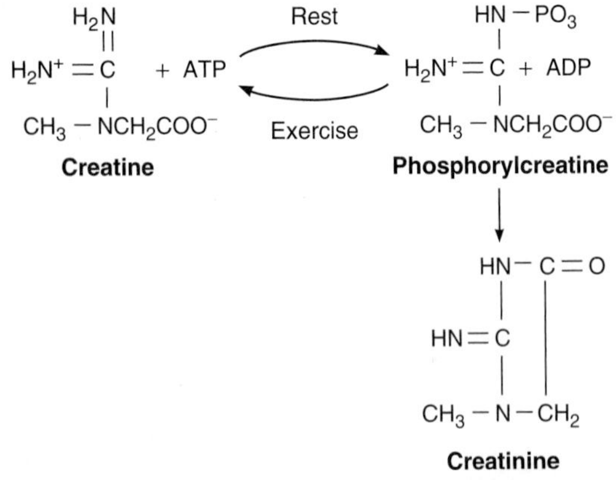

Creatine

Phosphorylcreatine

Creatinine

Figure 17–21. Creatine, phosphorylcreatine, and creatinine.

Normally, 98% of the filtered uric acid is reabsorbed and the remaining 2% makes up approximately 20% of the amount excreted. The remaining 80% comes from the tubular secretion. The uric acid excretion on a purine-free diet is about 0.5 g/24 h and on a regular diet about 1 g/24 h.

"Primary" & "Secondary" Gout

Gout is a disease characterized by recurrent attacks of arthritis; urate deposits in the joints, kidneys, and other tissues; and elevated blood and urine uric acid levels. The joint most commonly affected initially is the metatarsophalangeal joint of the great toe. There are two forms of "primary" gout. In one, uric acid production is increased because of various enzyme abnormalities. In the other, there is a selective deficit in renal tubular transport of uric acid. In "secondary" gout, the uric acid levels in the body fluids are elevated as a result of decreased excretion or increased production secondary to some other disease process. For example, excretion is decreased in patients treated with thiazide diuretics (see Chapter 38) and those with renal disease. Production is increased in leukemia and pneumonia because of increased breakdown of uric acid-rich white blood cells.

The treatment of gout is aimed at relieving the acute arthritis with drugs such as colchicine or nonsteroidal anti-inflammatory agents and decreasing the uric acid level in the blood. Colchicine does not affect uric acid metabolism, and it apparently relieves gouty attacks by inhibiting the phagocytosis of uric acid crystals by

Table 17–3. Purine- and pyrimidine-containing compounds.

Type of Compound	Components
Nucleoside	Purine or pyrimidine plus ribose or 2-deoxyribose
Nucleotide (mononucleotide)	Nucleoside plus phosphoric acid residue
Nucleic acid	Many nucleotides forming double-helical structures of two polynucleotide chains
Nucleoprotein	Nucleic acid plus one or more simple basic proteins
Contain ribose	Ribonucleic acids (RNA)
Contain 2-deoxyribose	Deoxyribonucleic acids (DNA)

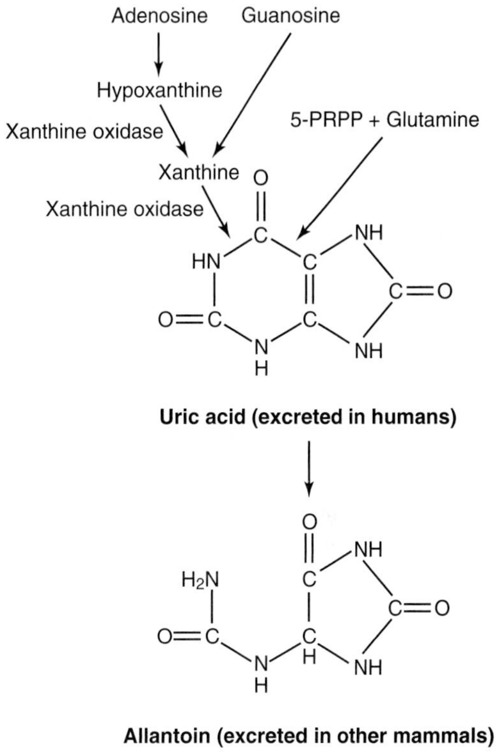

Figure 17–23. Synthesis and breakdown of uric acid. Adenosine is converted to hypoxanthine, which is then converted to xanthine, and xanthine is converted to uric acid. The latter two reactions are both catalyzed by xanthine oxidase. Guanosine is converted directly to xanthine. Xanthine oxidase is inhibited by allopurinol, one of the drugs used to treat gout.

leukocytes, a process that in some way produces the joint symptoms. Phenylbutazone and probenecid inhibit uric acid reabsorption in the renal tubules. Allopurinol, which inhibits xanthine oxidase (Figure 17–23), is one of the drugs used to decrease uric acid production.

Nitrogen Balance

A moderate daily protein intake is necessary to replace protein and amino acid losses. This requirement is not for protein itself but for the constituent amino acids, and the need can be met by feeding pure amino acids. Loss of protein and its derivatives in the stools is normally very small. Consequently, the amount of nitrogen in the urine is a reliable indicator of the amount of irreversible protein and amino acid breakdown. When the amount of nitrogen in the urine is equal to the nitrogen content of the protein in the diet, the individual is said to be in **nitrogen balance.** If protein intake is increased in a normal individual, the extra amino acids are deaminated and urea excretion increases, maintaining nitrogen balance. However, in conditions in which secretion of the catabolic hormones of the adrenal cortex is elevated or that of insulin is decreased, and during starvation and forced immobilization, nitrogen losses exceed intake and the nitrogen balance is **negative.** During growth or recovery from severe illness, or following administration of anabolic steroids such as testosterone, nitrogen intake exceeds excretion and nitrogen balance is **positive.**

When any one of the nutritionally essential amino acids necessary for synthesis of a particular protein is unavailable, the protein is not synthesized. The other amino acids that would have gone into the protein are deaminated, like other excess amino acids, and their nitrogen is excreted as urea. This is probably why nitrogen balance becomes negative whenever a single essential amino acid is omitted from the diet.

Response to Starvation

When an individual eats a diet that is low in protein but calorically adequate, excretion of urea and inorganic and ethereal sulfates declines. Uric acid excretion falls by 50%. Creatine excretion is not affected. The creatine and about half of the uric acid in the urine must therefore be the result of "wear-and-tear" processes that are unaffected by the protein intake. Total nitrogen excretion fails to fall below 3.6 g/d during protein starvation when the diet is calorically adequate because of the negative nitrogen balance produced by essential amino acid deficiencies.

On a diet that is inadequate in calories as well, urea nitrogen excretion averages about 10 g/d as proteins are catabolized for energy. Small amounts of glucose counteract this catabolism to a marked degree (**protein-sparing effect** of glucose). This protein-sparing effect is probably due for the most part to the increased insulin secretion produced by the glucose. The insulin in turn inhibits the breakdown of protein in muscle. Intravenous injection of relatively small amounts of amino acids also exerts a considerable protein-sparing effect.

Fats also spare nitrogen. During prolonged starvation, keto acids derived from fats (see below) are used by the brain and other tissues. These substances share cofactors for metabolism in muscle with three branched-chain amino acids, leucine, isoleucine, and valine, and to the extent that the fat-derived keto acids are utilized, these amino acids are apparently spared. Infusion of the non-nitrogen-containing analogs of these amino acids produces protein sparing and decreases urea and ammonia formation in patients with renal and hepatic failure.

Most of the protein burned during total starvation comes from the liver, spleen, and muscles and relatively little from the heart and brain. The blood glucose falls somewhat after liver glycogen is depleted (see above), but is maintained above levels that produce hypoglycemic symptoms by gluconeogenesis. Ketosis is present, and neutral fat is rapidly catabolized. When fat stores are used up, protein catabolism increases even further, and death soon follows. An average 70-kg man has 0.1 kg of glycogen in his liver, 0.4 kg of glycogen in his muscles, and 12 kg of fat. The glycogen is enough fuel for about 1 day of starvation. In hospitalized obese patients given nothing except water and vitamins, weight loss was observed to be about 1 kg/d for the first 10 days. It then declined and stabilized at about 0.3 kg/d. The patients did quite well for a time, although postural hypotension and attacks of acute gouty arthritis were troublesome complications in some instances. In the Irish prisoners who starved themselves to death several years ago, the average time from the start of the fast to death was about 60 days.

FAT METABOLISM

Lipids

The biologically important lipids are the fatty acids and their derivatives, the neutral fats (triglycerides), the phospholipids and related compounds, and the sterols. The triglycerides are made up of three fatty acids

bound to glycerol (Table 17–4). Naturally occurring fatty acids contain an even number of carbon atoms. They may be saturated (no double bonds) or unsaturated (dehydrogenated, with various numbers of double bonds). The phospholipids are constituents of cell

Table 17–4. Lipids.

Typical fatty acids:

Palmitic acid: $CH_3(CH_2)_{14}-\overset{\overset{\displaystyle O}{\|}}{C}-OH$

Stearic acid: $CH_3(CH_2)_{16}-\overset{\overset{\displaystyle O}{\|}}{C}-OH$

Oleic acid: $CH_3(CH_2)_7CH=CH(CH_2)_7-\overset{\overset{\displaystyle O}{\|}}{C}-OH$
(Unsaturated)

Triglycerides (triacylglycerols): Esters of glycerol and three fatty acids.

$$
\begin{array}{l}
CH_2-O-\overset{\overset{\displaystyle O}{\|}}{C}-R \\
| \\
CH-O-\overset{\overset{\displaystyle O}{\|}}{C}-R + 3H_2O \rightleftharpoons CHOH + 3HO-\overset{\overset{\displaystyle O}{\|}}{C}-R \\
| \\
CH_2-O-\overset{\overset{\displaystyle O}{\|}}{C}-R \qquad\qquad CH_2OH \\
\text{Triglyceride} \qquad\qquad\qquad \text{Glycerol}
\end{array}
$$

R = Aliphatic chain of various lengths and degrees of saturation.

Phospholipids:
A. Esters of glycerol, two fatty acids, and
   1. Phosphate = phosphatidic acid
   2. Phosphate plus inositol = phosphatidylinositol
   3. Phosphate plus choline = phosphatidylcholine (lecithin)
   4. Phosphate plus ethanolamine = phosphatidylethanolamine (cephalin)
   5. Phosphate plus serine = phosphatidylserine
B. Other phosphate-containing derivatives of glycerol
C. Sphingomyelins: Esters of fatty acid, phosphate, choline, and the amino alcohol sphingosine.

Cerebrosides: Compounds containing galactose, fatty acid, and sphingosine.

Sterols: Cholesterol and its derivatives, including steroid hormones, bile acids, and various vitamins.

membranes. The sterols include the various steroid hormones and cholesterol.

Fatty Acid Oxidation & Synthesis

In the body, fatty acids are broken down to acetyl-CoA, which enters the citric acid cycle. The main breakdown occurs in the mitochondria by β-oxidation. Fatty acid oxidation begins with activation of the fatty acid (Figure 17–24), a reaction that occurs both inside and outside the mitochondria. Medium- and short-chain fatty acids can enter the mitochondria without difficulty, but long-chain fatty acids must be bound to **carnitine** in ester linkage before they can cross the inner mitochondrial membrane. Carnitine is β-hydroxy-γ-trimethylammonium butyrate, and it is synthesized in the body from lysine and methionine. A translocase moves the fatty acid-carnitine ester into the matrix space in exchange for free carnitine. In the matrix space, the ester is hydrolyzed, making the activated fatty acid molecule available for β-oxidation and providing free carnitine for further exchange. β-Oxidation proceeds by serial removal of two carbon fragments from the fatty acid (Figure 17–24). The energy yield of this process is large. For example, catabolism of 1 mol of a six-carbon fatty acid through the citric acid cycle to CO_2 and H_2O generates 44 mol of ATP, compared with the 38 mol generated by catabolism of 1 mol of the six-carbon carbohydrate glucose.

Deficient β-oxidation of fatty acids can be produced by carnitine deficiency or genetic defects in the translocase or other enzymes involved in the transfer of long-chain fatty acids into the mitochondria. This causes cardiomyopathy. In addition, it causes **hypoketonemic hypoglycemia** with coma, a serious and often fatal condition triggered by fasting, in which glucose stores are used up because of the lack of fatty acid oxidation to provide energy, and ketone bodies (see below) are not formed in normal amounts because of the lack of adequate CoA in the liver.

Many tissues can synthesize fatty acids from acetyl-CoA. Some synthesis of long-chain fatty acids from short-chain fatty acids occurs in the mitochondria by simple reversal of the reactions shown in Figure 17–24. However, most of the synthesis of fatty acids occurs de novo from acetyl-CoA via a different pathway located principally outside the mitochondria, in the microsomes. The steps in this pathway, which involves the multienzyme complex fatty acid synthase, are summarized in Figure 17–25.

For unknown reasons, fatty acid synthesis stops in practically all cells when the chain is 16 carbon atoms long. Only small amounts of 12- and 14-carbon fatty acids are formed, and none with more than 16 carbons. Particularly in fat depots, the fatty acids are combined

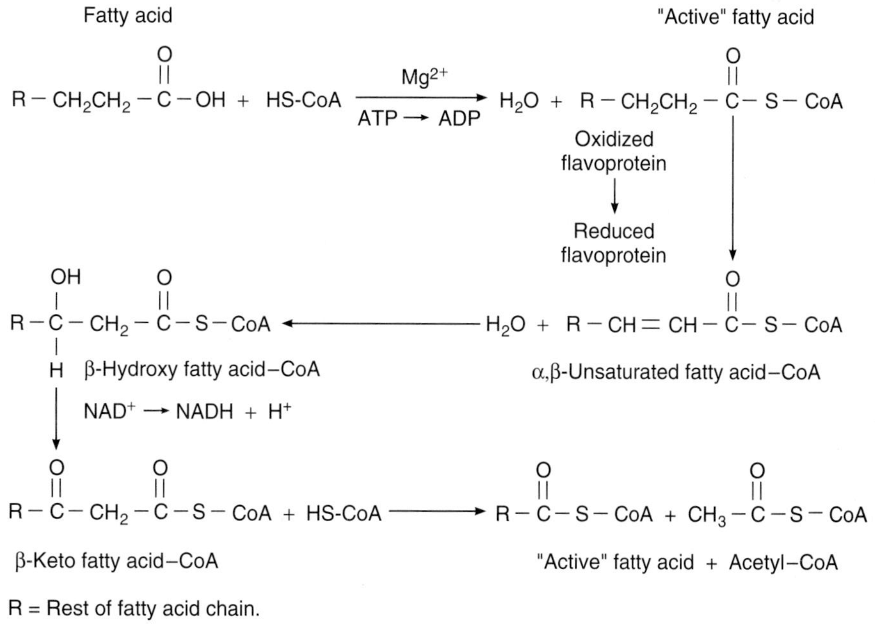

Figure 17–24. Fatty acid oxidation. This process, splitting off two carbon fragments at a time, is repeated to the end of the chain.

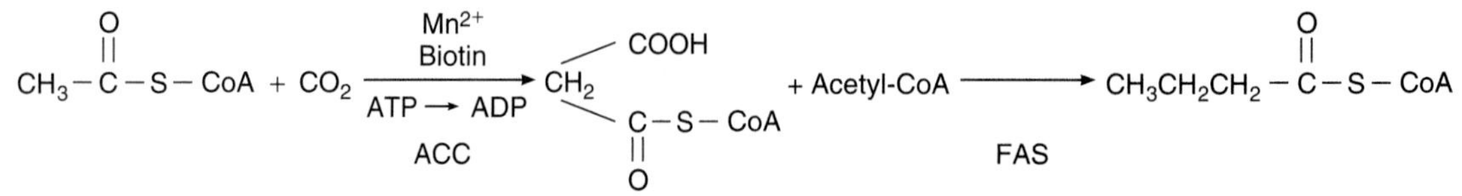

Figure 17–25. Fatty acid synthesis via the pathway found in microsomes. ACC, acetyl-CoA carboxylase; FAS, fatty acid synthase. In mammals, FAS is a dimer with multiple enzymatic functions. The reaction on the right is repeated, forming a six-carbon fatty acid, then an eight-carbon fatty acid, etc.

with glycerol to form neutral fats. This combination takes place in the mitochondria.

Ketone Bodies

In many tissues, acetyl-CoA units condense to form acetoacetyl-CoA (Figure 17–26). In the liver, which (unlike other tissues) contains a deacylase, free acetoacetate is formed. This β-keto acid is converted to β-hydroxybutyrate and acetone, and because these compounds are metabolized with difficulty in the liver, they diffuse into the circulation. Acetoacetate is also formed in the liver via the formation of 3-hydroxy-3-methylglutaryl-CoA (Figure 17–26), and this pathway is quantitatively more important than deacylation. Acetoacetate, β-hydroxybutyrate, and acetone are called **ketone bodies.** Tissues other than liver transfer CoA from succinyl-CoA to acetoacetate and metabolize the "active" acetoacetate to CO_2 and H_2O via the citric acid cycle. There are also other pathways whereby ketone bodies are metabolized. Acetone is discharged in the urine and expired air.

The normal blood ketone level in humans is low (about 1 mg/dL) and less than 1 mg is excreted per 24 hours, because the ketones are normally metabolized as rapidly as they are formed. However, if the entry of acetyl-CoA into the citric acid cycle is depressed because of a decreased supply of the products of glucose metabolism, or if the entry does not increase when the supply of acetyl-CoA increases, acetyl-CoA accumulates, the rate of condensation to acetoacetyl-CoA increases, and more acetoacetate is formed in the liver. The ability of the tissues to oxidize the ketones is soon exceeded, and they accumulate in the bloodstream **(ketosis).** Two of the three ketone bodies, acetoacetate and β-hydroxybutyrate, are anions of the moderately strong acids acetoacetic acid and β-hydroxybutyric acid. Many of their protons are buffered, reducing the decline in pH that would otherwise occur. However, the buffering capacity can be exceeded, and the metabolic acidosis that develops in conditions such as diabetic ketosis can be severe and even fatal.

Three conditions lead to deficient intracellular glucose supplies: starvation, diabetes mellitus, and a high-fat, low-carbohydrate diet. In diabetes, glucose entry into cells is impaired. When most of the caloric intake is supplied by fat, carbohydrate deficiency develops because there is no major pathway for converting fat to carbohydrate. The liver cells also become filled with fat, which damages them and displaces any glycogen that is formed. In all of these conditions, ketosis develops primarily because the supply of ketones is overabundant.

The acetone odor on the breath of children who have been vomiting is due to the ketosis of starvation. Parenteral administration of relatively small amounts of glucose abolishes the ketosis, and it is for this reason that carbohydrate is said to be **antiketogenic.**

Cellular Lipids

The lipids in cells are of two main types: **structural lipids,** which are an inherent part of the membranes and other parts of cells; and **neutral fat,** stored in the adipose cells of the fat depots. Neutral fat is mobilized during starvation, but structural lipid is preserved. The fat depots obviously vary in size, but in nonobese individuals they make up about 15% of body weight in men and 21% in women. They are not the inert lumps they were once thought to be but, rather, active dynamic tissues undergoing continuous breakdown and resynthesis. In the depots, glucose is metabolized to fatty acids, and neutral fats are synthesized. Neutral fat is also broken down, and free fatty acids are released into the circulation.

Brown Fat

A third, special type of lipid is **brown fat,** which makes up a small percentage of total body fat. Brown fat, which is somewhat more abundant in infants but is present in adults as well, is located between the scapulas, at the nape of the neck, along the great vessels in the thorax and abdomen, and in other scattered locations in the body. In brown fat depots, the fat cells as well as the blood vessels have an extensive sympathetic

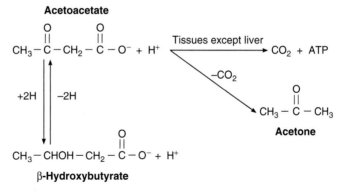

$$CH_3-\overset{\overset{\displaystyle O}{\|}}{C}-S-CoA \;+\; CH_3-\overset{\overset{\displaystyle O}{\|}}{C}-S-CoA \;\underset{\xrightarrow{\hspace{1.5cm}}}{\overset{\text{β-Ketothiolase}}{\xleftarrow{\hspace{1.5cm}}}}\; CH_3-\overset{\overset{\displaystyle O}{\|}}{C}-CH_2-\overset{\overset{\displaystyle O}{\|}}{C}-S-CoA \;+\; HS\text{-}CoA$$

2 Acetyl-CoA   **Acetoacetyl-CoA**

$$CH_3-\overset{\overset{\displaystyle O}{\|}}{C}-CH_2-\overset{\overset{\displaystyle O}{\|}}{C}-S-CoA \;+\; H_2O \;\xrightarrow[\text{(liver only)}]{\text{Deacylase}}\; CH_3-\overset{\overset{\displaystyle O}{\|}}{C}-CH_2-\overset{\overset{\displaystyle O}{\|}}{C}-O^- \;+\; H^+ \;+\; HS\text{-}CoA$$

Acetoacetyl-CoA   **Acetoacetate**

$$\text{Acetyl-CoA} \;+\; \text{Acetoacetyl-CoA} \;\longrightarrow\; CH_3-\underset{\underset{\displaystyle CH_2-COO^-}{|}}{\overset{\overset{\displaystyle OH}{|}}{C}}-CH_2-\overset{\overset{\displaystyle O}{\|}}{C}-S-CoA \;+\; H^+$$

3-Hydroxy-3-methylglutaryl-CoA
(HMG-CoA)

$$\text{HMG-CoA} \;\longrightarrow\; \text{Acetoacetate} \;+\; H^+ \;+\; \text{Acetyl-CoA}$$

Acetoacetate

$$CH_3-\overset{\overset{\displaystyle O}{\|}}{C}-CH_2-\overset{\overset{\displaystyle O}{\|}}{C}-O^- \;+\; H^+ \;\xrightarrow{\text{Tissues except liver}}\; CO_2 \;+\; ATP$$

$$\downarrow{\scriptstyle -CO_2}$$

$$CH_3-\overset{\overset{\displaystyle O}{\|}}{C}-CH_3$$

Acetone

$$+2H \updownarrow -2H$$

$$CH_3-CHOH-CH_2-\overset{\overset{\displaystyle O}{\|}}{C}-O^- \;+\; H^+$$

β-Hydroxybutyrate

Figure 17–26. Formation and metabolism of ketone bodies. Note that there are two pathways for the formation of acetoacetate.

innervation. This is in contrast to white fat depots, in which there may be innervation of some fat cells but the principal sympathetic innervation is solely on blood vessels. In addition, ordinary lipocytes have only a single large droplet of white fat, whereas brown fat cells contain several small droplets of fat. Brown fat cells also contain many mitochondria. In these mitochondria, there is the usual inward proton conductance that generates ATP (oxidative phosphorylation; see above), but there is in addition a second proton conductance that does not generate ATP. This "short-circuit" conductance depends on a 32-kDa uncoupling protein (UCP), now called UCP 1. It causes uncoupling of metabolism and generation of ATP, so that more heat is produced (Figure 17–27). Two additional uncoupling proteins, UCP2 and UCP3, have been characterized. However,

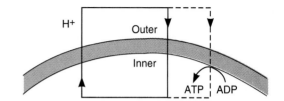

Figure 17–27. Proton transport across the mitochondrial membrane in brown fat. Protons are transported outward by the electron transport system, as in other mitochondria, but in addition to the inward proton movement that generates ATP, there is an inward proton "leak" that does not generate ATP. Consequently, metabolism of fat and generation of ATP are partially uncoupled. Compare with Figure 17–7.

they are distributed more widely than UCP1, which is found only in brown fat, and their function is unsettled. Stimulation of the sympathetic innervation to brown fat releases norepinephrine, which acts via β_3-adrenergic receptors to increase lipolysis, and increased fatty acid oxidation in the mitochondria increases heat production. Thus, variations in the activity in nerves to brown fat produce variations in the efficiency with which food is utilized and energy produced; ie, they provide a mechanism for varying the weight gained per unit of food ingested. Variations in the expression of the UCPs can also alter the efficiency of food utilization.

There is evidence that brown fat functions in this way in animals and presumably in humans adapted to cold, so that the rate of heat production in brown fat is increased. There is also a marked increase in blood flow. Nerve discharge to brown fat is also increased after eating, so that heat production is increased. Note that there are two components to the heat production after eating: the prompt specific dynamic action (SDA; see above) due to assimilation of food, and a second, somewhat slower increase in heat produced by brown fat.

Plasma Lipids & Lipid Transport

The major lipids are relatively insoluble in aqueous solutions and do not circulate in the free form. **Free fatty acids** (variously called FFA, UFA, or NEFA) are bound to albumin, whereas cholesterol, triglycerides, and phospholipids are transported in the form of lipoprotein complexes. The complexes greatly increase the solubility of the lipids. There are six families of lipoproteins (Table 17–5), which are graded in size and lipid content. The density of these lipoproteins (and consequently the speed at which they sediment in the ultracentrifuge) is inversely proportionate to their lipid content. In general, the lipoproteins consist of a hydrophobic core of triglycerides and cholesteryl esters surrounded by phospholipids and protein (Figure 17–28). The way these lipoproteins are organized into an **exogenous pathway,** which transports lipids from the intestine to the liver, and an **endogenous pathway,** which transports lipids to and from the tissues, is summarized in Figure 17–29.

The protein constituents of the lipoproteins are called **apoproteins.** The major apoproteins are called APO E, APO C, and APO B (Figure 17–29). There are two forms of APO B, a low-molecular-weight form called APO B-48, which is characteristic of the exogenous system that transports exogenous ingested lipids (see below), and a high-molecular-weight form called APO B-100, which is characteristic of the endogenous system.

Chylomicrons are formed in the intestinal mucosa during the absorption of the products of fat digestion (see Chapter 25). They are very large lipoprotein complexes that enter the circulation via the lymphatic ducts. After meals, there are so many of these particles

Table 17–5. The principal lipoproteins. The plasma lipids include these components plus free fatty acids from adipose tissue, which circulate bound to albumin.

| Lipoprotein | Size (nm) | Composition (%) | | | | | Origin |
		Protein	Free Cholesterol	Cholesteryl Esters	Triglyceride	Phospholipid	
Chylomicrons	75–1000	2	2	3	90	3	Intestine
Chylomicron remnants	30–80	. . .	. . .	. . .	. . .	. . .	Capillaries
Very low density lipoproteins (VLDL)	30–80	8	4	16	55	17	Liver and intestine
Intermediate-density lipoproteins (IDL)	25–40	10	5	25	40	20	VLDL
Low-density lipoproteins (LDL)	20	20	7	46	6	21	IDL
High-density lipoproteins (HDL)	7.5–10	50	4	16	5	25	Liver and intestine

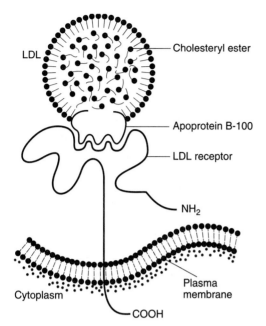

Figure 17–28. Diagrammatic representation of the structure of low-density lipoprotein (LDL), the LDL receptor, and the binding of the LDL to the receptor via APO B-100.

in the blood that the plasma may have a milky appearance **(lipemia).** The chylomicrons are cleared from the circulation by the action of **lipoprotein lipase,** which is located on the surface of the endothelium of the capillaries. The enzyme catalyzes the breakdown of the triglyceride in the chylomicrons to FFA and glycerol, which then enter adipose cells and are reesterified. Alternatively, the FFA remain in the circulation bound to albumin. Lipoprotein lipase, which requires heparin as a cofactor, also removes triglycerides from circulating **very low density lipoproteins (VLDL)** (see below). Chylomicrons and VLDL contain APO C, a complex of proteins that separates from them in the capillaries. One component of the complex, apolipoprotein C-II, activates lipoprotein lipase.

Chylomicrons depleted of their triglyceride remain in the circulation as cholesterol-rich lipoproteins called **chylomicron remnants,** which are 30–80 nm in diameter. The remnants are carried to the liver, where they bind to chylomicron remnant and LDL receptors. They are immediately internalized by receptor-mediated endocytosis (see Chapter 1), and are degraded in lysosomes.

The chylomicrons and their remnants constitute a transport system for ingested exogenous lipids (Figure 17–29). There is also an endogenous system made up

of VLDL, **intermediate-density lipoproteins (IDL), low-density lipoproteins (LDL),** and **high-density lipoproteins (HDL),** which transports triglycerides and cholesterol throughout the body. VLDL are formed in the liver and transport triglycerides formed from fatty acids and carbohydrates in the liver to extrahepatic tissues. After their triglyceride is largely removed by the action of lipoprotein lipase, they become IDL. The IDL give up phospholipids and, through the action of the plasma enzyme **lecithin-cholesterol acyl-transferase (LCAT;** Figure 17–29), pick up cholesteryl esters formed from cholesterol in the HDL. Some IDL are taken up by the liver. The remaining IDL then lose more triglyceride and protein, probably in the sinusoids of the liver, and become LDL. During this conversion, they lose APO E, but APO B-100 remains.

LDL provide cholesterol to the tissues. The cholesterol is an essential constituent in cell membranes and is used by gland cells to make steroid hormones. In the liver and most extrahepatic tissues, LDL are taken up by receptor-mediated endocytosis in coated pits (see Chapter 1). The receptors recognize the APO B-100 component of the LDL (Figure 17–28). They also bind APO E but do not bind APO B-48.

The human LDL receptor is one member of a family of receptors specialized for transport of macromolecules into cells via endocytosis in clathrin-coated pits (see Chapter 1). It is a large, complex molecule made up of a cysteine-rich region of 292 amino acid residues that binds LDL; a region of about 400 amino acid residues that is homologous to the precursor for epidermal growth factor; a 58-amino-acid region that is rich in serine and threonine and is the site of glycosylation; a stretch of 22 hydrophobic amino acid residues that spans the cell membrane; and a portion of 50 amino acid residues that projects into the cytoplasm (Figure 17–28). The gene for this protein contains 18 exons, 13 of which encode protein sequences homologous to sequences in other proteins. Thus, it appears that the LDL receptor is a mosaic protein formed by exons which code for parts of other proteins.

In the process of receptor-mediated endocytosis, each coated pit is pinched off to form a coated vesicle and then an endosome. Protein pumps in the membranes of the endosomes lower the pH in this organelle. In the case of the LDL receptor, but not the chylomicron remnant receptor, this triggers release of the LDL receptors, which recycle to the cell membrane (Figure 17–30). The endosome then fuses with a lysosome, where cholesterol formed from the cholesteryl esters by the acid lipase in the lysosomes becomes available to meet the cell's needs (Figure 17–30). The cholesterol in the cells also inhibits intracellular synthesis of cholesterol by inhibiting HMG-CoA reductase (see below), stimulates esterification of any excess cho-

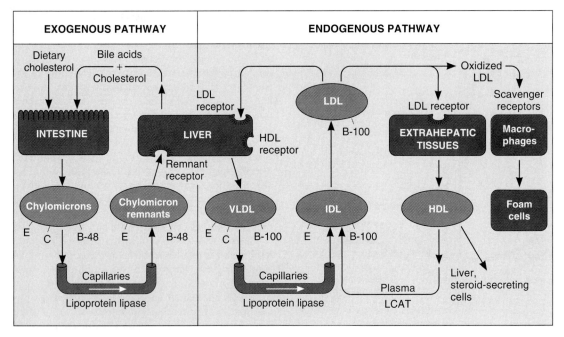

Figure 17–29. Simplified diagram of lipoprotein systems for transporting lipids in humans. In the exogenous system, chylomicrons rich in triglycerides of dietary origin are converted to chylomicron remnants rich in cholesteryl esters by the action of lipoprotein lipase. In the endogenous system, VLDL rich in triglycerides are secreted by the liver and converted to IDL and then to LDL rich in cholesteryl esters. Some of the LDL enter the subendothelial space of arteries and are oxidized, then taken up by macrophages, which become foam cells. LCAT, lecithin-cholesterol acyltransferase. The letters on the chylomicron remnants—VLDL, IDL, and LDL—identify primary apoproteins found in them. (Reproduced, with permission, from McPhee SJ, Lingappa VR, Ganong WF [editors]: *Pathophysiology of Disease*, 4th ed. McGraw-Hill, 2003.)

lesterol that is released, and inhibits a synthesis of new LDL receptors. All of these reactions provide feedback control of the amount of cholesterol in the cell.

LDL are also taken up by a lower-affinity system in the macrophages and some other cells. In addition, macrophages preferentially take up LDL that have been modified by oxidation. Oxidation can also occur in macrophages. Large doses of antioxidants such as vitamin E appear to slow the progress of atherosclerosis in experimental animals, but to date, results in humans have been disappointing. The LDL receptor on macrophages and related cells is called the **scavenger receptor.** It is different from the receptor on other cells and has a greater affinity for altered LDL. When the macrophages become overloaded with oxidized LDL, they become the "foam cells" that are seen in early atherosclerotic lesions.

In the steady state, cholesterol leaves as well as enters cells. Cholesterol appears to leave cells via one of the ABC cassette proteins (see Chapter 1), and this cholesterol is taken up by HDL. These lipoproteins are synthesized in the liver and the intestine. A separate HDL receptor has now been identified and cloned. It is found primarily in endocrine glands that make steroid hormones and in the liver. The HDL system transfers cholesterol to the liver, which is then excreted in the bile. In this way, it lowers plasma cholesterol.

APO E is synthesized by cells in the brain, spleen, lung, adrenal, ovary, and kidney, as well as the liver. Its concentration is greatly increased in injured nerves, where it appears to play a role in nerve regeneration. The apolipoprotein E gene is present in the general population in three alleles: *APO-2*, *APO-3*, and *APO-4*. *APO-4* is less common than *APO-2* and *APO-3* but is overrepresented in patients with Alzheimer's disease (see Chapter 16) and seems to predispose to this disease.

Free Fatty Acid Metabolism

Free fatty acids (FFA) are provided to fat cells and other tissues by chylomicrons and VLDL (see above). They

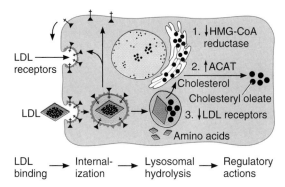

Figure 17–30. Cellular uptake and metabolism of cholesterol. LDL bind to receptors and are internalized by receptor-mediated endocytosis into endosomes with a low pH. Receptors are freed and recycle to the membrane. The cholesteryl esters enter lysosomes, where free cholesterol is released and is used for cellular processes. Cholesterol also (1) inhibits HMG-CoA reductase, (2) is then processed in part to other cholesteryl esters by the enzyme acetyl-CoA:cholesterol acyltransferase (ACAT), and (3) inhibits the formation of LDL receptors. (Courtesy of MS Brown.)

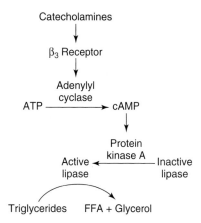

Figure 17–31. Mechanism by which catecholamines increase the activity of the hormone-sensitive lipase in adipose tissue.

are also synthesized in the fat depots in which they are stored. They circulate bound to albumin and are a major source of energy for many organs. They are used extensively in the heart, but probably all tissues, including the brain, can oxidize FFA to CO_2 and H_2O.

The supply of FFA to the tissues is regulated by two lipases. As noted above, lipoprotein lipase on the surface of the endothelium of the capillaries hydrolyzes the triglycerides in chylomicrons and VLDL, providing FFA and glycerol, which are reassembled into new triglycerides in the fat cells. The intracellular **hormone-sensitive lipase** of adipose tissue catalyzes the breakdown of stored triglycerides into glycerol and fatty acids, with the latter entering the circulation.

The hormone-sensitive lipase is converted from an inactive to an active form by cAMP via protein kinase A (Figure 17–31). The adenylyl cyclase in adipose cells is in turn activated by glucagon. It is also activated by the catecholamines norepinephrine and epinephrine via a β_3-adrenergic receptor that differs from the β_1- and β_2-adrenergic receptors. Growth hormone, glucocorticoids, and thyroid hormones increase the activity of the hormone-sensitive lipase, but they do it by a slower process that requires synthesis of new protein. Growth hormone appears to produce a protein that increases the ability of catecholamines to activate cAMP, whereas cortisol produces a protein that increases the action of cAMP. On the other hand, insulin and prostaglandin E

decrease the activity of the hormone-sensitive lipase, possibly by inhibiting the formation of cAMP.

Given the hormonal effects described in the preceding paragraph, it is not surprising that the activity of the hormone-sensitive lipase is increased by fasting and stress and decreased by feeding and insulin. Conversely, feeding increases and fasting and stress decrease the activity of lipoprotein lipase.

Cholesterol Metabolism

Cholesterol is the precursor of the steroid hormones and bile acids and is an essential constituent of cell membranes (see Chapter 1). It is found only in animals. Related sterols occur in plants, but plant sterols are not normally absorbed from the gastrointestinal tract. Most of the dietary cholesterol is contained in egg yolks and animal fat.

Cholesterol is absorbed from the intestine and incorporated into the chylomicrons formed in the intestinal mucosa. After the chylomicrons discharge their triglyceride in adipose tissue, the chylomicron remnants bring cholesterol to the liver. The liver and other tissues also synthesize cholesterol. Some of the cholesterol in the liver is excreted in the bile, both in the free form and as bile acids. Some of the biliary cholesterol is reabsorbed from the intestine. Most of the cholesterol in the liver is incorporated into VLDL and circulates in lipoprotein complexes (see above).

The biosynthesis of cholesterol from acetate is summarized in Figure 17–32. Cholesterol feeds back to inhibit its own synthesis by inhibiting **HMG-CoA reductase,** the enzyme that converts 3-hydroxy-3-methylglutaryl-coenzyme A (HMG-CoA) to mevalonic

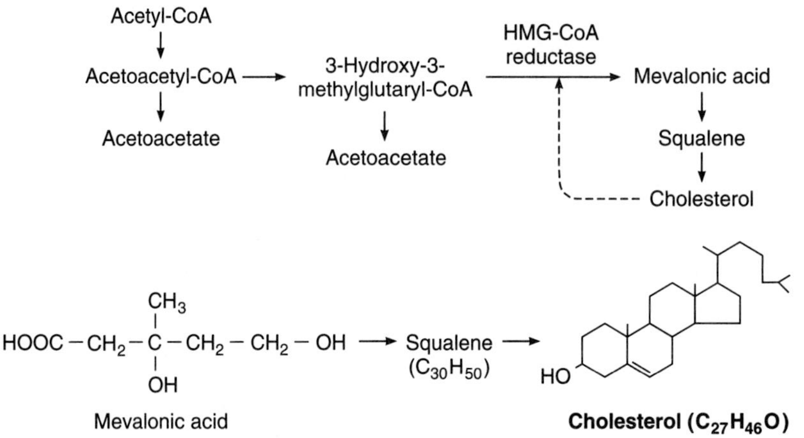

Figure 17–32. Biosynthesis of cholesterol. Six mevalonic acid molecules condense to form squalene, which is then hydroxylated and converted to cholesterol. The dashed arrow indicates feedback inhibition by cholesterol of HMG-CoA reductase, the enzyme that catalyzes mevalonic acid formation.

acid. Thus, when dietary cholesterol intake is high, hepatic cholesterol synthesis is decreased, and vice versa. However, the feedback compensation is incomplete, because a diet that is low in cholesterol and saturated fat leads to only a modest decline in circulating plasma cholesterol.

The plasma cholesterol level is decreased by thyroid hormones and estrogens, both of which increase the number of LDL receptors in the liver. Estrogens also increase plasma HDL levels. Drugs that increase the number of hepatic LDL receptors are currently being tested in animals. Plasma cholesterol is elevated by biliary obstruction and in untreated diabetes mellitus. If bile acid reabsorption in the intestine is decreased by resins such as colestipol, more cholesterol is diverted to bile acid formation. However, the drop in plasma cholesterol is relatively small because there is a compensatory increase in cholesterol synthesis. Another drug commonly used to lower plasma cholesterol is the vitamin niacin, which in large doses inhibits mobilization of free fatty acids from peripheral fat deposits, and thus reduces VLDL synthesis in the liver. However, the most effective and most commonly used cholesterol-lowering drugs are lovastatin and other **statins,** which reduce cholesterol synthesis by inhibiting HMG-CoA (Figure 17–32).

Relation to Atherosclerosis

The interest in cholesterol-lowering drugs stems from the role of cholesterol in the etiology and course of **atherosclerosis.** This extremely widespread disease predisposes to myocardial infarction, cerebral thrombosis, is-

chemic gangrene of the extremities, and other serious illnesses. It is characterized by infiltration of cholesterol into macrophages, converting them into foam cells in lesions of the arterial walls. This is followed by a complex sequence of changes involving platelets, macrophages, smooth muscle cells, growth factors, and inflammatory mediators that produces proliferative lesions which eventually ulcerate and may calcify. The lesions distort the vessels and make them rigid. Laymen often refer to this condition as arteriosclerosis, but technically, arteriosclerosis is a more general term that means loss of elasticity or hardening of the arteries from any cause. In individuals with elevated plasma cholesterol levels, there is an increased incidence of atherosclerosis and its complications. The normal range for plasma cholesterol is said to be 120–200 mg/dL, but in men, there is a clear, tight, positive correlation between the death rate from ischemic heart disease and plasma cholesterol levels above 180 mg/dL. Furthermore, it is now clear that lowering plasma cholesterol by diet and drugs slows and may even reverse the progression of atherosclerotic lesions and the complications they cause.

In evaluating plasma cholesterol levels in relation to atherosclerosis, it is important to analyze the LDL and HDL levels as well. LDL deliver cholesterol to peripheral tissues, including atheromatous lesions, and the LDL plasma concentration correlates positively with myocardial infarctions and ischemic strokes. On the other hand, as noted above, HDL pick up cholesterol from peripheral tissues and transport it to the liver, thus lowering plasma cholesterol. It is interesting that women, who have a lower incidence of myocardial infarction than men, have higher HDL levels. In addition, HDL levels are increased in individuals who exercise and those who drink one or two alcoholic drinks per day, whereas they are decreased in individuals who smoke, are obese, or live sedentary lives. Moderate drinking decreases the incidence of myocardial infarction, and obesity and smoking are risk factors that increase it. Plasma cholesterol and the incidence of cardiovascular diseases are increased in **familial hypercholesterolemia,** due to various loss-of-function mutations in the LDL receptors. Other factors predisposing to atherosclerosis are discussed in Chapter 32.

Essential Fatty Acids

Animals fed a fat-free diet fail to grow, develop skin and kidney lesions, and become infertile. Adding linolenic, linoleic, and arachidonic acids to the diet cures all the deficiency symptoms. These three acids are polyunsaturated fatty acids and because of their action are called **essential fatty acids.** Similar deficiency symptoms have not been unequivocally demonstrated in humans, but there is reason to believe that some unsaturated fats are essential dietary constituents, especially in children. Dehydrogenation of fats is known to occur in the body, but there does not appear to be any synthesis of carbon chains with the arrangement of double bonds found in the essential fatty acids.

Eicosanoids

One of the reasons, and possibly the only reason, that essential fatty acids are necessary for health is that they are the precursors of prostaglandins, prostacyclin, thromboxanes, lipoxins, leukotrienes, and related compounds. These substances are called **eicosanoids,** reflecting their origin from the 20-carbon (eicosa-) polyunsaturated fatty acid **arachidonic acid (arachidonate)** and the 20-carbon derivatives of linoleic and linolenic acids. They are produced from arachidonic acid by three separate groups of enzymes (Table 17–6).

The **prostaglandins** are a series of 20-carbon unsaturated fatty acids containing a cyclopentane ring. They were first isolated from semen but are now known to be synthesized in most and possibly in all organs in the body. The structures of some of them are shown in Figure 17–33. The prostaglandins are divided into groups—PGE and PGF, for example—on the basis of the configuration of the cyclopentane ring. The number of double bonds in the side chains is indicated by subscript numbers; for example, the E series prostaglandin shown in Figure 17–33 is PGE_2.

Prostaglandin H_2 (PGH_2) is the precursor for various other prostaglandins, for thromboxanes, and for prostacyclin. Arachidonic acid is formed from tissue phospholipids by phospholipase A_2 (Figure 17–33). It is converted to prostaglandin H_2 (PGH_2) by **prostaglandin G/H synthases** 1 and 2. These are bifunctional enzymes that have both cyclooxygenase and peroxidase activity, but they are more commonly known by the names cyclooxygenase 1 (**COX 1**) and cyclooxygenase 2 (**COX 2**). Their structures are very similar, but COX 1 is constitutive whereas COX 2 is induced by growth factors, cytokines, and tumor promoters. PGH_2 is converted to prostacyclin, thromboxanes, and prostaglandins by various tissue isomerases (Figure 17–33).

The actions of the derivatives of PGH_2 are mediated at least in large part by the serpentine receptors coupled to G proteins that are listed in Table 17–7. The table also lists examples of tissues in which the effects of the PGH_2 derivatives are predominant.

The effects of prostaglandins are multitudinous and varied. Many of them are discussed in the chapters on the systems in which they play an important role. They are particularly important in the female reproductive cycle, in parturition, in the cardiovascular system, in inflammatory responses, and in the causation of pain. New evidence indicates that they are also involved in carcinogenesis and the regulation of apoptosis and angiogenesis.

Thromboxane A_2 is synthesized by platelets and promotes vasoconstriction and platelet aggregation, whereas prostacyclin is produced in the endothelium and produces vasodilation. The important balance between thromboxane A and prostacyclin in hemostasis is discussed in Chapter 31.

Arachidonic acid is also converted to 5-hydroperoxy-eicosatetraenoic acid (5-HPETE; Figure 17–34). The 5-lipoxygenase that is involved is activated by a 5-lipoxygenase-activating protein (FLAP). 5-HPETE is converted to the **leukotrienes.** Four of the leukotrienes are **aminolipids,** which contain amino acids; leukotriene C_4 (LTC_4) contains the tripeptide glutathione, LTD_4 contains glycine and cysteine, LTE_4 contains cysteine, and LTF_4 contains cysteine and glutamic acid (Figure 17–34). In addition, arachidonic acid is converted to **lipoxins** via 15-HPETE (Figure 17–34).

The leukotrienes, thromboxanes, lipoxins, and prostaglandins have been called local hormones. They have short half-lives and are inactivated in many different tissues. They undoubtedly act mainly in the tissues at sites in which they are produced.

The leukotrienes are mediators of allergic responses and inflammation. Their release is provoked when specific allergens combine with IgE antibodies on the surfaces of mast cells. They produce bronchoconstriction, constrict arterioles, increase vascular permeability, and attract neutrophils and eosinophils to inflammatory

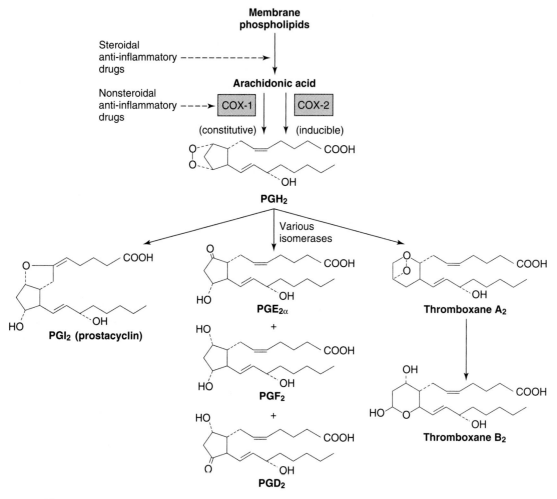

Figure 17–33. Formation of prostaglandins and thromboxanes from membrane phospholipids.

sites. In knockout mice in which the 5-lipoxygenase gene has been disrupted, development and general health are good but there is resistance to certain forms of inflammation. Diseases in which they may be involved include asthma, psoriasis, adult respiratory distress syndrome, allergic rhinitis, rheumatoid arthritis, Crohn's disease, and ulcerative colitis.

Two receptors for the leukotrienes containing cysteine, $CysLT_1$ and $CysLT_2$, have been characterized pharmacologically, though their structures are not yet known. The receptor for leukotriene B_4, BLT, is a serpentine receptor coupled to a G protein. The $CysLT_1$ receptor mediates bronchoconstriction, chemotaxis, and increased vascular permeability. The $CysLT_2$ receptor mediates constriction of pulmonary vascular smooth muscle, and the BLT receptor predominantly

mediates chemotaxis. The relation of these receptors to asthma is discussed in Chapter 37.

Lipoxin A dilates the microvasculature, and lipoxin A and lipoxin B both inhibit the cytotoxic effects of natural killer cells (see Chapter 27). However, their physiologic role is still uncertain.

12-HETE, several dihydroxy derivatives of eicosatetraenoic acid (DHTs), and several epoxyeicosatrienoic acids (EETs) are produced from arachidonic acid by **cytochrome P450 (CYP) monooxygenases** (Table 17–6). The role of these products is still unsettled, but DHTs and EETs have effects on renal excretion of salt and water that could be physiologically important. The P450s are a remarkable group of more than 300 enzymes that catalyze oxidations, epoxidations, aliphatic hydroxylations, and other reactions. In mammals, they

Table 17–6. Groups of enzymes involved in the metabolism of arachidonic acid and their eicosanoid products.

Enzymes	Products
Cyclooxygenases	Prostaglandins, prostacyclin, thromboxanes
Lipoxygenases	5-HETE, 12-HETE, 15-HETE, lipoxins, leukotrienes
CYP monooxygenases	12-HETE, EETs, DHTs

are involved not only in eicosanoid metabolism but also in steroid hormone synthesis, drug metabolism, and fatty acid oxidation. They have been divided into families and subfamilies on the basis of sequence homology (CYP1, CYP2, CYP3, etc). Twelve families have been identified in humans.

Pharmacology of Prostaglandins

Because prostaglandins play a prominent role in the genesis of pain, inflammation, and fever, pharmacologists have long sought drugs to inhibit their synthesis. Glucocorticoids inhibit phospholipase A_2 and thus inhibit the formation of all eicosanoids. A variety of nonsteroidal anti-inflammatory drugs (NSAIDs) inhibit both isomers of cyclooxygenases, inhibiting the production of PGH_2 and its derivatives (Figure 17–33). Aspirin is the best-known of these, but ibuprofen, indomethacin, and others are also used. However, there is evidence that prostaglandins synthesized by COX 2 are more involved in the production of pain and inflammation, and prostaglandins synthesized by COX 1 are more involved in protecting the gastrointestinal mucosa from ulceration (see Chapter 26). Drugs such as celecoxib (Celebrex) and rofecoxib (Vioxx) that selectively inhibit COX 2 have been developed, and in clinical use they do relieve pain and inflammation with a significantly lower incidence of gastrointestinal ulcerations and its complications than is seen with nonspecific NSAIDs.

Obesity

Obesity is the most common and most expensive nutritional problem in the USA. A convenient and reliable indicator of body fat is the **body mass index** (BMI), which is the body weight (in kilograms) divided by the square of the height (in meters). Values above 25 are abnormal. Individuals with values of 25–30 are overweight, and those with values > 30 are obese. In the USA, 55% of the population are overweight and 22% are obese. The incidence of obesity is also increasing in other countries. Indeed, the Worldwatch Institute has estimated that although starvation continues to be a problem in many parts of the world, the number of overweight people in the world is now as great as the number of underfed.

Obesity is a problem because of its complications. It is associated with accelerated atherosclerosis and an increased incidence of gallbladder and other diseases. Its association with type 2 diabetes is especially striking. As weight increases, insulin resistance increases and frank diabetes appears. At least in some cases, glucose tolerance is restored when weight is lost. This relation is discussed in more detail in Chapter 19.

The causes of the high incidence of obesity in the general population are probably multiple. Studies of twins raised apart show that there is a definite genetic component. It has been pointed out that through much of human evolution, famines were common, and mechanisms that permitted increased energy storage as fat had survival value. Now, however, food is plentiful in many countries, and the ability to gain and retain fat has become a liability. As noted above, the fundamental cause of obesity is still excess of energy intake in food over energy expenditure. If human volunteers are fed a fixed high-calorie diet, some gain weight more rapidly than others, but the slower weight gain is due to increased energy expenditure in the form of small, fidgety movements (**nonexercise activity thermogenesis; NEAT**).

Body weight generally increases at a slow but steady rate throughout adult life. Decreased physical activity is undoubtedly a factor in this increase, but decreased sensitivity to leptin may also play a role.

NUTRITION

The aim of the science of nutrition is the determination of the kinds and amounts of foods that promote health and well-being. This includes not only the problems of undernutrition but those of overnutrition, taste, and availability. However, certain substances are essential constituents of any human diet. Many of these compounds have been mentioned in previous sections of this chapter, and a brief summary of the essential and desirable dietary components is presented below.

Essential Dietary Components

An optimal diet includes, in addition to sufficient water (see Chapter 38), adequate calories, protein, fat, minerals, and vitamins (Table 17–8).

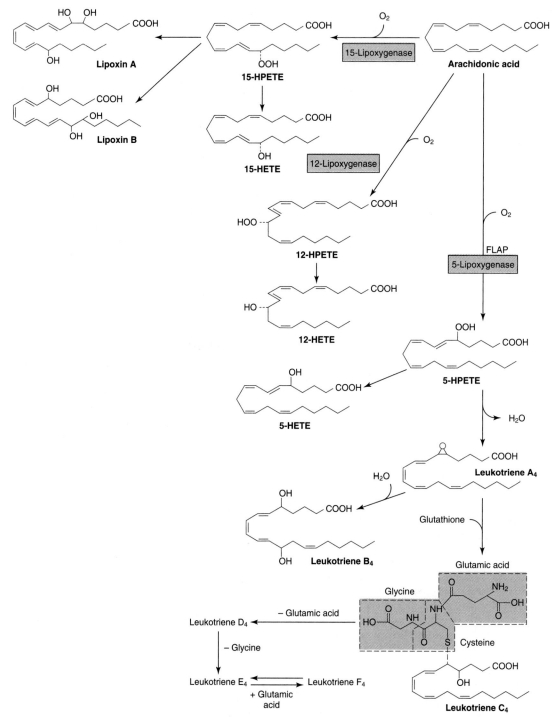

Figure 17–34. Metabolism of arachidonic acid by lipoxygenases. FLAP, 5-lipoxygenase-activating protein.

Table 17–7. Derivatives of PGH_2, their receptors, and tissues in which their effects are prominent.

Derivatives	Receptors	Tissues
Prostacyclin	IP	Endothelium, kidneys, platelets, brain
Thromboxanes	TP_α, TP_β	Platelets, vascular smooth muscle, macrophages, kidneys
Prostaglandin D_2	DP_1, DP_2	Most cells, brain, airways
Prostaglandin E_2	EP_1, EP_2, EP_3, EP_4	Brain, kidneys, vascular smooth muscle, platelets
Prostaglandin $F_{2\alpha}$	FP_α, FP_β	Uterus, vascular smooth muscle, airways

Caloric Intake & Distribution

As noted above, the caloric value of the dietary intake must be approximately equal to the energy expended if body weight is to be maintained. In addition to the 2000 kcal/d necessary to meet basal needs, 500–2500 kcal/d (or more) are required to meet the energy demands of daily activities.

The distribution of the calories among carbohydrate, protein, and fat is determined partly by physiologic factors and partly by taste and economic considerations. A daily protein intake of 1 g/kg body weight to supply the eight nutritionally essential amino acids and other amino acids is desirable. The source of the protein is also important. **Grade I proteins,** the animal proteins of meat, fish, and eggs, contain amino acids in approximately the proportions required for protein synthesis and other uses. Some of the plant proteins are also grade I, but most are **grade II** because they supply different proportions of amino acid and some lack one or more of the essential amino acids. Protein needs can be met with a mixture of grade II proteins, but the intake must be large because of the amino acid wastage.

Fat is the most compact form of food, since it supplies 9.3 kcal/g. However, often it is also the most expensive. Indeed, internationally there is a reasonably good positive correlation between fat intake and standard of living. In the past, Western diets have contained large amounts (100 g/d or more). The evidence indicating that a high unsaturated/saturated fat ratio in the diet is of value in the prevention of atherosclerosis and the current interest in preventing obesity may change this. In Central and South American Indian communities where corn (carbohydrate) is the dietary staple, adults live without ill effects for years on a very low fat intake. Therefore, provided that the needs for essential fatty acids are met, a low fat intake does not seem to be harmful, and a diet low in saturated fats is desirable.

Carbohydrate is the cheapest source of calories and provides 50% or more of the calories in most diets. In the average middle-class American diet, approximately 50% of the calories come from carbohydrate, 15% from protein, and 35% from fat. When calculating dietary needs, it is usual to meet the protein requirement first and then split the remaining calories between fat and carbohydrate, depending upon taste, income, and other factors. For example, a 65-kg man who is moderately active needs about 2800 kcal/d. He should eat at least 65 g of protein daily, supplying 267 (65×4.1) kcal. Some of this should be grade I protein. A reasonable figure for fat intake is 50–60 g. The rest of the caloric requirement can be met by supplying carbohydrate.

Mineral Requirements

A number of minerals must be ingested daily for the maintenance of health. Besides those for which recommended daily dietary allowances have been set (Table 17–8), a variety of different trace elements should be included. Trace elements are defined as elements found in tissues in minute amounts. Those believed to be essential for life, at least in experimental animals, are listed in Table 17–9. In humans, iron deficiency causes anemia (see Chapter 26). Cobalt is part of the vitamin B_{12} molecule, and vitamin B_{12} deficiency leads to megaloblastic anemia (see Chapter 25). Iodine deficiency causes thyroid disorders (see Chapter 18). Zinc deficiency causes skin ulcers, depressed immune responses, and hypogonadal dwarfism. Copper deficiency causes anemia and changes in ossification. Chromium deficiency causes insulin resistance. Fluorine deficiency increases the incidence of dental caries.

Conversely, some minerals can be toxic when present in the body in excess. For example, severe iron overload causes hemochromatosis (see Chapter 25), copper excess causes brain damage (Wilson's disease), and aluminum poisoning in patients with renal failure who are receiving dialysis treatment causes a rapidly progressive dementia that resembles Alzheimer's disease (see Chapter 16).

Sodium and potassium are also essential minerals, but listing them is academic, because it is very difficult to prepare a sodium-free or potassium-free diet. A low-salt diet is well tolerated for prolonged periods because of the compensatory mechanisms that conserve Na^+.

Table 17–8. Food and Nutrition Board, National Academy of Sciences—National Research Council recommended dietary allowances, revised 1989.[1,2]

Category	Age (years) or Condition	Weight[3] kg	Weight[3] lb	Height[3] cm	Height[3] in	Protein (g)	Fat-Soluble Vitamins				Water-Soluble Vitamins							Minerals						
							Vitamin A (µg of RE)[4]	Vitamin D (µg)[5]	Vitamin E (mg of α-TE)[6]	Vitamin K (µg)	Vitamin C (mg)	Thiamine (mg)	Riboflavin (mg)	Niacin (mg NE)[7]	Vitamin B$_6$ (mg)	Folate (µg)	Vitamin B$_{12}$ (µg)	Calcium (mg)[8]	Phosphorus (mg)	Magnesium (mg)	Iron (mg)	Zinc (mg)	Iodine (µg)	Selenium (µg)
Infants	0.0–0.5	6	13	60	24	13	375	7.5	3	5	30	0.3	0.4	5	0.3	25	0.3	400	300	40	6	5	40	10
	0.5–1.0	9	20	71	28	14	375	10	4	10	35	0.4	0.5	6	0.6	35	0.5	600	500	60	10	5	50	15
Children	1–3	13	29	90	35	16	400	10	6	15	40	0.7	0.8	9	1.0	50	0.7	800	800	80	10	10	70	20
	4–6	20	44	112	44	24	500	10	7	20	45	0.9	1.1	12	1.1	75	1.0	800	800	120	10	10	90	20
	7–10	28	62	132	52	28	700	10	7	30	45	1.0	1.2	13	1.4	100	1.4	800	800	170	10	10	120	30
Males	11–14	45	99	157	62	45	1000	10	10	45	50	1.3	1.5	17	1.7	150	2.0	1200	1200	270	12	15	150	40
	15–18	66	145	176	69	59	1000	10	10	65	60	1.5	1.8	20	2.0	200	2.0	1200	1200	400	12	15	150	50
	19–24	72	160	177	70	58	1000	10	10	70	60	1.5	1.7	19	2.0	200	2.0	1200	1200	350	10	15	150	70
	25–50	79	174	176	70	63	1000	5	10	80	60	1.5	1.7	19	2.0	200	2.0	800	800	350	10	15	150	70
	51+	77	170	173	68	63	1000	5	10	80	60	1.2	1.4	15	2.0	200	2.0	800	800	350	10	15	150	70
Females	11–14	46	101	157	62	46	800	10	8	45	50	1.1	1.3	15	1.4	150	2.0	1200	1200	280	15	12	150	45
	15–18	55	120	163	64	44	800	10	8	55	60	1.1	1.3	15	1.5	180	2.0	1200	1200	300	15	12	150	50
	19–24	58	128	164	65	46	800	10	8	60	60	1.1	1.3	15	1.6	180	2.0	1200	1200	280	15	12	150	55
	25–50	63	138	163	64	50	800	5	8	65	60	1.1	1.3	15	1.6	180	2.0	800	800	280	15	12	150	55
	51+	65	143	160	63	50	800	5	8	65	60	1.0	1.2	13	1.6	180	2.0	1200	800	280	10	12	150	55
Pregnant						60	800	10	10	65	70	1.5	1.6	17	2.2	400	2.2	1200	1200	320	30	15	175	65
Lactating	1st 6 months					65	1300	10	12	65	95	1.6	1.8	20	2.1	280	2.6	1200	1200	355	15	19	200	75
	2nd 6 months					62	1200	10	11	65	90	1.6	1.7	20	2.1	260	2.6	1200	1200	340	15	16	200	75

[1]Modified and reproduced, with permission, from *Recommended Dietary Allowances*, 10th ed, National Academy Press, 1989. Copyright © 1989 by the National Academy of Sciences. Courtesy of the National Academy Press, Washington, D.C.

[2]The allowances, expressed as average daily intakes over time, are intended to provide for individual variations among most normal persons as they live in the United States under usual environmental stresses. Diets should be based on a variety of common foods to provide other nutrients for which human requirements have been less well defined.

[3]Weights and heights of Reference Adults are actual medians for the U.S. population of the designated age. The median weights and heights of those under 19 years of age are not necessarily the ideal values.

[4]Retinol equivalents. 1 retinol equivalent = 1 mg of retinol or 6 µg of β-carotere.

[5]As cholecalciferol. 10 µg of cholecalciferol = 400 IU of vitamin D.

[6]α-Tocopherol equivalents. 1 mg of *d*-α tocopherol = 1 α-TE.

[7]1 NE (niacin equivalent) is equal to 1 mg of niacin or 60 mg of dietary tryptophan.

[8]Calcium values increased after age 50.

Table 17–9. Trace elements believed essential for life.

Arsenic	Manganese
Chromium	Molybdenum
Cobalt	Nickel
Copper	Selenium
Fluorine	Silicon
Iodine	Vanadium
Iron	Zinc

Vitamins

Vitamins were discovered when it was observed that diets adequate in calories, essential amino acids, fats, and minerals failed to maintain health. The term **vitamin** has now come to refer to any organic dietary constituent necessary for life, health, and growth that does not function by supplying energy.

Because there are minor differences in metabolism between mammalian species, some substances are vitamins in one species and not in another. The sources and functions of the major vitamins in humans are listed in Table 17–10 and the recommended daily dietary allowances in Table 17–8. Most vitamins have important functions in intermediary metabolism or the special metabolism of the various organ systems. Those that are water-soluble (vitamin B complex, vitamin C) are easily absorbed, but the fat-soluble vitamins (vitamins A, D, E, and K) are poorly absorbed in the absence of bile or pancreatic lipase. Some dietary fat intake is necessary for their absorption, and in obstructive jaundice or disease of the exocrine pancreas, deficiencies of the fat-soluble vitamins can develop even if their intake is adequate (see Chapter 26). Vitamin A and vitamin D are bound to transfer proteins in the circulation. The α-tocopherol form of vitamin E is normally bound to chylomicrons and then transferred in the liver to VLDL for distribution to tissues by an α-tocopherol transfer protein. When this protein is abnormal due to mutation of its gene in humans, there is cellular deficiency of vitamin E and the development of a condition resembling Friedreich's ataxia. Two Na^+-dependent L-ascorbic acid transporters have recently been isolated. One is found in the kidneys, intestines, and liver and the other in the brain and eyes.

The diseases caused by deficiency of each of the vitamins are listed in Table 17–10. It is worth remembering, however, particularly in view of the advertising campaigns for vitamin pills and supplements, that very large doses of the fat-soluble vitamins are definitely toxic. **Hypervitaminosis A** is characterized by anorexia, headache, hepatosplenomegaly, irritability, scaly dermatitis, patchy loss of hair, bone pain, and hyperostosis. Acute vitamin A intoxication was first described by Arctic explorers, who developed headache, diarrhea, and dizziness after eating polar bear liver. The liver of this animal is particularly rich in vitamin A. **Hypervitaminosis D** is associated with weight loss, calcification of many soft tissues, and eventual renal failure. **Hypervitaminosis K** is characterized by gastrointestinal disturbances and anemia. Large doses of water-soluble vitamins have been thought to be less likely to cause problems because they can be rapidly cleared from the body. However, it has now been demonstrated that ingestion of megadoses of pyridoxine (vitamin B_6) can produce peripheral neuropathy.

Table 17–10. Vitamins essential or probably essential to human nutrition. Choline is synthesized in the body in small amounts, but it has recently been added to the list of essential nutrients.

Vitamin	Action	Deficiency Symptoms	Sources	Chemistry
A (A_1, A_2)	Constituents of visual pigments (see Chapter 8); necessary for fetal development and for cell development throughout life	Night blindness, dry skin	Yellow vegetables and fruit	Vitamin A_1 alcohol (retinol).

(continues)

Table 17-10. Vitamins essential or probably essential to human nutrition. Choline is synthesized in the body in small amounts, but it has recently been added to the list of essential nutrients. *(continued)*

Vitamin	Action	Deficiency Symptoms	Sources	Chemistry
B complex Thiamin (vitamin B_1)	Cofactor in de-carboxylations	Beriberi, neuritis	Liver, un-refined cereal grains	
Riboflavin (vitamin B_2)	Constituent of flavoproteins	Glossitis, cheilosis	Liver, milk	
Niacin	Constituent of NAD^+ and $NADP^+$	Pellagra	Yeast, lean meat, liver	Can be synthesized in body from tryptophan.
Pyridoxine (vitamin B_6)	Forms pros-thetic group of certain decar-boxylases and transaminases. Converted in body into pyri-doxal phos-phate and pyridoxamine phosphate	Convul-sions, hyper-irritability	Yeast, wheat, corn, liver	
Pantothenic acid	Constituent of CoA	Dermatitis, enteritis, alopecia, adrenal in-sufficiency	Eggs, liver, yeast	
Biotin	Catalyzes CO_2 "fixation" (in fatty acid synthesis, etc)	Dermatitis, enteritis	Egg yolk, liver, to-matoes	

Table 17–10. Vitamins essential or probably essential to human nutrition. Choline is synthesized in the body in small amounts, but it has recently been added to the list of essential nutrients. *(continued)*

Vitamin	Action	Deficiency Symptoms	Sources	Chemistry
Folates (folic acid) and related compounds	Coenzymes for "1-carbon" transfer; involved in methylating reactions	Sprue, anemia. Neural tube defects in children born to folate-deficient women	Leafy green vegetables	Folic acid
Cyanocobalamin (vitamin B_{12})	Coenzyme in amino acid metabolism. Stimulates erythropoiesis	Pernicious anemia (see Chapter 26)	Liver, meat, eggs, milk	Complex of four substituted pyrrole rings around a cobalt atom (see Chapter 26)
C	Maintains prosthetic metal ions in their reduced form; scavenges free radicals	Scurvy	Citrus fruits, leafy green vegetables	Ascorbic acid (synthesized in most mammals except guinea pigs and primates, including humans).
D group	Increase intestinal absorption of calcium and phosphate (see Chapter 21)	Rickets	Fish liver	Family of sterols (see Chapter 21)
E group	Antioxidants; cofactors in electron transport in cytochrome chain?	Ataxia and other symptoms and signs of spinocerebellar dysfunction	Milk, eggs, meat, leafy vegetables	α-Tocopherol; β- and γ-tocopherol also active.
K group	Catalyze γ carboxylation of glutamic acid residues on various proteins concerned with blood clotting	Hemorrhagic phenomena	Leafy green vegetables	Vitamin K_3; a large number of similar compounds have biological activity.

The Thyroid Gland

18

INTRODUCTION

The thyroid gland maintains the level of metabolism in the tissues that is optimal for their normal function. Thyroid hormones stimulate the O_2 consumption of most of the cells in the body, help regulate lipid and carbohydrate metabolism, and are necessary for normal growth and maturation. The thyroid gland is not essential for life, but its absence causes mental and physical slowing, poor resistance to cold, and, in children, mental retardation and dwarfism. Conversely, excess thyroid secretion leads to body wasting, nervousness, tachycardia, tremor, and excess heat production. Thyroid function is controlled by the thyroid-stimulating hormone (TSH, thyrotropin) of the anterior pituitary. The secretion of this tropic hormone is in turn regulated in part by thyrotropin-releasing hormone (TRH) from the hypothalamus and is subject to negative feedback control by high circulating levels of thyroid hormones acting on the anterior pituitary and the hypothalamus. In this way, changes in the internal and external environment bring about appropriate adjustments in the rate of thyroid secretion.

In mammals, the thyroid gland also secretes calcitonin, a calcium-lowering hormone. This hormone is discussed in Chapter 21.

ANATOMIC CONSIDERATIONS

Thyroid tissue is present in all vertebrates. In mammals, the thyroid originates from an evagination of the floor of the pharynx, and a **thyroglossal duct** marking the path of the thyroid from the tongue to the neck sometimes persists in the adult. The two lobes of the human thyroid are connected by a bridge of tissue, the **thyroid isthmus,** and there is sometimes a **pyramidal lobe** arising from the isthmus in front of the larynx (Figure 18–1). The gland is well vascularized, and the thyroid has one of the highest rates of blood flow per gram of tissue of any organ in the body.

The thyroid is made up of multiple **acini (follicles).** Each spherical follicle is surrounded by a single layer of cells and filled with pink-staining proteinaceous material called **colloid.** When the gland is inactive, the colloid is abundant, the follicles are large, and the cells lining them are flat. When the gland is active, the follicles are small, the cells are cuboid or columnar, and the edge of the colloid is scalloped, forming many small "reabsorption lacunae" (Figure 18–2).

Microvilli project into the colloid from the apexes of the thyroid cells, and canaliculi extend into them. There is a prominent endoplasmic reticulum, a feature common to most glandular cells, and secretory droplets of thyroglobulin are seen (Figure 18–3). The individual thyroid cells rest on a basal lamina that separates them from the adjacent capillaries. The capillaries are fenestrated (Figure 18–3), like those of other endocrine glands (see Chapter 30).

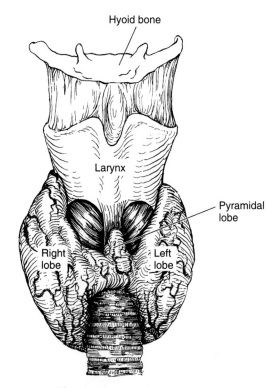

Figure 18–1. The human thyroid.

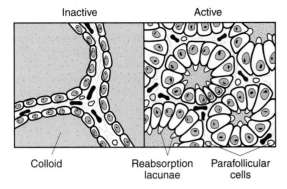

Inactive   Active

Colloid   Reabsorption   Parafollicular
            lacunae   cells

Figure 18–2. Thyroid histology. Note the small, punched-out "reabsorption lacunae" in the colloid next to the cells in the active gland.

FORMATION & SECRETION OF THYROID HORMONES

Chemistry

The principal hormones secreted by the thyroid are **thyroxine (T_4)** and **triiodothyronine (T_3)**. T_3 is also formed in the peripheral tissues by deiodination of T_4 (see below). Both hormones are iodine-containing amino acids (Figure 18–4). Small amounts of reverse triiodothyronine (3,3',5'-triiodothyronine, RT_3) and other compounds are also found in thyroid venous blood. T_3 is more active than T_4, whereas RT_3 is inactive. The naturally occurring forms of T_4 and its congeners with an asymmetric carbon atom are the L isomers. D-Thyroxine has only a small fraction of the activity of the L form.

Iodine Metabolism

Iodine is a raw material essential for thyroid hormone synthesis. Ingested iodine is converted to iodide and absorbed. The fate of the absorbed I^- is summarized in Figure 18–5. The minimum daily iodine intake that will maintain normal thyroid function is 150 µg in adults (see Table 17–8), but in the United States the average dietary intake is approximately 500 µg/d. The normal plasma I^- level is about 0.3 µg/dL, and I^- is distributed in a "space" of approximately 25 L (35% of body weight). The principal organs that take up the I^- are the thyroid, which uses it to make thyroid hormones, and the kidneys, which excrete it in the urine. About 120 µg/d enter the thyroid at normal rates of thyroid hormone synthesis and secretion. The thyroid secretes 80 µg/d as iodine in T_3 and T_4. Forty micrograms of I^- per day diffuses into the ECF. The secreted

T_3 and T_4 are metabolized in the liver and other tissues, with the release of 60 µg of I^- per day into the ECF. Some thyroid hormone derivatives are excreted in the bile, and some of the iodine in them is reabsorbed (enterohepatic circulation), but there is a net loss of I^- in the stool of approximately 20 µg/d. The total amount of I^- entering the ECF is thus 500 + 40 + 60, or 600 µg/d; 20% of this I^- enters the thyroid, whereas 80% is excreted in the urine.

The Na$^+$/I$^-$ Symporter

The thyroid cell membranes facing the capillaries contain a **symporter,** or iodide pump, that transports Na$^+$ and I^- into the cells against the electrochemical gradient for I^-. This Na$^+$/I$^-$ symporter (NIS) is capable of producing intracellular I^- concentrations that are 20–40 times as great as the concentration in plasma. The process involved is secondary active transport (see Chapter 1), with the energy provided by transport of

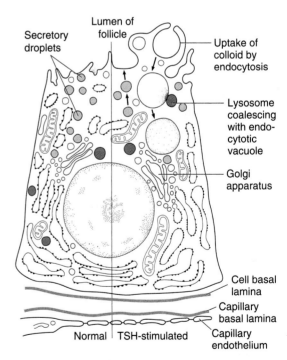

Figure 18–3. Thyroid cell. **Left:** Normal pattern. **Right:** After TSH stimulation. The arrows on the left show the secretion of thyroglobulin into the colloid. On the right, endocytosis of the colloid and merging of a colloid-containing vacuole with a lysosome are shown. The cell rests on a capillary with gaps (fenestrations) in the endothelial wall.

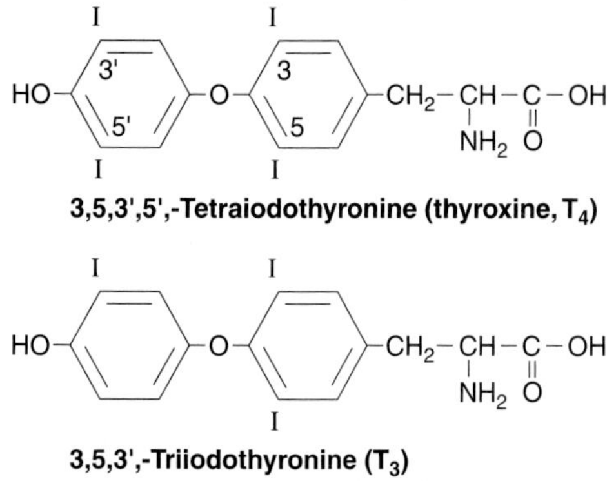

I I
HO—[3' / 5']—O—[3 / 5]—CH$_2$—CH—C—OH
 I I NH$_2$ O
 I I

3,5,3',5',-Tetraiodothyronine (thyroxine, T$_4$)

I I
HO—[ ]—O—[ ]—CH$_2$—CH—C—OH
 I NH$_2$ O

3,5,3',-Triiodothyronine (T$_3$)

Figure 18–4. Thyroid hormones. The numbers in the rings in the T$_4$ formula indicate the number of positions in the molecule. RT$_3$ is 3,3',5'-triiodothyronine.

Na$^+$ out of thyroid cells by Na$^+$-K$^+$ ATPase. I$^-$ moves by diffusion, possibly along an I$^-$ channel, to the colloid. The NIS protein has 12 transmembrane domains, and its amino and carboxyl terminals are inside the cell.

The relation of thyroid function to iodide is unique; as discussed in more detail below, iodide is essential for normal thyroid function, but iodide deficiency and iodide excess both inhibit thyroid function.

The salivary glands, the gastric mucosa, the placenta, the ciliary body of the eye, the choroid plexus, and the mammary glands also contain the NIS and transport iodide against a concentration gradient, but their uptake is not affected by TSH. The mammary glands also bind the iodine; diiodotyrosine is formed in mammary tissue, but T$_4$ and T$_3$ are not. The physiologic significance of all these extrathyroidal iodide-concentrating mechanisms is obscure.

Thyroid Hormone Synthesis

In the thyroid gland, iodide is oxidized to iodine and bound to the 3 position of tyrosine residues that are part of the thyroglobulin molecule in the colloid (Figure 18–6). **Thyroglobulin** is a glycoprotein made up of two subunits and has a molecular weight of 660,000. It contains 10% carbohydrate by weight. It also contains 123 tyrosine residues, but only four to eight of these are normally incorporated into thyroid hormones. Thyroglobulin is synthesized in the thyroid cells and secreted into the colloid by exocytosis of granules that also contain **thyroid peroxidase,** the enzyme that catalyzes the oxidation of I$^-$ and its binding. The thyroid hormones remain part of the thyroglobulin molecules until secreted. When they are secreted, colloid is ingested by the thyroid cells, the peptide

bonds are hydrolyzed, and free T$_4$ and T$_3$ are discharged into the capillaries (see below). The thyroid cells thus have three functions: They collect and transport iodine; they synthesize thyroglobulin and secrete it into the colloid; and they remove the thyroid hormones from thyroglobulin and secrete them into the circulation.

In the process of hormone synthesis, the first product is monoiodotyrosine (MIT). MIT is next iodinated in the 5 position to form diiodotyrosine (DIT). Two DIT molecules then undergo an oxidative condensation to form T$_4$ with the elimination of the alanine side chain from the molecule that forms the outer ring. There are two theories of how this **coupling reaction** occurs. One holds that the coupling occurs with both DIT molecules attached to thyroglobulin (intramolecular coupling). The other holds that the DIT that forms the outer ring is first detached from thyroglobulin (intermolecular coupling). In any case, thyroid peroxidase is probably involved in coupling as well as iodination. T$_3$ is probably formed by condensation of MIT with DIT. A small amount of RT$_3$ is also formed, probably by condensation of DIT with MIT. In the normal human thyroid, the average distribution of iodinated compounds is 23% MIT, 33% DIT, 35% T$_4$, and 7% T$_3$. Only traces of RT$_3$ and other components are present.

Secretion

The human thyroid secretes about 80 μg (103 nmol) of T$_4$, 4 μg (7 nmol) of T$_3$, and 2 μg (3.5 nmol) of RT$_3$ per day (Figure 18–7). However, MIT and DIT are not secreted. The thyroid cells ingest colloid by endocytosis

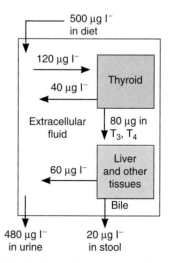

Figure 18–5. Iodine metabolism.

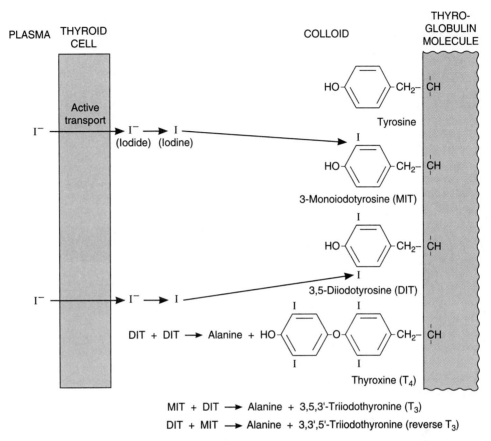

Figure 18–6. Outline of thyroid hormone biosynthesis. Iodination of tyrosine takes place at the apical border of the thyroid cells while the molecules are bound in peptide linkage in thyroglobulin.

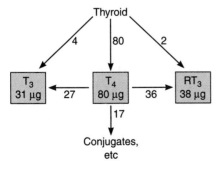

Figure 18–7. Secretion and interconversion of thyroid hormones in normal adult humans. Figures are in micrograms per day. Note that most of the T_3 and RT_3 are formed from T_4 deiodination in the tissues, and only small amounts are secreted by the thyroid.

(see Chapter 1). This chewing away at the edge of the colloid produces the reabsorption lacunae seen in active glands (Figure 18–2). In the cells, the globules of colloid merge with lysosomes (Figure 18–3). The peptide bonds between the iodinated residues and the thyroglobulin are broken by proteases in the lysosomes, and T_4, T_3, DIT, and MIT are liberated into the cytoplasm. The iodinated tyrosines are deiodinated by a microsomal **iodotyrosine deiodinase.** This enzyme does not attack iodinated thyronines, and T_4 and T_3 pass into the circulation. The iodine liberated by deiodination of MIT and DIT is reutilized in the gland and normally provides about twice as much iodide for hormone synthesis as the iodide pump does. In patients with congenital absence of the iodotyrosine deiodinase, MIT and DIT appear in the urine and there are symptoms of iodine deficiency (see below).

TRANSPORT & METABOLISM OF THYROID HORMONES

Protein Binding

The normal total **plasma T_4** level in adults is approximately 8 μg/dL (103 nmol/L), and the **plasma T_3** level is approximately 0.15 μg/dL (2.3 nmol/L). Large amounts of both are bound to plasma proteins. Both are measured by radioimmunoassay.

The free thyroid hormones in plasma are in equilibrium with the protein-bound thyroid hormones in plasma and in tissues (Figure 18–8). Free thyroid hormones are added to the circulating pool by the thyroid. It is the free thyroid hormones in plasma that are physiologically active and that inhibit pituitary secretion of TSH.

Many other hormones are bound to plasma proteins, and there is an equilibrium between their free active forms and their bound inactive forms in the circulation. The function of protein-binding appears to be maintenance of a large pool of readily available free hormone. In addition, at least for T_3, hormone binding prevents excess uptake by the first cells encountered and promotes uniform tissue distribution.

Capacity & Affinity of Plasma Proteins for Thyroid Hormones

The plasma proteins that bind thyroid hormones are **albumin;** a prealbumin formerly called **thyroxine-binding prealbumin (TBPA)** and now called **transthyretin;** and a globulin with an electrophoretic mobility between those of α_1- and α_2-globulin, **thyroxine-binding globulin (TBG).** Of the three proteins, albumin has the largest **capacity** to bind T_4—ie, it can bind the most T_4 before becoming saturated—and TBG has the smallest capacity. However, the

affinities of the proteins for T_4—ie, the avidity with which they bind T_4 under physiologic conditions—are such that most of the circulating T_4 is bound to TBG (Table 18–1), with over a third of the binding sites on the protein occupied. Smaller amounts of T_4 are bound to transthyretin and albumin. The half-life of transthyretin is 2 days, that of TBG is 5 days, and that of albumin 13 days.

Normally, 99.98% of the T_4 in plasma is bound; the free T_4 level is only about 2 ng/dL. There is very little T_4 in the urine. Its biologic half-life is long (about 6–7 days), and its volume of distribution is less than that of ECF (10 L, or about 15% of body weight). All of these properties are characteristic of a substance that is strongly bound to protein.

T_3 is not bound to quite as great an extent; of the 0.15 μg/dL normally found in plasma, 0.2% (0.3 ng/dL) is free. The remaining 99.8% is protein-bound, 46% to TBG and most of the remainder to albumin, with very little binding to transthyretin (Table 18–1). The lesser binding of T_3 correlates with the facts that T_3 has a shorter half-life than T_4 and that its action on the tissues is much more rapid. RT_3 also binds to TBG.

Until recently, it was difficult to measure free T_4 and free T_3 in blood directly, so indexes were used for both of them. A **free thyroxine index (FT_4I)** can be calculated by measuring total T_4 and multiplying it by the percentage of labeled T_4 taken up by a resin or charcoal added to the plasma specimen; resin and charcoal bind the free T_4 in the plasma. A **free triiodothyronine index (FT_3I)** can be calculated in a similar fashion. Methods are now available for direct assay of free T_4 and free T_3 that appear to be accurate, and these assays are replacing indexes.

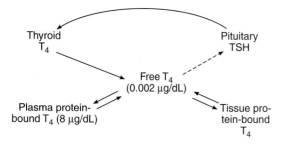

Figure 18–8. Distribution of T_4 in the body. The distribution of T_3 is similar. The dashed arrow indicates inhibition of TSH secretion by increases in the free T_4 level in ECF. Approximate concentrations in human blood are shown in parentheses.

Table 18–1. Binding of thyroid hormones to plasma proteins in normal adult humans.

Protein	Plasma Concentration (mg/dL)	Amount of Circulating Hormone Bound (%)	
		T_4	T_3
Thyroxine-binding globulin (TBG)	2	67	46
Transthyretin (thyroxine-binding prealbumin, TBPA)	15	20	1
Albumin	3500	13	53

Fluctuations in Binding

When there is a sudden, sustained increase in the concentration of thyroid-binding proteins in the plasma, the concentration of free thyroid hormones falls. This change is temporary, however, because the decrease in the concentration of free thyroid hormones in the circulation stimulates TSH secretion, which in turn causes an increase in the production of free thyroid hormones. A new equilibrium is eventually reached at which the total quantity of thyroid hormones in the blood is elevated but the concentration of free hormones, the rate of their metabolism, and the rate of TSH secretion are normal. Corresponding changes in the opposite direction occur when the concentration of thyroid-binding protein is reduced. Consequently, patients with elevated or decreased concentrations of binding proteins, particularly TBG, are neither hyper- nor hypothyroid; ie, they are **euthyroid.**

TBG levels are elevated in estrogen-treated patients and during pregnancy, as well as after treatment with various drugs (Table 18–2). They are depressed by glucocorticoids, androgens, the weak androgen danazol, and the cancer chemotherapeutic agent L-asparaginase. A number of other drugs, including salicylates, the anticonvulsant phenytoin, and the cancer chemotherapeutic agents mitotane (*o,p'*-DDD) and 5-fluorouracil inhibit binding of T_4 and T_3 to TBG and consequently produce changes similar to those produced by a decrease in TBG concentration. Changes in total plasma T_4 and T_3 can also be produced by changes in plasma concentrations of albumin and prealbumin.

Metabolism of Thyroid Hormones

T_4 and T_3 are deiodinated in the liver, the kidneys, and many other tissues. One-third of the circulating T_4 is normally converted to T_3 in adult humans, and 45% is converted to RT_3. As shown in Figure 18–7, only about 13% of the circulating T_3 is secreted by the thyroid and 87% is formed by deiodination of T_4; similarly, only 5% of the circulating RT_3 is secreted by the thyroid and 95% is formed by deiodination of T_4. It should be noted as well that there are marked differences in the ratio of T_3 to T_4 in various tissues. Two tissues that have very high T_3/T_4 ratios are the pituitary and the cerebral cortex.

Three different deiodinases act on thyroid hormones: D_1, D_2, and D_3. All are unique in that they contain the rare amino acid selenocysteine, with selenium in place of sulfur (see Table 17–2), and selenium is essential for their enzymic activity. D_1 is present in high concentration in the liver, kidneys, thyroid, and pituitary. It appears to be primarily responsible for monitoring the formation of T_3 from T_4 in the periphery. D_2 is present in the brain, pituitary, and brown fat. It also contributes to the formation of T_3. In the brain, it is located in astroglia and produces a supply of T_3 to neurons. D_3 is also present in the brain and in reproductive tissues. It acts only on the 5 position at T_4 and T_3 and is probably the main source of rT_3 in the blood and tissues. Overall, the deiodinases appear to be responsible for maintaining the differences in T_3/T_4 ratios in the various tissues in the body.

Some of the T_4 and T_3 is further converted to deiodotyrosines by deiodinases. T_4 and T_3 are also conjugated in the liver to form sulfates and glucuronides. These conjugates enter the bile and pass into the intestine. The thyroid conjugates are hydrolyzed, and some are reabsorbed (enterohepatic circulation), but some are excreted in the stool. In addition, some T_4 and T_3 pass directly from the circulation to the intestinal lumen. The iodide lost by these routes amounts to about 4% of the total daily iodide loss.

Table 18–2. Effect of variations in the concentrations of thyroid hormone-binding proteins in the plasma on various parameters of thyroid function after equilibrium has been reached.

Condition	Concentrations of Binding Proteins	Total Plasma T_4, T_3, RT_3	Free Plasma T_4, T_3, RT_3	Plasma TSH	Clinical State
Hyperthyroidism	Normal	High	High	Low	Hyperthyroid
Hypothyroidism	Normal	Low	Low	High	Hypothyroid
Estrogens, methadone, heroin, major tranquilizers, clofibrate	High	High	Normal	Normal	Euthyroid
Glucocorticoids, androgens, danazol, asparaginase	Low	Low	Normal	Normal	Euthyroid

Fluctuations in Deiodination

Much more RT_3 and much less T_3 are formed during fetal life, and the ratio shifts to that of adults about 6 weeks after birth. Various drugs inhibit deiodinases, producing a fall in the plasma T_3 level and a rise in the plasma RT_3 level. Selenium deficiency has the same effect. A wide variety of nonthyroidal illnesses also depress deiodinases. These include burns, trauma, advanced cancer, cirrhosis, renal failure, myocardial infarction, and febrile states. The low-T_3 state produced by these conditions disappears with recovery. It is difficult to decide whether individuals with the low-T_3 state produced by drugs and illness have mild hypothyroidism.

Diet also has a clear-cut effect on conversion of T_4 to T_3. In fasted individuals, plasma T_3 is reduced 10–20% in 24 hours and about 50% in 3–7 days, with a corresponding rise in RT_3 (Figure 18–9). Free and bound T_4 levels remain normal. During more prolonged starvation, RT_3 returns to normal but T_3 remains depressed. At the same time, the BMR falls and urinary nitrogen excretion, an index of protein breakdown, is decreased. Thus, the decline in T_3 conserves

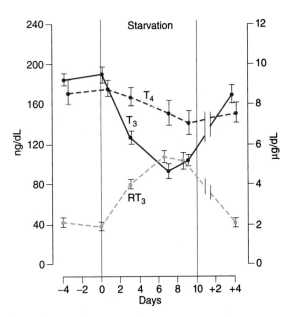

Figure 18–9. Effect of starvation on plasma levels of T_4, T_3, and RT_3 in humans. Similar changes occur in wasting diseases. The scale for T_3 and RT_3 is on the left and the scale for T_4 is on the right. (Reproduced, with permission, from Burger AG: New aspects of the peripheral action of thyroid hormones. Triangle, Sandoz J Med Sci 1983;22:175. Copyright © Sandoz Ltd., Basel, Switzerland.)

calories and protein. Conversely, overfeeding increases T_3 and reduces RT_3.

EFFECTS OF THYROID HORMONES

Some of the widespread effects of thyroid hormones in the body are secondary to stimulation of O_2 consumption (**calorigenic action**), although the hormones also affect growth and development in mammals, help regulate lipid metabolism, and increase the absorption of carbohydrates from the intestine (Table 18–3). They also increase the dissociation of oxygen from hemoglobin by increasing red cell 2,3-diphosphoglycerate (DPG) (see Chapter 35).

Mechanism of Action

Thyroid hormones enter cells, and T_3 binds to thyroid receptors (TR) in the nuclei. T_4 can also bind, but not as avidly. The hormone-receptor complex then binds to DNA via zinc fingers and increases or in some cases decreases the expression of a variety of different genes that code for enzymes which regulate cell function (see Chapter 1). Thus, the nuclear receptors for thyroid hormones are members of the superfamily of hormone-sensitive nuclear transcription factors.

There are two human TR genes: an α receptor gene on chromosome 17 and a β receptor gene on chromosome 3. By alternative splicing, each forms at least two different mRNAs and therefore two different receptor proteins. TRβ2 is found only in the brain, but TRα1, TRα2, and TRβ1 are widely distributed. TRα2 differs from the other three in that it does not bind T_3 and its function is unsettled. TRs bind to DNA as monomers, homodimers, and heterodimers with other nuclear receptors, particularly the retinoid X receptor (**RXR**). This heterodimer does not bind 9-*cis* retinoic acid, the usual ligand for RXR, but the TR binding to DNA is greatly enhanced. There are also coactivator and corepressor proteins that affect the actions of the TRs. Presumably, this complexity permits thyroid hormones to produce their many different effects in the body, but the overall physiologic significance of the complexity is still largely unknown.

In most of its actions, T_3 acts more rapidly and is three to five times more potent than T_4 (Figure 18–10). This is because it is less tightly bound to plasma proteins but binds more avidly to thyroid hormone receptors. RT_3 is inert.

Calorigenic Action

T_4 and T_3 increase the O_2 consumption of almost all metabolically active tissues. The exceptions are the adult brain, testes, uterus, lymph nodes, spleen, and anterior

Table 18–3. Physiologic effects of thyroid hormones.[1]

Target Tissue	Effect	Mechanism
Heart	Chronotropic Inotropic	Increase number of β-adrenergic receptors. Enhance responses to circulating catecholamines. Increase proportion of a myosin heavy chain (with higher ATPase activity).
Adipose tissue	Catabolic	Stimulate lipolysis.
Muscle	Catabolic	Increase protein breakdown.
Bone	Developmental	Promote normal growth and skeletal development.
Nervous system	Developmental	Promote normal brain development.
Gut	Metabolic	Increase rate of carbohydrate absorption.
Lipoprotein	Metabolic	Stimulate formation of LDL receptors.
Other	Calorigenic	Stimulate oxygen consumption by metabolically active tissues (exceptions: testes, uterus, lymph nodes, spleen, anterior pituitary). Increase metabolic rate.

[1]Modified and reproduced, with permission, from McPhee SJ, Lingarra VR, Ganong WF (editors): *Pathophysiology of Disease.* 4th ed, McGraw-Hill, 2003.

pituitary. T_4 actually depresses the O_2 consumption of the anterior pituitary, presumably because it inhibits TSH secretion. The increase in metabolic rate produced by a single dose of T_4 becomes measurable after a latent period of several hours and lasts 6 days or more.

Some of the calorigenic effect of thyroid hormones is due to metabolism of the fatty acids they mobilize. In addition, thyroid hormones increase the activity of the membrane-bound Na^+-K^+ ATPase in many tissues.

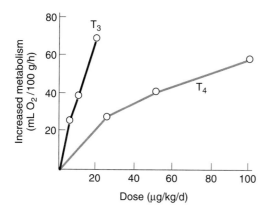

Figure 18–10. Calorigenic responses of thyroidectomized rats to subcutaneous injections of T_4 and T_3. (Redrawn and reproduced, with permission, from Barker SB: Peripheral actions of thyroid hormones. Fed Proc 1962;21:635.)

Effects Secondary to Calorigenesis

When the metabolic rate is increased by T_4 and T_3 in adults, nitrogen excretion is increased; if food intake is not increased, endogenous protein and fat stores are catabolized and weight is lost. In hypothyroid children, small doses of thyroid hormones cause a positive nitrogen balance because they stimulate growth, but large doses cause protein catabolism similar to that produced in the adult. The potassium liberated during protein catabolism appears in the urine, and there is an increase in urinary hexosamine and uric acid excretion.

When the metabolic rate is increased, the need for all vitamins is increased and vitamin deficiency syndromes may be precipitated. Thyroid hormones are necessary for hepatic conversion of carotene to vitamin A, and the accumulation of carotene in the bloodstream (**carotenemia**) in hypothyroidism is responsible for the yellowish tint of the skin. Carotenemia can be distinguished from jaundice because in the former condition the scleras are not yellow.

The skin normally contains a variety of proteins combined with polysaccharides, hyaluronic acid, and chondroitin sulfuric acid. In hypothyroidism, these complexes accumulate, promoting water retention and the characteristic puffiness of the skin (myxedema). When thyroid hormones are administered, the proteins are metabolized, and diuresis continues until the myxedema is cleared.

Milk secretion is decreased in hypothyroidism and stimulated by thyroid hormones, a fact sometimes put to practical use in the dairy industry. Thyroid hormones do not stimulate the metabolism of the uterus but are essential for normal menstrual cycles and fertility.

Effects on the Cardiovascular System

Large doses of thyroid hormones cause enough extra heat production to lead to a slight rise in body temperatures (Chapter 14), which in turn activates heat-dissipating mechanisms. Peripheral resistance decreases because of cutaneous vasodilation, and this increases levels of renal Na^+ and water absorption, expanding blood volume. Cardiac output is increased by direct action of thyroid hormones and catecholamines on the heart, so that pulse pressure and cardiac rate are increased and circulation time is shortened.

T_3 is not formed from T_4 in myocytes to any degree, but circulatory T_3 enters the myocytes, combines with its receptors, and enters the nucleus, where it promotes the expression of some genes and inhibits the expression of others. Those that are enhanced include the genes for α-myosin heavy chain, sarcoplasmic reticulum Ca^{2+} ATPase, β-adrenergic receptors, G proteins, Na^+-K^+ ATPase, and certain K^+ channels. Those that are inhibited include the genes for β-myosin heavy chain, phospholamban, two types of adenylyl cyclase, T_3 nuclear receptors, and the Na^+-Ca^{2+} exchanger. The net result is increased heart rate and force of contraction.

The heart contains two myosin heavy chain (MHC) isoforms, α-MHC and β-MHC. They are encoded by two highly homologous genes located in tandem in humans on the short arm of chromosome 17. Each myosin molecule consists of two heavy chains and two pairs of light chains (see Chapter 3). The myosin containing β-MHC has less ATPase activity than the myosin containing α-MHC. α-MHC predominates in the atria in adults, and its level is increased by treatment with thyroid hormone. This increases the speed of cardiac contraction. Conversely, expression of the α-MHC gene is depressed and that of the β-MHC gene is enhanced in hypothyroidism.

Effects on the Nervous System

In hypothyroidism, mentation is slow and the CSF protein level elevated. Thyroid hormones reverse these changes, and large doses cause rapid mentation, irritability, and restlessness. Overall cerebral blood flow and glucose and O_2 consumption by the brain are normal in adult hypo- and hyperthyroidism. However, thyroid hormones enter the brain in adults and are found in gray matter in numerous different locations. In addition, astrocytes in the brain convert T_4 to T_3,

and there is a sharp increase in brain D_2 activity after thyroidectomy that is reversed within 4 hours by a single intravenous dose of T_3. Some of the effects of thyroid hormones on the brain are probably secondary to increased responsiveness to catecholamines, with consequent increased activation of the reticular activating system (see Chapter 11). In addition, thyroid hormones have marked effects on brain development. The parts of the CNS most affected are the cerebral cortex and the basal ganglia. In addition, the cochlea is also affected. Consequently, thyroid hormone deficiency during development causes mental retardation, motor rigidity, and deaf-mutism.

Thyroid hormones also exert effects on reflexes. The reaction time of stretch reflexes (see Chapter 6) is shortened in hyperthyroidism and prolonged in hypothyroidism. Measurement of the reaction time of the ankle jerk (Achilles reflex) has attracted attention as a clinical test for evaluating thyroid function, but the reaction time is also affected by other diseases.

Relation to Catecholamines

The actions of thyroid hormones and the catecholamines norepinephrine and epinephrine are intimately interrelated. Epinephrine increases the metabolic rate, stimulates the nervous system, and produces cardiovascular effects similar to those of thyroid hormones, although the duration of these actions is brief. Norepinephrine has generally similar actions. The toxicity of the catecholamines is markedly increased in rats treated with T_4. Although plasma catecholamine levels are normal in hyperthyroidism, the cardiovascular effects, tremulousness, and sweating produced by thyroid hormones can be reduced or abolished by sympathectomy. They can also be reduced by drugs such as propranolol that block β-adrenergic receptors. Indeed, propranolol and other β blockers are used extensively in the treatment of thyrotoxicosis and in the treatment of the severe exacerbations of hyperthyroidism called **thyroid storms.** However, even though β blockers are weak inhibitors of extrathyroidal conversion of T_4 to T_3, and consequently may produce a small fall in plasma T_3, they have little effect on the other actions of thyroid hormones.

Effects on Skeletal Muscle

Muscle weakness occurs in most patients with hyperthyroidism (**thyrotoxic myopathy),** and when the hyperthyroidism is severe and prolonged, the myopathy may be severe. The muscle weakness may be due in part to increased protein catabolism. Thyroid hormones affect the expression of the MHC genes in skeletal as well as cardiac muscle (see Chapter 3). However, the effects produced are complex and their relation to the myopathy is not established. Hypothyroidism is also associated with muscle weakness, cramps, and stiffness.

Effects on Carbohydrate Metabolism

Thyroid hormones increase the rate of absorption of carbohydrate from the gastrointestinal tract, an action that is probably independent of their calorigenic action. In hyperthyroidism, therefore, the plasma glucose level rises rapidly after a carbohydrate meal, sometimes exceeding the renal threshold. However, it falls again at a rapid rate.

Effects on Cholesterol Metabolism

Thyroid hormones lower circulating cholesterol levels. The plasma cholesterol level drops before the metabolic rate rises, which indicates that this action is independent of the stimulation of O_2 consumption. As noted in Chapter 17, the decrease in plasma cholesterol concentration is due to increased formation of LDL receptors in the liver, resulting in increased hepatic removal of cholesterol from the circulation. Despite considerable effort, it has not been possible to produce a clinically useful thyroid hormone analog that lowers plasma cholesterol without increasing metabolism.

Effects on Growth

Thyroid hormones are essential for normal growth and skeletal maturation (see Chapter 22). In hypothyroid children, bone growth is slowed and epiphysial closure delayed. In the absence of thyroid hormones, growth hormone secretion is also depressed, and thyroid hormones potentiate the effect of growth hormone on the tissues.

Another example of the role of thyroid hormones in growth and maturation is their effect on amphibian metamorphosis. Tadpoles treated with T_4 and T_3 metamorphose early into dwarf frogs, whereas hypothyroid tadpoles never become frogs.

REGULATION OF THYROID SECRETION

Thyroid function is regulated primarily by variations in the circulating level of pituitary TSH. TSH secretion is increased by the hypophysiotropic hormone thyrotropin-releasing hormone (TRH; see Chapter 14) and inhibited in a negative feedback fashion by circulating free T_4 and T_3. The effect of T_4 is enhanced by production of T_3 in the cytoplasm of the pituitary cells by the 5'-DII they contain. TSH secretion is also inhibited by stress, and in experimental animals it is increased by cold and decreased by warmth.

Chemistry & Metabolism of TSH

Human TSH is a glycoprotein that contains 211 amino acid residues, plus hexoses, hexosamines, and sialic acid. It is made up of two subunits, designated α and β. The α subunit is encoded by a gene on chromosome 6 and the β subunit by a gene on chromosome 1. The α and β subunits become noncovalently linked in the thyrotropes. TSH-α is identical to the α subunit of LH, FSH, and hCG-α (see Chapters 22 and 23). The functional specificity of TSH is conferred by the β unit. The structure of TSH varies from species to species, but other mammalian TSHs are biologically active in humans.

The biologic half-life of human TSH is about 60 minutes. TSH is degraded for the most part in the kidneys and to a lesser extent in the liver. Secretion is pulsatile, and mean output starts to rise at about 9 PM, peaks at midnight, and then declines during the day. The normal secretion rate is about 110 μg/d. The average plasma level is about 2 μU/mL (Figure 18–11).

Since the α subunit in hCG is the same as that in TSH, large amounts of hCG can activate thyroid receptors. In some patients with benign or malignant tumors of placental origin, plasma hCG levels can rise so high that they produce mild hyperthyroidism.

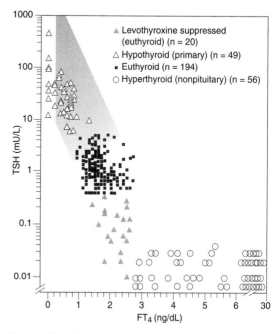

Figure 18–11. Relation between plasma TSH, measured by a highly sensitive radioimmunoassay, and plasma free T_4, measured by dialysis (FT$_4$). Note that the TSH scale is a log scale.

Effects of TSH on the Thyroid

When the pituitary is removed, thyroid function is depressed and the gland atrophies; when TSH is administered, thyroid function is stimulated. Within a few minutes after the injection of TSH, there are increases in iodide binding; synthesis of T_3, T_4, and iodotyrosines; secretion of thyroglobulin into the colloid; and endocytosis of colloid. Iodide trapping is increased in a few hours; blood flow increases; and, with chronic TSH treatment, the cells hypertrophy and the weight of the gland increases.

Whenever TSH stimulation is prolonged, the thyroid becomes detectably enlarged. Enlargement of the thyroid is called **goiter.**

TSH Receptors

The TSH receptor is a typical serpentine receptor that activates adenylyl cyclase through G_s. It also activates PLC. Like other glycoprotein hormone receptors, it has an extended, glycosylated extracellular domain.

Other Factors Affecting Thyroid Growth

In addition to TSH receptors, the thyroid cells contain receptors for IGF-I, EGF, and other growth factors. IGF-I and EGF promote growth whereas γ interferon and tumor necrosis factor α inhibit growth. The exact physiologic role of these factors in the thyroid has not been established.

Control Mechanisms

The mechanisms regulating thyroid secretion are summarized in Figure 18–12. The negative feedback effect of thyroid hormones on TSH secretion is exerted in part at the hypothalamic level, but it is also due in large part to an action on the pituitary, since T_4 and T_3 block the increase in TSH secretion produced by TRH. The relation between plasma TSH and the FT_4I index is shown in Figure 18–11. Infusion of T_4 as well as T_3 reduces the circulating level of TSH, which declines measurably within 1 hour. In experimental animals, there is an initial rise in pituitary TSH content before the decline, indicating that thyroid hormones inhibit secretion before they inhibit synthesis. The effects on secretion and synthesis of TSH both appear to depend on protein synthesis, even though the former is relatively rapid.

The day-to-day maintenance of thyroid secretion depends on the feedback interplay of thyroid hormones with TSH and TRH (Figure 18–12). The adjustments that appear to be mediated via TRH include the increased secretion of thyroid hormones produced by cold and, presumably, the decrease produced by heat. It

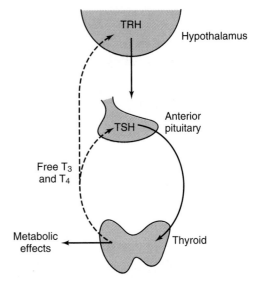

Figure 18–12. Feedback control of thyroid secretion. The dashed arrows indicate inhibitory effects, and the solid arrows indicate stimulatory effects. Compare with Figures 20–21, 22–10, 23–22, and 23–35.

is worth noting that although cold produces clear-cut increases in circulating TSH in experimental animals and human infants, the rise produced by cold in adult humans is negligible. Consequently, in adults, increased heat production due to increased thyroid hormone secretion (**thyroid hormone thermogenesis**) plays little if any role in the response to cold. Stress has an inhibitory effect on TRH secretion. Dopamine and somatostatin act at the pituitary level to inhibit TSH secretion, but it is not known whether they play a physiologic role in the regulation of TSH secretion. Glucocorticoids also inhibit TSH secretion.

The amount of thyroid hormone necessary to maintain normal cellular function in thyroidectomized individuals used to be defined as the amount necessary to normalize the BMR, but it is now defined as the amount necessary to return plasma TSH to normal. Indeed, with the accuracy and sensitivity of modern assays for TSH and the marked inverse correlation between plasma free thyroid hormone levels and plasma TSH, measurement of TSH is now widely regarded as one of the best tests of thyroid function. The amount of T_4 that normalizes plasma TSH in athyreotic individuals averages 112 μg of T_4 by mouth per day in adults. About 80% of this dose is absorbed from the gastrointestinal tract. It produces a slightly greater than normal FT_4I but a normal FT_3I, indicating that in humans, unlike some experimental animals, it is circulating T_3 rather than T_4

that is the principal feedback regulator of TSH secretion.

CLINICAL CORRELATES

The signs, symptoms, and complications of hypothyroidism and hyperthyroidism in humans are predictable consequences of the physiologic effects of thyroid hormones discussed above.

Hypothyroidism

The syndrome of adult hypothyroidism is generally called **myxedema,** although this term is also used to refer specifically to the skin changes in this syndrome. Hypothyroidism may be the end result of a number of diseases of the thyroid gland, or it may be secondary to pituitary failure (pituitary hypothyroidism) or hypothalamic failure (hypothalamic hypothyroidism). In the latter two conditions, unlike the first, the thyroid responds to a test dose of TSH, and at least in theory, hypothalamic hypothyroidism can be distinguished from pituitary hypothyroidism by the presence in the former of a rise in plasma TSH following a test dose of TRH. The TSH response to TRH is usually normal in hypothalamic hypothyroidism, while it is increased in hypothyroidism caused by thyroid disease and decreased in hyperthyroidism because of the feedback of thyroid hormones on the pituitary gland.

In completely athyreotic humans, the BMR falls to about 40. The hair is coarse and sparse, the skin is dry and yellowish (carotenemia), and cold is poorly tolerated. The voice is husky and slow, the basis of the aphorism that "myxedema is the one disease that can be diagnosed over the telephone." Mentation is slow, memory is poor, and in some patients there are severe mental symptoms ("myxedema madness"). Plasma cholesterol is elevated.

Cretinism

Children who are hypothyroid from birth or before are called cretins. They are dwarfed and mentally retarded and have potbellies and enlarged, protruding tongues (Figure 18–13). Worldwide, congenital hypothyroidism is one of the most common causes of preventable mental retardation. The main causes are listed in Table 18–4. They include not only maternal iodine deficiency and various congenital abnormalities of the fetal hypothalamo-pituitary-thyroid axis but also maternal antithyroid antibodies that cross the placenta and damage the fetal thyroid. T_4 crosses the placenta, and unless the mother is hypothyroid, growth and development are normal until birth. If treatment is started at birth, the prognosis for normal growth and development is

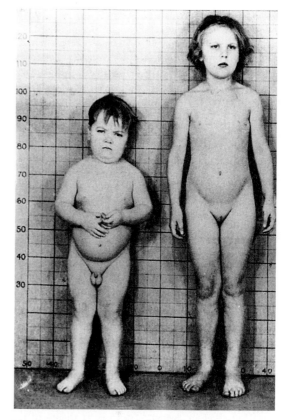

Figure 18–13. Fraternal twins, age 8 years. The boy has congenital hypothyroidism. (Reproduced, with permission, from Wilkins L in: *Clinical Endocrinology I.* Astwood EB, Cassidy CE [editors]. Grune & Stratton, 1960.)

good, and mental retardation can generally be avoided; for this reason, screening tests for congenital hypothyroidism are routine in all states of the USA and most other developed countries. When the mother is hypothyroid as well, as in the case of iodine deficiency, the mental deficiency is more severe and less responsive to treatment after birth. In addition, there may be deafmutism and rigidity (see above). Increased use of

Table 18–4. Causes of congenital hypothyroidism.

Maternal iodine deficiency
Fetal thyroid dysgenesis
Inborn errors of thyroid hormone synthesis
Maternal antithyroid antibodies that cross the placenta
Fetal hypopituitary hypothyroidism

iodized salt has now reduced the incidence of maternal iodine deficiency. However, it is still common in many parts of the world, and it has been estimated that 20 million people in the world now have various degrees of brain damage caused by iodine deficiency in utero.

Hyperthyroidism

Hyperthyroidism is characterized by nervousness; weight loss; hyperphagia; heat intolerance; increased pulse pressure; a fine tremor of the outstretched fingers; a warm, soft skin; sweating; and a BMR from +10 to as high as +100. It has various causes (Table 18–5). However, the most common cause is **Graves' disease (Graves' hyperthyroidism),** which accounts for 60–80% of the cases. The condition—which for unknown reasons is much more common in women—is an autoimmune disease in which antibodies to the TSH receptor stimulate the receptor. This produces marked T_4 and T_3 secretion and enlargement of the thyroid gland (goiter). However, due to the feedback effects of T_4 and T_3, plasma TSH is low, not high.

Another hallmark of Graves' disease is the occurrence of swelling of tissues in the orbits, producing protrusion of the eyeballs (**exophthalmos;** Figure 18–14). This occurs in 50% of patients and often precedes the development of obvious hyperthyroidism. A subpopulation of fibroblasts in the orbits ultimately develop into adipocytes, and these preadipocyte fibroblasts contain TSH receptor protein. The current theory of the development of exophthalmos is that when stimulated by the TSH receptor-stimulating antibodies in the circulation, these cells release cytokines that promote inflammation and edema.

Other antithyroid antibodies are present in Graves' disease, including antibodies to thyroglobulin and thyroid peroxidase. In Hashimoto's thyroiditis, autoimmune antibodies destroy the thyroid with little if any stimulation, producing hypothyroidism.

Table 18–5. Causes of hyperthyroidism.

Thyroid overactivity
Graves' disease
Solitary toxic adenoma
Toxic multinodular goiter
TSH-secreting pituitary tumor
Thyroiditis
Mutations causing constitutive activation of TSH receptor

Extrathyroidal
Administration of T_3 or T_4 (factitious or iatrogenic hyperthyroidism)
Ectopic thyroid tissue

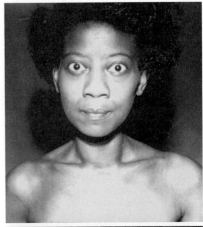
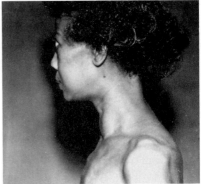

Figure 18–14. Graves' disease. Note the goiter and the exophthalmos. (Courtesy of PH Forsham.)

Thyroid Hormone Resistance

Some mutations in the gene that codes for hTRβ are associated with resistance to the effects of T_3 and T_4. Most commonly, there is resistance to thyroid hormones in the peripheral tissues and the anterior pituitary gland. Patients with this abnormality are usually not clinically hypothyroid, because they maintain plasma levels of T_3 and T_4 that are high enough to overcome the resistance, and hTRα is unaffected. The plasma TSH level is inappropriately high for the high circulating T_3 and T_4 levels and is difficult to suppress with exogenous thyroid hormone. Some patients have thyroid hormone resistance only in the pituitary. They have hypermetabolism and elevated plasma T_3 and T_4 levels with normal, nonsuppressible levels of TSH. A few patients apparently have peripheral resistance with normal pituitary sensitivity. They have hypometabolism despite normal plasma levels of T_3, T_4, and TSH, and they require large doses of thyroid hormones to increase their metabolic rate.

An interesting finding is that **attention deficit hyperactivity disorder,** a condition frequently diagnosed in children who are overactive and impulsive, is much more common in individuals with thyroid hormone resistance than in the general population. This suggests that hTRβ may play a special role in brain development (see above).

Iodine Deficiency

When the dietary iodine intake falls below 50 μg/d, thyroid hormone synthesis is inadequate and secretion declines. As a result of increased TSH secretion, the thyroid hypertrophies, producing an **iodine deficiency goiter** that may become very large. Such "endemic goiters" have been known since ancient times. Before the practice of adding iodide to table salt became widespread, they were very common in Central Europe and the area around the Great Lakes in the United States, the inland "goiter belts" where iodine has been leached out of the soil by rain so that food grown in the soil is iodine-deficient.

Radioactive Iodine Uptake

Iodine uptake is an index of thyroid function that can be easily measured by using tracer doses of radioactive isotopes of iodine that have no known deleterious effect on the thyroid. The tracer is administered orally and the thyroid uptake determined by placing a gamma ray counter over the neck. An area such as the thigh is also counted, and counts in this region are subtracted from the neck counts to correct for nonthyroidal radioactivity in the neck. The isotope of iodine that is most commonly used is ^{123}I because it has a half-life of only 0.55 day, compared with ^{131}I, which has a half-life of 8.1 days, and ^{125}I, which has a half-life of 60 days. Diagnostic use of radioactive iodine uptake has become rare, because of the general availability of methods for measuring T_4, T_3, and TSH in plasma. In addition, the use of iodized salt is widespread, and this causes uptake to be low because the iodide pool is so large that the tracer is excessively diluted. However, an analysis of radioactive iodine uptake is helpful in understanding the physiology of the thyroid gland. The uptake in a normal subject is plotted in Figure 18–15. In hyperthyroidism, iodide is rapidly incorporated into T_4 and T_3, and these hormones are released at an accelerated rate. Therefore, the amount of radioactivity in the thyroid rises sharply, but it then levels off and may start to decline within 24 hours, at a time when the uptake in normal subjects is still rising. In hypothyroidism, the uptake is low.

Large amounts of radioactive iodine destroy thyroid tissue because the radiation kills the cells. Radioiodine

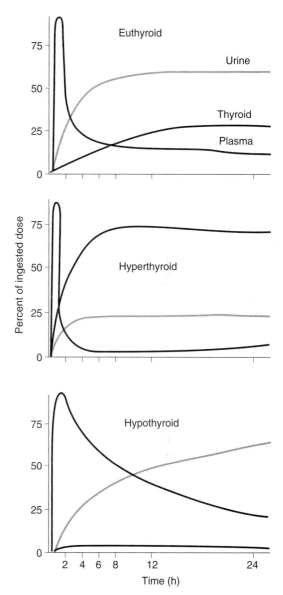

Figure 18–15. Distribution of radioactive iodine in individuals on a relatively low-iodine diet. Percentages are plotted against time after an oral dose of radioactive iodine. In hyperthyroidism, plasma radioactivity falls rapidly and then rises again as a result of release of labeled T_4 and T_3 from the thyroid.

therapy is useful in some cases of Graves' disease and some cases of thyroid cancer.

Radioactive isotopes of iodine are major products of nuclear fission, and if fission products are released into the atmosphere as a result of an accident at a nuclear power plant or explosion of a nuclear bomb, the isotopes spread for considerable distances in the atmosphere because they are more volatile than the other products. Treatment with potassium iodide is regularly instituted in fallout areas to enlarge the iodide pool and depress thyroid uptake to low levels.

Antithyroid Drugs

Most of the drugs that inhibit thyroid function act either by interfering with the iodide-trapping mechanism or by blocking the organic binding of iodine. In either case, TSH secretion is stimulated by the decline in circulating thyroid hormones, and goiter is produced. A number of monovalent anions compete with iodide for transport into the thyroid via the Na^+/I^- symporter. The anions include chlorate, pertechnetate, periodate, biiodate, nitrate, and perchlorate. Thiocyanate, another monovalent anion, inhibits iodide transport but is not itself concentrated within the gland. The activity of perchlorate is about ten times that of thiocyanate.

The **thiourylenes,** a group of compounds related to thiourea, inhibit the iodination of monoiodotyrosine (organic binding of iodide) and block the coupling reaction. The two used clinically are propylthiouracil and methimazole (Figure 18–16). Iodination of tyrosine is inhibited because propylthiouracil and methimazole compete with tyrosine residues for iodine and become iodinated. In addition, propylthiouracil but not methimazole inhibits D_2 deiodinase, reducing the conversion of T_4 to T_3 in many extrathyroidal tissues. Both drugs may also ameliorate hyperthyroidism by suppressing the immune system and thereby depressing the formation of stimulatory antibodies.

Another substance that inhibits thyroid function under certain conditions is iodide itself. The position of iodide in thyroid physiology is thus unique in that while some iodide is needed for normal thyroid function, too little iodide and too much both cause abnormal thyroid function. In normal individuals, large doses of iodides act directly on the thyroid to produce a mild and transient inhibition of organic binding of iodide and hence of hormone synthesis. This inhibition is known as the **Wolff-Chaikoff effect.** The Wolff-Chaikoff effect is greater and more prolonged when io-

dide transport is increased, and this is why patients with hyperthyroidism are more responsive to iodide than normal individuals. There are at least two additional mechanisms by which excess I^- inhibits thyroid function. It reduces the effect of TSH on the gland by reducing the cAMP response to this hormone, and it inhibits proteolysis of thyroglobulin.

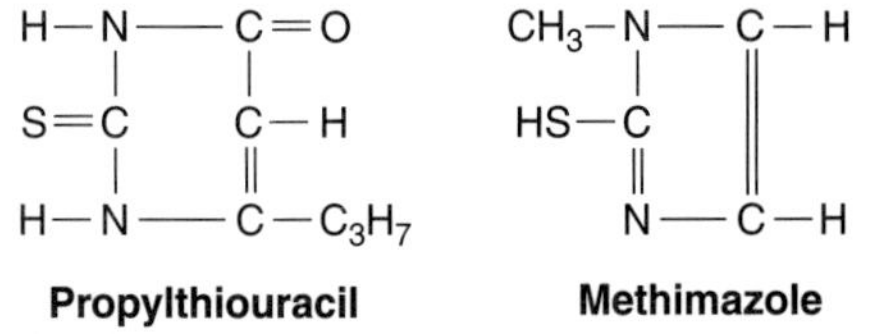

Naturally Occurring Goitrogens

Thiocyanates are sometimes ingested with food, and there are relatively large amounts of naturally occurring goitrogens in some foods. Vegetables of the Brassicaceae family, particularly rutabagas, cabbage, and turnips, contain **progoitrin** and a substance that converts this compound into **goitrin,** an active antithyroid agent (Figure 18–17). The progoitrin activator in vegetables is heat-labile, but because there are activators in the intestine (presumably of bacterial origin), goitrin is formed even if the vegetables are cooked. The goitrin intake on a normal mixed diet is usually not great enough to be harmful, but in vegetarians and food faddists, "cabbage goiters" do occur. Other as yet unidentified plant goitrogens probably exist and may be responsible for the occasional small "goiter epidemics" reported from various parts of the world.

Use of Thyroid Hormones in Nonthyroidal Diseases

When the pituitary-thyroid axis is normal, doses of exogenous thyroid hormone that provide less than the amount secreted endogenously have no significant effect on metabolism because there is a compensatory decline in endogenous secretion resulting from inhibition of TSH secretion. In euthyroid humans, the oral dose of T_4 that merely suppresses endogenous thyroid function can be estimated from the dose that normalizes plasma TSH in thyroidectomized adults, and this is 100–125 μg/d. Suppression of TSH secretion by exoge-

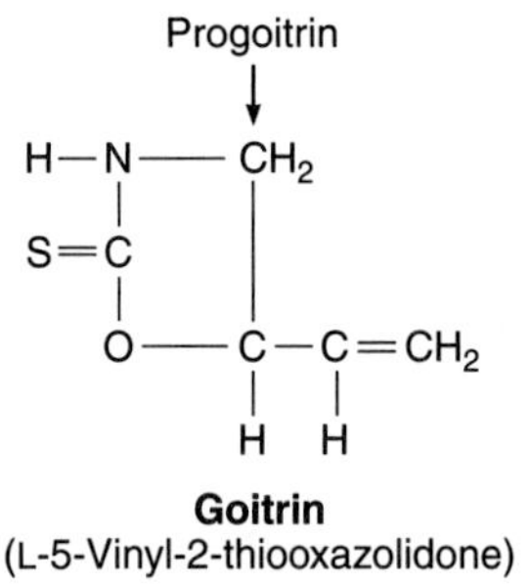

Figure 18–17. The naturally occurring goitrogen in vegetables of the family Brassicaceae.

Figure 18–16. Commonly used thiourylenes.

nous T_4 or pituitary disease leads eventually to thyroid atrophy. An atrophic gland initially responds sluggishly to TSH, and if the TSH suppression has been prolonged, it may take some time for normal thyroid responsiveness to return. The adrenal cortex and some other endocrine glands respond in an analogous fashion; when they are deprived of the support of their tropic hormones for some time, they become atrophic and only sluggishly responsive to their tropic hormone until the hormone has had some time to act on the gland.

Use of thyroid hormones to promote weight loss is of value only if the patient pays the price of some nervousness and heat intolerance. In addition, appetite must be curbed so that there is no compensatory increase in caloric intake.

Endocrine Functions of the Pancreas & Regulation of Carbohydrate Metabolism

19

INTRODUCTION

At least four polypeptides with hormonal activity are secreted by the islets of Langerhans in the pancreas. Two of these hormones, **insulin** and **glucagon,** have important functions in the regulation of the intermediary metabolism of carbohydrates, proteins, and fats. The third hormone, **somatostatin,** plays a role in the regulation of islet cell secretion, and the fourth, **pancreatic polypeptide,** is probably concerned primarily with gastrointestinal function. Glucagon, somatostatin, and possibly pancreatic polypeptide are also secreted by cells in the mucosa of the gastrointestinal tract.

Insulin is anabolic, increasing the storage of glucose, fatty acids, and amino acids. Glucagon is catabolic, mobilizing glucose, fatty acids, and the amino acids from stores into the bloodstream. The two hormones are thus reciprocal in their overall action and are reciprocally secreted in most circumstances. Insulin excess causes hypoglycemia, which leads to convulsions and coma. Insulin deficiency, either absolute or relative, causes diabetes mellitus, a complex and debilitating disease that if untreated is eventually fatal. Glucagon deficiency can cause hypoglycemia, and glucagon excess makes diabetes worse. Excess pancreatic production of somatostatin causes hyperglycemia and other manifestations of diabetes.

A variety of other hormones also have important roles in the regulation of carbohydrate metabolism.

ISLET CELL STRUCTURE

The islets of Langerhans (Figure 19–1) are ovoid, 76-× 175-μm collections of cells scattered throughout the pancreas, although they are more plentiful in the tail than in the body and head. They make up about 2% of the volume of the gland, whereas the exocrine portion of the pancreas makes up 80% and ducts and blood vessels make up the rest. In humans, there are 1–2 million islets. Each has a copious blood supply; and blood from the islets, like that from the gastrointestinal tract but unlike that from any other endocrine organs, drains into the hepatic portal vein.

The cells in the islets can be divided into types on the basis of their staining properties and morphology. There are at least four distinct cell types in humans: A, B, D, and F cells. A, B, and D cells are also called α, β, and δ cells. However, this leads to confusion in view of the use of Greek letters to refer to other structures in the body, particularly adrenergic receptors (see Chapter 4). The A cells secrete glucagon, the B cells secrete insulin, the D cells secrete somatostatin, and the F cells secrete pancreatic polypeptide. The B cells, which are the most common and account for 60–75% of the cells in the islets, are generally located in the center of each

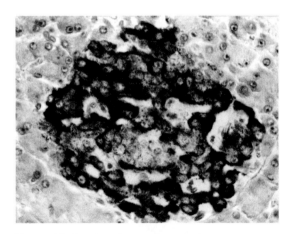

Figure 19–1. Islet of Langerhans in the rat pancreas. Darkly stained cells are B cells. Surrounding pancreatic acinar tissue is light-colored. (×400) (Courtesy of LL Bennett.)

islet. They tend to be surrounded by the A cells, which make up 20% of the total, and the less common D and F cells. The islets in the tail, the body, and the anterior and superior part of the head of the human pancreas have many A cells and few if any F cells in the outer rim, whereas in rats and probably in humans, the islets in the posterior part of the head of the pancreas have a relatively large number of F cells and few A cells. The A-cell-rich (glucagon-rich) islets arise embryologically from the dorsal pancreatic bud, and the F-cell-rich (pancreatic polypeptide-rich) islets arise from the ventral pancreatic bud. These buds arise separately from the duodenum.

The B cell granules are packets of insulin in the cell cytoplasm. Each packet is contained in a membrane-lined vesicle (Figure 19–2), and, characteristically, there is a clear space (halo) between the wall of the vesicle and the packet. The shape of the packets varies from species to species; in humans, some are round whereas others are rectangular. In the B cells, the insulin molecule forms polymers and also complexes with zinc. The differences in the shape of the packets are probably due to differences in the size of polymers or zinc aggregates of insulin. The A granules, which contain glucagon, are relatively uniform from species to species (Figure 19–3). The D cells also contain large numbers of relatively homogeneous granules.

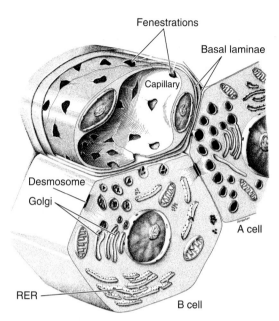

Figure 19–3. A and B cells, showing their relation to blood vessels. RER, rough endoplasmic reticulum. Insulin from the B cell and glucagon from the A cell are secreted by exocytosis and cross the basal lamina of the cell and the basal lamina of the capillary before entering the lumen of the fenestrated capillary. (Reproduced, with permission, from Junqueira IC, Carneiro J, Kelley RO: *Basic Histology: Text and Atlas,* 10th ed. McGraw-Hill, 2003.)

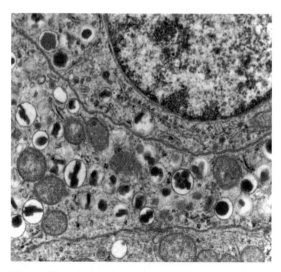

Figure 19–2. Electron micrograph of two adjoining B cells in the human pancreas. The B granules are the membrane-lined vesicles containing crystals that vary in shape from rhombic to round. (×26,000) (Courtesy of A Like. Reproduced, with permission, from Fawcett DW: *Bloom and Fawcett, A Textbook of Histology,* 11th ed. Saunders, 1986.)

STRUCTURE, BIOSYNTHESIS, & SECRETION OF INSULIN

Structure & Species Specificity

Insulin is a polypeptide containing two chains of amino acids linked by disulfide bridges (Table 19–1). There are minor differences in the amino acid composition of the molecule from species to species. The differences are generally not sufficient to affect the biologic activity of a particular insulin in heterologous species but are sufficient to make the insulin antigenic. If insulin of one species is injected for a prolonged period into another species, the anti-insulin antibodies formed inhibit the injected insulin. Almost all humans who have received commercial beef insulin for more than 2 months have antibodies against beef insulin, but the titer is usually low. Pork insulin differs from human insulin by only one amino acid residue and has low antigenicity. Human insulin produced in bacteria by recombinant DNA technology is now widely used to avoid antibody formation.

Table 19–1. Structure of human insulin (molecular weight 5808) and (below) variations in this structure in other mammalian species. In the rat, the islet cells secrete two slightly different insulins, and in certain fish four different chains are found.

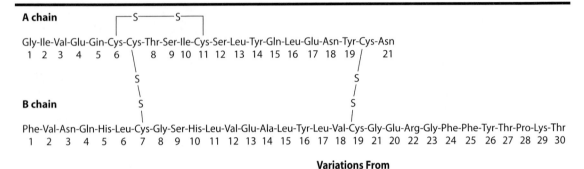

	Variations From Human Amino Acid Sequence	
	---	---
Species	A Chain Position 8 9 10	B Chain Position 30
Pig, dog, sperm whale	Thr-Ser-Ile	Ala
Rabbit	Thr-Ser-Ile	Ser
Cattle, goat	Ala-Ser-Val	Ala
Sheep	Ala-Gly-Val	Ala
Horse	Thr-Gly-Ile	Ala
Sei whale	Ala-Ser-Thr	Ala

Biosynthesis & Secretion

Insulin is synthesized in the rough endoplasmic reticulum of the B cells (Figure 19–3). It is then transported to the Golgi apparatus, where it is packaged in membrane-bound granules. These granules move to the plasma membrane by a process involving microtubules, and their contents are expelled by exocytosis (see Chapter 1). The insulin then crosses the basal laminas of the B cell and a neighboring capillary and the fenestrated endothelium of the capillary to reach the bloodstream. The fenestrations are discussed in detail in Chapter 30.

Like other polypeptide hormones and related proteins that enter the endoplasmic reticulum, insulin is synthesized as part of a larger preprohormone (see Chapter 1). The gene for insulin is located on the short arm of chromosome 11 in humans. It has two introns and three exons (Figure 19–4). **Preproinsulin** has a 23-amino-acid signal peptide removed as it enters the endoplasmic reticulum. The remainder of the molecule is then folded, and the disulfide bonds are formed to make **proinsulin.** The peptide segment connecting the A and B chains, the **connecting peptide (C peptide),** facilitates the folding and then is detached in the granules before secretion. Two proteases are involved in processing the proinsulin; it has no other established physi-

ologic activity. Normally, 90–97% of the product released from the B cells is insulin along with equimolar amounts of C peptide. The rest is mostly proinsulin. C peptide can be measured by radioimmunoassay, and its level provides an index of B cell function in patients receiving exogenous insulin.

FATE OF SECRETED INSULIN

Insulin & Insulin-Like Activity in Blood

Plasma contains a number of substances with insulin-like activity in addition to insulin (Table 19–2). The activity that is not suppressed by anti-insulin antibodies has been called **nonsuppressible insulin-like activity (NSILA).** Most, if not all, of this activity persists after pancreatectomy and is due to the insulin-like growth factors **IGF-I** and **IGF-II** (see Chapter 22). These IGFs are polypeptides. Small amounts are free in the plasma (low-molecular-weight fraction), but large amounts are bound to proteins (high-molecular-weight fraction).

One may well ask why pancreatectomy causes diabetes mellitus (see below) when the NSILA persists in the plasma. However, the insulin-like activities of IGF-I and IGF-II are weak compared to that of insulin.

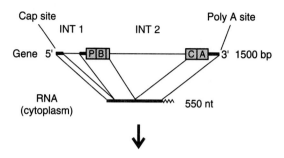

Chromosomal DNA (nucleus)

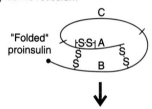

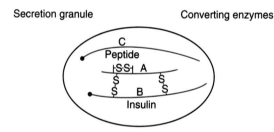

Figure 19–4. Biosynthesis of insulin. The three exons of the insulin gene (top) are separated by two introns (INT 1 and INT 2). Exons 1 and 2 code for an untranslated part of the mRNA, exon 2 codes for the signal peptide and the B chain (B), exons 2 and 3 code for the C peptide (C), and exon 3 codes for the A chain (A) plus an untranslated part of the mRNA; bp, base pairs; nt, nucleotides. The signal peptide guides the polypeptide chain into the endoplasmic reticulum and is then removed. The molecule is next folded, with formation of the disulfide bonds. The C peptide is separated by converting enzymes in the secretory granule.

Metabolism

The half-life of insulin in the circulation in humans is about 5 minutes. Insulin binds to insulin receptors, and some is internalized (see below). It is destroyed by proteases in the endosomes formed by the endocytotic process.

Table 19–2. Substances with insulin-like activity in human plasma.

Insulin
Proinsulin
Nonsuppressible insulin-like activity (NSILA)
  Low-molecular-weight fraction
    IGF-I
    IGF-II
  High-molecular-weight fraction (mostly IGF bound to protein)

EFFECTS OF INSULIN

The physiologic effects of insulin are far-reaching and complex. They are conveniently divided into rapid, intermediate, and delayed actions, as listed in Table 19–3. The best known is the hypoglycemic effect, but there are additional effects on amino acid and electrolyte transport, many enzymes, and growth (see below). The net effect of the hormone is storage of carbohydrate, protein, and fat. Therefore, insulin is appropriately called the "hormone of abundance."

The actions of insulin on adipose tissue; skeletal, cardiac, and smooth muscle; and the liver are summarized in Table 19–4.

Glucose Transporters

Glucose enters cells by **facilitated diffusion** (see Chapter 1) or, in the intestine and kidneys, by secondary active transport with Na⁺. In muscle, fat and some other tissues, insulin facilitates glucose entry into cells by increasing the number of glucose transporters in the cell membranes.

The glucose transporters that are responsible for facilitated diffusion of glucose across cell membranes are

Table 19–3. Principal actions of insulin.[1]

Rapid (seconds)
  Increased transport of glucose, amino acids, and K⁺ into insulin-sensitive cells

Intermediate (minutes)
  Stimulation of protein synthesis
  Inhibition of protein degradation
  Activation of glycolytic enzymes and glycogen synthase
  Inhibition of phosphorylase and gluconeogenic enzymes

Delayed (hours)
  Increase in mRNAs for lipogenic and other enzymes

[1]Courtesy of ID Goldfine.

Table 19–4. Effects of insulin on various tissues.

Adipose tissue
Increased glucose entry
Increased fatty acid synthesis
Increased glycerol phosphate synthesis
Increased triglyceride deposition
Activation of lipoprotein lipase
Inhibition of hormone-sensitive lipase
Increased K^+ uptake

Muscle
Increased glucose entry
Increased glycogen synthesis
Increased amino acid uptake
Increased protein synthesis in ribosomes
Decreased protein catabolism
Decreased release of gluconeogenic amino acids
Increased ketone uptake
Increased K^+ uptake

Liver
Decreased ketogenesis
Increased protein synthesis
Increased lipid synthesis
Decreased glucose output due to decreased gluconeogenesis, increased glycogen synthesis, and increased glycolysis

General
Increased cell growth

a family of closely related proteins that cross the cell membrane 12 times and have their amino and carboxyl terminals inside the cell. They differ from and have no homology with the sodium-dependent glucose transporters, SGLT 1 and SGLT 2, responsible for the secondary active transport of glucose out of the intestine (see Chapter 25) and renal tubules (see Chapter 38), although the SGLTs also have 12 transmembrane domains. Particularly in transmembrane helical segments 3, 5, 7, and 11, the amino acids of the facilitative transporters appear to surround channels that glucose can enter. Presumably, conformation then changes and glucose is released inside the cell.

Seven different glucose transporters, called in order of discovery GLUT 1–7, have been characterized (Table 19–5). They contain 492–524 amino acid residues, and their affinity for glucose varies. Each transporter appears to have evolved for special tasks. GLUT 4 is the transporter in muscle and adipose tissue that is stimulated by insulin. A pool of GLUT 4 molecules is maintained in vesicles in the cytoplasm of insulin-sensitive cells. When the insulin receptors of these cells are activated, the vesicles move rapidly to the cell membrane and fuse with it, inserting the transporters

into the cell membrane (Figure 19–5). When insulin action ceases, the transporter-containing patches of membrane are endocytosed, and the vesicles are ready for the next exposure to insulin. Activation of the insulin receptor brings about the movement of the vesicles to the cell membrane by activating phosphoinositol-3 kinase (Figure 19–5), but how this activation triggers vesicle movement is still unsettled. Most of the other GLUT transporters that are not insulin-sensitive appear to stay in the cell membrane.

In the tissues in which insulin increases the number of glucose transporters in the cell membranes, the rate of phosphorylation of the glucose, once it has entered the cells, is regulated by other hormones. Growth hormone and cortisol both inhibit phosphorylation in certain tissues. The process is normally so rapid that it is not a rate-limiting step in glucose metabolism. However, it is rate-limiting in the B cells (see below).

Insulin also increases the entry of glucose into liver cells, but it does not exert this effect by increasing the number of GLUT 4 transporters (see below) in the cell membranes. Instead, it induces glucokinase, and this increases the phosphorylation of glucose, so that the intracellular free glucose concentration stays low, facilitating the entry of glucose into the cell.

Insulin-sensitive tissues also contain a population of GLUT 4 vesicles that move into the cell membrane in response to exercise and are independent of the action of insulin. This is why exercise lowers blood sugar (see below). A 5'-AMP-activated kinase may be responsible for the insertion of these vesicles in the cell membrane.

Insulin Preparations

The maximal decline in plasma glucose occurs 30 minutes after intravenous injection of crystalline insulin. After subcutaneous administration, the maximal fall occurs in 2–3 hours. A wide variety of insulin preparations are now available commercially. These include insulins that have been complexed with protamine and other polypeptides to delay absorption and synthetic insulins in which there have been changes in amino acid residues. In general, they fall into three categories: rapid, intermediate-acting, and long-acting (24–36 hours).

Relation to Potassium

Insulin causes K^+ to enter cells, with a resultant lowering of the extracellular K^+ concentration. Infusions of insulin and glucose significantly lower the plasma K^+ level in normal individuals and are very effective for the temporary relief of hyperkalemia in patients with renal failure. Hypokalemia often develops when patients with diabetic acidosis are treated with insulin. The rea-

Table 19–5. Glucose transporters in mammals.[1]

	Function	K_m (mM)[2]	Major Sites of Expression
Secondary active transport (Na⁺-glucose cotransport)			
SGLT 1	Absorption of glucose	0.1–1.0	Small intestine, renal tubules
SGLT 2	Absorption of glucose	1.6	Renal tubules
Facilitated diffusion			
GLUT 1	Basal glucose uptake	1–2	Placenta, blood-brain barrier, brain, red cells, kidneys, colon, many other organs
GLUT 2	B cell glucose sensor; transport out of intestinal and renal epithelial cells	12–20	B cells of islets, liver, epithelial cells of small intestine, kidneys
GLUT 3	Basal glucose uptake	<1	Brain, placenta, kidneys, many other organs
GLUT 4	Insulin-stimulated glucose uptake	5	Skeletal and cardiac muscle, adipose tissue, other tissues
GLUT 5	Fructose transport	1–2	Jejunum, sperm
GLUT 6	None	—	Pseudogene
GLUT 7	Glucose 6-phosphate transporter in endoplasmic reticulum	—	Liver, ? other tissues

[1]Modified from Stephens JM, Pilch PF: The metabolic regulation and vesicular transport of GLUT 4, the major insulin-responsive glucose transporter. Endocr Rev 1995;16:529.

[2]The K_m is the glucose concentration at which transport is half-maximal.

son for the intracellular migration of K⁺ is still uncertain. However, insulin increases the activity of Na⁺-K⁺ ATPase in cell membranes, so that more K⁺ is pumped into cells.

Other Actions

The hypoglycemic and other effects of insulin are summarized in temporal terms in Table 19–3, and the net effects on various tissues are summarized in Table 19–4. The action on glycogen synthase fosters glycogen storage, and the actions on glycolytic enzymes favor glucose metabolism to two carbon fragments (see Figure 17–9), with resulting promotion of lipogenesis. Stimulation of protein synthesis from amino acids entering the cells and inhibition of protein degradation foster growth.

The anabolic effect of insulin is aided by the protein-sparing action of adequate intracellular glucose supplies. Failure to grow is a symptom of diabetes in

children, and insulin stimulates the growth of immature hypophysectomized rats to almost the same degree as growth hormone. Maximum insulin-induced growth is present, however, only when the protein-sparing action of glucose is fostered by feeding a high-carbohydrate diet.

MECHANISM OF ACTION

Insulin Receptors

Insulin receptors are found on many different cells in the body, including cells in which insulin does not increase glucose uptake. Selective knockout of the insulin receptors in the CNS produces mild hyperphagia and defects in reproduction in mice.

The insulin receptor, which has a molecular weight of approximately 340,000, is a tetramer made up of two α and two β glycoprotein subunits (Figure 19–6). All these are synthesized on a single mRNA and then proteolytically separated and bound to each other by

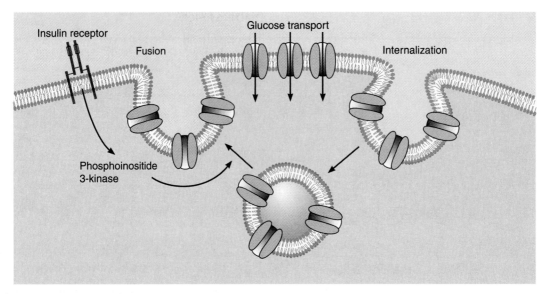

Figure 19–5. Cycling of GLUT 4 transporters through endosomes in insulin-sensitive tissues. Activation of the insulin receptor causes activation of phosphoinositide 3-kinase, which speeds translocation of the GLUT 4-containing endosomes into the cell membrane. The GLUT 4 transporters then mediate glucose transport into the cell.

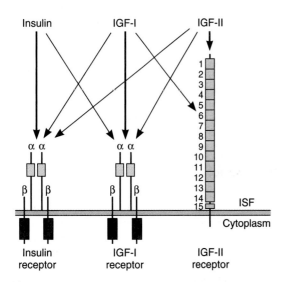

Figure 19–6. Insulin, IGF-I, and IGF-II receptors. Each hormone binds primarily to its own receptor, but insulin also binds to the IGF-I receptor, and IGF-I and IGF-II bind to all three. The dark-colored boxes are intracellular tyrosine kinase domains. Note the marked similarity between the insulin receptor and the IGF-I receptor; also note the 15 repeat sequences in the extracellular portion of the IGF-II receptor.

disulfide bonds. The gene for the insulin receptor has 22 exons and in humans is located on chromosome 19. The α subunits bind insulin and are extracellular, whereas the β subunits span the membrane. The intracellular portions of the β subunits have tyrosine kinase activity. The α and β subunits are both glycosylated, with sugar residues extending into the interstitial fluid.

Binding of insulin triggers the tyrosine kinase activity of the β subunits, producing autophosphorylation of the β subunits on tyrosine residues. The autophosphorylation, which is necessary for insulin to exert its biologic effects, triggers phosphorylation of some cytoplasmic proteins and dephosphorylation of others, mostly on serine and threonine residues. (Figure 19–7). Four related insulin receptor substrate (IRS) proteins in cells have been described: IRS-1, IRS-2, IRS-3, and IRS-4. IRS-1 has received the most attention. However, mice in which the insulin receptor gene is knocked out show marked growth retardation in utero, have abnormalities of the CNS and skin, and die at birth of respiratory failure, whereas IRS-1 knockouts show only moderate growth retardation in utero, survive, and are insulin-resistant but otherwise nearly normal. Thus, intracellular pathways in addition to IRS-1 play important roles in the actions of insulin.

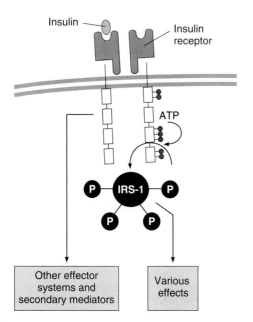

Figure 19–7. Intracellular responses triggered by insulin binding to the insulin receptor. Black balls and balls labeled P represent phosphate groups. IRS-1, insulin receptor substrate-1.

It is interesting to compare the insulin receptor with other related receptors. It is very similar to the receptor for IGF-I but different from the receptor for IGF-II (Figure 19–6). Other receptors for growth factors and receptors for various oncogenes are also tyrosine kinases. However, the amino acid composition of these receptors is quite different.

When insulin binds to its receptors, they aggregate in patches and are taken into the cell by receptor-mediated endocytosis (see Chapter 1). Eventually, the insulin-receptor complexes enter lysosomes, where the receptors are broken down or recycled. The half-life of the insulin receptor is about 7 hours.

The number or the affinity, or both, of insulin receptors is affected by insulin and other hormones, exercise, food, and other factors. Exposure to increased amounts of insulin decreases receptor concentration (down regulation), and exposure to decreased insulin levels increases the number of receptors.

CONSEQUENCES OF INSULIN DEFICIENCY

The far-reaching physiologic effects of insulin are highlighted by a consideration of the extensive and serious consequences of insulin deficiency.

Diabetes Mellitus

In humans, insulin deficiency is a common pathologic condition. In animals, it can be produced by pancreatectomy; by administration of alloxan, streptozocin, or other toxins that in appropriate doses cause selective destruction of the B cells of the pancreatic islets; by administration of drugs that inhibit insulin secretion; and by administration of anti-insulin antibodies. Strains of mice, rats, hamsters, guinea pigs, miniature swine, and monkeys that have a high incidence of spontaneous diabetes mellitus have also been described.

The constellation of abnormalities caused by insulin deficiency is called **diabetes mellitus.** Greek and Roman physicians used the term "diabetes" to refer to conditions in which the cardinal finding was a large urine volume, and two types were distinguished: "diabetes mellitus," in which the urine tasted sweet; and "diabetes insipidus," in which the urine had little taste. Today, the term diabetes insipidus is reserved for conditions in which there is a deficiency of the production or action of vasopressin (see Chapter 14), and the unmodified word diabetes is generally used as a synonym for diabetes mellitus.

Diabetes is characterized by polyuria, polydipsia, weight loss in spite of polyphagia (increased appetite), hyperglycemia, glycosuria, ketosis, acidosis, and coma. There are widespread biochemical abnormalities, but the fundamental defects to which most of the abnormalities can be traced are (1) reduced entry of glucose into various "peripheral" tissues and (2) increased liberation of glucose into the circulation from the liver. There is therefore an extracellular glucose excess and, in many cells, an intracellular glucose deficiency, a situation that has been called "starvation in the midst of plenty." There is also a decrease in the entry of amino acids into muscle and an increase in lipolysis.

Glucose Tolerance

In diabetes, glucose piles up in the bloodstream, especially after meals. If a glucose load is given to a diabetic, the plasma glucose rises higher and returns to the baseline more slowly than it does in normal individuals. The response to a standard oral test dose of glucose, the **oral glucose tolerance test,** is used in the clinical diagnosis of diabetes (Figure 19–8).

Impaired glucose tolerance in diabetes is due in part to reduced entry of glucose into cells (**decreased peripheral utilization**). In the absence of insulin, the entry of glucose into skeletal, cardiac, and smooth muscle and other tissues is decreased (Figure 19–9). Glucose uptake by the liver is also reduced, but the effect is indirect. Intestinal absorption of glucose is unaffected, as is its reabsorption from the urine by the cells of the

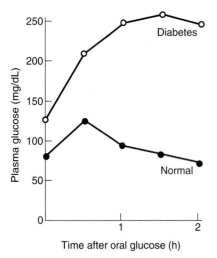

Figure 19–8. Oral glucose tolerance test. Adults are given 75 g of glucose in 300 mL of water. In normal individuals, the fasting venous plasma glucose is less than 115 mg/dL, the 2-hour value is less than 140 mg/dL, and no value is greater than 200 mg/dL. Diabetes mellitus is present if the 2-hour value and one other value are greater than 200 mg/dL. Impaired glucose tolerance is diagnosed when the values are above the upper limits of normal but below the values diagnostic of diabetes.

proximal tubules of the kidneys. Glucose uptake by most of the brain and the red blood cells is also normal.

The second and the major cause of hyperglycemia in diabetes is derangement of the glucostatic function of the liver (see Chapter 17). The liver takes up glucose

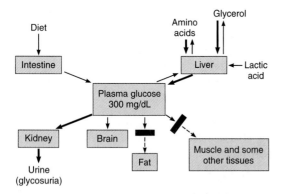

Figure 19–9. Disordered plasma glucose homeostasis in insulin deficiency. Compare with Figure 17–14. The heavy arrows indicate reactions that are accentuated. The rectangles across arrows indicate reactions that are blocked.

from the bloodstream and stores it as glycogen, but because the liver contains glucose 6-phosphatase it also discharges glucose into the bloodstream. Indeed, Claude Bernard spoke of the liver as an endocrine gland that secreted glucose. Insulin facilitates glycogen synthesis and inhibits hepatic glucose output. When the plasma glucose is high, insulin secretion is normally increased and hepatic glucogenesis is decreased. This effect is missing in diabetes. Glucagon also contributes to the hyperglycemia, and glucose output by the liver is facilitated by catecholamines, cortisol, and growth hormone when the stress of illness is severe.

Effects of Hyperglycemia

Hyperglycemia by itself can cause symptoms resulting from the hyperosmolality of the blood. In addition, there is glycosuria because the renal capacity for glucose reabsorption is exceeded. Excretion of the osmotically active glucose molecules entails the loss of large amounts of water (osmotic diuresis; see Chapter 38). The resultant dehydration activates the mechanisms regulating water intake, leading to polydipsia. There is an appreciable urinary loss of Na^+ and K^+ as well. For every gram of glucose excreted, 4.1 kcal is lost from the body. Increasing the oral caloric intake to cover this loss simply raises the plasma glucose further and increases the glycosuria, so mobilization of endogenous protein and fat stores and weight loss are not prevented.

When plasma glucose is episodically elevated over time, small amounts of hemoglobin A are nonenzymatically glycated to form **HbA$_{Ic}$** (see Chapter 27). Careful control of the diabetes with insulin reduces the amount formed and consequently HbA$_{Ic}$ concentration is measured clinically as an integrated index of diabetic control for the 4- to 6-week period before the measurement.

The role of chronic hyperglycemia in production of the long-term complications of diabetes is discussed below.

Effects of Intracellular Glucose Deficiency

The plethora of glucose outside the cells in diabetes contrasts with the intracellular deficit. Glucose catabolism is normally a major source of energy for cellular processes, and in diabetes, energy requirements can be met only by drawing on protein and fat reserves. Mechanisms are activated that greatly increase the catabolism of protein and fat, and one of the consequences of increased fat catabolism is ketosis.

Deficient glucose utilization in the cells of the hypothalamic ventromedial nuclei is probably the cause of the hyperphagia in diabetes. When the activity of the satiety area of this nucleus is decreased in response to decreased glucose utilization in its cells, the lateral appetite area operates unopposed, and food intake is increased (see Chapter 14). However, other explanations have also been proposed.

Glycogen depletion is a common consequence of intracellular glucose deficit, and the glycogen content of liver and skeletal muscle in diabetic animals is usually reduced.

Changes in Protein Metabolism

In diabetes, the rate at which amino acids are catabolized to CO_2 and H_2O is increased. In addition, more amino acids are converted to glucose in the liver. An idea of the rate of gluconeogenesis in fasting diabetic animals is obtained by measuring the ratio of glucose (dextrose) to nitrogen in the urine (**D/N ratio**). It can be calculated that the amount of carbon in the protein represented by 1 g of urinary nitrogen is sufficient to form 8.3 g of glucose. Consequently, the D/N ratio of approximately 3 seen in diabetes indicates the conversion to glucose of about 33% of the carbon of the protein metabolized.

The increased gluconeogenesis has many causes. Glucagon stimulates gluconeogenesis, and hyperglucagonemia is generally present in diabetes. Adrenal glucocorticoids also contribute to increased gluconeogenesis when they are elevated in severely ill diabetics. There is an increased supply of amino acids for gluconeogenesis because, in the absence of insulin, less protein synthesis occurs in muscle and hence blood amino acid levels rise. Alanine is particularly easily converted to glucose. In addition, the activity of the enzymes that catalyze the conversion of pyruvate and other two-carbon metabolic fragments to glucose is increased. These include phosphoenolpyruvate carboxykinase, which facilitates the conversion of oxaloacetate to phosphoenolpyruvate (see Chapter 17). They also include fructose 1,6-diphosphatase, which catalyzes the conversion of fructose diphosphate to fructose 6-phosphate, and glucose 6-phosphatase, which controls the entry of glucose into the circulation from the liver. Increased acetyl-CoA increases pyruvate carboxylase activity, and insulin deficiency increases the supply of acetyl-CoA because lipogenesis is decreased. Pyruvate carboxylase catalyzes the conversion of pyruvate to oxaloacetate (see Figure 17–9).

In diabetes, the net effect of accelerated protein conversion to CO_2, H_2O, and glucose, plus diminished protein synthesis, is a **negative nitrogen balance,** protein depletion, and wasting. Protein depletion from any cause is associated with poor "resistance" to infections.

Fat Metabolism in Diabetes

The principal abnormalities of fat metabolism in diabetes are acceleration of lipid catabolism, with increased formation of ketone bodies, and decreased synthesis of fatty acids and triglycerides. The manifestations of the disordered lipid metabolism are so prominent that diabetes has been called "more a disease of lipid than of carbohydrate metabolism."

Fifty percent of an ingested glucose load is normally burned to CO_2 and H_2O; 5% is converted to glycogen; and 30–40% is converted to fat in the fat depots. In diabetes, less than 5% is converted to fat even though the amount burned to CO_2 and H_2O is also decreased and the amount converted to glycogen is not increased. Therefore, glucose accumulates in the bloodstream and spills over into the urine.

The role of lipoprotein lipase and hormone-sensitive lipase in the regulation of the metabolism of fat depots is discussed in Chapter 17. In diabetes, there is decreased conversion of glucose to fatty acids in the depots because of the intracellular glucose deficiency. Insulin inhibits the hormone-sensitive lipase in adipose tissue, and, in the absence of this hormone, the plasma level of **free fatty acids** (NEFA, UFA, FFA) is more than doubled. The increased glucagon also contributes to the mobilization of FFA. Thus, the FFA level parallels the plasma glucose level in diabetes and in some ways is a better indicator of the severity of the diabetic state. In the liver and other tissues, the fatty acids are catabolized to acetyl-CoA. Some of the acetyl-CoA is burned along with amino acid residues to yield CO_2 and H_2O in the citric acid cycle. However, the supply exceeds the capacity of the tissues to catabolize the acetyl-CoA.

In addition to the previously mentioned increase in gluconeogenesis and marked outpouring of glucose into the circulation, there is a marked impairment of the conversion of acetyl-CoA to malonyl-CoA and thence to fatty acids. This is due to a deficiency of acetyl-CoA carboxylase, the enzyme that catalyzes the conversion. The excess acetyl-CoA is converted to ketone bodies.

In uncontrolled diabetes, there is an increase in the plasma concentration of triglycerides and chylomicrons as well as FFA, and the plasma is often lipemic. The rise in these constituents is due mainly to decreased removal of triglycerides into the fat depots. The decreased activity of lipoprotein lipase contributes to this decreased removal.

Ketosis

When there is excess acetyl-CoA in the body, some of it is converted to acetoacetyl-CoA and then, in the liver, to acetoacetate. Acetoacetate and its derivatives, acetone and β-hydroxybutyrate, enter the circulation in large quantities (see Chapter 17).

These circulating ketone bodies are an important source of energy in fasting. Half of the metabolic rate in fasted normal dogs is said to be due to metabolism of ketones. The rate of ketone utilization in diabetics is also appreciable. It has been calculated that the maximal rate at which fat can be catabolized without significant ketosis is 2.5 g/kg body weight/d in diabetic humans. In untreated diabetes, production is much greater than this, and ketone bodies pile up in the bloodstream.

Acidosis

As noted in Chapter 17, acetoacetate and β-hydroxybutyrate are anions of the fairly strong acids acetoacetic acid and β-hydroxybutyric acids. The hydrogen ions from these acids are buffered, but the buffering capacity is soon exceeded if production is increased. The resulting acidosis stimulates respiration, producing the rapid, deep respiration described by Kussmaul as "air hunger" and named, for him, **Kussmaul breathing.** The urine becomes acidic. However, when the ability of the kidneys to replace the plasma cations accompanying the organic anions with H^+ and NH_4^+ is exceeded, Na^+ and K^+ are lost in the urine. The electrolyte and water losses lead to dehydration, hypovolemia, and hypotension. Finally, the acidosis and dehydration depress consciousness to the point of coma. Diabetic acidosis is a medical emergency. Now that the infections which used to complicate the disease can be controlled with antibiotics, acidosis is the commonest cause of early death in clinical diabetes.

In severe acidosis, total body Na^+ is markedly depleted, and when sodium loss exceeds water loss, plasma Na^+ may also be low. Total body potassium is also low, but the plasma K^+ is usually normal, partly because ECF volume is reduced and partly because K^+ moves from cells to ECF when the ECF H^+ concentration is high. Another factor tending to maintain the plasma K^+ is the lack of insulin-induced entry of K^+ into cells.

The degree to which ketoacidosis complicates experimental diabetes varies in different species. The size of the body fat stores is also a factor conditioning the response to diabetes. Before the isolation of insulin by Banting and Best in 1921, the principal treatment of human diabetes was a starvation diet **(Allen regimen).** This not only lowered the plasma glucose level but also reduced depot fat stores to the point where there was little fat to mobilize.

Coma

Coma in diabetes can be due to acidosis and dehydration. However, the plasma glucose can be elevated to such a degree that independent of plasma pH, the hyperosmolarity of the plasma causes unconsciousness **(hyperosmolar coma).** Accumulation of lactate in the blood **(lactic acidosis)** may also complicate diabetic ketoacidosis if the tissues become hypoxic (see Chapter 33), and lactic acidosis may itself cause coma. Brain edema occurs in about 1% of children with ketoacidosis, and it can cause coma. Its cause is unsettled, but it is a serious complication, with a mortality rate of about 25%.

Cholesterol Metabolism

In diabetes, the plasma cholesterol level is usually elevated, and this plays a role in the accelerated development of the atherosclerotic vascular disease that is a major long-term complication of diabetes in humans. The rise in plasma cholesterol level is due to an increase in the plasma concentration of VLDL and LDL (see Chapter 17), which may be due to increased hepatic production of VLDL or decreased removal of VLDL and LDL from the circulation.

Summary

Because of the complexities of the metabolic abnormalities in diabetes, a summary is in order. One of the key features of insulin deficiency (Figure 19–10) is decreased entry of glucose into many tissues (decreased peripheral utilization). There is also increased net release of glucose from the liver (increased production), due in part to glucagon excess. The resultant hyperglycemia leads to glycosuria and a dehydrating osmotic

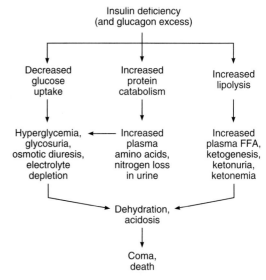

Figure 19–10. Effects of insulin deficiency. (Courtesy of RJ Havel.)

diuresis. Dehydration leads to polydipsia. In the face of intracellular glucose deficiency, appetite is stimulated, glucose is formed from protein (gluconeogenesis), and energy supplies are maintained by metabolism of proteins and fats. Weight loss, debilitating protein deficiency, and inanition are the result.

Fat catabolism is increased, and the system is flooded with triglycerides and FFA. Fat synthesis is inhibited, and the overloaded catabolic pathways cannot handle the excess acetyl-CoA that is formed. In the liver, the acetyl-CoA is converted to ketone bodies. Two of these are organic acids, and metabolic acidosis develops as ketones accumulate. Na^+ and K^+ depletion is added to the acidosis because these plasma cations are excreted with the organic anions not covered by the H^+ and NH_4^+ secreted by the kidneys. Finally, the acidotic, hypovolemic, hypotensive, depleted animal or patient becomes comatose because of the toxic effects of acidosis, dehydration, and hyperosmolarity on the nervous system and dies if treatment is not instituted.

All of these abnormalities are corrected by administration of insulin. Although emergency treatment of acidosis also includes administration of alkali to combat the acidosis and parenteral water, Na^+, and K^+ to replenish body stores, only insulin repairs the fundamental defects in a way that permits a return to normal.

INSULIN EXCESS

Symptoms

All the known consequences of insulin excess are manifestations, directly or indirectly, of the effects of hypoglycemia on the nervous system. Except in individuals who have been fasting for some time, glucose is the only fuel used in appreciable quantities by the brain. The carbohydrate reserves in neural tissue are very limited, and normal function depends upon a continuous glucose supply. As the plasma glucose level falls, the first symptoms are palpitations, sweating, and nervousness due to autonomic discharge. At lower plasma glucose levels, so-called **neuroglycopenic symptoms** begin to appear. These include hunger as well as confusion and the other cognitive abnormalities. At even lower plasma glucose levels, lethargy, coma, convulsions, and eventually death occur. The thresholds for these abnormalities are shown in Figure 19–11. Obviously, the onset of hypoglycemic symptoms calls for prompt treatment with glucose or glucose-containing drinks such as orange juice. Although a dramatic disappearance of symptoms is the usual response, abnormalities ranging from transient intellectual dulling to permanent coma may persist if the hypoglycemia was severe or prolonged.

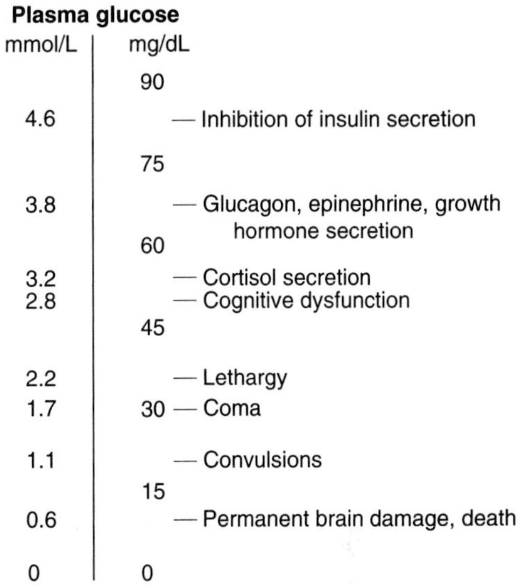

Plasma glucose

mmol/L	mg/dL	
	90	
4.6		— Inhibition of insulin secretion
	75	
3.8		— Glucagon, epinephrine, growth hormone secretion
	60	
3.2		— Cortisol secretion
2.8		— Cognitive dysfunction
	45	
2.2		— Lethargy
1.7	30	— Coma
1.1		— Convulsions
	15	
0.6		— Permanent brain damage, death
0	0	

Figure 19–11. Plasma glucose levels in arterialized venous blood at which various effects of hypoglycemia appear.

Compensatory Mechanisms

One important compensation for hypoglycemia is cessation of the secretion of endogenous insulin. Inhibition of insulin secretion is complete at a plasma glucose level of about 80 mg/dL (Figures 19–11 and 19–12). In addition, hypoglycemia triggers increased secretion of at least four counterregulatory hormones: glucagon, epinephrine, growth hormone, and cortisol. The epinephrine response is reduced during sleep. Glucagon and epinephrine increase the hepatic output of glucose by increasing glycogenolysis. Growth hormone decreases the utilization of glucose in various peripheral tissues, and cortisol has a similar action. The keys to counterregulation appear to be epinephrine and glucagon: if the plasma concentration of either increases, the decline in the plasma glucose level is reversed; but if both fail to increase, there is little if any compensatory rise in the plasma glucose level. The actions of the other hormones are supplementary.

Note that the autonomic discharge and release of counterregulatory hormones normally occurs at a higher plasma glucose level than the cognitive deficits and other more serious CNS changes (Figure 19–11). The symptoms caused by the autonomic discharge serve as a warning to seek glucose replacement that diabetics treated with insulin soon learn to heed. However, particularly in long-term diabetics who have been tightly regulated, the autonomic symptoms may not

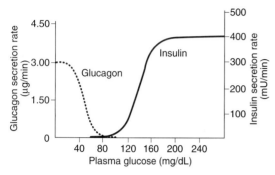

Figure 19-12. Mean rates of insulin and glucagon delivery from an artificial pancreas at various plasma glucose levels. The device was programmed to establish and maintain various plasma glucose levels in insulin-requiring diabetic humans, and the values for hormone output approximate the output of the normal human pancreas. The shape of the insulin curve also resembles the insulin response of incubated B cells to graded concentrations of glucose. (Reproduced, with permission, from Marliss EB et al: Normalization of glycemia in diabetics during meals with insulin and glucagon delivery by the artificial pancreas. Diabetes 1977;26:663.)

occur, and the resulting **hypoglycemia unawareness** can be a clinical problem of some magnitude.

REGULATION OF INSULIN SECRETION

The normal concentration of insulin measured by radioimmunoassay in the peripheral venous plasma of fasting normal humans is 0–70 µU/mL (0–502 pmol/L). The amount of insulin secreted in the basal state is about 1 U/h, with a fivefold to tenfold increase following ingestion of food. Therefore, the average amount secreted per day in a normal human is about 40 U (287 nmol).

Factors that stimulate and inhibit insulin secretion are summarized in Table 19–6.

Effects of the Plasma Glucose Level

It has been known for many years that glucose acts directly on pancreatic B cells to increase insulin secretion. The response to glucose is biphasic; there is a rapid but short-lived increase in secretion followed by a more slowly developing prolonged increase (Figure 19–13).

Glucose enters the B cells via GLUT 2 transporters and is phosphorylated by glucokinase, then metabolized to pyruvate in the cytoplasm (Figure 19–14). The pyruvate enters the mitochondria and is metabolized to CO_2 and H_2O via the citric acid cycle with the forma-

Table 19–6. Factors affecting insulin secretion.

Stimulators	Inhibitors
Glucose	Somatostatin
Mannose	2-Deoxyglucose
Amino acids (leucine,	Mannoheptulose
arginine, others)	α-Adrenergic stimulators
Intestinal hormones (GIP,	(norepinephrine,
GLP-1 [7–36], gastrin,	epinephrine)
secretin, CCK; others?)	β-Adrenergic blockers
β-Keto acids	(propranolol)
Acetylcholine	Galanin
Glucagon	Diazoxide
Cyclic AMP and various	Thiazide diuretics
cAMP-generating	K+ depletion
substances	Phenytoin
β-Adrenergic stimulators	Alloxan
Theophylline	Microtubule inhibitors
Sulfonylureas	Insulin

tion of ATP by oxidative phosphorylation. The ATP enters the cytoplasm, where it inhibits ATP-sensitive K+ channels, reducing K+ efflux. This depolarizes the B cell, and Ca^{2+} enters the voltage-gated Ca^{2+} channels. The Ca^{2+} influx causes exocytosis of a readily releasable

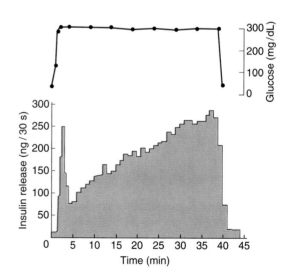

Figure 19–13. Insulin secretion from perfused rat pancreas in response to sustained glucose infusion. Values are means of three preparations. The top record shows the glucose concentration in the effluent perfusion mixture. (Reproduced, with permission, from Curry DL, Bennett LL, Grodsky GM: Dynamics of insulin secretion by the perfused rat pancreas. Endocrinology 1968;83:572.)

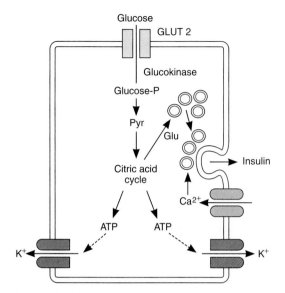

Figure 19–14. Insulin secretion. Glucose enters B cells by GLUT 2 transporters. It is phosphorylated and metabolized to pyruvate (Pyr) in the cytoplasm. The Pyr enters the mitochondria and is metabolized via the citric acid cycle. The ATP formed by oxidative phosphorylation inhibits ATP-sensitive K^+ channels, reducing K^+ efflux. This depolarizes the B cell, and Ca^{2+} influx is increased. The Ca^{2+} stimulates release of insulin by exocytosis. Glutamate (Glu) is also formed, and this primes secretory granules, preparing them for exocytosis.

pool of insulin-containing secretory granules, producing the initial spike of insulin secretion.

Metabolism of pyruvate via the citric acid cycle also causes an increase in intracellular glutamate. The glutamate appears to act on a second pool of secretory granules, committing them to the releasable form. The action of glutamate may be to decrease the pH in the secretory granules, a necessary step in their maturation. The release of these granules then produces the prolonged second phase of the insulin response to glucose. Thus, glutamate appears to act as an intracellular second messenger that primes secretory granules for secretion.

The feedback control of plasma glucose on insulin secretion normally operates with great precision, so that plasma glucose and insulin levels parallel each other with remarkable consistency.

Protein & Fat Derivatives

Insulin stimulates the incorporation of amino acids into proteins and combats the fat catabolism that pro-

duces the β-keto acids. Therefore, it is not surprising that arginine, leucine, and certain other amino acids stimulate insulin secretion, as do β-keto acids such as acetoacetate. Like glucose, these compounds generate ATP when metabolized, and this closes ATP-sensitive K^+ channels in the B cells. In addition, L-arginine is the precursor of NO, and NO stimulates insulin secretion.

Oral Hypoglycemic Agents

Tolbutamide and other sulfonylurea derivatives such as acetohexamide, tolazamide, glipizide, and glyburide are orally active hypoglycemic agents that lower blood glucose by increasing the secretion of insulin. They only work in patients with some remaining B cells and are ineffective after pancreatectomy or in type 1 diabetes. They bind to the ATP-inhibited K^+ channels in the B cell membranes and inhibit channel activity, depolarizing the B cell membrane and increasing Ca^{2+} influx and hence insulin release.

Persistent hyperinsulinemic hypoglycemia of infancy is a rare condition in which plasma insulin is elevated despite the hypoglycemia. The condition is caused by inactivating mutations in the ATP-inhibited K^+ channels. Treatment consists of administration of diazoxide, a drug that increases the activity of the K^+ channels—or, in more severe cases, subtotal pancreatectomy.

The biguanides phenformin and metformin are oral hypoglycemic agents that act in the absence of insulin. Phenformin caused lactic acidosis in an unacceptably large number of patients, and because of the seriousness of this side effect, it was withdrawn from the United States market. Metformin can also cause lactic acidosis, but the incidence is only 5–10% of the incidence associated with phenformin. Metformin acts primarily by reducing gluconeogenesis and therefore decreasing hepatic glucose output. It is sometimes combined with a sulfonylurea in the treatment of type 2 diabetes.

Troglitazone (Rezulin) and related **thiazolidinediones** are also used in the treatment of diabetes because they increase insulin-mediated peripheral glucose disposal, thus reducing insulin resistance. They bind to and activate peroxisome proliferator-activated receptor γ (PPARγ) in the nucleus of cells (see Chapter 1). Activation of this receptor, which is a member of the superfamily of hormone-sensitive nuclear transcription factors (see Chapter 1), has a unique ability to normalize a variety of metabolic functions.

Cyclic AMP & Insulin Secretion

Stimuli that increase cAMP levels in B cells increase insulin secretion, probably by increasing intracellular Ca^{2+}. These include β-adrenergic agonists, glucagon, and phosphodiesterase inhibitors such as theophylline.

Catecholamines have a dual effect on insulin secretion; they inhibit insulin secretion via α_2-adrenergic receptors and stimulate insulin secretion via β-adrenergic receptors. The net effect of epinephrine and norepinephrine is usually inhibition. However, if catecholamines are infused after administration of α-adrenergic blocking drugs, the inhibition is converted to stimulation.

Effect of Autonomic Nerves

Branches of the right vagus nerve innervate the pancreatic islets, and stimulation of the right vagus causes increased insulin secretion via M_4 receptors (see Table 4–2). Atropine blocks the response, and acetylcholine stimulates insulin secretion. The effect of acetylcholine, like that of glucose, is due to increased cytoplasmic Ca^{2+}, but acetylcholine activates phospholipase C, with the released IP_3 releasing the Ca^{2+} from the endoplasmic reticulum.

Stimulation of the sympathetic nerves to the pancreas inhibits insulin secretion. The inhibition is produced by released norepinephrine acting on α_2-adrenergic receptors. However, if α-adrenergic receptors are blocked, stimulation of the sympathetic nerves causes increased insulin secretion mediated by β_2-adrenergic receptors. The polypeptide galanin is found in some of the autonomic nerves innervating the islets, and galanin inhibits insulin secretion by activating the K^+ channels that are inhibited by ATP. Thus, although the denervated pancreas responds to glucose, the autonomic innervation of the pancreas is involved in the overall regulation of insulin secretion.

Intestinal Hormones

Orally administered glucose exerts a greater insulin-stimulating effect than intravenously administered glucose, and orally administered amino acids also produce a greater insulin response than intravenous amino acids. These observations led to exploration of the possibility that a substance secreted by the gastrointestinal mucosa stimulated insulin secretion. Glucagon, glucagon derivatives, secretin, cholecystokinin (CCK), gastrin, and gastric inhibitory peptide (GIP) all have such an action (see Chapter 26), and CCK potentiates the insulin-stimulating effects of amino acids. However, GIP is the only one of these peptides that produces stimulation when administered in doses that produce blood GIP levels comparable to those produced by oral glucose.

Recently, attention has focused on glucagon-like polypeptide 1 (7–36) (GLP-1 [7–36]) as an additional gut factor that stimulates insulin secretion. This polypeptide is a product of preproglucagon (see below).

There are GLP-1 (7–36) receptors as well as GIP receptors on B cells, and GLP-1 (7–36) is a more potent insulinotropic hormone than GIP. GIP and GLP-1 (7–36) both appear to act by increasing Ca^{2+} influx through voltage-gated Ca^{2+} channels.

The possible roles of pancreatic somatostatin and glucagon in the regulation of insulin secretion are discussed below.

Effects of K^+ Depletion

K^+ depletion decreases insulin secretion, and K^+-depleted patients, eg, patients with primary hyperaldosteronism (see Chapter 20), develop diabetic glucose tolerance curves. These curves are restored to normal by K^+ repletion. The thiazide diuretics, which cause loss of K^+ as well as Na^+ in the urine (see Chapter 38), decrease glucose tolerance and make diabetes worse. They apparently exert this effect primarily because of their K^+-depleting effects, although some of them also cause pancreatic islet cell damage.

Long-Term Changes in B Cell Responses

The magnitude of the insulin response to a given stimulus is determined in part by the secretory history of the B cells. Individuals fed a high-carbohydrate diet for several weeks not only have higher fasting plasma insulin levels but also show a greater secretory response to a glucose load than individuals fed an isocaloric low-carbohydrate diet.

Although the B cells respond to stimulation with hypertrophy like other endocrine cells, they become exhausted and stop secreting (**B cell exhaustion**) when the stimulation is marked or prolonged. The pancreatic reserve is large, and it is difficult to produce B cell exhaustion in normal animals; but if the pancreatic reserve is reduced by partial pancreatectomy or small doses of alloxan, exhaustion of the remaining B cells can be produced by any procedure that chronically raises the plasma glucose level. For example, diabetes can be produced in animals with limited pancreatic reserves by anterior pituitary extracts, growth hormone, thyroid hormones, or the prolonged continuous infusion of glucose alone. The diabetes precipitated by hormones in animals is at first reversible, but with prolonged treatment it becomes permanent. The transient diabetes is usually named for the agent producing it, eg, "hypophysial diabetes," "thyroid diabetes." Permanent diabetes persisting after treatment has been discontinued is indicated by the prefix meta-, eg, **"metahypophysial diabetes"** or **"metathyroid diabetes."** When insulin is administered along with the diabetogenic hormones, the B cells are protected, probably because the plasma glucose is lowered, and diabetes does not develop.

It is interesting in this regard that genetic factors may be involved in the control of B cell reserve. In mice in which the gene for IRS-1 has been knocked out (see above), there is a robust compensatory B cell response. However, in IRS-2 knockouts, the compensation is reduced and a more severe diabetic phenotype is produced.

Obesity, the Metabolic Syndrome, & Type 2 Diabetes

Obesity is discussed in relation to the regulation of food intake and energy balance in Chapter 14 and in terms of overall nutrition in Chapter 17. It deserves additional consideration in this chapter because of its special relation to disordered carbohydrate metabolism and diabetes. As body weight increases, there is increasing insulin resistance, ie, a decreased ability of insulin to move glucose into fat and muscle and to shut off glucose release from the liver. Weight reduction decreases insulin resistance. Associated with obesity there is hyperinsulinemia, dyslipidemia—characterized by high circulating triglycerides and low HDL—and accelerated development of atherosclerosis. This combination of findings is commonly called the **metabolic syndrome,** or **syndrome X.** Some of the patients with the syndrome are prediabetic, whereas others have type 2 diabetes. It has not been proved but it is logical to assume that the hyperinsulinemia is a compensatory response to the increased insulin resistance and that frank diabetes develops in individuals with reduced B cell reserves.

These observations and other data strongly suggest that fat produces a chemical signal or signals that act on muscles and the liver to increase insulin resistance. Evidence for this includes the recent observation that when glucose transporters are selectively knocked out in adipose tissue, there is an associated decrease in glucose transport in muscle in vivo, but when the muscles of those animals are tested in vitro, their transport is normal.

One possible signal is the circulating free fatty acid level, which is elevated in many insulin-resistant states. Other possibilities are peptides and proteins secreted by fat cells. It is now clear that white fat depots are not inert lumps but are actually endocrine tissues that secrete not only leptin (see Chapter 14) but also other hormones that affect fat metabolism. The most intensively studied of these **adipokines** are listed in Table 19–7. Some of the adipokines increase rather than decrease insulin sensitivity. Leptin, for example, increases insulin sensitivity and so does **adiponectin,** whereas TNFα and a newly described peptide called **resistin** decrease

Table 19–7. Adipokines.

Agent	Effect on Insulin Resistance
Leptin	Decreases
TNFα	Increases
Adiponectin	Decreases
Resistin	Increases

it. Further complicating the situation, there is marked insulin resistance in the rare metabolic disease **congenital lipodystrophy,** in which fat depots fail to develop. This resistance is reduced by leptin and adiponectin. Finally, a variety of knockouts of intracellular second messengers have been reported to increase insulin resistance. It is not clear how—or indeed if—these findings fit together to provide an explanation of the relation of obesity to insulin tolerance, but the topic is obviously an important one in which additional important results can be expected in the near future.

GLUCAGON

Chemistry

Human glucagon, a linear polypeptide with a molecular weight of 3485, is produced by the A cells of the pancreatic islets and the upper gastrointestinal tract. It contains 29 amino acid residues (see Table 26–2). All mammalian glucagons appear to have the same structure. Human preproglucagon (Figure 19–15) is a 179-

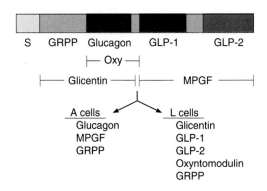

Figure 19–15. Posttranslational processing of preproglucagon in A and L cells. S, signal peptide; MPGF, major proglucagon fragment; Oxy, oxyntomodulin; GRPP, glicentin-related polypeptide; GLP, glucagon-like peptide. (Modified from Drucker, DJ: Glucagon and glucagon-like peptides. Pancreas 1990;5:484.)

amino-acid polypeptide that is found in pancreatic A cells, in L cells in the lower gastrointestinal tract, and in the brain. It is the product of a single mRNA, but it is processed differently in different tissues. In A cells, it is processed primarily to glucagon and **major proglucagon fragment (MPGF).** In L cells, it is processed primarily to **glicentin,** a polypeptide that consists of glucagon extended by additional amino acid residues at either end, plus **glucagon-like polypeptides 1** and **2 (GLP-1** and **GLP-2).** Some **oxyntomodulin** is also formed, and in both A and L cells, residual **glicentin-related polypeptide (GRPP)** is left. Glicentin has some glucagon activity. GLP-1 and GLP-2 have no definite biologic activity by themselves. However, GLP-1 is processed further by removal of its seven amino terminal amino acid residues, and the product, **GLP-1 (7–36),** is a potent stimulator of insulin secretion that also increases glucose utilization (see above). GLP-1 and GLP-2 are also produced in the brain. The function of GLP-1 in this location is uncertain, but GLP-2 appears to be the mediator in a pathway from the nucleus tractus solitarius (NTS) to the dorsomedial nuclei of the hypothalamus, and injection of GLP-2 inhibits food intake. Oxyntomodulin inhibits gastric acid secretion, though its physiologic role is unsettled, and GRPP does not have any established physiologic effects.

Action

Glucagon is glycogenolytic, gluconeogenic, lipolytic, and ketogenic. It acts on serpentine receptors with a molecular weight of about 190,000. In the liver, it acts via G_s to activate adenylyl cyclase and increase intracellular cAMP. This leads via protein kinase A to activation of phosphorylase and therefore to increased breakdown of glycogen and an increase in plasma glucose. However, glucagon acts on different glucagon receptors located on the same hepatic cells to activate phospholipase C, and the resulting increase in cytoplasmic Ca^{2+} also stimulates glycogenolysis. Protein kinase A also decreases the metabolism of glucose 6-phosphate (Figure 19–16) by inhibiting the conversion of phosphoenolpyruvate to pyruvate. It also decreases the concentration of fructose 2,6-diphosphate, and this in turn inhibits the conversion of fructose 6-phosphate to fructose 1,6-diphosphate. The resultant buildup of glucose 6-phosphate leads to increased release to glucose.

Glucagon does not cause glycogenolysis in muscle. It increases gluconeogenesis from available amino acids in the liver and elevates the metabolic rate. It increases ketone body formation by decreasing malonyl-CoA levels in the liver (see Chapter 17). Its lipolytic activity, which leads in turn to increased ketogenesis, is discussed in Chapter 17. The calorigenic action of

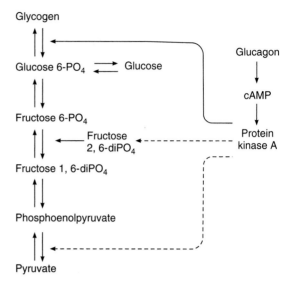

Figure 19–16. Mechanisms by which glucagon increases glucose output from the liver. Solid arrows indicate facilitation, and dashed arrows indicate inhibition.

glucagon is not due to the hyperglycemia per se but probably to the increased hepatic deamination of amino acids.

Large doses of exogenous glucagon exert a positively inotropic effect on the heart (see Chapter 29) without producing increased myocardial excitability, presumably because they increase myocardial cAMP. Use of this hormone in the treatment of heart disease has been advocated, but there is no evidence for a physiologic role of glucagon in the regulation of cardiac function. Glucagon also stimulates the secretion of growth hormone, insulin, and pancreatic somatostatin.

Metabolism

Glucagon has a half-life in the circulation of 5–10 minutes. It is degraded by many tissues but particularly by the liver. Since glucagon is secreted into the portal vein and reaches the liver before it reaches the peripheral circulation, peripheral blood levels are relatively low. The rise in peripheral blood glucagon levels produced by excitatory stimuli (see below) is exaggerated in patients with cirrhosis, presumably because of decreased hepatic degradation of the hormone.

Regulation of Secretion

The principal factors known to affect glucagon secretion are summarized in Table 19–8. Secretion is increased by hypoglycemia and decreased by a rise in

Table 19–8. Factors affecting glucagon secretion.

Stimulators	Inhibitors
Amino acids (particularly the glucogenic amino acids: alanine, serine, glycine, cysteine, and threonine)	Glucose
	Somatostatin
	Secretin
	FFA
CCK, gastrin	Ketones
Cortisol	Insulin
Exercise	Phenytoin
Infections	α-Adrenergic stimulators
Other stresses	GABA
β-Adrenergic stimulators	
Theophylline	
Acetylcholine	

plasma glucose. The B cells contain GABA, and there is evidence that coincident with the increased insulin secretion produced by hyperglycemia, GABA is released and acts on the A cells to inhibit glucagon secretion by activating $GABA_A$ receptors. The $GABA_A$ receptors are Cl^- channels, and the resulting Cl^- influx hyperpolarizes the A cells.

Secretion is also increased by stimulation of the sympathetic nerves to the pancreas, and this sympathetic effect is mediated via β-adrenergic receptors and cAMP. It appears that the A cells are like the B cells in that stimulation of β-adrenergic receptors increases secretion and stimulation of α-adrenergic receptors inhibits secretion (see above). However, the pancreatic response to sympathetic stimulation in the absence of blocking drugs is increased secretion of glucagon, so the effect of β-receptors predominates in the glucagon-secreting cells. The stimulatory effects of various stresses and possibly of exercise and infection are mediated at least in part via the sympathetic nervous system. Vagal stimulation also increases glucagon secretion.

A protein meal and infusion of various amino acids increase glucagon secretion. It seems appropriate that the glucogenic amino acids are particularly potent in this regard, since these are the amino acids that are converted to glucose in the liver under the influence of glucagon. The increase in glucagon secretion following a protein meal is also valuable, since the amino acids stimulate insulin secretion, and the secreted glucagon prevents the development of hypoglycemia while the insulin promotes storage of the absorbed carbohydrates and lipids. Glucagon secretion increases during starvation. It reaches a peak on the third day of a fast, at the time of maximal gluconeogenesis. Thereafter, the plasma glucagon level declines as fatty acids and ketones become the major sources of energy.

During exercise, there is an increase in glucose utilization (see below) that is balanced by an increase in glucose production caused by an increase in circulatory glucagon levels.

The glucagon response to oral administration of amino acids is greater than the response to intravenous infusion of amino acids, suggesting that a glucagon-stimulating factor is secreted from the gastrointestinal mucosa. CCK and gastrin increase glucagon secretion, whereas secretin inhibits it. Since CCK and gastrin secretion are both increased by a protein meal, either hormone could be the gastrointestinal mediator of the glucagon response. The inhibition produced by somatostatin is discussed below.

Glucagon secretion is also inhibited by FFA and ketones. However, this inhibition can be overridden, since plasma glucagon levels are high in diabetic ketoacidosis.

Insulin-Glucagon Molar Ratios

As noted above, insulin is glycogenic, antigluconeogenetic, antilipolytic, and antiketotic in its actions. It thus favors storage of absorbed nutrients and is a "hormone of energy storage." Glucagon, on the other hand, is glycogenolytic, gluconeogenetic, lipolytic, and ketogenic. It mobilizes energy stores and is a "hormone of energy release." Because of their opposite effects, the blood levels of both hormones must be considered in any given situation. It is convenient to think in terms of the molar ratios of these hormones.

The insulin-glucagon molar ratios fluctuate markedly because the secretion of glucagon and insulin are both modified by the conditions that preceded the application of any given stimulus (Table 19–9). Thus,

Table 19–9. Insulin-glucagon molar ratios (I/G) in blood in various conditions.[1]

Condition	Hepatic Glucose Storage (S) or Production (P)[2]	I/G
Glucose availability		
Large carbohydrate meal	4+ (S)	70
Intravenous glucose	2+ (S)	25
Small meal	1+ (S)	7
Glucose need		
Overnight fast	1+ (P)	2.3
Low-carbohydrate diet	2+ (P)	1.8
Starvation	4+ (P)	0.4

[1]Courtesy of RH Unger.

[2]1+ to 4+ indicate relative magnitude.

for example, the insulin-glucagon molar ratio on a balanced diet is approximately 2.3. An infusion of arginine increases the secretion of both hormones and raises the ratio to 3.0. After 3 days of starvation, the ratio falls to 0.4, and an infusion of arginine in this state lowers the ratio to 0.3. Conversely, the ratio is 25 in individuals receiving a constant infusion of glucose and rises to 170 on ingestion of a protein meal during the infusion. The rise occurs because insulin secretion rises sharply, while the usual glucagon response to a protein meal is abolished. Thus, when energy is needed during starvation, the insulin-glucagon molar ratio is low, favoring glycogen breakdown and gluconeogenesis; conversely, when the need for energy mobilization is low, the ratio is high, favoring the deposition of glycogen, protein, and fat.

OTHER ISLET CELL HORMONES

In addition to insulin and glucagon, the pancreatic islets secrete somatostatin and pancreatic polypeptide into the bloodstream. In addition, somatostatin may be involved in regulatory processes within the islets that adjust the pattern of hormones secreted in response to various stimuli.

Somatostatin

Somatostatin and its receptors are discussed in Chapter 4. Somatostatin 14 (SS 14) and its amino terminal-extended form somatostatin 28 (SS 28) are found in the D cells of pancreatic islets. Both forms inhibit the secretion of insulin, glucagon, and pancreatic polypeptide and may act locally within the pancreatic islets in a paracrine fashion. SS 28 is more active than SS 14 in inhibiting insulin secretion, and it apparently acts via the SSTR5 receptor (see Chapter 4). Patients with somatostatin-secreting pancreatic tumors (**somatostatinomas**) develop hyperglycemia and other manifestations of diabetes that disappear when the tumor is removed. They also develop dyspepsia due to slow gastric emptying and decreased gastric acid secretion, and gallstones, which are precipitated by decreased gallbladder contraction due to inhibition of CCK secretion. The secretion of pancreatic somatostatin is increased by several of the same stimuli that increase insulin secretion, ie, glucose and amino acids, particularly arginine and leucine. It is also increased by CCK. Somatostatin is released from the pancreas and the gastrointestinal tract into the peripheral blood.

Pancreatic Polypeptide

Human pancreatic polypeptide is a linear polypeptide that contains 36 amino acid residues and is produced by F cells in the islets. It is closely related to two other 36-amino-acid polypeptides: **polypeptide YY (PYY),** which is found in the intestine and may be a gastrointestinal hormone, and **neuropeptide Y,** which is found in the brain and the autonomic nervous system (see Chapter 4). All end in tyrosine and are amidated at their carboxyl terminal. Pancreatic polypeptide can be measured by radioimmunoassay in peripheral blood. At least in part, its secretion is under cholinergic control; plasma levels fall after administration of atropine. Its secretion is increased by a meal containing protein and by fasting, exercise, and acute hypoglycemia. Secretion is decreased by somatostatin and intravenous glucose. Infusions of leucine, arginine, and alanine do not affect it, so the stimulatory effect of a protein meal may be mediated indirectly. Pancreatic polypeptide slows the absorption of food in humans, and it may smooth out the peaks and valleys of absorption. However, its exact physiologic function is still uncertain.

Organization of the Pancreatic Islets

The presence in the pancreatic islets of hormones that affect the secretion of other islet hormones suggests that the islets function as secretory units in the regulation of nutrient homeostasis. Somatostatin inhibits the secretion of insulin, glucagon, and pancreatic polypeptide (Figure 19–17); insulin inhibits the secretion of glucagon; and glucagon stimulates the secretion of insulin and somatostatin. As noted above, A and D cells and pancreatic polypeptide-secreting cells are generally located around the periphery of the islets, with the B cells in the center. There are clearly two types of islets, glucagon-rich islets and pancreatic polypeptide-rich islets, but the functional significance of this separation is not known. The islet cell hormones released into the ECF probably diffuse to other islet cells and influence their function (paracrine communication; see Chapter 1). It has been demonstrated that there are gap junctions between A, B, and D cells and that these permit the passage of ions and other small molecules from one

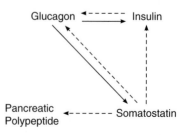

Figure 19–17. Effects of islet cell hormones on the secretion of other islet cell hormones. Solid arrows indicate stimulation; dashed arrows indicate inhibition.

cell to another, which could coordinate their secretory functions.

EFFECTS OF OTHER HORMONES & EXERCISE ON CARBOHYDRATE METABOLISM

Exercise has direct effects on carbohydrate metabolism. Many hormones in addition to insulin, IGF-I, IGF-II, glucagon, and somatostatin also have important roles in the regulation of carbohydrate metabolism. They include epinephrine, thyroid hormones, glucocorticoids, and growth hormone. The other functions of these hormones are considered elsewhere, but it seems wise to summarize their effects on carbohydrate metabolism in the context of the present chapter. The "directional flow valves" in metabolism at which these hormones act are discussed in Chapter 17.

Exercise

The entry of glucose into skeletal muscle is increased during exercise in the absence of insulin by causing an insulin-independent increase in the number of GLUT 4 transporters in muscle cell membranes (see above). This increase in glucose entry persists for several hours after exercise, and regular exercise training can produce prolonged increases in insulin sensitivity. Exercise can precipitate hypoglycemia in diabetics not only because of the increase in muscle uptake of glucose but also because absorption of injected insulin is more rapid during exercise. Patients with diabetes should take in extra calories or reduce their insulin dosage when they exercise.

Catecholamines

The activation of phosphorylase in liver by catecholamines is discussed in Chapters 17 and 20. Activation occurs via β-adrenergic receptors, which increase intracellular cAMP, and α-adrenergic receptors, which increase intracellular Ca^{2+}. Hepatic glucose output is increased, producing hyperglycemia. In muscle, the phosphorylase is also activated via cAMP and presumably via Ca^{2+}, but the glucose 6-phosphate formed can be catabolized only to pyruvate because of the absence of glucose 6-phosphatase. For reasons that are not entirely clear, large amounts of pyruvate are converted to lactate, which diffuses from the muscle into the circulation (Figure 19–18). The lactate is oxidized in the liver to pyruvate and converted to glycogen. Therefore, the response to an injection of epinephrine is an initial glycogenolysis followed by a rise in hepatic glycogen content. Lactate oxidation may be responsible for the calorigenic effect of epinephrine (see Chapter 20). Epi-

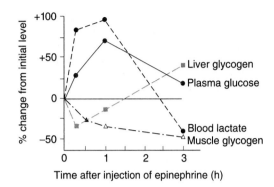

Figure 19–18. Effect of epinephrine on tissue glycogen, plasma glucose, and blood lactate levels in fed rats. (Reproduced, with permission, from Ruch TC, Patton HD [editors]: *Physiology and Biophysics*, 20th ed. Vol. 3. Saunders, 1973.)

nephrine and norepinephrine also liberate FFA into the circulation, and epinephrine decreases peripheral utilization of glucose.

Thyroid Hormones

Thyroid hormones make experimental diabetes worse; thyrotoxicosis aggravates clinical diabetes; and metathyroid diabetes can be produced in animals with decreased pancreatic reserve. The principal diabetogenic effect of thyroid hormones is to increase absorption of glucose from the intestine, but the hormones also cause (probably by potentiating the effects of catecholamines) some degree of hepatic glycogen depletion. Glycogen-depleted liver cells are easily damaged. When the liver is damaged, the glucose tolerance curve is diabetic because the liver takes up less of the absorbed glucose. Thyroid hormones may also accelerate the degradation of insulin. All these actions have a hyperglycemic effect and, if the pancreatic reserve is low, may lead to B cell exhaustion.

Adrenal Glucocorticoids

Glucocorticoids from the adrenal cortex (see Chapter 20) elevate blood glucose and produce a diabetic type of glucose tolerance curve. In humans, this effect may occur only in individuals with a genetic predisposition to diabetes. Glucose tolerance is reduced in 80% of patients with Cushing's syndrome (see Chapter 20), and 20% of these patients have frank diabetes. The glucocorticoids are necessary for glucagon to exert its gluconeogenic action during fasting. They are gluconeogenic themselves, but their role is mainly permissive. In adrenal insufficiency, the blood glucose is normal as long as food intake is maintained, but fasting precipitates hypoglycemia and collapse. The plasma-glucose-lowering effect of insulin is greatly enhanced in patients with adrenal insufficiency. In animals with experimental diabetes, adrenalectomy markedly ameliorates the diabetes. The major diabetogenic effects are an increase in protein catabolism with increased gluconeogenesis in the liver; increased hepatic glycogenesis and ketogenesis; and a decrease in peripheral glucose utilization relative to the blood insulin level that may be due to inhibition of glucose phosphorylation (see below).

Growth Hormone

Human growth hormone makes clinical diabetes worse, and 25% of patients with growth hormone-secreting tumors of the anterior pituitary have diabetes. Hypophysectomy ameliorates diabetes and increases sensitivity to insulin even more than adrenalectomy, whereas growth hormone treatment decreases insulin responsiveness.

The effects of growth hormone are partly direct and partly mediated via IGF-I (see Chapter 22). Growth hormone mobilizes FFA from adipose tissue, thus favoring ketogenesis. It decreases glucose uptake into some tissues ("anti-insulin action"), increases hepatic glucose output, and may decrease tissue binding of insulin. Indeed, it has been suggested that the ketosis and decreased glucose tolerance produced by starvation are due to hypersecretion of growth hormone. Growth hormone does not stimulate insulin secretion directly, but the hyperglycemia it produces secondarily stimulates the pancreas and may eventually exhaust the B cells.

HYPOGLYCEMIA & DIABETES MELLITUS IN HUMANS

Hypoglycemia

"Insulin reactions" are common in type 1 diabetics, and occasional hypoglycemic episodes are the price of good diabetic control in most diabetics. There is an increase in glucose uptake by skeletal muscle and an increased absorption of injected insulin during exercise (see above).

Symptomatic hypoglycemia also occurs in nondiabetics, and a review of some of the more important causes serves to emphasize the variables affecting blood glucose homeostasis. Chronic mild hypoglycemia can cause incoordination and slurred speech, and the condition can be mistaken for drunkenness. Mental aberrations and convulsions in the absence of frank coma also occur. When the level of insulin secretion is chronically elevated by an **insulinoma,** a rare, insulin-secreting tumor of the pancreas, symptoms are most common in

the morning. This is because a night of fasting has depleted hepatic glycogen reserves. However, symptoms can develop at any time, and in such patients, the diagnosis may be missed. Some cases of insulinoma have been erroneously diagnosed as epilepsy or psychosis. Hypoglycemia also occurs in some patients with large malignant tumors that do not involve the pancreatic islets, and the hypoglycemia in these cases is apparently due to excess secretion of IGF-II.

As noted above, the autonomic discharge that produces shakiness, sweating, anxiety, and hunger normally occurs at higher plasma glucose levels than cognitive dysfunction and serves as a warning to ingest sugar. However, in some individuals, these warning symptoms fail to occur before symptoms due to cerebral dysfunction, and this **hypoglycemia unawareness** is potentially dangerous. The condition is prone to develop in patients with insulinomas and in diabetics receiving intensive insulin therapy, so it appears that repeated bouts of hypoglycemia cause the eventual development of hypoglycemia unawareness. If blood sugar rises again for some time, the warning symptoms again appear at a higher plasma glucose level than cognitive abnormalities and coma. The reason why prolonged hypoglycemia causes loss of the warning symptoms is unsettled.

In liver disease, the glucose tolerance curve is diabetic but the fasting plasma glucose level is low (Figure 19–19). In **functional hypoglycemia,** the plasma glucose rise is normal after a test dose of glucose but the subsequent fall overshoots to hypoglycemic levels, pro-

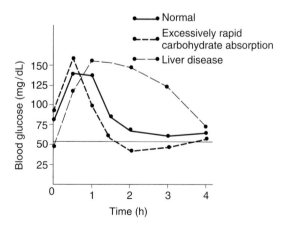

Figure 19–19. Typical glucose tolerance curves after an oral glucose load in liver disease and in conditions causing excessively rapid absorption of glucose from the intestine. The horizontal line is the approximate plasma glucose level at which hypoglycemic symptoms may appear.

ducing symptoms 3–4 hours after meals. This pattern is sometimes seen in individuals who later develop diabetes. Patients with this syndrome should be distinguished from the more numerous patients with similar symptoms due to psychologic or other problems who do not have hypoglycemia when blood is drawn during the symptomatic episode. It has been postulated that the overshoot of the plasma glucose is due to insulin secretion stimulated by impulses in the right vagus, but cholinergic blocking agents do not routinely correct the abnormality. In some thyrotoxic patients and in patients who have had gastrectomies or other operations that speed the passage of food into the intestine, glucose absorption is abnormally rapid. The plasma glucose rises to a high, early peak, but it then falls rapidly to hypoglycemic levels because the wave of hyperglycemia evokes a greater than normal rise in insulin secretion. Symptoms characteristically occur about 2 hours after meals.

Infants born to diabetic mothers often have high birth weights and large organs (**macrosomia**). This condition is caused by excess circulating insulin in the fetus, which in turn is caused in part by stimulation of the fetal pancreas by glucose and amino acids from the blood of the mother. Free insulin in maternal blood is destroyed by proteases in the placenta, but antibody-bound insulin is protected, so it reaches the fetus. Therefore, fetal macrosomia also occurs in women who develop antibodies against various animal insulins and then continue to receive the animal insulin during pregnancy.

Infants with **GLUT 1 deficiency** have defective transport of glucose across the blood-brain barrier. They have low CSF glucose in the presence of normal plasma glucose, seizures, and developmental delay.

Diabetes Mellitus

The incidence of diabetes mellitus in the human population has reached epidemic proportions worldwide, and it is increasing at a rapid rate. In 2000, there were an estimated 150 million cases in the world, and this number is projected to increase to 221 million by 2010. Ninety percent of the present cases are type 2 diabetes (see below), and most of the increase will be in type 2, paralleling the increase in the incidence of obesity.

Diabetes is sometimes complicated by acidosis and coma, and in long-standing diabetes there are additional complications. These include microvascular, macrovascular, and neuropathic disease. The microvascular abnormalities are proliferative scarring of the retina (**diabetic retinopathy**), leading to blindness; and renal disease (**diabetic nephropathy**), leading to renal failure. The macrovascular abnormalities are due

to accelerated atherosclerosis, which is secondary to increased plasma LDL. The result is an increased incidence of stroke and myocardial infarction. The neuropathic abnormalities (**diabetic neuropathy**) involve the autonomic nervous system and peripheral nerves. The neuropathy plus the atherosclerotic circulatory insufficiency in the extremities and reduced resistance to infection can lead to chronic ulceration and gangrene, particularly in the feet.

The ultimate cause of the microvascular and neuropathic complications is chronic hyperglycemia, and tight control of the diabetes reduces their incidence. Intracellular hyperglycemia activates the enzyme aldose reductase. This increases the formation of sorbitol in cells, which in turn reduces cellular Na^+-K^+ ATPase. In addition, intracellular glucose can be converted to so-called "Amadori products," and these in turn can form **advanced glycosylation end products (AGEs),** which cross-link matrix proteins. This damages blood vessels. The AGEs also interfere with the leukocyte responses to infection.

Types of Diabetes

The cause of clinical diabetes is always a deficiency of the effects of insulin at the tissue level, but the deficiency may be relative. One of the common forms, **type 1,** or **insulin-dependent diabetes mellitus (IDDM),** is due to insulin deficiency caused by autoimmune destruction of the B cells in the pancreatic islets; the A, D, and F cells remain intact. The second common form, **type 2,** or **non-insulin-dependent diabetes mellitus (NIDDM),** is characterized by insulin resistance and impaired insulin secretion. It is uncertain which comes first, but a case can be made for insulin resistance raising plasma glucose, which in turn stimulates insulin secretion until the B cell reserve is exceeded. In this situation, the plasma insulin level is generally elevated, rather than depressed, but it is not as high as it would be at that plasma glucose level under normal conditions.

There are in addition cases of diabetes due to other diseases or conditions such as chronic pancreatitis, total pancreatectomy, Cushing's syndrome (see Chapter 20), and acromegaly (see Chapter 22). These make up 5% of the total cases and are sometimes classified as **secondary diabetes.**

Type 1 diabetes usually develops before the age of 40 and hence is called **juvenile diabetes.** Patients with this disease are not obese, and they have a high incidence of ketosis and acidosis. Various anti-B cell antibodies are present in plasma, but the current thinking is that type 1 diabetes is primarily a T lymphocyte-mediated disease. There is a definite genetic susceptibility as well; if one identical twin develops the disease, there is a one-in-three chance that the other twin will do so also. In other words, the **concordance rate** is about 33%. The main genetic abnormality is in the major histocompatibility complex on chromosome 6, making individuals with certain types of histocompatibility antigens (see Chapter 27) much more prone to develop the disease. Other genes are also involved.

Immunosuppression with drugs such as cyclosporine ameliorate type 1 diabetes if given early in the disease before all B cells are lost. Attempts have been made to treat type 1 diabetes by transplanting pancreatic tissue or isolated islet cells, but results to date have been poor, largely because B cells are easily damaged and it is difficult to transplant enough of them to normalize glucose responses.

As mentioned above, type 2 is the most common type of diabetes and is usually associated with obesity. It usually develops after age 40 and is not associated with total loss of the ability to secrete insulin. It has an insidious onset, is rarely associated with ketosis, and is usually associated with normal B cell morphology and insulin content if the B cells have not become ex-hausted. There is a genetic component that is actually stronger than the genetic component in type 1 diabetes; in identical twins, the concordance rate is higher, ranging in some studies to nearly 100%.

In some patients with type 2 diabetes, there are genetic defects in glucokinase (about 1% of the cases), the insulin molecule itself (about 0.5% of the cases), the insulin receptor (about 1% of the cases), GLUT 4 (about 1% of the cases), or IRS-1 (about 15% of the cases). In maturity-onset diabetes occurring in young individuals (MODY), which accounts for about 1% of the cases of type 2 diabetes, loss-of-function mutations have been described in six different genes. Five code for transcription factors affecting the production of enzymes involved in glucose metabolism. The sixth is the gene for glucokinase (Figure 19–14), the enzyme that controls the rate of glucose phoshorylation and hence its metabolism in the B cells. However, the vast majority of cases of type 2 diabetes are almost certainly polygenic in origin, and the actual genes involved are still unknown.

The Adrenal Medulla & Adrenal Cortex

20

INTRODUCTION

There are two endocrine organs in the adrenal gland, one surrounding the other. The main secretions of the inner **adrenal medulla** (Figure 20–1) are the catecholamines **epinephrine, norepinephrine,** and **dopamine;** the outer **adrenal cortex** secretes steroid hormones.

The adrenal medulla is in effect a sympathetic ganglion in which the postganglionic neurons have lost

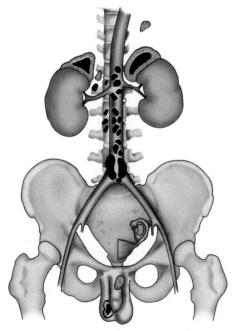

Figure 20–1. Human adrenal glands. Adrenocortical tissue is stippled; adrenal medullary tissue is black. Note the location of the adrenals at the superior pole of each kidney. Also shown are extra-adrenal sites at which cortical and medullary tissue is sometimes found. (Reproduced, with permission, from Forsham PH: The adrenal cortex. In: *Textbook of Endocrinology,* 4th ed. Williams RH [editor]. Saunders, 1968.)

their axons and become secretory cells. The cells secrete when stimulated by the preganglionic nerve fibers that reach the gland via the splanchnic nerves. Adrenal medullary hormones are not essential for life, but they help to prepare the individual to deal with emergencies.

Factors that stimulate and inhibit insulin secretion are summarized in Table 19–6. On the other hand, the adrenal cortex is essential for life. It secretes **glucocorticoids,** steroids with widespread effects on the metabolism of carbohydrate and protein; a **mineralocorticoid** essential to the maintenance of a sodium balance and ECF volume; and **sex hormones** that exert minor effects on reproductive function. Of these, the mineralocorticoids and the glucocorticoids are necessary for survival. Adrenocortical secretion is controlled primarily by ACTH from the anterior pituitary, but mineralocorticoid secretion is also subject to independent control by circulating factors, of which the most important is angiotensin II, a peptide formed in the bloodstream. The formation of angiotensin is in turn dependent on renin secreted by the kidneys.

ADRENAL MORPHOLOGY

The adrenal medulla, which constitutes 28% of the mass of the adrenal gland, is made up of interlacing cords of densely innervated granule-containing cells that abut on venous sinuses. Two cell types can be distinguished morphologically: an epinephrine-secreting type that has larger, less dense granules; and a norepinephrine-secreting type in which smaller, very dense granules fail to fill the vesicles in which they are contained (Figure 20–2). In humans, 90% of the cells are the epinephrine-secreting type and 10% are the norepinephrine-secreting type. The type of cell that secretes dopamine is unknown. **Paraganglia,** small groups of cells resembling those in the adrenal medulla, are found near the thoracic and abdominal sympathetic ganglia (Figure 20–1).

In adult mammals, the adrenal cortex is divided into three zones of variable distinctness (Figure 20–3). The

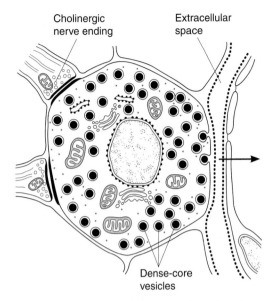

Figure 20–2. Norepinephrine-secreting adrenal medullary cell. The granules are released by exocytosis, and the granule contents enter the bloodstream (arrow). (Modified from Poirier J, Dumas JLR: *Review of Medical Histology.* Saunders, 1977.)

outer **zona glomerulosa** is made up of whorls of cells that are continuous with the columns of cells which form the **zona fasciculata.** These columns are separated by venous sinuses. The inner portion of the zona fasciculata merges into the **zona reticularis,** where the cell columns become interlaced in a network. The zona glomerulosa makes up 15% of the mass of the adrenal gland, the zona fasciculata 50%, and the zona reticularis 7%. The adrenal cells contain abundant lipid, especially in the outer portion of the zona fasciculata. All three cortical zones secrete corticosterone (see below), but the active enzymatic mechanism for aldosterone biosynthesis is limited to the zona glomerulosa, whereas the enzymatic mechanisms for forming cortisol and sex hormones is found in the two inner zones. Furthermore, there is subspecialization within the inner two zones, the zona fasciculata secreting mostly glucocorticoids and the zona reticularis mainly sex hormones.

Arterial blood reaches the adrenal from many small branches of the phrenic and renal arteries and the aorta. From a plexus in the capsule, blood flows to the sinusoids of the medulla. The medulla is also supplied by a few arterioles that pass directly to it from the capsule. Blood from the medulla flows into a central adrenal vein. In most species, including humans, there is a single large adrenal vein. The blood flow through the adrenal is large, as it is in most endocrine glands.

During fetal life, the human adrenal is large and under pituitary control, but the three zones of the permanent cortex represent only 20% of the gland. The remaining 80% is the large **fetal adrenal cortex,** which undergoes rapid degeneration at the time of birth. A major function of this fetal adrenal is synthesis and secretion of sulfate conjugates of androgens that are converted in the placenta to estrogens (see Chapter 23). There is no structure comparable to the human fetal adrenal in laboratory animals.

An important function of the zona glomerulosa, in addition to aldosterone biosynthesis, is the formation of new cortical cells. Like other tissues of neural origin, the adrenal medulla does not regenerate; but when the inner two zones of the cortex are removed, a new zona fasciculata and zona reticularis regenerate from glomerular cells attached to the capsule. Small capsular remnants will regrow large pieces of adrenocortical tissue. Immediately after hypophysectomy, the zona fasciculata and zona reticularis begin to atrophy whereas the zona glomerulosa is unchanged (Figure 20–3) because of the action of angiotensin II on this zone. The ability to secrete aldosterone and conserve Na^+ is normal for some time, but in long-standing hypopituitarism, aldosterone deficiency may develop, apparently because of the absence of a pituitary factor that maintains the responsiveness of the zona glomerulosa (see below). In-

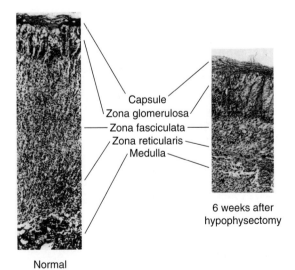

Figure 20–3. Effect of hypophysectomy on the morphology of the adrenal cortex of the dog. Note that the atrophy does not involve the zona glomerulosa. The morphology of the human adrenal is similar.

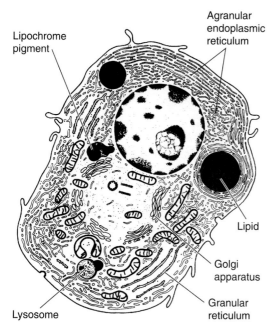

Figure 20–4. Diagrammatic representation of the cytologic features of steroid-secreting cells. Note the abundant agranular endoplasmic reticulum, the pleomorphic mitochondria, and the lipid droplets. (Reproduced, with permission, from Fawcett DW, Long JA, Jones AL: The ultrastructure of endocrine glands. Recent Prog Horm Res 1969;25:315.)

jections of ACTH and stimuli that cause endogenous ACTH secretion produce hypertrophy of the zona fasciculata and zona reticularis but actually decrease, rather than increase, the size of the zona glomerulosa.

The cells of the adrenal cortex contain large amounts of smooth endoplasmic reticulum, which is involved in the steroid-forming process. Other steps in steroid biosynthesis occur in the mitochondria. The structure of steroid-secreting cells is very similar throughout the body. The typical features of such cells are shown in Figure 20–4.

ADRENAL MEDULLA

STRUCTURE & FUNCTION OF MEDULLARY HORMONES

Catecholamines

Norepinephrine, epinephrine, and dopamine are secreted by the adrenal medulla. Cats and some other species secrete mainly norepinephrine, but in dogs and humans, most of the catecholamine output in the adrenal vein is epinephrine. Norepinephrine also enters the circulation from noradrenergic nerve endings.

The structures of norepinephrine, epinephrine, and dopamine and the pathways for their biosynthesis and metabolism are shown in Figures 4–19, 4–20, and 4–21. Norepinephrine is formed by hydroxylation and decarboxylation of tyrosine, and epinephrine by methylation of norepinephrine. Phenylethanolamine-*N*-methyltransferase (PNMT), the enzyme that catalyzes the formation of epinephrine from norepinephrine, is found in appreciable quantities only in the brain and the adrenal medulla. Adrenal medullary PNMT is induced by glucocorticoids. Although relatively large amounts are required, the glucocorticoid concentration is high in the blood draining from the cortex to the medulla. After hypophysectomy, the glucocorticoid concentration of this blood falls and epinephrine synthesis is decreased. In addition, glucocorticoids are apparently necessary for the normal development of the adrenal medulla; in 21β-hydroxylase deficiency (see below), glucocorticoid secretion is reduced during fetal life and the adrenal medulla is dysplastic. In untreated 21β-hydroxylase deficiency, after birth, circulating catecholamines are low.

In plasma, about 95% of the dopamine and 70% of the norepinephrine and epinephrine are conjugated to sulfate. Sulfate conjugates are inactive, and their function is unsettled. In recumbent humans, the normal plasma level of free norepinephrine is about 300 pg/mL (1.8 nmol/L). There is a 50–100% increase upon standing (Figure 20–5). The plasma norepinephrine level is generally unchanged after adrenalectomy, but the free epinephrine level, which is normally about 30 pg/mL (0.16 nmol/L), falls to essentially zero. The epinephrine found in tissues other than the adrenal medulla and the brain is for the most part absorbed from the bloodstream rather than synthesized in situ. Interestingly, low levels of epinephrine reappear in the blood some time after bilateral adrenalectomy, and these levels are regulated like those secreted by the adrenal medulla. They may come from cells such as the ICA cells (see Chapter 13), but their exact source is unknown.

The plasma free dopamine level is about 35 pg/mL (0.23 nmol/L), and there are appreciable quantities of dopamine in the urine. Half the plasma dopamine comes from the adrenal medulla, whereas the remaining half presumably comes from the sympathetic ganglia or other components of the autonomic nervous system.

The catecholamines have a half-life of about 2 minutes in the circulation. For the most part, they are methoxylated and then oxidized to 3-methoxy-4-hydroxymandelic acid (vanillylmandelic acid, VMA; see Figure 4–20). About 50% of the secreted catecholamines appear

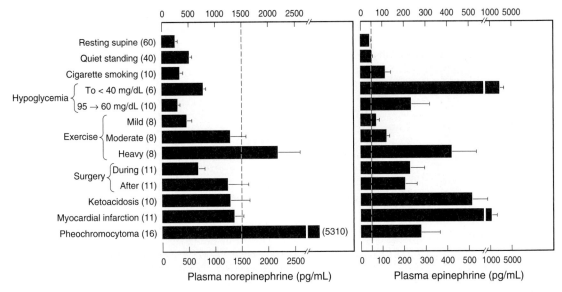

Figure 20–5. Norepinephrine and epinephrine levels in human venous blood in various physiologic and pathologic states. Note that the horizontal scales are different. The numbers in parentheses are the numbers of subjects tested. In each case, the vertical dashed line identifies the threshold plasma concentration at which detectable physiologic changes are observed. (Modified and reproduced, with permission, from Cryer PE: Physiology and pathophysiology of the human sympathoadrenal neuroendocrine system. N Engl J Med 1980;303:436.)

in the urine as free or conjugated metanephrine and normetanephrine, and 35% as VMA. Only small amounts of free norepinephrine and epinephrine are excreted. In normal humans, about 30 μg of norepinephrine, 6 μg of epinephrine, and 700 μg of VMA are excreted per day.

Other Substances Secreted by the Adrenal Medulla

In the medulla, norepinephrine and epinephrine are stored in granules with ATP. The granules also contain chromogranin A (see Chapter 4). Secretion is initiated by acetylcholine released from the preganglionic neurons that innervate the secretory cells. The acetylcholine opens cation channels, and the Ca^{2+} that enters the cells from the ECF triggers exocytosis (see Chapter 1). In this fashion, the catecholamines, ATP, and proteins in the granules are all released together.

Epinephrine-containing cells of the medulla also contain and secrete opioid peptides (see Chapter 4). The precursor molecule is preproenkephalin (see Table 4–4). Most of the circulating metenkephalin comes from the adrenal medulla. The circulating opioid peptides do not cross the blood-brain barrier to any degree, and their function in the blood is unknown.

Adrenomedullin, a vasodepressor polypeptide found in the adrenal medulla, is discussed in Chapter 31.

Effects of Epinephrine & Norepinephrine

In addition to mimicking the effects of noradrenergic nervous discharge, norepinephrine and epinephrine exert metabolic effects that include glycogenolysis in liver and skeletal muscle, mobilization of FFA, increased plasma lactate, and stimulation of the metabolic rate. The effects of norepinephrine and epinephrine are brought about by actions on two classes of receptors, α- and β-adrenergic receptors. Alpha receptors are subdivided into two groups, α_1 and α_2 receptors, and β receptors into β_1, β_2, and β_3 receptors, as outlined in Chapter 4. There are three subtypes of α_1 receptors and three subtypes of α_2 receptors (see Table 4–2).

Norepinephrine and epinephrine both increase the force and rate of contraction of the isolated heart. These responses are mediated by β_1 receptors. The catecholamines also increase myocardial excitability, causing extrasystoles and, occasionally, more serious cardiac arrhythmias. Norepinephrine produces vasoconstriction in most if not all organs via α_1 receptors, but epinephrine dilates the blood vessels in skeletal muscle and the liver via β_2 receptors. This usually overbalances the vasoconstriction produced by epinephrine elsewhere, and the total peripheral resistance drops. When norepinephrine is infused slowly in normal animals or humans, the systolic and diastolic blood pressures rise. The hypertension stimulates the carotid and aortic baroreceptors, producing reflex bradycardia that overrides the direct cardioacceleratory effect of norepinephrine. Consequently, cardiac output per minute falls. Epinephrine causes a widening of the pulse pressure, but because baroreceptor stimulation is insufficient to obscure the direct effect of the hormone on the heart, cardiac rate and output increase. These changes are summarized in Figure 20–6.

Catecholamines increase alertness (see Chapter 11). Epinephrine and norepinephrine are equally potent in this regard, although in humans epinephrine usually evokes more anxiety and fear.

The catecholamines have several different actions that affect blood glucose. Epinephrine and norepinephrine both cause glycogenolysis. They produce this effect via β-adrenergic receptors that increase cAMP, with activation of phosphorylase, and via α-adrenergic receptors that increase intracellular Ca^{2+} (see Chapter 17). In addition, the catecholamines increase the secretion of insulin and glucagon via β-adrenergic mechanisms and inhibit the secretion of these hormones via α-adrenergic mechanisms.

Norepinephrine and epinephrine also produce a prompt rise in the metabolic rate that is independent of the liver and a smaller, delayed rise that is abolished by hepatectomy and coincides with the rise in blood lactate concentration. The initial rise in metabolic rate may be due to cutaneous vasoconstriction, which decreases heat loss and leads to a rise in body temperature, or to increased muscular activity, or both. The second rise is probably due to oxidation of lactate in the liver. Mice unable to make norepinephrine or epinephrine because their dopamine β-hydroxylase gene is knocked out are intolerant to cold, but surprisingly, their basal metabolic rate is elevated. The cause of this elevation is unknown.

When injected, epinephrine and norepinephrine cause an initial rise in plasma K^+ because of release of K^+ from the liver and then a prolonged fall in plasma K^+ because of an increased entry of K^+ into skeletal muscle that is mediated by $β_2$-adrenergic receptors. There is some evidence that activation of α receptors opposes this effect. Thus, the catecholamines may play a significant role in regulating the ratio between extracellular and intracellular K^+.

The increases in plasma norepinephrine and epinephrine that are needed to produce the various effects listed above have been determined by infusion of catecholamines in resting humans. In general, the threshold for the cardiovascular and the metabolic effects of norepinephrine is about 1500 pg/mL, ie, about five times the resting value (Figure 20–5). Epinephrine, on the other hand, produces tachycardia when the plasma level is about 50 pg/mL, ie, about twice the resting value. The threshold for increased systolic blood pressure and lipolysis is about 75 pg/mL; the threshold for hyperglycemia, increased plasma lactate, and decreased diastolic blood pressure is about 150 pg/mL; and the threshold for the α-mediated decrease in insulin secretion is about 400 pg/mL. Plasma epinephrine often exceeds these thresholds. On the other hand, plasma norepinephrine rarely exceeds the threshold for its cardiovascular and metabolic effects, and most of its effects are due to its local release from postganglionic sympathetic neurons. Most adrenal medullary tumors (**pheochromocytomas**) secrete norepinephrine, or epinephrine, or both, and produce sustained hypertension. However, 15% of epinephrine-secreting tumors secrete this catecholamine episodically, producing intermittent bouts of palpitations, headache, glycosuria, and extreme systolic hypertension. These same symptoms are produced by intravenous injection of a large dose of epinephrine.

Effects of Dopamine

The physiologic function of the dopamine in the circulation is unknown. However, injected dopamine produces renal vasodilation, probably by acting on a specific dopaminergic receptor. It also produces vasodilation in the mesentery. Elsewhere, it produces vasoconstriction, probably by releasing norepinephrine, and it has a positively inotropic effect on the heart by an action on $β_1$-adrenergic receptors. The net effect of moderate

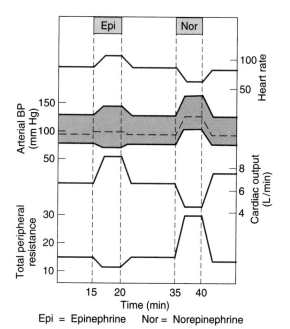

Figure 20–6. Circulatory changes produced in humans by the slow intravenous infusion of epinephrine and norepinephrine.

doses of dopamine is an increase in systolic pressure and no change in diastolic pressure. Because of these actions, dopamine is useful in the treatment of traumatic and cardiogenic shock (see Chapter 33).

Dopamine is made in the renal cortex, and there are appreciable amounts in the urine. It causes natriuresis and may exert this effect by inhibiting renal Na^+-K^+ ATPase.

REGULATION OF ADRENAL MEDULLARY SECRETION

Neural Control

Certain drugs act directly on the adrenal medulla, but physiologic stimuli affect medullary secretion through the nervous system. Catecholamine secretion is low in basal states, but the secretion of epinephrine and, to a lesser extent, that of norepinephrine is reduced even further during sleep.

Increased adrenal medullary secretion is part of the diffuse sympathetic discharge provoked in emergency situations, which Cannon called the "emergency function of the sympathoadrenal system." The ways in which this discharge prepares the individual for flight or fight are described in Chapter 13, and the increases in plasma catecholamines under various conditions are shown in Figure 20–5.

The metabolic effects of circulating catecholamines are probably important, especially in certain situations. The calorigenic action of catecholamines in animals exposed to cold is an example, and so is the glycogenolytic effect (see Chapter 19) in combating hypoglycemia.

Selective Secretion

When adrenal medullary secretion is increased, the ratio of norepinephrine to epinephrine in the adrenal effluent is generally unchanged. However, there is a tendency for norepinephrine secretion to be selectively increased by emotional stresses with which the individual is familiar, whereas epinephrine secretion rises selectively in situations in which the individual does not know what to expect.

ADRENAL CORTEX

STRUCTURE & BIOSYNTHESIS OF ADRENOCORTICAL HORMONES

Classification & Structure

The hormones of the adrenal cortex are derivatives of cholesterol. Like cholesterol, bile acids, vitamin D,

and ovarian and testicular steroids, they contain the **cyclopentanoperhydrophenanthrene nucleus** (Figure 20–7). Gonadal and adrenocortical steroids are of three types: C_{21} steroids, which have a two-carbon side chain at position 17; C_{19} steroids, which have a keto or hydroxyl group at position 17; and C_{18} steroids, which, in addition to a 17-keto or hydroxyl group, have no angular methyl group attached to position 10. The adrenal cortex secretes primarily C_{21} and C_{19} steroids. Most of the C_{19} steroids have a keto group at position 17 and are therefore called **17-ketosteroids.** The C_{21} steroids that have a hydroxyl group at the 17 position in addition to the side chain are often called 17-hydroxycorticoids or 17-hydroxycorticosteroids.

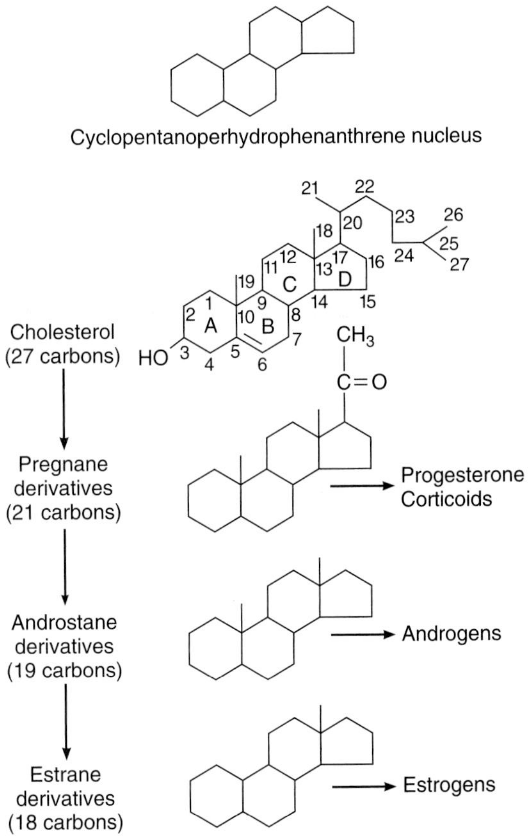

Figure 20–7. Basic structure of adrenocortical and gonadal steroids. The letters in the formula for cholesterol identify the four basic rings, and the numbers identify the positions in the molecule. As shown here, the angular methyl groups (positions 18 and 19) are usually indicated simply by straight lines.

The C_{19} steroids have androgenic activity. The C_{21} steroids are classified, using Selye's terminology, as mineralocorticoids or glucocorticoids. All secreted C_{21} steroids have both mineralocorticoid and glucocorticoid activity; **mineralocorticoids** are those in which effects on Na^+ and K^+ excretion predominate, and **glucocorticoids** are those in which effects on glucose and protein metabolism predominate.

The details of steroid nomenclature and isomerism can be found elsewhere. However, it is pertinent to mention that the Greek letter Δ indicates a double bond and that the groups which lie above the plane of each of the steroid rings are indicated by the Greek letter β and a solid line (–OH) whereas those which lie below the plane are indicated by α and a dashed line (- - -OH). Thus, the C_{21} steroids secreted by the adrenal have a Δ^4-3-keto configuration in the A ring. In most naturally occurring adrenal steroids, 17-hydroxy groups are in the α configuration, whereas 3-, 11-, and 21-hydroxy groups are in the β configuration. The 18-aldehyde configuration on naturally occurring aldosterone is the D form. L-Aldosterone is physiologically inactive.

Secreted Steroids

Innumerable steroids have been isolated from adrenal tissue, but the only steroids normally secreted in physiologically significant amounts are the mineralocorticoid **aldosterone,** the glucocorticoids **cortisol** and **corticosterone,** and the androgens **dehydroepiandrosterone (DHEA)** and **androstenedione.** The structures of these steroids are shown in Figures 20–8 and 20–9. **Deoxycorticosterone** is a mineralocorticoid that is normally secreted in about the same amount as aldosterone (Table 20–1) but has only 3% of the mineralocorticoid activity of aldosterone. Its effect on mineral metabolism is usually negligible, but in diseases in which its secretion is increased, its effect can be appreciable. Most of the estrogens that are not formed in the ovaries are produced in the circulation from adrenal androstenedione. Almost all the dehydroepiandrosterone is secreted conjugated with sulfate, although most if not all of the other steroids are secreted in the free, unconjugated form.

The secretion rate for individual steroids can be determined by injecting a very small dose of isotopically

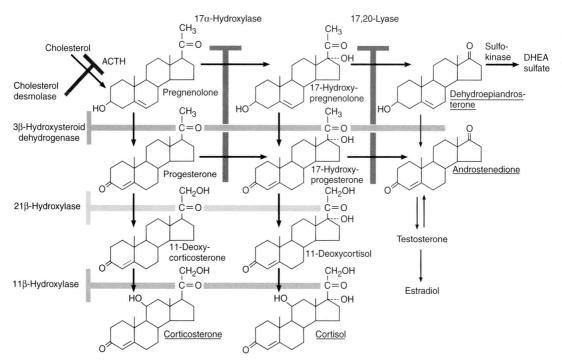

Figure 20–8. Outline of hormone biosynthesis in the zona fasciculata and zona reticularis of the adrenal cortex. The major secretory products are underlined. The enzymes for the reactions are shown on the left and at the top of the chart. When a particular enzyme is deficient, hormone production is blocked at the points indicated by the shaded bars.

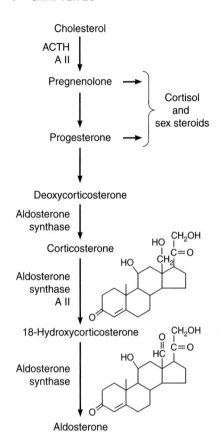

Cholesterol

ACTH
A II

Pregnenolone →

⎱ Cortisol
⎰ and
⎱ sex steroids

Progesterone →

Deoxycorticosterone

Aldosterone
synthase

Corticosterone

Aldosterone
synthase
A II

18-Hydroxycorticosterone

Aldosterone
synthase

Aldosterone

Figure 20–9. Hormone synthesis in the zona glomerulosa. The zona glomerulosa lacks 17α-hydroxylase activity, and only the zona glomerulosa can convert corticosterone to aldosterone because it is the only zone that normally contains aldosterone synthase. A II, angiotensin II.

labeled steroid and determining the degree to which the radioactive steroid excreted in the urine is diluted by unlabeled secreted hormone. This technique is used to measure the output of many different hormones.

Species Differences

In all species from amphibia to humans, the major C_{21} steroid hormones secreted by adrenocortical tissue appear to be aldosterone, cortisol, and corticosterone, although the ratio of cortisol to corticosterone varies. Birds, mice, and rats secrete corticosterone almost exclusively; dogs secrete approximately equal amounts of the two glucocorticoids; and cats, sheep, monkeys, and humans secrete predominantly cortisol. In humans, the ratio of secreted cortisol to corticosterone is approximately 7:1.

Synthetic Steroids

As with many other naturally occurring substances, the activity of adrenocortical steroids can be increased by altering their structure. A number of synthetic steroids are now available that have many times the activity of cortisol. The relative glucocorticoid and mineralocorticoid potencies of the natural steroids are compared with those of the synthetic steroids 9α-fluorocortisol, prednisolone, and dexamethasone in Table 20–2. The potency of dexamethasone is due to its high affinity for glucocorticoid receptors and its long half-life (see below). Prednisolone also has a long half-life.

Steroid Biosynthesis

The major paths by which the naturally occurring adrenocortical hormones are synthesized in the body are summarized in Figures 20–8 and Figure 20–9. The precursor of all steroids is cholesterol. Some of the cholesterol is synthesized from acetate, but most of it is taken up from LDL in the circulation (see Chapter 17). LDL receptors are especially abundant in adrenocortical cells. The cholesterol is esterified and stored in lipid droplets. **Cholesterol ester hydrolase** catalyzes the formation of free cholesterol in the lipid droplets (Figure 20–10). The cholesterol is transported to mitochondria by a sterol carrier protein. In the mitochondria, it is

Table 20–1. Principal adrenocortical hormones in adult humans.[1]

Name	Synonyms	Average Plasma Concentration (Free and Bound)[1] (μg/dL)	Average Amount Secreted (mg/24 h)
Cortisol	Compound F, hydrocortisone	13.9	10
Corticosterone	Compound B	0.4	3
Aldosterone		0.006	0.15
Deoxycorticosterone	DOC	0.006	0.20
Dehydroepiandrosterone sulfate	DHEAS	175.0	20

[1]All plasma concentration values except DHEAS are fasting morning values after overnight recumbency.

Table 20–2. Relative potencies of corticosteroids compared with cortisol.[1]

Steroid	Glucocorticoid Activity	Mineralocorticoid Activity
Cortisol	1.0	1.0
Corticosterone	0.3	15
Aldosterone	0.3	3000
Deoxycorti- costerone	0.2	100
Cortisone	0.7	0.8
Prednisolone	4	0.8
9a-Fluorocortisol	10	125
Dexamethasone	25	~0

[1]Values are approximations based on liver glycogen deposition or anti-inflammatory assays for glucocorticoid activity, and effect on urinary Na^+/K^+ or maintenance of adrenalectomized animals for mineralocorticoid activity. The last three steroids listed are synthetic compounds that do not occur naturally. (Data from various sources.)

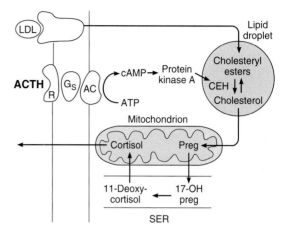

Figure 20–10. Mechanism of action of ACTH on cortisol-secreting cells in the inner two zones of the adrenal cortex. When ACTH binds to its receptor (R), adenylyl cyclase (AC) is activated via G_s. The resulting increase in cAMP activates protein kinase A, and the kinase phosphorylates cholesteryl ester hydrolase (CEH), increasing its activity. Consequently, more free cholesterol is formed and converted to pregnenolone in the mitochondria. Note that in the subsequent steps in steroid biosynthesis, products are shuttled between the mitochondria and the smooth endoplasmic reticulum (SER). Corticosterone is also synthesized and secreted.

converted to pregnenolone in a reaction catalyzed by an enzyme known as **cholesterol desmolase** or **side-chain cleavage enzyme.** This enzyme, like most of the enzymes involved in steroid biosynthesis, is a member of the cytochrome P450 superfamily and is also known as **P450scc** or **CYP11A1.** The basis of the new CYP terminology for cytochrome P450s is discussed in Chapter 17. For convenience, the various names of the enzymes involved in adrenocortical steroid biosynthesis are summarized in Table 20–3.

Pregnenolone moves to the smooth endoplasmic reticulum, where some of it is dehydrogenated to form progesterone in a reaction catalyzed by **3β-hydroxysteroid dehydrogenase.** This enzyme has a molecular weight of 46,000 and is not a cytochrome P450. It also catalyzes the conversion of 17α-hydroxypregnenolone

to 17α-hydroxyprogesterone, and dehydroepiandrosterone to androstenedione (Figure 20–8) in the smooth endoplasmic reticulum. The 17α-hydroxypregnenolone and the 17α-hydroxyprogesterone are formed from pregnenolone and progesterone, respectively (Figure 20–8) by the action of **17α-hydroxylase.** This is another cytochrome P450, and it is also known as **P450c17** or **CYP17.** Located in another part of the

Table 20–3. Nomenclature for adrenal steroidogenic enzymes and their location in adrenal cells.

Trivial Name	P450	CYP	Location
Cholesterol desmolase; side-chain cleavage enzyme	P450scc	CYP11A1	Mitochondria
3β-Hydroxysteroid dehydrogenase	...	...	SER
17α-Hydroxylase, 17,20-lyase	P450c17	CYP17	SER
21β-Hydroxylase	P450c21	CYP21A2	SER
11β-Hydroxylase	P450c11	CYP11B1	Mitochondria
Aldosterone synthase	P450c11AS	CYP11B2	Mitochondria

SER = smooth endoplasmic reticulum.

same enzyme is **17,20-lyase** activity that breaks the 17,20 bond, converting 17α-pregnenolone and 17α-progesterone to the C19 steroids dehydroepiandrosterone and androstenedione.

Hydroxylation of progesterone to 11-deoxycorticosterone and of 17α-hydroxyprogesterone to 11-deoxycortisol also occurs in the smooth endoplasmic reticulum. These reactions are catalyzed by 21β-hydroxylase, a cytochrome P450 that is also known as **P450c21** or **CYP21A2.**

11-Deoxycorticosterone and the 11-deoxycortisol move back to the mitochondria, where they are 11-hydroxylated to form corticosterone and cortisol. These reactions occur in the zona fasciculata and zona reticularis and are catalyzed by 11β-hydroxylase, a cytochrome P450 also known as **P450c11** or **CYP11B1.**

In the zona glomerulosa, there is no 11β-hydroxylase but there is a closely related enzyme called **aldosterone synthase.** This cytochrome P450 is 95% identical to 11β-hydroxylase and is also known as **P450c11AS** or **CYP11B2.** The genes that code CYP11B1 and CYP11B2 are both located on chromosome 8. However, aldosterone synthase is normally found only in the zona glomerulosa. The zona glomerulosa also lacks 17α-hydroxylase. This is why the zona glomerulosa makes aldosterone but fails to make 17-hydroxysteroids or sex hormones.

Furthermore, there is subspecialization within the inner two zones. The zona fasciculata has more 3β-hydroxysteroid dehydrogenase activity then the zona reticularis, and the zona reticularis has more of the cofactors required for the expression of the 17,20-lyase activity of 17α-hydroxylase. Therefore, the zona fasciculata makes more cortisol and corticosterone, and the zona reticularis makes more androgens. Most of the dehydroepiandrosterone that is formed is converted to dehydroepiandrosterone sulfate by **adrenal sulfokinase,** and this enzyme is localized in the zona reticularis as well.

Action of ACTH

ACTH binds to high-affinity receptors on the plasma membrane of adrenocortical cells. This activates adenylyl cyclase via G_s (see Chapter 1), and the resultant increase in intracellular cAMP activates protein kinase A. Protein kinase A phosphorylates cholesteryl ester hydrolase, increasing its activity, and conversion of cholesteryl esters to free cholesterol is increased (Figure 20–10). This, in turn, leads to a prompt increase in the formation of pregnenolone and its derivatives. Over longer periods, ACTH also increases the synthesis of the P450s involved in the synthesis of glucocorticoids.

Actions of Angiotensin II

Angiotensin II binds to AT_1 receptors (see Chapter 24) in the zona glomerulosa which act via a G protein to activate phospholipase C (see Chapter 1). The resulting increase in protein kinase C fosters the conversion of cholesterol to pregnenolone (Figure 20–9) and facilitates the formation of 18-hydroxycorticosterone, which, in turn, facilitates the production of aldosterone.

Enzyme Deficiencies

The consequences of inhibiting any of the enzyme systems involved in steroid biosynthesis can be predicted from Figures 20–8 and 20–9. Congenital defects in the enzymes lead to deficient cortisol secretion and the syndrome of **congenital adrenal hyperplasia.** The hyperplasia is due to increased ACTH secretion. Cholesterol desmolase deficiency is fatal in utero because it prevents the placenta from making the progesterone necessary for pregnancy to continue. A cause of severe congenital adrenal hyperplasia in newborns is a loss of function mutation of the gene for the **steroidogenic acute regulatory (StAR) protein.** This protein is essential in the adrenals and gonads but not in the placenta for the normal movement of cholesterol into the mitochondria to reach cholesterol desmolase, which is located on the matrix space side of the internal mitochondrial membrane. In its absence, only small amounts of steroids are formed. The degree of ACTH stimulation is marked, resulting eventually in accumulation of large numbers of lipoid droplets in the adrenal. For this reason, the condition is called **congenital lipoid adrenal hyperplasia.** Since androgens are not formed, female genitalia develop regardless of genetic sex (see Chapter 23). In 3β-hydroxysteroid dehydrogenase deficiency, another rare condition, DHEA secretion is increased. This steroid is a weak androgen that can cause some masculinization in females with the disease, but it is not adequate to produce full masculinization of the genitalia in genetic males. Consequently, hypospadias is common. In fully developed 17α-hydroxylase deficiency, a third rare condition due to a mutated gene for **CYP17,** no sex hormones are produced, so there are female external genitalia. However, the pathway leading to corticosterone and aldosterone is intact, and elevated levels of 11-deoxycorticosterone and other mineralocorticoids produce hypertension and hypokalemia. Cortisol is deficient, but this is partially compensated by the glucocorticoid activity of corticosterone. Variants of this syndrome have been reported in which the lack of 17,20-lyase activity prevents the production of normal amounts of sex hormones but 17α-hydroxylase activity is adequate to produce cortisol. These variants are apparently explained by deficient availability of cofactors for the lyase activity of the 17α-hydroxylase enzyme.

Unlike the defects discussed in the preceding paragraph, 21β-hydroxylase and 11β-hydroxylase deficiencies are common; the former accounts for 90–95% of the cases of congenital adrenal hyperplasia, and the latter accounts for all but 1% of the remaining cases. Both are characterized by **virilization,** because the increase in ACTH secretion causes steroids to pile up behind the blockades and to be shunted to the production of androgens. The characteristic pattern that develops in females in the absence of treatment is the **adrenogenital syndrome** (Figure 20–11). The gene for 21β-hydroxylase is on the short arm of chromosome 6, closely linked to the HLA major histocompatibility complex (see Chapter 19), and various defects can produce degrees of 21β-hydroxylase deficiency that range from mild to severe. Enough glucocorticoids and mineralocorticoids are usually formed to sustain life. In severe cases, the genitalia of genetic females are masculinized (**female pseudohermaphroditism;** see Chapter 23). However, masculinization may not be marked until later in life, and mild cases can be detected only by laboratory tests. Many of the patients with 21β-hydroxylase deficiency lose Na+ to an excessive degree (**salt-losing form** of congenital virilizing adrenal hyperplasia). The Na+ loss is due both to the mineralocorticoid deficiency and to the aldosterone-antagonizing effect of some of the steroids being secreted in excess amounts. In 11β-hydroxylase deficiency, there is virilization plus excess secretion of 11-deoxycortisol and 11-deoxycorticosterone. Since the latter is an active mineralocorticoid, patients with this condition also have salt and water retention and, in two-thirds of the cases, hypertension (**hypertensive form** of congenital virilizing adrenal hyperplasia).

Glucocorticoid treatment is indicated in all of the virilizing forms of congenital adrenal hyperplasia because it repairs the glucocorticoid deficit and inhibits ACTH secretion, reducing the abnormal secretion of androgens and other steroids. However, doses that just restore glucocorticoid activity to normal do not completely eliminate excess androgen production because the intrinsic abnormality in steroid biosynthesis is still present and some endogenous substrate is diverted to the androgen pathway. Some clinicians get around this problem by giving supernormal amounts of glucocorticoids. Others suggest combining glucocorticoid treatment with drugs that prevent the effects of excess sex hormones.

Expression of the cytochrome P450 enzymes responsible for steroid hormone biosynthesis depends on **steroid factor-1 (SF-1),** an orphan nuclear receptor. If *Ft2-F1,* the gene for SF-1, is knocked out, gonads as well as adrenals fail to develop and there are additional abnormalities at the pituitary and hypothalamic level.

TRANSPORT, METABOLISM, & EXCRETION OF ADRENOCORTICAL HORMONES

Glucocorticoid Binding

Cortisol is bound in the circulation to an α globulin called **transcortin** or **corticosteroid-binding globulin (CBG).** There is also a minor degree of binding to albumin (see Table 23–5). Corticosterone is similarly bound, but to a lesser degree. The half-life of cortisol in the circulation is therefore longer (about 60–90 minutes) than that of corticosterone (50 minutes). Bound steroids are physiologically inactive. Because of protein binding, there is relatively little free cortisol and corticosterone in the urine.

The equilibrium between cortisol and its binding protein and the implications of binding in terms of tissue supplies and ACTH secretion are summarized in Figure 20–12. The bound cortisol functions as a circulating reservoir of hormone that keeps a supply of free cortisol available to the tissues. The relationship is similar to that of T$_4$ and its binding protein (see Figure 18–8). At normal levels of total plasma cortisol (13.5 μg/dL, or 375 nmol/L), there is very little free cortisol in the plasma, but the binding sites on CBG become saturated when the total plasma cortisol exceeds 20 μg/dL. At higher plasma levels, there is some increased binding to albumin but the main increase is in the unbound fraction.

CBG is synthesized in the liver, and its production is increased by estrogen. CBG levels are elevated during

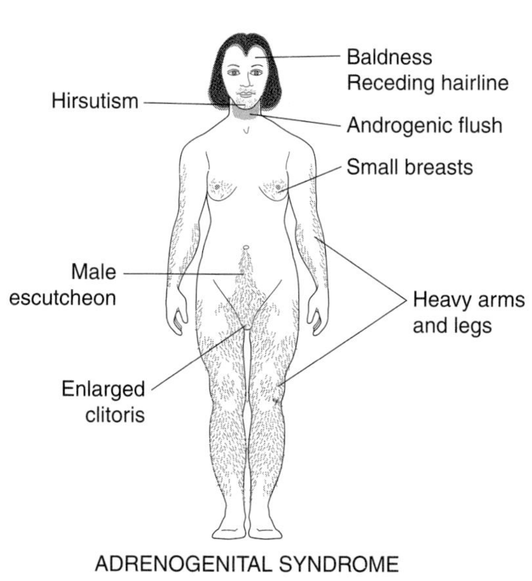

Figure 20–11. Typical findings in the adrenogenital syndrome in a postpubertal woman. (Reproduced, with permission, from Forsham PH, Di Raimondo VC: *Traumatic Medicine and Surgery for the Attorney.* Butterworth, 1960.)

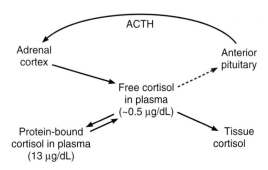

Figure 20–12. The interrelationships of free and bound cortisol. The dashed arrow indicates that cortisol inhibits ACTH secretion. The value for free cortisol is an approximation; in most studies, it is calculated by subtracting the protein-bound cortisol from the total plasma cortisol.

pregnancy and depressed in cirrhosis, nephrosis, and multiple myeloma. When the CBG level rises, more cortisol is bound, and initially there is a drop in the free cortisol level. This stimulates ACTH secretion, and more cortisol is secreted until a new equilibrium is reached at which the bound cortisol is elevated but the free cortisol is normal. Changes in the opposite direction occur when the CBG level falls. This explains why pregnant women have high total plasma cortisol levels without symptoms of glucocorticoid excess and, conversely, why some patients with nephrosis have low total plasma cortisol without symptoms of glucocorticoid deficiency.

Metabolism & Excretion of Glucocorticoids

Cortisol is metabolized in the liver, which is the principal site of glucocorticoid catabolism. Most of the cortisol is reduced to dihydrocortisol and then to tetrahydrocortisol, which is conjugated to glucuronic acid (Figure 20–13). The glucuronyl transferase system responsible for this conversion also catalyzes the formation of the glucuronides of bilirubin (see Chapter 26) and a number of hormones and drugs. There is competitive inhibition between these substrates for the enzyme system.

The liver and other tissues contain the enzyme 11β-hydroxysteroid dehydrogenase (see below). There are at least two forms of this enzyme. Type 1 catalyzes both the conversion of cortisol to cortisone and the reverse

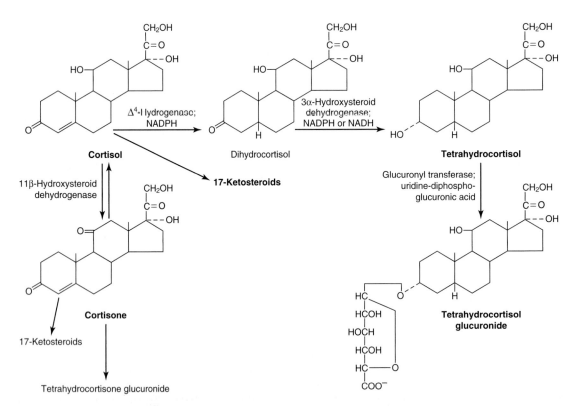

Figure 20–13. Outline of hepatic metabolism of cortisol.

reaction, though it functions primarily as a reductase, forming cortisol from corticosterone. Type 2 catalyzes almost exclusively the one-way conversion of cortisol to cortisone. Cortisone is an active glucocorticoid because it is converted to cortisol, and it is well known because of its extensive use in medicine. It is not secreted in appreciable quantities by the adrenal glands. Little if any of the cortisone formed in the liver enters the circulation, because it is promptly reduced and conjugated to form tetrahydrocortisone glucuronide. The tetrahydroglucuronide derivatives ("conjugates") of cortisol and corticosterone are freely soluble. They enter the circulation, where they do not become bound to protein. They are rapidly excreted in the urine, in part by tubular secretion.

About 10% of the secreted cortisol is converted in the liver to the 17-ketosteroid derivatives of cortisol and cortisone. The ketosteroids are conjugated for the most part to sulfate and then excreted in the urine. Other metabolites, including 20-hydroxy derivatives, are formed. There is an enterohepatic circulation of glucocorticoids, and about 15% of the secreted cortisol is excreted in the stool. The metabolism of corticosterone is similar to that of cortisol, except that it does not form a 17-ketosteroid derivative.

Variations in the Rate of Hepatic Metabolism

The rate of hepatic inactivation of glucocorticoids is depressed in liver disease and, interestingly, during surgery and other stresses. Thus, in stressed humans, the plasma free cortisol level rises higher than it does with maximal ACTH stimulation in the absence of stress.

Aldosterone

Aldosterone is bound to protein to only a slight extent, and its half-life is short (about 20 minutes). The amount secreted is small (Table 20–1), and the total plasma aldosterone level in humans is normally about 0.006 µg/dL (0.17 nmol/L), compared with a cortisol level (bound and free) of about 13.5 µg/dL (375 nmol/L). Much of the aldosterone is converted in the liver to the tetrahydroglucuronide derivative, but some is changed in the liver and in the kidneys to an 18-glucuronide. This glucuronide, which is unlike the breakdown products of other steroids, is converted to free aldosterone by hydrolysis at pH 1.0, and it is therefore often referred to as the "acid-labile conjugate." Less than 1% of the secreted aldosterone appears in the urine in the free form. Another 5% is in the form of the acid-labile conjugate, and up to 40% is in the form of the tetrahydroglucuronide.

17-Ketosteroids

The major adrenal androgen is the 17-ketosteroid dehydroepiandrosterone, although androstenedione is also secreted. The 11-hydroxy derivative of androstenedione and the 17-ketosteroids formed from cortisol and cortisone by side chain cleavage in the liver are the only 17-ketosteroids that have an =O or an –OH group in the 11 position ("11-oxy-17-ketosteroids"). Testosterone is also converted to 17-ketosteroids. Since the daily 17-ketosteroid excretion in normal adults is 15 mg in men and 10 mg in women, about two-thirds of the urinary ketosteroids in men are secreted by the adrenal or formed from cortisol in the liver and about one-third are of testicular origin.

Etiocholanolone, one of the metabolites of the adrenal androgens and testosterone, can cause fever when it is unconjugated (see Chapter 14). Certain individuals have episodic bouts of fever due to periodic accumulation in the blood of unconjugated etiocholanolone ("etiocholanolone fever").

EFFECTS OF ADRENAL ANDROGENS & ESTROGENS

Androgens

Androgens are the hormones that exert masculinizing effects, and they promote protein anabolism and growth (see Chapter 23). Testosterone from the testes is the most active androgen, and the adrenal androgens have less than 20% of its activity. Secretion of the adrenal androgens is controlled acutely by ACTH and not by gonadotropins. However, the concentration of dehydroepiandrosterone sulfate (DHEAS) increases until it peaks at about 225 mg/dL in the early twenties, then falls to very low values in old age (Figure 20–14). These long-term changes are not due to changes in ACTH secretion and appear to be due instead to a rise and then a gradual fall in the lyase activity of 17α-hydroxylase.

All but about 0.3% of the circulating DHEA is conjugated to sulfate (DHEAS). The secretion of adrenal androgens is nearly as great in castrated males and females as it is in normal males, so it is clear that these hormones exert very little masculinizing effect when secreted in normal amounts. However, they can produce appreciable masculinization when secreted in excessive amounts. In adult males, excess adrenal androgens merely accentuate existing characteristics; but in prepubertal boys they can cause precocious development of the secondary sex characteristics without testicular growth (**precocious pseudopuberty**). In females they cause female pseudohermaphroditism and the adrenogenital syndrome (Figure 20–11). Some health practitioners recommend injections of dehydroepiandrosterone to combat the effects

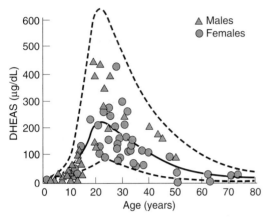

Figure 20–14. Change in serum dehydroepiandrosterone sulfate (DHEAS) with age. The middle line is the mean, and the dashed lines identify ± 1.96 standard deviations. (Reproduced, with permission, from Smith MR et al: A radioimmunoassay for the estimation of serum dehydroepiandrosterone sulfate in normal and pathological sera. Clin Chim Acta 1975;65:5.)

of aging (see Chapter 1), but results to date are controversial at best.

Estrogens

The adrenal androgen androstenedione is converted to testosterone and to estrogens (aromatized) in fat and other peripheral tissues. This is an important source of estrogens in men and postmenopausal women (see Chapter 23).

PHYSIOLOGIC EFFECTS OF GLUCOCORTICOIDS

Adrenal Insufficiency

In untreated adrenal insufficiency, there is Na$^+$ loss and shock due to the lack of mineralocorticoid activity, and there are abnormalities of water, carbohydrate, protein, and fat metabolism due to the lack of glucocorticoids. These metabolic abnormalities are eventually fatal despite mineralocorticoid treatment. Small amounts of glucocorticoids correct the metabolic abnormalities, in part directly and in part by permitting other reactions to occur. It is important to separate these physiologic actions of glucocorticoids from the quite different effects produced by large amounts of the hormones.

Mechanism of Action

The multiple effects of glucocorticoids are triggered by binding to glucocorticoid receptors, and the steroid-

receptor complexes act as transcription factors that promote the transcription of certain segments of DNA (see Chapter 1). This, in turn, leads via the appropriate mRNAs to synthesis of enzymes that alter cell function. In addition, it seems likely that there are nongenomic actions of glucocorticoids (see Chapter 1).

Effects on Intermediary Metabolism

The actions of glucocorticoids on the intermediary metabolism of carbohydrate, protein, and fat are discussed in Chapter 19. They include increased protein catabolism and increased hepatic glycogenesis and gluconeogenesis. Glucose 6-phosphatase activity is increased, and the plasma glucose level rises. Glucocorticoids exert an anti-insulin action in peripheral tissues and make diabetes worse. However, the brain and the heart are spared, so the increase in plasma glucose provides extra glucose to these vital organs. In diabetics, glucocorticoids raise plasma lipid levels and increase ketone body formation, but in normal individuals, the increase in insulin secretion provoked by the rise in plasma glucose obscures these actions. In adrenal insufficiency, the plasma glucose level is normal as long as an adequate caloric intake is maintained, but fasting causes hypoglycemia that can be fatal. The adrenal cortex is not essential for the ketogenic response to fasting.

Permissive Action

Small amounts of glucocorticoids must be present for a number of metabolic reactions to occur, although the glucocorticoids do not produce the reactions by themselves. This effect is called their **permissive action.** Permissive effects include the requirement for glucocorticoids to be present for glucagon and catecholamines to exert their calorigenic effects (see above and Chapter 19), for catecholamines to exert their lipolytic effects, and for catecholamines to produce pressor responses and bronchodilation.

Effects on ACTH Secretion

Glucocorticoids inhibit ACTH secretion, and ACTH secretion is increased in adrenalectomized animals. The consequences of the feedback action of cortisol on ACTH secretion are discussed below in the section on regulation of glucocorticoid secretion.

Vascular Reactivity

In adrenally insufficient animals, vascular smooth muscle becomes unresponsive to norepinephrine and epinephrine. The capillaries dilate and, terminally, become permeable to colloidal dyes. Failure to respond to the norepinephrine liberated at noradrenergic nerve endings probably impairs vascular compensation for the hypovolemia of adrenal insufficiency and promotes vascular collapse. Glucocorticoids restore vascular reactivity.

Effects on the Nervous System

Changes in the nervous system in adrenal insufficiency that are reversed only by glucocorticoids include the appearance of electroencephalographic waves slower than the normal α rhythm and personality changes. The latter, which are mild, include irritability, apprehension, and inability to concentrate.

Effects on Water Metabolism

Adrenal insufficiency is characterized by inability to excrete a water load. The load is eventually excreted, but the excretion is so slow that there is danger of water intoxication. Only glucocorticoids repair this deficit. In patients with adrenal insufficiency who have not received glucocorticoids, glucose infusion may cause high fever ("glucose fever") followed by collapse and death. Presumably, the glucose is metabolized, the water dilutes the plasma, and the resultant osmotic gradient between the plasma and the cells causes the cells of the thermoregulatory centers in the hypothalamus to swell to such an extent that their function is disrupted.

The cause of defective water excretion in adrenal insufficiency is unsettled. Plasma vasopressin levels are elevated in adrenal insufficiency and reduced by glucocorticoid treatment. The glomerular filtration rate is low, and this probably contributes to the deficiency in water excretion. The selective effect of glucocorticoids on the abnormal water excretion is consistent with this possibility, because even though the mineralocorticoids improve filtration by restoring plasma volume, the glucocorticoids raise the glomerular filtration rate to a much greater degree.

Effects on the Blood Cells & Lymphatic Organs

Glucocorticoids decrease the number of circulating eosinophils by increasing their sequestration in the spleen and lungs. Glucocorticoids also lower the number of basophils in the circulation and increase the number of neutrophils, platelets, and red blood cells (Table 20–4).

Glucocorticoids decrease the circulating lymphocyte count and the size of the lymph nodes and thymus by inhibiting lymphocyte mitotic activity. Their ability to reduce secretion of cytokines by inhibiting the effect of NF-$_\kappa$B on the nucleus is discussed in Chapter 33. The reduced secretion of the cytokine IL-2 leads to reduced

Table 20–4. Typical effects of cortisol on the white and red blood cell counts in humans (cells/μL).

Cell	Normal	Cortisol-Treated
White blood cells		
Total	9000	10,000
PMNs	5760	8330
Lymphocytes	2370	1080
Eosinophils	270	20
Basophils	60	30
Monocytes	450	540
Red blood cells	5 million	5.2 million

proliferation of lymphocytes (see Chapter 27), and these cells undergo apoptosis.

Resistance to "Stress"

When an animal or human is exposed to any of an immense variety of noxious or potentially noxious stimuli, there is an increased secretion of ACTH and, consequently, a rise in the circulating glucocorticoid level. This rise is essential for survival. Hypophysectomized animals, or adrenalectomized animals treated with maintenance doses of glucocorticoids, die when exposed to the same noxious stimuli.

Selye defined noxious stimuli that increase ACTH secretion as "stressors," and it is fashionable today to lump these stimuli together under the term "stress." This word is a short, emotionally charged word for something that otherwise takes many words to say, and it is a convenient term to use as long as it is understood that, in the context of this chapter, it denotes only those stimuli that have been proved to increase ACTH secretion in normal animals and humans.

The reason an elevated circulating glucocorticoid level is essential for resisting stress remains for the most part unknown. Most of the stressful stimuli that increase ACTH secretion also activate the sympathetic nervous system, and part of the function of circulating glucocorticoids may be maintenance of vascular reactivity to catecholamines. Glucocorticoids are also necessary for the catecholamines to exert their full FFA-mobilizing action, and the FFAs are an important emergency energy supply. However, sympathectomized animals tolerate a variety of stresses with relative impunity. Another theory holds that glucocorticoids prevent other stress-induced changes from becoming excessive. At present, all that can be said is that stress causes increases in plasma glucocorticoids to high "pharmacologic" levels that in the short run are life-saving but in the long run are definitely harmful and disruptive.

PHARMACOLOGIC & PATHOLOGIC EFFECTS OF GLUCOCORTICOIDS

Cushing's Syndrome

The clinical picture produced by prolonged increases in plasma glucocorticoids was described by Harvey Cushing, and is called **Cushing's syndrome** (Figure 20–15). It may be **ACTH-independent** or **ACTH-dependent.** The causes of ACTH-independent Cushing's syndrome include glucocorticoid-secreting adrenal tumors, adrenal hyperplasia, and prolonged administration of exogenous glucocorticoids for diseases such as rheumatoid arthritis. Rare but interesting ACTH-independent cases have been reported in which adrenocortical cells abnormally express receptors for GIP (see Chapter 26), vasopressin, β-adrenergic agonists, IL-1, or GnRH (see Chapter 14), causing these peptides to increase glucocorticoid secretion. The causes of ACTH-dependent Cushing's syndrome include ACTH-secreting tumors of the anterior pituitary gland and tumors of other organs, usually the lungs, that secrete ACTH (ectopic ACTH syndrome) or CRH. Cushing's syndrome due to anterior pituitary tumors is often called **Cushing's disease** because these tumors were the cause of the cases described by Cushing. However, it is confusing to speak of Cushing's disease as a subtype of Cushing's syndrome, and the distinction seems to be of little more than historical value.

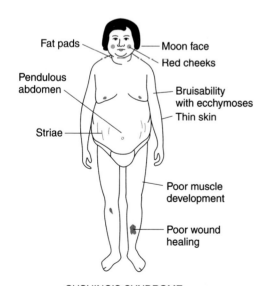

Fat pads
Moon face
Red cheeks
Pendulous abdomen
Bruisability with ecchymoses
Thin skin
Striae
Poor muscle development
Poor wound healing

CUSHING'S SYNDROME

Figure 20–15. Typical findings in Cushing's syndrome. (Reproduced, with permission, from Forsham PH, Di Raimondo VC: *Traumatic Medicine and Surgery for the Attorney.* Butterworth, 1960.)

Patients with Cushing's syndrome are protein-depleted as a result of excess protein catabolism. The skin and subcutaneous tissues are therefore thin, and the muscles are poorly developed. Wounds heal poorly, and minor injuries cause bruises and ecchymoses. The hair is thin and scraggly. Many patients with the disease have some increase in facial hair and acne, but this is caused by the increased secretion of adrenal androgens and often accompanies the increase in glucocorticoid secretion.

Body fat is redistributed in a characteristic way. The extremities are thin, but fat collects in the abdominal wall, face, and upper back, where it produces a "buffalo hump." As the thin skin of the abdomen is stretched by the increased subcutaneous fat depots, the subdermal tissues rupture to form prominent reddish-purple **striae.** These scars are seen normally whenever there is a rapid stretching of skin (eg, around the breasts of girls at puberty or in the abdominal wall during pregnancy), but in normal individuals, the striae are usually inconspicuous and lack the intense purplish color.

Many of the amino acids liberated from catabolized proteins are converted into glucose in the liver, and the resultant hyperglycemia and decreased peripheral utilization of glucose may be sufficient to precipitate insulin-resistant diabetes mellitus, especially in patients genetically predisposed to diabetes. Hyperlipemia and ketosis are associated with the diabetes, but acidosis is usually not severe.

The glucocorticoids are present in such large amounts in Cushing's syndrome that they may exert a significant mineralocorticoid action. Deoxycorticosterone secretion is also elevated in cases due to ACTH hypersecretion. The salt and water retention plus the facial obesity cause the characteristic plethoric, rounded "moon-faced" appearance, and there may be significant K^+ depletion and weakness. About 85% of patients with Cushing's syndrome are hypertensive. The hypertension may be due to increased deoxycorticosterone secretion, increased angiotensinogen secretion, or a direct glucocorticoid effect on blood vessels (see Chapter 33).

Glucocorticoid excess leads to bone dissolution by decreasing bone formation and increasing bone resorption. This leads to **osteoporosis,** a loss of bone mass that leads eventually to collapse of vertebral bodies and other fractures. The mechanisms by which glucocorticoids produce their effects on bone are discussed in Chapter 21.

Glucocorticoids in excess accelerate the basic electroencephalographic rhythms and produce mental aberrations ranging from increased appetite, insomnia, and euphoria to frank toxic psychoses. As noted above, glucocorticoid deficiency is also associated with mental symptoms, but the symptoms produced by glucocorticoid excess are more severe.

Anti-inflammatory & Antiallergic Effects of Glucocorticoids

Glucocorticoids inhibit the inflammatory response to tissue injury. The glucocorticoids also suppress manifestations of allergic disease that are due to the release of histamine from tissues. Both of these effects require high levels of circulating glucocorticoids and cannot be produced by administering steroids without producing the other manifestations of glucocorticoid excess. Furthermore, large doses of exogenous glucocorticoids inhibit ACTH secretion to the point that severe adrenal insufficiency can be a dangerous problem when therapy is stopped. However, local administration of glucocorticoids—eg, by injection into an inflamed joint or near an irritated nerve—produces a high local concentration of the steroid, often without enough systemic absorption to cause serious side effects.

The actions of glucocorticoids in patients with bacterial infections are dramatic but dangerous. For example, in pneumococcal pneumonia or active tuberculosis, the febrile reaction, the toxicity, and the lung symptoms disappear; but unless antibiotics are given at the same time, the bacteria spread throughout the body. It is important to remember that the symptoms are the warning that disease is present; when these symptoms are masked by treatment with glucocorticoids, there may be serious and even fatal delays in diagnosis and the institution of treatment with antimicrobial drugs.

The role of NF-$_\kappa$B in the anti-inflammatory and antiallergic effects of glucocorticoids has been mentioned above and is discussed in Chapter 33. An additional action that combats local inflammation is inhibition of phospholipase A$_2$. This reduces the release of arachidonic acid from tissue phospholipids and consequently reduces the formation of leukotrienes, thromboxanes, prostaglandins, and prostacyclin (see Figure 17–33).

Other Effects

Large doses of glucocorticoids inhibit growth, decrease growth hormone secretion (see Chapter 22), induce PNMT, and decrease TSH secretion. During fetal life, glucocorticoids accelerate the maturation of surfactant in the lungs (see Chapter 34).

REGULATION OF GLUCOCORTICOID SECRETION

Role of ACTH

Both basal secretion of glucocorticoids and the increased secretion provoked by stress are dependent upon ACTH from the anterior pituitary. Angiotensin II also stimulates the adrenal cortex, but its effect is mainly on aldosterone secretion. Large doses of a num-

ber of other naturally occurring substances, including vasopressin, serotonin, and VIP, are capable of stimulating the adrenal directly, but there is no evidence that these agents play any role in the physiologic regulation of glucocorticoid secretion.

Chemistry & Metabolism of ACTH

ACTH is a single-chain polypeptide containing 39 amino acids. Its origin from pro-opiomelanocortin (POMC) in the pituitary is discussed in Chapter 22. The first 23 amino acids in the chain generally constitute the active "core" of the molecule. Amino acids 24–39 constitute a "tail" that stabilizes the molecule and varies slightly in composition from species to species (Figure 20–16). The ACTHs that have been isolated are generally active in all species but antigenic in heterologous species.

ACTH is inactivated in blood in vitro more slowly than in vivo; its half-life in the circulation in humans is about 10 minutes. A large part of an injected dose of ACTH is found in the kidneys, but neither nephrectomy nor evisceration appreciably enhances its in vivo activity, and the site of its inactivation is not known.

Effect of ACTH on the Adrenal

After hypophysectomy, glucocorticoid synthesis and output decline within 1 hour to very low levels, al-

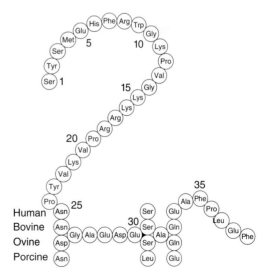

Figure 20–16. Structure of ACTH. In the species shown, the amino acid composition varies only at positions 25, 31, and 33. (Reproduced, with permission, from Li CH: Adrenocorticotropin 45: Revised amino acid sequences for sheep and bovine hormones. Biochem Biophys Res Commun 1972;49:835.)

though some hormone is still secreted. Within a short time after an injection of ACTH (in dogs, less than 2 minutes), glucocorticoid output rises (Figure 20–17). With low doses of ACTH, there is a linear relationship between the log of the dose and the increase in glucocorticoid secretion. However, the maximal rate at which glucocorticoids can be secreted is rapidly reached; and in dogs, doses larger than 10 mU only prolong the period of maximal secretion. A similar "ceiling on output" exists in humans. The effects of ACTH on adrenal morphology and the mechanism by which it increases steroid secretion are discussed above.

Adrenal Responsiveness

ACTH not only produces prompt increases in glucocorticoid secretion but also increases the sensitivity of the adrenal to subsequent doses of ACTH. Conversely, single doses of ACTH do not increase glucocorticoid secretion in chronically hypophysectomized animals and patients with hypopituitarism, and repeated injections or prolonged infusions of ACTH are necessary to restore normal adrenal responses to ACTH. Decreased responsiveness is also produced by doses of glucocorticoids that inhibit ACTH secretion. The decreased adrenal responsiveness to ACTH is detectable within 24 hours after hypophysectomy and increases progressively with time (Figure 20–18). It is marked when the adrenal is atrophic but develops before there are visible changes in adrenal size or morphology.

Circadian Rhythm

ACTH is secreted in irregular bursts throughout the day, and plasma cortisol tends to rise and fall in response to these bursts (Figure 20–19). In humans, the

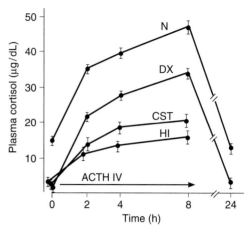

Figure 20–18. Loss of ACTH responsiveness when ACTH secretion is decreased in humans. The 1–24-amino-acid sequence of ACTH was infused intravenously (IV) in a dose of 250 μg over 8 hours. N, normal subjects; DX, dexamethasone 0.75 mg every 8 hours for 3 days; CST, long-term corticosteroid therapy; HI, anterior pituitary insufficiency. (Reproduced, with permission, from Kolanowski J et al: Adrenocortical response upon repeated stimulation with corticotropin in patients lacking endogenous corticotropin secretion. Acta Endocrinol [Kbh] 1977;85:595.)

bursts are most frequent in the early morning, and about 75% of the daily production of cortisol occurs between 4 AM and 10 AM. The bursts are least frequent in the evening. This **diurnal (circadian) rhythm** in ACTH secretion is present in patients with adrenal insufficiency receiving constant doses of glucocorticoids. It is not due to the stress of getting up in the morning, traumatic as that may be, because the increased ACTH secretion occurs before waking up. If the "day" is lengthened experimentally to more than 24 hours—ie, if the individual is isolated and the day's activities are spread over more than 24 hours—the adrenal cycle also lengthens, but the increase in ACTH secretion still occurs during the period of sleep. The biologic clock responsible for the diurnal ACTH rhythm is located in the suprachiasmatic nuclei of the hypothalamus (see Chapter 14).

The Response to Stress

The morning plasma ACTH concentration in a healthy resting human is about 25 pg/mL (5.5 pmol/L). ACTH and cortisol values in various abnormal conditions are summarized in Figure 20–20. During severe stress, the amount of ACTH secreted exceeds the amount necessary to produce maximal glucocorticoid

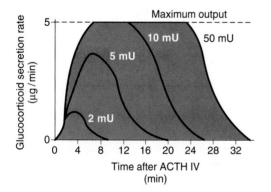

Figure 20–17. Changes in glucocorticoid output from the adrenal in hypophysectomized dogs following the intravenous (IV) administration of various doses of ACTH.

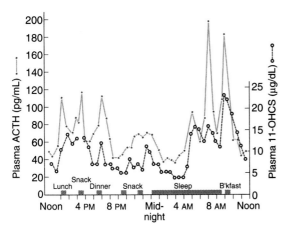

Figure 20–19. Fluctuations in plasma ACTH and glucocorticoids throughout the day in a normal girl (age 16). The ACTH was measured by immunoassay and the glucocorticoids as 11-oxysteroids (11-OHCS). Note the greater ACTH and glucocorticoid rises in the morning, before awakening. (Reproduced, with permission, from Krieger DT et al: Characterization of the normal temporal pattern of plasma corticosteroid levels. J Clin Endocrinol Metab 1971;32:266.)

the hypothalamus via release of CRH. This polypeptide is produced by neurons in the paraventricular nuclei. It is secreted in the median eminence and transported in the portal-hypophysial vessels to the anterior pituitary, where it stimulates ACTH secretion (see Chapter 14). If the median eminence is destroyed, increased secretion in response to many different stresses is blocked. Afferent nerve pathways from many parts of the brain converge on the paraventricular nuclei. Fibers from the amygdaloid nuclei mediate responses to emotional stresses, and fear, anxiety, and apprehension cause marked increases in ACTH secretion. Input from the suprachiasmatic nuclei provides the drive for the diurnal rhythm. Impulses ascending to the hypothalamus via the nociceptive pathways and the reticular formation trigger ACTH secretion in response to injury (Figure 20–20). There is an inhibitory input from the baroreceptors via the nucleus of the tractus solitarius. Contrary to previously proposed theories, circulating epinephrine and norepinephrine do not increase ACTH secretion in humans, and adrenocortical and adrenal medullary secretion are independently regulated by the hypothalamus.

Glucocorticoid Feedback

Free glucocorticoids inhibit ACTH secretion, and the degree of pituitary inhibition is proportionate to the circulating glucocorticoid level. The inhibitory effect is exerted at both the pituitary and the hypothalamic levels. The inhibition is due primarily to an action on DNA, and maximal inhibition takes several hours to

output. However, prolonged exposure to ACTH in conditions such as the ectopic ACTH syndrome increases the adrenal maximum.

Increases in ACTH secretion to meet emergency situations are mediated almost exclusively through

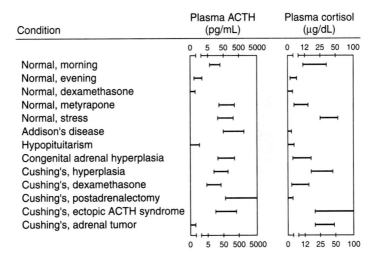

Figure 20–20. Plasma concentrations of ACTH and cortisol in various clinical states. (Reproduced, with permission, from Liddle G: The adrenal cortex. In: *Textbook of Endocrinology*, 5th ed. Williams RH [editor]. Saunders, 1974.)

develop, although there is in addition more rapid "fast feedback." The ACTH-inhibiting activity of the various steroids parallels their glucocorticoid potency. A drop in resting corticoid levels stimulates ACTH secretion, and in chronic adrenal insufficiency there is a marked increase in the rate of ACTH synthesis and secretion.

Thus, the rate of ACTH secretion is determined by two opposing forces: the sum of the neural and possibly other stimuli converging through the hypothalamus to increase ACTH secretion, and the magnitude of the braking action of glucocorticoids on ACTH secretion, which is proportionate to their level in the circulating blood (Figure 20–21).

The dangers involved when prolonged treatment with anti-inflammatory doses of glucocorticoids is stopped deserve emphasis. Not only is the adrenal atrophic and unresponsive after such treatment, but even if its responsiveness is restored by injecting ACTH, the pituitary may be unable to secrete normal amounts of ACTH for as long as a month. The cause of the deficiency is presumably diminished ACTH synthesis. Thereafter, there is a slow rise in ACTH to supranormal levels. These in turn stimulate the adrenal, and

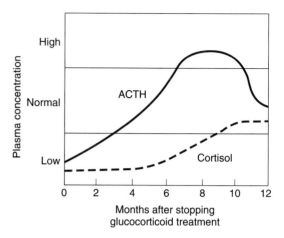

Figure 20–22. Pattern of plasma ACTH and cortisol values in patients recovering from prior long-term daily treatment with large doses of glucocorticoids. (Courtesy of R Ney.)

glucocorticoid output rises, with feedback inhibition gradually reducing the elevated ACTH levels to normal (Figure 20–22). The complications of sudden cessation of steroid therapy can usually be avoided by slowly decreasing the steroid dose over a long period of time.

EFFECTS OF MINERALOCORTICOIDS

Actions

Aldosterone and other steroids with mineralocorticoid activity increase the reabsorption of Na^+ from the urine, sweat, saliva, and the contents of the colon. Thus, mineralocorticoids cause retention of Na^+ in the ECF. This expands ECF volume. In the kidneys, they act primarily on the **principal cells (P cells)** of the collecting ducts (see Chapter 38). Under the influence of aldosterone, increased amounts of Na^+ are in effect exchanged for K^+ and H^+ in the renal tubules, producing a K^+ diuresis (Figure 20–23) and an increase in urine acidity.

Mechanism of Action

Like many other steroids, aldosterone binds to a cytoplasm receptor, and the receptor-hormone complex moves to the nucleus where it alters the transcription of mRNAs (see Figure 1–35). This in turn increases the production of proteins that alter cell function. The aldosterone-stimulated proteins have two effects—a rapid effect, to increase the activity of epithelial sodium channels (ENaCs) by increasing the insertion of these channels into the cell membrane from a cytoplasmic

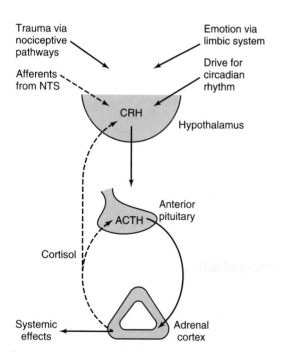

Figure 20–21. Feedback control of the secretion of cortisol and other glucocorticoids. The dashed arrows indicate inhibitory effects, and the solid arrows indicate stimulatory effects. NTS, nucleus tractus solitarius. Compare with Figures 18–12, 22–10, 23–22, and 23–35.

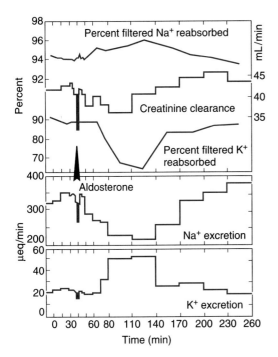

Figure 20–23. Effect of aldosterone (5 μg as a single dose injected into the aorta) on electrolyte excretion in an adrenalectomized dog. The scale for creatinine clearance is on the right.

pool; and a slower effect to increase the synthesis of ENaCs. Among the genes activated by aldosterone is the gene for **serum- and glucocorticoid-regulated kinase (sgk),** a serine-threonine protein kinase. The gene for sgk is an early response gene, and sgk increases ENaC activity. Aldosterone also increases the mRNAs for the three subunits that make up ENaCs. The fact that sgk is activated by glucocorticoids as well as aldosterone is not a problem because glucocorticoids are inactivated at mineralocorticoid receptor sites (see below). However, aldosterone activates the genes for other proteins in addition to sgk and ENaCs and inhibits others. Therefore, the exact mechanism by which aldosterone-induced proteins increase Na$^+$ reabsorption is still unsettled.

Evidence is accumulating that aldosterone also binds to the cell membrane and by a rapid, nongenomic action increases the activity of membrane Na$^+$-K$^+$ exchangers. This produces an increase in intracellular Na$^+$, and the second messenger involved is probably IP$_3$. In any case, the principal effect of aldosterone on Na$^+$ transport takes 10–30 min to develop and peaks even later (Figure 20–23), indicating that it depends on the synthesis of new protein by the genomic mechanism.

Relation of Mineralocorticoid to Glucocorticoid Receptors

It is intriguing that in vitro, the mineralocorticoid receptor has an appreciably higher affinity for glucocorticoids than the glucocorticoid receptor does, and glucocorticoids are present in large amounts in vivo. This raises the question of why glucocorticoids do not bind to the mineralocorticoid receptors in the kidneys and other locations and produce mineralocorticoid effects. At least in part, the answer is that the kidneys and other mineralocorticoid-sensitive tissues also contain the enzyme **11β-hydroxysteroid dehydrogenase type 2.** This enzyme leaves aldosterone untouched, but it converts cortisol to cortisone (Figure 20–13) and corticosterone to its 11-oxy derivative. These 11-oxy derivatives do not bind to the receptor.

Apparent Mineralocorticoid Excess

If 11β-hydroxysteroid dehydrogenase type 2 is inhibited or absent, cortisol has marked mineralocorticoid effects. The resulting syndrome is called **apparent mineralocorticoid excess (AME).** Patients with this condition have the clinical picture of hyperaldosteronism, but their plasma aldosterone level as well as their plasma renin activity is low. The condition can be due to congenital absence of the enzyme or to prolonged ingestion of licorice, because licorice contains glycyrrhetinic acid, which inhibits the enzyme.

Other Steroids That Affect Na$^+$ Excretion

Aldosterone is the principal mineralocorticoid secreted by the adrenal, although corticosterone is secreted in sufficient amounts to exert a minor mineralocorticoid effect (Tables 20–1 and 20–2). Deoxycorticosterone, which is secreted in appreciable amounts only in abnormal situations, has about 3% of the activity of aldosterone. Large amounts of progesterone and some other steroids cause natriuresis, but there is little evidence that they play any normal role in the control of Na$^+$ excretion.

Effect of Adrenalectomy

In adrenal insufficiency, Na$^+$ is lost in the urine. K$^+$ is retained, and the plasma K$^+$ rises. When adrenal insufficiency develops rapidly, the amount of Na$^+$ lost from the ECF exceeds the amount excreted in the urine, indicating that Na$^+$ also must be entering cells. When the posterior pituitary is intact, salt loss exceeds water loss, and the plasma Na$^+$ falls (Table 20–5). However, the plasma volume also is reduced, resulting in hypotension, circulatory insufficiency, and, eventually, fatal shock. These changes can be prevented to a degree by

Table 20–5. Typical plasma electrolyte levels in normal humans and patients with adrenocortical diseases.

State	Plasma Electrolytes (meq/L)			
	Na$^+$	K$^+$	Cl$^-$	HCO$_3^-$
Normal	142	4.5	105	25
Adrenal insufficiency	120	6.7	85	25
Primary hyper-aldosteronism	145	2.4	96	41

increasing the dietary NaCl intake. Rats survive indefinitely on extra salt alone, but in dogs and most humans, the amount of supplementary salt needed is so large that it is almost impossible to prevent eventual

collapse and death unless mineralocorticoid treatment is also instituted.

Secondary Effects of Excess Mineralocorticoids

K$^+$ depletion due to prolonged K$^+$ diuresis is a prominent feature of prolonged mineralocorticoid excess (Table 20–5). H$^+$ is also lost in the urine. Na$^+$ is retained initially, but the plasma Na$^+$ is elevated only slightly if at all, because water is retained with the osmotically active sodium ions. Consequently, ECF volume is expanded and the blood pressure rises. When the ECF expansion passes a certain point, Na$^+$ excretion is usually increased in spite of the continued action of mineralocorticoids on the renal tubules. This **escape phenomenon** (Figure 20–24) is probably due to increased secretion of ANP (see Chapter 24). Because of increased excretion of Na$^+$ when the ECF volume is expanded, mineralocorticoids do not produce edema in

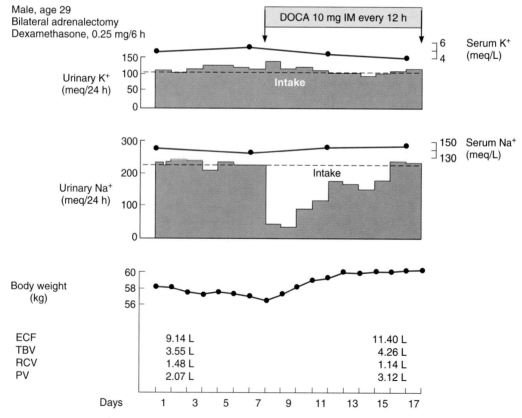

Figure 20–24. "Escape" from the sodium-retaining effect of desoxycorticosterone acetate (DOCA) in an adrenalectomized patient. ECF, extracellular fluid volume; TBV, total blood volume; RCV, red cell volume; PV, plasma volume. (Courtesy of EG Biglieri.)

normal individuals and patients with hyperaldosteronism. However, escape may not occur in certain disease states, and in these situations, continued expansion of ECF volume leads to edema (see Chapter 38).

REGULATION OF ALDOSTERONE SECRETION

Stimuli

The principal stimuli that increase aldosterone secretion are summarized in Table 20–6. Some of them also increase glucocorticoid secretion; others selectively affect the output of aldosterone. The primary regulatory factors involved are ACTH from the pituitary, renin from the kidney via angiotensin II, and a direct stimulatory effect of a rise in plasma K^+ concentration on the adrenal cortex.

Effect of ACTH

When first administered, ACTH stimulates the output of aldosterone as well as that of glucocorticoids and sex hormones. Although the amount of ACTH required to increase aldosterone output is somewhat greater than the amount that stimulates maximal glucocorticoid secretion (Figure 20–25), it is well within the range of endogenous ACTH secretion. The effect is transient, and even if ACTH secretion remains elevated, aldosterone output declines in 1 or 2 days. On the other hand, the output of the mineralocorticoid deoxycorticosterone remains elevated. The decline in aldosterone output is partly due to decreased renin secretion secondary to hypervolemia (see below), but it is possible that some other factor also decreases the conversion of corticosterone to aldosterone. After hypophysectomy, the basal rate of aldosterone secretion is normal. The increase normally produced by surgi-

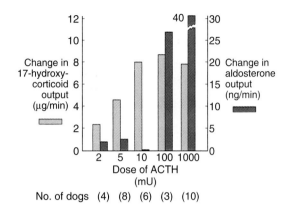

Figure 20–25. Changes in adrenal venous output of steroids produced by ACTH in nephrectomized hypophysectomized dogs.

cal and other stresses is absent, but the increase produced by dietary salt restriction is unaffected for some time. Later on, atrophy of the zona glomerulosa complicates the picture in long-standing hypopituitarism, and this may lead to salt loss and hypoaldosteronism.

Normally, glucocorticoid treatment does not suppress aldosterone secretion. However, an interesting recently described syndrome is **glucocorticoid-remediable aldosteronism (GRA).** This is an autosomal dominant disorder in which the increase in aldosterone secretion produced by ACTH is no longer transient. The hypersecretion of aldosterone and the accompanying hypertension are remedied when ACTH secretion is suppressed by administering glucocorticoids. The genes encoding aldosterone synthase and 11β-hydroxylase are 95% identical and are close together on chromosome 8. In individuals with GRA, there is unequal crossing over so that the 5' regulatory region of the 11β-hydroxylase gene is fused to the coding region of the aldosterone synthase. This creates an ACTH-sensitive aldosterone synthase hybrid gene.

Effects of Angiotensin II & Renin

The octapeptide angiotensin II is formed in the body from angiotensin I, which is liberated by the action of renin on circulating angiotensinogen (see Chapter 24). Injections of angiotensin II stimulate adrenocortical secretion and, in small doses, affect primarily the secretion of aldosterone (Figure 20–26). The sites of action of angiotensin II are both early and late in the steroid biosynthetic pathway. The early action is on the conversion of cholesterol to pregnenolone, and the late action is on the conversion of corticosterone to aldosterone (Figure 20–9). Angiotensin II does not increase

Table 20–6. Representative stimuli that increase aldosterone secretion.

Glucocorticoid secretion also increased
Surgery
Anxiety
Physical trauma
Hemorrhage
Glucocorticoid secretion unaffected
High potassium intake
Low sodium intake
Constriction of inferior vena cava in thorax
Standing
Secondary hyperaldosteronism (in some cases of congestive heart failure, cirrhosis, and nephrosis)

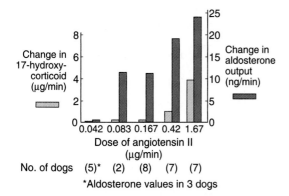

No. of dogs   (5)*   (2)   (8)   (7)   (7)

*Aldosterone values in 3 dogs

Figure 20–26. Changes in adrenal venous output of steroids produced by angiotensin II in nephrectomized hypophysectomized dogs.

the secretion of deoxycorticosterone, which is controlled by ACTH.

Renin is secreted from the juxtaglomerular cells that surround the renal afferent arterioles as they enter the glomeruli (see Chapter 24). Aldosterone secretion is regulated via the renin-angiotensin system in a feedback fashion (Figure 20–27). A drop in ECF volume or intra-arterial vascular volume leads to a reflex increase in renal nerve discharge and decreases renal arterial pressure. Both changes increase renin secretion, and the angiotensin II formed by the action of the renin increases the rate of secretion of aldosterone. The aldosterone causes Na^+ and, secondarily, water retention, expanding ECF volume and shutting off the stimulus that initiated increased renin secretion.

Hemorrhage stimulates ACTH and renin secretion. Like hemorrhage, standing and constriction of the thoracic inferior vena cava decrease intra-arterial vascular volume. Dietary sodium restriction also increases aldosterone secretion via the renin-angiotensin system (Figure 20–28). Such restriction reduces ECF volume, but aldosterone and renin secretion are increased before there is any consistent decrease in blood pressure. Consequently, the initial increase in renin secretion produced by dietary sodium restriction is probably due to reflex increases in the activity of the renal nerves. The increase in circulating angiotensin II produced by salt depletion up-regulates the angiotensin II receptors in the adrenal cortex and hence increases the response to angiotensin II, whereas it down-regulates the angiotensin II receptors in the blood vessels.

Electrolytes & Other Factors

An acute decline in plasma Na^+ of about 20 meq/L stimulates aldosterone secretion, but changes of this magnitude are rare. However, the plasma K^+ level need increase only 1 meq/L to stimulate aldosterone secretion, and transient increases of this magnitude may occur after a meal, particularly if it is rich in K^+. Like angiotensin II, K^+ stimulates the conversion of cholesterol to pregnenolone and the conversion of corticosterone to aldosterone. It appears to act by depolarizing the cell, which opens voltage-gated Ca^{2+} channels, increasing intracellular Ca^{2+}. The sensitivity of the zona glomerulosa to angiotensin II and consequently to a low-sodium diet is decreased by a low-potassium diet.

In normal individuals, there is an increase in plasma aldosterone concentration during the portion of the

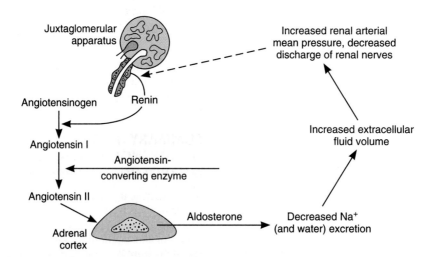

Figure 20–27. Feedback mechanism regulating aldosterone secretion. The dashed arrow indicates inhibition.

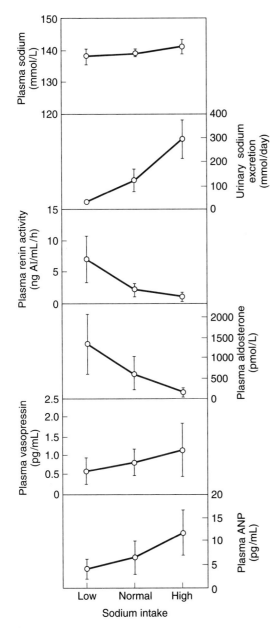

Figure 20–28. Effect of low-, normal-, and high-sodium diets on sodium metabolism and plasma renin activity, aldosterone, vasopressin, and ANP in normal humans. (Data from Sagnella GA et al: Plasma atrial natriuretic peptide: Its relationship to changes in sodium intake, plasma renin activity, and aldosterone in man. Clin Sci 1987;72:25.)

Table 20–7. Second messengers involved in the regulation of aldosterone secretion.

Secretagogue	Intracellular mediator
ACTH	Cyclic AMP, protein kinase A
Angiotensin II	Diacylglycerol, protein kinase C
K^+	Ca^{2+} via voltage-gated Ca^{2+} channels

day that the individual is carrying on activities in the upright position. This increase is due to a decrease in the rate of removal of aldosterone from the circulation by the liver and an increase in aldosterone secretion due to a postural increase in renin secretion. Individuals who are confined to bed show a circadian rhythm of aldosterone and renin secretion, with the highest values in the early morning before awakening.

ANP inhibits renin secretion and decreases the responsiveness of the zona glomerulosa to angiotensin II (see Chapter 24).

The mechanisms by which ACTH, angiotensin II, and K^+ stimulate aldosterone secretion are summarized in Table 20–7.

ROLE OF MINERALOCORTICOIDS IN THE REGULATION OF SALT BALANCE

Variation in aldosterone secretion is only one of many factors affecting Na^+ excretion. Other major factors include the glomerular filtration rate, ANP, the presence or absence of osmotic diuresis, and changes in tubular reabsorption of Na^+ independent of aldosterone. It takes some time for aldosterone to act. Thus, when one rises from the supine to the standing position, aldosterone secretion is increased and there is Na^+ retention; but the decrease in Na^+ excretion develops too rapidly to be explained solely by increased aldosterone secretion. The primary function of the aldosterone-secreting mechanism is the defense of intravascular volume, but it is only one of the homeostatic mechanisms involved.

SUMMARY OF THE EFFECTS OF ADRENOCORTICAL HYPER- & HYPOFUNCTION IN HUMANS

Recapitulating the manifestations of excess and deficiency of the adrenocortical hormones in humans is a convenient way to summarize the multiple and complex actions of these steroids. A characteristic clinical syndrome is associated with excess secretion of each of the types of hormones.

Excess androgen secretion causes masculinization (**adrenogenital syndrome**) and precocious pseudopuberty or female pseudohermaphroditism.

Excess glucocorticoid secretion produces a moonfaced, plethoric appearance, with trunk obesity, purple abdominal striae, hypertension, osteoporosis, protein depletion, mental abnormalities, and, frequently, diabetes mellitus (**Cushing's syndrome**). The causes of Cushing's syndrome are discussed above.

Excess mineralocorticoid secretion leads to K$^+$ depletion and Na$^+$ retention, usually without edema but with weakness, hypertension, tetany, polyuria, and hypokalemic alkalosis (**hyperaldosteronism**). This condition may be due to primary adrenal disease (**primary hyperaldosteronism; Conn's syndrome**) such as an adenoma of the zona glomerulosa, unilateral or bilateral adrenal hyperplasia, adrenal carcinoma, or GRA. In patients with primary hyperaldosteronism, renin secretion is depressed. **Secondary hyperaldosteronism** with high plasma renin activity is caused by cirrhosis, heart failure, and nephrosis. Increased renin secretion is also found in individuals with the salt-losing form of the adrenogenital syndrome (see above), because their ECF volume is low. In patients with elevated renin secretion due to renal artery constriction, aldosterone secretion is increased; in those in whom renin secretion is not elevated, aldosterone secretion is normal. The relationship of aldosterone to hypertension is discussed in Chapter 33.

Primary adrenal insufficiency due to disease processes that destroy the adrenal cortex is called **Addison's disease.** The condition used to be a relatively common complication of tuberculosis, and now it is usually due to autoimmune inflammation of the adrenal. Patients lose weight, are tired, and become chronically hypotensive. They have small hearts, probably because the hypotension decreases the work of the heart. Eventually they develop severe hypotension and shock (**addisonian crisis**). This is due not only to mineralocorticoid deficiency but to glucocorticoid deficiency as well. Fasting causes fatal hypoglycemia, and any stress causes collapse. Water is retained, and there is always the danger of water intoxication. Circulating ACTH levels are elevated. The diffuse tanning of the skin and the spotty pigmentation characteristic of chronic glucocorticoid deficiency (Figure 20–29) are due, at least in part, to the MSH activity of the ACTH in the blood (see Chapter 22). Minor menstrual abnormalities occur in women, but the deficiency of adrenal

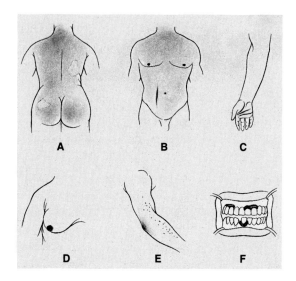

Figure 20–29. Pigmentation in Addison's disease. **A:** Tan and vitiligo. **B:** Pigmentation of scars from lesions that occurred after the development of the disease. **C:** Pigmentation of skin creases. **D:** Darkening of areolas. **E:** Pigmentation of pressure points. **F:** Pigmentation of the gums. (Reproduced, with permission, from Forsham PH, Di Raimondo VC: *Traumatic Medicine and Surgery for the Attorney.* Butterworth, 1960.)

sex hormones usually has little effect in the presence of normal testes or ovaries.

Secondary adrenal insufficiency is caused by pituitary diseases that decrease ACTH secretion, and **tertiary adrenal insufficiency** is caused by hypothalamic disorders disrupting CRH secretion. Both are usually milder than primary adrenal insufficiency because electrolyte metabolism is affected to a lesser degree. In addition, there is no pigmentation because in both of these conditions, plasma ACTH is low, not high.

Cases of isolated aldosterone deficiency have also been reported in patients with renal disease and a low circulating renin level (**hyporeninemic hypoaldosteronism**). In addition, **pseudohypoaldosteronism** is produced when there is resistance to the action of aldosterone. Patients with these syndromes have marked hyperkalemia, salt wasting, and hypotension, and they may develop metabolic acidosis.

Hormonal Control of Calcium Metabolism & the Physiology of Bone

21

INTRODUCTION

Three hormones are primarily concerned with the regulation of calcium metabolism. **1,25-Dihydroxycholecalciferol** is a steroid hormone formed from vitamin D by successive hydroxylations in the liver and kidneys. Its primary action is to increase calcium absorption from the intestine. **Parathyroid hormone (PTH)** is secreted by the parathyroid glands. Its main action is to mobilize calcium from bone and increase urinary phosphate excretion. **Calcitonin,** a calcium-lowering hormone that in mammals is secreted primarily by cells in the thyroid gland, inhibits bone resorption. Although the role of calcitonin seems to be relatively minor, all three hormones probably operate in concert to maintain the constancy of the Ca^{2+} level in the body fluids. A fourth local hormone, **parathyroid hormone-related protein (PTHrP),** acts on one of the PTH receptors and is important in skeletal development in utero. Glucocorticoids, growth hormone, estrogens, and various growth factors also affect calcium metabolism.

CALCIUM & PHOSPHORUS METABOLISM

Calcium

The body of a young adult human contains about 1100 g (27.5 mol) of calcium. Ninety-nine percent of the calcium is in the skeleton. The plasma calcium, normally about 10 mg/dL (5 meq/L, 2.5 mmol/L), is partly bound to protein and partly diffusible (Table 21–1). The distribution of calcium inside cells is discussed in Chapter 1.

It is the free, ionized calcium in the body fluids that is a vital second messenger (see Chapter 1) and is necessary for blood coagulation, muscle contraction, and nerve function. A decrease in extracellular Ca^{2+} exerts a net excitatory effect on nerve and muscle cells in vivo (see Chapter 2). The result is **hypocalcemic tetany,** which is characterized by extensive spasms of skeletal muscle, involving especially the muscles of the extremities and the larynx. Laryngospasm becomes so severe that the airway is obstructed and fatal asphyxia is produced. Ca^{2+} plays an important role in clotting (see Chapter 27); in vivo, however, the level of plasma Ca^{2+} at which fatal tetany occurs is still above the level at which clotting defects would occur.

Since the extent of Ca^{2+} binding by plasma proteins is proportionate to the plasma protein level, it is important to know the plasma protein level when evaluating the total plasma calcium. Plasma ionized calcium can be measured by use of a calcium-sensitive electrode. Other electrolytes and pH affect the Ca^{2+} level. Thus, for example, symptoms of tetany appear at much higher total calcium levels if the patient hyperventilates, increasing plasma pH. Plasma proteins are more ionized when the pH is high, providing more protein anion to bind with Ca^{2+}.

The calcium in bone is of two types: a readily exchangeable reservoir and a much larger pool of stable calcium that is only slowly exchangeable. There are two independent but interacting homeostatic systems af-

Table 21–1. Distribution (mmol/L) of calcium in normal human plasma.

Total diffusible	**1.34**
Ionized (Ca^{2+})	1.18
Complexed to HCO_3^-, citrate, etc	0.16
Total nondiffusible (protein-bound)	**1.16**
Bound to albumin	0.92
Bound to globulin	0.24
Total plasma calcium	**2.50**

fecting the calcium in bone. One is the system that regulates plasma Ca^{2+}, and in the operation of this system, about 500 mmol of Ca^{2+} per day moves into and out of the readily exchangeable pool in the bone (Figure 21–1). The other system is the one concerned with bone remodeling by the constant interplay of bone resorption and deposition (see below). However, the Ca^{2+} interchange between plasma and this stable pool of bone calcium is only about 7.5 mmol/d.

A large amount of calcium is filtered in the kidneys, but 98–99% of the filtered calcium is reabsorbed. About 60% of the reabsorption occurs in the proximal tubules and the remainder in the ascending limb of the loop of Henle and the distal tubule. Distal tubular reabsorption is regulated by parathyroid hormone.

The absorption of Ca^{2+} from the gastrointestinal tract is discussed in Chapter 25. Ca^{2+} is actively transported out of the intestine by a system in the brush border of the epithelial cells that involves a calcium-dependent ATPase, and this process is regulated by 1,25-dihydroxycholecalciferol (see below). There is also some absorption by passive diffusion. When Ca^{2+} intake is high, 1,25-dihydroxycholecalciferol levels fall because of the increased plasma Ca^{2+}. Consequently, Ca^{2+} absorption undergoes adaptation; ie, it is high when the calcium intake is low and decreased when the calcium intake is high. Calcium absorption is also decreased by substances that form insoluble salts with Ca^{2+} (eg, phosphates and oxalates) or by alkalis, which favor formation of insoluble calcium soaps. A high-protein diet increases absorption in adults.

Phosphorus

Phosphate is found in ATP, cAMP, 2,3-diphosphoglycerate, many proteins, and other vital compounds in the body. Phosphorylation and dephosphorylation of proteins are involved in the regulation of cell function (see Chapter 1). Therefore, it is not surprising that phosphate metabolism is closely regulated. Total body phosphorus is 500–800 g (16.1–25.8 mol), 85–90% of which is in the skeleton. Total plasma phosphorus is about 12 mg/dL, with two-thirds of this total in organic compounds and the remaining inorganic phosphorus (Pi) mostly in PO_4^{3-}, HPO_4^{2-}, and $H_2PO_4^{-}$.

The amount of phosphorus normally entering bone is about 3 mg (97 μmol)/kg/d, with an equal amount leaving via reabsorption.

The Pi in the plasma is filtered in the glomeruli, and 85–90% of the filtered Pi is reabsorbed. Active transport in the proximal tubule accounts for most of the reabsorption, and this active transport process is powerfully inhibited by parathyroid hormone (see below).

Pi is absorbed in the duodenum and small intestine by both active transport and passive diffusion. However, unlike the absorption of Ca^{2+}, the absorption of Pi is linearly proportionate to dietary intake. Many stimuli that increase Ca^{2+} absorption, including 1,25-dihydroxycholecalciferol, also increase Pi absorption.

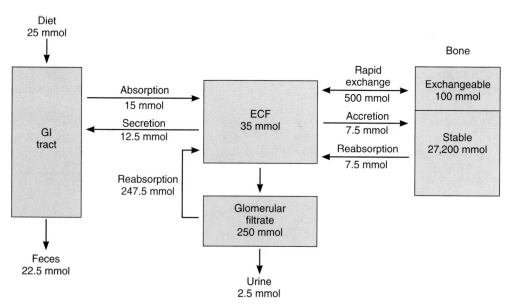

Figure 21–1. Calcium metabolism in an adult human ingesting 25 mmol (1000 mg) of calcium per day.

BONE PHYSIOLOGY

Bone is a special form of connective tissue with a collagen framework impregnated with Ca^{2+} and PO_4^{3-} salts, particularly **hydroxyapatites,** which have the general formula $Ca_{10}(PO_4)_6(OH)_2$. Bone is involved in overall Ca^{2+} and PO_4^{3-} homeostasis. It protects vital organs, and the rigidity it provides permits locomotion and the support of loads against gravity. Old bone is constantly being resorbed and new bone formed (see below), permitting remodeling that allows it to respond to the stresses and strains that are put upon it. It is a living tissue that is well vascularized and has a total blood flow of 200–400 mL/min in adult humans.

Structure

Bone in children and adults is of two types: **compact** or **cortical bone,** which makes up the outer layer of most bones (Figure 21–2) and accounts for 80% of the bone in the body; and **trabecular** or **spongy bones** inside the cortical bone, which make up the remaining 20% of bone in the body. In compact bone, the surface-to-volume ratio is low, and bone cells lie in lacunae. They receive nutrients by way of canaliculi that ramify throughout the compact bone (Figure 21–2). Trabecular bone is made up of spicules or plates, with a high surface to volume ratio and many cells sitting on the surface of the plates. Nutrients diffuse from bone ECF into the trabeculae, but in compact bone, nutrients are provided via **Haversian canals** (Figure 21–2), which contain blood vessels. Around each Haversian canal, collagen is arranged in concentric layers, forming cylinders called **osteons** or **Haversian systems.**

The protein in bone matrix is over 90% type I collagen, which is also the major structural protein in tendons and skin. This collagen, which weight for weight is as strong as steel, is made up of a triple helix of three polypeptides bound tightly together. Two of these are identical α_1 polypeptides encoded by one gene, and one is an α_2 polypeptide encoded by a different gene. Type I and other collagens make up a family of structurally related proteins that maintain the integrity of many different organs. Fifteen different types have been identified, and these are encoded by more than 20 different genes.

Bone Growth

During fetal development, most of the bones are modeled in cartilage and then transformed into bone by ossification **(enchondral bone formation).** The exceptions are the clavicles, the mandibles, and certain bones of the skull in which mesenchymal cells form bone directly **(intramembranous bone formation).**

During growth, specialized areas at the ends of each long bone **(epiphyses)** are separated from the shaft of the bone by a plate of actively proliferating cartilage, the **epiphysial plate** (Figure 21–3). The bone increases in length as this plate lays down new bone on the end of the shaft. The width of the epiphysial plate is proportionate to the rate of growth. The width is affected by a number of hormones but most markedly by the pituitary growth hormone and IGF-I (see Chapter 22).

Linear bone growth can occur as long as the epiphyses are separated from the shaft of the bone, but such growth ceases after the epiphyses unite with the shaft **(epiphysial closure).** The cartilage cells stop proliferating, become hypertrophic, and secrete VEGF, leading to vascularization and ossification. The epiphyses of the various bones close in an orderly temporal sequence, the last epiphyses closing after puberty. The normal age at which each of the epiphyses closes is known, and the "bone age" of a young individual can be determined by x-raying the skeleton and noting which epiphyses are open and which are closed.

Bone Formation & Resorption

The cells responsible for bone formation are **osteoblasts** and the cells responsible for bone resorption are **osteoclasts.**

Osteoblasts are modified fibroblasts. Their early development from the mesenchyme is the same as that of fibroblasts, and the same large number of growth factors is involved. Later, ossification-specific factors begin to appear. One of the most interesting is the transcription factor Cbfa1. Mice in which the gene for Cbfa1 is knocked out develop to term with their skeletons made exclusively of cartilage; no ossification occurs. Normal osteoblasts are able to lay down type 1 collagen and form new bone.

Osteoclasts, on the other hand, are members of the monocyte family. Stromal cells in the bone marrow, osteoblasts, and T lymphocytes all express a molecule called RANKL (RANK ligand) on their surface, and when they come in contact with appropriate monocytes they bind to RANKL receptors (RANK) on the surfaces of the monocytes. They also secrete M-CSF (see Chapter 27), and it binds to a receptor, c-fins, on the monocytes. The combination converts the monocytes into osteoclasts. The precursor cells also secrete **osteoprotegrin (OPG),** which checks the conversion of the monocytes by competing with RANK for binding of RANKL.

Osteoclasts erode and absorb previously formed bone. They become attached to bone via integrins in a membrane extension called the **sealing zone.** This creates an isolated area between the bone and a portion of the osteoclast (Figure 21–4). Proton pumps, which are H^+-dependent ATPases, then move from endosomes into the cell membrane apposed to the isolated area,

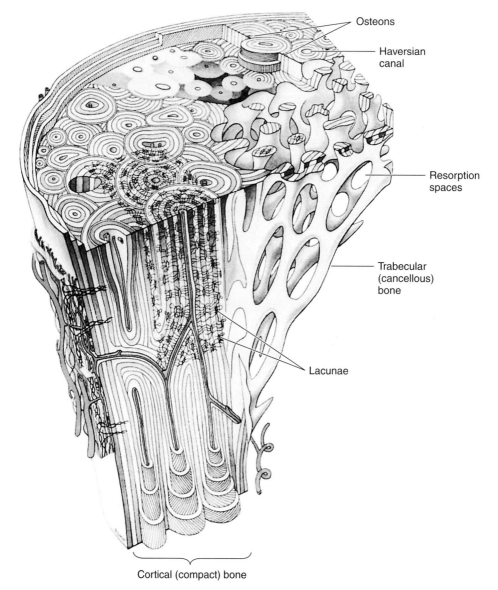

Figure 21–2. Structure of compact and trabecular bone. The compact bone is shown in horizontal section **(top)** and vertical section **(left)**. (Reproduced, with permission, from Williams PL et al (editors): *Gray's Anatomy,* 37th edition, Churchill Livingstone, 1989.)

and they acidify the area to approximately pH 4.0. Similar proton pumps are found in the endosomes and lysosomes of all eukaryotic cells (see Chapter 1), but in only a few other instances do they move into the cell membrane. Note in this regard that the sealed-off space formed by the osteoclast resembles a large lysosome. The acidic pH dissolves hydroxyapatite, and acid proteases secreted by the cell break down collagen, forming a shallow depression in the bone. The products of digestion are then endocytosed and move across the osteoclast by transcytosis (see Chapter 1), with release into the interstitial fluid. The collagen breakdown products have pyridinoline structures, and pyridinolines can be measured in the urine as an index of the rate of bone resorption.

Throughout life, bone is being constantly resorbed

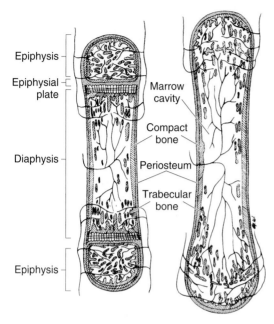

Figure 21–3. Structure of a typical long bone before **(left)** and after **(right)** epiphysial closure.

and new bone is being formed. The calcium in bone turns over at a rate of 100% per year in infants and 18% per year in adults. Bone remodeling is mainly a local process carried out in small areas by populations of cells called bone-remodeling units. First, osteoclasts resorb bone, and then osteoblasts lay down new bone in the same general area. This cycle takes about 100 days. Modeling drifts also occur in which the shapes of bones change as bone is resorbed in one location and added in another. Osteoclasts tunnel into cortical bone followed by osteoblasts, whereas in trabecular bone remodeling occurs on the surface of the trabeculas. About 5% of the bone mass is being remodeled by about 2 million bone-remodeling units in the human skeleton at any one time. The renewal rate for bone is about 4% per year for compact bone and 20% per year for trabecular bone. The remodeling is related in part to the stresses and strains imposed on the skeleton by gravity

How is the close pairing of the osteoclast and osteoblast activity regulated? Osteoblasts regulate osteoclast formation via the RANKL-RANK and the M-CSF-OPG mechanism, but there is no known direct feedback of osteoclasts on osteoblasts. Instead, the whole bone remodeling process is primarily under endocrine control. Parathyroid hormone accelerates bone resorption (see below), and estrogens slow bone resorption by inhibiting the production of bone-eroding cy-

tokines. An interesting new observation is that intracerebroventricular but not intravenous leptin decreases bone formation. This finding is consistent with the observations that obesity protects against bone loss and that most obese humans are resistant to the effects of leptin on appetite (see Chapter 14). Thus, there may be a neuroendocrine control of bone mass via leptin.

Bone Disease

The diseases produced by selective abnormalities of the cells and processes discussed above illustrate the interplay of factors that maintain normal bone function.

In **osteopetrosis,** a rare and often severe disease, the osteoclasts are defective and are unable to resorb bone in their usual fashion so the osteoblasts operate unopposed. The result is a steady increase in bone density, neurologic defects due to narrowing and distortion of foramina through which nerves normally pass, and hematologic abnormalities due to crowding out of the marrow cavities. Mice lacking the protein encoded by the immediate-early gene c-*fos* develop osteopetrosis, and osteopetrosis also occurs in mice lacking the PU.1 transcription factor. This suggests that all these factors are involved in normal osteoclast development and function.

On the other hand, **osteoporosis** is caused by a relative excess of osteoclastic function. Loss of bone matrix in this condition (Figure 21–5) is marked, and the incidence of fractures is increased. Fractures are particularly common in the distal forearm (Colles' fracture), verte-

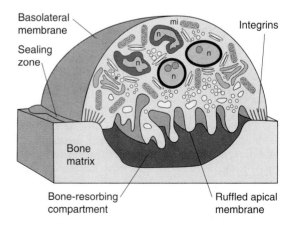

Figure 21–4. Osteoclast resorbing bone. The edges of the cell are tightly sealed to bone, permitting secretion of acid from the ruffled apical membrane and consequent erosion of the bone underneath the cell. Note the multiple nuclei (n) and mitochondria (mi). (Courtesy of R Baron.)

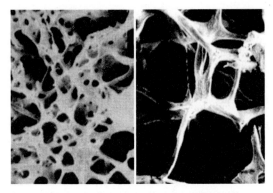

Figure 21–5. Normal trabecular bone **(left)** compared with trabecular bone from a patient with osteoporosis **(right).**

bral body, and hip. All of these areas have a high content of trabecular bone, and since trabecular bone is more active metabolically, it is lost more rapidly. Fractures of the vertebrae with compression cause kyphosis, with the production of a typical "widow's hump" that is common in elderly women with osteoporosis. Fractures of the hip in elderly individuals are associated with a mortality rate of 12–20%, and half of those who survive require prolonged expensive care.

Osteoporosis has multiple causes, but by far the commonest form is **involutional osteoporosis.** All normal humans gain bone early in life, during growth. After a plateau, they begin to lose bone as they grow older (Figure 21–6). When this loss is accelerated or exaggerated, it leads to osteoporosis.

Adult women have less bone mass than adult men, and after menopause they initially lose it more rapidly than men of comparable age do (Figure 21–6). Consequently, they are more prone to development of serious osteoporosis. The cause of the bone loss after menopause is primarily estrogen deficiency, and estrogen treatment arrests the progress of the disease. Estrogens inhibit secretion of cytokines such as IL-1, IL-6, and TNFα, and these cytokines foster the development of osteoclasts. Estrogen also stimulates production of TGF-β, and this cytokine increases apoptosis of osteoclasts. Large doses of estrogens may increase the incidence of myocardial infarction and stroke, but small doses that are effective in slowing bone loss may protect against cardiovascular disease. Estrogens alone increase the incidence of endometrial cancer, but it appears that this is avoided if the estrogen is given with a progestin. On the other hand, prolonged treatment with estrogens may increase the incidence of breast cancer. Therefore, the decision to treat a postmenopausal woman with estrogens depends on a careful weighing of the risk-benefit ratio.

Increased intake of calcium, particularly from natural sources such as milk, and moderate exercise may also help prevent or slow the progress of osteoporosis, although their effects are not great. Bisphosphonates such as etidronate that inhibit osteoclastic activity, increase the mineral content of bone when administered in a cyclic fashion and decrease the rate of new vertebral fractures. Fluoride stimulates osteoblasts, making bone more dense, but it has proved to be of little value in the treatment of the disease.

In patients who are immobilized for any reason, and during space flight (see Chapter 33), bone resorption exceeds bone formation and disuse osteoporosis develops. The plasma calcium level is not markedly elevated, but plasma concentrations of parathyroid hormone and 1,25-dihydroxycholecalciferol fall and large amounts of calcium are lost in the urine. Osteoporosis also occurs in patients with excess glucocorticoid secretion (Cushing's syndrome; see Chapter 20).

VITAMIN D & THE HYDROXYCHOLECALCIFEROLS

Chemistry

The active transport of Ca^{2+} and PO_4^{3-} from the intestine is increased by a metabolite of **vitamin D.** The term "vitamin D" is used to refer to a group of closely

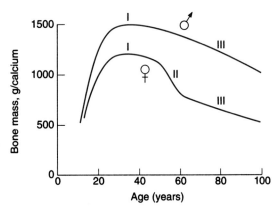

Figure 21–6. Total body calcium (g/calcium), an index of bone mass, at various ages in men and women. Note the rapid increase to young adult levels (phase I) followed by the steady loss of bone with advancing age in both sexes (phase III) and the superimposed rapid loss in women after menopause (phase II). (Reproduced, by permission of Oxford University Press, from Riggs BL, Melton LJ III: Involutional osteoporosis. In Evans TG, Williams TF (editors): *Oxford Textbook of Geriatric Medicine.* Oxford Univ Press, London, 1992.)

related sterols produced by the action of ultraviolet light on certain provitamins (Figure 21–7). Vitamin D_3, which is also called cholecalciferol, is produced in the skin of mammals from 7-dehydrocholesterol by the action of sunlight. The reaction involves the rapid formation of previtamin D_3, which is then converted more slowly to vitamin D_3 (cholecalciferol). Vitamin D_3 and its hydroxylated derivatives are transported in the plasma bound to a globulin vitamin D-binding protein (DBP), which is also known as G_c protein and binds G-actin (see Chapter 1). In addition, it enhances the stimulation of chemotactic activity of neutrophils induced by complement. The affinity of DBP for previtamin D_3 is low, but the affinity for vitamin D_3 is high, so DBP moves vitamin D_3 from the skin into the circulation. Vitamin D_3 is also ingested in the diet.

Vitamin D_3 is metabolized by enzymes that are members of the cytochrome P450 (CYP) superfamily (see Chapters 17 and 20). In the liver, vitamin D_3 is converted to **25-hydroxycholecalciferol** (calcidiol, 25-OHD_3). The 25-hydroxycholecalciferol is converted in the cells of the proximal tubules of the kidneys to the more active metabolite **1,25-dihydroxycholecalciferol,** which is also called calcitriol or 1,25-$(OH)_2D_3$. 1,25-Dihydroxycholecalciferol is also made in the placenta, in keratinocytes in the skin, and in macrophages. In patients with sarcoidosis, pulmonary alveolar macrophages also produce 1,25-dihydroxycholecalciferol, apparently upon stimulation by γ-interferon. The normal plasma level of 25-hydroxycholecalciferol is about 30 ng/mL, and that of 1,25-dihydroxycholecalciferol is about 0.03 ng/mL (approximately 100 pmol/L). The less active metabolite 24,25-dihydroxycholecalciferol is also formed in the kidneys (Figure 21–7).

Vitamin D_3 and its derivatives are **secosteroids;** ie, they are steroids in which one of the rings has been opened. In this case, it is the B ring (Figure 21–7). 1,25-Dihydroxycholecalciferol is a hormone because it is produced in the body and transported in the bloodstream to produce effects in target cells.

Mechanism of Action

Since 1,25-dihydroxycholecalciferol is a steroid, it is not surprising that it acts via a receptor which is one of the superfamily of receptors by which steroids, thyroid hormones, and a number of other substances trigger

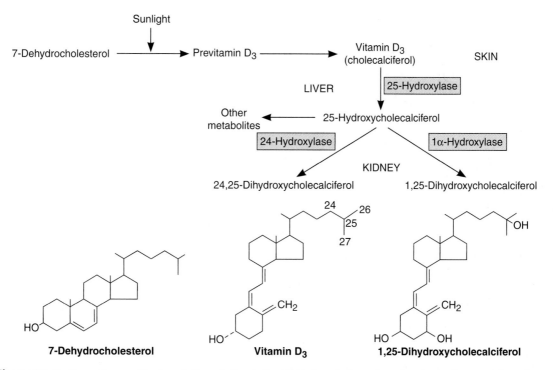

Figure 21–7. Formation and hydroxylation of vitamin D_3. 25-Hydroxylation takes place in the liver, and the other hydroxylations occur primarily in the kidneys. The formulas of 7-dehydrocholesterol, vitamin D_3, and 1,25-dihydroxycholecalciferol are also shown.

changes in gene expression (see Chapter 1). Binding of the steroid to the receptor exposes a DNA-binding region, and the result in this case is increased transcription of some mRNAs and inhibition of the transcription of others.

Actions

The mRNAs that are produced in response to 1,25-dihydroxycholecalciferol dictate the formation of a family of **calbindin-D** proteins. These are members of the troponin C superfamily of Ca^{2+}-binding proteins that also includes calmodulin (see Chapter 1). Calbindin-Ds are found in human intestine, brain, and kidneys and in many different tissues in rats. In the intestinal epithelium and many other tissues, two calbindins are induced: calbindin-D_{9K}, which has a molecular weight of 9000 and binds 2 Ca^{2+}; and calbindin-D_{28K}, which has a molecular weight of 28,000 and normally binds 4 Ca^{2+} even though it has 6 Ca^{2+}-binding sites. In the intestine, increases in calbindin-D_{9K} and calbindin-D_{28K} levels are correlated with increased Ca^{2+} transport, but the precise way they facilitate Ca^{2+} movement across the intestinal epithelium is still uncertain. There is also evidence that 1,25-dihydroxycholecalciferol increases the number of Ca^{2+}-H^+ ATPase molecules in the intestinal cells; these are needed to pump Ca^{2+} into the interstitium.

In addition to increasing Ca^{2+} absorption from the intestine, 1,25-dihydroxycholecalciferol facilitates Ca^{2+} reabsorption in the kidneys, increases the synthetic activity of osteoblasts, and is necessary for normal calcification of matrix. The stimulation of osteoblasts brings about a secondary increase in the activity of osteoclasts (see above).

1,25-Dihydroxycholecalciferol receptors are found in many issues other than the intestine, kidneys, and bones. Among them are the skin, lymphocytes, monocytes, skeletal and cardiac muscle, breast, and anterior pituitary gland. Evidence is accumulating that 1,25-dihydroxycholecalciferol stimulates the differentiation of immune cells and keratinocytes in the skin. It is interesting in this regard that there is an increased incidence of infections in patients with vitamin D deficiency and that 1,25-dihydroxycholecalciferol has shown promise in the treatment of psoriasis. In other tissues, 1,25-dihydroxycholecalciferol appears to be involved in the regulation of growth and the production of growth factors. However, its exact role remains to be determined.

Regulation of Synthesis

The formation of 25-hydroxycholecalciferol does not appear to be stringently regulated. However, the formation of 1,25-dihydroxycholecalciferol in the kidneys, which is catalyzed by 1α-hydroxylase, is regulated in a feedback fashion by plasma Ca^{2+} and PO_4^{3+} (Figure 21–8). Its formation is facilitated by PTH, and when the plasma Ca^{2+} level is low, PTH secretion is increased. When the plasma Ca^{2+} level is high, little 1,25-dihydroxycholecalciferol is produced, and the kidneys produce the relatively inactive metabolite 24,25-dihydroxycholecalciferol instead. This effect of Ca^{2+} on production of 1,25-dihydroxycholecalciferol is the mechanism that brings about adaptation of Ca^{2+} absorption from the intestine (see above). The production of 1,25-dihydroxycholecalciferol is also increased by low and inhibited by high plasma PO_4^{3+} levels, by a direct inhibitory effect of PO_4^{3+} on 1α-hydroxylase. Additional control of 1,25-dihydroxycholecalciferol formation is exerted by a direct negative feedback effect of the metabolite on 1α-hydroxylase, a positive feedback action on the formation of 24,25-dihydroxycholecalciferol, and a direct action on the parathyroid gland to inhibit the production of mRNA for PTH. Prolactin increases the activity of 1α-hydroxylase, and circulating 1,25-dihydroxycholecalciferol is increased during lactation. Estrogen increases total circulating 1,25-dihydroxycholecalciferol, but this is probably due to an increase in the secretion of its binding protein without any steady-state change in free 1,25-dihydroxycholecalciferol. Hyperthyroidism is associated with decreased circulating 1,25-dihydroxycholecalciferol and an increased incidence of osteoporosis. 1,25-Dihydroxycholecalciferol production is depressed by metabolic acidosis. Growth hormone, hCS, and calcitonin also stimulate 1,25-dihydroxycholecalciferol formation.

Rickets & Osteomalacia

Vitamin D deficiency causes defective calcification of bone matrix and the disease called **rickets** in children and **osteomalacia** in adults. Even though 1,25-dihy-

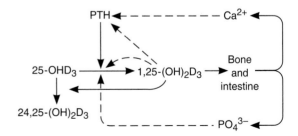

Figure 21–8. Feedback control of the formation of 1,25-dihydroxycholecalciferol (1,25-[OH]$_2$D$_3$) from 25-hydroxycholecalciferol (25-OHD$_3$) in the kidneys. Solid arrows indicate stimulation, and dashed arrows indicate inhibition.

droxycholecalciferol is necessary for normal mineralization of bone matrix, the main defect in this condition is failure to deliver adequate amounts of Ca^{2+} and PO_4^{3-} to the sites of mineralization. The full-blown condition in children is characterized by weakness and bowing of weight-bearing bones, dental defects, and hypocalcemia. In adults, the condition is less obvious. It used to be most commonly due to inadequate exposure to the sun in smoggy cities, but now it is more commonly due to inadequate intake of the provitamins on which the sun acts in the skin. These cases respond to administration of vitamin D. The condition can also be caused by inactivating mutations of the gene for renal 1α-hydroxylase, in which case there is no response to vitamin D but a normal response to 1,25-dihydroxycholecalciferol (**type I vitamin D-resistant rickets**). In rare instances, it can be due to inactivating mutations of the gene for the 1,25-dihydroxycholecalciferol receptor (**type II vitamin D-resistant rickets**), in which case there is a deficient response to both vitamin D and 1,25-dihydroxycholecalciferol.

THE PARATHYROID GLANDS

Anatomy

In humans, there are usually four parathyroid glands: two embedded in the superior poles of the thyroid and two in its inferior poles (Figure 21–9). However, the lo-

cations of the individual parathyroids and their number can vary considerably. Parathyroid tissue is sometimes found in the mediastinum.

Each parathyroid gland is a richly vascularized disk, about $3 \times 6 \times 2$ mm, containing two distinct types of cells. The abundant **chief cells,** which contain a prominent Golgi apparatus plus endoplasmic reticulum and secretory granules (Figure 21–10), synthesize and secrete **parathyroid hormone (PTH).** The less abundant and larger **oxyphil cells** contain oxyphil granules and large numbers of mitochondria in their cytoplasm. In humans, few are seen before puberty, and thereafter they increase in number with age. Their function is unknown, although some have argued that they are degenerated chief cells.

Synthesis & Metabolism of PTH

Human PTH is a linear polypeptide with a molecular weight of 9500 that contains 84 amino acid residues (Figure 21–11). Its structure is very similar to that of bovine and porcine PTH. It is synthesized as part of a larger molecule containing 115 amino acid residues (**preproPTH**). Upon entry of preproPTH into the endoplasmic reticulum, a leader sequence is removed from the amino terminal to form the 90-amino-acid polypeptide **proPTH.** Six additional amino acid residues are removed from the amino terminal of proPTH in the Golgi apparatus, and the 84-amino-acid polypeptide PTH is packaged in secretory granules and released as the main secretory product of the chief cells.

The normal plasma level of intact PTH is 10–55 pg/mL. The half-life of PTH is approximately 10 min-

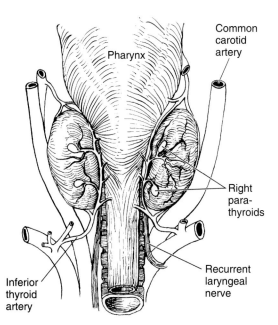

Figure 21–9. The human parathyroid glands, viewed from behind.

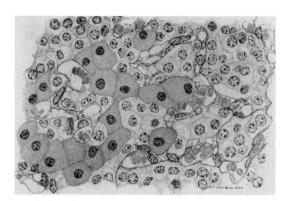

Figure 21–10. Section of human parathyroid. (Reduced 50% from × 960.) Small cells are chief cells; large stippled cells (especially prominent in the lower left of picture) are oxyphil cells. (Reproduced, with permission, from Fawcett DW: *Bloom and Fawcett, A Textbook of Histology,* 11th ed. Saunders, 1986.)

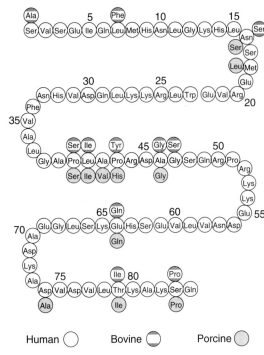

Figure 21–11. Parathyroid hormone. The symbols above and below the human structure show where amino acid residues are different in bovine and porcine PTH. (Reproduced, with permission, from Keutmann HT et al: Complete amino acid sequence of human parathyroid hormone. Biochemistry 1978;17:5723. Copyright © 1978 by the American Chemical Society.)

utes, and the secreted polypeptide is rapidly cleaved by the Kupffer cells in the liver into midregion and carboxyl terminal fragments that are probably biologically inactive. PTH and these fragments are then cleared by the kidneys. It is interesting in this regard that a synthetic polypeptide containing the amino terminal 34 amino acid residues of PTH has all the known biologic effects of the full molecule.

Since many of the old radioimmunoassays for PTH used antibodies against the midregion of the molecule, they measured the fragments as well as the intact hormone and hence gave values that were falsely high. This was particularly true in patients with renal failure, in whom the fragments were not cleared. To get around this problem, two-site immunoassays have been developed, using one antibody against the amino terminal and another against the carboxyl terminal. Typically, plasma is first reacted with a radiolabeled amino terminal antibody. It is next added to beads to which an unlabeled carboxyl terminal antibody has been bound. The beads are then washed, and the radioactivity bound to the beads is measured. Only intact PTH is recognized by both antibodies, and hence an accurate measurement of circulating PTH (1–84) is obtained.

Actions

PTH acts directly on bone to increase bone resorption and mobilize Ca^{2+}. In addition to increasing the plasma Ca^{2+} and depressing the plasma phosphate, PTH increases phosphate excretion in the urine. This **phosphaturic action** is due to a decrease in reabsorption of phosphate in the proximal tubules. PTH also increases reabsorption of Ca^{2+} in the distal tubules, although Ca^{2+} excretion is often increased in hyperparathyroidism because the increase in the amount filtered overwhelms the effect on reabsorption. PTH also increases the formation of 1,25-dihydroxycholecalciferol, and this increases Ca^{2+} absorption from the intestine.

On a longer timescale, PTH stimulates both osteoblasts and osteoclasts. The net effect varies, but with mildly elevated plasma PTH levels it is usually anabolic.

Mechanism of Action

It now appears that there are at least three different PTH receptors. One also binds parathyroid hormone-related protein (PTHrP; see below) and is known as the hPTH/PTHrP receptor. A second receptor, PTH2 (hPTH2-R), does not bind PTHrP and is found in the brain, placenta, and pancreas. In addition, there is evidence for a third receptor, CPTH, which reacts with the carboxyl terminal rather than the amino terminal of PTH. The first two are serpentine receptors coupled to G_s, and via this heterotrimeric G protein they activate adenylyl cyclase, increasing intracellular cAMP. The hPTH/PTHrP receptor also activates PLC via G_q, increasing intracellular Ca^{2+} and activating protein kinase C (Figure 21–12). However, the way these second messengers affect Ca^{2+} in bone is unsettled.

In the disease called **pseudohypoparathyroidism,** the signs and symptoms of hypoparathyroidism develop but the circulating level of PTH is normal or elevated. Since the tissues fail to respond to the hormone, this is a receptor disease. There are two forms. In the more common form, there is a congenital 50% reduction of the activity of G_s, and PTH fails to produce a normal increase in cAMP concentration. In a different, less common form, the cAMP response is normal but the phosphaturic action of the hormone is defective.

Regulation of Secretion

Circulating ionized calcium acts directly on the parathyroid glands in a negative feedback fashion to

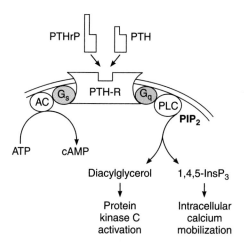

Figure 21–12. Signal transduction pathways activated by PTH or PTHrP binding to the hPTH/hPTHrP receptor. Intracellular cAMP is increased via G_s and adenylyl cyclase (AC). Diacylglycerol and IP_3 (1,4,5-InsP$_3$) are increased via G_q and phospholipase C (PLC). (Modified and reproduced, with permission, from Shoback DM, Strewler GJ: Disorders of the parathyroids and calcium metabolism. In: McPhee SJ, Lingappa VR, Ganong WF [editors]: *Pathophysiology of Disease*, 4th ed. McGraw-Hill, 2003.)

regulate the secretion of PTH (Figure 21–13). The key to this regulation is a cell membrane Ca^{2+} receptor. This serpentine receptor is coupled via a G protein to phosphoinositide turnover and is found in many tissues. In the parathyroid, its activation inhibits PTH secretion. In this way, when the plasma Ca^{2+} level is high, PTH secretion is inhibited and the Ca^{2+} is deposited in the bones. When it is low, secretion is increased and Ca^{2+} is mobilized from the bones.

1,25-Dihydroxycholecalciferol acts directly on the parathyroid glands to decrease preproPTH mRNA. Increased plasma phosphate stimulates PTH secretion by lowering plasma Ca^{2+} and inhibiting the formation of 1,25-dihydroxycholecalciferol. Magnesium is required to maintain normal parathyroid secretory responses. Impaired PTH release along with diminished target organ responses to PTH account for the hypocalcemia that occasionally occurs in magnesium deficiency.

Effects of Parathyroidectomy

PTH is essential for life. After parathyroidectomy, there is a steady decline in the plasma Ca^{2+} level. Signs of neuromuscular hyperexcitability appear, followed by full-blown hypocalcemic tetany (see above). Plasma phosphate levels usually rise as the plasma calcium level

falls after parathyroidectomy, but the rise does not always occur.

In humans, tetany is most often due to inadvertent parathyroidectomy during thyroid surgery. Symptoms usually develop 2–3 days postoperatively but may not appear for several weeks or more. In rats on a low-calcium diet, tetany develops much more rapidly, and death occurs 6–10 hours after parathyroidectomy. Injections of PTH correct the chemical abnormalities, and the symptoms disappear. Injections of Ca^{2+} salts give temporary relief.

The signs of tetany in humans include **Chvostek's sign,** a quick contraction of the ipsilateral facial muscles elicited by tapping over the facial nerve at the angle of the jaw; and **Trousseau's sign,** a spasm of the muscles of the upper extremity that causes flexion of the wrist and thumb with extension of the fingers (Figure 21–14). In individuals with mild tetany in whom spasm is not evident, Trousseau's sign can sometimes be produced by occluding the circulation for a few minutes with a blood pressure cuff.

Parathyroid Hormone Excess

Hyperparathyroidism due to injections of parathyroid extract in animals or hypersecretion of a functioning parathyroid tumor in humans is characterized by hypercalcemia and hypophosphatemia. Humans with PTH-secreting adenomas are usually asymptomatic, with the condition detected when plasma Ca^{2+} is measured in conjunction with a routine physical examina-

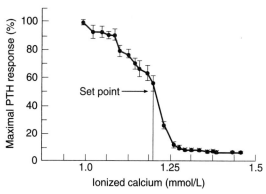

Figure 21–13. Relation between plasma Ca^{2+} concentration and PTH response in humans. The set point is the plasma Ca^{2+} at which half the maximal response occurred. (Modified and reproduced, with permission, from Brown E: Extracellular Ca^{2+} sensing, regulation of parathyroid cell functions, and role of Ca^{2+} and other ions as extracellular (first) messengers. Physiol Rev 1991;71:371.)

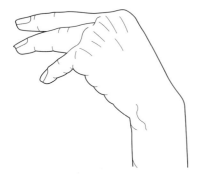

Figure 21–14. Position of the hand in hypocalcemic tetany (Trousseau's sign).

tion. However, there may be minor changes in personality, and calcium-containing kidney stones occasionally form.

Secondary Hyperparathyroidism

In conditions such as chronic renal disease and rickets, in which the plasma Ca^{2+} level is chronically low, stimulation of the parathyroid glands causes compensatory parathyroid hypertrophy and **secondary hyperparathyroidism.** The plasma Ca^{2+} level is low in chronic renal disease primarily because the diseased kidneys lose the ability to form 1,25-dihydroxycholecalciferol.

Familial Hypercalcemia & Hypocalcemia

Mutations in the gene for the Ca^{2+} receptor cause predictable long-term changes in plasma Ca^{2+}. Individuals heterozygous for inactivating mutations have **familial benign hypocalciuric hypercalcemia,** a condition in which there is a chronic moderate elevation in plasma Ca^{2+} because the feedback inhibition of PTH secretion by Ca^{2+} is reduced. Plasma PTH levels are normal or even elevated. However, children who are homozygous for inactivating mutations develop **neonatal severe primary hyperparathyroidism.** Individuals with activating mutations of the gene for the Ca^{2+} receptor develop **familial hypercalciuric hypocalcemia** due to increased sensitivity of the parathyroid glands to plasma Ca^{2+}.

PTHrP

A protein with PTH activity, **parathyroid hormone-related protein (PTHrP),** is produced by many different tissues in the body. It has 140 amino acid residues, compared with 84 in PTH, and is encoded by a gene on human chromosome 12, whereas PTH is encoded

by a gene on chromosome 11. However, PTHrP and PTH have marked homology at their amino terminal ends, with 8 of the first 13 amino acid residues in the same positions, and they both bind to the hPTH/PTHrP receptor. Yet their physiologic effects are very different. How is this possible when they bind to the same receptor? For one thing, PTHrP is primarily a tissue hormone or factor, acting where it is produced. It may be that circulating PTH cannot reach these sites. Another possibility is action of one or the other hormone on other, more selective receptors.

In any case, PTHrP has a marked effect on the growth and development of cartilage in utero. Mice in which both alleles of the PTHrP gene are knocked out (ie, those homozygous for the defect) have severe skeletal deformities and die soon after birth. In utero, when PTHrP is released, cartilage is stimulated by a protein called **Indian hedgehog.** The PTHrP-stimulated cartilage cells proliferate and their terminal differentiation is inhibited. PTHrP is also expressed in the brain, where there is evidence that it inhibits excitotoxic damage to developing neurons. In addition, there is evidence that it is involved in Ca^{2+} transport in the placenta.

Although homozygous PTHrP knockouts are lethal, it is possible to genetically engineer mice that survive because they have a local increase in PTHrP in their cartilage. In these animals, breasts fail to develop, a finding that is consistent with the observation that in normal lactating mice there is abundant PTHrP in the milk. However, the plasma level does not increase during lactation, and the function of PTHrP in milk is unknown. PTHrP is also found in keratinocytes in the skin, in smooth muscle, and in the teeth, where it is present in the enamel epithelium that caps each tooth. In the absence of PTHrP, teeth cannot erupt.

Hypercalcemia of Malignancy

Hypercalcemia is a common metabolic complication of cancer. About 20% of hypercalcemic patients have bone metastases that produce the hypercalcemia by eroding bone (**local osteolytic hypercalcemia).** There is evidence that this erosion is produced by prostaglandins such as PGE from the tumor. The hypercalcemia in the remaining 80% of the patients is due to elevated circulating levels of PTHrP (**humoral hypercalcemia of malignancy).** The tumors responsible for the hypersecretion include cancers of the breast, kidney, ovary, and skin.

CALCITONIN

Origin

In dogs, perfusion of the thyroparathyroid region with solutions containing high concentrations of Ca^{2+} leads to a fall in peripheral plasma calcium, and after damage to this region, Ca^{2+} infusions cause a greater increase in plasma Ca^{2+} than they do in control animals. These and other observations led to the discovery that a Ca^{2+}-lowering as well as a Ca^{2+}-elevating hormone was secreted by structures in the neck. The Ca^{2+}-lowering hormone has been named **calcitonin.** In nonmammalian vertebrates, the source of calcitonin is the **ultimobranchial bodies,** a pair of glands derived embryologically from the fifth branchial arches. In mammals, these bodies have for the most part become incorporated into the thyroid gland, where the ultimobranchial tissue is distributed around the follicles as the **parafollicular cells,** which are also known as the clear or C cells (Figures 18–2 and 21–15).

Structure

Human calcitonin has a molecular weight of 3500 and contains 32 amino acid residues (Figure 21–16). Much of the mRNA transcribed from the calcitonin gene is processed to a different mRNA in the nervous system, so that **calcitonin gene-related peptide (CGRP)** is formed rather than calcitonin (see Chapter 4). The calcitonins of the other species that have been studied also contain 32 amino acid residues, but the amino acid composition varies considerably. Salmon calcitonin is of interest because it is more than 20 times more active in humans than human calcitonin is.

Secretion & Metabolism

Secretion of calcitonin is increased when the thyroid gland is perfused with solutions containing a high Ca^{2+} concentration. Measurement of circulating calcitonin by immunoassay indicates that it is not secreted until the plasma calcium level reaches approximately 9.5 mg/dL and that above this calcium level, plasma calcitonin is directly proportionate to plasma calcium. β-Adrenergic agonists, dopamine, and estrogens also stimulate calcitonin secretion. Gastrin, CCK, glucagon, and secretin have all been reported to stimulate calcitonin secretion, with gastrin being the most potent stimulus (see Chapter 26). The plasma calcitonin level

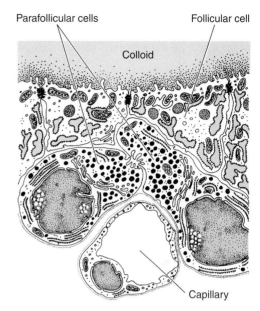

Figure 21–15. Parafollicular cells in the thyroid. (Modified from Poirier J, Dumas JLR: *Review of Medical Histology.* Saunders, 1977.)

is elevated in Zollinger-Ellison syndrome (see Chapter 26) and in pernicious anemia, in which the plasma gastrin level is also elevated. However, the dose of gastrin needed to stimulate calcitonin secretion produces an increase in plasma gastrin concentration greater than that produced by food, so it is premature to conclude that calcium in the intestine initiates secretion of a calcium-lowering hormone before the calcium is absorbed.

Human calcitonin has a half-life of less than 10 minutes.

Actions

Serpentine receptors for calcitonin are found in bones and the kidneys. Calcitonin lowers the circulating calcium and phosphate levels. It exerts its calcium-

Cys-Gly-Asn-Leu-Ser-Thr-Cys-Met-Leu-Gly-Thr-Tyr-Thr-Gln-Asp-Phe-Asn-
1 2 3 4 5 6 7 8 9 10 11 12 13 14 15 16 17

Lys-Phe-His-Thr-Phe-Pro-Gln-Thr-Ala-Ile-Gly-Val-Gly-Ala-Pro-NH$_2$
18 19 20 21 22 23 24 25 26 27 28 29 30 31 32

Figure 21–16. Human calcitonin.

lowering effect by inhibiting bone resorption. This action is direct, and calcitonin inhibits the activity of osteoclasts in vitro. It also increases Ca^{2+} excretion in the urine.

The exact physiologic role of calcitonin is uncertain. The calcitonin content of the human thyroid is low, and after thyroidectomy, bone density and plasma Ca^{2+} level are normal as long as the parathyroid glands are intact. In addition, there are only transient abnormalities of Ca^{2+} metabolism when a Ca^{2+} load is injected after thyroidectomy. This may be explained in part by secretion of calcitonin from tissues other than the thyroid. However, there is general agreement that the hormone has little long-term effect on the plasma Ca^{2+} level in adult animals and humans. Moreover, patients with medullary carcinoma of the thyroid have a very high circulating calcitonin level but no symptoms directly attributable to the hormone, and their bones are essentially normal. No syndrome due to calcitonin deficiency has been described. More hormone is secreted in young individuals, and it may play a role in skeletal development. It may protect against postprandial hypercalcemia. In addition, it may protect the bones of the mother from excess calcium loss during pregnancy. Bone formation in the infant and lactation are major drains on calcium stores, and plasma concentrations of 1,25-dihydroxycholecalciferol are elevated in pregnancy. They would cause bone loss in the mother if bone resorption were not simultaneously inhibited by an increase in the plasma calcitonin level.

Clinical Correlates

Calcitonin is useful in the treatment of Paget's disease, a condition in which increased osteoclastic activity triggers compensatory formation of disorganized new bone. It also has beneficial effects in severe hypercalcemia, but the hormone must be injected and its effect generally wears off.

Summary

The actions of the three principal hormones that regulate the plasma concentration of Ca^{2+} can now be sum-marized. PTH increases plasma Ca^{2+} by mobilizing this ion from bone. It increases Ca^{2+} reabsorption in the kidney, but this may be offset by the increase in filtered Ca^{2+}. It also increases the formation of 1,25-dihydroxycholecalciferol. 1,25-Dihydroxycholecalciferol increases Ca^{2+} absorption from the intestine and increases Ca^{2+} reabsorption in the kidneys. Calcitonin inhibits bone resorption and increases the amount of Ca^{2+} in the urine.

EFFECTS OF OTHER HORMONES & HUMORAL AGENTS ON CALCIUM METABOLISM

Calcium metabolism is affected by various hormones in addition to 1,25-dihydroxycholecalciferol, PTH, PTHrP, and calcitonin. Glucocorticoids lower plasma Ca^{2+} levels by inhibiting osteoclast formation and activity, but over long periods they cause osteoporosis by decreasing bone formation and increasing bone resorption. They decrease bone formation by inhibiting protein synthesis in osteoblasts. They also decrease the absorption of Ca^{2+} and PO_4^{3-} from the intestine and increase the renal excretion of these ions. This is why they depress the hypercalcemia of vitamin D intoxication. The decrease in plasma Ca^{2+} concentration increases the secretion of PTH, and bone resorption is facilitated. **Growth hormone** increases calcium excretion in the urine, but it also increases intestinal absorption of calcium, and this effect may be greater than the effect on excretion, with a resultant positive calcium balance. **IGF-I** generated by the action of growth hormone stimulates protein synthesis in bone. As noted above, **thyroid hormones** may cause hypercalcemia, hypercalciuria, and, in some instances, osteoporosis. **Estrogens** prevent osteoporosis by inhibiting the stimulatory effects of certain cytokines on osteoclasts (see above). **Insulin** increases bone formation, and there is significant bone loss in untreated diabetes.

The Pituitary Gland

22

INTRODUCTION

The anterior, intermediate, and posterior lobes of the pituitary gland are actually three more or less separate endocrine organs that, at least in some species, contain 14 or more hormonally active substances. The intermediate lobe is rudimentary in humans. The six established hormones that are secreted by the anterior pituitary are **thyroid-stimulating hormone (TSH, thyrotropin), adrenocorticotropic hormone (ACTH), luteinizing hormone (LH), follicle-stimulating hormone (FSH), prolactin,** and **growth hormone** (Table 22–1). ACTH, prolactin, and growth hormone are simple polypeptides or proteins, whereas TSH, LH, and FSH are glycoproteins. Prolactin acts on the breast (see Figure 14–16). The remaining five are, at least in part, **tropic hormones;** ie, they stimulate secretion of hormonally active substances by other endocrine glands or, in the case of growth hormone, the liver and other tissues (see below). In addition, the anterior lobe of the pituitary secretes **β-lipotropin (β-LPH).** This linear polypeptide contains 91 amino acid residues, and although its physiologic role is uncertain, it contains the amino acid sequences of endorphins and enkephalins, peptides that bind to opiate receptors (see Chapter 4). In addition, the anterior and intermediate lobes con-

Table 22–1. Pituitary hormones in mammals.[1]

Name and Source	Principal Actions
Anterior lobe	
Thyroid-stimulating hormone (TSH, thyrotropin)	Stimulates thyroid secretion and growth of thyroid gland.
Adrenocorticotropic hormone (ACTH, corticotropin)	Stimulates secretion and growth of zona fasciculata and zona reticularis of adrenal cortex.
Growth hormone (GH, somatotropin, STH)	Accelerates body growth; stimulates secretion of IGF-I.
Follicle-stimulating hormone (FSH)	Stimulates ovarian follicle growth in female and spermatogenesis in male.
Luteinizing hormone (LH)	Stimulates ovulation and luteinization of ovarian follicles in female and testosterone secretion in male.
Prolactin (PRL)	Stimulates secretion of milk and maternal behavior.
β-Lipotropin (β-LPH)	?
Intermediate lobe	
α-, β-, and γ-melanocyte-stimulating hormones (α-, β-, and γ-MSH; referred to collectively as melanotropin or intermedin)	Expand melanophores in fish, amphibians, and reptiles; stimulate melanin synthesis in melanocytes in humans.
γ-Lipotropin (γ-LPH), corticotropin-like intermediate lobe peptide (CLIP), other fragments of pro-opiomelanocortin	?
Posterior lobe	
Vasopressin (antidiuretic hormone, ADH)	Promotes water retention.
Oxytocin	Causes milk ejection, contraction of pregnant uterus.

[1]In addition, a variety of gastrointestinal and other polypeptides are found in one or more lobes of the pituitary gland. These include CCK, gastrin, renin, angiotensin II, and calcitonin gene-related peptide (CGRP).

tain other hormonally active derivatives of the pro-opiomelanocortin molecule (see below). The hormones tropic to a particular endocrine gland are discussed in the chapter on that gland TSH in Chapter 18; ACTH in Chapter 20; and the gonadotropins FSH and LH in Chapter 23, along with prolactin. The hormones secreted by the posterior pituitary in mammals (**oxytocin** and **vasopressin**) and the neural regulation of anterior and posterior pituitary secretion are discussed in Chapter 14. Growth hormone and the melanocyte-stimulating hormones of the intermediate lobe of the pituitary, α-MSH and β-MSH, are the subject of this chapter, along with a number of general considerations about the pituitary.

MORPHOLOGY

Gross Anatomy

The anatomy of the pituitary gland is summarized in Figure 22–1 and discussed in detail in Chapter 14. The posterior pituitary is made up largely of the endings on blood vessels of axons from the supraoptic and paraventricular nuclei of the hypothalamus, whereas the anterior pituitary has a special vascular connection with the brain, ie, the portal hypophysial vessels. The intermediate lobe is formed in the embryo from the dorsal half of **Rathke's pouch,** an evagination of the roof of the pharynx, but is closely adherent to the posterior lobe in the adult. It is separated from the anterior lobe by the remains of Rathke's pouch, the **residual cleft.**

Histology

In the posterior lobe, the endings of the supraoptic and paraventricular axons can be observed in close relation to blood vessels. In some species, the endings are rod-like palisaded structures. There are also **pituicytes—** stellate cells which are modified astrocytes.

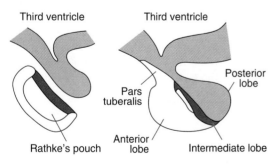

Figure 22–1. Diagrammatic outline of the formation of the pituitary and the various parts of the organ in the adult.

The intermediate lobe is rudimentary in humans and a few other mammalian species. In these species, most of its cells are agranular, although there are often a few basophilic elements that resemble anterior lobe cells. Along the residual cleft are small thyroid-like follicles, some containing a little colloid. The function of the colloid, if any, is unknown.

The anterior pituitary is made up of interlacing cell cords and an extensive network of sinusoidal capillaries. The endothelium of the capillaries is fenestrated, like that in other endocrine organs. The cells contain granules of stored hormone that are extruded from the cells by exocytosis. The granules presumably break down in the pericapillary space, and their contents enter the capillaries.

Cell Types in the Anterior Pituitary

Human anterior pituitary cells have traditionally been divided on the basis of their staining reactions into chromophobes and chromophils. The chromophilic cells are subdivided into acidophils, which stain with acidic dyes, and basophils, which stain with basic dyes. Many of the chromophobic cells are secretory cells that are inactive and have few secretory granules. The others are probably folliculostellate cells (see below). Five types of chromophilic secretory cells have been identified by immunocytochemistry and electron microscopy. They are the somatotropes, which secrete growth hormone; the lactotropes (also called mammotropes), which secrete prolactin; the corticotropes, which secrete ACTH and β-LPH; the thyrotropes, which secrete TSH; and the gonadotropes, which secrete FSH and LH. The characteristics of these cells are summarized in Table 22–2. Their appearance and granule content can vary with changes in hormone status. Note that in this classification, two or more polypeptide hormones were secreted by several types of cells.

The anterior pituitary also contains folliculostellate cells, which send processes between established secretory cells. Current evidence indicates that these cells contain and secrete the cytokine IL-6 (see Chapter 27), but their physiologic role remains unsettled.

Two-Unit Structure of FSH, LH, & TSH

The three pituitary glycoprotein hormones, FSH, LH, and TSH, are each made up of two subunits. The subunits, which have been designated α and β, have some activity but must be combined for maximal physiologic activity. In addition, the placental glycoprotein gonadotropin human chorionic gonadotropin (hCG) has α and β subunits (see Chapter 23). All of the α subunits of these hormones are products of a single gene

Table 22–2. Hormone-secreting cells of the human anterior pituitary gland.

Cell Type	Hormones Secreted	% of Total Secretory Cells	Stain Affinity	Diameter of Secretory Granules (nm)
Somatotrope	Growth hormone	50	Acidophilic	300–400
Lactotrope	Prolactin	10–30	Acidophilic	200
Corticotrope	ACTH, β-LPH	10	Basophilic	400–550
Thyrotrope	TSH	5	Basophilic	120–200
Gonadotrope	FSH, LH	20	Basophilic	250–400

and have the same amino acid composition, although their carbohydrate residues vary. The β subunits, which are produced by separate genes and differ in structure, confer hormonal specificity. The α subunits are remarkably interchangeable, and hybrid molecules can be created. However, the physiologic and evolutionary significance of the unique two-unit structure of these glycoprotein hormones remains to be determined.

INTERMEDIATE-LOBE HORMONES

Pro-opiomelanocortin

Intermediate-lobe cells and corticotropes of the anterior lobe both synthesize a large precursor protein that is cleaved to form a family of hormones. After removal of the signal peptide, this prohormone is known as **pro-opiomelanocortin (POMC).** This molecule is also synthesized in the hypothalamus and other parts of the nervous system, the lungs, the gastrointestinal tract, and the placenta (see Chapter 4). Its structure is shown in Figure 22–2. In the corticotropes, it is hydrolyzed to ACTH and β-LPH plus a small amount of β-endorphin, and these substances are secreted. In the intermediate lobe cells, POMC is further hydrolyzed to corticotropin-like intermediate-lobe peptide (CLIP), γ-LPH, and appreciable quantities of β-endorphin. The functions, if any, of CLIP and γ-LPH are unknown, whereas β-endorphin is an opioid peptide (see Chapter 4) that has the five amino acid residues of met-enkephalin at its amino terminal end. The **melanotropins** α- and β-MSH are also formed. However, the intermediate lobe in humans is rudimentary, and it appears that neither α-MSH nor β-MSH is secreted in adults.

Control of Skin Coloration

Fish, reptiles, and amphibia change the color of their skin for thermoregulation, camouflage, and behavioral displays. They do this in part by moving black or brown granules into or out of the periphery of pigment cells called **melanophores.** The granules are made up of **melanins,** which are synthesized from dopa (see Chapter 4) and dopaquinone. The movement of these granules is controlled by a variety of hormones and neurotransmitters, including α- and β-MSH, melanin-concentrating hormone, melatonin, and catecholamines.

In mammals, there are no melanophores containing pigment granules that disperse and aggregate, but there are **melanocytes,** which have multiple processes containing melanin granules. The melanocytes contain **melanotropin-1** receptors, one of several types of melanotropin receptors that have been cloned. Treatment with MSHs accelerates melanin synthesis, causing readily detectable darkening of the skin in humans in 24 hours. As noted above, α- and β-MSH do not circulate in adult humans, and their function is unknown. However, ACTH binds to melanotropin-1 receptors.

Pigment Abnormalities in Humans

The pigmentary changes in several endocrine diseases are due to changes in circulating ACTH. For example, abnormal pallor is a hallmark of hypopituitarism. Hyperpigmentation occurs in patients with adrenal insufficiency due to primary adrenal disease. Indeed, the presence of hyperpigmentation in association with adrenal insufficiency rules out the possibility that the insufficiency is secondary to pituitary or hypothalamic disease because in these conditions, plasma ACTH is not increased (see Chapter 20).

Albinos have a congenital inability to synthesize melanin. Albinism occurs in humans and many other mammalian species. It can be due to a variety of different genetic defects in the pathways for melanin synthesis. **Piebaldism** is characterized by patches of skin that lack melanin as a result of congenital defects in the migration of pigment cell precursors from the neural crest

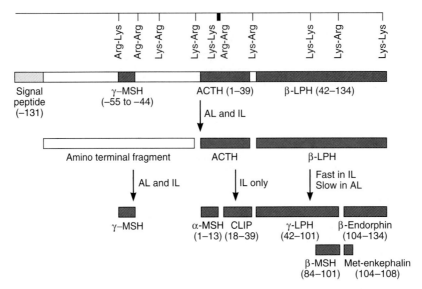

Figure 22–2. Schematic representation of the prepro-opiomelanocortin molecule formed in pituitary cells, neurons, and other tissues. The numbers in parentheses identify the amino acid sequences in each of the polypeptide fragments. For convenience, the amino acid sequences are numbered from the amino terminal of ACTH and read toward the carboxyl terminal portion of the parent molecule, whereas the amino acid sequences in the other portion of the molecule read to the left to –131, the amino terminal of the parent molecule. The locations of Lys-Arg and other pairs of basic amino acids residues are also indicated; these are the sites of proteolytic cleavage in the formation of the smaller fragments of the parent molecule. AL, anterior lobe; IL, intermediate lobe.

during embryonic development. Not only the condition but also the precise pattern of the loss is passed from one generation to the next. **Vitiligo** is due to a similar patchy loss of melanin, but the loss develops after birth and is progressive.

GROWTH HORMONE

Biosynthesis & Chemistry

The long arm of human chromosome 17 contains the growth hormone-hCS cluster that contains five genes: one, *hGH-N* (for normal), codes for the most abundant form of growth hormone; a second, *hGH-V* (for variant), codes for the variant form of growth hormone (see below); two code for hCS (see Chapter 23); and the fifth is probably an hCS pseudogene.

"Normal" human growth hormone, the product of *hGH-N,* accounts for 75% of the circulating hGH. Because of its molecular weight of 22,000, it is also known as 22 K hGH. Figure 22–3 shows its structure and compares it with the structure of hCS, which also has 191 amino acid residues and differs by only 29 residues. The *hGH-V* gene, which is expressed primarily in the placenta, produces a 191-amino-acid growth

hormone that differs from normal human growth hormone by 13 amino acids scattered throughout the polypeptide chain. Appreciable amounts appear in the circulation only in pregnancy. The mRNA produced by the *hGH-N* gene undergoes alternative splicing to produce a growth hormone identical to 22 K hGH except that amino acid residues 32–46 are deleted. This smaller form, 20 K hGH, is also biologically active and makes up about 10% of the circulating growth hormone. The physiologic significance of having the two principal forms, 22 K and 20 K, is unknown.

Species Specificity

Not surprisingly, there is considerable variation in the structure of growth hormone from one species to another. Porcine and simian growth hormones have only a transient effect in the guinea pig, probably because they stimulate the rapid formation of anti-growth-hormone antibodies. In monkeys and humans, bovine and porcine growth hormones do not even have a significant transient effect on growth, although monkey and human growth hormones are fully active in both monkeys and humans. Human growth hormone has intrinsic lactogenic activity.

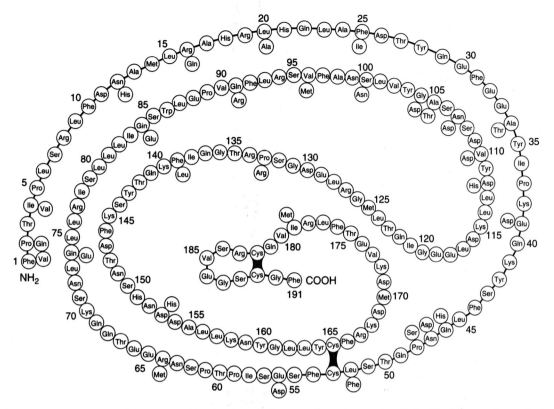

Figure 22–3. Structure of the principal human growth hormone (continuous chain). The black bars indicate disulfide bridges. The 29 residues alongside the chain identify residues that differ in human chorionic somatomammotropin (hCS; see Chapter 23). All the other residues in hCS are the same, and hCS also has 191 amino acid residues. (Reproduced, with permission, from Parsons JA [editor]: *Peptide Hormones.* University Park Press, 1976.)

Plasma Levels, Binding, & Metabolism

Growth hormone is bound to a protein in plasma that is a large fragment of the extracellular domain of the growth hormone receptor (see below). It appears to be produced by cleavage of receptors in humans, and its concentration is an index of the number of growth hormone receptors in the tissues. About half the growth hormone activity is bound, providing a reservoir of the hormone to compensate for the wide fluctuations that occur in secretion (see below).

Growth hormone is metabolized rapidly, probably at least in part in the liver. The half-life of circulating growth hormone in humans is 6–20 minutes, and the daily growth hormone output has been calculated to be 0.2–1.0 mg/d in adults.

The basal plasma growth hormone level measured by radioimmunoassay in adult humans is normally less than 3 ng/mL. This is mostly 22 K hGH, since the affinity of 20 K hGH for most antibodies is about 30% of the affinity of 22 K hGH. Both the protein-bound and free forms are measured by radioimmunoassays, since the antibodies that are employed have a higher affinity for the hormone than for the binding proteins.

Growth Hormone Receptors

The growth hormone receptor is a 620-amino-acid protein with a large extracellular portion, a transmembrane domain, and a large cytoplasmic portion. It is a member of the cytokine receptor superfamily, which is discussed in Chapter 27. Growth hormone has two binding sites for receptors, and when it binds to one of the receptor subunits, the other binding site attracts another subunit, producing a homodimer (Figure 22–4). Dimerization is essential for receptor activation.

Growth hormone has widespread effects in the body (see below), so even though it is not yet possible to correlate intracellular and whole body effects, it is not surprising that, like insulin, growth hormone activates many different intracellular enzyme cascades (Figure 22–4). Of particular note is its activation of the JAK2-STAT pathway. JAK2 is a member of the Janus family of cytoplasmic tyrosine kinases. STATs (for signal transducers and activators of transcription) are a family of inactive cytoplasmic transcription factors that upon phosphorylation by JAK kinases migrate to the nucleus and activate various genes. The JAK-STAT pathways also mediate the effects of prolactin and various growth factors.

Effects on Growth

In young animals in which the epiphyses have not yet fused to the long bones (see Chapter 21), growth is inhibited by hypophysectomy (Figure 22–5) and stimulated by growth hormone. Chondrogenesis is ac-

celerated, and as the cartilaginous epiphysial plates widen, they lay down more bone matrix at the ends of long bones. In this way, stature is increased, and prolonged treatment with growth hormone leads to gigantism.

When the epiphyses are closed, linear growth is no longer possible, and growth hormone produces the pattern of bone and soft tissue deformities known in humans as **acromegaly** (Figure 22–6). The sizes of most of the viscera are increased. The protein content of the body is increased, and the fat content is decreased.

Effects on Protein & Electrolyte Metabolism

Growth hormone is a protein anabolic hormone and produces a positive nitrogen and phosphorus balance, a

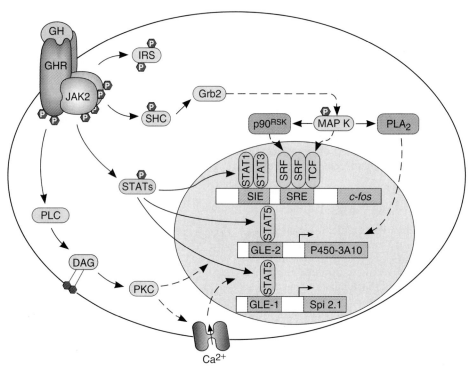

Figure 22–4. Some of the principal signaling pathways activated by the dimerized growth hormone receptor (GHR). Solid arrows indicate established pathways; dashed arrows indicate probable pathways. The details of the PLC pathway and the pathway from Grb2 to MAP K are shown in Chapter 1. GLE-1 and GLE-2, interferon γ-activated response elements; IRS, insulin receptor substrate; p90^{RSK}, an S6 kinase; PLA$_2$, phospholipase A$_2$; SIE, Sis-induced element; SRE, serum response element; SRF, serum response factor; TCF, ternary complex factor. The genes being activated are those for C-fos (see Chapter 1), P450-3A10, a cytochrome, and Spi 2.1, a serine protease inhibitor. (Modified and reproduced, with permission, from Argetsinger LS, Carter-Su C: Mechanism of signalling by growth hormone receptor. Physiol Rev 1996;76:1089.)

Figure 22–5. Effect of hypophysectomy on growth of the immature rhesus monkey. Both monkeys were the same size and weight 2 years previously, when the one on the left was hypophysectomized. (Reproduced, with permission, from Knobil E in: *Growth in Living Systems.* Zarrow MX [editor]. Basic Books, 1961.)

rise in plasma phosphorus, and a fall in the blood urea nitrogen and amino acid levels. In adults with growth hormone deficiency, recombinantly produced human growth hormone produces an increase in lean body mass and a decrease in body fat, along with an increase in metabolic rate and a fall in plasma cholesterol. Gastrointestinal absorption of Ca^{2+} is increased. Na^+ and K^+ excretion is reduced by an action independent of the adrenal glands, probably because these electrolytes are diverted from the kidneys to the growing tissues. Excretion of the amino acid 4-hydroxyproline is increased during growth and in acromegaly, but it is also increased in a number of other diseases. Much of the excreted hydroxyproline comes from collagen, and hydroxyproline excretion is increased in diseases associated with increased collagen destruction. However, it is also increased when synthesis of soluble collagen is increased, and growth hormone stimulates the synthesis of soluble collagen.

Effects on Carbohydrate & Fat Metabolism

The actions of growth hormone on carbohydrate metabolism are discussed in Chapter 19. Growth hormone is diabetogenic because it increases hepatic glucose output and exerts an anti-insulin effect in muscle. It is ketogenic because it increases circulating FFA levels. The increase in plasma FFA, which takes several hours to develop, provides a ready source of energy for the tissues during hypoglycemia, fasting, and stressful stimuli. Growth hormone does not stimulate B cells of the pancreas directly, but it increases the ability of the pancreas to respond to insulinogenic stimuli such as arginine and glucose. This is an additional way growth hormone promotes growth, since insulin has a protein anabolic effect (see Chapter 19).

Somatomedins

The effects of growth hormone on growth, cartilage, and protein metabolism depend on an interaction between growth hormone and **somatomedins,** which are polypeptide growth factors secreted by the liver and other tissues. The first of these factors isolated was called sulfation factor because it stimulated the incorporation of sulfate into cartilage. However, it also stimulated collagen formation, and its name was changed to somatomedin. It then became clear that there are a variety of different somatomedins and that they are members of an increasingly large family of **growth factors** that affect many different tissues and organs.

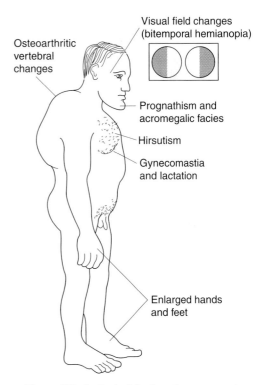

Figure 22–6. Typical findings in acromegaly.

The principal (and in humans probably the only) circulating somatomedins are **insulin-like growth factor I (IGF-I, somatomedin C)** and **insulin-like growth factor II (IGF-II).** These factors are closely related to insulin, except that their C chains are not separated (Figure 22–7) and they have an extension of the A chain called the D domain. The hormone relaxin (see Chapter 23) is also a member of this family; there are two related forms of relaxin in humans, and both resemble IGF-II. In humans a variant form of IGF-I lacking three amino terminal amino acid residues has been found in the brain, and there are several variant forms of human IGF-II (Figure 22–8). The mRNAs for IGF-I and IGF-II are found in the liver, in cartilage, and in many other tissues, indicating that they are synthesized in these tissues.

The properties of IGF-I, IGF-II, and insulin are compared in Table 22–3. Both are tightly bound to proteins in the plasma, and, at least for IGF-I, this prolongs the IGF half-life in the circulation. Six different IGF-binding proteins, with different patterns of distribution in various tissues, have been identified. All are present in plasma, with IGF-binding protein-3 (IGFBP-3) accounting for 95% of the binding in the circulation. The contribution of the IGFs to the insulin-like activity in blood is discussed in Chapter 19. The IGF-I receptor is very similar to the insulin receptor and probably uses much of the same intracellular machinery. The IGF-II receptor is a mannose-6-phosphate receptor (see Figure 19–6) which is involved in the intracellular targeting of acid hydrolases and other proteins to intracellular organelles. Secretion of IGF-I is independent of growth hormone before birth but is stimulated by growth hormone after birth, and it has pronounced growth-stimulating activity. Its concentration in plasma rises during childhood and peaks at the time of puberty, then declines to low levels in old age. IGF-II is largely independent of growth hormone and plays a role in the growth of the fetus before birth. In human fetuses in which it is overexpressed, there is dis-

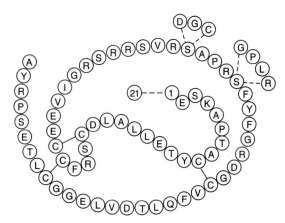

Figure 22–8. Primary structure of human IGF-II and three identified variants: a 21-amino-acid extension of the carboxyl terminal, a tetrapeptide substitution at Ser-29, and a tripeptide substitution of Ser-33. Single-letter codes are used for amino acid residues. (Reproduced, with permission, from Sara VR, Hall K: Insulin-like growth factors and their binding proteins. Physiol Rev 1990;70:591.)

proportionate growth of organs, especially the tongue, other muscles, kidneys, heart, and liver. In adults, the gene for IGF-II is expressed only in the choroid plexus and meninges.

Direct & Indirect Action of Growth Hormone

Ideas about the mechanism of action of growth hormone have undergone a series of changes as new information has become available. Growth hormone was originally thought to produce growth by a direct action on tissues, and later it was believed to act solely

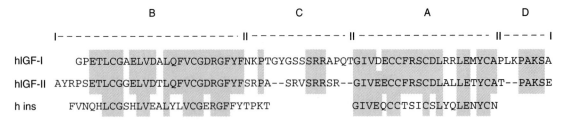

Figure 22–7. Structure of human IGF-I, IGF-II, and insulin (ins), with amino acid residues aligned to show the homologies identified by the colored areas. Single-letter codes are used for amino acid residues. Note that in insulin the C peptide is removed, whereas in IGF-I and IGF-II the A and B chains (domains) remain connected by the C peptide. In addition, there are D domains. (Modified and reproduced, with permission, from Sara VR, Hall K: Insulin-like growth factors and their binding proteins. Physiol Rev 1990;70:591.)

Table 22–3. Comparison of insulin and the insulin-like growth factors.

	Insulin	IGF-I	IGF-II
Other names	...	Somatomedin C	Multiplication-stimulating activity (MSA)
Number of amino acids	51	70	67
Source	Pancreatic B cells	Liver and other tissues	Diverse tissues
Level regulated by	Glucose	Growth hormone after birth, nutritional status	Unknown
Plasma levels	0.3–2 ng/mL	10–700 ng/mL; peaks at puberty	300–800 ng/mL
Plasma-binding proteins	No	Yes	Yes
Major physiologic role	Control of metabolism	Skeletal and cartilage growth	Growth during fetal development

through somatomedins. However, if growth hormone is injected into one proximal tibial epiphysis, a unilateral increase in cartilage width is produced, and cartilage, like other tissues, makes IGF-I. A current hypothesis to explain these results holds that growth hormone acts on cartilage to convert stem cells into cells that respond to IGF-I and then locally produced and circulating IGF-I makes the cartilage grow. However, the role of circulating IGF-I remains important, since infusion of IGF-I to hypophysectomized rats restores bone and body growth.

Figure 22–9 is a summary of current views of the other actions of growth hormone and IGF-I. However, growth hormone probably combines with circulating and locally produced IGF-I in various proportions to produce at least some of these effects.

In preliminary studies, a growth hormone receptor-blocking drug has been found to reduce plasma IGF-I and produce clinical improvement in acromegaly.

Hypothalamic Control of Growth Hormone Secretion

The secretion of growth hormone undergoes marked and rapid spontaneous fluctuation in children and young adults before it declines in old age. Therefore, it is not surprising that it is under hypothalamic control. One factor is growth hormone-releasing hormone (GHRH); a second is somatostatin, the growth hormone release-inhibiting factor (see Chapter 14). A third is probably **ghrelin.** Several years ago, it was discovered that in addition to GHRH and somatostatin receptors on anterior pituitary somatotropes, there was a third type of G protein-linked receptor that mediated increases in growth hormone secretion in response to various synthetic hexapeptides (growth hormone secretagogues; GHSs). The search for the endogenous ligand of these GHS receptors led to the discovery of ghrelin, a 28-amino-acid polypeptide. The main site of

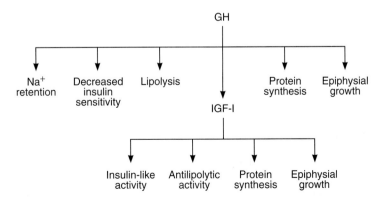

Figure 22–9. Current view of actions mediated by growth hormone (GH) and IGF-I. (Courtesy of R Clark and N Gesundheit.)

ghrelin synthesis and secretion is the stomach, but it is also produced in the hypothalamus and has marked growth hormone-stimulating activity. In addition, it appears to be involved in the regulation of food intake (see Chapter 14).

Growth hormone secretion is under feedback control, like the secretion of other anterior pituitary hormones. Growth hormone increases circulating IGF-I, and IGF-I in turn exerts a direct inhibitory action on growth hormone secretion from the pituitary. It also stimulates somatostatin secretion (Figure 22–10).

Stimuli Affecting Growth Hormone Secretion

The basal plasma growth hormone concentration ranges from 0 to 3 ng/mL in normal adults. It is not significantly higher in children. However, secretory rates cannot be estimated from single values, because there are irregular "spikes" of secretion throughout the day (see below). Growth hormone secretion declines in old age, and there has been considerable interest in injecting growth hormone to counterbalance the effects of aging (see Chapter 1). The hormone increases lean body mass and decreases body fat, but it does not produce statistically significant increases in muscle strength or mental status.

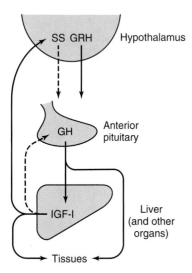

Figure 22–10. Feedback control of growth hormone secretion. The dashed arrows indicate inhibitory effects and the solid arrows stimulatory effects. Note that IGF-I stimulates the secretion of somatostatin (SS) from the hypothalamus and acts directly on the pituitary to inhibit growth hormone (GH) secretion. Compare with Figures 18–12, 20–21, 23–22, and 23–35.

Table 22–4. Stimuli that affect growth hormone secretion in humans.

Stimuli that increase secretion
Deficiency of energy substrate
Hypoglycemia
2-Deoxyglucose
Exercise
Fasting
Increase in circulating levels of certain amino acids
Protein meal
Infusion of arginine and some other amino acids
Glucagon
Stressful stimuli
Pyrogen
Lysine vasopressin
Various psychologic stresses
Going to sleep
L-Dopa and α-adrenergic agonists that penetrate the brain
Apomorphine and other dopamine receptor agonists
Estrogens and androgens
Stimuli that decrease secretion
REM sleep
Glucose
Cortisol
FFA
Medroxyprogesterone
Growth hormone

The stimuli that increase growth hormone secretion are summarized in Table 22–4. Most of them fall into three general categories: (1) conditions such as hypoglycemia and fasting in which there is an actual or threatened decrease in the substrate for energy production in the cells; (2) conditions in which there are increased amounts of certain amino acids in the plasma; and (3) stressful stimuli. The response to glucagon has been used as a test of growth hormone reserve. A spike in growth hormone secretion occurs with considerable regularity upon going to sleep, but the significance of the association between growth hormone and sleep is an enigma. Growth hormone secretion is increased in subjects deprived of REM sleep (see Chapter 11) and inhibited during normal REM sleep.

Glucose infusions lower plasma growth hormone levels and inhibit the response to exercise. The increase produced by 2-deoxyglucose is presumably due to intracellular glucose deficiency, since this compound blocks the catabolism of glucose 6-phosphate. Sex hormones, particularly estrogens, increase growth hormone responses to provocative stimuli such as arginine and insulin. Growth hormone secretion is inhibited by cortisol, FFA, and medroxyprogesterone.

Growth hormone secretion is increased by L-dopa, which increases the release of dopamine and norepinephrine in the brain, and by the dopamine receptor agonist apomorphine.

PHYSIOLOGY OF GROWTH

Growth is a complex phenomenon that is affected not only by growth hormone and somatomedins but also by thyroid hormones, androgens, estrogens, glucocorticoids, and insulin. It is also affected, of course, by genetic factors, and it depends on adequate nutrition. It is normally accompanied by an orderly sequence of maturational changes, and it involves accretion of protein and increase in length and size, not just an increase in weight, which may be due to the formation of fat or retention of salt and water.

Role of Nutrition

The food supply is the most important extrinsic factor affecting growth. The diet must be adequate not only in protein content but also in essential vitamins and minerals (see Chapter 17) and in calories, so that ingested protein is not burned for energy. However, the age at which a dietary deficiency occurs appears to be an important consideration. For example, once the pubertal growth spurt has commenced, considerable lin-

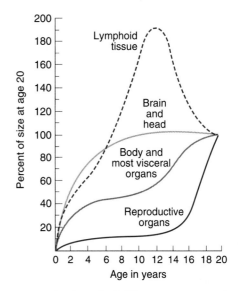

Figure 22–12. Growth of different tissues at various ages as a percentage of size at age 20. The curves are composites that include data for both boys and girls.

ear growth continues even if caloric intake is reduced. Injury and disease stunt growth because they increase protein catabolism.

Growth Periods

Patterns of growth vary somewhat from species to species. Rats continue to grow, although at a declining rate, throughout life. In humans, there are two periods of rapid growth (Figure 22–11), the first in infancy and the second in late puberty just before growth stops. The first period of accelerated growth is partly a continuation of the fetal growth period. The second growth spurt, at the time of puberty, is due to growth hormone, androgens, and estrogens, and the subsequent cessation of growth is due in large part to closure of the epiphyses by estrogens (see Chapter 23). Since girls mature earlier than boys, this growth spurt appears earlier in girls. Of course, in both sexes the rate of growth of individual tissues varies (Figure 22–12).

It is interesting that at least during infancy, growth is not a continuous process but is episodic or saltatory. Increases in length of human infants of 0.5–2.5 cm in a few days are separated by periods of 2–63 days during which no measurable growth can be detected. The cause of the episodic growth is unknown.

Hormonal Effects

The contributions of hormones to growth after birth are shown diagrammatically in Figure 22–13. In labo-

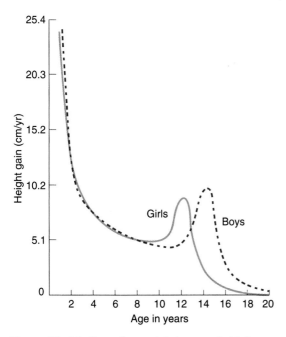

Figure 22–11. Rate of growth in boys and girls from birth to age 20.

ratory animals and in humans, growth in utero is independent of fetal growth hormone.

Plasma growth hormone is elevated in newborns. Subsequently, average resting levels fall but the spikes of growth hormone secretion are larger, especially during puberty, so the mean plasma level over 24 hours is increased; it is 2–4 ng/mL in normal adults but 5–8 ng/mL in children. One of the factors stimulating IGF-I secretion is growth hormone, and plasma IGF-I levels rise during childhood, reaching a peak at 13–17 years of age. In contrast, IGF-II levels are constant throughout postnatal growth.

The growth spurt that occurs at the time of puberty (Figure 22–11) is due in part to the protein anabolic effect of androgens, and the secretion of adrenal androgens increases at this time in both sexes. However, it is also due to an interaction between sex steroids, growth hormone, and IGF-I. Treatment with estrogens and androgens increases the growth hormone responses to stimuli such as insulin and arginine. Sex steroids also increase plasma IGF-I but fail to produce this increase in individuals with growth hormone deficiency. Thus, it appears that the sex hormones produce an increase in the amplitude of the spikes in growth hormone secretion that increases IGF-I secretion, and this in turn causes growth.

Although androgens and estrogens initially stimulate growth, estrogens ultimately terminate growth by causing the epiphyses to fuse to the long bones (epiphysial closure). Once the epiphyses have closed, linear growth ceases (see Chapter 21). This is why patients with sexual precocity are apt to be dwarfed. On the other hand, men who were castrated before puberty tend to be tall because their estrogen production is decreased and their epiphyses remain open so that some growth continues past the normal age of puberty.

When growth hormone is administered to hypophysectomized animals, the animals do not grow as rapidly as they do when treated with growth hormone plus thyroid hormones. Thyroid hormones alone have no effect on growth in this situation. Their action is therefore permissive to that of growth hormone, possibly via potentiation of the actions of somatomedins. Thyroid hormones also appear to be necessary for a completely normal rate of growth hormone secretion; basal growth hormone levels are normal in hypothyroidism, but the response to hypoglycemia is frequently subnormal in hypothyroid children. Thyroid hormones have widespread effects on the ossification of cartilage, the growth of teeth, the contours of the face, and the proportions of the body. Cretins are therefore dwarfed and have infantile features (Figure 22–14). Patients who are dwarfed because of panhypopituitarism have features consistent with their chronologic age until puberty, but since they do not mature sexually, they have juvenile features in adulthood.

The effect of insulin on growth is discussed in Chapter 19. Diabetic animals fail to grow, and insulin causes growth in hypophysectomized animals. However, the growth is appreciable only when large amounts of carbohydrate and protein are supplied with the insulin.

Adrenocortical hormones other than androgens exert a permissive action on growth in the sense that adrenalectomized animals fail to grow unless their blood pressures and circulations are maintained by replacement therapy. On the other hand, glucocorticoids are potent inhibitors of growth because of their direct action on cells, and treatment of children with pharmacologic doses of steroids slows or stops growth for as long as the treatment is continued.

Catch-Up Growth

Following illness or starvation in children, there is a period of **catch-up growth** (Figure 22–15) during which the growth rate is greater than normal. The accelerated growth usually continues until the previous growth curve is reached, then slows to normal. The mechanisms that bring about and control catch-up growth are unknown.

Dwarfism

Short stature can be due to GRH deficiency, growth hormone deficiency, deficient secretion of IGF-I, or other causes. Isolated growth hormone deficiency is often due to GRH deficiency, and in these instances, the growth hormone response to GRH is normal. However, some patients with isolated growth hormone defi-

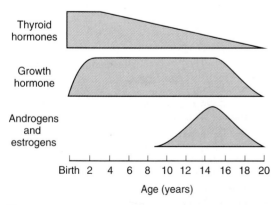

Figure 22–13. Relative importance of hormones in human growth at various ages. (Courtesy of DA Fisher.)

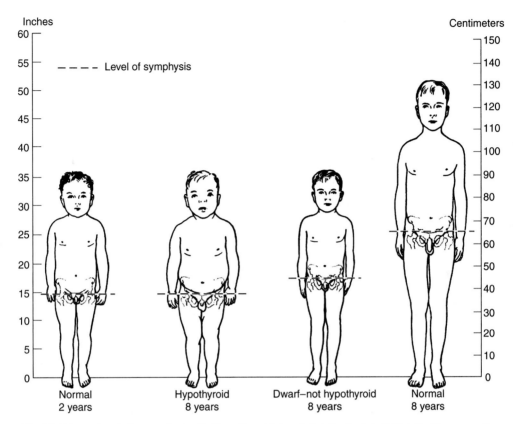

Figure 22–14. Normal and abnormal growth. Hypothyroid dwarfs (cretins) retain their infantile proportions, whereas dwarfs of the constitutional type and, to a lesser extent, of the hypopituitary type have proportions characteristic of their chronologic age. (Reproduced, with permission, from Wilkins L: *The Diagnosis and Treatment of Endocrine Disorders in Childhood and Adolescence,* 3rd ed. Thomas, 1966.)

ciency have abnormalities of their growth hormone-secreting cells.

In another group of dwarfed children, the plasma growth hormone concentration is normal or elevated but their growth hormone receptors are unresponsive as a result of loss-of-function mutations of the gene for the receptors. The resulting condition is known as **growth hormone insensitivity** or **Laron dwarfism.** Plasma IGF-I is markedly reduced, along with IGFBP-3 (see above), which is also growth hormone-dependent.

African pygmies have normal plasma growth hormone levels and a modest reduction in the plasma level of growth hormone-binding protein. Their plasma IGF-I concentration fails to increase at the time of puberty. However, there is less growth than in nonpygmy controls throughout the prepubertal period. Thus, the explanation for the short stature of pygmies is still unsettled.

As noted above, short stature is characteristic of cretinism and occurs in patients with precocious puberty. It is also part of the syndrome of **gonadal dysgenesis** seen in patients who have an XO chromosomal pattern instead of an XX or XY pattern (see Chapter 23). Various bone and metabolic diseases also cause stunted growth, and in many cases there is no known cause ("constitutional delayed growth"). Chronic abuse and neglect can also cause dwarfism in children. This condition is known as **psychosocial dwarfism** or the **Kaspar Hauser syndrome,** named for the patient with the first reported case.

Achondroplasia, the most common form of dwarfism in humans, is characterized by short limbs with a normal trunk. It is an autosomal dominant condition caused by a mutation in the gene that codes for **fibroblast growth factor receptor 3 (FGFR3).** This member of the fibroblast growth receptor family is normally expressed in cartilage and the brain.

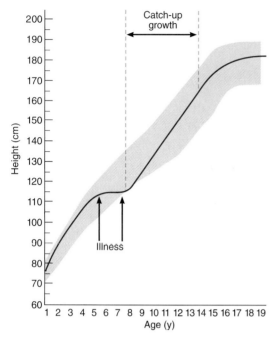

Figure 22–15. Growth curve for a normal boy who had an illness beginning at age 5 and ending at age 7. Catch-up growth eventually returned his height to his previous normal growth curve. (Modified from Boersma B, Wit JM: Catch-up growth. Endocr Rev 1997;18:646.)

PITUITARY INSUFFICIENCY

Changes in Other Endocrine Glands

The widespread changes that develop when the pituitary is removed surgically or destroyed by disease in humans or animals are predictable in terms of the known hormonal functions of the gland. In hypopituitarism, the adrenal cortex atrophies, and the secretion of adrenal glucocorticoids and sex hormones falls to low levels, although some secretion persists. Stress-induced increases in aldosterone secretion are absent, but basal aldosterone secretion and increases induced by salt depletion are normal, at least for some time. Since there is no mineralocorticoid deficiency, salt loss and hypovolemic shock do not develop, but the inability to increase glucocorticoid secretion makes patients with pituitary insufficiency sensitive to stress. The development of salt loss in long-standing hypopituitarism is discussed in Chapter 20. Growth is inhibited (see above). Thyroid function is depressed to low levels, and cold is tolerated poorly. The gonads atrophy, sexual cycles stop, and some of the secondary sex characteristics disappear.

Insulin Sensitivity

Hypophysectomized animals have a tendency to become hypoglycemic, especially when fasted. In some species, but not in humans, fatal hypoglycemic reactions are fairly common. Hypophysectomy ameliorates diabetes mellitus (see Chapter 19) and markedly increases the hypoglycemic effect of insulin. This is due in part to the deficiency of adrenocortical hormones, but hypophysectomized animals are more sensitive to insulin than adrenalectomized animals because they also lack the anti-insulin effect of growth hormone.

Water Metabolism

Although selective destruction of the supraoptic-posterior pituitary mechanism causes diabetes insipidus (see Chapter 14), removal of both the anterior and posterior pituitary usually causes no more than a transient polyuria. In the past, there was speculation that the anterior pituitary secreted a "diuretic hormone," but the amelioration of the diabetes insipidus is explained by a decrease in the osmotic load presented for excretion. Osmotically active particles hold water in the renal tubules (see Chapter 38). Because of the ACTH deficiency, the rate of protein catabolism is decreased in hypophysectomized animals. Because of the TSH deficiency, the metabolic rate is low. Consequently, fewer osmotically active products of catabolism are filtered and urine volume declines, even in the absence of vasopressin. Growth hormone deficiency contributes to the depression of the glomerular filtration rate in hypophysectomized animals, and growth hormone increases the glomerular filtration rate and renal plasma flow in humans. Finally, because of the glucocorticoid deficiency, there is the same defective excretion of a water load that is seen in adrenalectomized animals. The "diuretic" activity of the anterior pituitary can thus be explained in terms of the actions of ACTH, TSH, and growth hormone.

Other Defects

Most patients with growth hormone deficiency developing in adulthood have deficiencies in other anterior pituitary hormones as well, though there is evidence that growth hormone deficiency causes loss of body protein and increased body fat along with decreased vitality. The deficiency of ACTH and other pituitary hormones with MSH activity may be responsible for the pallor of the skin in patients with hypopituitarism. There may be some loss of protein in adults, but wasting is not a feature of hypopituitarism in humans, and most patients with pituitary insufficiency are well nourished (Figure 22–16). It used to be thought that

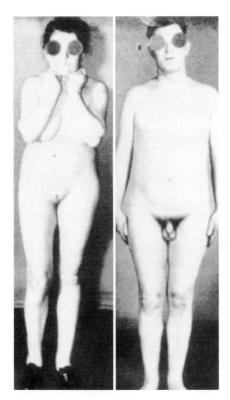

Figure 22–16. Typical picture of hypopituitarism in adults. Note the well-nourished appearance and pallor. (Reproduced, with permission, from Daughaday WH: The adenohypophysis. In: *Textbook of Endocrinology,* 5th ed. Williams RH [editor]. Saunders, 1974.)

cachexia was part of the clinical picture, but it is now generally accepted that emaciated patients described in the older literature had anorexia nervosa rather than hypopituitarism.

Causes of Pituitary Insufficiency in Humans

Tumors of the anterior pituitary cause pituitary insufficiency. Suprasellar cysts, remnants of Rathke's pouch that enlarge and compress the pituitary, are another cause of hypopituitarism. In women who have an episode of shock due to postpartum hemorrhage, the pituitary may become infarcted, with the subsequent development of postpartum necrosis (**Sheehan's syndrome**). The blood supply to the anterior lobe is vulnerable because it descends on the pituitary stalk through the rigid diaphragma sellae, and during pregnancy the pituitary is enlarged. Pituitary infarction is usually extremely rare in men, but it was fairly com-

mon in soldiers who contracted hemorrhagic fever in Korea. In hemorrhagic fever, there was a diffuse vasculitis that probably caused pituitary enlargement as a result of edema, and the patients who developed pituitary infarction were those who went into shock in the course of their disease.

PITUITARY HYPERFUNCTION IN HUMANS

Acromegaly

Tumors of the somatotropes of the anterior pituitary secrete large amounts of growth hormone, leading in children to **gigantism** and in adults to **acromegaly.** Hypersecretion of growth hormone is accompanied by hypersecretion of prolactin in 20–40% of patients with acromegaly. In addition, secretion of glycoprotein hormone α subunits is said to occur in up to 37% of patients. Acromegaly can be caused by extrapituitary as well as intrapituitary growth hormone-secreting tumors and by hypothalamic tumors that secrete GRH, but these are rare.

The principal findings in acromegaly are those related to the local effects of the tumor (enlargement of the sella turcica, headache, visual disturbances) and those due to growth hormone secretion. As mentioned above, there is enlargement of the hands and feet (**acral** parts; hence the term acromegaly) and a protrusion of the lower jaw (**prognathism**) (Figure 22–6). Overgrowth of the malar, frontal, and facial bones combines with prognathism to produce the coarse facial features called **acromegalic facies.** Body hair is increased in amount. The skeletal changes predispose to osteoarthritis. About 25% of patients have abnormal glucose tolerance tests, and 4% develop lactation in the absence of pregnancy.

Cushing's Syndrome

The clinical picture of Cushing's syndrome and its various causes are described in Chapter 20. Many patients with bilaterally hyperplastic adrenals have small ACTH-secreting pituitary tumors (microadenomas) that are difficult to detect. However, a significant percentage of the patients who have bilaterally hyperplastic adrenals removed develop rapidly growing ACTH-secreting pituitary tumors (**Nelson's syndrome**). These tumors cause hyperpigmentation of the skin and neurologic signs due to pressure on structures in the sellar region. Some are malignant. Blood ACTH levels are extremely high, and the intrinsic MSH activity of the ACTH probably accounts for the cutaneous pigmentation. It is difficult to say whether these patients had undetected tumors to start with or developed neoplastic changes in the pituitary when the feedback check on ACTH secretion was removed.

Other Hormone-Secreting Tumors

Animals sometimes develop TSH-secreting tumors after thyroidectomy and gonadotropin-secreting tumors after gonadectomy. TSH-secreting tumors are rare in humans. On the other hand, prolactin-secreting tumors are common (see Chapter 23). In addition, many apparently nonsecretory pituitary tumors in women secrete gonadotropins. In other women, gonadotropin α subunits, β subunits, or both are secreted. Secretion of those substances does not cause clinical abnormalities, but elevated concentrations in the blood can be detected by suitable laboratory tests. Most if not all anterior pituitary tumors are monoclonal, ie, they arise from a single abnormal cell.

The Gonads: Development & Function of the Reproductive System

23

INTRODUCTION

Modern genetics and experimental embryology make it clear that, in most species of mammals, the multiple differences between the male and the female depend primarily on a single chromosome (the Y chromosome) and a single pair of endocrine structures, the testes in the male and the ovaries in the female. The differentiation of the primitive gonads into testes or ovaries in utero is genetically determined in humans, but the formation of male genitalia depends upon the presence of a functional, secreting testis; in the absence of testicular tissue, development is female. There is evidence that male sexual behavior and, in some species, the male pattern of gonadotropin secretion are due to the action of male hormones on the brain in early development. After birth, the gonads remain quiescent until adolescence, when they are activated by gonadotropins from the anterior pituitary. Hormones secreted by the gonads at this time cause the appearance of features typical of the adult male or female and the onset of the sexual cycle in the female. In human females, ovarian function regresses after a number of years and sexual cycles cease (the menopause). In males, there is a slow decline in gonadal function with advancing age, but the ability to father children persists.

In both sexes, the gonads have a dual function: the production of germ cells (**gametogenesis**) and the secretion of **sex hormones.** The **androgens** are the steroid sex hormones that are masculinizing in their action; the **estrogens** are those that are feminizing. Both types of hormones are normally secreted in both sexes. The testes secrete large amounts of androgens, principally **testosterone,** but they also secrete small amounts of estrogens. The ovaries secrete large amounts of estrogens and small amounts of androgens. Androgens are secreted from the adrenal cortex in both sexes, and some of the androgens are converted to estrogens in fat and other extragonadal and extraadrenal tissues. The ovaries also secrete **progesterone,** a steroid that has spe-

cial functions in preparing the uterus for pregnancy. Particularly during pregnancy, the ovaries secrete the polypeptide hormone **relaxin,** which loosens the ligaments of the pubic symphysis and softens the cervix, facilitating delivery of the fetus. In both sexes, the gonads secrete other polypeptides, including **inhibin B,** a polypeptide that inhibits FSH secretion.

The secretory and gametogenic functions of the gonads are both dependent upon the secretion of the anterior pituitary gonadotropins, FSH, and LH. The sex hormones and inhibin B feed back to inhibit gonadotropin secretion. In males, gonadotropin secretion is noncyclic; but in postpubertal females an orderly, sequential secretion of gonadotropins is necessary for the occurrence of menstruation, pregnancy, and lactation.

SEX DIFFERENTIATION & DEVELOPMENT

CHROMOSOMAL SEX

The Sex Chromosomes

Sex is determined genetically by two chromosomes, called the **sex chromosomes** to distinguish them from the **somatic chromosomes (autosomes).** In humans and many other mammals, the sex chromosomes are called X and Y chromosomes. The Y chromosome is necessary and sufficient for the production of testes, and the testis-determining gene product is called SRY (for sex-determining region of the Y chromosome). SRY is a DNA-binding regulatory protein. It bends the DNA and acts as a transcription factor that initiates transcription of a cascade of genes necessary for testicular differentiation, including the gene for MIS (see below). The gene for SRY is located near the tip of the short arm of the human Y chromosome. Male cells with the diploid number of chromosomes contain an X and a Y chromosome (XY pattern), whereas female cells contain two X chromosomes (XX pattern). As a consequence of meiosis during gametogenesis, each normal ovum contains a single X chromosome, but half the normal sperms contain an X chromosome and half contain a Y chromosome (Figure 23–1). When a sperm containing a Y chromosome fertilizes an ovum, an XY pattern results and the zygote develops into a **genetic male.** When fertilization occurs with an X-containing sperm, an XX pattern and a **genetic female** result. Cell division and the chemical nature of chromosomes are discussed in Chapter 1.

Human Chromosomes

Human chromosomes can be studied in detail. Human cells are grown in tissue culture; treated with the drug colchicine, which arrests mitosis at the metaphase; exposed to a hypotonic solution that makes the chromosomes swell and disperse; and then "squashed" onto slides. Fluorescent and other staining techniques make it possible to identify the individual chromosomes and study them in detail (Figure 23–2). There are 46 chromosomes: in males, 22 pairs of autosomes plus an X chromosome and a Y chromosome; in females, 22 pairs of autosomes plus two X chromosomes. The individual

chromosomes are usually arranged in an arbitrary pattern **(karyotype).** The individual autosome pairs are identified by the numbers 1–22 on the basis of their morphologic characteristics. The human Y chromosome is smaller than the X chromosome, and it has been hypothesized that sperms containing the Y chromosome are lighter and able to "swim" faster up the female genital tract, thus reaching the ovum more rapidly. This supposedly accounts for the fact that the number of males born is slightly greater than the number of females.

Sex Chromatin

Soon after cell division has started during embryonic development, one or the other of the two X chromosomes of the somatic cells in normal females becomes functionally inactive. In abnormal individuals with more than two X chromosomes, only one remains active. The process that is normally responsible for inactivation is initiated in an X-inactivation center in the chromosome, probably via the trans-activating factor CTCF, which is also induced in gene imprinting. However, the details of the inactivation process are still incompletely understood. The choice of which X chromosome remains active is random, so normally one X chromosome remains active in approximately half of

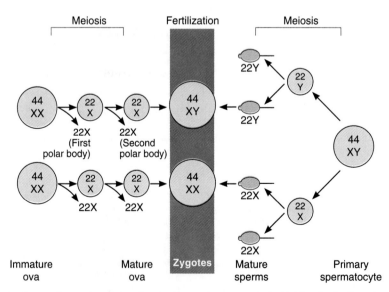

Figure 23–1. Basis of genetic sex determination. In the two-stage meiotic division in the female, only one cell survives as the mature ovum. In the male, the meiotic division results in the formation of four sperms, two containing the X and two the Y chromosome. Fertilization thus produces a male zygote with 22 pairs of autosomes plus an X and a Y or a female zygote with 22 pairs of autosomes and two X chromosomes. Note that for clarity, this figure and Figures 23–5 and 23–8 differ from the current international nomenclature for karyotypes, which lists the total number of chromosomes followed by the sex chromosome pattern. Thus, XO is 45,X, XY is 46,XY, XXY is 47,XXY, etc.

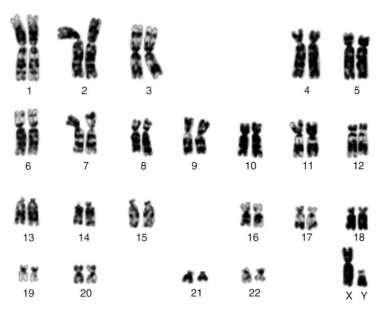

Figure 23–2. Karyotype of chromosomes from a normal male. The chromosomes have been stained with Giemsa's stain, which produces a characteristic banding pattern. (Reproduced, with permission, from Lingappa VJ, Farey K: *Physiological Medicine*. McGraw-Hill, 2000.)

the cells and the other X chromosome is active in the other half. The selection persists through subsequent divisions of these cells, and consequently some of the somatic cells in adult females contain an active X chromosome of paternal origin and some contain an active X chromosome of maternal origin.

In normal cells, the inactive X chromosome condenses and can be seen in various types of cells, usually near the nuclear membrane, as the **Barr body,** also called sex chromatin (Figure 23–3). Thus, there is a Barr body for each X chromosome in excess of one in the cell. The inactive X chromosome is also visible as a small "drumstick" of chromatin projecting from the

nuclei of 1–15% of the polymorphonuclear leukocytes in females but not in males (Figure 23–3).

EMBRYOLOGY OF THE HUMAN REPRODUCTIVE SYSTEM

Development of the Gonads

On each side of the embryo, a primitive gonad arises from the genital ridge, a condensation of tissue near the adrenal gland. The gonad develops a **cortex** and a **medulla.** Until the sixth week of development, these structures are identical in both sexes. In genetic males,

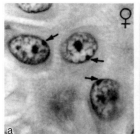
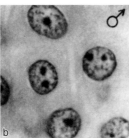
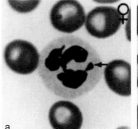
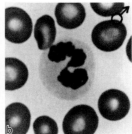

Figure 23–3. **Left:** Barr body (arrows) in the epidermal spinous cell layer. **Right:** Nuclear appendage ("drumstick") identified by arrow in white blood cells. (Reproduced, with permission, from Grumbach MM, Barr ML: Cytologic tests of chromosomal sex in relation to sex anomalies in man. Recent Prog Horm Res 1958;14:255.)

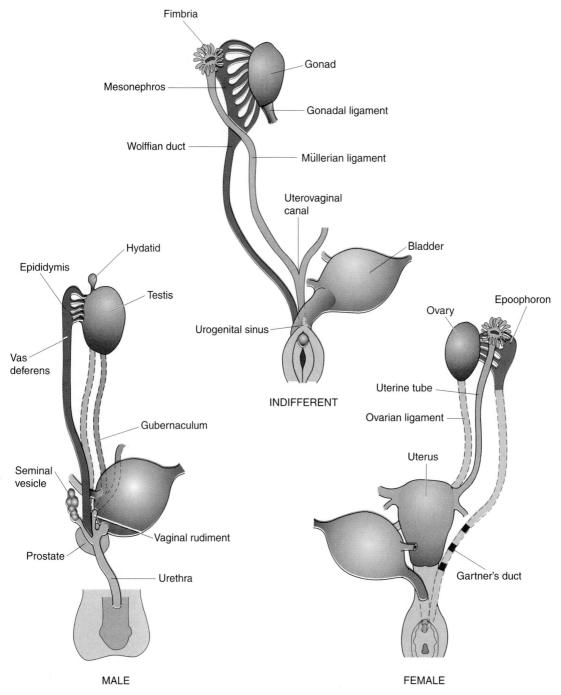

Figure 23–4. Embryonic differentiation of male and female internal genitalia (genital ducts) from wolffian (male) and müllerian (female) primordia. (After Corning HK, Wilkins L. Redrawn and reproduced, with permission, from Grumbach M, Conte FA, in: *Williams Textbook of Endocrinology,* 7th ed. Wilson JD, Foster DW [editors]. Saunders, 1985.)

the medulla develops during the seventh and eighth weeks into a testis, and the cortex regresses. Leydig and Sertoli cells appear, and testosterone and müllerian inhibiting substance are secreted. In genetic females, the cortex develops into an ovary and the medulla regresses. The embryonic ovary does not secrete hormones. Hormonal treatment of the mother has no effect on gonadal (as opposed to ductal and genital) differentiation in humans, although it does in some experimental animals.

Embryology of the Genitalia

In the seventh week of gestation, the embryo has both male and female primordial genital ducts (Figure 23–4). In a normal female fetus, the müllerian duct system

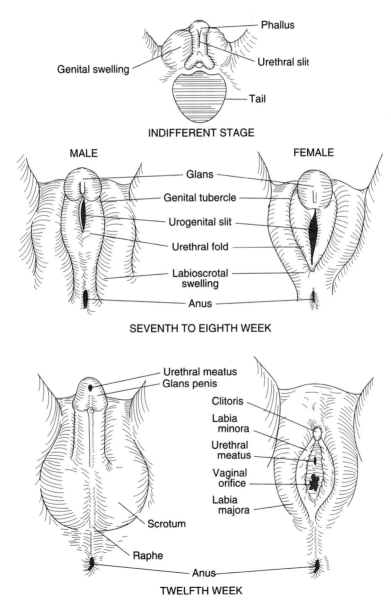

Figure 23–5. Differentiation of male and female external genitalia from indifferent primordial structures in the embryo.

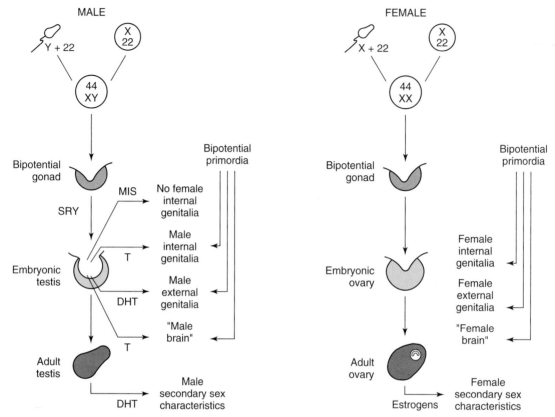

Figure 23–6. Diagrammatic summary of normal sex determination, differentiation, and development in humans. MIS, müllerian inhibiting substance; T, testosterone; DHT, dihydrotestosterone.

then develops into uterine tubes (oviducts) and a uterus. In the normal male fetus, the wolffian duct system on each side develops into the epididymis and vas deferens. The external genitalia are similarly bipotential until the eighth week (Figure 23–5). Thereafter, the urogenital slit disappears and male genitalia form, or, alternatively, it remains open and female genitalia form.

When there are functional testes in the embryo, male internal and external genitalia develop. The Leydig cells of the fetal testis secrete testosterone, and the Sertoli cells secrete **müllerian inhibiting substance** (**MIS**; also called müllerian regression factor, or MRF). MIS is a 536-amino-acid homodimer that is a member of the TGFβ superfamily of growth factors which includes inhibins and activins (see below). In their effects on the internal as opposed to the external genitalia, MIS and testosterone act unilaterally. MIS causes regression of the müllerian ducts by apoptosis on the side on which it is secreted, and testosterone fosters the development of the vas deferens and related structures from the wolffian ducts. The testosterone metabolite dihydrotestosterone (see below) induces the formation

of male external genitalia and male secondary sex characteristics (Figure 23–6).

MIS continues to be secreted by the Sertoli cells, and it reaches mean values of 48 ng/mL in plasma in 1- to 2-year-old boys. Thereafter, it declines to low levels by the time of puberty and persists at low but detectable levels throughout life. In girls, MIS is produced by granulosa cells in small follicles in the ovaries, but plasma levels are very low or undetectable until puberty. Thereafter, plasma MIS is about the same as in adult men, ie, about 2 ng/mL. The functions of MIS after early embryonic life are unsettled, but it is probably involved in germ cell maturation in both sexes and in control of testicular descent in boys (see below).

Development of the Brain

At least in some species, the development of the brain as well as the external genitalia is affected by androgens early in life. In rats, a brief exposure to androgens during the first few days of life causes the male pattern of sexual behavior and the male pattern of hypothalamic control of gonadotropin secretion to develop after puberty. In the absence of androgens, female patterns develop (see Chapter 15). In monkeys, similar effects on sexual behavior are produced by exposure to androgens in utero, but the pattern of gonadotropin secretion remains cyclic. Early exposure of female human fetuses to androgens also appears to cause subtle but significant masculinizing effects on behavior. However, women with adrenogenital syndrome due to congenital adrenocortical enzyme deficiency (see Chapter 20) develop normal menstrual cycles when treated with cortisol. Thus, the human, like the monkey, appears to retain the cyclic pattern of gonadotropin secretion despite the exposure to androgens in utero.

ABERRANT SEXUAL DIFFERENTIATION

Chromosomal Abnormalities

From the preceding discussion, it might be expected that abnormalities of sexual development could be caused by genetic or hormonal abnormalities as well as by other nonspecific teratogenic influences, and this is indeed the case. The major classes of abnormalities are listed in Table 23–1.

An established defect in gametogenesis is **nondisjunction,** a phenomenon in which a pair of chromo-

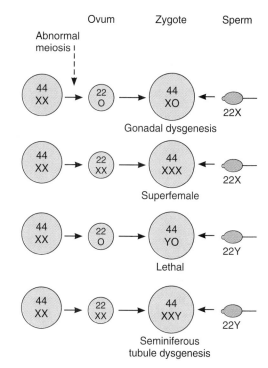

Figure 23–7. Summary of four possible defects produced by maternal nondisjunction of the sex chromosomes at the time of meiosis. The YO combination is believed to be lethal, and the fetus dies in utero.

Table 23–1. Classification of the major disorders of sex differentiation in humans. Many of these syndromes can have great variation in degree and, consequently, in manifestations.

Chromosomal disorders
Gonadal dysgenesis (XO and variants)
"Superfemales" (XXX)
Seminiferous tubule dysgenesis (XXY and variants)
True hermaphroditism

Developmental disorders
Female pseudohermaphroditism
  Congenital virilizing adrenal hyperplasia of fetus
  Maternal androgen excess
    Virilizing ovarian tumor
    Iatrogenic: Treatment with androgens or certain
      synthetic progestational drugs
Male pseudohermaphroditism
  Androgen resistance
  Defective testicular development
Congenital 17α-hydroxylase deficiency
Congenital adrenal hyperplasia due to blockade of pregnenolone formation
Various nonhormonal anomalies

---

somes fail to separate, so that both go to one of the daughter cells during meiosis. Four of the abnormal zygotes that can form as a result of nondisjunction of one of the X chromosomes during oogenesis are shown in Figure 23–7. In individuals with the XO chromosomal pattern, the gonads are rudimentary or absent, so that female external genitalia develop. Stature is short, other congenital abnormalities are often present, and no maturation occurs at puberty. This syndrome is called **gonadal dysgenesis** or, alternatively, **ovarian agenesis** or **Turner's syndrome.** Individuals with the XXY pattern, the most common sex chromosome disorder, have the genitalia of a normal male. Testosterone secretion at puberty is often great enough for the development of male characteristics. However, the seminiferous tubules are abnormal, and there is a higher than normal incidence of mental retardation. This syndrome is known as **seminiferous tubule dysgenesis** or **Klinefelter's syndrome.** The XXX ("superfemale") pattern is second in frequency only to the XXY pattern and may be even more common in the general population, since it does not seem to be associated with any characteristic abnormalities. The YO combination is probably lethal.

Meiosis is a two-stage process, and although nondisjunction usually occurs during the first meiotic division, it can occur in the second, producing more complex chromosomal abnormalities. In addition, nondisjunction or simple loss of a sex chromosome can occur during the early mitotic divisions after fertilization. The result of faulty mitoses in the early zygote is the production of a **mosaic,** an individual with two or more populations of cells with different chromosome complements. **True hermaphroditism,** the condition in which the individual has both ovaries and testes, is probably due to XX/XY mosaicism and related mosaic patterns, although other genetic aberrations are possible.

Chromosomal abnormalities also include transposition of parts of chromosomes to other chromosomes. Rarely, genetic males are found to have the XX karyotype because the short arm of their father's Y chromosome was transposed to their father's X chromosome during meiosis and they received that X chromosome along with their mother's. Similarly, deletion of the small portion of the Y chromosome containing SRY produces females with the XY karyotype.

Sex chromosome abnormalities are, of course, not the only abnormalities associated with disease states; nondisjunction of several different autosomal chromosomes is known to occur. For example, nondisjunction of chromosome 21 produces **trisomy 21,** the chromosomal abnormality associated with **Down's syndrome** (mongolism). The additional chromosome 21 is normal, so Down's syndrome is a pure case of gene excess causing abnormalities. In most instances, nondisjunction occurs in the ovary rather than the testis and the incidence of Down's syndrome increases with advancing age of the mother.

There are many other chromosomal abnormalities as well as numerous diseases due to defects in single genes. These conditions are generally diagnosed in utero by analysis of fetal cells in a sample of amniotic fluid collected by inserting a needle through the abdominal wall (**amniocentesis**) or, earlier in pregnancy, by examining fetal cells obtained by a needle biopsy of chorionic villi (**chorionic villus sampling**).

Hormonal Abnormalities

Development of the male external genitalia occurs normally in genetic males in response to androgen secreted by the embryonic testes, but male genital development may also occur in genetic females exposed to androgens from some other source during the eighth to the thirteenth weeks of gestation. The syndrome that results is **female pseudohermaphroditism.** A pseudohermaphrodite is an individual with the genetic constitution and gonads of one sex and the genitalia of the other. After the thirteenth week, the genitalia are fully formed, but exposure to androgens can cause hypertrophy of the clitoris. Female pseudohermaphroditism may be due to congenital virilizing adrenal hyperplasia (see Chapter 20), or it may be caused by androgens administered to the mother. Conversely, development of female external genitalia in genetic males (**male pseudohermaphroditism**) occurs when the embryonic testes are defective. Because the testes also secrete MIS, genetic males with defective testes have female internal genitalia.

Another cause of male pseudohermaphroditism is **androgen resistance,** in which, as a result of various congenital abnormalities, male hormones cannot exert their full effects on the tissues. One form of androgen resistance is a **5α-reductase deficiency,** in which the enzyme responsible for the formation of dihydrotestosterone, the active form of testosterone, is decreased. The consequences of this deficiency are discussed in the section on the male reproductive system. Other forms of androgen resistance are due to various mutations in the androgen receptor gene, and the resulting defects in receptor function range from minor to severe. Mild defects cause infertility with or without gynecomastia (see below). When the loss of receptor function is complete, the **testicular feminizing syndrome,** now known as **complete androgen resistance syndrome,** results. In this condition, MIS is present and testosterone is secreted at normal or even elevated rates. The external genitalia are female, but the vagina ends blindly because there are no female internal genitalia. Individuals with this syndrome develop enlarged breasts at puberty and usually are considered to be normal women until they are diagnosed when they seek medical advice because of lack of menstruation.

It is worth noting that genetic males with congenital blockage of the formation of pregnenolone are pseudohermaphrodites because testicular as well as adrenal androgens are normally formed from pregnenolone. Male pseudohermaphroditism also occurs when there is a congenital deficiency of 17α-hydroxylase (see Chapter 20).

PUBERTY

As noted above, there is a burst of testosterone secretion in male fetuses before birth (Figure 23–8). In the neonatal period there is another burst, whose function is unknown, but thereafter the Leydig cells become quiescent. There follows in all mammals a period in which the gonads of both sexes are quiescent until they are activated by gonadotropins from the pituitary to bring about the final maturation of the reproductive system. This period of final maturation is known as **adolescence.** It is often also called **puberty,** although puberty, strictly defined, is the period when the endocrine and

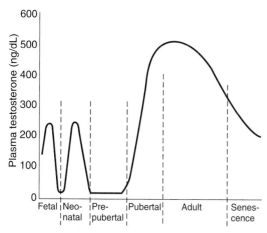

Figure 23–8. Plasma testosterone levels at various ages in human males.

and their hypothalami contain GnRH (see Chapter 14). However, their gonadotropins are not secreted. In immature monkeys, normal menstrual cycles can be brought on by pulsatile injection of GnRH, and they persist as long as the pulsatile injection is continued. Thus, it seems clear that pulsatile secretion of GnRH brings on puberty. During the period from birth to puberty, a neural mechanism is operating to prevent the normal pulsatile release of GnRH. The nature of the mechanism inhibiting the GnRH pulse generator is unknown.

gametogenic functions of the gonads have first developed to the point where reproduction is possible. In girls, the first event is **thelarche,** the development of breasts, followed by **pubarche,** the development of axillary and pubic hair, and then by **menarche,** the first menstrual period. The initial periods are generally anovulatory, and regular ovulation appears about a year later. In contrast to the situation in adulthood, removal of the gonads during the period from soon after birth to puberty causes little or no increase in gonadotropin secretion, so gonadotropin secretion is not being held in check by the gonadal hormones. In children between the ages of 7 and 10, a slow increase in estrogen and androgen secretion precedes the more rapid rise in the early teens (Figure 23–9).

The age at the time of puberty is variable. In Europe and the United States, it has been declining at the rate of 1–3 months per decade for more than 175 years. In the United States in recent years, puberty generally occurs between the ages of 8 and 13 in girls and 9 and 14 in boys.

Another event that occurs in humans at the time of puberty is an increase in the secretion of adrenal androgens (see Figure 20–14). The onset of this increase is called **adrenarche.** It occurs at age 8–10 years in girls and age 10–12 years in boys. DHEA values peak at about 25 years of age in females and slightly later than that in males. They then decline slowly to low values in old age. The rise appears to be due to an increase in the lyase activity of 17α-hydroxylase.

Control of the Onset of Puberty

The gonads of children can be stimulated by gonadotropins; their pituitaries contain gonadotropins;

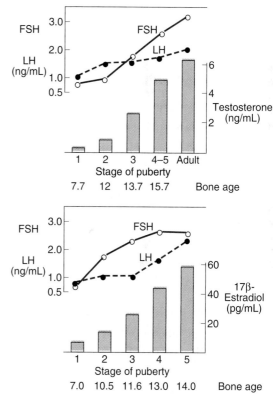

Figure 23–9. Changes in plasma hormone concentrations during puberty in boys **(top)** and girls **(bottom).** Stage 1 of puberty is preadolescence in both sexes. In boys, stage 2 is characterized by beginning enlargement of the testes, stage 3 by penile enlargement, stage 4 by growth of the glans penis, and stage 5 by adult genitalia. In girls, stage 2 is characterized by breast buds, stage 3 by elevation and enlargement of the breasts, stage 4 by projection of the areolas, and stage 5 by adult breasts. (Modified and reproduced, with permission, from Grumbach MM: Onset of puberty. In: *Puberty: Biologic and Psychosocial Components.* Berenberg SR [editor]. HE Stenfoert Kroese BV, 1975.)

Relation to Leptin

It has been argued for some time that normally there is a critical body weight that must be reached for puberty to occur. Thus, for example, young women who engage in strenuous athletics lose weight and stop menstruating. So do girls with anorexia nervosa. If these girls start to eat and gain weight, they menstruate again, ie, they "go back through puberty." It now appears that leptin, the satiety-producing hormone secreted by fat cells (see Chapter 14), may be the link between body weight and puberty. Obese ob/ob mice that cannot make leptin are infertile, and their fertility is restored by injections of leptin. Leptin treatment also induces precocious puberty in immature female mice. However, the way that leptin fits into the overall control of puberty remains to be determined.

PRECOCIOUS & DELAYED PUBERTY

Sexual Precocity

The major causes of precocious sexual development in humans are listed in Table 23–2. Early development of secondary sexual characteristics without gametogenesis is caused by abnormal exposure of immature males to androgen or females to estrogen. This syndrome should be called **precocious pseudopuberty** to distinguish it from **true precocious puberty** due to an early but otherwise normal pubertal pattern of gonadotropin secretion from the pituitary (Figure 23–10).

Constitutional precocious puberty—ie, precocious puberty in which no cause can be determined—is more

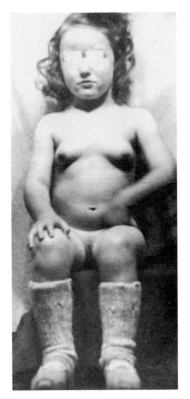

Figure 23–10. Constitutional precocious puberty in a 3-year-old girl. The patient developed pubic hair and started to menstruate at the age of 17 months. (Reproduced, with permission, from Jolly H: *Sexual Precocity.* Thomas, 1955.)

Table 23–2. Classification of the causes of precocious sexual development in humans.

True precocious puberty
  Constitutional
  Cerebral: Disorders involving posterior hypothalamus
    Tumors
    Infections
    Developmental abnormalities
  Gonadotropin-independent precocity

Precocious pseudopuberty (no spermatogenesis or ovarian development)
  Adrenal
    Congenital virilizing adrenal hyperplasia
    Androgen-secreting tumors (in males)
    Estrogen-secreting tumors (in females)
  Gonadal
    Leydig cell tumors of testis
    Granulosa cell tumors of ovary
  Miscellaneous

---

common in girls than in boys. In both sexes, tumors or infections involving the hypothalamus cause precocious puberty. Indeed, in one large series of cases, precocious puberty was the most common endocrine symptom of hypothalamic disease. In experimental animals, precocious puberty can be produced by hypothalamic lesions. However, it is not clear whether hypothalamic disease and lesions in humans stimulate GnRH secretion by itself or interrupt a pathway that normally holds GnRH secretion in check. Pineal tumors are sometimes associated with precocious puberty, but there is evidence that these tumors are associated with precocity only when there is secondary damage to the hypothalamus.

Precocious gametogenesis and steroidogenesis can occur without the pubertal pattern of gonadotropin secretion (gonadotropin-independent precocity). At least in some cases of this condition, the sensitivity of LH receptors to gonadotropins is increased because of an activating mutation in the G protein that couples the receptors to adenylyl cyclase.

Delayed or Absent Puberty

The normal variation in the age at which adolescent changes occur is so wide that puberty cannot be considered to be pathologically delayed until the menarche has failed to occur by the age of 17 or testicular development by the age of 20. Failure of maturation due to panhypopituitarism is associated with dwarfing and evidence of other endocrine abnormalities. Patients with the XO chromosomal pattern and gonadal dysgenesis are also dwarfed. In some individuals, puberty is delayed even though the gonads are present and other endocrine functions are normal. In males, this clinical picture is called **eunuchoidism.** In females, it is called **primary amenorrhea** (see below).

MENOPAUSE

The human ovaries become unresponsive to gonadotropins with advancing age, and their function declines, so that sexual cycles disappear **(menopause).** This unresponsiveness is associated with and probably caused by a decline in the number of primordial follicles, which becomes precipitous at the time of menopause (Figure 23–11). The ovaries no longer secrete progesterone and 17β-estradiol in appreciable quantities, and estrogen is formed only in small amounts by aromatization of androstenedione in peripheral tissues (see Chapter 20). The uterus and the vagina gradually become atrophic. As the negative feedback effect of estrogens and progesterone is reduced, secretion of FSH and LH is increased, and plasma FSH and LH increase to high levels. Old female mice and rats have long periods of diestrus and increased levels of gonadotropin secretion, but a clear-cut "menopause" has only been described in women.

In women, the menses usually become irregular and cease between the ages of 45 and 55. The average age at onset of the menopause has been increasing since the end of the 19th century and is currently 52 years.

Sensations of warmth spreading from the trunk to the face (hot flushes; also called hot flashes), night sweats, and various psychic symptoms are common after ovarian function has ceased. Hot flushes are said to occur in 75% of menopausal women and may continue intermittently for as long as 40 years. They also occur when early menopause is produced by bilateral ovariectomy, and they are prevented by estrogen treatment. In addition, they occur after castration in men. Their cause is unknown. However, they coincide with surges of LH secretion. LH is secreted in episodic bursts at intervals of 30–60 minutes or more **(circhoral secretion),** and in the absence of gonadal hormones these bursts are large. Each hot flush begins with the start of a burst. However, LH itself is not responsible for the symptoms, because they can continue after removal of the pituitary. Instead, it appears that some estrogen-sensitive event in the hypothalamus initiates both the release of LH and the episode of flushing.

Although the function of the testes tends to decline slowly with advancing age, the evidence is clear that there is no "male menopause" **(andropause)** similar to that occurring in women.

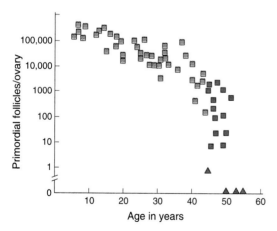

Figure 23–11. Number of primordial follicles per ovary in women at various ages. Colored squares, premenopausal women (regular menses); black squares, perimenopausal women (irregular menses for at least 1 year); black triangles, postmenopausal women (no menses for at least 1 year). Note that the vertical scale is a log scale and that the values are from one rather than two ovaries. (Redrawn by PM Wise and reproduced, with permission, from Richardson SJ, Senikas V, Nelson JF: Follicular depletion during the menopausal transition: Evidence for accelerated loss and ultimate exhaustion. J Clin Endocrinol Metab 1987;65:1231.)

PITUITARY GONADOTROPINS & PROLACTIN

Chemistry

FSH and LH are each made up of an α and a β subunit whose nature is discussed in Chapter 22. They are glycoproteins that contain the hexoses mannose and galactose, the hexosamines N-acetylgalactosamine and N-acetylglycosamine, and the methylpentose fucose. They also contain sialic acid. The carbohydrate in the gonadotropin molecules increases their potency by markedly slowing their metabolism. The half-life of human FSH is about 170 minutes; the half-life of LH is about 60 minutes. Documented mutations in the β subunit of FSH have now been reported in a man with hypogonadism and a woman with delayed puberty and hypogonadism.

Human pituitary prolactin contains 199 amino acid residues and three disulfide bridges (Figure 23–12) and has considerable structural similarity to human growth hormone and hCS. The half-life of prolactin, like that of growth hormone, is about 20 minutes. Structurally similar prolactins are secreted by the endometrium and by the placenta (see below).

Receptors

The receptors for FSH and LH are serpentine receptors coupled to adenylyl cyclase through G_s (see Chapter 1). In addition, each has an extended, glycosylated extracellular domain.

The human prolactin receptor resembles the growth hormone receptor and is one of the superfamily of receptors that includes the growth hormone receptor and receptors for many cytokines and hematopoietic growth factors (see Chapters 1, 22, 24, and 27). It dimerizes and activates the JAK-Stat and other intracellular enzyme cascades.

Actions

The testes and ovaries become atrophic when the pituitary is removed or destroyed. The actions of prolactin and the gonadotropins FSH and LH, as well as those of the gonadotropin secreted by the placenta, are described in detail in succeeding sections of this chapter. In brief, FSH helps maintain the spermatogenic epithelium by stimulating Sertoli cells in the male and is responsible for the early growth of ovarian follicles in the female. LH is tropic to the Leydig cells and, in females, is responsible for the final maturation of the ovarian follicles and estrogen secretion from them. It is also responsible for ovulation, the initial formation of the corpus luteum, and secretion of progesterone.

Prolactin causes milk secretion from the breast after estrogen and progesterone priming. Its effect on the breast involves increased action of mRNA and increased production of casein and lactalbumin. However, the action of the hormone is not exerted on the cell nucleus and is prevented by inhibitors of microtubules. Prolactin also inhibits the effects of gonadotropins, possibly by an action at the level of the ovary. Its role in preventing ovulation in lactating women is discussed below. The function of prolactin in

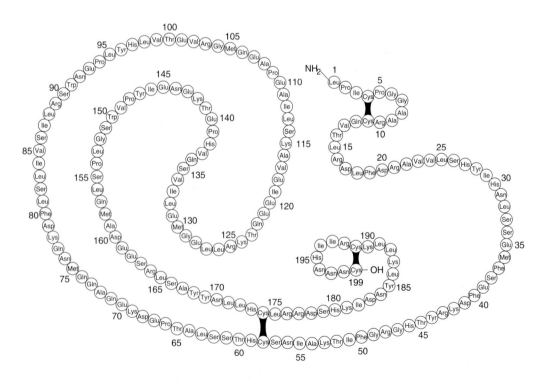

Figure 23–12. Structure of human prolactin.

normal males is unsettled, but excess prolactin secreted by tumors causes impotence. An action of prolactin that has been used as the basis for bioassay of this hormone is stimulation of the growth and "secretion" of the crop sacs in pigeons and other birds. The paired crop sacs are outpouchings of the esophagus which form, by desquamation of their inner cell layers, a nutritious material ("milk") that the birds feed to their young. However, prolactin, FSH, and LH are now regularly measured by radioimmunoassay.

Regulation of Prolactin Secretion

The normal plasma prolactin concentration is approximately 5 ng/mL in men and 8 ng/mL in women. Secretion is tonically inhibited by the hypothalamus, and section of the pituitary stalk leads to an increase in circulating prolactin. Thus, the effect of the hypothalamic prolactin-inhibiting hormone (PIH) dopamine is normally greater than the effects of the various hypothalamic peptides with prolactin-releasing activity. In humans, prolactin secretion is increased by exercise, surgical and psychologic stresses, and stimulation of the nipple (Table 23–3). The plasma prolactin level rises during sleep, the rise starting after the onset of sleep and persisting throughout the sleep period. Secretion is increased during pregnancy, reaching a peak at the time of parturition. After delivery, the plasma concentration falls to nonpregnant levels in about 8 days. Suckling produces a prompt increase in secretion, but the magnitude of this rise gradually declines after a woman has been nursing for more than 3 months. With prolonged lactation, milk secretion occurs with prolactin levels that are in the normal range.

L-Dopa decreases prolactin secretion by increasing the formation of dopamine, and bromocriptine and other dopamine agonists inhibit secretion because they stimulate dopamine receptors. Chlorpromazine and related drugs that block dopamine receptors increase prolactin secretion. TRH stimulates the secretion of prolactin in addition to TSH, and there are additional prolactin-releasing polypeptides in hypothalamic tissue. Estrogens produce a slowly developing increase in prolactin secretion as a result of a direct action on the lactotropes.

It has now been established that prolactin facilitates the secretion of dopamine in the median eminence. Thus, prolactin acts in the hypothalamus in a negative feedback fashion to inhibit its own secretion.

Hyperprolactinemia

Up to 70% of the patients with chromophobe adenomas of the anterior pituitary have elevated plasma pro-

Table 23–3. Factors affecting the secretion of human prolactin and growth hormone.

Factor	Prolactin[1]	Growth Hormone[1]
Sleep	I+	I+
Nursing	I++	N
Breast stimulation in nonlactating women	I	N
Stress	I+	I+
Hypoglycemia	I	I+
Strenuous exercise	I	I
Sexual intercourse in women	I	N
Pregnancy	I++	N
Estrogens	I	I
Hypothyroidism	I	N
TRH	I+	N
Phenothiazines, butyrophenones	I+	N
Opioids	I	I
Glucose	N	D
Somatostatin	N	D+
L-Dopa	D+	I+
Apomorphine	D+	I+
Bromocriptine and related ergot derivatives	D+	I

[1]I, moderate increase; I+, marked increase; I++, very marked increase; N, no change; D, moderate decrease; D+, marked decrease.

lactin levels. In some instances, the elevation may be due to damage to the pituitary stalk, but in most cases, the tumor cells are actually secreting the hormone. The hyperprolactinemia may cause galactorrhea, but in many individuals there are no demonstrable endocrine abnormalities. Conversely, most women with galactorrhea have normal prolactin levels; definite elevations are found in less than a third of patients with this condition.

Another interesting observation is that 15–20% of women with secondary amenorrhea have elevated prolactin levels, and when prolactin secretion is reduced, normal menstrual cycles and fertility return. It appears that the prolactin may produce amenorrhea by blocking the action of gonadotropins on the ovaries, but definitive proof of this hypothesis must await further research. The hypogonadism produced by prolactinomas is associated with osteoporosis due to estrogen deficiency.

As noted above, hyperprolactinemia in men is associated with impotence and hypogonadism that disappear when prolactin secretion is reduced.

THE MALE REPRODUCTIVE SYSTEM

STRUCTURE

The testes are made up of loops of convoluted **seminiferous tubules,** in the walls of which the spermatozoa are formed from the primitive germ cells (**spermatogenesis).** Both ends of each loop drain into a network of ducts in the head of the **epididymis.** From there, spermatozoa pass through the tail of the epididymis into the **vas deferens.** They enter through the **ejaculatory ducts** into the urethra in the body of the **prostate** at the time of ejaculation (Figure 23–13). Between the tubules in the testes are nests of cells containing lipid granules, the **interstitial cells of Leydig** (Figures 23–14 and 23–15), which secrete testosterone into the bloodstream. The spermatic arteries to the testes are tortuous, and blood in them runs parallel but in the opposite direction to blood in the pampiniform plexus of spermatic veins. This anatomic arrangement may permit countercurrent exchange of heat and testosterone. The principles of countercurrent exchange are considered in detail in relation to the kidney in Chapter 38.

GAMETOGENESIS & EJACULATION

Blood-Testis Barrier

The walls of the seminiferous tubules are lined by primitive germ cells and **Sertoli cells**—large, complex glycogen-containing cells that stretch from the basal lamina of the tubule to the lumen (Figure 23–15).

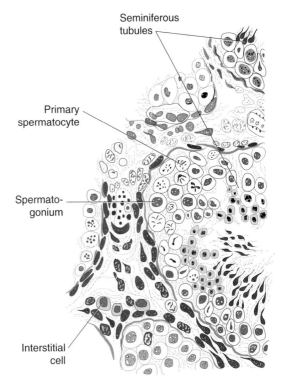

Figure 23–14. Section of human testis.

Germ cells must stay in contact with Sertoli cells to survive, and this contact is maintained by bridges of carbohydrate molecules. Tight junctions between adjacent Sertoli cells near the basal lamina form a **blood-testis barrier** that prevents many large molecules from passing from the interstitial tissue and the

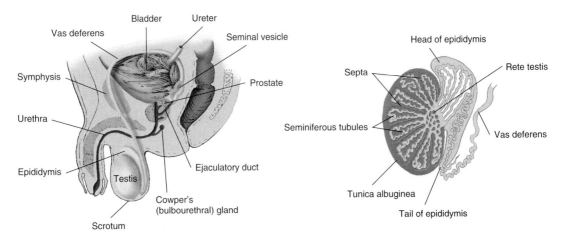

Figure 23–13. **Left:** Male reproductive system. **Right:** Duct system of the testis.

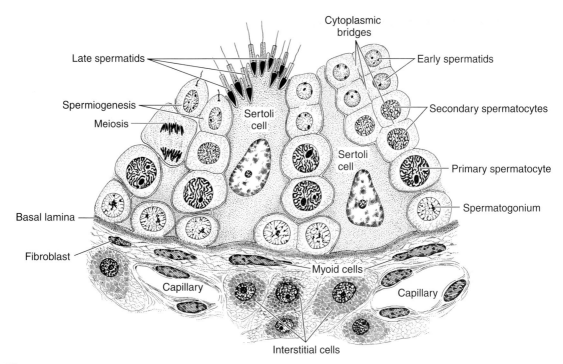

Figure 23–15. Seminiferous epithelium. Note that maturing germ cells remain connected by cytoplasmic bridges through the early spermatid stage and that these cells are closely invested by Sertoli cell cytoplasm as they move from the basal lamina to the lumen. (Reproduced, with permission, from Junqueira LC, Carneiro J: *Basic Histology: Text and Atlas,* 10th ed. McGraw-Hill, 2003.)

part of the tubule near the basal lamina (basal compartment) to the region near the tubular lumen (adluminal compartment) and the lumen. However, steroids penetrate this barrier with ease, and there is evidence that some proteins pass from the Sertoli cells to the Leydig cells and vice versa in a paracrine fashion. In addition, maturing germ cells must pass through the barrier as they move to the lumen. This appears to occur without disruption of the barrier by progressive breakdown of the tight junctions above the germ cells, with concomitant formation of new tight junctions below them.

The fluid in the lumen of the seminiferous tubules is quite different from plasma; it contains very little protein and glucose but is rich in androgens, estrogens, K^+, inositol, and glutamic and aspartic acids. Maintenance of its composition presumably depends on the blood-testis barrier. The barrier also protects the germ cells from blood-borne noxious agents, prevents antigenic products of germ cell division and maturation from entering the circulation and generating an autoimmune response, and may help establish an osmotic gradient that facilitates movement of fluid into the tubular lumen.

Spermatogenesis

The **spermatogonia,** the primitive germ cells next to the basal lamina of the seminiferous tubules, mature into **primary spermatocytes** (Figure 23–15). This process begins during adolescence. The primary spermatocytes undergo meiotic division, reducing the number of chromosomes. In this two-stage process, they divide into **secondary spermatocytes** and then into **spermatids,** which contain the haploid number of 23 chromosomes. The spermatids mature into **spermatozoa (sperms).** As a single spermatogonium divides and matures, its descendants remain tied together by cytoplasmic bridges until the late spermatid stage. This apparently ensures synchrony of the differentiation of each clone of germ cells. The estimated number of spermatids formed from a single spermatogonium is 512. In humans, it takes an average of 74 days to form a mature sperm from a primitive germ cell by this orderly process of spermatogenesis.

Each sperm is an intricate motile cell, rich in DNA, with a head that is made up mostly of chromosomal material (Figure 23–16). Covering the head like a cap is the **acrosome,** a lysosome-like organelle rich in enzymes involved in sperm penetration of the ovum and other events involved in fertilization. The motile tail of the sperm is wrapped in its proximal portion by a sheath holding numerous mitochondria. The membranes of late spermatids and spermatozoa contain a special small form of angiotensin-converting enzyme called **germinal angiotensin II-converting enzyme** (see Chapter 24). The function of this enzyme in the sperms is unknown, although male mice in which the function of the angiotensin-converting enzyme gene has been disrupted have reduced fertility.

The spermatids mature into spermatozoa in deep folds of the cytoplasm of the Sertoli cells (Figure 23–15). Mature spermatozoa are released from the Sertoli cells and become free in the lumens of the tubules. The Sertoli cells secrete **androgen-binding protein (ABP), inhibin,** and **MIS.** They do not synthesize androgens, but they contain **aromatase (CYP19),** the enzyme responsible for conversion of androgens to estrogens, and they can produce estrogens. ABP probably functions to maintain a high, stable supply of androgen in the tubular fluid. Inhibin inhibits FSH secretion (see below). MIS causes regression of the müllerian ducts in males during fetal life (see above).

FSH and androgens maintain the gametogenic function of the testis. After hypophysectomy, injection of LH produces a high local concentration of androgen in the testes, and this maintains spermatogenesis. The stages from spermatogonia to spermatids appear to be androgen-independent. However, the maturation from spermatids to spermatozoa depends on androgen acting on the Sertoli cells in which the developing spermatozoa are embedded. FSH acts on the Sertoli cells to facilitate the last stages of spermatid maturation. In addition, it promotes the production of ABP.

An interesting observation is that the estrogen content of the fluid in the rete testis (Figure 23–13) is high, and the walls of the rete contain numerous ERα estrogen receptors. In this region, fluid is reabsorbed and the spermatozoa are concentrated. If this does not occur, the sperm entering the epididymis are diluted in a large volume of fluid, and infertility results.

Further Development of Spermatozoa

Spermatozoa leaving the testes are not fully mobile. They continue their maturation and acquire motility during their passage through the epididymis. Motility appears to be the important factor, since fertilization occurs if an immotile spermatozoon from the head of the epididymis is microinjected directly into an ovum. The ability to move forward **(progressive motility),** which is acquired in the epididymis, involves activation of a unique protein called **CatSper,** which is localized to the principal piece of the sperm tail. This protein appears to be a Ca^{2+} ion channel that permits cAMP-generalized Ca^{2+} influx.

Ejaculation of the spermatozoon (see below) involves contractions of the vas deferens mediated in part by P2X receptors for ATP (see Chapter 4), and fertility is reduced in mice in which these receptors are knocked out.

Once ejaculated into the female, the spermatozoa move up the uterus to the isthmus of the uterine tubes, where they slow down and undergo **capacitation.** This further maturation process involves two components: increasing the motility of the spermatozoa and facilitating their preparation for the acrosome reaction. However, the role of capacitation appears to be facilitatory rather than obligatory, because fertilization is readily produced in vitro. From the isthmuses the capacitated spermatozoa move rapidly to the tubal ampullas, where fertilization takes place (see below).

Effect of Temperature

Spermatogenesis requires a temperature considerably lower than that of the interior of the body. The testes are normally maintained at a temperature of about 32 °C. They are kept cool by air circulating around the

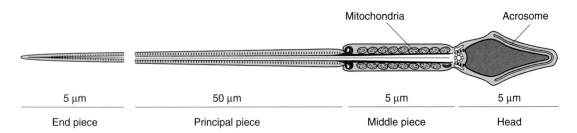

Figure 23–16. Human spermatozoon, profile view. Note the acrosome, an organelle that covers half the sperm head inside the plasma membrane of the sperm. (Reproduced, with permission, from Junqueira LC, Carneiro J: *Basic Histology: Text and Atlas,* 10th ed. McGraw-Hill, 2003.)

scrotum and probably by heat exchange in a counter-current fashion between the spermatic arteries and veins. When the testes are retained in the abdomen or when, in experimental animals, they are held close to the body by tight cloth binders, degeneration of the tubular walls and sterility result. Hot baths (43–45 °C for 30 minutes per day) and insulated athletic supporters reduce the sperm count in humans, in some cases by 90%. However, the reductions produced in this manner are not consistent enough to make the procedures reliable forms of male contraception. In addition, there is evidence suggesting a seasonal effect in men, with sperm counts being greater in the winter regardless of the temperature to which the scrotum is exposed.

Semen

The fluid that is ejaculated at the time of orgasm, the **semen,** contains sperms and the secretions of the seminal vesicles, prostate, Cowper's glands, and, probably, the urethral glands (Table 23–4). An average volume per ejaculate is 2.5–3.5 mL after several days of conti-

Table 23–4. Composition of human semen.

Color: White, opalescent	
Specific gravity: 1.028	
pH: 7.35–7.50	
Sperm count: Average about 100 million/mL, with fewer than 20% abnormal forms	
Other components:	
Fructose (1.5–6.5 mg/mL) Phosphorylcholine Ergothioneine Ascorbic acid Flavins Prostaglandins	From seminal vesicles (contributes 60% of total volume)
Spermine Citric acid Cholesterol, phospholipids Fibrinolysin, fibrinogenase Zinc Acid phosphatase	From prostate (contributes 20% of total volume)
Phosphate Bicarbonate	Buffers
Hyaluronidase	

nence. The volume of semen and the sperm count decrease rapidly with repeated ejaculation. Even though it takes only one sperm to fertilize the ovum, there are normally about 100 million sperms per milliliter of semen. Fifty percent of men with counts of 20–40 million/mL and essentially all of those with counts under 20 million/mL are sterile. The presence of many morphologically abnormal or immotile spermatozoa also correlates with infertility. The **prostaglandins** in semen, which actually come from the seminal vesicles, are in high concentration, but the function of these fatty acid derivatives in semen is not known. Their structure and their multiple actions in other parts of the body are discussed in Chapter 17.

Human sperms move at a speed of about 3 mm/min through the female genital tract. Sperms reach the uterine tubes 30–60 minutes after copulation. In some species, contractions of the female organs facilitate the transport of the sperms to the uterine tubes, but it is not known if such contractions occur in humans.

Erection

Erection is initiated by dilation of the arterioles of the penis. As the erectile tissue of the penis fills with blood, the veins are compressed, blocking outflow and adding to the turgor of the organ. The integrating centers in the lumbar segments of the spinal cord are activated by impulses in afferents from the genitalia and descending tracts that mediate erection in response to erotic psychic stimuli. The efferent parasympathetic fibers are in the pelvic splanchnic nerves **(nervi erigentes).** The fibers presumably release acetylcholine and the vasodilator VIP as cotransmitters (see Chapter 4).

There are also nonadrenergic noncholinergic fibers in the nervi erigentes, and these contain large amounts of **NO synthase,** the enzyme that catalyzes the formation of nitric oxide (NO; see Chapter 31). NO activates guanylyl cyclase, resulting in increased production of cGMP, and cGMP is a potent vasodilator. Injection of inhibitors of NO synthase prevents the erection normally produced by stimulation of the pelvic nerve in experimental animals. Thus, it seems clear that NO plays a prominent role in the production of erection. Sildenafil (Viagra) inhibits the breakdown of cGMP by phosphodiesterases and has gained worldwide fame for the treatment of impotence. The multiple phosphodiesterases (PDEs) in the body have been divided into seven isoenzyme families, and sildenafil is most active against PDE5, the type of phosphodiesterase found in the corpora cavernosa. It is worth noting, however, that sildenafil also produces significant inhibition of PDE6. This is the type of phosphodiesterase found in the retina, and one of the side effects of sildenafil is transient loss of the ability to discriminate between blue and green (see Chapter 8).

Normally, erection is terminated by sympathetic vasoconstrictor impulses to the penile arterioles.

Ejaculation

Ejaculation is a two-part spinal reflex that involves **emission,** the movement of the semen into the urethra; and **ejaculation** proper, the propulsion of the semen out of the urethra at the time of orgasm. The afferent pathways are mostly fibers from touch receptors in the glans penis that reach the spinal cord through the internal pudendal nerves. Emission is a sympathetic response, integrated in the upper lumbar segments of the spinal cord and effected by contraction of the smooth muscle of the vasa deferentia and seminal vesicles in response to stimuli in the hypogastric nerves. The semen is propelled out of the urethra by contraction of the bulbocavernosus muscle, a skeletal muscle. The spinal reflex centers for this part of the reflex are in the upper sacral and lowest lumbar segments of the spinal cord, and the motor pathways traverse the first to third sacral roots and the internal pudendal nerves. Carbon monoxide may be involved in the control of ejaculation, since HO2, the enzyme that catalyzes its production in the nervous system (see Chapter 4), is abundant in the pathways concerned with ejaculation, and ejaculatory performance is diminished when the gene for HO2 is knocked out.

PSA

The prostate produces and secretes into the semen and the bloodstream a 30-kDa serine protease generally called **prostate-specific antigen (PSA).** The gene for PSA has two androgen response elements. PSA hydrolyzes the sperm motility inhibitor semenogelin in semen, and it has several substrates in plasma, but its precise function in the circulation is unknown. An elevated plasma PSA occurs in prostate cancer and is widely used as a screening test for this disease, though PSA is also elevated in benign prostatic hyperplasia and prostatitis.

Vasectomy

Bilateral ligation of the vas deferens (vasectomy) has proved to be a relatively safe and convenient contraceptive procedure. However, it has proven difficult to restore the patency of the vas in those wishing to restore fertility, and the current success rate for such operations, as measured by the subsequent production of pregnancy, is about 50%. Half of the men who have been vasectomized develop antibodies against sperma-

tozoa, and in monkeys, the presence of such antibodies is associated with a higher incidence of infertility after restoration of the patency of the vas. However, there do not appear to be any other adverse effects of the antisperm antibodies.

ENDOCRINE FUNCTION OF THE TESTES

Chemistry & Biosynthesis of Testosterone

Testosterone, the principal hormone of the testes, is a C_{19} steroid (see Chapter 20) with an $-OH$ group in the 17 position (Figure 23–17). It is synthesized from cholesterol in the Leydig cells and is also formed from androstenedione secreted by the adrenal cortex. The biosynthetic pathways in all endocrine organs that form steroid hormones are similar, the organs differing only in the enzyme systems they contain. In the Leydig cells, the 11- and 21-hydroxylases found in the adrenal cortex (see Figure 20–8) are absent, but 17α-hydroxylase is present. Pregnenolone is therefore hydroxylated in the 17 position and then subjected to side chain cleavage to form dehydroepiandrosterone. Androstenedione is also formed via progesterone and 17-hydroxyprogesterone, but this pathway is less prominent in humans. Dehydroepiandrosterone and androstenedione are then converted to testosterone.

The secretion of testosterone is under the control of LH, and the mechanism by which LH stimulates the Leydig cells involves increased formation of cAMP via the serpentine LH receptor and G_s. Cyclic AMP increases the formation of cholesterol from cholesteryl esters and the conversion of cholesterol to pregnenolone via the activation of protein kinase A.

Secretion

The testosterone secretion rate is 4–9 mg/d (13.9–31.33 µmol/d) in normal adult males. Small amounts of testosterone are also secreted in females, probably from the ovary but possibly from the adrenal as well.

Transport & Metabolism

Ninety-eight percent of the testosterone in plasma is bound to protein: 65% is bound to a β-globulin called **gonadal steroid-binding globulin (GBG)** or **sex steroid-binding globulin,** and 33% to albumin (Table 23–5). GBG also binds estradiol. The plasma testosterone level (free and bound) is 300–1000 ng/dL (10.4–34.7 nmol/L) in adult men (Figure 23–8) and 30–70 ng/dL (1.04–2.43 nmol/L) in adult women. It declines somewhat with age in males.

A small amount of circulating testosterone is converted to estradiol (see below), but most of the testosterone is converted to 17-ketosteroids, principally an-

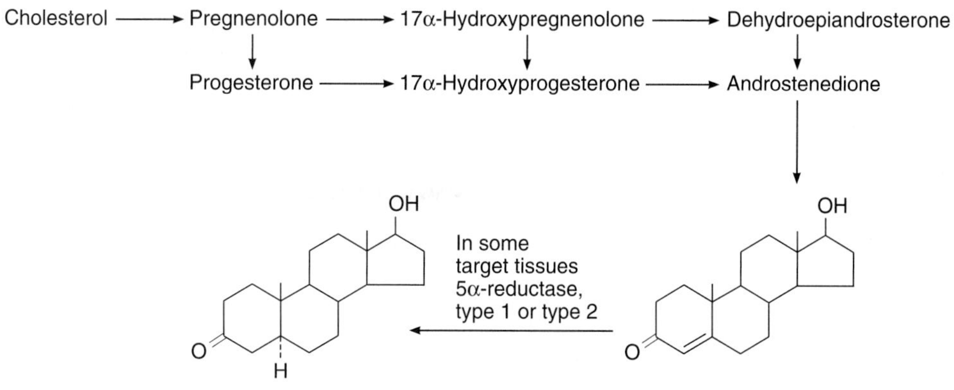

Dihydrotestosterone **Testosterone**

Figure 23–17. Biosynthesis of testosterone. The formulas of the precursor steroids are shown in Figure 20–8. Although the main secretory product of the Leydig cells is testosterone, some of the precursors also enter the circulation.

drosterone and its isomer etiocholanolone (Figure 23–18), and excreted in the urine. About two-thirds of the urinary 17-ketosteroids are of adrenal origin, and one-third are of testicular origin. Although most of the 17-ketosteroids are weak androgens (they have 20% or less the potency of testosterone), it is worth emphasizing that not all 17-ketosteroids are androgens and not all androgens are 17-ketosteroids. Etiocholanolone, for example, has no androgenic activity, and testosterone itself is not a 17-ketosteroid.

Actions

In addition to their actions during development, testosterone and other androgens exert an inhibitory feedback effect on pituitary LH secretion; develop and maintain the male secondary sex characteristics; exert an important protein-anabolic, growth-promoting effect; and, along with FSH, maintain spermatogenesis (see above).

Secondary Sex Characteristics

The widespread changes in hair distribution, body configuration, and genital size that develop in boys at puberty—the male **secondary sex characteristics**—are summarized in Table 23–6. The prostate and seminal vesicles enlarge, and the seminal vesicles begin to secrete fructose. This sugar appears to function as the main nutritional supply for the spermatozoa. The psy-

Table 23–5. Distribution of gonadal steroids and cortisol in plasma.[1]

		% Bound to		
Steroid	**% Free**	**CBG**	**GBG**	**Albumin**
Testosterone	2	0	65	33
Androstenedione	7	0	8	85
Estradiol	2	0	38	60
Progesterone	2	18	0	80
Cortisol	4	90	0	6

[1]CBG, corticosteroid-binding globulin; GBG, gonadal steroid-binding globulin. (Courtesy of S Munroe.)

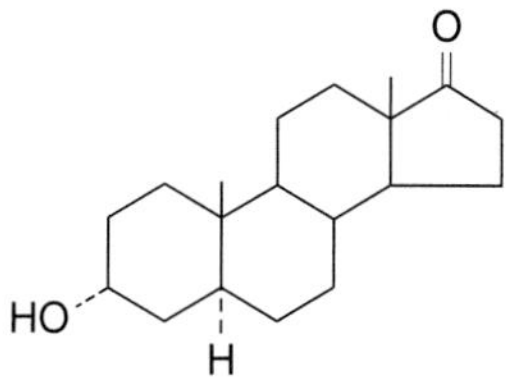

Androsterone

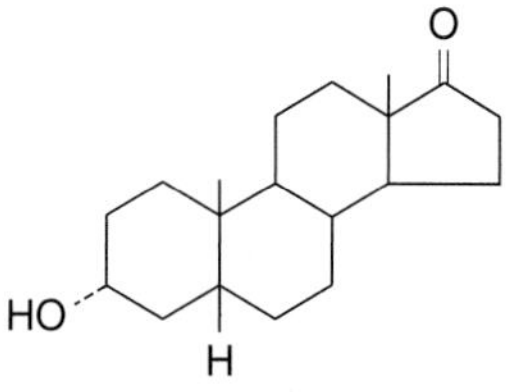

Etiocholanolone

Figure 23–18. Two 17-ketosteroid metabolites of testosterone.

Table 23–6. Body changes at puberty in boys (male secondary sex characteristics).

External genitalia: Penis increases in length and width. Scrotum becomes pigmented and rugose.

Internal genitalia: Seminal vesicles enlarge and secrete and begin to form fructose. Prostate and bulbourethral glands enlarge and secrete.

Voice: Larynx enlarges, vocal cords increase in length and thickness, and voice becomes deeper.

Hair growth: Beard appears. Hairline on scalp recedes anterolaterally. Pubic hair grows with male (triangle with apex up) pattern. Hair appears in axillas, on chest, and around anus; general body hair increases.

Mental: More aggressive, active attitude. Interest in opposite sex develops.

Body conformation: Shoulders broaden, muscles enlarge.

Skin: Sebaceous gland secretion thickens and increases (predisposing to acne).

---

chic effects of testosterone are difficult to define in humans, but in experimental animals, androgens provoke boisterous and aggressive play. The effects of androgens and estrogens on sexual behavior are considered in detail in Chapter 15. Although body hair is increased by androgens, scalp hair is decreased (Figure 23–19). Hereditary baldness often fails to develop unless dihydrotestosterone is present.

Anabolic Effects

Androgens increase the synthesis and decrease the breakdown of protein, leading to an increase in the rate of growth. It used to be argued that they cause the epiphyses to fuse to the long bones, thus eventually stopping growth, but it now appears that epiphysial closure is due to estrogens (see below and Chapter 22). Secondary to their anabolic effects, androgens cause mod-

erate sodium, potassium, water, calcium, sulfate, and phosphate retention; and they also increase the size of the kidneys. Doses of exogenous testosterone that exert significant anabolic effects are also masculinizing and increase libido, which limits the usefulness of the hormone as an anabolic agent in patients with wasting diseases. Attempts to develop synthetic steroids in which the anabolic action is divorced from the androgenic action have not been successful.

Mechanism of Action

Like other steroids (see Chapter 1), testosterone binds to an intracellular receptor, and the receptor-steroid complex then binds to DNA in the nucleus, facilitating transcription of various genes. In addition, testosterone is converted to **dihydrotestosterone (DHT)** by 5α-reductase in some target cells (Figure 23–17), and DHT binds to the same intracellular receptor as testosterone. DHT also circulates, with a plasma level that is about 10% of the testosterone level. Testosterone-receptor complexes are less stable than DHT-receptor complexes in target cells, and they conform less well to the DNA-binding state. Thus, DHT formation is a way of amplifying the action of testosterone in target tissues. There are two 5α-reductases in humans, encoded by different genes. Type 1 5α-reductase is present in skin throughout the body and is the dominant enzyme in the scalp. Type 2 5α-reductase is present in genital skin, the prostate, and other genital tissues.

Testosterone-receptor complexes are responsible for the maturation of wolffian duct structures and consequently for the formation of male internal genitalia during development, but DHT-receptor complexes are needed to form male external genitalia (Figure 23–20). DHT-receptor complexes are also primarily responsible for enlargement of the prostate and probably of the penis at the time of puberty, as well as for the facial hair, the acne, and the temporal recession of the hairline. On

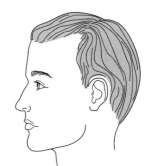

Figure 23–19. Hairline in children and adults. The hairline of the woman is like that of the child, whereas that of the man is indented in the lateral frontal region.

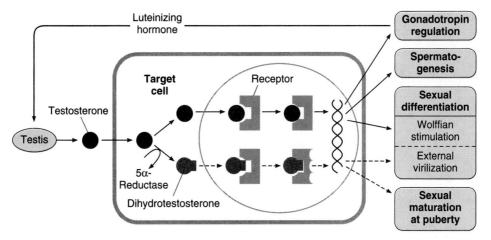

Figure 23–20. Schematic diagram of the actions of testosterone (solid arrows) and dihydrotestosterone (dashed arrows). Note that they both bind to the same receptor, but DHT binds more effectively. (Reproduced, with permission, from Wilson JD, Griffin JE, Russell W: Steroid 5a-reductase 2 deficiency. Endocr Rev 1993;14:577. Copyright © 1993 by The Endocrine Society.)

the other hand, the increase in muscle mass and the development of male sex drive and libido depend primarily on testosterone rather than DHT.

Congenital 5α-reductase deficiency, in which the gene for type 2 5α-reductase is mutated, is common in certain parts of the Dominican Republic. It produces an interesting form of male pseudohermaphroditism. Individuals with this syndrome are born with male internal genitalia including testes, but they have female external genitalia and are usually raised as girls. However, when they reach puberty, LH secretion and circulating testosterone levels are increased. Consequently, they develop male body contours and male libido. At this point, they usually change their gender identities and "become boys." Their clitorises enlarge ("penis-at-12 syndrome") to the point that some of them can have intercourse with women. This enlargement probably occurs because with the high LH, there is enough testosterone to overcome the need for DHT amplification in the genitalia.

5α-Reductase-inhibiting drugs are now being used clinically to treat benign prostatic hyperplasia, and finasteride, the most extensively used drug, has its greatest effect on type 2 5α-reductase.

Testicular Production of Estrogens

Over 80% of the estradiol and 95% of the estrone in the plasma of adult men is formed by extragonadal and extra-adrenal aromatization of circulating testosterone and androstenedione. The remainder comes from the testes. Some of the estradiol in testicular venous blood comes from the Leydig cells, but some is also produced by aromatization of androgens in Sertoli cells. In men, the plasma estradiol level is 20–50 pg/mL (73–184 pmol/L) and the total production rate is approximately 50 μg/d (184 nmol/d). In contrast to the situation in women, there is a moderate increase in estrogen production with advancing age in men.

CONTROL OF TESTICULAR FUNCTION

FSH is tropic to the Sertoli cells, and FSH and androgens maintain the gametogenic function of the testes. FSH also stimulates the secretion of ABP and inhibin. Inhibin feeds back to inhibit FSH secretion. LH is tropic to the Leydig cells and stimulates the secretion of testosterone, which in turn feeds back to inhibit LH secretion. Hypothalamic lesions in animals and hypothalamic disease in humans lead to atrophy of the testes and loss of their function.

Inhibins

Testosterone reduces plasma LH, but except in large doses, it has no effect on plasma FSH. Plasma FSH is elevated in patients who have atrophy of the seminiferous tubules but normal levels of testosterone and LH secretion. These observations led to the search for **inhibin,** a factor of testicular origin that inhibits FSH secretion. There are two inhibins in extracts of testes in men and in antral fluid from ovarian follicles in women. They are formed from three polypeptide subunits: a glycosylated α subunit with a molecular weight of 18,000, and two nonglycosylated β subunits, β_A and β_B, each with a molecular weight of 14,000. The subunits are formed from precursor proteins (Figure 23–21). The α subunit combines with β_A to form a heterodimer and with β_B to form another heterodimer, with the subunits linked by disulfide bonds. Both $\alpha\beta_A$ (inhibin A) and $\alpha\beta_B$ (inhibin B) inhibit FSH secretion by a direct action on the pituitary, though it now appears that it is inhibin B that is the FSH-regulating inhibin in adult men and women. Inhibins are produced by Sertoli cells in males and granulosa cells in females.

The heterodimer $\beta_A\beta_B$ and the homodimers $\beta_A\beta_A$ and $\beta_B\beta_B$ are also formed. They stimulate rather than inhibit FSH secretion and consequently are called **activins.** Their function in reproduction is unsettled. However, the inhibins and activins are members of the TGFβ superfamily of dimeric growth factors that also includes MIS (see above). Two **activin receptors** have been cloned, and both appear to be serine kinases. Inhibins and activins are found not only in the gonads but also in the brain and many other tissues. In the bone marrow, activins are involved in the development of white blood cells. In embryonic life, activins are involved in the formation of mesoderm. All mice with a targeted deletion of the α-inhibin subunit gene initially grow in a normal fashion but then develop gonadal stromal tumors, so the gene is a tumor suppressor gene.

In plasma, α_2-macroglobulin binds activins and inhibins. In tissues, activins bind to a family of four glycoproteins called **follistatins.** Binding of the activins inactivates their biologic activity, but the relation of follistatins to inhibin and their physiologic function remain unsettled.

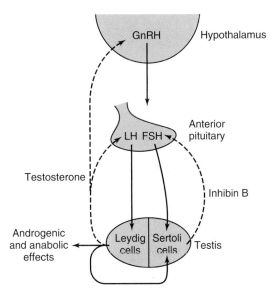

Figure 23–22. Postulated interrelationships between the hypothalamus, anterior pituitary, and testes. Solid arrows indicate excitatory effects; dashed arrows indicate inhibitory effects. Compare with Figures 18–12, 20–21, 22–10, and 23–35.

Steroid Feedback

A current "working hypothesis" of the way the functions of the testes are regulated is shown in Figure 23–22. Castration is followed by a rise in the pituitary content and secretion of FSH and LH, and hypothalamic lesions prevent this rise. Testosterone inhibits LH secretion by acting directly on the anterior pituitary and by inhibiting the secretion of GnRH from the hypothalamus. Inhibin acts directly on the anterior pituitary to inhibit FSH secretion.

In response to LH, some of the testosterone secreted from the Leydig cells bathes the seminiferous epithelium and provides the high local concentration of androgen to the Sertoli cells that is necessary for normal spermatogenesis. Systemically administered testosterone does not raise the androgen level in the testes to as great a degree, and it inhibits LH secretion. Consequently, the net effect of systemically administered testosterone is generally a decrease in sperm count. Testosterone therapy has been suggested as a means of male contraception. However, the dose of testosterone needed to suppress spermatogenesis causes sodium and water retention. The possible use of inhibins as male contraceptives is now being explored.

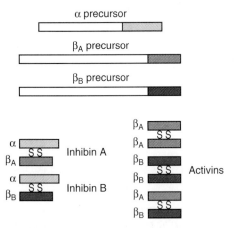

Figure 23–21. Inhibin precursor proteins and the various inhibins and activins that are formed from the carboxyl terminal regions of these precursors. SS, disulfide bonds.

ABNORMALITIES OF TESTICULAR FUNCTION

Cryptorchidism

The testes develop in the abdominal cavity and normally migrate to the scrotum during fetal development. **Testicular descent** to the inguinal region depends on MIS, and descent from the inguinal region to the scrotum depends on other factors. Descent is incomplete on one or, less commonly, both sides in 10% of newborn males, the testes remaining in the abdominal cavity or inguinal canal. Gonadotropic hormone treatment speeds descent in some cases, or the defect can be corrected surgically. Spontaneous descent of the testes is the rule, and the proportion of boys with undescended testes (**cryptorchidism**) falls to 2% at age 1 year and 0.3% after puberty. However, early treatment is now recommended despite these figures because there is a higher incidence of malignant tumors in undescended than in scrotal testes and because after puberty the higher temperature in the abdomen eventually causes irreversible damage to the spermatogenic epithelium.

Male Hypogonadism

The clinical picture of male hypogonadism depends upon whether testicular deficiency develops before or after puberty. In adults, if it is due to testicular disease, circulating gonadotropin levels are elevated (**hypergonadotropic hypogonadism**); if it is secondary to disorders of the pituitary or the hypothalamus (eg, Kallmann's syndrome; see Chapter 14), circulating gonadotropin levels are depressed (**hypogonadotropic hypogonadism**). If the endocrine function of the testes is lost in adulthood, the secondary sex characteristics regress slowly because it takes very little androgen to maintain them once they are established. The growth of the larynx during adolescence is permanent, and the voice remains deep. Men castrated in adulthood suffer some loss of libido, although the ability to copulate persists for some time. They occasionally have hot flushes and are generally more irritable, passive, and depressed than men with intact testes. When the Leydig cell deficiency dates from childhood, the clinical picture is that of **eunuchoidism.** Eunuchoid individuals over the age of 20 are characteristically tall, although not as tall as hyperpituitary giants, because their epiphyses remain open and some growth continues past the normal age of puberty. They have narrow shoulders and small muscles, a body configuration resembling that of the adult female. The genitalia are small and the voice high-pitched. Pubic hair and axillary hair are present because of adrenocortical androgen secretion. However, the hair is sparse, and the pubic hair has the

female "triangle with the base up" distribution rather than the "triangle with the base down" pattern (male escutcheon) seen in normal males.

Androgen-Secreting Tumors

"Hyperfunction" of the testes in the absence of tumor formation is not a recognized entity. Androgen-secreting Leydig cell tumors are rare and cause detectable endocrine symptoms only in prepubertal boys, who develop precocious pseudopuberty (Table 23–2).

THE FEMALE REPRODUCTIVE SYSTEM

THE MENSTRUAL CYCLE

The reproductive system of women (Figure 23–23), unlike that of men, shows regular cyclic changes that teleologically may be regarded as periodic preparations for fertilization and pregnancy. In humans and other primates, the cycle is a **menstrual** cycle, and its most conspicuous feature is the periodic vaginal bleeding that occurs with the shedding of the uterine mucosa (**menstruation**). The length of the cycle is notoriously variable in women, but an average figure is 28 days from the start of one menstrual period to the start of the next. By common usage, the days of the cycle are identified by number, starting with the first day of menstruation.

Ovarian Cycle

From the time of birth, there are many **primordial follicles** under the ovarian capsule. Each contains an immature ovum (Figure 23–24). At the start of each cycle, several of these follicles enlarge, and a cavity forms around the ovum (**antrum formation**). This cavity is filled with follicular fluid. In humans, one of the follicles in one ovary starts to grow rapidly on about the sixth day and becomes the **dominant follicle,** while the others regress, forming **atretic follicles.** The atretic process involves apoptosis. It is uncertain how one follicle is selected to be the dominant follicle in this **follicular phase** of the menstrual cycle, but it seems to be related to the ability of the follicle to secrete the estrogen inside it that is needed for final maturation. When women are given highly purified human pituitary gonadotropin preparations by injection, many follicles develop simultaneously.

    The structure of a maturing ovarian (**graafian**) follicle is shown in Figure 23–24. The cells of the **theca interna** of the follicle are the primary source of circulat-

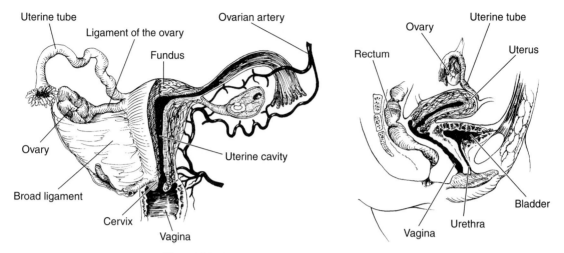

Figure 23–23. The female reproductive system.

ing estrogens. However, the follicular fluid has a high estrogen content, and much of this estrogen comes from the granulosa cells (see below).

At about the 14th day of the cycle, the distended follicle ruptures, and the ovum is extruded into the abdominal cavity. This is the process of **ovulation.** The ovum is picked up by the fimbriated ends of the uterine tubes (oviducts). It is transported to the uterus and, unless fertilization occurs, on out through the vagina.

The follicle that ruptures at the time of ovulation promptly fills with blood, forming what is sometimes called a **corpus hemorrhagicum.** Minor bleeding from the follicle into the abdominal cavity may cause peritoneal irritation and fleeting lower abdominal pain ("mittelschmerz"). The granulosa and theca cells of the follicle lining promptly begin to proliferate, and the clotted blood is rapidly replaced with yellowish, lipid-rich **luteal cells,** forming the **corpus luteum.** This initiates the **luteal phase** of the menstrual cycle, during which the luteal cells secrete estrogens and progesterone. Growth of the corpus luteum depends on its developing an adequate blood supply, and there is evidence that VEGF (see Chapter 30) is essential for this process.

If pregnancy occurs, the corpus luteum persists and there are usually no more periods until after delivery. If there is no pregnancy, the corpus luteum begins to degenerate about 4 days before the next menses (24th day of the cycle) and is eventually replaced by scar tissue, forming a **corpus albicans.**

The ovarian cycle in other mammals is similar, except that in many species more than one follicle ovulates and multiple births are the rule. Corpora lutea form in some submammalian species but not in others.

In humans, no new ova are formed after birth. During fetal development, the ovaries contain over 7 million primordial follicles. However, many undergo atresia (involution) before birth and others are lost after birth. At the time of birth, there are 2 million ova, but 50% of these are atretic. The million that are normal undergo the first part of the first meiotic division at about this time and enter a stage of arrest in prophase in which those that survive persist until adulthood. There is continuing atresia during development, and the number of ova in both of the ovaries at the time of puberty is less than 300,000 (Figure 23–11). Only one of these ova per cycle (or about 500 in the course of a normal reproductive life) normally reaches maturity; the remainder degenerate. Just before ovulation, the first meiotic division is completed. One of the daughter cells, the **secondary oocyte,** receives most of the cytoplasm, while the other, the **first polar body,** fragments and disappears. The secondary oocyte immediately begins the second meiotic division, but this division stops at metaphase and is completed only when a sperm penetrates the oocyte. At that time, the **second polar body** is cast off and the fertilized ovum proceeds to form a new individual. The arrest in metaphase is due, at least in some species, to formation in the ovum of the protein **pp39**^{mos}, which is encoded by the **c-mos** proto-oncogene. When fertilization occurs, the pp39^{mos} is destroyed within 30 minutes by **calpain,** a calcium-dependent cysteine protease.

Uterine Cycle

At the end of menstruation, all but the deep layers of the endometrium have sloughed. Under the influence

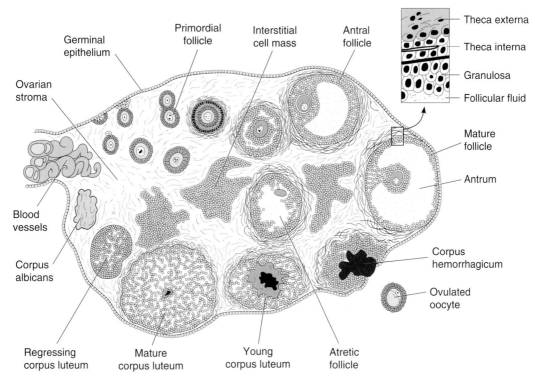

Figure 23–24. Diagram of a mammalian ovary, showing the sequential development of a follicle, formation of a corpus luteum, and, in the center, follicular atresia. A section of the wall of a mature follicle is enlarged at the upper right. The interstitial cell mass is not prominent in primates. (Reproduced, with permission, from Gorbman A, Bern H: *Textbook of Comparative Endocrinology.* Wiley, 1962.)

of estrogens from the developing follicle, the endometrium increases rapidly in thickness from the fifth to the fourteenth days of the menstrual cycle. As the thickness increases, the uterine glands are drawn out so that they lengthen (Figure 23–25), but they do not become convoluted or secrete to any degree. These endometrial changes are called proliferative, and this part of the menstrual cycle is sometimes called the **proliferative phase.** It is also called the preovulatory or follicular phase of the cycle. After ovulation, the endometrium becomes more highly vascularized and slightly edematous under the influence of estrogen and progesterone from the corpus luteum. The glands become coiled and tortuous (Figure 23–25), and they begin to secrete a clear fluid. Consequently, this phase of the cycle is called the **secretory** or **luteal phase.** Late in the luteal phase, the endometrium, like the anterior pituitary, produces prolactin, but the function of this endometrial prolactin is unknown.

The endometrium is supplied by two types of arteries. The superficial two-thirds of the endometrium that is shed during menstruation, the **stratum functionale,** is supplied by long, coiled **spiral arteries** (Figure 23–26), whereas the deep layer that is not shed, the **stratum basale,** is supplied by short, straight **basilar arteries.**

When the corpus luteum regresses, hormonal support for the endometrium is withdrawn. The endometrium becomes thinner, which adds to the coiling of the spiral arteries. Foci of necrosis appear in the endometrium, and these coalesce. There is in addition spasm and then necrosis of the walls of the spiral arteries, leading to spotty hemorrhages that become confluent and produce the menstrual flow.

The vasospasm is probably produced by locally released prostaglandins. There are large quantities of prostaglandins in the secretory endometrium and in menstrual blood, and infusions of $PGF_{2\alpha}$ produce endometrial necrosis and bleeding. One theory of the onset of menstruation holds that in necrotic endometrial cells, lysosomal membranes break down, with the release of enzymes that foster the formation of prostaglandins from cellular phospholipids.

From the point of view of endometrial function, the proliferative phase of the menstrual cycle represents

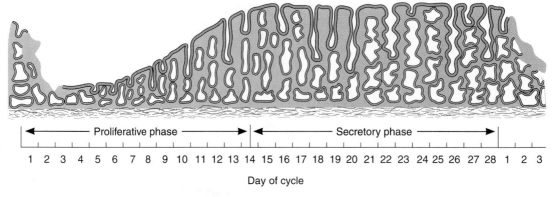

Figure 23–25. Changes in the endometrium during the menstrual cycle.

restoration of the epithelium from the preceding menstruation, and the secretory phase represents preparation of the uterus for implantation of the fertilized ovum. The length of the secretory phase is remarkably constant at about 14 days, and the variations seen in the length of the menstrual cycle are due for the most part to variations in the length of the proliferative phase. When fertilization fails to occur during the secretory phase, the endometrium is shed and a new cycle starts.

Normal Menstruation

Menstrual blood is predominantly arterial, with only 25% of the blood being of venous origin. It contains tissue debris, prostaglandins, and relatively large amounts of fibrinolysin from endometrial tissue. The fibrinolysin lyses clot, so that menstrual blood does not normally contain clots unless the flow is excessive.

The usual duration of the menstrual flow is 3–5 days, but flows as short as 1 day and as long as 8 days can occur in normal women. The amount of blood lost may range normally from slight spotting to 80 mL; the average amount lost is 30 mL. Loss of more than 80 mL is abnormal. Obviously, the amount of flow can be affected by various factors, including the thickness of the endometrium, medication, and diseases that affect the clotting mechanism. After menstruation, a new endometrium regenerates from the stratum basale.

Anovulatory Cycles

In some instances, ovulation fails to occur during the menstrual cycle. Such anovulatory cycles are common for the first 12–18 months after menarche and again before the onset of the menopause. When ovulation does not occur, no corpus luteum is formed and the effects of progesterone on the endometrium are absent. Estrogens continue to cause growth, however, and the proliferative endometrium becomes thick enough to

break down and begins to slough. The time it takes for bleeding to occur is variable, but it usually occurs in less than 28 days from the last menstrual period. The flow is also variable and ranges from scanty to relatively profuse.

Cyclic Changes in the Uterine Cervix

Although it is continuous with the body of the uterus, the cervix of the uterus is different in a number of ways. The mucosa of the uterine cervix does not undergo cyclic desquamation, but there are regular changes in the cervical mucus. Estrogen makes the mucus thinner and more alkaline, changes that promote the survival and transport of sperms. Progesterone makes it thick, tenacious, and cellular. The mucus is thinnest at the time of ovulation, and its elasticity, or **spinnbarkeit,** increases so that by midcycle, a drop can be stretched into a long, thin thread that may be 8–12 cm or more in length. In addition, it dries in an arborizing, fern-like pattern (Figure 23–27) when a thin layer is spread on a slide. After ovulation and during pregnancy, it becomes thick and fails to form the fern pattern.

Vaginal Cycle

Under the influence of estrogens, the vaginal epithelium becomes cornified, and cornified epithelial cells can be identified in the vaginal smear. Under the influence of progesterone, a thick mucus is secreted, and the epithelium proliferates and becomes infiltrated with leukocytes. The cyclic changes in the vaginal smear in rats are relatively marked. The changes in humans and other species are similar but not so clear-cut.

Cyclic Changes in the Breasts

Although lactation normally does not occur until the end of pregnancy, there are cyclic changes in the breasts

Uterine lumen

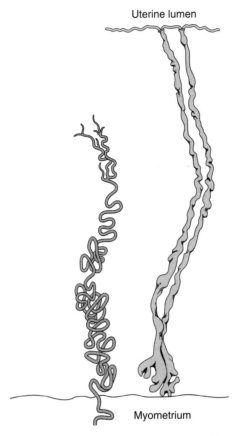

Figure 23–26. Spiral artery of endometrium. Drawing of a spiral artery **(left)** and two uterine glands **(right)** from the endometrium of a rhesus monkey; early secretory phase. (Reproduced, with permission, from Daron GH: The arterial pattern of the tunica mucosa of the uterus in the *Macacus rhesus. Am J Anat* 1936;58:349.)

during the menstrual cycle. Estrogens cause proliferation of mammary ducts, whereas progesterone causes growth of lobules and alveoli. The breast swelling, tenderness, and pain experienced by many women during the 10 days preceding menstruation are probably due to distention of the ducts, hyperemia, and edema of the interstitial tissue of the breast. All these changes regress, along with the symptoms, during menstruation.

Changes During Intercourse

During sexual excitement in women, fluid is secreted onto the vaginal walls, probably because of release of VIP from vaginal nerves. A lubricating mucus is also secreted by the vestibular glands. The upper part of the vagina is sensitive to stretch, while tactile stimulation

from the labia minora and clitoris adds to the sexual excitement. These stimuli are reinforced by tactile stimuli from the breasts and, as in men, by visual, auditory, and olfactory stimuli, which may build to the crescendo known as orgasm. During orgasm, there are autonomically mediated rhythmic contractions of the vaginal walls. Impulses also travel via the pudendal nerves and produce rhythmic contraction of the bulbocavernosus and ischiocavernosus muscles. The vaginal contractions may aid sperm transport but are not essential for it, since fertilization of the ovum is not dependent on orgasm.

Indicators of Ovulation

Knowing when during the menstrual cycle ovulation occurs is important in increasing fertility or, conversely, in family planning. A convenient and reasonably reliable indicator of the time of ovulation is a change—usually a rise—in the basal body temperature (Figure 23–28). Women interested in obtaining an accurate temperature chart should use a thermometer with wide gradations and take their temperatures (oral or rectal)

Normal cycle, 14th day

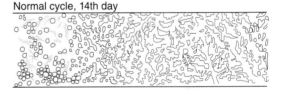

Midluteal phase, normal cycle

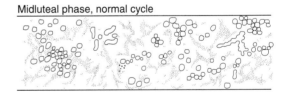

Anovulatory cycle with estrogen present

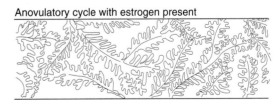

Figure 23–27. Patterns formed when cervical mucus is smeared on a slide, permitted to dry, and examined under the microscope. Progesterone makes the mucus thick and cellular. In the smear from a patient who failed to ovulate **(bottom),** there is no progesterone to inhibit the estrogen-induced fern pattern.

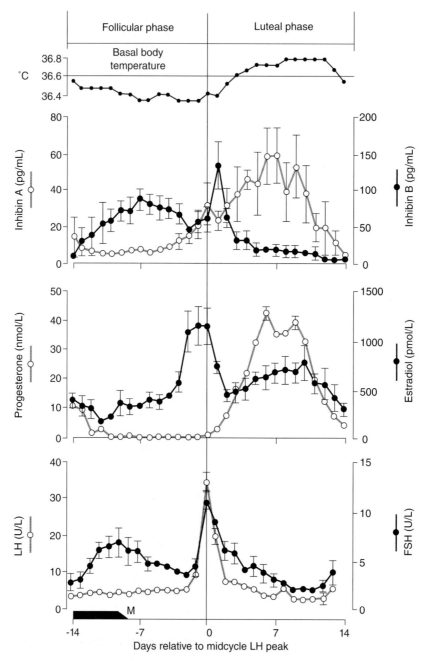

Figure 23–28. Basal body temperature and plasma hormone concentrations (mean ± standard error) during the normal human menstrual cycle. Values are aligned with respect to the day of the midcycle LH peak. (Hormone values from Chabbert Buffet N et al: Regulation of the human menstrual cycle. Front Neuroendocrinol 1998;19:151.)

in the morning before getting out of bed. The cause of the temperature change at the time of ovulation is probably the increase in progesterone secretion, since progesterone is thermogenic (see Chapter 14).

There is a surge in LH secretion that triggers ovulation (see below), and ovulation normally occurs about 9 hours after the peak of the LH surge at midcycle (Figure 23–28). The ovum lives for approximately 72 hours after it is extruded from the follicle, but it is fertilizable for a much shorter time than this. In a study of the relation of isolated intercourse to pregnancy, 36% of women had a detected pregnancy following intercourse on the day of ovulation, but with intercourse on days after ovulation, the percentage was zero. Isolated intercourse on the first and second day before ovulation also led to pregnancy in about 36% of women. A few pregnancies resulted from isolated intercourse on day 3, 4, or 5 before ovulation, although the percentage was much lower, eg, 8% on day 5 before ovulation. Thus, some sperms can survive in the female genital tract and produce fertilization for up to 120 hours before ovulation, but the most fertile period is clearly the 48 hours before ovulation. However, for those interested in the "rhythm method" of contraception, it should be noted that there are rare but documented cases in the literature of pregnancy resulting from isolated coitus on every day of the cycle.

The Estrous Cycle

Mammals other than primates do not menstruate, and their sexual cycle is called an **estrous cycle.** It is named for the conspicuous period of "heat" (**estrus**) at the time of ovulation, normally the only time during which the sexual interest of the female is aroused (see Chapter 15). In spontaneously ovulating species with estrous cycles, such as the rat, there is no episodic vaginal bleeding but the underlying endocrine events are essentially the same as those in the menstrual cycle. In other species, ovulation is induced by copulation (reflex ovulation).

OVARIAN HORMONES

Chemistry, Biosynthesis, & Metabolism of Estrogens

The naturally occurring estrogens are **17β-estradiol, estrone,** and **estriol** (Figure 23–29). They are C_{18} steroids (see Figure 20–7); they do not have an angular methyl group attached to the 10 position or a Δ^4-3-keto configuration in the A ring. They are secreted primarily by the granulosa cells of the ovarian follicles, the corpus luteum, and the placenta. The biosynthetic pathway involves their formation from androgens.

They are also formed by aromatization of androstenedione in fat and other tissues. **Aromatase (CYP19)** is the enzyme that catalyzes the conversion of androstenedione to estrone and the conversion of testosterone to estradiol (Figure 23–29).

Theca interna cells have many LH receptors, and LH acts via cAMP to increase conversion of cholesterol to androstenedione. Some of the androstenedione is converted to estradiol, which enters the circulation. The theca interna cells also supply androstenedione to the granulosa cells. The granulosa cells make estradiol when provided with androgens (Figure 23–30), and it appears that the estradiol they form in primates is secreted into the follicular fluid. Granulosa cells have many FSH receptors, and FSH facilitates their secretion of estradiol by acting via cAMP to increase their aromatase activity. Mature granulosa cells also acquire LH receptors, and LH also stimulates estradiol production.

Two percent of the circulating estradiol is free, and the remainder is bound to protein: 60% to albumin and 38% to the same gonadal steroid-binding globulin (GBG) that binds testosterone (Table 23–5).

In the liver, estradiol, estrone, and estriol are converted to glucuronide and sulfate conjugates. All these compounds, along with other metabolites, are excreted in the urine. Appreciable amounts are secreted in the bile and reabsorbed into the bloodstream (enterohepatic circulation).

Secretion

The concentration of estradiol in the plasma during the menstrual cycle is shown in Figure 23–28. Almost all of this estrogen comes from the ovary, and there are two peaks of secretion: one just before ovulation and one during the midluteal phase. The estradiol secretion rate is 36 μg/d (133 nmol/d) in the early follicular phase, 380 μg/d just before ovulation, and 250 μg/d during the midluteal phase (Table 23–7). After menopause, estrogen secretion declines to low levels.

As noted above, the estradiol production rate in men is about 50 μg/d (184 nmol/d).

Effects on the Female Genitalia

Estrogens facilitate the growth of the ovarian follicles and increase the motility of the uterine tubes. Their role in the cyclic changes in the endometrium, cervix, and vagina is discussed above. They increase uterine blood flow and have important effects on the smooth muscle of the uterus. In immature and castrate females, the uterus is small and the myometrium atrophic and inactive. Estrogens increase the amount of uterine muscle and its content of contractile proteins. Under the influ-

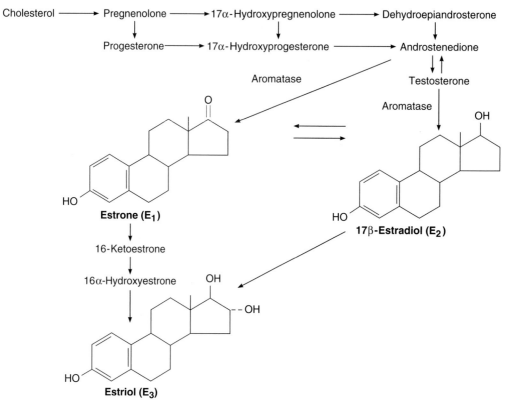

Figure 23–29. Biosynthesis and metabolism of estrogens. The formulas of the precursor steroids are shown in Figure 20–8.

ence of estrogens, the muscle becomes more active and excitable, and action potentials in the individual fibers become more frequent (see Chapter 3). The "estrogen-dominated" uterus is also more sensitive to oxytocin.

Chronic treatment with estrogens causes the endometrium to hypertrophy. When estrogen therapy is discontinued, there is sloughing with **withdrawal bleeding.** Some "breakthrough" bleeding may occur during treatment when estrogens are given for long periods.

Effects on Endocrine Organs

Estrogens decrease FSH secretion. Under some circumstances, they inhibit LH secretion (negative feedback); in other circumstances, they increase LH secretion (positive feedback; see below). Women are sometimes given large doses of estrogens for 4–6 days to prevent conception after coitus during the fertile period (post-coital or "morning-after" contraception). However, in

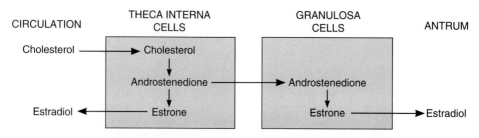

Figure 23–30. Interactions between theca and granulosa cells in estradiol synthesis and secretion.

Table 23–7. Twenty-four-hour production rates of sex steroids in women at different stages of the menstrual cycle.[1]

Sex Steroids	Early Follicular	Preovulatory	Midluteal
Progesterone (mg)	1.0	4.0	25.0
17-Hydroxyprogesterone (mg)	0.5	4.0	4.0
Dehydroepiandrosterone (mg)	7.0	7.0	7.0
Androstenedione (mg)	2.6	4.7	3.4
Testosterone (μg)	144.0	171.0	126.0
Estrone (μg)	50.0	350.0	250.0
Estradiol (μg)	36.0	380.0	250.0

[1]Modified and reproduced, with permission, from Yen SSC, Jaffe RB, Barbieri RL: *Reproductive Endocrinology,* 4th ed. Saunders, 1999.

this instance, pregnancy is probably prevented by interference with implantation of the fertilized ovum rather than changes in gonadotropin secretion.

Estrogens cause increased secretion of angiotensinogen (see Chapter 24) and thyroid-binding globulin (see Chapter 18). They exert an important protein anabolic effect in chickens and cattle, possibly by stimulating the secretion of androgens from the adrenal, and estrogen treatment has been used commercially to increase the weight of domestic animals. They cause epiphysial closure in humans.

Effects on the CNS

The estrogens are responsible for estrous behavior in animals, and they increase libido in humans. They apparently exert this action by a direct effect on certain neurons in the hypothalamus (Figure 23–31). The relation of estrogens, progesterone, and androgens to sexual behavior is discussed in Chapter 15.

Estrogens increase the proliferation of dendrites on neurons and the number of synaptic knobs in rats. In humans, they have been reported to slow the progression of Alzheimer's disease (see Chapter 16), but this is controversial.

Effects on the Breasts

Estrogens produce duct growth in the breasts and are largely responsible for breast enlargement at puberty in girls; they have been called the growth hormones of the breast. Breast enlargement that occurs when estrogen-containing skin creams are applied locally is due primarily to systemic absorption of the estrogen, although a slight local effect is also produced. Estrogens are responsible for the pigmentation of the areolas, although pigmentation usually becomes more intense during the first pregnancy than it does at puberty. The role of the estrogens in the overall control of breast growth and lactation is discussed below.

Female Secondary Sex Characteristics

The body changes that develop in girls at puberty—in addition to enlargement of breasts, uterus, and vagina—are due in part to estrogens, which are the "feminizing hormones," and in part simply to the absence of testicular androgens. Women have narrow shoulders and broad hips, thighs that converge, and arms that diverge (wide **carrying angle**). This body configuration, plus the female distribution of fat in the breasts and buttocks, is seen also in castrate males. In women, the larynx retains its prepubertal proportions and the voice stays high-pitched. There is less body hair and more scalp hair, and the pubic hair generally has a characteristic flat-topped pattern (female escutcheon). However, growth of pubic and axillary hair in both sexes is due primarily to androgens rather than estrogens.

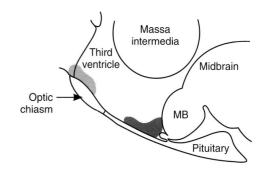

Figure 23–31. Loci where implantations of estrogen in the hypothalamus affect ovarian weight and sexual behavior in rats, projected on a sagittal section of the hypothalamus. The implants that stimulate sex behavior are located in the suprachiasmatic area above the optic chiasm (gray area), whereas ovarian atrophy is produced by implants in the arcuate nucleus and surrounding ventral hypothalamus (colored area). MB, mamillary body.

Other Actions

Normal women retain salt and water and gain weight just before menstruation. Estrogens cause some degree of salt and water retention. However, aldosterone secretion is slightly elevated in the luteal phase, and this also contributes to the premenstrual fluid retention.

Estrogens are said to make sebaceous gland secretions more fluid and thus to counter the effect of testosterone and inhibit formation of **comedones** ("blackheads") and acne. The liver palms, spider angiomas, and slight breast enlargement seen in advanced liver disease are due to increased circulating estrogens. The increase appears to be due to decreased hepatic metabolism of androstenedione, making more of this androgen available for conversion to estrogens.

Estrogens have a significant plasma cholesterol-lowering action (see Chapter 17) and they rapidly produce vasodilation by increasing the local production of NO (see Chapter 31). These actions inhibit atherogenesis and contribute to the low incidence of myocardial infarction and other complications of atherosclerotic vascular disease in premenopausal women. There is a debate about the cardiovascular effects of estrogen treatment in postmenopausal women. Large doses of orally active estrogens promote thrombosis, apparently because they reach the liver in high concentrations in the portal blood and alter hepatic production of clotting factors. Smaller doses still cause some clotting. However, considerable epidemiologic evidence indicates that estrogen therapy is still cardioprotective after menopause.

Mechanism of Action

Like other steroids (see Chapter 1), estrogens combine with protein receptors in the nucleus, and the complexes bind to DNA, promoting formation of mRNAs that in turn direct the formation of new proteins which modify cell function. Two nuclear estrogen receptors have been cloned: estrogen receptor α (ERα) encoded by a gene on chromosome 6; and estrogen receptor β (ERβ), encoded by a gene on chromosome 14. Although there is overlap, the distribution of these receptors is different. ERα expression is moderate to high in the uterus, testis, pituitary, kidney, epididymis, and adrenal, whereas ERβ expression is high in the ovary, prostate, lung, bladder, brain, and bone. It has been suggested that the regulation of ovarian function by the pituitary-ovarian axis is primarily ERα-mediated, whereas the estrogens secreted into the ovarian follicles act primarily via ERβs. Male and female mice in which the gene for ERα has been knocked out are sterile, develop osteoporosis, and continue to grow because their epiphyses do not close.

ERβ female knockouts are infertile, but ERβ male knockouts are fertile even though they have hyperplastic prostates and loss of fat. However, much additional research is needed to sort out the detailed effects of ERα and ERβ and their interactions.

Most of the actions of estrogens are genomic, ie, mediated via ERα and ERβ. However, a few effects are so rapid that it is difficult to believe they are mediated via production of mRNAs. These include effects on neuronal discharge in the brain and, possibly, feedback effects on gonadotropin secretion. Evidence is accumulating that these effects are mediated by cell membrane receptors that appear to be structurally related to the nuclear receptors and produce their effects by intracellular mitogen-activated protein kinase pathways. Similar rapid effects of progesterone, testosterone, glucocorticoids, aldosterone, and 1,25-dihydroxycholecalciferol may also be produced by membrane receptors.

Synthetic Estrogens

The ethinyl derivative of estradiol is a potent estrogen and—unlike the naturally occurring estrogens—is relatively active when given by mouth, because it is resistant to hepatic metabolism. The activity of the naturally occurring hormones is low when they are administered by mouth because the portal venous drainage of the intestine carries them to the liver, where they are inactivated before they can reach the general circulation. Some nonsteroidal substances and a few compounds found in plants have estrogenic activity. The plant estrogens are rarely a problem in human nutrition, but they may cause undesirable effects in farm animals.

Estradiol reduces the hot flushes and other symptoms of the menopause, and it also prevents the development of osteoporosis. There is debate about whether it reduces the progression of atherosclerosis and the incidence of heart attacks and strokes. It also stimulates growth of the endometrium and the breast and can lead to cancer of the uterus and, probably, the breast. Therefore, there has been an active search for "tailor-made" estrogens that have the bone and cardiovascular effects of estradiol but lack its growth-stimulating effects on the uterus and breast. Two compounds, **tamoxifen** and **raloxifene,** show promise in this regard. Neither combats the symptoms of the menopause, but both have the bone-preserving and cardiovascular effects of estradiol. In addition, tamoxifen does not stimulate the breast, and raloxifene does not stimulate the breast or the uterus. The way the selective effects are brought about by these selective estrogen receptor modulators **(SERMs)** is related to differences in the way the receptor-ligand complexes they form bind to DNA.

Chemistry, Biosynthesis, & Metabolism of Progesterone

Progesterone is a C_{21} steroid (Figure 23–32) secreted by the corpus luteum, the placenta, and (in small amounts) the follicle. It is an important intermediate in steroid biosynthesis in all tissues that secrete steroid hormones, and small amounts apparently enter the circulation from the testes and adrenal cortex. About 2% of the circulating progesterone is free (Table 23–5), whereas 80% is bound to albumin and 18% is bound to corticosteroid-binding globulin (see Chapter 20). Progesterone has a short half-life and is converted in the liver to pregnanediol, which is conjugated to glucuronic acid and excreted in the urine (Figure 23–32).

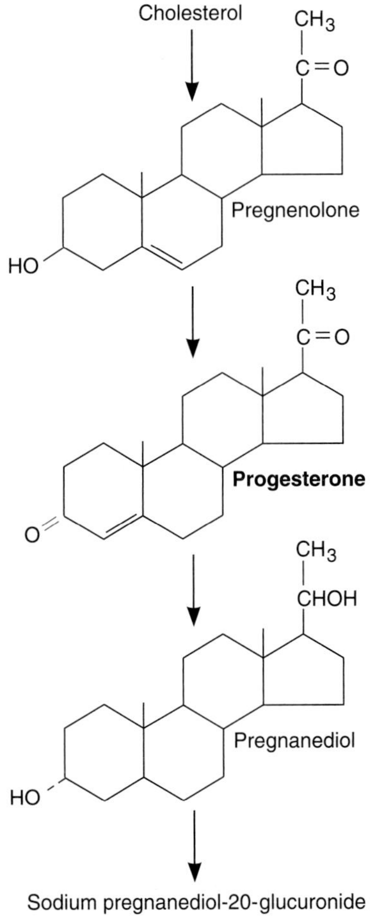

Figure 23–32. Biosynthesis of progesterone and major pathway for its metabolism. Other metabolites are also formed.

Cholesterol

Pregnenolone

Progesterone

Pregnanediol

Sodium pregnanediol-20-glucuronide

Secretion

In men, the plasma progesterone level is approximately 0.3 ng/mL (1 nmol/L). In women, the level is approximately 0.9 ng/mL (3 nmol/L) during the follicular phase of the menstrual cycle (Figure 23–28). The difference is due to secretion of small amounts of progesterone by cells in the ovarian follicles; theca cells provide pregnenolone to the granulosa cells, which convert it to progesterone. Late in the follicular phase, progesterone secretion begins to increase. During the luteal phase, the corpus luteum produces large quantities of progesterone (Table 23–7) and there is a marked increase in plasma progesterone to a peak value of approximately 18 ng/mL (60 nmol/L).

The stimulating effect of LH on progesterone secretion by the corpus luteum is due to activation of adenylyl cyclase and involves a subsequent step that is dependent on protein synthesis.

Actions

The principal target organs of progesterone are the uterus, the breasts, and the brain. Progesterone is responsible for the progestational changes in the endometrium and the cyclic changes in the cervix and vagina described above. It has an antiestrogenic effect on the myometrial cells, decreasing their excitability, their sensitivity to oxytocin, and their spontaneous electrical activity while increasing their membrane potential. It also decreases the number of estrogen receptors in the endometrium and increases the rate of conversion of 17β-estradiol to less active estrogens.

In the breast, progesterone stimulates the development of lobules and alveoli. It induces differentiation of estrogen-prepared ductal tissue and supports the secretory function of the breast during lactation.

The feedback effects of progesterone are complex and are exerted at both the hypothalamic and pituitary levels. Large doses of progesterone inhibit LH secretion and potentiate the inhibitory effect of estrogens, preventing ovulation.

Progesterone is thermogenic and is probably responsible for the rise in basal body temperature at the time of ovulation. It stimulates respiration, and the alveolar P_{CO_2} (P_{CO_2}; see Chapter 34) in women during the luteal phase of the menstrual cycle is lower than that in men. In pregnancy, the P_{CO_2} falls as progesterone secretion rises. However, the physiologic significance of this respiratory response is unknown.

Large doses of progesterone produce natriuresis, probably by blocking the action of aldosterone on the kidney. The hormone does not have a significant anabolic effect.

Mechanism of Action

The effects of progesterone, like those of other steroids, are brought about by an action on DNA to initiate synthesis of new mRNA. As noted in Chapter 1, the progesterone receptor is bound to a heat shock protein in the absence of the steroid, and progesterone binding releases the heat shock protein, exposing the DNA-binding domain of the receptor. The synthetic steroid **mifepristone (RU 486)** binds to the receptor but does not release the heat shock protein, and it blocks the binding of progesterone. Since the maintenance of early pregnancy depends on the stimulatory effect of progesterone on endometrial growth and its inhibition of uterine contractility, mifepristone causes abortion. In some countries, mifepristone combined with a prostaglandin is used to produce elective abortions.

There are two isoforms of the progesterone receptor—PR_A and PR_B—produced by differential processing from a single gene. PR_A is a truncated form, but it is likely that both isoforms mediate unique subsets of progesterone action.

Substances that mimic the action of progesterone are sometimes called **progestational agents, gestagens,** or **progestins.** They are used along with synthetic estrogens as oral contraceptive agents (see below).

Relaxin

Relaxin is a polypeptide hormone that is produced in the corpus luteum, uterus, placenta, and mammary glands in women and in the prostate gland in men. During pregnancy, it relaxes the pubic symphysis and other pelvic joints and softens and dilates the uterine cervix. Thus, it facilitates delivery. It also inhibits uterine contractions and may play a role in the development of the mammary glands. In nonpregnant women, relaxin is found in the corpus luteum and the endometrium during the secretory but not the proliferative phase of the menstrual cycle. Its function in nonpregnant women is unknown. In men, it is found in semen, where it may help maintain sperm motility and aid in sperm penetration of the ovum.

In most species there is only one relaxin gene, but in humans there are two genes on chromosome 9 that code for two structurally different polypeptides which both have relaxin activity. However, only one of these genes is active in the ovary and the prostate. The structure of the polypeptide produced in these two tissues is shown in Figure 23–33.

CONTROL OF OVARIAN FUNCTION

FSH from the pituitary is responsible for the early maturation of the ovarian follicles, and FSH and LH together are responsible for their final maturation. A

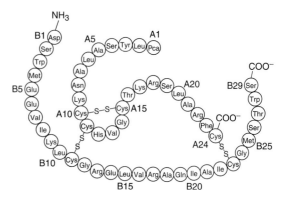

Figure 23–33. Structure of human luteal and seminal relaxin. Note the similarity to the structure of insulin, IGF-I, and IGF-II (see Figure 22–9). Pca, pyroglutamic acid. (Modified and reproduced, with permission, from Winslow JW et al: Human seminal relaxin is a product of the same gene as human luteal relaxin. Endocrinology 1992;130:2660. Copyright © 1992 by The Endocrine Society.)

burst of LH secretion (Figure 23–28) is responsible for ovulation and the initial formation of the corpus luteum. There is also a smaller midcycle burst of FSH secretion, whose significance is uncertain. LH stimulates the secretion of estrogen and progesterone from the corpus luteum.

Hypothalamic Components

The hypothalamus occupies a key position in the control of gonadotropin secretion. Hypothalamic control is exerted by GnRH secreted into the portal hypophysial vessels (see Chapter 14). GnRH stimulates the secretion of FSH as well as LH, and it is unlikely that there is an additional separate FRH.

GnRH is normally secreted in episodic bursts, and these bursts produce the circhoral peaks of LH secretion. They are essential for normal secretion of gonadotropins. If GnRH is administered by constant infusion, the GnRH receptors in the anterior pituitary down-regulate (see Chapter 1) and LH secretion declines to zero. However, if GnRH is administered episodically at a rate of one pulse per hour, LH secretion is stimulated. This is true even when endogenous GnRH secretion has been prevented by a lesion of the ventral hypothalamus.

It is now clear not only that episodic secretion of GnRH is a general phenomenon but also that fluctuations in the frequency and amplitude of the GnRH bursts are important in generating the other hormonal changes that are responsible for the menstrual cycle.

Frequency is increased by estrogens and decreased by progesterone and testosterone. The frequency increases late in the follicular phase of the cycle, culminating in the LH surge. During the secretory phase, the frequency decreases as a result of the action of progesterone (Figure 23–34), but when estrogen and progesterone secretion decrease at the end of the cycle, the frequency once again increases.

At the time of the midcycle LH surge, the sensitivity of the gonadotropes to GnRH is greatly increased because of their exposure to GnRH pulses of the frequency that exist at this time. This self-priming effect of GnRH is important in producing a maximum LH response.

The nature and the exact location of the GnRH pulse generator in the hypothalamus are still unsettled. However, it is known in a general way that norepinephrine and possibly epinephrine in the hypothalamus in-

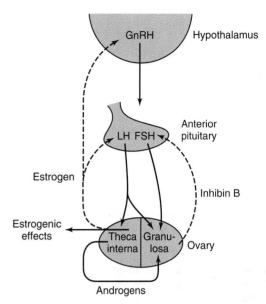

Figure 23–35. Feedback regulation of ovarian function. The cells of the theca interna provide androgens to the granulosa cells, and theca cells also produce the circulating estrogens that inhibit the secretion of GnRH, LH, and FSH. Inhibin from the granulosa cells inhibits FSH secretion. LH regulates the thecal cells, whereas the granulosa cells are regulated by both LH and FSH. The dashed arrows indicate inhibitory effects and the solid arrows stimulatory effects. Compare with Figures 18–12, 20–21, 22–10, and 23–22.

crease GnRH pulse frequencies. Conversely, opioid peptides such as the enkephalins and β-endorphin reduce the frequency of GnRH pulses.

The down-regulation of pituitary receptors and the consequent decrease in LH secretion produced by constantly elevated levels of GnRH has led to the use of long-acting GnRH analogs to inhibit LH secretion in precocious puberty and in cancer of the prostate (see below).

Feedback Effects

Changes in plasma LH, FSH, sex steroids, and inhibin during the menstrual cycle are shown in Figure 23–28, and their feedback relations are diagrammed in Figure 23–35. During the early part of the follicular phase, inhibin B is low and FSH is modestly elevated, fostering follicular growth. LH secretion is held in check by the negative feedback effect of the rising plasma estrogen level. At 36–48 hours before ovulation, the estrogen feedback effect becomes positive, and this initiates the burst of LH secretion (LH surge) that produces ovula-

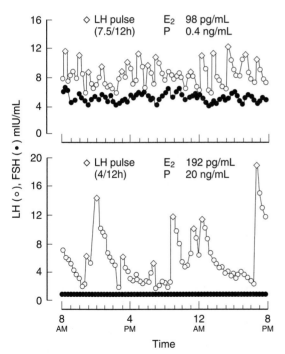

Figure 23–34. Episodic secretion of LH (s) and FSH (d) during the follicular stage **(top)** and the luteal stage **(bottom)** of the menstrual cycle. The numbers above each graph indicate the numbers of LH pulses per 12 hours and the plasma estradiol (E$_2$) and progesterone (P) concentrations at these two times of the cycle. (Reproduced, with permission, from Marshall JC, Kelch RO: Gonadotropin-releasing hormone: role of pulsatile secretion in the regulation of reproduction. N Engl J Med 1986;315:1459.)

tion. Ovulation occurs about 9 hours after the LH peak. FSH secretion also peaks, despite a small rise in inhibin, probably because of the strong stimulation of gonadotropes by GnRH. During the luteal phase, the secretion of LH and FSH is low because of the elevated levels of estrogen, progesterone, and inhibin.

It should be emphasized that a moderate, constant level of circulating estrogen exerts a negative feedback effect on LH secretion, whereas during the cycle, an elevated estrogen level exerts a positive feedback effect and stimulates LH secretion. It has been demonstrated that in monkeys there is also a minimum time that estrogens must be elevated to produce positive feedback. When circulating estrogen was increased about 300% for 24 hours, only negative feedback was seen; but when it was increased about 300% for 36 hours or more, a brief decline in secretion was followed by a burst of LH secretion that resembled the midcycle surge. When circulating levels of progesterone were high, the positive feedback effect of estrogen was inhibited. There is evidence that in primates, both the negative and the positive feedback effects of estrogen are exerted in the mediobasal hypothalamus, but exactly how negative feedback is switched to positive feedback and then back to negative feedback in the luteal phase remains unknown.

Control of the Cycle

In an important sense, regression of the corpus luteum (**luteolysis**) starting 3–4 days before menses is the key to the menstrual cycle. $PGF_{2\alpha}$ appears to be a physiologic luteolysin, but this prostaglandin is only active when endothelial cells producing ET-1 (see Chapter 31) are present. Therefore it appears that at least in some species luteolysis is produced by the combined action of $PGF2_{\alpha}$ and ET-1. In some domestic animals, oxytocin secreted by the corpus luteum appears to exert a local luteolytic effect, possibly by causing the release of prostaglandins. Once luteolysis begins, the estrogen and progesterone levels fall and the secretion of FSH and LH increases. A new crop of follicles develops, and then a single dominant follicle matures as a result of the action of FSH and LH. Near midcycle, there is a rise in estrogen secretion from the follicle. This rise augments the responsiveness of the pituitary to GnRH and triggers a burst of LH secretion. The resulting ovulation is followed by formation of a corpus luteum. There is a drop in estrogen secretion, but progesterone and estrogen levels then rise together, along with inhibin B. The elevated levels inhibit FSH and LH secretion for a while, but luteolysis again occurs and a new cycle starts.

Reflex Ovulation

Female cats, rabbits, mink, and some other animals have long periods of estrus, during which they ovulate only after copulation. Such **reflex ovulation** is brought about by afferent impulses from the genitalia and the eyes, ears, and nose that converge on the ventral hypothalamus and provoke an ovulation-inducing release of LH from the pituitary. In species such as rats, monkeys, and humans, ovulation is a spontaneous periodic phenomenon, but neural mechanisms are also involved. Ovulation can be delayed 24 hours in rats by administering pentobarbital or various other neurally active drugs 12 hours before the expected time of follicle rupture. In women, menstrual cycles may be markedly influenced by emotional stimuli.

Contraception

Methods commonly used to prevent conception are listed in Table 23–8, along with their failure rates. Once conception has occurred, abortion can be produced by progesterone antagonists such as mifepristone (see above).

Implantation of foreign bodies in the uterus causes changes in the duration of the sexual cycle in a number of mammalian species. In humans, such foreign bodies do not alter the menstrual cycle, but they act as effective contraceptive devices. Intrauterine implantation of pieces of metal or plastic (**intrauterine devices, IUDs**) has been used in programs aimed at controlling population growth. Although the mechanism of action of

Table 23–8. Relative effectiveness of frequently used contraceptive methods.[1]

Method	Failures per 100 Woman-Years
Vasectomy	0.02
Tubal ligation and similar procedures	0.13
Oral contraceptive	
> 50 μg estrogen and progestin	0.32
< 50 μg estrogen and progestin	0.27
Progestin only	1.2
IUD	
Copper 7	1.5
Loop D	1.3
Diaphragm	1.9
Condom	3.6
Withdrawal	6.7
Spermicide	11.9
Rhythm	15.5

[1]Data from Vessey M, Lawless M, Yeates D: Efficacy of different contraceptive methods. Lancet 1982;1:841. Reproduced with permission.

IUDs is still unsettled, they seem in general to prevent sperms from fertilizing ova. Those containing copper appear to exert a spermatocidal effect. IUDs that slowly release progesterone or synthetic progestins have the additional effect of thickening cervical mucus so that entry of sperms into the uterus is impeded. IUDs can cause intrauterine infections, but these usually occur in the first month after insertion and in women exposed to sexually transmitted diseases.

Women undergoing long-term treatment with relatively large doses of estrogen do not ovulate, probably because they have depressed FSH levels and multiple irregular bursts of LH secretion rather than a single midcycle peak. Women treated with similar doses of estrogen plus a progestational agent do not ovulate because the secretion of both gonadotropins is suppressed. In addition, the progestin makes the cervical mucus thick and unfavorable to sperm migration, and it may also interfere with implantation. For contraception, an orally active estrogen such as ethinyl estradiol is often combined with a synthetic progestin such as norethindrone. The pills are administered for 21 days, then withdrawn for 5–7 days to permit menstrual flow, and started again. Like ethinyl estradiol, norethindrone has an ethinyl group on position 17 of the steroid nucleus, so it is resistant to hepatic metabolism and consequently is effective by mouth. In addition to being a progestin, it is partly metabolized to ethinyl estradiol, and for this reason it also has estrogenic activity. It is now clear that small as well as large doses of estrogen are effective (Table 23–8); the use of small doses reduces the risk of thromboses or other complications. Progestins alone can be used for contraception, although they are more effective when combined with estrogens.

Implants made up primarily of progestins such as levonorgestrel are now seeing increased use in some parts of the world. These are inserted under the skin and can prevent pregnancy for up to 5 years. They often produce amenorrhea, but otherwise they appear to be effective and well tolerated.

ABNORMALITIES OF OVARIAN FUNCTION

Menstrual Abnormalities

Some women who are infertile have **anovulatory cycles;** they fail to ovulate but have menstrual periods at fairly regular intervals. As noted above, anovulatory cycles are the rule for the first 1–2 years after menarche and again before the menopause. **Amenorrhea** is the absence of menstrual periods. If menstrual bleeding has never occurred, the condition is called **primary amenorrhea.** Some women with primary amenorrhea have small breasts and other signs of failure to mature sexually. Cessation of cycles in a woman with previously normal periods is called **secondary amenorrhea.** The commonest cause of secondary amenorrhea is pregnancy, and the old clinical maxim that "secondary amenorrhea should be considered to be due to pregnancy until proved otherwise" has considerable merit. Other causes of amenorrhea include emotional stimuli and changes in the environment, hypothalamic diseases, pituitary disorders, primary ovarian disorders, and various systemic diseases. There is evidence that in some women with hypothalamic amenorrhea, the frequency of GnRH pulses is slowed as a result of excess opioid activity in the hypothalamus. In encouraging preliminary studies, the frequency of GnRH pulses has been increased by administration of the orally active opioid blocker naltrexone.

The terms **hypomenorrhea** and **menorrhagia** refer to scanty and abnormally profuse flow, respectively, during regular periods. **Metrorrhagia** is bleeding from the uterus between periods, and **oligomenorrhea** is reduced frequency of periods. **Dysmenorrhea** is painful menstruation. The severe menstrual cramps that are common in young women quite often disappear after the first pregnancy. Most of the symptoms of dysmenorrhea are due to accumulation of prostaglandins in the uterus, and symptomatic relief has been obtained by treatment with inhibitors of prostaglandin synthesis (see Chapter 17).

Some women develop symptoms such as irritability, bloating, edema, emotional lability, decreased ability to concentrate, depression, headache, and constipation during the last 7–10 days of their menstrual cycles. These symptoms of the **premenstrual syndrome (PMS)** have been attributed to salt and water retention. However, it seems unlikely that this or any of the other hormonal alterations that occur in the late luteal phase are responsible because the time course and severity of the symptoms are not modified if the luteal phase is terminated early and menstruation produced by administration of mifepristone. The antidepressant fluoxetine (Prozac), which is a serotonin reuptake inhibitor, and the benzodiazepine alprazolam produce symptomatic relief, and so do GnRH-releasing agonists in doses that suppress the pituitary-ovarian axis. How these diverse clinical observations fit together to produce a picture of the pathophysiology of PMS is still unknown.

Genetic Defects

A number of single-gene mutations cause reproductive abnormalities when they occur in women. Examples include (1) Kallmann's syndrome, which causes hypogonadotropic hypogonadism (see above); (2) GnRH resistance, FSH resistance, and LH resistance, which are due to defects in the GnRH, FSH, or LH receptors, respectively; and (3) aromatase deficiency, which prevents the formation of estrogens. These are all caused by loss-of-function mutations. An interesting gain-of-function mutation causes the **McCune-Albright syndrome,** in which $G_s\alpha$ becomes constitutively active in certain cells but not others (mosaicism) because a somatic mutation after initial cell division has occurred in the embryo (see Chapter 1). It is associated with multiple endocrine abnormalities, including precocious puberty and amenorrhea with galactorrhea.

PREGNANCY

Fertilization & Implantation

In humans, **fertilization** of the ovum by the sperm usually occurs in the ampulla of the uterine tube. Fertilization involves (1) chemoattraction of the sperm to the ovum by substances produced by the ovum; (2) adherence to the **zona pellucida,** the membranous structure surrounding the ovum; (3) penetration of the zona pellucida and the acrosome reaction; and (4) adherence of the sperm head to the cell membrane of the ovum, with breakdown of the area of fusion and release of the sperm nucleus into the cytoplasm of the ovum (Figure 23–36). Millions of sperms are deposited in the vagina during intercourse. Eventually, 50–100 sperms reach the ovum, and many of them contact the zona pellucida. Sperms bind to a sperm receptor called ZP3 in the zona, and this is followed by the **acrosomal reaction,** ie, the breakdown of the acrosome, the lysosome-like organelle on the head of the sperm (Figure 23–18). Various enzymes are released, including the trypsin-like protease **acrosin.** Acrosin facilitates but is not required for the penetration of the sperm through the zona pellucida. When one sperm reaches the membrane of the ovum, fusion to the ovum membrane is mediated by **fertilin,** a protein on the surface of the sperm head that resembles the viral fusion proteins which permit viruses to attack cells. The fusion provides the signal that initiates development. In addition, the fusion sets off a reduction in the membrane potential of the ovum that prevents polyspermy, the fertilization of the ovum by more than one sperm. This transient potential change is followed by a structural change in the zona pellucida that provides protection against polyspermy on a more long-term basis.

The developing embryo, now called a **blastocyst,** moves down the tube into the uterus. This journey takes about 3 days, during which the blastocyst reaches the 8- or 16-cell stage. Once in contact with the en-

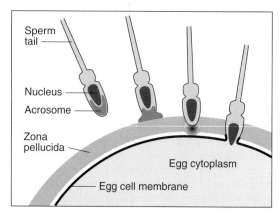

Figure 23–36. Sequential events in fertilization in mammals. Sperms are attracted to the ovum, bind to the zona pellucida, release acrosomal enzymes, penetrate the zona pellucida, and fuse with the membrane of the ovum, releasing the sperm nucleus into its cytoplasm. Current evidence indicates that the side—rather than the tip—of the sperm head fuses with the egg cell membrane. (Modified from Vacquier VD: Evolution of gamete recognition proteins. Science 1999;281:1995.)

dometrium, the blastocyst becomes surrounded by an outer layer of **syncytiotrophoblast,** a multinucleate mass with no discernible cell boundaries, and an inner layer of **cytotrophoblast** made up of individual cells. The syncytiotrophoblast erodes the endometrium, and the blastocyst burrows into it **(implantation).** The implantation site is usually on the dorsal wall of the uterus. A placenta then develops, and the trophoblast remains associated with it.

Failure to Reject the "Fetal Graft"

It should be noted that the fetus and the mother are two genetically distinct individuals, and the fetus is in effect a transplant of foreign tissue in the mother. However, the transplant is tolerated, and the rejection reaction that is characteristically produced when other foreign tissues are transplanted (see Chapter 27) fails to occur. The way the "fetal graft" is protected is unknown. However, one explanation may be that the placental trophoblast, which separates maternal and fetal tissues, does not express the polymorphic class I and class II MHC genes and instead expresses **HLA-G,** a nonpolymorphic gene. Therefore, antibodies against the fetal proteins do not develop. In addition, there is a Fas ligand on the surface of the placenta, and this bonds to T cells, causing them to undergo apoptosis (see Chapter 1). There is also some decrease in circulating antibodies in the mother. For example, the level of maternal antithyroid antibodies in the circulation of women with Graves' disease decreases during pregnancy, and they often become euthyroid until the pregnancy is over.

Infertility

The vexing clinical problem of infertility often requires extensive investigation before a cause is found. In 30% of cases the problem is in the man; in 45%, the problem is in the woman; in 20%, there are problems in both partners; and in 5% no cause can be found. **In vitro fertilization,** ie, removing mature ova, fertilizing them with sperms, and implanting one or more of them in the uterus at the four-cell stage is of some value in these cases. It has a 5–10% chance of producing a live birth.

Endocrine Changes

In all mammals, the corpus luteum in the ovary at the time of fertilization fails to regress and instead enlarges in response to stimulation by gonadotropic hormones secreted by the placenta. The placental gonadotropin in humans is called **human chorionic gonadotropin (hCG).** The enlarged **corpus luteum of pregnancy** secretes estrogens, progesterone, and relaxin. The relaxin helps maintain pregnancy by inhibiting myometrial contractions. In most species, removal of the ovaries at any time during pregnancy precipitates abortion. In humans, however, the placenta produces sufficient estrogen and progesterone from maternal and fetal precursors to take over the function of the corpus luteum after the sixth week of pregnancy. Ovariectomy before the sixth week leads to abortion, but ovariectomy thereafter has no effect on the pregnancy. The function of the corpus luteum begins to decline after 8 weeks of pregnancy, but it persists throughout pregnancy. hCG secretion decreases after an initial marked rise, but estrogen and progesterone secretion increase until just before parturition (Table 23–9).

hCG

hCG is a glycoprotein that contains galactose and hexosamine. It is produced by the syncytiotrophoblast. Like the pituitary glycoprotein hormones, it is made up of α and β subunits. hCG-α is identical to the α subunit of LH, FSH, and TSH. The molecular weight of hCG-α is 18,000, and that of hCG-β is 28,000. hCG is primarily luteinizing and luteotropic and has little FSH activity. It can be measured by radioimmunoassay and detected in the blood as early as 6 days after conception. Its presence in the urine in early pregnancy is the basis of the various laboratory tests for pregnancy, and it can sometimes be detected in the urine as early

Table 23–9. Hormone levels in human maternal blood during normal pregnancy.

Hormone	Approximate Peak Value	Time of Peak Secretion
hCG	5 mg/mL	First trimester
Relaxin	1 ng/mL	First trimester
hCS	15 mg/mL	Term
Estradiol	16 ng/mL	Term
Estriol	14 ng/mL	Term
Progesterone	190 ng/mL	Term
Prolactin	200 ng/mL	Term

as 14 days after conception. It appears to act on the same receptor as LH.

hCG is not absolutely specific for pregnancy. Small amounts are secreted by a variety of gastrointestinal and other tumors in both sexes, and hCG has been measured in individuals with suspected tumors as a "tumor marker." It also appears that the fetal liver and kidney normally produce small amounts of hCG.

hCS

The syncytiotrophoblast also secretes large amounts of a protein hormone that is lactogenic and has a small amount of growth-stimulating activity. This hormone has been called **chorionic growth hormone-prolactin (CGP)** and **human placental lactogen (hPL),** but it is now generally called **human chorionic somatomammotropin (hCS).** The structure of hCS is very similar to that of human growth hormone (see Figure 22–3), and it appears that these two hormones and prolactin evolved from a common progenitor hormone. Large quantities of hCS are found in maternal blood, but very little reaches the fetus. Secretion of growth hormone from the maternal pituitary is not increased during pregnancy and may actually be decreased by hCS. However, hCS has most of the actions of growth hormone and apparently functions as a "maternal growth hormone of pregnancy" to bring about the nitrogen, potassium, and calcium retention, lipolysis, and decreased glucose utilization seen in this state. These latter two actions divert glucose to the fetus. The amount of hCS secreted is proportionate to the size of the placenta, which normally weighs about one-sixth as much as the fetus, and low hCS levels are a sign of placental insufficiency.

Other Placental Hormones

In addition to hCG, hCS, progesterone, and estrogens, the placenta secretes other hormones. Human placental fragments probably produce POMC. In culture, they release CRH, β-endorphin, α-MSH, and dynorphin A, all of which appear to be identical to their hypothalamic counterparts. They also secrete GnRH and inhibin, and since GnRH stimulates and inhibin inhibits hCG secretion, locally produced GnRH and inhibin may act in a paracrine fashion to regulate hCG secretion. The trophoblast cells and amnion cells also secrete leptin (see Chapter 14), and moderate amounts of this satiety hormone enter the maternal circulation. Some also enters the amniotic fluid. Its function in pregnancy is unknown. The placenta also secretes prolactin in a number of variant forms.

Finally, the placenta secretes the α subunits of hCG, and the plasma concentration of free α subunits rises throughout pregnancy. These α subunits acquire a carbohydrate composition that makes them unable to combine with β subunits, and their prominence suggests that they have a function of their own. It is interesting in this regard that the secretion of the prolactin produced by the endometrium also appears to increase throughout pregnancy, and it may be that the circulating α subunits stimulate endometrial prolactin secretion.

The cytotrophoblast of the human chorion contains prorenin (see Chapter 24). There are large amounts of prorenin in amniotic fluid, but its function in this location is unknown.

Fetoplacental Unit

The fetus and the placenta interact in the formation of steroid hormones. The placenta synthesizes pregnenolone and progesterone from cholesterol. Some of the progesterone enters the fetal circulation and provides the substrate for the formation of cortisol and corticosterone in the fetal adrenal glands (Figure 23–37). Some of the pregnenolone enters the fetus and, along with pregnenolone synthesized in the fetal liver, is the substrate for the formation of dehydroepiandrosterone sulfate (DHEAS) and 16-hydroxydehydroepiandrosterone sulfate (16-OHDHEAS) in the fetal adrenal. Some 16-hydroxylation also occurs in the fetal liver. DHEAS and 16-OHDHEAS are transported back to the placenta, where DHEAS forms estradiol and 16-OHDHEAS forms estriol. The principal estrogen formed is estriol, and since fetal 16-OHDHEAS is the principal substrate for the estrogens, the urinary estriol excretion of the mother can be monitored as an index of the state of the fetus.

Parturition

The duration of pregnancy in humans averages 270 days from fertilization (284 days from the first day of

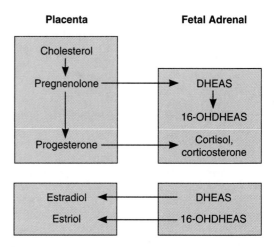

Figure 23–37. Interactions between the placenta and the fetal adrenal cortex in the production of steroids.

the menstrual period preceding conception). Irregular uterine contractions increase in frequency in the last month of pregnancy.

The difference between the body of the uterus and the cervix becomes evident at the time of delivery. The cervix, which is firm in the nonpregnant state and throughout pregnancy until near the time of delivery, softens and dilates, while the body of the uterus contracts and expels the fetus.

There is still considerable uncertainty about the mechanisms responsible for the onset of labor. One factor is the increase in circulating estrogens produced by increased circulating DHEAS. This makes the uterus more excitable, increases the number of gap junctions between myometrial cells, and causes production of more prostaglandins, which in turn cause uterine contractions. Progesterone has a quieting effect on the uterus, and in nonprimate mammals the circulating progesterone level also drops near term. However, this drop does not occur in primates. In humans, CRH secretion by the fetal hypothalamus increases and is supplemented by increased placental production of CRH. This increases circulating ACTH in the fetus, and the resulting increase in cortisol hastens the maturation of the respiratory system. Thus, in a sense, the fetus picks the time to be born by increasing CRH secretion.

The number of oxytocin receptors in the myometrium and the decidua (the endometrium of pregnancy) increases more than 100-fold during pregnancy and reaches a peak during early labor. Estrogens increase the number of oxytocin receptors, and uterine distention late in pregnancy may also increase their for-

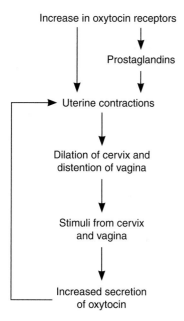

Figure 23–38. Role of oxytocin in parturition.

mation. In early labor, the oxytocin concentration in maternal plasma is not elevated from the prelabor value of about 25 pg/mL. It is possible that the marked increase in oxytocin receptors causes the uterus to respond to normal plasma oxytocin concentrations. However, at least in rats, the amount of oxytocin mRNA in the uterus increases, reaching a peak at term; this suggests that locally produced oxytocin also participates in the process.

Once labor is started, the uterine contractions dilate the cervix, and this dilation in turn sets up signals in afferent nerves that increase oxytocin secretion (Figure 23–38). The plasma oxytocin level rises, and more oxytocin becomes available to act on the uterus. Thus, a positive feedback loop is established that aids delivery and terminates on expulsion of the products of conception. Oxytocin increases uterine contractions in two ways: (1) It acts directly on uterine smooth muscle cells to make them contract, and (2) it stimulates the formation of prostaglandins in the decidua. The prostaglandins enhance the oxytocin-induced contractions.

During labor, spinal reflexes and voluntary contractions of the abdominal muscles ("bearing down") also aid in delivery. However, delivery can occur without bearing down and without a reflex increase in secretion of oxytocin from the posterior pituitary gland, since paraplegic women can go into labor and deliver.

LACTATION

Development of the Breasts

Many hormones are necessary for full mammary development. In general, estrogens are primarily responsible for proliferation of the mammary ducts and progesterone for the development of the lobules. In rats, some prolactin is also needed for development of the glands at puberty, but it is not known if prolactin is necessary in humans. In hypophysectomized rats, glucocorticoids, insulin, and growth hormone are necessary for mammary development in response to other hormones, but they do not by themselves cause growth of the breasts (Figure 23–39). During pregnancy, prolactin levels increase steadily until term, and under the influence of this hormone plus the high levels of estrogens and progesterone, full lobuloalveolar development of the breasts takes place.

Secretion & Ejection of Milk

The composition of human and cows' milk is shown in Table 23–10. In estrogen- and progesterone-primed rodents, injections of prolactin cause the formation of milk droplets and their secretion into the ducts. Oxytocin causes contraction of the myoepithelial cells lining the duct walls, with consequent ejection of the milk through the nipple (Figure 23–39). The reflex release of oxytocin initiated by touching the nipples and areolas (milk ejection reflex) is discussed in Chapter 14. Oxytocin is not essential for milk ejection in some species, but it is in humans.

The other hormonal relations in humans are generally similar to those in rats, although normal breast growth and lactation can occur in dwarfs with congenital growth hormone deficiency.

The transfer of antibodies to the infant by colostrum is discussed in Chapter 26.

Initiation of Lactation After Delivery

The breasts enlarge during pregnancy in response to high circulating levels of estrogens, progesterone, prolactin, and possibly hCG. Some milk is secreted into the ducts as early as the fifth month, but the amounts are small compared with the surge of milk secretion that follows delivery. In most animals, milk is secreted within an hour after delivery, but in women it takes 1–3 days for the milk to "come in."

After expulsion of the placenta at parturition, there is an abrupt decline in circulating estrogens and progesterone. The drop in circulating estrogen initiates lactation. Prolactin and estrogen synergize in producing

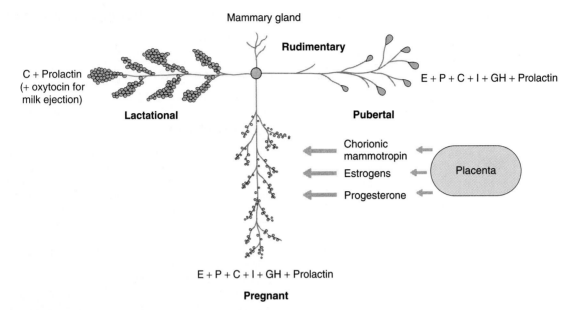

Figure 23–39. Hormonal control of breast development and lactation in rats. Estrogens (E) plus some progesterone (P) and some prolactin in the presence of glucocorticoids (C), insulin (I), and growth hormone (GH) cause duct proliferation and growth at puberty **(right)**. During pregnancy, all of these hormones bring about full alveolar development and some milk secretion **(below)**. After delivery, increased secretion of prolactin and a decline in estrogen and progesterone levels bring about copious milk secretion and, in the presence of oxytocin, ejection of milk **(left)**. Chorionic mammotropin is the lactogenic hormone presumably secreted by the placenta in rats and is analogous to hCS. It supplements the action of prolactin.

breast growth, but estrogen antagonizes the milk-producing effect of prolactin on the breast. Indeed, in women who do not wish to nurse their babies, estrogens may be administered to stop lactation.

Suckling not only evokes reflex oxytocin release and milk ejection; it also maintains and augments the secretion of milk because of the stimulation of prolactin secretion produced by suckling (see above).

Effect of Lactation on Menstrual Cycles

Women who do not nurse their infants usually have their first menstrual period 6 weeks after delivery. However, women who nurse regularly have amenorrhea for 25–30 weeks. Nursing stimulates prolactin secretion, and there is evidence that prolactin inhibits GnRH secretion, inhibits the action of GnRH on the pituitary, and antagonizes the action of gonadotropins on the ovaries. Ovulation is inhibited, and the ovaries are inactive, so estrogen and progesterone output falls to low levels. Consequently, only 5–10% of women become pregnant again during the suckling period, and nursing has long been known to be an important if only partly effective method of birth control. Further-

more, almost 50% of the cycles in the first 6 months after resumption of menses are anovulatory.

Chiari-Frommel Syndrome

An interesting although rare condition is persistence of lactation **(galactorrhea)** and amenorrhea in women who do not nurse after delivery. This condition, called the **Chiari-Frommel syndrome,** may be associated with some genital atrophy and is due to persistent prolactin secretion without the secretion of the FSH and LH necessary to produce maturation of new follicles and ovulation. A similar pattern of galactorrhea and amenorrhea with high circulating prolactin levels is seen in nonpregnant women with chromophobe pituitary tumors and in women in whom the pituitary stalk has been sectioned in treatment of cancer.

Gynecomastia

Breast development in the male is called **gynecomastia.** It may be unilateral but is more commonly bilateral. It is common, occurring in about 75% of newborns because of transplacental passage of maternal estrogens. It

Table 23–10. Composition of colostrum and milk.[1] (Units are weight per deciliter.)

Component	Human Colostrum	Human Milk	Cows' Milk
Water, g	...	88	88
Lactose, g	5.3	6.8	5.0
Protein, g	2.7	1.2	3.3
Casein:lactalbumin ratio	...	1:2	3:1
Fat, g	2.9	3.8	3.7
Linoleic acid	...	8.3% of fat	1.6% of fat
Sodium, mg	92	15	58
Potassium, mg	55	55	138
Chloride, mg	117	43	103
Calcium, mg	31	33	125
Magnesium, mg	4	4	12
Phosphorus, mg	14	15	100
Iron, mg	0.09[2]	0.15[2]	0.10[2]
Vit A, μg	89	53	34
Vit D, μg	...	0.03[2]	0.06[2]
Thiamine, μg	15	16	42
Riboflavin, μg	30	43	157
Nicotinic acid, μg	75	172	85
Ascorbic acid, mg	4.4[2]	4.3[2]	1.6[2]

[1]Reproduced, with permission, from Findlay ALR: Lactation. Res Reprod (Nov) 1974;6(6).
[2]Poor source.

also occurs in mild, transient form in 70% of normal boys at the time of puberty and in many men over the age of 50. It occurs in androgen resistance. It is a complication of estrogen therapy and is seen in patients with estrogen-secreting tumors. It is found in a wide variety of seemingly unrelated conditions, including eunuchoidism, hyperthyroidism, and cirrhosis of the liver. Digitalis can produce it, apparently because cardiac glycosides are weakly estrogenic. It can also be caused by many other drugs. It has been seen in malnourished prisoners of war, but only after they were liberated and eating an adequate diet. A feature common to many and perhaps all cases of gynecomastia is an increase in the plasma estrogen:androgen ratio due to either increased circulating estrogens or decreased circulating androgens.

Hormones & Cancer

About 35% of carcinomas of the breast in women of childbearing age are **estrogen-dependent;** their continued growth depends upon the presence of estrogens in the circulation. The tumors are not cured by decreasing estrogen secretion, but symptoms are dramatically relieved, and the tumor regresses for months or years before recurring. Women with estrogen-dependent tumors often have a remission when their ovaries are removed. The incidence of a favorable response is greater when the tumor contains estrogen receptors and greatest when the tumor contains both estrogen and progesterone receptors, because estrogen stimulates the formation of progesterone receptors and their presence indicates that estrogen is not only binding to but acting on the tumor cells. However, a few women with neither type of receptor still respond to this type of endocrine therapy. When the disease recurs, another remission follows bilateral adrenalectomy. Since ovarian and adrenal estrogen secretion are both inhibited by hypophysectomy, this operation has been performed in cancer patients. There also is some evidence that growth hormone and prolactin stimulate the growth of breast carcinomas, and hypophysectomy removes these stimuli.

Some carcinomas of the prostate are **androgen-dependent** and regress temporarily after the removal of the testes or treatment with GnRH agonists in doses that are sufficient to produce down-regulation of the GnRH receptors on gonadotropes and decrease LH secretion. The formation of pituitary tumors after removal of the target endocrine glands controlled by pituitary tropic hormones is discussed in Chapter 22.

Endocrine Functions of the Kidneys, Heart, & Pineal Gland

24

INTRODUCTION

The organs with endocrine functions include numerous structures in addition to the posterior, intermediate, and anterior lobes of the pituitary; the thyroid; the parathyroids; the pancreas; the adrenal cortex; the adrenal medulla; and the gonads. Hormones that stimulate or inhibit the secretion of anterior pituitary hormones are secreted by the hypothalamus (see Chapter 14), and a number of hormones are secreted by the mucosa of the gastrointestinal tract (see Chapter 26). Many different cells produce cytokines, interleukins, and growth factors (see Chapters 1, 22, and 27). The kidneys produce three hormones: 1,25-dihydroxycholecalciferol (see Chapter 21), renin, and erythropoietin. Natriuretic peptides, substances secreted by the heart and other tissues, increase excretion of sodium by the kidneys, and an additional natriuretic hormone inhibits Na^+-K^+ ATPase. The pineal gland secretes melatonin, and this indole may have an endocrine function. The endocrine functions of the kidneys, heart, and pineal gland are considered in this chapter.

THE RENIN-ANGIOTENSIN SYSTEM

Renin

The rise in blood pressure produced by injection of kidney extracts is due to **renin,** an acid protease secreted by the kidneys into the bloodstream. This enzyme acts in concert with angiotensin-converting enzyme to form angiotensin II (Figure 24–1). It is a glycoprotein with a molecular weight of 37,326 in humans. The molecule is made up of two lobes, or domains, between which the active site of the enzyme is located in a deep cleft. Two aspartic acid residues, one at position 104 and one at position 292 (residue numbers from human preprorenin), are juxtaposed in the cleft and are essential for activity. Thus, renin is an aspartyl protease.

Like other hormones, renin is synthesized as a large preprohormone. Human **preprorenin** contains 406 amino acid residues. The **prorenin** that remains after removal of a leader sequence of 23 amino acid residues from the amino terminal contains 383 amino acid

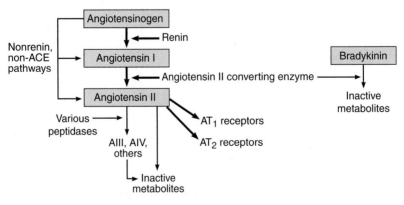

Figure 24–1. Formation and metabolism of circulating angiotensins.

residues, and after removal of the pro sequence from the amino terminal of prorenin, active **renin** contains 340 amino acid residues. Prorenin has little if any biologic activity.

Some prorenin is converted to renin in the kidneys, and some is secreted. Prorenin is secreted by other organs, including the ovaries. After nephrectomy, the prorenin level in the circulation is usually only moderately reduced and may actually rise, but the active-renin level falls to essentially zero. Thus, there is very little conversion of prorenin to renin in the circulation, and active renin is a product primarily if not exclusively of the kidneys. Prorenin is secreted constitutively, whereas active renin is formed in the secretory granules of the juxtaglomerular cells, the cells in the kidneys that produce renin (see below). Active renin has a half-life in the circulation of 80 minutes or less. Its only known function is to split the decapeptide **angiotensin I** from the amino terminal end of **angiotensinogen (renin substrate).**

Angiotensinogen

Circulating angiotensinogen is found in the α_2-globulin fraction of the plasma (Figure 24–1). It contains about 13% carbohydrate and is made up of 453 amino acid residues. It is synthesized in the liver with a 32-amino-acid signal sequence that is removed in the endoplasmic reticulum. Its circulating level is increased by glucocorticoids, thyroid hormones, estrogens, several cytokines, and angiotensin II.

Angiotensin-Converting Enzyme & Angiotensin II

Angiotensin-converting enzyme (ACE) is a dipeptidyl-carboxypeptidase that splits off histidyl-leucine from the physiologically inactive angiotensin I, forming the octapeptide **angiotensin II** (Figure 24–2). The same enzyme inactivates bradykinin (see Chapter 31). Increased tissue bradykinin produced when ACE is inhibited acts on B_2 receptors to produce the cough that is an annoying side effect in up to 20% of patients treated with ACE inhibitors. Most of the converting enzyme that forms angiotensin II in the circulation is located in endothelial cells. Much of the conversion occurs as the blood passes through the lungs, but there is also conversion in many other parts of the body.

ACE is an ectoenzyme that exists in two forms, a **somatic** form found throughout the body and a **germinal** form found solely in postmeiotic spermatogenic cells and spermatozoa (see Chapter 23). Both ACEs have a single transmembrane domain and a short cytoplasmic tail. However, somatic ACE is a 170-kDa protein with two homologous extracellular domains, each containing an active site (Figure 24–3). Germinal ACE is a 90-kDa protein that has only one extracellular domain and active site. Both enzymes are formed from a single gene. However, the gene has two different promoters, producing two different mRNAs. In male mice in which the ACE gene has been knocked out, blood pressure is lower than normal, but in females it is nor-

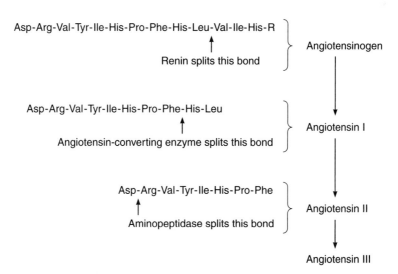

Figure 24–2. Structure of the amino terminal end of angiotensinogen and angiotensins I, II, and III in humans. R, remainder of protein. After removal of a 24-amino-acid leader sequence, angiotensinogen contains 453 amino acid residues. The structure of angiotensin II in dogs, rats, and many other mammals is the same as that in humans. Bovine and ovine angiotensin II have valine instead of isoleucine at position 5.

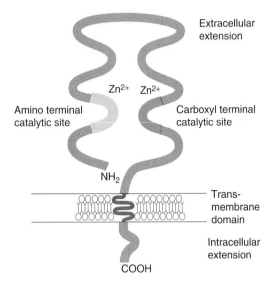

Figure 24–3. Diagrammatic representation of the structure of the somatic form of angiotensin-converting enzyme. Note the short cytoplasmic tail of the molecule and the two extracellular catalytic sites, each of which binds a zinc ion (Zn^{2+}). (Reproduced, with permission, from Johnston CI: Tissue angiotensin-converting enzyme in cardiac and vascular hypertrophy, repair, and remodeling. Hypertension 1994;23:258. Copyright © 1994 by The American Heart Association.)

mal. In addition, fertility is reduced in males but not in females.

Metabolism of Angiotensin II

Angiotensin II is metabolized rapidly, its half-life in the circulation in humans being 1–2 minutes. It is metabolized by various peptidases. An aminopeptidase removes the Asp residue from the amino terminal of the peptide. The resulting heptapeptide has physiologic activity and is sometimes called **angiotensin III** (see below). Removal of a second amino terminal residue from angiotensin III produces the hexapeptide sometimes called angiotensin IV, which is also said to have some activity. Most, if not all, of the other peptide fragments that are formed are inactive. In addition, aminopeptidase can act on angiotensin I to produce (des-Asp[1]) angiotensin I, and this compound can be converted directly to angiotensin III by the action of ACE. Angiotensin-metabolizing activity is found in red blood cells and many tissues. In addition, angiotensin II appears to be removed from the circulation by some sort of trapping mechanism in the vascular beds of tissues other than the lungs.

Renin is usually measured by incubating the sample to be assayed and measuring by immunoassay the amount of angiotensin I generated. This measures the **plasma renin activity (PRA)** of the sample. Deficiency of angiotensinogen as well as renin can cause low PRA values, and to avoid this problem, exogenous angiotensinogen is often added, so that **plasma renin concentration (PRC)** rather than PRA is measured. The normal PRA in supine subjects eating a normal amount of sodium is approximately 1 ng of angiotensin I generated per mL per hour. The plasma angiotensin II concentration in such subjects is about 25 pg/mL (approximately 25 pmol/L).

Actions of Angiotensins

Angiotensin I appears to function solely as the precursor of angiotensin II and does not have any other established action.

Angiotensin II—previously called hypertensin or angiotonin—produces arteriolar constriction and a rise in systolic and diastolic blood pressure. It is one of the most potent vasoconstrictors known, being four to eight times as active as norepinephrine on a weight basis in normal individuals. However, its pressor activity is decreased in sodium-depleted individuals and in patients with cirrhosis and some other diseases. In these conditions, circulating angiotensin II is increased, and this down regulates the angiotensin receptors in vascular smooth muscle. Consequently, there is less response to injected angiotensin II.

Angiotensin II also acts directly on the adrenal cortex to increase the secretion of aldosterone, and the renin-angiotensin system is a major regulator of aldosterone secretion (see Chapter 20). Additional actions of angiotensin II include facilitation of the release of norepinephrine by a direct action on postganglionic sympathetic neurons, contraction of mesangial cells with a resultant decrease in glomerular filtration rate (see Chapter 38), and a direct effect on the renal tubules to increase Na^+ reabsorption.

Angiotensin II also acts on the brain to decrease the sensitivity of the baroreflex (see Chapter 31), and this potentiates the pressor effect of angiotensin II. In addition, it acts on the brain to increase water intake (see Chapter 14), and increase the secretion of vasopressin and ACTH. It does not penetrate the blood-brain barrier, but it triggers these responses by acting on the circumventricular organs, four small structures in the brain that are outside the blood-brain barrier (see Chapter 32). One of these structures, the area postrema, is primarily responsible for the pressor potentiation, whereas two of the others, the subfornical organ (SFO) and the organum vasculosum of the lamina terminalis (OVLT), are responsible for the increase in water intake (dipsogenic effect). It is not certain which of the circumventricular organs are responsible for the increases in vasopressin and ACTH secretion.

Angiotensin III [(des-Asp^1) angiotensin II] has about 40% of the pressor activity of angiotensin II but 100% of the aldosterone-stimulating activity. It has been suggested that angiotensin III is the natural aldosterone-stimulating peptide whereas angiotensin II is the blood-pressure-regulating peptide. However, this appears not to be the case, and instead angiotensin III is simply a breakdown product with some biologic activity. The same is probably true of angiotensin IV, though some investigators have argued that it has unique effects in the brain.

Tissue Renin-Angiotensin Systems

In addition to the system that generates circulating angiotensin II, many different tissues contain independent renin-angiotensin systems that generate angiotensin II, apparently for local use. Components of the renin-angiotensin system are found in the walls of blood vessels and in the uterus, the placenta, and the fetal membranes. There is a high concentration of prorenin in amniotic fluid. In addition, there are tissue renin-angiotensin systems, or at least several components of the renin-angiotensin system, in the eyes, exocrine portion of the pancreas, heart, fat, adrenal cortex, testis, ovary, anterior and intermediate lobes of the pituitary, pineal, and brain. Tissue renin contributes very little to the circulating renin pool, since plasma renin activity falls to undetectable levels after the kidneys are removed. The functions of these tissue renin-angiotensin systems are unsettled, though evidence is accumulating that angiotensin II is a significant growth factor in the heart and blood vessels. As noted in Chapter 33, ACE inhibitors or AT$_1$ receptor blockers are now the treatment of choice for congestive heart failure, and part of their value may be due to inhibition of the growth effects of angiotensin II.

Angiotensin II Receptors

There are at least two classes of angiotensin II receptors (Figure 24–1). AT$_1$ receptors are serpentine receptors coupled by a G protein (G$_q$) to phospholipase C, and angiotensin II increases the cytosolic free Ca^{2+} level. It also activates numerous tyrosine kinases. In vascular smooth muscle, AT$_1$ receptors are associated with caveolae (see Chapter 1), and AII increases production of caveolin-1, one of the three isoforms of the protein that is characteristic of caveolae. In rodents, there are two different but closely related AT$_1$ subtypes, AT$_{1A}$ and AT$_{1B}$, coded by two separate genes. The AT$_{1A}$ subtype

is found in blood vessel walls, the brain, and many other organs. It mediates most of the known effects of angiotensin II. The AT$_{1B}$ subtype is found in the anterior pituitary and the adrenal cortex. In humans, there is an AT$_1$ receptor gene on chromosome 3. There may be a second AT$_1$ type, but it is still unsettled whether there are distinct AT$_{1A}$ and AT$_{1B}$ subtypes.

There are also AT$_2$ receptors, which are coded in humans by a gene on the X chromosome. Like the AT$_1$ receptors, they have seven transmembrane domains, but their actions are different. They act via a G protein to activate various phosphatases which in turn antagonize growth effects and open K$^+$ channels. In addition, AT$_2$ receptor activation increases the production of NO and therefore increases intracellular cGMP. The overall physiologic consequences of these second-messenger effects are unsettled. AT$_2$ receptors are more plentiful in fetal and neonatal life, but they persist in the brain and other organs in adults.

The AT$_1$ receptors in the arterioles and the AT$_1$ receptors in the adrenal cortex are regulated in opposite ways: an excess of angiotensin II down regulates the vascular receptors, but it up regulates the adrenocortical receptors, making the gland more sensitive to the aldosterone-stimulating effect of the peptide.

The Juxtaglomerular Apparatus

The renin in kidney extracts and the bloodstream is produced by the **juxtaglomerular cells (JG cells).** These epithelioid cells are located in the media of the afferent arterioles as they enter the glomeruli (Figure 24–4). The membrane-lined secretory granules in them have been shown to contain renin. Renin is also found in agranular **lacis cells** that are located in the junction between the afferent and efferent arterioles, but its significance in this location is unknown.

At the point where the afferent arteriole enters the glomerulus and the efferent arteriole leaves it, the tubule of the nephron touches the arterioles of the glomerulus from which it arose. At this point, which marks the start of the distal convolution, there is a modified region of tubular epithelium called the macula densa (Figure 24–4). The macula densa is in close proximity to the JG cells. The lacis cells, the JG cells, and the macula densa constitute the **juxtaglomerular apparatus.**

Regulation of Renin Secretion

Several different factors regulate renin secretion (Table 24–1), and the rate of renin secretion at any given time is determined by the summed activity of these factors. One factor is an intrarenal baroreceptor mechanism that causes renin secretion to decrease

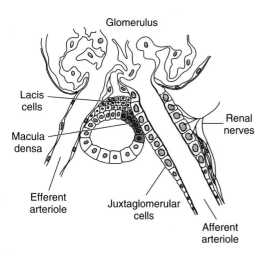

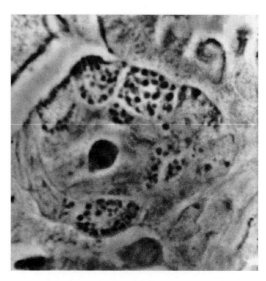

Figure 24–4. **Left:** Diagram of glomerulus, showing the juxtaglomerular apparatus. **Right:** Phase contrast photomicrograph of afferent arteriole in an unstained, freeze-dried preparation of the kidney of a mouse. Note the red blood cell in the lumen of the arteriole and the granulated juxtaglomerular cells in the wall. (Courtesy of C Peil.)

when arteriolar pressure at the level of the JG cells increases and to increase when arteriolar pressure at this level falls. Another renin-regulating sensor is in the macula densa. Renin secretion is inversely proportionate to the rate of transport of Na^+ and Cl^- across this portion of the tubule. The rate of transport depends not only on the mechanisms in the macula densa cells but also on the amount of electrolyte reaching the macula densa. Therefore, decreased delivery of Na^+ and Cl^- to the distal tubules is associated with increased renin secretion. Prostaglandins, especially prostacyclin (see Chapter 17), stimulate renin secretion, apparently by a direct action on the JG cells. There is some evidence that the effects of the macula densa on renin secretion are mediated by NO (see Chapter 31). Renin secretion also varies inversely with the plasma K^+ level, but the effect of K^+ appears to be

mediated by the changes it produces in Na^+ and Cl^- delivery to the macula densa.

Angiotensin II feeds back to inhibit renin secretion by a direct action on the JG cells. Vasopressin also inhibits renin secretion in vitro and in vivo, although there is some debate about whether its in vivo effect is direct or indirect.

Finally, increased activity of the sympathetic nervous system increases renin secretion. The increase is mediated both by increased circulating catecholamines and by norepinephrine secreted by postganglionic renal sympathetic nerves. The catecholamines act mainly on β_1-adrenergic receptors on the JG cells and the increases in renin are mediated by increased intracellular cAMP.

The principal conditions that are associated with increased renin secretion in humans are listed in Table 24–2. Most of them decrease central venous pressure, and this triggers increased sympathetic activity as well as a potential decrease in renal arteriolar pressure. Renal artery constriction and constriction of the aorta proximal to the renal arteries produce a decrease in renal arteriolar pressure. Psychologic stimuli increase the activity of the renal nerves.

Table 24–1. Factors that affect renin secretion.

Stimulatory

  Increased sympathetic activity via renal nerves
  Increased circulating catecholamines
  Prostaglandins

Inhibitory

  Increased Na^+ and Cl^- reabsorption across macula densa
  Increased afferent arteriolar pressure
  Angiotensin II
  Vasopressin

Pharmacologic Manipulation of the Renin-Angiotensin System

It is now possible to inhibit the secretion or the effects of renin in a variety of ways. Inhibitors of prostaglandin synthesis such as **indomethacin** and β-adrenergic

Table 24–2. Conditions that increase renin secretion.

Sodium depletion
Diuretics
Hypotension
Hemorrhage
Upright posture
Dehydration
Cardiac failure
Cirrhosis
Constriction of renal artery or aorta
Various psychologic stimuli

blocking drugs such as **propranolol** reduce renin secretion. The peptide **pepstatin** and newly developed renin inhibitors such as **enalkiren** prevent renin from generating angiotensin I. Angiotensin-converting enzyme inhibitors (ACE inhibitors) such as **captopril** and **enalapril** prevent conversion of angiotensin I to angiotensin II. **Saralasin** and several other analogs of angiotensin II are competitive inhibitors of the action of angiotensin II on both AT_1 and AT_2 receptors. **Losartan** (DuP-753) selectively blocks AT_1 receptors, and PD-123177 and several other drugs selectively block AT_2 receptors.

Role of Renin in Clinical Hypertension

Constriction of one renal artery causes a prompt increase in renin secretion and the development of sustained hypertension (**renal** or **Goldblatt hypertension**). Removal of the ischemic kidney or the arterial constriction cures the hypertension if it has not persisted too long. In general, the hypertension produced by constricting one renal artery with the other kidney intact (one-clip, two-kidney Goldblatt hypertension; see Table 33–5) is associated with increased circulating renin. The clinical counterpart of this condition is **renal hypertension** due to atheromatous narrowing of one renal artery or other abnormalities of the renal circulation. However, plasma renin activity is usually normal in one-clip one-kidney Goldblatt hypertension. The explanation of the hypertension in this situation is unsettled. However, many patients with hypertension respond to treatment with ACE inhibitors or losartan even when their renal circulation appears to be normal and they have normal or even low plasma renin activity.

The role of renin in a feedback mechanism that helps maintain the constancy of ECF volume through regulation of aldosterone secretion has been described in Chapter 20.

ERYTHROPOIETIN

Structure & Function

When an individual is bled or becomes hypoxic, hemoglobin synthesis is enhanced, and production and release of red blood cells from the bone marrow (**erythropoiesis**) are increased (see Chapter 27). Conversely, when the red cell volume is increased above normal by transfusion, the erythropoietic activity of the bone marrow decreases. These adjustments are brought about by changes in the circulating level of **erythropoietin,** a circulating glycoprotein that contains 165 amino acid residues and four oligosaccharide chains that are necessary for its activity in vivo. Its blood level is markedly increased in anemia (Figure 24–5).

Erythropoietin increases the number of erythropoietin-sensitive committed stem cells in the bone marrow that are converted to red blood cell precursors and subsequently to mature erythrocytes (see Figure 27–2). The receptor for erythropoietin is a linear protein with a single transmembrane domain that is a member of the cytokine receptor superfamily (see Chapter 1). The receptor has tyrosine kinase activity, and it activates a cascade of serine and threonine kinases, resulting in growth and development of its target cells. When erythropoietin levels are low, erythroid stem cells show DNA cleavage followed by programmed cell death

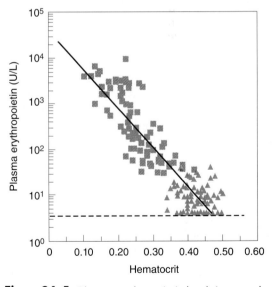

Figure 24–5. Plasma erythropoietin levels in normal blood donors (triangles) and patients with various forms of anemia (squares). (Reproduced, with permission, from Erslev AJ: Erythropoietin. N Engl J Med 1991;324:1339.)

(apoptosis). Apoptosis of various types of cells is now known to be part of normal development in many different tissues. In the erythroid stem cells, erythropoietin reduces the DNA cleavage and causes the cells to survive.

The principal site of inactivation of erythropoietin is the liver, and the hormone has a half-life in the circulation of about 5 hours. However, the increase in circulating red cells that it triggers takes 2–3 days to appear, since red cell maturation is a relatively slow process. Loss of even a small portion of the sialic acid residues in the carbohydrate moieties that are part of the erythropoietin molecule shortens its half-life to 5 minutes, making it biologically ineffective.

Sources

In adults, about 85% of the erythropoietin comes from the kidneys and 15% from the liver. Both these organs contain the mRNA for erythropoietin. Erythropoietin can also be extracted from the spleen and salivary glands, but these tissues do not contain the mRNA and consequently do not appear to manufacture the hormone. During fetal and neonatal life, the major site of erythropoiesis is the liver, and it is also the major site of erythropoietin production before erythropoiesis is taken over by the bone marrow and erythropoietin production by the kidneys. When renal mass is reduced in adults by renal disease or nephrectomy, the liver cannot compensate and anemia develops.

In adults, erythropoietin is produced by interstitial cells in the peritubular capillary bed of the kidneys and by perivenous hepatocytes in the liver. It is also produced in the brain, where it exerts a protective effect against excitotoxic damage triggered by hypoxia; and in the uterus and oviducts, where it is induced by estrogen and appears to mediate estrogen-dependent angiogenesis.

The gene for the hormone has been cloned, and recombinant erythropoietin produced in animal cells is available for clinical use as epoetin alfa. The recombinant erythropoietin is of value in the treatment of the anemia associated with renal failure; 90% of the patients with end-stage renal failure who are on dialysis are anemic as a result of erythropoietin deficiency. Erythropoietin is also used to stimulate red cell production in individuals who are banking a supply of their own blood in preparation for autologous transfusions during elective surgery (see Chapter 27).

Regulation of Secretion

The usual stimulus for erythropoietin secretion is hypoxia, but secretion of the hormone can also be stimulated by cobalt salts and androgens. Recent evidence suggests that the O_2 sensor regulating erythropoietin secretion in the kidneys and the liver is a heme protein that in the dioxy form stimulates and in the oxy form inhibits transcription of the erythropoietin gene to form erythropoietin mRNA. Secretion of the hormone is facilitated by the alkalosis that develops at high altitudes. Like renin secretion, erythropoietin secretion is facilitated by catecholamines via a β-adrenergic mechanism, although the renin-angiotensin system is totally separate from the erythropoietin system.

HORMONES OF THE HEART & OTHER NATRIURETIC FACTORS

Structure

The existence of various **natriuretic hormones** has been postulated for some time. Two of these are secreted by the heart. The muscle cells in the atria and, to a much lesser extent, in the ventricles contain secretory granules (Figure 24–6) that increase in number when sodium chloride intake is increased and extracellular fluid expanded, and extracts of atrial tissue cause natriuresis.

The first natriuretic hormone isolated from the heart was **atrial natriuretic peptide (ANP),** a polypeptide with a characteristic 17-amino-acid ring formed by a disulfide bond between two cysteines. The circulating form of this polypeptide has 28 amino acid residues (Figure 24–7). It is formed from a large precursor molecule that contains 151 amino acid residues, including a 24-amino-acid signal peptide. ANP was subsequently isolated from other tissues, including the brain, where

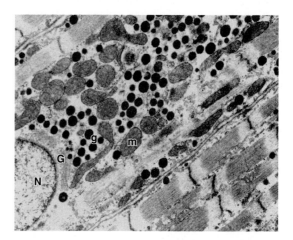

Figure 24–6. ANP granules (g) interspersed between mitochondria (m) in rat atrial muscle cell. G, Golgi complex; N, nucleus. The granules in human atrial cells are similar. × 17,640. (Courtesy of M Cantin.)

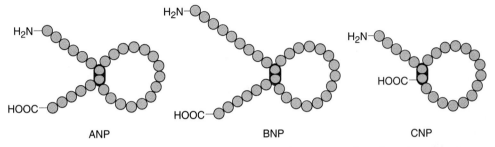

Figure 24–7. Human ANP, BNP, and CNP. **Top:** Single-letter codes for amino acid residues aligned to show common sequences (colored). **Bottom:** Shape of molecules. Note that one cysteine is the carboxyl terminal amino acid residue in CNP, so there is no carboxyl terminal extension from the 17-member ring. (Modified from Imura H, Nakao K, Itoh H: The natriuretic peptide system in the brain: Implication in the central control of cardiovascular and neuroendocrine functions. Front Neuroendocrinol 1992;13:217.)

it exists in two forms that are smaller than circulating ANP. A second natriuretic polypeptide was isolated from porcine brain and named **brain natriuretic peptide (BNP).** It is also present in the brain in humans, but more is present in the human heart, including the ventricles. The circulating form of this hormone contains 32 amino acid residues. It has the same 17-member ring as ANP, though some of the amino acid residues in the ring are different (Figure 24–7). A third member of this family has been named **C-type natriuretic peptide (CNP)** because it was the third in the sequence to be isolated. It contains 22 amino acid residues (Figure 24–7), and there is also a larger 53-amino-acid form. CNP is present in the brain, the pituitary, the kidneys, and vascular endothelial cells. However, there is very little in the heart and the circulation, and it appears to be primarily a paracrine mediator.

Actions

ANP and BNP in the circulation act on the kidneys to increase Na+ excretion, and injected CNP has a similar effect. They appear to produce this effect by dilating afferent arterioles and relaxing mesangial cells. Both of these actions increase glomerular filtration (see Chapter 38). In addition, they act on the renal tubules to inhibit Na+ reabsorption. Other actions include an increase in capillary permeability, leading to extravasation

of fluid and a decline in blood pressure. In addition, they relax vascular smooth muscle in arterioles and venules. CNP has a greater dilator effect on veins than ANP and BNP. These peptides also inhibit renin secretion and counteract the pressor effects of catecholamines and angiotensin II.

In the brain, ANP is present in neurons, and an ANP-containing neural pathway projects from the anteromedial part of the hypothalamus to the areas in the lower brain stem that are concerned with neural regulation of the cardiovascular system. In general, the effects of ANP in the brain are opposite to those of angiotensin II, and ANP-containing neural circuits appear to be involved in lowering blood pressure and promoting natriuresis. CNP and BNP in the brain probably have functions similar to those of ANP, but detailed information is not available.

Natriuretic Peptide Receptors

Three different natriuretic peptide receptors (NPR) have been isolated and characterized (Figure 24–8). The NPR-A and NPR-B receptors both span the cell membrane and have cytoplasmic domains that are guanylyl cyclases. ANP has the greatest affinity for the NPR-A receptor, and CNP has the greatest affinity for the NPR-B receptor. The third receptor, NPR-C, binds all three natriuretic peptides but has a markedly truncated cytoplasmic domain. There is some evidence that

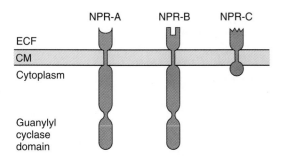

Figure 24–8. Diagrammatic representation of natriuretic peptide receptors. The NPR-A and NPR-B receptor molecules have intracellular guanylyl cyclase domains, whereas the clearance receptor, NPR-C, has only a small cytoplasmic domain. CM, cell membrane.

it acts via G proteins to activate phospholipase C and inhibit adenylyl cyclase. However, it has also been argued that this receptor does not trigger any intracellular change and is instead a **clearance receptor** which removes natriuretic peptides from the bloodstream and then releases them later, helping to maintain a steady blood level of the hormones.

Secretion & Metabolism

The concentration of ANP in plasma is about 5 fmol/mL in normal humans ingesting moderate amounts of sodium. ANP secretion is increased when the ECF volume is increased by infusion of isotonic saline or ingestion of a high-sodium diet. It is also increased by immersion in water up to the neck (Figure 24–9), a procedure that counteracts the effect of gravity on the circulation and increases central venous and consequently atrial pressure. Note that immersion also decreases the secretion of renin and aldosterone. Conversely, there is a small but measurable decrease in plasma ANP in association with a decrease in central venous pressure on rising from the supine to the standing position. Strips of atrial muscle release ANP when stretched in vitro. Thus, it appears likely that the atria respond directly to stretch in vivo and that the rate of ANP secretion is proportionate to the degree to which the atria are stretched by increases in central venous pressure.

Circulating ANP has a short half-life. It is metabolized by neutral endopeptidase (NEP), which is inhibited by thiorphan. Therefore, administration of thiorphan increases circulating ANP.

Less is known about the cardiac secretion of BNP and its metabolism, but they are presumably similar to those of ANP.

Na$^+$-K$^+$ ATPase-Inhibiting Factor

There is another natriuretic factor in blood. This factor produces natriuresis by inhibiting Na$^+$-K$^+$ ATPase and raises rather than lowers blood pressure. Current evidence indicates that it is the digitalis-like steroid **ouabain** and that it comes from the adrenal glands. However, its physiologic significance is not yet known.

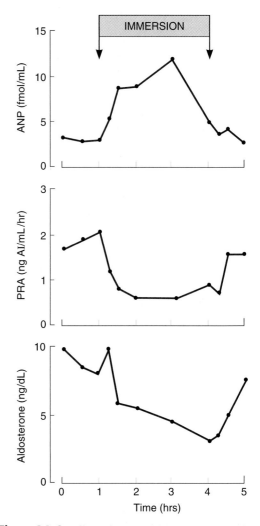

Figure 24–9. Effect of immersion in water up to the neck for 3 hours on plasma concentrations of ANP, PRA, and aldosterone. (Modified and reproduced, with permission, from Epstein M et al: Increases in circulating atrial natriuretic factor during immersion-induced central hypervolaemia in normal humans. Hypertension 1986;4[Suppl 2]:593.)

PINEAL GLAND

The **pineal gland** (epiphysis), believed by Descartes to be the seat of the soul, has at one time or another been regarded as having a wide variety of functions. It is now known to secrete melatonin, and it may function as a timing device to keep internal events synchronized with the light-dark cycle in the environment.

Anatomy

The pineal arises from the roof of the third ventricle under the posterior end of the corpus callosum and is connected by a stalk to the posterior commissure and habenular commissure. There are nerve fibers in the stalk, but they apparently do not reach the gland. The pineal stroma contains neuroglia and parenchymal cells with features suggesting that they have a secretory function (Figure 24–10). Like other endocrine glands, the pineal has highly permeable fenestrated capillaries. In young animals and infants, the pineal is large, and the cells tend to be arranged in alveoli. It begins to involute before puberty, and, in humans, small concretions of calcium phosphate and carbonate (**pineal sand**) appear in the tissue. Because the concretions are radiopaque, the normal pineal is often visible on x-ray films of the skull in adults. Displacement of a calcified pineal from its normal position indicates the presence of a space-occupying lesion such as a tumor in the brain.

Melatonin

The amphibian pineal contains an indole, *N*-acetyl-5-methoxytryptamine, named **melatonin** because it lightens the skin of tadpoles by an action on melanophores. However, it does not appear to play a physiologic role in the regulation of skin color, and it is present in mammals. Melatonin and the enzymes responsible for its synthesis from serotonin by N-acetylation and O-methylation (Figure 24–11) are present in humans. Melatonin is synthesized by pineal parenchymal cells and secreted by them into the blood and the cerebrospinal fluid. It is also synthesized in other organs.

Two melatonin-binding sites have been characterized: a high-affinity ML1 site and a low affinity ML2 site. Two subtypes of the ML1 receptor have been cloned: Mel 1a and Mel 1b. All the receptors are coupled to G proteins, with ML1 receptors inhibiting adenylyl cyclase and ML2 receptors stimulating phosphoinositide hydrolysis. However, the functions of each remain to be determined.

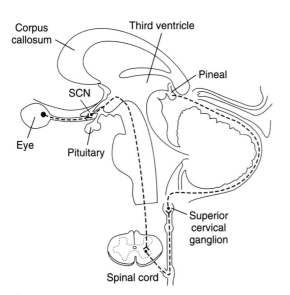

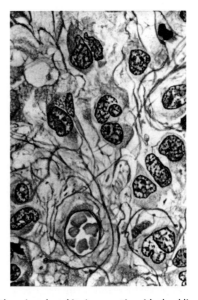

Figure 24–10. **Left:** Sagittal section of human brain stem showing the pineal and its innervation (dashed lines). Retino-hypothalamic fibers synapse in the suprachiasmatic nuclei (SCN), and there are connections from the SCN to the intermediolateral gray column in the spinal cord. Preganglionic neurons pass from the spinal cord to the superior cervical ganglion, and the postganglionic neurons project from this ganglion to the pineal in the nervi conarii. **Right:** Histology of pineal. Drawing of hematoxylin-and-eosin-stained section. (Reproduced, with permission, from Fawcett DW: Bloom and Fawcett, *A Textbook of Histology,* 11th ed. Saunders, 1986.)

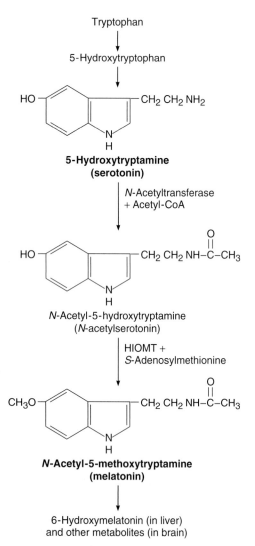

Figure 24–11. Formation and metabolism of melatonin. HIOMT, hydroxyindole-O-methyltransferase. For details of the synthesis and metabolism of serotonin, see Figure 4–23.

Regulation of Secretion

In humans and all other species studied to date, melatonin synthesis and secretion are increased during the dark period of the day and maintained at a low level during the daylight hours (Figure 24–12). This remarkable diurnal variation in secretion is brought about by norepinephrine secreted by the postganglionic sympathetic nerves (nervi conarii) that innervate the pineal (Figure 24–10). The norepinephrine acts via β-adrenergic receptors in the pineal to increase intracellular

cAMP, and the cAMP in turn produces a marked increase in N-acetyltransferase activity. This results in increased melatonin synthesis and secretion.

The discharge of the sympathetic nerves to the pineal is entrained to the light–dark cycle in the environment via the retinohypothalamic nerve fibers to the suprachiasmatic nuclei. The way these bring about entrainment of circadian rhythms is discussed in Chapter 14. From the hypothalamus, descending pathways converge on the intermediolateral gray column of the thoracic spinal cord and end on the preganglionic sympathetic neurons that in turn innervate the superior cervical ganglion, the site of origin of the postganglionic neurons to the pineal.

Circulating melatonin is rapidly metabolized in the liver by 6-hydroxylation followed by conjugation, and over 90% of the melatonin that appears in the urine is in the form of 6-hydroxy conjugates and 6-sulfatoxymelatonin. The pathway by which the brain metabolizes melatonin is unsettled but may involve cleavage of the indole nucleus.

Function of the Pineal

Injected melatonin has some effects on the gonads, but these effects vary markedly from species to species and also depend on the time of injection. In some situations, melatonin inhibits gonadal function, and in others, its effect is facilitatory. This variability has led to the hypothesis that it is not the melatonin per se but the diurnal change in melatonin secretion that functions as some sort of timing signal which coordinates internal events with the light–dark cycle in the environ-

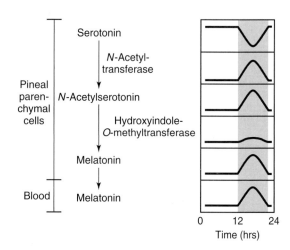

Figure 24–12. Diurnal rhythms of various compounds in the pineal and melatonin in blood. The shaded area represents the hours of darkness during the 24-hour day.

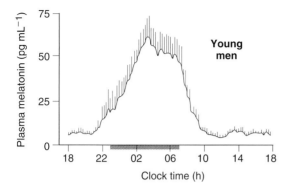

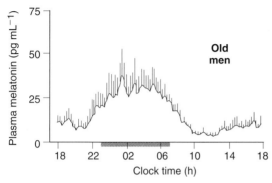

Figure 24-13. Daily plasma melatonin values (means ± SE) in men aged 20–27 years and men aged 67–84 years. The black bar on the horizontal axis indicates time in bed. (Reproduced, with permission, from Turek F: Melatonin hype hard to swallow. Nature 1996;379:295. Copyright © 1996 by Macmillan Magazines Ltd.)

ment. Evidence supporting this timing function of melatonin is the observation that in blind people with free-running circadian rhythms (see Chapter 14), melatonin injections entrain the rhythms.

It has been argued that the pineal normally inhibits the onset of puberty in humans, because pineal tumors are sometimes associated with sexual precocity. However, as noted in Chapter 23, it appears that pineal tumors produce precocity only when they produce hypothalamic damage. Nocturnal plasma melatonin concentrations are much higher in children than adults, and they decline with age. In children 1–3 years of age, they average about 250 pg/mL (1080 pmol/L); in adolescents 8–15 years of age, they average about 120 pg/mL; in young men 20–27 years of age, they average about 70 pg/mL; and in old men 67–84 years of age they average about 30 pg/mL (Figure 24–13). However, the decline is gradual throughout life, with no abrupt change at puberty, and daytime plasma melatonin concentrations average about 7 pg/mL at all ages.

REFERENCES FOR SECTION IV: ENDOCRINOLOGY, METABOLISM, & REPRODUCTIVE FUNCTION

Adashi EY, Hennebold JD: Single gene mutations resulting in reproductive dysfunction in women. N Engl J Med 1999;340:709.

Adrogue HJ, Madias NE: Hypernatremia. N Engl J Med 2000;342:1493.

Adrogue HJ, Madias NE: Hyponatremia. N Engl J Med 2000;342:101.

Bole-Feysot C et al: Prolactin (PRL) and its receptor: Actions, signal transduction pathways, and phenotypes observed in PRL receptor knockout mice. Endocr Rev 1998;19:225.

Bouillon R: The many faces of rickets. N Engl J Med 1998;338:681.

Braddock M et al: Born again bone: tissue engineering for bone repair. News Physiol Sci 2001;16:208.

Brzezinski A: Melatonin in humans. N Engl J Med 1997;336:186.

Corvol P, Jeunemaitre X: Molecular genetics of human hypertension: Role of angiotensinogen. Endocr Rev 1997;18:662.

de la Vieja A et al: Molecular aspects of the sodium-iodide symporter: impact on thyroid and extrathyroid pathophysiology. Physiol Rev 2000;80:1083.

Diaz MN et al: Antioxidants and atherosclerotic heart disease. N Engl J Med 1997;337:408.

Dunne MI, Aynsley-Green A, Lindley KI: Nature's KATP-channel knockout. News Physiol Sci 1997;12:197.

Eastell R: Drug therapy: Treatment of postmenopausal osteoporosis. N Engl J Med 1998;338:736.

Falkner F, Tanner JM (editors): *Human Growth,* 2nd ed. 3 vols. Plenum, 1986.

Fisher JW: Erythropoietin: Physiologic and pharmacologic aspects. Proc Soc Exper Biol Med 1997;216:358.

FitzGerald GA, Patrono CP: The coxibs, selective inhibitors of cyclooxygenase-2. N Engl J Med 2001;345:433.

Francomano CA: Clinical implications of basic research: The genetic basis of dwarfism. N Engl J Med 1995;332:58.

Gehlert DR: Multiple receptors for the pancreatic polypeptide (PP-fold) family: physiological implications. Proc Soc Exper Biol Med 1998;218:7.

Goldstein JL, Brown MS: The cholesterol quartet. Science 2001;292:1510.

Goodman HM (editor): *Handbook of Physiology,* Section 7: *The Endocrine System.* Oxford Univ Press, 2000.

Gregoire FM: Adipocyte differentiation: from fibroblast to endocrine cell. Exp Biol Med 2001;226:997.

Gruber CJ et al: Production and actions of estrogens. N Engl J Med 2002;346:340.

Inagami T: A memorial to Robert Tiegerstedt: The centennial of renin discovery. Hypertension 1998;32:953.

Kiberghs P, Smith O, Norman C (editors): Bone health in the balance. Science 2001;289:1457.

Kim S, Iwao H: Molecular and cellular mechanisms of angiotensin II-mediated cardiovascular and renal diseases. Pharmacol Rev 2000;52:11.

Kjos SL, Buchanan TA: Gestational diabetes mellitus. N Engl J Med 1999;341:1749.

Klein I, Qjamaa K: Thyroid hormone and the cardiovascular system. N Engl J Med 2001;344:501.

Knopp RH: Drug treatment of lipid disorders. N Engl J Med 1999;341:498.

Komers R, Vrana A: Thiazolidinediones—tools for the research of metabolic syndrome X. Physiol Res 1998;47:215.

Kuiper GGJM et al: The estrogen receptor β: a novel mediator of estrogen action in neuroendocrine systems. Front Neuroendocrinol 1998;19:253.

Lazar MA: Recent progress in understanding thyroid hormone action. Thyroid Today 1997 (Oct);20:1.

LeRoith D: Insulin-like growth factors. N Engl J Med 1997;336:633.

Levin ER, Gardner DG, Samson WK: Natriuretic peptides. N Engl J Med 1998;339.

Marcus R, Hoffman AR: Growth hormone as treatment in older men and women. Annu Rev Pharmacol Toxicol 1998;38:45.

Mather JP, Moore A, Li R-H: Activins, inhibins, and follistatins: Further thoughts on a growing family of regulators. Proc Soc Exper Biol Med 1997;215:209.

Melmed S (editor): *The Pituitary.* Blackwell, 1994.

Mendelsohn ME, Kavas RH: The protective effects of estrogen on the cardiovascular system. N Engl J Med 1999;340:1801.

Money J: *The Kaspar Hauser Syndrome of "Psychosocial Dwarfism": Deficient Structural, Intellectual, and Social Growth Induced by Child Abuse.* Prometheus Books, 1994.

Naz RK (editor): *Endocrine Disruptors.* CRC Press, 1998.

Norwitz ER, Robinson JN, Challis JRG: The control of labor. N Engl J Med 1999;341:660.

Pool R *Fat: Fighting the Fat Epidemic.* Oxford Univ Press, 2001.

Primakoff P, Nyles DG: Penetration, adhesion, and fusion in mammalian sperm-egg interaction. Science 2002;296:2183.

Refetoff S, Weiss RE, Usala SJ: The syndromes of resistance to thyroid hormone. Endocr Rev 1993;14:348.

Reppert SM, Weaver DR: Coordination of circadian timing in mammals. Nature 2002;418:935.

Rolfe DFS, Brown GC: Cellular energy utilization and molecular origin of standard metabolic rate. Physiol Rev 1997;77:731.

Rosenbaum M, Leibel RL, Hirsch J: Obesity. N Engl J Med 1997;337:396.

Sabaté J (editor): *Vegetarian Nutrition.* CRC Press, 2001.

Sebastian S, Bullum SK: Complex organization of the regulating region of the human CYP 19 (aromatase) gene revealed by the human genome project. J Clin Endocrinol Metab 2001;86:4000.

Stocco DM: A review of the characteristics of the protein required for the acute regulation of steroid hormone biosynthesis: the case for the steroidogenic acute regulatory (StAR) protein. Proc Soc Exp Biol Med 1998;217:123.

Strewler GL: The physiology of parathyroid hormone-related protein. N Engl J Med 2000;342:177.

Strugnell SA, Deluca HF: The vitamin D receptor-structure and transcriptional activation. Proc Soc Exper Biol Med 1997;215:223.

Surridge C, Narn D (editors): Diabetes. Nature 2001;414:781.

Veenstra TD, Pittleikow MR, Kumar R: Regulation of cellular growth by 1,25-dihydroxy vitamin D_3-mediated growth factor expression. News Physiol Sci 1999;14:37.

Volpé R: *Autoimmune Endocrinopathies.* Humana Press, 1999.

Waterman MR, Bischof LJ: Diversity of ACTH (cAMP)-dependent transcription of bovine steroid hydroxylase genes. FASEB J 1997;11:419.

Weetman AP: Graves' disease. N Engl J Med 2000;343:1236.

Wehling M: Specific, nongenomic actions of steroid hormones. Annu Rev Physiol 1997;59:365.

Welch GN: Mechanisms of disease: Homocysteine and atherothrombosis. N Engl J Med 1998;338:1042.

White PC: Disorders of aldosterone biosynthesis and action. N Engl J Med 1994;331:250.

Willett WC, Dietz WH, Colditz GA: Guidelines for healthy weight. N Engl J Med 1999;341:427.

Wilson JD et al (editors): *Williams Textbook of Endocrinology,* 9th ed. Saunders, 1998.

Yanovski SZ, Yanovski LA: Drug therapy: obesity. N Engl J Med 2002;346:591.

Yen P: Physiological and molecular basis of thyroid hormone action. Physiol Rev 2001;81;1097.

Yen SSC, Jaffe RB, Barbieri RL: *Reproductive Endocrinology: Physiology, Pathophysiology, and Clinical Management,* 4th ed. Saunders, 1999.

SECTION V
Gastrointestinal Function

Digestion & Absorption

INTRODUCTION

The gastrointestinal system is the portal through which nutritive substances, vitamins, minerals, and fluids enter the body. Proteins, fats, and complex carbohydrates are broken down into absorbable units **(digested)**, principally in the small intestine. The products of digestion and the vitamins, minerals, and water cross the mucosa and enter the lymph or the blood **(absorption).** The digestive and absorptive processes are the subject of this chapter. The details of the functions of the various parts of the gastrointestinal system are considered in Chapter 26.

Digestion of the major foodstuffs is an orderly process involving the action of a large number of **digestive enzymes** (Table 25–1). Enzymes from the salivary and lingual glands attack carbohydrates and fats; enzymes from the stomach attack proteins and fats; and enzymes from the exocrine portion of the pancreas attack carbohydrates, proteins, lipids, DNA, and RNA. Other enzymes that complete the digestive process are found in the luminal membranes and the cytoplasm of the cells that line the small intestine. The action of the enzymes is aided by the hydrochloric acid secreted by the stomach and the bile secreted by the liver.

The mucosal cells in the small intestine are called **enterocytes.** In the small intestine they have a **brush border** made up of numerous microvilli lining their apical surface (see Figure 26–28). This border is rich in enzymes. It is lined on its luminal side by a layer that is rich in neutral and amino sugars, the **glycocalyx.** The membranes of the mucosal cells contain glycoprotein enzymes that hydrolyze carbohydrates and peptides, and the glycocalyx is made up in part of the carbohydrate portions of these glycoproteins that extend into the intestinal lumen. Next to the brush border and glycocalyx is an **unstirred layer** similar to the layer adjacent to other biologic membranes (see Chapter 1). Solutes must diffuse across this layer to reach the mucosal cells. The mucous coat overlying the cells also constitutes a significant barrier to diffusion.

Most substances pass from the intestinal lumen into the enterocytes and then out of the enterocytes to the interstitial fluid. The processes responsible for movement across the luminal cell membrane are often quite different from those responsible for movement across the basal and lateral cell membranes to the interstitial fluid. The dynamics of transport in all parts of the body are considered in Chapter 1.

CARBOHYDRATES

Digestion

The principal dietary carbohydrates are polysaccharides, disaccharides, and monosaccharides. Starches (glucose polymers) and their derivatives are the only polysaccharides that are digested to any degree in the human gastrointestinal tract. In glycogen, the glucose molecules are mostly in long chains (glucose molecules in 1:4α linkage), but there is some chain-branching (produced by 1:6α linkages; see Figure 17–12). Amylopectin, which constitutes 80–90% of dietary starch, is similar but less branched, whereas amylose is a straight chain with only 1:4α linkages. Glycogen is found in animals, whereas amylose and amylopectin are of plant origin. The disaccharides

Table 25–1. Principal digestive enzymes. The corresponding proenzymes are shown in parentheses.

Source	Enzyme	Activator	Substrate	Catalytic Function or Products
Salivary glands	Salivary α-amylase	Cl⁻	Starch	Hydrolyzes 1:4α linkages, producing α-limit dextrins, maltotriose, and maltose
Lingual glands	Lingual lipase		Triglycerides	Fatty acids plus 1,2-diacylglycerols
Stomach	Pepsins (pepsinogens)	HCl	Proteins and polypeptides	Cleave peptide bonds adjacent to aromatic amino acids
	Gastric lipase		Triglycerides	Fatty acids and glycerol
Exocrine pancreas	Trypsin (trypsinogen)	Enteropeptidase	Proteins and polypeptides	Cleave peptide bonds on carboxyl side of basic amino acids (arginine or lysine)
	Chymotrypsins (chymotrypsinogens)	Trypsin	Proteins and polypeptides	Cleaves peptide bonds on carboxyl side of aromatic amino acids
	Elastase (proelastase)	Trypsin	Elastin, some other proteins	Cleaves bonds on carboxyl side of aliphatic amino acids
	Carboxypeptidase A (procarboxypeptidase A)	Trypsin	Proteins and polypeptides	Cleaves carboxyl terminal amino acids that have aromatic or branched aliphatic side chains
	Carboxypeptidase B (procarboxypeptidase B)	Trypsin	Proteins and polypeptides	Cleaves carboxyl terminal amino acids that have basic side chains
	Colipase (procolipase)	Trypsin	Fat droplets	Facilitates exposure of active site of pancreatic lipase
	Pancreatic lipase	...	Triglycerides	Monoglycerides and fatty acids
	Cholesteryl ester hydrolase	...	Cholesteryl esters	Cholesterol
	Pancreatic α-amylase	Cl⁻	Starch	Same as salivary α-amylase
	Ribonuclease	...	RNA	Nucleotides
	Deoxyribonuclease	...	DNA	Nucleotides
	Phospholipase A_2 (prophospholipase A_2)	Trypsin	Phospholipids	Fatty acids, lysophospholipids
Intestinal mucosa	Enteropeptidase	...	Trypsinogen	Trypsin
	Aminopeptidases	...	Polypeptides	Cleave amino terminal amino acid from peptide
	Carboxypeptidases	...	Polypeptides	Cleave carboxyl terminal amino acid from peptide
	Endopeptidases	...	Polypeptides	Cleave between residues in midportion of peptide
	Dipeptidases	...	Dipeptides	Two amino acids
	Maltase	...	Maltose, maltotriose, α-dextrins	Glucose

Table 25–1. Principal digestive enzymes. The corresponding proenzymes are shown in parentheses. *(continued)*

Source	Enzyme	Activator	Substrate	Catalytic Function or Products
Intestinal mucosa *(continued)*	Lactase	…	Lactose	Galactose and glucose
	Sucrase[1]	…	Sucrose; also maltotriose and maltose	Fructose and glucose
	α-Dextrinase[1]	…	α-Dextrins, maltose, maltotriose	Glucose
	Trehalase	…	Trehalose	Glucose
	Nuclease and related enzymes	…	Nucleic acids	Pentoses and purine and pyrimidine bases
Cytoplasm of mucosal cells	Various peptidases	…	Di-, tri-, and tetrapeptides	Amino acids

[1]Sucrase and α-dextrinase are separate subunits of a single protein.

lactose (milk sugar) and **sucrose** (table sugar) are also ingested, along with the monosaccharides fructose and glucose.

In the mouth, starch is attacked by salivary α-amylase. However, the optimal pH for this enzyme is 6.7, and its action is inhibited by the acidic gastric juice when food enters the stomach. In the small intestine, both the salivary and the pancreatic α-amylase also acts on the ingested polysaccharides. Both the salivary and the pancreatic α-amylases hydrolyze 1:4α linkages but spare 1:6α linkages, terminal 1:4α linkages, and the 1:4α linkages next to branching points. Consequently, the end products of α-amylase digestion are oligosaccharides: the disaccharide **maltose;** the trisaccharide **maltotriose;** some slightly larger polymers with glucose in 1:4α linkage; and α-**dextrins,** polymers of glucose containing an average of about eight glucose molecules with 1:6α linkages (Figure 25–1).

The oligosaccharidases responsible for the further digestion of the starch derivatives are located in the outer portion of the brush border, the membrane of the microvilli of the small intestine (Figure 25–2). Some of these enzymes have more than one substrate. α-**Dextrinase,** which is also known as **isomaltase,** is mainly responsible for hydrolysis of 1:6α linkages. Along with **maltase** and **sucrase,** it also breaks down maltotriose and maltose. Sucrase and α-dextrinase are initially synthesized as a single glycoprotein chain which is inserted into the brush border membrane. It is then hydrolyzed by pancreatic proteases into sucrase and isomaltase sub-

units, but the subunits reassociate noncovalently at the intestinal surface.

Sucrase hydrolyzes sucrose into a molecule of glucose and a molecule of fructose. In addition, there are two disaccharidases in the brush border: **lactase,** which hydrolyzes lactose to glucose and galactose, and **treha-**

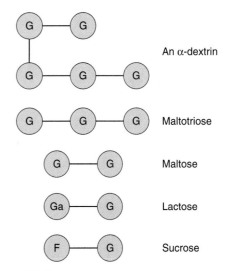

Figure 25–1. Principal end products of carbohydrate digestion in the intestinal lumen. Each circle represents a hexose molecule. G, glucose; F, fructose; Ga, galactose.

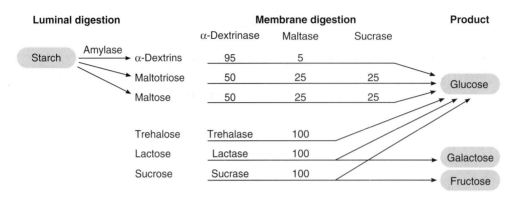

Figure 25–2. Substrate specificities of the enzymes involved in carbohydrate digestion, and the hexoses that are the final products. Numbers are percentages of each substrate cleaved by a particular enzyme. Note that trehalase, lactase, and sucrase are solely responsible for the breakdown of trehalose, lactose, and sucrose respectively, but that α-dextrins, maltotriose, and maltose are substrates for several enzymes. (Reproduced, with permission, from Johnson LR [editor]: *Essential Medical Physiology,* Raven, 1992.)

lase, which hydrolyzes trehalose, a 1:1α-linked dimer of glucose, into two glucose molecules.

Deficiency of one or more of the brush border oligosaccharidases may cause diarrhea, bloating, and flatulence after ingestion of sugar. The diarrhea is due to the increased number of osmotically active oligosaccharide molecules that remain in the intestinal lumen, causing the volume of the intestinal contents to increase. In the colon, bacteria break down some of the oligosaccharides, further increasing the number of osmotically active particles. The bloating and flatulence are due to the production of gas (CO_2 and H_2) from disaccharide residues in the lower small intestine and colon.

Lactase is of interest because, in most mammals and in many races of humans, intestinal lactase activity is high at birth, then declines to low levels during childhood and adulthood. The low lactase levels are associated with intolerance to milk **(lactose intolerance).** Most Europeans and their American descendants retain their intestinal lactase activity in adulthood; the incidence of lactase deficiency in northern and western Europeans is only about 15%. However, the incidence in blacks, American Indians, Orientals, and Mediterranean populations is 70–100%. Milk intolerance can be ameliorated by administration of commercial lactase preparations, but this is expensive. Yogurt is better tolerated than milk in intolerant individuals because it contains its own bacterial lactase.

Absorption

Hexoses and pentoses are rapidly absorbed across the wall of the small intestine (Table 25–2). Essentially all of the hexoses are removed before the remains of a meal

reach the terminal part of the ileum. The sugar molecules pass from the mucosal cells to the blood in the capillaries draining into the portal vein.

The transport of most hexoses is uniquely affected by the amount of Na^+ in the intestinal lumen; a high concentration of Na^+ on the mucosal surface of the cells facilitates and a low concentration inhibits sugar influx into the epithelial cells. This is because glucose and Na^+ share the same **cotransporter,** or **symport,** the **sodium-dependent glucose transporter** (SGLT, Na^+-glucose cotransporter). The members of this family of transporters, SGLT 1 and SGLT 2, resemble the glucose transporters responsible for facilitated diffusion (see Chapter 19) in that they cross the cell membrane 12 times and have their −COOH and −NH$_2$ terminals on the cytoplasmic side of the membrane. However, there is no homology to the GLUT series of transporters. SGLT 1 and SGLT 2 are also responsible for glucose transport out of the renal tubules (see Chapter 38).

Since the intracellular Na^+ concentration is low in intestinal cells as it is in other cells, Na^+ moves into the cell along its concentration gradient. Glucose moves with the Na^+ and is released in the cell (Figure 25–3). The Na^+ is transported into the lateral intercellular spaces, and the glucose is transported by GLUT 2 into the interstitium and thence to the capillaries. Thus, glucose transport is an example of secondary active transport (see Chapter 1); the energy for glucose transport is provided indirectly, by the active transport of Na^+ out of the cell. This maintains the concentration gradient across the luminal border of the cell, so that more Na^+ and consequently more glucose enter. When the Na^+/glucose cotransporter is congenitally defective, the resulting **glucose/galactose malabsorption** causes severe diarrhea that is often fatal if glucose and galac-

Table 25–2. Normal transport of substances by the intestine and location of maximum absorption or secretion.[1]

Absorption of:	Small Intestine			Colon
	Upper[2]	Mid	Lower	
Sugars (glucose, galactose, etc)	++	+++	++	0
Amino acids	++	+++	++	0
Water-soluble and fat-soluble vitamins except vitamin B_{12}	+++	++	0	0
Betaine, dimethylglycine, sarcosine	+	++	++	?
Antibodies in newborns	+	++	+++	?
Pyrimidines (thymine and uracil)	+	+	?	?
Long-chain fatty acid absorption and conversion to triglyceride	+++	++	+	0
Bile salts	+	+	+++	
Vitamin B_{12}	0	+	+++	0
Na^+	+++	++	+++	+++
K^+	+	+	+	Sec
Ca^{2+}	+++	++	+	?
Fe^{2+}	+++	+	+	?
Cl^-	+++	++	+	+
SO_4^{2-}	++	+	0	?

[1]Amount of absorption is graded + to +++. Sec, secreted when luminal K^+ is low.

[2]Upper small intestine refers primarily to jejunum, although the duodenum is similar in most cases studied (with the notable exception that the duodenum secretes HCO_3^- and shows little net absorption or secretion of NaCl).

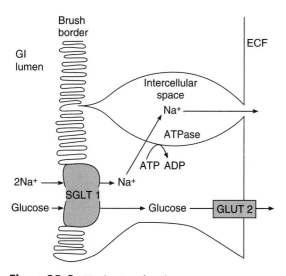

Figure 25–3. Mechanism for glucose transport across intestinal epithelium. Glucose transport into the intestinal cell is coupled to Na^+ transport, utilizing the cotransporter SGLT 1. Na^+ is then actively transported out of the cell, and glucose enters the interstitium by facilitated diffusion via GLUT 2. From there, it diffuses into the blood.

tose are not promptly removed from the diet. The use of glucose and its polymers to retain Na^+ in diarrheal disease is discussed below.

The glucose mechanism also transports galactose. Fructose utilizes a different mechanism. Its absorption is independent of Na^+ or the transport of glucose and galactose; it is transported instead by facilitated diffusion from the intestinal lumen into the enterocytes by GLUT 5 and out of the enterocytes into the interstitium by GLUT 2. Some fructose is converted to glucose in the mucosal cells. Pentoses are absorbed by simple diffusion.

Insulin has little effect on intestinal transport of sugars. In this respect, intestinal absorption resembles glucose reabsorption in the proximal convoluted tubules of the kidneys (see Chapter 38); neither process requires phosphorylation, and both are essentially normal in diabetes but are depressed by the drug phlorhizin. The maximal rate of glucose absorption from the intestine is about 120 g/h.

PROTEINS & NUCLEIC ACIDS

Protein Digestion

Protein digestion begins in the stomach, where pepsins cleave some of the peptide linkages. Like many of the other enzymes concerned with protein digestion, pepsins are secreted in the form of inactive precursors (**proenzymes**) and activated in the gastrointestinal tract. The pepsin precursors are called pepsinogens and are activated by gastric hydrochloric acid. Human gastric mucosa contains a number of related pepsinogens, which can be divided into two immunohistochemically distinct groups, pepsinogen I and pepsinogen II. Pepsinogen I is found only in acid-secreting regions, whereas pepsinogen II is also found in the pyloric region. Maximal acid secretion correlates with pepsinogen I levels.

Pepsins hydrolyze the bonds between aromatic amino acids such as phenylalanine or tyrosine and a second amino acid, so the products of peptic digestion are polypeptides of very diverse sizes. A **gelatinase** that liquefies gelatin is also found in the stomach. **Chymosin,** a milk-clotting gastric enzyme also known as **rennin,** is found in the stomachs of young animals but is probably absent in humans.

Because pepsins have a pH optimum of 1.6–3.2, their action is terminated when the gastric contents are mixed with the alkaline pancreatic juice in the duodenum and jejunum. The pH of the intestinal contents in the duodenal cap is 2.0–4.0, but in the rest of the duodenum it is about 6.5.

In the small intestine, the polypeptides formed by digestion in the stomach are further digested by the powerful proteolytic enzymes of the pancreas and intestinal mucosa. Trypsin, the chymotrypsins, and elastase act at interior peptide bonds in the peptide molecules and are called **endopeptidases.** The formation of the active endopeptidases from their inactive precursors is discussed in Chapter 26. The carboxypeptidases of the pancreas are **exopeptidases** that hydrolyze the amino acids at the carboxyl and amino ends of the polypeptides. Some free amino acids are liberated in the intestinal lumen, but others are liberated at the cell surface by the aminopeptidases, carboxypeptidases, endopeptidases, and dipeptidases in the brush border of the mucosal cells. Some di- and tripeptides are actively transported into the intestinal cells and hydrolyzed by intracellular peptidases, with the amino acids entering the bloodstream. Thus, the final digestion to amino acids occurs in three locations: the intestinal lumen, the brush border, and the cytoplasm of the mucosal cells.

Absorption

At least seven different transport systems transport amino acids into enterocytes. Five of these require Na^+ and cotransport amino acids and Na^+ in a fashion similar to the cotransport of Na^+ and glucose (Figure 25–3). Two of these five also require Cl^-. In two systems, transport is independent of Na^+.

Di- and tripeptides are transported into enterocytes by a system that requires H^+ instead of Na^+. There is very little absorption of larger peptides. In the enterocytes, amino acids released from the peptides by intracellular hydrolysis plus the amino acids absorbed from the intestinal lumen and brush border are transported out of the enterocytes along their basolateral borders by at least five transport systems. From there, they enter the hepatic portal blood. Two of these systems are dependent on Na^+, and three are not. Significant amounts of small peptides also enter the portal blood.

Absorption of amino acids is rapid in the duodenum and jejunum but slow in the ileum. Approximately 50% of the digested protein comes from ingested food, 25% from proteins in digestive juices, and 25% from desquamated mucosal cells. Only 2–5% of the protein in the small intestine escapes digestion and absorption. Some of the ingested protein enters the colon and is eventually digested by bacterial action. The protein in the stools is not of dietary origin but comes from bacteria and cellular debris. There is evidence that the peptidase activities of the brush border and the mucosal cell cytoplasm are increased by resection of part of the ileum and that they are independently altered in starvation. Thus, these enzymes appear to be subject to homeostatic regulation. In humans, a congenital defect in the mechanism that transports neutral amino acids in the intestine and renal tubules causes **Hartnup disease.** A congenital defect in the transport of basic amino acids causes **cystinuria.**

In infants, moderate amounts of undigested proteins are also absorbed. The protein antibodies in maternal colostrum are largely secretory immunoglobulins (IgAs), the production of which is increased in the breast in late pregnancy. They cross the mammary epithelium by transcytosis and enter the circulation of the infant from the intestine, providing passive immunity against infections. Absorption is by endocytosis and subsequent exocytosis.

Protein absorption declines with age, but adults still absorb small quantities. Foreign proteins that enter the circulation provoke the formation of antibodies, and the antigen-antibody reaction occurring upon subsequent entry of more of the same protein may cause allergic symptoms. Thus, absorption of proteins from the intestine may explain the occurrence of allergic symptoms after eating certain foods. The incidence of food allergy in children is said to be as high as 8%. Certain foods are more allergenic than others. Crustaceans, mollusks, and fish are common offenders, and allergic responses to legumes, cows' milk, and egg white are also relatively frequent.

Absorption of protein antigens, particularly bacterial and viral proteins, takes place in large **microfold cells** or **M cells,** specialized intestinal epithelial cells that overlie aggregates of lymphoid tissue (Peyer's patches). These cells pass the antigens to the lymphoid cells, and lymphocytes are activated. The activated lymphoblasts enter the circulation, but they later return to the intestinal mucosa and other epithelia, where they secrete IgA in response to subsequent exposures to the same antigen. This **secretory immunity** is an important defense mechanism; it is discussed in more detail in Chapter 27.

Nucleic Acids

Nucleic acids are split into nucleotides in the intestine by the pancreatic nucleases, and the nucleotides are split into the nucleosides and phosphoric acid by enzymes that appear to be located on the luminal surfaces of the mucosal cells. The nucleosides are then split into their constituent sugars and purine and pyrimidine bases. The bases are absorbed by active transport.

LIPIDS

Fat Digestion

A lingual lipase is secreted by Ebner's glands on the dorsal surface of the tongue, and the stomach also secretes a lipase (Table 25–1). The gastric lipase is of little importance except in pancreatic insufficiency, but lingual lipase is active in the stomach and can digest as much as 30% of dietary triglyceride.

Most fat digestion begins in the duodenum, pancreatic lipase being one of the most important enzymes involved. This enzyme hydrolyzes the 1- and 3-bonds of the triglycerides (triacylglycerols) with relative ease but acts on the 2-bonds at a very low rate, so the principal products of its action are free fatty acids and 2-monoglycerides (2-monoacylglycerols). It acts on fats that have been emulsified. Its activity is facilitated when an amphipathic helix that covers the active site like a lid is bent back. **Colipase,** a protein with a molecular weight of about 11,000, is also secreted in the pancreatic juice, and when this molecule binds to the –COOH-terminal domain of the pancreatic lipase, opening of the lid is facilitated. Colipase is secreted in an inactive proform (Table 25–1) and is activated in the intestinal lumen by trypsin.

Another pancreatic lipase that is activated by bile salts has been characterized. This 100,000-kDa **bile salt-activated lipase** represents about 4% of the total protein in pancreatic juice. In adults, pancreatic lipase is 10–60 times more active, but unlike pancreatic lipase, bile salt-activated lipase catalyzes the hydrolysis of cholesterol esters, esters of fat-soluble vitamins, and phospholipids, as well as triglycerides. A very similar enzyme is found in human milk.

Most of the dietary cholesterol is in the form of cholesteryl esters, and cholesteryl ester hydrolase also hydrolyzes these esters in the intestinal lumen.

Fats are finely emulsified in the small intestine by the detergent action of bile salts, lecithin, and monoglycerides. When the concentration of bile salts in the intestine is high, as it is after contraction of the gallbladder, lipids and bile salts interact spontaneously to form **micelles** (Figure 25–4). These cylindrical aggregates take up lipids (see Chapter 26), and although their lipid concentration varies, they generally contain fatty acids, monoglycerides, and cholesterol in their hydrophobic centers. Micellar formation further solubilizes the lipids and provides a mechanism for their transport to the enterocytes. Thus, the micelles move down their concentration gradient through the unstirred layer to the brush border of the mucosal cells. The lipids diffuse out of the micelles, and a saturated aqueous solution of the lipids is maintained in contact with the brush border of the mucosal cells (Figure 25–4).

Steatorrhea

Pancreatectomized animals and patients with diseases that destroy the exocrine portion of the pancreas have

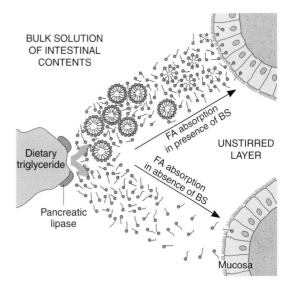

Figure 25–4. Lipid digestion and passage to intestinal mucosa. Fatty acids (FA) are liberated by the action of pancreatic lipase on dietary triglycerides and, in the presence of bile salts (BS), form micelles (the circular structures), which diffuse through the unstirred layer to the mucosal surface. (Reproduced, with permission, from Thomson ABR: Intestinal absorption of lipids: Influence of the unstirred water layer and bile acid micelle. In: *Disturbances in Lipid and Lipoprotein Metabolism.* Dietschy JM, Gotto AM Jr, Ontko JA [editors]. American Physiological Society, 1978.)

fatty, bulky, clay-colored stools (**steatorrhea**) because of the impaired digestion and absorption of fat. The steatorrhea is due mostly to the lipase deficiency—but in addition, in the absence of the bicarbonate that is secreted from the pancreas, the relatively acidic milieu in the duodenum precipitates some bile salts. Acid also inhibits pancreatic lipase, and this is why patients with excess secretion of gastric acid and a consequent low duodenal pH caused by a gastrin-secreting tumor (gastrinoma) may develop steatorrhea. Another cause of steatorrhea is defective reabsorption of bile salts in the distal ileum (see Chapter 26).

Fat Absorption

Traditionally, lipids were thought to enter the enterocytes by passive diffusion, but there is some evidence that carriers are involved. Inside the cells, the lipids are rapidly esterified, maintaining a favorable concentration gradient from the lumen into the cells (Figure 25–5).

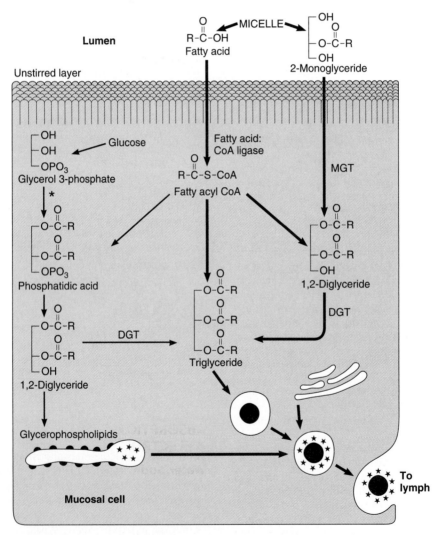

Figure 25–5. Lipid absorption. Triglycerides are formed in the mucosal cells from monoglycerides and fatty acids. Some of the glycerides also come from glucose via phosphatidic acid. The triglycerides are then converted to chylomicrons and released by exocytosis. From the extracellular space, they enter the lymph. Heavy arrows indicate major pathways. *, reaction inhibited by monoglyceride; MGT, monoacylglycerol acyltransferase; DGT, diacylglycerol acyltransferase.

Unlike the ileal mucosa, the rate of uptake of bile salts by the jejunal mucosa is low, and for the most part the bile salts remain in the intestinal lumen, where they are available for the formation of new micelles.

The fate of the fatty acids in enterocytes depends on their size. Fatty acids containing less than 10–12 carbon atoms pass from the mucosal cells directly into the portal blood, where they are transported as free (unesterified) fatty acids. The fatty acids containing more than 10–12 carbon atoms are reesterified to triglycerides in the mucosal cells. In addition, some of the absorbed cholesterol is esterified. The triglycerides and cholesteryl esters are then coated with a layer of protein, cholesterol, and phospholipid to form chylomicrons. These leave the cell and enter the lymphatics (Figure 25–5).

In mucosal cells, most of the triglyceride is formed by the acylation of the absorbed 2-monoglycerides, primarily in the smooth endoplasmic reticulum. However, some of the triglyceride is formed from glycerophosphate, which in turn is a product of glucose catabolism. Glycerophosphate is also converted into glycerophospholipids that participate in chylomicron formation. The acylation of glycerophosphate and the formation of lipoproteins occur in the rough endoplasmic reticulum. Carbohydrate moieties are added to the proteins in the Golgi apparatus, and the finished chylomicrons are extruded by exocytosis from the basal or lateral aspects of the cell.

Absorption of long-chain fatty acids is greatest in the upper parts of the small intestine, but appreciable amounts are also absorbed in the ileum (Figure 25–6). On a moderate fat intake, 95% or more of the ingested fat is absorbed. The processes involved in fat absorption are not fully mature at birth, and infants fail to absorb 10–15% of ingested fat. Thus, they are more susceptible to the ill effects of disease processes that reduce fat absorption.

Short-Chain Fatty Acids

Increasing attention is being focused on short-chain fatty acids (SCFAs) that are produced in the colon and absorbed from it. SCFAs are two- to five-carbon weak acids that have an average normal concentration of about 80 mmol/L in the lumen. About 60% of this total is acetate, 25% propionate, and 15% butyrate. They are formed by the action of colonic bacteria (see Chapter 26) on complex carbohydrates, resistant starches, and other components of the dietary fiber, ie, the material that escapes digestion in the upper gastrointestinal tract and enters the colon.

Absorbed SCFAs are metabolized and make a significant contribution to the total caloric intake. In addition, they exert a trophic effect on the colonic epithelial

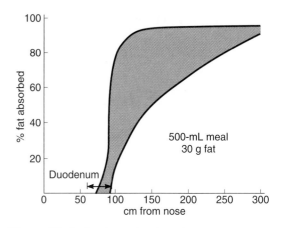

Figure 25–6. Fat absorption, based on measurement after a fat meal in humans. The double-headed arrow identifies the duodenum. (Redrawn and reproduced, with permission, from Davenport HW: *Physiology of the Digestive Tract,* 2nd ed. Year Book, 1966.)

cells, combat inflammation, and are absorbed in part by exchange for H^+, helping to maintain acid-base equilibrium. There are a family of anion exchangers in the colonic epithelial cells. SCFAs also promote the absorption of Na^+, although the exact mechanism for coupled Na^+-SCFA absorption is unsettled.

Absorption of Cholesterol & Other Sterols

Cholesterol is readily absorbed from the small intestine if bile, fatty acids, and pancreatic juice are present. Closely related sterols of plant origin are poorly absorbed. Almost all the absorbed cholesterol is incorporated into chylomicrons that enter the circulation via the lymphatics, as noted above. Nonabsorbable plant sterols such as those found in soybeans reduce the absorption of cholesterol, probably by competing with cholesterol for esterification with fatty acids.

ABSORPTION OF WATER & ELECTROLYTES

Water, Sodium, Potassium, & Chloride

Overall water balance in the gastrointestinal tract is summarized in Table 25–3. The intestines are presented each day with about 2000 mL of ingested fluid plus 7000 mL of secretions from the mucosa of the gastrointestinal tract and associated glands. Ninety-eight percent of this fluid is reabsorbed, with a daily fluid loss of only 200 mL in the stools. Only small amounts of water move across the gastric mucosa, but water moves in both directions across the mucosa of the small

Table 25–3. Daily water turnover (mL) in the gastrointestinal tract[1]

Ingested		2000
Endogenous secretions		7000
Salivary glands	1500	
Stomach	2500	
Bile	500	
Pancreas	1500	
Intestine	1000	
	7000	
Total Input		9000
Reabsorbed		8800
Jejunum	5500	
Ileum	2000	
Colon	1300	
	8800	
Balance in stool		200

[1]Data from Moore EW: *Physiology of Intestinal Water and Electrolyte Absorption,* American Gastroenterological Society, 1976.

and large intestines in response to osmotic gradients. Some Na+ diffuses into or out of the small intestine depending on the concentration gradient. Because the luminal membranes of all enterocytes in the small intestine and colon are permeable to Na+ and their basolateral membranes contain Na+-K+ ATPase, Na+ is also actively absorbed throughout the small and large intestines.

In the small intestine, secondary active transport of Na+ is important in bringing about absorption of glucose, some amino acids (see above), and other substances. Conversely, the presence of glucose in the intestinal lumen facilitates the reabsorption of Na+. This is the physiologic basis for the treatment of Na+ and water loss in diarrhea by oral administration of solutions containing NaCl and glucose. Cereals containing carbohydrates are also useful in the treatment of diarrhea. This type of treatment has even proved to be beneficial in the treatment of cholera, a disease associated with very severe and, if untreated, frequently fatal diarrhea.

Cl− normally enters enterocytes from the interstitial fluid via Na+-K+-2Cl− cotransporters in their basolateral membranes (Figure 25–7), and the Cl− is then secreted into the intestinal lumen via channels that are regulated by various protein kinases. One of these is activated by protein kinase A and hence by cAMP. The cAMP concentration is increased in cholera. The cholera vibrio stays in the intestinal lumen, but it pro-

duces a toxin which binds to GM-1 ganglioside receptors, and this permits part of the A subunit (A[1] peptide) of the toxin to enter the cell. The A[1] peptide binds adenosine diphosphate ribose to the α subunit of G[s], inhibiting its GTPase activity (see Chapter 1). Therefore, the constitutively activated G protein produces prolonged stimulation of adenylyl cyclase and a marked increase in the intracellular cAMP concentration. In addition to increased Cl− secretion, the function of the mucosal carrier for Na+ is reduced, thus reducing NaCl absorption. The resultant increase in electrolyte and water content of the intestinal contents causes the diarrhea. However, Na+-K+ ATPase and the Na+/glucose cotransporter are unaffected, so coupled reabsorption of glucose and Na+ bypasses the defect.

Water moves into or out of the intestine until the osmotic pressure of the intestinal contents equals that of the plasma. The osmolality of the duodenal contents may be hypertonic or hypotonic, depending on the meal ingested, but by the time the meal enters the jejunum, its osmolality is close to that of plasma. This osmolality is maintained throughout the rest of the small intestine; the osmotically active particles produced by digestion are removed by absorption, and water moves passively out of the gut along the osmotic gradient thus generated. In the colon, Na+ is pumped out and water moves passively with it, again along the osmotic gradient. **Saline cathartics** such as magnesium sulfate are poorly absorbed salts that retain their osmotic equivalent of water in the intestine, thus increas-

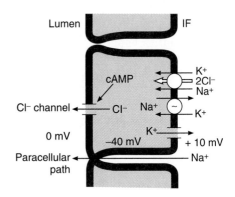

Figure 25–7. Movement of ions across enterocytes in the small intestine. Cl− enters the cell via the Na+-K+-2Cl− cotransporter on its basolateral surface and is secreted into the intestinal lumen via Cl− channels, some of which are activated by cyclic AMP. K+ recycles to the interstitial fluid (IF) via basolateral K+ channels. (Reproduced, with permission, from Field M, Roa MC, Chang EB: Intestinal electrolyte transport and diarrheal disease. N Engl J Med 1989;321:800.)

ing intestinal volume and consequently exerting a laxative effect.

There is some secretion of K^+ into the intestinal lumen, especially as a component of mucus, but for the most part, the movement of K^+ across the gastrointestinal mucosa is due to diffusion. On the other hand, there are K^+ channels in the luminal as well as the basolateral membrane of the enterocytes of the colon, so K^+ is secreted into the colon. In addition, K^+ moves passively down its electrochemical gradient. The accumulation of K^+ in the colon is partially offset by H^+-K^+ ATPase in the luminal membrane of cells in the distal colon, with resulting active transport of K^+ into the cells. Nevertheless, loss of ileal or colonic fluids in chronic diarrhea can lead to severe hypokalemia.

When the dietary intake of K^+ is high for a prolonged period, aldosterone secretion is increased and more K^+ enters the colon. This is due in part to the appearance of more Na^+-K^+ ATPase pumps in the basolateral membranes of the cells, with a consequent increase in intracellular K^+ and K^+ diffusion across the luminal membranes of the cells.

ABSORPTION OF VITAMINS & MINERALS

Vitamins

Absorption of the fat-soluble vitamins A, D, E, and K is deficient if fat absorption is depressed because of lack of pancreatic enzymes or if bile is excluded from the intestine by obstruction of the bile duct. Most vitamins are absorbed in the upper small intestine, but vitamin B_{12} is absorbed in the ileum. This vitamin binds to intrinsic factor, a protein secreted by the stomach, and the complex is absorbed across the ileal mucosa (see Chapter 26).

Vitamin B_{12} absorption and folate absorption are Na^+ independent, but all seven of the remaining water-soluble vitamins—thiamin, riboflavin, niacin, pyridoxine, pantothenate, biotin, and ascorbic acid—are absorbed by carriers that are Na^+ cotransporters.

Calcium

From 30% to 80% of ingested calcium is absorbed. The absorptive process and its relation to 1,25-dihydroxycholecalciferol are discussed in Chapter 21. Through this vitamin D derivative, Ca^{2+} absorption is adjusted to body needs; absorption is increased in the presence of Ca^{2+} deficiency and decreased in the presence of Ca^{2+} excess. Ca^{2+} absorption is also facilitated by protein. It is inhibited by phosphates and oxalates because these anions form insoluble salts with Ca^{2+} in the intestine. Magnesium absorption is facilitated by protein.

Iron

In adults, the amount of iron lost from the body is relatively small. The losses are generally unregulated, and total body stores of iron are regulated by changes in the rate at which it is absorbed from the intestine. Men lose about 0.6 mg/d, largely in the stools. Women have a variable, larger loss averaging about twice this value because of the additional iron lost in the blood shed during menstruation. The average daily iron intake in the United States and Europe is about 20 mg, but the amount absorbed is equal only to the losses. Thus, the amount of iron absorbed ranges normally from about 3 to 6% of the amount ingested. Various dietary factors affect the availability of iron for absorption; for example, the phytic acid found in cereals reacts with iron to form insoluble compounds in the intestine. So do phosphates and oxalates.

Most of the iron in the diet is in the ferric (Fe^{3+}) form, whereas it is the ferrous (Fe^{2+}) form that is absorbed. There is Fe^{3+} reductase activity associated with the iron transporter in the brush borders of the enterocytes (Figure 25–8).

No more than a trace of iron is absorbed in the stomach, but the gastric secretions dissolve the iron and permit it to form soluble complexes with ascorbic acid and other substances that aid its reduction to the Fe^{2+} form. The importance of this function in humans is indicated by the fact that iron deficiency anemia is a troublesome and relatively frequent complication of partial gastrectomy.

Almost all iron absorption occurs in the duodenum. Transport of Fe^{2+} into the enterocytes occurs via **DMT1** (Figure 25–8). Some is stored in ferritin, and the remainder is transported out of the enterocytes by a basolateral transporter named **ferroportin 1.** A protein called **hephaestin (Hp)** is associated with ferroportin 1. It is not a transporter itself, but it facilitates basolateral transport. In the plasma, Fe^{2+} is converted to Fe^{3+} and bound to the iron transport protein **transferrin.** This protein has two iron-binding sites. Normally, transferrin is about 35% saturated with iron, and the normal plasma iron level is about 130 µg/dL (23 µmol/L) in men and 110 µg/dL (19 µmol/L) in women.

Heme (see Chapter 27) binds to an apical transport protein in enterocytes that has not yet been identified and is carried into the cytoplasm where a subtype of heme oxygenase (HO2) removes Fe^{2+} from the porphyrin and adds it to the intracellular Fe^{2+} pool.

Seventy percent of the iron in the body is in hemoglobin, 3% in myoglobin, and the rest in ferritin, which is present not only in enterocytes but also in many other cells. Apoferritin is a globular protein made up of 24 subunits. Iron forms a micelle of ferric hy-

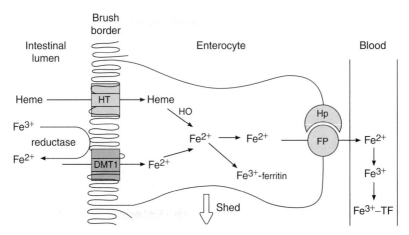

Figure 25–8. Absorption of iron. Fe^{3+} is converted to Fe^{2+} by ferric reductase, and Fe^{2+} is transported into the enterocyte by the apical membrane iron transporter DMT1. Heme is transported into the enterocyte by a separate heme transporter (HT), and heme oxidase (HO) releases Fe^{2+} from the heme. Some of the intracellular Fe^{2+} is converted to Fe^{3+} and bound to ferritin. The rest binds to the basolateral Fe^{2+} transporter ferroportin (FP) and is transported to the interstitial fluid. The transport is aided by hephaestin (Hp). In plasma, Fe^{2+} is converted to Fe^{3+} and bound to the iron transport protein transferrin (TF).

droxyphosphate, and in ferritin, the subunits surround this micelle. The ferritin micelle can contain as many as 4500 atoms of iron. Ferritin is readily visible under the electron microscope and has been used as a tracer in studies of phagocytosis and related phenomena. Ferritin molecules in lysosomal membranes may aggregate in deposits that contain as much as 50% iron. These deposits are called **hemosiderin.**

Intestinal absorption of iron is regulated by three factors: recent dietary intake of iron, the state of the iron stores in the body, and the state of erythropoiesis in the bone marrow. However, the ways these factors signal the absorptive apparatus are still unsettled.

The normal operation of the factors that maintain iron balance is essential for health. Iron deficiency causes anemia. Conversely, iron overload causes hemosiderin to accumulate in the tissues, producing **hemosiderosis.** Large amounts of **hemosiderin** can damage tissues, causing hemochromatosis. This syndrome is

characterized by pigmentation of the skin, pancreatic damage with diabetes ("bronze diabetes"), cirrhosis of the liver, a high incidence of hepatic carcinoma, and gonadal atrophy. Hemochromatosis may be hereditary or acquired. The most common cause of the hereditary forms is a mutated *HFE* gene that is common in the Caucasian population. It is located on the short arm of chromosome 6 and is closely linked to the HLA-A locus. It is still unknown how mutations in *HFE* cause hemochromatosis, but individuals who are homogenous for *HFE* mutations absorb excess amounts of iron. If the abnormality is diagnosed before there is excessive iron deposition in the tissues, life expectancy can be prolonged by repeated withdrawal of blood. Acquired hemochromatosis occurs when the iron-regulating system is overwhelmed by excess iron loads due to chronic destruction of red blood cells, liver disease, or repeated transfusions in diseases such as intractable anemia.

Regulation of Gastrointestinal Function

26

INTRODUCTION

The digestive and absorptive functions of the gastrointestinal system outlined in the previous chapter depend upon a variety of mechanisms that soften the food, propel it through the gastrointestinal tract, and mix it with hepatic bile stored in the gallbladder and digestive enzymes secreted by the salivary glands and pancreas. Some of these mechanisms depend upon intrinsic properties of the intestinal smooth muscle. Others involve the operation of reflexes involving the neurons intrinsic to the gut, reflexes involving the CNS, paracrine effects of chemical messengers, and gastrointestinal hormones. The hormones are humoral agents secreted by cells in the mucosa and transported in the circulation to influence the functions of the stomach, the intestines, the pancreas, and the gallbladder. They also act in a paracrine fashion.

GENERAL CONSIDERATIONS

Organization

The organization of the structures that make up the wall of the gastrointestinal tract from the posterior pharynx to the anus is shown in Figure 26–1. There is some local variation, but in general there are four layers from the lumen outward: the mucosa, the submucosa, the muscularis, and the serosa. There are smooth muscle fibers in the submucosa (muscularis mucosae) and two layers of smooth muscle in the muscularis, an outer longitudinal and an inner circular layer. The wall is lined by mucosa throughout and, except in the case of the esophagus and distal rectum, is covered by serosa. The serosa continues onto the mesentery, which contains the nerves, lymphatics, and blood vessels supplying the tract.

Gastrointestinal Circulation

The blood flow to the stomach, intestines, pancreas, and liver is arranged in a series of parallel circuits, with all the blood from the intestines and pancreas draining via the portal vein to the liver. The physiology of this

important portion of the circulation is discussed in Chapter 32.

The Enteric Nervous System

There are two major networks of nerve fibers that are intrinsic to the gastrointestinal tract: the **myenteric plexus** (Auerbach's plexus), between the outer longitudinal and middle circular muscle layers, and the **sub-**

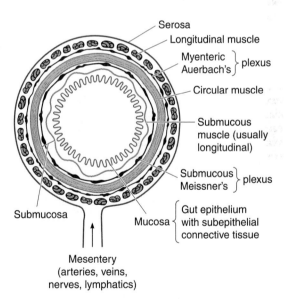

Figure 26–1. Diagrammatic representation of the layers of the wall of the stomach, small intestine, and colon. The structure of the esophagus and the distal rectum is similar, except that they have no serosa or mesentery. In addition, the muscle in the upper quarter of the esophagus is striated, and there is a transitional zone of mixed smooth and striated muscle before the muscle becomes solely smooth in the distal esophagus. (Reproduced, with permission, from Bell GH, Emslie-Smith D, Paterson CR: *Textbook of Physiology and Biochemistry,* 9th ed. Churchill Livingstone, 1976.)

mucous plexus (Meissner's plexus), between the middle circular layer and the mucosa (Figure 26–1). Collectively, these neurons constitute the **enteric nervous system.** The system contains about 100 million sensory neurons, interneurons, and motor neurons in humans—as many as are found in the whole spinal cord—and the system is probably best viewed as a displaced part of the CNS that is concerned with the regulation of gastrointestinal function. It is connected to the CNS by parasympathetic and sympathetic fibers but can function autonomously without these connections (see below). The myenteric plexus innervates the longitudinal and circular smooth muscle layers and is concerned primarily with motor control, whereas the submucous plexus innervates the glandular epithelium, intestinal endocrine cells, and submucosal blood vessels and is primarily involved in the control of intestinal secretion. The neurotransmitters in the system include acetylcholine, the amines norepinephrine and serotonin, the amino acid GABA, the purine ATP, the gases NO and CO, and many different peptides and polypeptides (Table 26–1). Some of these peptides also act in a paracrine fashion, and some enter the blood stream, becoming hormones. Not surprisingly, most of them are also found in the brain.

Extrinsic Innervation

The intestine receives a dual extrinsic innervation from the autonomic nervous system, with parasympathetic cholinergic activity generally increasing the activity of intestinal smooth muscle and sympathetic noradrenergic activity generally decreasing it while causing sphincters to contract. The preganglionic parasympathetic fibers consist of about 2000 vagal efferents and other efferents in the sacral nerves. They generally end on cholinergic nerve cells of the myenteric and submucous plexuses. The sympathetic fibers are postganglionic, but many of them end on postganglionic cholinergic neurons, where the norepinephrine they secrete inhibits acetylcholine secretion by activating α_2 presynaptic receptors. Other sympathetic fibers appear to end directly on intestinal smooth muscle cells. The electrical properties of intestinal smooth muscle are discussed in Chapter 3. Still other fibers innervate blood vessels, where they produce vasoconstriction. It appears that the intestinal blood vessels have a dual innervation; they have an extrinsic noradrenergic innervation and an intrinsic innervation by fibers of the enteric nervous system. VIP and NO are among the mediators in the intrinsic innervation, which seems among other things to be responsible for the hyperemia that accompanies digestion of food. It is unsettled whether there is an additional cholinergic innervation of the blood vessels.

Peristalsis

Peristalsis is a reflex response that is initiated when the gut wall is stretched by the contents of the lumen, and it occurs in all parts of the gastrointestinal tract from the esophagus to the rectum. The stretch initiates a circular contraction behind the stimulus and an area of relaxation in front of it. The wave of contraction then moves in an oral to caudal direction, propelling the contents of the lumen forward at rates that vary from 2 to 25 cm/s. Peristaltic activity can be increased or decreased by the autonomic input to the gut, but its occurrence is independent of the extrinsic innervation. Indeed, progression of the contents is not blocked by removal and resuture of a segment of intestine in its original position and is blocked only if the segment is reversed before it is sewn back into place. Peristalsis is an excellent example of the integrated activity of the enteric nervous system. It appears that local stretch releases serotonin, which activates sensory neurons that activate the myenteric plexus. Cholinergic neurons passing in a retrograde direction in this plexus activate neurons that release substance P and acetylcholine, causing smooth muscle contraction. At the same time, cholinergic neurons passing in an anterograde direction activate neurons that secrete NO, VIP, and ATP, producing the relaxation ahead of the stimulus.

Basic Electrical Activity & Regulation of Motility

Except in the esophagus and the proximal portion of the stomach, the smooth muscle of the gastrointestinal tract has spontaneous rhythmic fluctuations in membrane potential between about −65 and −45 mV. This **basic electrical rhythm (BER)** is initiated by the **in-**

Table 26–1. Principal peptides found in the enteric nervous system.

CGRP
CCK
Endothelin-2
Enkephalins
Galanin
GRP
Neuropeptide Y
Neurotensin
Peptide YY
PACAP
Somatostatin
Substance P
TRH
VIP

terstitial cells of Cajal, stellate mesenchymal pacemaker cells with smooth muscle-like features that send long multiply branched processes into the intestinal smooth muscle. In the stomach and the small intestine, these cells are located in the outer circular muscle layer near the myenteric plexus; in the colon, they are at the submucosal border of the circular muscle layer. In the stomach and small intestine, there is a descending gradient in pacemaker frequency, and as in the heart, the pacemaker with the highest frequency usually dominates.

The BER itself rarely causes muscle contraction, but **spike potentials** superimposed on the most depolarizing portions of the BER waves do increase muscle tension (Figure 26–2). The depolarizing portion of each spike is due to Ca^{2+} influx, and the repolarizing portion is due to K^+ efflux. Many polypeptides and neurotransmitters affect the BER. For example, acetylcholine increases the number of spikes and the tension of the smooth muscle, whereas epinephrine decreases the number of spikes and the tension. The rate of the BER is about 4/min in the stomach. It is about 12/min in the duodenum and falls to about 8/min in the distal ileum. In the colon, the BER rate rises from about 9/min at the cecum to about 16/min at the sigmoid. The function of the BER is to coordinate peristaltic and other motor activity; contractions occur only during the depolarizing part of the waves. After vagotomy or transection of the stomach wall, for example, peristalsis in the stomach becomes irregular and chaotic.

Migrating Motor Complex

During fasting between periods of digestion, the pattern of electrical and motor activity in gastrointestinal smooth muscle becomes modified so that cycles of motor activity migrate from the stomach to the distal ileum. Each cycle, or **migrating motor complex (MMC),** starts with a quiescent period (phase I), continues with a period of irregular electrical and mechanical activity (phase II), and ends with a burst of regular activity (phase III) (Figure 26–3). The MMCs migrate aborally at a rate of about 5 cm/min, and they occur at intervals of approximately 90 minutes. Their function is unsettled, although gastric secretion, bile flow, and pancreatic secretion increase during each MMC. They

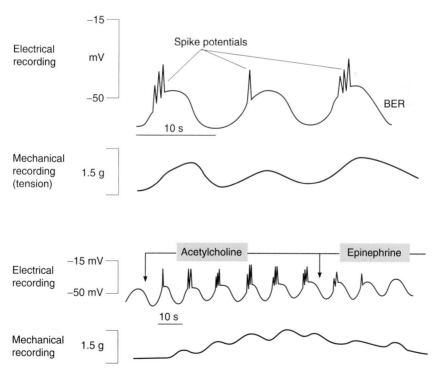

Figure 26–2. Basic electrical rhythm (BER) of gastrointestinal smooth muscle. **Top:** Morphology, and relation to muscle contraction. **Bottom:** Stimulatory effect of acetylcholine and inhibitory effect of epinephrine. (Modified and reproduced, with permission, from Chang EB, Sitrin MD, Black DD: *Gastrointestinal, Hepatobiliary, and Nutritional Physiology.* Lippincott-Raven, 1996.)

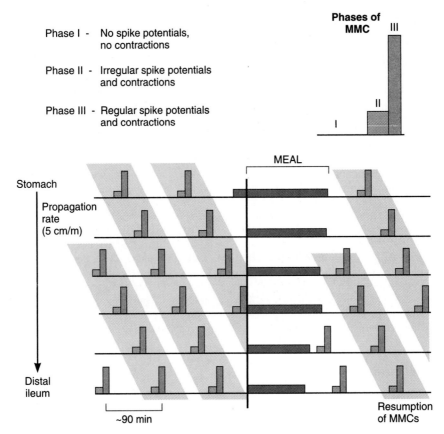

Phase I - No spike potentials,
no contractions

Phase II - Irregular spike potentials
and contractions

Phase III - Regular spike potentials
and contractions

Figure 26–3. Migrating motor complexes (MMCs). Note that the complexes move down the gastrointestinal tract at a regular rate during fasting, that they are completely inhibited by a meal, and that they resume 90–120 minutes after the meal. (Reproduced, with permission, from Chang EB, Sitrin MD, Black DD: *Gastrointestinal, Hepatobiliary, and Nutritional Physiology.* Lippincott-Raven, 1996.)

may clear the stomach and small intestine of luminal contents in preparation for the next meal. They are immediately stopped by ingestion of food, with a return to peristalsis and the other forms of BER and spike potentials.

Other aspects of muscle contractions in the gut are unique to specific regions and are discussed in the sections on those regions.

GASTROINTESTINAL HORMONES

Biologically active polypeptides that are secreted by nerve cells and gland cells in the mucosa act in a paracrine fashion, but they also enter the circulation. Experiments with them and measurement of their concentrations in blood by radioimmunoassay have identified the roles these **gastrointestinal hormones** play in the regulation of gastrointestinal secretion and motility.

When large doses of the hormones are given, their actions overlap. However, their physiologic effects appear to be relatively discrete. On the basis of structural similarity (Table 26–2) and, to a degree, similarity of function, some of the hormones fall into one of two families: the gastrin family, the primary members of which are gastrin and cholecystokinin (CCK); and the secretin family, the primary members of which are secretin, glucagon, glicentin (GLI), VIP, and gastric inhibitory polypeptide (GIP). There are other hormones that do not fall readily into these families.

Enteroendocrine Cells

More than 15 types of hormone-secreting **enteroendocrine cells** have been identified in the mucosa of the stomach, small intestine, and colon. Many of these secrete only one hormone and are identified by letters (G cells, S cells, etc). Some, but not all, manufacture sero-

Table 26–2. Structures of some of the hormonally active polypeptides secreted by cells in the human gastrointestinal tract.[1]

Gastrin Family		Secretin Family				Other Polypeptides			
CCK 39	Gastrin 34	GIP	Glucagon	Secretin	VIP	Motilin	Substance P	GRP	Guanylin
Tyr		Tyr	His	His	His	Phe	Arg	Val	Pro
Ile		Ala	Ser	Ser	Ser	Val	Pro	Pro	Asn
Gln		Glu	Gln	Asp	Asp	Pro	Lys	Leu	Thr
Gln		Gly	Gly	Gly	Ala	Ile	Pro	Pro	Cys
Ala		Thr	Thr	Thr	Val	Phe	Gln	Ala	Glu
Arg	(pyro)Glu	Phe	Phe	Phe	Phe	Thr	Gln	Gly	Ile
→Lys	Leu	Ile	Thr	Thr	Thr	Tyr	Phe	Gly	Cys
Ala	Gly	Ser	Ser	Ser	Asp	Gly	Phe	Gly	Ala
Pro	Pro	Asp	Asp	Glu	Asn	Glu	Gly	Thr	Tyr
Ser	Gln	Tyr	Tyr	Leu	Tyr	Leu	Leu	Val	Ala
Gly	Gly	Ser	Ser	Ser	Thr	Gln	Met-NH2	Leu	Ala
Arg	Pro	Ile	Lys	Arg	Arg	Arg		Thr	Cys
Met	Pro	Ala	Tyr	Leu	Leu	Met		Lys	Thr
Ser	His	Met	Leu	Arg	Arg	Gln		Met	Gly
Ile	Leu	Asp	Asp	Glu	Lys	Glu		Tyr	Cys
Val	Val	Lys	Ser	Gly	Gln	Lys		Pro	
Lys	Ala	Ile	Arg	Ala	Met	Glu		Arg	
Asn	Asp	His	Arg	Arg	Ala	Arg		Gly	
Leu	Pro	Gln	Ala	Leu	Val	Asn		Asn	
Gln	Ser	Gln	Gln	Gln	Lys	Lys		His	
Asn	Lys	Asp	Asp	Arg	Lys	Gly		Trp	
Leu	→Lys	Phe	Phe	Leu	Tyr	Gln		Ala	
Asp	Gln	Val	Val	Leu	Leu			Val	
Pro	Gly	Asn	Gln	Leu	Asn			Gly	
Ser	→Pro	Trp	Gln	Gln	Ser			His	
His	Trp	Leu	Trp	Gly	Ile			Leu	
→Arg	Leu	Leu	Leu	Leu	Leu			Met-NH2	
Ile	Glu	Leu	Met	Val-NH2	Asn-NH2				
Ser	Glu	Ala	Asn						
Asp	Glu	Gly	Thr						
→Arg	Glu	Glu							
Asp	Glu	Lys							
Tys	Ala	Gly							
Met	Tys	Lys							
→Gly	→Gly	Lys							
Trp	Trp	Asn							
Met	Met	Asp							
Asp	Asp	Trp							
Phe-NH2	Phe-NH2	Lys							
		His							
		Asn							
		Ile							
		Thr							
		Gln							

[1]Homologous amino acid residues are enclosed by the lines that generally cross from one polypeptide to another. Arrows indicate points of cleavage to form smaller variants. Tys, tyrosine sulfate. All gastrins occur in unsulfated (gastrin I) and sulfated (gastrin II) forms. Glicentin, an additional member of the secretin family, is a C-terminally extended relative of glucagon (see Chapter 19).

tonin as well and are called **enterochromaffin cells.** Cells that manufacture amines in addition to polypeptides are sometimes called **APUD cells** (for *a*mine *p*recursor *u*ptake and *d*ecarboxylase) or **neuroendocrine cells** and are found in the lungs and other organs in addition to the gastrointestinal tract. They are the cells that form carcinoid tumors.

Gastrin

Gastrin is produced by cells called G cells in the lateral walls of the glands in the antral portion of the gastric mucosa (Figure 26–4). G cells are flask-shaped, with a broad base containing many gastrin granules and a narrow apex that reaches the mucosal surface. Microvilli project from the apical end into the lumen. Receptors mediating gastrin responses to changes in gastric contents are present on the microvilli. Other cells in the gastrointestinal tract that secrete hormones have a similar morphology.

Gastrin is also found in the pancreatic islets in fetal life. Gastrin-secreting tumors, called **gastrinomas,** occur in the pancreas, but it is uncertain whether any gastrin is present in the pancreas in normal adults. In addition, gastrin is found in the anterior and intermediate lobes of the pituitary gland, in the hypothalamus and medulla oblongata, and in the vagus and sciatic nerves.

Gastrin is typical of a number of polypeptide hormones in that it shows both **macroheterogeneity** and **microheterogeneity.** Macroheterogeneity refers to the occurrence in tissues and body fluids of peptide chains of various lengths; microheterogeneity refers to differences in molecular structure due to derivatization of single amino acid residues. Preprogastrin is processed into fragments of various sizes. Three main fragments contain 34, 17, and 14 amino acid residues. All have the same carboxyl terminal configuration (Table 26–2). These forms are also known as G 34, G 17, and G 14 gastrins, respectively. Another form is the carboxyl terminal tetrapeptide, and there is also a large form that is extended at the amino terminal and contains more than 45 amino acid residues. One form of derivatization is sulfation of the tyrosine that is the sixth amino acid residue from the carboxyl terminal. There are approximately equal amounts of nonsulfated and sulfated forms in blood and tissues, and they are equally active. Another derivatization is amidation of the carboxyl terminal phenylalanine.

What is the physiologic significance of this marked heterogeneity? There are some differences in activity between the various components, and the proportions of the components also differ in the various tissues in which gastrin is found. This suggests that different forms are tailored for different actions. However, all

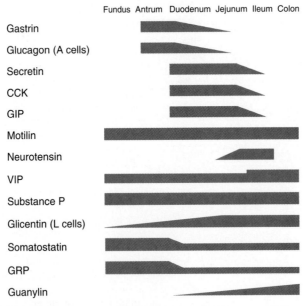

Figure 26–4. Distribution of gastrointestinal peptides along the gastrointestinal tract. The thickness of each bar is roughly proportionate to the concentration of the peptide in the mucosa. Preproglucagon is processed primarily to glucagon in A cells in the upper gastrointestinal tract and to glicentin, GLP-1, GLP-2, and other derivatives in L cells in the lower gastrointestinal tract.

that can be concluded at present is that G 17 is the principal form with respect to gastric acid secretion. The carboxyl terminal tetrapeptide has all the activities of gastrin but only 10% of the strength of G 17.

G 14 and G 17 have half-lives of 2–3 minutes in the circulation, whereas G 34 has a half-life of 15 minutes. Gastrins are inactivated primarily in the kidney and small intestine.

In large doses, gastrin has a variety of actions, but its principal physiologic actions are stimulation of gastric acid and pepsin secretion and stimulation of the growth of the mucosa of the stomach and small and large intestines **(trophic action).** Stimulation of gastric motility is probably a physiologic action as well. Gastrin stimulates insulin secretion; however, only after a protein meal, and not after a carbohydrate meal, does circulating endogenous gastrin reach the level necessary to increase insulin secretion. The functions of gastrin in the pituitary gland, brain, and peripheral nerves are unknown.

Gastrin secretion is affected by the contents of the stomach, the rate of discharge of the vagus nerves, and blood-borne factors (Table 26–3). Atropine does not inhibit the gastrin response to a test meal in humans, because the transmitter secreted by the postganglionic vagal fibers that innervate the G cells is gastrin-releasing polypeptide (GRP; see below) rather than acetylcholine. Gastrin secretion is also increased by the presence of the products of protein digestion in the stomach, particularly amino acids, which act directly on the G cells. Phenylalanine and tryptophan are particularly effective.

Acid in the antrum inhibits gastrin secretion, partly by a direct action on G cells and partly by release of somatostatin, a relatively potent inhibitor of gastrin secretion. The effect of acid is the basis of a negative feedback loop regulating gastrin secretion. Increased secretion of the hormone increases acid secretion, but the acid then feeds back to inhibit further gastrin secretion.

The role of gastrin in the pathophysiology of duodenal ulcers is discussed below. In conditions such as pernicious anemia in which the acid-secreting cells of the stomach are damaged, gastrin secretion is chronically elevated.

Cholecystokinin-Pancreozymin

It was formerly thought that a hormone called cholecystokinin produced contraction of the gallbladder whereas a separate hormone called pancreozymin increased the secretion of pancreatic juice rich in enzymes. It is now clear that a single hormone secreted by cells in the mucosa of the upper small intestine has both activities, and the hormone has therefore been named **cholecystokinin-pancreozymin.** It is also called **CCK-PZ** or, most commonly, **CCK.**

Like gastrin, CCK shows both macroheterogeneity and microheterogeneity. Prepro-CCK is processed into many fragments. A large CCK contains 58 amino acid residues (CCK 58). There are, in addition, CCK peptides containing 39 amino acid residues (CCK 39) and 33 amino acid residues (CCK 33), several forms containing 12 (CCK 12) or slightly more amino acid residues, and a form containing 8 amino acid residues (CCK 8). All of these forms have the same 5 amino acids at the carboxyl terminal as gastrin (Table 26–2). The carboxyl terminal tetrapeptide (CCK 4) also exists in tissues. The carboxyl terminal is amidated, and the tyrosine that is the seventh amino acid residue from the carboxyl terminal is sulfated. Unlike gastrin, the nonsulfated form of CCK has not been found in tissues. However, derivatization of other amino acid residues in CCK can occur. The half-life of circulating CCK is about 5 minutes, but little is known about its metabolism.

In addition to its secretion by endocrine cells, the I cells in the upper intestine, CCK is found in nerves in the distal ileum and colon. It is also found in neurons in the brain, especially the cerebral cortex, and in nerves in many parts of the body (see Chapter 4). In the brain, it may be involved in the regulation of food intake (see Chapter 14), and it appears to be related to the production of anxiety and analgesia. The CCK secreted in the duodenum and jejunum is probably mostly CCK 8 and CCK 12, although CCK 58 is also present in the intestine and circulating blood in some species. The enteric and pancreatic nerves contain primarily CCK 4. CCK 58 and CCK 8 are found in the brain.

In addition to causing contraction of the gallbladder and secretion of a pancreatic juice rich in enzymes,

Table 26–3. Stimuli that affect gastrin secretion.

Stimuli that increase gastrin secretion
Luminal
 Peptides and amino acids
 Distention
Neural
 Increased vagal discharge via GRP
Blood-borne
 Calcium
 Epinephrine

Stimuli that inhibit gastrin secretion
Luminal
 Acid
 Somatostatin
Blood-borne
 Secretin, GIP, VIP, glucagon, calcitonin

CCK augments the action of secretin in producing secretion of an alkaline pancreatic juice. It also inhibits gastric emptying, exerts a trophic effect on the pancreas, increases the secretion of enterokinase, and may enhance the motility of the small intestine and colon. There is some evidence that, along with secretin, it augments the contraction of the pyloric sphincter, thus preventing the reflux of duodenal contents into the stomach. Gastrin and CCK stimulate glucagon secretion, and since the secretion of both gastrointestinal hormones is increased by a protein meal, either or both may be the "gut factor" that stimulates glucagon secretion (see Chapter 19). As noted in Chapters 4 and 14, two CCK receptors have been identified. CCK-A receptors are primarily located in the periphery, whereas both CCK-A and CCK-B receptors are found in the brain. Both activate PLC, causing increased production of IP_3 and DAG (see Chapter 1).

The secretion of CCK is increased by contact of the intestinal mucosa with the products of digestion, particularly peptides and amino acids, and also by the presence in the duodenum of fatty acids containing more than 10 carbon atoms. Since the bile and pancreatic juice that enter the duodenum in response to CCK further the digestion of protein and fat and the products of this digestion stimulate further CCK secretion, a sort of positive feedback operates in the control of the secretion of this hormone. The positive feedback is terminated when the products of digestion move on to the lower portions of the gastrointestinal tract.

Secretin

Secretin occupies a unique position in the history of physiology. In 1902, Bayliss and Starling first demonstrated that the excitatory effect of duodenal stimulation on pancreatic secretion was due to a blood-borne factor. Their research led to the identification of secretin. They also suggested that many chemical agents might be secreted by cells in the body and pass in the circulation to affect organs some distance away. Starling introduced the term **hormone** to categorize such "chemical messengers." Modern endocrinology is the proof of the correctness of this hypothesis.

Secretin is secreted by S cells that are located deep in the glands of the mucosa of the upper portion of the small intestine. The structure of secretin (Table 26–2) is different from that of CCK and gastrin but very similar to that of glucagon, GLI, VIP, and GIP. Only one form of secretin has been isolated, and the fragments of the molecule that have been tested to date are inactive. Its half-life is about 5 minutes, but little is known about its metabolism.

Secretin increases the secretion of bicarbonate by the duct cells of the pancreas and biliary tract. It thus causes the secretion of a watery, alkaline pancreatic juice. Its action on pancreatic duct cells is mediated via cAMP. It also augments the action of CCK in producing pancreatic secretion of digestive enzymes. It decreases gastric acid secretion and may cause contraction of the pyloric sphincter.

The secretion of secretin is increased by the products of protein digestion and by acid bathing the mucosa of the upper small intestine. The release of secretin by acid is another example of feedback control: secretin causes alkaline pancreatic juice to flood into the duodenum, neutralizing the acid from the stomach and thus stopping further secretion of the hormone.

GIP

GIP contains 43 amino acid residues (Table 26–2) and is produced by K cells in the mucosa of the duodenum and jejunum. Its secretion is stimulated by glucose and fat in the duodenum, and because in large doses it inhibits gastric secretion and motility, it was named gastric inhibitory peptide. However, it now appears that it does not have significant gastric inhibiting activity when administered in smaller doses that raise the blood level to that seen after a meal. In the meantime, it was found that GIP stimulates insulin secretion. Gastrin, CCK, secretin, and glucagon also have this effect, but GIP is the only one of these that stimulates insulin secretion when administered in doses that produce blood levels comparable to those produced by oral glucose. For this reason, it is often called **glucose-dependent insulinotropic polypeptide.** The glucagon derivative GLP-1 (7–36) (see Chapter 19) also stimulates insulin secretion and is said to be more potent in this regard than GIP. Therefore, it may also be a physiologic B cell-stimulating hormone of the gastrointestinal tract.

The integrated action of gastrin, CCK, secretin, and GIP in facilitating digestion and utilization of absorbed nutrients is summarized in Figure 26–5.

VIP

VIP contains 28 amino acid residues (Table 26–2). It is found in nerves in the gastrointestinal tract. Prepro-VIP contains both VIP and a closely related polypeptide (**PHM-27** in humans, PHI-27 in other species). VIP is also found in blood, in which it has a half-life of about 2 minutes. In the intestine, it markedly stimulates intestinal secretion of electrolytes and hence of water. Its other actions include relaxation of intestinal smooth muscle, including sphincters; dilation of peripheral blood vessels; and inhibition of gastric acid secretion. It is also found in the brain and many autonomic nerves (see Chapter 4), where it often occurs in the same neurons as acetylcholine. It potentiates the ac-

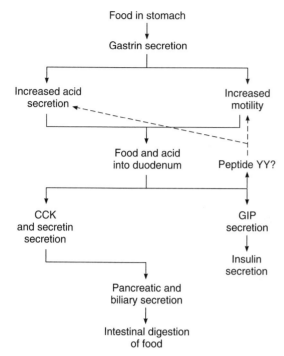

Food in stomach
↓
Gastrin secretion

Increased acid secretion

Increased motility

Food and acid into duodenum

Peptide YY?

CCK and secretin secretion

GIP secretion

Insulin secretion

Pancreatic and biliary secretion
↓
Intestinal digestion of food

Figure 26–5. Integrated action of gastrointestinal hormones in regulating digestion and utilization of absorbed nutrients. The dashed arrows indicate inhibition. The exact identity of the hormonal factor or factors from the intestine that inhibit(s) gastric acid secretion and motility is unsettled, but it may be peptide YY.

tion of acetylcholine in salivary glands. However, VIP and acetylcholine do not coexist in neurons that innervate other parts of the gastrointestinal tract. VIP-secreting tumors (VIPomas) have been described in patients with severe diarrhea. The relation of GIP and VIP to **enterogastrone,** a putative hormone that inhibits gastric acid secretion and motility, is unsettled. However, peptide YY (see Chapter 19) is also a good candidate to be enterogastrone. Fat causes its release from the jejunum, and it is an effective inhibitor of gastrin-stimulated acid secretion.

Motilin

Motilin is a polypeptide containing 22 amino acid residues that is secreted by enterochromaffin cells and Mo cells in the stomach, small intestine, and colon. It acts on G protein-coupled receptors on enteric neurons in the duodenum and colon and upon injection produces contraction of smooth muscle in the stomach and intestines. Its circulating level increases at intervals of approximately 100 minutes in the interdigestive

state, and it is a major regulator of the MICs (Figure 26–3) that control gastrointestinal motility between meals. The antibiotic erythromycin binds to motilin receptors, and derivatives of this compound may be of value in treating patients in whom gastrointestinal motility is decreased.

Other Gastrointestinal Hormones

Neurotensin, a 13-amino-acid polypeptide, is produced by neurons and cells that are abundant in the mucosa of the ileum. Its release is stimulated by fatty acids, and it inhibits gastrointestinal motility and increases ileal blood flow. **Substance P** (Table 26–2) is found in endocrine and nerve cells in the gastrointestinal tract and may enter the circulation. It increases the motility of the small intestine. **GRP** contains 27 amino acid residues, and the 10 amino acid residues at its carboxyl terminal are almost identical to those of amphibian **bombesin.** It is present in the vagal nerve endings that terminate on G cells and is the neurotransmitter producing vagally mediated increases in gastrin secretion. It may enter the circulation when secreted in very large amounts. **Somatostatin,** the growth-hormone-inhibiting hormone originally isolated from the hypothalamus, is secreted into the circulation by D cells in the pancreatic islets (see Chapter 19) and by similar D cells in the gastrointestinal mucosa. It exists in tissues in two forms, somatostatin 14 and somatostatin 28 (see Figure 14–18), and both are secreted. Somatostatin inhibits the secretion of gastrin, VIP, GIP, secretin, and motilin. Like several other gastrointestinal hormones, somatostatin is secreted in larger amounts into the gastric lumen than into the bloodstream. Its secretion is stimulated by acid in the lumen, and it probably acts in a paracrine fashion via the gastric juice to mediate the inhibition of gastrin secretion produced by acid. It also inhibits pancreatic exocrine secretion; gastric acid secretion and motility; gallbladder contraction; and the absorption of glucose, amino acids, and triglycerides. **Glucagon** from the gastrointestinal tract may be responsible (at least in part) for the hyperglycemia after pancreatectomy. The products formed from preproglucagon in the upper and lower intestine are discussed in Chapter 19.

The polypeptide **ghrelin** is secreted by the stomach, with secretion decreased by feeding and increased by fasting. It appears to be involved in the control of food intake (see Chapter 14) and growth hormone secretion (see Chapter 22).

Guanylin is a gastrointestinal polypeptide that binds to guanylyl cyclase. It is made up of 15 amino acid residues (Table 26–2) and is secreted by cells of the intestinal mucosa. Stimulation of guanylyl cyclase increases the concentration of intracellular cGMP, and this in turn causes increased activity of the cystic fibrosis-regulated Cl⁻ channel and increased secretion of Cl⁻ into the intestinal lumen. Most of the guanylin appears to act in a paracrine fashion, and it is produced in cells from the pylorus to the rectum. In an interesting example of molecular mimicry, the heat-stable enterotoxin of certain diarrhea-producing strains of *E coli* has a structure very similar to guanylin and activates guanylin receptors in the intestine.

Guanylin receptors are also found in the kidneys, the liver, and the female reproductive tract, and guanylin may act in an endocrine fashion to regulate fluid movement in these tissues as well.

A substance called **urogastrone,** which was originally isolated from urine and was found to aid the healing of ulcers, is now known to be epidermal growth factor (see Chapter 22).

The cells that secrete gastrointestinal polypeptides can form tumors. Fifty percent of these tumors are gastrinomas and 25% are glucagonomas, but VIPomas, neurotensinomas, and others have also been described.

MOUTH & ESOPHAGUS

In the mouth, food is mixed with saliva and propelled into the esophagus. Peristaltic waves in the esophagus move the food into the stomach.

Mastication

Chewing (**mastication**) breaks up large food particles and mixes the food with the secretions of the salivary glands. This wetting and homogenizing action aids swallowing and subsequent digestion. Large food particles can be digested, but they cause strong and often painful contractions of the esophageal musculature. Particles that are small tend to disperse in the absence of saliva and also make swallowing difficult because they do not form a bolus. The number of chews that is optimal depends on the food, but usually ranges from 20 to 25.

Edentulous patients are generally restricted to a soft diet and have considerable difficulty eating dry food.

Salivary Glands & Saliva

In the salivary glands, the secretory (**zymogen**) granules containing the salivary enzymes are discharged from the acinar cells into the ducts (Figure 26–6). The characteristics of each of the three pairs of salivary glands in humans are summarized in Table 26–4.

About 1500 mL of saliva is secreted per day. The pH of saliva from resting glands is slightly less than 7.0, but during active secretion, it approaches 8.0. Saliva contains two digestive enzymes: **lingual lipase,** secreted by glands on the tongue, and **salivary α-amylase,** se-

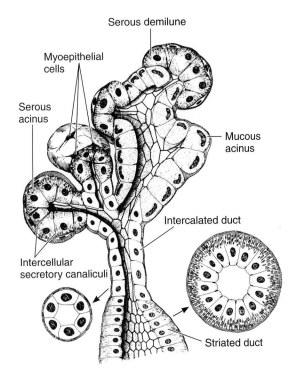

Figure 26–6. Structure of the submandibular gland (also known as the submaxillary gland). Note that the cells in the mucous acini have flattened basal nuclei, whereas the cells in the serous acini have round nuclei and collections of zymogen secretory granules at their apexes. The intercalated ducts drain into the striated ducts, where the cells are specialized for ion transport.

creted by the salivary glands. The functions of these enzymes are discussed in Chapter 25. Saliva also contains **mucins,** glycoproteins that lubricate the food, bind bacteria, and protect the oral mucosa. It also contains the secretory immune globulin IgA (see Chapter 27); lysozyme, which attacks the walls of bacteria; lactoferrin, which binds iron and is bacteriostatic; and proline-rich proteins that protect tooth enamel and bind toxic tannins.

Saliva performs a number of important functions. It facilitates swallowing, keeps the mouth moist, serves as a solvent for the molecules that stimulate the taste buds, aids speech by facilitating movements of the lips and tongue, and keeps the mouth and teeth clean. The saliva also has some antibacterial action, and patients with deficient salivation (**xerostomia**) have a higher than normal incidence of dental caries. The buffers in saliva help maintain the oral pH at about 7.0. They also help neutralize gastric acid and relieve heartburn when gastric juice is regurgitated into the esophagus.

Table 26–4. Characteristics of each of the pairs of salivary glands in humans.

Gland	Histologic Type	Secretion[1]	Percentage of Total Saliva in Humans[2] (1.5 L/d)
Parotid	Serous	Watery	20
Submandibular (submaxillary)	Mixed	Moderately viscous	70
Sublingual	Mucous	Viscous	5

[1]Serous cells secrete ptyalin; mucous cells secrete mucin.

[2]The remaining 5% of salivary volume is contributed by lingual and other minor glands in the oral cavity.

Ionic Composition of Saliva

There is considerable variation in the ionic composition of saliva from species to species and from gland to gland. In general, however, saliva secreted in the acini is probably isotonic, with concentrations of Na^+, K^+, Cl^-, and HCO_3^- that are close to those in plasma. The excretory ducts and probably the intercalated ducts that drain into them modify the composition of the saliva by extracting Na^+ and Cl^- and adding K^+ and HCO_3^-. The ducts are relatively impermeable to water. Therefore, at low salivary flows, the saliva that reaches the mouth is hypotonic, slightly acidic, and rich in K^+ but relatively depleted of Na^+ and Cl^-. When salivary flow is rapid, there is less time for ionic composition to change in the ducts. Consequently, although still hypotonic in humans, saliva is closer to isotonic, with higher concentrations of Na^+ and Cl^-. Aldosterone increases the K^+ concentration and reduces the Na^+ concentration of saliva in an action analogous to its action on the kidneys (see Chapters 20 and 38), and a high salivary Na^+/K^+ ratio is seen when aldosterone is deficient in Addison's disease.

Control of Salivary Secretion

Salivary secretion is under neural control. Stimulation of the parasympathetic nerve supply causes profuse secretion of watery saliva with a relatively low content of organic material. Associated with this secretion is a pronounced vasodilation in the gland, which appears to be due to the local release of VIP. This polypeptide is a cotransmitter with acetylcholine in some of the postganglionic parasympathetic neurons. Atropine and other cholinergic blocking agents reduce salivary secretion. Stimulation of the sympathetic nerve supply causes vasoconstriction and, in humans, secretion of small amounts of saliva rich in organic constituents from the submandibular glands.

Food in the mouth causes reflex secretion of saliva, and so does stimulation of the vagal afferent fibers at the gastric end of the esophagus. Salivary secretion is

easily conditioned, as shown in Pavlov's original experiments (see Chapter 16). In humans, the sight, smell, and even thought of food causes salivary secretion ("makes the mouth water").

Swallowing

Swallowing (deglutition) is a reflex response that is triggered by afferent impulses in the trigeminal, glossopharyngeal, and vagus nerves. These impulses are integrated in the nucleus of the tractus solitarius and the nucleus ambiguus. The efferent fibers pass to the pharyngeal musculature and the tongue via the trigeminal, facial, and hypoglossal nerves. Swallowing is initiated by the voluntary action of collecting the oral contents on the tongue and propelling them backward into the pharynx. This starts a wave of involuntary contraction in the pharyngeal muscles that pushes the material into the esophagus. Inhibition of respiration and glottic closure are part of the reflex response. A peristaltic ring contraction of the esophageal muscle forms behind the material, which is then swept down the esophagus at a speed of approximately 4 cm/s. When humans are in an upright position, liquids and semisolid foods generally fall by gravity to the lower esophagus ahead of the peristaltic wave.

Swallowing is difficult if not impossible when the mouth is open, as anyone who has spent time in the dentist's chair feeling saliva collect in the throat is well aware. A normal adult swallows frequently while eating, but swallowing also continues between meals. The total number of swallows per day is about 600: 200 while eating and drinking, 350 while awake without food, and 50 while sleeping.

Lower Esophageal Sphincter

Unlike the rest of the esophagus, the musculature of the gastroesophageal junction (**lower esophageal sphincter; LES**) is tonically active but relaxes upon swallowing. The tonic activity of the LES between meals prevents reflux of gastric contents into the esophagus. The

LES is made up of three components (Figure 26–7). The esophageal smooth muscle is more prominent at the junction with the stomach (intrinsic sphincter). Fibers of the crural portion of the diaphragm, a skeletal muscle, surround the esophagus at this point (extrinsic sphincter) and exert a pinchcock-like action on the esophagus. In addition, the oblique or sling fibers of the stomach wall create a flap valve that helps close off the esophagogastric junction and prevent regurgitation when intragastric pressure rises.

The tone of the LES is under neural control. Release of acetylcholine from vagal endings causes the intrinsic sphincter to contract, and release of NO and VIP from interneurons innervated by other vagal fibers causes it to relax. Contraction of the crural portion of the diaphragm, which is innervated by the phrenic nerves, is coordinated with respiration and contractions of chest and abdominal muscles. Thus, the intrinsic and extrinsic sphincters operate together to permit orderly flow of food into the stomach and to prevent reflux of gastric contents into the esophagus.

Motor Disorders of the Esophagus

Achalasia is a condition in which food accumulates in the esophagus and the organ becomes massively dilated. It is due to increased resting LES tone and incomplete relaxation upon swallowing. The myenteric plexus of the esophagus is deficient at the LES in this condition, and there is defective release of NO and VIP. It can be treated by pneumatic dilation of the sphincter or incision of the esophageal muscle (myotomy). Inhibition of acetylcholine release by injection of botulinum toxin into the LES is also effective and produces relief that lasts for several months.

The opposite condition is LES incompetence, which permits reflux of acid gastric contents into the esophagus (**gastroesophageal reflux disease**). This common condition causes heartburn and esophagitis and can lead to ulceration and stricture of the esophagus due to scarring. In severe cases, there is weakness of the intrinsic sphincter, the extrinsic sphincter, or both, but less severe cases are caused by intermittent periods of poorly understood decreases in the neural drive to both sphincters. The condition can be treated by inhibition of acid secretion with H_2 receptor blockers or omeprazole (see below). Surgical treatment in which a portion of the fundus of the stomach is wrapped around the lower esophagus so that the LES is inside a short tunnel of stomach (**fundoplication**) is also effective.

Aerophagia & Intestinal Gas

Some air is unavoidably swallowed in the process of eating and drinking (**aerophagia**). Some of the swal-

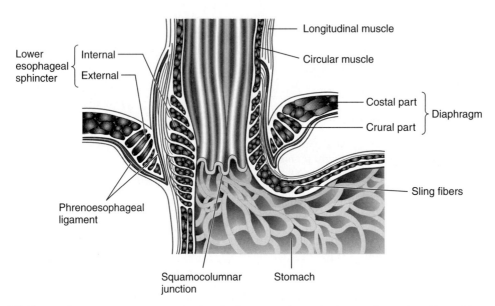

Figure 26–7. Esophagogastric junction. Note that the lower esophageal sphincter (intrinsic sphincter) is supplemented by the crural portion of the diaphragm (extrinsic sphincter), and that the two are anchored to each other by the phrenoesophageal ligament. (Reproduced, with permission, from Mittal RK, Balaban DH: The esophagogastric junction. N Engl J Med 1997;336:924. Copyright 1997 by Massachusetts Medical Society. All rights reserved.)

lowed air is regurgitated (belching), and some of the gases it contains are absorbed, but much of it passes on to the colon. Here, some of the oxygen is absorbed, and hydrogen, hydrogen sulfide, carbon dioxide, and methane formed by the colonic bacteria from carbohydrates and other substances are added to it. It is then expelled as **flatus.** The smell is largely due to sulfides. The volume of gas normally found in the human gastrointestinal tract is about 200 mL, and the daily production is 500–1500 mL. In some individuals, gas in the intestines causes cramps, **borborygmi** (rumbling noises), and abdominal discomfort.

STOMACH

Food is stored in the stomach; mixed with acid, mucus, and pepsin; and released at a controlled, steady rate into the duodenum.

Anatomic Considerations

The gross anatomy of the stomach is shown in Figure 26–8. The gastric mucosa contains many deep glands. In the pyloric and cardiac regions, the glands secrete mucus. In the body of the stomach, including the fundus, the glands contain **parietal (oxyntic) cells,** which secrete hydrochloric acid and intrinsic factor, and **chief (zymogen, peptic) cells,** which secrete pepsinogens (Figure 26–9). These secretions mix with mucus se-

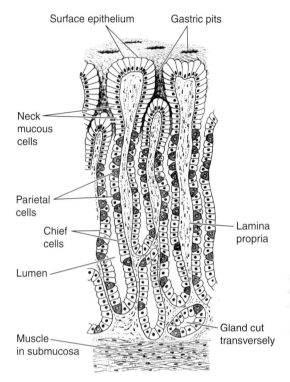

Figure 26–9. Glands in the mucosa of the body of the human stomach. (Reproduced, with permission, from Bell GH, Davidson N, Scarborough G: *Textbook of Physiology and Biochemistry,* 6th ed. Livingstone, 1965.)

creted by the cells in the necks of the glands. Several of the glands open on a common chamber (**gastric pit**) that opens in turn on the surface of the mucosa. Mucus is also secreted along with HCO_3^- by mucus cells on the surface of the epithelium between glands.

The stomach has a very rich blood and lymphatic supply. Its parasympathetic nerve supply comes from the vagi and its sympathetic supply from the celiac plexus.

Gastric Secretion

The cells of the gastric glands secrete about 2500 mL of **gastric juice** daily. This juice contains a variety of substances (Table 26–5). The gastric enzymes are discussed in Chapter 25. The **hydrochloric acid** secreted by the glands in the body of the stomach kills many ingested bacteria, provides the necessary pH for pepsin to start protein digestion, and stimulates the flow of bile.

Mucosal Barrier

The acid in gastric juice is concentrated enough to cause tissue damage. Normally, damage fails to occur

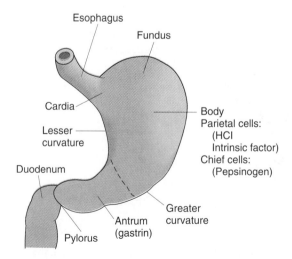

Figure 26–8. Anatomy of the stomach. The principal secretions are listed in parentheses under the labels indicating the locations where they are produced. In addition, mucus is secreted in all parts of the stomach. The dashed line marks the border between the body and the antrum.

Table 26–5. Contents of normal gastric juice (fasting state).

Cations: Na^+, K^+, Mg^{2+}, H^+ (pH approximately 1.0)
Anions: Cl^-, HPO_4^{2-}, SO_4^{2-}
Pepsins
Lipase
Mucus
Intrinsic factor

because a mucosal barrier is produced by mucus and secreted HCO_3^-. Mucus, which is secreted by neck cells of the gastric glands and surface mucosal cells, is made up of glycoproteins called mucins that form a flexible gel coating the mucosa. The surface mucosal cells also secrete HCO_3^-. Much of this is trapped in the mucus gel, so that a pH gradient is established that ranges from pH 1.0–2.0 at the luminal side to pH 6.0–7.0 at the surface of the epithelial cells. HCl secreted by the parietal cells in the gastric glands crosses this barrier in finger-like channels, leaving the rest of the gel layer intact.

Mucus and HCO_3^- secreted by mucosal cells also play an important role in protecting the duodenum from damage when acid-rich gastric juice is secreted into it. Prostaglandins stimulate mucus secretion. HCO_3^- secretion is also stimulated by prostaglandins and by local reflexes.

Some of the resistance of the mucosa of the gastrointestinal tract to autodigestion is also provided by **trefoil peptides** in the mucosa. These are of several types and are acid-resistant. They are also found in the hypothalamus and pituitary, and more generally in rapidly proliferating tissues. They are characterized by a three-loop structure that looks like a three-leaf clover. In mice in which the gene for one of these peptides has been knocked out, the gastric and intestinal mucosa are histologically abnormal and there is a high incidence of benign and malignant mucosal tumors.

Pepsinogen Secretion

The chief cells that secrete pepsinogens, the inactive precursors of the pepsins in gastric juice (see Chapter 25), contain zymogen granules. The secretory process is similar to that involved in the secretion of trypsinogen and the other pancreatic enzymes by the pancreas. Pepsinogen activity can be detected in the plasma and in the urine, where it is called **uropepsinogen.**

Hydrochloric Acid Secretion

The purest specimens of parietal cell secretion that have been obtained contain approximately 0.17 N HCl,

with pHs as low as 0.87. Therefore, parietal cell secretion is probably an isotonic solution of HCl that contains 150 meq of Cl^- and 150 meq of H^+ per liter. Yet the pH of the cytoplasm of the parietal cells, like that of other cells, is 7.0–7.2, and the comparable concentrations per liter of plasma are about 100 meq of Cl^- and 0.00004 meq of H^+.

It is H^+-K^+ ATPase in the apical membrane of the parietal cells that pumps H^+ against a concentration gradient of this magnitude (Figure 26–10). The parietal cells are polarized, with an apical membrane facing the lumen of the gastric glands and a basolateral membrane in contact with the interstitial fluid. Canaliculi extend from the apical surface into the cell (Figure 26–11). At rest, the cells contain abundant tubulovesicular structures with H^+-K^+ ATPase molecules in their walls. When the parietal cells are stimulated, the tubulovesicular structures move to the apical membrane and fuse with it, thus inserting many more H^+-K^+ ATPase molecules into the membrane. These ATPase molecules are now exposed to the K^+ in the ECF, and H^+-K^+ exchange begins.

Cl^- is also extruded down its electrochemical gradient through channels that are activated by cAMP. The H^+ that is extruded comes from H_2CO_3, and H_2CO_3 in turn is formed by the hydration of CO_2 (Figure 26–10). This latter reaction is catalyzed by carbonic anhydrase, and the parietal cells are particularly rich in

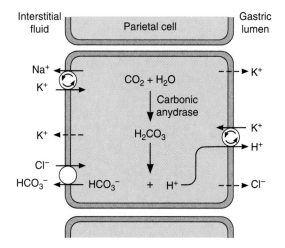

Figure 26–10. HCl secretion by parietal cells in the stomach. Active transport by ATPase is indicated by arrows in circles. H^+ is secreted into the gastric lumen in exchange for K^+ by H^+-K^+ ATPase. HCO_3^- is exchanged for Cl^- in the interstitial fluid by an antiport, and Na^+-K^+ ATPase keeps intracellular Na^+ low. Dashed arrows indicate diffusion. Compare with Figure 38–20.

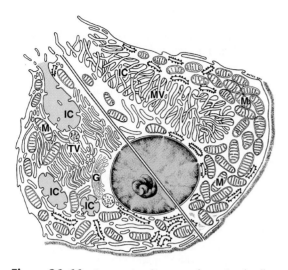

Figure 26–11. Composite diagram of a parietal cell, showing the resting state **(lower left)** and the active state **(upper right).** The resting cell has intracellular canaliculi (IC), which open on the apical membrane of the cell, and many tubulovesicular structures (TV) in the cytoplasm. When the cell is activated, the TVs fuse with the cell membrane and microvilli (MV) project into the canaliculi, so the area of cell membrane in contact with gastric lumen is greatly increased. M, mitochondrion; G, Golgi apparatus. (Reproduced, with permission, from Junqueira LC, Carneiro J: *Basic Histology: Text and Atlas,* 10th ed. McGraw-Hill, 2003.)

particularly those of the E series, inhibit acid secretion by activating G_i.

ECL Cells

Gastrin also acts by stimulating the secretion of histamine from **enterochromaffin-like cells (ECL cells).** These are vesicle- and granule-containing cells that are the predominant endocrine cell type in the acid-secreting portion of the stomach. They have acetylcholine receptors in addition to gastrin receptors, but the relative importance of acetylcholine in stimulating their secretion is unsettled. They are inhibited by somato-

this enzyme. The HCO_3^- formed by dissociation of H_2CO_3 is extruded by an antiport in the basolateral membrane of the parietal cells that exchanges HCO_3^- for another anion, and since Cl^- is the most abundant anion in interstitial fluid, the exchange is mainly for Cl^-. Because of the efflux of HCO_3^- into the blood, the stomach has a negative respiratory quotient (RQ)—ie, the amount of CO_2 in arterial blood is greater than the amount in gastric venous blood. When gastric acid secretion is elevated after a meal, sufficient H^+ may be secreted to raise the pH of systemic blood and make the urine alkaline (**postprandial alkaline tide).**

Acid secretion is stimulated by histamine via H_2 receptors and by acetylcholine via M_3 muscarinic receptors. Gastrin probably acts directly as well (Figure 26–12), although the main stimulation by gastrin is via ECL cells (see below). The H_2 receptors increase intracellular cAMP via G_s, whereas the muscarinic receptors and the gastrin receptors exert their effects by increasing intracellular free Ca^{2+}. The intracellular events interact so that activation of one receptor type potentiates the response of another to stimulation. Prostaglandins,

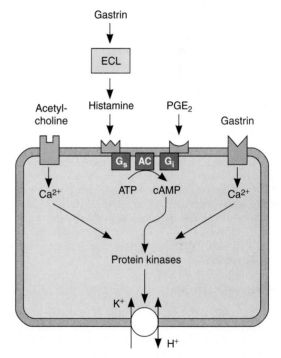

Figure 26–12. Regulation of gastric acid secretion by the parietal cell. Acid secretion is increased by acetylcholine acting on M_3 muscarinic receptors to increase intracellular Ca^{2+} and by gastrin acting on gastrin receptors to increase intracellular Ca^{2+}. In addition, gastrin stimulates histamine secretion by enterochromaffin-like (ECL) cells, and this is the principal way in which gastrin stimulates H^+ secretion. Histamine binds to H_2 receptors, and via G_s, this increases adenylyl cyclase (AC) activity and intracellular cAMP. PGE_2 acts via G_i to decrease adenylyl cyclase activity and intracellular cAMP. Cyclic AMP and Ca^{2+} act via protein kinases to increase the transport of H^+ into the gastric lumen by H^+-K^+ ATPase.

statin. They undergo hypertrophy when gastric acid secretion is suppressed for prolonged periods.

Gastric Motility & Emptying

When food enters the stomach, the fundus and upper portion of the body relax and accommodate the food with little if any increase in pressure (**receptive relaxation**). Peristalsis then begins in the lower portion of the body, mixing and grinding the food and permitting small, semiliquid portions of it to pass through the pylorus and enter the duodenum.

Receptive relaxation is vagally mediated and triggered by movement of the pharynx and esophagus. Peristaltic waves controlled by the gastric BER begin soon thereafter and sweep toward the pylorus. The contraction of the distal stomach caused by each wave is sometimes called **antral systole** and can last up to 10 seconds. Waves occur three to four times per minute.

In the regulation of gastric emptying, the antrum, pylorus, and upper duodenum apparently function as a unit. Contraction of the antrum is followed by sequential contraction of the pyloric region and the duodenum. In the antrum, partial contraction ahead of the advancing gastric contents prevents solid masses from entering the duodenum, and they are mixed and crushed instead. The more liquid gastric contents are squirted a bit at a time into the small intestine. Normally, regurgitation from the duodenum does not occur, because the contraction of the pyloric segment tends to persist slightly longer than that of the duodenum. The prevention of regurgitation may also be due to the stimulating action of CCK and secretin on the pyloric sphincter.

Hunger Contractions

Gastric contractions between meals, presumably associated with the MMCs, can sometimes be felt and may even be mildly painful. These **hunger contractions** are associated with the sensation of hunger and were once thought to be an important regulator of appetite. However, food intake is normal in animals after denervation of the stomach and intestines (see Chapter 14).

Regulation of Gastric Secretion

Gastric motility and secretion are regulated by neural and humoral mechanisms. The neural components are local autonomic reflexes, involving cholinergic neurons, and impulses from the CNS by way of the vagus nerves. The humoral components are the hormones discussed above. Vagal stimulation increases gastrin secretion by release of gastrin-releasing peptide (see above). Other vagal fibers release acetylcholine, which acts directly on the cells in the glands in the body and the fundus to increase acid and pepsin secretion. Stimulation of the vagus nerve in the chest or neck increases acid and pepsin secretion, but vagotomy does not abolish the secretory response to local stimuli.

For convenience, the physiologic regulation of gastric secretion is usually discussed in terms of cephalic, gastric, and intestinal influences, although these overlap. The **cephalic** influences are vagally mediated responses induced by activity in the CNS. The **gastric** influences are primarily local reflex responses and responses to gastrin. The **intestinal** influences are the reflex and hormonal feedback effects on gastric secretion initiated from the mucosa of the small intestine.

Cephalic Influences

The presence of food in the mouth reflexly stimulates gastric secretion. The efferent fibers for this reflex are in the vagus nerves. Vagally mediated increases in gastric secretion are easily conditioned. In humans, for example, the sight, smell, and thought of food increase gastric secretion. These increases are due to alimentary conditioned reflexes that become established early in life. Cephalic influences are responsible for one-third to one-half of the acid secreted in response to a normal meal.

Psychic states have effects on gastric secretion and motility that are mediated principally via the vagi. Among his famous observations on Alexis St. Martin, the Canadian with a permanent gastric fistula resulting from a gunshot wound, William Beaumont noted that anger and hostility were associated with turgor, hyperemia, and hypersecretion of the gastric mucosa. Similar observations have been made on other patients with gastric fistulas. Fear and depression decrease gastric secretion and blood flow and inhibit gastric motility.

Gastric Influences

Food in the stomach accelerates the increase in gastric secretion produced by the sight and smell of food and the presence of food in the mouth (Figure 26–13). Receptors in the wall of the stomach and the mucosa respond to stretch and chemical stimuli, mainly amino acids and related products of digestion. The fibers from the receptors enter the submucous plexus, where the cell bodies of the receptor neurons are located. They synapse on postganglionic parasympathetic neurons that end on parietal cells and stimulate acid secretion. Thus, the acid responses are produced by local reflexes in which the reflex arc is totally within the wall of the stomach. The postganglionic neurons in the local reflex arc are the same ones innervated by the descending vagal preganglionic neurons from the brain that mediate the cephalic phase of secretion. The products of protein

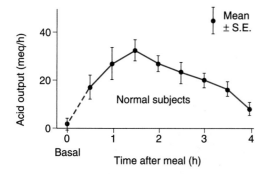

Figure 26–13. Human gastric acid secretion after a steak meal. (Modified from Brooks FP: Integrative lecture: Response of the GI tract to a meal. *Undergraduate Teaching Project.* American Gastroenterological Association, 1974.)

digestion also bring about increased secretion of gastrin, and this augments the flow of acid.

Intestinal Influences

Although there are gastrin-containing cells in the mucosa of the small intestine as well as in the stomach, instillation of amino acids directly into the duodenum does not increase circulating gastrin levels. Fats, carbohydrates, and acid in the duodenum inhibit gastric acid and pepsin secretion and gastric motility via neural and hormonal mechanisms. The identity of enterogastrone, the putative intestinal hormone responsible for the inhibition, is unsettled, although it may be peptide YY (see above). Gastric acid secretion is increased following removal of large parts of the small intestine. The hypersecretion, which is roughly proportionate in degree to the amount of intestine removed, may be due in part to removal of the source of hormones that inhibit acid secretion.

Other Influences

Hypoglycemia acts via the brain and vagal efferents to stimulate acid and pepsin secretion. Other stimulants include **alcohol** and **caffeine,** both of which act directly on the mucosa. The beneficial effects of moderate amounts of alcohol on appetite and digestion, as a result of this stimulatory effect on gastric secretion, have been known since ancient times.

Regulation of Gastric Motility & Emptying

The rate at which the stomach empties into the duodenum depends on the type of food ingested. Food rich in carbohydrate leaves the stomach in a few hours. Protein-rich food leaves more slowly, and emptying is

slowest after a meal containing fat (Figure 26–14). The rate of emptying also depends on the osmotic pressure of the material entering the duodenum. Hyperosmolality of the duodenal contents is sensed by "duodenal osmoreceptors" that initiate a decrease in gastric emptying which is probably neural in origin.

Products of protein digestion and hydrogen ions bathing the duodenal mucosa initiate a neurally mediated decrease in gastric motility, the **enterogastric reflex.** Distention of the duodenum also initiates this reflex. Peptide YY inhibits gastric motility. Cutting the vagus nerves slows gastric emptying. In humans, vagotomy may cause relatively severe gastric atony and distention. Excitement is said to hasten gastric emptying, and fear to slow it.

Since fats are particularly effective in inhibiting gastric emptying, some people drink milk, cream, or even olive oil before a cocktail party. The fat keeps the alcohol in the stomach for a long time, where its absorption is slower than in the small intestine, and the intoxicant enters the small intestine in a slow, steady stream so that—theoretically, at least—a sudden rise of the blood alcohol to a high level and consequent embarrassing intoxication are avoided.

Peptic Ulcer

Gastric and duodenal ulceration in humans is related primarily to a breakdown of the barrier that normally prevents irritation and autodigestion of the mucosa by

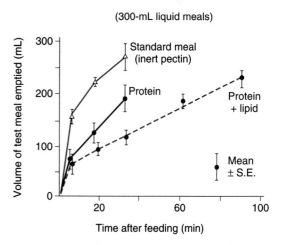

Figure 26–14. Effect of protein and fat on the rate of emptying of the human stomach. Subjects were fed 300-mL liquid meals. (Reproduced, with permission, from Brooks FP: Integrative lecture. Response of the GI tract to a meal. *Undergraduate Teaching Project.* American Gastroenterological Association, 1974.)

the gastric secretions. Infection with the bacterium *Helicobacter pylori* disrupts this barrier. So do aspirin and other nonsteroidal anti-inflammatory drugs (NSAIDs), which inhibit the production of prostaglandins and consequently decrease mucus and HCO_3^- secretion (see above). The NSAIDs are widely used to combat pain and treat arthritis. An additional cause of ulceration is prolonged excess secretion of acid. An example of this is the ulcers that occur in the **Zollinger-Ellison syndrome.** This syndrome is seen in patients with gastrinomas. These tumors can occur in the stomach and duodenum, but most of them are found in the pancreas. The gastrin causes prolonged hypersecretion of acid, and severe ulcers are produced.

Gastric and duodenal ulcers can be given a chance to heal by inhibition of acid secretion with drugs such as cimetidine that block the H_2 histamine receptors on parietal cells (Figure 26–12) or omeprazole and related drugs that inhibit H^+-K^+ ATPase. *Helicobacter pylori* can be eradicated with antibiotics, and NSAID-induced ulcers can be treated by stopping the NSAID or, when this is not advisable, by treatment with the prostaglandin agonist misoprostol. Gastrinomas can be removed surgically.

Other Functions of the Stomach

The stomach has a number of functions in addition to the storage of food and the control of its release into the duodenum.

In addition to hydrochloric acid, the parietal cells in the gastric mucosa secrete **intrinsic factor,** a 49-kDa glycoprotein that binds to **cyanocobalamin** (vitamin B_{12}) and is necessary for its absorption from the small intestine.

Cyanocobalamin (Figure 26–15) is a complex cobalt-containing vitamin. Deficiency of this vitamin causes an anemia characterized by the appearance in the bloodstream of large red blood cells called megaloblasts and deterioration of certain sensory pathways in the CNS. A complete remission of the deficiency syndrome occurs when cyanocobalamin is injected parenterally but not when it is administered by mouth unless the intrinsic factor secreted by the gastric mucosa is present. Deficiency due to an inadequate dietary intake of cyanocobalamin is very rare, apparently because the minimum daily requirements are quite low and the vitamin is found in most foods of animal origin.

Vitamin B_{12} normally binds to intrinsic factor, and the complex is taken up by **cubilin,** a high-affinity apolipoprotein in receptors in the distal ileum. This triggers absorption of the complex by endocytosis. In the ileal enterocytes, the cyanocobalamin is transferred to **transcobalamin II,** a cyancobalamin transport protein that transports the vitamin in plasma.

Cyanocobalamin deficiency is produced by gastrectomy, with removal of the intrinsic factor-secreting tissue, and by **pernicious anemia,** a disease in which there is autoimmune destruction of the parietal cells. Vitamin B_{12} deficiency can also be produced by diseases of the distal ileum.

In totally gastrectomized patients, the intrinsic factor deficiency must be circumvented by parenteral injection of cyanocobalamin. Protein digestion is normal in the absence of pepsin, and nutrition can be maintained. However, these patients are prone to develop iron deficiency anemia (see Chapter 25) and other abnormalities, and they must eat frequent small meals. Because of rapid absorption of glucose from the intestine and the resultant hyperglycemia and abrupt rise in insulin secretion, gastrectomized patients sometimes develop hypoglycemic symptoms about 2 hours after meals (see Chapter 19). Weakness, dizziness, and sweating after meals, due in part to hypoglycemia, are part of the picture of the **"dumping syndrome,"** a distressing syndrome that develops in patients in whom portions of the stomach have been removed or the jejunum has

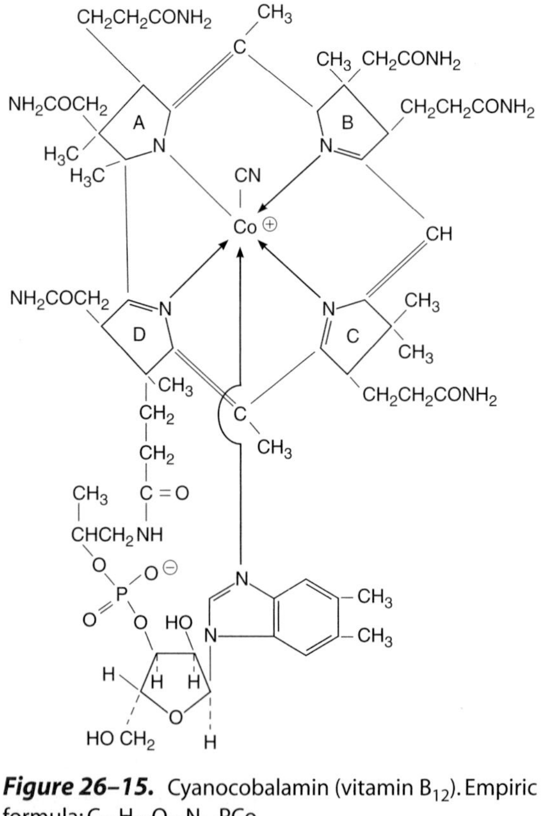

Figure 26–15. Cyanocobalamin (vitamin B_{12}). Empiric formula: $C_{63}H_{88}O_{14}N_{14}PCo$.

been anastomosed to the stomach. Another cause of the symptoms is rapid entry of hypertonic meals into the intestine; this provokes the movement of so much water into the gut that significant hypovolemia and hypotension are produced.

EXOCRINE PORTION OF THE PANCREAS

The pancreatic juice contains enzymes that are of major importance in digestion (see Table 25–1). Its secretion is controlled in part by a reflex mechanism and in part by the gastrointestinal hormones secretin and CCK.

Anatomic Considerations

The portion of the pancreas that secretes pancreatic juice is a compound alveolar gland resembling the salivary glands. Granules containing the digestive enzymes **(zymogen granules)** are formed in the cell and discharged by exocytosis (see Chapter 1) from the apexes of the cells into the lumens of the pancreatic ducts (Figure 26–16). The small duct radicles coalesce into a single duct (pancreatic duct of Wirsung), which usually joins the common bile duct to form the ampulla of Vater (Figure 26–17). The ampulla opens through the duodenal papilla, and its orifice is encircled by the sphincter of Oddi. In some individuals, there is an accessory pancreatic duct (duct of Santorini) that enters the duodenum more proximally.

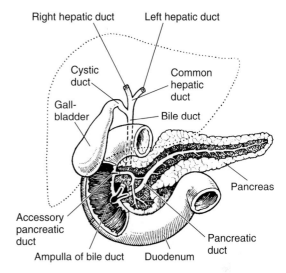

Figure 26–17. Connections of the ducts of the gallbladder, liver, and pancreas. (Reproduced, with permission, from Bell GH, Emslie-Smith D, Paterson CR: *Textbook of Physiology and Biochemistry*, 9th ed. Churchill Livingstone, 1976.)

Composition of Pancreatic Juice

The pancreatic juice is alkaline (Table 26–6) and has a high HCO_3^- content (approximately 113 meq/L versus

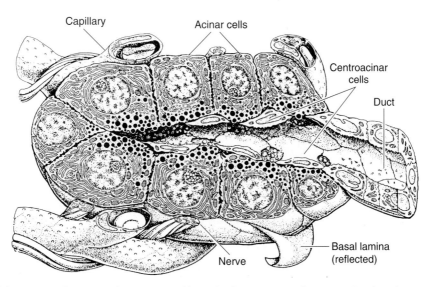

Figure 26–16. Acinar cells surrounding a terminal branch of a pancreatic duct. Note the abundant rough endoplasmic reticulum and the zymogen granules concentrated at the apexes of the cells. (Reproduced, with permission, from Krstic RV: *Die Gewebe des Menschen und der Säugetiere*. Springer, 1978.)

Table 26–6. Composition of normal human pancreatic juice.

Cations: Na$^+$, K$^+$, Ca^{2+}, Mg^{2+} (pH approximately 8.0)
Anions: HCO$_3^-$, Cl$^-$, SO$_4^{2-}$, HPO$_4^{2-}$
Digestive enzymes (see Table 25–1; 95% of protein in juice)
Other proteins

24 meq/L in plasma). About 1500 mL of pancreatic juice is secreted per day. Bile and intestinal juices are also neutral or alkaline, and these three secretions neutralize the gastric acid, raising the pH of the duodenal contents to 6.0–7.0. By the time the chyme reaches the jejunum, its reaction is nearly neutral, but the intestinal contents are rarely alkaline.

The powerful protein-splitting enzymes of the pancreatic juice are secreted as inactive proenzymes. Trypsinogen is converted to the active enzyme trypsin by the brush border enzyme **enteropeptidase (enterokinase)** when the pancreatic juice enters the duodenum. Enteropeptidase contains 41% polysaccharide, and this high polysaccharide content apparently prevents it from being digested itself before it can exert its effect. Trypsin converts chymotrypsinogens into chymotrypsins and other proenzymes into active enzymes (Figure 26–18). Trypsin can also activate trypsinogen; therefore, once some trypsin is formed, there is an autocatalytic chain reaction. Enteropeptidase deficiency occurs as a congenital abnormality and leads to protein malnutrition.

The potential danger of the release into the pancreas of a small amount of trypsin is apparent; the resulting chain reaction would produce active enzymes that could digest the pancreas. It is therefore not surprising that the pancreas normally contains a trypsin inhibitor.

Another enzyme activated by trypsin is phospholipase A$_2$. This enzyme splits a fatty acid off lecithin, forming lysolecithin. Lysolecithin damages cell membranes. It has been hypothesized that in **acute pancreatitis,** a severe and sometimes fatal disease, phospholipase A$_2$ is activated in the pancreatic ducts, with the formation of lysolecithin from the lecithin that is a normal constituent of bile. This causes disruption of pancreatic tissue and necrosis of surrounding fat.

Small amounts of pancreatic digestive enzymes normally leak into the circulation, but in acute pancreatitis, the circulating levels of the digestive enzymes rise markedly. Measurement of the plasma amylase or lipase concentration is therefore of value in diagnosing the disease.

Regulation of the Secretion of Pancreatic Juice

Secretion of pancreatic juice is primarily under hormonal control. Secretin acts on the pancreatic ducts to cause copious secretion of a very alkaline pancreatic juice that is rich in HCO$_3^-$ and poor in enzymes. The effect on duct cells is due to an increase in intracellular cAMP. Secretin also stimulates bile secretion. CCK acts on the acinar cells to cause the release of zymogen granules and production of pancreatic juice rich in enzymes but low in volume. Its effect is mediated by phospholipase C (see Chapter 1).

The response to intravenous secretin is shown in Figure 26–19. Note that as the volume of pancreatic secretion increases, its Cl$^-$ concentration falls and its HCO$_3^-$ concentration increases. Although HCO$_3^-$ is secreted in the small ducts, it is reabsorbed in the large ducts in exchange for Cl$^-$. The magnitude of the exchange is inversely proportionate to the rate of flow.

Like CCK, acetylcholine acts on acinar cells via phospholipase C to cause discharge of zymogen granules, and stimulation of the vagi causes secretion of a small amount of pancreatic juice rich in enzymes. There is evidence for vagally mediated conditioned reflex secretion of pancreatic juice in response to the sight or smell of food.

LIVER & BILIARY SYSTEM

Bile is secreted by the cells of the liver into the bile duct, which drains into the duodenum. Between meals, the duodenal orifice of this duct is closed and bile flows into the gallbladder, where it is stored. When food enters the mouth, the sphincter around the orifice relaxes; when the gastric contents enter the duodenum, the hormone CCK from the intestinal mucosa causes the gallbladder to contract.

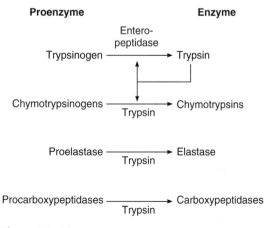

Figure 26–18. Activation of the pancreatic proteases in the duodenal lumen.

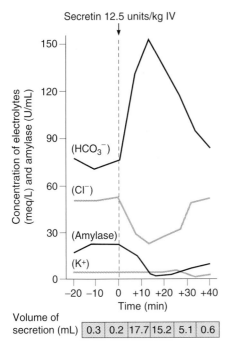

Secretin 12.5 units/kg IV

Volume of secretion (mL)	0.3	0.2	17.7	15.2	5.1	0.6

Figure 26–19. Effect of a single dose of secretin on the composition and volume of the pancreatic juice in humans.

Anatomic Considerations

The liver is organized in lobules, within which blood flows past hepatic cells via sinusoids from branches of the portal vein to the central vein of each lobule. There are large gaps between endothelial cells, and plasma is in close contact with liver cells (Figure 26–20). There is usually only one layer of hepatocytes between sinusoids, so the total area of contact between liver cells and plasma is very large. Hepatic artery blood also enters the sinusoids. The central veins coalesce to form the hepatic veins, which drain into the inferior vena cava. Hepatic blood flow is discussed in more detail in Chapter 32. The average transit time for blood across the liver lobule from the portal venule to the central hepatic vein is about 8.4 seconds. Numerous macrophages (**Kupffer cells**) are anchored to the endothelium of the sinusoids and project into the lumen. The functions of these phagocytic cells are discussed in Chapter 27.

Another way of looking at the organization of the liver that has functional implications is its division into hepatic acini. The center of each acinus is a vascular stalk containing terminal branches of portal veins, hepatic arteries, and bile ducts (see Figure 32–16). Blood flows from the vascular stalk to terminal hepatic venules located on the outside of the acinus. This way,

the cells closest to the vascular stalk receive the best oxygenated blood and the cells on the periphery of the acinus are least well oxygenated and hence most subject to anoxic injury.

Each liver cell is also apposed to several bile canaliculi (Figure 26–21). The canaliculi drain into intralobular bile ducts, and these coalesce via interlobular bile ducts to form the right and left hepatic ducts. These ducts join outside the liver to form the common hepatic duct. The cystic duct drains the gallbladder. The hepatic duct unites with the cystic duct to form the common bile duct. The common bile duct enters the duodenum at the duodenal papilla. Its orifice is surrounded by the sphincter of Oddi, and it usually unites with the main pancreatic duct just before entering the duodenum (Figure 26–17).

The walls of the extrahepatic biliary ducts and the gallbladder contain fibrous tissue and smooth muscle. The mucous membrane contains mucous glands and is lined by a layer of columnar cells. In the gallbladder, the mucous membrane is extensively folded; this increases its surface area and gives the interior of the gallbladder a honeycombed appearance. In primates, the mucous membranes of the cystic duct are also folded to form the so-called spiral valves.

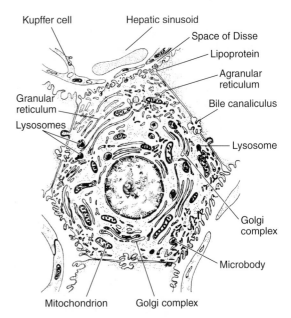

Figure 26–20. Hepatocyte. Note the relation of the cell to bile canaliculi and sinusoids. Note also the wide openings between the endothelial cells next to the hepatocyte. (Reproduced, with permission, from Fawcett DW: *Bloom and Fawcett, A Textbook of Histology,* 11th ed. Saunders, 1986.)

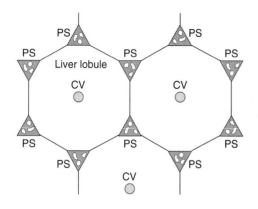

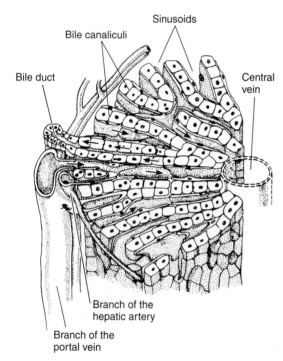

Figure 26–21. Top: Organization of the liver. CV, central vein. PS, portal space containing branches of bile duct, portal vein, and hepatic artery. **Bottom:** Arrangement of plates of liver cells, sinusoids, and bile ducts in a liver lobule, showing centripetal flow of blood in sinusoids to central vein and centrifugal flow of bile in bile canaliculi to bile ducts. (Reproduced, with permission, from Fawcett DW: *Bloom and Fawcett, A Textbook of Histology*, 11th ed. Saunders, 1986.)

Functions of the Liver

The liver, the largest gland in the body, has many complex functions. These are summarized in Table 26–7. Discussion of these functions in one place requires that they be taken out of their proper context at considerable cost in terms of clarity and integration of function. Therefore, while some of them are discussed in this chapter, others are discussed individually in the chapters on the systems of which each is a part.

Synthesis of Plasma Proteins

The principal proteins synthesized by the liver are listed in Table 27–9. Many of them are **acute-phase proteins,** proteins synthesized and secreted into the plasma upon exposure to stressful stimuli. Others are proteins that transport steroids and other hormones in the plasma, and still others are clotting factors.

Bile

Bile is made up of the bile salts, bile pigments, and other substances dissolved in an alkaline electrolyte solution that resembles pancreatic juice (Table 26–8). About 500 mL is secreted per day. Some of the components of the bile are reabsorbed in the intestine and then excreted again by the liver (**enterohepatic circulation**).

Table 26–7. Principal functions of the liver.[1]

Formation and secretion of bile (26)
Nutrient and vitamin metabolism (17, 27)
Glucose and other sugars
Amino acids
Lipids
Fatty acids
Cholesterol
Lipoproteins
Fat-soluble vitamins
Water-soluble vitamins
Inactivation of various substances
Toxins (17)
Steroids (20, 23, 26)
Other hormones (14, 18)
Synthesis of plasma proteins (26, 27; see Table 27–9)
Acute-phase proteins
Albumin
Clotting factors
Steroid-binding and other hormone-binding proteins
Immunity (27)
Kupffer cells

[1]Numbers in parentheses are chapters in this book in which the functions are considered.

Table 26–8. Composition of human hepatic duct bile.

Water	97.0%
Bile salts	0.7%
Bile pigments	0.2%
Cholesterol	0.06%
Inorganic salts	0.7%
Fatty acids	0.15%
Lecithin	0.1%
Fat	0.1%
Alkaline phosphatase	…

The glucuronides of the **bile pigments,** bilirubin and biliverdin, are responsible for the golden-yellow color of bile. The formation of these breakdown products of hemoglobin is discussed in detail in Chapter 27, and their excretion is discussed below.

The **bile salts** are sodium and potassium salts of **bile acids,** and all those secreted into the bile are conjugated to glycine or taurine, a derivative of cysteine. The bile acids are synthesized from cholesterol. The four found in humans are listed in Figure 26–22. In common with vitamin D, cholesterol, a variety of steroid hormones, and the digitalis glycosides, the bile acids contain the cyclopentanoperhydrophenanthrene nucleus (see Chapter 20). The two principal (primary) bile acids formed in the liver are cholic acid and chenodeoxycholic acid. In the colon, bacteria convert cholic acid to deoxycholic acid and chenodeoxycholic acid to lithocholic acid. Since they are formed by bacterial action, deoxycholic acid and lithocholic acid are called secondary bile acids.

The bile salts have a number of important actions. They reduce surface tension and, in conjunction with phospholipids and monoglycerides, are responsible for the emulsification of fat preparatory to its digestion and absorption in the small intestine (see Chapter 25). They are **amphipathic,** ie, they have both hydrophilic and hydrophobic domains; one surface of the molecule is hydrophilic because the polar peptide bond and the carboxyl and hydroxyl groups are on that surface, whereas the other surface is hydrophobic. Therefore, the bile salts tend to form cylindrical disks called **micelles,** with the hydrophilic surfaces facing out and the hydrophobic surfaces facing in (Figure 26–23). Above a certain concentration called the **critical micelle concentration,** all bile salts added to a solution form micelles. Lipids collect in the micelles, with cholesterol in the hydrophobic center and amphipathic phospholipids and monoglycerides lined up with their hydrophilic heads on the outside and their hydrophobic tails in the center. The micelles play an important role in keeping lipids in solution and transporting them to the brush border of the intestinal epithelial cells, where they are absorbed (see Chapter 25).

Ninety to 95 percent of the bile salts are absorbed from the small intestine. Some are absorbed by nonionic diffusion, but most are absorbed from the terminal ileum (Figure 26–24) by an extremely efficient Na^+-bile salt cotransport system powered by basolateral Na^+-K^+ ATPase. One of the Na^+-bile salt cotransporters involved in this secondary active transport system has been cloned, and there is evidence that at least one more exists.

The remaining 5–10% of the bile salts enter the colon and are converted to the salts of deoxycholic acid

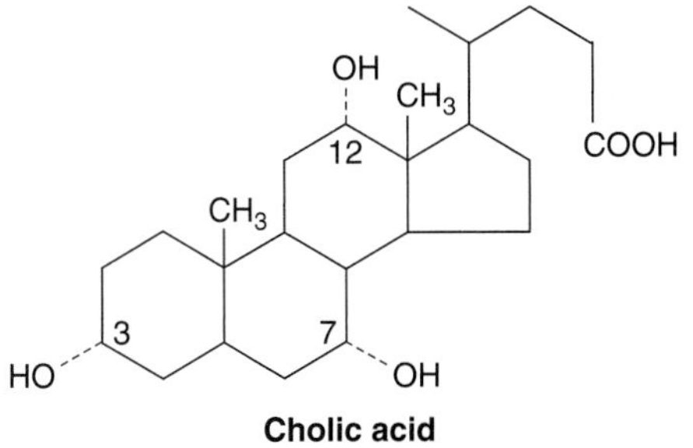

Cholic acid

	Group at position			Percent in human bile
	3	7	12	
Cholic acid	OH	OH	OH	50
Chenodeoxycholic acid	OH	OH	H	30
Deoxycholic acid	OH	H	OH	15
Lithocholic acid	OH	H	H	5

Figure 26–22. Human bile acids. The numbers in the formula for cholic acid refer to the positions in the steroid ring.

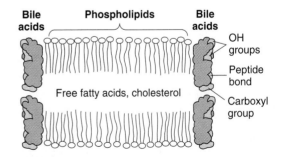

Figure 26–23. Cross section of disk-shaped bile acid-lipid mixed micelle with free fatty acids and cholesterol in its hydrophobic interior. The surface of each bile acid that faces outward is hydrophilic because of the polar peptide bond and the carboxyl and OH groups.

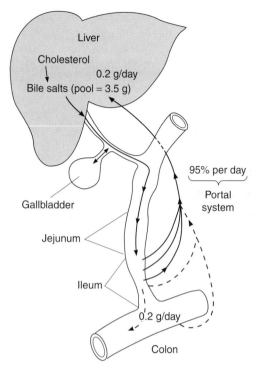

Figure 26–24. Enterohepatic circulation of bile salts. The solid lines entering the portal system represent bile salts of hepatic origin, whereas the dashed lines represent bile salts resulting from bacterial action.

and lithocholic acid. Lithocholate is relatively insoluble and is mostly excreted in the stools; only 1% is absorbed. However, deoxycholate is absorbed.

The absorbed bile salts are transported back to the liver in the portal vein and reexcreted in the bile (enterohepatic circulation). Those lost in the stool are replaced by synthesis in the liver; the normal rate of bile salt synthesis is 0.2–0.4 g/d. The total bile salt pool of approximately 3.5 g recycles repeatedly via the enterohepatic circulation; it has been calculated that the entire pool recycles twice per meal and six to eight times per day. When bile is excluded from the intestine, up to 50% of ingested fat appears in the feces. There is also severe malabsorption of fat-soluble vitamins. When bile salt reabsorption is prevented by resection of the terminal ileum or by disease in this portion of the small intestine, the amount of fat in the stools is also increased because when the enterohepatic circulation is interrupted, the liver cannot increase the rate of bile salt production to a sufficient degree to compensate for the loss. The other effects of resection of the terminal ileum are discussed in the section on malabsorption.

Bilirubin Metabolism & Excretion

Most of the bilirubin in the body is formed in the tissues by the breakdown of hemoglobin (see Chapter 27). The bilirubin is bound to albumin in the circulation. Some of it is tightly bound, but most of it can dissociate in the liver, and free bilirubin enters liver cells, where it is bound to cytoplasmic proteins (Figure 26–25). It is next conjugated to glucuronic acid in a reaction catalyzed by the enzyme **glucuronyl transferase** (UDP-glucuronosyltransferase). This enzyme is located primarily in the smooth endoplasmic reticulum. Each bilirubin molecule reacts with two uridine diphosphoglucuronic acid (UDPGA) molecules to form bilirubin diglucuronide. This glucuronide, which is more water-soluble than the free bilirubin, is then transported against a concentration gradient by a presumably active process into the bile canaliculi. A small amount of the bilirubin glucuronide escapes into the blood, where it is bound less tightly to albumin than is free bilirubin, and is excreted in the urine. Thus, the total plasma bilirubin normally includes free bilirubin plus a small amount of conjugated bilirubin. Most of the bilirubin glucuronide passes via the bile ducts to the intestine.

The intestinal mucosa is relatively impermeable to conjugated bilirubin but is permeable to unconjugated bilirubin and to urobilinogens, a series of colorless derivatives of bilirubin formed by the action of bacteria in the intestine. Consequently, some of the bile pigments and urobilinogens are reabsorbed in the portal circulation. Some of the reabsorbed substances are again excreted by the liver (enterohepatic circulation), but small amounts of urobilinogens enter the general circulation and are excreted in the urine.

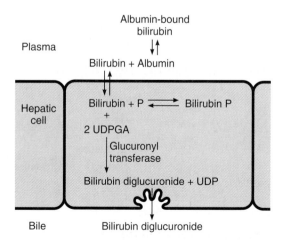

Figure 26–25. Metabolism of bilirubin in the liver. P, intracellular binding proteins; UDPGA, uridine diphosphoglucuronic acid; UDP, uridine diphosphate.

Jaundice

When free or conjugated bilirubin accumulates in the blood, the skin, scleras, and mucous membranes turn yellow. This yellowness is known as **jaundice** (icterus) and is usually detectable when the total plasma bilirubin is greater than 2 mg/dL (34 μmol/L). Hyperbilirubinemia may be due to (1) excess production of bilirubin (hemolytic anemia, etc; see Chapter 27); (2) decreased uptake of bilirubin into hepatic cells; (3) disturbed intracellular protein binding or conjugation; (4) disturbed secretion of conjugated bilirubin into the bile canaliculi; or (5) intrahepatic or extrahepatic bile duct obstruction. When it is due to one of the first three processes, the free bilirubin rises. When it is due to disturbed secretion of conjugated bilirubin or bile duct obstruction, bilirubin glucuronide regurgitates into the blood, and it is predominantly the conjugated bilirubin in the plasma that is elevated.

Other Substances Conjugated by Glucuronyl Transferase

The glucuronyl transferase system in the smooth endoplasmic reticulum catalyzes the formation of the glucuronides of a variety of substances in addition to bilirubin. The list includes steroids (see Chapters 20 and 23) and various drugs. These other compounds can compete with bilirubin for the enzyme system when they are present in appreciable amounts. In addition, several barbiturates, antihistamines, anticonvulsants, and other compounds cause marked proliferation of the smooth endoplasmic reticulum in the hepatic cells, with a concurrent increase in hepatic glucuronyl transferase activity. Phenobarbital has been used successfully for the treatment of a congenital disease in which there is a relative deficiency of glucuronyl transferase (type 2 UDP-glucuronosyltransferase deficiency).

Other Substances Excreted in the Bile

Cholesterol and alkaline phosphatase are excreted in the bile. In patients with jaundice due to intra- or extrahepatic obstruction of the bile duct, the blood levels of these two substances usually rise; a much smaller rise is generally seen when the jaundice is due to nonobstructive hepatocellular disease. Adrenocortical and other steroid hormones and a number of drugs are excreted in the bile and subsequently reabsorbed (enterohepatic circulation).

Functions of the Gallbladder

In normal individuals, bile flows into the gallbladder when the sphincter of Oddi is closed. In the gallbladder, the bile is concentrated by absorption of water.

The degree of this concentration is shown by the increase in the concentration of solids (Table 26–9); liver bile is 97% water, whereas the average water content of gallbladder bile is 89%. When the bile duct and cystic duct are clamped, the intrabiliary pressure rises to about 320 mm of bile in 30 minutes, and bile secretion stops. However, when the bile duct is clamped and the cystic duct is left open, water is reabsorbed in the gallbladder, and the intrabiliary pressure rises only to about 100 mm of bile in several hours. Acidification of the bile is another function of the gallbladder (Table 26–9).

Regulation of Biliary Secretion

When food enters the mouth, the resistance of the sphincter of Oddi decreases. Fatty acids and amino acids in the duodenum release CCK, which causes gallbladder contraction. Substances that cause contraction of the gallbladder are called **cholagogues.**

The production of bile is increased by stimulation of the vagus nerves and by the hormone secretin, which increases the water and HCO_3^- content of bile. Substances that increase the secretion of bile are known as **choleretics.** Bile salts themselves are among the most important physiologic choleretics.

Effects of Cholecystectomy

The periodic discharge of bile from the gallbladder aids digestion but is not essential for it. Cholecystectomized patients maintain good health and nutrition with a constant slow discharge of bile into the duodenum, although eventually the bile duct becomes somewhat dilated, and more bile tends to enter the duodenum after meals than at other times. Cholecystectomized patients can even tolerate fried foods, although they generally must avoid foods that are particularly high in fat content.

Visualizing the Gallbladder

Exploration of the right upper quadrant with an ultrasonic beam (**ultrasonography**) and computed tomog-

Table 26–9. Comparison of human hepatic duct bile and gallbladder bile.

	Hepatic Duct Bile	Gallbladder Bile
Percentage of solids	2–4	10–12
Bile salts (mmol/L)	10–20	50–200
pH	7.8–8.6	7.0–7.4

raphy (CT) have become the most widely used methods for visualizing the gallbladder and detecting gallstones. A third method of diagnosing gallbladder disease is **nuclear cholescintigraphy.** When administered intravenously, technetium-99m-labeled derivatives of iminodiacetic acid are excreted in the bile and provide excellent gamma camera images of the gallbladder and bile ducts. The response of the gallbladder to CCK can then be observed following intravenous administration of the hormone.

Gallstones

Cholelithiasis, ie, the presence of gallstones, is a common condition. Its incidence increases with age, so that in the United States, for example, 20% of the women and 5% of the men between the ages of 50 and 65 have gallstones. The stones are of two types: calcium bilirubinate stones and cholesterol stones. In the United States and Europe, 85% of the stones are cholesterol stones.

Three factors appear to be involved in the formation of cholesterol stones. One is bile stasis; stones form in the bile that is sequestrated in the gallbladder rather than the bile that is flowing in the bile ducts. A second is supersaturation of the bile with cholesterol. Cholesterol is very insoluble in bile, and it is maintained in solution in micelles only at certain concentrations of bile salts and lecithin (Figure 26–26). At concentrations above line ABC in Figure 26–26, the bile is supersaturated and contains small crystals of cholesterol in addition to micelles. However, many normal individuals who do not develop gallstones also have supersaturated bile. The third factor is a mix of nucleation factors that favors formation of stones from the supersaturated bile. Outside the body, bile from patients with cholelithiasis forms stones in 2–3 days, whereas it takes more than 2 weeks for stones to form in bile from normal individuals. The exact nature of the nucleation factors is unsettled, although glycoproteins in gallbladder mucus have been implicated. In addition, it is unsettled whether stones form as a result of excess production of components that favor nucleation or decreased production of antinucleation components that prevent stones from forming in normal individuals.

SMALL INTESTINE

In the small intestine, the intestinal contents are mixed with the secretions of the mucosal cells and with pancreatic juice and bile. Digestion, which begins in the mouth and stomach, is completed in the lumen and mucosal cells of the small intestine, and the products of digestion are absorbed, along with most of the vitamins and fluid. The small intestine is presented with about 9

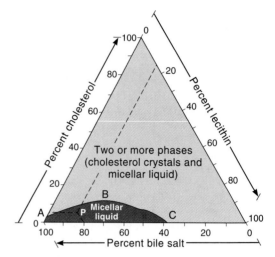

Figure 26–26. Cholesterol solubility in bile as a function of the proportions of lecithin, bile salts, and cholesterol. In bile that has a composition described by any point below line ABC (eg, point P), cholesterol is solely in micellar solution; points above line ABC describe bile in which there are cholesterol crystals as well. (Reproduced, with permission, from Small DM: Gallstones. N Engl J Med 1968;279:588.)

L of fluid per day—2 L from dietary sources and 7 L of gastrointestinal secretions (see Table 25–3); however, only 1–2 L pass into the colon.

Anatomic Considerations

The general arrangement of the muscular layers, nerve plexuses, and mucosa in the small intestine is shown in Figure 26–1. The first portion of the duodenum is sometimes called the duodenal cap or bulb. It is the region struck by the acid gastric contents squirted through the pylorus and is a common site of peptic ulceration. At the ligament of Treitz, the duodenum becomes the jejunum. Arbitrarily, the upper 40% of the small intestine below the duodenum is called the jejunum and the lower 60% the ileum, although there is no sharp anatomic boundary between the two. The ileocecal valve marks the point where the ileum ends in the colon.

The small intestine is shorter during life than it is in cadavers; it relaxes and elongates after death. The distance from the pylorus to the ileocecal valve in living humans has been reported to be 285 cm (Table 26–10), but at autopsy, this distance is about 700 cm.

The mucosa of the small intestine contains **solitary lymphatic nodules** and, especially in the ileum, **aggregated lymphatic nodules** (Peyer's patches) along the

Table 26–10. Mean lengths of various segments of the gastrointestinal tract as measured by intubation in living humans.[1]

Segment	Length (cm)
Pharynx, esophagus, and stomach	65
Duodenum	25
Jejunum and ileum	260
Colon	110

[1]Data from Hirsch JE, Ahrens EH Jr, Blankenhorn DH: Measurement of human intestinal length in vivo and some causes of variation. Gastroenterology 1956;31:274.

antimesenteric border. Throughout the small intestine there are simple tubular **intestinal glands** (crypts of Lieberkühn). In the duodenum there are in addition the small, coiled acinotubular **duodenal glands** (Brunner's glands). As noted above, the epithelium of the small intestine contains various kinds of enteroendocrine cells, and there are many valve-like folds **(valvulae conniventes)** in the mucous membrane.

Throughout the length of the small intestine, the mucous membrane is covered by **villi** (Figure 26–27). There are 20–40 villi per square millimeter of mucosa. Each intestinal villus is a finger-like projection, 0.5–1 mm long, covered by a single layer of columnar epithelium and containing a network of capillaries and a lymphatic vessel **(lacteal).** Fine extensions of the smooth muscle of the submucosa run longitudinally up each villus to its tip. The free edges of the cells of the epithelium of the villi are divided into minute **microvilli.** These in turn are covered by the **glycocalyx,** an amorphous layer rich in neutral and amino sugars. The microvilli make up the **brush border** (Figure 26–28). The cells are connected to each other by tight junctions. The outer layer of the cell membrane of the mucosal cells contains many of the enzymes involved in the digestive processes initiated by salivary, gastric, and pancreatic enzymes. Enzymes found in this membrane include various disaccharidases, peptidases, and enzymes involved in the breakdown of nucleic acids (see Chapter 25).

The absorptive surface of the small intestine is increased about 600-fold by the valvulae conniventes, villi, and microvilli. It has been estimated that the inner surface area of a mucosal cylinder the size of the small intestine would be about 3300 cm^2, that the valvulae increase the surface area to 10,000 cm^2, that the villi increase it to 100,000 cm^2, and that the microvilli increase it to 2 million cm^2.

The enterocytes in the small intestine are formed from mitotically active undifferentiated cells in the crypts of Lieberkühn. They migrate up to the tips of the villi, where they undergo apoptosis and are sloughed into the intestinal lumen in large numbers (Figure 26–27). The average life of these cells is 2–5 days, depending on the species. The number of cells shed per day has been calculated to be about 17 billion in humans, and the amount of protein "secreted" in this fashion is about 30 g/d. Mucosal cells are also rapidly sloughed and replaced in the stomach. The crypts are also the site of cAMP-mediated secretion of water and electrolytes (see Chapter 25).

Paneth cells—endocrine cells located in the depths of the crypts of Lieberkühn—secrete **defensins,** naturally occurring peptide antibiotics that are also secreted elsewhere in the body (see Chapter 27). The migrating

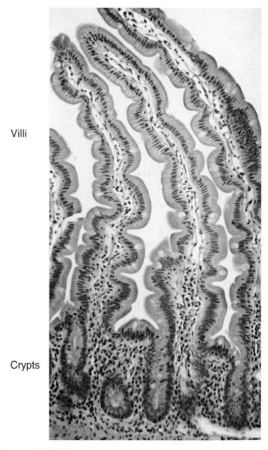

Villi

Crypts

Figure 26–27. Mucosa of the small intestine, showing crypts and villi lined with enterocytes. The enterocytes originate in the crypts, migrate up the villi, and are shed into the intestinal lumen at the tips of the villi. (Modified from Chandrasoma P, Taylor CR: *Concise Pathology,* 3rd ed. Originally published by Appleton & Lange. Copyright © 1998 by the McGraw-Hill Companiies, Inc.)

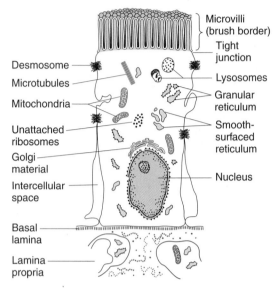

Figure 26–28. Diagram of enterocyte from human small intestine. Note the microvilli, the tight connections to other cells at the mucosal edge, and the space between cells at the base (intercellular space).

enterocytes are exposed to a high concentrate of the defensins, and this may protect them as they move to the tops of the villi. Paneth cells may also secret guanylin (see above).

Intestinal Mucus

As noted above, mucus is secreted by surface epithelial cells throughout the gastrointestinal tract, by Brunner's glands in the duodenum, and by characteristic goblet cells in the mucosa of the small and large intestine. In addition to its protective function on the surface of the mucosa, the mucus gel lubricates the food and holds immunoglobulins in place. Mucin secretion is accelerated by cholinergic stimulation and by chemical and physical irritation. Somewhat different mucins are secreted by different goblet cells. The composition of mucins is altered in individuals with tumors of the small intestine and in patients with ulcerative colitis.

Intestinal Motility

The MMCs that pass along the intestine at regular intervals in the fasting state and their replacement by peristaltic and other contractions controlled by the BER are described above. In the small intestine, there are an average of 12 BER cycles/min in the proximal jejunum, declining to 8/min in the distal ileum. There are three types of smooth muscle contractions: peri-

staltic waves, segmentation contractions, and tonic contractions. **Peristalsis** is described above. It propels the intestinal contents (**chyme**) toward the large intestines. **Segmentation contractions** are ring-like contractions that appear at fairly regular intervals along the gut and then disappear and are replaced by another set of ring contractions in the segments between the previous contractions (Figure 26–29). They move the chyme to and fro and increase its exposure to the mucosal surface. They are initiated by focal increases in Ca^{2+} influx with waves of increased Ca^{2+} concentration spreading from each focus. **Tonic contractions** are relatively prolonged contractions that in effect isolate one segment of the intestine from another. Note that these last two types of contraction slow transit in the small intestine to the point that the transit time is actually longer in the fed than in the fasted state. This permits longer contact of the chyme with the enterocytes and fosters absorption.

Very intense peristaltic waves called **peristaltic rushes** are not seen in normal individuals, but they do occur when the intestine is obstructed. Weak antiperistalsis is sometimes seen in the colon, but most waves pass regularly in an oral-caudal direction.

Regulation of Intestinal Secretion

The intestinal glands secrete an isotonic fluid. Most of the enzymes usually found in this secretion are in desquamated mucosal cells; cell-free intestinal juice probably contains few if any enzymes. Gastrointestinal hormones and other polypeptides such as VIP (see above) stimulate the secretion of intestinal juice.

The Malabsorption Syndrome

The digestive and absorptive functions of the small intestine are essential for life. Removal of short segments of the jejunum or ileum generally does not cause severe

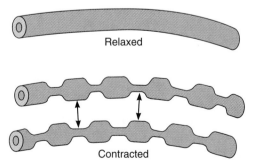

Relaxed

Contracted

Figure 26–29. Diagram of segmentation contractions of the intestine. Arrows indicate how areas of relaxation become areas of constriction and vice versa.

symptoms, and there is compensatory hypertrophy and hyperplasia of the remaining mucosa, with gradual return of the absorptive function toward normal (**intestinal adaptation**). This adaptation is partly due to a direct effect of nutrients in the intestinal lumen on the mucosa and partly due to circulating factors such as gastrointestinal hormones. However, when more than 50% of the small intestine is resected or bypassed, the absorption of nutrients and vitamins is so compromised that it is very difficult to prevent malnutrition and wasting (**malabsorption**).

The increased gastric acid secretion produced by intestinal resection has been mentioned above. Resection of the ileum prevents the absorption of bile acids, and this leads in turn to deficient fat absorption. It also causes diarrhea because the unabsorbed bile salts enter the colon, where they increase adenylyl cyclase activity, thus increasing intestinal secretion (see Chapter 25) and the entry of small molecules into the lumen. Because the capacity of the jejunum to adapt is lower than that of the ileum, distal small bowel resection causes a greater degree of malabsorption than removal of a comparable length of proximal small bowel. Other complications of intestinal resection or bypass include hypocalcemia, arthritis, hyperuricemia, and possibly fatty infiltration of the liver, followed by cirrhosis. Operations in which segments of the small intestine are bypassed have been recommended for the treatment of obesity. However, bypass operations should not be undertaken lightly in view of their complications and dangers. Various disease processes also impair absorption (Table 26–11). The pattern of deficiencies that results is sometimes called the **malabsorption syndrome**. This pattern varies somewhat with the cause, but it can include deficient absorption of amino acids, with marked body wasting and, eventually, hypoproteinemia and edema. Carbohydrate and fat absorption are also depressed. Because of the defective fat absorption, the fat-soluble vitamins (vitamins A, D, E, and K)

are not absorbed in adequate amounts. The amount of fat and protein in the stools is increased, and the stools become bulky, pale, foul-smelling, and greasy (**steatorrhea**).

The defective intestinal function in tropical sprue (Table 26–11) is apparently due in part to folic acid deficiency. However, the intestinal changes in experimentally produced folic acid deficiency are less intense than those in tropical sprue, and patients with tropical sprue respond to treatment with antibiotics such as tetracycline. Celiac disease occurs in genetically predisposed individuals who have the MHC class II antigens HLA-DR3 and HLA-DQW2. In these individuals, the gliadin portion of gluten, the protein found in wheat, rye, barley, and to a lesser extent in oats—but not in rice or corn—causes intestinal T cells to mount an inflammatory allergic response that disrupts and flattens the intestinal mucosa. When grains containing gluten are omitted from the diet, bowel function is generally restored to normal.

Adynamic Ileus

When the intestines are traumatized, there is a direct inhibition of smooth muscle, which causes a decrease in intestinal motility. It is due in part to activation of opioid receptors and is relieved by opioid-blocking drugs. When the peritoneum is irritated, there is reflex inhibition due to increased discharge of noradrenergic fibers in the splanchnic nerves. Both types of inhibition operate to cause **paralytic (adynamic) ileus** after abdominal operations. Because of the diffuse decrease in peristaltic activity in the small intestine, its contents are not propelled into the colon, and it becomes irregularly distended by pockets of gas and fluid. Intestinal peristalsis returns in 6–8 hours, followed by gastric peristalsis, but colonic activity takes 2–3 days to return. Adynamic ileus can be relieved by passing a tube through the nose down to the small intestine and aspirating the fluid and gas for a few days until peristalsis returns.

Mechanical Obstruction of the Small Intestine

Localized mechanical obstruction of the small intestine causes severe cramping pain (**intestinal colic**), whereas adynamic ileus is often painless. The segment of intestine above the point of mechanical obstruction dilates and becomes filled with fluid and gas. The pressure in the segment rises, and the blood vessels in its wall are compressed, causing local ischemia. Activity in visceral afferent nerve fibers from the distended segment causes sweating, a drop in blood pressure, and severe vomiting, with resultant metabolic alkalosis and dehydration. If the obstruction is not relieved, the condition is fatal.

Table 26–11. Disease processes associated with malabsorption.

Abnormalities of digestion in the intestinal lumen
  Glucose/galactose malabsorption
  Inadequate lipolysis (eg, due to pancreatic insufficiency or excess secretion of gastric acid)
  Decreased conjugated bile salts (eg, due to ileal resection or bacterial overgrowth)

Abnormalities of mucosal cell transport
  Nonspecific (due to tropical sprue, celiac disease, etc)
  Specific (due to deficiency of various disaccharidases, etc)

Abnormalities of fat transport in intestinal lymphatics

COLON

The main function of the colon is absorption of water, Na^+, and other minerals. By removal of about 90% of the fluid, it converts the 1000–2000 mL of isotonic chyme that enters it each day from the ileum to about 200–250 mL of semisolid feces.

Anatomic Considerations

The diameter of the colon is greater than that of the small intestine. Its length is about 100 cm in living adults and about 150 cm at autopsy. The fibers of its external muscular layer are collected into three longitudinal bands, the **teniae coli.** Because these bands are shorter than the rest of the colon, the wall of the colon forms outpouchings (**haustra**) between the teniae (Figure 26–30). There are no villi on the mucosa. The colonic glands are short inward projections of the mucosa that secrete mucus. Solitary lymph follicles are present, especially in the cecum and appendix.

Motility & Secretion of the Colon

The portion of the ileum containing the ileocecal valve projects slightly into the cecum, so that increases in colonic pressure squeeze it shut whereas increases in ileal pressure open it. Therefore, it effectively prevents reflux of colonic contents into the ileum. It is normally closed. Each time a peristaltic wave reaches it, it opens briefly, permitting some of the ileal chyme to squirt into the cecum. If the valve is resected in experimental animals, the chyme enters the colon so rapidly that absorption in the small intestine is reduced; however, a significant reduction does not occur in humans. When food leaves the stomach, the cecum relaxes and the passage of chyme through the ileocecal valve increases (**gastroileal reflex).** This is presumably a vagal reflex, although there is some argument about whether vagal stimulation affects the ileocecal valve. Sympathetic stimulation increases the contraction of the valve.

The movements of the colon include segmentation contractions and peristaltic waves like those occurring in the small intestine. Segmentation contractions mix the contents of the colon and, by exposing more of the contents to the mucosa, facilitate absorption. Peristaltic waves propel the contents toward the rectum, although weak antiperistalsis is sometimes seen. A third type of contraction that occurs only in the colon is the **mass action contraction,** in which there is simultaneous contraction of the smooth muscle over large confluent areas. These contractions move material from one portion of the colon to another. They also move material into the rectum, and rectal distention initiates the defecation reflex (see below).

The movements of the colon are coordinated by the BER of the colon. The frequency of this wave, unlike the wave in the small intestine, increases along the colon, from about 2/min at the ileocecal valve to 6/min at the sigmoid.

Transit Time in the Small Intestine & Colon

The first part of a test meal reaches the cecum in about 4 hours, and all of the undigested portions have entered the colon in 8 or 9 hours. On average, the first remnants of the meal reach the hepatic flexure in 6 hours, the splenic flexure in 9 hours, and the pelvic colon in 12 hours. From the pelvic colon to the anus, transport is much slower. When small colored beads are fed with a meal, an average of 70% of them are recovered in the stool in 72 hours, but total recovery requires more than a week. Transit time, pressure fluctuations, and changes in pH in the gastrointestinal tract can be observed by monitoring the progress of a small pill that contains sensors and a miniature radio transmitter.

Absorption in the Colon

The absorptive capacity of the mucosa of the large intestine is great. Na^+ is actively transported out of the colon, and water follows along the osmotic gradient thus generated. Normally, there is net secretion of K^+ and HCO_3^- into the colon (see Chapter 25). The absorptive capacity of the colon makes rectal instillation a practical route for drug administration, especially in

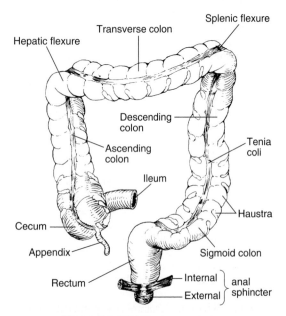

Figure 26–30. The human colon.

Splenic flexure
Transverse colon
Hepatic flexure
Descending colon
Ascending colon
Tenia coli
Ileum
Haustra
Cecum
Appendix
Sigmoid colon
Rectum
Internal } anal
External } sphincter

children. Many compounds, including anesthetics, sedatives, tranquilizers, and steroids, are absorbed rapidly by this route. Some of the water in an enema is absorbed, and if the volume of an enema is large, absorption may be rapid enough to cause water intoxication.

Feces

The stools contain inorganic material, undigested plant fibers, bacteria, and water. Their composition (Table 26–12) is relatively unaffected by variations in diet because a large fraction of the fecal mass is of nondietary origin. This is why appreciable amounts of feces continue to be passed during prolonged starvation.

Intestinal Bacteria

The chyme in the jejunum normally contains few if any bacteria. There are more microorganisms in the ileum, but it is only the colon that regularly contains large numbers of bacteria. The reason for the relative sterility of the jejunal contents is unsettled, although gastric acid and the comparatively rapid transit of the chyme through this region may inhibit bacterial growth.

The bacteria in the gastrointestinal tract can be divided into three types. Some are **pathogens** that cause disease; others are **symbionts** that benefit the host and vice versa; and most are **commensals,** which have no particular effect on the host and vice versa. They include in the colon bacilli, such as various strains of *Escherichia coli* and *Enterobacter aerogenes,* but also pleomorphic organisms such as *Bacteroides fragilis* and cocci of various types. Great masses of bacteria are passed in the stool. At birth, the colon is sterile, but the intestinal bacterial flora becomes established early in life.

Antibiotics improve growth rates in a variety of species, including humans; and small amounts of antibiotics are frequently added to the diets of domestic animals. Animals raised under sanitary but not germ-free conditions grow faster than controls. They assimilate food better and do not require certain amino acids that are essential dietary constituents in other animals. They also have larger litters and a lower neonatal death rate.

The reason for the improved growth is unsettled. Nutritionally important substances such as ascorbic acid, cyanocobalamin, and choline are utilized by some intestinal bacteria. On the other hand, some enteric microorganisms synthesize vitamin K and a number of B complex vitamins, and the folic acid produced by bacteria can be shown to be absorbed in significant amounts. In addition, short-chain fatty acids produced by the action of bacteria in the colon are of considerable physiologic importance (see Chapter 25).

Recent evidence indicates that nonpathogenic salmonella bacteria are able to block the ubiquination of $I_\kappa B\alpha$, the step which is necessary for the transcription factor NF-$_\kappa$B to initiate inflammation (see Chapter 33).

The brown color of the stools is due to pigments formed from the bile pigments by the intestinal bacteria. When bile fails to enter the intestine, the stools become white **(acholic stools).** Bacteria produce some of the gases in the flatus. Organic acids formed from carbohydrates by bacteria are responsible for the slightly acid reaction of the stools (pH 5.0–7.0). Amines formed by the intestinal bacteria—especially indole and skatole—contribute to the odor of the feces, as do sulfides.

Intestinal bacteria appear to play a role in cholesterol metabolism, since the poorly absorbed antibiotic neomycin that modifies the intestinal flora lowers plasma LDL and cholesterol levels.

When normal animals with the usual intestinal bacterial flora are exposed to ionizing radiation, the body defenses that prevent intestinal bacteria from invading the rest of the body break down, and a major cause of death in **radiation poisoning** is overwhelming sepsis. Germ-free animals have extremely hypoplastic lymphoid tissue and poorly developed immune mechanisms, probably because these mechanisms have never been challenged. However, they are much more resistant to radiation than animals with the usual intestinal flora because they have no intestinal bacteria to cause sepsis.

Comensal bacteria have recently been genetically engineered to produce the anti-inflammatory cytokine IL-

Table 26–12. Approximate composition of feces on an average diet.

Component	Percentage of Total Weight
Water	75
Solids	25

	Percentage of Total Solids
Cellulose and other indigestible fiber	Variable
Bacteria	30
Inorganic material (mostly calcium and phosphates)	15
Fat and fat derivatives	5
Also desquamated mucosal cells, mucus, and small amounts of digestive enzymes	

10, then fed to experimental animals with promising results in terms of relieving inflammatory diseases of the intestines. The potential of this technique for treating human diseases is obviously great.

Blind Loop Syndrome

Overgrowth of bacteria within the intestinal lumen can cause definite harmful effects. Such overgrowth occurs when there is stasis of the contents of the small intestine, and it causes macrocytic anemia and steatorrhea. Because the condition is prominent in patients with surgically created blind loops of small intestine, it has acquired the name **blind loop syndrome;** however, it can occur in any condition that promotes massive bacterial contamination of the small intestine. The cause of the anemia is malabsorption of cyanocobalamin. The steatorrhea is due to excessive hydrolysis of conjugated bile salts by the bacteria. The important role of bile salts in fat digestion is discussed in Chapter 25.

Dietary Fiber

Adequate nutrition in herbivorous animals depends upon the action of gastrointestinal microorganisms that break down cellulose and related plant carbohydrates. In humans, there is no appreciable digestion of these vegetable products. Cellulose, hemicellulose, and lignin in the diet are important components of the **dietary fiber,** which by definition is all ingested food that reaches the large intestine in an essentially unchanged state. Various gums, algal polysaccharides, and pectic substances also contribute to dietary fiber.

If the amount of dietary fiber is small, the diet is said to lack **bulk.** Since the amount of material in the colon is small, the colon is inactive and bowel movements are infrequent. In addition, starvation and parenteral nutrition lead to atrophy of the mucosa of the colon, and this is reversed when substances like pectin are placed in the colon. So-called bulk laxatives work by providing a larger volume of indigestible material to the colon.

There has been a recent upsurge of interest in dietary fiber because of epidemiologic evidence indicating that groups of people who live on a diet which contains large amounts of vegetable fiber have a low incidence of diverticulitis, cancer of the colon, diabetes mellitus, and coronary artery disease. However, the relationship between dietary fiber and the incidence of disease is still unsettled and needs further study.

Defecation

Distention of the rectum with feces initiates reflex contractions of its musculature and the desire to defecate. In humans, the sympathetic nerve supply to the internal (involuntary) anal sphincter is excitatory, whereas the parasympathetic supply is inhibitory. This sphincter relaxes when the rectum is distended. The nerve supply to the external anal sphincter, a skeletal muscle, comes from the pudendal nerve. The sphincter is maintained in a state of tonic contraction, and moderate distention of the rectum increases the force of its contraction (Figure 26–31). The urge to defecate first occurs when rectal pressure increases to about 18 mm Hg. When this pressure reaches 55 mm Hg, the external as well as the internal sphincter relaxes and the contents of the rectum are expelled. This is why there is reflex evacuation of the rectum in chronic spinal animals and humans. Before the pressure that relaxes the external anal sphincter is reached, voluntary defecation can be initiated by voluntarily relaxing the external sphincter and contracting the abdominal muscles (straining), thus aiding the reflex emptying of the distended rectum. Defecation is therefore a spinal reflex that can be voluntarily inhibited by keeping the external sphincter contracted or facilitated by relaxing the sphincter and contracting the abdominal muscles.

Distention of the stomach by food initiates contractions of the rectum and, frequently, a desire to defecate. The response is called the **gastrocolic reflex,** although there is some evidence that it is due to an action of gastrin on the colon and is not neurally mediated. Because of the response, defecation after meals is the rule in children. In adults, habit and cultural factors play a large role in determining when defecation occurs.

Effects of Colectomy

Humans can survive after total removal of the colon if fluid and electrolyte balance is maintained. When total colectomy is performed, the ileum is brought out through the abdominal wall **(ileostomy)** and the chyme expelled from the ileum is collected in a plastic bag fastened around the opening. If the diet is carefully regulated, the volume of ileal discharge decreases and its consistency increases over time. Care of an ileostomy used to be a time-consuming, difficult job, but when ileostomies are constructed by modern techniques, they can be relatively trouble-free and patients with them can lead essentially normal lives.

Constipation

In bowel-conscious America, the amount of misinformation and undue apprehension about constipation probably exceeds that about any other health topic. Patients with persistent constipation, and particularly those with a recent change in bowel habits, should of course be examined carefully to rule out underlying organic disease. However, many normal humans defecate only once every 2–3 days, even though others defecate

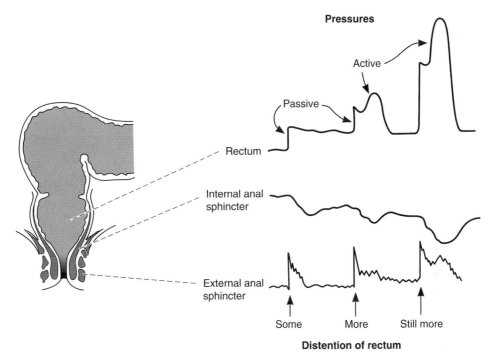

Figure 26–31. Responses to distention of the rectum by pressures less than 55 mm Hg. Distention produces passive tension due to stretching of the wall of the rectum, and additional active tension when the smooth muscle in the wall contracts. (Reproduced, with permission, from Davenport HW: *A Digest of Digestion,* 2nd ed. Year Book, 1978.)

once a day and some as often as three times a day. Furthermore, the only symptoms caused by constipation are slight anorexia and mild abdominal discomfort and distention. These symptoms are not due to absorption of "toxic substances," because they are promptly relieved by evacuating the rectum and can be reproduced by distending the rectum with inert material. Other symptoms attributed by the lay public to constipation are due to anxiety or other causes.

Megacolon

The lack of harmful effects of infrequent bowel movements is emphasized by the relative absence of symptoms other than abdominal distention, anorexia, and lassitude in children with **aganglionic megacolon** (Hirschsprung's disease). This disease is due to congenital absence of the ganglion cells in both the myenteric and submucous plexuses of a segment of the distal colon, probably as a result of failure of the normal cranial-to-caudal migration of neuroblasts during development. The absence of peristalsis causes feces to pass the aganglionic region with difficulty, and children with the disease may defecate as infrequently as once every 3 weeks. The condition can be relieved if the aganglionic

portion of the colon is resected and the portion of the colon above it anastomosed to the rectum.

Endothelins acting by the endothelin B receptor (see Chapter 31) are necessary for normal migration of certain neural crest cells, and knockout mice lacking endothelin B receptors develop megacolon. In addition, one cause of congenital aganglionic megacolon in humans appears to be a mutation in the endothelin B receptor gene.

Diarrhea

Diarrhea has many causes, some of which have been mentioned in particular contexts in preceding sections of this chapter and in other chapters. Severe diarrhea is caused by certain strains of *E coli* that produce toxins which stimulate secretion of Na^+ and water in the small intestine. Rotavirus appears to secrete a substance that stimulates the nerves to secretory cells in the colon. In any case, when large amounts of Na^+, K^+, and water are washed out of the colon and the small intestine in the diarrheal stools, there is dehydration, hypovolemia, and, eventually, shock and cardiovascular collapse. A more insidious complication of chronic diarrhea, if fluid balance is maintained, is severe hypokalemia.

Acute diarrhea is a major problem in developing countries. It is also common in travelers. Fortunately, fluid and electrolyte loss can be effectively reduced by oral administration of Na^+ and glucose so that Na^+ is absorbed via SGLT 1, the Na^+-glucose cotransporter (see Chapter 25). In addition, bismuth subsalicylate has been shown to be effective in reducing stool output, and compounds containing this material are generally available in the United States without prescription.

REFERENCES FOR SECTION V: GASTROINTESTINAL FUNCTION

Andrews NC: Disorders of iron metabolism. N Engl J Med 1999;341:1986.

Ankoma-Sey V: Hepatic regeneration—revising the myth of Prometheus. News Physiol Sci 1999;14:149.

Arias JM et al (editors): *The Liver: Biology and Pathology,* 3rd ed. Raven Press, 1994.

Baron TH, Morgan DE: Current concepts: acute necrotizing pancreatitis. N Engl J Med 1999;340:1412.

Bengmark S: Econutrition and health maintenance—a new concept to prevent GI inflammation, ulceration, and sepsis. Clin Nutr 1996;15:1.

Boyer JL, Graf J, Meier PJ: Hepatocyte transport systems regulating pH_i, cell volume, and bile secretion. Annu Rev Physiol 1992;54:415.

Chong L, Marx J (editors): Lipids in the limelight. Science 2001;294:1861.

Cohen S, Parkman HP: Heartburn—a serious symptom. N Engl J Med 1999;340:878.

Farrell RJ, Kelly CP: Celiac sprue. N Engl J Med 2002;346:180.

Field M, Rao MC, Chang EB: Intestinal electrolyte transport and diarrheal disease. N Engl J Med 1989;321:800.

Forte LR, Hamra FK: Guanylin and uroguanylin: Intestinal peptide hormones that regulate epithelial transport. News Physiol Sci 1996;11:17.

Go VLW et al: *The Pancreas: Biology, Pathobiology and Disease,* 2nd ed. Raven Press, 1993.

Hersey SJ, Sachs G: Gastric acid secretion. Physiol Rev 1995;75:155.

Hofmann AF: Bile acids: the good, the bad, and the ugly. News Physiol Sci 1999;14:24.

Hunt RH, Tytgat GN (editors): *Helicobacter pylori: Basic Mechanisms to Clinical Cure.* Kluwer Academic, 2000.

Itoh Z: Motilin and clinical application. Peptides 1997;18:593.

Johnston DE, Kaplan MM: Pathogenesis and treatment of gallstones. N Engl J Med 1993;328:412.

Kunzelmann K, Mall M: Electrolyte transport in the mammalian colon: mechanisms and implications for disease. Physiol Rev 2002;82:245.

Lamberts SWJ et al: Octreotide. N Engl J Med 1996;334:246.

Levitt MD, Bond JH: Volume, composition and source of intestinal gas. Gastroenterology 1970;59:921.

Lewis JH (editor): *A Pharmacological Approach to Gastrointestinal Disorders.* Williams & Wilkins, 1994.

Mackowiak PA: The normal microbial flora. N Engl J Med 1982;307:83.

Mann NS, Mann SK: Enterokinase. Proc Soc Exp Biol Med 1994;206:114.

Mayer EA, Sun XP, Willenbucher RF: Contraction coupling in colonic smooth muscle. Annu Rev Physiol 1992;54:395.

Meier PJ, Stieger B: Molecular mechanisms of bile formation. News Physiol Sci 2000;15:89.

Michalopoulos GK, DeFrances MC: Liver regeneration. Science 1997;276:60.

Mittal RK, Balaban DH: The esophagogastric junction. N Engl J Med 1997;336:924.

Montecucco C, Rappuoli R: Living dangerously: how *Helicobacter pylori* survives in the human stomach. Nat Rev Mol Cell Biol 2001;2:457.

Nakazato M: Guanylin family: new intestinal peptides regulating electrolyte and water homeostatis. J Gastroenterol 2001;36:219.

Nilson S et al: Mechanisms of estrogen action. Physiol Rev 2001;18:1535.

Perlman AL, Schulze-Delrieu KS (editors): *Deglutition and Its Disorders: Anatomy, Physiology, Clinical Diagnosis, and Management.* Singular Publication Group, 1997.

Rabon EC, Reuben MA: The mechanism and structure of the gastric H^+, K^+-ATPase. Annu Rev Physiol 1990;52:321.

Sachs G, Zeng N, Prinz C: Pathophysiology of isolated gastric endocrine cells. Annu Rev Physiol 1997;59:234.

Sanders KM, Warm SM: Nitric oxide as a mediator of noncholinergic neurotransmission. Am J Physiol 1992;262:G379.

Sellin JH: SCKAs: the enigma of weak electrolyte transport in the colon. News Physiol Sci 1999;14:58.

Sleisenger MH, Brandborg LL: *Malabsorption.* Saunders, 1977.

Specian RD, Oliver MG: Functional biology of intestinal goblet cells. Am J Med 1991;260:C183.

Topping DL, Clifton PM: Short-chain fatty acids and human colonic function: select resistant starch and nonstarch polysaccharides. Physiol Rev 2001;81:1031.

Trauner M, Meier PJ, Boyer JL: Molecular mechanisms of cholestasis. N Engl J Med 1998;339:1217.

Walsh JH (editor): *Gastrin.* Raven Press, 1993.

Williams JA, Blevins GT Jr: Cholecystokinin and regulation of pancreatic acinar cell function. Physiol Rev 1993;73:701.

Wolfe MM, Lichtenstein DR, Singh G: Gastrointestinal toxicity of nonsteroidal antiinflammatory drugs. N Engl J Med 1999;340:1888.

Wright EM: The intestinal Na^+/glucose cotransporter. Annu Rev Physiol 1993;55:575.

Young JA, van Lennep EW: *The Morphology of Salivary Glands.* Academic Press, 1978.

Circulating Body Fluids

27

INTRODUCTION

The **circulatory system** is the transport system that supplies O_2 and substances absorbed from the gastrointestinal tract to the tissues, returns CO_2 to the lungs and other products of metabolism to the kidneys, functions in the regulation of body temperature, and distributes hormones and other agents that regulate cell function. The blood, the carrier of these substances, is pumped through a closed system of blood vessels by the heart, which in mammals is really two pumps in series with each other. From the left ventricle, blood is pumped through the arteries and arterioles to the capillaries, where it equilibrates with the interstitial fluid. The capillaries drain through venules into the veins and back to the right atrium. This is the **major (systemic) circulation.** From the right atrium, blood flows to the right ventricle, which pumps it through the vessels of the lungs—the **lesser (pulmonary) circulation**—and the left atrium to the left ventricle. In the pulmonary capillaries, the blood equilibrates with the O_2 and CO_2 in the alveolar air. Some tissue fluids enter another system of closed vessels, the lymphatics, which drain lymph via the thoracic duct and the right lymphatic duct into the venous system (the **lymphatic circulation**). The circulation is controlled by multiple regulatory systems that function in general to maintain adequate capillary blood flow—when possible in all organs, but particularly in the heart and brain. This chapter is concerned with blood and lymph and with the multiple functions of the cells they contain.

BLOOD

The cellular elements of the blood—white blood cells, red blood cells, and platelets—are suspended in the plasma. The normal total circulating blood volume is about 8% of the body weight (5600 mL in a 70-kg man). About 55% of this volume is plasma.

BONE MARROW

In the adult, red blood cells, many white blood cells, and platelets are formed in the bone marrow. In the fetus, blood cells are also formed in the liver and spleen, and in adults such **extramedullary hematopoiesis** may occur in diseases in which the bone marrow becomes destroyed or fibrosed. In children, blood cells are actively produced in the marrow cavities of all the bones. By age 20, the marrow in the cavities of the long bones, except for the upper humerus and femur, has become inactive (Figure 27–1). Active cellular marrow is called **red marrow;** inactive marrow that is infiltrated with fat is called **yellow marrow.**

The bone marrow is actually one of the largest organs in the body, approaching the size and weight of the liver. It is also one of the most active. Normally, 75% of the cells in the marrow belong to the white blood cell-producing myeloid series and only 25% are maturing red cells, even though there are over 500 times as many red cells in the circulation as there are

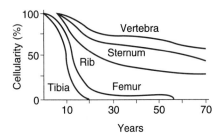

Figure 27–1 Changes in red bone marrow cellularity with age. 100% equals the degree of cellularity at birth. (Reproduced, with permission, from Whitby LEH, Britton CJC: *Disorders of the Blood,* 10th ed. Churchill Livingstone, 1969.)

white cells. This difference in the marrow reflects the fact that the average life span of white cells is short, whereas that of red cells is long.

The bone marrow contains multipotent uncommitted stem cells (**pluripotential stem cells**) that differentiate into one or another type of committed stem cells (**progenitor cells**). These in turn form the various differentiated types of blood cells. There are separate pools of progenitor cells for megakaryocytes, lymphocytes, erythrocytes, eosinophils, and basophils, whereas neutrophils and monocytes arise from a common precursor. The bone marrow stem cells are also the source of osteoclasts (see Chapter 21), Kupffer cells (see Chapter 26), mast cells, dendritic cells, and Langerhans cells (see below).

The pluripotential cells are few in number but are capable of completely replacing the bone marrow when injected into a host whose own bone marrow has been completely destroyed. The best current source for these hematopoietic stem cells is umbilical cord blood.

The pluripotential cells are derived from uncommitted, totipotent stem cells that at least in theory can be stimulated to form any cell in the body. There are a few of these in adults, but they are more readily obtained from the blastocysts of embryos. Totipotential cells from human embryos have now been cultured, and there is immense interest in stem cell research. However, there are ethical as well as scientific issues involved, and debate on these issues will undoubtedly continue.

WHITE BLOOD CELLS

There are normally 4000–11,000 white blood cells per microliter of human blood (Table 27–1). Of these, the **granulocytes (polymorphonuclear leukocytes, PMNs)** are the most numerous. Young granulocytes have horse-

shoe-shaped nuclei that become multilobed as the cells grow older (Figure 27–2). Most of them contain neutrophilic granules (**neutrophils**), but a few contain granules that stain with acidic dyes (**eosinophils**), and some have basophilic granules (**basophils**). The other two cell types found normally in peripheral blood are **lymphocytes,** which have large round nuclei and scanty cytoplasm, and **monocytes,** which have abundant agranular cytoplasm and kidney-shaped nuclei (Figure 27–2). Acting together, these cells provide the body with powerful defenses against tumors and viral, bacterial, and parasitic infections.

Granulocytes

All granulocytes have cytoplasmic granules that contain biologically active substances involved in inflammatory and allergic reactions.

The average half-life of a neutrophil in the circulation is 6 hours. To maintain the normal circulating blood level, it is therefore necessary to produce over 100 billion neutrophils per day. Many of the neutrophils enter the tissues. They are attracted to the endothelial surface by selectins, and they roll along it. They then bind firmly to neutrophil adhesion molecules of the integrin family. They next insinuate themselves through the walls of the capillaries between endothelial cells by a process called **diapedesis.** Many of those that leave the circulation enter the gastrointestinal tract and are lost from the body.

Invasion of the body by bacteria triggers the **inflammatory response.** The bone marrow is stimulated to

Table 27–1. Normal values for the cellular elements in human blood.

Cell	Cells/μL (average)	Approximate Normal Range	Percentage of Total White Cells
Total WBC	9000	4000–11,000	...
Granulocytes			
Neutrophils	5400	3000–6000	50–70
Eosinophils	275	150–300	1–4
Basophils	35	0–100	0.4
Lymphocytes	2750	1500–4000	20–40
Monocytes	540	300–600	2–8
Erythrocytes			
Females	4.8×10^6	...	...
Males	5.4×10^6	...	...
Platelets	300,000	200,000–500,000	...

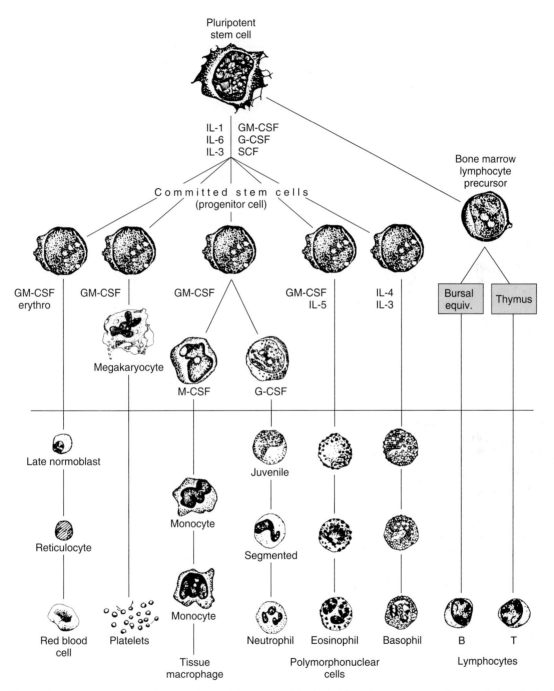

Figure 27–2 Development of various formed elements of the blood from bone marrow cells. Cells below the horizontal line are found in normal peripheral blood. The principal sites of action of erythropoietin (erythro) and the various colony-stimulating factors (CSF) that stimulate the differentiation of the components are indicated. G, granulocyte; M, macrophage; IL, interleukin; see Tables 27–2 and 27–3.

produce and release large numbers of neutrophils. Bacterial products interact with plasma factors and cells to produce agents that attract neutrophils to the infected area **(chemotaxis).** The chemotactic agents, which are part of a large and expanding family of **chemokines** (see below), include a component of the complement system (C5a), leukotrienes, and polypeptides from lymphocytes, mast cells, and basophils. The stimulatory effect of C5a on chemotactic activity is enhanced by G_c-globulin, and neutrophil membranes contain this protein, which also binds and transports vitamin D in the plasma (see Chapter 21). Other plasma factors act on the bacteria to make them "tasty" to the phagocytes **(opsonization).** The principal opsonins that coat the bacteria are immunoglobulins of a particular class (IgG) and complement proteins (see below). The coated bacteria then bind to receptors on the neutrophil cell membrane. This triggers, via heterotrimeric G protein-mediated responses, increased motor activity of the cell, exocytosis, and the so-called respiratory burst. The increased motor activity leads to prompt ingestion of the bacteria by endocytosis **(phagocytosis).** By **exocytosis,** neutrophil granules discharge their contents into the phagocytic vacuoles containing the bacteria and, to a degree, into the interstitial space **(degranulation).** The granules contain various proteases plus antimicrobial proteins called **defensins.** Two types of defensins, α and β, are found in mammals, but other types are found in invertebrates and plants. In addition, the cell membrane-bound enzyme **NADPH oxidase** is activated, with the production of toxic oxygen metabolites. The combination of the toxic oxygen metabolites and the proteolytic enzymes from the granules makes the neutrophil a very effective killing machine.

Activation of NADPH oxidase is associated with a sharp increase in O_2 uptake and metabolism in the neutrophil (the **respiratory burst**) and generation of O_2^- by the following reaction:

$$NADPH + H^+ + 2O_2 \rightarrow NADP^+ + 2H^+ + 2O_2^-$$

O_2^- is a **free radical** formed by the addition of one electron to O_2. Two O_2^- react with two H^+ to form H_2O_2 in a reaction catalyzed by the cytoplasmic form of superoxide dismutase (SOD):

$$O_2^- + O_2^- + H^+ + H^+ \xrightarrow{SOD} H_2O_2 + O_2$$

O_2^- and H_2O_2 are both oxidants that are effective bactericidal agents, but H_2O_2 is converted to H_2O and O_2 by the enzyme **catalase.** The cytoplasmic form of SOD contains both Zn and Cu. It is found in many parts of the body. It is defective as a result of genetic mutation in a familial form of **amyotrophic lateral sclerosis**

(ALS; see Chapter 16). Therefore, it may be that O_2^- accumulates in motor neurons and kills them in at least one form of this progressive, fatal disease. Two other forms of SOD encoded by at least one different gene are also found in humans.

Neutrophils also discharge the enzyme **myeloperoxidase,** which catalyzes the conversion of Cl^-, Br^-, I^-, and SCN^- to the corresponding acids (HOCl, HOBr, etc). These acids are also potent oxidants. Since Cl^- is present in greatest abundance in body fluids, the principal product is HOCl.

In addition to myeloperoxidase and defensins, neutrophil granules contain an elastase, two metalloproteinases that attack collagen, and a variety of other proteases that help destroy invading organisms. These enzymes act in a cooperative fashion with the O_2^-, H_2O_2, and HOCl formed by the action of the NADPH oxidase and myeloperoxidase to produce a killing zone around the activated neutrophil. This zone is effective in killing invading organisms, but in certain diseases, eg, rheumatoid arthritis, the neutrophils may also cause local destruction of host tissue.

The movements of the cell in phagocytosis, as well as migration to the site of infection, involve microtubules and microfilaments (see Chapter 1). Proper function of the microfilaments involves the interaction of the actin they contain with myosin-I on the inside of the cell membrane (see Chapter 1).

Like neutrophils, **eosinophils** have a short half-life in the circulation, are attracted to the surface of endothelial cells by selectins, bind to integrins which attach them to the vessel wall, and enter the tissues by diapedesis. Like neutrophils, they release proteins, cytokines, and chemokines that produce inflammation but are capable of killing invading organisms. However, there is some selectivity in the selectins and integrins to which they respond and in the killing molecules they secrete. Their maturation and activation in tissues is particularly stimulated by IL-3, IL-5, and GM-CSF (see below). They are especially abundant in the mucosa of the gastrointestinal tract, where they defend against parasites, and in the mucosa of the respiratory and urinary tracts. Circulating eosinophils are increased in allergic diseases such as asthma and in various other respiratory and gastrointestinal diseases.

Basophils also enter tissues and release proteins and cytokines. They resemble but are not identical to mast cells, and like mast cells they contain histamine and heparin (see below). They release histamine and other inflammatory mediators when activated by a histamine-releasing factor secreted by T lymphocytes (see below) and are essential for immediate-type hypersensitivity reactions. These range from mild urticaria and rhinitis to severe anaphylactic shock.

Mast Cells

Mast cells are heavily granulated wandering cells that are found in areas rich in connective tissue, and they are abundant beneath epithelial surfaces. Their granules contain heparin, histamine, and many proteases. The heparin appears to play a role in granule formation. They have IgE receptors on their cell membranes, and, like basophils, they degranulate when IgE-coated antigens bind to their surface. They are involved in inflammatory responses initiated by immunoglobulins IgE and IgG (see below). The inflammation combats invading parasites. In addition to this involvement in acquired immunity, they release TNF-α in response to bacterial products by an antibody-independent mechanism, thus participating in the nonspecific **natural immunity** that combats infections (see below). Marked mast cell degranulation produces clinical manifestations of allergy up to and including anaphylaxis.

Monocytes

Monocytes enter the blood from the bone marrow and circulate for about 72 hours. They then enter the tissues and become **tissue macrophages** (Figure 27–3). Their life span in the tissues is unknown, but bone marrow transplantation data in humans suggest that they persist for about 3 months. It appears that they do not reenter the circulation. Some of them end up as the multinucleated giant cells seen in chronic inflammatory diseases such as tuberculosis. The tissue macrophages

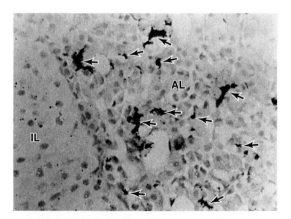

Figure 27–3 Tissue macrophages in the pituitary gland. The macrophages, which are identified by arrows, have been stained immunocytochemically with a monoclonal antibody that is specific for these cells. AL, anterior lobe; IL, intermediate lobe. (Courtesy of S Gordon.)

include the Kupffer cells of the liver, pulmonary alveolar macrophages (see Chapter 34), and microglia in the brain, all of which come from the circulation. In the past, they have been called the **reticuloendothelial system,** but the general term **tissue macrophage system** seems more appropriate.

The macrophages become activated by lymphokines from T lymphocytes. The activated macrophages migrate in response to chemotactic stimuli and engulf and kill bacteria by processes generally similar to those occurring in neutrophils. They play a key role in immunity (see below). They also secrete up to 100 different substances, including factors that affect lymphocytes and other cells, prostaglandins of the E series, and clot-promoting factors.

Granulocyte & Macrophage Colony-Stimulating Factors

The production of red and white blood cells is regulated with great precision in healthy individuals, and the production of granulocytes is rapidly and dramatically increased in infections. The proliferation and self-renewal of the pluripotential cells depend on **stem cell factor (SCF).** Other factors are also involved. The proliferation and maturation of the cells that enter the blood from the marrow are regulated by glycoprotein growth factors or hormones that cause cells in one or more of the committed cell lines to proliferate and mature (Figure 27–2, Table 27–2). The regulation of erythrocyte production by **erythropoietin** is discussed in Chapter 24. Three additional factors are called **colony-stimulating factors (CSFs),** because they cause appropriate single stem cells to proliferate in soft agar, forming colonies in this culture medium. The factors stimulating the production of committed stem cells include **granulocyte-macrophage CSF (GM-CSF), granulocyte CSF (G-CSF),** and **macrophage CSF (M-CSF).** Interleukins **IL-1** and **IL-6** followed by **IL-3** (Table 27–2) act in sequence to convert pluripotential uncommitted stem cells to committed progenitor cells (Figure 27–2). IL-3 is also known as **multi-CSF.** Each of the CSFs has a predominant action, but all the CSFs and interleukins also have other overlapping actions. In addition, they activate and sustain mature blood cells. It is interesting in this regard that the genes for many of these factors are located together on the long arm of chromosome 5 and may have originated by duplication of an ancestral gene. It is also interesting that basal hematopoiesis is normal in mice in which the GM-CSF gene is knocked out, indicating that loss of one factor can be compensated for by others. On the other hand, the absence of GM-CSF causes accumulation of surfactant in the lungs (see Chapter 34).

Table 27–2. Cytokines that regulate hematopoiesis.

Cytokine	Cell Lines Stimulated	Cytokine Source
IL-1	Erythrocyte Granulocyte Megakaryocyte Monocyte	Multiple cell types
IL-3	Erythrocyte Granulocyte Megakaryocyte Monocyte	T lymphocytes
IL-4	Basophil	T lymphocytes
IL-5	Eosinophil	T lymphocytes
IL-6	Erythrocyte Granulocyte Megakaryocyte Monocyte	Endothelial cells Fibroblasts Macrophages
IL-11	Erythrocyte Granulocyte Megakaryocyte	Fibroblasts Osteoblasts
Erythropoietin	Erythrocyte	Kidney Kupffer cells of liver
SCF	Erythrocyte Granulocyte Megakaryocyte Monocyte	Multiple cell types
G-CSF	Granulocyte	Endothelial cells Fibroblasts Monocytes
GM-CSF	Erythrocyte Granulocyte Megakaryocyte	Endothelial cells Fibroblasts Monocytes T lymphocytes
M-CSF	Monocyte	Endothelial cells Fibroblasts Monocytes
Thrombopoietin	Megakaryocyte	Liver, kidney

Key: IL = interleukin; CSF = colony stimulating factor; G = granulocyte; M = macrophage; SCF = stem cell factor
Reproduced with permission, from McPhee SJ, Lingappa VR, Ganong WF (editors): *Pathophysiology of Disease,* 4th ed, McGraw-Hill, 2003.

As noted in Chapter 24, erythropoietin is produced in part by kidney cells and is a circulating hormone. The other factors are produced by macrophages, activated T cells, fibroblasts, and endothelial cells. For the most part, the factors act locally in the bone marrow.

Disorders of Phagocytic Function

More than 15 primary defects in neutrophil function have been described, along with at least 30 other con-

ditions in which there is a secondary depression of the function of the neutrophils. Patients with these diseases are prone to infections that are relatively mild when only the neutrophil system is involved but severe when the monocyte-tissue macrophage system is also involved. In one syndrome (neutrophil hypomotility), actin in the neutrophils does not polymerize normally, and the neutrophils move slowly. In another, there is a congenital deficiency of leukocyte integrins. In a more serious disease (chronic granulomatous disease), there is a failure to generate O_2^- in both the neutrophils and monocytes and consequent inability to kill many phagocytosed bacteria. In severe congenital glucose 6-phosphate dehydrogenase deficiency, there are multiple infections because of failure to generate the NADPH necessary for O_2^- production. In congenital myeloperoxidase deficiency, microbial killing power is reduced because hypohalite ions are not formed.

Lymphocytes

Lymphocytes are key elements in the production of immunity (see below). After birth, some lymphocytes are formed in the bone marrow. However, most are formed in the lymph nodes (Figure 27–4), thymus, and spleen from precursor cells that originally came from the bone marrow and were processed in the thymus or bursal equivalent (see below). Lymphocytes enter the bloodstream for the most part via the lymphatics. At any given time, only about 2% of the body lymphocytes are in the peripheral blood. Most of the rest are in the lymphoid organs. It has been calculated that in humans,

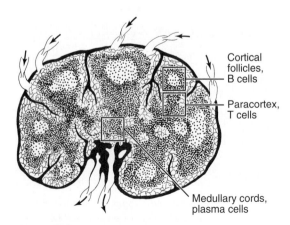

Cortical follicles, B cells

Paracortex, T cells

Medullary cords, plasma cells

Figure 27–4 Anatomy of a normal lymph node. (After Chandrasoma. Reproduced, with permission, from McPhee SJ, Lingappa VR, Ganong WF [editors]: *Pathophysiology of Disease,* 4th ed. McGraw-Hill, 2003.)

3.5×10^{10} lymphocytes per day enter the circulation via the thoracic duct alone; however, this count includes cells that reenter the lymphatics and thus traverse the thoracic duct more than once. The effects of adrenocortical hormones on the lymphoid organs, the circulating lymphocytes, and the granulocytes are discussed in Chapter 20.

IMMUNITY

Overview

Insects and other invertebrates have **innate immunity.** The key to this system is receptors that bind sequences of sugars, fats, or amino acids in common bacteria and activate various defense mechanisms. The receptors are coded in the germ line, and their fundamental structure is not modified by exposure to antigen. The activated defenses include, in various species, release of interferons, phagocytosis, production of antibacterial peptides, activation of the complement system, and several proteolytic cascades. Even plants release antibacterial peptides in response to infection. In vertebrates, innate immunity is complemented by **acquired immunity,** a system in which T and B lymphocytes are activated by very specific antigens. In both innate and acquired immunity, the receptors recognize the shape of antigens, not their specific chemical composition. In acquired immunity, the activated lymphocytes form clones that produce more antibodies which attack foreign proteins. After the invasion is repelled, small numbers persist as memory cells so that a second exposure to the same antigen provokes a prompt and magnified immune attack. The genetic event that led to acquired immunity occurred 450 million years ago in the ancestors of jawed vertebrates and was probably insertion of a transposon into the genome in a way that made possible the generation of the immense repertoire of T cells that are present in the body.

In vertebrates, including humans, innate immunity provides the first line of defense against infections, but it also triggers the slower but more specific acquired immune response (Figure 27–5). In vertebrates, natural and acquired immune mechanisms also attack tumors and tissue transplanted from other animals.

Once activated, immune cells communicate by means of cytokines and chemokines. They kill viruses, bacteria, and other foreign cells by secreting other cytokines and activating the complement system.

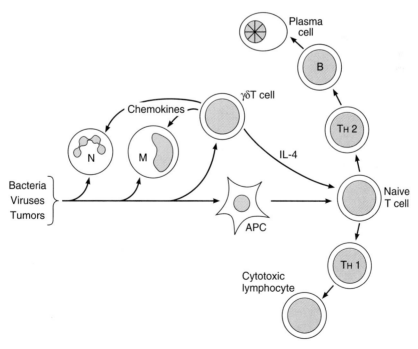

Figure 27–5 How bacteria, viruses, and tumors trigger innate immunity and initiate the acquired immune response. APC, antigen-presenting cell; M, monocyte; N, neutrophil; TH1 and TH2, helper T cells type 1 and type 2, respectively.

There have been rapid advances in immunology in recent years, and the field is large and complex. Only a summary of the fundamentals can be presented in this book.

Cytokines

Cytokines are hormone-like molecules that act, generally in a paracrine fashion, to regulate immune responses. They are secreted not only by lymphocytes and macrophages but by endothelial cells, neurons, glial cells, and other types of cells. Most of the cytokines are initially named for their actions, eg, B cell-differentiating factor, B cell-stimulating factor 2. However, there is a convention that once the amino acid sequence of a factor in humans is known, its name is changed to **interleukin.** Thus, for example, the name of B cell-differentiating factor was changed to interleukin-4. In a

field that is moving this fast, it is very difficult to construct a comprehensive list of factors, but most of the principal cytokines are listed in Table 27–3.

Many of the receptors for the cytokines and the hematopoietic growth factors (see above), as well as the receptors for prolactin (see Chapter 23), and growth hormone (see Chapter 22) are members of a cytokine-receptor superfamily that has three subfamilies (Figure 27–6). The members of subfamily 1, which includes the receptors for IL-4 and IL-7, are homodimers. The members of subfamily 2, which includes the receptors for IL-3, IL-5, and IL-6, are heterodimers. The receptor for IL-2 and several other cytokines is unique in that it consists of a heterodimer and an unrelated protein, the so-called Tac antigen. The other members of subfamily 3 have the same γ chain as IL-2R. The extracellular domain of the homodimer and heterodimer subunits all contain four conserved cysteine residues

Table 27–3. Examples of cytokines and their clinical relevance.[1]

Cytokine	Cellular Sources	Major Activities	Clinical Relevance
Interleukin-1	Macrophages	Activation of T cells and macrophages; promotion of inflammation	Implicated in the pathogenesis of septic shock, rheumatoid arthritis, and atheroscelerosis
Interleukin-2	Type 1 (TH1) helper T cells	Activation of lymphocyes, natural killer cells, and macrophages	Used to induce lymphokine-activated killer cells; used in the treatment of metastatic renal-cell carcinoma, melanoma, and various other tumors
Interleukin-4	Type 2 (TH2) helper T cells, mast cells, basophils, and eosinophils	Activation of lymphocytes, monocytes, and IgE class switching	As a result of its ability to stimulate IgE production, plays a part in mast-cell sensitization and thus in allergy and in defense against nematode infections
Interleukin-5	Type 2 (TH2) helper T cells, mast cells, and eosinophils	Differentiation of eosinophils	Monoclonal antibody against interleukin-5 used to inhibit the antigen-induced late-phase eosinophilia in animal models of allergy
Interleukin-6	Type 2 (TH2) helper T cells and macrophages	Activation of lymphocytes; differentiation of B cells; stimulation of the production of acute-phase proteins	Overproduced in Castleman's disease; acts as an autocrine growth factor in myeloma and in mesangial proliferative glomerulonephritis
Interleukin-8	T cells and macrophages	Chemotaxis of neutrophils, basophils, and T cells	Levels are increased in diseases accompanied by neutrophilia, making it a potentially useful marker of disease activity
Interleukin-11	Bone marrow stromal cells	Stimulation of the production of acute-phase proteins	Used to reduce chemotherapy-induced thrombocytopenia in patients with cancer

Table 27–3. Examples of cytokines and their clinical relevance.[1] (continued)

Cytokine	Cellular Sources	Major Activities	Clinical Relevance
Interleukin-12	Macrophages and B cells	Stimulation of the production of interferon γ by type 1 (TH1) helper T cells and by natural killer cells; induction of type 1 (TH1) helper T cells	May be useful as an adjuvant for vaccines
Tumor necrosis factor α	Macrophages, natural killer cells, T cells, B cells, and mast cells	Promotion of inflammation	Treatment with antibodies against tumor necrosis factor α beneficial in rheumatoid arthritis
Lymphotoxin (tumor necrosis factor β	Type 1 (TH1) helper T cells and B cells	Promotion of inflammation	Implicated in the pathogenesis of multiple sclerosis and insulin-dependent diabetes mellitus
Transforming growth factor β	T cells, macrophages, B cells, and mast cells	Immunosuppression	May be useful therapeutic agent in multiple sclerosis and myasthenia gravis
Granulocyte-macrophage colony-stimulating factor	T cells, macrophages, natural killer cells, and B cells	Promotion of the growth of granulocytes and monocytes	Used to reduce neutropenia after chemotherapy for tumors and in ganciclovir-treated patients with AIDS; used to stimulate cell production after bone marrow transplantation
Interferon-α	Virally infected cells	Induction of resistance of cells to viral infection	Used to treat AIDS-related Kaposi's sarcoma, melanoma, chronic hepatitis B infection, and chronic hepatitis C infection
Interferon-β	Virally infected cells	Induction of resistance of cells to viral infection	Used to reduce the frequency and severity of relapses in multiple sclerosis
Interferon-γ	Type 1 (TH1) helper T cells and natural killer cells	Activation of macrophages; inhibition of type 2 (TH2) helper T cells	Used to enhance the killing of phagocytosed bacteria in chronic granulomatous disease

[1]Reproduced with permission from Delves PJ, Roitt IM: The immune system. First of two parts. N Engl J Med 2000;343:37.

plus a conserved Trp-Ser-X-Trp-Ser domain, and although the intracellular portions do not contain tyrosine kinase catalytic domains, they activate cytoplasmic tyrosine kinases when ligand binds to the receptors.

The effects of the principal cytokines are listed in Table 27–3. Some of them have systemic as well as local paracrine effects. For example, IL-1, IL-6, and tumor necrosis factor α cause fever, and IL-1 increases slow-wave sleep and reduces appetite.

Another superfamily of cytokines is the **chemokine** family. Chemokines are substances that attract neutrophils (see above) and other white blood cells to areas of inflammation or immune response. Over 40 have now been identified, and it is clear that they also play a role in the regulation of cell growth and angiogenesis. The chemokine receptors are serpentine receptors that act via heterotrimeric G proteins to cause, among other things, extension of pseudopodia with migration of the cell toward the source of the chemokine.

The Complement System

The cell-killing effects of innate and acquired immunity are mediated in part by a system of plasma enzymes originally named the **complement system** because they "complemented" the effects of antibodies.

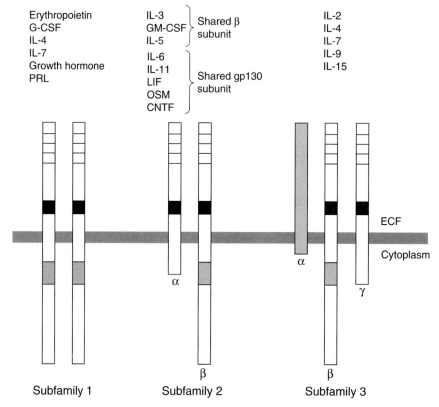

Figure 27–6 Members of one of the cytokine receptor superfamilies, showing shared structural elements. Note that all the subunits except the α subunit in subfamily 3 have four conserved cysteine residues (open boxes at top) and a Trp-Ser-X-Trp-Ser motif (dark color). Many subunits also contain a critical regulatory domain in their cytoplasmic portions (light color). IL-6, LIF, OSM, and CNTF share gp130 instead of a β subunit. CNTF, ciliary neurotrophic factor; LIF, leukemia inhibitory factor; OSM, oncostatin M. (Modified from D'Andrea AD: Cytokine receptors in congenital hematopoietic disease. N Engl J Med 1994;330:839.)

Nomenclature for the over 30 proteins in the system is confusing because it is a mixture of letters and numbers: examples include C1q, C3, and C3b. Three different pathways or enzyme cascades activate the system: the **classic pathway,** triggered by immune complexes; the **mannose-binding lectin pathway,** triggered when this lectin binds mannose groups in bacteria; and the **alternative** or **properdin** pathway, triggered by contact with various viruses, bacteria, fungi, and tumor cells. The proteins that are produced have three functions: They help kill invading organisms by opsonization, chemotaxis, and eventual lysis of the cells; they serve in part as a bridge from innate to acquired immunity by activating B cells and aiding immune memory; and they help dispose of waste products after apoptosis. Cell lysis, one of the principal ways the complement system kills cells, is brought about by inserting proteins called **perforins** into their cell membranes. The holes produced in this fashion permit free flow of ions, with disruption of membrane polarity.

Innate Immunity

The cells that mediate innate immunity include neutrophils, macrophages, and **natural killer (NK) cells,** large lymphocytes that are not T cells but are cytotoxic. All these cells respond to lipid and carbohydrate sequences unique to bacterial cell walls and to other substances characteristic of tumor and transplant cells. They exert their effects by way of the complement and other systems, with the cells they attack frequently dying by osmotic lysis or apoptosis. Their cytokines also activate cells of the acquired immune system.

An important link in innate immunity in **Drosophila** is a receptor protein named **toll,** which binds fungal antigens and triggers activation of genes coding for antifungal proteins. An expanding list of toll-like receptors (TLRs) have now been identified in humans. One of these, TLR4, binds bacterial lipopolysaccharide and a protein called CD14, and this initiates a cascade of intracellular events that activate transcription of genes for a variety of proteins involved in innate immune responses. This is important because bacterial lipopolysaccharide produced by gram-negative organisms is the cause of septic shock. TLR2 mediates the response to microbial lipoproteins; TLR6 cooperates with TLR2 in recognizing certain peptidoglycans; and TLR9 recognizes the DNA of certain bacteria.

Acquired Immunity

As noted above, the key to acquired immunity is the ability of lymphocytes to produce antibodies that are specific for one of the many millions of foreign agents that may invade the body. The antigens stimulating antibody production are usually proteins and polypeptides, but antibodies can also be formed against nucleic acids and lipids if these are presented as nucleoproteins and lipoproteins, and antibodies to smaller molecules can be produced experimentally when the molecules are bound to protein. Acquired immunity has two components: humoral immunity and cellular immunity. **Humoral immunity** is mediated by circulating immunoglobulin antibodies in the γ-globulin fraction of the plasma proteins. Immunoglobulins are produced by B lymphocytes, and they activate the complement system and attack and neutralize antigens. Humoral immunity is a major defense against bacterial infections. **Cellular immunity** is mediated by T lymphocytes. It is responsible for delayed allergic reactions and rejection of transplants of foreign tissue. Cytotoxic T cells attack and destroy cells that have the antigen which activated them. They kill by inserting perforins

(see above) and by initiating apoptosis. Cellular immunity constitutes a major defense against infections due to viruses, fungi, and a few bacteria such as the tubercle bacillus. It also helps defend against tumors.

Development of the Immune System

During fetal development, lymphocyte precursors come from the bone marrow. Those that populate the thymus (Figure 27–7) become transformed by the environment in this organ into the lymphocytes responsible for cellular immunity (**T lymphocytes**). In birds, the precursors that populate the bursa of Fabricius, a lymphoid structure near the cloaca, become transformed into the lymphocytes responsible for humoral immunity (**B lymphocytes**). There is no bursa in mammals, and the transformation to B lymphocytes occurs in **bursal equivalents,** ie, the fetal liver and, after birth, the bone marrow. After residence in the thymus or liver, many of the T and B lymphocytes migrate to the lymph nodes and bone marrow. Most of the processing occurs during fetal and neonatal life. However, there is also a slow, continuous production of new lymphocytes from stem cells in adults.

T and B lymphocytes are morphologically indistinguishable but can be identified by markers on their cell membranes. B cells differentiate into **plasma cells** and **memory B cells.** There are three major types of T cells: **cytotoxic T cells, helper T cells,** and **memory T cells.** There are two subtypes of helper T cells: T helper 1 (TH1) cells secrete IL-2 and γ-interferon and are concerned primarily with cellular immunity; T helper 2 (TH2) cells secrete IL-4 and IL-5 and interact primarily with B cells in relation to humoral immunity. Cytotoxic T cells destroy transplanted and other foreign cells, with their development aided and directed by helper T cells. Markers on the surface of lymphocytes

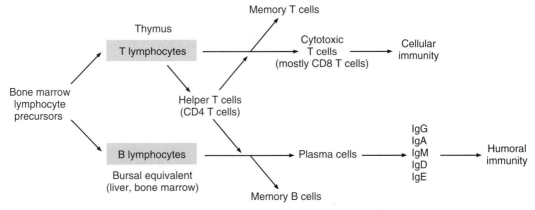

Figure 27–7 Development of the system mediating acquired immunity.

are assigned CD (clusters of differentiation) numbers on the basis of their reactions to a panel of monoclonal antibodies. Most cytotoxic T cells display the glycoprotein CD8, and helper T cells display the glycoprotein CD4. These proteins are closely associated with the T cell receptors and may function as coreceptors. On the basis of differences in their receptors and functions, cytotoxic T cells are divided into $\alpha\beta$ and $\gamma\delta$ types (see below). Natural killer cells (see above) are also cytotoxic lymphocytes, though they are not T cells. Thus, there are three main types of cytotoxic lymphocytes in the body: $\alpha\beta$ T cells, $\gamma\delta$ T cells, and NK cells.

Memory B Cells & T Cells

After exposure to a given antigen, a small number of activated B and T cells persist as memory B and T cells. These cells are readily converted to effector cells by a later encounter with the same antigen. This ability to produce an accelerated response to a second exposure to an antigen is a key characteristic of acquired immunity. The ability persists for long periods of time, and in some instances (eg, immunity to measles) it can be lifelong.

After activation in lymph nodes, lymphocytes disperse widely throughout the body and are especially plentiful in areas where invading organisms enter the body, eg, the mucosa of the respiratory and gastrointestinal tracts. This puts memory cells close to sites of reinfection and may account in part for the rapidity and strength of their response. Chemokines are involved in guiding activated lymphocytes to these locations.

It had been argued that the long life of memory cells involves their repeated exposure to small amounts of antigen. However, memory cells persist when infused into mice in which the ability to process the antigen to which they are sensitive has been abolished by gene knockout. It may be that they avoid apoptosis by taking up nerve growth factor in the peripheral tissues.

Antigen Recognition

The number of different antigens recognized by lymphocytes in the body is extremely large. The recognition ability is innate and develops without exposure to the antigen. Stem cells differentiate into many million different T and B lymphocytes, each with the ability to respond to a particular antigen. When the antigen first enters the body, it can bind directly to the appropriate receptors on B cells. However, a full antibody response requires that the B cells contact helper T cells. In the case of T cells, the antigen is taken up by an antigen-presenting cell and partially digested. A peptide fragment of it is presented to the appropriate receptors on

T cells. In either case, the cells are stimulated to divide, forming **clones** of cells that respond to this antigen **(clonal selection).**

Antigen Presentation

Antigen-presenting cells (APCs) include specialized cells called **dendritic cells** in the lymph nodes and spleen and the Langerhans dendritic cells in the skin. Macrophages and B cells themselves can also function as APCs. In APCs, polypeptide products of antigen digestion are coupled to protein products of the **major histocompatibility complex (MHC)** genes and presented on the surface of the cell. The products of the MHC genes are called human leukocyte antigens (HLA).

The genes of the MHC, which are located on the short arm of human chromosome 6, encode glycoproteins and are divided into two classes on the basis of structure and function. Class I antigens are composed of a 45-kDa heavy chain associated noncovalently with β_2-microglobulin encoded by a gene outside the MHC (Figure 27–8). They are found on all nucleated cells. Class II antigens are heterodimers made up of a 29- to 34-kDa α chain associated noncovalently with a 25- to 28-kDa β chain. They are present in antigen-presenting cells, including B cells, and in activated T cells.

The class I MHC proteins (MHC-I proteins) are coupled primarily to peptide fragments generated from proteins synthesized within cells. The peptides to which the host is not tolerant, eg, those from mutant or viral proteins, are recognized by T cells. The digestion of these proteins occurs in **proteasomes,** complexes of proteolytic enzymes that may be produced by genes in the MHC group, and the peptide fragments appear to bind to MHC proteins in the endoplasmic reticulum. The class II MHC proteins (MHC-II proteins) are concerned primarily with peptide products of extracellular antigens, such as bacteria, that enter the cell by endocytosis and are digested in the late endosomes.

T Cell Receptors

The MHC protein-peptide complexes on the surface of the antigen-presenting cells bind to appropriate T cells. Therefore, receptors on the T cells must recognize a very wide variety of complexes. Most of the receptors on circulating T cells are made up of two polypeptide units designated α and β. They form heterodimers that recognize the MHC proteins and the antigen fragments with which they are combined (Figure 27–9). These cells are called $\alpha\beta$ T cells. About 10% of the circulating T cells have two different polypeptides designated γ and δ in their receptors, and they are called $\gamma\delta$ T cells. These T cells are prominent in the mucosa of the gastrointestinal

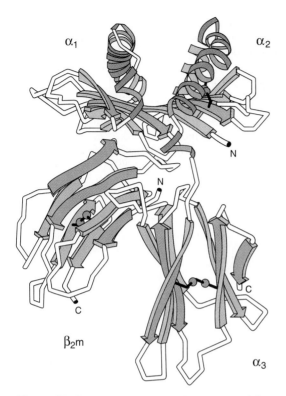

Figure 27–8 Structure of human histocompatibility antigen HLA-A2. The antigen-binding pocket is at the top and is formed by the α_1 and α_2 parts of the molecule. The α_3 portion and the associated β_2-microglobulin (β_2m) are close to the membrane. The extension of the C̄ terminal from α_3 that provides the transmembrane domain and the small cytoplasmic portion of the molecule have been omitted. (Reproduced, with permission, from Bjorkman PJ et al: Structure of the human histocompatibility antigen HLA-A2. Nature 1987;329:506.)

tract, and there is evidence that they form a link between the innate and acquired immune systems by way of the cytokines they secrete (Figure 27–5).

CD8 occurs on the surface of cytotoxic T cells that bind MHC-I proteins, and CD4 occurs on the surface of helper T cells that bind MHC-II proteins (Figure 27–10). The CD8 and CD4 proteins facilitate the binding of the MHC proteins to the T cell receptors, and they also foster lymphocyte development, but how they produce these effects is unsettled. The activated CD8 cytotoxic T cells kill their targets directly, whereas the activated CD4 helper T cells secrete cytokines that activate other lymphocytes.

The T cell receptors are surrounded by adhesion molecules and proteins that bind to complementary proteins in the antigen-presenting cell when the two cells transiently join to form the "immunologic synapse" that permits T cell activation to occur. It is now generally accepted that two signals are necessary to produce activation. One is produced by the binding of the digested antigen to the T cell receptor. The other is produced by the joining of the surrounding proteins in the "synapse." If the first signal occurs but the second does not, the T cell is inactivated and becomes unresponsive.

B Cells

As noted above, B cells can bind antigens directly, but they must contact helper T cells to produce full activation and antibody formation. It is the TH2 subtype that is mainly involved. Helper T cells are pushed along the TH2 line by the cytokine IL-4 (see below). On the other hand, IL-12 pushes helper T cells along the TH1 line. IL-2 acts in an autocrine fashion to cause activated T cells to proliferate. The role of various cytokines in B cell and T cell activation is summarized in Figure 27–11.

The activated B cells proliferate and transform into **memory B cells** (see above) and **plasma cells.** The plasma cells secrete large quantities of antibodies into

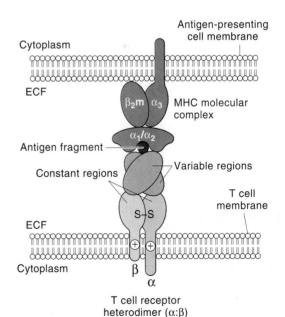

Figure 27–9 Interaction between antigen-presenting cell **(top)** and T lymphocyte **(bottom).** The MHC proteins (in this case, MHC-I) and their peptide antigen fragment bind to the α and β units that combine to form the T cell receptor.

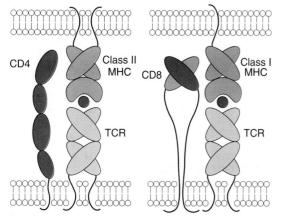

Figure 27–10 Diagrammatic summary of the structure of CD4 and CD8, and their relation to MHC-I and MHC-II proteins. Note that CD4 is a single protein, whereas CD8 is a heterodimer. (Reproduced, with permission, from Leahy DJ: A structural view of CD4 and CD8. FASEB J 1995;9:17.)

the general circulation. The antibodies circulate in the globulin fraction of the plasma (see below) and, like antibodies elsewhere, are called **immunoglobulins.** The immunoglobulins are actually the secreted form of antigen-binding receptors on the B cell membrane.

Immunoglobulins

Circulating antibodies protect their host by binding to and neutralizing some protein toxins, by blocking the attachment of some viruses and bacteria to cells, by opsonizing bacteria (see above) and by activating complement. Five general types of immunoglobulin antibodies are produced by the lymphocyte-plasma cell system. The basic component of each is a symmetric unit containing four polypeptide chains (Figure 27–12). The two long chains are called **heavy chains,** whereas the two short chains are called **light chains.** There are two types of light chains, κ and λ, and eight types of heavy chains. The chains are joined by disulfide bridges that permit mobility, and there are intrachain disulfide bridges as well. In addition, the heavy chains are flexible in a region called the hinge. Each heavy chain has a variable (V) segment in which the amino acid sequence is highly variable, a diversity (D) segment in which the amino acid segment is also highly variable, a joining (J) segment in which it is moderately variable, and a constant (C) segment in which the sequence is constant. Each light chain has a V, a J, and a C segment. The V segments form part of the antigen-binding sites (Fab portion of the molecule

[Figure 27–12]). The Fc portion of the molecule is the effector portion, which mediates the reactions initiated by antibodies.

Two of the classes of immunoglobulins contain additional polypeptide components (Table 27–4). In IgMs, five of the basic immunoglobulin units join around a polypeptide called the J chain to form a pentamer. In IgAs, the **secretory immunoglobulins,** the immunoglobulin units form dimers and trimers around a J chain and a polypeptide that comes from epithelial cells, the secretory component (SC).

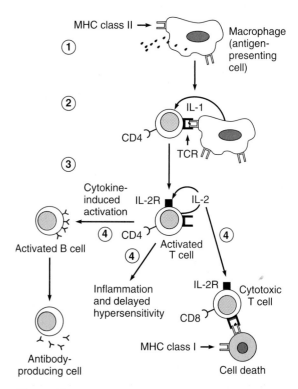

Figure 27–11 Summary of acquired immunity. **(1)** An antigen-presenting cell ingests and partially digests an antigen, then presents part of the antigen along with MHC peptides (in this case, MHC II peptides on the cell surface). **(2)** An "immune synapse" forms with a naive CD4 T cell, which is activated to produce IL-2. **(3)** IL-2 acts in an autocrine fashion to cause the cell to multiply, forming a clone. **(4)** The activated CD4 cell may promote B cell activation and production of plasma cells or it may activate a cytotoxic CD8 cell. The CD8 cell can also be activated by forming a synapse with an MCH I antigen-presenting cell. (Reproduced, with permission, from McPhee SJ, Lingappa VR, Ganong WF [editors]: *Pathophysiology of Disease,* 4th ed. McGraw-Hill, 2003.)

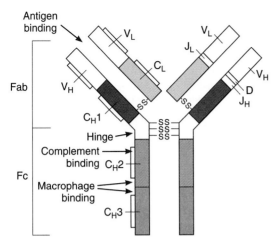

Figure 27–12 Typical immunoglobulin G molecule. Fab, portion of the molecule that is concerned with antigen binding; Fc, effector portion of the molecule. The constant regions are colored, and the variable regions are clear. The constant segment of the heavy chain is subdivided into C_H1, C_H2, and C_H3. The lines indicating the presence of intrasegmental disulfide bonds are omitted on the right side, and the J_H, D, V_H, J_L, and V_L segments are labeled on that side.

In the intestine, bacterial and viral antigens are taken up by M cells (see Chapter 25) and passed on to underlying aggregates of lymphoid tissue **(Peyer's patches),** where they activate naive T cells. These lymphocytes then form B cells which infiltrate mucosa of the gastrointestinal, respiratory, genitourinary, and female reproductive tracts and the breast. There they se-

crete large amounts of IgAs when exposed again to the antigen that initially stimulated them. The epithelial cells produce the SC, which acts as a receptor for and binds the IgA. The resulting secretory immunoglobulin passes through the epithelial cell and is secreted by exocytosis. This system of **secretory immunity** is an important and effective defense mechanism. Some immune cells contain inducible NOS (see Chapter 31), and NO appears to enhance secretory immunity in the gastrointestinal tract (see Chapter 26).

Monoclonal Antibodies

Large quantities of the immunoglobulin produced by a single plasma cell can be obtained by fusing the cell with a tumor cell, producing an antibody "factory." In practice, animals are immunized with a particular antigen or cell preparation. They are then sacrificed, and the antibody-producing cells are extracted from their spleens and fused to myeloma cells. Myelomas are B lymphocyte tumors that readily fuse with plasma cells to form antibody-producing **hybridomas,** which grow and reproduce very well. The fused cells are separated by standard techniques, and each starts a **clone** of cells descended from a single cell.

Genetic Basis of Diversity in the Immune System

The genetic mechanism for the production of the immensely large number of different configurations of immunoglobulins produced by human B cells is a fascinating biologic problem. Diversity is brought about in part by the fact that in immune globulin molecules there are two kinds of light chains and eight kinds of heavy chains. As noted above, there are areas of great

Table 27–4. Human immunoglobulins. In all instances, the light chains are κ or λ.

Immuno-globulin	Function	Heavy Chain	Additional Chain	Structure	Plasma Concentration (mg/dL)
IgG	Complement activation.	γ_1, γ_2, γ_3, γ_4		Monomer.	1000
IgA	Localized protection in external secretions (tears, intestinal secretions, etc).	α_1, α_2	J, SC	Monomer; dimer with J or SC chain; trimer with J chain.	200
IgM	Complement activation.	μ	J	Pentamer with J chain.	120
IgD	Antigen recognition by B cells.	δ		Monomer.	3
IgE	Reagin activity; releases histamine from basophils and mast cells.	ε		Monomer.	0.05

variability **(hypervariable regions)** in each chain. The variable portion of the heavy chains consists of the V, D, and J segments. In the gene family responsible for this region, there are several hundred different coding regions for the V segment, about 20 for the D segment, and 4 for the J segment. During B cell development, one V coding region, one D coding region, and one J coding region are selected at random and recombined to form the gene that produces that particular variable portion. There is similar variable recombination in the coding regions responsible for the two variable segments (V and J) in the light chain. In addition, the J segments are variable because the gene segments join in an imprecise and variable fashion (junctional site diversity) and nucleotides are sometimes added (junctional insertion diversity). It has been calculated that these mechanisms permit the production of about 10^{15} different immunoglobulin molecules. Additional variability is added by somatic mutation.

Similar gene rearrangement and joining mechanisms operate to produce the diversity in T cell receptors. In humans, the α subunit has a V region encoded by one of about 50 different genes and a J region encoded by 1 of another 50 different genes. The β subunits have a V region encoded by 1 of about 50 genes, a D region encoded by 1 of 2 genes, and a J region encoded by 1 of 13 genes. These variable regions permit the generation of up to an estimated 10^{15} different T cell receptors.

Recognition of Self

A key question is why T and B cells do not form antibodies against and destroy the cells and organs of the individual in which they develop. Current evidence indicates that self antigens are presented along with nonself antigens but are then eliminated during development (tolerance). Central tolerance occurs in the thymus for T cells and the bone marrow for B cells. This is supplemented by peripheral tolerance occurring in the lymph nodes and elsewhere in the body. A number of different mechanisms are involved—exactly which ones and the details of their operation are unsettled questions.

Autoimmunity

Sometimes the processes that eliminate antibodies against self antigens fail, and a variety of different **autoimmune diseases** are produced. These can be B cell- or T cell-mediated and can be organ-specific or systemic. They include type 1 diabetes mellitus (antibodies against pancreatic islet B cells), myasthenia gravis (antibodies against nicotinic cholinergic receptors), and multiple sclerosis (antibodies against myelin basic protein and several other components of myelin). In some

instances, the antibodies are against receptors and are capable of activating receptors; for example, antibodies against TSH receptors increase thyroid activity and cause Graves' disease (see Chapter 18). Other conditions are due to the production of antibodies against invading organisms that cross-react with normal body constituents **(molecular mimicry).** An example is rheumatic fever following a streptococcal infection; a portion of cardiac myosin resembles a portion of the streptococcal M protein, and antibodies induced by the latter attack the former and damage the heart. Some conditions may be due to **bystander effects,** in which inflammation sensitizes T cells in the neighborhood, causing them to become activated when otherwise they would not respond. However, much is still uncertain about the pathogenesis of autoimmune disease.

Tissue Transplantation

The T lymphocyte system is responsible for the rejection of transplanted tissue. When tissues such as skin and kidneys are transplanted from a donor to a recipient of the same species, the transplants "take" and function for a while but then become necrotic and are "rejected" because the recipient develops an immune response to the transplanted tissue. This is generally true even if the donor and recipient are close relatives, and the only transplants that are never rejected are those from an identical twin.

A number of treatments have been developed to overcome the rejection of transplanted organs in humans. The goal of treatment is to stop rejection without leaving the patient vulnerable to massive infections. One approach is to kill T lymphocytes by killing all rapidly dividing cells with drugs such as azathioprine, a purine antimetabolite, but this makes patients susceptible to infections and cancer. Another is to administer glucocorticoids, which inhibit cytotoxic T cell proliferation by inhibiting production of IL-2, but these cause osteoporosis, mental changes, and the other stigmas of Cushing's syndrome (see Chapter 20). A third is treatment with **cyclosporine** or **tacrolimus (FK-506).** Activation of the T cell receptor normally increases intracellular Ca^{2+}, which acts via calmodulin to activate calcineurin (Figure 27–13). Calcineurin dephosphorylates the transcription factor NF-AT, which moves to the nucleus and increases the activity of genes coding for IL-2 and related stimulatory cytokines. Cyclosporine and tacrolimus prevent the dephosphorylation of NF-AT. However, these drugs inhibit all T cell-mediated immune responses, and cyclosporine causes kidney damage and cancer. A new and promising approach to transplant rejection is the production of T cell unresponsiveness by using drugs that block the second signal costimulation that is required for normal ac-

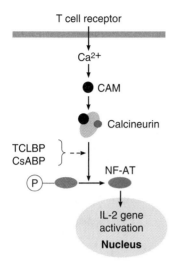

Figure 27–13 Action of cyclosporine (CsA) and tacrolimus (TCL) in lymphocytes. Activation of the T cell receptor permits Ca$^+$ influx, which, via calmodulin (CAM), activates calcineurin. The calcineurin dephosphorylates the transcription factor NF-AT, which then enters the nucleus and activates the IL-2 gene. When bound to their respective immunophilins (BP), cyclosporine and tacrolimus inhibit the action of calcineurin on NF-AT and thus prevent increased production of IL-2.

tivation (see above). Clinically effective drugs that act in this fashion could be of great value to transplant surgeons.

Other Clinical Correlates

As knowledge about the immune system has increased, over 50 immunodeficiency syndromes due to abnormalities in the function of immune cells have been described. These produce abnormalities ranging from a moderate increase in the incidence of infections to severe, usually fatal conditions. Figure 27–14 illustrates how blockade at various points along the B and T cell maturation pathway produces predictable abnormalities. There are also various forms of complement deficiency.

Malignant transformation can occur at various stages of lymphocyte development. Most if not all cases of chronic lymphocytic leukemia are due to uncontrolled proliferation of B lymphocytes, whereas multiple myeloma is due to malignant proliferation of clones of mature plasma cells. Some cases of acute lymphocytic leukemia are T lymphocyte cancers.

Acquired immune deficiency syndrome (AIDS), a disease that is currently a major worldwide problem, is unique in that **HIV (human immunodeficiency virus),** the retrovirus that causes many cases of it, binds to CD4 and produces a decrease in the number of CD4 helper T cells. The loss of helper lymphocytes leads in turn to failure of proliferation of CD8 and B cells, with eventual loss of immune function and death from infections due to normally nonpathogenic bacteria or cancer.

PLATELETS

The platelets are small, granulated bodies that aggregate at sites of vascular injury. They lack nuclei and are 2–4 μm in diameter (Figure 27–2). There are about 300,000/μL of circulating blood, and they normally have a half-life of about 4 days. The **megakaryocytes,** giant cells in the bone marrow, form platelets by pinching off bits of cytoplasm and extruding them into the circulation. Between 60% and 75% of the platelets that have been extruded from the bone marrow are in the circulating blood, and the remainder are mostly in the spleen. Splenectomy causes an increase in the platelet count **(thrombocytosis).**

Platelets have a ring of microtubules around their periphery and an extensively invaginated membrane with an intricate canalicular system in contact with the ECF. Their membranes contain receptors for collagen, ADP, vessel wall von Willebrand factor (see below), and fibrinogen. Their cytoplasm contains actin, myosin, glycogen, lysosomes, and two types of granules: (1) dense granules, which contain the nonprotein substances that are secreted in response to platelet activation, including serotonin, ADP and other adenine nucleotides, and (2) α-granules, which contain secreted proteins other than the hydrolases in lysosomes. These proteins include clotting factors and **platelet-derived growth factor (PDGF).** PDGF is also produced by macrophages and endothelial cells. It is a dimer made up of A and B subunit polypeptides. Homodimers (AA and BB), as well as the heterodimer (AB), are produced. PDGF stimulates wound healing and is a potent mitogen for vascular smooth muscle. Blood vessel walls as well as platelets contain von Willebrand factor, which, in addition to its role in adhesion, regulates circulating levels of factor VIII (see below).

When a blood vessel wall is injured, platelets adhere to the exposed collagen and von Willebrand factor in the wall via the receptors on the platelet membrane. Binding produces platelet activations which release the contents of their granules. The released ADP acts on the ADP receptors in the platelet membranes to produce further accumulation of more platelets **(platelet aggregation).** There are at least three different types of platelet **ADP receptors** in humans; **P2Y$_1$, P2Y$_2$,** and **P2X$_1$.** These are obviously attractive targets for drug

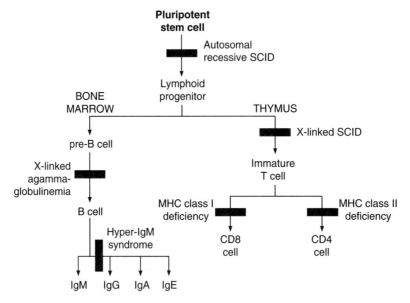

Figure 27–14 Sites of congenital blockade of B and T lymphocyte maturation in various immunodeficiency states. SCID, severe combined immune deficiency. (Modified from Rosen FS, Cooper MD, Wedgwood RJP: The primary immunodeficiencies. N Engl J Med 1995;333:431.)

development, and several new inhibitors have shown promise in the prevention of heart attacks and strokes. Aggregation is also fostered by **platelet-activating factor (PAF)**, a cytokine secreted by neutrophils and monocytes as well as platelets. This compound also has inflammatory activity. It is an ether phospholipid, 1-alkyl-2-acetylglyceryl-3-phosphorylcholine, which is produced from membrane lipids. It acts via a G protein-coupled receptor to increase the production of arachidonic acid derivatives, including thromboxane A_2. The role of this compound in the balance between clotting and anticlotting activity at the site of vascular injury is discussed in Chapter 31.

Platelet production is regulated by the colony-stimulating factors that control the production of megakaryocytes (Figure 27–2), plus **thrombopoietin**, a circulating protein factor. This factor, which facilitates megakaryocyte maturation, is produced constitutively by the liver and kidneys, and there are thrombopoietin receptors on platelets. Consequently, when the number of platelets is low, less is bound and more is available to stimulate production of platelets. Conversely, when the number of platelets is high, more is bound and less is available, producing a form of feedback control of platelet production. The amino terminal portion of the thrombopoietin molecule has the platelet-stimulating activity, whereas the carboxyl terminal portion contains many carbohydrate residues and is concerned with the bioavailability of the molecule.

When the platelet count is low, clot retraction is deficient and there is poor constriction of ruptured vessels. The resulting clinical syndrome (**thrombocytopenic purpura**) is characterized by easy bruisability and multiple subcutaneous hemorrhages. Purpura may also occur when the platelet count is normal, and in some of these cases, the circulating platelets are abnormal (**thrombasthenic purpura**). Individuals with thrombocytosis are predisposed to thrombotic events.

RED BLOOD CELLS

The red blood cells (**erythrocytes**) carry hemoglobin in the circulation. They are biconcave disks (Figure 27–15) that are manufactured in the bone marrow. In mammals, they lose their nuclei before entering the circulation. In humans, they survive in the circulation for an average of 120 days. The average normal red blood cell count is 5.4 million/μL in men and 4.8 million/μL in women. Each human red blood cell is about 7.5 μm in diameter and 2 μm thick, and each contains approximately 29 pg of hemoglobin (Table 27–5). There are thus about 3×10^{13} red blood cells and about 900 g of hemoglobin in the circulating blood of an adult man (Figure 27–16).

The feedback control of erythropoiesis by erythropoietin is discussed in Chapter 24, and the role of IL-1, IL-3, IL-6, and GM-CSF in development of the relevant erythroid stem cells is shown in Figure 27–2.

Figure 27–15 Human red blood cells and fibrin fibrils. Blood was placed on a polyvinyl chloride surface, fixed, and photographed with a scanning electron microscope. Reduced from ×2590. (Courtesy of NF Rodman.)

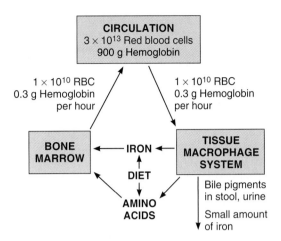

Figure 27–16 Red cell formation and destruction. RBC, red blood cells.

Red Cell Fragility

Red blood cells, like other cells, shrink in solutions with an osmotic pressure greater than that of normal plasma. In solutions with a lower osmotic pressure they swell, becoming spherical rather than disk-shaped, and eventually lose their hemoglobin (**hemolysis**). The hemoglobin of hemolyzed red cells dissolves in the plasma, coloring it red. A 0.9% sodium chloride solution is isotonic with plasma. When **osmotic fragility** is normal, red cells begin to hemolyze when suspended in 0.5% saline; 50% lysis occurs in 0.40–0.42% saline, and lysis is complete in 0.35% saline. In **hereditary spherocytosis** (congenital hemolytic icterus), the cells are spherocytic in normal plasma and hemolyze more readily than normal cells in hypotonic sodium chloride solutions. Spherocytes are

Table 27–5. Characteristics of human red cells.[1]

		Male	Female
Hematocrit (Hct)(%)		47	42
Red blood cells (RBC) (10^6/μL)		5.4	4.8
Hemoglobin (Hb) (g/dL)		16	14
Mean corpuscular volume (MCV) (fL)	$= \dfrac{Hct \times 10}{RBC\ (10^6/\mu L)}$	87	87
Mean corpuscular hemoglobin (MCH) (pg)	$= \dfrac{Hb \times 10}{RBC\ (10^6/\mu L)}$	29	29
Mean corpuscular hemoglobin concentration (MCHC) (g/dL)	$= \dfrac{Hb \times 100}{Hct}$	34	34
Mean cell diameter (MCD) (μm)	$= \dfrac{\text{Mean diameter of 500}}{\text{cells in smear}}$	7.5	7.5

[1]Cells with MCVs > 95 fL are called macrocytes; cells with MCVs < 80 fL are called microcytes; cells with MCHs < 25 g/dL are called hypochromic.

also removed by the spleen (see below). Consequently, hereditary spherocytosis is one of the most common causes of **hereditary hemolytic anemia.** The spherocytosis is caused by abnormalities of the protein network that maintains the shape and flexibility of the red cell membrane. The membrane skeleton is made up in part of **spectrin** and is anchored to the transmembrane protein band 3 by the protein **ankyrin** (Figure 1–10). Band 3 is also an important anion exchanger (see Chapter 35). Defects in band 3, spectrin, and ankyrin have all been reported.

Red cells can also be lysed by drugs and infections. The susceptibility of red cells to hemolysis by these agents is increased by deficiency of the enzyme glucose 6-phosphate dehydrogenase (G6PD), which catalyzes the initial step in the oxidation of glucose via the hexose monophosphate pathway (see Chapter 17). This pathway generates NADPH, which is needed in some way for the maintenance of normal red cell fragility. Severe G6PD deficiency also inhibits the killing of bacteria by granulocytes and predisposes to severe infections (see above).

Role of the Spleen

The spleen is an important blood filter that removes spherocytes and other abnormal red cells. It also contains many platelets and plays a significant role in the immune system. Abnormal red cells are removed if they are not as flexible as normal red cells and consequently are unable to squeeze through the slits between the endothelial cells that line the splenic sinuses.

Hemoglobin

The red, oxygen-carrying pigment in the red blood cells of vertebrates is **hemoglobin,** a protein with a molecular weight of 64,450. Hemoglobin is a globular molecule made up of 4 subunits (Figure 27–17). Each subunit contains a **heme** moiety conjugated to a polypeptide. Heme is an iron-containing porphyrin derivative (Figure 27–18). The polypeptides are referred to collectively as the **globin** portion of the hemoglobin molecule. There are two pairs of polypeptides in each hemoglobin molecule. In normal adult human hemoglobin **(hemoglobin A),** the two types of polypeptide are called the α chains, each of which contains 141 amino acid residues, and the β chains, each of which contains 146 amino acid residues. Thus, hemoglobin A is designated $\alpha_2\beta_2$. Not all the hemoglobin in the blood of normal adults is hemoglobin A. About 2.5% of the hemoglobin is hemoglobin A_2, in which β chains are replaced by δ chains ($\alpha_2\delta_2$). The δ chains also contain 146 amino acid residues, but 10 individual residues differ from those in the β chains.

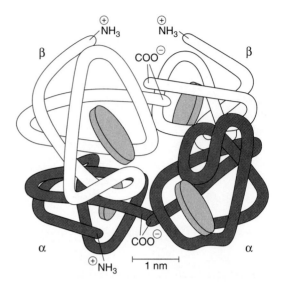

Figure 27–17 Diagrammatic representation of a molecule of hemoglobin A, showing the four subunits. There are two α and two β polypeptide chains, each containing a heme moiety. These moieties are represented by the disks. (Reproduced, with permission, from Harper HA et al: *Physiologische Chemie.* Springer, 1975.)

There are small amounts of hemoglobin A derivatives closely associated with hemoglobin A that represent glycated hemoglobins. One of these, **hemoglobin A_{1c} (HbA$_{1c}$),** has a glucose attached to the terminal valine in each β chain and is of special interest because the quantity in the blood increases in poorly controlled diabetes mellitus (see Chapter 19).

Reactions of Hemoglobin

Hemoglobin binds O_2 to form **oxyhemoglobin,** O_2 attaching to the Fe^{2+} in the heme. The affinity of hemoglobin for O_2 is affected by pH, temperature, and the concentration in the red cells of 2,3-diphosphoglycerate (2,3-DPG). 2,3-DPG and H^+ compete with O_2 for binding to deoxygenated hemoglobin, decreasing the affinity of hemoglobin for O_2 by shifting the positions of the four peptide chains (quaternary structure). The details of the oxygenation and deoxygenation of hemoglobin and the physiologic role of these reactions in O_2 transport are discussed in Chapter 35.

When blood is exposed to various drugs and other oxidizing agents in vitro or in vivo, the ferrous iron (Fe^{2+}) that is normally in the molecule is converted to ferric iron (Fe^{3+}), forming **methemoglobin.** Methemoglobin is dark-colored, and when it is present in large quantities in the circulation, it causes a dusky discol-

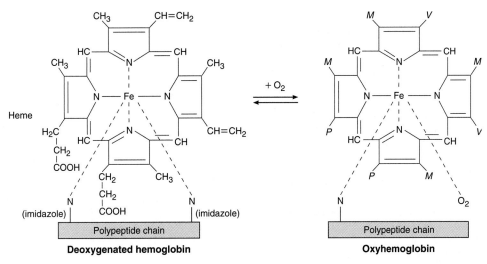

Deoxygenated hemoglobin   **Oxyhemoglobin**

Figure 27–18 Reaction of heme with O_2. The abbreviations M, V, and P stand for the groups shown on the molecule on the left.

oration of the skin resembling cyanosis (see Chapter 37). Some oxidation of hemoglobin to methemoglobin occurs normally, but an enzyme system in the red cells, the NADH-methemoglobin reductase system, converts methemoglobin back to hemoglobin. Congenital absence of this system is one cause of hereditary methemoglobinemia.

Carbon monoxide reacts with hemoglobin to form **carbon monoxyhemoglobin (carboxyhemoglobin).** The affinity of hemoglobin for O_2 is much lower than its affinity for carbon monoxide, which consequently displaces O_2 on hemoglobin, reducing the oxygen-carrying capacity of blood (see Chapter 37).

Heme is also part of the structure of **myoglobin,** an oxygen-binding pigment found in red (slow) muscles (see Chapter 35). In addition, **neuroglobin,** an oxygen-binding globin, is found in the brain. It appears to help supply O_2 to neurons. There is heme in the respiratory chain enzyme **cytochrome c** (see Chapter 17). Porphyrins other than that found in heme play a role in the pathogenesis of a number of metabolic diseases (congenital and acquired porphyria, etc).

Hemoglobin in the Fetus

The blood of the human fetus normally contains **fetal hemoglobin (hemoglobin F).** Its structure is similar to that of hemoglobin A except that the β chains are replaced by γ chains; ie, hemoglobin F is $\alpha_2\gamma_2$. The γ chains also contain 146 amino acid residues but have 37 that differ from those in the β chain. Fetal hemoglobin is normally replaced by adult hemoglobin soon after birth (Figure 27–19). In certain individuals, it fails

to disappear and persists throughout life. In the body, its O_2 content at a given P_{O_2} is greater than that of adult hemoglobin because it binds 2,3-DPG less avidly. This facilitates movement of O_2 from the maternal to the fetal circulation (see Chapter 32). In young embryos there are, in addition, ζ and ε chains, forming Gower 1 hemoglobin ($\zeta_2\varepsilon_2$) and Gower 2 hemoglobin ($\alpha_2\varepsilon_2$). There are two copies of the α globin gene on human chromosome 16. In addition, there are five globin genes in tandem on chromosome 11 that encode β, γ, and δ globin chains and the two chains normally found only during fetal life. Switching from one form of hemoglobin to another during development seems

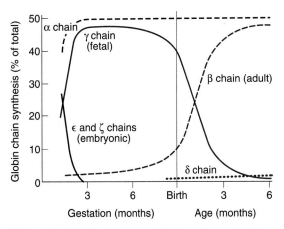

Figure 27–19 Development of human hemoglobin chains.

to be regulated by a locus control region (LCR) in the DNA upstream from the human gene for ε hemoglobin, but the details of the switching remain to be determined.

Abnormalities of Hemoglobin Production

The amino acid sequences in the polypeptide chains of hemoglobin are determined by globin genes.

There are two major types of inherited disorders of hemoglobin in humans: the **hemoglobinopathies,** in which abnormal polypeptide chains are produced, and the **thalassemias** and related disorders, in which the chains are normal in structure but produced in decreased amounts or absent because of defects in the regulatory portion of the globin genes. The α and β thalassemias are defined by decreased or absent α and β polypeptides, respectively.

Mutant genes that cause the production of abnormal hemoglobins are widespread, and about 1000 abnormal hemoglobins have been described in humans. They are usually identified by letter—hemoglobin C, E, I, J, S, etc. In most instances, the abnormal hemoglobins differ from normal hemoglobin A in the structure of the polypeptide chains. For example, in hemoglobin S, the α chains are normal but the β chains are abnormal, because among the 146 amino acid residues in each β polypeptide chain, one glutamic acid residue has been replaced by a valine residue (Table 27–6).

When an abnormal gene inherited from one parent dictates formation of an abnormal hemoglobin—ie, when the individual is heterozygous—half the circulating hemoglobin is abnormal and half is normal. When identical abnormal genes are inherited from both parents, the individual is homozygous and all of the hemoglobin is abnormal. It is theoretically possible to inherit two different abnormal hemoglobins, one from the father and one from the mother. Studies of the inheritance and geographic distribution of abnormal hemoglobins have made it possible in some cases to decide where the mutant gene originated and approximately how long ago the mutation occurred. In general, harmful mutations tend to die out, but mutant genes that confer traits with survival value persist and spread in the population.

Many of the abnormal hemoglobins are harmless. However, some have abnormal O_2 equilibriums. Others cause anemia. For example, hemoglobin S polymerizes at low O_2 tensions, and this causes the red cells to become sickle-shaped, hemolyze, and form aggregates that block blood vessels. The result is the severe hemolytic anemia known as **sickle cell anemia.** Heterozygous individuals have the **sickle cell trait** and rarely have severe symptoms, but homozygous individuals develop the full-blown disease. The sickle cell gene is an example of a gene that has persisted and spread in the population. It originated in the black population in Africa, and it confers resistance to one type of malaria. This is an important benefit in Africa, and in some parts of Africa 40% of the population have the sickle cell trait. In the United States black population its incidence is about 10%.

Table 27–6. Partial amino acid composition of normal human β chain, and some hemoglobins with abnormal β chains.[1]

Hemoglobin	Positions on β Polypeptide Chain of Hemoglobin						
	1 2 3	6 7	26	63	67	121	146
A (normal)	Val-His-Leu	Glu-Glu	Glu	His	Val	Glu	His
S (sickle cell)		Val					
C		Lys					
G$_{San Jose}$		Gly					
E			Lys				
M$_{Saskatoon}$				Tyr			
M$_{Milwaukee}$					Glu		
O$_{Arabia}$						Lys	

[1]Other hemoglobins have abnormal α chains. Abnormal hemoglobins that are very similar electrophoretically but differ slightly in composition are indicated by the same letter and a subscript indicating the geographic location where they were first discovered; hence, M$_{Saskatoon}$ and M$_{Milwaukee}$.

Hemoglobin F has the ability to decrease the polymerization of deoxygenated hemoglobin S, and hydroxyurea causes hemoglobin F to be produced in children and adults. It has proved to be a very valuable agent for the treatment of sickle cell disease. In patients with severe sickle cell disease, bone marrow transplantation has been carried out and the patients have generally done well, though more study is needed.

Synthesis of Hemoglobin

The average normal hemoglobin content of blood is 16 g/dL in men and 14 g/dL in women, all of it in red cells. In the body of a 70-kg man, there are about 900 g of hemoglobin, and 0.3 g of hemoglobin is destroyed and 0.3 g synthesized every hour (Figure 27–16). The heme portion of the hemoglobin molecule is synthesized from glycine and succinyl-CoA.

Catabolism of Hemoglobin

When old red blood cells are destroyed in the tissue macrophage system, the globin portion of the hemoglobin molecule is split off, and the heme is converted to **biliverdin.** The enzyme involved is a subtype of heme oxygenase (see Figure 4–33), and CO is formed in the process. CO may be an intercellular messenger, like NO (see Chapter 4).

In humans, most of the biliverdin is converted to **bilirubin** (Figure 27–20) and excreted in the bile (see Chapter 26). The iron from the heme is reused for hemoglobin synthesis.

Exposure of the skin to white light converts bilirubin to lumirubin, which has a shorter half-life than bilirubin. **Phototherapy** (exposure to light) is of value in treating infants with jaundice due to hemolysis. Iron is essential for hemoglobin synthesis; if blood is lost from the body and the iron deficiency is not corrected,

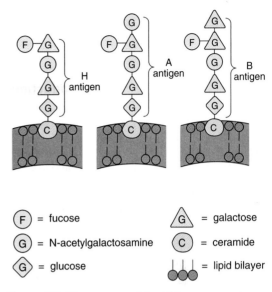

Figure 27–21 Antigens of the ABO system on the surface of red blood cells.

iron deficiency anemia results. The metabolism of iron is discussed in Chapter 25.

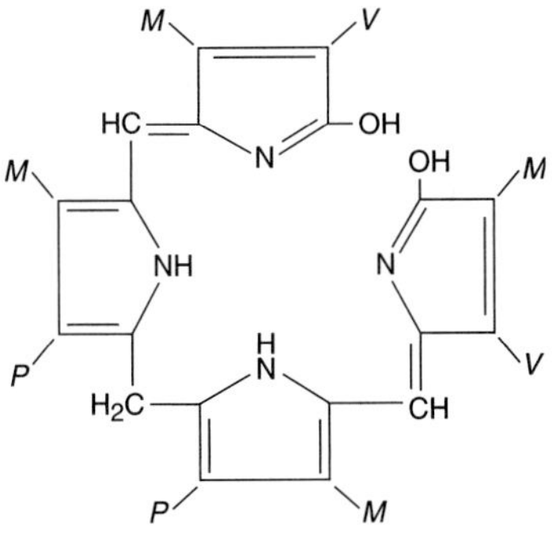

BLOOD TYPES

The membranes of human red cells contain a variety of **blood group antigens,** which are also called **agglutinogens.** The most important and best known of these are the A and B antigens, but there are many more.

The ABO System

The A and B antigens are inherited as mendelian dominants, and individuals are divided into four major **blood types** on this basis. Type A individuals have the A antigen, type B have the B, type AB have both, and type O have neither. These antigens are found in many tissues in addition to blood: these include salivary glands, saliva, pancreas, kidney, liver, lungs, testes, semen, and amniotic fluid.

The A and B antigens are actually complex oligosaccharides that differ in their terminal sugar. On red cells they are mostly glycosphingolipids, whereas in other tissues they are glycoproteins. An *H* gene codes for a fucose transferase that puts a fucose on the end of these glycolipids or glycoproteins, forming the H antigen that is usually present in individuals of all blood types (Figure 27–21). Individuals who are type A have a gene which codes for a transferase that catalyzes placement of a terminal *N*-acetylgalactosamine on the H antigen, whereas individuals who are type B have a gene which

Figure 27–20 Bilirubin. The abbreviations M, V, and P stand for the groups shown on the molecule on the left in Figure 27–18.

codes for a transferase that places a terminal galactose. Individuals who are type AB have both transferases. Individuals who are type O have neither, so the H antigen persists. It now appears that type O individuals have a single-base deletion in their corresponding gene. This creates an open reading frame, and consequently they produce a protein that has no transferase activity.

Subgroups of blood types A and B have been described, the most important being A_1 and A_2. However, the difference between A_1 and A_2 appears to be quantitative; each A_1 cell has about 1,000,000 copies of the A antigen on its surface, and each A_2 cell has about 250,000.

Antibodies against red cell agglutinogens are called **agglutinins.** Antigens very similar to A and B are common in intestinal bacteria and possibly in foods to which newborn individuals are exposed. Therefore, infants rapidly develop antibodies against the antigens not present in their own cells. Thus, type A individuals develop anti-B antibodies, type B individuals develop anti-A antibodies, type O individuals develop both, and type AB individuals develop neither (Table 27–7). When the plasma of a type A individual is mixed with type B red cells, the anti-B antibodies cause the type B red cells to clump (agglutinate), as shown in Figure 27–22. The other agglutination reactions produced by mismatched plasma and red cells are summarized in Table 27–7. **Blood typing** is performed by mixing an individual's red blood cells with antisera containing the various agglutinins on a slide and seeing whether agglutination occurs.

Transfusion Reactions

Dangerous **hemolytic transfusion reactions** occur when blood is transfused into an individual with an incompatible blood type, ie, an individual who has agglutinins against the red cells in the transfusion. The plasma in the transfusion is usually so diluted in the recipient that it rarely causes agglutination even when

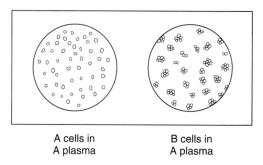

A cells in   B cells in
A plasma   A plasma

Figure 27–22 Red cell agglutination in incompatible plasma.

the titer of agglutinins against the recipient's cells is high. However, when the recipient's plasma has agglutinins against the donor's red cells, the cells agglutinate and hemolyze. Free hemoglobin is liberated into the plasma. The severity of the resulting transfusion reaction may vary from an asymptomatic minor rise in the plasma bilirubin level to severe jaundice and renal tubular damage (caused in some way by the products liberated from hemolyzed cells), with anuria and death.

Incompatibilities in the ABO blood group system are summarized in Table 27–7. Persons with type AB blood are "universal recipients" because they have no circulating agglutinins and can be given blood of any type without developing a transfusion reaction due to ABO incompatibility. Type O individuals are "universal donors" because they lack A and B antigens, and type O blood can be given to anyone without producing a transfusion reaction due to ABO incompatibility. This does not mean, however, that blood should ever be transfused without being cross-matched except in the most extreme emergencies, since the possibility of reactions or sensitization due to incompatibilities in systems other than ABO systems always exists. In cross-matching, donor red cells are mixed with recipient plasma on a slide and checked for agglutination. It is advisable to check the action of the donor's plasma on the recipient cells in addition, even though, as noted above, this is rarely a source of trouble.

A procedure that has recently become popular is to withdraw the patient's own blood in advance of elective surgery and then infuse this blood back (**autologous transfusion**) if a transfusion is needed during the surgery. With iron treatment, 1000–1500 mL can be withdrawn over a 3-week period. The popularity of banking one's own blood is due primarily to fear of transmission of AIDS by heterologous transfusions, but of course another advantage is elimination of the risk of transfusion reactions.

Table 27–7. Summary of ABO system.

Blood Type	Agglutinins in Plasma	Frequency in United States (%)	Plasma Agglutinates Red Cells of Type:
O	Anti-A, anti-B	45	A, B, AB
A	Anti-B	41	B, AB
B	Anti-A	10	A, AB
AB	None	4	None

Inheritance of A & B Antigens

The A and B antigens are inherited as mendelian allelomorphs, A and B being dominants. For example, an individual with type B blood may have inherited a B antigen from each parent or a B antigen from one parent and an O from the other; thus, an individual whose **phenotype** is B may have the **genotype** BB **(homozygous)** or BO **(heterozygous).**

When the blood types of the parents are known, the possible genotypes of their children can be stated. When both parents are type B, they could have children with genotype BB (B antigen from both parents), BO (B antigen from one parent, O from the other heterozygous parent), or OO (O antigen from both parents, both being heterozygous). When the blood types of a mother and her child are known, typing can prove that a man cannot be the father, although it cannot prove that he is the father. The predictive value is increased if the blood typing of the parties concerned includes identification of antigens other than the ABO agglutinogens. With the use of DNA fingerprinting (see Chapter 1), the exclusion rate for paternity rises to close to 100%.

Other Agglutinogens

In addition to the ABO system of antigens in human red cells, there are systems such as the Rh, MNSs, Lutheran, Kell, Kidd, and many others. There are over 500 billion possible known blood group phenotypes, and because undiscovered antigens undoubtedly exist, it has been calculated that the number of phenotypes is actually in the trillions.

The number of blood groups in animals is as large as it is in humans. An interesting question is why this degree of polymorphism developed and has persisted through evolution. Certain diseases are more common in individuals with one blood type or another, but the differences are not great. One, the Duffy antigen, is a chemokine receptor. Many of the others seem to be cell recognition molecules, but the significance of a recognition code of this complexity is unknown.

The Rh Group

Aside from the antigens of the ABO system, those of the Rh system are of the greatest clinical importance. The "Rh factor," named for the rhesus monkey because it was first studied using the blood of this animal, is a system composed primarily of the C, D, and E antigens, although it actually contains many more. Unlike the ABO antigens, the system has not been detected in tissues other than red cells. D is by far the most antigenic component, and the term "Rh-positive" as it is generally used means that the individual has agglutino-

gen D. The D protein is not glycosylated, and its function is unknown. The "Rh-negative" individual has no D antigen and forms the anti-D agglutinin when injected with D-positive cells. The Rh typing serum used in routine blood typing is anti-D serum. Eighty-five percent of Caucasians are D-positive and 15% are D-negative; over 99% of Asians are D-positive. Unlike the antibodies of the ABO system, anti-D antibodies do not develop without exposure of a D-negative individual to D-positive red cells by transfusion or entrance of fetal blood into the maternal circulation. However, D-negative individuals who have received a transfusion of D-positive blood (even years previously) can have appreciable anti-D titers and thus may develop transfusion reactions when transfused again with D-positive blood.

Hemolytic Disease of the Newborn

Another complication due to "Rh incompatibility" arises when an Rh-negative mother carries an Rh-positive fetus. Small amounts of fetal blood leak into the maternal circulation at the time of delivery, and some mothers develop significant titers of anti-Rh agglutinins during the postpartum period. During the next pregnancy, the mother's agglutinins cross the placenta to the fetus. In addition, there are some cases of fetal-maternal hemorrhage during pregnancy, and sensitization can occur during pregnancy. In any case, when anti-Rh agglutinins cross the placenta to an Rh-positive fetus, they can cause hemolysis and various forms of **hemolytic disease of the newborn (erythroblastosis fetalis).** If hemolysis in the fetus is severe, the infant may die in utero or may develop anemia, severe jaundice, and edema **(hydrops fetalis). Kernicterus,** a neurologic syndrome in which unconjugated bilirubin is deposited in the basal ganglia, may also develop, especially if birth is complicated by a period of hypoxia. Bilirubin rarely penetrates the brain in adults, but it does in infants with erythroblastosis, possibly in part because the blood-brain barrier is more permeable in infancy. However, the main reasons that the concentration of unconjugated bilirubin is very high in this condition are that production is increased and the bilirubin-conjugating system is not yet mature.

About 50% of Rh-negative individuals are sensitized (develop an anti-Rh titer) by transfusion of Rh-positive blood. Since sensitization of Rh-negative mothers by carrying an Rh-positive fetus generally occurs at birth, the first child is usually normal. However, hemolytic disease occurs in about 17% of the Rh-positive fetuses born to Rh-negative mothers who have previously been pregnant one or more times with Rh-positive fetuses. Fortunately, it is usually possible to prevent sensitization from occurring the first time by administering a single dose of anti-Rh antibodies in the form of Rh immune globulin during the postpartum period. Such passive immunization does not harm the mother and has been demonstrated to prevent active antibody formation by the mother. In obstetric clinics, the institution of such treatment on a routine basis to unsensitized Rh-negative women who have delivered an Rh-positive baby has reduced the overall incidence of hemolytic disease by more than 90%. In addition, fetal Rh typing with material obtained by amniocentesis or chorionic villus sampling is now possible, and treatment with a small dose of Rh immune serum will prevent sensitization during pregnancy.

PLASMA

The fluid portion of the blood, the **plasma,** is a remarkable solution containing an immense number of ions, inorganic molecules, and organic molecules that are in transit to various parts of the body or aid in the transport of other substances. The normal plasma volume is about 5% of body weight, or roughly 3500 mL in a 70-kg man. Plasma clots on standing, remaining

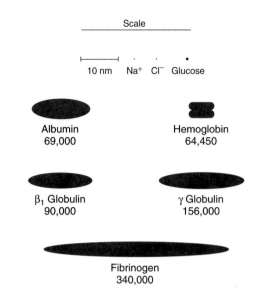

Figure 27–23 Relative dimensions and molecular masses of some of the proteins and other substances in plasma. (Modified and reproduced, with permission, from Murray RK et al: *Harper's Biochemistry,* 26th ed. McGraw-Hill, 2003.)

Table 27–8. System for naming blood-clotting factors.

Factor[1]	Names
I	Fibrinogen
II	Prothrombin
III	Thromboplastin
IV	Calcium
V	Proaccelerin, labile factor, accelerator globulin
VII	Proconvertin, SPCA, stable factor
VIII	Antihemophilic factor (AHF), antihemophilic factor A, antihemophilic globulin (AHG)
IX	Plasma thromboplastic component (PTC), Christmas factor, antihemophilic factor B
X	Stuart-Prower factor
XI	Plasma thromboplastin antecedent (PTA), antihemophilic factor C
XII	Hageman factor, glass factor
XIII	Fibrin-stabilizing factor, Laki-Lorand factor
HMW-K	High-molecular-weight kininogen, Fitzgerald factor
Pre-K$_a$	Prekallikrein, Fletcher factor
Ka	Kallikrein
PL	Platelet phospholipid

[1]Factor VI is not a separate entity and has been dropped.

fluid only if an anticoagulant is added. If whole blood is allowed to clot and the clot is removed, the remaining fluid is called **serum.** Serum has essentially the same composition as plasma except that its fibrinogen and clotting factors II, V, and VIII (Table 27–8) have been removed and it has a higher serotonin content because of the breakdown of platelets during clotting. The normal plasma levels of various substances are discussed in the chapters on the systems with which the substances are concerned and summarized on the inside back cover of this book.

Plasma Proteins

The plasma proteins consist of **albumin, globulin,** and **fibrinogen** fractions. The globulin fraction is subdivided into numerous components. One classification divides it into α_1, α_2, β_1, β_2, and γ globulins and fibrinogen. The molecular masses and overall configurations of several of the plasma proteins are shown in Figure 27–23. Of course, the actual structure of the proteins is more complicated than shown in the illustration, but it is helpful to see their approximate size and shape relative to each other and to molecules such as glucose.

The capillary walls are relatively impermeable to the proteins in plasma, and the proteins therefore exert an osmotic force of about 25 mm Hg across the capillary wall (**oncotic pressure;** see Chapter 1) that pulls water into the blood. The plasma proteins are also responsible for 15% of the buffering capacity of the blood (see Chapter 39), because of the weak ionization of their substituent COOH and NH_2 groups. At the normal plasma pH of 7.40, the proteins are mostly in the anionic form (see Chapter 1) and constitute a significant part of the anionic complement of plasma. Plasma proteins such as antibodies and the proteins concerned with blood clotting have specific functions. Some of the proteins function in the transport of thyroid, adrenocortical, gonadal, and other hormones. Binding keeps these hormones from being rapidly filtered through the glomeruli and provides a stable reservoir of hormones on which the tissues can draw. In addition, albumin serves as a carrier for metals, ions, fatty acids, amino acids, bilirubin, enzymes, and drugs.

Origin of Plasma Proteins

Circulating antibodies in the γ globulin fraction of the plasma proteins are manufactured in the plasma cells (see above). Most of the other plasma proteins are synthesized in the liver. These proteins and their principal functions are listed in Table 27–9.

Data on the turnover of albumin provide an indication of the role played by synthesis in the maintenance of normal albumin levels. In normal adult humans, the plasma albumin level is 3.5–5.0 g/dL, and the total exchangeable albumin pool is 4.0–5.0 g/kg body weight; 38–45% of this albumin is intravascular, and much of the rest of it is in the skin. Between 6 and 10% of the exchangeable pool is degraded per day, and the degraded albumin is replaced by hepatic synthesis of 200–400 mg/kg/d. The albumin is probably transported to the extravascular areas by vesicular transport across the walls of the capillaries (see Chapter 1). Albumin synthesis is carefully regulated. It is decreased during fasting and increased in conditions such as nephrosis in which there is excessive albumin loss.

Hypoproteinemia

Plasma protein levels are maintained during starvation until body protein stores are markedly depleted. However, in prolonged starvation and in the malabsorption syndrome due to intestinal diseases such as sprue, plasma protein levels are low (**hypoproteinemia**). They are also low in liver disease, because hepatic protein synthesis is depressed, and in nephrosis, because large amounts of albumin are lost in the urine. Because of the decrease in the plasma oncotic pressure, edema tends to develop (see Chapter 30). Rarely, there is congenital absence of one or another plasma protein fraction. An example of congenital protein deficiency is the congenital form of **afibrinogenemia,** characterized by defective blood clotting.

HEMOSTASIS

Hemostasis is the process of forming clots in the walls of damaged blood vessels and preventing blood loss while maintaining blood in a fluid state within the vascular system. A collection of complex interrelated systemic mechanisms operates to maintain this balance between coagulation and anticoagulation. In addition, the balance is affected by local factors in various different organs.

Response to Injury

When a small blood vessel is transected or damaged, the injury initiates a series of events (Figure 27–24) that leads to the formation of a clot (**hemostasis**). This seals off the damaged region and prevents further blood loss. The initial event is constriction of the vessel and formation of a temporary **hemostatic plug** of platelets that is triggered when platelets bind to collagen and aggregate. This is followed by conversion of the plug into the definitive clot.

The constriction of an injured arteriole or small artery may be so marked that its lumen is obliterated. The vasoconstriction is due to serotonin and other vasoconstrictors liberated from platelets that adhere to the walls of the damaged vessels. It is claimed that for a time after being divided transversely, arteries as large as the radial artery constrict and may stop bleeding. However, this is no excuse for delay in ligating the damaged vessel. Furthermore, arterial walls cut longitudinally or irregularly do not constrict in such a way that the lumen of the artery is occluded, and bleeding continues.

The Clotting Mechanism

The loose aggregation of platelets in the temporary plug is bound together and converted into the definitive clot by **fibrin.** The clotting mechanism responsible for the formation of fibrin involves a cascade of reactions in which inactive enzymes are activated, and the activated enzymes in turn activate other inactive enzymes. The complexity of the system has in the past been compounded by variations in nomenclature, but acceptance of a numbering system for most of the various clotting factors (Table 27–8) has simplified the situation.

The fundamental reaction in the clotting of blood is conversion of the soluble plasma protein fibrinogen to insoluble fibrin (Figure 27–25). The process involves the release of two pairs of polypeptides from each fibrinogen molecule. The remaining portion, **fibrin**

Table 27–9. Some of the proteins synthesized by the liver: Physiologic functions and properties.[1]

Name	Principal Function	Binding Characteristics	Serum or Plasma Concentration
Albumin	Binding and carrier protein; osmotic regulator	Hormones, amino acids, steroids, vitamins, fatty acids	4500–5000 mg/dL
Orosomucoid	Uncertain; may have a role in inflammation		Trace; rises in inflammation
α_1-Antiprotease	Trypsin and general protease inhibitor	Proteases in serum and tissue secretions	1.3–1.4 mg/dL
α-Fetoprotein	Osmotic regulation; binding and carrier protein[2]	Hormones, amino acids	Found normally in fetal blood
α_2-Macroglobulin	Inhibitor of serum endoproteases	Proteases	150–420 mg/dL
Antithrombin-III	Protease inhibitor of intrinsic coagulation system	1:1 binding to proteases	17–30 mg/dL
Ceruloplasmin	Transport of copper	Six atoms copper/mol	15–60 mg/dL
C-reactive protein	Uncertain; has role in tissue inflammation	Complement C1q	< 1 mg/dL; rises in inflammation
Fibrinogen	Precursor to fibrin in hemostasis		200–450 mg/dL
Haptoglobin	Binding, transport of cell-free hemoglobin	Hemoglobin 1:1 binding	40–180 mg/dL
Hemopexin	Binds to porphyrins, particularly heme for heme recycling	1:1 with heme	50–100 mg/dL
Transferrin	Transport of iron	Two atoms iron/mol	3.0–6.5 mg/dL
Apolipoprotein B	Assembly of lipoprotein particles	Lipid carrier	
Angiotensinogen	Precursor to pressor peptide angiotensin II		
Proteins, coagulation factors II, VII, IX, X	Blood clotting		20 mg/dL
Antithrombin C, protein C	Inhibition of blood clotting		
Insulin-like growth factor I	Mediator of anabolic effects of growth hormone	IGF-I receptor	
Steroid hormone-binding globulin	Carrier protein for steroids in bloodstream	Steroid hormones	3.3 mg/dL
Thyroxine-binding globulin	Carrier protein for thyroid hormone in bloodstream	Thyroid hormones	1.5 mg/dL
Transthyretin (thyroid-binding prealbumin)	Carrier protein for thyroid hormone in bloodstream	Thyroid hormones	25 mg/dL

[1]Reproduced, with permission, from McPhee SJ, Lingappa VR, Ganong WF (editors): *Pathophysiology of Disease,* 4th ed, McGraw-Hill, 2003.

[2]The function of alpha-fetoprotein is uncertain, but because of its structural homology to albumin it is often assigned these functions.

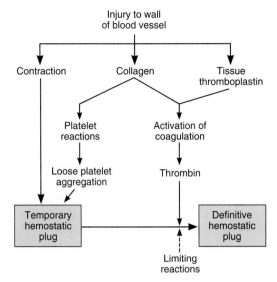

Figure 27–24 Summary of reactions involved in hemostasis. The dashed arrow indicates inhibition. (Modified and reproduced, with permission, from Deykin D: Thrombogenesis, N Engl J Med 1967;267:622.)

monomer, then polymerizes with other monomer molecules to form **fibrin.** The fibrin is initially a loose mesh of interlacing strands. It is converted by the formation of covalent cross-linkages to a dense, tight aggregate (stabilization). This latter reaction is catalyzed by activated factor XIII and requires Ca^{2+}.

The conversion of fibrinogen to fibrin is catalyzed by thrombin. Thrombin is a serine protease that is formed from its circulating precursor, prothrombin, by the action of activated factor X. It has additional actions, including activation of platelets, endothelial cells, and leukocytes via at least one G protein-coupled receptor.

Factor X can be activated by reactions in either of two systems, an intrinsic and an extrinsic system (Figure 27–25). The initial reaction in the **intrinsic system** is conversion of inactive factor XII to active factor XII (XIIa). This activation, which is catalyzed by high-molecular-weight kininogen and kallikrein (see Chapter 31), can be brought about in vitro by exposing the blood to electronegatively charged wettable surfaces such as glass and collagen fibers. Activation in vivo occurs when blood is exposed to the collagen fibers underlying the endothelium in the blood vessels. Active factor XII then activates factor XI, and active factor XI activates factor IX. Activated factor IX forms a complex with active factor VIII, which is activated when it is separated from von Willebrand factor. The complex of IXa and VIIIa activate factor X. Phospholipids from ag-

gregated platelets (PL) and Ca^{2+} are necessary for full activation of factor X. The **extrinsic system** is triggered by the release of tissue thromboplastin, a protein-phospholipid mixture that activates factor VII. The tissue thromboplastin and factor VII activate factors IX and X. In the presence of PL, Ca^{2+}, and factor V, activated factor X catalyzes the conversion of prothrombin to thrombin. The extrinsic pathway is inhibited by a **tissue factor pathway inhibitor** that forms a quaternary structure with TPL, factor VIIa, and factor Xa.

Anticlotting Mechanisms

The tendency of blood to clot is balanced in vivo by limiting reactions that tend to prevent clotting inside the blood vessels and to break down any clots that do form. These reactions include the interaction between the platelet-aggregating effect of thromboxane A_2 and

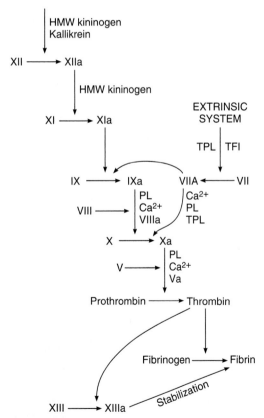

Figure 27–25 The clotting mechanism. a, active form of clotting factor; HMW, high-molecular-weight; PL, platelet phospholipid; TPL, tissue thromboplastin. TFI, tissue factor pathway inhibitor.

the antiaggregating effect of prostacyclin, which causes clots to form at the site when a blood vessel is injured but keeps the vessel lumen free of clot (see Chapter 31).

Antithrombin III is a circulating protease inhibitor that binds to the serine proteases in the coagulation system, blocking their activity as clotting factors. This binding is facilitated by **heparin,** a naturally occurring anticoagulant that is a mixture of sulfated polysaccharides with molecular weights averaging 15,000–18,000. The clotting factors that are inhibited are the active forms of factors IX, X, XI, and XII.

The endothelium of the blood vessels also plays an active role in preventing the extension of clots into blood vessels. All endothelial cells except those in the cerebral microcirculation produce **thrombomodulin,** a thrombin-binding protein, and express it on their surface. In the circulating blood, thrombin is a procoagulant that activates factors V and VIII, but when it binds to thrombomodulin, it becomes an anticoagulant in that the thrombomodulin-thrombin complex activates protein C (Figure 27–26). Activated protein C (APC), along with its cofactor protein S, inactivates factors V and VIII and inactivates an inhibitor of tissue plasminogen activator, increasing the formation of plasmin.

Plasmin (fibrinolysin) is the active component of the **plasminogen (fibrinolytic) system** (Figure 27–26). This enzyme lyses fibrin and fibrinogen, with the production of fibrinogen degradation products (FDP) that inhibit thrombin. Plasmin is formed from

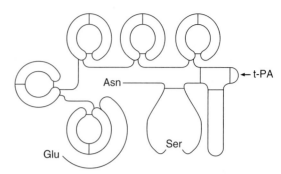

Figure 27–27 Structure of human plasminogen. Note the Glu at the amino terminal, the Asn at the carboxyl terminal, and five uniquely shaped loop structures (kringles). Hydrolysis by t-PA at the arrow separates the carboxyl terminal light chain from the amino terminal heavy chain but leaves the disulfide bonds intact. This activates the molecule. (Modified and reproduced, with permission, from Bachman F, in: *Thrombosis and Hemostasis.* Verstraete M et al [editors]. Leuven University Press, 1987.)

its inactive precursor, plasminogen, by the action of thrombin and **tissue-type plasminogen activator (t-PA).** It is also activated by **urokinase-type plasminogen activator (u-PA).** If the t-PA gene or the u-PA gene is knocked out in mice, there is some fibrin deposition and clot lysis is slowed. However when both are knocked out, there is extensive spontaneous fibrin deposition. Wound healing is delayed (see Chapter 33). There are also defects in growth and fertility, since the plasminogen system not only lyses clots but also plays a role in cell movement and in ovulation.

Human plasminogen consists of a 560-amino-acid heavy chain and a 241-amino-acid light chain. The heavy chain, with glutamate at its amino terminal, is folded into five loop structures, each held together by three disulfide bonds (Figure 27–27). These loops are called kringles because of their resemblance to a Danish pastry of the same name. The kringles are lysine-binding sites by which the molecule attaches to fibrin and other clot proteins, and they are also found in prothrombin. Plasminogen is converted to active plasmin when t-PA hydrolyzes the bond between Arg 560 and Val 561.

Plasminogen receptors are located on the surfaces of many different types of cells and are plentiful on endothelial cells. When plasminogen binds to its receptors, it becomes activated, so intact blood vessel walls are provided with a mechanism that discourages clot formation.

Human t-PA is now produced by recombinant DNA techniques and is available (as alteplase) for clini-

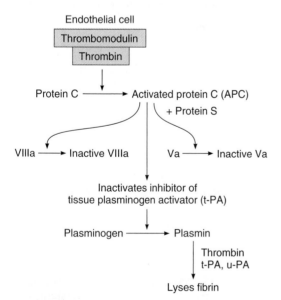

Figure 27–26 The fibrinolytic system and its regulation by protein C.

cal use. It lyses clots in the coronary arteries if given to patients soon after the onset of myocardial infarction. Streptokinase, a bacterial enzyme, is also fibrinolytic and is also used in the treatment of early myocardial infarction (see Chapter 32).

A group of homologous proteins now called **annexins** are associated with coagulation and fibrinolysis. Over 20 of these have been described, ten in mammals. One of these, annexin II, forms a platform on endothelial cells on which components of the fibrinolytic system interact, producing fibrinolysis. Another, annexin V, forms a shield around phospholipids involved in coagulation and exerts an antithrombotic effect. However, the exact physiologic roles of the various annexins remain to be determined.

Anticoagulants

Hemoglobin is a naturally occurring anticoagulant that facilitates the action of antithrombin III. It is also a cofactor for lipoprotein lipase (clearing factor; see Chapter 17). The highly basic protein protamine forms an irreversible complex with heparin and is used clinically to neutralize heparin. Low-molecular-weight fragments with an average molecular weight of 5000 have been produced by depolymerizing unfractionated heparin, and these low-molecular-weight heparins are seeing increased clinical use because they have a longer half-life

and produce a more predictable anticoagulant response than unfractionated heparin.

In vivo, a plasma Ca^{2+} level low enough to interfere with blood clotting is incompatible with life, but clotting can be prevented in vitro if Ca^{2+} is removed from the blood by the addition of substances such as oxalates, which form insoluble salts with Ca^{2+}, or **chelating agents,** which bind Ca^{2+}. Coumarin derivatives such as **dicumarol** and **warfarin** are also effective anticoagulants. They inhibit the action of vitamin K, and this vitamin is a necessary cofactor for the enzyme that catalyzes the conversion of glutamic acid residues to γ-carboxyglutamic acid residues. Six of the proteins involved in clotting require conversion of a number of glutamic acid residues to γ-carboxyglutamic acid residues before being released into the circulation, and hence all six are vitamin K-dependent. These proteins are factors II (prothrombin), VII, IX, and X, protein C, and protein S (see above).

Abnormalities of Hemostasis

In addition to clotting abnormalities due to platelet disorders (see above), hemorrhagic diseases can be produced by selective deficiencies of most of the clotting factors (Table 27–10). Hemophilia A, which is caused by factor VIII deficiency, is relatively common. The disease has been treated with factor VIII-rich prepara-

Table 27–10. Examples of diseases due to deficiency of clotting factors.

Deficiency of Factor:	Clinical Syndrome	Cause
I	Afibrinogenemia	Depletion during pregnancy with premature separation of placenta; also congenital (rare)
II	Hypoprothrombinemia (hemorrhagic tendency in liver disease)	Decreased hepatic synthesis, usually secondary to vitamin K deficiency
V	Parahemophilia	Congenital
VII	Hypoconvertinemia	Congenital
VIII	Hemophilia A (classic hemophilia)	Congenital defect due to various abnormalities of the gene on X chromosome that codes for factor VIII; disease is therefore inherited as sex-linked characteristic
IX	Hemophilia B (Christmas disease)	Congenital
X	Stuart-Prower factor deficiency	Congenital
XI	PTA deficiency	Congenital
XII	Hageman trait	Congenital

tions made from plasma, but unfortunately this has led to transmission of HIV to a significant number of patients. Highly purified preparations that have been treated to inactivate viruses are now available. So is factor VIII produced by recombinant DNA techniques. In addition to promoting platelet adherence (see above), **von Willebrand factor** forms a complex with factor VIII and regulates its plasma levels. Congenital deficiency of von Willebrand factor causes a bleeding disorder (von Willebrand's disease).

The absorption of the K vitamins, along with that of the other fat-soluble vitamins, is depressed in obstructive jaundice because of the lack of bile in the intestine and consequent depression of fat absorption (see Chapter 26). The resulting clotting factor deficiencies may cause the development of a significant bleeding tendency.

Formation of clots inside blood vessels is called **thrombosis** to distinguish it from the normal extravascular clotting of blood. Thromboses are a major medical problem. They are particularly prone to occur where blood flow is sluggish, eg, in the veins of the legs after operations and delivery, because the slow flow permits activated clotting factors to accumulate instead of being washed away. They also occur in vessels such as the coronary and cerebral arteries at sites where the intima is damaged by atherosclerotic plaques, and over areas of damage to the endocardium. They frequently occlude the arterial supply to the organs in which they form, and bits of thrombus (**emboli**) sometimes break off and travel in the bloodstream to distant sites, damaging other organs. Examples are obstruction of the pulmonary artery or its branches by thrombi from the leg veins (**pulmonary embolism**) and embolism of cerebral or leg vessels by bits of clot breaking off from a thrombus in the left ventricle (**mural thrombus**) overlying a myocardial infarct.

Congenital absence of protein C leads to uncontrolled intravascular coagulation and, in general, death in infancy. If this condition is diagnosed and treatment is instituted with blood concentrates rich in protein C, the coagulation defect disappears. Resistance to APC is another cause of thrombosis, and this condition is common. It is due to a point mutation in the gene for factor V, which prevents APC from inactivating the factor. Mutations in protein S and antithrombin III that increase the incidence of thrombosis have also been described but are less common.

Disseminated intravascular coagulation is a serious complication of septicemia, extensive tissue injury, and other diseases in which fibrin is deposited in the vascular system and many small- and medium-sized vessels are thrombosed. The increased consumption of platelets and coagulation factors causes bleeding to occur at the same time. The cause of the condition appears to be increased generation of thrombin due to increased TPL activity without adequate tissue factor inhibitory pathway activity.

LYMPH

Lymph is tissue fluid that enters the lymphatic vessels. It drains into the venous blood via the thoracic and right lymphatic ducts. It contains clotting factors and clots on standing in vitro. In most locations, it also contains proteins that traverse capillary walls and return to the blood via the lymph. Its protein content is generally lower than that of plasma, which contains about 7 g/dL, but lymph protein content varies with the region from which the lymph drains (Table 27–11). Water-insoluble fats are absorbed from the intestine into the lymphatics, and the lymph in the thoracic duct after a meal is milky because of its high fat content (see Chapter 25). Lymphocytes enter the circulation principally through the lymphatics, and there are appreciable numbers of lymphocytes in thoracic duct lymph.

Table 27–11. Probable approximate protein content of lymph in humans.[1]

Source of Lymph	Protein Content (g/dL)
Choroid plexus	0
Ciliary body	0
Skeletal muscle	2
Skin	2
Lung	4
Gastrointestinal tract	4.1
Heart	4.4
Liver	6.2

[1]Data largely from JN Diana.

Origin of the Heartbeat & the Electrical Activity of the Heart

28

INTRODUCTION

The parts of the heart normally beat in orderly sequence: Contraction of the atria (**atrial systole**) is followed by contraction of the ventricles (**ventricular systole**), and during **diastole** all four chambers are relaxed. The heartbeat originates in a specialized **cardiac conduction system** and spreads via this system to all parts of the myocardium. The structures that make up the conduction system (Figure 28–1) are the **sinoatrial node** (SA node), the **internodal atrial pathways,** the **atrioventricular node** (AV node), the **bundle of His** and its branches, and the **Purkinje**

system. The various parts of the conduction system and, under abnormal conditions, parts of the myocardium are capable of spontaneous discharge. However, the SA node normally discharges most rapidly, depolarization spreading from it to the other regions before they discharge spontaneously. The SA node is therefore the normal **cardiac pacemaker,** its rate of discharge determining the rate at which the heart beats. Impulses generated in the SA node pass through the atrial pathways to the AV node, through this node to the bundle of His, and through the branches of the bundle of His via the Purkinje system to the ventricular muscle.

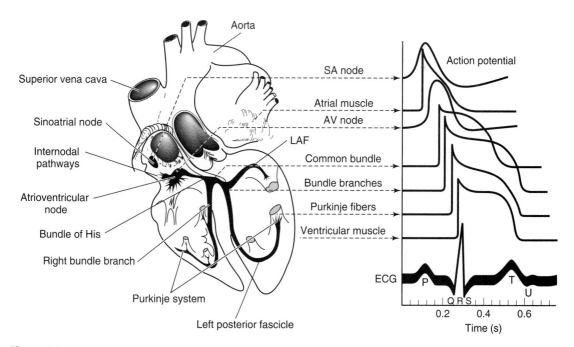

Figure 28–1. Conducting system of the heart. Typical transmembrane action potentials for the SA and AV nodes, other parts of the conduction system, and the atrial and ventricular muscles are shown along with the correlation to the extracellularly recorded electrical activity, ie, the electrocardiogram (ECG). The action potentials and ECG are plotted on the same time axis but with different zero points on the vertical scale. LAF, left anterior fascicle.

ORIGIN & SPREAD OF CARDIAC EXCITATION

Anatomic Considerations

In the human heart, the SA node is located at the junction of the superior vena cava with the right atrium. The AV node is located in the right posterior portion of the interatrial septum (Figure 28–1). There are three bundles of atrial fibers that contain Purkinje type fibers and connect the SA node to the AV node: the anterior internodal tract of Bachman, the middle internodal tract of Wenckebach, and the posterior internodal tract of Thorel. Conduction also occurs through atrial myocytes, but it is more rapid in these bundles. The AV node is normally the only conducting pathway between the atria and ventricles. It is continuous with the bundle of His, which gives off a left bundle branch at the top of the interventricular septum and continues as the right bundle branch. The left bundle branch divides into an anterior fascicle and a posterior fascicle. The branches and fascicles run subendocardially down either side of the septum and come into contact with the Purkinje system, whose fibers spread to all parts of the ventricular myocardium.

The histology of cardiac muscle is described in Chapter 3. The conduction system is composed for the most part of modified cardiac muscle that has fewer striations and indistinct boundaries. The SA node and, to a lesser extent, the AV node, also contain small round cells with few organelles, which are connected by gap junctions. These are probably the actual pacemaker cells, and therefore they are called **P cells.** The atrial muscle fibers are separated from those of the ventricles by a fibrous tissue ring, and normally the only conducting tissue between the atria and ventricles is the bundle of His.

The SA node develops from structures on the right side of the embryo and the AV node from structures on the left. This is why in the adult the right vagus is distributed mainly to the SA node and the left vagus mainly to the AV node. Similarly, the sympathetic innervation on the right side is distributed primarily to the SA node and the sympathetic innervation on the left side primarily to the AV node. On each side, most sympathetic fibers come from the stellate ganglion. Noradrenergic fibers are epicardial, whereas the vagal fibers are endocardial. However, connections exist for reciprocal inhibitory effects of the sympathetic and parasympathetic innervation of the heart on each other. Thus, acetylcholine acts presynaptically to reduce norepinephrine release from the sympathetic nerves, and conversely, neuropeptide Y released from noradrenergic endings may inhibit the release of acetylcholine.

Properties of Cardiac Muscle

The electrical responses of cardiac muscle and nodal tissue and the ionic fluxes that underlie them are discussed in detail in Chapter 3. As shown in Figure 3–15, myocardial fibers have a resting membrane potential of approximately −90 mV. The relation of the action potential to the contractile response is shown in Figure 3–14. The individual fibers are separated by membranes, but depolarization spreads radially through them as if they were a syncytium, because of the presence of gap junctions. The transmembrane action potential of single cardiac muscle cells is characterized by rapid depolarization, a plateau, and a slow repolarization process (see Figure 3–15). The initial depolarization is due to Na^+ influx through rapidly opening Na^+ channels (the Na^+ current, I_{Na}). Ca^{2+} influx through more slowly opening Ca^{2+} channels (the Ca^{2+} current, I_{Ca}) produces the plateau phase, and repolarization is due to net K^+ efflux through three types of K^+ channels (see Chapter 3). Recorded extracellularly, the summed electrical activity of all the cardiac muscle fibers is the ECG. The timing of the discharge of the individual units relative to the ECG is shown in Figure 28–1.

Pacemaker Potentials

Rhythmically discharging cells have a membrane potential that, after each impulse, declines to the firing level. Thus, this **prepotential** or **pacemaker potential** (Figure 28–2) triggers the next impulse. At the peak of each impulse, I_K begins and brings about repolarization. I_K then declines, and as K^+ efflux decreases, the membrane begins to depolarize, forming the first part of the prepotential. Ca^{2+} channels then open. These are of two types in the heart, the **T** (for transient) **channels** and the **L** (for long-lasting) **channels.** The calcium current (I_{Ca}) due to opening of T channels completes the prepotential, and I_{Ca} due to opening of L channels produces the impulse. Other ion channels are also involved, and there is evidence that local Ca^{2+} release from the sarcoplasmic reticulum (**Ca^{2+} sparks**) occurs during the prepotential.

The action potentials in the SA and AV nodes are largely due to Ca^{2+}, with little contribution by Na^+ influx. Consequently there is no sharp, rapid depolarizing spike before the plateau, as there is in other parts of the conduction system and the atrial and ventricular fibers (Figure 28–1). In addition, prepotentials are normally prominent only in the SA and AV nodes. However, there are "latent pacemakers" in other portions of the conduction system that can take over when the SA and AV nodes are depressed or conduction from them is blocked. Atrial and ventricular muscle fibers do not have prepotentials, and they discharge spontaneously only when injured or abnormal.

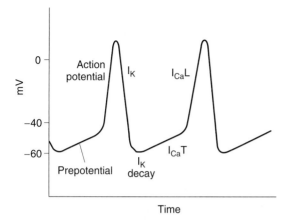

Figure 28–2. Diagram of the membrane potential of pacemaker tissue. The principal current responsible for each part of the potential is shown under or beside the component. L, long-lasting; T, transient. Other ion channels contribute to the electrical response. Note that the resting membrane potential of pacemaker tissue is somewhat lower than that of atrial and ventricular muscle.

When the cholinergic vagal fibers to nodal tissue are stimulated, the membrane becomes hyperpolarized and the slope of the prepotentials is decreased (Figure 28–3) because the acetylcholine released at the nerve endings increases the K+ conductance of nodal tissue. This action is mediated by M_2 muscarinic receptors, which, via the βγ subunit of a G protein, open a special set of K+ channels. The resulting I_{KAch} counters the decay of I_K. In addition, activation of the M_2 receptors decreases cAMP in the cells, and this slows the opening of the Ca^{2+} channels. The result is a decrease in firing rate. Strong va-

gal stimulation may abolish spontaneous discharge for some time.

Conversely, stimulation of the sympathetic cardiac nerves makes the membrane potential fall more rapidly, and the rate of spontaneous discharge increases (Figure 28–3). Norepinephrine secreted by the sympathetic endings binds to β_1 receptors, and the resulting increase in intracellular cAMP facilitates the opening of L channels, increasing I_{Ca} and the rapidity of the depolarization phase of the impulse.

Because of the "sidedness" of the cardiac innervation (see above), stimulation of the right vagus slows the heart by inhibiting the SA node, whereas stimulation of the left vagus mainly slows AV conduction. Similarly, stimulation of the right stellate ganglion accelerates the heart, whereas stimulation of the left stellate ganglion shortens the AV nodal conduction time and refractoriness.

The rate of discharge of the SA node and other nodal tissue is influenced by temperature and by drugs. The discharge frequency is increased when the temperature rises, and this may contribute to the tachycardia associated with fever. Digitalis depresses nodal tissue and exerts an effect like that of vagal stimulation, particularly on the AV node.

Spread of Cardiac Excitation

Depolarization initiated in the SA node spreads radially through the atria, then converges on the AV node. Atrial depolarization is complete in about 0.1 s. Because conduction in the AV node is slow (Table 28–1), there is a delay of about 0.1 s (**AV nodal delay**) before excitation spreads to the ventricles. This delay is shortened by stimulation of the sympathetic nerves to the heart and lengthened by stimulation of the vagi. From the top of the septum, the wave of depolarization spreads in the rapidly conducting Purkinje fibers to all parts of the ventricles in the 0.08–0.1 s. In humans, depolarization of the ventricular muscle starts at the left side of the interventricular septum and moves first to the right across the midportion of the septum. The wave of depolarization then spreads down the septum

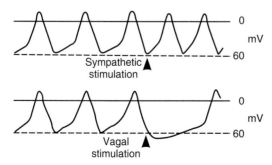

Figure 28–3. Effect of sympathetic (noradrenergic) and vagal (cholinergic) stimulation on the membrane potential of the SA node.

Table 28–1. Conduction speeds in cardiac tissue.

Tissue	Conduction Rate (m/s)
SA node	0.05
Atrial pathways	1
AV node	0.05
Bundle of His	1
Purkinje system	4
Ventricular muscle	1

to the apex of the heart. It returns along the ventricular walls to the AV groove, proceeding from the endocardial to the epicardial surface (Figure 28–4). The last parts of the heart to be depolarized are the posterobasal portion of the left ventricle, the pulmonary conus, and the uppermost portion of the septum.

THE ELECTROCARDIOGRAM

Because the body fluids are good conductors (ie, because the body is a **volume conductor**), fluctuations in potential that represent the algebraic sum of the action potentials of myocardial fibers can be recorded extracellularly. The record of these potential fluctuations during the cardiac cycle is the **electrocardiogram (ECG).** Most electrocardiograph machines record these fluctuations on a moving strip of paper.

The ECG may be recorded by using an **active or exploring electrode** connected to an indifferent electrode at zero potential (**unipolar recording**) or by using two active electrodes (**bipolar recording).** In a volume conductor, the sum of the potentials at the points of an equilateral triangle with a current source in the center is zero at all times. A triangle with the heart at its center (**Einthoven's triangle**) can be approximated by placing electrodes on both arms and on the left leg. These are the three **standard limb leads** used in electrocardiography. If these electrodes are connected to a common terminal, an indifferent electrode that stays near zero potential is obtained. Depolarization moving toward an active electrode in a volume conductor produces a positive deflection, whereas depolarization moving in the opposite direction produces a negative deflection.

The names of the various waves and segments of the ECG in humans are shown in Figure 28–5. By convention, an upward deflection is written when the active electrode becomes positive relative to the indifferent electrode, and a downward deflection is written when the active electrode becomes negative. The P wave is produced by atrial depolarization, the QRS complex by ventricular depolarization, and the ST segment and T wave by ventricular repolarization. The manifestations of atrial repolarization are not normally seen because they are obscured by the QRS complex. The U wave is an inconstant finding, believed to be due to slow repolarization of the papillary muscles. The intervals between the various waves of the ECG and the events in the heart that occur during these intervals are shown in Table 28–2.

Bipolar Leads

Bipolar leads were used before unipolar leads were developed. The **standard limb leads,** leads I, II, and III,

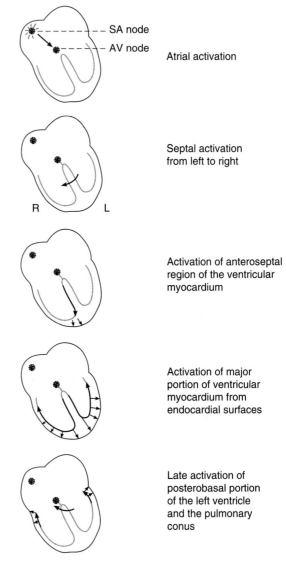

Figure 28–4. Normal spread of electrical activity in the heart. (Reproduced, with permission, from Goldman MJ: *Principles of Clinical Electrocardiography*, 12th ed. Originally published by Appleton & Lange. Copyright © 1986 by The McGraw-Hill Companies, Inc.)

each record the differences in potential between two limbs. Since current flows only in the body fluids, the records obtained are those that would be obtained if the electrodes were at the points of attachment of the limbs, no matter where on the limbs the electrodes are placed. In lead I, the electrodes are connected so that an upward deflection is inscribed when the left arm becomes positive relative to the right (left arm positive).

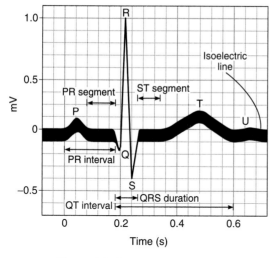

Figure 28–5. Waves of the ECG.

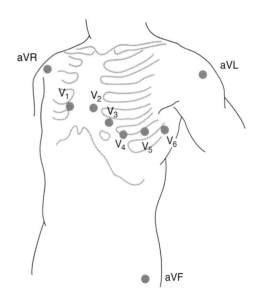

Figure 28–6. Unipolar electrocardiographic leads.

In lead II, the electrodes are on the right arm and left leg, with the leg positive; and in lead III, the electrodes are on the left arm and left leg, with the leg positive.

Unipolar (V) Leads

An additional nine unipolar leads, ie, leads that record the potential difference between an exploring electrode and an indifferent electrode, are commonly used in clinical electrocardiography. There are six unipolar chest leads (precordial leads) designated V_1–V_6 (Figure 28–6) and three unipolar limb leads: VR (right arm), VL (left arm), and VF (left foot). **Augmented limb leads,** designated by the letter *a* (aVR, aVL, aVF), are generally used. The augmented limb leads are recordings between one limb and the other two limbs. This increases the size of the potentials by 50% without any change in configuration from the nonaugmented record.

Unipolar leads can also be placed at the tips of catheters and inserted into the esophagus or heart.

Normal ECG

The ECG of a normal individual is shown in Figure 28–7. The sequence in which the parts of the heart are depolarized (Figure 28–4) and the position of the heart relative to the electrodes are the important considerations in interpreting the configurations of the waves in each lead. The atria are located posteriorly in the chest. The ventricles form the base and anterior surface of the heart, and the right ventricle is anterolateral to the left.

Table 28–2. ECG intervals.

Intervals	Normal Duration (s)		Events in the Heart During Interval
	Average	Range	
PR interval[1]	0.18[2]	0.12–0.20	Atrial depolarization and conduction through AV node
QRS duration	0.08	to 0.10	Ventricular depolarization and atrial repolarization
QT interval	0.40	to 0.43	Ventricular depolarization plus ventricular repolarization
ST interval (QT minus QRS)	0.32	...	Ventricular repolarization

[1]Measured from the beginning of the P wave to the beginning of the QRS complex.
[2]Shortens as heart rate increases from average of 0.18 s at a rate of 70 beats/min to 0.14 s at a rate of 130 beats/min.

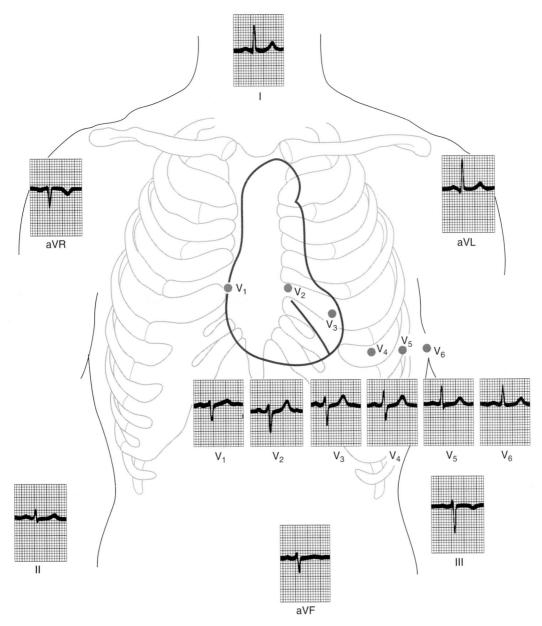

Figure 28–7. Normal ECG. (Reproduced, with permission, from Goldman MJ: *Principles of Clinical Electrocardiography*, 12th ed. Originally published by Appleton & Lange. Copyright © 1986 by The McGraw-Hill Companies, Inc.)

Thus, aVR "looks at" the cavities of the ventricles. Atrial depolarization, ventricular depolarization, and ventricular repolarization move away from the exploring electrode, and the P wave, QRS complex, and T wave are therefore all negative (downward) deflections; aVL and aVF look at the ventricles, and the deflections are therefore predominantly positive or biphasic. There

is no Q wave in V_1 and V_2, and the initial portion of the QRS complex is a small upward deflection because ventricular depolarization first moves across the mid-portion of the septum from left to right toward the exploring electrode. The wave of excitation then moves down the septum and into the left ventricle away from the electrode, producing a large S wave. Finally, it moves back along the ventricular wall toward the electrode, producing the return to the isoelectric line. Conversely, in the left ventricular leads ($V_4–V_6$) there may be an initial small Q wave (left to right septal depolarization), and there is a large R wave (septal and left ventricular depolarization) followed in V_4 and V_5 by a moderate S wave (late depolarization of the ventricular walls moving back toward the AV junction).

There is considerable variation in the position of the normal heart, and the position affects the configuration of the electrocardiographic complexes in the various leads.

Bipolar Limb Leads & the Cardiac Vector

Because the standard limb leads are records of the potential differences between two points, the deflection in each lead at any instant indicates the magnitude and direction in the axis of the lead of the electromotive force generated in the heart (**cardiac vector** or **axis**). The vector at any given moment in the two dimensions of the frontal plane can be calculated from any two standard limb leads (Figure 28–8) if it is assumed that the three electrode locations form the points of an equilateral triangle (Einthoven's triangle) and that the heart lies in the center of the triangle. These assumptions are not completely warranted, but calculated vectors are useful approximations. An approximate **mean QRS vector** ("electrical axis of the heart") is often plotted by using the average QRS deflection in each lead as shown in Figure 28–8. This is a **mean** vector as opposed to an **instantaneous** vector, and the average QRS deflections should be measured by integrating the QRS complexes. However, they can be approximated by measuring the net differences between the positive and negative peaks of the QRS. The normal direction of the mean QRS vector is generally said to be −30 to +110 degrees on the coordinate system shown in Figure 28–8. **Left** or **right axis deviation** is said to be present if the calculated axis falls to the left of −30 degrees or to the right of +110 degrees, respectively. Right axis deviation suggests right ventricular hypertrophy, and left axis deviation may be due to left ventricular hypertrophy, but there are better and more reliable electrocardiographic criteria for ventricular hypertrophy.

Vectorcardiography

If the tops of the arrows representing all of the instantaneous cardiac vectors in the frontal plane during the cardiac cycle are connected, from first to last, the line connecting them forms a series of three loops: one for

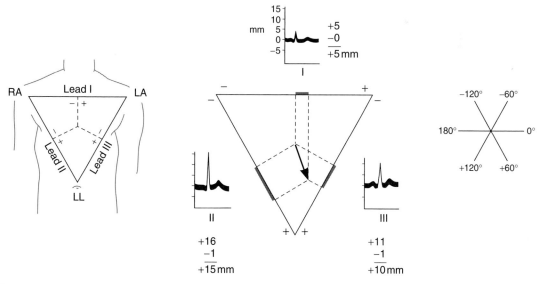

Figure 28–8. Cardiac vector. **Left:** Einthoven's triangle. Perpendiculars dropped from the midpoints of the sides of the equilateral triangle intersect at the center of electrical activity. RA, right arm; LA, left arm; LL, left leg. **Center:** Calculation of mean QRS vector. In each lead, distances equal to the height of the R wave minus the height of the largest negative deflection in the QRS complex are measured off from the midpoint of the side of the triangle representing that lead. An arrow drawn from the center of electrical activity to the point of intersection of perpendiculars extended from the distances measured off on the sides represents the magnitude and direction of the mean QRS vector. **Right:** Reference axes for determining the direction of the vector.

the P wave, one for the QRS complex, and one for the T wave. This can be done electronically and the loops, called **vectorcardiograms,** projected on the face of a cathode ray oscilloscope.

His Bundle Electrogram

In patients with heart block, the electrical events in the AV node, bundle of His, and Purkinje system are frequently studied with a catheter containing an electrode at its tip that is passed through a vein to the right side of the heart and manipulated into a position close to the tricuspid valve. Three or more standard electrocardiographic leads are recorded simultaneously. The record of the electrical activity obtained with the catheter (Figure 28–9) is the **His bundle electrogram (HBE).** It normally shows an A deflection when the AV node is activated, an H spike during transmission through the His bundle, and a V deflection during ventricular depolarization. With the HBE and the standard electrocardiographic leads, it is possible to accurately time three intervals: (1) the PA interval, the time from the first appearance of atrial depolarization to the A wave in the HBE, which represents conduction time from the SA node to the AV node; (2) the AH interval, from the A wave to the start of the H spike, which represents the AV nodal conduction time; and (3) the HV interval, the time from the start of the H spike to the start of the QRS deflection in the ECG, which represents conduction in the bundle of His and the bundle branches. The approximate normal values for these intervals in adults are PA, 27 ms; AH, 92 ms; and HV, 43 ms. These values illustrate the relative slowness of conduction in the AV node (Table 28–1).

Monitoring

The ECG is often recorded continuously in hospital coronary care units, with alarms arranged to sound at the onset of life-threatening arrhythmias. Using a small portable tape recorder **(Holter monitor),** it is also possible to record the ECG in ambulatory individuals as they go about their normal activities. The recording is later played back at high speed and analyzed. Long-term continuous records can be obtained. Transtelephonic ECG devices are also available with portable recorders that can be activated when a patient has symptoms, and the recording can then be forwarded to a doctor's office or a laboratory by telephone. Recordings obtained with monitors have proved valuable in the diagnosis of arrhythmias and in planning the treatment of patients recovering from myocardial infarctions.

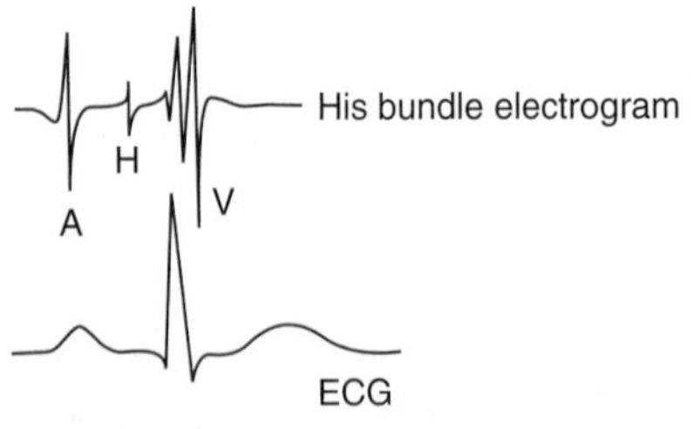

CARDIAC ARRHYTHMIAS

Normal Cardiac Rate

In the normal human heart, each beat originates in the SA node **(normal sinus rhythm, NSR).** The heart beats about 70 times a minute at rest. The rate is slowed **(bradycardia)** during sleep and accelerated **(tachycardia)** by emotion, exercise, fever, and many other stimuli. The control of heart rate is discussed in Chapter 31. In healthy young individuals breathing at a normal rate, the heart rate varies with the phases of respiration: it accelerates during inspiration and decelerates during expiration, especially if the depth of breathing is increased. This **sinus arrhythmia** (Figure 28–10) is a normal phenomenon and is due primarily

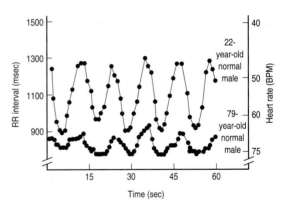

Figure 28–10. Sinus arrhythmia in a young man and an old man. The subjects breathed five times per minute, and with each inspiration the RR interval (the interval between R waves) declined, indicating an increase in heart rate. Note the marked reduction in the magnitude of the arrhythmia in the older man. These records were obtained after β-adrenergic blockade but would have been generally similar in its absence. (Reproduced, with permission, from Pfeifer MA et al: Differential changes of autonomic nervous system function with age in man. Am J Med 1983;75:249.)

Figure 28–9. Normal His bundle electrogram (HBE) with simultaneously recorded ECG.

to fluctuations in parasympathetic output to the heart. During inspiration, impulses in the vagi from the stretch receptors in the lungs inhibit the cardioinhibitory area in the medulla oblongata. The tonic vagal discharge that keeps the heart rate slow decreases, and the heart rate rises.

Disease processes affecting the sinus node lead to marked bradycardia accompanied by dizziness and syncope (**sick sinus syndrome**).

Abnormal Pacemakers

The AV node and other portions of the conduction system can in abnormal situations become the cardiac pacemaker. In addition, diseased atrial and ventricular muscle fibers can have their membrane potentials reduced and discharge repetitively.

As noted above, the discharge rate of the SA node is more rapid than that of the other parts of the conduction system, and this is why the SA node normally controls the heart rate. When conduction from the atria to the ventricles is completely interrupted, **complete (third-degree) heart block** results, and the ventricles beat at a low rate (**idioventricular rhythm**) indepen-

dently of the atria (Figure 28–11). The block may be due to disease in the AV node (**AV nodal block**) or in the conducting system below the node (**infranodal block**). In patients with AV nodal block, the remaining nodal tissue becomes the pacemaker and the rate of the idioventricular rhythm is approximately 45 beats/min. In patients with infranodal block due to disease in the bundle of His, the ventricular pacemaker is located more peripherally in the conduction system and the ventricular rate is lower; it averages 35 beats/min, but in individual cases it can be as low as 15 beats/min. In such individuals, there may also be periods of asystole lasting a minute or more. The resultant cerebral ischemia causes dizziness and fainting (**Stokes-Adams syndrome**). Causes of third-degree heart block include septal myocardial infarction and damage to the bundle of His during surgical correction of congenital interventricular septal defects.

When conduction between the atria and ventricles is slowed but not completely interrupted, **incomplete heart block** is present. In the form called **first-degree heart block**, all the atrial impulses reach the ventricles but the PR interval is abnormally long. In the form called **second-degree heart block**, not all atrial im-

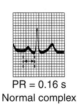

PR = 0.16 s
Normal complex

PR = 0.38 s
First-degree heart block

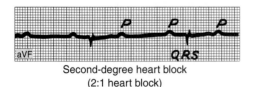

Second-degree heart block
(2:1 heart block)

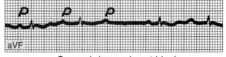

Second-degree heart block
(Wenckebach phenomenon)

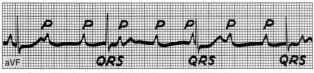

Complete heart block. Atrial rate, 107; ventricular rate, 43

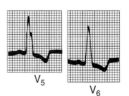

V₅   V₆

Two V leads in left
bundle branch block

Figure 28–11. Heart block.

pulses are conducted to the ventricles. There may be, for example, a ventricular beat following every second or every third atrial beat (2:1 block, 3:1 block, etc). In another form of incomplete heart block, there are repeated sequences of beats in which the PR interval lengthens progressively until a ventricular beat is dropped **(Wenckebach phenomenon).** The PR interval of the cardiac cycle that follows each dropped beat is usually normal or only slightly prolonged (Figure 28–11).

Sometimes one branch of the bundle of His is interrupted, causing **right** or **left bundle branch block.** In bundle branch block, excitation passes normally down the bundle on the intact side and then sweeps back through the muscle to activate the ventricle on the blocked side. The ventricular rate is therefore normal, but the QRS complexes are prolonged and deformed (Figure 28–11). Block can also occur in the anterior or posterior fascicle of the left bundle branch, producing the condition called **hemiblock** or **fascicular block.** Left anterior hemiblock produces abnormal left axis deviation in the ECG, whereas left posterior hemiblock produces abnormal right axis deviation. It is not uncommon to find combinations of fascicular and branch blocks bifascicular or trifascicular block). The His bundle electrogram permits detailed analysis of the site of block when there is a defect in the conduction system.

Implanted Pacemakers

When there is marked bradycardia in patients with sick sinus syndrome or third-degree heart block, an electronic pacemaker is frequently implanted. These devices, which have become sophisticated and reliable, are useful in patients with sinus node dysfunction, AV block, and bifascicular or trifascicular block. They are useful also in patients with severe neurogenic syncope in whom carotid sinus stimulation produces pauses of more than 3 seconds between heartbeats.

Ectopic Foci of Excitation

Normally, myocardial cells do not discharge spontaneously, and the possibility of spontaneous discharge of the His bundle and Purkinje system is low because the normal pacemaker discharge of the SA node is more rapid than their rate of spontaneous discharge. However, in abnormal conditions, the His-Purkinje fibers or the myocardial fibers may discharge spontaneously. In these conditions, **increased automaticity** of the heart is said to be present. If an irritable **ectopic focus** discharges once, the result is a beat that occurs before the expected next normal beat and transiently interrupts the cardiac rhythm (atrial, nodal, or ventricular **extrasystole** or **premature beat).** If the focus discharges repetitively at a rate higher than that of the SA node, it produces rapid, regular tachycardia (atrial, ventricular, or nodal **paroxysmal tachycardia** or **atrial flutter).**

Reentry

A more common cause of paroxysmal arrhythmias is a defect in conduction that permits a wave of excitation to propagate continuously within a closed circuit **(circus movement).** For example, if there is a transient block on one side of a portion of the conducting system, the impulse can go down the other side. If the block then wears off, the impulse may conduct in a retrograde direction in the previously blocked side back to the origin and then descend again, establishing a circus movement. An example of this in a ring of tissue is shown in Figure 28–12. If the reentry is in the AV node, the reentrant activity depolarizes the atrium, and the resulting atrial beat is called an echo beat. In addition, the reentrant activity in the node propagates back down to the ventricle, producing paroxysmal nodal tachycardia. Circus movements can also become established in the atrial or ventricular muscle fibers. In individuals with an abnormal extra bundle of conducting tissue connecting the atria to the ventricles (bundle of Kent; see below), the circus activity can pass in one direction through the AV node and in the other direction through the bundle, thus involving both the atria and the ventricles.

Atrial Arrhythmias

Excitation spreading from an independently discharging focus in the atria stimulates the AV node prematurely and is conducted to the ventricles. The P waves of atrial extrasystoles are abnormal, but the QRST configurations are usually normal (Figure 28–13). The excitation may depolarize the SA node, which must repolarize and

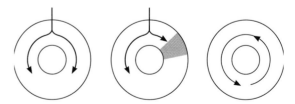

Figure 28–12. Depolarization of a ring of cardiac tissue. Normally, the impulse spreads in both directions in the ring **(left)** and the tissue immediately behind each branch of the impulse is refractory. When there is a transient block on one side **(center),** the impulse on the other side goes around the ring, and if the transient block has now worn off **(right),** the impulse passes this area and continues to circle indefinitely (circus movement).

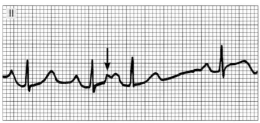

Atrial extrasystole

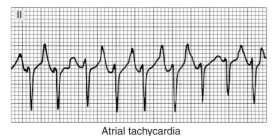

Atrial tachycardia

Atrial flutter

Atrial fibrillation

Figure 28–13. Atrial arrhythmias. The illustration shows an atrial premature beat with its P wave superimposed on the T wave of the preceding beat (arrow); atrial tachycardia; atrial flutter with 4:1 AV block; and atrial fibrillation with a totally irregular ventricular rate. (Tracings reproduced, with permission, from Goldschlager N, Goldman MJ: *Principles of Clinical Electrocardiography,* 13th ed. Originally published by Appleton & Lange. Copyright © 1989 by The McGraw-Hill Companies, Inc.)

ing atrial rates up to 220/min. Sometimes, especially in digitalized patients, some degree of atrioventricular block is associated with the tachycardia (**paroxysmal atrial tachycardia with block).**

In atrial flutter, the atrial rate is 200–350/min (Figure 28–13). In the most common form of this arrhythmia, there is large counterclockwise circus movement in the right atrium. This produces a characteristic saw-tooth pattern of flutter waves due to atrial contractions. It is almost always associated with 2:1 or greater AV block because in adults, the AV node cannot conduct more than about 230 impulses per minute.

In **atrial fibrillation,** the atria beat very rapidly (300–500/min) in a completely irregular and disorganized fashion. Because the AV node discharges at irregular intervals, the ventricles beat at a completely irregular rate, usually 80–160/min (Figure 28–13). The condition can be paroxysmal or chronic, and in some cases there appears to be a genetic predisposition. The cause of atrial fibrillation is still a matter of debate, but in most cases it appears to be due to multiple concurrently circulating reentrant excitation waves in both atria. However, some cases of paroxysmal atrial fibrillation seem to be produced by discharge of one or more ectopic foci. Many of these foci appear to be located in the pulmonary veins as much as 4 cm from the heart. Atrial muscle fibers extend along the pulmonary veins and are the origin of these discharges.

Consequences of Atrial Arrhythmias

Occasional atrial extrasystoles occur from time to time in most normal humans and have no pathologic significance. In paroxysmal atrial tachycardia and flutter, the ventricular rate may be so high that diastole is too short for adequate filling of the ventricles with blood between contractions. Consequently, cardiac output is reduced and symptoms of heart failure appear. The relationship between cardiac rate and cardiac output is discussed in detail in Chapter 29. Heart failure may also complicate atrial fibrillation when the ventricular rate is high. Acetylcholine liberated at vagal endings depresses conduction in the atrial musculature and AV node. This is why stimulating reflex vagal discharge by pressing on the eyeball (**oculocardiac reflex**) or massaging the carotid sinus often converts tachycardia and sometimes converts atrial flutter to normal sinus rhythm. Alternatively, vagal stimulation increases the degree of AV block, abruptly lowering the ventricular rate. Digitalis also depresses AV conduction and is used to lower a rapid ventricular rate in atrial fibrillation.

Ventricular Arrhythmias

Premature beats that originate in an ectopic ventricular focus usually have bizarrely shaped prolonged QRS

then depolarize to the firing level before it can initiate the next normal beat. Consequently, there is a pause between the extrasystole and the next normal beat that is usually equal in length to the interval between the normal beats preceding the extrasystole, and the rhythm is "reset" (see below). If the excitation does not reach the SA node before its next normal discharge, there is a pause equal to the time between two normal beats.

Atrial tachycardia occurs when an atrial focus discharges regularly or there is reentrant activity produc-

complexes (Figure 28–14) because of the slow spread of the impulse from the focus through the ventricular muscle to the rest of the ventricle. They are usually incapable of exciting the bundle of His, and retrograde conduction to the atria therefore does not occur. In the meantime, the next succeeding normal SA nodal impulse depolarizes the atria. The P wave is usually buried in the QRS of the extrasystole. If the normal impulse reaches the ventricles, they are still in the refractory period following depolarization from the ectopic focus. However, the second succeeding impulse from the SA node produces a normal beat. Thus, ventricular premature beats are followed by a **compensatory pause** that is often longer than the pause after an atrial extrasystole. Furthermore, ventricular premature beats do not interrupt the regular discharge of the SA node, whereas atrial premature beats often interrupt and "reset" the normal rhythm.

Atrial and ventricular premature beats are not strong enough to produce a pulse at the wrist if they occur early in diastole, when the ventricles have not had time to fill with blood and the ventricular musculature is still in its relatively refractory period. They may not even open the aortic and pulmonary valves, in which case there is, in addition, no second heart sound (see Chapter 29).

Paroxysmal ventricular tachycardia (Figure 28–14) is in effect a series of rapid, regular ventricular depolarizations usually due to a circus movement in-

volving the ventricles. **Torsade des pointes** is a form of ventricular tachycardia in which the QRS morphology varies (Figure 28–15). Tachycardias originating above the ventricles (supraventricular tachycardias such as paroxysmal nodal tachycardia) can be distinguished from paroxysmal ventricular tachycardia by use of the HBE; in supraventricular tachycardias, there is a His bundle H deflection, whereas in ventricular tachycardias, there is none. Ventricular premature beats are common and, in the absence of ischemic heart disease, usually benign. Ventricular tachycardia is more serious because cardiac output is decreased, and ventricular fibrillation is an occasional complication of ventricular tachycardia.

In **ventricular fibrillation** (Figure 28–15), the ventricular muscle fibers contract in a totally irregular and ineffective way because of the very rapid discharge of multiple ventricular ectopic foci or a circus movement. The fibrillating ventricles, like the fibrillating atria, look like a quivering "bag of worms." Ventricular fibrillation can be produced by an electric shock or an extrasystole during a critical interval, the **vulnerable period.** The vulnerable period coincides in time with the midportion of the T wave—ie, it occurs at a time when some of the ventricular myocardium is depolarized, some is incompletely repolarized, and some is completely repolarized. These are excellent conditions in which to establish reentry and a circus movement. The fibrillating ventricles cannot pump blood effectively, and circulation of the blood stops. Therefore, in the absence of emergency treatment, ventricular fibrillation that lasts more than a few minutes is fatal. The most frequent cause of sudden death in patients with myocardial infarcts is ventricular fibrillation.

Another indication of the vulnerability of the heart during repolarization is the fact that in patients in whom the QT interval is prolonged, there is an increased incidence of ventricular arrhythmias and sudden death. In three congenital forms of the **long QT syndrome,** genetic defects have been described. One produces blockade of one type of K⁺ channel (the HERG channel), slowing repolarization (Figure 28–16). Another blocks another kind of K⁺ channel, and is associated with deafness (see Chapter 9). A third produces constitutive activation of the cardiac Na⁺ channel, which also slows repolarization.

Although ventricular fibrillation may be produced by electrocution, it can often be stopped and converted to normal sinus rhythm by means of electrical shocks. Electronic defibrillators are now available not only in hospitals but also in emergency vehicles and commercial aircraft. They should be used as rapidly as possible. Their size has been reduced and their sophistication increased to the point that they can also be implanted surgically in patients at high risk for ventricular fibrillation and pro-

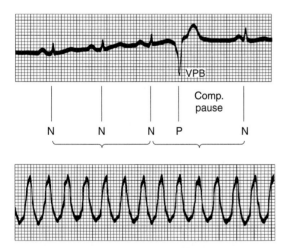

Figure 28–14. **Top:** Ventricular premature beats (VPB). The lines under the tracing illustrate the compensatory pause and show that the duration of the premature beat plus the preceding normal beat is equal to the duration of two normal beats. **Bottom:** Ventricular tachycardia.

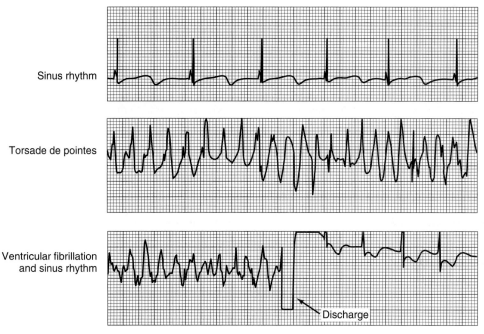

Sinus rhythm

Torsade de pointes

Ventricular fibrillation
and sinus rhythm

Discharge

Figure 28–15. Record obtained from an implanted cardioverter-defibrillator in a 12-year-old boy with congenital long QT syndrome who collapsed while answering a question in school. **Top:** Normal sinus rhythm with long QT interval. **Middle:** Torsade de pointes. **Bottom:** Ventricular fibrillation with discharge of defibrillator, as programmed 7.5 s after the start of ventricular tachycardia, converting the heart to normal sinus rhythm. The boy recovered consciousness in 2 minutes and had no neurologic sequelae. (Reproduced, with permission, from Moss AJ, Daubert JP: Images in clinical medicine. Internal ventricular fibrillation. N Engl J Med 2000;342:398.)

grammed to discharge automatically after 5–10 seconds of ventricular tachycardia or ventricular fibrillation.

Long QT Syndrome

Another indication of vulnerability of the heart during repolarization is the fact that in patients in whom the QT interval is prolonged, there is irregular cardiac repolarization and an increased incidence of ventricular arrhythmias and sudden death. The syndrome can be caused by a number of different drugs, by electrolyte abnormalities, and by myocardial ischemia. It can also be congenital (Figure 28–15). Three documented causes of the congenital form are mutations in *KVLQT-1*, *HERG* (Figure 28–16), and *minK*, three genes that encode parts of K+ channels. Mutations in *KVLQT-1* also cause deafness because the resulting K+ channel abnormality causes abnormal K+ movement across the stria vascularis (see Chapter 9). A mutation in the Na+ channel gene *SCN5A* also has been shown to produce congenital long QT syndrome.

Cardiopulmonary Resuscitation

In patients who are fibrillating or whose hearts have stopped, cardiac output and perfusion of the coronaries can be partially maintained by closed-chest **cardiac massage.** The person conducting external massage places the heel of one hand on the lower sternum above the xiphoid process and the heel of the other hand on top of the first (Figure 28–17). Pressure is applied straight down, depressing the sternum 4 or 5 cm toward the spine. This procedure is repeated 80–100 times per minute. It is worth remembering that when the heart stops suddenly, the pulmonary veins, left heart, and arteries are full of oxygenated blood, so attention should first be directed to the circulation. However, if breathing has also stopped and does not resume, full cardiopulmonary resuscitation (CPR) should be initiated; cardiac compression should be alternated with mouth-to-mouth breathing (see Chapter 37) at a rate of one ventilation to five chest compressions.

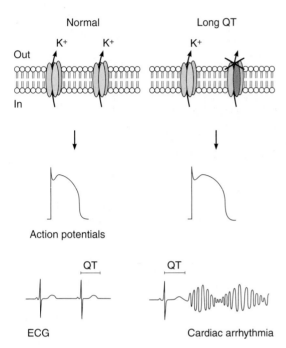

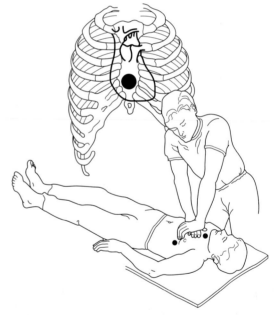

Figure 28–16. Long QT syndrome due to genetic abnormality that blocks HERG K+ channels. This predisposes to ventricular arrhythmias because it slows K+ efflux, prolonging the cardiac action potential and hence the QT interval. (Modified from Keating M, Sanguinetti MC: Molecular genetic insights into cardiovascular disease. Science 1996;272:681.)

Accelerated AV Conduction

An interesting condition seen in some otherwise normal individuals who are prone to attacks of paroxysmal atrial arrhythmias is **accelerated AV conduction (Wolff-Parkinson-White syndrome).** Normally, the only conducting pathway between the atria and the ventricles is the AV node. Individuals with Wolff-Parkinson-White syndrome have an additional aberrant muscular or nodal tissue connection (**bundle of Kent**) between the atria and ventricles. This conducts more rapidly than the slowly conducting AV node, and one ventricle is excited early. The manifestations of its activation merge with the normal QRS pattern, producing a short PR interval and a prolonged QRS deflection slurred on the upstroke (Figure 28–18), with a normal interval between the start of the P wave and the end of the QRS complex ("PJ interval"). The paroxysmal atrial tachycardias seen in this syndrome often follow an atrial premature beat. This beat conducts normally down the AV node but finds the aberrant bundle refractory, since the bundle has a longer refractory period than the AV node. However, when ventricular activation spreads to the ventricular end of the aberrant bundle, the bundle is no longer refractory and the impulse is transmitted retrograde to the atrium. A circus movement is thus established. Less commonly, an atrial premature beat finds the AV node refractory but reaches the ventricles via the bundle of Kent, setting up a circus movement in which the impulse passes from the ventricles to the atria via the AV node.

In some instances, the Wolff-Parkinson-White syndrome is familial. In two such families, there is a mutation in a gene that codes for an AMP-activated protein kinase. Presumably, this kinase is normally involved in suppressing abnormal atrioventricular pathways during fetal development.

Attacks of paroxysmal supraventricular tachycardia, usually nodal tachycardia, are seen in individuals with short PR intervals and normal QRS complexes (**Lown-Ganong-Levine syndrome**). In this condition, depolarization presumably passes from the atria to the ventricles via an aberrant bundle that bypasses the AV

Figure 28–17. Technique of external, closed-chest cardiac massage. The black circle on the diagram of the heart shows the area where force should be applied. Circles on the supine figure at the apex of the heart and just to the right of the upper portion of the sternum show where electrodes should be applied for external defibrillation. (Reproduced, with permission, from Schroeder SA et al [editors]: *Current Medical Diagnosis & Treatment*. Originally published by Appleton & Lange. Copyright © 1990 by The McGraw-Hill Companies, Inc.)

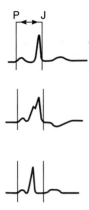

Figure 28–18. Accelerated AV conduction. **Top:** Normal sinus beat. **Middle:** Short PR interval; wide, slurred QRS complex; normal PJ interval (Wolff-Parkinson-White syndrome). **Bottom:** Short PR interval, normal QRS complex (Lown-Ganong-Levine syndrome). (Reproduced, with permission, from Goldschlager N, Goldman MJ: *Principles of Clinical Electrocardiography*, 13th ed. Originally published by Appleton & Lange. Copyright © 1989 by The McGraw-Hill Companies, Inc.)

node but enters the intraventricular conducting system distal to the node.

Antiarrhythmic Drugs

Many different drugs have been developed that are used in the treatment of arrhythmias because they slow conduction in the conduction system and the myocardium. This depresses ectopic activity and reduces the discrepancy between normal and reentrant paths so that reentry does not occur. However, it has now become clear that in some patients any of these drugs can be **proarrhythmic** rather than antiarrhythmic—ie, they can also cause various arrhythmias. Therefore, they are increasingly being replaced by radiofrequency catheter ablation for the treatment of arrhythmias.

Radiofrequency Catheter Ablation of Reentrant Pathways

Catheters with electrodes at the tip can now be inserted into the chambers of the heart and its environs and used to map the exact location of an ectopic focus or accessory bundle that is responsible for the production of reentry and supraventricular tachycardia. The pathway can then be ablated by passing radiofrequency current with the catheter tip placed close to the bundle or focus. In skilled hands, this form of treatment can be very effective and is associated with few complications. It is particularly useful in conditions that cause supraventricular tachycardias, including Wolf-Parkinson-White syndrome and atrial flutter. It has also been used with success to ablate foci in the pulmonary veins causing paroxysmal atrial fibrillation (see above).

ELECTROCARDIOGRAPHIC FINDINGS IN OTHER CARDIAC & SYSTEMIC DISEASES

Myocardial Infarction

When the blood supply to part of the myocardium is interrupted, there are profound changes in the myocardium that lead to irreversible changes and death of muscle cells (**myocardial infarction;** see Chapter 32). The ECG is very useful for diagnosing ischemia and locating areas of infarction. The underlying electrical events and the resulting electrocardiographic changes are complex, and only a brief review can be presented here.

The three major abnormalities that cause electrocardiographic changes in acute myocardial infarction are summarized in Table 28–3. The first change, abnormally rapid repolarization after discharge of the infarcted muscle fibers as a result of accelerated opening of K^+ channels, develops seconds after occlusion of a coronary artery in experimental animals. It lasts only a few minutes, but before it is over the resting membrane potential of the infarcted fibers declines because of the loss of intracellular K^+. Starting about 30 minutes later, the infarcted fibers also begin to depolarize more slowly than the surrounding normal fibers.

All three of these changes cause current flow that produces elevation of the ST segment in electrocardiographic leads recorded with electrodes over the infarcted area (Figure 28–19). Because of the rapid repolarization in the infarct, the membrane potential of the

Table 28–3. Summary of the three major abnormalities of membrane polarization associated with acute myocardial infarction.

Defect in Infarcted Cells	Current Flow	Resultant ECG Change in Leads Over Infarct
Rapid repolarization	Out of infarct	ST segment elevation
Decreased resting membrane potential	Into infarct	TQ segment depression (manifested as ST segment elevation)
Delayed depolarization	Out of infarct	ST segment elevation

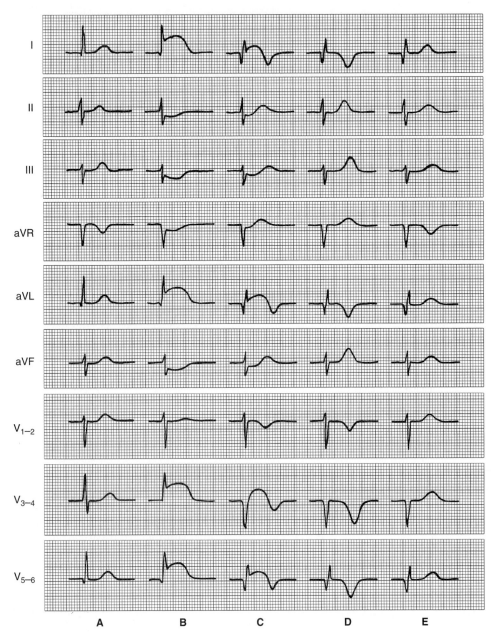

Figure 28–19. Diagrammatic illustration of serial electrocardiographic patterns in anterior infarction. **A:** Normal tracing. **B:** Very early pattern (hours after infarction): ST segment elevation in I, aVL, and V_{3-6}; reciprocal ST depression in II, III, and aVF. **C:** Later pattern (many hours to a few days): Q waves have appeared in I, aVL, and V_{5-6}. QS complexes are present in V_{3-4}. This indicates that the major transmural infarction is underlying the area recorded by V_{3-4}; ST segment changes persist but are of lesser degree, and the T waves are beginning to invert in the leads in which the ST segments are elevated. **D:** Late established pattern (many days to weeks): The Q waves and QS complexes persist, the ST segments are isoelectric, and the T waves are symmetric and deeply inverted in leads that had ST elevation and tall in leads that had ST depression. This pattern may persist for the remainder of the patient's life. **E:** Very late pattern: This may occur many months to years after the infarction. The abnormal Q waves and QS complexes persist. The T waves have gradually returned to normal. (Reproduced, with permission, from Goldschlager N, Goldman MJ: *Principles of Clinical Electrocardiography,* 13th ed. Originally published by Appleton & Lange. Copyright © 1989 by The McGraw-Hill Companies, Inc.)

area is greater than it is in the normal area during the latter part of repolarization, making the normal region negative relative to the infarct. Extracellularly, current therefore flows out of the infarct into the normal area (since, by convention, current flow is from positive to negative). This current flows toward electrodes over the injured area, causing increased positivity between the S and T waves of the ECG. Similarly, the delayed depolarization of the infarcted cells causes the infarcted area to be positive relative to the healthy tissue (Table 28–3) during the early part of repolarization, and the result is also ST segment elevation. The remaining change, the decline in resting membrane potential during diastole, causes a current flow into the infarct during ventricular diastole. The result of this current flow is a depression of the TQ segment of the ECG. However, the elec-

tronic arrangement in electrocardiographic recorders is such that a TQ segment depression is recorded as an ST segment elevation. Thus, the hallmark of acute myocardial infarction is elevation of the ST segments in the leads overlying the area of infarction (Figure 28–19). Leads on the opposite side of the heart show ST segment depression.

After some days or weeks, the ST segment abnormalities subside. The dead muscle and scar tissue become electrically silent. The infarcted area is therefore negative relative to the normal myocardium during systole, and it fails to contribute its share of positivity to the electrocardiographic complexes. The manifestations of this negativity are multiple and subtle. Common changes include the appearance of a Q wave in some of the leads in which it was not previously present and an

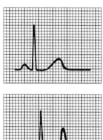

Normal tracing (plasma K+ 4–5.5 meq/L). PR interval = 0.16 s; QRS interval = 0.06 s; QT interval = 0.4 s (normal for an assumed heart rate of 60).

Hyperkalemia (plasma K+ ±7.0 meq/L). The PR and QRS intervals are within normal limits. Very tall, slender peaked T waves are now present.

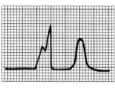

Hyperkalemia (plasma K+ ±8.5 meq/L). There is no evidence of atrial activity; the QRS complex is broad and slurred and the QRS interval has widened to 0.2 s. The T waves remain tall and slender. Further elevation of the plasma K+ level may result in ventricular tachycardia and ventricular fibrillation.

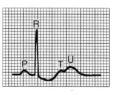

Hypokalemia (plasma K+ ±3.5 meq/L). PR interval = 0.2 s; QRS interval = 0.06 s; ST segment depression. A prominent U wave is now present immediately following the T. The actual QT interval remains 0.4 s. If the U wave is erroneously considered a part of the T, a falsely prolonged QT interval of 0.6 s will be measured.

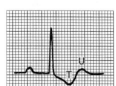

Hypokalemia (plasma K+ ±2.5 meq/L). The PR interval is lengthened to 0.32 s; the ST segment is depressed; the T wave is inverted; a prominent U wave is seen. The true QT interval remains normal.

Figure 28–20. Correlation of plasma K+ level and the ECG, assuming that the plasma Ca2+ level is normal. The diagrammed complexes are left ventricular epicardial leads. (Reproduced, with permission, from Goldman MJ: *Principles of Clinical Electrocardiography,* 12th ed. Originally published by Appleton & Lange. Copyright © 1989 by The McGraw-Hill Companies, Inc.)

increase in the size of the normal Q wave in some of the other leads, although so-called non-Q-wave infarcts are also seen. These infarcts tend to be less severe, but there is a high incidence of subsequent reinfarction. Another finding in infarction of the anterior left ventricle is "failure of progression of the R wave"; ie, the R wave fails to become successively larger in the precordial leads as the electrode is moved from right to left over the left ventricle. If the septum is infarcted, the conduction system may be damaged, causing bundle branch block or other forms of heart block.

Myocardial infarctions are often complicated by serious ventricular arrhythmias, with the threat of ventricular fibrillation and death. In experimental animals and presumably in humans, ventricular arrhythmias occur during three periods. During the first 30 minutes of an infarction, arrhythmias due to reentry are common. There follows a period relatively free from arrhythmias, but, starting 12 hours after infarction, arrhythmias occur as a result of increased automaticity. Arrhythmias occurring 3 days to several weeks after infarction are once again usually due to reentry. It is worth noting in this regard that infarcts that damage the epicardial portions of the myocardium interrupt sympathetic nerve fibers, producing denervation supersensitivity to catecholamines (see Chapter 4) in the area beyond the infarct. Alternatively, endocardial lesions can selectively interrupt vagal fibers (see above), leaving the actions of sympathetic fibers unopposed.

Effects of Changes in the Ionic Composition of the Blood

Changes in ECF Na^+ and K^+ concentration would be expected to affect the potentials of the myocardial fibers, because the electrical activity of the heart depends upon the distribution of these ions across the muscle cell membranes. Clinically, a fall in the plasma level of Na^+ may be associated with low-voltage electrocardiographic complexes, but changes in the plasma K^+ level produce severe cardiac abnormalities. Hyperkalemia is a very dangerous and potentially lethal condition because of its effects on the heart. As the plasma K^+ level rises, the first change in the ECG is the appearance of tall peaked T waves, a manifestation of altered repolarization (Figure 28–20). At higher K^+ levels, paralysis of the atria and prolongation of the QRS complexes occur. Ventricular arrhythmias may develop. The resting membrane potential of the muscle fibers decreases as the extracellular K^+ concentration increases. The fibers eventually become unexcitable, and the heart stops in diastole. Conversely, a decrease in the plasma K^+ level causes prolongation of the PR interval, prominent U waves, and, occasionally, late T wave inversion in the precordial leads. If the T and U waves merge, the apparent QT interval is often prolonged; if the T and U waves are separated, the true QT interval is seen to be of normal duration. Hypokalemia is a serious condition, but it is not as rapidly fatal as hyperkalemia.

Increases in extracellular Ca^{2+} concentration enhance myocardial contractility. When large amounts of Ca^{2+} are infused into experimental animals, the heart relaxes less during diastole and eventually stops in systole (**calcium rigor**). However, in clinical conditions associated with hypercalcemia, the plasma calcium level is rarely if ever high enough to affect the heart. Hypocalcemia causes prolongation of the ST segment and consequently of the QT interval, a change that is also produced by phenothiazines and tricyclic antidepressant drugs and by various diseases of the central nervous system.

The Heart as a Pump

29

INTRODUCTION

The orderly depolarization process described in the previous chapter triggers a wave of contraction that spreads through the myocardium. In single muscle fibers, contraction starts just after depolarization and lasts until about 50 ms after repolarization is completed (see Figure 3–15). Atrial systole starts after the P wave of the ECG; ventricular systole starts near the end of the R wave and ends just after the T wave. The contraction produces sequential changes in pressures and flows in the heart chambers and blood vessels. It should be noted that the term **systolic pressure** in the vascular system refers to the peak pressure reached during systole, not the mean pressure; similarly, the **diastolic pressure** refers to the lowest pressure during diastole.

MECHANICAL EVENTS OF THE CARDIAC CYCLE

Events in Late Diastole

Late in diastole, the mitral and tricuspid valves between the atria and ventricles are open and the aortic and pulmonary valves are closed. Blood flows into the heart throughout diastole, filling the atria and ventricles. The rate of filling declines as the ventricles become distended, and—especially when the heart rate is low—the cusps of the atrioventricular (AV) valves drift toward the closed position (Figure 29–1). The pressure in the ventricles remains low.

Atrial Systole

Contraction of the atria propels some additional blood into the ventricles, but about 70% of the ventricular filling occurs passively during diastole. Contraction of the atrial muscle that surrounds the orifices of the superior and inferior vena cava and pulmonary veins narrows their orifices, and the inertia of the blood moving toward the heart tends to keep blood in it; however, there is some regurgitation of blood into the veins during atrial systole.

Ventricular Systole

At the start of ventricular systole, the mitral and tricuspid (AV) valves close. Ventricular muscle initially shortens relatively little, but intraventricular pressure rises sharply as the myocardium presses on the blood in the ventricle (Figure 29–2). This period of **isovolumetric (isovolumic, isometric) ventricular contraction** lasts about 0.05 s, until the pressures in the left and right ventricles exceed the pressures in the aorta (80 mm Hg; 10.6 kPa) and pulmonary artery (10 mm Hg) and the aortic and pulmonary valves open. During isovolumetric contraction, the AV valves bulge into the atria, causing a small but sharp rise in atrial pressure (Figure 29–3).

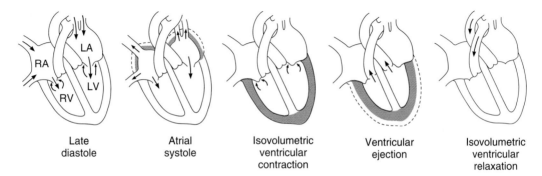

| Late diastole | Atrial systole | Isovolumetric ventricular contraction | Ventricular ejection | Isovolumetric ventricular relaxation |

Figure 29–1. Blood flow in the heart and great vessels during the cardiac cycle. The portions of the heart contracting in each phase are indicated in color. RA and LA, right and left atria; RV and LV, right and left ventricles.

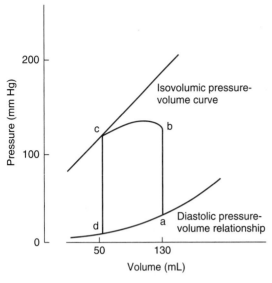

Figure 29–2. Pressure-volume loop of the left ventricle. During diastole, the ventricle fills and pressure increases from d to a. Pressure then rises sharply from a to b during isovolumetric contraction and from b to c during ventricular ejection. At c, the aortic valves close and pressure falls during isovolumetric relaxation from c back to d. (Reproduced, with permission, from McPhee SJ, Lingappa VR, Ganong WF [editors]: *Pathophysiology of Disease,* 4th ed. McGraw-Hill, 2003.)

When the aortic and pulmonary valves open, the phase of **ventricular ejection** begins. Ejection is rapid at first, slowing down as systole progresses. The intraventricular pressure rises to a maximum and then declines somewhat before ventricular systole ends. Peak left ventricular pressure is about 120 mm Hg, and peak right ventricular pressure is 25 mm Hg or less. Late in systole, the aortic pressure actually exceeds the ventricular, but for a short period momentum keeps the blood moving forward. The AV valves are pulled down by the contractions of the ventricular muscle, and atrial pressure drops. The amount of blood ejected by each ventricle per stroke at rest is 70–90 mL. The **end-diastolic ventricular volume** is about 130 mL. Thus, about 50 mL of blood remains in each ventricle at the end of systole **(end-systolic ventricular volume),** and the **ejection fraction,** the percent of the end-diastolic ventricular volume that is ejected with each stroke, is about 65%. The ejection fraction is a valuable index of ventricular function. It can be measured by injecting radionuclide-labeled red blood cells, imaging the cardiac blood pool at the end of diastole and the end of systole (equilibrium radionuclide angiocardiography), and then calculating the ejection fraction.

Early Diastole

Once the ventricular muscle is fully contracted, the already falling ventricular pressures drop more rapidly. This is the period of **protodiastole.** It lasts about 0.04 s. It ends when the momentum of the ejected blood is overcome and the aortic and pulmonary valves close, setting up transient vibrations in the blood and blood vessel walls. After the valves are closed, pressure continues to drop rapidly during the period of **isovolumetric ventricular relaxation.** Isovolumetric relaxation ends when the ventricular pressure falls below the atrial pressure and the AV valves open, permitting the ventricles to fill. Filling is rapid at first, then slows as the next cardiac contraction approaches. Atrial pressure continues to rise after the end of ventricular systole until the AV valves open, then drops and slowly rises again until the next atrial systole.

Pericardium

The heart is separated from the rest of the thoracic viscera by the pericardium. The myocardium itself is covered by the fibrous epicardium. The pericardial sac normally contains 5–30 mL of clear fluid, which lubricates the heart and permits it to contract with minimal friction.

Timing

Although events on the two sides of the heart are similar, they are somewhat asynchronous. Right atrial systole precedes left atrial systole, and contraction of the right ventricle starts after that of the left (see Chapter 28). However, since pulmonary arterial pressure is lower than aortic pressure, right ventricular ejection begins before left ventricular ejection. During expiration, the pulmonary and aortic valves close at the same time; but during inspiration, the aortic valve closes slightly before the pulmonary. The slower closure of the pulmonary valve is due to lower impedance of the pulmonary vascular tree. When measured over a period of minutes, the outputs of the two ventricles are, of course, equal, but transient differences in output during the respiratory cycle occur in normal individuals.

Length of Systole & Diastole

Cardiac muscle has the unique property of contracting and repolarizing faster when the heart rate is high (see Chapter 3), and the duration of systole decreases from 0.27 s at a heart rate of 65 to 0.16 s at a rate of 200 beats/min (Table 29–1). The shortening is due mainly to a decrease in the duration of systolic ejection. However, the duration of systole is much more fixed than

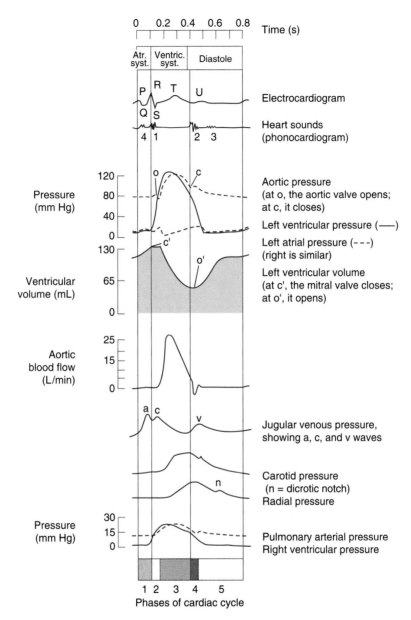

Figure 29–3. Events of the cardiac cycle at a heart rate of 75 beats/min. The phases of the cardiac cycle identified by the numbers at the bottom are as follows: 1, atrial systole; 2, isovolumetric ventricular contraction; 3, ventricular ejection; 4, isovolumetric ventricular relaxation; 5, ventricular filling. Note that late in systole, aortic pressure actually exceeds left ventricular pressure. However, the momentum of the blood keeps it flowing out of the ventricle for a short period. The pressure relationships in the right ventricle and pulmonary artery are similar. Abbreviations: Atr. syst., atrial systole; Ventric. syst., ventricular systole.

Table 29–1. Variation in length of action potential and associated phenomena with cardiac rate.[1]

	Heart Rate 75/min	Heart Rate 200/min	Skeletal Muscle
Duration, each cardiac cycle	0.80	0.30	...
Duration of systole	0.27	0.16	...
Duration of action potential	0.25	0.15	0.005
Duration of absolute refractory period	0.20	0.13	0.004
Duration of relative refractory period	0.05	0.02	0.003
Duration of diastole	0.53	0.14	...

[1]All values are in seconds. (Courtesy of AC Barger and GS Richardson.)

that of diastole, and when the heart rate is increased, diastole is shortened to a much greater degree. For example, at a heart rate of 65, the duration of diastole is 0.62 s, whereas at a heart rate of 200, it is only 0.14 s. This fact has important physiologic and clinical implications. It is during diastole that the heart muscle rests, and coronary blood flow to the subendocardial portions of the left ventricle occurs only during diastole (see Chapter 32). Furthermore, most of the ventricular filling occurs in diastole. At heart rates up to about 180, filling is adequate as long as there is ample venous return, and cardiac output per minute is increased by an increase in rate. However, at very high heart rates, filling may be compromised to such a degree that cardiac output per minute falls and symptoms of heart failure develop.

Because it has a prolonged action potential, cardiac muscle is in its refractory period and will not contract in response to a second stimulus until near the end of the initial contraction (see Figure 3–15). Therefore, cardiac muscle cannot be tetanized like skeletal muscle. The highest rate at which the ventricles can contract is theoretically about 400/min, but in adults the AV node will not conduct more than about 230 impulses/min because of its long refractory period. A ventricular rate of more than 230 is seen only in paroxysmal ventricular tachycardia (see Chapter 28).

Exact measurement of the duration of isovolumetric ventricular contraction is difficult in clinical situations, but it is relatively easy to measure the duration of **total electromechanical systole (QS$_2$)**, the **preejection period (PEP)**, and the **left ventricular ejection time (LVET)** by recording the ECG, phonocardiogram, and carotid pulse simultaneously. QS$_2$ is the period from the onset of the QRS complex to the closure of the aortic valves, as determined by the onset of the second heart sound. LVET is the period from the beginning of the carotid pressure rise to the dicrotic notch (see below). PEP is the difference between QS$_2$ and LVET and represents the time for the electrical as well as the mechanical events that precede systolic ejection. The ratio PEP/LVET is normally about 0.35, and it increases without a change in QS$_2$ when left ventricular performance is compromised in a variety of cardiac diseases.

Arterial Pulse

The blood forced into the aorta during systole not only moves the blood in the vessels forward but also sets up a pressure wave that travels along the arteries. The pressure wave expands the arterial walls as it travels, and the expansion is palpable as the **pulse.** The rate at which the wave travels, which is independent of and much higher than the velocity of blood flow, is about 4 m/s in the aorta, 8 m/s in the large arteries, and 16 m/s in the small arteries of young adults. Consequently, the pulse is felt in the radial artery at the wrist about 0.1 s after the peak of systolic ejection into the aorta (Figure 29–3). With advancing age, the arteries become more rigid, and the pulse wave moves faster.

The strength of the pulse is determined by the pulse pressure and bears little relation to the mean pressure. The pulse is weak ("thready") in shock. It is strong when stroke volume is large, eg, during exercise or after the administration of histamine. When the pulse pressure is high, the pulse waves may be large enough to be felt or even heard by the individual (palpitation, "pounding heart"). When the aortic valve is incompetent (aortic insufficiency), the pulse is particularly strong, and the force of systolic ejection may be sufficient to make the head nod with each heartbeat. The pulse in aortic insufficiency is called a **collapsing, Corrigan,** or **water-hammer pulse.** A water hammer is an evacuated glass tube half-filled with water that was a popular toy in the 19th century. When held in the hand and inverted, it delivers a short, hard knock.

The **dicrotic notch,** a small oscillation on the falling phase of the pulse wave caused by vibrations set up when the aortic valve snaps shut (Figure 29–3), is visible if the pressure wave is recorded but is not palpable at the wrist. There is also a dicrotic notch on the pulmonary artery pressure curve produced by the closure of the pulmonary valves.

Atrial Pressure Changes & the Jugular Pulse

Atrial pressure rises during atrial systole and continues to rise during isovolumetric ventricular contraction when the AV valves bulge into the atria. When the AV valves are pulled down by the contracting ventricular muscle, pressure falls rapidly and then rises as blood flows into the atria until the AV valves open early in diastole. The return of the AV valves to their relaxed position also contributes to this pressure rise by reducing atrial capacity. The atrial pressure changes are transmitted to the great veins, producing three characteristic waves in the record of jugular pressure (Figure 29–3). The **a wave** is due to atrial systole. As noted above, some blood regurgitates into the great veins when the atria contract, even though the orifices of the great veins are constricted. In addition, venous inflow stops, and the resultant rise in venous pressure contributes to the a wave. The **c wave** is the transmitted manifestation of the rise in atrial pressure produced by the bulging of the tricuspid valve into the atria during isovolumetric ventricular contraction. The **v wave** mirrors the rise in atrial pressure before the tricuspid valve opens during diastole. The jugular pulse waves are superimposed on the respiratory fluctuations in venous pressure. Venous pressure falls during inspiration as a result of the increased negative intrathoracic pressure and rises again during expiration.

Careful bedside inspection of the pulsations of the jugular veins may give clinical information of some importance. For example, in tricuspid insufficiency there is a giant c wave with each ventricular systole. In complete heart block, when the atria and ventricles are beating at different rates, the a waves that are not synchronous with the radial pulse can be made out, and there is a giant a wave ("cannon wave") whenever the atria contract while the tricuspid valve is closed. It may also be possible to distinguish atrial from ventricular extrasystoles by inspection of the jugular pulse, because atrial premature beats produce an a wave, whereas ventricular premature beats do not.

Heart Sounds

Two sounds are normally heard through a stethoscope during each cardiac cycle. The first is a low, slightly prolonged "lub" (**first sound**), caused by vibrations set up by the sudden closure of the mitral and tricuspid valves at the start of ventricular systole (Figure 29–3). The second is a shorter, high-pitched "dup" (**second sound**), caused by vibrations associated with closure of the aortic and pulmonary valves just after the end of ventricular systole. A soft, low-pitched **third sound** is heard about one-third of the way through diastole in many normal young individuals. It coincides with the period of rapid ventricular filling and is probably due to vibrations set up by the inrush of blood. A **fourth sound** can sometimes be heard immediately before the first sound when atrial pressure is high or the ventricle is stiff in conditions such as ventricular hypertrophy. It is due to ventricular filling and is rarely heard in normal adults.

The first sound has a duration of about 0.15 s and a frequency of 25–45 Hz. It is soft when the heart rate is low, because the ventricles are well filled with blood and the leaflets of the AV valves float together before systole. The second sound lasts about 0.12 s, with a frequency of 50 Hz. It is loud and sharp when the diastolic pressure in the aorta or pulmonary artery is elevated, causing the respective valves to shut briskly at the end of systole. The interval between aortic and pulmonary valve closure during inspiration is frequently long enough for the second sound to be reduplicated (physiologic splitting of the second sound). Splitting also occurs in various diseases. The third sound has a duration of 0.1 s.

Murmurs

Murmurs, or **bruits,** are abnormal sounds heard in various parts of the vascular system. The two terms are used interchangeably, though "murmur" is more commonly used to denote noise heard over the heart than over blood vessels. As discussed in detail in Chapter 30, blood flow is laminar and nonturbulent up to a critical velocity; above this velocity, and beyond an obstruction, blood flow is turbulent. Laminar flow is silent, but turbulent flow creates sounds. Blood flow speeds up when an artery or a heart valve is narrowed.

Examples of vascular sounds outside the heart are the bruit heard over a large, highly vascular goiter, the bruit heard over a carotid artery when its lumen is narrowed and distorted by atherosclerosis, and the murmurs heard over an aneurysmal dilation of one of the large arteries, an arteriovenous (A-V) fistula, or a patent ductus arteriosus.

The major, but certainly not the only, cause of cardiac murmurs is disease of the heart valves. When the orifice of a valve is narrowed (**stenosis**), blood flow through it in the normal direction is accelerated and turbulent. When a valve is incompetent, blood flows backward through it (**regurgitation** or **insufficiency**), again through a narrow orifice that accelerates flow. The timing (systolic or diastolic) of a murmur due to stenosis or insufficiency of any particular valve (Table 29–2) can be predicted from a knowledge of the me-

Table 29–2. Heart murmurs.

Valve	Abnormality	Timing of Murmur
Aortic or pulmonary	Stenosis	Systolic
	Insufficiency	Diastolic
Mitral or tricuspid	Stenosis	Diastolic
	Insufficiency	Systolic

chanical events of the cardiac cycle. Murmurs due to disease of a particular valve can generally be heard best when the stethoscope is over that particular valve; thus, murmurs due to disorders of the aortic and pulmonic valves are usually heard best at the base of the heart, and murmurs due to mitral disease are usually heard best at the apex. There are other aspects of the duration, character, accentuation, and transmission of the sound that help to locate its origin in one valve or the other. One of the loudest murmurs is that produced when blood flows backward in diastole through a hole in a cusp of the aortic valve. Most murmurs can be heard only with the aid of the stethoscope, but this high-pitched musical diastolic murmur is sometimes audible to the unaided ear several feet from the patient.

In patients with congenital interventricular septal defects, flow from the left to the right ventricle causes a systolic murmur. Soft murmurs may also be heard in patients with interatrial septal defects, although they are not a constant finding.

Soft systolic murmurs are common in individuals, especially children, who have no cardiac disease. Systolic murmurs are also common in anemic patients as a result of the low viscosity of the blood and the rapid flow (see Chapter 30).

Echocardiography

Wall movement and other aspects of cardiac function can be evaluated by **echocardiography,** a noninvasive technique that does not involve injections or insertion of a catheter. In echocardiography, pulses of ultrasonic waves, commonly at a frequency of 2.25 MHz, are emitted from a transducer that also functions as a receiver to detect waves reflected back from various parts of the heart. Reflections occur wherever acoustic impedance changes, and a recording of the echoes displayed against time on an oscilloscope provides a record of the movements of the ventricular wall, septum, and valves during the cardiac cycle. When combined with Doppler techniques, echocardiography can be used to measure velocity and volume of flow through valves. It has considerable clinical usefulness, particularly in evaluating and planning therapy in patients with valvular lesions.

CARDIAC OUTPUT

Methods of Measurement

In experimental animals, cardiac output can be measured with an electromagnetic flow meter placed on the ascending aorta. Two methods of measuring output that are applicable to humans, in addition to Doppler combined with echocardiography, are the **direct Fick method** and the **indicator dilution method.**

The **Fick principle** states that the amount of a substance taken up by an organ (or by the whole body) per unit of time is equal to the arterial level of the substance minus the venous level (**A-V difference**) times the blood flow. This principle can be applied, of course, only in situations in which the arterial blood is the sole source of the substance taken up. The principle can be used to determine cardiac output by measuring the amount of O_2 consumed by the body in a given period and dividing this value by the A-V difference across the lungs. Because systemic arterial blood has the same O_2 content in all parts of the body, the arterial O_2 content can be measured in a sample obtained from any convenient artery. A sample of venous blood in the pulmonary artery is obtained by means of a cardiac catheter. Right atrial blood has been used in the past, but mixing of this blood may be incomplete, so that the sample is not representative of the whole body. An example of the calculation of cardiac output using a typical set of values is as follows:

$$\text{Output of left ventricle} = \frac{O_2 \text{ consumption (mL/min)}}{[A_{O_2}] - [V_{O_2}]}$$

$$= \frac{250 \text{ mL/min}}{\begin{array}{c}190 \text{ mL/L arterial blood} - \\ 140 \text{ mL/L venous blood in} \\ \text{pulmonary artery}\end{array}}$$

$$= \frac{250 \text{ mL/min}}{50 \text{ mL/L}}$$

$$= 5 \text{ L/min}$$

It has now become commonplace to insert a long catheter through a forearm vein and to guide its tip into the heart with the aid of a fluoroscope. The technique was initially developed by Forssmann, who catheterized himself but was summarily fired from his job when he sought permission to explore the use of the catheter in others for diagnostic purposes. However, the procedure is now known to be essentially harmless. Catheters can be inserted not only into the right atrium but also through the atrium and the right ventricle into the small branches of the pulmonary artery. Catheters can also be inserted in peripheral arteries and guided in a retrograde direction to the heart and into coronary or other arteries.

In the indicator dilution technique, a known amount of a substance such as a dye or, more commonly, a radioactive isotope is injected into an arm vein and the concentration of the indicator in serial samples of arterial blood is determined. The output of the heart is equal to the amount of indicator injected divided by its average concentration in arterial blood after a single circulation through the heart (Figure 29–4). The indicator must, of course, be a substance that stays in the bloodstream during the test and has no harmful or hemodynamic effects. In practice, the log of the indicator concentration in the serial arterial samples is plotted against time as the concentration rises, falls, and then rises again as the indicator recirculates. The initial decline in concentration, linear on a semilog plot, is extrapolated to the abscissa, giving the time for first passage of the indicator through the circulation. The cardiac output for that period is calculated (Figure 29–4) and then converted to output per minute.

A popular indicator dilution technique is **thermodilution,** in which the indicator used is cold saline. The saline is injected into the right atrium through one side of a double-lumen catheter, and the temperature change in the blood is recorded in the pulmonary artery, using a thermistor in the other, longer side of the catheter. The temperature change is inversely proportionate to the amount of blood flowing through the pulmonary artery, ie, to the extent that the cold saline is diluted by blood. This technique has two important advantages: (1) the saline is completely innocuous; and (2) the cold is dissipated in the tissues so recirculation is not a problem, and it is easy to make repeated determinations.

Cardiac output can also be measured by Doppler techniques combined with echocardiography (see above).

Cardiac Output in Various Conditions

The amount of blood pumped out of each ventricle per beat, the **stroke volume,** is about 70 mL in a resting man of average size in the supine position (70 mL from the left ventricle and 70 mL from the right, with the two ventricular pumps in series). The output of the heart per unit time is the **cardiac output.** In a resting, supine man, it averages about 5.0 L/min (70 mL × 72 beats/min). There is a correlation between resting cardiac output and body surface area. The output per minute per square meter of body surface (the **cardiac index**) averages 3.2 L. The effects of various conditions on cardiac output are summarized in Table 29–3.

Factors Controlling Cardiac Output

Variations in cardiac output can be produced by changes in cardiac rate or stroke volume (Figure 29–5).

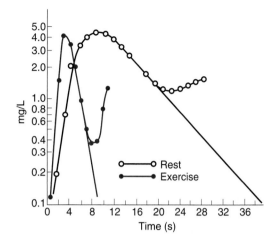

$$F = \frac{E}{\int_0^\alpha C\,dt}$$

F = flow
E = amount of indicator injected
C = instantaneous concentration of indicator in arterial blood

In the **rest** example above,

$$\frac{\text{Flow in 39 s}}{\text{(time of first passage)}} = \frac{5 \text{ mg injection}}{\frac{1.6 \text{ mg/L}}{\text{(avg concentration)}}}$$

Flow = 3.1 L in 39 s
Flow (cardiac output)/min = $3.1 \times \dfrac{60}{39}$ = 4.7 L

For the **exercise** example,

Flow in 9 s = $\dfrac{5 \text{ mg}}{1.51 \text{ mg/L}}$ = 3.3 L

Flow/min = $3.3 \times \dfrac{60}{9}$ = 22.0 L

Figure 29–4. Determination of cardiac output by indicator (dye) dilution.

The cardiac rate is controlled primarily by the cardiac innervation, sympathetic stimulation increasing the rate and parasympathetic stimulation decreasing it (see Chapter 28). The stroke volume is also determined in part by neural input, sympathetic stimuli making the myocardial muscle fibers contract with greater strength at any given length and parasympathetic stimuli having the opposite effect. When the strength of contraction increases without an increase in fiber length, more of

Table 29–3. Effect of various conditions on cardiac output.

	Condition or Factor[1]
No change	Sleep
	Moderate changes in environmental temperature
Increase	Anxiety and excitement (50–100%)
	Eating (30%)
	Exercise (up to 700%)
	High environmental temperature
	Pregnancy
	Epinephrine
Decrease	Sitting or standing from lying position (20–30%)
	Rapid arrhythmias
	Heart disease

[1]Approximate percent changes are shown in parentheses.

the blood that normally remains in the ventricles is expelled; ie, the ejection fraction increases and the end-systolic ventricular blood volume falls. The cardiac accelerator action of the catecholamines liberated by sympathetic stimulation is referred to as their **chronotropic action,** whereas their effect on the strength of cardiac contraction is called their **inotropic action.** Factors that increase the strength of cardiac contraction are said to be positively inotropic; those that decrease it are said to be negatively inotropic.

The force of contraction of cardiac muscle is dependent upon its preloading and its afterloading. These factors are illustrated in Figure 29–6, in which a muscle strip is stretched by a load (the **preload**) that rests on a platform. The initial phase of the contraction is isometric; the elastic component in series with the contractile element is stretched, and tension increases until it is sufficient to lift the load. The tension at which the load is lifted is the **afterload.** The muscle then contracts isotonically without developing further tension. In vivo, the preload is the degree to which the myocardium is stretched before it contracts and the afterload is the resistance against which blood is expelled.

Relation of Tension to Length in Cardiac Muscle

The length-tension relationship in cardiac muscle (see Figure 3–17) is similar to that in skeletal muscle (see Figure 3–11); as the muscle is stretched, the developed tension increases to a maximum and then declines as stretch becomes more extreme. Starling pointed this out when he stated that the "energy of contraction is proportional to the initial length of the cardiac muscle fiber." This pronouncement has come to be known as **Starling's law of the heart** or the **Frank-Starling law.** For the heart, the length of the muscle fibers (ie, the extent of the preload) is proportionate to the end-diastolic volume. The relation between ventricular stroke

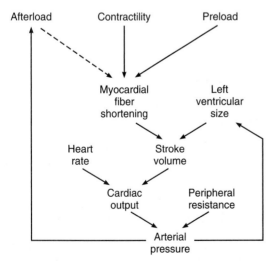

Figure 29–5. Interactions between the components that regulate cardiac output and arterial pressure. Solid lines indicate increases, and the dashed line indicates a decrease.

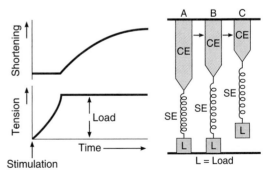

Figure 29–6. Model for contraction of afterloaded muscles. **A:** Rest. **B:** Partial contraction of the contractile element of the muscle (CE), with stretching of the series elastic element (SE) but no shortening. **C:** Complete contraction, with shortening. (Reproduced, with permission, from Sonnenblick EH in: *The Myocardial Cell: Structure, Function and Modification.* Briller SA, Conn HL [editors]. Univ Pennsylvania Press, 1966.)

volume and end-diastolic volume is called the Frank-Starling curve.

Regulation of cardiac output as a result of changes in cardiac muscle fiber length is sometimes called **heterometric regulation,** whereas regulation due to changes in contractility independent of length is sometimes called **homometric regulation.**

Heart-Lung Preparation

The effect of changes in peripheral resistance can be demonstrated in the **heart-lung preparation.** In this preparation, the heart and lungs of an anesthetized experimental animal are cannulated in such a way that blood flows from the aorta through a system of tubing and a reservoir to the right atrium, and from there through the animal's heart and lungs back to the aorta. The heart is functionally denervated, and so its rate varies little if at all. When the tubing is constricted (afterload is increased), the heart dilates and beats more forcefully. If the reservoir is elevated, venous pressure and venous return are increased (preload is increased), and again the heart dilates and puts out the blood returned to it.

Factors Affecting End-Diastolic Volume

Alterations in systolic and diastolic function have different effects on the heart. When systolic contractions are reduced, there is a primary reduction in stroke volume. Diastolic function also affects stroke volume, but in a different way.

The factors that normally operate to regulate end-diastolic volume, ie, the degree to which cardiac muscle is stretched, are listed in Table 29–4. An increase in intrapericardial pressure limits the extent to which the ventricle can fill. So does a decrease in ventricular compliance, ie, an increase in ventricular stiffness produced

Table 29–4. Factors that normally increase or decrease the length of ventricular cardiac muscle fibers.

Increase
  Stronger atrial contractions
  Increased total blood volume
  Increased venous tone
  Increased pumping action of skeletal muscle
  Increased negative intrathoracic pressure

Decrease
  Standing
  Increased intrapericardial pressure
  Decreased ventricular compliance

by myocardial infarction, infiltrative disease, and other abnormalities. Atrial contractions aid ventricular filling. The other factors affect the amount of blood returning to the heart and hence the degree of cardiac filling during diastole. An increase in total blood volume increases venous return. Constriction of the veins reduces the size of the venous reservoirs, decreasing venous pooling and thus increasing venous return. An increase in the normal negative intrathoracic pressure increases the pressure gradient along which blood flows to the heart, whereas a decrease impedes venous return. Standing decreases venous return, and muscular activity increases it as a result of the pumping action of skeletal muscle.

The effects of systolic and diastolic dysfunction on the pressure-volume loop of the left ventricle are summarized in Figure 29–7.

Myocardial Contractility

The contractility of the myocardium exerts a major influence on stroke volume. When the sympathetic nerves to the heart are stimulated, the whole length-tension curve shifts upward and to the left (Figure 29–8). The positively inotropic effect of the norepinephrine liberated at the nerve endings is augmented by circulating norepinephrine, and epinephrine has a similar effect. There is a negatively inotropic effect of vagal stimulation on the atrial muscle and a small negatively inotropic effect on the ventricular muscle.

Changes in cardiac rate and rhythm also affect myocardial contractility (force-frequency relation, Figure 29–8). Ventricular extrasystoles condition the myocardium in such a way that the next succeeding contraction is stronger than the preceding normal contraction. This **postextrasystolic potentiation** is independent of ventricular filling, since it occurs in isolated cardiac muscle, and is due to increased availability of intracellular Ca^{2+}. A sustained increment in contractility can be produced by delivering paired electrical stimuli to the heart in such a way that the second stimulus is delivered shortly after the refractory period of the first. It has also been shown that myocardial contractility increases as the heart rate increases, although this effect is relatively small.

The catecholamines exert their inotropic effect via an action on cardiac β_1-adrenergic receptors and G_s, with resultant activation of adenylyl cyclase and increased intracellular cAMP. Xanthines such as caffeine and theophylline that inhibit the breakdown of cAMP are positively inotropic. Glucagon, which increases the formation of cAMP, is positively inotropic, and it has been recommended for use in the treatment of some heart diseases. The positively inotropic effect of digitalis and related drugs (Figure 29–8) is due to their in-

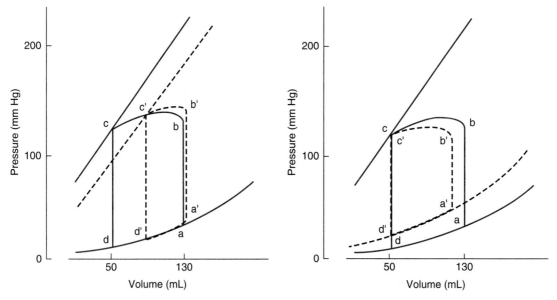

Figure 29–7. Effect of systolic and diastolic dysfunction on the pressure-volume loop of the left ventricle. **Left:** Systolic dysfunction shifts the isovolumic pressure-volume curve (see Figure 29–2) to the right, decreasing the stroke volume from b–c to b′–c′. **Right:** Diastolic dysfunction increases end-diastolic volume and shifts the diastolic pressure-volume relationship upward and to the left. This reduces the stroke volume from b–c to b′–c′. (Reproduced, with permission, from McPhee SJ, Lingappa VR, Ganong WF [editors]: *Pathophysiology of Disease,* 4th ed. McGraw-Hill, 2003.)

hibitory effect on the Na⁺-K⁺ ATPase in the myocardium (see Chapter 3). Hypercapnia, hypoxia, acidosis, and drugs such as quinidine, procainamide, and barbiturates depress myocardial contractility. The contractility of the myocardium is also reduced in heart failure (intrinsic depression). The cause of this depression is not known.

Integrated Control of Cardiac Output

In intact animals and humans, the mechanisms listed above operate in an integrated way to maintain cardiac output. During muscular exercise, there is increased sympathetic discharge, so that myocardial contractility is increased and the heart rate rises. The increase in heart rate is particularly prominent in normal individuals, and there is only a modest increase in stroke volume (Table 29–5). However, patients with transplanted hearts are able to increase their cardiac output during exercise in the absence of cardiac innervation through the operation of the Frank-Starling mechanism (Figure 29–9). Circulating catecholamines also contribute. The increase seen in these patients is not as rapid, and their maximal increase is smaller than in normal individuals but is still appreciable. If venous return increases and there is no change in sympathetic tone, venous pressure

rises, diastolic inflow is greater, ventricular end-diastolic pressure increases, and the heart muscle contracts more forcefully. During muscular exercise, venous return is increased by the pumping action of the muscles and the increase in respiration (see Chapter 33). In addition, because of vasodilation in the contracting muscles, peripheral resistance and consequently afterload are decreased. The end result in both normal and transplanted hearts is thus a prompt and marked increase in cardiac output.

Normal hearts probably never dilate to the point that they are on the "descending limb" of the Frank-Starling curve (Figure 29–8), ie, at a point where further stretch decreases rather than increases stroke volume. If this did occur, an eventually fatal vicious cycle would be set up in which further increases in cardiac filling would lead to decreased output, which would in turn further increase cardiac filling.

One of the differences between untrained individuals and trained athletes is that the athletes have lower heart rates, greater end-systolic ventricular volumes, and greater stroke volumes at rest. Therefore, they can potentially achieve a given increase in cardiac output by further increases in stroke volume without increasing their heart rate to as great a degree as an untrained individual.

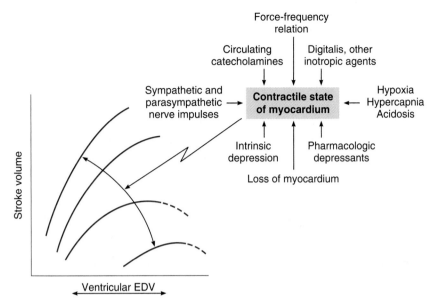

Figure 29–8. Effect of changes in myocardial contractility on the Frank-Starling curve. The curve shifts downward and to the right as contractility is decreased. The major factors influencing contractility are summarized on the right. The dashed lines indicate portions of the ventricular function curves where maximum contractility has been exceeded; ie, they identify points on the "descending limb" of the Frank-Starling curve. EDV, end-diastolic volume. (Reproduced, with permission, from Braunwald E, Ross J, Sonnenblick EH: Mechanisms of contraction of the normal and failing heart. N Engl J Med 1967;277:794. Courtesy of Little, Brown, Inc.)

Oxygen Consumption by the Heart

The basal O_2 consumption by the myocardium, which can be determined by stopping the heart while artificially maintaining the coronary circulation, is about 2 mL/100 g/min. This value is considerably higher than that of resting skeletal muscle. O_2 consumption by the beating heart is about 9 mL/100 g/min at rest. Increases occur during exercise and in a number of different states. Cardiac venous O_2 tension is low, and little additional O_2 can be extracted from the blood in the coronaries, so increases in O_2 consumption require increases in coronary blood flow. The regulation of coronary flow is discussed in Chapter 32.

The O_2 consumption by the heart is determined primarily by the intramyocardial tension, the contractile state of the myocardium, and the heart rate. Ventricular work per beat correlates with O_2 consumption. The work is the product of stroke volume and mean arterial pressure in the pulmonary artery (for the right

Table 29–5. Changes in cardiac function with exercise. Note that stroke volume levels off, then falls somewhat (as a result of the shortening of diastole) when the heart rate rises to high values.[1]

Work (kg-m/min)	O₂ Usage (mL/min)	Pulse Rate (per min)	Cardiac Output (L/min)	Stroke Volume (mL)	A-V O₂ Difference (mL/dL)
Rest	267	64	6.4	100	4.3
288	910	104	13.1	126	7.0
540	1430	122	15.2	125	9.4
900	2143	161	17.8	110	12.3
1260	3007	173	20.9	120	14.5

[1]Reproduced, with permission, from Asmussen E, Nielsen M: The cardiac output in rest and work determined by the acetylene and the dye injection methods. Acta Physiol Scand 1952;27:217.

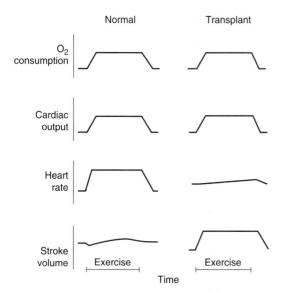

Figure 29–9. Cardiac responses to moderate supine exercise in normal humans and patients with transplanted and hence denervated hearts. (Reproduced, with permission, from Kent KM, Cooper T: The denervated heart. N Engl J Med 1974;291:1017.)

ventricle) or the aorta (for the left ventricle). Since aortic pressure is 7 times greater than pulmonary artery pressure, the stroke work of the left ventricle is approximately 7 times the stroke work of the right ventricle. In theory, a 25% increase in stroke volume without a

change in arterial pressure should produce the same increase in O_2 consumption as a 25% increase in arterial pressure without a change in stroke volume. However, for reasons that are incompletely understood, pressure work produces a greater increase in O_2 consumption than volume work. In other words, an increase in afterload causes a greater increase in cardiac O_2 consumption than an increase in preload does. This is why angina pectoris due to deficient delivery of O_2 to the myocardium is more common in aortic stenosis than in aortic insufficiency. In aortic stenosis, intraventricular pressure must be increased to force blood through the stenotic valve, whereas in aortic insufficiency, regurgitation of blood produces an increase in stroke volume with little change in aortic impedance.

It is worth noting that the increase in O_2 consumption produced by increased stroke volume when the myocardial fibers are stretched is an example of the operation of the law of Laplace. This law, which is discussed in detail in Chapter 30, states that the tension developed in the wall of a hollow viscus is proportionate to the radius of the viscus, and the radius of a dilated heart is increased. O_2 consumption per unit time increases when the heart rate is increased by sympathetic stimulation because of the increased number of beats and the increased velocity and strength of each contraction. However, this is somewhat offset by the decrease in end-systolic volume and hence in the radius of the heart.

Dynamics of Blood & Lymph Flow   30

INTRODUCTION

The blood vessels are a closed system of conduits that carry blood from the heart to the tissues and back to the heart. Some of the interstitial fluid enters the lymphatics and passes via these vessels to the vascular system. Blood flows through the vessels primarily because of the forward motion imparted to it by the pumping of the heart, although in the case of the systemic circulation, diastolic recoil of the walls of the arteries, compression of the veins by skeletal muscles during exercise, and the negative pressure in the thorax during inspiration also move the blood forward. The resistance to flow depends to a minor degree upon the viscosity of the blood but mostly upon the diameter of the vessels, principally the arterioles. The blood flow to each tissue is regulated by local chemical and general neural and humoral mechanisms that dilate or constrict the vessels of the tissue. All of the blood flows through the lungs, but the systemic circulation is made up of numerous different circuits in parallel (Figure 30–1), an arrangement that permits wide variations in regional blood flow without changing total systemic flow.

This chapter is concerned with the general principles that apply to all parts of the circulation and with pressure and flow in the systemic circulation. The homeostatic mechanisms operating to adjust flow are the subject of Chapter 31. The special characteristics of pulmonary and renal circulation are discussed in Chapters 34 and 38 and the unique features of the circulation to other organs in Chapter 32.

FUNCTIONAL MORPHOLOGY

Arteries & Arterioles

The characteristics of the various types of blood vessels are listed in Table 30–1. The walls of all arteries are made up of an outer layer of connective tissue, the adventitia; a middle layer of smooth muscle, the media; and an inner layer, the intima, made up of the endothelium and underlying connective tissue (Figure 30–2). The walls of the aorta and other arteries of large diameter contain a relatively large amount of elastic tissue, primarily located in the inner and external elastic laminas. They are stretched during systole and recoil on the blood during diastole. The walls of the arterioles contain less elastic tissue but much more smooth muscle. The muscle is innervated by noradrenergic nerve fibers, which are constrictor in function, and in some instances by cholinergic fibers, which dilate the vessels. The arterioles are the major site of the resistance to blood flow, and small changes in their caliber cause large changes in the total peripheral resistance.

Capillaries

The arterioles divide into smaller muscle-walled vessels, sometimes called **metarterioles,** and these in turn feed into capillaries (Figure 30–3). In some of the vascular beds that have been studied in detail, a metarteriole is connected directly with a venule by a capillary **thoroughfare vessel,** and the true capillaries are an anastomosing network of side branches of this thoroughfare vessel. The openings of the true capillaries are surrounded on the upstream side by minute

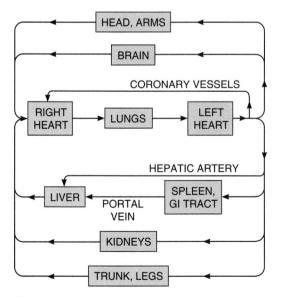

Figure 30–1. Diagram of the circulation in the adult.

Table 30–1. Characteristics of various types of blood vessels in humans.

Vessel	Lumen Diameter	Wall Thickness	All Vessels of Each Type	
			Approximate Total Cross-Sectional Area (cm²)	Percentage of Blood Volume Contained[1]
Aorta	2.5 cm	2 mm	4.5	2
Artery	0.4 cm	1 mm	20	8
Arteriole	30 µm	20 µm	400	1
Capillary	5 µm	1 µm	4500	5
Venule	20 µm	2 µm	4000 ⎤	
Vein	0.5 cm	0.5 mm	40 ⎬	54
Vena cava	3 cm	1.5 mm	18 ⎦	

[1]In systemic vessels. There is an additional 12% in the heart and 18% in the pulmonary circulation.

smooth muscle **precapillary sphincters.** It is unsettled whether the metarterioles are innervated, and it appears that the precapillary sphincters are not. However, they can of course respond to local or circulating vasoconstrictor substances. The true capillaries are about 5 µm in diameter at the arterial end and 9 µm in diameter at the venous end. When the sphincters are dilated, the diameter of the capillaries is just sufficient to permit red blood cells to squeeze through in "single file." As they pass through the capillaries, the red cells become thimble- or parachute-shaped, with the flow pushing the center ahead of the edges. This configuration appears to be due simply to the pressure in the center of the vessel whether or not the edges of the red blood cell are in contact with the capillary walls.

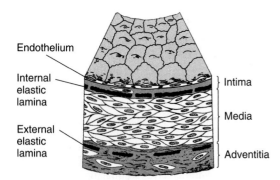

Endothelium

Internal elastic lamina

External elastic lamina

Intima

Media

Adventitia

Figure 30–2. Structure of normal muscle artery. (Reproduced, with permission, from Ross R, Glomset JA: The pathogenesis of atherosclerosis. N Engl J Med 1976;295:369.)

The total area of all the capillary walls in the body exceeds 6300 m² in the adult. The walls, which are about 1 µm thick, are made up of a single layer of endothelial cells. The structure of the walls varies from organ to organ. In many beds, including those in skeletal, cardiac, and smooth muscle, the junctions between the endothelial cells (Figure 30–4) permit the passage of molecules up to 10 nm in diameter. It also appears that plasma and its dissolved proteins are taken up by endocytosis, transported across the endothelial cells, and discharged by exocytosis (**vesicular transport;** see Chapter 1). However, this process can account for only a small portion of the transport across the endothelium. In the brain, the capillaries resemble the capillaries in muscle, but the junctions between endothelial cells are tighter, and transport across them is largely limited to small molecules. In most endocrine glands, the intestinal villi, and parts of the kidneys, the cytoplasm of the endothelial cells is attenuated to form gaps called **fenestrations.** These fenestrations are 20–100 nm in diameter. They permit the passage of relatively large molecules and make the capillaries porous. Except in the renal glomeruli, they appear to be closed by a thin membrane. However, in a number of different tissues, the membrane can be shown by a rapid freeze-fracture technique to be discontinuous, consisting of a central hub joined by spokes of membrane to the edges of the fenestration (Figure 30–5). In the liver, where the sinusoidal capillaries are extremely porous, the endothelium is discontinuous and there are large gaps between endothelial cells that are not closed by membranes (see Figure 26–20). Some of the gaps are 600 nm in diameter, and others may be as large as 3000 nm. The permeabilities of capillaries in various parts of the body, expressed in

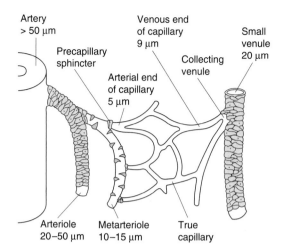

Figure 30–3. The microcirculation. Arterioles give rise to metarterioles, which give rise to capillaries. The capillaries drain via short collecting venules to the venules. The walls of the arteries, arterioles, and small venules contain relatively large amounts of smooth muscle. There are scattered smooth muscle cells in the walls of the metarterioles, and the openings of the capillaries are guarded by muscular precapillary sphincters. The diameters of the various vessels are also shown. (Courtesy of JN Diana.)

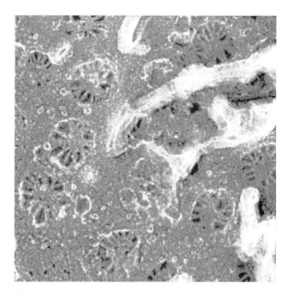

Figure 30–5. Fenestrations in capillaries in pancreatic islets. The fenestrations are the rosette-shaped structures in this quick-frozen deep-etched face view (× 64,000). (Reproduced, with permission, from Orci L: The insulin cell: Its cellular environment and how it processes (pro)insulin. Diabetes Metab Rev 1986;2:71.)

terms of their hydraulic conductivity, are summarized in Table 30–2.

Capillaries and postcapillary venules have **pericytes** outside the endothelial cells (Figure 30–4). These cells have long processes that wrap around the vessels. They

are contractile and release a wide variety of vasoactive agents. They also synthesize and release constituents of the basement membrane and extracellular matrix. One of their physiologic functions appears to be regulation of flow through the junctions between endothelial cells,

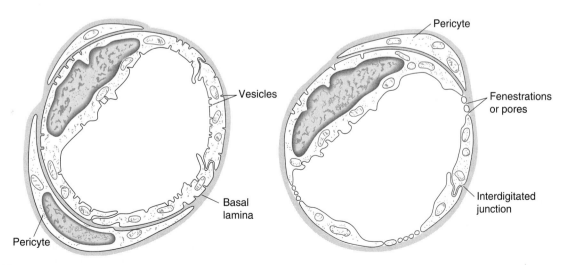

Figure 30–4. Cross-sections of capillaries. **Left:** Continuous type of capillary found in muscle. **Right:** Fenestrated type of capillary. (Reproduced, with permission, from Fawcett DW: *Bloom and Fawcett, Textbook of Histology*, 11th ed. Saunders, 1986.)

Table 30–2. Hydraulic conductivity of capillaries in various parts of the body.[1]

Organ	Conductivity[2]	Type of Endothelium
Brain (excluding circumventricular organs)	3	
Skin	100	Continuous
Skeletal muscle	250	
Lung	340	
Heart	860	
Gastrointestinal tract (intestinal mucosa)	13,000	Fenestrated
Glomerulus in kidney	15,000	

[1]Data courtesy of JN Diana.

[2]Units of conductivity are 10^{-13} cm^3 s^{-1} dyne^{-1}.

particularly in the presence of inflammation. They are closely related to the mesangial cells in the renal glomeruli (see Chapter 38).

Lymphatics

The lymphatics drain from the lungs and from the rest of the body tissues via a system of vessels that coalesce and eventually enter the right and left subclavian veins at their junctions with the respective internal jugular veins. The lymph vessels contain valves and regularly traverse lymph nodes along their course. The ultrastructure of the small lymph vessels differs from that of the capillaries in several details: There are no visible fenestrations in the lymphatic endothelium; there is very little if any basal lamina under the endothelium; and the junctions between endothelial cells are open, with no tight intercellular connections.

Arteriovenous Anastomoses

In the fingers, palms, and ear lobes of humans and the paws, ears, and other tissues of some animals, there are short channels that connect arterioles to venules, bypassing the capillaries. These **arteriovenous (A-V) anastomoses,** or **shunts,** have thick, muscular walls and are abundantly innervated, presumably by vasoconstrictor nerve fibers.

Venules & Veins

The walls of the venules are only slightly thicker than those of the capillaries. The walls of the veins are also thin and easily distended. They contain relatively little smooth muscle, but considerable venoconstriction is produced by activity in the noradrenergic nerves to the veins and by circulating vasoconstrictors such as endothelins. Anyone who has had trouble making venipunctures has observed the marked local venospasm produced in superficial forearm veins by injury. Variations in venous tone are important in circulatory adjustments.

The intima of the limb veins is folded at intervals to form **venous valves** that prevent retrograde flow. The way these valves function was first demonstrated by William Harvey in the 17th century. There are no valves in the very small veins, the great veins, or the veins from the brain and viscera.

Endothelium

Located between the circulating blood and the media and adventitia of the blood vessels, the endothelial cells constitute a large and important organ. They respond to flow changes, stretch, a variety of circulating substances, and inflammatory mediators. They secrete growth regulators and vasoactive substances (see below and Chapter 31).

Vascular Smooth Muscle

The smooth muscle in blood vessel walls has been one of the most-studied forms of visceral smooth muscle because of its importance in the regulation of blood pressure and hypertension. The membranes of the muscle cells contain various types of K$^+$, Ca^{2+}, and Cl$^-$ channels. Contraction is produced primarily by the myosin light chain mechanism described in Chapter 3 and illustrated in Figure 30–6. However, vascular smooth muscle also undergoes the prolonged contractions that determine vascular tone. These may be due in part to the latch-bridge mechanism (see Chapter 3), but other factors also play a role. Some of the molecular mechanisms that appear to be involved in contraction and relaxation are shown in Figure 30–6.

Vascular smooth muscle cells provide an interesting example of the way high and low cytosolic Ca^{2+} can have different and even opposite effects (see Chapter 1). In these cells, influx of Ca^{2+} via voltage-gated Ca^{2+} channels produces a diffuse increase in cytosolic Ca^{2+} that initiates contraction. However, the Ca^{2+} influx also initiates Ca^{2+} release from the sarcoplasmic reticulum via ryanodine receptors (see Chapter 3), and the high local Ca^{2+} concentration produced by these Ca^{2+} sparks increases the activity of **Ca^{2+}**-activated K$^+$ channels in the cell membrane. These are also known as big K or **BK channels** because K$^+$ flows through them at a

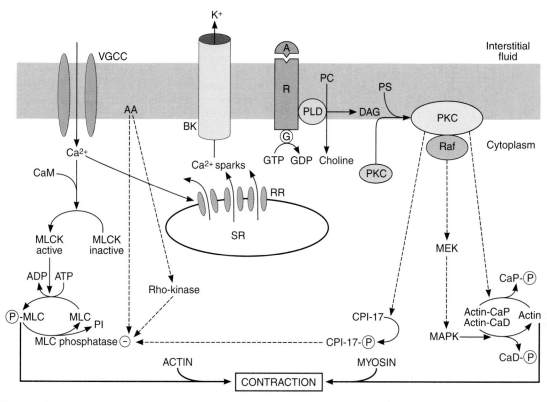

Figure 30–6. Some of the established and postulated mechanisms involved in the contraction and relaxation of vascular smooth muscle. A, agonist; AA, arachidonic acid; BK, Ca⁺-activated K⁺ channel; G, heterotrimeric G protein; MLC, myosin light chain; MLCK, myosin light chain kinase; PLD, phospholipase D; R, receptor; SF, sarcoplasmic reticulum; VGCC, voltage-gated Ca²⁺ channel; RR, ryanodine receptors. For other abbreviations, see Chapter 1 and Appendix. (Modified from Khahl R: Mechanisms of vascular smooth muscle contraction. Council for High Blood Pressure Newsletter, Spring 2001.)

high rate. The increased K⁺ efflux increases the membrane potential, shutting off voltage-gated Ca²⁺ channels and producing relaxation. The site of action of the Ca²⁺ sparks is the β_1-subunit of the BK channel, and mice in which this subunit is knocked out develop increased vascular tone and blood pressure. Obviously, therefore, the sensitivity of the β_1 subunit to Ca²⁺ sparks plays an important role in the control of vascular tone.

Angiogenesis

When tissues grow, blood vessels must proliferate if the tissue is to maintain a normal blood supply. Therefore, angiogenesis, the formation of new blood vessels, is important during fetal life and growth to adulthood. It is also important in adulthood for processes such as wound-healing, formation of the corpus luteum after

ovulation, and formation of new endometrium after menstruation. Abnormally, it is important in tumor growth; if tumors do not develop a blood supply, they do not grow.

During embryonic development, a network of leaky capillaries is formed in tissues from angioblasts: this process is sometimes called **vasculogenesis.** Vessels then branch off of nearby vessels, hook up with the capillaries, and provide them with smooth muscle, which brings about their maturation. Angiogenesis in adults is presumably similar.

Many factors are involved in angiogenesis. A key compound is the protein growth factor **vascular endothelial growth factor (VEGF).** This factor exists in multiple isoforms, and there are at least three VEGF receptors, two of which are known to be tyrosine kinases. VEGF appears to be primarily responsible for vasculogenesis, whereas the budding of vessels which connect to the immature capillary network is regulated by other as yet unidentified factors.

Some of the VEGF isoforms and some of its receptors seem to be concerned primarily with the formation of lymphatic vessels (**lymphangiogenesis**) rather than blood vessels. Most of the vascular growth factors act on all tissues. However, a VEGF produced by endocrine glands and acting only on the blood vessels of these glands has recently been described. The actions of VEGF and the other factors are complex, and there is much still to be learned about the factors involved in angiogenesis and the way they work together to regulate blood and lymph vessel formation in living animals.

BIOPHYSICAL CONSIDERATIONS

Flow, Pressure, & Resistance

Blood always flows, of course, from areas of high pressure to areas of low pressure, except in certain situations when momentum transiently sustains flow (see Figure 29–3). The relationship between mean flow, mean pressure, and resistance in the blood vessels is analogous in a general way to the relationship between the current, electromotive force, and resistance in an electrical circuit expressed in Ohm's law:

$$\text{Current (I)} = \frac{\text{Electromotive force (E)}}{\text{Resistance (R)}}$$

$$\text{Flow (F)} = \frac{\text{Pressure (P)}}{\text{Resistance (R)}}$$

Flow in any portion of the vascular system is equal to the **effective perfusion pressure** in that portion divided by the **resistance.** The effective perfusion pressure is the mean intraluminal pressure at the arterial end minus the mean pressure at the venous end. The units of resistance (pressure divided by flow) are dyne·s/cm^5. To avoid dealing with such complex units, resistance in the cardiovascular system is sometimes expressed in **R units,** which are obtained by dividing pressure in mm Hg by flow in mL/s (see also Table 32–1). Thus, for example, when the mean aortic pressure is 90 mm Hg and the left ventricular output is 90 mL/s, the total peripheral resistance is

$$\frac{90 \text{ mm Hg}}{90 \text{ mL/s}} = 1 \text{ R unit}$$

Methods for Measuring Blood Flow

Blood flow can be measured by cannulating a blood vessel, but this has obvious limitations. Various devices have been developed to measure flow in a blood vessel

without opening it. **Electromagnetic flow meters** depend on the principle that a voltage is generated in a conductor moving through a magnetic field and that the magnitude of the voltage is proportionate to the speed of movement. Since blood is a conductor, a magnet is placed around the vessel, and the voltage, which is proportionate to the volume flow, is measured with an appropriately placed electrode on the surface of the vessel. Blood flow velocity can be measured with **Doppler flow meters.** Ultrasonic waves are sent into a vessel diagonally from one crystal, and the waves reflected from the red and white blood cells are picked up by a second, downstream crystal. The frequency of the reflected waves is higher by an amount that is proportionate to the rate of flow toward the second crystal because of the Doppler effect.

Indirect methods for measuring the blood flow of various organs in humans include adaptations of the Fick and indicator dilution techniques described in Chapter 29. One example is the use of the Kety N_2O method for measuring cerebral blood flow (see Chapter 32). Another is determination of the renal blood flow by measuring the clearance of para-aminohippuric acid (see Chapter 38). A considerable amount of data on blood flow in the extremities has been obtained by **plethysmography** (Figure 30–7). The forearm, for example, is sealed in a watertight chamber (**plethysmograph**). Changes in the volume of the forearm, reflecting changes in the amount of blood and interstitial fluid it contains, displace the water, and this displacement is measured with a volume recorder. When the venous drainage of the forearm is occluded, the rate of increase

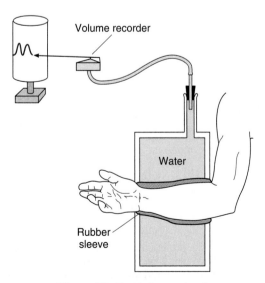

Figure 30–7. Plethysmography.

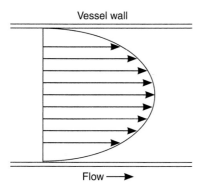

Figure 30–8. Diagram of the velocities of concentric laminas of a viscous fluid flowing in a tube, illustrating the parabolic distribution of velocities (streamline flow).

in the volume of the forearm is a function of the arterial blood flow (**venous occlusion plethysmography**).

Applicability of Physical Principles to Flow in Blood Vessels

Physical principles and equations that are applicable to the description of the behavior of perfect fluids in rigid tubes have often been used indiscriminately to explain the behavior of blood in blood vessels. Blood vessels are not rigid tubes, and the blood is not a perfect fluid but a two-phase system of liquid and cells. Therefore, the behavior of the circulation deviates, sometimes markedly, from that predicted by these principles. However, the physical principles are of value when used as an aid to understanding what goes on in the body rather than as an end in themselves or as a test of the memorizing ability of students.

Laminar Flow

The flow of blood in straight blood vessels, like the flow of liquids in narrow rigid tubes, is normally **laminar (streamline).** Within the blood vessels, an infinitely thin layer of blood in contact with the wall of the vessel does not move. The next layer within the vessel has a low velocity, the next a higher velocity, and so forth, velocity being greatest in the center of the stream (Figure 30–8). Laminar flow occurs at velocities up to a certain **critical velocity.** At or above this velocity, flow is turbulent. Streamline flow is silent, but turbulent flow creates sounds.

The probability of turbulence is also related to the diameter of the vessel and the viscosity of the blood. This probability can be expressed by the ratio of inertial to viscous forces as follows:

$$Re = \frac{\rho D \dot{V}}{\eta}$$

where Re is the Reynolds number, named for the man who described the relationship; ρ is the density of the fluid; D is the diameter of the tube under consideration; $\dot{V}$ is the velocity of the flow; and η is the viscosity of the fluid. The higher the value of Re, the greater the probability of turbulence. When D is in cm, $\dot{V}$ is in cm/s^{-1}, and η is in poises, flow is usually not turbulent if Re is less than 2000. When Re is more than 3000, turbulence is almost always present. Laminar flow is disturbed at branching of arteries, but normally not to the point that turbulence is produced. Constriction of an artery increases the velocity of blood flow through the constriction, producing turbulence, and consequently sound, beyond the constriction (Figure 30–9). Examples are bruits heard over arteries constricted by atherosclerotic plaques and the sounds of Korotkoff heard when measuring blood pressure (see below).

In humans, the critical velocity is sometimes exceeded in the ascending aorta at the peak of systolic ejection, but it is usually exceeded only when an artery is constricted. Turbulence occurs more frequently in anemia because the viscosity of the blood is lower. This may be the explanation of the systolic murmurs that are common in anemia.

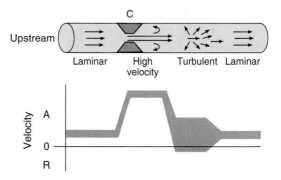

Figure 30–9. Top: Effect of constriction (C) on the profile of velocities in a blood vessel. The arrows indicate direction of velocity components, and their length is proportionate to their magnitude. **Bottom:** Range of velocities at each point along the vessel. In the area of turbulence, there are many different anterograde (A) and some retrograde (R) velocities. (Modified and reproduced, with permission, from Richards KE: Doppler echocardiography in diagnosis and quantification of vascular disease. Mod Concepts Cardiovasc Dis 1987;56:43. By permission of the American Heart Association, Inc.)

Shear Stress & Gene Activation

Flowing blood creates a force on the endothelium that is parallel to the long axis of the vessel. This **shear stress** (γ) is proportionate to viscosity (η) times the shear rate (dy/dr), which is the rate at which the axial velocity increases from the vessel wall toward the lumen.

$$\gamma = \eta \, (dy/dr)$$

Change in shear stress and other physical variables such as cyclic strain and stretch produce marked changes in the expression of genes in the endothelial cells that are related to cardiovascular function. The receptors are probably integrins attached to the cytoskeleton of the cells. The second messengers are IP_3, DAG, and components of the MAP kinase pathways (see Chapter 1), and the genes that are activated are those that produce growth factors, integrins, and related molecules (Table 30–3). Over 15 endothelial cell genes have been shown to be activated by various physical forces.

Average Velocity

When considering flow in a system of tubes, it is important to distinguish between velocity, which is displacement per unit time (eg, cm/s), and flow, which is volume per unit time (eg, cm³/s). Velocity ($\dot{V}$) is proportionate to flow (Q) divided by the area of the conduit (A):

$$\dot{V} = \frac{Q}{A}$$

Therefore, $Q = A\dot{V}$, and if flow stays constant, velocity increases in direct proportion to any decrease in A (Figure 30–9).

The average velocity of fluid movement at any point in a system of tubes in parallel is inversely proportionate to the *total* cross-sectional area at that point. Therefore, the average velocity of the blood is high in the aorta, declines steadily in the smaller vessels, and is lowest in the capillaries, which have 1000 times the *total* cross-sectional area of the aorta (Table 30–1). The average velocity of blood flow increases again as the blood enters the veins and is relatively high in the vena cava, although not so high as in the aorta. Clinically, the velocity of the circulation can be measured by injecting a bile salt preparation into an arm vein and timing the first appearance of the bitter taste it produces (Figure 30–10). The average normal arm-to-tongue **circulation time** is 15 seconds.

Poiseuille-Hagen Formula

The relation between the flow in a long narrow tube, the viscosity of the fluid, and the radius of the tube is

Table 30–3. Genes in human, bovine, and rabbit endothelial cells that are affected by shear stress, and transcription factors involved.[1,2]

Gene	Transcription Factors
Endothelin-1	AP-1
VCAM-1	AP-1, NF-$_\kappa$B
ACE	SSRE, AP-1, Egr-1
Tissue factor	SP1
Tissue factor	Egr-1
TM	AP-1
PDGF-α	SSRE, Egr-1
PDGF-β	SSRE
ICAM-1	SSRE, AP-1, NF-$_\kappa$B
TGF-β	SSRE, AP-1, NF-$_\kappa$B
Egr-1	SREs
c-fos	SSRE
c-jun	SSRE, AP-1
NOS 3	SSRE, AP-1, NF-$_\kappa$B
MCP-1	SSRE, AP-1, NF-$_\kappa$B

[1]Modified from Braddock M et al: Fluid shear stress modulation of gene expression in endothelial cells. News Physiol Sci 1998;13:241.

[2]Acronyms are expanded in the Appendix.

expressed mathematically in the **Poiseuille-Hagen formula:**

$$F = (P_A - P_B) \times \left(\frac{\pi}{8}\right) \times \left(\frac{1}{\eta}\right) \times \left(\frac{r^4}{L}\right)$$

where

$$
\begin{aligned}
F &= \text{flow} \\
P_A - P_B &= \text{pressure difference between the two ends} \\
&\quad \text{of the tube} \\
\eta &= \text{viscosity} \\
r &= \text{radius of tube} \\
L &= \text{length of tube}
\end{aligned}
$$

Since flow is equal to pressure difference divided by resistance (R),

$$R = \frac{8\eta L}{\pi r^4}$$

Since flow varies directly and resistance inversely with the fourth power of the radius, blood flow and resistance in vivo are markedly affected by small changes in the caliber of the vessels. Thus, for example, flow through a vessel is doubled by an increase of only 19% in its radius; and when the radius is doubled, resistance is reduced to 6% of its previous value. This is why organ blood flow is so effectively regulated by small changes in the caliber of the arterioles and why varia-

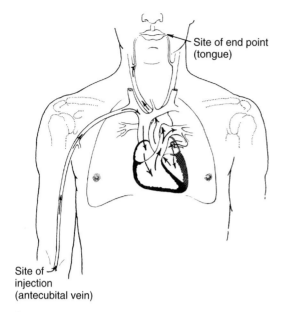

Figure 30–10. Pathway traversed by the injected material when the arm-to-tongue circulation time is measured.

Viscosity & Resistance

The resistance to blood flow is determined not only by the radius of the blood vessels (**vascular hindrance**) but also by the viscosity of the blood. Plasma is about 1.8 times as viscous as water, whereas whole blood is 3–4 times as viscous as water. Thus viscosity depends for the most part on the **hematocrit,** ie, the percentage of the volume of blood occupied by red blood cells. The effect of viscosity in vivo deviates from that predicted by the Poiseuille-Hagen formula. In large vessels, increases in hematocrit cause appreciable increases in viscosity. However, in vessels smaller than 100 μm in diameter, ie, in arterioles, capillaries, and venules, the viscosity change per unit change in hematocrit is much less than it is in large-bore vessels. This is due to a difference in the nature of flow through the small vessels. Therefore, the net change in viscosity per unit change in hematocrit is considerably smaller in the body than it is in vitro (Figure 30–11). This is why hematocrit changes have relatively little effect on the peripheral resistance except when the changes are large. In severe polycythemia, the increase in resistance does increase the work of the heart. Conversely, in anemia, peripheral resistance is decreased, in part because of the de-

cline in viscosity. Of course, the decrease in hemoglobin decreases the O_2-carrying ability of the blood, but the improved blood flow due to the decrease in viscosity partially compensates for this.

Viscosity is also affected by the composition of the plasma and the resistance of the cells to deformation. Clinically significant increases in viscosity are seen in diseases in which plasma proteins such as the immunoglobulins are markedly elevated and in diseases such as hereditary spherocytosis, in which the red blood cells are abnormally rigid.

Critical Closing Pressure

In rigid tubes the relation between pressure and flow of homogeneous fluids is linear, but in thin-walled blood vessels in vivo it is not. When the pressure in a small blood vessel is reduced, a point is reached at which there is no flow of blood even though the pressure is not zero (Figure 30–12). The vessels are surrounded by tissues that exert a small but definite pressure on the vessels, and when the intraluminal pressure falls below the tissue pressure, the vessels collapse. In inactive tissues, for example, the pressure in many capillaries is low because the precapillary sphincters and metarterioles are constricted, and many of these capillaries are

tions in arteriolar diameter have such a pronounced effect on systemic arterial pressure.

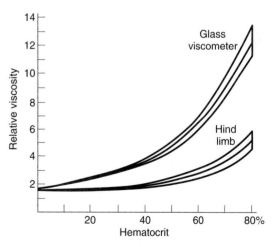

Figure 30–11. Effect of changes in hematocrit on the relative viscosity of blood measured in a glass viscometer and in the hind leg of a dog. In each case, the middle line represents the mean and the upper and lower lines the standard deviation. (Reproduced, with permission, from Whittaker SRF, Winton FR: The apparent viscosity of blood flowing in the isolated hind limb of the dog, and its variation with corpuscular concentration. J Physiol [Lond] 1933;78:338.)

collapsed. The pressure at which flow ceases is called the **critical closing pressure.**

Law of Laplace

The relation between distending pressure and tension is shown diagrammatically in Figure 30–13. It is perhaps surprising that structures as thin-walled and delicate as the capillaries are not more prone to rupture. The principal reason for their relative invulnerability is their small diameter. The protective effect of small size in this case is an example of the operation of the **law of Laplace,** an important physical principle with several other applications in physiology. This law states that tension in the wall of a cylinder (T) is equal to the product of the transmural pressure (P) and the radius (r) divided by the wall thickness (w).

$$T = Pr/w$$

The **transmural pressure** is the pressure inside the cylinder minus the pressure outside the cylinder, but since tissue pressure in the body is low, it can generally be ignored and P equated to the pressure inside the viscus. In a thin-walled viscus, w is very small and it too can be ignored, but it becomes a significant factor in vessels such as arteries. Therefore, in a thin-walled viscus, P = T divided by the two principal radii of curvature of the viscus.

$$P = T\left(\frac{1}{r_1} + \frac{1}{r_2}\right)$$

In a sphere, $r_1 = r_2$, so

$$P = \frac{2T}{r}$$

In a cylinder such as a blood vessel, one radius is infinite, so

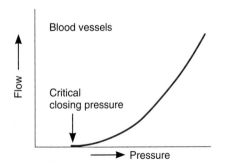

Figure 30–12. Relation of pressure to flow in thin-walled blood vessel.

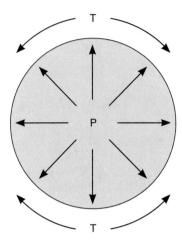

Figure 30–13. Relation between distending pressure (P) and wall tension (T) in a hollow viscus.

$$P = \frac{T}{r}$$

Consequently, the smaller the radius of a blood vessel, the lower the tension in the wall necessary to balance the distending pressure. In the human aorta, for example, the tension at normal pressures is about 170,000 dynes/cm, and in the vena cava it is about 21,000 dynes/cm; but in the capillaries, it is approximately 16 dynes/cm.

The law of Laplace also makes clear a disadvantage faced by dilated hearts. When the radius of a cardiac chamber is increased, a greater tension must be developed in the myocardium to produce any given pressure; consequently, a dilated heart must do more work than a nondilated heart. In the lungs, the radii of curvature of the alveoli become smaller during expiration, and these structures would tend to collapse because of the pull of surface tension if the tension were not reduced by the surface-tension-lowering agent, surfactant (see Chapter 34). Another example of the operation of this law is seen in the urinary bladder (see Chapter 38).

Resistance & Capacitance Vessels

When a segment of the vena cava or another large distensible vein is filled with blood, the pressure does not rise rapidly until large volumes of fluid are injected (Figure 30–12). In vivo, the veins are an important blood reservoir. Normally they are partially collapsed and oval in cross section. A large amount of blood can be added to the venous system before the veins become distended to the point where further increments in volume produce a large rise in venous pressure. The veins are therefore called **capacitance vessels.** The small arteries and arterioles are referred to as **resistance vessels** because they are the principal site of the peripheral resistance (see below).

At rest, at least 50% of the circulating blood volume is in the systemic veins. Twelve percent is in the heart cavities, and 18% is in the low-pressure pulmonary circulation. Only 2% is in the aorta, 8% in the arteries, 1% in the arterioles, and 5% in the capillaries (Table 30–1). When extra blood is administered by transfusion, less than 1% of it is distributed in the arterial system (the **"high-pressure system"**), and all the rest is found in the systemic veins, pulmonary circulation, and heart chambers other than the left ventricle (the **"low-pressure system"**).

ARTERIAL & ARTERIOLAR CIRCULATION

The pressure and velocities of the blood in the various parts of the systemic circulation are summarized in Figure 30–14. The general relationships in the pulmonary circulation are similar, but the pressure in the pulmonary artery is 25/10 mm Hg or less.

Velocity & Flow of Blood

Although the mean velocity of the blood in the proximal portion of the aorta is 40 cm/s, the flow is phasic, and velocity ranges from 120 cm/s during systole to a negative value at the time of the transient backflow before the aortic valve closes in diastole. In the distal portions of the aorta and in the large arteries, velocity is also greater in systole than it is in diastole. However, the vessels are elastic, and forward flow is continuous because of the recoil during diastole of the vessel walls that have been stretched during systole (Figure 30–15). This recoil effect is sometimes called the **Windkessel effect,** and the vessels are called Windkessel vessels; Windkessel is the German word for an elastic reservoir. Pulsatile flow appears, in some poorly understood way, to maintain optimal function of the tissues. If an organ is perfused with a pump that delivers a nonpulsatile flow, there is a gradual rise in vascular resistance, and tissue perfusion fails.

Arterial Pressure

The pressure in the aorta and in the brachial and other large arteries in a young adult human rises to a peak value **(systolic pressure)** of about 120 mm Hg during each heart cycle and falls to a minimum value **(diastolic pressure)** of about 70 mm Hg. The arterial pressure is conventionally written as systolic pressure over diastolic pressure—eg, 120/70 mm Hg. One millime-

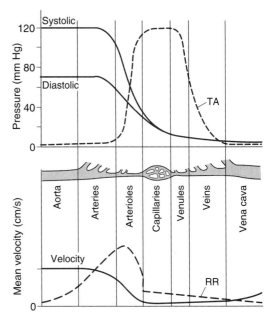

Figure 30–14. Diagram of the changes in pressure and velocity as blood flows through the systemic circulation. TA, total cross-sectional area of the vessels, which increases from 4.5 cm² in the aorta to 4500 cm² in the capillaries (Table 30–1). RR, relative resistance, which is highest in the arterioles.

ter of mercury equals 0.133 kPa, so in SI units (see Appendix) this value is 16.0/9.3 kPa. The **pulse pressure,** the difference between the systolic and diastolic pressures, is normally about 50 mm Hg. The **mean pressure** is the average pressure throughout the cardiac cycle. Because systole is shorter than diastole, the mean pressure is slightly less than the value halfway between systolic and diastolic pressure. It can actually be determined only by integrating the area of the pressure curve (Figure 30–16); however, as an approximation, mean pressure equals the diastolic pressure plus one-third of the pulse pressure.

The pressure falls very slightly in the large and medium-sized arteries because their resistance to flow is small, but it falls rapidly in the small arteries and arterioles, which are the main sites of the peripheral resistance against which the heart pumps. The mean pressure at the end of the arterioles is 30–38 mm Hg. Pulse pressure also declines rapidly to about 5 mm Hg at the ends of the arterioles (Figure 30–13). The magnitude of the pressure drop along the arterioles varies considerably depending upon whether they are constricted or dilated.

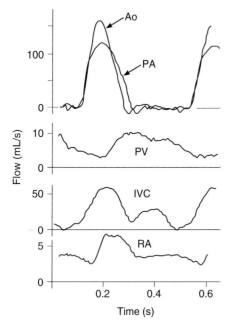

Figure 30–15. Changes in blood flow during the cardiac cycle in the dog. Diastole is followed by systole starting at 0.1 and again at 0.5 s. Flow patterns in humans are similar. Ao, aorta; PA, pulmonary artery; PV, pulmonary vein; IVC, inferior vena cava; RA, renal artery. (Reproduced, with permission, from Milnor WR: Pulsatile blood flow. N Engl J Med 1972;287:27.)

Effect of Gravity

The pressures in Figure 30–14 are those in blood vessels at heart level. The pressure in any vessel below heart level is increased and that in any vessel above heart level is decreased by the effect of gravity. The magnitude of the gravitational effect—the product of the density of the blood, the acceleration due to gravity (980 cm/s/s), and the vertical distance above or below the heart—is 0.77 mm Hg/cm at the density of normal blood. Thus, in an adult human in the upright position, when the mean arterial pressure at heart level is 100 mm Hg, the mean pressure in a large artery in the head (50 cm above the heart) is 62 mm Hg (100 − [0.77 × 50]) and the pressure in a large artery in the foot (105 cm below the heart) is 180 mm Hg (100 + [0.77 × 105]) (Figure 30–17). The effect of gravity on venous pressure is similar (see below).

Methods of Measuring Blood Pressure

If a cannula is inserted into an artery, the arterial pressure can be measured directly with a mercury manometer or a suitably calibrated strain gauge and an oscillograph arranged to write directly on a moving strip of paper. When an artery is tied off beyond the point at which the cannula is inserted, an **end pressure** is recorded. Flow in the artery is interrupted, and all the kinetic energy of flow is converted into pressure energy. If, alternatively, a T tube is inserted into a vessel and the pressure is measured in the side arm of the tube, the recorded **side pressure** under conditions where pressure drop due to resistance is negligible is lower than the end pressure by the kinetic energy of flow. This is because in a tube or a blood vessel the total energy—the sum of the kinetic energy of flow and the pressure energy—is constant (**Bernoulli's principle**).

It is worth noting that the pressure drop in any segment of the arterial system is due both to resistance and to conversion of potential into kinetic energy. The pressure drop due to energy lost in overcoming resistance is irreversible, since the energy is dissipated as heat; but the pressure drop due to conversion of potential to kinetic energy as a vessel narrows is reversed when the vessel widens out again (Figure 30–18).

Bernoulli's principle also has a significant application in pathophysiology. According to the principle, the greater the velocity of flow in a vessel, the lower the lateral pressure distending its walls. When a vessel is narrowed, the velocity of flow in the narrowed portion increases and the distending pressure decreases. Therefore, when a vessel is narrowed by a pathologic process such as an atherosclerotic plaque, the lateral pressure at the constriction is decreased and the narrowing tends to maintain itself.

Auscultatory Method

The arterial blood pressure in humans is routinely measured by the **auscultatory method.** An inflatable cuff (**Riva-Rocci cuff**) attached to a mercury manometer

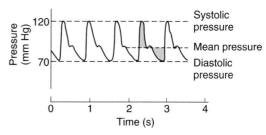

Figure 30–16. Brachial artery pressure curve of a normal young human, showing the relation of systolic and diastolic pressure to mean pressure. The shaded area above the mean pressure line is equal to the shaded area below it.

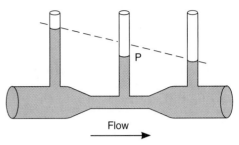

Figure 30–18. Bernoulli's principle. When fluid flows through the narrow portion of the tube, the kinetic energy of flow is increased as the velocity increases, and the potential energy is reduced. Consequently, the measured pressure (P) is lower than it would have been at that point if the tube had not been narrowed. The dashed line indicates what the pressure drop due to frictional forces would have been if the tube had been of uniform diameter.

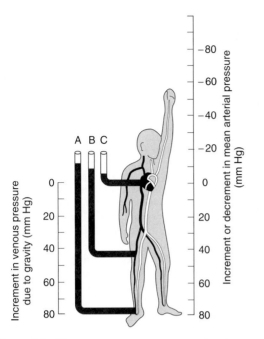

Figure 30–17. Effects of gravity on arterial and venous pressure. The scale on the right indicates the increment (or decrement) in mean pressure in a large artery at each level. The mean pressure in all large arteries is approximately 100 mm Hg when they are at the level of the left ventricle. The scale on the left indicates the increment in venous pressure at each level due to gravity. The manometers on the left of the figure indicate the height to which a column of blood in a tube would rise if connected to an ankle vein (A), the femoral vein (B), or the right atrium (C), with the subject in the standing position. The approximate pressures in these locations in the recumbent position—ie, when the ankle, thigh, and right atrium are at the same level—are A, 10 mm Hg; B, 7.5 mm Hg; and C, 4.6 mm Hg.

the sounds become louder, then dull and muffled. Finally, in most individuals, they disappear. These are the **sounds of Korotkoff.** When direct and indirect blood pressure measurements are made simultaneously, the diastolic pressure in resting adults correlates best with the pressure at which the sound disappears. However, in adults after exercise and in children, the diastolic pressure correlates best with the pressure at which the sounds become muffled. This is also true in diseases such as hyperthyroidism and aortic insufficiency.

(**sphygmomanometer**) is wrapped around the arm and a stethoscope is placed over the brachial artery at the elbow (Figure 30–19). The cuff is rapidly inflated until the pressure in it is well above the expected systolic pressure in the brachial artery. The artery is occluded by the cuff, and no sound is heard with the stethoscope. The pressure in the cuff is then lowered slowly. At the point at which systolic pressure in the artery just exceeds the cuff pressure, a spurt of blood passes through with each heartbeat and, synchronously with each beat, a tapping sound is heard below the cuff. The cuff pressure at which the sounds are first heard is the systolic pressure. As the cuff pressure is lowered further,

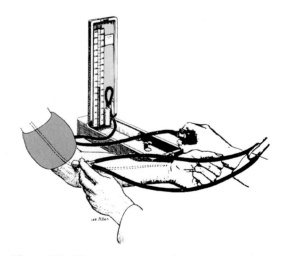

Figure 30–19. Determination of blood pressure by the auscultatory method. (Reproduced, with permission, from Schottelius BA, Schottelius D: Textbook of Physiology, 18th ed. Mosby, 1978.)

The sounds of Korotkoff are produced by turbulent flow in the brachial artery. The streamline flow in the unconstricted artery is silent, but when the artery is narrowed, the velocity of flow through the constriction exceeds the **critical velocity** and turbulent flow results (Figure 30–9). At cuff pressures just below the systolic pressure, flow through the artery occurs only at the peak of systole, and the intermittent turbulence produces a tapping sound. As long as the pressure in the cuff is above the diastolic pressure in the artery, flow is interrupted at least during part of diastole, and the intermittent sounds have a staccato quality. When the cuff pressure is near the arterial diastolic pressure, the vessel is still constricted, but the turbulent flow is continuous. Continuous sounds have a muffled rather than a staccato quality.

The auscultatory method is accurate when used properly, but a number of precautions must be observed. The cuff must be at heart level to obtain a pressure that is uninfluenced by gravity. The blood pressure in the thighs can be measured with the cuff around the thigh and the stethoscope over the popliteal artery, but there is more tissue between the cuff and the artery in the leg than there is in the arm, and some of the cuff pressure is dissipated. Therefore, pressures obtained by using the standard arm cuff are falsely high. The same thing is true when brachial arterial pressures are measured in individuals with obese arms, because the blanket of fat dissipates some of the cuff pressure. In both situations, accurate pressures can be obtained by using a cuff that is wider than the standard arm cuff. If the cuff is left inflated for some time, the discomfort may cause generalized reflex vasoconstriction, raising the blood pressure. It is always wise to compare the blood pressure in both arms when examining an individual for the first time. Persistent major differences between the pressure on the two sides indicate the presence of vascular obstruction.

Automated machines employing the auscultatory or other methods are now routinely used for continuous monitoring of blood pressure in hospitals and in the home.

Palpation Method

The systolic pressure can be determined by inflating an arm cuff and then letting the pressure fall and determining the pressure at which the radial pulse first becomes palpable. Because of the difficulty in determining exactly when the first beat is felt, pressures obtained by this **palpation method** are usually 2–5 mm Hg lower than those measured by the auscultatory method.

It is wise to form a habit of palpating the radial pulse while inflating the blood pressure cuff during measurement of the blood pressure by the auscultatory method. When the cuff pressure is lowered, the sounds of Korotkoff sometimes disappear at pressures well above diastolic pressure, then reappear at lower pressures ("auscultatory gap"). If the cuff is initially inflated until the radial pulse disappears, the examiner can be sure that the cuff pressure is above systolic pressure, and falsely low pressure values will be avoided.

Normal Arterial Blood Pressure

The blood pressure in the brachial artery in young adults in the sitting or lying position at rest is approximately 120/70 mm Hg (Figure 30–20). Since the arterial pressure is the product of the cardiac output and the peripheral resistance, it is affected by conditions that affect either or both of these factors. Emotion increases the cardiac output and peripheral resistance, and about 20% of hypertensive patients have blood pressures that are higher in the doctor's office than at home, going about their regular daily activities ("white coat hypertension"). Blood pressure normally falls up to 20 mg Hg during sleep. This fall is reduced or absent in hypertension. Consequently, normals are sometimes called "dippers" and hypertensives "nondippers."

There is general agreement that blood pressure rises with advancing age, but there has been uncertainty about the magnitude of this rise because hypertension is a common disease and its incidence increases with advancing age. One way to mitigate the problem is to select groups of young people with various systolic blood pressures and follow them prospectively over a period of years. The results of such a study are shown in Figure 30–20. Four groups of individuals were followed: those with initial systolic blood pressures < 120 mm Hg, those with pressures of 120–139 mm Hg, those with pressures of 140–159 mm Hg, and those with pressures > 160 mm Hg. There was a steady rise in systolic pressure in all groups, whether or not their pressures rose to hypertensive levels. Diastolic pressures also rose until 50–60 years of age, but after that they fell. Consequently, there was a marked increase in pulse pressure in elderly individuals. The increase in systolic pressure and pulse pressure in these individuals is due mainly to increased stiffness of the arteries.

It is interesting that systolic and diastolic blood pressures are lower in young women than in young men until age 55–65, after which they become comparable. Since there is a positive correlation between blood pressure and the incidence of heart attacks and strokes (see below), the lower blood pressure before menopause in women may be one reason that, on average, they live longer than men.

CAPILLARY CIRCULATION

At any one time, only 5% of the circulating blood is in the capillaries, but this 5% is in a sense the most important part of the blood volume because it is across the systemic capillary walls that O_2 and nutrients enter the interstitial fluid and CO_2 and waste products enter the bloodstream. The exchange across the capillary walls is essential to the survival of the tissues.

Methods of Study

It is difficult to obtain accurate measurements of capillary pressures and flows. The capillaries in the mesentery of experimental animals and the fingernail beds of humans are readily visible under the dissecting microscope, and observations on flow patterns under various conditions have been made in these and in some other tissues. Capillary pressure has been estimated by determining the amount of external pressure necessary to occlude the capillaries or the amount of pressure necessary to make saline start to flow through a micropipette inserted so that its tip faces the arteriolar end of the capillary.

Capillary Pressure & Flow

Capillary pressures vary considerably, but typical values in human nail bed capillaries are 32 mm Hg at the arteriolar end and 15 mm Hg at the venous end. The pulse pressure is approximately 5 mm Hg at the arteriolar end and zero at the venous end. The capillaries are short, but blood moves slowly (about 0.07 cm/s) because the total cross-sectional area of the capillary bed is large. Transit time from the arteriolar to the venular end of an average-sized capillary is 1–2 seconds.

Equilibration With Interstitial Fluid

As noted above, the capillary wall is a thin membrane made up of endothelial cells. Substances pass through

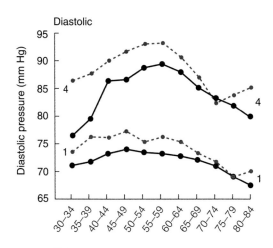
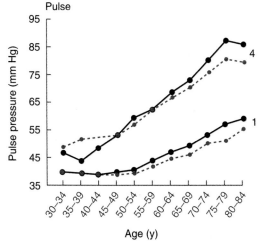

Figure 30–20. Effects of age and sex on arterial pressure components in humans. Four groups of individuals were selected on the basis of their systolic blood pressure and followed prospectively (see text). The values shown are those from group 1 (initial systolic pressure < 120 mm Hg) and group 4 (initial systolic pressure > 160 mm Hg). The solid black line shows the values for women, and the dashed colored line shows the values for men. (Modified and reproduced, with permission, from Franklin SS et al: Hemodynamic patterns of age-related changes in blood pressure: the Framingham Heart Study. Circulation 1997;96:308.)

the junctions between endothelial cells and through fenestrations, when they are present. Some also pass through the cells by vesicular transport or, in the case of lipid-soluble substances, through the cytoplasm.

The factors other than vesicular transport that are responsible for transport across the capillary wall are diffusion and filtration (see Chapter 1). Diffusion is quantitatively much more important in terms of the exchange of nutrients and waste materials between blood and tissue. O_2 and glucose are in higher concentration in the bloodstream than in the interstitial fluid and diffuse into the interstitial fluid, whereas CO_2 diffuses in the opposite direction.

The rate of filtration at any point along a capillary depends upon a balance of forces sometimes called the **Starling forces** after the physiologist who first described their operation in detail. One of these forces is the **hydrostatic pressure gradient** (the hydrostatic pressure in the capillary minus the hydrostatic pressure of the interstitial fluid) at that point. The interstitial fluid pressure varies from one organ to another, and there is considerable evidence that it is subatmospheric—about −2 mm Hg—in subcutaneous tissue. It is positive in the liver and kidneys and is as high as 6 mm Hg in the brain. The other force is the **osmotic pressure gradient** across the capillary wall (colloid osmotic pressure of plasma minus colloid osmotic pressure of interstitial fluid). This component is directed inward.

Thus:

$$\text{Fluid movement} = k[(P_c + \pi_i) - (P_i + \pi_c)]$$

where

$$
\begin{aligned}
k &= \text{capillary filtration coefficient} \\
P_c &= \text{capillary hydrostatic pressure} \\
P_i &= \text{interstitial hydrostatic pressure} \\
\pi_c &= \text{capillary colloid osmotic pressure} \\
\pi_i &= \text{interstitial colloid osmotic pressure}
\end{aligned}
$$

π_i is usually negligible, so the osmotic pressure gradient $(\pi_c - \pi_i)$ usually equals the oncotic pressure. The capillary filtration coefficient takes into account, and is proportionate to, the permeability of the capillary wall and the area available for filtration. The magnitude of the Starling forces along a typical muscle capillary is shown in Figure 30–21. Fluid moves into the interstitial space at the arteriolar end of the capillary, where the filtration pressure across its wall exceeds the oncotic pressure, and into the capillary at the venular end, where the oncotic pressure exceeds the filtration pressure. In other capillaries, the balance of Starling forces is different and, for example, fluid moves out of almost the entire length of the capillaries in the renal glomeruli. On the

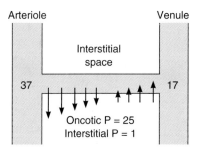

Figure 30–21. Schematic representation of pressure gradients across the wall of a muscle capillary. The numbers at the arteriolar and venular ends of the capillary are the hydrostatic pressures in mm Hg at these locations. The arrows indicate the approximate magnitude and direction of fluid movement. In this example, the pressure differential at the arteriolar end of the capillary is 11 mm Hg ([37 − 1] − 25) outward; at the opposite end, it is 9 mm Hg (25 − [17 − 1]) inward.

other hand, fluid moves into the capillaries through almost their entire length in the intestines.

It is worth noting that small molecules often equilibrate with the tissues near the arteriolar end of each capillary. In this situation, total diffusion can be increased by increasing blood flow; ie, exchange is **flow-limited** (Figure 30–22). Conversely, transfer of substances that do not reach equilibrium with the tissues during their passage through the capillaries is said to be **diffusion-limited.**

It has been estimated that about 24 L of fluid are filtered through the capillaries per day. This is about 0.3% of the cardiac output. About 85% of the filtered fluid is reabsorbed into the capillaries, and the remainder returns to the circulation via the lymphatics.

Active & Inactive Capillaries

In resting tissues, most of the capillaries are collapsed, and blood flows for the most part through the thoroughfare vessels from the arterioles to the venules. In active tissues, the metarterioles and the precapillary sphincters dilate. The intracapillary pressure rises, overcoming the critical closing pressure of the vessels, and blood flows through all the capillaries. Relaxation of the smooth muscle of the metarterioles and precapillary sphincters is due to the action of vasodilator metabolites formed in active tissue (see Chapter 31) and possibly also to a decrease in the activity of the sympathetic vasoconstrictor nerves that innervate the smooth muscle.

After noxious stimulation, substance P released by the axon reflex (see Chapter 32) increases capillary permeability. Bradykinin and histamine also increase capil-

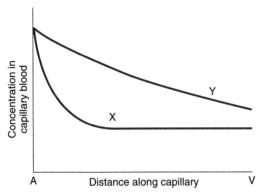

Figure 30–22. Flow-limited and diffusion-limited exchange across capillary walls. A and V indicate the arteriolar and venular ends of the capillary. Substance X equilibrates with the tissues (movement into the tissues equals movement out) well before the blood leaves the capillary, whereas substance Y does not equilibrate. If other factors stay constant, the amount of X entering the tissues can be increased only by increasing blood flow; ie, it is flow-limited. The movement of Y is diffusion-limited.

lary permeability. When capillaries are stimulated mechanically, they empty (white reaction; see Chapter 32), probably due to contraction of the precapillary sphincters.

LYMPHATIC CIRCULATION & INTERSTITIAL FLUID VOLUME

Lymphatic Circulation

Fluid efflux normally exceeds influx across the capillary walls, but the extra fluid enters the lymphatics and drains through them back into the blood. This keeps the interstitial fluid pressure from rising and promotes the turnover of tissue fluid. The normal 24-hour lymph flow is 2–4 L. The composition of lymph is discussed in Chapter 27.

Lymphatic vessels can be divided into two types: initial lymphatics and collecting lymphatics (Figure 30–23). The former lack valves and smooth muscle in their walls, and they are found in regions such as the intestine or skeletal muscle. Tissue fluid appears to enter them through loose junctions between the endothelial cells that form their walls. The fluid in them apparently is massaged by muscle contractions of the organs and contraction of arterioles and venules, with which they are often associated. They drain into the collecting lymphatics, which have valves and smooth muscle in their walls and contract in a peristaltic fashion, pro-

pelling the lymph along the vessels. Flow in the collecting lymphatics is further aided by movements of skeletal muscle, the negative intrathoracic pressure during inspiration, and the suction effect of high-velocity flow of blood in the veins in which the lymphatics terminate. However, the contractions are the principal factor propelling the lymph.

Agents that increase lymph flow are called **lymphagogues**. They include a variety of agents that increase capillary permeability. Agents that cause contraction of smooth muscle also increase lymph flow from the intestines.

Other Functions of the Lymphatic System

Appreciable quantities of protein enter the interstitial fluid in the liver and intestine, and smaller quantities enter from the blood in other tissues. The macromolecules enter the lymphatics, presumably at the junctions between the endothelial cells, and the proteins are returned to the bloodstream via the lymphatics. The amount of protein returned in this fashion in 1 day is equal to 25–50% of the total circulating plasma protein. The transport of absorbed long-chain fatty acids and cholesterol from the intestine via the lymphatics has been discussed in Chapter 25.

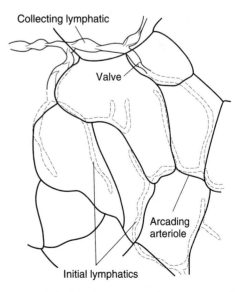

Figure 30–23. Initial lymphatics draining into collecting lymphatics in the mesentery. Note the close association with arcading arterioles, indicated by the single black lines. (Reproduced, with permission, from Schmid-Schönbein GW, Zeifach BW: Fluid pump mechanisms in initial lymphatics. News Physiol Sci 1994;9:67.)

Interstitial Fluid Volume

The amount of fluid in the interstitial spaces depends upon the capillary pressure, the interstitial fluid pressure, the oncotic pressure, the capillary filtration coefficient, the number of active capillaries, the lymph flow, and the total ECF volume. The ratio of precapillary to postcapillary venular resistance is also important. Precapillary constriction lowers filtration pressure, whereas postcapillary constriction raises it. Changes in any of these variables lead to changes in the volume of interstitial fluid. Factors promoting an increase in this volume are summarized in Table 30–4. **Edema** is the accumulation of interstitial fluid in abnormally large amounts.

In active tissues, capillary pressure rises, often to the point where it exceeds the oncotic pressure throughout the length of the capillary. In addition, osmotically active metabolites may temporarily accumulate in the interstitial fluid because they cannot be washed away as rapidly as they are formed. To the extent that they accumulate, they exert an osmotic effect that decreases the magnitude of the osmotic gradient due to the oncotic pressure. The amount of fluid leaving the capillaries is therefore markedly increased and the amount entering them reduced. Lymph flow is increased, decreasing the degree to which the fluid would otherwise accumulate, but exercising muscle, for example, still increases in volume by as much as 25%.

Interstitial fluid tends to accumulate in dependent parts because of the effect of gravity. In the upright position, the capillaries in the legs are protected from the high arterial pressure by the arterioles, but the high venous pressure is transmitted to them through the venules. Skeletal muscle contractions keep the venous pressure low by pumping blood toward the heart (see below) when the individual moves about; however, if one stands still for long periods, fluid accumulates and edema eventually develops. The ankles also swell during long trips when travelers sit for prolonged periods with their feet in a dependent position. Venous obstruction may contribute to the edema in these situations.

Whenever there is abnormal retention of salt in the body, water is also retained. The salt and water are distributed throughout the ECF, and since the interstitial fluid volume is therefore increased, there is a predisposition to edema. Salt and water retention is a factor in the edema seen in heart failure, nephrosis, and cirrhosis, but there are also variations in the mechanisms that govern fluid movement across the capillary walls in these diseases. In congestive heart failure, for example, there is usually an elevation in venous pressure, and capillary pressure is consequently elevated. In cirrhosis of the liver, oncotic pressure is low because hepatic synthesis of plasma proteins is depressed; and in nephrosis, oncotic pressure is low because large amounts of protein are lost in the urine.

Another cause of edema is inadequate lymphatic drainage. Edema caused by lymphatic obstruction is called **lymphedema,** and the edema fluid has a high protein content. If it persists, it causes a chronic inflammatory condition that leads to fibrosis of the interstitial tissue. One cause of lymphedema is radical mastectomy, an operation for cancer of the breast in which removal of the axillary lymph nodes on one side reduces lymph drainage. Edema of the arm on that side occurs in 10–30% of the patients. In filariasis, parasitic worms migrate into the lymphatics and obstruct them. Fluid accumulation plus tissue reaction lead in time to massive swelling, usually of the legs or scrotum (**elephantiasis**). The extent of the reaction is perhaps most graphically illustrated by the remarkable account of the man with elephantiasis whose scrotum was so edematous that he had to place it in a wheelbarrow and wheel it along with him when he walked.

An interesting and beneficial treatment for lymphedema is administration of benzopyrones. These drugs increase proteolysis by tissue macrophages, and reduction of the protein content of the edema fluid reduces inflammatory reactions and permits the fluid to be reabsorbed.

VENOUS CIRCULATION

Blood flows through the blood vessels, including the veins, primarily because of the pumping action of the

Table 30–4. Causes of increased interstitial fluid volume and edema.

Increased filtration pressure
  Arteriolar dilation
  Venular constriction
  Increased venous pressure (heart failure, incompetent valves, venous obstruction, increased total ECF volume, effect of gravity, etc)

Decreased osmotic pressure gradient across capillary
  Decreased plasma protein level
  Accumulation of osmotically active substances in interstitial space

Increased capillary permeability
  Substance P
  Histamine and related substances
  Kinins, etc

Inadequate lymph flow

heart. However, venous flow is aided by the heartbeat, the increase in the negative intrathoracic pressure during each inspiration, and contractions of skeletal muscles that compress the veins (**muscle pump**).

Venous Pressure & Flow

The pressure in the venules is 12–18 mm Hg. It falls steadily in the larger veins to about 5.5 mm Hg in the great veins outside the thorax. The pressure in the great veins at their entrance into the right atrium (**central venous pressure**) averages 4.6 mm Hg but fluctuates with respiration and heart action.

Peripheral venous pressure, like arterial pressure, is affected by gravity. It is increased by 0.77 mm Hg for each cm below the right atrium and decreased by a like amount for each 1 cm above the right atrium the pressure is measured (Figure 30–17).

When blood flows from the venules to the large veins, its average velocity increases as the total cross-sectional area of the vessels decreases. In the great veins, the velocity of blood is about one-fourth that in the aorta, averaging about 10 cm/s.

Thoracic Pump

During inspiration, the intrapleural pressure falls from −2.5 to −6 mm Hg. This negative pressure is transmitted to the great veins and, to a lesser extent, the atria, so that central venous pressure fluctuates from about 6 mm Hg during expiration to approximately 2 mm Hg during quiet inspiration. The drop in venous pressure during inspiration aids venous return. When the diaphragm descends during inspiration, intra-abdominal pressure rises, and this also squeezes blood toward the heart because backflow into the leg veins is prevented by the venous valves.

Effects of Heartbeat

The variations in atrial pressure are transmitted to the great veins, producing the **a, c,** and **v waves** of the venous pressure-pulse curve (see Chapter 29). Atrial pressure drops sharply during the ejection phase of ventricular systole because the atrioventricular valves are pulled downward, increasing the capacity of the atria. This action sucks blood into the atria from the great veins. The sucking of the blood into the atria during systole contributes appreciably to the venous return, especially at rapid heart rates.

Close to the heart, venous flow becomes pulsatile. When the heart rate is slow, two periods of peak flow are detectable, one during ventricular systole, due to pulling down of the atrioventricular valves, and one in early diastole, during the period of rapid ventricular filling (Figure 30–15).

Muscle Pump

In the limbs, the veins are surrounded by skeletal muscles, and contraction of these muscles during activity compresses the veins. Pulsations of nearby arteries may also compress veins. Since the venous valves prevent reverse flow, the blood moves toward the heart. During quiet standing, when the full effect of gravity is manifest, venous pressure at the ankle is 85–90 mm Hg (Figure 30–17). Pooling of blood in the leg veins reduces venous return, with the result that cardiac output is reduced, sometimes to the point where fainting occurs. Rhythmic contractions of the leg muscles while the person is standing serve to lower the venous pressure in the legs to less than 30 mm Hg by propelling blood toward the heart. This heartward movement of the blood is decreased in patients with **varicose veins** because their valves are incompetent. These patients may develop stasis and ankle edema. However, even when the valves are incompetent, muscle contractions continue to produce a basic heartward movement of the blood because the resistance of the larger veins in the direction of the heart is less than the resistance of the small vessels away from the heart.

Venous Pressure in the Head

In the upright position, the venous pressure in the parts of the body above the heart is decreased by the force of gravity. The neck veins collapse above the point where the venous pressure is close to zero, and the pressure all along the collapsed segments is close to zero rather than subatmospheric. However, the dural sinuses have rigid walls and cannot collapse. The pressure in them in the standing or sitting position is therefore subatmospheric. The magnitude of the negative pressure is proportionate to the vertical distance above the top of the collapsed neck veins, and in the superior sagittal sinus may be as much as −10 mm Hg. This fact must be kept in mind by neurosurgeons. Neurosurgical procedures are sometimes performed with the patient seated. If one of the sinuses is opened during such a procedure, it sucks air, causing **air embolism.**

Air Embolism

Because air, unlike fluid, is compressible, its presence in the circulation has serious consequences. The forward movement of the blood depends upon the fact that blood is incompressible. Large amounts of air fill the heart and effectively stop the circulation, causing sudden death, because most of the air is compressed by the contracting ventricles rather than propelled into the arteries. Small amounts of air are swept through the heart with the blood, but the bubbles lodge in the small blood vessels. The surface capillarity of the bubbles markedly increases the resistance to blood flow, and flow is reduced or abolished. Blockage of small vessels in the brain leads to serious and even fatal neurologic abnormalities. Treatment with hyperbaric oxygen (see Chapter 37) is of value because the pressure reduces the size of the gas emboli. In experimental animals, the amount of air that produces fatal air embolism varies considerably, depending in part upon the rate at which it enters the veins. Sometimes as much as 100 mL can be injected without ill effects, whereas at other times as little as 5 mL is lethal.

Measuring Venous Pressure

Central venous pressure can be measured directly by inserting a catheter into the thoracic great veins. **Peripheral venous pressure** correlates well with central venous pressure in most conditions. To measure peripheral venous pressure, a needle attached to a manometer containing sterile saline is inserted into an arm vein. The peripheral vein should be at the level of the right atrium (a point 10 cm or half the chest diameter from the back in the supine position). The values obtained in mm of saline can be converted into mm Hg by dividing by 13.6 (the density of mercury). The amount by which peripheral venous pressure exceeds central venous pressure increases with the distance from the heart along the veins. The mean pressure in the antecubital vein is normally 7.1 mm Hg, compared with a mean pressure of 4.6 mm Hg in the central veins.

A fairly accurate estimate of central venous pressure can be made without any equipment by simply noting the height to which the external jugular veins are distended when the subject lies with the head slightly above the heart. The vertical distance between the right atrium and the place the vein collapses (the place where the pressure in it is zero) is the venous pressure in mm of blood.

Central venous pressure is decreased during negative pressure breathing and shock. It is increased by positive pressure breathing, straining, expansion of the blood volume, and heart failure. In advanced congestive heart failure or obstruction of the superior vena cava, the pressure in the antecubital vein may reach values of 20 mm Hg or more.

Cardiovascular Regulatory Mechanisms

31

INTRODUCTION

In humans and other mammals, multiple cardiovascular regulatory mechanisms have evolved. These mechanisms increase the blood supply to active tissues and increase or decrease heat loss from the body by redistributing the blood. In the face of challenges such as hemorrhage, they maintain the blood flow to the heart and brain. When the challenge faced is severe, flow to these vital organs is maintained at the expense of the circulation to the rest of the body.

Circulatory adjustments are effected by altering the output of the pump (the heart), changing the diameter of the resistance vessels (primarily the arterioles), or altering the amount of blood pooled in the capacitance vessels (the veins). Regulation of cardiac output is discussed in Chapter 29. The caliber of the arterioles is adjusted in part by autoregulation. It is also increased in active tissues by locally produced vasodilator metabolites, is affected by substances secreted by the endothelium, and is regulated systemically by circulating vasoactive substances and the nerves that innervate the arterioles. The caliber of the capacitance vessels is also affected by circulating vasoactive substances and by vasomotor nerves. The systemic regulatory mechanisms synergize with the local mechanisms and adjust vascular responses throughout the body.

The terms **vasoconstriction** and **vasodilation** are generally used to refer to constriction and dilation of the resistance vessels. Changes in the caliber of the veins are referred to specifically as **venoconstriction** or **venodilation.**

LOCAL REGULATION

Autoregulation

The capacity of tissues to regulate their own blood flow is referred to as **autoregulation.** Most vascular beds have an intrinsic capacity to compensate for moderate changes in perfusion pressure by changes in vascular resistance, so that blood flow remains relatively constant. This capacity is well developed in the kidneys (see Chapter 38), but it has also been observed in the mesentery, skeletal muscle, brain, liver, and myocardium. It is probably due in part to the intrinsic contractile response of smooth muscle to stretch (**myogenic theory of autoregulation**). As the pressure rises, the blood vessels are distended and the vascular smooth muscle fibers that surround the vessels contract. If it is postulated that the muscle responds to the tension in the vessel wall, this theory could explain the greater degree of contraction at higher pressures; the wall tension is proportionate to the distending pressure times the radius of the vessel (law of Laplace; see Chapter 30), and the maintenance of a given wall tension as the pressure rises would require a decrease in radius. Vasodilator substances tend to accumulate in active tissues, and these "metabolites" also contribute to autoregulation (**metabolic theory of autoregulation**). When blood flow decreases, they accumulate and the vessels dilate; when blood flow increases, they tend to be washed away.

Vasodilator Metabolites

The metabolic changes that produce vasodilation include, in most tissues, decreases in O_2 tension and pH. These changes cause relaxation of the arterioles and precapillary sphincters. Increases in CO_2 tension and osmolality also dilate the vessels. The direct dilator action of CO_2 is most pronounced in the skin and brain. The neurally mediated vasoconstrictor effects of systemic as opposed to local hypoxia and hypercapnia are discussed below. A rise in temperature exerts a direct vasodilator effect, and the temperature rise in active tissues (due to the heat of metabolism) may contribute to the vasodilation. K^+ is another substance that accumulates locally, has demonstrated dilator activity, and probably plays a role in the dilation that occurs in skeletal muscle. Lactate may also contribute to the dilation. In injured tissues, histamine released from damaged cells increases capillary permeability. Thus, it is probably responsible for some of the swelling in areas of inflammation. Adenosine may play a vasodilator role in cardiac muscle but not in skeletal muscle. It also inhibits the release of norepinephrine.

Localized Vasoconstriction

Injured arteries and arterioles constrict strongly. The constriction appears to be due in part to the local liberation of serotonin from platelets that stick to the vessel wall in the injured area (see Chapter 27). Injured veins also constrict.

A drop in tissue temperature causes vasoconstriction, and this local response to cold plays a part in temperature regulation (see Chapter 14).

SUBSTANCES SECRETED BY THE ENDOTHELIUM

Endothelial Cells

As noted in Chapter 30, the endothelial cells make up a large and important organ. This organ secretes many growth factors and vasoactive substances. The vasoactive substances include prostaglandins and thromboxanes, nitric oxide, and endothelins.

Prostacyclin & Thromboxane A_2

Prostacyclin is produced by endothelial cells and thromboxane A_2 by platelets from their common precursor arachidonic acid via the cyclooxygenase pathway (see Figure 17–33). Thromboxane A_2 promotes platelet aggregation and vasoconstriction, whereas prostacyclin inhibits platelet aggregation and promotes vasodilation. The balance between platelet thromboxane A_2 and prostacyclin fosters localized platelet aggregation and consequent clot formation (see Chapter 27) while preventing excessive extension of the clot and maintaining blood flow around it.

The thromboxane A_2-prostacyclin balance can be shifted toward prostacyclin by administration of low doses of aspirin. Aspirin produces irreversible inhibition of cyclooxygenase by acetylating a serine residue in its active site. Obviously, this reduces production of both thromboxane A_2 and prostacyclin. However, endothelial cells produce new cyclooxygenase in a matter of hours whereas platelets cannot manufacture the enzyme, and the level rises only as new platelets enter the circulation. This is a slow process because platelets have a half-life of about 4 days. Therefore, administration of small amounts of aspirin for prolonged periods reduces clot formation and has been shown to be of value in preventing myocardial infarctions, unstable angina, transient ischemic attacks, and stroke.

Endothelium-Derived Relaxing Factor

A chance observation 2 decades ago led to the discovery that the endothelium plays a key role in vasodilation. Many different stimuli act on the endothelial cells to produce **endothelium-derived relaxing factor (EDRF)**, a substance that is now known to be **nitric oxide (NO)**. NO is synthesized from arginine (Figure 31–1) in a reaction catalyzed by nitric oxide synthase (NO synthase, NOS). Three isoforms of NOS have been identified: NOS 1, found in the nervous system; NOS 2, found in macrophages and other immune cells; and NOS 3, found in endothelial cells. NOS 1 and NOS 3 are activated by agents that increase intracellular Ca^{2+} concentration, including the vasodilators acetylcholine and bradykinin. The NOS in immune cells is not induced by Ca^{2+} but is activated by cytokines. The NO that is formed in the endothelium diffuses to smooth muscle cells, where it activates soluble guanylyl cyclase, producing cGMP (Figure 31–1), which in turn mediates the relaxation of vascular smooth muscle. NO is inactivated by hemoglobin.

Adenosine, ANP, and histamine via H_2 receptors produce relaxation of vascular smooth muscle that is independent of the endothelium. However, acetylcholine, histamine via H_1 receptors, bradykinin, VIP, substance P, and some other polypeptides act via the endothelium, and various vasoconstrictors that act directly on vascular smooth muscle would produce much greater constriction if they did not simultaneously

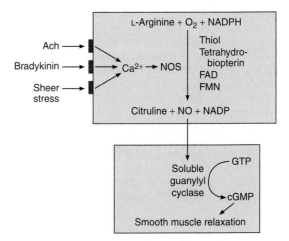

Figure 31–1. Synthesis of NO from arginine in endothelial cells and its action via stimulation of soluble guanylyl cyclase and generation of cGMP to produce relaxation in vascular smooth muscle cells. The endothelial form of nitric oxide synthase (NOS) is activated by increased intracellular Ca^{2+} concentration, and an increase is produced by acetylcholine (Ach), bradykinin, or shear stress acting on the cell membrane. Thiol, tetrahydrobiopterin, FAD, and FMN are requisite cofactors.

cause the release of NO. When flow to a tissue is suddenly increased by arteriolar dilation, the large arteries to the tissue also dilate. This flow-induced dilation is due to local release of NO. Products of platelet aggregation also cause release of NO, and the resulting vasodilation helps keep blood vessels with an intact endothelium patent. This is in contrast to injured blood vessels, where the endothelium is damaged at the site of injury and platelets therefore aggregate and produce vasoconstriction (see Chapter 27).

Further evidence for a physiologic role of NO is the observation that when various derivatives of arginine that inhibit NO synthase are administered to experimental animals, there is a prompt rise in blood pressure. This suggests that tonic release of NO is necessary to maintain normal blood pressure.

NO is also involved in vascular remodeling and angiogenesis, and NO may be involved in the pathogenesis of atherosclerosis. It is interesting in this regard that some patients with heart transplants develop an accelerated form of atherosclerosis in the vessels of the transplant, and there is reason to believe that this is triggered by endothelial damage. Nitroglycerin and other nitrovasodilators that are of great value in the treatment of angina act by stimulating guanylyl cyclase in the same manner as NO does.

There is good evidence that penile erection is produced by release of NO, with consequent vasodilation and engorgement of the corpora cavernosa (see Chapter 23).

Other Functions of NO

It has become almost commonplace to discover a compound that plays an important role in cardiovascular regulation and then learn that it is produced in other systems and has additional diverse functions. This is true, for example, of angiotensin II (see Chapter 24) and the endothelins (see below). It is also true of NO. NO is present in the brain, and, acting via cGMP, it is important in brain function (see Chapter 4). It is necessary for the cytotoxic activity of macrophages, including their ability to kill cancer cells. In the gastrointestinal tract, it is a major dilator of smooth muscle. Other functions of NO are mentioned in other parts of this book.

CO

The production of CO from heme is shown in Figure 4–33. HO2, the enzyme that catalyzes the reaction, is present in cardiovascular tissues, and there is evidence that CO as well as NO produces local dilation in blood vessels.

Endothelins

Endothelial cells also produce **endothelin-1**, one of the most potent vasoconstrictor agents yet isolated. Endothelin-1 (ET-1), endothelin-2 (ET-2), and endothelin-3 (ET-3) are the members of a family of three similar 21-amino-acid polypeptides (Figure 31–2). Each is encoded by a different gene. The unique structure of the endothelins resembles that of the sarafotoxins, polypeptides found in the venom of a snake, the Israeli burrowing asp.

Endothelin-1

In endothelial cells, the product of the endothelin-1 gene is processed to a 39-amino-acid prohormone, **big endothelin-1,** which has about 1% of the activity of endothelin-1. The prohormone is cleaved at a Trp-Val bond to form endothelin-1 by **endothelin-converting**

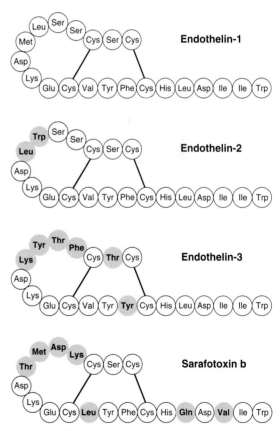

Figure 31–2. Structure of human endothelins and one of the snake venom sarafotoxins. The amino acid residues that differ from endothelin-1 are indicated in color.

enzyme. There is a family of these enzymes, apparently related to the cleavage of big endothelin-2 and big endothelin-3 as well as big endothelin-1. Small amounts of big endothelin-1 and endothelin-1 are secreted into the blood, but for the most part, they are secreted into the media of blood vessels and act in a paracrine fashion.

Two different endothelin receptors have been cloned, both of which are coupled via G proteins to phospholipase C (see Chapter 1). The ET_A receptor, which is specific for endothelin-1, is found in many tissues and mediates the vasoconstriction produced by endothelin-1. The ET_B receptor responds to all three endothelins, and is coupled to G_i. It may mediate vasodilation, and it appears to mediate the developmental effects of the endothelins (see below).

Regulation of Secretion

Endothelin-1 is not stored in secretory granules, and most regulatory factors alter the transcription of its gene, with changes in secretion occurring promptly thereafter. Factors activating and inhibiting the gene are summarized in Table 31–1.

Cardiovascular Functions

As noted above, endothelin-1 appears to be primarily a local, paracrine regulator of vascular tone. Big endothelin-1 and endothelin-1 are both present in the circulation. However, they are not increased in hypertension, and in mice in which one allele of the endothelin-1 gene is knocked out, blood pressure is actually elevated rather than reduced. The concentration of circulating endothelin-1 is elevated in congestive heart failure and after myocardial infarction, so it may play a role in the pathophysiology of these diseases.

Other Functions of Endothelins

Endothelin-1 is found in the brain and kidneys as well as the endothelial cells. Endothelin-2 is produced primarily in the kidneys and intestine. Endothelin-3 is present in the blood and is found in high concentrations in the brain. It is also found in the kidneys and gastrointestinal tract. The actions listed in Table 31–2 suggest functions for endothelins in these various locations. In the brain, endothelins are abundant and, in early life, are produced by both astrocytes and neurons.

Table 31–1. Regulation of endothelin-1 secretion via transcription of its gene.

Stimulators
Angiotensin II
Catecholamines
Growth factors
Hypoxia
Insulin
Oxidized LDL
HDL
Shear stress
Thrombin
Inhibitors
NO
ANP
PGE_2
Prostacyclin

Table 31–2. Biologic actions of endothelins.[1]

Hemodynamic effects
Contracts vascular smooth muscle; veins possibly more sensitive than arteries
Causes initial depressor response followed by sustained pressor effect (regional differences in vasoconstriction exist)

Cardiac effects
Evokes positive inotropic and chronotropic effects on myocardium
Stimulates intense vasoconstriction of coronary arteries

Neuroendocrine effects
Increases plasma levels of ANP, renin, aldosterone, and catecholamines
Modulates synaptic transmission

Various effects on endocrine glands

Renal effects
Increases renal vascular resistance
Decreases glomerular filtration rate, renal blood flow, and the glomerular ultrafiltration coefficient
Increases Na^+ reabsorption through hemodynamic actions
Decreases Na^+ reabsorption through inhibition of Na^+-K^+ ATPase

Pulmonary effects
Produces bronchoconstriction

Gastrointestinal effects
Enhances gluconeogenesis
Regulates gastrointestinal blood flow

Promitogenic effects
Stimulates cell growth in numerous cell lines

[1]Modified from Thomas CP, Simonson MS, Dunn MJ: Endothelin: Receptors and transmembrane signals. News Physiol Sci 1992;7:207.

They are found in the dorsal root ganglia, ventral horn cells, the cortex, the hypothalamus, and cerebellar Purkinje cells. They also play a role in regulating transport across the blood-brain barrier. There are endothelin receptors on mesangial cells (see Chapter 38), and the polypeptide presumably produces mesangial cell-mediated decreases in the glomerular filtration rate.

Mice that have both alleles of the endothelin-1 gene deleted have severe craniofacial abnormalities and die of respiratory failure at birth. They also have megacolon (Hirschsprung's disease), apparently because the cells that normally form the myenteric plexus fail to migrate to the distal colon. In addition, endothelins play a role in closing the ductus arteriosus at birth.

SYSTEMIC REGULATION BY HORMONES

Many circulating hormones affect the vascular system. The vasodilator hormones include kinins, VIP, and ANP. Circulating vasoconstrictor hormones include vasopressin, norepinephrine, epinephrine, and angiotensin II.

Kinins

Two related vasodilator peptides called **kinins** are found in the body. One is the nonapeptide **bradykinin,** and the other is the decapeptide **lysylbradykinin,** also known as **kallidin** (Figure 31–3). Lysylbradykinin can be converted to bradykinin by aminopeptidase. Both peptides are metabolized to inactive fragments by **kininase I,** a carboxypeptidase that removes the carboxyl terminal Arg. In addition, the dipeptidylcarboxypeptidase **kininase II** inactivates bradykinin and lysylbradykinin by removing Phe-Arg from the carboxyl terminal. Kininase II is the same enzyme as **angiotensin-converting enzyme** (see Chapter 24), which removes His-Leu from the carboxyl terminal end of angiotensin I.

Bradykinin and lysylbradykinin are formed from two precursor proteins, **high-molecular-weight kininogen** and **low-molecular-weight kininogen** (Figure 31–4). They are formed by alternative splicing of a single gene

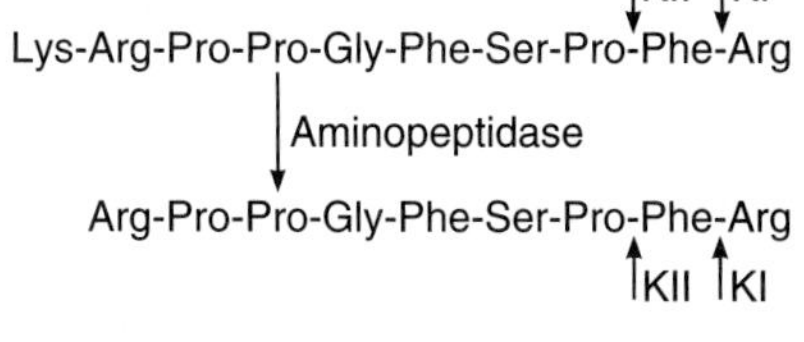

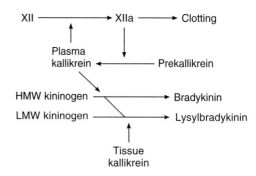

Figure 31–4. Formation of kinins from high-molecular-weight (HMW) and low-molecular-weight (LMW) kininogens.

located on chromosome 3. The biologic activities of bradykinin and lysylbradykinin are generally similar, and it is not known why two types are produced.

Proteases called **kallikreins** release the peptides from their precursors. They are produced in humans by a family of three genes located on chromosome 19. There are two types of kallikreins: **plasma kallikrein,** which circulates in an inactive form, and **tissue kallikrein,** which appears to be located primarily on the apical membranes of cells concerned with transcellular electrolyte transport. Tissue kallikrein is found in many tissues, including sweat and salivary glands, the pancreas, the prostate, the intestine, and the kidneys. Tissue kallikrein acts on high-molecular-weight kininogen and low-molecular-weight kininogen to form lysylbradykinin. When activated, plasma kallikrein acts on high-molecular-weight kininogen to form bradykinin.

Inactive plasma kallikrein **(prekallikrein)** is converted to the active form, kallikrein, by active factor XII, the factor which initiates the intrinsic blood clotting cascade. Kallikrein also activates factor XII in a positive feedback loop, and high-molecular-weight kininogen has a factor XII-activating action (see Figure 27–25).

The actions of the kinins resemble those of histamine. They are primarily tissue hormones, although small amounts are also found in the circulating blood. They cause contraction of visceral smooth muscle, but they relax vascular smooth muscle via NO, lowering blood pressure. They also increase capillary permeability, attract leukocytes, and cause pain upon injection under the skin. They are formed during active secretion in sweat glands, salivary glands, and the exocrine portion of the pancreas (see Chapter 26), and they are probably responsible for the increase in blood flow when these tissues are actively secreting their products. They are present in the kidneys, where their function is uncertain.

Lys-Arg-Pro-Pro-Gly-Phe-Ser-Pro-Phe-Arg

↓ Aminopeptidase

Arg-Pro-Pro-Gly-Phe-Ser-Pro-Phe-Arg

Figure 31–3. Kinins. Lysylbradykinin **(top)** can be converted to bradykinin **(bottom)** by aminopeptidase. The peptides are inactivated by kininase I (KI) or kininase II (KII) at the sites indicated by the short arrows.

Two bradykinin receptors, B_1 and B_2, have been identified. Their amino acid residues are 36% identical, and both are serpentine receptors coupled to G proteins. The B_1 receptor may mediate the pain-producing effects of the kinins, but little is known about its distribution and function. The B_2 receptor has strong homology to the H_2 receptor and is found in many different tissues.

Adrenomedullin

Adrenomedullin (AM) is a depressor polypeptide first isolated from pheochromocytoma cells. Its prohormone is also the source of another depressor polypeptide, proadrenomedullin amino terminal 20 peptide **(PAMP).** AM also inhibits aldosterone secretion in salt-depleted animals and appears to produce its depressor effect by increasing production of NO. PAMP appears to act by inhibiting peripheral sympathetic nerve activity. Both AM and PAMP are found in plasma and in many tissues in addition to the adrenal medulla, including the kidney and the brain. However, the role, if any, of AM and PAMP in cardiovascular control is still unknown.

Natriuretic Hormones

The atrial natriuretic peptide (ANP) secreted by the heart (see Chapter 24) antagonizes the action of various vasoconstrictor agents and lowers blood pressure, but its exact role in the regulation of the circulation is still unsettled. The natriuretic Na^+-K^+ ATPase inhibitor, which is now thought to be endogenously produced ouabain, apparently raises rather than lowers blood pressure.

Circulating Vasoconstrictors

Vasopressin is a potent vasoconstrictor, but when it is injected in normal individuals, there is a compensating decrease in cardiac output, so that there is little change in blood pressure. Its role in blood pressure regulation is discussed in Chapter 14.

Norepinephrine has a generalized vasoconstrictor action, whereas epinephrine dilates the vessels in skeletal muscle and the liver. The relative unimportance of circulating norepinephrine, as opposed to norepinephrine released from vasomotor nerves, is pointed out in Chapter 20, where the cardiovascular actions of catecholamines are discussed in detail.

The octapeptide angiotensin II has a generalized vasoconstrictor action. It is formed from angiotensin I liberated by the action of renin from the kidney on circulating angiotensinogen (see Chapter 24). Its formation is increased because renin secretion is increased when the blood pressure falls or ECF volume is reduced, and it helps maintain blood pressure. An-

giotensin II also increases water intake and stimulates aldosterone secretion, and increased formation of angiotensin II is part of a homeostatic mechanism that operates to maintain ECF volume (see Chapter 20). In addition, there are renin-angiotensin systems in many different organs, and there may be one in the walls of blood vessels. Angiotensin II produced in blood vessel walls could be important in some forms of clinical hypertension.

Urotensin-II, a polypeptide first isolated from the spinal cord of fish, is present in human cardiac and vascular tissue. It is one of the most potent mammalian vasoconstrictors known, but its physiologic role is still uncertain.

SYSTEMIC REGULATION BY THE NERVOUS SYSTEM

Neural Regulatory Mechanisms

Although the arterioles and the other resistance vessels are most densely innervated, all blood vessels except capillaries and venules contain smooth muscle and receive motor nerve fibers from the sympathetic division of the autonomic nervous system. The fibers to the resistance vessels regulate tissue blood flow and arterial pressure. The fibers to the venous capacitance vessels vary the volume of blood "stored" in the veins. The innervation of most veins is sparse, but the splanchnic veins are well innervated. Venoconstriction is produced by stimuli that also activate the vasoconstrictor nerves to the arterioles. The resultant decrease in venous capacity increases venous return, shifting blood to the arterial side of the circulation.

Innervation of the Blood Vessels

Noradrenergic fibers end on vessels in all parts of the body (Figure 31–5). The noradrenergic fibers are vasoconstrictor in function. In addition to their vasoconstrictor innervation, the resistance vessels of the skeletal muscles are innervated by vasodilator fibers, which, although they travel with the sympathetic nerves, are cholinergic (the **sympathetic vasodilator system.**) There is some evidence that blood vessels in the heart, lungs, kidneys, and uterus also receive a cholinergic innervation. Bundles of noradrenergic and cholinergic fibers form a plexus on the adventitia of the arterioles. Fibers with multiple varicosities extend from this plexus to the media and end primarily on the outer surface of the smooth muscle of the media without penetrating it. Transmitters reach the inner portions of the media by diffusion, and current spreads from one smooth muscle cell to another via gap junctions.

There is no tonic discharge in the vasodilator fibers, but the vasoconstrictor fibers to most vascular beds have

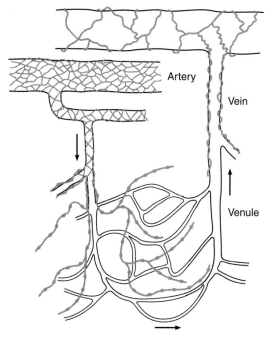

Figure 31–5. Relation of noradrenergic nerve fibers to blood vessels in the mesentery of the rat. Arrows indicate direction of flow. (Reproduced, with permission, from Furness JB, Marshall JM: Correlation of the directly observed responses of mesenteric vessels of the rat to nerve stimulation and noradrenaline with the distribution of adrenergic nerves. J Physiol 1974;75:239.)

Table 31–3. Summary of factors affecting the caliber of the arterioles.

Constriction	Dilation
Local factors	
Decreased local temperature	Increased CO_2 and decreased O_2
Autoregulation	Increased K^+, adenosine, lactate, etc.
	Decreased local pH
	Increased local temperature
Endothelial products	
Endothelin-1	NO
Locally released platelet serotonin	Kinins
Thromboxane A_2	Prostacyclin
Circulating hormones	
Epinephrine (except in skeletal muscle and liver)	Epinephrine in skeletal muscle and liver
Norepinephrine	CGRPα
AVP	Substance P
Angiotensin II	Histamine
Circulating Na^+-K^+ ATPase inhibitor	ANP
Neuropeptide Y	VIP
Neural factors	
Increased discharge of noradrenergic vasomotor nerves	Decreased discharge of noradrenergic vasomotor nerves
	Activation of cholinergic dilator fibers to skeletal muscle

some tonic activity. When the sympathetic nerves are cut **(sympathectomy),** the blood vessels dilate. In most tissues, vasodilation is produced by decreasing the rate of tonic discharge in the vasoconstrictor nerves, although in skeletal muscles it can also be produced by activating the sympathetic vasodilator system (Table 31–3).

Nerves containing polypeptides are found on many blood vessels. The cholinergic nerves also contain VIP, which produces vasodilation. The noradrenergic postganglionic sympathetic nerves also contain neuropeptide Y, which is a vasoconstrictor. Substance P and CGRPα, which produce vasodilation, are found in sensory nerves near blood vessels.

Afferent impulses in sensory nerves from the skin are relayed antidromically down branches of the sensory nerves that innervate blood vessels, and these impulses cause release of substance P from the nerve endings. Substance P causes vasodilation and increased capillary permeability. This local neural mechanism is called the **axon reflex** (see Figure 32–17). Other cardiovascular reflexes are integrated in the central nervous system.

Cardiac Innervation

Impulses in the noradrenergic sympathetic nerves to the heart increase the cardiac rate (chronotropic effect) and the force of cardiac contraction (inotropic effect). They also inhibit the effects of vagal stimulation, probably by release of neuropeptide Y, which is a cotransmitter in the sympathetic endings. Impulses in the cholinergic vagal cardiac fibers decrease the heart rate. There is a moderate amount of tonic discharge in the cardiac sympathetic nerves at rest, but there is a good deal of tonic vagal discharge **(vagal tone)** in humans and other large animals. When the vagi are cut in experimental animals, the heart rate rises, and after the administration of parasympatholytic drugs such as atropine, the heart rate in humans increases from 70, its normal resting value, to 150–180 beats/min because the sympathetic tone is unopposed. In humans in whom both noradrenergic and cholinergic systems are blocked, the heart rate is approximately 100.

Vasomotor Control

The sympathetic nerves that constrict arterioles and veins and increase heart rate and stroke volume discharge in a tonic fashion, and blood pressure is adjusted by variations in the rate of this tonic discharge (Figure 31–6). Spinal reflex activity affects blood pressure, but the main control of blood pressure is exerted by groups of neurons in the medulla oblongata that are sometimes called collectively the **vasomotor area** or **vasomotor center.** Neurons that mediate increased sympathetic discharge to blood vessels and the heart (Figure 31–7) project directly to sympathetic preganglionic neurons in the intermediolateral gray column (IML) of the spinal cord. On each side, the cell bodies of these neurons are located near the pial surface of the medulla in the rostral ventrolateral medulla (RVLM). Their axons course dorsally and medially and then descend in the lateral column of the spinal cord to the IML. They contain PNMT (see Chapter 4), but it appears that the excitatory transmitter they secrete is glutamate rather than epinephrine.

Impulses reaching the medulla also affect the heart rate via vagal discharge to the heart. The neurons from which the vagal fibers arise are in the dorsal motor nucleus of the vagus and the nucleus ambiguus (Figure 31–8).

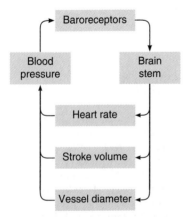

Figure 31–6. Feedback control of blood pressure.

When vasoconstrictor discharge is increased, there is increased arteriolar constriction and a rise in blood pressure. Venoconstriction and a decrease in the stores of blood in the venous reservoirs usually accompany these changes, although changes in the capacitance vessels do not always parallel changes in the resistance vessels. Heart rate and stroke volume are increased because of activity in the sympathetic nerves to the heart, and cardiac output is increased. There is usually an associated decrease in the tonic activity of vagal fibers to the heart. Conversely, a decrease in vasomotor discharge causes vasodilation, a fall in blood pressure, and an increase in the storage of blood in the venous reservoirs. There is usually a concomitant decrease in heart rate, but this is mostly due to stimulation of the vagal innervation of the heart.

Afferents to the Vasomotor Area

The afferents that converge on the vasomotor area are summarized in Table 31–4. They include not only the very important fibers from arterial and venous baroreceptors but also fibers from other parts of the nervous system and from the carotid and aortic chemoreceptors. In addition, some stimuli act directly on the vasomotor area.

There are descending tracts to the vasomotor area from the cerebral cortex (particularly the limbic cortex) that relay in the hypothalamus. These fibers are responsible for the blood pressure rise and tachycardia produced by emotions such as sexual excitement and anger. The connections between the hypothalamus and the vasomotor area are reciprocal, with afferents from the brain stem closing the loop.

Inflation of the lungs causes vasodilation and a decrease in blood pressure. This response is mediated via vagal afferents from the lungs that inhibit vasomotor discharge. Pain usually causes a rise in blood pressure via afferent impulses in the reticular formation converging on the vasomotor area. However, prolonged severe pain may cause vasodilation and fainting.

Somatosympathetic Reflex

Pain causes increased arterial pressure, and activity in afferents from exercising muscles probably exerts a similar pressor effect via the C1 neurons in the rostral ventrolateral medulla. The pressor response to stimulation of somatic afferent nerves is called the **somatosympathetic reflex.**

Baroreceptors

The **baroreceptors** are stretch receptors in the walls of the heart and blood vessels. The **carotid sinus** and

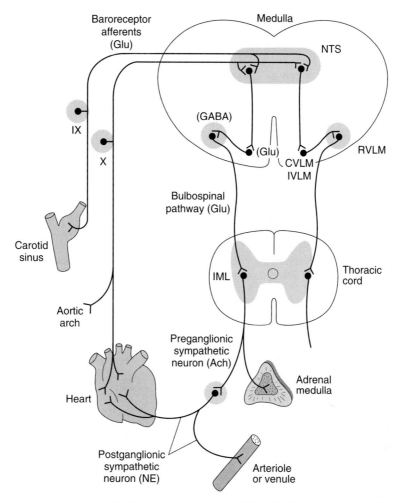

Figure 31–7. Basic pathways involved in the medullary control of blood pressure. The vagal efferent pathways that slow the heart are not shown. The putative neurotransmitters in the pathways are indicated in parentheses. Glu, glutamate; GABA, γ-aminobutyric acid; Ach, acetylcholine; NE, norepinephrine; IML, intermediolateral gray column; NTS, nucleus of the tractus solitarius; CVLM, IVLM, RVLM, caudal, intermediate, and rostral ventrolateral medulla; IX and X, glossopharyngeal and vagus nerves.

aortic arch receptors monitor the arterial circulation. Receptors are also located in the walls of the right and left atria at the entrance of the superior and inferior venae cavae and the pulmonary veins, as well as in the pulmonary circulation. These receptors in the low-pressure part of the circulation are referred to collectively as the cardiopulmonary receptors. The baroreceptors are stimulated by distention of the structures in which they are located, and so they discharge at an increased rate when the pressure in these structures rises. Their afferent fibers pass via the glossopharyngeal and vagus nerves to the medulla. Most of them end in

the nucleus of the tractus solitarius (NTS), and the excitatory transmitter they secrete is probably glutamate. There are excitatory, presumably glutaminergic, projections from the NTS to the caudal and intermediate ventrolateral medulla, where they apparently stimulate GABA-secreting inhibitory neurons that project to the rostral ventrolateral medulla. There are also excitatory projections, probably polyneuronal, from the NTS to the vagal motor neurons in the dorsal motor nucleus and the nucleus ambiguus. Thus, increased baroreceptor discharge *inhibits* the tonic discharge of the vasoconstrictor nerves and *excites* the vagal innervation of

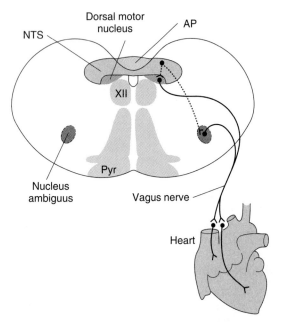

Figure 31–8. Basic pathways involved in the medullary control of heart rate by the vagus nerves. NTS neurons (dashed lines) project to and inhibit preganglionic parasympathetic neurons in the dorsal motor nucleus of the vagus and the nucleus ambiguus. Postganglionic cholinergic neurons innervate the atria and the ventricles. Pyr, pyramid; XII, hypoglossal nucleus.

the heart, producing vasodilation, venodilation, a drop in blood pressure, bradycardia, and a decrease in cardiac output.

Carotid Sinus & Aortic Arch

The carotid sinus is a small dilation of the internal carotid artery just above the bifurcation of the common carotid into external and internal carotid branches (Figure 31–9). Baroreceptors are located in this dilation. They are also found in the wall of the arch of the aorta. The receptors are located in the adventitia of the vessels. They are extensively branched, knobby, coiled, and intertwined ends of myelinated nerve fibers that resemble Golgi tendon organs (see Figure 6–5). Similar receptors have been found in various other parts of the large arteries of the thorax and neck in some species. The afferent nerve fibers from the carotid sinus and carotid body form a distinct branch of the glossopharyngeal nerve, the **carotid sinus nerve,** but the fibers from the aortic arch form a separate distinct branch of the vagus only in the rabbit. The carotid sinus nerves

Table 31–4. Factors affecting the activity of the vasomotor area in the medulla.

Direct stimulation
CO_2
Hypoxia
Excitatory inputs
From cortex via hypothalamus
From pain pathways and muscles
From carotid and aortic chemoreceptors
Inhibitory inputs
From cortex via hypothalamus
From lungs
From carotid, aortic, and cardiopulmonary baroreceptors

and vagal fibers from the aortic arch are commonly called the **buffer nerves.**

Buffer Nerve Activity

At normal blood pressure levels, the fibers of the buffer nerves discharge at a low rate (Figure 31–10). When the pressure in the sinus and aortic arch rises, the dis-

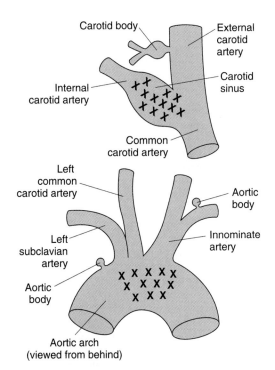

Figure 31–9. Baroreceptor areas in the carotid sinus and aortic arch. X, sites where receptors are located.

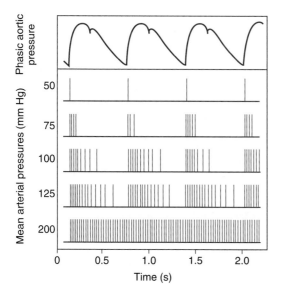

Figure 31–10. Discharges (vertical lines) in a single afferent nerve fiber from the carotid sinus at various arterial pressures, plotted against changes in aortic pressure with time. (Reproduced, with permission, from Berne RM, Levy MN: *Cardiovascular Physiology*, 3rd ed. Mosby, 1977.)

charge rate increases; and when the pressure falls, the rate declines.

When one carotid sinus of a monkey is isolated and perfused and the other baroreceptors are denervated, there is no discharge in the afferent fibers from the perfused sinus and no drop in the animal's arterial pressure or heart rate when the perfusion pressure is below 30 mm Hg. At perfusion pressures of 70–110 mm Hg, there is an essentially linear relation between the perfusion pressure and the fall in blood pressure and heart rate produced in the monkey. At perfusion pressures above 150 mm Hg there is no further increase in response (Figure 31–12), presumably because the rate of baroreceptor discharge and the degree of inhibition of the vasomotor center are maximal.

The carotid receptors respond both to sustained pressure and to pulse pressure. A decline in carotid pulse pressure without any change in mean pressure decreases the rate of baroreceptor discharge and provokes a rise in blood pressure and tachycardia. The receptors also respond to changes in pressure as well as steady pressure; when the pressure is fluctuating, they sometimes discharge during the rises and are silent during the falls (Figure 31–10) at mean pressures at which if there were no fluctuations, there would be a steady discharge.

The aortic receptors have not been studied in such great detail, but there is no reason to believe that their responses differ significantly from those of the receptors in the carotid sinus.

From the foregoing discussion, it is apparent that the baroreceptors on the arterial side of the circulation, their afferent connections to the vasomotor and cardioinhibitory areas, and the efferent pathways from these areas constitute a reflex feedback mechanism that operates to stabilize the blood pressure and heart rate. Any drop in systemic arterial pressure decreases the inhibitory discharge in the buffer nerves, and there is a compensatory rise in blood pressure and cardiac output. Any rise in pressure produces dilation of the arterioles and decreases cardiac output until the blood pressure returns to its previous normal level.

Baroreceptor Resetting

In chronic hypertension, the baroreceptor reflex mechanism is "reset" to maintain an elevated rather than a normal blood pressure. In perfusion studies on hypertensive experimental animals, raising the pressure in the isolated carotid sinus lowers the elevated systemic pressure, and decreasing the perfusion pressure raises the elevated pressure (Figure 31–11). Little is known about how and why this occurs, but resetting occurs rapidly in experimental animals. It is also rapidly reversible, both in experimental animals and in clinical situations.

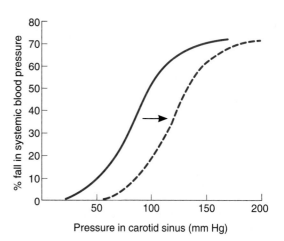

Figure 31–11. Solid line: Fall in systemic blood pressure produced by raising the pressure in the isolated carotid sinus of a normal monkey to various values. **Dashed line:** Response in a hypertensive monkey, demonstrating baroreceptor resetting (arrow).

Effect of Carotid Clamping & Buffer Nerve Section

Bilateral clamping of the carotid arteries proximal to the carotid sinuses elevates the blood pressure and heart rate because the procedure lowers the pressure in the sinuses. Cutting the carotid sinus nerves on each side has the same effect. The pressor response following these two procedures is moderate, because the aortic baroreceptors are still functioning normally, and they buffer the rise. If baroreceptor afferents in the vagi are also interrupted, blood pressure rises to 300/200 mm Hg or higher and is unstable. Bilateral lesions of the NTS, the site of termination of the baroreceptor afferents, cause severe hypertension that can be fatal. These forms of experimental hypertension are called **neurogenic hypertension.**

Atrial Stretch Receptors

The stretch receptors in the atria are of two types: those that discharge primarily during atrial systole (type A), and those that discharge primarily late in diastole, at the time of peak atrial filling (type B). The discharge of type B baroreceptors is increased when venous return is increased and decreased by positive-pressure breathing, indicating that these baroreceptors respond primarily to distention of the atrial walls. The reflex circulatory adjustments initiated by increased discharge from most if not all of these receptors include vasodilation and a fall in blood pressure. However, the heart rate is increased rather than decreased.

Role of Baroreceptors in Endocrine Defense of ECF Volume

When the ECF volume falls, central venous pressure declines, and the decreased firing of the atrial stretch receptors leads to increased secretion of vasopressin (see Chapter 14). Sympathetic activity is increased, and this leads to increased secretion of renin. The increase in renin secretion increases the secretion of aldosterone. When loss of volume is more severe, arterial pressure falls and decreased discharge of the carotid and aortic baroreceptors also contributes to the increases in hormone secretion. Other factors contributing to the increase in vasopressin and renin secretion are discussed in Chapters 20, 24, and 39. The net result is a retention of water and sodium that helps restore the ECF volume.

Bainbridge Reflex

Rapid infusion of blood or saline in anesthetized animals sometimes produces a rise in heart rate if the initial heart rate is low. This effect was described by Bainbridge in 1915, and since then it has been known as the **Bainbridge reflex.** It appears to be a true reflex rather than a local response to stretch, since it is abolished by bilateral vagotomy, and infusion of fluids in animals with transplanted hearts increases the rate of the recipient's atrial remnant but fails to affect the rate of the transplanted heart. The receptors may be the tachycardia-producing atrial receptors mentioned above. The reflex competes with the baroreceptor-mediated decrease in heart rate produced by volume expansion and is diminished or absent when the initial heart rate is high. There has been much debate about its significance, and its physiologic role remains unsettled.

Left Ventricular Receptors

When the left ventricle is distended in experimental animals, there is a fall in systemic arterial pressure and heart rate. It takes considerable ventricular distention to produce this response, and its physiologic significance is uncertain. However, left ventricular stretch receptors may play a role in the maintenance of the vagal tone that keeps the heart rate low at rest.

In experimental animals, injections of serotonin, veratridine, capsaicin, phenyldiguanide, and some other drugs into the coronary arteries supplying the left ventricle cause apnea followed by rapid breathing, hypotension, and bradycardia (the **coronary chemoreflex** or **Bezold-Jarisch reflex**). The receptors are probably C fiber endings, and the afferents are vagal. The response is not produced by injections into the blood supply of the atria or the right ventricle. Its physiologic role is uncertain, but in patients with myocardial infarcts, substances released from the infarcted tissue may stimulate ventricular receptors, contributing to the hypotension that is not infrequently a stubborn complication of this disease.

Pulmonary Receptors

Injections of serotonin, capsaicin, veratridine, and related drugs into the pulmonary artery activate C fiber endings close to capillaries in the lungs and produce apnea followed by rapid breathing, hypotension, and bradycardia **(pulmonary chemoreflex).** This response, which is discussed in Chapter 36, is blocked by vagotomy and is essentially the same as the coronary chemoreflex caused by injection of drugs into the arterial supply of the left ventricle. However, the response is too rapid to be caused by the drugs reaching the left ventricular receptors.

Clinical Testing & Stimulation

The changes in pulse rate and blood pressure that occur in humans on standing up or lying down (see Chapter

33) are due for the most part to baroreceptor reflexes. The function of the receptors can be tested by monitoring changes in heart rate as a function of increasing arterial pressure during infusion of the α-adrenergic agonist phenylephrine. A normal response is shown in Figure 31–12; from a systolic pressure of about 120–150 mm Hg, there is a linear relation between pressure and lowering of the heart rate (greater RR interval).

The function of the receptors can also be tested by monitoring the changes in pulse and blood pressure that occur in response to brief periods of straining (forced expiration against a closed glottis: the **Valsalva maneuver**). The blood pressure rises at the onset of straining (Figure 31–13) because the increase in intrathoracic pressure is added to the pressure of the blood in the aorta. It then falls because the high intrathoracic pressure compresses the veins, decreasing venous return and cardiac output. The decreases in arterial pressure and pulse pressure inhibit the baroreceptors, causing tachycardia and a rise in peripheral resistance. When the glottis is opened and the intrathoracic pressure returns to normal, cardiac output is restored but the peripheral vessels are constricted. The blood pressure therefore rises above normal, and this stimulates the baroreceptors, causing bradycardia and a drop in pressure to normal levels.

In sympathectomized patients, heart rate changes still occur because the baroreceptors and the vagi are intact. However, in patients with autonomic insufficiency, a syndrome in which there is widespread disruption of autonomic function, the heart rate changes are absent. For reasons that are still obscure, patients with primary hyperaldosteronism also fail to show the heart rate changes and the blood pressure rise when the in-

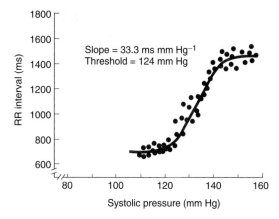

Figure 31–12. Baroreflex-mediated lowering of the heart rate during infusion of phenylephrine in a human subject. Note that the values for the RR interval of the ECG, which are plotted on the vertical axis, are inversely proportionate to the heart rate. (Reproduced, with permission, from Kotrly K et al: Effects of fentanyl-diazepam-nitrous oxide anaesthesia on arterial baroreflex control of heart rate in man. Br J Anaesth 1986;58:406.)

trathoracic pressure returns to normal. Their response to the Valsalva maneuver returns to normal after removal of the aldosterone-secreting tumor.

Effects of Chemoreceptor Stimulation on the Vasomotor Area

Afferents from the chemoreceptors in the carotid and aortic bodies exert their main effect on respiration, and their function is discussed in Chapter 36. How-

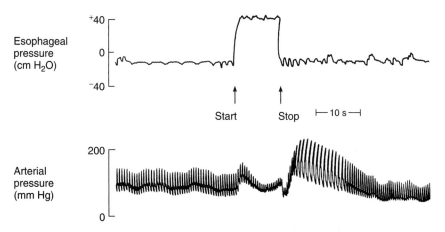

Figure 31–13. Diagram of the response to straining (the Valsalva maneuver) in a normal man, recorded with a needle in the brachial artery. (Courtesy of M McIlroy.)

ever, they also converge on the vasomotor area. The cardiovascular response to chemoreceptor stimulation consists of peripheral vasoconstriction and bradycardia. However, hypoxia also produces hyperpnea and increased catecholamine secretion from the adrenal medulla, both of which produce tachycardia and an increase in cardiac output. Hemorrhage that produces hypotension leads to chemoreceptor stimulation. This is due to decreased blood flow to the chemoreceptors and consequent stagnant anoxia of these organs (see Chapter 37). In hypotensive animals, baroreceptor discharge is low (see below), and section of the glossopharyngeal and vagus nerves leads to a fall rather than a rise in blood pressure, because the chemoreceptor drive to the vasomotor area is removed. Chemoreceptor discharge may also contribute to the production of **Mayer waves.** These should not be confused with **Traube-Hering waves,** which are fluctuations in blood pressure synchronized with respiration. The Mayer waves are slow regular oscillations in arterial pressure that occur at the rate of about one per 20–40 seconds during hypotension. Under these conditions, hypoxia stimulates the chemoreceptors. The stimulation raises the blood pressure, which improves the blood flow in the receptor organs and eliminates the stimulus to the chemoreceptors, so that the pressure falls and a new cycle is initiated. However, Mayer waves are reduced but not abolished by chemoreceptor denervation and are sometimes present in spinal animals, so oscillation in spinal vasopressor reflexes is also involved.

Direct Effects on the Vasomotor Area

Hypoxia and hypercapnia both stimulate the vasomotor area directly and presumably act on the RVLM, although the direct effect of hypoxia is small. When intracranial pressure is increased, the blood supply to the vasomotor area is compromised, and the local hypoxia and hypercapnia increase its discharge. The resultant rise in systemic arterial pressure **(Cushing reflex)** tends to restore the blood flow to the medulla. The rise in blood pressure causes a reflex decrease in heart rate via the arterial baroreceptors (see below), and this is why bradycardia rather than tachycardia is characteristically seen in patients with increased intracranial pressure.

A rise in arterial P_{CO_2} stimulates the vasomotor area, but the direct peripheral effect of hypercapnia is vasodilation. Therefore, the peripheral and central actions tend to cancel each other. Moderate hyperventilation, which significantly lowers the CO_2 tension of the blood, causes cutaneous and cerebral vasoconstriction in humans, but there is little change in blood pressure. Exposure to high concentrations of CO_2 is associated with marked cutaneous and cerebral vasodilation, but

Table 31–5. Factors affecting heart rate.[1]

Heart rate accelerated by:
Decreased activity of baroreceptors in the arteries, left ventricle, and pulmonary circulation
Increased activity of atrial stretch receptors
Inspiration
Excitement
Anger
Most painful stimuli
Hypoxia
Exercise
Epinephrine
Thyroid hormones
Fever
Bainbridge reflex

Heart rate slowed by:
Norepinephrine[1]
Increased activity of baroreceptors in the arteries, left ventricle, and pulmonary circulation
Expiration
Fear
Grief
Stimulation of pain fibers in trigeminal nerve
Increased intracranial pressure

[1]Norepinephrine has a direct chronotropic effect on the heart, but in the intact animal its pressor action stimulates the baroreceptors, leading to enough reflex increase in vagal tone to overcome the direct effect and produce bradycardia.

there is vasoconstriction elsewhere and usually a slow rise in blood pressure.

Sympathetic Vasodilator System

The cholinergic sympathetic vasodilator fibers are part of a regulatory system that originates in the cerebral cortex, relays in the hypothalamus and mesencephalon, and passes through the medulla without interruption to the intermediolateral gray column of the spinal cord. The preganglionic neurons which are part of this system activate postganglionic neurons to blood vessels in skeletal muscle that are anatomically sympathetic but secrete acetylcholine. Stimulation of this system produces vasodilation in skeletal muscle, but the resultant increase in blood flow is associated with a decrease rather than an increase in muscle O_2 consumption. This suggests that the blood is being diverted through thoroughfare channels rather than capillaries. Adrenal medullary secretion of norepinephrine and epinephrine is apparently increased when this system is stimulated, the epinephrine probably reinforcing the dilation of muscle blood vessels. In cats and dogs, the system discharges in response to emotional stimuli such as fear, apprehension, and rage. Its role in humans is uncertain, but it has been suggested that the sympathetic vasodilator system is responsible for fainting in emotional situations. In addition, there is direct evidence for cholinergically mediated vasodilation in muscle at or even before the start of muscular exercise (see Chapter 33). However, it has been argued that the vasodilation before the start of exercise is not a constant or marked phenomenon.

Control of Heart Rate

The sympathetic and parasympathetic nerves to the heart and baroreceptor-mediated reflex changes in heart rate have been considered in detail in preceding sections of this chapter. However, Table 31–5 is a convenient summary of conditions that affect the heart rate. In general, stimuli that increase the heart rate also increase the blood pressure, whereas those that decrease the heart rate lower the blood pressure. However, there are exceptions such as the production of hypotension and tachycardia by stimulation of atrial stretch receptors and the production of hypertension and bradycardia by increased intracranial pressure (see above).

There is evidence that there is a positive correlation between the resting heart rate and the incidence of death due to cardiac disease, suggesting that increased vagal tone is beneficial. However, this view is not supported by all investigators.

Circulation Through Special Regions

32

INTRODUCTION

The distribution of the cardiac output to various parts of the body at rest in a normal man is shown in Table 32–1. The general principles described in preceding chapters apply to the circulation of all these regions, but the vascular supplies of most organs have additional special features. The portal circulation of the anterior pituitary is discussed in Chapter 14, the pulmonary circulation in Chapter 34, and the renal circulation in Chapter 38. The circulation of skeletal muscle is discussed with the physiology of exercise in Chapter 33. This chapter is concerned with the circulation of the brain, the heart, the splanchnic area, the skin, the placenta, and the fetus.

CEREBRAL CIRCULATION

ANATOMIC CONSIDERATIONS

Vessels

The principal arterial inflow to the brain in humans is via four arteries: two internal carotids and two vertebrals. The vertebral arteries unite to form the basilar artery, and the basilar artery and the carotids form the **circle of Willis** below the hypothalamus. The circle of Willis is the origin of the six large vessels supplying the cerebral cortex. In some animals the vertebrals are large and the internal carotids small, but in humans a relatively small fraction of the total arterial flow is carried by the vertebral arteries. Substances injected into one carotid artery are distributed almost exclusively to the cerebral hemisphere on that side. There is normally no crossing over, probably because the pressure is equal on both sides. Even when it is not, the anastomotic channels in the circle do not permit a very large flow. Occlusion of one carotid artery, particularly in older patients, often causes serious symptoms of cerebral ischemia. There are precapillary anastomoses between the cerebral arterioles in humans and some other

species, but flow through these channels is generally insufficient to maintain the circulation and prevent infarction when a cerebral artery is occluded.

Venous drainage from the brain by way of the deep veins and dural sinuses empties principally into the internal jugular veins in humans, although a small amount of venous blood drains through the ophthalmic and pterygoid venous plexuses, through emissary veins to the scalp, and down the system of paravertebral veins in the spinal canal. In other species, the internal jugular veins are small, and the venous blood from the brain mixes with blood from other structures.

The cerebral vessels have a number of unique anatomic features. In the choroid plexuses there are gaps between the endothelial cells of the capillary wall, but the choroid epithelial cells that separate them from the cerebrospinal fluid (CSF) are connected to one another by tight junctions. The capillaries in the brain substance resemble nonfenestrated capillaries in muscle (see Chapter 30), but there are tight junctions between the endothelial cells that limit the passage of substances through the junctions. In addition, there are relatively few vesicles in the endothelial cytoplasm, and presumably there is little vesicular transport. The brain capillaries are surrounded by the endfeet of astrocytes (Figure 32–1). These endfeet are closely applied to the basal lamina of the capillaries, but they do not cover the entire capillary wall, and there are gaps of about 20 nm between endfeet (Figure 32–2). However, the endfeet induce the tight junctions in the capillaries (see Chapter 2). The protoplasm of astrocytes is also found around synapses, where it appears to isolate the synapses in the brain from one another.

Innervation

Three systems of nerves innervate the cerebral blood vessels. Postganglionic sympathetic neurons have their cell bodies in the superior cervical ganglia, and their endings contain norepinephrine and neuropeptide Y. Cholinergic neurons that probably originate in the sphenopalatine ganglia also innervate the cerebral vessels, and the postganglionic cholinergic neurons on the

Table 32–1. Resting blood flow and O_2 consumption of various organs in a 63-kg adult man with a mean arterial blood pressure of 90 mm Hg and an O_2 consumption of 250 mL/min.[1]

Region	Mass (kg)	Blood Flow		Arteriovenous Oxygen Difference (mL/L)	Oxygen Consumption		Resistance (R units)[2]		Percentage of Total	
		mL/ min	mL/100 g/min		mL/ min	mL/100 g/min	Absolute	per kg	Cardiac Output	Oxygen Consumption
Liver	2.6	1500	57.7	34	51	2.0	3.6	9.4	27.8	20.4
Kidneys	0.3	1260	420.0	14	18	6.0	4.3	1.3	23.3	7.2
Brain	1.4	750	54.0	62	46	3.3	7.2	10.1	13.9	18.4
Skin	3.6	462	12.8	25	12	0.3	11.7	42.1	8.6	4.8
Skeletal muscle	31.0	840	2.7	60	50	0.2	6.4	198.4	15.6	20.0
Heart muscle	0.3	250	84.0	114	29	9.7	21.4	6.4	4.7	11.6
Rest of body	23.8	336	1.4	129	44	0.2	16.1	383.2	6.2	17.6
Whole body	63.0	5400	8.6	46	250	0.4	1.0	63.0	100.0	100.0

[1]Reproduced, with permission, from Bard P (editor): *Medical Physiology,* 11th ed. Mosby, 1961.
[2]R units are pressure (mm Hg) divided by blood flow (mL/s).

blood vessels contain acetylcholine, VIP, and PHM-27 (see Chapter 26). These nerves end primarily on large arteries. Sensory nerves are found on more distal arteries. They have their cell bodies in the trigeminal ganglia and contain substance P, neurokinin A, and CGRP. Substance P, CGRP, VIP, and PHM-27 cause vasodilation, whereas neuropeptide Y is a vasoconstric-tor. Touching or pulling on the cerebral vessels causes pain.

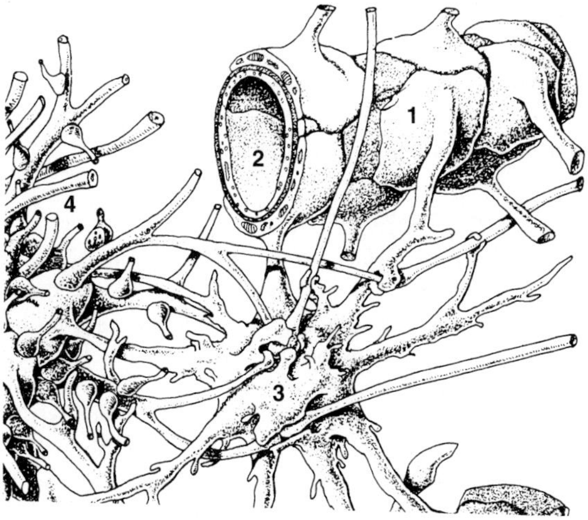

CEREBROSPINAL FLUID

Formation & Absorption

CSF fills the ventricles and subarachnoid space. In humans, the volume of CSF is about 150 mL and the rate of CSF production is about 550 mL/d. Thus the CSF turns over about 3.7 times a day. In experiments on animals, it has been estimated that 50–70% of the CSF is formed in the choroid plexuses and the remainder is formed around blood vessels and along ventricular walls. Presumably, the situation in humans is similar. The CSF in the ventricles flows through the foramens of Magendie and Luschka to the subarachnoid space and is absorbed through the **arachnoid villi** into veins, primarily the cerebral venous sinuses. The villi consist of projections of the fused arachnoid membrane and endothelium of the sinuses into the venous sinuses. There are similar, smaller villi projecting into veins around spinal nerve routes. In a poorly understood way, these projections act as valves which permit **bulk flow** (direct flow) of CSF into venous blood. Bulk flow through these villi is about 500 mL/d, with additional small amounts of CSF being absorbed by diffusion into cerebral blood vessels.

The composition of CSF (Table 32–2) is essentially the same as that of brain ECF, which in living humans makes up 15% of the brain volume. In adults, there appears to be free communication between the brain interstitial fluid and CSF, although the diffusion distances from some parts of the brain to the CSF are

Figure 32–1. Relation of fibrous astrocyte (3) to a capillary (2) and neuron (4) in the brain. The endfeet of the astrocyte processes form a discontinuous membrane around the capillary (1). Astrocyte processes also envelop the neuron. (Reproduced, with permission, from Krstic RV: *Die Gewebe des Menschen und der Säugetiere.* Springer, 1978.)

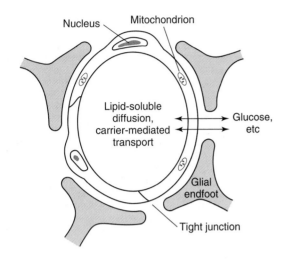

Table 32-2. Concentration of various substances in human CSF and plasma.

Substance		CSF	Plasma	Ratio CSF/Plasma
Na$^+$	(meq/kg H$_2$O)	147.0	150.0	0.98
K$^+$	(meq/kg H$_2$O)	2.9	4.6	0.62
Mg^{2+}	(meq/kg H$_2$O)	2.2	1.6	1.39
Ca^{2+}	(meq/kg H$_2$O)	2.3	4.7	0.49
Cl$^-$	(meq/kg H$_2$O)	113.0	99.0	1.14
HCO$_3^-$	(meq/L)	25.1	24.8	1.01
P$_{CO_2}$	(mm Hg)	50.2	39.5	1.28
pH		7.33	7.40	...
Osmolality	(mosm/kg H$_2$O)	289.0	289.0	1.00
Protein	(mg/dL)	20.0	6000.0	0.003
Glucose	(mg/dL)	64.0	100.0	0.64
Inorganic P	(mg/dL)	3.4	4.7	0.73
Urea	(mg/dL)	12.0	15.0	0.80
Creatinine	(mg/dL)	1.5	1.2	1.25
Uric acid	(mg/dL)	1.5	5.0	0.30
Cholesterol	(mg/dL)	0.2	175.0	0.001

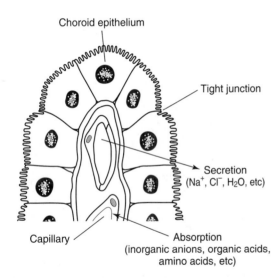

Figure 32–2. Top: Transport across cerebral capillaries. **Bottom:** Transport across cells of the choroid plexus. (Modified from Fishman RA: *Cerebrospinal Fluid in Diseases of the Nervous System,* 2nd ed. Saunders, 1992.)

appreciable. Consequently, equilibration may take some time to occur, and local areas of the brain may have extracellular microenvironments that are transiently different from CSF.

Lumbar CSF pressure is normally 70–180 mm CSF. Up to pressures well above this range, the rate of CSF formation is independent of intraventricular pressure. However, absorption, which takes place largely by bulk flow, is proportionate to the pressure (Figure 32–3). At a pressure of 112 mm CSF, which is the average normal CSF pressure, filtration and absorption are equal. Below a pressure of approximately 68 mm CSF, absorption stops. Large amounts of fluid accumulate when the reabsorptive capacity of the arachnoid villi is decreased (**external hydrocephalus, communicating hydrocephalus**). Fluid also accumulates proximal to the block and distends the ventricles when the foramens of Luschka and Magendie are blocked or there is obstruction within the ventricular system (**internal hydrocephalus, noncommunicating hydrocephalus**).

Protective Function

The meninges and the CSF protect the brain. The dura is attached firmly to bone. There is normally no "subdural space," the arachnoid being held to the dura by the surface tension of the thin layer of fluid between the two membranes. As shown in Figure 32–4, the brain itself is supported within the arachnoid by the blood vessels and nerve roots and by the multiple fine fibrous **arachnoid trabeculae.** The brain weighs about 1400 g in air, but in its "water bath" of CSF it has a net weight of only 50 g. The buoyancy of the brain in the CSF permits its relatively flimsy attachments to suspend it very effectively. When the head receives a blow, the arachnoid slides on the dura and the brain moves, but its motion is gently checked by the CSF cushion and by the arachnoid trabeculae.

The pain produced by spinal fluid deficiency illustrates the importance of CSF in supporting the brain.

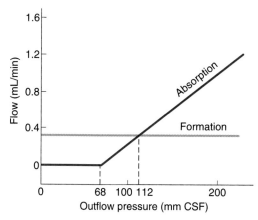

Figure 32–3. CSF formation and absorption in humans at various CSF pressures. Note that at 112 mm CSF, formation and absorption are equal, and at 68 mm CSF, absorption is zero. (Modified and reproduced, with permission, from Cutler RWP et al: Formation and absorption of cerebrospinal fluid in man. Brain 1968;91:707.)

Removal of CSF during lumbar puncture can cause a severe headache after the fluid is removed, because the brain hangs on the vessels and nerve roots, and traction on them stimulates pain fibers. The pain can be relieved by intrathecal injection of sterile isotonic saline.

Head Injuries

Without the protection of the spinal fluid and the meninges, the brain would probably be unable to withstand even the minor traumas of everyday living; but with the protection afforded, it takes a fairly severe blow to produce cerebral damage. The brain is damaged most commonly when the skull is fractured and bone is driven into neural tissue (depressed skull fracture), when the brain moves far enough to tear the delicate bridging veins from the cortex to the bone, or when the brain is accelerated by a blow on the head and is driven against the skull or the tentorium at a point opposite where the blow was struck (**contrecoup injury**).

THE BLOOD-BRAIN BARRIER

The tight junctions between capillary endothelial cells in the brain and between the epithelial cells in the choroid plexus effectively prevent proteins from entering the brain in adults and slow the penetration of smaller molecules. An example is the slow penetration of urea (Figure 32–5). This uniquely limited exchange of substances into the brain is referred to as the **blood-brain barrier.** Some physiologists use this term to refer

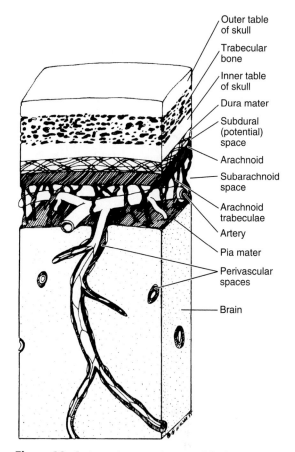

Figure 32–4. Investing membranes of the brain, showing their relation to the skull and to brain tissue. (Reproduced, with permission, from Wheater PR et al, *Functional Histology.* Churchill Livingstone, 1979.)

to the barrier in the capillary walls and the term **blood-CSF barrier** to refer to the barrier in the choroid epithelium. However, the barriers are similar, and it seems more appropriate to use the term blood-brain barrier to refer to exchange across both barriers. Passive diffusion across the tight cerebral capillaries is very limited, and there is little vesicular transport. However, there are numerous carrier-mediated and active transport systems in the cerebral capillaries. These move substances out of as well as into the brain, though movement out of the brain is generally more free than movement into it because of bulk flow of CSF into venous blood via the arachnoid villi (see above).

Penetration of Substances Into the Brain

Water, CO_2, and O_2 penetrate the brain with ease. So do the lipid-soluble free forms of steroid hormones,

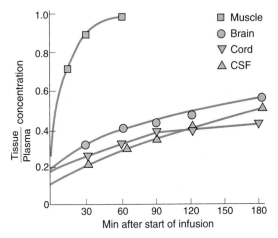

Figure 32–5. Penetration of urea into muscle, brain, spinal cord, and CSF. Urea was administered by constant infusion. (Modified and reproduced, with permission, from Kleeman CR, Davson H, Levin E: Urea transport in the central nervous system. Am J Physiol 1962;203:739.)

whereas their protein-bound forms and, in general, all proteins and polypeptides do not. The easy penetration of CO_2 contrasts with the slow penetration of H^+ and HCO_3^- and has physiologic significance in the regulation of respiration (see Chapter 36).

Glucose is the major ultimate source of energy for nerve cells. Its passive penetration of the blood-brain barrier is slow, but it is transported across the walls of brain capillaries by the glucose transporter GLUT 1 (see Table 19–5). The brain contains two forms of GLUT 1, GLUT 1 55K and GLUT 1 45K. Both are encoded by the same gene, but they differ in the extent to which they are glycosylated. GLUT 1 55K is present in high concentration in brain capillaries (Figure 32–6). Infants with congenital GLUT 1 deficiency develop low CSF glucose concentrations in the presence of normal plasma glucose, and they have seizures and delayed development.

Another transporter in the cerebral capillaries is a unique Na^+-K^+-$2Cl^-$ cotransporter that is stimulated by ET-1 and ET-3 and apparently induced by a humoral factor from astrocytes. It may help keep the brain K^+ concentration low. There are in addition transporters for thyroid hormones, several organic acids, choline, nucleic acid precursors, and neutral, basic, and acidic amino acids.

It has recently been shown that a variety of drugs and peptides actually cross the cerebral capillaries but are promptly transported back into the blood by a multidrug nonspecific transporter in the apical membranes

of the endothelial cells. This **P-glycoprotein** is a member of the family of ATP-binding cassettes that transport various proteins and lipids across cell membranes (see Chapter 1). In mice in which the function of this cassette has been disrupted by gene inactivation, much larger proportions of systemically administered doses of various chemotherapeutic drugs, analgesics, and opioid peptides are found in the brain than in controls. If pharmacologic agents that inhibit this transporter can be developed, they could be of value in the treatment of brain tumors and other CNS diseases in which it is difficult to introduce adequate amounts of therapeutic agents into the brain.

Circumventricular Organs

When dyes that bind to proteins in the plasma are injected, they stain many tissues but spare most of the brain. It was this observation, made many years ago, that led to the concept of the blood-brain barrier. However, four small areas in or near the brain stem stain like the tissues outside the brain. These areas are (1) the **posterior pituitary** (neurohypophysis) and the adjacent ventral part of the **median eminence** of the hypothalamus, (2) the **area postrema,** (3) the **organum vasculosum of the lamina terminalis** (**OVLT,** supraoptic crest), and (4) the **subfornical organ** (**SFO**).

These areas are referred to collectively as the **circumventricular organs** (Figure 32–7). All have fenestrated capillaries, and because of their permeability they

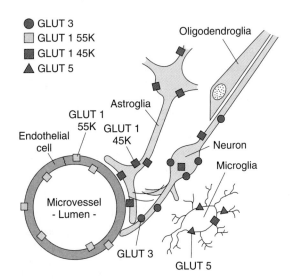

Figure 32–6. Localization of the various GLUT transporters in the brain. (Reproduced, with permission, from Maher F, Vannucci SJ, Simpson IA: Glucose transporter proteins in brain. FASEB J 1994;8:1003.)

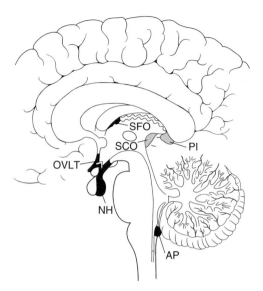

Figure 32–7. Circumventricular organs. The neurohypophysis (NH), organum vasculosum of the lamina terminalis (OVLT, supraoptic crest), subfornical organ (SFO), and area postrema (AP) are shown projected on a sagittal section of the human brain. SCO, subcommissural organ; PI, pineal.

are said to be "outside the blood-brain barrier." Some of them function as **neurohemal organs,** ie, areas in which polypeptides secreted by neurons enter the circulation; for example, oxytocin and vasopressin enter the general circulation in the posterior pituitary, and hypothalamic hypophysiotropic hormones enter the portal hypophysial circulation in the median eminence. Other circumventricular organs contain receptors for many different peptides and other substances, and they function as chemoreceptor zones, ie, areas in which substances in the circulating blood can act to trigger changes in brain function without penetrating the blood-brain barrier. The area postrema is a chemoreceptor trigger zone that initiates vomiting in response to chemical changes in the plasma (see Chapter 14). It is also concerned with cardiovascular control, and in many species circulating angiotensin II acts on the area postrema to trigger an increase in blood pressure. Angiotensin II also acts on the SFO and possibly on the OVLT to increase water intake. In addition, it appears that the OVLT is the site of the osmoreceptor controlling vasopressin secretion (see Chapter 14), and there is evidence that circulating IL-1 produces fever by acting on this circumventricular organ.

The subcommissural organ (Figure 32–7) is closely associated with the pineal and histologically resembles the circumventricular organs. However, it does not

have fenestrated capillaries, is less permeable, and has no established function. The pineal and the anterior pituitary have fenestrated capillaries and are outside the blood-brain barrier, but both are endocrine glands and are not part of the brain.

Function of the Blood-Brain Barrier

The blood-brain barrier probably maintains the constancy of the environment of the neurons in the central nervous system. These neurons are so dependent upon the concentrations of K^+, Ca^{2+}, Mg^{2+}, H^+, and other ions in the fluid bathing them that even minor variations have far-reaching consequences. The constancy of the composition of the ECF in all parts of the body is maintained by multiple homeostatic mechanisms (see Chapters 1 and 39), but because of the sensitivity of the cortical neurons to ionic change, it is not surprising that an additional defense has evolved to protect them. Similarly, a blood-testis barrier protects the composition of the fluid surrounding the germinal epithelium and a placental barrier protects the composition of the body fluids of the fetus.

Other suggested functions for the blood-brain barrier are protection of the brain from endogenous and exogenous toxins in the blood and prevention of the escape of neurotransmitters into the general circulation.

Development of the Blood-Brain Barrier

In experimental animals, many small molecules penetrate the brain more readily during the fetal and neonatal period than they do in the adult. On this basis, it is often stated that the blood-brain barrier is immature at birth. Humans are more mature at birth than rats and various other experimental animals, and detailed data on passive permeability of the human blood-brain barrier are not available. However, in severely jaundiced infants with high plasma levels of free bilirubin and an immature hepatic bilirubin-conjugating system, free bilirubin enters the brain and, in the presence of asphyxia, damages the basal ganglia **(kernicterus).** The counterpart of this situation in later life is the Crigler-Najjar syndrome, in which there is a congenital deficiency of glucuronyl transferase. These individuals can have very high free bilirubin levels in the blood and develop encephalopathy. In other conditions, free bilirubin levels are generally not high enough to produce brain damage.

Clinical Implications

Physicians must know the degree to which drugs penetrate the brain in order to treat diseases of the nervous system intelligently. For example, it is clinically relevant that the amines dopamine and serotonin penetrate brain tissue to a very limited degree but their corresponding acid precursors, L-dopa and 5-hydroxytryptophan, respectively, enter with relative ease (see Chapters 12 and 15).

Another important clinical consideration is the fact that the blood-brain barrier tends to break down in areas of infection or injury. Tumors develop new blood vessels, and the capillaries that are formed lack contact with normal astrocytes. Therefore, there are no tight junctions, and the vessels may even be fenestrated. The lack of a barrier helps in identifying the location of tumors; substances such as radioactive iodine-labeled albumin penetrate normal brain tissue very slowly, but they enter tumor tissue, making the tumor stand out as an island of radioactivity in the surrounding normal brain. The blood-brain barrier can also be temporarily disrupted by sudden marked increases in blood pressure or by intravenous injection of hypertonic fluids.

CEREBRAL BLOOD FLOW & ITS REGULATION

Kety Method

According to the **Fick principle** (see Chapter 29), the blood flow of any organ can be measured by determining the amount of a given substance (Q_x) removed from the bloodstream by the organ per unit of time and dividing that value by the difference between the concentration of the substance in arterial blood and the concentration in the venous blood from the organ ($[A_x] - [V_x]$). Thus:

$$\text{Cerebral blood flow (CBF)} = \frac{Q_x}{[A_x] - [V_x]}$$

This can be applied clinically, using inhaled nitrous oxide (N_2O) **(Kety method).** The average cerebral blood flow in young adults is 54 mL/100 g/min. The average adult brain weighs about 1400 g, so the flow for the whole brain is about 756 mL/min. Note that the Kety method provides an average value for perfused areas of brain; that it gives no information about regional differences in blood flow; and that since it depends on N_2O uptake, it measures flow to perfused parts of the brain only. If the blood flow to a portion of the brain is occluded, there is no change in the measured flow, because the nonperfused area does not take up any N_2O.

In spite of the marked local fluctuations in brain blood flow with neural activity, the cerebral circulation is regulated in such a way that total blood flow remains relatively constant. The factors involved in regulating the flow are summarized in Figure 32–8.

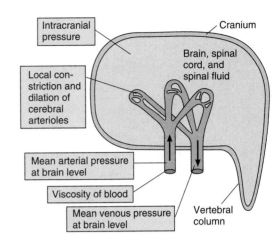

Figure 32–8. Diagrammatic summary of the factors affecting overall cerebral blood flow.

Role of Intracranial Pressure

In adults, the brain, spinal cord, and spinal fluid are encased, along with the cerebral vessels, in a rigid bony enclosure. The cranial cavity normally contains a brain weighing approximately 1400 g, 75 mL of blood, and 75 mL of spinal fluid. Because brain tissue and spinal fluid are essentially incompressible, the volume of blood, spinal fluid, and brain in the cranium at any time must be relatively constant **(Monro-Kellie doctrine).** More importantly, the cerebral vessels are compressed whenever the intracranial pressure rises. Any change in venous pressure promptly causes a similar change in intracranial pressure. Thus, a rise in venous pressure decreases cerebral blood flow both by decreasing the effective perfusion pressure and by compressing the cerebral vessels. This relationship helps to compensate for changes in arterial blood pressure at the level of the head. For example, if the body is accelerated upward (positive *g*), blood moves toward the feet and arterial pressure at the level of the head decreases. However, venous pressure also falls and intracranial pressure falls, so that the pressure on the vessels decreases and blood flow is much less severely compromised than it would otherwise be. Conversely, during acceleration downward, force acting toward the head (negative *g*) increases arterial pressure at head level, but intracranial pressure also rises, so that the vessels are supported and do not rupture. The cerebral vessels are protected during the straining associated with defecation or delivery in the same way.

Effect of Intracranial Pressure Changes on Systemic Blood Pressure

When intracranial pressure is elevated to more than 33 mm Hg over a short period, cerebral blood flow is significantly reduced. The resultant ischemia stimulates the vasomotor area (see Chapter 31), and systemic blood pressure rises. Stimulation of vagal outflow produces bradycardia, and respiration is slowed. The blood pressure rise, which was described by Cushing and is sometimes called the **Cushing reflex,** helps to maintain the cerebral blood flow. Over a considerable range, the rise in systemic blood pressure is proportionate to the rise in intracranial pressure, although eventually a point is reached where the intracranial pressure exceeds the arterial pressure and cerebral circulation ceases.

Autoregulation

Autoregulation is prominent in the brain (Figure 32–9). This process, by which the flow to many tissues is maintained at relatively constant levels despite variations in perfusion pressure, is discussed in Chapter 31. In the brain, autoregulation maintains a normal cerebral blood flow at arterial pressures of 65–140 mm Hg.

Role of Vasomotor & Sensory Nerves

The innervation of large cerebral blood vessels by postganglionic sympathetic and parasympathetic nerves and the additional distal innervation by sensory nerves are described above. The role of these nerves remains a matter of debate. It has been argued that noradrenergic discharge occurs when the blood pressure is markedly elevated. This reduces the resultant passive increase in blood flow and helps protect the blood-brain barrier from the disruption that could otherwise occur (see above). Thus, vasomotor discharges affect autoregulation. With sympathetic stimulation, the constant-flow, or plateau, part of the pressure-flow curve is extended to the right (Figure 32–9); ie, greater increases in pressure can occur without an increase in flow. On the other hand, the vasodilator

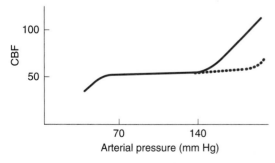

Figure 32–9. Autoregulation of cerebral blood flow (CBF) during steady-state conditions. The dotted line shows the alteration produced by sympathetic stimulation during autoregulation.

hydralazine and the ACE inhibitor captopril reduce the length of the plateau.

Blood Flow in Various Parts of the Brain

A major advance in recent decades has been the development of techniques for monitoring regional blood flow in living, conscious humans. Among the most valuable methods are **positron emission tomography (PET)** and related techniques in which a short-lived radioisotope is used to label a compound and the compound is injected. The arrival and clearance of the tracer are monitored by a battery of scintillation detectors placed over the head. The output from the detectors is processed in a computer and can be displayed on a color television screen in such a way that the color corresponding to the location of each detector is proportionate to the flow it is detecting. Since blood flow is tightly coupled to brain metabolism, local uptake of 2-deoxyglucose is also a good index of blood flow (see below and Chapter 8). If the 2-deoxyglucose is labeled with a short-half-life positron emitter such as ^{18}F, ^{11}O, or ^{15}O, its concentration in any part of the brain can be monitored.

Another valuable technique involves magnetic resonance imaging (MRI). MRI is based on detecting resonant signals from different tissues in a magnetic field. **Functional magnetic resonance imaging (fMRI)** measures the amount of blood in a tissue area. When neurons become active, there is an increase in the spike output of the cells and the local field potential, which measures current flow in and around the neurons, including nearby glia involved in glutamate metabolism. A still unsettled mechanism triggers an increase in local blood flow and of oxygen in excess of what is needed. The increase in oxygenated blood is detected by fMRI. It is now known that the increased blood flow correlates best with the local field potential rather than spike output. The resolution of fMRI is now better than that of PET.

In resting humans, the average blood flow in gray matter is 69 mL/100 g/min compared with 28 mL/100 g/min in white matter. A striking feature of cerebral function is the marked variation in local blood flow with changes in brain activity. Examples are shown in Figures 16–1 and 32–10. In subjects who are awake but at rest, blood flow is greatest in the premotor and frontal regions. This is the part of the brain that is believed to be concerned with decoding and analyzing afferent input and with intellectual activity. During voluntary clenching of the right hand, flow is increased in the hand area of the left motor cortex and the corresponding sensory areas in the postcentral gyrus. Especially when the movements being performed are sequential, the flow is also increased in the supplementary

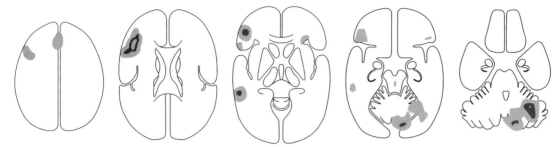

Figure 32–10. Activity in the human brain at five different horizontal levels while a subject generates a verb which is appropriate for each noun presented by an examiner. This mental task activates the frontal cortex (slices 1–4), anterior cingulate gyrus (slice 1), and posterior temporal lobe (slice 3) on the left side and the cerebellum (slices 4 and 5) on the right side. Light color, moderate activation; dark color, marked activation. (Based on PET scans in Posner MI, Raichle ME: *Images of Mind.* Scientific American Library, 1994.)

motor area. When subjects talk, there is a bilateral increase in blood flow in the face, tongue, and mouth-sensory and motor areas and the upper premotor cortex in the categorical (usually the left) hemisphere. When the speech is stereotyped, Broca's and Wernicke's areas do not show increased flow, but when the speech is creative, ie, when it involves ideas, there are flow increases in both these areas. Reading produces widespread increases in blood flow. Problem-solving, reasoning, and motor ideation without movement produce increases in selected areas of the premotor and frontal cortex. In anticipation of a cognitive task, many of the brain areas that will be activated during the task are activated beforehand, as if the brain produces an internal model of the expected task. In right-handed individuals, blood flow to the left hemisphere is greater when a verbal task is being performed and blood flow to the right hemisphere is greater when a spatial task is being performed.

PET scanning and fMRI have been applied to the study of various diseases. Epileptic foci are hyperemic during seizures, whereas flow is reduced in other parts of the brain. Between seizures, flow is sometimes reduced in the foci that generate the seizures. There is decreased parieto-occipital flow in patients with symptoms of agnosia (see Chapter 16). In Alzheimer's disease, the earliest change is decreased metabolism and blood flow in the superior parietal cortex, with later spread to the temporal and finally the frontal cortex. There is relative sparing of the pre- and postcentral gyri, basal ganglia, thalamus, brain stem, and cerebellum. In Huntington's disease, there is a bilateral reduction in the blood flow to the caudate nucleus, and this alteration in flow occurs early in the disease. In manic depressives but not in patients with unipolar depression, there is a general decrease in cortical blood flow when the patients are depressed. In schizophrenia there is some evidence for decreased blood flow in the frontal

lobes, temporal lobes, and basal ganglia. During the aura in patients with migraine, a bilateral decrease in blood flow starts in the occipital cortex and spreads anteriorly to the temporal and parietal lobes.

BRAIN METABOLISM & OXYGEN REQUIREMENTS

Uptake & Release of Substances by the Brain

If the cerebral blood flow is known, it is possible to calculate the consumption or production by the brain of O_2, CO_2, glucose, or any other substance present in the bloodstream by multiplying the cerebral blood flow by the difference between the concentration of the substance in arterial blood and its concentration in cerebral venous blood (Table 32–3). When calculated in this fashion, a negative value indicates that the brain is producing the substance.

Oxygen Consumption

O_2 consumption by the human brain (**cerebral metabolic rate for O_2**, CMRO$_2$) averages about 3.5 mL/100 g of brain/min (49 mL/min for the whole brain) in an adult. This figure represents approximately 20% of the total body resting O_2 consumption (Table 32–1). The brain is extremely sensitive to hypoxia, and occlusion of its blood supply produces unconsciousness in as short a period as 10 seconds. The vegetative structures in the brain stem are more resistant to hypoxia than the cerebral cortex, and patients may recover from accidents such as cardiac arrest and other conditions causing fairly prolonged hypoxia with normal vegetative functions but severe, permanent intellectual deficiencies. The basal ganglia use O_2 at a very high rate,

Table 32–3. Utilization and production of substances by the adult human brain in vivo.

Substance	Uptake (+) or Output (−) per 100 g of Brain/min	Total/min
Substances utilized		
Oxygen	+3.5 mL	+49 mL
Glucose	+5.5 mg	+77 mg
Glutamate	+0.4 mg	+5.6 mg
Substances produced		
Carbon dioxide	−3.5 mL	−49 mL
Glutamine	−0.6 mL	−8.4 mg

Substances not used or produced in the fed state: lactate, pyruvate, total ketones, α-ketoglutarate.

and symptoms of Parkinson's disease as well as intellectual deficits can be produced by chronic hypoxia. The thalamus and the inferior colliculus are also very susceptible to hypoxic damage.

Energy Sources

Glucose is the major ultimate source of energy for the brain; under normal conditions, 90% of the energy needed to maintain ion gradients across cell membranes and transmit electrical impulses comes from this source. Glucose enters the brain via GLUT 1 in cerebral capillaries (see above). Other transporters then distribute it to neurons and glial cells.

Glucose is taken up from the blood in large amounts, and the RQ (respiratory quotient; see Chapter 17) of cerebral tissue is 0.95–0.99 in normal individuals. In general, glucose utilization at rest parallels blood flow and O_2 consumption. This does not mean that the total source of energy is always glucose. During prolonged starvation, there is appreciable utilization of other substances. Indeed, there is evidence that as much as 30% of the glucose taken up under normal conditions is converted to amino acids, lipids, and proteins and that substances other than glucose are metabolized for energy during convulsions. There may also be some utilization of amino acids from the circulation even though the amino acid arteriovenous difference across the brain is normally minute. Insulin is not required for most cerebral cells to utilize glucose.

The consequences of hypoglycemia in terms of neural function are discussed in Chapter 19.

Glutamate & Ammonia Removal

The brain's uptake of glutamate is approximately balanced by its output of glutamine. Glutamate entering the brain takes up ammonia and leaves as glutamine (see Chapter 17). The glutamate-glutamine conversion in the brain—the opposite of the reaction in the kidney that produces some of the ammonia entering the tubules—serves as a detoxifying mechanism to keep the brain free of ammonia. Ammonia is very toxic to nerve cells, and ammonia intoxication is believed to be a major cause of the bizarre neurologic symptoms in hepatic coma.

Stroke

When the blood supply to a part of the brain is interrupted, ischemia damages or kills the cells in the area, producing the signs and symptoms of a stroke. There are two general types of strokes: hemorrhagic and ischemic. Hemorrhagic stroke occurs when a cerebral artery or arteriole ruptures, sometimes but not always at the site of a small aneurysm. Ischemic stroke occurs when flow in a vessel is compromised by atherosclerotic plaques on which thrombi form. Thrombi may also be produced elsewhere (eg, in the atria in patients with atrial fibrillation) and pass to the brain as emboli.

Until recently, there was little that could be done to modify the course of a stroke and its consequences. However, it has now become clear that in the penumbra, the area surrounding the most severe brain damage, ischemia reduces glutamate uptake and the increase in local glutamate causes excitotoxic damage and death to neurons (see Chapter 4). In experimental animals and perhaps in humans, drugs that prevent this excitotoxic damage significantly reduce the effects of strokes. In addition, clot-lysing drugs such as t-PA (see Chapter 27) are of benefit. Both antiexcitotoxic treatment and t-PA must be given early in the course of a stroke to be of maximum benefit, and this is why stroke has become a condition in which rapid diagnosis and treatment have become important. In addition, of course, it is important to determine if a stroke is thrombotic or hemorrhagic, since clot lysis is contraindicated in the latter.

CORONARY CIRCULATION

Anatomic Considerations

The two coronary arteries that supply the myocardium arise from the sinuses behind two of the cusps of the aortic valve at the root of the aorta (Figure 32–11). Eddy currents keep the valves away from the orifices of the arteries, and they are patent throughout the cardiac cycle. The right coronary artery has a greater flow in 50% of individuals, the left has a greater flow in 20%,

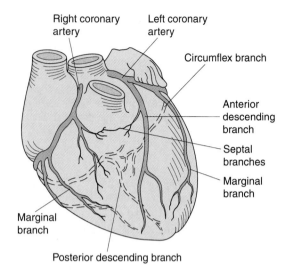

Figure 32–11. Coronary arteries and their principal branches in humans. (Reproduced, with permission, from Ross G: The cardiovascular system. In: *Essentials of Human Physiology.* Ross G [editor]. Copyright © 1978 by Year Book Medical Publishers, Inc.)

and the flow is equal in 30%. Most of the venous blood returns to the heart through the coronary sinus and anterior cardiac veins (Figure 32–12), which drain into the right atrium. In addition, there are other vessels that empty directly into the heart chambers. These include **arteriosinusoidal vessels,** sinusoidal capillary-like vessels that connect arterioles to the chambers; **thebesian veins** that connect capillaries to the chambers; and a few **arterioluminal vessels** that are small arteries draining directly into the chambers. There are a

few anastomoses between the coronary arterioles and extracardiac arterioles, especially around the mouths of the great veins. Anastomoses between coronary arterioles in humans only pass particles less than 40 μm in diameter, but there is evidence that these channels enlarge and increase in number in patients with coronary artery disease.

Pressure Gradients & Flow in the Coronary Vessels

The heart is a muscle that, like skeletal muscle, compresses its blood vessels when it contracts. The pressure inside the left ventricle is slightly higher than in the aorta during systole (Table 32–4). Consequently, flow occurs in the arteries supplying the subendocardial portion of the left ventricle only during diastole, although the force is sufficiently dissipated in the more superficial portions of the left ventricular myocardium to permit some flow in this region throughout the cardiac cycle. Since diastole is shorter when the heart rate is high, left ventricular coronary flow is reduced during tachycardia. On the other hand, the pressure differential between the aorta and the right ventricle, and the differential between the aorta and the atria, are somewhat greater during systole than during diastole. Consequently, coronary flow in those parts of the heart is not appreciably reduced during systole. Flow in the right and left coronary arteries is shown in Figure 32–13. Because there is no blood flow during systole in the subendocardial portion of the left ventricle, this region is prone to ischemic damage and is the most common site of myocardial infarction. Blood flow to the left ventricle is decreased in patients with stenotic aortic valves because in aortic stenosis the pressure in the left ventricle must be much higher than that in the aorta to eject the blood. Consequently, the coronary vessels are severely compressed during systole. Patients with this disease are particularly prone to develop symptoms of my-

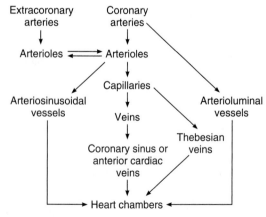

Figure 32–12. Diagram of the coronary circulation.

Table 32–4. Pressure in aorta and left and right ventricles (vent) in systole and diastole.

	Pressure (mm Hg) in			Pressure Differential (mm Hg) Between Aorta and	
	Aorta	Left Vent	Right Vent	Left Vent	Right Vent
Systole	120	121	25	−1	95
Diastole	80	0	0	80	80

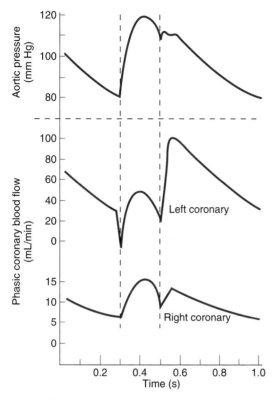

Figure 32–13. Blood flow in the left and right coronary arteries during various phases of the cardiac cycle. Systole occurs between the two vertical dashed lines. (Reproduced, with permission, from Berne RM, Levy MN: *Physiology.* Mosby, 1983.)

ocardial ischemia, in part because of this compression and in part because the myocardium requires more O_2 to expel blood through the stenotic aortic valve. Coronary flow is also decreased when the aortic diastolic pressure is low. The rise in venous pressure in conditions such as congestive heart failure reduces coronary flow because it decreases effective coronary perfusion pressure.

Coronary blood flow has been measured by inserting a catheter into the coronary sinus and applying the Kety method to the heart on the assumption that the N_2O content of coronary venous blood is typical of the entire myocardial effluent. Coronary flow at rest in humans is about 250 mL/min (5% of the cardiac output). A number of techniques utilizing **radionuclides,** radioactive tracers that can be detected with radiation detectors over the chest, have been used to study regional blood flow in the heart and to detect areas of ischemia and infarct as well as to evaluate ventricular function. Radionuclides such as thallium 201 (^{201}T1) are

pumped into cardiac muscle cells by Na^+-K^+ ATPase and equilibrate with the intracellular K^+ pool. For the first 10–15 minutes after intravenous injection, ^{201}T1 distribution is directly proportionate to myocardial blood flow, and areas of ischemia can be detected by their low uptake. The uptake of this isotope is often determined soon after exercise and again several hours later to bring out areas in which exertion leads to compromised flow. Conversely, radiopharmaceuticals such as technetium 99m stannous pyrophosphate (^{99m}Tc-PYP) are selectively taken up by infarcted tissue by an incompletely understood mechanism and make infarcts stand out as "hot spots" on scintiscans of the chest. Coronary angiography can be combined with measurement of ^{133}Xe washout (see above) to provide detailed analysis of coronary blood flow. Radiopaque contrast medium is first injected into the coronary arteries, x-rays being used to outline their distribution. The angiographic camera is then replaced with a multiple-crystal scintillation camera, and ^{133}Xe washout is measured. An example of normal flow distribution after injection in a left coronary artery is shown in Figure 32–14.

Variations in Coronary Flow

At rest, the heart extracts 70–80% of the O_2 from each unit of blood delivered to it (Table 32–1). O_2 consumption can be increased significantly only by increasing blood flow. Therefore, it is not surprising that blood flow increases when the metabolism of the myocardium is increased. The caliber of the coronary vessels, and consequently the rate of coronary blood flow, is influenced not only by pressure changes in the aorta but also by chemical and neural factors. The coronary circulation shows considerable autoregulation.

Chemical Factors

The close relationship between coronary blood flow and myocardial O_2 consumption indicates that one or more of the products of metabolism cause coronary vasodilation. Factors suspected of playing this role include O_2 lack and increased local concentrations of CO_2, H^+, K^+, lactate, prostaglandins, adenine nucleotides, and adenosine. More than one of these vasodilator metabolites could be involved. Asphyxia, hypoxia, and intracoronary injections of cyanide all increase coronary blood flow 200–300% in denervated as well as intact hearts, and the feature common to these three stimuli is hypoxia of the myocardial fibers. A similar increase in flow is produced in the area supplied by a coronary artery if the artery is occluded and then released. This **reactive hyperemia** is similar to that seen in the skin (see below). There is evidence that in the heart it is due to release of adenosine. The

Mean LV flow (65 mL/100 g/min)

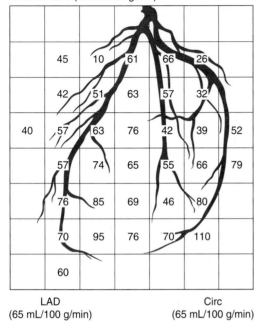

LAD
(65 mL/100 g/min)

Circ
(65 mL/100 g/min)

Figure 32–14. Normal human myocardial perfusion pattern following injection of ^{133}Xe into the left main coronary artery. The branches of the artery are shown in color, and the numbers in the squares are the flow values (in mL/100 g/min) for the regions under each scintillation detector. Circ, circumflex artery; LAD, left anterior descending artery; LV, left ventricle. (Reproduced, with permission, from Cannon PJ et al: Evaluation of myocardial circulation with radionuclides. Cardiovasc Med 1978;2:371.)

adenosine appears in addition to ameliorate the reperfusion-induced injury that occurs when blood flow is reestablished (see Chapter 33).

Neural Factors

The coronary arterioles contain α-adrenergic receptors, which mediate vasoconstriction, and β-adrenergic receptors, which mediate vasodilation. Activity in the noradrenergic nerves to the heart and injections of norepinephrine cause coronary vasodilation. However, norepinephrine increases the heart rate and the force of cardiac contraction, and the vasodilation is due to production of vasodilator metabolites in the myocardium secondary to the increase in its activity. When the inotropic and chronotropic effects of noradrenergic discharge are blocked by a β-adrenergic blocking drug, stimulation of the noradrenergic nerves or injection of norepinephrine in unanesthetized animals elicits coronary vasoconstriction. Thus, the direct effect of noradrenergic stimulation is constriction rather than dilation of the coronary vessels. Stimulation of vagal fibers to the heart dilates the coronaries.

When the systemic blood pressure falls, the overall effect of the reflex increase in noradrenergic discharge is increased coronary blood flow secondary to the metabolic changes in the myocardium at a time when the cutaneous, renal, and splanchnic vessels are constricted. In this way the circulation of the heart, like that of the brain, is preserved when flow to other organs is compromised.

Coronary Artery Disease

When flow through a coronary artery is reduced to the point that the myocardium it supplies becomes hypoxic, "P factor" accumulates and **angina pectoris** develops (see Chapter 7). If the myocardial ischemia is severe and prolonged, irreversible changes occur in the muscle, and the result is **myocardial infarction.** Many individuals have angina only on exertion, and blood flow is normal at rest. Others have more severe restriction of blood flow and have anginal pain at rest as well. Partially occluded coronary arteries can be constricted further by vasospasm, producing myocardial infarction. However, it is now clear that the most common cause of myocardial infarction is rupture of an atherosclerotic **plaque,** or hemorrhage into it, which triggers the formation of a coronary occluding clot at the site of the plaque.

The electrocardiographic changes in myocardial infarction are discussed in Chapter 28. When myocardial cells actually die, they leak enzymes into the circulation, and measuring the rises in serum enzymes and isoenzymes produced by infarcted myocardial cells also plays an important role in the diagnosis of myocardial infarction. The enzymes most commonly measured today are the MB isomer of creatine kinase (CK-MB), troponin T, and troponin I.

Myocardial infarction is a very common cause of death in developed countries because of the widespread occurrence of atherosclerosis. The relation of dietary cholesterol and hypercholesterolemia to the artherosclerosis that causes plaques in coronary and other arteries is discussed in Chapter 17. In addition, there is a relation between atherosclerosis and circulating levels of **lipoprotein(a) (Lp[a]).** Lp(a) interferes with fibrinolysis by down-regulating plasmin generation (see Chapter 27). There is also a strong positive correlation to circulating levels of homocysteine. This substance damages endothelial cells. It is converted to nontoxic methionine in the presence of folate and vitamin B_{12}, and clinical trials are under way to determine whether supplements of folate and B_{12} lower the incidence of coronary disease.

It now appears that there is an important inflammatory component in atherosclerosis as well. The lesions of the disease contain inflammatory cells, and there is a positive correlation between increased levels of C-reactive protein and other **inflammatory markers** in the circulation and subsequent myocardial infarction. Much more work needs to be done in this area, but it is interesting that there is a correlation between circulating antibodies to *Chlamydia pneumoniae* and atherosclerotic plaques. This microorganism has an amino acid sequence in its plasma membrane that resembles a sequence in heart α myosin heavy chains, and injection of this sequence causes autoimmune inflammation and fibrosis of coronary arteries in mice, an example of **molecular mimicry** (see Chapter 27).

Treatment of myocardial infarction has become very complex and is beyond the scope of this book except to note that it should be started as promptly as possible.

SPLANCHNIC CIRCULATION

The blood from the intestines, pancreas, and spleen drains via the hepatic portal vein to the liver and from the liver via the hepatic veins to the inferior vena cava. The viscera and the liver receive about 30% of the cardiac output via the celiac, superior mesenteric, and inferior mesenteric arteries (Figure 32–15). The liver receives about 1000 mL/min from the portal vein and 500 mL/min from the hepatic artery.

Intestinal Circulation

The intestines are supplied by a series of parallel circulations via the branches of the superior and inferior mesenteric arteries (Figure 32–15). There are extensive anastomoses between these vessels, but blockage of a large intestinal artery still leads to infarction of the bowel. The blood flow to the mucosa is greater than that to the rest of the intestinal wall, and it responds to changes in metabolic activity. Thus, blood flow to the small intestine (and hence blood flow in the portal vein) doubles after a meal, and the increase lasts up to 3 hours. The intestinal circulation is capable of extensive autoregulation.

Hepatic Circulation

There are large gaps between endothelial cells in the walls of hepatic sinusoids, and the sinusoids are highly permeable. The way the intrahepatic branches of the hepatic artery and portal vein converge on the sinu-

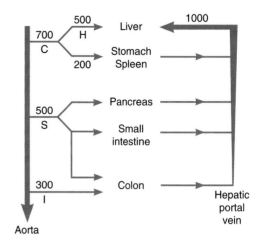

Figure 32–15. Splanchnic circulation. Note that most of the viscera are supplied by a series of parallel circuits, whereas the liver receives blood from the hepatic artery and the portal vein. The figures are average blood flows (mL/min). C, celiac axis; S, superior mesenteric artery; I, inferior mesenteric artery.

soids and drain into the central lobular veins of the liver is shown in Figures 26–20 and 32–16. The functional unit of the liver is the acinus. Each acinus is at the end of a vascular stalk containing terminal branches of portal veins, hepatic arteries, and bile ducts. Blood flows from the center of this functional unit to the terminal branches of the hepatic veins at the periphery (Figure 32–16). This is why the central portion of the acinus, sometimes called zone 1, is well oxygenated, the intermediate zone (zone 2) is moderately well oxygenated, and the peripheral zone (zone 3) is least well oxygenated and most susceptible to anoxic injury. The hepatic veins drain into the inferior vena cava. The acini have been likened to grapes or berries, each on a vascular stem. There are about 100,000 acini in the human liver.

Portal venous pressure is normally about 10 mm Hg in humans, and hepatic venous pressure is approximately 5 mm Hg. The mean pressure in the hepatic artery branches that converge on the sinusoids is about 90 mm Hg, but the pressure in the sinusoids is lower than the portal venous pressure, so there is a marked pressure drop along the hepatic arterioles. This pressure drop is adjusted so that there is an inverse relationship between hepatic arterial and portal venous blood flow. This inverse relationship may be maintained in part by the rate at which adenosine is removed from the region around the arterioles. According to this hypothesis, adenosine is produced by metabolism at a constant rate. When portal flow is reduced, it is washed away

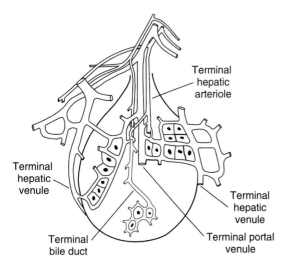

Figure 32–16. Concept of the acinus as the functional unit of the liver. In each acinus, blood in the portal venule and hepatic arteriole enters the center of the acinus and flows outward to the hepatic venule. (Reproduced, with permission, from Lautt WW, Greenway CV: Conceptual review of the hepatic vascular bed. Hepatology 1987;7:952. Copyright © by The American Association for the Study of Liver Diseases.)

more slowly, and the local accumulation of adenosine dilates the terminal arterioles.

The intrahepatic portal vein radicles have smooth muscle in their walls that is innervated by noradrenergic vasoconstrictor nerve fibers reaching the liver via the third to eleventh thoracic ventral roots and the splanchnic nerves. The vasoconstrictor innervation of the hepatic artery comes from the hepatic sympathetic plexus. There are no known vasodilator fibers reaching the liver. At rest, circulation in the peripheral portions of the liver is sluggish, and only a portion of the organ is actively perfused. When systemic venous pressure rises, the portal vein radicles are dilated passively and the amount of blood in the liver increases. In congestive heart failure, this hepatic venous congestion may be extreme. Conversely, when there is diffuse noradrenergic discharge in response to a drop in systemic blood pressure, the intrahepatic portal radicles constrict, portal pressure rises, and blood flow through the liver is brisk, bypassing most of the organ. Most of the blood in the liver enters the systemic circulation. Constriction of the hepatic arterioles diverts blood from the liver, and constriction of the mesenteric arterioles reduces portal inflow. In severe shock, hepatic blood flow may be reduced to such a degree that there is patchy necrosis of the liver.

Reservoir Function of the Splanchnic Circulation

In dogs and other carnivores, there is a large amount of smooth muscle in the capsule of the spleen. The spleen traps blood, and rhythmic contractions of its capsule pump plasma into the lymphatics. The spleen therefore contains a reservoir of blood rich in cells. Noradrenergic nerve discharge and epinephrine make the spleen contract strongly, discharging the blood into the circulation. This function of the spleen is quantitatively unimportant in humans. However, the reservoir function of the whole visceral circulation is important. For example, 25–30% of the volume of the liver is accounted for by blood. Contraction of the capacitance vessels in the viscera can pump a liter of blood into the arterial circulation in less than a minute.

Other blood reservoirs that contain a large volume of blood at rest are the skin and lungs. During vigorous exercise, constriction of the vessels in these organs and decreased blood "storage" in the liver and other portions of the splanchnic bed, the skin, and the lungs may increase the volume of actively circulating blood perfusing the muscles by as much as 30%.

CUTANEOUS CIRCULATION

The amount of heat lost from the body is regulated to a large extent by varying the amount of blood flowing through the skin (see Chapter 14). The fingers, toes, palms, and earlobes contain well-innervated anastomotic connections between arterioles and venules (arteriovenous anastomoses; see Chapter 30). Blood flow in response to thermoregulatory stimuli can vary from 1 to as much as 150 mL/100 g of skin/min, and it has been postulated that these variations are possible because blood can be shunted through the anastomoses. The subdermal capillary and venous plexus is a blood reservoir of some importance, and the skin is one of the few places where the reactions of blood vessels can be observed visually.

White Reaction

When a pointed object is drawn lightly over the skin, the stroke lines become pale (**white reaction**). The mechanical stimulus apparently initiates contraction of the precapillary sphincters, and blood drains out of the capillaries and small veins. The response appears in about 15 seconds.

Triple Response

When the skin is stroked more firmly with a pointed instrument, instead of the white reaction there is reddening at the site that appears in about 10 seconds (**red reaction**). This is followed in a few minutes by local swelling and diffuse, mottled reddening around the injury. The initial redness is due to capillary dilation, a direct response of the capillaries to pressure. The swelling (**wheal**) is local edema due to increased permeability of the capillaries and postcapillary venules, with consequent extravasation of fluid. The redness spreading out from the injury (**flare**) is due to arteriolar dilation. This three-part response, the red reaction, wheal, and flare, is called the **triple response** and is part of the normal reaction to injury (see Chapter 20). It is present after total sympathectomy. The flare is absent in locally anesthetized skin and in denervated skin after the sensory nerves have degenerated, but it is present immediately after nerve block or section above the site of the injury. This plus other evidence indicates that it is due to an **axon reflex,** a response in which impulses initiated in sensory nerves by the injury are relayed antidromically down other branches of the sensory nerve fibers (Figure 32–17). This is the one situation in the body where there is substantial evidence for a physiologic effect due to antidromic conduction. The transmitter released at the central termination of the sensory C fiber neurons is substance P (see Chapter 4), and substance P and CGRP are present in all parts of the neurons. Both dilate arterioles, and, in addition, sub-stance P causes extravasation of fluid. Effective nonpeptide antagonists to substance P have now been developed, and they reduce the extravasation. Thus, it appears that these peptides produce the wheal.

Reactive Hyperemia

A response of the blood vessels that occurs in many organs but is visible in the skin is **reactive hyperemia,** an increase in the amount of blood in a region when its circulation is reestablished after a period of occlusion. When the blood supply to a limb is occluded, the cutaneous arterioles below the occlusion dilate. When the circulation is reestablished, blood flowing into the dilated vessels makes the skin become fiery red. O_2 diffuses a short distance through the skin, and reactive hyperemia is prevented if the circulation of the limb is occluded in an atmosphere of 100% O_2. Therefore, the arteriolar dilation is apparently due to a local effect of hypoxia.

Generalized Responses

Noradrenergic nerve stimulation and circulating epinephrine and norepinephrine constrict cutaneous blood vessels. There are no known vasodilator nerve fibers to the cutaneous vessels, and vasodilation is brought about by a decrease in constrictor tone as well as the local production of bradykinin in sweat glands and vasodilator metabolites. Skin color and temperature also depend on the state of the capillaries and venules. A cold blue or gray skin is one in which the arterioles are constricted and the capillaries dilated; a warm red skin is one in which both are dilated.

Because painful stimuli cause diffuse noradrenergic discharge, a painful injury causes generalized cutaneous vasoconstriction in addition to the local triple response. When the body temperature rises during exercise, the cutaneous blood vessels dilate in spite of continuing noradrenergic discharge in other parts of the body. Dilation of cutaneous vessels in response to a rise in hypothalamic temperature (see Chapter 14) is a prepotent reflex response that overcomes other reflex activity. Cold causes cutaneous vasoconstriction; however, with severe cold, superficial vasodilation may supervene. This vasodilation is the cause of the ruddy complexion seen on a cold day.

Shock is more profound in patients with elevated temperatures because of the cutaneous vasodilation, and patients in shock should not be warmed to the point that their body temperature rises. This is sometimes a problem because well-meaning laymen have read in first-aid books that "injured patients should be kept warm," and they pile blankets on accident victims who are in shock.

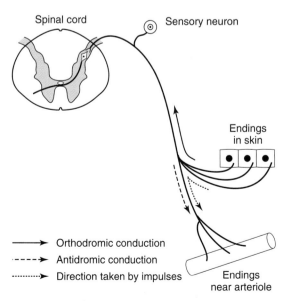

Figure 32–17. Axon reflex.

PLACENTAL & FETAL CIRCULATION

Uterine Circulation

The blood flow of the uterus parallels the metabolic activity of the myometrium and endometrium and undergoes cyclic fluctuations that correlate well with the menstrual cycle in nonpregnant women. The function of the spiral and basilar arteries of the endometrium in menstruation is discussed in Chapter 23. During pregnancy, blood flow increases rapidly as the uterus increases in size (Figure 32–18). Vasodilator metabolites are undoubtedly produced in the uterus, as they are in other active tissues. In early pregnancy the arteriovenous O_2 difference across the uterus is small, and it has been suggested that estrogens act on the blood vessels to increase uterine blood flow in excess of tissue O_2 needs. However, even though uterine blood flow increases 20-fold during pregnancy, the size of the conceptus increases much more, changing from a single cell to a fetus plus a placenta that weighs 4–5 kg at term in humans. Consequently, more O_2 is extracted from

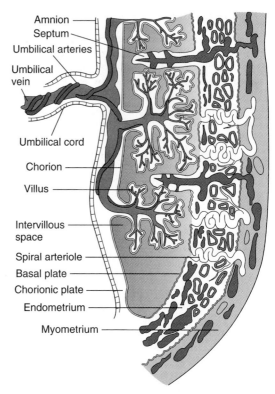

Figure 32–19. Diagram of a section through the human placenta, showing the way the fetal villi project into the maternal sinuses. (Reproduced, with permission, from Benson RC: *Handbook of Obstetrics and Gynecology*, 8th ed. Originally published by Appleton & Lange. Copyright © 1983 by The McGraw-Hill Companies, Inc.

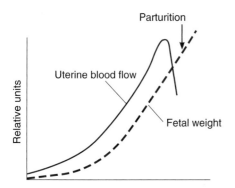

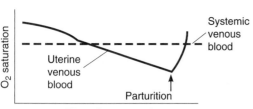

Figure 32–18. Changes in uterine blood flow and the amount of O_2 in uterine venous blood during pregnancy. (After Barcroft H. Modified and redrawn, with permission, from Keele CA, Neil E: *Samson Wright's Applied Physiology*, 12th ed. Oxford Univ Press, 1971.)

the uterine blood during the latter part of pregnancy, and the O_2 saturation of uterine blood falls. Just before parturition there is a sharp decline in uterine blood flow, but the significance of this is not clear.

Placenta

The placenta is the "fetal lung." Its maternal portion is in effect a large blood sinus. Into this "lake" project the villi of the fetal portion containing the small branches of the fetal umbilical arteries and vein (Figure 32–19). O_2 is taken up by the fetal blood and CO_2 is discharged into the maternal circulation across the walls of the villi in a fashion analogous to O_2 and CO_2 exchange in the lungs (see Chapter 34). However, the cellular layers covering the villi are thicker and less permeable than the alveolar membranes in the lungs, and exchange is much less efficient. The placenta is also the route by which all nutritive materials enter the fetus and by which fetal wastes are discharged to the maternal blood.

Fetal Circulation

The arrangement of the circulation in the fetus is shown diagrammatically in Figure 32–20. Fifty-five percent of the fetal cardiac output goes through the placenta. The blood in the umbilical vein in humans is believed to be about 80% saturated with O_2, compared with 98% saturation in the arterial circulation of the adult. The **ductus venosus** (Figure 32–21) diverts some of this blood directly to the inferior vena cava, and the remainder mixes with the portal blood of the fetus. The portal and systemic venous blood of the fetus is only 26% saturated, and the saturation of the mixed blood in the inferior vena cava is approximately 67%. Most of the blood entering the heart through the inferior vena cava is diverted directly to the left atrium via the patent foramen ovale. Most of the blood from the superior vena cava enters the right ventricle and is expelled into the pulmonary artery. The resistance of the collapsed lungs is high, and the pressure in the pulmonary artery is several mm Hg higher than it is in the aorta, so that most of the blood in the pulmonary artery passes through the **ductus arteriosus** to the aorta. In this fashion, the relatively unsaturated blood from the right ventricle is diverted to the trunk and lower body of the fetus, while the head of the fetus receives the better-oxygenated blood from the left ventricle. From the aorta, some of the blood is pumped into the umbilical arteries and back to the placenta. The O_2 saturation of the blood in the lower aorta and umbilical arteries of the fetus is approximately 60%.

Fetal Respiration

The tissues of fetal and newborn mammals have a remarkable but poorly understood resistance to hypoxia. However, the O_2 saturation of the maternal blood in the placenta is so low that the fetus might suffer hypoxic damage if fetal red cells did not have a greater O_2 affinity than adult red cells (Figure 32–22). The fetal red cells contain fetal hemoglobin (hemoglobin F), whereas the adult cells contain adult hemoglobin (hemoglobin A). The cause of the difference in O_2 affinity between the two is that hemoglobin F binds 2,3-DPG less effectively than hemoglobin A does. The decrease in O_2 affinity due to the binding of 2,3-DPG is discussed in Chapter 35.

There is some hemoglobin A in blood during fetal life (see Figure 27–19). After birth no more hemoglo-

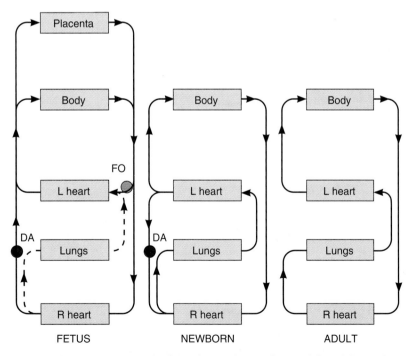

Figure 32–20. Diagram of the circulation in the fetus, the newborn infant, and the adult. DA, ductus arteriosus; FO, foramen ovale. (Redrawn and reproduced, with permission, from Born GVR et al: Changes in the heart and lungs at birth. Cold Spring Harbor Symp Quant Biol 1954;19:102.)

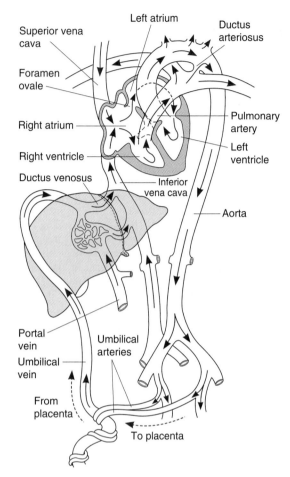

Figure 32–21. Circulation in the fetus. Most of the oxygenated blood reaching the heart via the umbilical vein and inferior vena cava is diverted through the foramen ovale and pumped out the aorta to the head, while the deoxygenated blood returned via the superior vena cava is mostly pumped through the pulmonary artery and ductus arteriosus to the feet and the umbilical arteries.

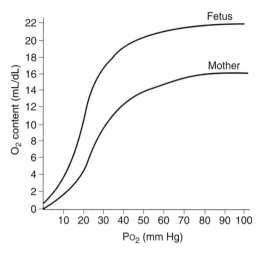

Figure 32–22. Dissociation curves of hemoglobin in human maternal and fetal blood.

bin F is normally formed, and by the age of 4 months 90% of the circulating hemoglobin is hemoglobin A.

Changes in Fetal Circulation & Respiration at Birth

Because of the patent ductus arteriosus and foramen ovale (Figure 32–21), the left heart and right heart pump in parallel in the fetus rather than in series as they do in the adult. At birth, the placental circulation is cut off and the peripheral resistance suddenly

rises. The pressure in the aorta rises until it exceeds that in the pulmonary artery. Meanwhile, because the placental circulation has been cut off, the infant becomes increasingly asphyxial. Finally, the infant gasps several times, and the lungs expand. The markedly negative intrapleural pressure (−30 to −50 mm Hg) during the gasps contributes to the expansion of the lungs, but other poorly understood factors are also involved. The sucking action of the first breath plus constriction of the umbilical veins squeezes as much as 100 mL of blood from the placenta (the "placental transfusion").

Once the lungs are expanded, the pulmonary vascular resistance falls to less than 20% of the in utero value, and pulmonary blood flow increases markedly. Blood returning from the lungs raises the pressure in the left atrium, closing the foramen ovale by pushing the valve that guards it against the interatrial septum. The ductus arteriosus constricts within a few hours after birth, producing functional closure, and permanent anatomic closure follows in the next 24–48 hours due to extensive intimal thickening. The mechanism producing the initial constriction is not completely understood, but the increase in arterial O_2 tension plays an important role. There are relatively high concentrations of vasodilators in the ductus in utero—especially prostaglandin $E_{2\alpha}$—and synthesis of these prostaglandins is inhibited by inhibition of cyclooxygenase at birth.

In many premature infants the ductus fails to close spontaneously, but closure can be produced by infusion of drugs that inhibit cyclooxygenase. Best results have been obtained with drugs that inhibit both COX-1 and COX-2.

Cardiovascular Homeostasis in Health & Disease

33

INTRODUCTION

The compensatory adjustments of the cardiovascular system to the challenges faced by the circulation normally in everyday life and abnormally in disease illustrate the integrated operation of the cardiovascular regulatory mechanisms described in the preceding chapters. The adjustments to gravity, exercise, inflammation, wound healing, shock, fainting, hypertension, and heart failure are considered in this chapter.

COMPENSATIONS FOR GRAVITATIONAL EFFECTS

In the standing position, as a result of the effect of gravity on the blood (see Chapter 30), the mean arterial blood pressure in the feet of a normal adult is 180–200 mm Hg and venous pressure is 85–90 mm Hg. The arterial pressure at head level is 60–75 mm Hg, and the venous pressure is zero. If the individual does not move, 300–500 mL of blood pools in the venous capacitance vessels of the lower extremities, fluid begins to accumulate in the interstitial spaces because of increased hydrostatic pressure in the capillaries, and stroke volume is decreased. Symptoms of cerebral ischemia develop when the cerebral blood flow decreases to less than about 60% of the flow in the recumbent position. If there were no compensatory cardiovascular changes, the reduction in cardiac output due to pooling on standing would lead to a reduction of cerebral flow of this magnitude, and consciousness would be lost.

The major compensations on assuming the upright position are triggered by the drop in blood pressure in the carotid sinus and aortic arch. The heart rate increases, helping to maintain cardiac output. There is relatively little venoconstriction in the periphery, but there is a prompt increase in the circulating levels of renin and aldosterone. The arterioles constrict, helping to maintain blood pressure. The actual blood pressure change at heart level is variable, depending upon the balance between the degree of arteriolar constriction and the drop in cardiac output (Figure 33–1).

In the cerebral circulation, there are additional compensatory changes. The arterial pressure at head level

drops 20–40 mm Hg, but jugular venous pressure falls 5–8 mm Hg, reducing the drop in perfusion pressure (arterial pressure minus venous pressure). Cerebral vascular resistance is reduced because intracranial pressure falls as venous pressure falls, decreasing the pressure on

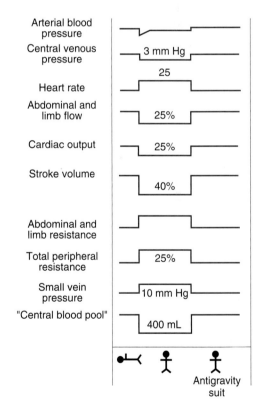

Figure 33–1. Effect on the cardiovascular system of rising from the supine to the upright position. Figures shown are average changes. Changes in abdominal and limb resistance and in blood pressure are variable from individual to individual. (Redrawn and reproduced, with permission, from Brobeck JR [editor]: *Best and Taylor's Physiological Basis of Medical Practice*, 9th ed. Williams & Wilkins, 1973.)

the cerebral vessels. The decline in cerebral blood flow increases the partial pressure of CO_2 (PCO_2) and decreases the PO_2 and the pH in brain tissue, further actively dilating the cerebral vessels. Because of the operation of these autoregulatory mechanisms, cerebral blood flow declines only 20% on standing. In addition, the amount of O_2 extracted from each unit of blood increases, and the net effect is that cerebral O_2 consumption is about the same in the supine and the upright positions.

Prolonged standing presents an additional problem because of increasing interstitial fluid volume in the lower extremities. As long as the individual moves about, the operation of the muscle pump (see Chapter 30) keeps the venous pressure below 30 mm Hg in the feet, and venous return is adequate. However, with prolonged quiet standing (eg, in military personnel standing at attention for long periods), fainting may result. In a sense, the fainting is a "homeostatic mechanism," because falling to the horizontal position promptly restores venous return, cardiac output, and cerebral blood flow to adequate levels.

The effects of gravity on the circulation in humans depend in part upon the blood volume. When the blood volume is low, these effects are marked; when it is high, they are minimal.

The compensatory mechanisms that operate on assumption of the erect posture are better developed in humans than in quadrupeds even though these animals have sensitive carotid sinus mechanisms. Quadrupeds tolerate tilting to the upright position poorly. Of course, giraffes are an exception. These long-legged animals do not develop ankle edema despite the very large increment in vascular pressure in their legs due to gravity because they have tight skin and fascia in the lower legs—in a sense a built-in antigravity suit (see below)—and a very effective muscle pump. Perfusion in the head is maintained by a high mean arterial pressure. When giraffes lower their heads to drink, blood is pumped up their jugular veins to the chest, presumably by rhythmic contractions of the muscles of the jaws.

Postural Hypotension

In some individuals, sudden standing causes a fall in blood pressure, dizziness, dimness of vision, and even fainting. The causes of this **orthostatic (postural) hypotension** are multiple. It is common in patients receiving sympatholytic drugs. It also occurs in diseases such as diabetes and syphilis, in which there is damage to the sympathetic nervous system. This underscores the importance of the sympathetic vasoconstrictor fibers in compensating for the effects of gravity on the circulation. Another cause of postural hypotension is

primary autonomic failure (Table 33–1). Autonomic failure occurs in a variety of diseases. One form is caused by a congenital deficiency of dopamine β-hydroxylase (see Chapter 4) with little or no production of norepinephrine and epinephrine. Baroreceptor reflexes are also abnormal in patients with primary hyperaldosteronism. However, these patients generally do not have postural hypotension, because their blood volumes are expanded sufficiently to maintain cardiac output in spite of changes in position. Indeed, mineralocorticoids are used to treat patients with postural hypotension.

Effects of Acceleration

The effects of gravity on the circulation are multiplied during acceleration or deceleration in vehicles that in modern civilization range from elevators to rockets. Force acting on the body as a result of acceleration is commonly expressed in *g* units, 1 *g* being the force of

Table 33–1. Major forms of primary autonomic failure.[1]

Bradbury-Eggleston syndrome (idiopathic orthostatic hypotension)
  Onset late in life
  Sympathetic and parasympathetic failure
  Absent or minimal other neurologic involvement
  Plasma norepinephrine/dopamine ratio greater than 1

Shy-Drager syndrome (multiple system atrophy)
  Onset in mid to late life
  Sympathetic and parasympathetic failure
  Other neurologic involvement (extrapyramidal, cerebellar, etc)
  Plasma norepinephrine/dopamine ratio greater than 1

Riley-Day syndrome (familial dysautonomia)
  Congenital onset and premature mortality
  Ashkenazi Jewish extraction
  Sympathetic and parasympathetic involvement
  Emotional lability
  Plasma norepinephrine/dopamine ratio greater than 1

Dopamine β-hydroxylase deficiency
  Congenital onset
  Sympathoadrenomedullary failure (orthostatic hypotension)
  Intact sweating
  Parasympathetic sparing
  Plasma norepinephrine/dopamine ratio much less than 1

[1]Reproduced, with permission, from Robertson D et al: Dopamine β-hydroxylase deficiency: A genetic disorder of cardiovascular regulation. Hypertension 1991;18:1. By permission of the American Heart Association, Inc.

gravity on the earth's surface. "Positive *g*" is force due to acceleration acting in the long axis of the body, from head to foot; "negative *g*" is force due to acceleration acting in the opposite direction. During exposure to positive *g,* blood is "thrown" into the lower part of the body. Arterial pressure in the head is reduced, but so are venous pressure and intracranial pressure, and this reduces the decrease in arterial blood flow that would otherwise occur (see Chapter 32). Cardiac output is maintained for a time because blood is drawn from the pulmonary venous reservoir and because the force of cardiac contraction is increased. At accelerations producing more than 5 *g,* however, vision fails ("blackout") in about 5 seconds and unconsciousness follows almost immediately thereafter. The effects of positive *g* are effectively cushioned by the use of antigravity "*g* suits," double-walled pressure suits containing water or compressed air and regulated in such a way that they compress the abdomen and legs with a force proportionate to the positive *g.* This decreases venous pooling and helps maintain venous return (Figure 33–1).

Negative *g* causes increased cardiac output, a rise in cerebral arterial pressure, intense congestion of the head and neck vessels, ecchymoses around the eyes, severe throbbing head pain, and, eventually, mental confusion ("redout"). In spite of the great rise in cerebral arterial pressure, the vessels in the brain do not rupture, because generally there is an increase in intracranial pressure and their walls are supported (see Chapter 32). The tolerance for *g* forces exerted across the body is much greater than it is for axial *g.* Humans tolerate 11 *g* acting in a back-to-chest direction for 3 minutes and 17 *g* acting in a chest-to-back direction for 4 minutes. Astronauts are therefore positioned to take the *g* forces of rocket flight in the chest-to-back direction. The tolerances in this position are sufficiently large to permit acceleration to orbital or escape velocity and deceleration back into the earth's atmosphere without ill effects.

Effects of Zero Gravity on the Cardiovascular System

From the data available to date, cardiovascular function is maintained for up to 14 months of weightlessness, though there is some disuse atrophy of the mechanisms that withstand gravity on earth. Upon return to earth, astronauts have postural hypotension, but this disappears and readaptation to gravity appears to be complete in 4–7 weeks. Of course, longer exposure to weightlessness might be a bigger problem.

Other Effects of Zero Gravity

Muscular effort is much reduced when objects to be moved are weightless, and the decrease in the extensive

normal proprioceptive input due to the action of gravity on the body leads to flaccidity and atrophy of skeletal muscles. A program of regular exercises against resistance, eg, pushing against a wall or stretching a heavy rubber band, appears to decrease the loss of muscle. However, the compensation is incomplete.

Other changes produced by exposure to space flight include space motion sickness (see Chapter 9), a problem that has proved to be of greater magnitude than initially expected; loss of plasma volume, probably because of headward shift of body fluids, with subsequent diuresis; loss of muscle mass; steady loss of bone mineral, with increased Ca^{2+} excretion; loss of red-cell mass; and alterations in plasma lymphocytes. The loss of body Ca^{2+} is equivalent to 0.4% of the total body Ca^{2+} per month, and although there is some evidence that the loss tapers off during prolonged space flight, loss at this rate might create problems of appreciable magnitude if continued for more than 14 months. A high-calcium diet helps overcome this problem, but no totally effective treatment has as yet been developed. The psychological problems associated with the isolation and monotony of prolonged space flight are also a matter of concern.

EXERCISE

Exercise is associated with very extensive alterations in the circulatory and respiratory systems. For convenience, the circulatory adjustments are considered in this chapter and the respiratory adjustments in Chapter 37. However, it should be emphasized that they occur together in an integrated fashion as part of the homeostatic responses that make moderate to severe exercise possible.

Muscle Blood Flow

The blood flow of resting skeletal muscle is low (2–4 mL/100 g/min). When a muscle contracts, it compresses the vessels in it if it develops more than 10% of its maximal tension (Figure 33–2); when it develops more than 70% of its maximal tension, blood flow is completely stopped. Between contractions, however, flow is so greatly increased that blood flow per unit of time in a rhythmically contracting muscle is increased as much as 30-fold. Blood flow sometimes increases at or even before the start of exercise, so the initial rise is probably a neurally mediated response. Impulses in the sympathetic vasodilator system (see Chapter 31) may be involved. The blood flow in resting muscle doubles after sympathectomy, so some decrease in tonic vasoconstrictor discharge may also be involved. However, once exercise has started, local mechanisms maintain the high blood flow, and there

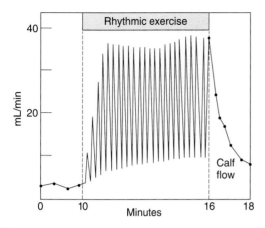

Figure 33–2. Blood flow through a portion of the calf muscles during rhythmic contraction. (Reproduced, with permission, from Barcroft H, Swann HJC: *Sympathetic Control of Human Blood Vessels.* Arnold, 1953.)

is no difference in flow in normal and sympathectomized animals.

Local mechanisms maintaining a high blood flow in exercising muscle include a fall in tissue P_{O_2}, a rise in tissue P_{CO_2}, and accumulation of K^+ and other vasodilator metabolites (see Chapter 31). The temperature rises in active muscle, and this further dilates the vessels. Dilation of the arterioles and precapillary sphincters causes a 10 to 100-fold increase in the number of open capillaries. The average distance between the blood and the active cells—and the distance O_2 and metabolic products must diffuse—is thus greatly decreased. The dilation increases the cross-sectional area of the vascular bed, and the velocity of flow therefore decreases. The capillary pressure increases until it exceeds the oncotic pressure throughout the length of the capillaries. In addition, the accumulation of osmotically active metabolites more rapidly than they can be carried away decreases the osmotic gradient across the capillary walls. Therefore, fluid transudation into the interstitial spaces is tremendously increased. Lymph flow is also greatly increased, limiting the accumulation of interstitial fluid and in effect greatly increasing its turnover. The decreased pH and increased temperature shift the dissociation curve for hemoglobin to the right, so that more O_2 is given up by the blood. The concentration of 2,3-DPG in the red blood cells is increased, and this further decreases the O_2 affinity of hemoglobin (see Chapters 27 and 35). The net result is an up to threefold increase in the arteriovenous O_2 difference, and the transport of CO_2 out of the tissue is also facilitated. All of these changes combine to make it possible for the O_2 consumption of skeletal muscle to increase

100-fold during exercise. An even greater increase in energy output is possible for short periods during which the energy stores are replenished by anaerobic metabolism of glucose and the muscle incurs an O_2 debt (see Chapter 3). The overall changes in intermediary metabolism during exercise are discussed in Chapter 17.

K^+ dilates arterioles in exercising muscle, particularly during the early part of exercise. Muscle blood flow increases to a lesser degree during exercise in K^+-depleted individuals, and there is a greater tendency for severe disintegration of muscle (**exertional rhabdomyolysis**) to occur.

Systemic Circulatory Changes

The systemic cardiovascular response to exercise depends on whether the muscle contractions are primarily isometric or primarily isotonic with the performance of external work. With the start of an isometric muscle contraction, the heart rate rises. This increase still occurs if the muscle contraction is prevented by local infusion of a neuromuscular blocking drug. It also occurs with just the thought of performing a muscle contraction, so it is probably the result of psychic stimuli acting on the medulla oblongata. The increase is largely due to decreased vagal tone, although increased discharge of the cardiac sympathetic nerves plays some role. Within a few seconds of the onset of an isometric muscle contraction, systolic and diastolic blood pressures rise sharply. Stroke volume changes relatively little, and blood flow to the steadily contracting muscles is reduced as a result of compression of their blood vessels.

The response to exercise involving isotonic muscle contraction is similar in that there is a prompt increase in heart rate but different in that there is a marked increase in stroke volume. In addition, there is a net fall in total peripheral resistance (Figure 33–3) due to vasodilation in exercising muscles (Table 33–2). Consequently, systolic blood pressure rises only moderately, whereas diastolic pressure usually remains unchanged or falls. The difference in response to isometric and isotonic exercise is explained in part by the fact that the active muscles are tonically contracted during isometric exercise and consequently contribute to increased total peripheral resistance. In addition, there is a general increase in muscle sympathetic nerve activity, apparently because of a signal from the contracted muscle, but since there is cholinergic sympathetic vasodilation in the inactive skeletal muscles, the significance of this increase is unclear.

Cardiac output is increased during isotonic exercise to values that may exceed 35 L/min, the amount being proportionate to the increase in O_2 consumption. The

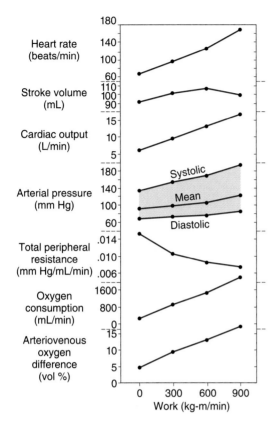

Figure 33–3. Effects of different levels of isotonic exercise on cardiovascular function. (Reproduced, with permission, from Berne RM, Levy MN: *Cardiovascular Physiology,* 5th ed. Mosby, 1986.)

mechanisms responsible for this increase are discussed above and in Chapter 29. The maximal heart rate achieved during exercise decreases with age. In children, it rises to 200 or more beats per minute; in adults it rarely exceeds 195 beats per minute, and in elderly individuals the rise is even smaller.

There is a great increase in venous return, although the increase in venous return is not the primary cause of the increase in cardiac output. Venous return is increased by the great increase in the activity of the muscle and thoracic pumps; by mobilization of blood from the viscera; by increased pressure transmitted through the dilated arterioles to the veins; and by noradrenergically mediated venoconstriction, which decreases the volume of blood in the veins. The amount of blood mobilized from the splanchnic area and other reservoirs may increase the amount of blood in the arterial portion of the circulation by as much as 30% during strenuous exercise.

After exercise, the blood pressure may transiently drop to subnormal levels, presumably because accumulated metabolites keep the muscle vessels dilated for a short period. However, the blood pressure soon returns to the preexercise level. The heart rate returns to normal more slowly.

Temperature Regulation

The quantitative aspects of heat dissipation during exercise are summarized in Figure 33–4. In many locations, the skin is supplied by branches of muscle arteries, so that some of the blood warmed in the muscles is transported directly to the skin, where some of the heat is radiated to the environment. There is a marked increase in ventilation (see Chapter 37), and some heat is lost in the expired air. The body temperature rises, and the hypothalamic centers that control heat-dissipating mechanisms are activated. The temperature increase is due at least in part to the inability of the heat-dissipating mechanism to handle the great increase in heat production. Sweat secretion is greatly increased, and vaporization of this sweat is the major path for heat loss. The cutaneous vessels also dilate. This dilation is primarily due to inhibition of vasoconstrictor tone, although local release of vasodilator polypeptides may also contribute (see Chapter 31).

Training

Both at rest and at any given level of exercise, trained athletes have a larger stroke volume and lower heart rate than untrained individuals (see Chapter 29), and they tend to have larger hearts. Training increases the maximal oxygen consumption (VO_{2max}) that can be produced by exercise in an individual. VO_{2max} averages

Table 33–2. Cardiac output and regional blood flow in a sedentary man.[1]

	Quiet Standing	Exercise
Cardiac output	5900	24,000
Blood flow to:		
Heart	250	1000
Brain	750	750
Active skeletal muscle	650	20,850
Inactive skeletal muscle	650	300
Skin	500	500
Kidney, liver, gastrointestinal tract, etc.	3100	600

[1]Values are mL/min at rest and during isotonic exercise at maximal oxygen uptake.

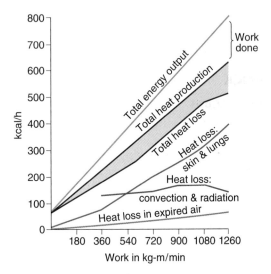

Figure 33–4. Energy exchange in muscular exercise. The shaded area represents the excess of heat production over heat loss. The total energy output equals the heat production plus the work done. (Redrawn and reproduced, with permission, from Nielsen M: *Die Regulation der Korpertemperatur bei Muskelarbeit.* Scand Arch Physiol 1938;79:193.)

about 38 mL/kg/min in active healthy men and about 29 mL/kg/min in active healthy women. It is lower in sedentary individuals. VO_{2max} is the product of maximal cardiac output and maximal O_2 extraction by the tissues, and both increase with training.

The changes that occur in skeletal muscles with training include increases in the number of mitochondria and the enzymes involved in oxidative metabolism. There is an increase in the number of capillaries, with better distribution of blood to the muscle fibers. The net effect is more complete extraction of O_2 and consequently, for a given work load, less increase in lactate production. There is less increase in blood flow to muscles as well and, because of this, less increase in heart rate and cardiac output than in an untrained individual. This is one of the reasons that exercise is of benefit to patients with heart disease.

Relation to Cardiovascular Disease

It is said that the internist's mantra for cardiovascular health is, "Stop smoking, lose weight, and get more exercise." The beneficial effects of a program of regular isotonic exercise are well established in terms of helping patients to feel better, have less severe heart attacks when they have them, and avoid heart attacks in the first place. Regular exercise improves coronary perfu-

sion apparently because the exercise through sheer stress improves the production of prostacyclin and NO by the endothelium of the coronary vessels.

On the other hand, it is also true that the incidence of heart attacks increases during and up to 30 minutes after heavy exercise, particularly in individuals leading sedentary lives. The cause of the increase is unknown but may be related to increased rupture of atherosclerotic plaques. The long-term benefits of exercise probably outweigh the short-term dangers, but it is important to start an exercise program gradually and not let it become too strenuous.

INFLAMMATION & WOUND HEALING

Local Injury

Inflammation is a complex localized response to foreign substances such as bacteria or in some instances to internally produced substances. It includes a sequence of reactions initially involving cytokines, neutrophils, adhesion molecules, complement, and IgG. PAF, an agent with potent inflammatory effects (see Chapter 27), also plays a role. Later, monocytes and lymphocytes are involved. Arterioles in the inflamed area dilate, and capillary permeability is increased (see Chapters 31 and 32). When the inflammation occurs in or just under the skin (Figure 33–5), it is characterized by redness, swelling, tenderness, and pain. Elsewhere, it is a key component of asthma, ulcerative colitis, and many other diseases.

Evidence is accumulating that a transcription factor, **nuclear factor-$_\kappa$B,** plays a key role in the inflammatory response. NF-$_\kappa$B is a heterodimer that normally exists in the cytoplasm of cells bound to I$_\kappa$B$_\alpha$, which renders it inactive. Stimuli such as cytokines, viruses, and oxidants separate NF-$_\kappa$B from I$_\kappa$B$_\alpha$, which is then degraded. NF-$_\kappa$B moves to the nucleus, where it binds to the DNA of the genes for numerous inflammatory mediators, resulting in their increased production and secretion. Glucocorticoids inhibit the activation of NF-$_\kappa$B by increasing the production of I$_\kappa$B$_\alpha$, and this is probably the main basis of their anti-inflammatory action (see Chapter 20).

Systemic Response to Injury

Cytokines produced in response to inflammation and other injuries also produce systemic responses. These include alterations in plasma **acute phase proteins,** defined as proteins whose concentration is increased or decreased by at least 25% following injury. Many of the proteins are of hepatic origin and are listed in Table 27–10. A number of them are shown in Figure 33–6. The causes of the changes in concentration are incompletely understood, but it can be said that many of the

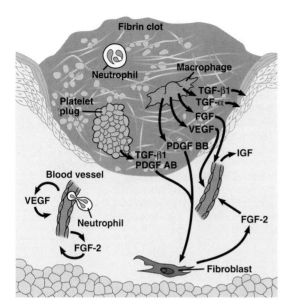

Figure 33–5. Cutaneous wound 3 days after injury, showing the multiple cytokines and growth factors affecting the repair process. VEGF, vascular endothelial growth factor. For other abbreviations, see Appendix. Note the epidermis growing down under the fibrin clot to restore skin continuity. (Modified from Singer AJ, Clark RAF: Cutaneous wound healing. N Engl J Med 1999;341:738.)

changes make homeostatic sense. Thus, for example, an increase in C-reactive protein activates monocytes and causes further production of cytokines.

Other changes that occur in response to injury include somnolence, negative nitrogen balance, and fever.

Wound Healing

When tissue is damaged, platelets adhere to exposed matrix via integrins that bind to collagen and laminin (Figure 33–5). Blood coagulation produces thrombin, which promotes platelet aggregation and granule release. The platelet granules generate an inflammatory response. White blood cells are attracted by selectins and bind to integrins on endothelial cells, leading to their extravasation through the blood vessel walls. Cytokines released by the white blood cells and platelets up-regulate integrins on macrophages, which migrate to the area of injury, and on fibroblasts and epithelial cells, which mediate wound healing and scar formation. Plasmin aids healing by removing excess fibrin. This aids the migration of keratinocytes into the wound to restore the epithelium under the scab. Collagen proliferates, producing the scar. Wounds gain 20%

of their ultimate strength in 3 weeks and later gain more strength, but they never reach more than about 70% of the strength of normal skin.

SHOCK

General Considerations

Shock is a syndrome about which there has been a great deal of confusion and controversy. Part of the difficulty lies in the loose use of the term by physiologists and physicians as well as laymen. Electric shock and spinal shock, for example, bear no relation to the condition produced by hemorrhage and related cardiovascular abnormalities. Shock in the restricted sense of "circulatory shock" is still a collection of different entities that share certain common features. However, the feature that is common to all the entities is inadequate tissue perfusion with a relatively or absolutely inadequate cardiac output. The cardiac output may be inadequate because the amount of fluid in the vascular system is inadequate to fill it (**hypovolemic shock**). Alternatively, it may be inadequate in the relative sense because the size of the vascular system is increased by vasodilation even though the blood volume is normal (**distributive, vasogenic,** or **low-resistance shock**). Shock may also be caused by inade-

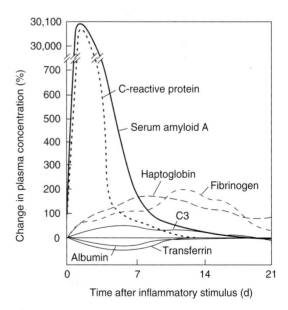

Figure 33–6. Time course of changes in some major acute phase proteins. C3, C3 component of complement. (Modified and reproduced with permission, from Gitlin JD, Colten HR: Molecular biology of acute phase plasma proteins. In Pick F et al [editors]. *Lymphokines,* vol 14, pages 123–153. Academic Press, 1987.)

quate pumping action of the heart as a result of myocardial abnormalities (**cardiogenic shock**), and by inadequate cardiac output as a result of obstruction of blood flow in the lungs or heart (**obstructive shock**). These forms of shock are listed in Table 33–3, along with examples of the disease processes that can cause them.

Hypovolemic Shock

Hypovolemic shock is also called "cold shock." It is characterized by hypotension; a rapid, thready pulse; a cold, pale, clammy skin; intense thirst; rapid respiration; and restlessness or, alternatively, torpor. None of these findings, however, are invariably present. The hypotension may be relative. A hypertensive patient whose blood pressure is regularly 240/140, for example, may be in severe shock when the blood pressure is 120/90.

Hypovolemic shock is commonly subdivided into categories on the basis of cause. The use of terms such as hemorrhagic shock, traumatic shock, surgical shock, and burn shock is of some benefit because, although there are similarities between these various forms of shock, there are important features that are unique to each.

Hemorrhagic Shock

It is useful to consider the effects of hemorrhage in some detail because they illustrate the features of a ma-

Table 33–3. Types of shock, with examples of conditions or diseases that can cause each type.

Hypovolemic shock (decreased blood volume)
  Hemorrhage
  Trauma
  Surgery
  Burns
  Fluid loss due to vomiting or diarrhea

Distributive shock (marked vasodilation; also called vasogenic or low-resistance shock)
  Fainting (neurogenic shock)
  Anaphylaxis
  Sepsis (also causes hypovolemia due to increased capillary permeability with loss of fluid into tissues)

Cardiogenic shock (inadequate output by a diseased heart)
  Myocardial infarction
  Congestive heart failure
  Arrhythmias

Obstructive shock (obstruction of blood flow)
  Tension pneumothorax
  Pulmonary embolism
  Cardiac tumor
  Cardiac tamponade

Table 33–4. Compensatory reactions activated by hemorrhage.

Vasoconstriction
Tachycardia
Venoconstriction
Tachypnea → increased thoracic pumping
Restlessness → increased skeletal muscle pumping (in some cases)
Increased movement of interstitial fluid into capillaries
Increased secretion of norepinephrine and epinephrine
Increased secretion of vasopressin
Increased secretion of glucocorticoids
Increased secretion of renin and aldosterone
Increased secretion of erythropoietin
Increased plasma protein synthesis

---

jor form of hypovolemic shock and the multiple compensatory reactions that come into play to defend ECF volume. The principal reactions are listed in Table 33–4.

The decline in blood volume produced by bleeding decreases venous return, and cardiac output falls. The heart rate is increased, and with severe hemorrhage, there is always a fall in blood pressure. With moderate hemorrhage (5–15 mL/kg body weight), pulse pressure is reduced but mean arterial pressure may be normal. The blood pressure changes vary from individual to individual, even when exactly the same amount of blood is lost. The skin is cool and pale and may have a grayish tinge because of stasis in the capillaries and a small amount of cyanosis. Respiration is rapid, and in patients whose consciousness is not obtunded, intense thirst is a prominent symptom.

In hypovolemic and other forms of shock, the inadequate perfusion of the tissue leads to increased anaerobic glycolysis, with the production of large amounts of lactic acid. In severe cases, the blood lactate level rises from the normal value of about 1 mmol/L to 9 mmol/L or more. The resulting **lactic acidosis** depresses the myocardium, decreases peripheral vascular responsiveness to catecholamines, and may be severe enough to cause coma.

Rapid Compensatory Reactions

When blood volume is reduced and venous return is decreased, the arterial baroreceptors are stretched to a lesser degree and sympathetic output is increased. Even if there is no drop in mean arterial pressure, the decrease in pulse pressure decreases the rate of discharge in the arterial baroreceptors, and reflex tachycardia and vasoconstriction result. It is interesting that with more severe blood loss, tachycardia is replaced by bradycardia; this occurs while shock is still reversible (see below). With even greater hemorrhage, the heart rate rises again. The bradycardia is presumably due to unmasking a vagally mediated depressor reflex, and the response may have evolved as a mechanism for stopping further blood loss.

The vasoconstriction is generalized, sparing only the vessels of the brain and heart. The vasoconstrictor innervation of the cerebral arterioles is probably insignificant from a functional point of view, and the coronary vessels are dilated because of the increased myocardial metabolism secondary to the increase in heart rate (see Chapter 32). Vasoconstriction is most marked in the skin, where it accounts for the coolness and pallor, and in the kidneys and viscera.

Hemorrhage also evokes a widespread reflex venoconstriction that helps maintain the filling pressure of the heart, although the receptors that initiate the venoconstriction are unsettled. The intense vasoconstriction in the splanchnic area shifts blood from the visceral reservoir into the systemic circulation. Blood is also shifted out of the subcutaneous and pulmonary veins. Contraction of the spleen discharges more "stored" blood into the circulation, although the volume mobilized in this way in humans is small.

In the kidneys, both afferent and efferent arterioles are constricted, but the efferent vessels are constricted to a greater degree. The glomerular filtration rate is depressed, but renal plasma flow is decreased to a greater extent, so that the filtration fraction (glomerular filtration rate divided by renal plasma flow) increases. Na^+ retention is marked, and there is retention of the nitrogenous products of metabolism in the blood (**azotemia** or **uremia**). Especially when the hypotension is prolonged, there may be severe renal tubular damage (**acute renal failure).**

Hemorrhage is a potent stimulus to adrenal medullary secretion (see Chapter 20). Circulating norepinephrine is also increased because of the increased discharge of sympathetic noradrenergic neurons. The increase in circulating catecholamines probably contributes relatively little to the generalized vasoconstriction, but it may lead to stimulation of the reticular formation (see Chapter 11). Possibly because of such reticular stimulation, some patients in hemorrhagic shock are restless and apprehensive. Others are quiet and apathetic, and their sensorium is dulled, probably because of cerebral ischemia and acidosis. When restlessness is present, increased motor activity and increased respiratory movements increase the muscular and thoracic pumping of venous blood.

The loss of red cells decreases the O_2-carrying power of the blood, and the blood flow in the carotid and aortic bodies is reduced. The resultant anemia and stagnant hypoxia (see Chapter 37), as well as the acidosis,

stimulate the chemoreceptors. Increased activity in chemoreceptor afferents is probably the main cause of respiratory stimulation in shock. Chemoreceptor activity also excites the vasomotor areas in the medulla, increasing vasoconstrictor discharge. In fact, in hemorrhaged dogs with arterial pressures of less than 70 mm Hg, cutting the nerves to the carotid baroreceptors and chemoreceptors may cause a further fall in blood pressure rather than a rise. This paradoxic result occurs because there is no baroreceptor discharge at pressures below 70 mm Hg and activity in fibers from the carotid chemoreceptors is driving the vasomotor area beyond the maximal rate produced by release of baroreceptor inhibition.

The increase in the level of circulating angiotensin II produced by the increase in plasma renin activity during hemorrhage causes thirst by an action on the subfornical organ (see Chapter 32), and ingestion of fluid helps restore the ECF volume. The increase in angiotensin II also helps to maintain blood pressure. The blood pressure fall produced by removal of a given volume of blood is greater in animals infused with drugs that block angiotensin II receptors than it is in controls. Vasopressin also raises blood pressure when administered in large doses in normal animals, but infusion of doses that produce the same plasma vasopressin levels produced by hemorrhage causes only a small increase in blood pressure because there is a compensatory decrease in cardiac output (see Chapter 31). However, blood pressure falls when peptides that antagonize the effects of vasopressin are injected following hemorrhage. Thus, it appears that vasopressin also plays a significant role in maintaining blood pressure. The increases in circulating angiotensin II and ACTH levels increase aldosterone secretion, and the increased circulating levels of aldosterone and vasopressin cause retention of Na^+ and water, which helps reexpand the blood volume. However, aldosterone takes about 30 minutes to exert its effect, and the initial decline in urine volume and Na^+ excretion is certainly due for the most part to the hemodynamic alterations in the kidney.

When the arterioles constrict and the venous pressure falls because of the decrease in blood volume, there is a drop in capillary pressure. Fluid moves into the capillaries along most of their course, helping to maintain the circulating blood volume. This decreases interstitial fluid volume, and fluid moves out of the cells.

Long-Term Compensatory Reactions

After a moderate hemorrhage, the circulating plasma volume is restored in 12–72 hours (Figure 33–7). There is also a rapid entry of preformed albumin from extravascular stores, but most of the tissue fluids that are mobilized are protein-free. They dilute the plasma

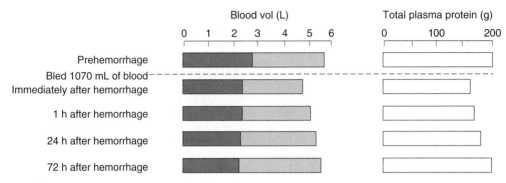

Figure 33–7. Changes in red cell volume (dark color), plasma volume (light color), and total plasma protein following hemorrhage in a normal human subject.

proteins and cells, but when whole blood is lost, the hematocrit may not fall for several hours after the onset of bleeding. After the initial influx of preformed albumin, the rest of the plasma protein losses are replaced, presumably by hepatic synthesis, over a period of 3–4 days. Erythropoietin appears in the circulation, and the reticulocyte count increases, reaching a peak in 10 days. The red cell mass is restored to normal in 4–8 weeks. However, a low hematocrit is remarkably well tolerated because of various compensatory mechanisms. One of these is an increase in the concentration of 2,3-DPG in the red blood cells, which causes hemoglobin to give more O_2 to the tissues (see Chapter 27). In long-standing anemia in otherwise healthy individuals, exertional dyspnea is not observed until the hemoglobin concentration is about 7.5 g/dL. Weakness becomes appreciable at about 6 g/dL; dyspnea at rest appears at about 3 g/dL; and the heart fails when the hemoglobin level falls to 2 g/dL.

Refractory Shock

Depending largely upon the amount of blood lost, some patients die soon after hemorrhage and others recover as the compensatory mechanisms, aided by appropriate treatment, gradually restore the circulation to normal. In an intermediate group of patients, shock persists for hours and gradually progresses to a state in which there is no longer any response to vasopressor drugs and in which, even if the blood volume is returned to normal, cardiac output remains depressed. This is known as **refractory shock.** The condition is not unique to hemorrhagic shock but occurs in other forms as well. It used to be called **irreversible shock,** and patients still do die despite vigorous treatment. However, more and more patients are saved as understanding of the pathophysiologic mechanisms increases and treatment is improved. Therefore, refractory shock seems to be a more appropriate term.

Various positive feedback mechanisms contribute to the production of refractory shock. For example, severe cerebral ischemia leads eventually to depression of the vasomotor and cardiac areas of the brain, causing vasodilation and reduction of the heart rate. These both make the blood pressure drop further, with a further reduction in cerebral blood flow and further depression of the vasomotor and cardiac areas.

Another important example of this type of positive feedback is myocardial depression. In severe shock, the coronary blood flow is reduced because of the hypotension and tachycardia (see Chapter 32), even though the coronary vessels are dilated. The myocardial failure makes the shock and the acidosis worse, and this in turn leads to further depression of myocardial function. If the reduction is marked and prolonged, the myocardium may be damaged to the point where cardiac output cannot be restored to normal in spite of reexpansion of the blood volume.

A late complication of shock that can be fatal is pulmonary damage with the production of **acute respiratory distress syndrome (ARDS, adult respiratory distress syndrome;** see Chapter 37). This syndrome is characterized by acute respiratory failure with a high mortality, and it can be triggered not only by shock but also by sepsis, lung contusion, other forms of trauma, and other serious conditions. The common feature seems to be damage to capillary endothelial cells and alveolar epithelial cells, with release of cytokines.

Other Forms of Hypovolemic Shock

Traumatic shock develops when there is severe damage to muscle and bone. This is the type of shock seen in battle casualties and automobile accident victims. Frank bleeding into the injured areas is the principal cause of the shock, although some plasma also enters the tissue. The amount of blood which can be lost into an injury that appears relatively minor is remarkable;

the thigh muscles can accommodate 1 L of extravasated blood, for example, with an increase in the diameter of the thigh of only 1 cm.

Breakdown of skeletal muscle (**rhabdomyolysis**) is a serious additional problem when shock is accompanied by extensive muscle crushing (**crush syndrome**). Kidney damage is also common in the crush syndrome. It is due to accumulation of myoglobin and other products from reperfused tissue in kidneys in which glomerular filtration is already reduced by shock. The products damage and clog the tubules, frequently causing anuria, which may be fatal.

Surgical shock is due to the combination in various proportions of external hemorrhage, bleeding into injured tissues, and dehydration.

In **burn shock**, the most apparent abnormality is loss of plasma as exudate from the burned surfaces. Since the loss in this situation is plasma rather than whole blood, the hematocrit rises and **hemoconcentration** is a prominent finding. Burns also cause complex, poorly understood metabolic changes in addition to fluid loss. For example, there is a 50% rise in metabolic rate of nonthyroidal origin, and some burned patients develop hemolytic anemia. Because of these complications, plus the severity of the shock and the problems of sepsis and kidney damage, the mortality rate when third-degree burns cover more than 75% of the body is still close to 100%.

Hypovolemic shock is a complication of various metabolic and infectious diseases. For example, although the mechanism is different in each case, adrenal insufficiency, diabetic ketoacidosis, and severe diarrhea are all characterized by loss of Na^+ from the circulation. The resultant decline in plasma volume may be severe enough to precipitate cardiovascular collapse.

Distributive Shock

As noted above, distributive shock occurs when the blood volume is normal but the capacity of the circulation is increased by marked vasodilation. It is also called "warm shock" because the skin is not cold and clammy, as it is in hypovolemic shock. A good example is **anaphylactic shock**, a rapidly developing, severe allergic reaction that sometimes occurs when an individual who has previously been sensitized to an antigen is reexposed to it. The resultant antigen-antibody reaction releases large quantities of histamine, causing increased capillary permeability and widespread dilation of arterioles and capillaries.

Septic Shock

Septic shock is a common and serious condition in which infections, usually due to gram-negative bacteria, cause shock which has both distributive and hypovolemic features. **Endotoxins,** the cell wall lipopolysaccharides produced by some bacteria, cause vasodilation and increased capillary permeability, with loss of plasma in the tissues. They also initiate a complex series of cytokine and coagulant reactions that can lead eventually to multiple organ failure. The mortality of the condition is 30–50%, and numerous drugs designed to inhibit the inflammatory response, including glucocorticoids have failed to lower this figure. However, promising results have recently been obtained with **activated protein C,** which has anticoagulant activity (see Chapter 27).

Fainting

A third type of distributive shock is neurogenic shock, in which there is sudden autonomic activity producing vasodilation, pooling of blood in the extremities, and fainting. These are called vasovagal attacks, and they are short-lived and benign. Other forms of syncope include **postural syncope,** fainting due to pooling of blood in the dependent parts of the body on standing. **Micturition syncope,** fainting during urination, occurs in patients with orthostatic hypotension. It is due to the combination of the orthostasis and reflex bradycardia induced by voiding in these patients. Pressure on the carotid sinus, produced, for example, by a tight collar, can cause such marked bradycardia and vasodilation that fainting results (**carotid sinus syncope**). Rarely, vasodilation and bradycardia may be precipitated by swallowing (**deglutition syncope**). **Cough syncope** occurs when the increase in intrathoracic pressure during straining or coughing is sufficient to block venous return. **Effort syncope** is fainting on exertion as a result of inability to increase cardiac output to meet the increased demands of the tissues and is particularly common in patients with aortic or pulmonary stenosis.

Syncope can also be due to more serious abnormalities. About 25% of syncopal episodes are of cardiac origin and are due to either transient obstruction of blood flow through the heart or sudden decreases in cardiac output owing to various cardiac arrhythmias. Fainting due to bradycardia, heart block, or sinus arrest is called **neurocardiogenic syncope.** In addition, fainting is the presenting symptom in 7% of patients with myocardial infarctions. Thus, all cases of syncope should be investigated to determine the cause.

Cardiogenic & Obstructive Shock

When the pumping function of the heart is impaired to the point that blood flow to the tissues is no longer adequate to meet resting metabolic demands, the condition that results is called cardiogenic shock. It is most commonly due to extensive infarction of the left ventricle, but it can also be caused by other diseases that severely compromise ventricular function. The symptoms are those of shock plus congestion of the lungs and viscera because the heart fails to put out all the venous blood returned to it. Consequently, the condition is sometimes called "congested shock." The incidence of this shock in patients with myocardial infarction is about 10%, and it has a mortality of 60–90%.

The picture of congested shock is also seen in obstructive shock. When the obstruction is due to tension pneumothorax with kinking of the great veins (see Chapter 37) or bleeding into the pericardium with external pressure on the heart **(cardiac tamponade),** prompt surgical intervention is required to prevent death.

Treatment of Shock

The treatment of shock should be aimed at correcting the cause and helping the physiologic compensatory mechanisms to restore an adequate level of tissue perfusion. In hemorrhagic, traumatic, and surgical shock, for example, the primary cause of the shock is blood loss, and the treatment should include early and rapid transfusion of adequate amounts of compatible whole blood. Saline is of limited temporary value. The immediate goal is restoration of an adequate circulating blood volume, and since saline is distributed in the ECF, only 25% of the amount administered stays in the vascular system. In burn shock and other conditions in which there is hemoconcentration, plasma is the treatment of choice to restore the fundamental defect, the loss of plasma. "Plasma expanders," solutions of sugars of high molecular weight and related substances that do not cross capillary walls, have some merit. Concentrated human serum albumin and other hypertonic solutions expand the blood volume by drawing fluid out of the interstitial spaces. They are valuable in emergency treatment but have the disadvantage of further dehydrating the tissues of an already dehydrated patient.

In anaphylactic shock, epinephrine has a highly beneficial and almost specific effect that must represent more than just constriction of the dilated vessels.

HYPERTENSION

Hypertension is a sustained elevation of the systemic arterial pressure. **Pulmonary hypertension** also occurs, but the pressure in the pulmonary artery (see Chapter 34) is relatively independent of that in the systemic arteries.

Experimental Hypertension

The arterial pressure is determined by the cardiac output and the peripheral resistance (pressure = flow × re-

sistance; see Chapter 30). The peripheral resistance is determined by the viscosity of the blood and, more importantly, by the caliber of the resistance vessels. Hypertension can be produced by elevating the cardiac output, but sustained hypertension is usually due to increased peripheral resistance. Some of the procedures that have been reported to produce sustained hypertension in experimental animals are listed in Table 33–5. For the most part, the procedures involve manipulation of the kidneys, the nervous system, or the adrenals. There are in addition a number of strains of rats that develop hypertension either spontaneously (SHR rats) or when fed a high-sodium diet (Dahl salt-sensitive rats).

The hypertension that follows constriction of the renal arterial blood supply or compression of the kidney is called **renal hypertension.** As noted in Chapter 24, some animals with renal hypertension have elevated plasma renin activity, whereas others do not. In general, one-clip, two-kidney Goldblatt hypertension (Table 33–5) is renin-dependent, whereas one-clip, one-kidney Goldblatt hypertension is not. An additional factor that probably contributes to renal hypertension is decreased ability of the constricted kidney to excrete Na^+.

Neurogenic hypertension is discussed in Chapter 31. Provided that salt intake is normal or high, deoxy-

Table 33–5. Procedures that produce sustained hypertension in experimental animals.

Interference with renal blood flow (renal hypertension)
  Constriction of one renal artery; other kidney removed (one-clip, one-kidney Goldblatt hypertension)
  Constriction of one renal artery; other kidney intact (one-clip, two-kidney Goldblatt hypertension)
  Constriction of aorta or both renal arteries (two-clip, two-kidney Goldblatt hypertension)
  Compression of kidney by rubber capsules, production of perinephritis, etc

---

Interruptions of afferent input from arterial baroreceptors (neurogenic hypertension)
  Denervation of carotid sinuses and aortic arch
  Bilateral lesions of nucleus of tractus solitarius

Treatment with corticosteroids
  Deoxycorticosterone and salt
  Other mineralocorticoids

Partial adrenalectomy (adrenal regeneration hypertension)

Genetic
  Spontaneous hypertension in various strains of rats
  Salt-induced hypertension in genetically sensitive rats
  Endothelial NOS gene knockout in mice
  Various types of transgenic animals

corticosterone causes hypertension which may persist after treatment is stopped. The hypertension is more severe in unilaterally nephrectomized animals.

Hypertension in Humans

Hypertension is a very common abnormality in humans. It can be produced by many diseases (Table 33–6). It causes a number of serious disorders. When the resistance against which the left ventricle must pump (afterload) is elevated for a long period, the cardiac muscle hypertrophies. The initial response is activation of immediate-early genes in the ventricular muscle, followed by activation of a series of genes involved in growth during fetal life. Left ventricular hypertrophy is associated with a poor prognosis. The total O_2 consumption of the heart, already increased by the work of expelling blood against a raised pressure (see Chapter 29), is increased further because there is more muscle. Therefore, any decrease in coronary blood flow has more serious consequences in hypertensive patients than it does in normal individuals, and degrees of coronary narrowing that do not produce symptoms when the size of the heart is normal may produce myocardial infarction when the heart is enlarged. There is an increased incidence of atherosclerosis in hypertension, and myocardial infarcts are common even when the heart is not enlarged. Eventually, the ability to compensate for the high peripheral resistance is exceeded, and the heart fails. Hypertensive individuals are also predis-

posed to thromboses of cerebral vessels and cerebral hemorrhage. An additional complication is renal failure. However, the incidence of heart failure, strokes, and renal failure can be markedly reduced by active treatment of hypertension, even when the hypertension is relatively mild.

Malignant Hypertension

Chronic hypertension can enter an accelerated phase in which necrotic arteriolar lesions develop and there is a rapid downhill course with papilledema, cerebral symptoms, and progressive renal failure. This syndrome is known as **malignant hypertension,** and without treatment it is fatal in less than 2 years. However, its progression can be stopped, and it can be reversed by appropriate antihypertensive therapy.

Essential Hypertension

In over 90% of patients with elevated blood pressure, the cause of the hypertension is unknown, and they are said to have **essential hypertension.**

The progression of untreated essential hypertension is variable. Particularly in women, it is often benign, with elevated pressure being the only finding for many years. However, it may also progress rapidly into the malignant phase.

At present, essential hypertension is treatable but not curable. Effective lowering of the blood pressure can be produced by drugs that block α-adrenergic receptors, either in the periphery or in the central nervous system; drugs that block β-adrenergic receptors; drugs that inhibit the activity of angiotensin-converting enzyme; and calcium channel blockers that relax vascular smooth muscle.

Other Types of Hypertension

Although the vast majority of patients have essential hypertension of unknown cause, a growing number of subsets of patients are having causes of their hypertension identified. This removes them from the essential hypertension category and makes the point that hypertension, like diabetes mellitus, is not a single disease but a syndrome with multiple causes. The known causes include in particular renal abnormalities, disorders of the adrenal glands, other endocrine abnormalities, and genetic disorders. They are summarized in Table 33–6. It is important to recognize them because some are not only treatable but a few are curable.

A subgroup of hypertensives appear to be **salt-sensitive,** like the Dahl salt-sensitive rats, whereas others, like the Dahl salt-resistant rats, have small increases in blood pressure when fed a high-salt diet.

Table 33–6. Estimated frequency of various forms of hypertension in the general hypertensive population.[1]

	Percentage of Population
Essential hypertension	88
Renal hypertension	
Renovascular	2
Parenchymal	3
Endocrine hypertension	
Primary aldosteronism	5
Cushing's syndrome	0.1
Pheochromocytoma	0.1
Other adrenal forms	0.2
Estrogen treatment ("pill hypertension")	1
Miscellaneous (Liddle's syndrome, coarctation of the aorta, etc)	0.6

[1]Reproduced, with permission, from McPhee SJ, Lingappa V, Ganong WF. *Pathophysiology of Disease,* 4th ed. McGraw-Hill, 2003.

In pregnant women, hypertension associated with **preeclampsia** and **eclampsia** may be due to a pressor peptide secreted by the placenta.

In humans, deoxycorticosterone and aldosterone both elevate the blood pressure, and hypertension is a prominent feature of primary hyperaldosteronism. It is also seen in patients who secrete excess deoxycorticosterone (see Chapter 20). It is especially prominent in glucocorticoid-remediable aldosteronism (GRA), the congenital disorder in which the aldosteronism is due to an ACTH-stimulatable aldosterone synthase (see Chapter 20).

Plasma renin activity is low in the forms of hypertension that are due to excess secretion of aldosterone and deoxycorticosterone. It has also been found to be low, for unknown reasons, in the presence of normal or low aldosterone and deoxycorticosterone secretion rates in 10–15% of patients with what otherwise appears to be essential hypertension (**low-renin hypertension**).

Hypertension is also seen in Cushing's syndrome, in which aldosterone secretion is usually normal. The cause of the hypertension in this syndrome is uncertain; it may be due to the increased secretion of deoxycorticosterone produced by increased circulating ACTH, increased secretion of angiotensinogen due to the increase in circulating glucocorticoids, a direct action of glucocorticoids on the arterioles, or a combination of these factors.

Pheochromocytomas—tumors of the adrenal medulla or of catecholamine-secreting tissue elsewhere in the body—also produce hypertension. They are discussed in Chapter 20.

Renal hypertension due to narrowing of the renal arteries is discussed above and in Chapter 24. In Liddle's syndrome, constitutive activation of epithelial Na^+ channels causes excess salt retention and hypertension (see Chapter 38). In another recently described form of hypertension, there is a mutation in the gene for the mineralocorticoid receptor that makes it constitutively active, causing Na^+ retention and expanded EDF volume. The mutant receptor is also activated by progesterone, so the hypertension is worse during pregnancy.

Coarctation of the aorta, a congenital narrowing of a segment of the thoracic aorta, increases the resistance to flow, producing severe hypertension in the upper part of the body. The blood pressure in the lower part of the body is usually normal but may be elevated as a result of increased renin secretion.

Chronic treatment with oral contraceptives containing estrogens produces significant hypertension in some women. The hypertension is due to an increase in circulating levels of angiotensinogen, the production of which is stimulated by estrogens (see Chapter 24). Normally, an increase in the circulating angiotensin II level produced by an increase in the circulating angiotensinogen level inhibits renin secretion, and the resulting decline in circulating renin restores the circulating angiotensin II level to normal. However, this feedback is incomplete in some women, and they develop "pill hypertension." Certainly, the occasional occurrence of hypertension is not reason enough for normotensive women to avoid oral contraceptives, but it would seem wise for them to have their blood pressures checked at 6-month intervals.

HEART FAILURE

Pathogenesis

Chronic hypertension increases the load on the myocardium of the left ventricle, and so do a number of other diseases. The response of the myocytes is activation of genes that make the myocytes hypertrophy; there is little if any hyperplasia because the capacity of the myocytes to divide is very limited. In some cases, there is also ventricular remodeling in response to distortions produced by the disease processes. Initially, these changes are compensatory, but the heart eventually fails, ie, if it does not put out enough blood to maintain tissue perfusion. At first, the cardiac output is only decreased during exercise and remains normal at rest, but with continued progression of disease processes it becomes inadequate at rest as well (Figure 33–8). Two processes are distinguished: **systolic heart**

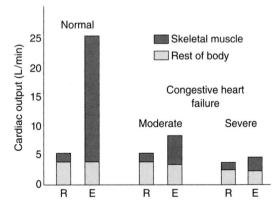

Figure 33–8. Decreased cardiac output in congestive heart failure. R, rest; E, maximal exercise. Note that with moderate failure, resting cardiac output is normal and only the portion going to skeletal muscle during exercise is reduced. As failure progresses, resting cardiac output is also reduced. (Modified and reproduced, with permission, from Zelis R et al: Vasoconstrictor mechanisms in congestive heart failure, Part I. Mod Concepts Cardiovasc Dis 1989;58:7. By permission of the American Heart Association, Inc.)

failure, in which stroke volume is reduced because ventricular contractions are weak; and **diastolic heart failure,** in which the elasticity of the heart is reduced, hindering cardiac filling. Systolic weakness causes an increase in the end-systolic ventricular volume, so that the **ejection fraction**— the fraction of the blood in the ventricle that is ejected during systole—falls from its normal value of 65% to as low as 20%.

Inadequate filling of the arterial system triggers baroreceptor-mediated increased sympathetic activity with renal vasoconstriction and increased secretion of renin and hence aldosterone (see Chapter 31). This causes marked Na^+ and water retention, which adds to the venous congestion and edema.

Heart failure can involve primarily the right ventricle **(cor pulmonale),** but it much more commonly involves the larger, thicker left ventricle. Furthermore, the decrease in cardiac output may be relative rather than absolute. When there is a large arteriovenous fistula— or in thyrotoxicosis and thiamin deficiency—output may be elevated in absolute terms. However, when it is inadequate relative to the needs of the tissues, heart failure is present **(high-output failure).**

Manifestations

The manifestations of heart failure range from sudden death (eg, in ventricular fibrillation or air embolism), through cardiogenic shock, to chronic **congestive heart failure,** depending upon the degree of circulatory inadequacy and the rapidity with which it develops. The principal symptoms and signs of congestive failure include cardiac enlargement and the signs and symptoms listed in Table 33–7. The terms "forward failure" and "backward failure" are sometimes used to refer to the manifestations produced primarily by systolic dysfunction and diastolic dysfunction, respectively. These terms are misleading because the entities occur together and are not due to separate disorders. However, the terms are useful in understanding the manifestations of heart failure, so they are retained in the table.

Treatment

Treatment of congestive heart failure is aimed at improving cardiac contractility, treating the symptoms, and decreasing the load on the heart. Currently, the most effective treatment in general use is inhibition of the production of angiotensin II with angiotensin-converting enzyme inhibitors. Blockade of the effects of angiotensin II on AT_1 receptors with nonpeptide antagonists is also of value. Angiotensin II appears to have direct effects on the heart, but these are controversial. Blocking the production of angiotensin II or its effects also reduces the circulating aldosterone level and decreases blood pressure, reducing the afterload against which the heart pumps. The effects of aldosterone can be further reduced by administering aldosterone recep-

Table 33–7. Simplified summary of pathogenesis of major findings in congestive heart failure.

Abnormality	Cause
Weakness, exercise intolerance	"Forward failure" of left ventricle; cardiac output inadequate to perfuse muscles; especially, failure of output to rise with exercise.
Ankle, sacral edema	"Backward failure" → increased peripheral venous pressure → increased fluid transudation.
Hepatomegaly	Increased peripheral venous pressure → increased resistance to portal flow.
Pulmonary congestion	"Backward failure" → increased pulmonary venous pressure → pulmonary venous distention and transudation of fluid into air spaces.
Dyspnea on exertion	Failure of left ventricular output to rise during exercise → increased pulmonary venous pressure.
Paroxysmal dyspnea, pulmonary edema	Probably sudden failure of left heart output to keep up with right heart output → acute rise in pulmonary venous and capillary pressure → transudation of fluid into air spaces.
Orthopnea	Normal pooling of blood in lungs in supine position added to already congested pulmonary vascular system; increased venous return not put out by left ventricle. (Relieved by sitting up, raising head of bed, lying on extra pillows.)
Cardiac dilation	Greater ventricular end-diastolic volume.

tor blockers, and these have shown promise in recent trials. Reducing venous tone with nitrates or hydralazine increases venous capacity so that the amount of blood returned to the heart is reduced, lowering the preload. Diuretics reduce the fluid overload. Drugs that block β-adrenergic receptors have been shown to decrease mortality and morbidity. Digitalis derivatives such as digoxin have classically been used to treat congestive failure because of their ability to increase intracellular Ca^{2+} and hence exert a positively inotropic effect (see Chapter 3), but they are now used in a secondary role to treat systolic dysfunction and slow the ventricular rate in patients with atrial fibrillation (see Chapter 28).

REFERENCES FOR SECTION VI: CIRCULATION

Bajetto A et al: Chemokines and their receptors in the central nervous system. Front Neuroendocrinol 2001;22:147.

Begley DJ, Bradbury MW, Kreater J (editors): *The Blood-Brain Barrier and Drug Delivery to the CNS*. Marcel Dekker, 2000.

Birmingham K (editor): The heart. Nature 2002;415:197.

Bouchama A, Knochel JP: Heat stroke. N Engl J Med 2002;346:1978.

Brott T, Bogousslavsky J: Treatment of acute ischemic stroke. N Engl J Med 2000;343:710.

Bunn HF: Mechanisms of disease: Pathogenesis and treatment of sickle cell anemia. N Engl J Med 1997;337:762.

Dampney RAL: Functional organization of central pathways regulating the cardiovascular system. Physiol Rev 1994;74:323.

Davidson A, Diamond B: Autoimmune disease. N Engl J Med 2001;345:340.

De Mello WC, Danser AH: Angiotensin II and the heart: On the intracrine renin-angiotensin system. Hypertension 2000;35:1183.

Delves PJ, Roitt IM: The immune system. (Two parts.) N Engl J Med 2000;343:37, 108.

Dennery PA, Seidman DS, Stevenson DK: Neonatal hyperbilirubinemia. N Engl J Med 2001;344:581.

Dhainaut J-K, Thijs LG, Park G (editors): *Septic Shock*. Saunders, 2000.

Ferrara N, Alitgalo K: Clinical applications of angiogenic growth factors and their inhibitors. Nat Med 1999;5:1359.

Ganz T: Defensins and host defense. Science 1999;286:420.

Garrett WE Jr, Kirkendall DT (editors): *Exercise and Sports Medicine*. Lippincott Williams & Wilkins, 2000.

Hibi M, Nakajima K, Hirano T: IL-6 cytokine family and signal transduction: A model of the cytokine system. J Mol Med 1996;74:1.

Hunter JJ, Chien KR: Signaling pathways for cardiac hypertrophy and failure. N Engl J Med 1999;341:1276.

Jackson WF: Ion channels and vascular tone. Hypertension 2000;35(Part 2):173.

Kamisago M et al: Mutations in sarcomere protein genes as a cause of dilated cardiomyopathy. N Engl J Med 2000;243:1688.

Levi M, ten Cate H: Disseminated intravascular coagulation. N Engl J Med 1999;341:586.

Lusis AJ: Atherosclerosis. Nature 2000;407:233.

Mendelsohn ME, Karas RH: The preventive effects of estrogen on the cardiovascular system. N Engl J Med 1999;300:1801.

Morndy F: Radiofrequency ablation as treatment for cardiac arrhythmias. N Engl J Med 1999;340:534.

Niemann JT: Cardiopulmonary resuscitation. N Engl J Med 1992;327:1075.

Opie LH: *The Heart: Physiology From Cell to Circulation,* 3rd ed. Lippincott-Raven, 1998.

Premack BA, Schall TJ: Chemokine receptors: Gateways to inflammation and infection. Nat Med 1996;2:1174.

Raichle ME: Cognitive neuroscience. Bold insights. Nature 2001;412:128.

Ross R: Atherosclerosis—an inflammatory disease. N Engl J Med 1999;340:115.

Rowell LB: *Human Cardiovascular Control*. Oxford Univ Press, 1993.

Samstein B, Emond JC: Liver transplant from living related donors. Annu Rev Med 2001;52:147.

Singer AJ, Clark RAF: Cutaneous wound healing. N Engl J Med 1999;341:738

Spahn DR, Pasch T: Physiological properties of blood substitutes. News Physiol Sci 2001;16:38.

Stamatoyannopoulos G et al (editors). *The Molecular Basis of Blood Disease,* 3rd ed. Saunders, 2001.

Swales JD (editor): *Textbook of Hypertension*. Blackwell, 1994.

Tedder TF et al: The selectins: Vascular adhesion molecules. FASEB J 1995;9:866.

Wagner GS: *Marriott's Practical Electrocardiography,* 10th ed. Lippincott Williams & Wilkins, 2000.

Walport MJ: Complement. (Two parts.) N Engl J Med 2001;344:1058, 1140.

Wright SJ: Human embryonic stem-cell research: science and ethics. Am Scientist 1999;87:352.

Zaret BL, Wackers FJ: Nuclear cardiology. N Engl J Med 1993;329:855.

Zhang F et al: Vasoregulatory function of the heme-heme oxygenase-carbon monoxide system. Am J Hypertens 2001;14(6 Part 2):62S.

SECTION VII
Respiration

Pulmonary Function

34

INTRODUCTION

Respiration, as the term is generally used, includes two processes: **external respiration,** the absorption of O_2 and removal of CO_2 from the body as a whole; and **internal respiration,** the utilization of O_2 and production of CO_2 by cells and the gaseous exchanges between the cells and their fluid medium. Details of the utilization of O_2 and the production of CO_2 by cells are considered in Chapter 17. This chapter is concerned with the functions of the respiratory system in external respiration, ie, the processes responsible for the uptake of O_2 and excretion of CO_2 in the lungs. Chapter 35 is concerned with the transport of O_2 and CO_2 to and from the tissues.

The respiratory system is made up of a gas-exchanging organ (the lungs) and a pump that ventilates the lungs. The pump consists of the chest wall; the respiratory muscles, which increase and decrease the size of the thoracic cavity; the areas in the brain that control the muscles; and the tracts and nerves that connect the brain to the muscles. At rest, a normal human breathes 12–15 times a minute. About 500 mL of air per breath, or 6–8 L/min, is inspired and expired. This air mixes with the gas in the alveoli, and, by simple diffusion, O_2 enters the blood in the pulmonary capillaries while CO_2 enters the alveoli. In this manner, 250 mL of O_2 enters the body per minute and 200 mL of CO_2 is excreted.

Traces of other gases such as methane from the intestines are also found in expired air. Alcohol and acetone are expired when present in appreciable quantities in the body. Indeed, over 250 different volatile substances have been identified in human breath.

PROPERTIES OF GASES

The pressure of a gas is proportionate to its temperature and the number of moles per volume

$$P = \frac{nRT}{V} \quad \text{(from equation of state of ideal gas)}$$

where

P = Pressure
n = Number of moles
R = Gas constant
T = Absolute temperature
V = Volume

Partial Pressures

Unlike liquids, gases expand to fill the volume available to them, and the volume occupied by a given number of gas molecules at a given temperature and pressure is (ideally) the same regardless of the composition of the gas. Therefore, the pressure exerted by any one gas in a mixture of gases (its **partial pressure**) is equal to the total pressure times the fraction of the total amount of gas it represents.

The composition of dry air is 20.98% O_2, 0.04% CO_2, 78.06% N_2, and 0.92% other inert constituents such as argon and helium. The barometric pressure (P_B) at sea level is 760 mm Hg (1 atmosphere). The partial pressure (indicated by the symbol P) of O_2 in dry air is therefore 0.21×760, or 160 mm Hg at sea level. The partial pressure of N_2 and the other inert gases is 0.79×760, or 600 mm Hg; and the P_{CO_2} is 0.0004×760, or 0.3 mm Hg. The water vapor in the air in most climates reduces these percentages, and therefore the partial pressures, to a slight degree. Air equilibrated with water is saturated with water vapor, and inspired air is saturated by the time it reaches the lungs. The P_{H_2O} at body temperature (37 °C) is 47 mm Hg. Therefore, the partial pressures at sea level of the other gases in the air reaching the lungs are P_{O_2}, 149 mm Hg; P_{CO_2}, 0.3 mm Hg; and P_{N_2} (including the other inert gases), 564 mm Hg.

Gas diffuses from areas of high pressure to areas of low pressure, the rate of diffusion depending upon the concentration gradient and the nature of the barrier between the two areas. When a mixture of gases is in contact with and permitted to equilibrate with a liquid, each gas in the mixture dissolves in the liquid to an extent determined by its partial pressure and its solubility in the fluid. The partial pressure of a gas in a liquid is that pressure which in the gaseous phase in equilibrium with the liquid would produce the concentration of gas molecules found in the liquid.

Methods of Quantitating Respiratory Phenomena

Respiratory excursions can be recorded by devices that measure chest expansion or by recording spirometers (see Figure 17–1), which also permit measurement of gas intake and output. Since gas volumes vary with temperature and pressure and since the amount of water vapor in them varies, it is important to correct respiratory measurements involving volume to a stated set of standard conditions. The four most commonly used standards and their abbreviations are shown in Table 34–1. Modern techniques for gas analysis make possible rapid, reliable measurements of the composition of gas mixtures and the gas content of body fluids. For example, O_2 and CO_2 electrodes, small probes sensitive to O_2 or CO_2, can be inserted into the airway or into blood vessels or tissues and the P_{O_2} and P_{CO_2} recorded

continuously. Chronic assessment of oxygenation is carried out noninvasively with a **pulse oximeter,** which is usually attached to the ear.

ANATOMY OF THE LUNGS

Air Passages

After passing through the nasal passages and pharynx, where it is warmed and takes up water vapor, the inspired air passes down the trachea and through the bronchioles, respiratory bronchioles, and alveolar ducts to the alveoli (Figure 34–1).

Between the trachea and the alveolar sacs, the airways divide 23 times. The first 16 generations of passages form the conducting zone of the airways that transports gas from and to the exterior. They are made up of bronchi, bronchioles, and terminal bronchioles. The remaining seven generations form the transitional and respiratory zones where gas exchange occurs and are made up of respiratory bronchioles, alveolar ducts, and alveoli. These multiple divisions greatly increase the total cross-sectional area of the airways, from 2.5 cm^2 in the trachea to 11,800 cm^2 in the alveoli (Figure 34–2). Consequently, the velocity of air flow in the small airways declines to very low values.

The alveoli are surrounded by pulmonary capillaries. In most areas, air and blood are separated only by the alveolar epithelium and the capillary endothelium, so they are about 0.5 μm apart (Figure 34–3). There are 300 million alveoli in humans, and the total area of the alveolar walls in contact with capillaries in both lungs is about 70 m^2.

The alveoli are lined by two types of epithelial cells. **Type I cells** are flat cells with large cytoplasmic extensions and are the primary lining cells. **Type II cells (granular pneumocytes)** are thicker and contain numerous lamellar inclusion bodies. These cells secrete surfactant (see below). There may be other special types of epithelial cells, and the lungs also contain pulmonary alveolar macrophages (PAMs), lymphocytes, plasma cells, APUD cells (see Chapter 26), and mast cells. The mast cells (see Chapter 27) contain heparin, various lipids, histamine, and various proteases that participate in allergic reactions.

The Bronchi & Their Innervation

The trachea and bronchi have cartilage in their walls but relatively little smooth muscle. They are lined by a ciliated epithelium that contains mucous and serous glands. Cilia are present as far as the respiratory bronchioles, but glands are absent from the epithelium of the bronchioles and terminal bronchioles, and their walls do not contain cartilage. However, their walls contain more smooth

Table 34–1. Standard conditions to which measurements involving gas volumes are corrected.

STPD	0 °C, 760 mm Hg, dry (standard temperature and pressure, dry)
BTPS	Body temperature and pressure, saturated with water vapor
ATPD	Ambient temperature and pressure, dry
ATPS	Ambient temperature and pressure, saturated with water vapor

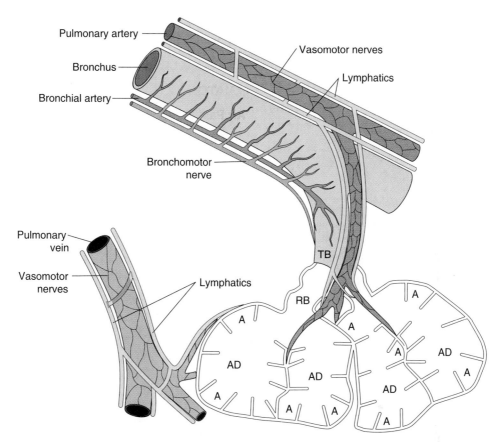

Figure 34–1. Structure of the lung. A, alveolus; AD, alveolar duct; RB, respiratory bronchiole; TB, terminal bronchiole. (Modified from Staub NC: The pathophysiology of pulmonary edema. Hum Pathol 1970;1:419.)

muscle, of which the largest amount relative to the thickness of the wall is present in the terminal bronchioles.

The walls of the bronchi and bronchioles are innervated by the autonomic nervous system. There are abundant muscarinic receptors, and cholinergic discharge causes bronchoconstriction. There are β_2-adrenergic receptors in the bronchial epithelium and smooth muscle. Many of these are not innervated. Some may be located on cholinergic endings, where they inhibit acetylcholine release. The β_2 receptors mediate bronchodilation. They increase bronchial secretion (see Table 13–2), while α_1 adrenergic receptors inhibit secretion. There is in addition a **noncholinergic, nonadrenergic innervation** of the bronchioles that produces bronchodilation, and there is evidence that VIP is the mediator responsible for the dilation.

Pulmonary Circulation

Almost all the blood in the body passes via the pulmonary artery to the pulmonary capillary bed, where it is oxygenated and returned to the left atrium via the pulmonary veins (Figure 34–5). The separate and much smaller bronchial arteries come from systemic arteries. They form capillaries, which drain into bronchial veins or anastomose with pulmonary capillaries or veins (Figure 34–5). The bronchial veins drain into the azygos vein. The bronchial circulation nourishes the bronchi and pleura. Lymphatic channels are more abundant in the lungs than in any other organ (Figure 34–1).

Further aspects of the pulmonary circulation are considered below.

MECHANICS OF RESPIRATION

Inspiration & Expiration

The lungs and the chest wall are elastic structures. Normally, there is no more than a thin layer of fluid between the lungs and the chest wall. The lungs slide easily on the chest wall but resist being pulled away from

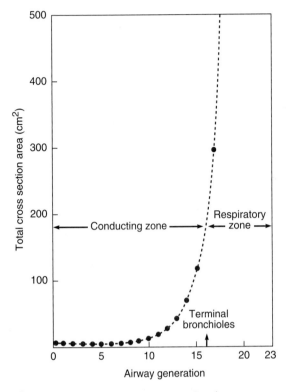

Figure 34–2. Total airway cross-sectional area as a function of airway generation. Note the extremely rapid increase in total cross-sectional area in the respiratory zone. As a result, forward velocity of gas during inspiration falls to a very low level in this zone. (Reproduced, with permission, from West JB: *Respiratory Physiology: The Essentials,* 4th ed. Williams & Wilkins, 1991.)

slightly negative, and air flows into the lungs (Figure 34–6). At the end of inspiration, the lung recoil begins to pull the chest back to the expiratory position, where the recoil pressures of the lungs and chest wall balance. The pressure in the airway becomes slightly positive, and air flows out of the lungs. Expiration during quiet breathing is passive in the sense that no muscles which decrease intrathoracic volume contract. However, there is some contraction of the inspiratory muscles in the early part of expiration. This contraction exerts a braking action on the recoil forces and slows expiration.

Strong inspiratory efforts reduce intrapleural pressure to values as low as −30 mm Hg, producing correspondingly greater degrees of lung inflation. When ventilation is increased, the extent of lung deflation is also increased by active contraction of expiratory muscles that decrease intrathoracic volume.

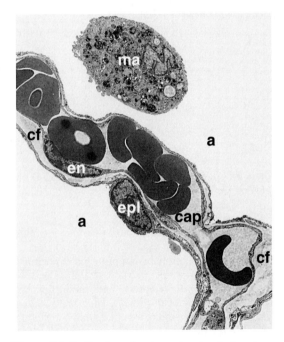

Figure 34–3. Portion of an interalveolar septum in the adult human lung. The pulmonary capillary (cap) in the septum contains plasma and red blood cells. Note the closely apposed endothelial wall and pulmonary epithelium, separated at places by connective tissue fibers (cf). en, nucleus of endothelial cell; epl, nucleus of type I alveolar epithelial cell; a, alveolar space; ma, alveolar macrophage. (Reproduced, with permission, from Burri PA: Development and growth of the human lung. In: *Handbook of Physiology,* Section 3, *The Respiratory System.* Fishman AP, Fisher AB [editors]. American Physiological Society, 1985.)

it in the same way that two moist pieces of glass slide on each other but resist separation. The pressure in the "space" between the lungs and chest wall (intrapleural pressure) is subatmospheric (Figure 34–4). The lungs are stretched when they are expanded at birth, and at the end of quiet expiration their tendency to recoil from the chest wall is just balanced by the tendency of the chest wall to recoil in the opposite direction. If the chest wall is opened, the lungs collapse; and if the lungs lose their elasticity, the chest expands and becomes barrel-shaped.

Inspiration is an active process. The contraction of the inspiratory muscles increases intrathoracic volume. The intrapleural pressure at the base of the lungs, which is normally about −2.5 mm Hg (relative to atmospheric) at the start of inspiration, decreases to about −6 mm Hg. The lungs are pulled into a more expanded position. The pressure in the airway becomes

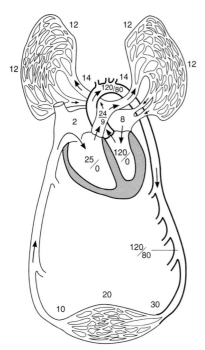

Figure 34–4. Blood pressures (mm Hg) in the pulmonary and systemic circulation. (Modified from Comroe JH Jr: *Physiology of Respiration,* 2nd ed. Year Book, 1974.)

Lung Volumes

The amount of air that moves into the lungs with each inspiration (or the amount that moves out with each expiration) is called the **tidal volume.** The air inspired with a maximal inspiratory effort in excess of the tidal volume is the **inspiratory reserve volume.** The volume expelled by an active expiratory effort after passive expiration is the **expiratory reserve volume,** and the air left in the lungs after a maximal expiratory effort is the **residual volume.** Normal values for these lung volumes, and names applied to combinations of them, are shown in Figure 34–7. The space in the conducting zone of the airways occupied by gas that does not exchange with blood in the pulmonary vessels is the **respiratory dead space.** The **vital capacity,** the largest amount of air that can be expired after a maximal inspiratory effort, is frequently measured clinically as an index of pulmonary function. It gives useful information about the strength of the respiratory muscles and other aspects of pulmonary function. The fraction of the vital capacity expired during the first second of a forced expiration (**FEV$_1$, timed vital capacity**) (Figure 34–8) gives additional information; the vital capacity

may be normal but the FEV$_1$ reduced in diseases such as asthma, in which airway resistance is increased because of bronchial constriction. The amount of air inspired per minute (**pulmonary ventilation, respiratory minute volume**) is normally about 6 L (500 mL/breath × 12 breaths/min). The **maximal voluntary ventilation (MVV),** or, as it was formerly called, the **maximal breathing capacity,** is the largest volume of gas that can be moved into and out of the lungs in 1 minute by voluntary effort. The normal MVV is 125–170 L/min.

Respiratory Muscles

Movement of the **diaphragm** accounts for 75% of the change in intrathoracic volume during quiet inspiration. Attached around the bottom of the thoracic cage, this muscle arches over the liver and moves downward like a piston when it contracts. The distance it moves ranges from 1.5 cm to as much as 7 cm with deep inspiration (Figure 34–9).

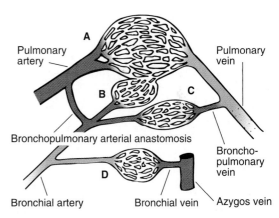

Figure 34–5. Relationship between the bronchial and pulmonary circulations. The pulmonary artery supplies pulmonary capillary network **A.** The bronchial artery supplies capillary networks B, C, and D. Network **B** represents the bronchial capillary supply to bronchioles that anastomoses with pulmonary capillaries and drains through pulmonary veins. Network **C** represents the bronchial capillary supply to most bronchi; these vessels form bronchopulmonary veins that empty into pulmonary veins. Network **D** represents the bronchial capillary supply to lobar and segmental bronchi; these vessels form true bronchial veins that drain into the azygos, hemiazygos, or intercostal veins. Dark-colored areas represent blood of low O$_2$ content. (Reproduced, with permission, from Murray JF: *The Normal Lung.* Saunders, 1986.)

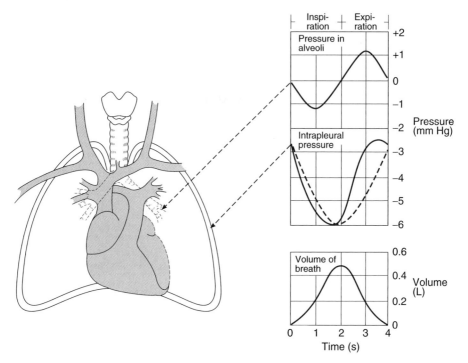

Figure 34–6. Changes in intrapleural (intrathoracic) and intrapulmonary pressure relative to atmospheric pressure during inspiration and expiration. Note that if there were no airway and tissue resistance, intrapleural pressure would follow the dashed line and that the actual pressure curve is skewed to the left by those resistances.

The diaphragm has three parts: the costal portion, made up of muscle fibers that are attached to the ribs around the bottom of the thoracic cage; the crural portion, made up of fibers that are attached to the ligaments along the vertebras; and the central tendon, into which the costal and the crural fibers insert. The central tendon is also the inferior part of the pericardium. The crural fibers pass on either side of the esophagus and can compress it when they contract. The costal and crural portions are innervated by different parts of the phrenic nerve and can contract separately. For example, during vomiting and eructation, intra-abdominal pressure is increased by contraction of the costal fibers but the crural fibers remain relaxed, allowing material to pass from the stomach into the esophagus. The role of the crural fibers in swallowing is discussed in Chapter 26.

The other important **inspiratory muscles** are the **external intercostal muscles,** which run obliquely downward and forward from rib to rib. The ribs pivot as if hinged at the back, so that when the external intercostals contract they elevate the lower ribs. This pushes the sternum outward and increases the anteroposterior diameter of the chest. The transverse diameter also in-

creases, but to a lesser degree. Either the diaphragm or the external intercostal muscles alone can maintain adequate ventilation at rest. Transection of the spinal cord above the third cervical segment is fatal without artificial respiration, but transection below the fifth cervical segment is not, because it leaves the phrenic nerves that innervate the diaphragm intact; the phrenic nerves arise from cervical segments 3–5. Conversely, in patients with bilateral phrenic nerve palsy but intact innervation of their intercostal muscles, respiration is somewhat labored but adequate to maintain life. The scalene and sternocleidomastoid muscles in the neck are accessory inspiratory muscles that help to elevate the thoracic cage during deep labored respiration.

A decrease in intrathoracic volume and forced expiration result when the **expiratory muscles** contract. The internal intercostals have this action because they pass obliquely downward and posteriorly from rib to rib and therefore pull the rib cage downward when they contract. Contractions of the muscles of the anterior abdominal wall also aid expiration by pulling the rib cage downward and inward and by increasing the intra-abdominal pressure, which pushes the diaphragm upward.

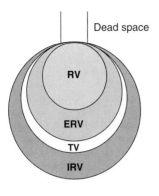
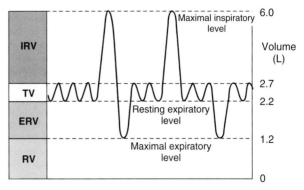

IRV = Inspiratory reserve volume   TV = Tidal volume
ERV = Expiratory reserve volume   RV = Residual volume

Volume (L)			
	Men	**Women**	
Vital capacity IRV	3.3	1.9	Inspiratory capacity
TV	0.5	0.5	
ERV	1.0	0.7	Functional residual capacity
RV	1.2	1.1	
Total lung capacity	6.0	4.2	

Respiratory minute volume (rest): 6 L/min
Alveolar ventilation (rest): 4.2 L/min
Maximal voluntary ventilation (BTPS): 125–170 L/min

Timed vital capacity: 83% of total in 1 s; 97% in 3 s
Work of quiet breathing: 0.5 kg-m/min
Maximal work of breathing: 10 kg-m/breath

Figure 34–7. Lung volumes and some measurements related to the mechanics of breathing. The diagram at the upper right represents the excursions of a spirometer plotted against time.

Glottis

The abductor muscles in the larynx contract early in inspiration, pulling the vocal cords apart and opening the glottis. During swallowing or gagging, there is reflex contraction of the adductor muscles that closes the glottis and prevents aspiration of food, fluid, or vomitus into the lungs. In unconscious or anesthetized patients, glottic closure may be incomplete and vomitus may enter the trachea, causing an inflammatory reaction in the lung (**aspiration pneumonia**).

The laryngeal muscles are supplied by the vagus nerves. When the abductors are paralyzed, there is inspiratory stridor. When the adductors are paralyzed, food and fluid enter the trachea, causing aspiration pneumonia and edema. Bilateral cervical vagotomy in animals causes the slow development of fatal pulmonary congestion and edema. The edema is due at least in part to aspiration, although some edema develops even if a tracheostomy is performed before the vagotomy.

Bronchial Tone

In general, the smooth muscle in the bronchial walls aids respiration. The bronchi dilate during inspiration and constrict during expiration. Dilation is produced by sympathetic discharge and constriction by parasympathetic discharge. Stimulation of sensory receptors in the airways by irritants and chemicals such as sulfur dioxide produces reflex bronchoconstriction that is mediated via cholinergic pathways. Cool air also causes bronchoconstriction, and so does exercise, possibly because the increased respiration associated with it cools the airways. In addition, the bronchial muscles protect the bronchi during coughing. There is a circadian rhythm in bronchial tone, with maximal constriction at about 6 AM and maximal dilation at about 6 PM.

As noted above, VIP produces bronchodilation. On the other hand, substance P causes bronchoconstriction. So does adenosine acting via its A_1 receptor (see Chapter 4). and many other cytokines and inflammatory modulators. The relation of these substances to asthma

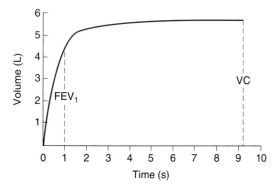

Figure 34–8. Volume of gas expired by a normal adult man during a forced expiration, demonstrating the FEV$_1$ and the total vital capacity (VC). (Reproduced, with permission, from Crapo RO: Pulmonary-function testing. N Engl J Med 1994;331:25. Copyright © 1994, Massachusetts Medical Society.)

is discussed in Chapter 37. Their role in the physiologic regulation of bronchial tone is still unsettled.

Compliance of the Lungs & Chest Wall

The interaction between the recoil of the lungs and recoil of the chest can be demonstrated in living subjects.

The nostrils are clipped shut, and the subject breathes through a spirometer that has a valve just beyond the mouthpiece. The mouthpiece contains a pressure-measuring device. After the subject inhales a given amount, the valve is shut, closing off the airway. The respiratory muscles are then relaxed while the pressure in the airway is recorded. The procedure is repeated after inhaling or actively exhaling various volumes. The curve of airway pressure obtained in this way, plotted against volume, is the **relaxation pressure curve** of the total respiratory system (Figure 34–10). The pressure is zero at a lung volume that corresponds to the volume of gas in the lungs at the end of quiet expiration (**relaxation volume,** which equals the functional residual capacity). It is positive at greater volumes and negative at smaller volumes. The change in lung volume per unit change in airway pressure ($\Delta V/\Delta P$) is the stretchability (**compliance**) of the lungs and chest wall. It is normally measured in the pressure range where the relaxation pressure curve is steepest, and the normal value is approximately 0.2 L/cm H$_2$O. However, compliance depends on lung volume; an individual with only one lung has approximately half the ΔV for a given ΔP. Compliance is also slightly greater when measured during deflation than when measured during inflation. Consequently, it is more informative to examine the whole pressure-volume curve. The curve is shifted downward and to the right (compliance is decreased)

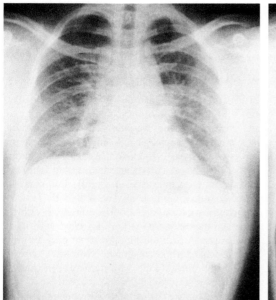

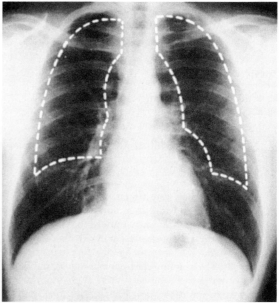

Figure 34–9. X-ray of chest in full expiration (**left**) and full inspiration (**right**). The dashed white line on the right is an outline of the lungs in full expiration. (Reproduced, with permission, from Comroe JH Jr: *Physiology of Respiration,* 2nd ed. Year Book, 1974.)

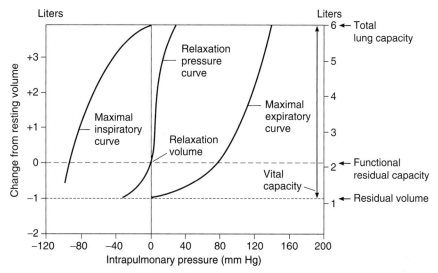

Figure 34–10. Relation between intrapulmonary pressure and volume. The middle curve is the relaxation pressure curve of the total respiratory system, ie, the static pressure curve of values obtained when the lungs are inflated or deflated by various amounts and the intrapulmonary pressure (elastic recoil pressure) is measured with the airway closed. The relaxation volume is the point where the recoil of the chest and the recoil of the lungs balance. The slope of the curve is the compliance of the lungs and chest wall. The maximal inspiratory and expiratory curves are the airway pressures that can be developed during maximal inspiratory and expiratory efforts.

by pulmonary congestion and interstitial pulmonary fibrosis (Figure 34–11); pulmonary fibrosis is a progressive disease of unknown cause in which there is stiffening and scarring of the lung. The curve is shifted upward and to the left (compliance is increased) in emphysema (see Chapter 37). It should be noted that compliance is a static measure of lung and chest recoil. The **resistance** of the lung and chest is the pressure difference required for a unit of air flow, and this measurement, which is dynamic rather than static, also takes into account the resistance to air flow in the airways.

Alveolar Surface Tension

An important factor affecting the compliance of the lungs is the surface tension of the film of fluid that lines the alveoli. The magnitude of this component at various lung volumes can be measured by removing the lungs of an experimental animal from the body and distending them alternately with saline and with air while measuring the intrapulmonary pressure. Because saline reduces the surface tension to nearly zero, the pressure-volume curve obtained with saline measures only the tissue elasticity (Figure 34–12), whereas the curve obtained with air measures both tissue elasticity and surface tension. The difference between the two curves, the elasticity due to surface tension, is much smaller at small than at large lung volumes. The surface

tension is also much lower than the expected surface tension at a water-air interface of the same dimensions.

Surfactant

The low surface tension when the alveoli are small is due to the presence in the fluid lining the alveoli of **surfactant,** a lipid surface-tension-lowering agent. Surfactant is a mixture of dipalmitoylphosphatidylcholine (DPPC), other lipids, and proteins (Table 34–2). If the surface tension is not kept low when the alveoli become smaller during expiration, they collapse in accordance with the law of Laplace (see Chapter 30). In spherical structures like the alveoli, the distending pressure equals 2 times the tension divided by the radius ($P = 2T/r$); if T is not reduced as r is reduced, the tension overcomes the distending pressure. Surfactant also helps to prevent pulmonary edema. It has been calculated that if it were not present, the unopposed surface tension in the alveoli would produce a 20 mm Hg force favoring transudation of fluid from the blood into the alveoli.

Phospholipids, which have a hydrophilic "head" and two parallel hydrophobic fatty acids "tails" (see Chapter 1), line up in the alveoli with their tails facing the alveolar lumen (Figure 34–13), and surface tension is inversely proportionate to their concentration per unit area. They move farther apart as the alveoli enlarge dur-

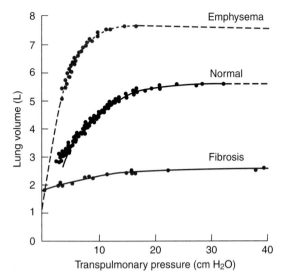

Figure 34–11. Static expiratory pressure-volume curves of lungs in normal subjects and subjects with severe emphysema and pulmonary fibrosis. (Modified and reproduced, with permission, from Pride NB, Macklem PT: Lung mechanics in disease. In: *Handbook of Physiology.* Section 3, *The Respiratory System.* Vol III, part 2. Fishman AP [editor]. American Physiological Society, 1986.)

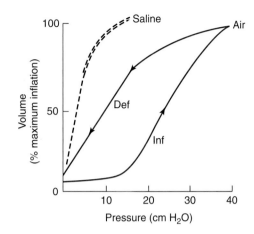

Figure 34–12. Pressure-volume relations in the lungs of a cat after removal from the body. **Air:** lungs inflated (Inf) and deflated (Def) with air. **Saline:** lungs inflated and deflated with saline. (Reproduced, with permission, from Morgan TE: Pulmonary surfactant. N Engl J Med 1971;284:1185.)

ing inspiration, and surface tension increases, whereas it decreases when they move closer together during expiration.

Surfactant is produced by type II alveolar epithelial cells (Figure 34–13). Typical **lamellar bodies,** membrane-bound organelles containing whorls of phospholipid, are formed in these cells and secreted into the alveolar lumen by exocytosis. Tubes of lipid called **tubular myelin** form from the extruded bodies, and the tubular myelin in turn forms the phospholipid film. Some of the protein-lipid complexes in surfactant are taken up by endocytosis in type II alveolar cells and recycled.

Formation of the phospholipid film is greatly facilitated by the proteins in surfactant. This material contains four unique proteins, SP-A, SP-B, SP-C, and SP-D. SP-A is a large glycoprotein and has a collagen-like domain within its structure. It probably has multiple functions, including regulation of the feedback uptake of surfactant by the type II alveolar epithelial cells that secrete it. SP-B and SP-C are smaller proteins, which facilitate formation of the monomolecular film of phospholipid. A mutation of the gene for SP-C has been reported to be associated with familial interstitial lung disease. Like SP-A, SP-D is a glycoprotein. Its function is uncertain. However, SP-A and SP-D are members of

the collectin family of proteins that are involved in innate immunity (see Chapter 27) in other parts of the body.

Surfactant is important at birth. The fetus makes respiratory movements in utero, but the lungs remain collapsed until birth. After birth, the infant makes several strong inspiratory movements and the lungs expand. Surfactant keeps them from collapsing again. Surfactant deficiency is an important cause of **infant respiratory distress syndrome (IRDS; hyaline membrane disease),** the serious pulmonary disease that develops in infants born before their surfactant system is functional. Surface tension in the lungs of these infants is high, and there are many areas in which the alveoli are collapsed (atelectasis). An additional factor in IRDS is retention of fluid in the lungs. During fetal life, Cl⁻

Table 34–2. Approximate composition of surfactant.

Component	Percentage Composition
Dipalmitoylphosphatidylcholine	62
Phosphatidylglycerol	5
Other phospholipids	10
Neutral lipids	13
Proteins	8
Carbohydrate	2

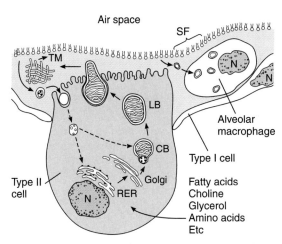

Figure 34–13. Formation and metabolism of surfactant. Lamellar bodies (LB) are formed in type II alveolar epithelial cells and secreted by exocytosis. The released lamellar body material is converted to tubular myelin (TM), and the TM is probably the source of the phospholipid surface film (SF). Some surfactant is taken up by alveolar macrophages, but more is taken up by endocytosis in type II epithelial cells. N, nucleus; RER, rough endoplasmic reticulum; CB, composite body. (Reproduced, with permission, from Wright JR: Metabolism and turnover of lung surfactant. Am Rev Respir Dis 1987;136:426.)

is secreted with fluid by the pulmonary epithelial cells. At birth, there is a shift to Na^+ absorption by these cells via the epithelial Na^+ channels (ENaCs), and fluid is absorbed with the Na^+. Prolonged immaturity of the ENaCs contributes to the pulmonary abnormalities in IRDS.

Administration of phospholipid alone by inhalation has little value in the treatment of IRDS. However, a synthetic surfactant and a surfactant preparation derived from bovine lungs are available for use by inhalation. Used prophylactically at birth and as replacement therapy, they decrease the severity of IRDS and the severity but not the incidence of chronic lung disease in survivors.

Maturation of surfactant in the lungs is accelerated by glucocorticoid hormones. There is an increase in fetal and maternal cortisol near term, and the lungs are rich in glucocorticoid receptors.

Patchy atelectasis is also associated with surfactant deficiency in patients who have undergone cardiac surgery involving use of a pump oxygenator and interruption of the pulmonary circulation. In addition, surfactant deficiency may play a role in some of the abnormalities that develop following occlusion of a main

bronchus, occlusion of one pulmonary artery, or long-term inhalation of 100% O_2. There is a decrease in surfactant in the lungs of cigarette smokers.

An interesting recent finding is the presence of excess surfactant lipids and proteins in mice with the GM-CSF gene knocked out. The role of GM-CSF in hematopoiesis is discussed in Chapter 27. The pathologic findings in the lungs of the knockout mice resemble those in the lungs of humans with **pulmonary alveolar proteinosis.**

Work of Breathing

Work is performed by the respiratory muscles in stretching the elastic tissues of the chest wall and lungs (elastic work), moving inelastic tissues (viscous resistance), and moving air through the respiratory passages (Table 34–3). Since pressure times volume ($g/cm^2 \times cm^3 = g \times cm$) has the same dimensions as work (force × distance), the work of breathing can be calculated from the relaxation pressure curve (Figures 34–10 and 34–14). In Figure 34–11, the total elastic work required for inspiration is area ABCA. Note that the relaxation pressure curve of the total respiratory system differs from that of the lungs alone. The actual elastic work required to increase the volume of the lungs alone is area ABDEA. The amount of elastic work required to inflate the whole respiratory system is less than the amount required to inflate the lungs alone because part of the work comes from elastic energy stored in the thorax. The elastic energy lost from the thorax (area AFGBA) is equal to that gained by the lungs (area AEDCA).

The frictional resistance to air movement is relatively small during quiet breathing, but it does cause the intrapleural pressure changes to lead the lung volume changes during inspiration and expiration (Figure 34–6), producing a **hysteresis loop** rather than a straight line when pressure is plotted against volume (Figure 34–15). In this diagram, area AXBYA represents the work done to overcome airway resistance and lung viscosity. If the air flow becomes turbulent during rapid respiration, the energy required to move the air is greater than when the flow is laminar.

Table 34–3. Components that make up the work of breathing during quiet inspiration, and the percent contribution of each.

Nonelastic work
Viscous resistance (7%)
Airway resistance (28%)
Elastic work (65%)

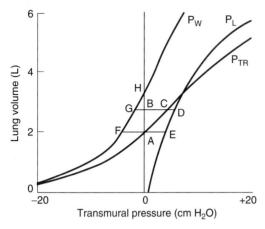

Figure 34–14. Relaxation pressure curve of the total respiratory system (P_{TR}), and the relaxation pressure curves of the lungs (P_L) and the chest (P_W). The transmural pressure is intrapulmonary pressure minus intrapleural pressure in the case of the lungs, intrapleural pressure minus outside (barometric) pressure in the case of the chest wall, and intrapulmonary pressure minus barometric pressure in the case of the total respiratory system. (Modified from Mines AH: *Respiratory Physiology,* 3rd ed. Raven Press, 1993.)

Estimates of the total work of quiet breathing range from 0.3 up to 0.8 kg-m/min. The value rises markedly during exercise, but the energy cost of breathing in normal individuals represents less than 3% of the total energy expenditure during exercise. The work of breathing is greatly increased in diseases such as emphysema, asthma, and congestive heart failure with dyspnea and orthopnea. The respiratory muscles have length-tension relations like those of other skeletal and cardiac muscles, and when they are severely stretched, they contract with less strength. They can also become fatigued and fail (pump failure), leading to inadequate ventilation (see Chapter 37). For unknown reasons, aminophylline increases the force of contraction of the human diaphragm and is useful in the treatment of pump failure.

Differences in Ventilation & Blood Flow in Different Parts of the Lung

In the upright position, ventilation per unit lung volume is greater at the base of the lung than at the apex. The reason for this is that at the start of inspiration, intrapleural pressure is less negative at the base than at the apex (Figure 34–16), and since the intrapulmonary-intrapleural pressure difference is less than at the apex, the lung is less expanded. Conversely, at the apex, the

lung is more expanded; ie, the percentage of maximum lung volume is greater. Because of the stiffness of the lung, the increase in lung volume per unit increase in pressure is smaller when the lung is initially more expanded, and ventilation is consequently greater at the base. Blood flow is also greater at the base than the apex (see below). The relative change in blood flow from the apex to the base is greater than the relative change in ventilation, so the ventilation/perfusion ratio is low at the base and high at the apex.

The ventilation and perfusion differences from the apex to the base of the lung have usually been attributed to gravity; they tend to disappear in the supine position, and the weight of the lung would be expected to make the intrapleural pressure lower at the base in the upright position. However, the inequalities of ventilation and blood flow in humans were found to persist to a remarkable degree in the weightlessness of space. Therefore, other as yet unknown factors apparently also play a role in producing the inequalities.

It should be noted that at very low lung volumes such as those after forced expiration, intrapleural pressure at the bases of the lungs can actually exceed the atmospheric pressure in the airways, and the small airways such as respiratory bronchioles collapse (**airway closure**). In older people and in those with chronic lung disease, some of the elastic recoil is lost, with a resulting decrease in intrapleural pressure. Consequently, airway closure may occur in the bases of the lungs in the upright position without forced expiration, at volumes as high as the functional residual capacity.

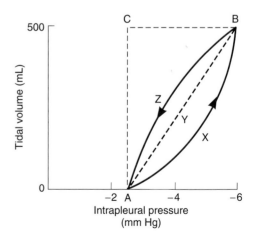

Figure 34–15. Diagrammatic representation of pressure and volume changes during quiet inspiration (line AXB) and expiration (line BZA). Line AYB is the compliance line.

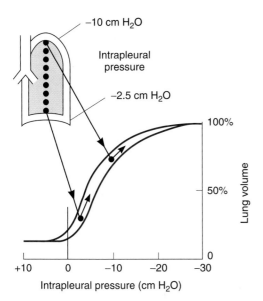

Figure 34–16. Intrapleural pressures in the upright position and their effect on ventilation. Note that because intrapulmonary pressure is atmospheric, the more negative intrapleural pressure at the apex holds the lung in a more expanded position at the start of inspiration. Further increases in volume per unit increase in intrapleural pressure are smaller than at the base because the expanded lung is stiffer. (Reproduced, with permission, from West JB: *Ventilation/Blood Flow and Gas Exchange*, 3rd ed. Blackwell, 1977.)

Dead Space & Uneven Ventilation

Since gaseous exchange in the respiratory system occurs only in the terminal portions of the airways, the gas that occupies the rest of the respiratory system is not available for gas exchange with pulmonary capillary blood. Normally, the volume of this dead space is approximately equal to the body weight in pounds. Thus, in a man who weighs 150 lb (68 kg), only the first 350 mL of the 500 mL inspired with each breath at rest mixes with the air in the alveoli. Conversely, with each expiration, the first 150 mL expired is gas that occupied the dead space, and only the last 350 mL is gas from the alveoli. Consequently, the **alveolar ventilation**, ie, the amount of air reaching the alveoli per minute, is less than the respiratory minute volume. Note in addition that because of the dead space, rapid shallow breathing produces much less alveolar ventilation than slow deep breathing at the same respiratory minute volume (Table 34–4).

It is important to distinguish between the **anatomic dead space** (respiratory system volume exclusive of alve-

Table 34–4. Effect of variations in respiratory rate and depth on alveolar ventilation.

Respiratory rate	30/min	10/min
Tidal volume	200 mL	600 mL
Minute volume	6 L	6 L
Alveolar ventilation	$(200 - 150) \times 30$ $= 1500$ mL	$(600 - 150) \times 10$ $= 4500$ mL

oli) and the **total (physiologic) dead space** (volume of gas not equilibrating with blood, ie, wasted ventilation). In healthy individuals, the two dead spaces are identical; but in disease states, there may be no exchange between the gas in some of the alveoli and the blood, and some of the alveoli may be overventilated. The volume of gas in nonperfused alveoli and any volume of air in the alveoli in excess of that necessary to arterialize the blood in the alveolar capillaries is part of the dead space (nonequilibrating) gas volume. The anatomic dead space can be measured by analysis of the single-breath N_2 curves (Figure 34–17). From mid inspiration, the subject takes as deep a breath as possible of pure O_2, then exhales steadily while the N_2 content of the expired gas is continuously measured. The initial gas exhaled (phase I) is the gas that filled the dead space and that consequently contains no N_2. This is followed by a mixture of dead space and alveolar gas (phase II) and then by alveolar gas (phase III). The volume of the dead space is the volume of the gas expired from peak inspiration to the midportion of phase II (Figure 34–17).

Phase III of the single-breath N_2 curve terminates at the **closing volume (CV)** and is followed by phase IV,

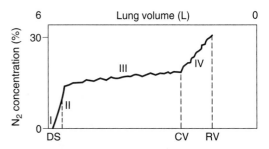

Figure 34–17. Single-breath N_2 curve. From mid inspiration, the subject takes a deep breath of pure O_2, then exhales steadily. The changes in the N_2 concentration of expired gas during expiration are shown, with the various phases of the curve indicated by roman numerals. DS, dead space, CV, closing volume; RV, residual volume.

during which the N_2 content of the expired gas is increased. The CV is the lung volume above residual volume at which airways in the lower, dependent parts of the lungs begin to close off because of the lesser transmural pressure in these areas (see above). The gas in the upper portions of the lungs is richer in N_2 than the gas in the lower, dependent portions because the alveoli in the upper portions are more distended at the start of the inspiration of O_2 (see above) and, consequently, the N_2 in them is less diluted with O_2. It is also worth noting that in most normal individuals, phase III has a slight positive slope even before phase IV is reached. This indicates that even during phase III there is a gradual increase in the proportion of the expired gas coming from the relatively N_2-rich upper portions of the lungs.

The pattern of ventilation in the lungs can be assessed by having the subject inhale a radioactive isotope of the inert gas xenon (^{113}Xe) while the chest is monitored with a battery of radiation detectors. Areas that show little radioactivity are poorly ventilated.

The total dead space can be calculated from the PCO_2 of expired air, the PCO_2 of arterial blood, and the tidal volume. The tidal volume (V_T) times the PCO_2 of the expired gas ($PECO_2$) equals the arterial PCO_2 ($PaCO_2$) times the difference between the tidal volume and the dead space (V_D) plus the PCO_2 of inspired air ($PICO_2$) times V_D (**Bohr's equation**):

$$PECO_2 \times V_T = PaCO_2 \times (V_T - V_D) + PICO_2 \times V_D$$

The term $PICO_2 \times V_D$ is so small that it can be ignored and the equation solved for V_D. If, for example,

$$PECO_2 = 28 \text{ mm Hg}$$
$$PaCO_2 = 40 \text{ mm Hg}$$
$$VT = 500 \text{ mL}$$

then,

$$V_D = 150 \text{ mL}$$

The equation can also be used to measure the anatomic dead space if one replaces $PaCO_2$ with alveolar PCO_2 (PCO_2), which is the PCO_2 of the last 10 mL of expired gas (see below). PCO_2 is an average of gas from different alveoli in proportion to their ventilation regardless of whether they are perfused. This is in contrast to $PaCO_2$, which is gas equilibrated only with perfused alveoli, and consequently, in individuals with unperfused alveoli, is greater than PCO_2.

Although it is possible to stand under water and breathe through a tube that projects above the surface, it may be worth noting that such a tube is in effect an extension of the respiratory dead space. For each milliliter of tube volume, the depth of inspiration would have to be increased 1 mL to supply the same volume

of air to the alveoli. Thus, if the volume of the tube were at all large, breathing would become very laborious. Additional effort is also required to expand the chest against the pressure of the surrounding water.

GAS EXCHANGE IN THE LUNGS

Sampling Alveolar Air

Theoretically, all but the first 150 mL expired with each expiration is the gas that was in the alveoli (**alveolar air**), but there is always some mixing at the interface between the dead-space gas and the alveolar air (Figure 34–17). A later portion of expired air is therefore the portion taken for analysis. Using modern apparatus with a suitable automatic valve, it is possible to collect the last 10 mL expired during quiet breathing. The composition of alveolar gas is compared with that of inspired and expired air in Figure 34–18.

PAO_2 can also be calculated from the **alveolar gas equation:**

$$PAO_2 = PIO_2 - PACO_2 \left(FIO_2 + \frac{1 - FIO_2}{R} \right)$$

where FIO_2 is the fraction of O_2 molecules in the dry gas, PIO_2 is the inspired PO_2, and R is the respiratory

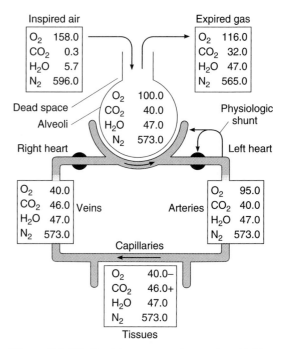

Figure 34–18. Partial pressures of gases (mm Hg) in various parts of the respiratory system and in the circulatory system.

exchange ratio (see Chapter 17), ie, the flow of CO_2 molecules across the alveolar membrane per minute divided by the flow of O_2 molecules across the membrane per minute.

Composition of Alveolar Air

Oxygen continuously diffuses out of the gas in the alveoli into the bloodstream, and CO_2 continuously diffuses into the alveoli from the blood. In the steady state, inspired air mixes with the alveolar gas, replacing the O_2 that has entered the blood and diluting the CO_2 that has entered the alveoli. Part of this mixture is expired. The O_2 content of the alveolar gas then falls and its CO_2 content rises until the next inspiration. Since the volume of gas in the alveoli is about 2 L at the end of expiration (functional residual capacity; Figure 34–7), each 350-mL increment of inspired and expired air has relatively little effect on PO_2 and PCO_2. Indeed, the composition of alveolar gas remains remarkably constant, not only at rest but also under a variety of other conditions (see Chapter 36).

Diffusion Across the Alveolocapillary Membrane

Gases diffuse from the alveoli to the blood in the pulmonary capillaries or vice versa across the thin alveolocapillary membrane made up of the pulmonary epithelium, the capillary endothelium, and their fused basement membranes (Figure 34–3). Whether or not substances passing from the alveoli to the capillary blood reach equilibrium in the 0.75 s that blood takes to traverse the pulmonary capillaries at rest depends on their reaction with substances in the blood. Thus, for example, the anesthetic gas nitrous oxide does not react, and N_2O reaches equilibrium in about 0.1 s (Figure 34–19). In this situation, the amount of N_2O taken up is not limited by diffusion but by the amount of blood flowing through the pulmonary capillaries; ie, it is **flow-limited.** On the other hand, carbon monoxide is taken up by the hemoglobin in the red blood cells at such a high rate that the partial pressure of CO in the capillaries stays very low and equilibrium is not reached in the 0.75 s the blood is in the pulmonary capillaries. Therefore, the transfer of CO is not limited by perfusion at rest and instead is **diffusion-limited.** O_2 is intermediate between N_2O and CO; it is taken up by hemoglobin, but much less avidly than CO, and it reaches equilibrium with capillary blood in about 0.3 s. Thus, its uptake is also perfusion-limited.

The **diffusing capacity** of the lung for a given gas is directly proportionate to the surface area of the alveolocapillary membrane and inversely proportionate to its thickness. The diffusing capacity for CO ($DLCO$) is

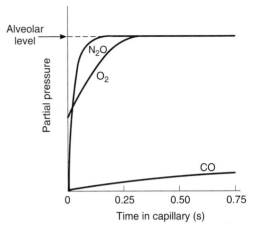

Figure 34–19. Uptake of various substances during the 0.75 s they are in transit through a pulmonary capillary. N_2O is not bound in blood, so its partial pressure in blood rises rapidly to its partial pressure in the alveoli. Conversely, CO is avidly taken up by red blood cells, so its partial pressure reaches only a fraction of its partial pressure in the alveoli. O_2 is intermediate between the two.

measured as an index of diffusing capacity because its uptake is diffusion-limited. $DLCO$ is proportionate to the amount of CO entering the blood (VCO) divided by the partial pressure of CO in the alveoli minus the partial pressure of CO in the blood entering the pulmonary capillaries. Except in habitual cigarette smokers, this latter term is close to zero, so it can be ignored and the equation becomes

$$DLCO = \frac{\dot{V}CO}{PACO}$$

The normal value of $DLCO$ at rest is about 25 mL/min/mm Hg. It increases up to threefold during exercise because of capillary dilation and an increase in the number of active capillaries.

The PO_2 of alveolar air is normally 100 mm Hg (Figure 34–18), and the PO_2 of the blood entering the pulmonary capillaries is 40 mm Hg. The diffusing capacity for O_2, like that for CO at rest, is about 25 mL/min/mm Hg, and the PO_2 of blood is raised to 97 mm Hg, a value just under the alveolar PO_2. This falls to 95 mm Hg in the aorta because of the physiologic shunt (see below). DLO_2 increases to 65 mL/min/mm Hg or more during exercise and is reduced in diseases such as sarcoidosis and beryllium poisoning (berylliosis) that cause fibrosis of the alveolar walls. Another cause of pulmonary fibrosis is excess secretion of PDGF (see Chapter 27) by alveolar macrophages, with resulting stimulation of neighboring mesenchymal cells.

The PCO_2 of venous blood is 46 mm Hg, whereas that of alveolar air is 40 mm Hg, and CO_2 diffuses from the blood into the alveoli along this gradient. The PCO_2 of blood leaving the lungs is 40 mm Hg. CO_2 passes through all biological membranes with ease, and the diffusing capacity of the lung for CO_2 is much greater than the capacity for O_2. It is for this reason that CO_2 retention is rarely a problem in patients with alveolar fibrosis even when the reduction in diffusing capacity for O_2 is severe.

PULMONARY CIRCULATION

Pulmonary Blood Vessels

The pulmonary vascular bed resembles the systemic (see Chapter 30), except that the walls of the pulmonary artery and its large branches are about 30% as thick as the wall of the aorta, and the small arterial vessels, unlike the systemic arterioles, are endothelial tubes with relatively little muscle in their walls. There is also some smooth muscle in the walls of the postcapillary vessels. The pulmonary capillaries are large, and there are multiple anastomoses, so that each alveolus sits in a capillary basket.

Pressure, Volume, & Flow

With two quantitatively minor exceptions, the blood put out by the left ventricle returns to the right atrium and is ejected by the right ventricle, making the pulmonary vasculature unique in that it accommodates a blood flow that is almost equal to that of all the other organs in the body. One of the exceptions is part of the bronchial blood flow. As noted above, there are anastomoses between the bronchial capillaries and the pulmonary capillaries and veins, and although some of the bronchial blood enters the bronchial veins, some enters the pulmonary capillaries and veins, bypassing the right ventricle. The other exception is blood that flows from the coronary arteries into the chambers of the left side of the heart (see Chapter 32). Because of the small **physiologic shunt** created by those two exceptions, the blood in systemic arteries has a PO_2 about 2 mm Hg lower than that of blood that has equilibrated with alveolar air, and the saturation of hemoglobin is 0.5% less (see Chapter 35).

The entire pulmonary vascular system is a distensible low-pressure system. The pulmonary arterial pressure is about 24/9 mm Hg, and the mean pressure is about 15 mm Hg. The pressure in the left atrium is about 8 mm Hg during diastole, so that the pressure gradient in the pulmonary system is about 7 mm Hg, compared with a gradient of about 90 mm Hg in the

systemic circulation (Figure 34–4). It is interesting that the pressure fall from the pulmonary artery to the capillaries is relatively small and that there is an appreciable pressure drop in the veins.

The volume of blood in the pulmonary vessels at any one time is about 1 L, of which less than 100 mL is in the capillaries. The mean velocity of the blood in the root of the pulmonary artery is the same as that in the aorta (about 40 cm/s). It falls off rapidly, then rises slightly again in the larger pulmonary veins. It takes a red cell about 0.75 s to traverse the pulmonary capillaries at rest and 0.3 s or less during exercise.

Capillary Pressure

Pulmonary capillary pressure is about 10 mm Hg, whereas the oncotic pressure is 25 mm Hg, so that there is an inward-directed pressure gradient of about 15 mm Hg which keeps the alveoli free of fluid. When the pulmonary capillary pressure is more than 25 mm Hg—as it may be, for example, when there is "backward failure" of the left ventricle—pulmonary congestion and edema result. Patients with mitral stenosis also have a chronic, progressive rise in pulmonary capillary pressure and extensive fibrotic changes in the pulmonary vessels.

Effect of Gravity

Gravity has a relatively marked effect on the pulmonary circulation. In the upright position, the upper portions of the lungs are well above the level of the heart, and the bases are at or below it. Consequently, there is a relatively marked pressure gradient in the pulmonary arteries from the top to the bottom of the lungs, because of the effect of gravity, and a resulting linear increase in pulmonary blood flow from the apices to the bases of the lungs (Figure 34–20). The pressure in the capillaries at the top of the lungs is close to the atmospheric pressure in the alveoli. Pulmonary arterial pressure is normally just sufficient to maintain perfusion, but if it is reduced or if alveolar pressure is increased, some of the capillaries collapse. Under these circumstances, there is no gas exchange in the affected alveoli and they become part of the physiologic dead space.

In the middle portions of the lungs, the pulmonary arterial and capillary pressure exceeds alveolar pressure, but the pressure in the pulmonary venules may be lower than alveolar pressure during normal expiration, so they are collapsed. Under these circumstances, blood flow is determined by the pulmonary artery-alveolar pressure difference rather than the pulmonary artery-pulmonary vein difference. Beyond the constriction, blood "falls" into the pulmonary veins, which are compliant and take whatever amount of blood the constric-

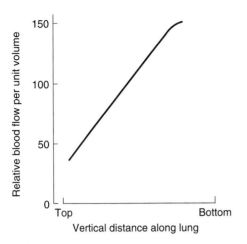

Figure 34–20. Relative blood flow from the top to the bottom of the lung in the upright position. The values for lung blood flow are scaled so that if flow were uniform, the value would be 100 throughout.

tion lets flow into them. This has been called the **waterfall effect.** Obviously, the degree of constriction decreases and pulmonary blood flow increases as the arterial pressure increases toward the base of the lung.

In the lower portions of the lungs, alveolar pressure is lower than the pressure in all parts of the pulmonary circulation and blood flow is determined by the arterial-venous pressure difference.

Ventilation/Perfusion Ratios

The ratio of pulmonary ventilation to pulmonary blood flow for the whole lung at rest is about 0.8 (4.2 L/min ventilation divided by 5.5 L/min blood flow). However, there are relatively marked differences in this **ventilation/perfusion ratio** in various parts of the normal lung as a result of the effect of gravity, and local changes in the ventilation/perfusion ratio are common in disease. If the ventilation to an alveolus is reduced relative to its perfusion, the PO_2 in the alveolus falls because less O_2 is delivered to it and the PCO_2 rises because less CO_2 is expired. Conversely, if perfusion is reduced relative to ventilation, the PCO_2 falls because less CO_2 is delivered and the PO_2 rises because less O_2 enters the blood. These effects are summarized in Figure 34–21.

As noted above, ventilation, as well as perfusion, in the upright position declines in a linear fashion from the bases to the apices of the lungs. However, the ventilation/perfusion ratios are high in the upper portions of the lungs. It is said that the high ventilation/perfusion ratios at the apices account for the predilection of tu-

berculosis for this area because the relatively high alveolar PO_2 that results provides a favorable environment for the growth of the tuberculosis bacteria.

When widespread, nonuniformity of ventilation and perfusion in the lungs can cause CO_2 retention and declines in systemic arterial PO_2. The consequences of nonuniformity in disease states are discussed in Chapter 37.

Pulmonary Reservoir

Because of their distensibility, the pulmonary veins are an important blood reservoir. When a normal individual lies down, the pulmonary blood volume increases by up to 400 mL, and when the person stands up this blood is discharged into the general circulation. This shift is the cause of the decrease in vital capacity in the supine position and is responsible for the occurrence of orthopnea in heart failure (see Chapter 33).

Regulation of Pulmonary Blood Flow

It is unsettled whether pulmonary veins and pulmonary arteries are regulated separately, although constriction of the veins increases pulmonary capillary pressure and constriction of pulmonary arteries increases the load on the right side of the heart. Small arteries several hundred micrometers in diameter are the major site of vascular resistance.

Pulmonary blood flow is affected by both active and passive factors. There is an extensive autonomic inner-

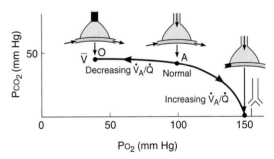

Figure 34–21. Effects of decreasing or increasing the ventilation/perfusion ratio ($\dot{V}_A/\dot{Q}$) on the PCO_2 and PO_2 in an alveolus. The drawings above the curve represent an alveolus and a pulmonary capillary, and the black areas indicate sites of blockage. With complete obstruction of the airway to the alveolus, PCO_2 and PO_2 approximate the values in mixed venous blood ($\bar{V}$). With complete block of perfusion, PCO_2 and PO_2 approximate the values in inspired air. (Reproduced, with permission, from West JB: *Ventilation/Blood Flow and Gas Exchange,* 3rd ed. Blackwell, 1977.)

vation of the pulmonary vessels, and stimulation of the cervical sympathetic ganglia reduces pulmonary blood flow by as much as 30%. The vessels also respond to circulating humoral agents. The receptors involved and the responses produced are summarized in Table 34–5. Many of the dilator responses are endothelium-dependent and presumably operate via release of NO (see Chapter 31).

Passive factors such as cardiac output and gravitational forces also have significant effects on pulmonary blood flow. Local adjustments of perfusion to ventilation are determined by local effects of O_2 or its lack. With ex-

Table 34–5. Receptors affecting smooth muscle in pulmonary arteries and veins.[1]

Receptor	Subtype	Response	Endothelium Dependency
Autonomic			
Adrenergic	α_1	Contraction	No
	α_2	Relaxation	Yes
	β_2	Relaxation	Yes
Muscarinic	M_3	Relaxation	Yes
Purinergic	P_{2x}	Contraction	No
	P_{2y}	Relaxation	Yes
Tachykinin	NK_1	Relaxation	Yes
	NK_2	Contraction	No
VIP	?	Relaxation	?
CGRP	?	Relaxation	No
Humoral			
Adenosine	A_1	Contraction	No
	A_2	Relaxation	No
Angiotensin II	AT_1	Contraction	No
ANP	ANP_A	Relaxation	No
	ANP_B	Relaxation	No
Bradykinin	B_1?	Relaxation	Yes
	B_2	Relaxation	Yes
Endothelin	ET_A	Contraction	No
	ET_B	Relaxation	Yes
Histamine	H_1	Relaxation	Yes
	H_2	Relaxation	No
5-HT	$5\text{-}HT_1$	Contraction	No
	$5\text{-}HT_{1C}$	Relaxation	Yes
Thromboxane	TP	Contraction	No
Vasopressin	V_1	Relaxation	Yes

[1]Modified and reproduced, with permission, from Barnes PJ, Lin SF: Regulation of pulmonary vascular tone. Pharmacol Rev 1995;47:88.

ercise, cardiac output increases and pulmonary arterial pressure rises proportionately with little or no vasodilation. More red cells move through the lungs without any reduction in the O_2 saturation of the hemoglobin in them, and consequently, the total amount of O_2 delivered to the systemic circulation is increased. Capillaries dilate, and previously underperfused capillaries are "recruited" to carry blood. The net effect is a marked increase in pulmonary blood flow with few if any alterations in autonomic outflow to the pulmonary vessels.

^{133}Xe can be used to survey local pulmonary blood flow by injecting a saline solution of the gas intravenously while monitoring the chest. The gas rapidly enters the alveoli that are perfused normally but fails to enter those that are not perfused. Another technique for locating poorly perfused areas is injection of macroaggregates of albumin labeled with radioactive iodine. These aggregates are large enough to block capillaries and small arterioles, and they lodge only in vessels in which blood was flowing when they reached the lungs. Although it seems paradoxic to study patients with defective pulmonary blood flow by producing vascular obstruction, the technique is safe because relatively few particles are injected. The particles block only a small number of pulmonary vessels and are rapidly removed by the body.

When a bronchus or a bronchiole is obstructed, hypoxia develops in the underventilated alveoli beyond the obstruction. The O_2 deficiency apparently acts directly on vascular smooth muscle in the area to produce constriction, shunting blood away from the hypoxic area. Accumulation of CO_2 leads to a drop in pH in the area, and a decline in pH also produces vasoconstriction in the lungs, as opposed to the vasodilation it produces in other tissues. Conversely, reduction of the blood flow to a portion of the lung lowers the alveolar P_{CO_2} in that area, and this leads to constriction of the bronchi supplying it, shifting ventilation away from the poorly perfused area.

Systemic hypoxia also causes the pulmonary arterioles to constrict, with a resultant increase in pulmonary arterial pressure.

OTHER FUNCTIONS OF THE RESPIRATORY SYSTEM

Lung Defense Mechanisms

The respiratory passages that lead from the exterior to the alveoli do more than serve as gas conduits. They humidify and cool or warm the inspired air so that even very hot or very cold air is at or near body temperature by the time it reaches the alveoli. Bronchial secretions contain secretory immunoglobulins (IgA; see Chapter 27) and other substances that help resist infections and maintain the integrity of the mucosa. In addition, the epithelium of the paranasal sinuses appears to produce NO, which is bacteriostatic and helps prevent infections.

The pulmonary epithelium contains an interesting group of protease-activated receptors (PARs) that when activated trigger release of PGE_2, which in turn protects the epithelial cells. These receptors, which are also present in the gastrointestinal tract, are activated when thrombin or trypsin partially digests ligands tethered to them. The PAR2 isoform is the form of the receptor in the respiratory tract.

The pulmonary alveolar macrophages (PAMs, "dust cells") are another important component of the pulmonary defense mechanisms. Like other macrophages (see Chapter 27), these cells come originally from the bone marrow. They are actively phagocytic and ingest inhaled bacteria and small particles. They also help process inhaled antigens for immunologic attack, and they secrete substances that attract granulocytes to the lungs as well as substances that stimulate granulocyte and monocyte formation in the bone marrow. Their role in the pathogenesis of emphysema is discussed in Chapter 37. When the macrophages ingest large amounts of the substances in cigarette smoke, they may also release lysosomal products into the extracellular space. This causes inflammation. Silica and asbestos particles also cause extracellular release of lysosomal enzymes.

Various mechanisms operate to prevent foreign matter from reaching the alveoli. The hairs in the nostrils strain out many particles larger than 10 μm in diameter. Most of the remaining particles of this size settle on mucous membranes in the nose and pharynx; because of their momentum, they do not follow the airstream as it curves downward into the lungs, and they impact on or near the **tonsils** and **adenoids,** large collections of immunologically active lymphoid tissue in the back of the pharynx. Particles 2–10 μm in diameter generally fall on the walls of the bronchi as the air flow slows in the smaller passages. There they initiate reflex bronchial constriction and coughing (see Chapter 14). They are also moved away from the lungs by the "ciliary escalator." The epithelium of the respiratory passages from the anterior third of the nose to the beginning of the respiratory bronchioles is ciliated, and the cilia, which are covered with mucus, beat in a coordinated fashion at a frequency of 1000–1500 cycles per minute. The ciliary mechanism is capable of moving particles away from the lungs at a rate of at least 16 mm/min. Particles less than 2 μm in diameter generally reach the alveoli, where they are ingested by the macrophages. The importance of these defense mechanisms is evident when one remembers that in modern cities, each liter of air may contain several million particles of dust and irritants.

When ciliary motility is defective, mucus transport is virtually absent. This leads to chronic sinusitis, recurrent lung infections, and bronchiectasis. Ciliary immotility may be produced by various air pollutants, or it may be congenital. One congenital form is **Kartagener's syndrome,** in which the axonemal dynein, the ATPase molecular motor that produces ciliary beating (see Chapter 1), is absent. Patients with this condition also are infertile because they lack motile sperm, and they often have situs inversus, presumably because the cilia necessary for rotating the viscera are nonfunctional during embryonic development.

Metabolic & Endocrine Functions of the Lungs

In addition to their functions in gas exchange, the lungs have a number of metabolic functions. They manufacture surfactant for local use as noted above. They also contain a fibrinolytic system that lyses clots in the pulmonary vessels. They release a variety of substances that enter the systemic arterial blood (Table 34–6), and they remove other substances from the systemic venous blood that reaches them via the pulmonary artery. Prostaglandins are removed from the circulation, but they are also synthesized in the lungs and released into the blood when lung tissue is stretched.

The lungs also activate one hormone; the physiologically inactive decapeptide angiotensin I is converted to the pressor, aldosterone-stimulating octapeptide angiotensin II in the pulmonary circulation (see Chapter 24). Large amounts of the angiotensin-converting

Table 34–6. Biologically active substances metabolized by the lungs.

Synthesized and used in the lungs
Surfactant
Synthesized or stored and released into the blood
Prostaglandins
Histamine
Kallikrein
Partially removed from the blood
Prostaglandins
Bradykinin
Adenine nucleotides
Serotonin
Norepinephrine
Acetylcholine
Activated in the lungs
Angiotensin I → angiotensin II

enzyme responsible for this activation are located on the surface of the endothelial cells of the pulmonary capillaries. The converting enzyme also inactivates bradykinin. Circulation time through the pulmonary capillaries is less than 1 s, yet 70% of the angiotensin I reaching the lungs is converted to angiotensin II in a single trip through the capillaries. Four other peptidases have been identified on the surface of the pulmonary endothelial cells, but their physiologic role is unsettled.

Removal of serotonin and norepinephrine reduces the amounts of these vasoactive substances reaching the systemic circulation. However, many other vasoactive hormones pass through the lungs without being metabolized. These include epinephrine, dopamine, oxytocin, vasopressin, and angiotensin II. In addition, as noted in Chapter 26, various amines and polypeptides are secreted by neuroendocrine cells in the lungs.

Gas Transport Between the Lungs & the Tissues

35

INTRODUCTION

The partial pressure gradients for O_2 and CO_2 plotted in graphic form in Figure 35–1 emphasize that they are the key to gas movement and that O_2 "flows downhill" from the air through the alveoli and blood into the tissues whereas CO_2 "flows downhill" from the tissues to the alveoli. However, the amount of both of these gases transported to and from the tissues would be grossly inadequate if it were not that about 99% of the O_2 which dissolves in the blood combines with the O_2-carrying protein hemoglobin and that about 94.5% of the CO_2 which dissolves enters into a series of reversible chemical reactions that convert it into other compounds. Thus, the presence of hemoglobin increases the O_2-carrying capacity of the blood 70-fold, and the reactions of CO_2 increase the blood CO_2 content 17-fold. This chapter reviews the mechanisms involved in O_2 and CO_2 transport.

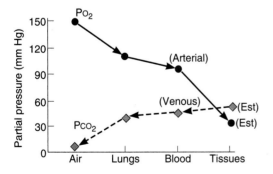

Figure 35–1. Summary of P_{O_2} and P_{CO_2} values in air, lungs, blood, and tissues, graphed to emphasize the fact that both O_2 and CO_2 diffuse "downhill" along gradients of decreasing partial pressure. (Redrawn and reproduced, with permission, from Kinney JM: Transport of carbon dioxide in blood. Anesthesiology 1960;21:615.)

OXYGEN TRANSPORT

Oxygen Delivery to the Tissues

The O_2 delivery system in the body consists of the lungs and the cardiovascular system. O_2 delivery to a particular tissue depends on the amount of O_2 entering the lungs, the adequacy of pulmonary gas exchange, the blood flow to the tissue, and the capacity of the blood to carry O_2. The blood flow depends on the degree of constriction of the vascular bed in the tissue and the cardiac output. The amount of O_2 in the blood is determined by the amount of dissolved O_2, the amount of hemoglobin in the blood, and the affinity of the hemoglobin for O_2.

Reaction of Hemoglobin & Oxygen

The dynamics of the reaction of hemoglobin with O_2 make it a particularly suitable O_2 carrier. Hemoglobin (see Chapter 27) is a protein made up of four subunits, each of which contains a **heme** moiety attached to a polypeptide chain. In normal adults, most of the hemoglobin molecules contain two α and two β chains. Heme (see Figure 27–17) is a complex made up of a porphyrin and one atom of ferrous iron. Each of the four iron atoms can bind reversibly one O_2 molecule. The iron stays in the ferrous state, so that the reaction is an **oxygenation,** not an oxidation. It has been customary to write the reaction of hemoglobin with O_2 as $Hb + O_2 \rightleftarrows HbO_2$. Since it contains four Hb units, the hemoglobin molecule can also be represented as Hb_4, and it actually reacts with four molecules of O_2 to form Hb_4O_8.

$$Hb_4 + O_2 \rightleftarrows Hb_4O_2$$
$$Hb_4O_2 + O_2 \rightleftarrows Hb_4O_4$$
$$Hb_4O_4 + O_2 \rightleftarrows Hb_4O_6$$
$$Hb_4O_6 + O_2 \rightleftarrows Hb_4O_8$$

The reaction is rapid, requiring less than 0.01 s. The deoxygenation (reduction) of Hb_4O_8 is also very rapid.

The quaternary structure of hemoglobin determines its affinity for O_2. In deoxyhemoglobin, the globin units are tightly bound in a **tense (T) configuration** which reduces the affinity of the molecule for O_2. When O_2 is first bound, the bonds holding the globin units are released, producing a **relaxed (R) configuration** which exposes more O_2 binding sites. The net result is a 500-fold increase in O_2 affinity. In the tissues, these reactions are reversed, releasing O_2. The transition from one state to another has been calculated to occur about 10^8 times in the life of a red blood cell.

The **oxygen-hemoglobin dissociation curve,** the curve relating percentage saturation of the O_2-carrying power of hemoglobin to the PO_2 (Figure 35–2), has a characteristic sigmoid shape due to the T–R interconversion. Combination of the first heme in the Hb molecule with O_2 increases the affinity of the second heme for O_2, and oxygenation of the second increases the affinity of the third, etc, so that the affinity of Hb for the fourth O_2 molecule is many times that for the first.

When blood is equilibrated with 100% O_2 (PO_2 = 760 mm Hg), the normal hemoglobin becomes 100% saturated. When fully saturated, each gram of normal hemoglobin contains 1.39 mL of O_2. However, blood normally contains small amounts of inactive hemoglobin derivatives, and the measured value in vivo is lower. The traditional figure is 1.34 mL of O_2. The hemoglobin concentration in normal blood is about 15 g/dL

Table 35–1. Gas content of blood.

	mL/dL of Blood Containing 15 g of Hemoglobin			
	Arterial Blood (PO_2 95 mm Hg; PCO_2 40 mm Hg; Hb 97% Saturated)		Venous Blood (PO_2 40 mm Hg; PCO_2 46 mm Hg; Hb 75% Saturated)	
Gas	Dissolved	Combined	Dissolved	Combined
O_2	0.29	19.5	0.12	15.1
CO_2	2.62	46.4	2.98	49.7
N_2	0.98	0	0.98	0

(14 g/dL in women and 16 g/dL in men; see Chapter 27). Therefore, 1 dL of blood contains 20.1 mL (1.34 mL × 15) of O_2 bound to hemoglobin when the hemoglobin is 100% saturated. The amount of dissolved O_2 is a linear function of the PO_2 (0.003 mL/dL blood/mm Hg PO_2).

In vivo, the hemoglobin in the blood at the ends of the pulmonary capillaries is about 97.5% saturated with O_2 (PO_2 = 97 mm Hg). Because of a slight admixture with venous blood that bypasses the pulmonary capillaries (physiologic shunt), the hemoglobin in systemic arterial blood is only 97% saturated. The arterial blood therefore contains a total of about 19.8 mL of O_2 per dL: 0.29 mL in solution and 19.5 mL bound to hemoglobin. In venous blood at rest, the hemoglobin is 75% saturated and the total O_2 content is about 15.2 mL/dL: 0.12 mL in solution and 15.1 mL bound to hemoglobin. Thus, at rest the tissues remove about 4.6 mL of O_2 from each deciliter of blood passing through them (Table 35–1); 0.17 mL of this total represents O_2 that was in solution in the blood, and the remainder represents O_2 that was liberated from hemoglobin. In this way, 250 mL of O_2 per minute is transported from the blood to the tissues at rest.

Factors Affecting the Affinity of Hemoglobin for Oxygen

Three important conditions affect the oxygen-hemoglobin dissociation curve: the **pH**, the **temperature,** and the concentration of **2,3-diphosphoglycerate (DPG; 2,3-DPG).** A rise in temperature or a fall in pH shifts the curve to the right (Figure 35–3). When the curve is shifted in this direction, a higher PO_2 is required for hemoglobin to bind a given amount of O_2. Conversely, a fall in temperature or a rise in pH shifts the curve to the left, and a lower PO_2 is required to bind a given amount of O_2. A convenient index of such

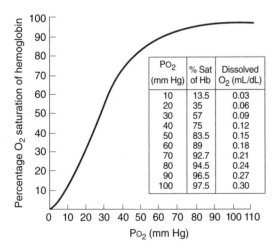

Figure 35–2. Oxygen-hemoglobin dissociation curve. pH 7.40, temperature 38 °C. (Redrawn and reproduced, with permission, from Comroe JH Jr et al: *The Lung: Clinical Physiology and Pulmonary Function Tests,* 2nd ed. Year Book, 1962.)

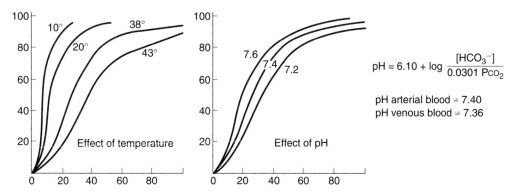

Figure 35–3. Effect of temperature and pH on the oxygen-hemoglobin dissociation curve. Ordinates and abscissas are as in Figure 35–2. (Redrawn and reproduced, with permission, from Comroe JH Jr et al: *The Lung: Clinical Physiology and Pulmonary Function Tests,* 2nd ed. Year Book, 1962.)

shifts is the P_{50}, the P_{O_2} at which hemoglobin is half saturated with O_2. The higher the P_{50}, the lower the affinity of hemoglobin for O_2.

The decrease in O_2 affinity of hemoglobin when the pH of blood falls is called the **Bohr effect** and is closely related to the fact that deoxygenated hemoglobin (deoxyhemoglobin) binds H^+ more actively than does oxyhemoglobin. The pH of blood falls as its CO_2 content increases (see below), so that when the P_{CO_2} rises, the curve shifts to the right and the P_{50} rises. Most of the unsaturation of hemoglobin that occurs in the tissues is secondary to the decline in the P_{O_2}, but an extra 1–2% unsaturation is due to the rise in P_{CO_2} and consequent shift of the dissociation curve to the right.

2,3-DPG is very plentiful in red cells. It is formed (Figure 35–4) from 3-phosphoglyceraldehyde, which is a product of glycolysis via the Embden-Meyerhof pathway (see Chapter 17). It is a highly charged anion that binds to the β chains of deoxyhemoglobin. One mole of deoxyhemoglobin binds 1 mol of 2,3-DPG. In effect,

$$HbO_2 + 2,3\text{-}DPG \rightleftarrows Hb - 2,3\text{-}DPG + O_2$$

In this equilibrium, an increase in the concentration of 2,3-DPG shifts the reaction to the right, causing more O_2 to be liberated. ATP binds to deoxyhemoglobin to a lesser extent, and some other organic phosphates bind to a minor degree.

Factors affecting the concentration of 2,3-DPG in the red cells include pH. Because acidosis inhibits red cell glycolysis, the 2,3-DPG concentration falls when the pH is low. Thyroid hormones, growth hormone, and androgens increase the concentration of 2,3-DPG and the P_{50}.

Exercise has been reported to produce an increase in 2,3-DPG within 60 minutes, although the rise may not occur in trained athletes. The P_{50} is also increased during exercise, because the temperature rises in active tissues and CO_2 and metabolites accumulate, lowering the pH. In addition, much more O_2 is removed from each unit of blood flowing through active tissues because the tissues P_{O_2} declines. Finally, at low P_{O_2} values, the oxygen-hemoglobin dissociation curve is steep,

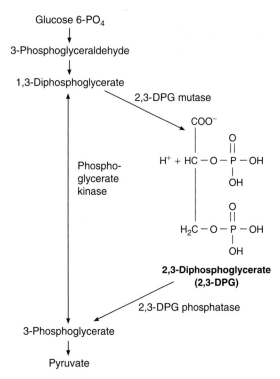

Figure 35–4. Formation and catabolism of 2,3-DPG.

and large amounts of O_2 are liberated per unit drop in Po_2.

Ascent to high altitude triggers a substantial rise in 2,3-DPG concentration in red cells, with a consequent increase in P_{50} and increase in the availability of O_2 to tissues. The rise in 2,3-DPG, which has a half-life of 6 hours, is secondary to the rise in blood pH (see Chapter 37). 2,3-DPG levels drop to normal upon return to sea level.

The greater affinity of fetal hemoglobin (hemoglobin F) than adult hemoglobin (hemoglobin A) for O_2 facilitates the movement of O_2 from the mother to the fetus (see Chapters 27 and 32). The cause of this greater affinity is the poor binding of 2,3-DPG by the γ polypeptide chains that replace β chains in fetal hemoglobin. Some abnormal hemoglobins in adults have low P_{50} values, and the resulting high O_2 affinity of the hemoglobin causes enough tissue hypoxia to stimulate increased red cell formation, with resulting polycythemia (see Chapter 24). It is interesting to speculate that these hemoglobins may not bind 2,3-DPG.

Red cell 2,3-DPG concentration is increased in anemia and in a variety of diseases in which there is chronic hypoxia. This facilitates the delivery of O_2 to the tissues by raising the Po_2 at which O_2 is released in peripheral capillaries. In bank blood that is stored, the 2,3-DPG level falls and the ability of this blood to release O_2 to the tissues is reduced. This decrease, which obviously limits the benefit of the blood if it is transfused into a hypoxic patient, is less if the blood is stored in citrate-phosphate-dextrose solution rather than the usual acid-citrate-dextrose solution.

Nitric Oxide Transport by Hemoglobin

The ferrous O_2 binding sites in hemoglobin also bind nitric oxide (NO), and there is an additional NO binding site on the β chains. The affinity of this second site is increased by O_2, so hemoglobin binds NO in the lungs and releases it in the tissues, where it promotes vasodilation (see Chapter 31)

Thus, hemoglobin has four functions: it facilitates O_2 transport; it facilitates CO_2 transport; it has an important role as a buffer (see Chapter 39); and it transports NO. Other aspects of the chemistry of hemoglobin are discussed in Chapter 27. Fetal hemoglobin and transplacental O_2 exchange are discussed in Chapter 32.

Myoglobin

Myoglobin is an iron-containing pigment found in skeletal muscle. It resembles hemoglobin but binds 1 rather than 4 mol of O_2 per mole. Its dissociation curve is a rectangular hyperbola rather than a sigmoid curve. Because its curve is to the left of the hemoglobin curve (Figure 35–5), it takes up O_2 from hemoglobin in the blood. It releases O_2 only at low Po_2 values, but the Po_2 in exercising muscle is close to zero. The myoglobin content is greatest in muscles specialized for sustained contraction. The muscle blood supply is compressed during such contractions, and myoglobin may provide O_2 when blood flow is cut off. There is also evidence that myoglobin facilitates the diffusion of O_2 from the blood to the mitochondria, where the oxidative reactions occur. However, mice in which myoglobin synthesis is prevented by gene knockout have normal exercise capacity, so further research on the physiologic role of myoglobin is needed.

CARBON DIOXIDE TRANSPORT

Buffers

Since CO_2 forms carbonic acid in the blood, an understanding of buffering in the body is required to understand CO_2 transport. Buffers are described in Chapter 1 and considered in detail in Chapter 39.

Fate of Carbon Dioxide in Blood

The solubility of CO_2 in blood is about 20 times that of O_2, so that there is considerably more CO_2 than O_2 in simple solution at equal partial pressures. The CO_2 that diffuses into red blood cells is rapidly hydrated to H_2CO_3 because of the presence of carbonic anhydrase. The H_2CO_3 dissociates to H^+ and HCO_3^-, and the H^+ is buffered, primarily by hemoglobin, while the HCO_3^- enters the plasma. Some of the CO_2 in the red cells reacts with the amino groups of

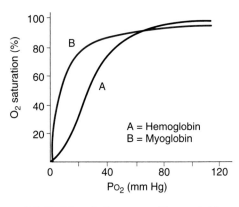

Figure 35–5. Dissociation curve of hemoglobin and myoglobin.

hemoglobin and other proteins (R), forming **carbamino compounds:**

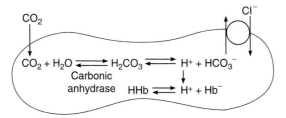

Since deoxygenated hemoglobin binds more H^+ than oxyhemoglobin does and forms carbamino compounds more readily, binding of O_2 to hemoglobin reduces its affinity for CO_2 **(Haldane effect).** Consequently, venous blood carries more CO_2 than arterial blood, and CO_2 uptake is facilitated in the tissues and CO_2 release is facilitated in the lungs. About 11% of the CO_2 added to the blood in the systemic capillaries is carried to the lungs as carbamino-CO_2.

In the plasma, CO_2 reacts with plasma proteins to form small amounts of carbamino compounds, and small amounts of CO_2 are hydrated; but the hydration reaction is slow in the absence of carbonic anhydrase.

Chloride Shift

Since the rise in the HCO_3^- content of red cells is much greater than that in plasma as the blood passes through the capillaries, about 70% of the HCO_3^- formed in the red cells enters the plasma. The excess HCO_3^- leaves the red cells in exchange for Cl^- (Figure 35–6), a process mediated by **Band 3,** a major membrane protein. This exchange is called the **chloride shift.** Because of it, the Cl^- content of the red cells in venous blood is therefore significantly greater than in arterial blood. The chloride shift occurs rapidly and is essentially complete in 1 second.

Note that for each CO_2 molecule added to a red cell, there is an increase of one osmotically active particle—either an HCO_3^- or a Cl^-—in the red cell (Figure 35–6). Consequently, the red cells take up water and

Table 35–2. Fate of CO_2 in blood.

In plasma

1. Dissolved

2. Formation of carbamino compounds with plasma protein

3. Hydration, H^+ buffered, HCO_3^- in plasma

In red blood cells

1. Dissolved

2. Formation of carbamino-Hb

3. Hydration, H^+ buffered, 70% of HCO_3^- enters the plasma

4. Cl^- shifts into cells; mosm in cells increases

increase in size. For this reason, plus the fact that a small amount of fluid in the arterial blood returns via the lymphatics rather than the veins, the hematocrit of venous blood is normally 3% greater than that of the arterial blood. In the lungs, the Cl^- moves out of the cells and they shrink.

Summary of Carbon Dioxide Transport

For convenience, the various fates of CO_2 in the plasma and red cells are summarized in Table 35–2.

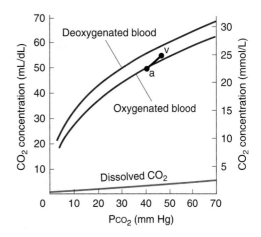

Figure 35–7. CO_2 dissociation curves. The arterial point (a) and the venous point (v) indicate the total CO_2 content found in arterial blood and venous blood of normal resting humans. (Modified and reproduced, with permission, from Schmidt RF, Thews G [editors]: *Human Physiology.* Springer, 1983.)

Figure 35–6. Summary of changes that occur in a red cell upon addition of CO_2 to blood. Note that for each CO_2 molecule that enters the red cell, there is an additional HCO_3^- or Cl^- ion in the cell.

The extent to which they increase the capacity of the blood to carry CO_2 is indicated by the difference between the lines indicating the dissolved CO_2 and the total CO_2 in the dissociation curves for CO_2 shown in Figure 35–7.

Of the approximately 49 mL of CO_2 in each deciliter of arterial blood (Table 35–1), 2.6 mL is dissolved, 2.6 mL is in carbamino compounds, and 43.8 mL is in HCO_3^-. In the tissues, 3.7 mL of CO_2 per deciliter of blood is added; 0.4 mL stays in solution, 0.8 mL forms carbamino compounds, and 2.5 mL forms HCO_3^-. The pH of the blood drops from 7.40 to 7.36. In the lungs, the processes are reversed, and the 3.7 mL of CO_2 is discharged into the alveoli. In this fashion, 200 mL of CO_2 per minute at rest and much larger amounts during exercise are transported from the tissues to the lungs and excreted. It is worth noting that this amount of CO_2 is equivalent in 24 hours to over 12,500 meq of H^+.

Regulation of Respiration

36

INTRODUCTION

Spontaneous respiration is produced by rhythmic discharge of motor neurons that innervate the respiratory muscles. This discharge is totally dependent on nerve impulses from the brain; breathing stops if the spinal cord is transected above the origin of the phrenic nerves.

The rhythmic discharges from the brain that produce spontaneous respiration are regulated by alternations in arterial PO_2, PCO_2, and H^+ concentration, and this chemical control of breathing is supplemented by a number of nonchemical influences.

NEURAL CONTROL OF BREATHING

Control Systems

Two separate neural mechanisms regulate respiration. One is responsible for voluntary control and the other for automatic control. The voluntary system is located in the cerebral cortex and sends impulses to the respiratory motor neurons via the corticospinal tracts. The automatic system is located in the pons and medulla, and the efferent output from this system to the respiratory motor neurons is located in the white matter of the spinal cord between the lateral and ventral corticospinal tracts. The nerve fibers mediating inspiration converge on the phrenic motor neurons located in the ventral horns from C3 to C5 and the external intercostal motor neurons in the ventral horns throughout the thoracic cord. The fibers concerned with expiration converge primarily on the internal intercostal motor neurons in the thoracic cord.

The motor neurons to the expiratory muscles are inhibited when those supplying the inspiratory muscles are active, and vice versa. Although spinal reflexes contribute to this **reciprocal innervation** (see Chapter 6), it is due primarily to activity in descending pathways. Impulses in these descending pathways excite agonists and inhibit antagonists. The one exception to the reciprocal inhibition is a small amount of activity in phrenic axons for a short period after inspiration. The function of this postinspiratory output appears to be to brake the lung's elastic recoil and make respiration smooth.

Medullary Systems

Rhythmic discharge of neurons in the medulla and pons produces automatic respiration; transection of the brain stem below the medulla (Section D in Figure 36–1) stops respiration, and automatic respiration is normal after transection rostral to the pons (Section A in Figure 36–1).

The main components of the **respiratory control pattern generator** responsible for automatic respiration are located in the medulla, since spontaneous respiration continues, albeit somewhat irregular and gasping, after transection of the brain stem at the inferior border of the pons (Section C in Figure 36–1).

It now seems clear that rhythmic respiration is initiated by a small group of synaptically coupled pacemaker cells in the **pre-Bötzinger complex** on either side of the medulla between the nucleus ambiguus and the lateral reticular nucleus (Figure 36–2). These neurons discharge rhythmically, and they produce rhythmic discharges in phrenic motor neurons that are abolished by sections between the pre-Bötzinger complex and these motor neurons. They also contact the hypoglossal nuclei, and the tongue is involved in the regulation of airway resistance. Neurons in the pre-Bötzinger complex discharge rhythmically in brain slice preparations in vitro, and if the slices become hypoxic, discharge changes to one associated with gasping. Addition of cadmium to the slices causes occasional sigh-like discharge patterns. There are NK1 receptors and μ-opioid receptors on these neurons, and—in vivo—substance P stimulates and opioids inhibit respiration.

There are in addition dorsal and ventral groups of respiratory neurons in the medulla (Figure 36–1). However, lesions of these neurons do not abolish respiratory activity, and they apparently project to the pre-Bötzinger pacemaker neurons.

Pontine & Vagal Influences

Although the rhythmic discharge of medullary neurons concerned with respiration is spontaneous, it is modified by neurons in the pons and afferents in the vagus from receptors in the airways and lungs. An area known as the **pneumotaxic center** in the medial parabrachial and Kölliker-Fuse nuclei of the dorsolat-

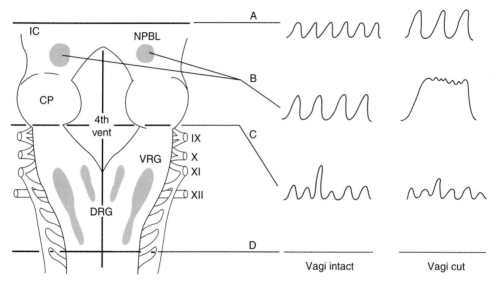

Figure 36–1. Respiratory neurons in the brain stem. Dorsal view of brain stem; cerebellum removed. The effects of various lesions and brain stem transections are also shown. The spirometer tracings at the right indicate the depth and rate of breathing. DRG, dorsal group of respiratory neurons; VRG, ventral group of respiratory neurons; NPBL, nucleus parabrachialis (pneumotaxic center); 4th vent, fourth ventricle; IC, inferior colliculus; CP, middle cerebellar peduncle. The roman numerals identify cranial nerves. (Modified and reproduced, with permission, from Mitchell RA, Berger A: State of the art: Review of neural regulation of respiration. Am Rev Respir Dis 1975;111:206.)

eral pons contains neurons active during inspiration and neurons active during expiration. When this area is damaged, respiration becomes slower and tidal volume greater, and when the vagi are also cut in anesthetized animals, there are prolonged inspiratory spasms that resemble breath holding (**apneusis;** Section B in Figure 36–1). The normal function of the pneumotaxic center is unknown, but it may play a role in switching between inspiration and expiration.

Stretching of the lungs during inspiration initiates impulses in afferent pulmonary vagal fibers. These impulses inhibit inspiratory discharge. This is why the depth of inspiration is increased after vagotomy (Figure 36–1) and apneusis develops if the vagi are cut after damage to the pneumotaxic center. As shown in Figure 36–3, the vagal feedback activity does not alter the rate of rise of the neural activity in respiratory motor neurons, but in its absence the activity is prolonged.

When the activity of the inspiratory neurons is increased in intact animals, the rate and the depth of breathing are increased. The depth of respiration is increased because the lungs are stretched to a greater degree before the amount of vagal and pneumotaxic center inhibitory activity is sufficient to overcome the more intense inspiratory neuron discharge. The respiratory rate is increased because the after-discharge in the

vagal and possibly the pneumotaxic afferents to the medulla is rapidly overcome.

REGULATION OF RESPIRATORY ACTIVITY

A rise in the P_{CO_2} or H^+ concentration of arterial blood or a drop in its P_{O_2} increases the level of respiratory neuron activity in the medulla, and changes in the opposite direction have a slight inhibitory effect. The effects of variations in blood chemistry on ventilation are mediated via respiratory **chemoreceptors**—the carotid and aortic bodies and collections of cells in the medulla and elsewhere that are sensitive to changes in the chemistry of the blood. They initiate impulses that stimulate the respiratory center. Superimposed on this basic **chemical control of respiration,** other afferents provide nonchemical controls that affect breathing in particular situations (Table 36–1).

CHEMICAL CONTROL OF BREATHING

The chemical regulatory mechanisms adjust ventilation in such a way that the alveolar P_{CO_2} is normally held constant, the effects of excess H^+ in the blood are combated, and the P_{O_2} is raised when it falls to a potentially dangerous level. The respiratory minute volume is

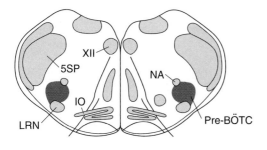

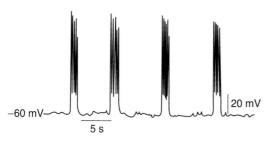

Figure 36–2. Rhythmic discharge (tracing below) of neurons in the pre-Bötzinger complex (pre-BÖTC) in a brain slice from a neonatal rat. IO, inferior olive; LRN, lateral reticular nucleus; NA, nucleus ambiguus; XII, nucleus of 12th cranial nerve; 5SP, spinal nucleus of trigeminal nerve. (Modified from Feldman JC, Gray PA: Sighs and gasps in a dish. Nat Neurosci 2000;3:531.)

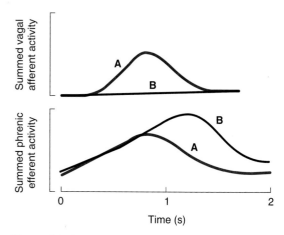

Figure 36–3. Superimposed records of two breaths: with **(A)** and without **(B)** feedback vagal afferent activity from stretch receptors in the lungs. Note that the rate of rise in phrenic nerve activity to the diaphragm is unaffected but the discharge is prolonged in the absence of vagal input.

proportionate to the metabolic rate, but the link between metabolism and ventilation is CO_2, not O_2. The receptors in the carotid and aortic bodies are stimulated by a rise in the P_{CO_2} or H^+ concentration of arterial blood or a decline in its P_{O_2}. After denervation of the carotid chemoreceptors, the response to a drop in P_{O_2} is abolished; the predominant effect of hypoxia after denervation of the carotid bodies is a direct depression of the respiratory center. The response to changes in arterial blood H^+ concentration in the pH 7.3–7.5 range is also abolished, although larger changes exert some effect. The response to changes in arterial P_{CO_2}, on the other hand, is affected only slightly; it is reduced no more than 30–35%.

Carotid & Aortic Bodies

There is a carotid body near the carotid bifurcation on each side, and there are usually two or more aortic bodies near the arch of the aorta (Figure 36–4). Each carotid and aortic body **(glomus)** contains islands of two types of cells, type I and type II cells, surrounded by fenestrated sinusoidal capillaries. The type I or **glomus cells** are closely associated with cup-like endings of the afferent nerves (Figure 36–5). The glomus cells resemble adrenal chromaffin cells and have dense-core granules containing catecholamines that are released upon exposure to hypoxia and cyanide (see below). The cells are excited by hypoxia, and the principal transmitter appears to be dopamine, which excites the nerve endings by way of D_2 receptors. The type II cells are glia-like, and each surrounds four to six type I cells. Their function is probably sustentacular.

Table 36–1. Stimuli affecting the respiratory center.

Chemical control
CO_2 (via CSF and brain interstitial fluid H^+ concentration)
O_2 ⎫
H^+ ⎭ (via carotid and aortic bodies)

Nonchemical control
Vagal afferents from receptors in the airways and lungs
Afferents from the pons, hypothalamus, and limbic system
Afferents from proprioceptors
Afferents from baroreceptors: arterial, atrial, ventricular, pulmonary

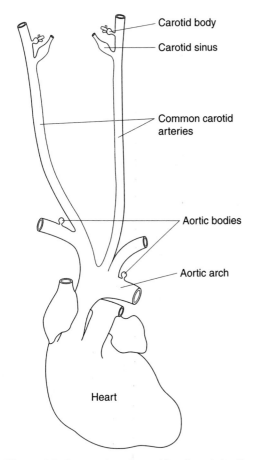

Figure 36–4. Location of carotid and aortic bodies.

tains similar O_2-sensitive K^+ channels, which mediate the vasoconstriction caused by hypoxia (see Chapter 37). This is in contrast to systemic arteries, which contain ATP-dependent K^+ channels that permit more K^+ efflux with hypoxia and consequently cause vasodilation instead of vasoconstriction.

The blood flow in each 2-mg carotid body is about 0.04 mL/min, or 2000 mL/100 g of tissue/min compared with a blood flow per 100 g/min of 54 mL in the brain and 420 mL in the kidneys (see Table 32–1). Because the blood flow per unit of tissue is so enormous, the O_2 needs of the cells can be met largely by dissolved O_2 alone. Therefore, the receptors are not stimulated in conditions such as anemia and carbon monoxide poisoning, in which the amount of dissolved O_2 in the blood reaching the receptors is generally normal even though the combined O_2 in the blood is markedly decreased. The receptors are stimulated when the arterial P_{O_2} is low or when, because of vascular stasis, the amount of O_2 delivered to the receptors per unit time is decreased. Powerful stimulation is also produced by

Outside the capsule of each body, the nerve fibers acquire a myelin sheath; however, they are only 2–5 µm in diameter and conduct at the relatively low rate of 7–12 m/s. Afferents from the carotid bodies ascend to the medulla via the carotid sinus and glossopharyngeal nerves, and fibers from the aortic bodies ascend in the vagi. Studies in which one carotid body has been isolated and perfused while recordings are being taken from its afferent nerve fibers show that there is a graded increase in impulse traffic in these afferent fibers as the P_{O_2} of the perfusing blood is lowered (Figure 36–6) or the P_{CO_2} raised.

Type I glomus cells have O_2-sensitive K^+ channels, whose conductance is reduced in proportion to the degree of hypoxia to which they are exposed. This reduces the K^+ efflux, depolarizing the cell and causing Ca^{2+} influx, primarily via L-type Ca^{2+} channels. The Ca^{2+} influx triggers action potentials and transmitter release, with consequent excitation of the afferent nerve endings. The smooth muscle of pulmonary arteries con-

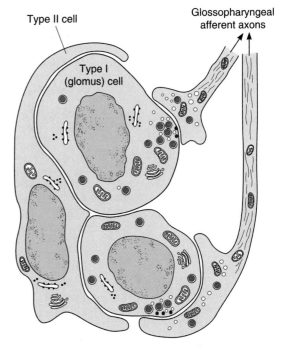

Figure 36–5. Organization of the carotid body. Type I (glomus) cells contain catecholamines. When exposed to hypoxia, they release their catecholamines, which stimulate the cup-like endings of the carotid sinus nerve fibers in the glossopharyngeal nerve. The glia-like type II cells surround the type I cells and probably have a sustentacular function.

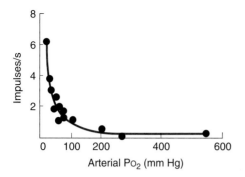

Figure 36–6. Change in the rate of discharge of a single afferent fiber from the carotid body when arterial P_{O_2} is reduced. (Courtesy of S Sampson.)

drugs such as cyanide, which prevent O_2 utilization at the tissue level. In sufficient doses, nicotine and lobeline activate the chemoreceptors. It has also been reported that infusion of K^+ increases the discharge rate in chemoreceptor afferents, and since the plasma K^+ level is increased during exercise, the increase may contribute to exercise-induced hyperpnea.

Because of their anatomic location, the aortic bodies have not been studied in as great detail as the carotid bodies. Their responses are probably similar but of lesser magnitude. In humans in whom both carotid bodies have been removed but the aortic bodies left intact, the responses are essentially the same as those following denervation of both carotid and aortic bodies in animals; there is little change in ventilation at rest, but the ventilatory response to hypoxia is lost and there is a 30% reduction in the ventilatory responses to CO_2.

Neuroepithelial bodies composed of innervated clusters of amine-containing cells are found in the airways of humans and animals. These cells have an outward K^+ current that is reduced by hypoxia, and this would be expected to produce depolarization. However, the function of these hypoxia-sensitive cells is uncertain because, as noted above, removal of the carotid bodies alone abolishes the respiratory response to hypoxia.

Chemoreceptors in the Brain Stem

The chemoreceptors that mediate the hyperventilation produced by increases in arterial P_{CO_2} after the carotid and aortic bodies are denervated are located in the medulla oblongata and consequently are called **medullary chemoreceptors.** They are separate from the dorsal and ventral respiratory neurons and are located on the ventral surface of the medulla (Figure 36–7). Recent evidence indicates that additional chemoreceptors are located in the vicinity of the solitary tract nuclei, the locus ceruleus, and the hypothalamus.

The chemoreceptors monitor the H^+ concentration of CSF, including the brain interstitial fluid. CO_2 readily penetrates membranes, including the blood-brain barrier, whereas H^+ and HCO_3^- penetrate slowly. The CO_2 that enters the brain and CSF is promptly hydrated. The H_2CO_3 dissociates, so that the local H^+ concentration rises. The H^+ concentration in brain interstitial fluid parallels the arterial P_{CO_2}. Experimentally produced changes in the P_{CO_2} of CSF have minor, variable effects on respiration as long as the H^+ concentration is held constant, but any increase in spinal fluid H^+ concentration stimulates respiration. The magnitude of the stimulation is proportionate to the rise in H^+ concentration. Thus, the effects of CO_2 on respiration are mainly due to its movement into the CSF and brain interstitial fluid, where it increases the H^+ concentration and stimulates receptors sensitive to H^+.

Ventilatory Responses to Changes in Acid-Base Balance

In metabolic acidosis due, for example, to the accumulation of the acid ketone bodies in the circulation in diabetes mellitus, there is pronounced respiratory stimulation (Kussmaul breathing; see Chapter 19). The hyperventilation decreases alveolar P_{CO_2} ("blows off CO_2") and thus produces a compensatory fall in blood H^+ concentration (see Chapter 39). Conversely, in metabolic alkalosis due, for example, to protracted vomiting with loss of HCl from the body, ventilation is depressed and the arterial P_{CO_2} rises, raising the H^+ concentration toward normal (see Chapter 39). If there

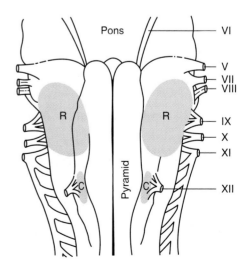

Figure 36–7. Rostral (R) and caudal (C) chemosensitive areas on the ventral surface of the medulla.

is an increase in ventilation that is not secondary to a rise in arterial H^+ concentration, the drop in P_{CO_2} lowers the H^+ concentration below normal (**respiratory alkalosis**); conversely, hypoventilation that is not secondary to a fall in plasma H^+ concentration causes **respiratory acidosis.**

Ventilatory Responses to CO_2

The arterial P_{CO_2} is normally maintained at 40 mm Hg. When there is a rise in arterial P_{CO_2} as a result of increased tissue metabolism, ventilation is stimulated and the rate of pulmonary excretion of CO_2 is increased until the arterial P_{CO_2} falls to normal, shutting off the stimulus. The operation of this feedback mechanism keeps CO_2 excretion and production in balance.

When a gas mixture containing CO_2 is inhaled, the alveolar P_{CO_2} rises, elevating the arterial P_{CO_2} and stimulating ventilation as soon as the blood that contains more CO_2 reaches the medulla. CO_2 elimination is increased, and the alveolar P_{CO_2} drops toward normal. This is why relatively large increments in the P_{CO_2} of inspired air (eg, 15 mm Hg) produce relatively slight increments in alveolar P_{CO_2} (eg, 3 mm Hg). However, the P_{CO_2} does not drop to normal, and a new equilibrium is reached at which the alveolar P_{CO_2} is slightly elevated and the hyperventilation persists as long as CO_2 is inhaled. The essentially linear relationship between respiratory minute volume and the alveolar P_{CO_2} is shown in Figure 36–8.

There is, of course, an upper limit to this linearity. When the P_{CO_2} of the inspired gas is close to the alveolar P_{CO_2}, elimination of CO_2 becomes difficult. When the CO_2 content of the inspired gas is more than 7%, the alveolar and arterial P_{CO_2} begin to rise abruptly in spite of hyperventilation. The resultant accumulation of CO_2 in the body (**hypercapnia**) depresses the central nervous system, including the respiratory center, and produces headache, confusion, and eventually coma (**CO_2 narcosis**).

Ventilatory Response to Oxygen Lack

When the O_2 content of the inspired air is decreased, there is an increase in respiratory minute volume. The stimulation is slight when the P_{O_2} of the inspired air is more than 60 mm Hg, and marked stimulation of respiration occurs only at lower P_{O_2} values (Figure 36–9). However, any decline in arterial P_{O_2} below 100 mm Hg produces increased discharge in the nerves from the carotid and aortic chemoreceptors. There are two reasons why in normal individuals this increase in impulse traffic does not increase ventilation to any extent until the P_{O_2} is less than 60 mm Hg. Because Hb is a weaker acid than HbO_2 (see Chapter 35), there is a slight de-

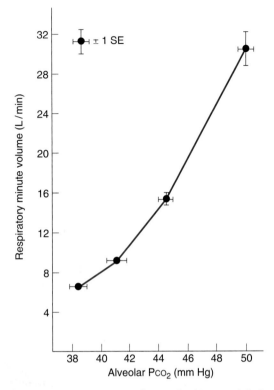

Figure 36–8. Responses of normal subjects to inhaling O_2 and approximately 2, 4, and 6% CO_2. The increase in respiratory minute volume is due to an increase in both the depth and rate of respiration. (Reproduced, with permission, from Lambertsen CJ in: *Medical Physiology,* 13th ed. Mountcastle VB [editor]. Mosby, 1974.)

crease in the H^+ concentration of arterial blood when the arterial P_{O_2} falls and hemoglobin becomes less saturated with O_2. The fall in H^+ concentration tends to inhibit respiration. In addition, any increase in ventilation that does occur lowers the alveolar P_{CO_2}, and this also tends to inhibit respiration. Therefore, the stimulatory effects of hypoxia on ventilation are not clearly manifest until they become strong enough to override the counterbalancing inhibitory effects of a decline in arterial H^+ concentration and P_{CO_2}.

The effects on ventilation of decreasing the alveolar P_{O_2} while holding the alveolar P_{CO_2} constant are shown in Figure 36–10. When the alveolar P_{CO_2} is stabilized at a level 2–3 mm Hg above normal, there is an inverse relationship between ventilation and the alveolar P_{O_2} even in the 90–110 mm Hg range; but when the alveolar P_{CO_2} is fixed at lower than normal values, there is no stimulation of ventilation by hypoxia until the alveolar P_{O_2} falls below 60 mm Hg.

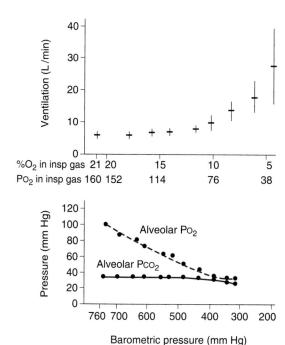

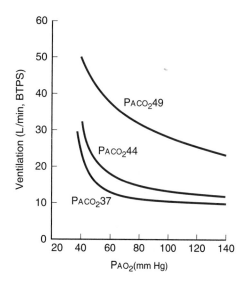

Figure 36–10. Ventilation at various alveolar P_{O_2} values when P_{CO_2} is held constant at 49, 44, or 37 mm Hg. (Data from Loeschke HH and Gertz KH.)

Figure 36–9. Top: Average respiratory minute volume during the first half hour of exposure to gases containing various amounts of O_2. The horizontal line in each case indicates the mean; the vertical bar indicates one standard deviation. **Bottom:** Alveolar P_{O_2} and P_{CO_2} values when breathing air at various barometric pressures. The two graphs are aligned so that the P_{O_2} of the inspired gas mixtures in the upper graph correspond to the P_{O_2} at the various barometric pressures in the lower graph. (Courtesy of RH Kellogg.)

Effect of H⁺ on the CO_2 Response

The stimulatory effects of H⁺ and CO_2 on respiration appear to be additive and not, like those of CO_2 and O_2, complexly interrelated. In metabolic acidosis, the CO_2 response curves are similar to those in Figure 36–11, except that they are shifted to the left. In other

Effects of Hypoxia on the CO_2 Response Curve

When the converse experiment is performed, ie, when the alveolar P_{O_2} is held constant while the response to varying amounts of inspired CO_2 is tested, a linear response is obtained (Figure 36–11). When the CO_2 response is tested at different fixed P_{O_2} values, the slope of the response curve changes, with the slope increased when alveolar P_{O_2} is decreased. In other words, hypoxia makes the individual more sensitive to increases in arterial P_{CO_2}. However, the alveolar P_{CO_2} level at which the curves in Figure 36–11 intersect is unaffected. In the normal individual, this threshold value is just below the normal alveolar P_{CO_2}, indicating that normally there is a very slight but definite "CO_2 drive" of the respiratory area.

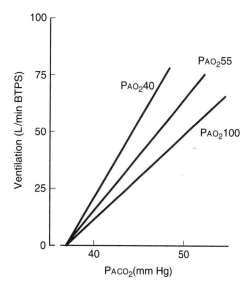

Figure 36–11. Fan of lines showing CO_2 response curves at various fixed values of alveolar P_{O_2}.

words, the same amount of respiratory stimulation is produced by lower arterial P_{CO_2} levels. It has been calculated that the CO_2 response curve shifts 0.8 mm Hg to the left for each nanomole rise in arterial H^+. About 40% of the ventilatory response to CO_2 is removed if the increase in arterial H^+ produced by CO_2 is prevented. As noted above, the remaining 60% is probably due to the effect of CO_2 on spinal fluid or brain interstitial fluid H^+ concentration.

Breath Holding

Respiration can be voluntarily inhibited for some time, but eventually the voluntary control is overridden. The point at which breathing can no longer be voluntarily inhibited is called the **breaking point.** Breaking is due to the rise in arterial P_{CO_2} and the fall in P_{O_2}. Individuals can hold their breath longer after removal of the carotid bodies. Breathing 100% oxygen before breath holding raises alveolar P_{O_2} initially, so that the breaking point is delayed. The same is true of hyperventilating room air, because CO_2 is blown off and arterial P_{CO_2} is lower at the start. Reflex or mechanical factors appear to influence the breaking point, since subjects who hold their breath as long as possible and then breathe a gas mixture low in O_2 and high in CO_2 can hold their breath for an additional 20 seconds or more. Psychological factors also play a role, and subjects can hold their breath longer when they are told their performance is very good than when they are not.

Hormonal Effects on Respiration

Ventilation is increased during the luteal phase of the menstrual cycle and during pregnancy (see Chapter 23). Experiments with animals indicate that this is due to activation of estrogen-dependent progesterone receptors in the hypothalamus. However, the physiologic significance of this increased ventilation is unknown.

NONCHEMICAL INFLUENCES ON RESPIRATION

Responses Mediated by Receptors in the Airways & Lungs

Receptors in the airways and lungs are innervated by myelinated and unmyelinated vagal fibers. The unmyelinated fibers are C fibers. The receptors innervated by myelinated fibers are commonly divided into **slowly adapting receptors** and **rapidly adapting receptors** on the basis of whether sustained stimulation leads to prolonged or transient discharge in their afferent nerve fibers (Table 36–2). The other group of receptors presumably consists of the endings of C fibers, and they are divided into pulmonary and bronchial subgroups on the basis of their location.

Table 36–2. Airway and lung receptors.[1]

Vagal Innervation	Type	Location in Interstitium	Stimulus	Response
Myelinated	Slowly adapting	Among airway smooth muscle cells(?)	Lung inflation	Inspiratory time shortening Hering-Breuer inflation and deflation reflexes Bronchodilation Tachycardia
	Rapidly adapting	Among airway epithelial cells	Lung hyperinflation Exogenous and endogenous substances (eg, histamine, prostaglandins)	Hyperpnea Cough Bronchoconstriction Mucus secretion
Unmyelinated C fibers	Pulmonary C fibers Bronchial C fibers	Close to blood vessels	Lung hyperinflation Exogenous and endogenous substances (eg, capsaicin, bradykinin, serotonin)	Apnea followed by rapid breathing Bronchoconstriction Bradycardia Hypotension Mucus secretion

1 Modified and reproduced, with permission, from Berger AJ, Hornbein TF: Control of respiration. In: *Textbook of Physiology*, 21st ed. Vol. 2. Patton HD et al (editors). Saunders, 1989.

The shortening of inspiration produced by vagal afferent activity (Figure 36–3) is mediated by slowly adapting receptors. So are the **Hering-Breuer reflexes.** The Hering-Breuer inflation reflex is an increase in the duration of expiration produced by steady lung inflation, and the Hering-Breuer deflation reflex is a decrease in the duration of expiration produced by marked deflation of the lung. Because the rapidly adapting receptors are stimulated by chemicals such as histamine, they have been called **irritant receptors.** Activation of rapidly adapting receptors in the trachea causes coughing, bronchoconstriction, and mucus secretion, and activation of rapidly adapting receptors in the lung may produce hyperpnea.

Because the C fiber endings are close to pulmonary vessels, they have been called J (juxtacapillary) receptors. They are stimulated by hyperinflation of the lung, but they respond as well to intravenous or intracardiac administration of chemicals such as capsaicin. The reflex response that is produced is apnea followed by rapid breathing, bradycardia, and hypotension (**pulmonary chemoreflex**). A similar response is produced by receptors in the heart (**Bezold-Jarisch reflex** or the **coronary chemoreflex;** see Chapter 31). The physiologic role of this reflex is uncertain, but it probably occurs in pathologic states such as pulmonary congestion or embolization, in which it is produced by endogenously released substances.

Coughing & Sneezing

Coughing begins with a deep inspiration followed by forced expiration against a closed glottis. This increases the intrapleural pressure to 100 mm Hg or more. The glottis is then suddenly opened, producing an explosive outflow of air at velocities up to 965 km (600 miles) per hour. Sneezing is a similar expiratory effort with a continuously open glottis (see Chapter 14). These reflexes help expel irritants and keep airways clear.

Responses in Patients With Heart-Lung Transplants

Transplantation of the heart and lungs is now an established treatment for severe pulmonary disease and some other conditions. In individuals with transplants, the recipient's right atrium is sutured to the donor heart, and the donor heart does not reinnervate, so the resting heart rate is elevated. The donor trachea is sutured to the recipient's just above the carina, and afferent fibers from the lungs do not regrow. Consequently, healthy patients with heart-lung transplants provide an opportunity to evaluate the role of lung innervation in normal physiology. Their cough responses to stimulation of the trachea are normal, because the trachea remains

innervated, but their cough responses to stimulation of the smaller airways are absent. Their bronchi tend to be dilated to a greater degree than normal. In addition, they have the normal number of yawns and sighs, indicating that these do not depend on innervation of the lungs. Finally, they lack Hering-Breuer reflexes, but their pattern of breathing at rest is normal, indicating that these reflexes do not play an important role in the regulation of resting respiration in humans.

Afferents From "Higher Centers"

Pain and emotional stimuli affect respiration, so there must also be afferents from the limbic system and hypothalamus to the respiratory neurons in the brain stem. In addition, even though breathing is not usually a conscious event, both inspiration and expiration are under voluntary control. The pathways for voluntary control pass from the neocortex to the motor neurons innervating the respiratory muscles, bypassing the medullary neurons.

Since voluntary and automatic control of respiration are separate, automatic control is sometimes disrupted without loss of voluntary control. The clinical condition that results has been called **Ondine's curse.** In German legend, Ondine was a water nymph who had an unfaithful mortal lover. The king of the water nymphs punished the lover by casting a curse upon him that took away all his automatic functions. In this state, he could stay alive only by staying awake and remembering to breathe. He eventually fell asleep from sheer exhaustion, and his respiration stopped. Patients with this intriguing condition generally have bulbar poliomyelitis or disease processes that compress the medulla.

Afferents From Proprioceptors

Carefully controlled experiments have shown that active and passive movements of joints stimulate respiration, presumably because impulses in afferent pathways from proprioceptors in muscles, tendons, and joints stimulate the inspiratory neurons. This effect probably helps increase ventilation during exercise.

Respiratory Components of Visceral Reflexes

The respiratory adjustments during vomiting, swallowing, and sneezing are discussed in Chapters 14 and 26. Inhibition of respiration and closure of the glottis during these activities not only prevent the aspiration of food or vomitus into the trachea but, in the case of vomiting, fix the chest so that contraction of the abdominal muscles increases the intra-abdominal pressure. Similar glottic closure and inhibition of respiration occur during voluntary and involuntary straining.

Hiccup is a spasmodic contraction of the diaphragm and other inspiratory muscles that produces an inspiration during which the glottis suddenly closes. The glottic closure is responsible for the characteristic sensation and sound. Hiccups occur in the fetus in utero as well as throughout extrauterine life. Their function is unknown. Most attacks of hiccups are of short duration, and they often respond to breath-holding or other measures that increase arterial P_{CO_2}. Intractable hiccups, which can be debilitating, sometimes respond to dopamine antagonists and perhaps to some centrally acting analgesic compounds.

Yawning is a peculiar "infectious" respiratory act whose physiologic basis and significance are uncertain. Like hiccuping, it occurs in utero, and it occurs in fish and tortoises as well as mammals. Underventilated alveoli have a tendency to collapse, and it has been suggested that the deep inspiration and stretching open them and prevent the development of atelectasis. Yawning also increases venous return to the heart. Sighing may have a similar function. However, in actual experiments, no atelectatic effect of yawning could be demonstrated. It has also been suggested that yawning is a nonverbal signal used for communication between animals in a group, and one could argue that on a different level, the same thing is true in humans.

Respiratory Effects of Baroreceptor Stimulation

Afferent fibers from the baroreceptors in the carotid sinuses, aortic arch, atria, and ventricles relay to the respiratory neurons as well as the vasomotor and cardioinhibitory neurons in the medulla. Impulses in them inhibit respiration, but the inhibitory effect is slight and of little physiologic importance. The hyperventilation in shock is due to chemoreceptor stimulation caused by acidosis and hypoxia secondary to local stagnation of blood flow and is not baroreceptor-mediated. The activity of inspiratory neurons affects blood pressure and heart rate (see Chapters 28 and 31), and activity in the vasomotor and cardiac areas in the medulla may have minor effects on respiration.

Effects of Sleep

Respiration is less rigorously controlled during sleep than in the waking state, and brief periods of apnea occur in normal sleeping adults. There are variable changes in the ventilatory response to hypoxia. If the P_{CO_2} falls during the waking state, various stimuli from proprioceptors and the environment maintain respiration, but during sleep, these stimuli are decreased and a decrease in P_{CO_2} can cause apnea. During REM sleep, breathing is irregular and the CO_2 response is highly variable.

Respiratory Adjustments in Health & Disease

37

INTRODUCTION

This chapter is concerned with the respiratory adjustments to exercise, hypoxia, including that produced by high altitude, hypercapnia, and respiratory disease. In the same way that the compensatory adjustments of the cardiovascular system to environmental changes and disease illustrate the integrated operation of the cardiovascular regulatory mechanisms (see Chapter 33), these respiratory adjustments highlight the operation of the respiratory regulatory mechanisms discussed in Chapters 34–36.

EFFECTS OF EXERCISE

Many cardiovascular and respiratory mechanisms must operate in an integrated fashion if the O_2 needs of the active tissue are to be met and the extra CO_2 and heat removed from the body during exercise. Circulatory changes increase muscle blood flow while maintaining adequate circulation in the rest of the body (see Chapter 33). There is in addition an increase in the extraction of O_2 from the blood in exercising muscles and an increase in ventilation that provides extra O_2, eliminates some of the heat, and excretes extra CO_2.

Heat dissipation during exercise is discussed in Chapter 33 and summarized in Figure 33–4. The changes in acid-base balance associated with respiration are reviewed in Chapter 39.

Changes in Ventilation

During exercise, the amount of O_2 entering the blood in the lungs is increased because the amount of O_2 added to each unit of blood and the pulmonary blood flow per minute are increased. The PO_2 of blood flowing into the pulmonary capillaries falls from 40 to 25 mm Hg or less, so that the alveolar-capillary PO_2 gradient is increased and more O_2 enters the blood. Blood flow per minute is increased from 5.5 L/min to as much as 20–35 L/min. The total amount of O_2 entering the blood therefore increases from 250 mL/min at rest to values as high as 4000 mL/min. The amount of

CO_2 removed from each unit of blood is increased, and CO_2 excretion increases from 200 mL/min to as much as 8000 mL/min. The increase in O_2 uptake is proportionate to work load up to a maximum. Above this maximum, O_2 consumption levels off and the blood lactate level continues to rise (Figure 37–1). The lactate comes from muscles in which aerobic resynthesis of energy stores cannot keep pace with their utilization and an **oxygen debt** is being incurred (see Chapter 3).

There is an abrupt increase in ventilation with the onset of exercise, followed after a brief pause by a further, more gradual increase (Figure 37–2). With moderate exercise, the increase is due mostly to an increase in the depth of respiration; this is accompanied by an increase in the respiratory rate when the exercise is more strenuous. There is an abrupt decrease in ventilation when exercise ceases, followed after a brief pause

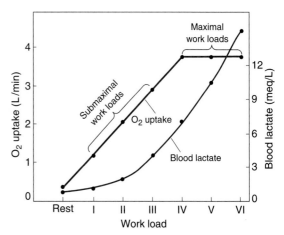

Figure 37–1. Relation between workload, blood lactate level, and O_2 uptake. I–VI, increasing work loads produced by increasing the speed and grade of a treadmill on which the subjects worked. (Reproduced, with permission, from Mitchell JH, Blomqvist G: Maximal oxygen uptake. N Engl J Med 1971;284:1018.)

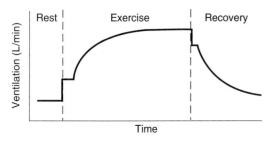

Figure 37–2. Diagrammatic representation of changes in ventilation during exercise.

by a more gradual decline to preexercise values. The abrupt increase at the start of exercise is presumably due to psychic stimuli and afferent impulses from proprioceptors in muscles, tendons, and joints. The more gradual increase is presumably humoral even though arterial pH, PCO_2, and PO_2 remain constant during moderate exercise. The increase in ventilation is proportionate to the increase in O_2 consumption, but the mechanisms responsible for the stimulation of respiration are still the subject of much debate. The increase in body temperature may play a role. As noted in Chapter 36, exercise increases the plasma K^+ level, and this increase may stimulate the peripheral chemoreceptors. In addition, it may be that the sensitivity of the neurons controlling the response to CO_2 is increased or that the respiratory fluctuations in arterial PCO_2 increase so that, even though the mean arterial PCO_2 does not rise, it is CO_2 that is responsible for the increase in ventilation. O_2 also seems to play some role despite the lack of a decrease in arterial PO_2, since during the performance of a given amount of work, the increase in ventilation while breathing 100% O_2 is 10–20% less than the increase while breathing air. Thus, it currently appears that a number of different factors combine to produce the increase in ventilation seen during moderate exercise.

When exercise becomes more vigorous, buffering of the increased amounts of lactic acid that are produced liberates more CO_2, and this further increases ventilation. The response to graded exercise is shown in Figure 37–3. With increased production of acid, the increases in ventilation and CO_2 production remain proportionate, so alveolar and arterial CO_2 change relatively little **(isocapnic buffering)**. Because of the hyperventilation, alveolar PO_2 increases. With further accumulation of lactic acid, the increase in ventilation outstrips CO_2 production and alveolar PCO_2 falls, as does arterial PCO_2. The decline in arterial PCO_2 provides respiratory compensation (see Chapter 39) for the metabolic acidosis produced by the additional lactic acid. The additional increase in ventilation produced by the acidosis

is dependent on the carotid bodies and does not occur if they are removed.

The respiratory rate after exercise does not reach basal levels until the O_2 debt is repaid. This may take as long as 90 minutes. The stimulus to ventilation after exercise is not the arterial PCO_2, which is normal or low, or the arterial PO_2, which is normal or high, but the elevated arterial H^+ concentration due to the lactic acidemia. The magnitude of the O_2 debt is the amount by which O_2 consumption exceeds basal consumption from the end of exertion until the O_2 consumption has returned to preexercise basal levels. During repayment of the O_2 debt, there is a small rise in the O_2 concentration in muscle myoglobin. ATP and phosphorylcreatine are resynthesized, and lactic acid is removed. Eighty percent of the lactic acid is converted to glycogen and 20% is metabolized to CO_2 and H_2O.

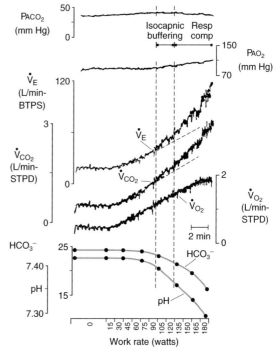

Figure 37–3. Changes in alveolar PCO_2, alveolar PO_2, ventilation ($\dot{V}_E$), CO_2 production ($\dot{V}_{CO_2}$), O_2 consumption ($\dot{V}_{O_2}$), arterial HCO_3^-, and arterial pH with graded increases in work on a bicycle ergometer. Resp comp, respiratory compensation. The subject was a normal adult male. (Reproduced, with permission, from Wasserman K, Whipp BJ, Casaburi R: Respiratory control during exercise. In: *Handbook of Physiology.* Section 3, *The Respiratory System.* Vol II, part 2. Fishman AP [editor]. American Physiological Society, 1986.)

Because of the extra CO_2 produced by the buffering of lactic acid during strenuous exercise, the R (see Chapter 17) rises, reaching 1.5–2.0. After exertion, while the O_2 debt is being repaid, the R falls to 0.5 or less.

Changes in the Tissues

Maximum O_2 uptake during exercise is limited by the maximum rate at which O_2 is transported to the mitochondria in the exercising muscle. However, this limitation is not normally due to deficient O_2 uptake in the lungs, and hemoglobin in arterial blood is saturated even during the most severe exercise.

During exercise, the contracting muscles use more O_2, and the tissue PO_2 and the PO_2 in venous blood from exercising muscle fall nearly to zero. More O_2 diffuses from the blood, the blood PO_2 of the blood in the muscles drops, and more O_2 is removed from hemoglobin. Because the capillary bed of contracting muscle is dilated and many previously closed capillaries are open, the mean distance from the blood to the tissue cells is greatly decreased; this facilitates the movement of O_2 from blood to cells. The oxygen-hemoglobin dissociation curve is steep in the PO_2 range below 60 mm Hg, and a relatively large amount of O_2 is supplied for each drop of 1 mm Hg in PO_2 (see Figure 35–2). Additional O_2 is supplied because, as a result of the accumulation of CO_2 and the rise in temperature in active tissues—and perhaps because of a rise in red blood cell 2,3-DPG—the dissociation curve shifts to the right (see Figure 35–3). The net effect is a threefold increase in O_2 extraction from each unit of blood. Since this increase is accompanied by a 30-fold or greater increase in blood flow, it permits the metabolic rate of muscle to rise as much as 100-fold during exercise (see Chapter 3).

Exercise Tolerance & Fatigue

What determines the maximum amount of exercise that can be performed by an individual? Obviously, exercise tolerance has a time as well as an intensity dimension. For example, a fit young man can produce a power output on a bicycle of about 700 watts for 1 minute, 300 watts for 5 minutes, and 200 watts for 40 minutes. It used to be argued that the limiting factors in exercise performance were the rate at which O_2 could be delivered to the tissues or the rate at which O_2 could enter the body in the lungs. These factors play a role, but it is clear that other factors also contribute and that exercise stops when the sensation of **fatigue** progresses to the sensation of exhaustion. Fatigue is produced in part by bombardment of the brain by neural impulses from muscles, and the decline in blood pH

produced by lactic acidosis also makes one feel tired. So do the rise in body temperature (see Chapter 33), dyspnea, and, perhaps, the uncomfortable sensations produced by activation of the J receptors in the lungs.

HYPOXIA

Hypoxia is O_2 deficiency at the tissue level. It is a more correct term than **anoxia,** there rarely being no O_2 at all left in the tissues.

Traditionally, hypoxia has been divided into four types. Numerous other classifications have been used, but the four-type system still has considerable utility if the definitions of the terms are kept clearly in mind. The four categories are (1) **hypoxic hypoxia (anoxic anoxia),** in which the PO_2 of the arterial blood is reduced; (2) **anemic hypoxia,** in which the arterial PO_2 is normal but the amount of hemoglobin available to carry O_2 is reduced; (3) **stagnant** or **ischemic hypoxia,** in which the blood flow to a tissue is so low that adequate O_2 is not delivered to it despite a normal PO_2 and hemoglobin concentration; and (4) **histotoxic hypoxia,** in which the amount of O_2 delivered to a tissue is adequate but, because of the action of a toxic agent, the tissue cells cannot make use of the O_2 supplied to them.

Effects on Cells

Hypoxia causes the production of transcription factors **(hypoxia-inducible factors; HIFs).** These are made up of α and β subunits. In normally oxygenated tissues, the α subunits are rapidly ubiquitinated (see Chapter 1) and destroyed. However, in hypoxic cells, the α factors dimerize with the β, and the dimers activate genes that produce angiogenic factors and erythropoietin. Many cancer cells are hypoxic, and there is considerable interest in the possibility of manipulating HIFs to kill cancer cells.

Effects on the Brain

The effects of stagnant hypoxia depend upon the tissue affected. In hypoxic hypoxia and the other generalized forms of hypoxia, the brain is affected first. A sudden drop in the inspired PO_2 to less than 20 mm Hg, which occurs, for example, when cabin pressure is suddenly lost in a plane flying above 16,000 m, causes loss of consciousness in 10–20 seconds (Figure 37–4) and death in 4–5 minutes. Less severe hypoxia causes a variety of mental aberrations not unlike those produced by alcohol: impaired judgment, drowsiness, dulled pain sensibility, excitement, disorientation, loss of time sense, and headache. Other symptoms include anorexia, nausea, vomiting, tachycardia, and, when the

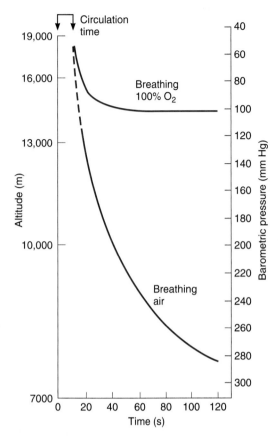

Figure 37–4. Duration of useful consciousness upon sudden exposure to the ambient pressure at various altitudes. Ten seconds is the approximate lung-to-brain circulation time.

hypoxia is severe, hypertension. The rate of ventilation is increased in proportion to the severity of the hypoxia of the carotid chemoreceptor cells.

Respiratory Stimulation

Dyspnea is by definition difficult or labored breathing in which the subject is conscious of shortness of breath; **hyperpnea** is the general term for an increase in the rate or depth of breathing regardless of the patient's subjective sensations. **Tachypnea** is rapid, shallow breathing. In general, a normal individual is not conscious of respiration until ventilation is doubled, and breathing is not uncomfortable until ventilation is tripled or quadrupled. Whether or not a given level of ventilation is uncomfortable also appears to depend on a variety of other factors. Hypercapnia and, to a lesser extent, hypoxia cause dyspnea. An additional factor is

the effort involved in moving the air in and out of the lungs (the work of breathing).

Cyanosis

Reduced hemoglobin has a dark color, and a dusky bluish discoloration of the tissues, called **cyanosis,** appears when the reduced hemoglobin concentration of the blood in the capillaries is more than 5 g/dL. Its occurrence depends upon the total amount of hemoglobin in the blood, the degree of hemoglobin unsaturation, and the state of the capillary circulation. Cyanosis is most easily seen in the nail beds and mucous membranes and in the earlobes, lips, and fingers, where the skin is thin. Cyanosis does not occur in anemic hypoxia, because the total hemoglobin content is low; in carbon monoxide poisoning, because the color of reduced hemoglobin is obscured by the cherry-red color of carbonmonoxyhemoglobin (see below); or in histotoxic hypoxia, because the blood gas content is normal. A discoloration of the skin and mucous membranes similar to cyanosis is produced by high circulating levels of methemoglobin (see Chapter 27).

HYPOXIC HYPOXIA

Hypoxic hypoxia is a problem in normal individuals at high altitudes and is a complication of pneumonia and a variety of other diseases of the respiratory system.

Effects of Decreased Barometric Pressure

The composition of air stays the same, but the total barometric pressure falls with increasing altitude (Figure 37–5). Therefore, the P_{O_2} also falls. At 3000 m (approximately 10,000 ft) above sea level, the alveolar P_{O_2} is about 60 mm Hg and there is enough hypoxic stimulation of the chemoreceptors to definitely increase ventilation. As one ascends higher, the alveolar P_{O_2} falls less rapidly and the alveolar P_{CO_2} declines somewhat because of the hyperventilation. The resulting fall in arterial P_{CO_2} produces respiratory alkalosis.

Hypoxic Symptoms Breathing Air

There are a number of compensatory mechanisms that operate over a period of time to increase altitude tolerance **(acclimatization),** but in unacclimatized subjects, mental symptoms such as irritability appear at about 3700 m. At 5500 m, the hypoxic symptoms are severe; and at altitudes above 6100 m (20,000 ft), consciousness is usually lost.

Hypoxic Symptoms Breathing Oxygen

The total atmospheric pressure becomes the limiting factor in altitude tolerance when breathing 100% O_2.

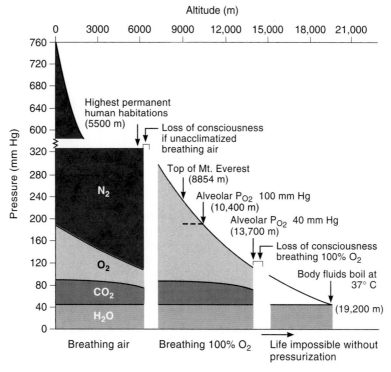

Figure 37–5. Composition of alveolar air in individuals breathing air (0–6100 m) and 100% O_2 (6100–13,700 m). The minimal alveolar P_{O_2} that an unacclimatized subject can tolerate without loss of consciousness is about 35–40 mm Hg. Note that with increasing altitude, the alveolar P_{CO_2} drops because of the hyperventilation due to hypoxic stimulation of the carotid and aortic chemoreceptors. The fall in barometric pressure with increasing altitude is not linear, because air is compressible.

The partial pressure of water vapor in the alveolar air is constant at 47 mm Hg, and that of CO_2 is normally 40 mm Hg, so that the lowest barometric pressure at which a normal alveolar P_{O_2} of 100 mm Hg is possible is 187 mm Hg, the pressure at about 10,400 m (34,000 ft). At greater altitudes, the increased ventilation due to the decline in alveolar P_{O_2} lowers the alveolar P_{CO_2} somewhat, but the maximum alveolar P_{O_2} that can be attained when breathing 100% O_2 at the ambient barometric pressure of 100 mm Hg at 13,700 m is about 40 mm Hg. At about 14,000 m, consciousness is lost in spite of the administration of 100% O_2. At 19,200 m, the barometric pressure is 47 mm Hg, and at or below this pressure the body fluids boil at body temperature. The point is largely academic, however, because any individual exposed to such a low pressure would be dead of hypoxia before the bubbles of steam could cause death.

Of course, an artificial atmosphere can be created around an individual; in a pressurized suit or cabin supplied with O_2 and a system to remove CO_2, it is possi-

ble to ascend to any altitude and to live in the vacuum of interplanetary space.

Delayed Effects of High Altitude

When they first arrive at a high altitude, many individuals develop transient "mountain sickness." This syndrome develops 8–24 hours after arrival at altitude and lasts 4–8 days. It is characterized by headache, irritability, insomnia, breathlessness, and nausea and vomiting. Its cause is unsettled, but it appears to be associated with cerebral edema. The low P_{O_2} at high altitude causes arteriolar dilation, and if cerebral autoregulation does not compensate, there is an increase in capillary pressure that favors increased transudation of fluid into brain tissue. Individuals who do not develop mountain sickness have a diuresis at high altitude, and urine volume is decreased in individuals who develop the condition.

High-altitude illness includes not only mountain sickness but also two more serious syndromes that complicate it: **high-altitude cerebral edema** and **high-**

altitude pulmonary edema. In high-altitude cerebral edema, the capillary leakage in mountain sickness progresses to frank brain swelling, with ataxia, disorientation, and in some cases coma and death due to herniation of the brain through the tentorium. High-altitude pulmonary edema is a patchy edema of the lungs that is related to the marked pulmonary hypertension that develops at high altitude. It has been argued that it occurs because not all pulmonary arteries have enough smooth muscle to constrict in response to hypoxia, and in the capillaries supplied by those arteries, the general rise in pulmonary arterial pressure causes a capillary pressure increase that disrupts their walls (stress failure).

All forms of high-altitude illness are benefited by descent to lower altitude and by treatment with the diuretic acetazolamide (see Chapter 38). This drug inhibits carbonic anhydrase, producing increased HCO_3^- excretion in the urine, stimulating respiration, increasing $PaCO_2$, and reducing the formation of CSF. When cerebral edema is marked, large doses of glucocorticoids are often administered as well. Their mechanism of action is unsettled. In high-altitude pulmonary edema, prompt treatment with O_2 is essential—and, if available, use of a hyperbaric chamber (see below). Portable hyperbaric chambers are now available in a number of mountain areas. Nifedipine, a Ca^{2+} channel blocker that lowers pulmonary artery pressure, is also useful.

Acclimatization

Acclimatization to altitude is due to the operation of a variety of compensatory mechanisms. The respiratory alkalosis produced by the hyperventilation shifts the oxygen-hemoglobin dissociation curve to the left, but there is a concomitant increase in red blood cell 2,3-DPG, which tends to decrease the O_2 affinity of hemoglobin. The net effect is a small increase in P_{50} (see Chapter 35). The decrease in O_2 affinity makes more O_2 available to the tissues. However, the value of the increase in P_{50} is limited because when the arterial PO_2 is markedly reduced, the decreased O_2 affinity also interferes with O_2 uptake by hemoglobin in the lungs.

The initial ventilatory response to increased altitude is relatively small, because the alkalosis tends to counteract the stimulating effect of hypoxia. However, there is a steady increase in ventilation over the next 4 days (Figure 37–6) because the active transport of H^+ into CSF, or possibly a developing lactic acidosis in the brain, causes a fall in CSF pH that increases the response to hypoxia. After 4 days, the ventilatory response begins to decline slowly, but it takes years of residence at higher altitudes for it to decline to the initial level. Associated with this decline is a gradual desensitization to the stimulatory effects of hypoxia.

Erythropoietin secretion increases promptly on ascent to high altitude (see Chapter 24) and then falls

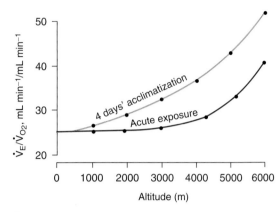

Figure 37–6. Effect of acclimatization on the ventilatory response at various altitudes. $\dot{V}_E / \dot{V}_{O_2}$ is the ventilatory equivalent, the ratio of expired minute volume ($\dot{V}_E$) to the O_2 consumption ($\dot{V}_{O_2}$). (Reproduced, with permission, from Lenfant C, Sullivan K: Adaptation to high altitude. N Engl J Med 1971;284:1298.)

somewhat over the following 4 days as the ventilatory response increases and the arterial PO_2 rises. The increase in circulating red blood cells triggered by the erythropoietin begins in 2–3 days and is sustained as long as the individual remains at high altitude.

There are also compensatory changes in the tissues. The mitochondria, which are the site of oxidative reactions, increase in number, and there is an increase in myoglobin (see Chapter 35) that facilitates the movement of O_2 in the tissues. There is also an increase in the tissue content of cytochrome oxidase.

The effectiveness of the acclimatization process is indicated by the fact that in the Andes and Himalayas there are permanent human habitations at elevations above 5500 m (18,000 ft). The natives who live in these villages are barrel-chested and markedly polycythemic. They have low alveolar PO_2 values, but in most other ways they are remarkably normal.

Diseases Causing Hypoxic Hypoxia

Hypoxic hypoxia is the most common form of hypoxia seen clinically. The diseases that cause it can be roughly divided into those in which the gas exchange apparatus fails, those such as congenital heart disease in which large amounts of blood are shunted from the venous to the arterial side of the circulation, and those in which the respiratory pump fails (Table 37–1). Lung failure occurs when conditions such as pulmonary fibrosis produce alveolar-capillary block or there is ventilation-perfusion imbalance. Pump failure can be due to fatigue of the respiratory muscles in conditions in which

Table 37–1. Disorders causing hypoxic hypoxia.

Lung failure (gas exchange failure)
  Pulmonary fibrosis
  Ventilation-perfusion imbalance

Shunt

Pump failure (ventilatory failure)
  Fatigue
  Mechanical defects
  Depression of respiratory controller in the brain

---

the work of breathing is increased or to a variety of mechanical defects such as pneumothorax or bronchial obstruction that limit ventilation. It can also be caused by abnormalities of the neural mechanisms that control ventilation, such as depression of the respiratory neurons in the medulla by morphine and other drugs.

Ventilation-Perfusion Imbalance

Patchy ventilation-perfusion imbalance is by far the most common cause of hypoxic hypoxia in clinical situations. The physiologic effects of ventilation-perfusion imbalance and their role in producing the alterations in alveolar gas due to gravity are discussed in Chapter 34.

In disease processes that prevent ventilation of some of the alveoli, the ventilation-blood flow ratios in different parts of the lung determine the extent to which systemic arterial P_{O_2} declines. If nonventilated alveoli are perfused, the nonventilated but perfused portion of the lung is in effect a right-to-left shunt, dumping unoxygenated blood into the left side of the heart. Lesser degrees of ventilation-perfusion imbalance are more common. In the example illustrated in Figure 37–7, the underventilated alveoli (B) have a low alveolar P_{O_2}, whereas the overventilated alveoli (A) have a high alveolar P_{O_2}. However, the unsaturation of the hemoglobin of the blood coming from B is not completely

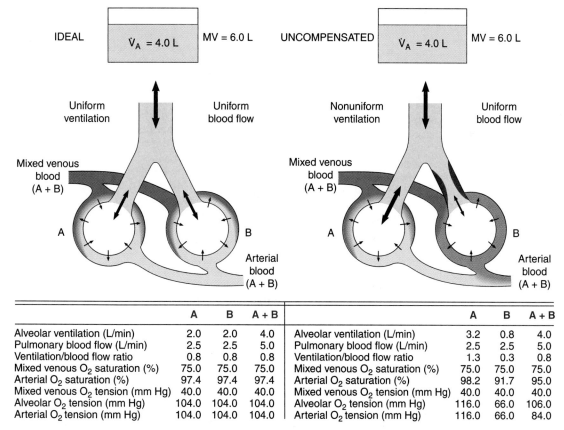

	A	B	A + B		A	B	A + B
Alveolar ventilation (L/min)	2.0	2.0	4.0	Alveolar ventilation (L/min)	3.2	0.8	4.0
Pulmonary blood flow (L/min)	2.5	2.5	5.0	Pulmonary blood flow (L/min)	2.5	2.5	5.0
Ventilation/blood flow ratio	0.8	0.8	0.8	Ventilation/blood flow ratio	1.3	0.3	0.8
Mixed venous O_2 saturation (%)	75.0	75.0	75.0	Mixed venous O_2 saturation (%)	75.0	75.0	75.0
Arterial O_2 saturation (%)	97.4	97.4	97.4	Arterial O_2 saturation (%)	98.2	91.7	95.0
Mixed venous O_2 tension (mm Hg)	40.0	40.0	40.0	Mixed venous O_2 tension (mm Hg)	40.0	40.0	40.0
Alveolar O_2 tension (mm Hg)	104.0	104.0	104.0	Alveolar O_2 tension (mm Hg)	116.0	66.0	106.0
Arterial O_2 tension (mm Hg)	104.0	104.0	104.0	Arterial O_2 tension (mm Hg)	116.0	66.0	84.0

Figure 37–7. **Left:** "Ideal" ventilation/blood flow relationship. **Right:** Nonuniform ventilation and uniform blood flow, uncompensated. $\dot{V}_A$, alveolar ventilation; MV, respiratory minute volume. (Reproduced, with permission, from Comroe JH Jr et al: *The Lung: Clinical Physiology and Pulmonary Function Tests,* 2nd ed. Year Book, 1962.)

compensated by the greater saturation of the blood coming from A, because hemoglobin is normally nearly saturated in the lungs and the higher alveolar P_{O_2} adds only a little more O_2 to the hemoglobin than it normally carries. Consequently, the arterial blood is unsaturated. On the other hand, the CO_2 content of the arterial blood is generally normal in such situations, since extra loss of CO_2 in overventilated regions can balance diminished loss in underventilated areas.

Venous-to-Arterial Shunts

When a cardiovascular abnormality such as an interatrial septal defect permits large amounts of unoxygenated venous blood to bypass the pulmonary capillaries and dilute the oxygenated blood in the systemic arteries ("right-to-left shunt"), chronic hypoxic hypoxia and cyanosis **(cyanotic congenital heart disease)** result. Administration of 100% O_2 raises the O_2 content of alveolar air and improves the hypoxia due to hypoventilation, impaired diffusion, or ventilation-perfusion imbalance (short of perfusion of totally unventilated segments) by increasing the amount of O_2 in the blood leaving the lungs. However, in patients with venous-to-arterial shunts and normal lungs, any beneficial effect of 100% O_2 is slight and is due solely to an increase in the amount of dissolved O_2 in the blood.

Collapse of the Lung

When a bronchus or bronchiole is obstructed, the gas in the alveoli beyond the obstruction is absorbed and the lung segment collapses. Collapse of alveoli is called **atelectasis.** The atelectatic area may range in size from a small patch to a whole lung. Some blood is diverted from the collapsed area to better ventilated portions of the lung, and this reduces the magnitude of the decline in arterial P_{O_2}.

When a large part of the lung is collapsed, there is an appreciable decrease in lung volume. The intrapleural pressure therefore becomes more negative and pulls the mediastinum, which in humans is a fairly flexible structure, to the affected side.

Another cause of atelectasis is absence or inactivation of surfactant, the surface-tension-depressing agent normally found in the thin fluid lining the alveoli (see Chapter 34). This abnormality is a major cause of failure of the lungs to expand normally at birth. Collapse of the lung may also be due to the presence in the pleural space of air **(pneumothorax)**, tissue fluids **(hydrothorax, chylothorax),** or blood **(hemothorax).**

Pneumothorax

When air is admitted to the pleural space, through either a rupture in the lung or a hole in the chest wall,

the lung on the affected side collapses because of its elastic recoil. Since the intrapleural pressure on the affected side is now atmospheric, the mediastinum shifts toward the normal side. If the communication between the pleural space and the exterior remains open **(open or sucking pneumothorax),** more air moves in and out of the pleural space each time the patient breathes. If the hole is large, the resistance to air flow into the pleural cavity is less than the resistance to air flow into the intact lung, and little air enters the lung. During inspiration, the mediastinum shifts farther to the intact side, kinking the great vessels until it flaps back during expiration. There is marked stimulation of respiration due to hypoxia, hypercapnia, and activation of pulmonary deflation receptors. Respiratory distress is severe.

If there is a flap of tissue over the hole in the lung or chest wall that acts as a flutter valve, permitting air to enter during inspiration but preventing its exit during expiration, the pressure in the pleural space rises above atmospheric pressure **(tension pneumothorax).** The hypoxic stimulus to respiration causes deeper inspiratory efforts, which further increase the pressure in the pleural cavity, kinking the great veins and causing further hypoxia and shock. Intrapleural pressure in such cases may rise to 20–30 mm Hg. The peripheral veins become distended, there is intense cyanosis, and the condition is potentially fatal if the pneumothorax is not decompressed by removing the air.

On the other hand, if the hole through which air enters the pleural space seals off **(closed pneumothorax),** respiratory distress is not great because, with each inspiration, air flows into the lung on the unaffected side rather than into the pleural space. Because the vascular resistance is increased in the collapsed lung, blood is diverted to the other lung. Consequently, unless the pneumothorax is very large, it does not cause much hypoxia.

The air in a closed pneumothorax is absorbed. Since it is at atmospheric pressure, its total pressure, P_{O_2}, and P_{N_2} are greater than the corresponding values in venous blood (compare values for air and venous blood in Figure 34–18). Gas diffuses down these gradients into the blood, and after 1–2 weeks all of the gas disappears.

Asthma

Asthma is characterized by episodic or chronic wheezing, cough, and a feeling of tightness in the chest as a result of bronchoconstriction. Its morbidity and mortality are increasing, and its fundamental cause is still unknown despite intensive research. However, three abnormalities are present: airway obstruction that is at least partially reversible, airway inflammation, and airway hyperresponsiveness to a variety of stimuli. A link to allergy has long been recognized, and plasma IgE levels are often elevated. Proteins released from eosinophils in the inflammatory reaction may damage the airway epithelium and contribute to the hyper-responsiveness. Leukotrienes (see Chapter 17) are released from eosinophils and mast cells, and leukotrienes cause bronchoconstriction. Numerous other amines, neuropeptides, chemokines, and interleukins have effects on bronchial smooth muscle or produce inflammation, and they may be involved in asthma.

Asthma attacks are more severe in the late-night and early-morning hours because, as noted above, this is the period of maximal constriction in the circadian rhythm of bronchial tone. Cool air and exercise, both of which normally cause bronchoconstriction, also trigger asthma attacks, and in about 5% of asthmatics attacks are triggered by aspirin. However, innumerable other substances have been found to trigger asthma attacks.

Beta$_2$-adrenergic agonists have long been the mainstay of treatment for mild to moderate asthma attacks because β_2-adrenergic receptors mediate bronchodilation. Inhaled and systemic steroids are used even in mild to moderate cases; they are very effective, but their side effects (see Chapter 20) are a problem. Agents that block synthesis of leukotrienes or their CysLT$_1$ receptor are useful in some cases. IgE, which stimulates mast cells and basophils, plays a prominent role in asthma, and a monoclonal antibody that blocks the action of IgE has produced promising therapeutic results in preliminary clinical trials.

Emphysema

In the degenerative and potentially fatal pulmonary disease called **emphysema,** the lungs lose their elasticity as a result of disruption of elastic tissue and the walls between the alveoli break down so that the alveoli are replaced by large air sacs. The physiologic dead space is greatly increased, and because of inadequate and uneven alveolar ventilation and perfusion of underventilated alveoli, severe hypoxia develops. Late in the disease, hypercapnia also develops. Inspiration and expiration are labored, and the work of breathing is greatly increased. The changes in the pressure-volume curve of the lungs are shown in Figure 34–11. The chest becomes enlarged and barrel-shaped because the chest wall expands as the opposing elastic recoil of the lungs declines. The hypoxia leads to polycythemia. Pulmonary hypertension develops, and the right side of the heart enlarges (**cor pulmonale**) and then fails.

The most common cause of emphysema is heavy cigarette smoking. The smoke causes an increase in the number of pulmonary alveolar macrophages, and these macrophages release a chemical substance that attracts leukocytes to the lungs. The leukocytes in turn release proteases including elastase, which attacks the elastic tissue in the lungs. At the same time, α_1-antitrypsin, a plasma protein that normally inactivates elastase and other proteases, is itself inhibited. The α_1-antitrypsin is inactivated by oxygen radicals, and these are released by the leukocytes. The final result is a protease-antiprotease imbalance with increased destruction of lung tissue.

In about 2% of cases of emphysema, there is a congenital deficiency of active α_1-antitrypsin. If individuals who are homozygous for this deficiency smoke, they develop crippling emphysema early in life and have a 20-year reduction in their life span. If they do not smoke, they may still develop emphysema, but they do much better and their life expectancy is much improved. Thus, α_1-antitrypsin deficiency provides an interesting example of the interaction between genetic factors and environmental factors in the production of disease.

Cystic Fibrosis

Cystic fibrosis is another condition that leads to repeated pulmonary infections, particularly with *Pseudomonas pneumoniae,* and progressive, eventually fatal destruction of the lungs. In this congenital recessive condition, the function of a Cl$^-$ channel, the CFTR channel, is depressed. One would expect Na$^+$ reabsorption to be depressed as well, and indeed in sweat glands it is. However, in the lungs, it is enhanced, so that the Na$^+$ and water move out of airways, leaving their other secretions inspissated and sticky. It appears that the cause of abnormally increased Na$^+$ absorption is excessive activation of the ENaCs (see Chapter 1). The mechanism responsible for this activation is unsettled.

Among Caucasians, cystic fibrosis is one of the most common genetic disorders: 5% of the population carry a defective gene, and the disease occurs in one of every 2000 births.

The gene that is abnormal in cystic fibrosis is located on the long arm of chromosome 7 and encodes a Cl$^-$ channel called the **cystic fibrosis transmembrane conductance regulator (CFTR),** which has 12 membrane-spanning domains, two ATP-binding sites, and a region containing phosphorylation sites for cAMP-dependent protein kinase (protein kinase A) (see Chapter 1). This channel is a member of a superfamily of transporters that mediate, among other things, export of the α-factor mating pheromone in yeast and probably secretion of proteins lacking a signal sequence in mammals. The number of reported mutations in the *CFTR* gene that cause cystic fibrosis is large, and the severity of the defect varies with the mutation; however this is not surprising in a gene encoding such a complex protein. In most cases, the mutations interfere with ATP-binding or the conformational change that is presumably produced by the binding.

In men with cystic fibrosis, inspissated secretions also occur in sperm ducts, and this may obstruct the passage of spermatozoa, causing sterility. Chronic pancreatitis also occurs in both sexes as a result of abnormal function of pancreatic ducts (see Chapter 26). As noted above, transport of Na^+ and Cl^- out of the lumens of sweat glands is defective, and a high sweat content of these electrolytes is usually present.

OTHER FORMS OF HYPOXIA

Anemic Hypoxia

Hypoxia due to anemia is not severe at rest unless the hemoglobin deficiency is marked, because red blood cell 2,3-DPG increases. However, anemic patients may have considerable difficulty during exercise because of limited ability to increase O_2 delivery to the active tissues (Figure 37–8).

Carbon Monoxide Poisoning

Small amounts of carbon monoxide (CO) are formed in the body, and this gas may function as a chemical messenger in the brain and elsewhere (see Chapters 4

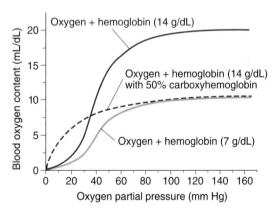

Figure 37–8. Normal oxyhemoglobin dissociation curve (top; hemoglobin concentration, 14 g/dL) compared to oxyhemoglobin dissociation curves in CO poisoning (50% carboxyhemoglobin) and anemia (hemoglobin concentration, 7 g/dL). Note that the CO-poisoning curve is shifted to the left of the anemia curve. (Reproduced, with permission from Leff AR, Schumacker PT: *Respiratory Physiology: Basics and Applications*. Saunders, 1993.)

and 27). In larger amounts, it is poisonous. Outside the body, it is formed by incomplete combustion of carbon. It was used by the Greeks and Romans to execute criminals, and today it causes more deaths than any other gas. CO poisoning has become less common in the United States, since natural gas, which does not contain CO, replaced artificial gases such as coal gas, which contains large amounts. However, the exhaust of gasoline engines is 6% or more CO.

CO is toxic because it reacts with hemoglobin to form **carbonmonoxyhemoglobin (carboxyhemoglobin, COHb),** and COHb cannot take up O_2 (Figure 37–8). Carbon monoxide poisoning is often listed as a form of anemic hypoxia because there is a deficiency of hemoglobin that can carry O_2, but the total hemoglobin content of the blood is unaffected by CO. The affinity of hemoglobin for CO is 210 times its affinity for O_2, and COHb liberates CO very slowly. An additional difficulty is that when COHb is present the dissociation curve of the remaining HbO_2 shifts to the left, decreasing the amount of O_2 released. This is why an anemic individual who has 50% of the normal amount of HbO_2 may be able to perform moderate work, whereas an individual whose HbO_2 is reduced to the same level because of the formation of COHb is seriously incapacitated.

Because of the affinity of CO for hemoglobin, there is progressive COHb formation when the alveolar P_{CO} is greater than 0.4 mm Hg. However, the amount of COHb formed depends upon the duration of exposure to CO as well as the concentration of CO in the inspired air and the alveolar ventilation.

CO is also toxic to the cytochromes in the tissues, but the amount of CO required to poison the cytochromes is 1000 times the lethal dose; tissue toxicity thus plays no role in clinical CO poisoning.

The symptoms of CO poisoning are those of any type of hypoxia, especially headache and nausea, but there is little stimulation of respiration, since in the arterial blood, P_{O_2} remains normal and the carotid and aortic chemoreceptors are not stimulated (see Chapter 36). The cherry-red color of COHb is visible in the skin, nail beds, and mucous membranes. Death results when about 70–80% of the circulating hemoglobin is converted to COHb. The symptoms produced by chronic exposure to sublethal concentrations of CO are those of progressive brain damage, including mental changes and, sometimes, a parkinsonism-like state (see Chapter 32).

Treatment of CO poisoning consists of immediate termination of the exposure and adequate ventilation, by artificial respiration if necessary. Ventilation with O_2 is preferable to ventilation with fresh air, since O_2 hastens the dissociation of COHb. Hyperbaric oxygenation (see below) is useful in this condition.

Stagnant Hypoxia

Hypoxia due to slow circulation is a problem in organs such as the kidneys and heart during shock (see Chapter 33). The liver and possibly the brain are damaged by stagnant hypoxia in congestive heart failure. The blood flow to the lung is normally very large, and it takes prolonged hypotension to produce significant damage. However, ARDS (see Chapter 33) can develop when there is prolonged circulatory collapse.

Histotoxic Hypoxia

Hypoxia due to inhibition of tissue oxidative processes is most commonly the result of cyanide poisoning. Cyanide inhibits cytochrome oxidase and possibly other enzymes. Methylene blue or nitrites are used to treat cyanide poisoning. They act by forming **methemoglobin,** which then reacts with cyanide to form **cyanmethemoglobin,** a nontoxic compound. The extent of treatment with these compounds is, of course, limited by the amount of methemoglobin that can be safely formed. Hyperbaric oxygenation may also be useful.

OXYGEN TREATMENT

Value

Administration of oxygen-rich gas mixtures is of very limited value in stagnant, anemic, and histotoxic hypoxia because all that can be accomplished in this way is an increase in the amount of dissolved O_2 in the arterial blood. This is also true in hypoxic hypoxia when it is due to shunting of unoxygenated venous blood past the lungs. In other forms of hypoxic hypoxia, O_2 is of great benefit. Treatment regimes that deliver less than 100% O_2 are of value both acutely and chronically, and administration of O_2 24 hours per day for 2 years in this fashion has been shown to significantly decrease the mortality of chronic obstructive pulmonary disease.

When 100% O_2 is first inhaled, there may be a slight decrease in respiration in normal individuals, suggesting that there is normally some hypoxic chemoreceptor drive. However, the effect is minor and can be demonstrated only by special techniques. In addition, it is offset by a slight accumulation of H^+ ions, since the concentration of deoxygenated hemoglobin in the blood is reduced and Hb is a better buffer than HbO_2 (see Chapter 35).

It should be remembered that in hypercapnic patients in severe pulmonary failure, the CO_2 level may be so high that it depresses rather than stimulates respiration. Some of these patients keep breathing only because the carotid and aortic chemoreceptors drive the respiratory center. If the hypoxic drive is withdrawn by administering O_2, breathing may stop. During the resultant apnea, the arterial PO_2 drops but breathing may not start again, because the increase in PCO_2 further depresses the respiratory center. Therefore, O_2 therapy in this situation must be started with care.

Oxygen Toxicity

It is interesting that while O_2 is necessary for life in aerobic organisms, it is also toxic. Indeed, 100% O_2 has been demonstrated to exert toxic effects not only in animals but also in bacteria, fungi, cultured animal cells, and plants. The toxicity seems to be due to the production of the superoxide anion (O_2^-), which is a free radical, and H_2O_2. When 80–100% O_2 is administered to humans for periods of 8 hours or more, the respiratory passages become irritated, causing substernal distress, nasal congestion, sore throat, and coughing.

Some infants treated with O_2 for respiratory distress syndrome develop a chronic condition characterized by lung cysts and densities (**bronchopulmonary dysplasia**). There is evidence that this syndrome is a manifestation of O_2 toxicity. Another complication in these infants is **retinopathy of prematurity (retrolental fibroplasia),** the formation of opaque vascular tissue in the eyes, which can lead to serious visual defects. The retinal receptors mature from the center to the periphery of the retina, and they use considerable O_2. This causes the retina to become vascularized in an orderly fashion. Oxygen treatment before maturation is complete provides the needed O_2 to the photoreceptors, and consequently the normal vascular pattern fails to develop. There is evidence that this condition can be prevented or ameliorated by treatment with vitamin E, which exerts an antioxidant effect, and, in animals, by growth hormone inhibitors.

Hyperbaric Oxygen Therapy

Administration of 100% O_2 at increased pressure accelerates the onset of O_2 toxicity, with the production not only of tracheobronchial irritation but also of muscle twitching, ringing in the ears, dizziness, convulsions, and coma. The speed with which these symptoms develop is proportionate to the pressure at which the O_2 is administered; eg, at 4 atmospheres, symptoms develop in half the subjects in 30 minutes, whereas at 6 atmospheres, convulsions develop in a few minutes. Administration of other gases at increased pressure also causes central nervous system symptoms (see below). Administration of O_2 at elevated pressures to rats decreases their brain GABA content (see Chapter 4) and their brain, liver, and kidney ATP content.

On the other hand, exposure to 100% O_2 at 2–3 atmospheres can increase dissolved O_2 in arterial blood to the point that arterial O_2 tension is greater than 2000 mm Hg and tissue O_2 tension is 400 mm Hg. If exposure is limited to 5 hours or less at these pressures, O_2 toxicity is not a problem. Therefore, **hyperbaric O_2** therapy in closed tanks is used to treat diseases in which improved oxygenation of tissues cannot be achieved in other ways. It is of demonstrated value in carbon monoxide poisoning, radiation-induced tissue injury, gas gangrene, very severe blood loss anemia, diabetic leg ulcers and other wounds that are slow to heal, and rescue of skin flaps and grafts in which the circulation is marginal. It is also the primary treatment for decompression sickness and air embolism (see below).

HYPERCAPNIA & HYPOCAPNIA

Hypercapnia

Retention of CO_2 in the body (hypercapnia) initially stimulates respiration. Retention of larger amounts produces symptoms due to depression of the central nervous system: confusion, diminished sensory acuity, and, eventually, coma with respiratory depression and death. In patients with these symptoms, the P_{CO_2} is markedly elevated, there is severe respiratory acidosis, and the plasma HCO_3^- may exceed 40 meq/L. Large amounts of HCO_3^- are excreted, but more HCO_3^- is reabsorbed, raising the plasma HCO_3^- and partially compensating for the acidosis (see Chapter 39).

CO_2 is so much more soluble than O_2 that hypercapnia is rarely a problem in patients with pulmonary fibrosis. However, it does occur in ventilation-perfusion inequality and when for any reason alveolar ventilation is inadequate in the various forms of pump failure (Table 37–1). It is exacerbated when CO_2 production is increased. For example, in febrile patients there is a 13% increase in CO_2 production for each 1 °C rise in temperature, and a high carbohydrate intake increases CO_2 production because of the increase in RQ (see Chapter 17). Normally, alveolar ventilation increases and the extra CO_2 is expired, but it accumulates when ventilation is compromised.

Hypocapnia

Hypocapnia is the result of hyperventilation. During voluntary hyperventilation, the arterial P_{CO_2} falls from 40 to as low as 15 mm Hg while the alveolar P_{O_2} rises to 120–140 mm Hg.

The more chronic effects of hypocapnia are seen in neurotic patients who chronically hyperventilate. Cerebral blood flow may be reduced 30% or more because of the direct constrictor effect of hypocapnia on the cerebral vessels (see Chapter 32). The cerebral ischemia causes light-headedness, dizziness, and paresthesias.

Hypocapnia also increases cardiac output. It has a direct constrictor effect on many peripheral vessels, but it depresses the vasomotor center, so that the blood pressure is usually unchanged or only slightly elevated.

Other consequences of hypocapnia are due to the associated respiratory alkalosis, the blood pH being increased to 7.5 or 7.6. The plasma HCO_3^- level is low, but HCO_3^- reabsorption is decreased because of the inhibition of renal acid secretion by the low P_{CO_2}. The plasma total calcium level does not change, but the plasma Ca^{2+} level falls and hypocapnic individuals develop carpopedal spasm, a positive Chvostek sign, and other signs of tetany (see Chapter 21).

OTHER RESPIRATORY ABNORMALITIES

Asphyxia

In asphyxia produced by occlusion of the airway, acute hypercapnia and hypoxia develop together. There is pronounced stimulation of respiration, with violent respiratory efforts. Blood pressure and heart rate rise sharply, catecholamine secretion is increased, and blood pH drops. Eventually the respiratory efforts cease, the blood pressure falls, and the heart slows. Asphyxiated animals can still be revived at this point by artificial respiration, although they are prone to ventricular fibrillation, probably because of the combination of hypoxic myocardial damage and high circulating catecholamine levels. If artificial respiration is not started, cardiac arrest occurs in 4–5 minutes.

Drowning

Drowning is asphyxia caused by immersion, usually in water. In about 10% of drownings, the first gasp of water after the losing struggle not to breathe triggers laryngospasm, and death results from asphyxia without any water in the lungs. In the remaining cases, the glottic muscles eventually relax and fluid enters the lungs. Fresh water is rapidly absorbed, diluting the plasma and causing intravascular hemolysis. Ocean water is markedly hypertonic and draws fluid from the vascular system into the lungs, decreasing plasma volume. The immediate goal in the treatment of drowning is, of course, resuscitation, but long-term treatment must also take into account the circulatory effects of the water in the lungs.

Periodic Breathing

The acute effects of voluntary hyperventilation demonstrate the interaction of the chemical respiratory regulating mechanisms. When a normal individual hyperventilates for 2–3 minutes, then stops and permits respiration to continue without exerting any voluntary control over it, there is a period of apnea. This is followed by a few shallow breaths and then by another period of apnea, followed again by a few breaths **(periodic breathing).** The cycles may last for some time before normal breathing is resumed (Figure 37–9). The apnea apparently is due to CO_2 lack because it does not occur following hyperventilation with gas mixtures containing 5% CO_2. During the apnea, the alveolar PO_2 falls and the PCO_2 rises. Breathing resumes because of hypoxic stimulation of the carotid and aortic chemoreceptors before the CO_2 level has returned to normal. A few breaths eliminate the hypoxic stimulus, and breathing stops until the alveolar PO_2 falls again. Gradually, however, the PCO_2 returns to normal, and normal breathing resumes.

Cheyne-Stokes Respiration

Periodic breathing occurs in various disease states and is often called **Cheyne-Stokes respiration.** It is seen most commonly in patients with congestive heart failure and uremia, but it occurs also in patients with brain disease and during sleep in some normal individuals (Figure 37–10). Some of the patients with Cheyne-Stokes respiration have increased sensitivity to CO_2. The increased response is apparently due to disruption of neural pathways that normally inhibit respiration. In these individuals, CO_2 causes relative hyperventilation,

Figure 37–10. Cheyne-Stokes breathing during sleep. Two periods of apnea are separated by an increase and then a smooth decrease in tidal volume (V_T). (Reproduced, with permission, from Cherniack NS: Respiratory dysrhythmias during sleep. N Engl J Med 1981;305:325.)

lowering the arterial PCO_2. During the resultant apnea, the arterial PCO_2 again rises to normal, but the respiratory mechanism again overresponds to CO_2. Breathing ceases, and the cycle repeats.

Another cause of periodic breathing in patients with cardiac disease is prolongation of the lung-to-brain circulation time, so that it takes longer for changes in arterial gas tensions to affect the respiratory area in the medulla. When individuals with a slower circulation hyperventilate, they lower the PCO_2 of the blood in their lungs, but it takes longer than normal for the blood with a low PCO_2 to reach the brain. During this time, the PCO_2 in the pulmonary capillary blood continues to be lowered, and when this blood reaches the brain, the low PCO_2 inhibits the respiratory area, producing apnea. In other words, the respiratory control system oscillates because the negative feedback loop from lungs to brain is abnormally long.

Sleep Apnea

Episodes of apnea during sleep can be central in origin, ie, due to failure of discharge in the nerves producing respiration, or they can be due to airway obstruction **(obstructive sleep apnea).** This can occur at any age and is produced when the pharyngeal muscles relax during sleep. In some cases, failure of the genioglossus muscles to contract during inspiration contributes to the blockage; these muscles pull the tongue forward, and when they do not contract the tongue falls back and obstructs the airway. After several increasingly strong respiratory efforts, the patient wakes up, takes a few normal breaths, and falls back to sleep. Not surprisingly, the apneic episodes are most common during REM sleep, when the muscles are most hypotonic (see Chapter 11). The symptoms are loud snoring, morning headaches, fatigue, and daytime sleepiness. When severe and prolonged, the condition apparently causes hypertension and its complications. In addition, the incidence of motor vehicle accidents in sleep apnea patients is seven times greater than it is in the general driving population.

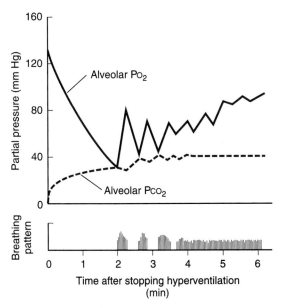

Figure 37–9. Changes in breathing and composition of alveolar air after forced hyperventilation for 2 minutes.

Sudden Infant Death Syndrome

It has been argued that sudden infant death syndrome (SIDS) may be a form of sleep apnea. This disorder, in which apparently healthy infants are found dead, often in their cribs, has attracted a great deal of attention. Apneic spells are common in premature infants. However, periods of prolonged apnea do not correlate with the subsequent occurrence of death, and none of the known tests of chemoresponsiveness reliably predict which infants will subsequently have difficulty. Some of the cases appear to be caused by cardiac arrhythmias complicating the congenital long QT syndrome (see Chapter 28). In addition, there is evidence that the incidence of SIDS is increased in infants of mothers who smoke. It is increased also by sleeping in the prone position, and teaching mothers to put their babies down on their backs has led to a significant reduction in incidence.

DISEASES AFFECTING THE PULMONARY CIRCULATION

Pulmonary Hypertension

Sustained primary pulmonary hypertension can occur at any age. Like systemic arterial hypertension (see Chapter 33), it is a syndrome with multiple causes. However, the causes are different from those causing systemic hypertension. They include hypoxia, inhalation of cocaine, treatment with dexfenfluramine and related appetite-suppressing drugs that increase extracellular serotonin, and systemic lupus erythematosus. Some cases are familial and appear to be related to mutations that increase the sensitivity of pulmonary vessels to growth factors or cause deformations in the pulmonary vascular system.

All these conditions lead to increased pulmonary vascular resistance. If appropriate therapy is not initiated, the increased right ventricular afterload leads eventually to right heart failure and death. Treatment with vasodilators such as prostacyclin and prostacyclin analogs is effective. Until recently, these had to be administered by continuous intravenous infusion, but aerosolized preparations that appear to be effective are now available.

Pulmonary Embolization

One of the normal functions of the lungs is to filter out small blood clots, and this occurs without any symptoms. When emboli block larger branches of the pulmonary artery, they provoke a rise in pulmonary arterial pressure and rapid, shallow respiration **(tachypnea).** The rise in pulmonary arterial pressure may be due to reflex vasoconstriction via the sympathetic nerve fibers, but reflex vasoconstriction appears to be absent when large branches of the pulmonary artery are blocked. The tachypnea is a reflex response to activation of vagally innervated pulmonary receptors close to the vessel walls (see Table 36–2) that appear to be activated at the site of the embolization.

EFFECTS OF INCREASED BAROMETRIC PRESSURE

The ambient pressure increases by 1 atmosphere for every 10 m of depth in seawater and every 10.4 m of depth in fresh water. Therefore, at a depth of 31 m (100 ft) in the ocean, a diver is exposed to a pressure of 4 atmospheres. Those who dig underwater tunnels are also exposed to the same hazards because the pressure in the chambers (caissons) in which they work is increased to keep out the water.

The hazards of exposure to increased barometric pressure used to be the concern largely of the specialists who cared for deep-sea divers and tunnel workers. However, the invention of SCUBA gear (self-contained underwater breathing apparatus, a tank-and-valve system carried by the diver) transformed diving from a business into a sport. The popularity of skin diving is so great that all physicians should be aware of its potential dangers.

Nitrogen Narcosis & the High-Pressure Nervous Syndrome

A diver must breathe air or other gases at increased pressure to equalize the increased pressure on the chest wall and abdomen. CO_2 is routinely removed to prevent its accumulation. At increased pressure, 100% O_2 causes central nervous system symptoms of oxygen toxicity (Table 37–2). Since the harmful effects of breathing O_2 (see above) are proportionate to the P_{O_2}, they can be prevented by decreasing the concentration of O_2 in the gas mixture to 20% or less.

If a diver breathes compressed air, the increased P_{N_2} can cause **nitrogen narcosis,** a condition also known as "rapture of the deep" (Table 37–2). At pressures of 4–5 atmospheres (ie, at depths of 30–40 m in the ocean), 80% N_2 produces definite euphoria. At greater pressures, the symptoms resemble alcohol intoxication. Manual dexterity is maintained, but intellectual functions are impaired.

The problem of nitrogen narcosis can be avoided by breathing mixtures of O_2 and helium, and deeper dives can be made. However, the **high-pressure nervous syndrome** (HPNS) develops during deep dives with such mixtures. This condition is characterized by tremors, drowsiness, and a depression of the α activity in the EEG. Unlike nitrogen narcosis, intellectual functions are not severely affected but manual dexterity is

Table 37–2. Potential problems associated with exposure to increased barometric pressure.

Oxygen toxicity
  Lung damage
  Convulsions

Nitrogen narcosis
  Euphoria
  Impaired performance

High-pressure nervous syndrome
  Tremors
  Somnolence

Decompression sickness
  Pain
  Paralyses

Air embolism
  Sudden death

---

impaired. The cause of HPNS is not settled, but it is worth noting that a variety of gases that are physiologically inert at atmospheric pressure are anesthetics at increased pressure. This is true of N_2 and also of xenon, krypton, argon, neon, and helium. Their anesthetic activity parallels their lipid solubility, and the anesthesia may be due to an action on nerve cell membranes.

Decompression Sickness

As a diver breathing 80% N_2 ascends from a dive, the elevated alveolar P_{N_2} falls. N_2 diffuses from the tissues into the lungs along the partial pressure gradient. If the return to atmospheric pressure (decompression) is gradual, no harmful effects are observed; but if the ascent is rapid, N_2 escapes from solution. Bubbles form in the tissues and blood, causing the symptoms of **decompression sickness (the bends, caisson disease).** Bubbles in the tissues cause severe pains, particularly around joints, and neurologic symptoms that include paresthesias and itching. Bubbles in the bloodstream, which occur in more severe cases, obstruct the arteries to the brain and spinal cord. Symptoms commonly appear 10–30 minutes after the diver resurfaces, and they progress. Abnormalities due to damage to the spinal cord are most common, but there can also be major paralyses and respiratory failure. Bubbles in the pulmonary capillaries are apparently responsible for the dyspnea that divers called **"the chokes,"** and bubbles in the coronary arteries may cause myocardial damage.

Treatment of this disease is prompt recompression in a pressure chamber, followed by slow decompression. Recompression is frequently lifesaving. Recovery is often complete, but there may be residual neurologic sequelae as a result of irreversible damage to the nervous system.

Ascent in an airplane is equivalent to ascent from a dive. The decompression during a climb from sea level to 8550 m in the unpressurized cabin of an airplane (pressure drop from 1 atm down to 0.33 atm) is the same as that during ascent to the surface after spending time in 20 m of seawater (pressure drop from 3 atm down to 1 atm). A rapid ascent in either situation can cause decompression sickness.

Air Embolism

If a diver breathing from a tank at increased pressure during a dive holds his or her breath and suddenly heads for the surface—as may occur if the diver gets into trouble and panics—the gas in the lungs may expand rapidly enough to rupture the pulmonary veins. This drives air into the vessels, causing air embolism. Fatal air embolism has occurred during rapid ascent from as shallow a depth as 5 m. The consequences of air in the circulatory system are discussed in Chapter 30. This cannot happen, of course, to an individual who takes a breath on the surface, dives, and returns to the surface still holding that breath, no matter how deep the dive.

Air embolism also occurs as a result of rapid expansion of the gas in the lungs when the external pressure is suddenly reduced from atmospheric to subatmospheric, as it is when the wall of the pressurized cabin of an airplane or rocket at high altitude is breached (**explosive decompression).**

ARTIFICIAL RESPIRATION

In acute asphyxia due to drowning, CO or other forms of gas poisoning, electrocution, anesthetic accidents, and other similar causes, artificial respiration after breathing has ceased may be lifesaving. It should always be attempted, because respiration stops before the heart stops. There are numerous methods of emergency artificial respiration, but the method presently recommended to produce adequate ventilation in all cases is mouth-to-mouth breathing. The advantages of mouth-to-mouth resuscitation lie not only in its simplicity but also in the fact that it works by expanding the lungs.

Mouth-to-Mouth Breathing

In this form of resuscitation, the operator first places the victim in the supine position and opens the airway by placing a hand under the neck and lifting, while keeping pressure with the other hand on the victim's forehead. This extends the neck and lifts the tongue away from the back of the throat. The victim's mouth is

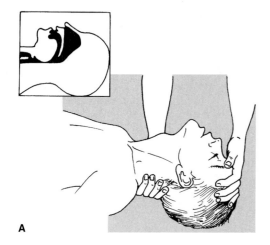

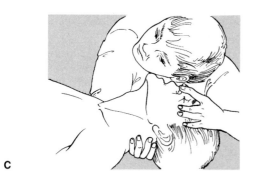

Figure 37–11. Proper performance of mouth-to-mouth resuscitation. **A:** Open airway by positioning neck anteriorly in extension. Inserts show airway obstructed when the neck is in resting flexed position and open when neck is extended. **B:** Close victim's nose with fingers, seal mouth around victim's mouth, and deliver breath by vigorous expiration. **C:** Allow victim to exhale passively by unsealing mouth and nose. Rescuer should listen and feel for expiratory air flow. Repeat B and C 12 times per minute. (Reproduced, with permission, from Schroeder SA, Krupp MA, Tierney LM Jr [editors]: *Current Medical Diagnosis & Treatment 1990.* Originally published by Appleton & Lange. Copyright © 1990 by The McGraw-Hill Companies, Inc.

---

can be expelled by applying upward pressure on the abdomen from time to time. In apneic individuals in whom there is no detectable heartbeat, mouth-to-mouth breathing should be alternated with cardiac massage (see Chapter 28).

Mechanical Ventilation

For treatment of chronic weakness of the respiratory muscles, airtight devices that cover the chest are available. By means of a motor, negative pressure is applied to the chest at intervals, drawing air into the lungs.

In acute respiratory failure and other conditions in which alveolar-capillary exchange is compromised, patients are intubated and pulses of air or mixtures of respiratory gases are delivered by machines. Various pressure settings are used, and it is common to maintain a positive end-expiratory pressure (PEEP) to aid movement of O_2 into the blood and prevent atelectasis. However, excessive pressure can rupture the alveoli. For this and other reasons, mechanical ventilation should be discontinued as soon as possible. For reasons that are incompletely understood, removing patients from a respirator (**weaning**) is difficult, and in 25% of cases patients have respiratory distress that is severe enough to require resumption of mechanical ventilation before final discontinuation.

REFERENCES FOR SECTION VII: RESPIRATION

Barnes PJ: Chronic obstructive pulmonary disease. N Engl J Med 2000;343:269.

Bartecchi CE, MacKenzie TD, Schner RW: The human costs of tobacco use. N Engl J Med 1994;930:907.

Beckett WS: Occupational respiratory diseases. N Engl J Med 2000;342:406.

---

covered by the operator's mouth while the fingers of the hand already on the forehead occlude the nostrils (Figure 37–11). About 12 times a minute, the operator blows into the victim's mouth a volume about twice the tidal volume, then permits the elastic recoil of the victim's lungs to produce passive expiration. The victim's neck is kept extended. Any gas blown into the stomach

Bove AA, Davis JC (editors): *Diving Medicine,* 2nd ed. Williams & Wilkins, 1990.

Busse WW, Lemanske RF: Asthma. N Engl J Med 2001;344:350.

Carrell RW, Lomas DA: Alpha$_1$-antitrypsin deficiency—a model of conformational diseases. N Engl J Med 2002;346:45.

Crapo RO: Pulmonary function testing. N Engl J Med 1994;331:25.

Fishman AP et al (editors): *Fishman's Pulmonary Diseases and Disorders,* 3rd ed. McGraw-Hill, 1998.

Gross I: Regulation of fetal lung maturation. Am J Physiol 1990;259:L337.

Gross TJ, Hunninghake GW: Idiopathic pulmonary fibrosis. N Engl J Med 2001;345:517.

Hackett PH, Roach RC: High-altitude illness. N Engl J Med 2001;345:107.

Higenbottam T, Otulana BA, Wallwork J: The physiology of heart-lung transplantation in humans. News Physiol Sci 1990;5:71.

Jones NL, Killian KJ: Exercise limitation in health and disease. N Engl J Med 2000;343:632.

Laffey JG, Kavanagh BP: Hypocapnia. N Engl J Med 2002;347:43.

Launois S et al: Hiccup in adults: An overview. Eur Respir J 1993;6:563.

Macklem PT: The act of breathing. News Physiol Sci 1990;5:233.

Maggi CA et al: Neuropeptides as regulators of airway function: Vasoactive intestinal peptide and the tachykinins. Physiol Rev 1995;75:277.

Marin MG: Update: Pharmacology of airway secretion. Pharmacol Rev 1994;46:36.

Modell JH: Drowning. N Engl J Med 1993;328:253.

Moffat K, Deatherage JF, Seybert DW: A structural model for the kinetic behavior of hemoglobin. Science 1979;206:1035.

Moon RE, Vann RD, Bennett PB: The physiology of decompression illness. Sci Am (Aug) 1995;273:70.

Murray JF: *The Normal Lung: The Basis for Diagnosis and Treatment of Pulmonary Disease,* 2nd ed. Saunders, 1985.

Prisk GK, Paiva M, West JB (editors): *Gravity and the Lung: Lessons From Micrography.* Marcel Dekker, 2001.

Putnam RW, Dean JB, Ballantyne D (editors): Central chemosensitivity. Respir Physiol 2001;129:1.

Rekling JC, Feldman JL: Pre-Bötzinger complex and pacemaker neurons: hypothesized site and kernel for respiratory rhythm generation. Annu Rev Physiol 1998;60:385.

Reynolds HY: Immunologic system in the respiratory tract. Physiol Rev 1991;71:1117.

Rigatto H: Control of ventilation in the newborn. Annu Rev Physiol 1984;46:661.

Rooney SA, Young SL, Mendelsohn CR: Molecular and cellular processing of lung surfactant. FASEB J 1994;8:957.

Rubin LJ: Primary pulmonary hypertension. N Engl J Med 1997;336:111.

Spengler CM, Boutellier U: Breathless legs? Consider training your respiration. News Physiol Sci 2000;15:101.

Sugar O: In search of Ondine's curse. JAMA 1978;240:236.

Tibbles PM, Edelsberg JS: Hyperbaric-oxygen therapy. N Engl J Med 1996;334:1642.

Tobin MJ: Advances in mechanical ventilation. N Engl J Med 2001;344:1986.

Voelkel NF: High-altitude pulmonary edema. N Engl J Med 2002;346:1607.

Ware LB, Matthay MA: The acute respiratory distress syndrome. N Engl J Med 2000;342:1334.

Wasserman K et al: *Principles of Exercise Testing and Interpretation,* 3rd ed. Lippincott Williams & Wilkins, 1999.

Wasserman K: Coupling of external to cellular respiration during exercise: The wisdom of the body revisited. Am J Physiol 1994;266:E519.

Weinacker AB, Vaszar LT: Acute respiratory distress syndrome: physiology and new management strategies. Annu Rev Med 2001;52:221.

Weir EK, Archer SL: The mechanism of acute hypoxic pulmonary vasoconstriction: The tale of two channels. FASEB J 1995;9:183.

Weir EK, Reeves JT (editors): *Pulmonary Edema.* Futura, 1997.

West JB et al: Pathogenesis of high-altitude pulmonary oedema: Direct evidence of stress failure of pulmonary capillaries. Eur Respir J 1995;8:523.

West JB: *Respiratory Physiology: The Essentials,* 4th ed. Williams & Wilkins, 1991.

Wright JR: Immunomodulatory functions of surfactant. Physiol Rev 1997;77:931.

INTRODUCTION

In the kidneys, a fluid that resembles plasma is filtered through the glomerular capillaries into the renal tubules **(glomerular filtration).** As this glomerular filtrate passes down the tubules, its volume is reduced and its composition altered by the processes of **tubular reabsorption** (removal of water and solutes from the tubular fluid) and **tubular secretion** (secretion of solutes into the tubular fluid) to form the urine that enters the renal pelvis. A comparison of the composition of the plasma and an average urine specimen illustrates the magnitude of some of these changes (Table 38–1) and emphasizes the manner in which wastes are eliminated while water and important electrolytes and metabolites are conserved. Furthermore, the composition of the urine can be varied, and many homeostatic regulatory mechanisms minimize or prevent changes in the composition of the ECF by changing the amount of water and various specific solutes in the urine. From

Table 38–1. Urinary and plasma concentrations of some physiologically important substances.

Substance	Concentration in		
	Urine (U)	Plasma (P)	U/P Ratio
Glucose (mg/dL)	0	100	0
Na⁺ (meq/L)	90	140	0.6
Urea (mg/dL)	900	15	60
Creatinine (mg/dL)	150	1	150

the renal pelvis, the urine passes to the bladder and is expelled to the exterior by the process of urination, or **micturition.** The kidneys are also endocrine organs, making kinins (see Chapter 31) and 1,25-dihydroxy-cholecalciferol (see Chapter 21) and making and secreting renin (see Chapter 24).

FUNCTIONAL ANATOMY

The Nephron

Each individual renal tubule and its glomerulus is a unit **(nephron).** The size of the kidneys in various species is determined largely by the number of nephrons they contain. There are approximately 1.3 million nephrons in each human kidney. The parts of the nephron are shown in diagrammatic fashion in Figure 38–1. On the right side of Figure 38–2, the scale drawing shows the long, thin nature of the nephrons.

The glomerulus, which is about 200 μm in diameter, is formed by the invagination of a tuft of capillaries into the dilated, blind end of the nephron **(Bowman's capsule).** The capillaries are supplied by an **afferent arteriole** and drained by a slightly smaller **efferent arteriole** (Figures 38–2 and 38–3). There are two cellular layers separating the blood from the glomerular filtrate in Bowman's capsule: the capillary endothelium and the specialized epithelium of the capsule that is made up of **podocytes** (see below) overlying the glomerular capillaries (Figure 38–3). These layers are separated by a basal lamina. Stellate cells called **mesangial cells** are located between the basal lamina and the endothelium. They are similar to cells called **pericytes,** which are found in the walls of capillaries elsewhere in the body.

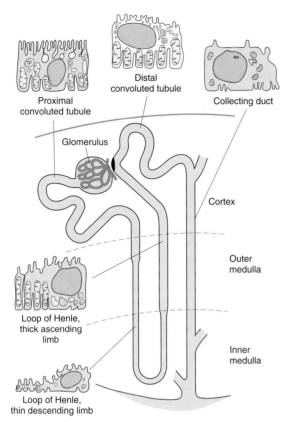

Figure 38–1. Diagram of a juxtamedullary nephron. The main histologic features of the cells that make up each portion of the tubule are also shown.

Mesangial cells are especially common between two neighboring capillaries, and in these locations the basal membrane forms a sheath shared by both capillaries (Figure 38–3). The mesangial cells are contractile and play a role in the regulation of glomerular filtration (see below). They also secrete various substances, take up immune complexes, and are involved in the production of glomerular disease.

The endothelium of the glomerular capillaries is fenestrated, with pores that are 70–90 nm in diameter. The cells of the epithelium (**podocytes**) have numerous pseudopodia that interdigitate (Figure 38–3) to form **filtration slits** along the capillary wall. The slits are approximately 25 nm wide, and each is closed by a thin membrane. The basal lamina does not contain visible gaps or pores.

Functionally, the glomerular membrane permits the free passage of neutral substances up to 4 nm in diameter and almost totally excludes those with diameters greater than 8 nm. However, the charges on molecules

as well as their diameters affect their passage into Bowman's capsule (see below). The total area of glomerular capillary endothelium across which filtration occurs in humans is about 0.8 m^2.

The general features of the cells that make up the walls of the tubules are shown in Figure 38–1. However, there are cell subtypes in all segments, and the anatomic differences between them correlate with differences in function (see below).

The human **proximal convoluted tubule** is about 15 mm long and 55 μm in diameter. Its wall is made up of a single layer of cells that interdigitate with one another and are united by apical tight junctions. Between the bases of the cells, there are extensions of the extracellular space called the **lateral intercellular spaces.** The luminal edges of the cells have a striate **brush border** due to the presence of innumerable 1 × 0.7 μm microvilli.

The convoluted portion of the proximal tubule (pars convoluta) drains into the straight portion (pars recta), which forms the first part of the **loop of Henle** (Figures 38–1 and 38–2). The proximal tubule terminates in the thin segment of the descending limb of the loop of Henle, which has an epithelium made up of attenuated, flat cells. The nephrons with glomeruli in the outer portions of the renal cortex have short loops of Henle (**cortical nephrons**), whereas those with glomeruli in the juxtamedullary region of the cortex (**juxtamedullary nephrons**) have long loops extending down into the medullary pyramids. In humans, only 15% of the nephrons have long loops. The total length of the thin segment of the loop varies from 2 to 14 mm. It ends in the thick segment of the ascending limb, which is about 12 mm in length. The cells of the thick ascending limb are cuboid. They have numerous mitochondria, and the basilar portions of their cell membranes are extensively invaginated.

The thick ascending limb of the loop of Henle reaches the glomerulus of the nephron from which the tubule arose and passes close to its afferent arteriole and efferent arteriole. The walls of the afferent arterioles contain the renin-secreting juxtaglomerular cells. At this point, the tubular epithelium is modified histologically to form the **macula densa.** The juxtaglomerular cells, the macula densa, and the lacis cells near them are known collectively as the **juxtaglomerular apparatus** (see Figure 24–4).

The **distal convoluted tubule** is about 5 mm long. Its epithelium is lower than that of the proximal tubule, and although there are a few microvilli, there is no distinct brush border. The distal tubules coalesce to form **collecting ducts** that are about 20 mm long and pass through the renal cortex and medulla to empty into the pelvis of the kidney at the apexes of the medullary pyramids. The epithelium of the collecting

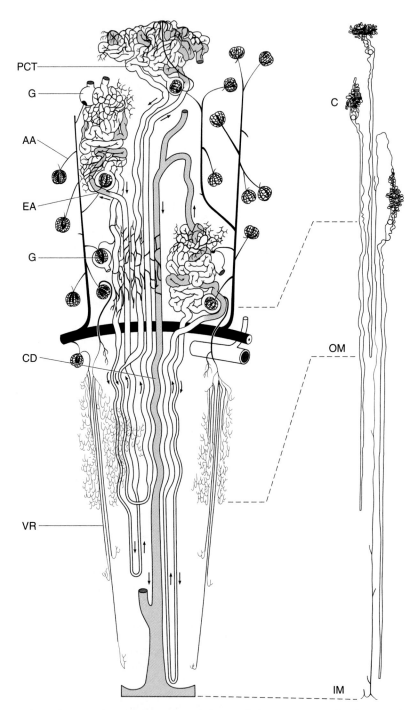

Figure 38–2. Renal circulation in the dog. The pattern in human kidneys is essentially identical. The vascular structures are simplified, and on the left the horizontal scale is expanded to show details. The same nephrons are shown undistorted on the right. AA, afferent arteriole; C, cortex; CD, collecting duct; EA, efferent arteriole; G, glomerulus; IM, inner medulla; OM, outer medulla; PCT, proximal convoluted tubule; VR, vasa recta. The distal convoluted tubules and collecting ducts are colored. (Reproduced, with permission, from Beeuwkes R III: The vascular organization of the kidney. Annu Rev Physiol 1980;42:531. Copyright © 1980 by Annual Reviews, Inc.)

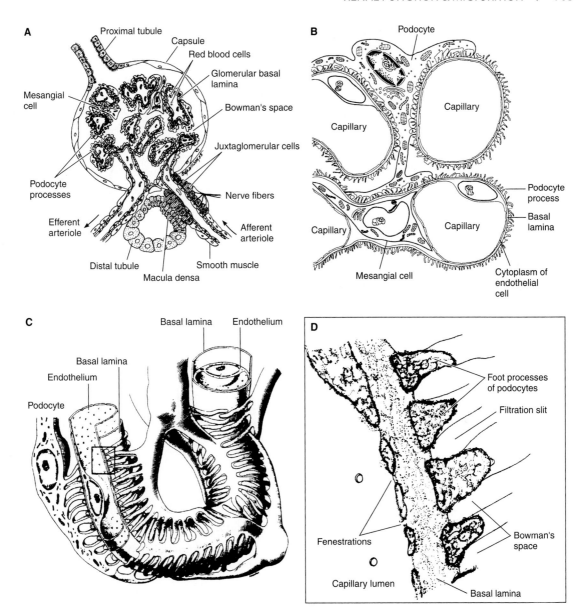

Figure 38–3. Structural details of glomerulus. **A:** Section through vascular pole, showing capillary loops. **B:** Relation of mesangial cells and podocytes to glomerular capillaries. **C:** Detail of the way podocytes form filtration slits on the basal lamina, and the relation of the lamina to the capillary endothelium. **D:** Enlargement of the rectangle in **C** to show the podocyte processes. The fuzzy material on their surfaces is glomerular polyanion.

ducts is made up of **principal cells (P cells)** and **intercalated cells (I cells).** The P cells, which predominate, are relatively tall and have few organelles. They are involved in Na^+ reabsorption and vasopressin-stimulated water reabsorption. The I cells, which are present in smaller numbers and are also found in the distal tubules, have more microvilli, cytoplasmic vesicles, and

mitochondria. They are concerned with acid secretion and HCO_3^- transport. The total length of the nephrons, including the collecting ducts, ranges from 45 to 65 mm.

Cells in the kidneys that appear to have a secretory function include not only the juxtaglomerular cells but also some of the cells in the interstitial tissue of the medulla. These cells are called **type I medullary interstitial cells.** They contain lipid droplets and probably secrete prostaglandins, predominantly PGE_2 (see Chapter 17). PGE_2 is also secreted by the cells in the collecting ducts, and prostacyclin (PGI_2), as well as other prostaglandins are secreted by the arterioles and glomeruli.

Blood Vessels

The renal circulation is diagrammed in Figure 38–2. The afferent arterioles are short, straight branches of the interlobular arteries. Each divides into multiple capillary branches to form the tuft of vessels in the glomerulus. The capillaries coalesce to form the efferent arteriole, which in turn breaks up into capillaries that supply the tubules (**peritubular capillaries)** before draining into the interlobular veins. The arterial segments between glomeruli and tubules are thus technically a portal system, and the glomerular capillaries are the only capillaries in the body that drain into arterioles. However, there is relatively little smooth muscle in the efferent arterioles.

The capillaries draining the tubules of the cortical nephrons form a peritubular network, whereas the efferent arterioles from the juxtamedullary glomeruli drain not only into a peritubular network but also into vessels that form hairpin loops (the **vasa recta).** These loops dip into the medullary pyramids alongside the loops of Henle (Figure 38–2). The descending vasa recta have a nonfenestrated endothelium that contains a facilitated transporter for urea, and the ascending vasa recta have a fenestrated endothelium, consistent with their function in conserving solute (see below).

The efferent arteriole from each glomerulus breaks up into capillaries that supply a number of different nephrons. Thus, the tubule of each nephron does not necessarily receive blood solely from the efferent arteriole of that nephron. In humans, the total surface of the renal capillaries is approximately equal to the total surface area of the tubules, both being about 12 m². The volume of blood in the renal capillaries at any given time is 30–40 mL.

Lymphatics

The kidneys have an abundant lymphatic supply that drains via the thoracic duct into the venous circulation in the thorax.

Capsule

The renal capsule is thin but tough. If the kidney becomes edematous, the capsule limits the swelling, and the tissue pressure (**renal interstitial pressure)** rises. This decreases the glomerular filtration rate and is claimed to enhance and prolong the anuria in acute renal failure (see Chapter 33).

Innervation of the Renal Vessels

The renal nerves travel along the renal blood vessels as they enter the kidney. They contain many postganglionic sympathetic efferent fibers and a few afferent fibers. There also appears to be a cholinergic innervation via the vagus nerve, but its function is uncertain. The sympathetic preganglionic innervation comes primarily from the lower thoracic and upper lumbar segments of the spinal cord, and the cell bodies of the postganglionic neurons are in the sympathetic ganglion chain, in the superior mesenteric ganglion, and along the renal artery. The sympathetic fibers are distributed primarily to the afferent and efferent arterioles, the proximal and distal tubules, and the juxtaglomerular cells (see Chapter 24). In addition, there is a dense noradrenergic innervation of the thick ascending limb of the loop of Henle.

Nociceptive afferents that mediate pain in kidney disease parallel the sympathetic efferents and enter the spinal cord in the thoracic and upper lumbar dorsal roots. Other renal afferents presumably mediate a **renorenal reflex** by which an increase in ureteral pressure in one kidney leads to a decrease in efferent nerve activity to the contralateral kidney, and this decrease permits an increase in its excretion of Na^+ and water.

RENAL CIRCULATION

Blood Flow

In a resting adult, the kidneys receive 1.2–1.3 L of blood per minute, or just under 25% of the cardiac output (see Table 32–1). Renal blood flow can be measured with electromagnetic or other types of flow meters, or it can be determined by applying the Fick principle (see Chapter 29) to the kidney—ie, by measuring the amount of a given substance taken up per unit of time and dividing this value by the arteriovenous difference for the substance across the kidney. Since the kidney filters plasma, the **renal plasma flow** equals the amount of a substance excreted per unit of time divided by the renal arteriovenous difference as long as the amount in the red cells is unaltered during passage through the kidney. Any excreted substance can be used if its concentration in arterial and renal venous plasma can be measured and if it is not metabolized, stored, or produced by the kidney and does not itself affect blood flow.

Renal plasma flow can be measured by infusing *p*-aminohippuric acid (PAH) and determining its urine and plasma concentrations. PAH is filtered by the glomeruli and secreted by the tubular cells, so that its **extraction ratio** (arterial concentration minus renal venous concentration divided by arterial concentration) is high. For example, when PAH is infused at low doses, 90% of the PAH in arterial blood is removed in a single circulation through the kidney. It has therefore become commonplace to calculate the "renal plasma flow" by dividing the amount of PAH in the urine by the plasma PAH level, ignoring the level in renal venous blood. Peripheral venous plasma can be used because its PAH concentration is essentially identical to that in the arterial plasma reaching the kidney. The value obtained should be called the **effective renal plasma flow (ERPF)** to indicate that the level in renal venous plasma was not measured. In humans, ERPF averages about 625 mL/min.

Effective renal plasma flow (ERPF) =

$$\frac{U_{PAH}\dot{V}}{P_{PAH}} = \text{Clearance of PAH (}C_{PAH}\text{)}$$

Example:

Concentration of PAH in urine (U_{PAH}): 14 mg/mL
Urine flow ($\dot{V}$): 0.9 mL/min
Concentration of PAH in plasma (P_{PAH}): 0.02 mg/mL

$$ERPF = \frac{14 \times 0.9}{0.02} = 630 \text{ mL/min}$$

It should be noted that the ERPF determined in this way is the **clearance** of PAH. The concept of clearance is discussed in detail below.

ERPF can be converted to actual renal plasma flow (RPF):

Average PAH extraction ratio: 0.9

$$\frac{ERP}{\text{Extraction ratio}} = \frac{630}{0.9} = \text{Actual RPF} = 700 \text{ mL/min}$$

From the renal plasma flow, the renal blood flow can be calculated by dividing by 1 minus the hematocrit:

Hematocrit (Hct): 45%

$$\text{Renal blood flow} = RPF \times \frac{1}{1 - Hct}$$

$$= 700 \times \frac{1}{0.55}$$

$$= 1273 \text{ mL/min}$$

Pressure in Renal Vessels

The pressure in the glomerular capillaries has been measured directly in rats and has been found to be considerably lower than predicted on the basis of indirect measurements. When the mean systemic arterial pressure is 100 mm Hg, the glomerular capillary pressure is about 45 mm Hg. The pressure drop across the glomerulus is only 1–3 mm Hg, but there is a further drop in the efferent arteriole so that the pressure in the peritubular capillaries is about 8 mm Hg. The pressure in the renal vein is about 4 mm Hg. Pressure gradients are similar in the squirrel monkey and presumably in humans, with a glomerular capillary pressure that is about 40% of systemic arterial pressure.

Regulation of the Renal Blood Flow

Norepinephrine constricts the renal vessels, with the greatest effect of injected norepinephrine being exerted on the interlobular arteries and the afferent arterioles. Dopamine is made in the kidney and causes renal vasodilation and natriuresis. Angiotensin II exerts a greater constrictor effect on the efferent arterioles than on the afferent. Prostaglandins increase blood flow in the renal cortex and decrease blood flow in the renal medulla. Acetylcholine also produces renal vasodilation. A high-protein diet raises glomerular capillary pressure and increases renal blood flow.

Functions of the Renal Nerves

Stimulation of the renal nerves increases renin secretion by a direct action of released norepinephrine on β_1-adrenergic receptors on the juxtaglomerular cells (see Chapter 24) and it increases Na^+ reabsorption, probably by a direct action of norepinephrine on renal tubular cells. The proximal and distal tubules and the thick ascending limb of the loop of Henle are richly innervated. When the renal nerves are stimulated at increasing strengths in experimental animals, the first response is an increase in the sensitivity of the juxtaglomerular cells (Table 38–2), followed by increased renin secretion, then increased Na^+ reabsorption, and finally, at the highest threshold, renal vasoconstriction with decreased glomerular filtration and renal blood flow. It is still unsettled whether the effect on Na^+ reabsorption is mediated via α- or β-adrenergic receptors, and it may be mediated by both. The physiologic role of the renal nerves in Na^+ metabolism is also unsettled, in part because most renal functions appear to be normal in patients with transplanted kidneys, and it takes some time for transplanted kidneys to acquire a functional innervation.

Strong stimulation of the sympathetic noradrenergic nerves to the kidneys causes a marked decrease in renal

Table 38–2. Renal responses to graded renal nerve stimulation.[1]

Renal Nerve Stimulation Frequency (Hz)	RSR[2]	$U_{Na}\dot{V}$[2]	GFR	RBF[2]
0.25	No effect on basal values; augments RSR mediated by nonneural stimuli.	0	0	0
0.50	Increased without changing $U_{Na}\dot{V}$, GFR, or RBF.	0	0	0
1.0	Increased with decreased $U_{Na}\dot{V}$ without changing GFR or RBF.	↓	0	0
2.50	Increased with decreased $U_{Na}\dot{V}$, GFR, and RBF.	↓	↓	↓

[1]Reproduced from DiBona GF: Neural control of renal function: Cardiovascular implications. Hypertension 1989;13: 539. By permission of the American Heart Association, Inc.

[2]RSR, renin secretion rate; $U_{Na}\dot{V}$, urinary sodium excretion; RBF, renal blood flow.

blood flow. This effect is mediated by α_1-adrenergic receptors and to a lesser extent by postsynaptic α_2-adrenergic receptors. There is some tonic discharge in the renal nerves at rest in animals and humans. When systemic blood pressure falls, the vasoconstrictor response produced by decreased discharge in the baroreceptor nerves includes renal vasoconstriction. Renal blood flow is decreased during exercise and, to a lesser extent, on rising from the supine position.

Autoregulation of Renal Blood Flow

When the kidney is perfused at moderate pressures (90–220 mm Hg in the dog), the renal vascular resistance varies with the pressure so that renal blood flow is relatively constant (Figure 38–4). Autoregulation of this type occurs in other organs, and several factors contribute to it (see Chapter 31). Renal autoregulation is present in denervated and in isolated, perfused kidneys but is prevented by the administration of drugs that paralyze vascular smooth muscle. It is probably produced in part by a direct contractile response of the

smooth muscle of the afferent arteriole to stretch. NO may also be involved. At low perfusion pressures, angiotensin II also appears to play a role by constricting the efferent arterioles, thus maintaining the glomerular filtration rate. This is believed to be the explanation of the renal failure that sometimes develops in patients with poor renal perfusion who are treated with drugs which inhibit angiotensin-converting enzyme.

Regional Blood Flow & Oxygen Consumption

The main function of the renal cortex is filtration of large volumes of blood through the glomeruli, so it is not surprising that the renal cortical blood flow is relatively great and little oxygen is extracted from the blood. Cortical blood flow is about 5 mL/g of kidney tissue/min (compared with 0.5 mL/g/min in the brain), and the arteriovenous oxygen difference for the whole kidney is only 14 mL/L of blood, compared with 62 mL/L for the brain and 114 mL/L for the heart (see Table 32–1). The P_{O_2} of the cortex is about 50 mm Hg. On the other hand, maintenance of the osmotic gradient in the medulla (see below) requires a relatively low blood flow. It is not surprising, therefore, that the blood flow is about 2.5 mL/g/min in the outer medulla and 0.6 mL/g/min in the inner medulla. However, metabolic work is being done, particularly to reabsorb Na^+ in the thick ascending limb of Henle (see below), so relatively large amounts of O_2 are extracted from the blood in the medulla. The P_{O_2} of the medulla is about 15 mm Hg. This makes the medulla vulnerable to hypoxia if flow is reduced further. NO, prostaglandins, and many cardiovascular peptides in this region function in a paracrine fashion to maintain the balance between low blood flow and metabolic needs.

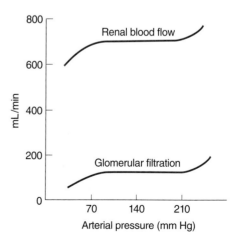

Figure 38–4. Autoregulation in the kidneys.

GLOMERULAR FILTRATION

Measuring GFR

The **glomerular filtration rate (GFR)** can be measured in intact animals and humans by measuring the excretion and plasma level of a substance that is freely filtered through the glomeruli and neither secreted nor reabsorbed by the tubules. The amount of such a substance in the urine per unit of time must have been provided by filtering exactly the number of milliliters of plasma that contained this amount. Therefore, if the substance is designated by the letter X, the GFR is equal to the concentration of X in urine (U_X) times the **urine flow** per unit of time ($\dot{V}$) divided by the **arterial plasma level** of X (P_X), or $U_X\dot{V}/P_X$. This value is called the clearance of X (C_X). P_X is, of course, the same in all parts of the arterial circulation, and if X is not metabolized to any extent in the tissues, the level of X in peripheral venous plasma can be substituted for the arterial plasma level.

Substances Used to Measure GFR

In addition to the requirement that it be freely filtered and neither reabsorbed nor secreted in the tubules, a substance suitable for measuring the GFR should be nontoxic and not metabolized by the body. Inulin, a polymer of fructose with a molecular weight of 5200 that is found in dahlia tubers, meets these criteria in humans and most animals and is extensively used to measure GFR. Radioisotopes such as ^{51}Cr-EDTA are also used, but inulin remains the standard substance. In practice, a loading dose of inulin is administered intravenously, followed by a sustaining infusion to keep the arterial plasma level constant. After the inulin has equilibrated with body fluids, an accurately timed urine specimen is collected and a plasma sample obtained halfway through the collection. Plasma and urinary inulin concentrations are determined and the clearance calculated.

Example:

$$U_{IN} = 35 \text{ mg/mL}$$

$$\dot{V} = 0.9 \text{ mL/min}$$

$$P_{IN} = 0.25 \text{ mg/mL}$$

$$C_{IN} = \frac{U_{IN}\dot{V}}{P_{IN}} = \frac{35 \times 0.9}{0.25}$$

$$C_{IN} = 126 \text{ mL/min}$$

In dogs, cats, rabbits, and a number of other mammalian species, clearance of creatinine (C_{Cr}) can also be used to determine the GFR, but in primates, including humans, some creatinine is secreted by the tubules and some may be reabsorbed. In addition, plasma creatinine determinations are inaccurate at low creatinine levels because the method for determining creatinine measures small amounts of other plasma constituents. In spite of this, the clearance of endogenous creatinine is frequently measured in patients. The values agree quite well with the GFR values measured with inulin because, although the value for $U_{Cr}\dot{V}$ is high as a result of tubular secretion, the value for P_{Cr} is also high as a result of nonspecific chromogens, and the errors thus tend to cancel. Endogenous creatinine clearance is easy to measure and is a worthwhile index of renal function, but when precise measurements of GFR are needed it seems unwise to rely on a method that owes what accuracy it has to compensating errors.

Normal GFR

The GFR in an average-sized normal man is approximately 125 mL/min. Its magnitude correlates fairly well with surface area, but values in women are 10% lower than those in men even after correction for surface area. A rate of 125 mL/min is 7.5 L/h, or 180 L/d, whereas the normal urine volume is about 1 L/d. Thus, 99% or more of the filtrate is normally reabsorbed. At the rate of 125 mL/min, the kidneys filter in 1 day an amount of fluid equal to 4 times the total body water, 15 times the ECF volume, and 60 times the plasma volume.

Control of GFR

The factors governing filtration across the glomerular capillaries are the same as those governing filtration across all other capillaries (see Chapter 30), ie, the size of the capillary bed, the permeability of the capillaries, and the hydrostatic and osmotic pressure gradients across the capillary wall. For each nephron:

$$GFR = K_f [(P_{GC} - P_T) - (\pi_{GC} - \pi_T)]$$

K_f, the glomerular ultrafiltration coefficient, is the product of the glomerular capillary wall hydraulic conductivity (ie, its permeability) and the effective filtration surface area. P_{GC} is the mean hydrostatic pressure in the glomerular capillaries, P_T the mean hydrostatic pressure in the tubule, π_{GC} the osmotic pressure of the plasma in the glomerular capillaries, and π_T the osmotic pressure of the filtrate in the tubule.

Permeability

The permeability of the glomerular capillaries is about 50 times that of the capillaries in skeletal muscle. Neutral substances with effective molecular diameters of less than 4 nm are freely filtered, and the filtration of neutral substances with diameters of more than 8 nm approaches zero (Figure 38–5). Between these values, filtration is inversely proportionate to diameter. However, sialoproteins in the glomerular capillary wall are negatively charged, and studies with anionically charged and cationically charged dextrans indicate that the negative charges repel negatively charged substances in blood, with the result that filtration of anionic substances 4 nm in diameter is less than half that of neutral substances of the same size. This probably explains why albumin, with an effective molecular diameter of approximately 7 nm, normally has a glomerular concentration only 0.2% of its plasma concentration rather than the higher concentration that would be expected on the basis of diameter alone; circulating albumin is negatively charged. Filtration of cationic substances is greater than that of neutral substances.

The amount of protein in the urine is normally less than 100 mg/d, and most of this is not filtered but comes from shed tubular cells. The presence of significant amounts of albumin in the urine is called **albuminuria.** In nephritis, the negative charges in the glomerular wall are dissipated, and albuminuria can occur for this reason without an increase in the size of the "pores" in the membrane.

Size of the Capillary Bed

K_f can be altered by the mesangial cells, contraction of these cells producing a decrease in K_f that is largely due

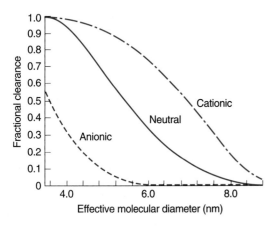

Figure 38–5. Effect of electrical charge on the fractional clearance of dextran molecules of various sizes in rats. The negative charges in the glomerular membrane retard the passage of negatively charged molecules (anionic dextran) and facilitate the passage of positively charged molecules (cationic dextran). (Reproduced, with permission, from Brenner BM, Beeuwkes R: The renal circulations. Hosp Pract [July] 1978;13:35.)

Table 38–3. Agents causing contraction or relaxation of mesangial cells.

Contraction	Relaxation
Endothelins	ANP
Angiotensin II	Dopamine
Vasopressin	PGE_2
Norepinephrine	cAMP
Platelet-activating factor	
Platelet-derived growth factor	
Thromboxane A_2	
PGF_2	
Leukotrienes C_4 and D_4	
Histamine	

to a reduction in the area available for filtration. Contraction of points where the capillary loops bifurcate probably shifts flow away from some of the loops, and elsewhere, contracted mesangial cells distort and encroach on the capillary lumen. Agents that have been shown to affect the mesangial cells are listed in Table 38–3. Angiotensin II is an important regulator of mesangial contraction, and there are angiotensin II receptors in the glomeruli. In addition, there is some evidence that mesangial cells make renin.

Hydrostatic & Osmotic Pressure

The pressure in the glomerular capillaries is higher than that in other capillary beds because the afferent arterioles are short, straight branches of the interlobular arteries. Furthermore, the vessels "downstream" from the glomeruli, the efferent arterioles, have a relatively high resistance. The capillary hydrostatic pressure is opposed by the hydrostatic pressure in Bowman's capsule. It is also opposed by the osmotic pressure gradient across the glomerular capillaries ($\pi_{GC} - \pi_T$). π_T is normally negligible, and the gradient is equal to the oncotic pressure of the plasma proteins.

The actual pressures in one strain of rats are shown in Figure 38–6. The net filtration pressure (P_{UF}) is 15 mm Hg at the afferent end of the glomerular capillaries, but it falls to zero—ie, filtration equilibrium is reached—proximal to the efferent end of the glomerular capillaries. This is because fluid leaves the plasma and the oncotic pressure rises as blood passes through the glomerular capillaries. The calculated change in $\Delta\pi$ along an idealized glomerular capillary is shown in Figure 38–6. It is apparent that portions of the glomerular capillaries do not normally contribute to the formation of the glomerular ultrafiltrate; ie, exchange across the glomerular capillaries is flow-limited rather than diffusion-limited (see Chapter 30). It is also apparent that a decrease in the rate of rise of the $\Delta\pi$ curve produced by an increase in renal plasma flow would increase filtration because it would increase the distance along the capillary in which filtration was taking place.

There is considerable species variation in whether filtration equilibrium is reached, and there are some uncertainties in the measurement of K_f. It is uncertain whether filtration equilibrium is reached in humans.

Changes in GFR

Variations in the factors discussed in the preceding paragraphs and listed in Table 38–4 have predictable effects on the GFR. Changes in renal vascular resistance as a result of autoregulation tend to stabilize filtration pressure, but when the mean systemic arterial pressure drops below 90 mm Hg, there is a sharp drop in GFR. The GFR tends to be maintained when efferent arteriolar constriction is greater than afferent constriction, but either type of constriction decreases blood flow to the tubules.

Filtration Fraction

The ratio of the GFR to the renal plasma flow (RPF), the **filtration fraction,** is normally 0.16–0.20. The

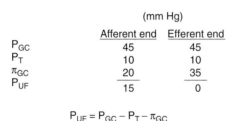

	(mm Hg)	
	Afferent end	Efferent end
P_{GC}	45	45
P_T	10	10
π_{GC}	20	35
P_{UF}	15	0

$$P_{UF} = P_{GC} - P_T - \pi_{GC}$$

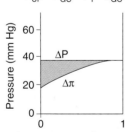

Dimensionless distance along idealized glomerular capillary

Figure 38–6. Hydrostatic pressure (P_{GC}) and osmotic pressure (π_{GC}) in a glomerular capillary in the rat. P_T, pressure in Bowman's capsule; P_{UF}, net filtration pressure. π_T is normally negligible, so $\Delta\pi = \pi_{GC}$. $\Delta P = P_{GC} - P_T$. (Reproduced, with permission, from Mercer PF, Maddox DA, Brenner BM: Current concepts of sodium chloride and water transport by the mammalian nephron. West J Med 1974;120:33.)

Table 38–4. Factors affecting the GFR.

Changes in renal blood flow
Changes in glomerular capillary hydrostatic pressure
  Changes in systemic blood pressure
  Afferent or efferent arteriolar constriction
Changes in hydrostatic pressure in Bowman's capsule
  Ureteral obstruction
  Edema of kidney inside tight renal capsule
Changes in concentration of plasma proteins:
  dehydration, hypoproteinemia, etc (minor factors)
Changes in K_f
  Changes in glomerular capillary permeability
  Changes in effective filtration surface area

GFR varies less than the RPF. When there is a fall in systemic blood pressure, the GFR falls less than the RPF because of efferent arteriolar constriction, and consequently the filtration fraction rises.

TUBULAR FUNCTION

General Considerations

The amount of any substance (X) that is filtered is the product of the GFR and the plasma level of the substance ($C_{In}P_X$). The tubular cells may add more of the substance to the filtrate (tubular secretion), may remove some or all of the substance from the filtrate (tubular reabsorption), or may do both. The amount of the substance excreted per unit time ($U_X\dot{V}$) equals the amount filtered plus the **net amount transferred** by the tubules. This latter quantity is conveniently indicated by the symbol T_X (Figure 38–7). The clearance of the substance equals the GFR if there is no net tubular secretion or reabsorption, exceeds the GFR if there is net tubular secretion, and is less than the GFR if there is net tubular reabsorption.

Much of our knowledge about glomerular filtration and tubular function has been obtained by using micropuncture techniques. Micropipettes can be inserted into the tubules of the living kidney and the composition of aspirated tubular fluid determined by the use of microchemical techniques. In addition, two pipettes can be inserted in a tubule and the tubule perfused in vivo. Alternatively, isolated perfused segments of tubules can be studied in vitro, and tubular cells can be grown and studied in culture.

Mechanisms of Tubular Reabsorption & Secretion

Small proteins and some peptide hormones are reabsorbed in the proximal tubules by endocytosis. Other

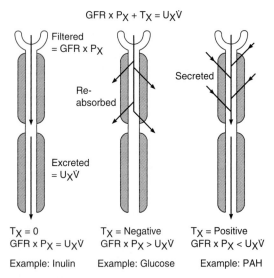

$$GFR \times P_X + T_X = U_X\dot{V}$$

Filtered
= GFR × P$_X$

Re-
absorbed

Secreted

Excreted
= U$_X$\dot{V}

$T_X = 0$
$GFR \times P_X = U_X\dot{V}$

Example: Inulin

T_X = Negative
$GFR \times P_X > U_X\dot{V}$

Example: Glucose

T_X = Positive
$GFR \times P_X < U_X\dot{V}$

Example: PAH

Figure 38–7. Tubular function. For explanation of symbols, see text.

substances are secreted or reabsorbed in the tubules by passive diffusion between cells and through cells by facilitated diffusion down chemical or electrical gradients or active transport against such gradients (see Chapter 1). Movement is by way of ion channels, exchangers, cotransporters, and pumps. Many of these have now been cloned, and their regulation is being studied. Mutations of individual genes for many of them cause specific syndromes such as Dent's disease, Bartter's syndrome, and Liddle's syndrome, and a large number of mutations have been described. An interesting example involves the proteins polycystin-1 (PKD-1) and polycystin-2 (PKD-2). PKD-1 appears to be a Ca^{2+} receptor that activates a nonspecific ion channel associated with PKD-2. The normal function of this apparent ion channel is unknown, but both proteins are abnormal in **autosomal dominant polycystic kidney disease,** in which renal parenchyma is progressively replaced by fluid-filled cysts until there is complete renal failure.

It is important to note that the pumps and other units in the luminal membrane are different from those in the basolateral membrane. It is this different distribution that makes possible net movement of solutes across the epithelia.

Like transport systems elsewhere, renal active transport systems have a maximal rate, or **transport maximum (Tm),** at which they can transport a particular solute. Thus, the amount of a particular solute transported is proportionate to the amount present up to the Tm for the solute, but at higher concentrations, the

transport mechanism is **saturated** and there is no appreciable increment in the amount transported. However, the Tms for some systems are high, and it is difficult to saturate them.

It should also be noted that the tubular epithelium, like that of the small intestine and gallbladder, is a **leaky epithelium** in that the tight junctions between cells (Figure 38–8) permit the passage of some water and electrolytes. The degree to which leakage by this **paracellular pathway** contributes to the net flux of fluid and solute into and out of the tubules is controversial since it is difficult to measure, but current evidence seems to suggest that it is a significant factor. One indication of this is that paracellin-1, a protein localized to tight junctions, is related to Mg^{2+} reabsorption, and a loss-of-function mutation of its gene causes severe Mg^{2+} and Ca^{2+} loss in the urine.

The effects of tubular reabsorption and secretion on substances of major physiologic interest are summarized in Table 38–5.

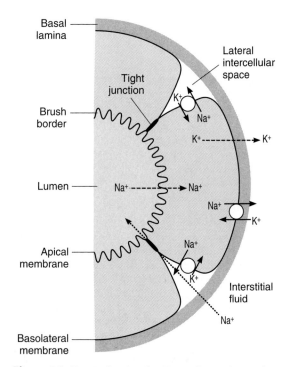

Basal lamina

Lateral intercellular space

Tight junction

Brush border

Lumen

Apical membrane

Basolateral membrane

Interstitial fluid

Figure 38–8. Mechanism for Na^+ reabsorption in the proximal tubule. Solid lines indicate active transport; dashed lines indicate cotransport; and the dotted line indicates passive diffusion. Note that Na^+ moves from the lumen into the cells by cotransport and that Na^+ and H_2O diffuse into the tubular lumen at the intercellular tight junctions.

Table 38–5. Renal handling of various plasma constituents in a normal adult human on an average diet.

Substance	Per 24 Hours				Percentage Reabsorbed	Location[1]
	Filtered	Reabsorbed	Secreted	Excreted		
Na$^+$ (meq)	26,000	25,850		150	99.4	P, L, D, C
K$^+$ (meq)	600	560[2]	50[2]	90	93.3	P, L, D, C
Cl$^-$ (meq)	18,000	17,850		150	99.2	P, L, D, C
HCO$_3^-$ (meq)	4,900	4,900		0	100	P, D
Urea (mmol)	870	460[3]		410	53	P, L, D, C
Creatinine (mmol)	12	1[4]	1[4]	12		
Uric acid (mmol)	50	49	4	5	98	P
Glucose (mmol)	800	800		0	100	P
Total solute (mosm)	54,000	53,400	100	700	98.9	P, L, D, C
Water (mL)	180,000	179,000		1000	99.4	P, L, D, C

[1]P, proximal tubules; L, loops of Henle; D, distal tubules; C, collecting ducts.
[2]K$^+$ is both reabsorbed and secreted.
[3]Urea moves into as well as out of some portions of the nephron.
[4]Variable secretion and probable reabsorption of creatinine in humans.

Na$^+$ Reabsorption

The reabsorption of Na$^+$ and Cl$^-$ plays a major role in body electrolyte and water metabolism. In addition, Na$^+$ transport is coupled to the movement of H$^+$, other electrolytes, glucose, amino acids, organic acids, phosphate, and other substances across the tubule walls. The principal cotransporters and exchangers in the various parts of the nephron are listed in Table 38–6. In the proximal tubules, the thick portion of the ascending limb of the loop of Henle, the distal tubules, and the collecting ducts, Na$^+$ moves by cotransport or exchange from the tubular lumen into the tubular epithelial cells down its concentration and electrical gradients and is actively pumped from these cells into the interstitial space. Thus, Na$^+$ is actively transported out of all parts of the renal tubule except the thin portions of the loop of Henle. Na$^+$ is pumped into the interstitium by Na$^+$-K$^+$ ATPase. The operation of this ubiquitous Na$^+$ pump is considered in detail in Chapter 1. It extrudes three Na$^+$ in exchange for two K$^+$ that are pumped into the cell.

The tubular cells are connected by tight junctions at their luminal edges, but there is space between the cells along the rest of their lateral borders. Much of the Na$^+$ is actively transported into these extensions of the interstitial space, the **lateral intercellular spaces** (Figure 38–8).

Proximal tubular reabsorbate is slightly hypertonic, and water moves passively along the osmotic gradient created by its absorption into tubular epithelial cells. It is now known that the apical membranes of proximal tubule cells contain water channels which aid the movement of water (see below). From the cells, the water moves into the lateral intercellular spaces. The rate at which solutes and water move into the capillaries from the lateral intercellular spaces and the rest of the interstitium is determined by the Starling forces determining movement across the walls of all capillaries, ie, the hydrostatic and osmotic pressures in the interstitium and the capillaries (see Chapter 30). Na$^+$ and H$_2$O leak back to the tubular lumen via the intercellular junctions, especially when the lateral intercellular spaces are distended.

Glucose Reabsorption

Glucose, amino acids, and bicarbonate are reabsorbed along with Na$^+$ in the early portion of the proximal tubule (Figure 38–9). Farther along the tubule, Na$^+$ is reabsorbed with Cl$^-$. Glucose is typical of substances removed from the urine by secondary active transport. It is filtered at a rate of approximately 100 mg/min (80 mg/dL of plasma × 125 mL/min). Essentially all of the glucose is reabsorbed, and no more than a few milligrams appear in the urine per 24 hours. The amount reabsorbed is proportionate to the amount filtered and

Table 38–6. Transport proteins involved in the movement of Na^+ and Cl^- across the apical membranes of renal tubular cells.[1]

Site	Apical Transporter	Function
Proximal tubule	Na^+/glucose CT	Na^+ uptake, glucose uptake
	Na^+/Pi CT	Na^+ uptake, Pi uptake
	Na^+ amino acid CT	Na^+ uptake, amino acid uptake
	Na^+/lactate CT	Na^+ uptake, lactate uptake
	Na^+/H^+ exchanger	Na^+ uptake, H extrusion
	Cl^-/base exchanger	Cl^- uptake
Thick ascending limb	Na^+, $2Cl^-$, K^+ CT	Na^+ uptake, Cl^- uptake, K^+ uptake
	Na^+/H^+ exchanger	Na^+ uptake, H^+ extrusion
	K^+ channels	K^+ extrusion (recycling)
Distal convoluted tubule	NaCl CT	Na^+ uptake, Cl^- uptake
Collecting duct	Na^+ channel (ENaC)	Na^+ uptake

[1]Uptake indicates movement from tubular lumen to cell interior, extrusion is movement from cell interior to tubular lumen. CT, cotransporter; Pi, inorganic phosphate. (Reproduced, with permission, from Schnermann JB, Sayegh EI: *Kidney Physiology.* Lippincott-Raven, 1998.)

hence to the plasma glucose level (P_G) times the GFR up to the transport maximum (Tm_G); but when the Tm_G is exceeded, the amount of glucose in the urine rises (Figure 38–10). The Tm_G is about 375 mg/min in men and 300 mg/min in women.

The **renal threshold** for glucose is the plasma level at which the glucose first appears in the urine in more than the normal minute amounts. One would predict that the renal threshold would be about 300 mg/dL— ie, 375 mg/min (Tm_G) divided by 125 mL/min (GFR). However, the actual renal threshold is about 200 mg/dL of arterial plasma, which corresponds to a venous level of about 180 mg/dL. Figure 38–10 shows why the actual renal threshold is less than the predicted threshold. The "ideal" curve shown in this diagram would be obtained if the Tm_G in all the tubules was identical and if all the glucose were removed from each tubule when the amount filtered was below the Tm_G. This is not the case, and in humans, for example, the actual curve is rounded and deviates considerably from the "ideal" curve. This deviation is called **splay**. The magnitude of the splay is inversely proportionate to the avidity with which the transport mechanism binds the substance it transports.

Glucose Transport Mechanism

Glucose reabsorption in the kidneys is similar to glucose reabsorption in the intestine (see Chapter 25). Glucose and Na^+ bind to the common carrier SGLT 2 in the luminal membrane (Figure 38–11), and glucose is carried into the cell as Na^+ moves down its electrical and chemical gradient. The Na^+ is then pumped out of the cell into the lateral intercellular spaces, and the glu-

cose is transported by GLUT 2 into the interstitial fluid. At least in the rat, there is some transport by SGLT 1 and GLUT 1 as well. Thus, glucose transport in the kidneys as well as in the intestine is an example of secondary active transport.

The common carrier specifically binds the d isomer of glucose, and the rate of transport of d-glucose is many times greater than that of l-glucose. Glucose transport in the kidneys is inhibited, as it is in the intestine, by the plant glucoside **phlorhizin,** which competes with d-glucose for binding to the carrier.

Additional Examples of Secondary Active Transport

A variety of other substances are transported by secondary active transport via symports, with the energy provided by active transport of Na^+ out of the renal tubular cells. These substances include some amino acids, lactate, inorganic phosphate (Pi), H^+, and Cl^-.

Like glucose reabsorption, amino acid reabsorption is most marked in the early portion of the proximal convoluted tubule. Absorption in this location resembles absorption in the intestine (see Chapter 25). The main carriers in the luminal membrane cotransport Na^+, whereas the carriers in the basolateral membranes are not Na^+-dependent. Na^+ is pumped out of the cells by Na^+-K^+ ATPase and the amino acids leave by passive or facilitated diffusion to the interstitial fluid.

Some Cl^- is reabsorbed with Na^+ and K^+ in the thick ascending limb of the loop of Henle (see below). In addition, two members of a family of Cl^- channels have been identified in the kidney. The family is characterized by 12 transmembrane domains, and members

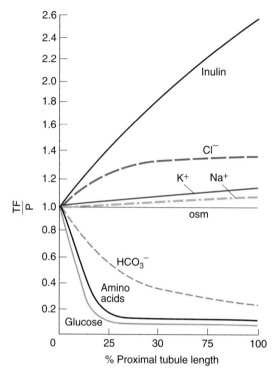

Figure 38–9. Reabsorption of various solutes in the proximal tubule. TF/P, tubular fluid:plasma concentration ratio. (Courtesy of FC Rector Jr.)

of it are also found in muscle and other tissues. Mutations in the gene for one of the renal channels is associated with Ca^{2+}-containing kidney stones and hypercalciuria **(Dent's disease),** but how tubular transport of Ca^{2+} and Cl^- are linked is still unsettled.

PAH Transport

The dynamics of PAH transport illustrate the operation of the active transport mechanisms that secrete substances into the tubular fluid. The filtered load of PAH is a linear function of the plasma level, but PAH secretion increases as P_{PAH} rises only until a maximal secretion rate (Tm_{PAH}) is reached (Figure 38–12). When P_{PAH} is low, C_{PAH} is high; but as P_{PAH} rises above Tm_{PAH}, C_{PAH} falls progressively. It eventually approaches the clearance of inulin (C_{In}) (Figure 38–13), because the amount of PAH secreted becomes a smaller and smaller fraction of the total amount excreted. Conversely, the clearance of glucose is essentially zero at P_G levels below the renal threshold; but above the threshold, C_G rises to approach C_{In} as P_G is raised.

The use of C_{PAH} to measure effective renal plasma flow (ERPF) is discussed above.

Other Substances Secreted by the Tubules

Derivatives of hippuric acid in addition to PAH, phenol red and other sulfonphthalein dyes, penicillin, and a variety of iodinated dyes are actively secreted into the tubular fluid. Substances that are normally produced in the body and secreted by the tubules include various ethereal sulfates, steroid and other glucuronides, and 5-hydroxyindoleacetic acid, the principal metabolite of serotonin (see Chapter 4).

Tubuloglomerular Feedback & Glomerulotubular Balance

Signals from the renal tubules feed back to affect glomerular filtration. As the rate of flow through the ascending limb of the loop of Henle and first part of the distal tubule increases, glomerular filtration in the same nephron decreases, and, conversely, a decrease in flow increases the GFR (Figure 38–14). This process, which is called **tubuloglomerular feedback,** tends to maintain the constancy of the load delivered to the distal tubule. The sensor for the response appears to be the macula densa, and GFR is adjusted by constriction or dilation of the afferent arteriole. Constriction may be mediated by thromboxane A_2.

Conversely, an increase in GFR causes an increase in the reabsorption of solutes, and consequently of water,

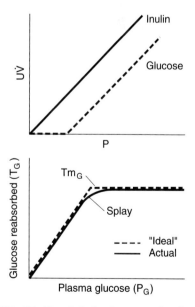

Figure 38–10. Top: Relation between the plasma level (P) and excretion (UV̇) of glucose and inulin. **Bottom:** Relation between the plasma glucose level (P_G) and amount of glucose reabsorbed (T_G).

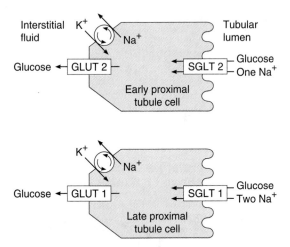

Figure 38–11. Glucose reabsorption. In the early portion of the proximal tubule, glucose and Na⁺ are cotransported from the tubular lumen by SGLT 2, Na⁺ is pumped out of the tubular cells by Na⁺-K⁺ ATPase in the basolateral membranes, and glucose is transported to the interstitium by GLUT 2. At least in the rat, some glucose is also removed by SGLT 1 and GLUT 1 in the terminal, straight portions of the proximal tubules. SGLT 1 transports two Na⁺ for each glucose, whereas SGLT 2 transports one Na⁺.

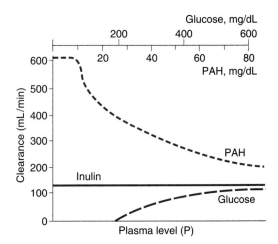

Figure 38–13. Clearance of inulin, glucose, and PAH at various plasma levels of each substance in humans.

there is a relatively large increase in the oncotic pressure of the plasma by the time it reaches the efferent arterioles and their capillary branches. This increases the reabsorption of Na⁺ from the tubule. However, other as yet unidentified intrarenal mechanisms are also involved.

primarily in the proximal tubule, so that in general the percentage of the solute reabsorbed is held constant. This process is called **glomerulotubular balance,** and it is particularly prominent for Na⁺. The change in Na⁺ reabsorption occurs within seconds after a change in filtration, so it seems unlikely that an extrarenal humoral factor is involved. One factor is the oncotic pressure in the peritubular capillaries. When the GFR is high,

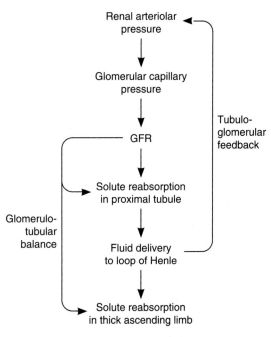

Figure 38–14. Mechanisms of glomerulotubular balance and tubuloglomerular feedback.

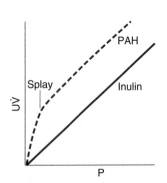

Figure 38–12. Relation between plasma levels (P) and excretion (U̇V) of PAH and inulin.

WATER EXCRETION

Normally, 180 L of fluid is filtered through the glomeruli each day, while the average daily urine volume is about 1 L. The same load of solute can be excreted per 24 hours in a urine volume of 500 mL with a concentration of 1400 mosm/kg or in a volume of 23.3 L with a concentration of 30 mosm/kg (Table 38–7). These figures demonstrate two important facts: first, that at least 87% of the filtered water is reabsorbed, even when the urine volume is 23 L; and second, that the reabsorption of the remainder of the filtered water can be varied without affecting total solute excretion. Therefore, when the urine is concentrated, water is retained in excess of solute; and when it is dilute, water is lost from the body in excess of solute. Both facts have great importance in the body economy and the regulation of the osmolality of the body fluids. A key regulator of water output is vasopressin acting on the collecting ducts.

Aquaporins

Research on mice, rats, and humans indicates that rapid diffusion of water across cell membranes depends on water channels made up of proteins called **aquaporins.** Four aquaporins—aquaporin-1, aquaporin-2, aquaporin-5, and aquaporin-9—have been characterized in humans, and additional aquaporins have been identified in rats. Most are found in the kidneys, though aquaporin-9 is found in human leukocytes, liver, lung, and spleen; and aquaporin-5 is found in human lacrimal glands. The key roles played by aquaporin-1 and aquaporin-2 in water excretion are discussed below.

Proximal Tubule

Many substances are actively transported out of the fluid in the proximal tubule, but fluid obtained by micropuncture remains essentially isosmotic to the end of

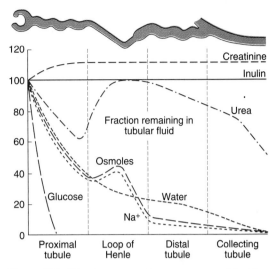

Figure 38–15. Changes in the percentage of the filtered amount of substances remaining in the tubular fluid along the length of the nephron in the presence of vasopressin. (Modified from Sullivan LP, Grantham JJ: *Physiology of the Kidney,* 2nd ed. Lea & Febiger, 1982.)

the proximal tubule (Figure 38–9). Therefore, in the proximal tubule, water moves passively out of the tubule along the osmotic gradients set up by active transport of solutes, and isotonicity is maintained. Since the ratio of the concentration in tubular fluid to the concentration in plasma (TF/P) of the nonreabsorbable substance inulin is 2.5–3.3 at the end of the proximal tubule, it follows that 60–70% of the filtered solute and 60–70% of the filtered water have been removed by the time the filtrate reaches this point (Figure 38–15).

Aquaporin-1 is localized in the proximal tubules. When it is knocked out in mice, proximal tubular water permeability is reduced 80%, and their plasma os-

Table 38–7. Alterations in water metabolism produced by vasopressin in humans. In each case, the osmotic load excreted is 700 mosm/d.

	GFR (mL/min)	Percentage of Filtered Water Reabsorbed	Urine Volume (L/d)	Urine Concentration (mosm/kg H_2O)	Gain or Loss of Water in Excess of Solute (L/d)
Urine isotonic to plasma	125	98.7	2.4	290	...
Vasopressin (maximal antidiuresis)	125	99.7	0.5	1400	1.9 gain
No vasopressin ("complete" diabetes insipidus)	125	87.1	23.3	30	20.9 loss

molality increases to 500 mosm/kg when the mice are subjected to dehydration even though their other aquaporins are intact. In humans with mutations that eliminate aquaporin-1 activity, the defect in water metabolism is not as severe, though their response to dehydration is defective.

Loop of Henle

As noted above, the loops of Henle of the juxtamedullary nephrons dip deeply into the medullary pyramids before draining into the distal convoluted tubules in the cortex, and all of the collecting ducts descend back through the medullary pyramids to drain at the tips of the pyramids into the renal pelvis. There is a graded increase in the osmolality of the interstitium of the pyramids, the osmolality at the tips of the papillae normally being about 1200 mosm/kg of H_2O, approximately four times that of plasma. The descending limb of the loop of Henle is permeable to water, but the ascending limb is impermeable (Table 38–8). Na^+, K^+,

Table 38–8. Permeability and transport in various segments of the nephron.[1]

	Permeability			Active Transport of Na^+
	H_2O	Urea	NaCl	
Loop of Henle				
Thin descending limb	4+	+	±	0
Thin ascending limb	0	+	4+	0
Thick ascending limb	0	±	±	4+
Distal convoluted tubule	±	±	±	3+
Collecting tubule				
Cortical portion	3+*	0	±	2+
Outer medullary portion	3+*	0	±	1+
Inner medullary portion	3+*	3+	±	1+

[1]Data are based on studies of rabbit and human kidneys. Values indicated by asterisks are in the presence of vasopressin. These values are 1+ in the absence of vasopressin. (Modified and reproduced, with permission, from Kokko JP: Renal concentrating and diluting mechanisms. Hosp Pract [Feb] 1979;110:14.)

and Cl^- are cotransported out of the thick segment of the ascending limb (see below). Therefore, the fluid in the descending limb of the loop of Henle becomes hypertonic as water moves into the hypertonic interstitium. In the ascending limb it becomes more dilute, and when it reaches the top it is hypotonic to plasma because of the movement of Na^+ and Cl^- out of the tubular lumen. In passing through the loop of Henle, another 15% of the filtered water is removed, so approximately 20% of the filtered water enters the distal tubule, and the TF/P of inulin at this point is about 5.

In the thick ascending limb, a carrier cotransports one Na^+, one K^+, and 2 Cl^- from the tubular lumen into the tubular cells. This is another example of secondary active transport; the Na^+ is actively transported from the cells into the interstitium by Na^+-K^+ ATPase in the basolateral membranes of the cells, keeping the intracellular Na^+ low. The Na^+-K^+-$2Cl^-$ transporter has 12 transmembrane domains with intracellular amino and carboxyl terminals. It is a member of a family of transporters found in many other locations, including salivary glands, the gastrointestinal tract, and the airways.

The K^+ diffuses back into the tubular lumen and back into the interstitium via ROMK and other K^+ channels. The Cl^- diffuses into the interstitium via ClC-Kb channels (Figure 38–16). This K^+ recycles across the luminal and the basolateral membrane, while there is no transport of Na^+ and Cl^- into the interstitium.

Bartter's Syndrome

Bartter's syndrome is a rare but interesting condition that is due to defective transport in the thick ascending limb. It is characterized by chronic Na^+ loss in the urine, with resultant hypovolemia causing stimulation of renin and aldosterone secretion without hypertension, plus hyperkalemia and alkalosis. The condition can be caused by loss-of-function mutations in the gene for any of four key proteins: the Na^+-K^+-$2Cl^-$ cotransporter, the ROMK K^+ channel, the ClC-Kb Cl^- channel, or **barttin**, a recently described integral membrane protein that is necessary for the normal function of ClC-Kb Cl^- channels.

The stria vascularis in the inner ear is responsible for maintaining the high K^+ concentration in the scala media that is essential for normal hearing (see Chapter 9). It contains both ClC-Kb and ClC-Ka Cl^- channels. Bartter's syndrome associated with mutated ClC Kb channels is not associated with deafness because the Clc-Ka channels can carry the load. However, both types of Cl^- channels are barttin-dependent, so patients with Bartter's syndrome due to mutated barttin are also deaf.

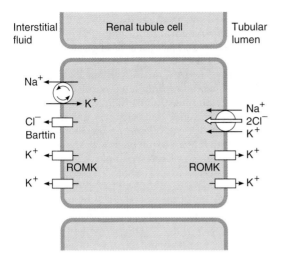

Figure 38–16. NaCl transport in the thick ascending limb of the loop of Henle. The Na+, K+, 2Cl– cotransporter moves these ions into the tubular cell by secondary active transport. Na+ is transported out of the cell into the interstitium by Na+-K+ ATPase in the basolateral membrane of the cell. Cl– exits in basolateral ClC-Kb Cl– channels. Barttin, a protein in the cell membrane, is essential for normal ClC-Kb function. K+ moves from the cell to the interstitium and the tubular lumen by ROMK and other K+ channels.

Distal Tubule

The distal tubule, particularly its first part, is in effect an extension of the thick segment of the ascending limb. It is relatively impermeable to water, and continued removal of the solute in excess of solvent further dilutes the tubular fluid. About 5% of the filtered water is removed in this segment.

Collecting Ducts

The collecting ducts have two portions: a cortical portion and a medullary portion. The changes in osmolality and volume in the collecting ducts depend on the amount of vasopressin acting on the ducts. This antidiuretic hormone from the posterior pituitary gland increases the permeability of the collecting ducts to water. The key to the action of vasopressin on the collecting ducts is aquaporin-2. Unlike the other aquaporins, this aquaporin is stored in vesicles in the cytoplasm of principal cells. Vasopressin causes rapid insertion of these vesicles into the apical membrane of cells. The effect is mediated via the vasopressin V_2 receptor, cyclic AMP, protein kinase A, and a molecular motor, one of the dyneins (see Chapter 1).

In the presence of enough vasopressin to produce maximal antidiuresis, water moves out of the hypotonic fluid entering the cortical collecting ducts into the interstitium of the cortex, and the tubular fluid becomes isotonic. In this fashion, as much as 10% of the filtered water is removed. The isotonic fluid then enters the medullary collecting ducts with a TF/P inulin of about 20. An additional 4.7% or more of the filtrate is reabsorbed into the hypertonic interstitium of the medulla, producing a concentrated urine with a TF/P inulin of over 300. In humans, the osmolality of urine may reach 1400 mosm/kg of H_2O, almost five times the osmolality of plasma, with a total of 99.7% of the filtered water being reabsorbed (Table 38–7). In other species, the ability to concentrate urine is even greater. Maximal urine osmolality is about 2500 mosm/kg in dogs, about 3200 mosm/kg in laboratory rats, and as high as 5000 mosm/kg in certain desert rodents.

When vasopressin is absent, the collecting duct epithelium is relatively impermeable to water. The fluid therefore remains hypotonic, and large amounts flow into the renal pelvis. In humans, the urine osmolality may be as low as 30 mosm/kg of H_2O. The impermeability of the distal portions of the nephron is not absolute; along with the salt that is pumped out of the collecting duct fluid, about 2% of the filtered water is reabsorbed in the absence of vasopressin. However, as much as 13% of the filtered water may be excreted, and urine flow may reach 15 mL/min or more.

The causes of diabetes insipidus, the condition caused by vasopressin deficiency or failure to respond to the hormone, are discussed in Chapter 14. In nephrogenic diabetes insipidus, the collecting ducts fail to respond to vasopressin. Two forms of this disease have been described. In one, the gene for the V_2 receptor is mutated, making the receptor unresponsive. The V_2 receptor gene is on the X chromosome, and the mode of inheritance is sex-linked recessive. In the other form, the autosomal gene for aquaporin-2 is mutated.

It is interesting that prolonged exposure to elevated levels of vasopressin can lead eventually to downregulation of the production of aquaporin-2. This permits urine flow to increase and plasma osmolality to fall despite exposure of the collecting ducts to elevated levels of the hormone—ie, the individual escapes from the renal effects of vasopressin.

The Countercurrent Mechanism

The concentrating mechanism depends upon the maintenance of a gradient of increasing osmolality along the medullary pyramids. This gradient is produced by the operation of the loops of Henle as **countercurrent multipliers** and maintained by the operation of the vasa recta as **countercurrent exchangers.** A countercurrent system is a system in which the inflow runs parallel to, counter to, and in close proximity to the outflow for some distance. This occurs for both the loops of Henle and the vasa recta in the renal medulla (Figure 38–2).

The operation of each loop of Henle as a countercurrent multiplier depends on the active transport of Na^+ and Cl^- out of its thick ascending limb (see above), the high permeability of its thin descending limb to water (Table 38–8), and the inflow of tubular fluid from the proximal tubule, with outflow into the distal tubule. The process is best understood in terms of hypothetical steps leading to the normal equilibrium condition, although of course the steps do not occur in vivo, and equilibrium is maintained unless the osmotic gradient is washed out. These steps are summarized in Figure 38–17 for a cortical nephron with no thin ascending limb. Assume first a condition in which osmolality is 300 mosm/kg of H_2O throughout the descending and ascending limbs and the medullary interstitium (Figure 38–17A). Assume in addition that the pumps in the thick ascending limb can pump 100 mosm/kg of Na^+ and Cl^- from the tubular fluid to the interstitium, increasing interstitial osmolality to 400 mosm/kg of H_2O. Water then moves out of the thin descending limb, and its contents equilibrate with the interstitium (Figure 38–17B). However, fluid containing 300 mosm/kg of H_2O is continuously entering this limb from the proximal tubule (Figure 38–17C), so the gradient against which the Na^+ and Cl^- are pumped is reduced and more enters the interstitium (Figure 38–17D). Meanwhile, hypotonic fluid flows into the

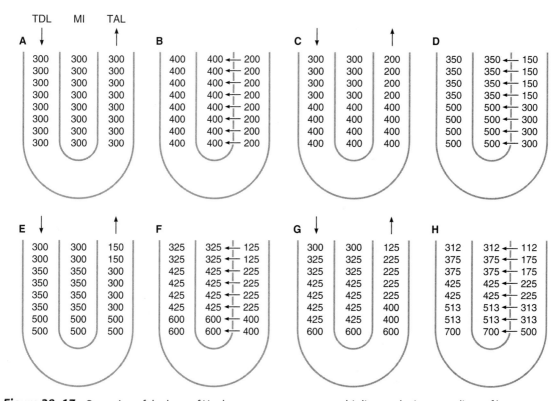

Figure 38–17. Operation of the loop of Henle as a countercurrent multiplier producing a gradient of hyperosmolarity in the medullary interstitium (MI). TDL, thin descending limb; TAL, thick ascending limb. The process of generation of the gradient is illustrated as occurring in hypothetical steps, starting at A, where osmolality in both limbs and the interstitium is 300 mosm/kg of water. The pumps in the thick ascending limb move Na^+ and Cl^- into the interstitium, increasing its osmolality to 400 mosm/kg, and this equilibrates with the fluid in the thin descending limb. However, isotonic fluid continues to flow into the thin descending limb and hypotonic fluid out of the thick ascending limb. Continued operation of the pumps makes the fluid leaving the thick ascending limb even more hypotonic, while hypertonicity accumulates at the apex of the loop. (Modified and reproduced, with permission, from Johnson LR [editor]. *Essential Medical Physiology*, Raven Press, 1992.)

distal tubule, and isotonic and subsequently hypertonic fluid flows into the ascending thick limb. The process keeps repeating, and the final result is a gradient of osmolality from the top to the bottom of the loop.

In juxtamedullary nephrons with longer loops and thin ascending limbs, the osmotic gradient is spread over a greater distance and the osmolality at the tip of the loop is greater. This is because the thin ascending limb is relatively impermeable to water but permeable to Na^+ and Cl^-. Therefore, Na^+ and Cl^- move down their concentration gradients into the interstitium, and there is additional passive countercurrent multiplication. The greater the length of the loop of Henle, the greater the osmolality that can be reached at the tip of the pyramid.

The osmotic gradient in the medullary pyramids would not last long if the Na^+ and urea in the interstitial spaces were removed by the circulation. These solutes remain in the pyramids primarily because the vasa recta operate as countercurrent exchangers (Figure 38–18). The solutes diffuse out of the vessels conducting blood toward the cortex and into the vessels descending into the pyramid. Conversely, water diffuses out of the descending vessels and into the fenestrated ascending vessels. Therefore, the solutes tend to recirculate in the medulla and water tends to bypass it, so that hypertonicity is maintained. The water removed from the collecting ducts in the pyramids is also removed by the vasa recta and enters the general circulation. Countercurrent exchange is a passive process; it depends upon movement of water and could not maintain the osmotic gradient along the pyramids if the process of countercurrent multiplication in the loops of Henle were to cease.

It is worth noting that there is a very large osmotic gradient in the loop of Henle and, in the presence of vasopressin, in the collecting ducts. It is the countercurrent system that makes this gradient possible by spreading it along a system of tubules 1 cm or more in length rather than across a single layer of cells that is only a few micrometers thick. There are other examples of the operation of countercurrent exchangers in animals. One is the heat exchange between the arteries and venae comitantes of the limbs. To a minor degree in humans but to a major degree in mammals living in cold water, heat is transferred from the arterial blood flowing into the limbs to the adjacent veins draining blood back into the body, making the tips of the limbs cold while conserving body heat.

Role of Urea

Urea contributes to the establishment of the osmotic gradient in the medullary pyramids and to the ability to form a concentrated urine in the collecting ducts. Urea transport is mediated by urea transporters, presumably by facilitated diffusion. There are at least four isoforms of the transport protein UT-A in the kidneys (UT-A1 to UT-A4). UT-B is found in erythrocytes. The amount of urea in the medullary interstitium and, consequently, in the urine varies with the amount of urea filtered, and this in turn varies with the dietary intake of protein. Therefore, a high-protein diet increases the ability of the kidneys to concentrate the urine.

Water Diuresis

The feedback mechanism controlling vasopressin secretion and the way vasopressin secretion is stimulated by a rise and inhibited by a drop in the effective osmotic pressure of the plasma are discussed in Chapter 14. The **water diuresis** produced by drinking large amounts of hypotonic fluid begins about 15 minutes after ingestion of a water load and reaches its maximum in about 40 minutes. The act of drinking produces a small decrease in vasopressin secretion before the water is absorbed, but most of the inhibition is produced by the

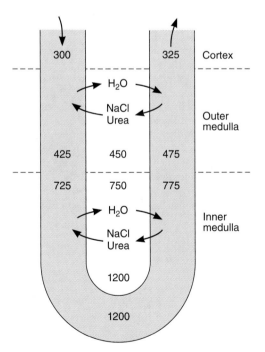

Figure 38–18. Operation of the vasa recta as countercurrent exchangers in the kidney. NaCl and urea diffuse out of the ascending limb of the vessel and into the descending limb, whereas water diffuses out of the descending and into the ascending limb of the vascular loop.

decrease in plasma osmolality after the water is absorbed.

Water Intoxication

During excretion of an average osmotic load, the maximal urine flow that can be produced during a water diuresis is about 16 mL/min. If water is ingested at a higher rate than this for any length of time, swelling of the cells because of the uptake of water from the hypotonic ECF becomes severe and, rarely, the symptoms of **water intoxication** may develop. Swelling of the cells in the brain causes convulsions and coma and leads eventually to death. Water intoxication can also occur when water intake is not reduced after administration of exogenous vasopressin or secretion of endogenous vasopressin in response to nonosmotic stimuli such as surgical trauma.

Osmotic Diuresis

The presence of large quantities of unreabsorbed solutes in the renal tubules causes an increase in urine volume called **osmotic diuresis.** Solutes that are not reabsorbed in the proximal tubules exert an appreciable osmotic effect as the volume of tubular fluid decreases and their concentration rises. Therefore, they "hold water in the tubules." In addition, there is a limit to the concentration gradient against which Na$^+$ can be pumped out of the proximal tubules. Normally, the movement of water out of the proximal tubule prevents any appreciable gradient from developing, but Na$^+$ concentration in the fluid falls when water reabsorption is decreased because of the presence in the tubular fluid of increased amounts of unreabsorbable solutes. The limiting concentration gradient is reached, and further proximal reabsorption of Na$^+$ is prevented; more Na$^+$ remains in the tubule, and water stays with it. The result is that the loop of Henle is presented with a greatly increased volume of isotonic fluid. This fluid has a decreased Na$^+$ concentration, but the total amount of Na$^+$ reaching the loop per unit time is increased. In the loop, reabsorption of water and Na$^+$ is decreased because the medullary hypertonicity is decreased. The decrease is due primarily to decreased reabsorption of Na$^+$, K$^+$, and Cl$^-$ in the ascending limb of the loop because the limiting concentration gradient for Na$^+$ reabsorption is reached. More fluid passes through the distal tubule, and because of the decrease in the osmotic gradient along the medullary pyramids, less water is reabsorbed in the collecting ducts. The result is a marked increase in urine volume and excretion of Na$^+$ and other electrolytes.

Osmotic diuresis is produced by the administration of compounds such as mannitol and related polysac-

charides that are filtered but not reabsorbed. It is also produced by naturally occurring substances when they are present in amounts exceeding the capacity of the tubules to reabsorb them. In diabetes mellitus, for example, the glucose that remains in the tubules when the filtered load exceeds the Tm$_G$ causes polyuria. Osmotic diuresis can also be produced by the infusion of large amounts of sodium chloride or urea.

It is important to recognize the difference between osmotic diuresis and water diuresis. In water diuresis, the amount of water reabsorbed in the proximal portions of the nephron is normal, and the maximal urine flow that can be produced is about 16 mL/min. In osmotic diuresis, increased urine flow is due to decreased water reabsorption in the proximal tubules and loops and very large urine flows can be produced. As the load of excreted solute is increased, the concentration of the urine approaches that of plasma (Figure 38–19) in spite of maximal vasopressin secretion, because an increasingly large fraction of the excreted urine is isotonic proximal tubular fluid. If osmotic diuresis is produced in an animal with diabetes insipidus, the urine concentration rises for the same reason.

Relation of Urine Concentration to GFR

The magnitude of the osmotic gradient along the medullary pyramids is increased when the rate of flow of fluid through the loops of Henle is decreased. A reduction in GFR such as that caused by dehydration produces a decrease in the volume of fluid presented to the countercurrent mechanism, so that the rate of flow in the loops declines and the urine becomes more concentrated. When the GFR is low, the urine can become quite concentrated in the absence of vasopressin. If one renal artery is constricted in an animal with diabetes insipidus, the urine excreted on the side of the constriction becomes hypertonic because of the reduction in GFR, whereas that excreted on the opposite side remains hypotonic.

"Free Water Clearance"

In order to quantitate the gain or loss of water by excretion of a concentrated or dilute urine, the "free water clearance" (C$_{H_2O}$) is sometimes calculated. This is the difference between the urine volume and the clearance of osmoles (C$_{osm}$):

$$C_{H_2O} = \dot{V} - \frac{U_{osm}\dot{V}}{P_{osm}}$$

where $\dot{V}$ is the urine flow rate and U$_{osm}$ and P$_{osm}$ the urine and plasma osmolality, respectively. C$_{osm}$ is the amount of water necessary to excrete the osmotic load in a urine that is isotonic with plasma. Therefore,

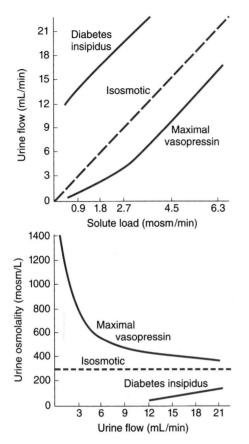

Figure 38–19. Approximate relationship between urine concentration and urine flow in osmotic diuresis in humans. The dashed line in the lower diagram indicates the concentration at which the urine is isosmotic with plasma. (Reproduced, with permission, from Berliner RW, Giebisch G in: *Best and Taylor's Physiological Basis of Medical Practice*, 9th ed. Brobeck JR [editor]. Williams & Wilkins, 1979.)

C_{H_2O} is negative when the urine is hypertonic and positive when the urine is hypotonic. In Table 38–7, for example, the values for C_{H_2O} are –1.3 mL/min (–1.9 L/d) during maximal antidiuresis and 14.5 mL/min (20.9 L/d) in the absence of vasopressin.

ACIDIFICATION OF THE URINE & BICARBONATE EXCRETION

H+ Secretion

The cells of the proximal and distal tubules, like the cells of the gastric glands, secrete hydrogen ions (see Chapter 26). Acidification also occurs in the collect-

ing ducts. The reaction that is primarily responsible for H+ secretion in the proximal tubules is Na+-H+ exchange (Figure 38–20). This is an example of secondary active transport; extrusion of Na+ from the cells into the interstitium by Na+-K+ ATPase lowers intracellular Na+, and this causes Na+ to enter the cell from the tubular lumen, with coupled extrusion of H+. The H+ comes from intracellular dissociation of H_2CO_3, and the HCO_3^- that is formed diffuses into the interstitial fluid. Thus, for each H+ ion secreted, one Na+ ion and one HCO_3^- ion enter the interstitial fluid.

Carbonic anhydrase catalyzes the formation of H_2CO_3, and drugs that inhibit carbonic anhydrase depress both secretion of acid by the proximal tubules and the reactions which depend on it.

There is some evidence that H+ is secreted in the proximal tubules by other types of pumps, but the evidence for these additional pumps is controversial, and in any case, their contribution is small relative to that of the Na+-H+ exchange mechanism. This is in contrast to what occurs in the distal tubules and collecting ducts, where H+ secretion is relatively independent of Na+ in the tubular lumen. In this part of the tubule, most H+ is secreted by an ATP-driven proton pump. Aldosterone acts on this pump to increase distal H+ secretion. The I cells in this part of the renal tubule secrete acid and, like the parietal cells in the stomach,

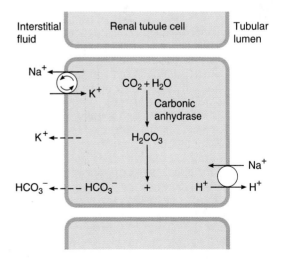

Figure 38–20. Secretion of acid by proximal tubular cells in the kidney. H+ is transported into the tubular lumen by an antiport in exchange for Na+. Active transport by Na+-K+ ATPase is indicated by arrows in the circle. Dashed arrows indicate diffusion. Compare with Figure 26–10.

contain abundant carbonic anhydrase and numerous tubulovesicular structures. There is evidence that the H^+-translocating ATPase which produces H^+ secretion is located in these vesicles as well as in the luminal cell membrane and that, in acidosis, the number of H^+ pumps is increased by insertion of these tubulovesicles into the luminal cell membrane. Some of the H^+ is also secreted by H^+-K^+ ATPase. The I cells contain **Band 3,** an anion exchange protein, in their basolateral cell membranes, and this protein may function as a Cl^--HCO_3^- exchanger for the transport of HCO_3^- to the interstitial fluid.

Fate of H^+ in the Urine

The amount of acid secreted depends upon the subsequent events in the tubular urine. The maximal H^+ gradient against which the transport mechanisms can secrete in humans corresponds to a urine pH of about 4.5, ie, an H^+ concentration in the urine that is 1000 times the concentration in plasma. pH 4.5 is thus the **limiting pH.** This is normally reached in the collecting ducts. If there were no buffers that "tied up" H^+ in the urine, this pH would be reached rapidly, and H^+ secretion would stop. However, three important reactions in the tubular fluid remove free H^+, permitting more acid to be secreted (Figure 38–21). These are the reactions with HCO_3^- to form CO_2 and H_2O, with HPO_4^{2-} to form $H_2PO_4^-$, and with NH_3 to form NH_4^+.

Reaction With Buffers

The dynamics of buffering are discussed in Chapters 1 and 39. The pK′ of the bicarbonate system is 6.1, that of the dibasic phosphate system is 6.8, and that of the ammonia system is 9.0. The concentration of HCO_3^- in the plasma, and consequently in the glomerular filtrate, is normally about 24 meq/L, whereas that of phosphate is only 1.5 meq/L. Therefore, in the proximal tubule, most of the secreted H^+ reacts with HCO_3^- to form H_2CO_3 (Figure 38–21). The H_2CO_3 breaks down to form CO_2 and H_2O. In the proximal (but not in the distal) tubule, there is carbonic anhydrase in the brush border of the cells; this facilitates the formation of CO_2 and H_2O in the tubular fluid. The CO_2, which diffuses readily across all biological membranes, enters the tubular cells, where it adds to the pool of CO_2 available to form H_2CO_3. Since most of the H^+ is removed from the tubule, the pH of the fluid is changed very little. This is the mechanism by which HCO_3^- is reabsorbed; for each mole of HCO_3^- removed from the tubular fluid, 1 mole of HCO_3^- diffuses from the tubular cells into the blood, even though it is not the same mole that disappeared from the tubular fluid.

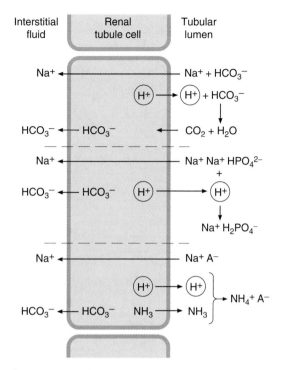

Figure 38–21. Fate of H^+ secreted into a tubule in exchange for Na^+. **Top:** Reabsorption of filtered bicarbonate via CO_2. **Middle:** Formation of monobasic phosphate. **Bottom:** Ammonium formation. Note that in each instance one Na^+ ion and one HCO_3^- ion enter the bloodstream for each H^+ ion secreted. A^-, anion.

Secreted H^+ also reacts with dibasic phosphate (HPO_4^{2-}) to form monobasic phosphate ($H_2PO_4^-$). This happens to the greatest extent in the distal tubules and collecting ducts, because it is here that the phosphate which escapes proximal reabsorption is greatly concentrated by the reabsorption of water. The reaction with NH_3 occurs in the proximal and distal tubules. H^+ also combines to a minor degree with other buffer anions.

Each H^+ ion that reacts with the buffers contributes to the urinary **titratable acidity,** which is measured by determining the amount of alkali that must be added to the urine to return its pH to 7.4, the pH of the glomerular filtrate. However, the titratable acidity obviously measures only a fraction of the acid secreted, since it does not account for the H_2CO_3 that has been converted to H_2O and CO_2. In addition, the pK′ of the ammonia system is 9.0, and the ammonia system is titrated only from the pH of the urine to pH 7.4, so it contributes very little to the titratable acidity.

Ammonia Secretion

Reactions in the renal tubular cells produce NH_4^+ and HCO_3^-. NH_4^+ is in equilibrium with $NH_3 + H^+$ in the cells. Since the pK' of this reaction is 9.0, the ratio of NH_3 to NH_4^+ at pH 7.0 is 1:100 (Figure 38–22). However, NH_3 is lipid-soluble and diffuses across the cell membranes down its concentration gradient into the interstitial fluid and tubular urine. In the urine it reacts with H^+ to form NH_4^+, and the NH_4^+ remains in the urine.

The principal reaction producing NH_4^+ in cells is conversion of glutamine to glutamate. This reaction is catalyzed by the enzyme **glutaminase,** which is abundant in renal tubular cells (Figure 38–22). **Glutamic dehydrogenase** catalyzes the conversion of glutamate to α-ketoglutarate, with the production of more NH_4^+. Subsequent metabolism of α-ketoglutarate utilizes $2H^+$, freeing $2HCO_3^-$.

In chronic acidosis, the amount of NH_4^+ excreted at any given urine pH also increases, because more NH_3 enters the tubular urine. The effect of this **adaptation** of NH_3 secretion, the cause of which is unsettled, is a further removal of H^+ from the tubular fluid and consequently a further enhancement of H^+ secretion.

The process by which NH_3 is secreted into the urine and then changed to NH_4^+, maintaining the concentration gradient for diffusion of NH_3, is called **nonionic diffusion** (see Chapter 1). Salicylates and a number of other drugs that are weak bases or weak acids are also secreted by nonionic diffusion. They diffuse into the tubular fluid at a rate that is dependent upon the pH of the urine, so the amount of each drug excreted varies with the pH of the urine.

pH Changes Along the Nephrons

There is a moderate drop in pH in the proximal tubular fluid, but, as noted above, most of the secreted H^+ has little effect on luminal pH because of the formation

$$NH_4^+ \rightleftharpoons NH_3 + H^+$$

$$pH = pK' + \log \frac{[NH_3]}{[NH_4^+]}$$

$$\text{Glutamine} \xrightarrow{\text{Glutaminase}} \text{Glutamate} + NH_4^+$$

$$\text{Glutamate} \xrightarrow{\substack{\text{Glutamic} \\ \text{dehydrogenase}}} \alpha\text{–Ketoglutarate} + NH_4^+$$

Figure 38–22. Major reactions involved in ammonia production in the kidneys. See also Chapter 17.

of CO_2 and H_2O from H_2CO_3. In contrast, the distal tubule has less capacity to secrete H^+, but secretion in this segment has a greater effect on urinary pH.

Factors Affecting Acid Secretion

Renal acid secretion is altered by changes in the intracellular P_{CO_2}, K^+ concentration, carbonic anhydrase level, and adrenocortical hormone concentration. When the P_{CO_2} is high (respiratory acidosis), more intracellular H_2CO_3 is available to buffer the hydroxyl ions and acid secretion is enhanced, whereas the reverse is true when the P_{CO_2} falls. K^+ depletion enhances acid secretion, apparently because the loss of K^+ causes intracellular acidosis even though the plasma pH may be elevated. Conversely, K^+ excess in the cells inhibits acid secretion. When carbonic anhydrase is inhibited, acid secretion is inhibited, because the formation of H_2CO_3 is decreased. Aldosterone and the other adrenocortical steroids that enhance tubular reabsorption of Na^+ also increase the secretion of H^+ and K^+.

Bicarbonate Excretion

Although the process of HCO_3^- reabsorption does not actually involve transport of this ion into the tubular cells, HCO_3^- reabsorption is proportionate to the amount filtered over a relatively wide range. There is no demonstrable Tm, but HCO_3^- reabsorption is decreased by an unknown mechanism when the ECF volume is expanded (Figure 38–23). When the plasma HCO_3^- concentration is low, all the filtered HCO_3^- is reabsorbed; but when the plasma HCO_3^- concentration is high, ie, above 26–28 meq/L (the renal threshold for HCO_3^-), HCO_3^- appears in the urine and the urine becomes alkaline. Conversely, when the plasma HCO_3^- falls below about 26 meq/L, the value at which all the secreted H^+ is being used to reabsorb HCO_3^-, more H^+ becomes available to combine with other buffer anions. Therefore, the lower the plasma HCO_3^- concentration drops, the more acidic the urine becomes and the greater its NH_4^+ content.

Implications of Urinary pH Changes

Depending upon the rates of the interrelated processes of acid secretion, NH_4^+ production, and HCO_3^- excretion, the pH of the urine in humans varies from 4.5 to 8.0. Excretion of a urine that is at a pH different from that of the body fluids has important implications for the body's electrolyte and acid-base economy which are discussed in detail in Chapter 39. Acids are buffered in the plasma and cells, the overall reaction being HA + $NaHCO_3 \rightarrow NaA + H_2CO_3$. The H_2CO_3 forms CO_2 and H_2O, and the CO_2 is expired, while the NaA ap-

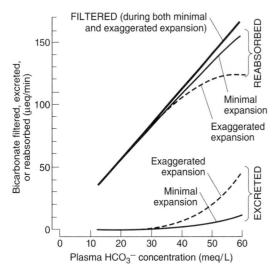

Figure 38–23. Effect of ECF volume on HCO_3^- filtration, reabsorption, and excretion in rats. The pattern of HCO_3^- excretion is similar in humans. The plasma HCO_3^- concentration is normally about 24 meq/L. (Reproduced, with permission, from Valtin H: *Renal Function*, 2nd ed. Little, Brown, 1983. Copyright © Little, Brown and Co., 1983.)

pears in the glomerular filtrate. To the extent that the Na^+ is replaced by H^+ in the urine, Na^+ is conserved in the body. Furthermore, for each H^+ ion excreted with phosphate or as NH_4^+, there is a net gain of one HCO_3^- ion in the blood, replenishing the supply of this important buffer anion. Conversely, when base is added to the body fluids, the OH^- ions are buffered, raising the plasma HCO_3^-. When the plasma level exceeds 28 meq/L, the urine becomes alkaline and the extra HCO_3^- is excreted in the urine. Because the rate of maximal H^+ secretion by the tubules varies directly with the arterial P_{CO_2}, HCO_3^- reabsorption also is affected by the P_{CO_2}. This relationship is discussed in more detail in Chapter 39.

REGULATION OF NA$^+$ & CL$^-$ EXCRETION

Na^+ is filtered in large amounts, but it is actively transported out of all portions of the tubule except the thin loop of Henle. Normally, 96% to well over 99% of the filtered Na^+ is reabsorbed. Most of the Na^+ is reabsorbed with Cl^- (Table 38–9), but some is reabsorbed in the processes by which one Na^+ ion enters the bloodstream for each H^+ ion secreted by the tubules, and in the distal tubules a small amount is actively reabsorbed in association with the secretion of K^+.

Regulation of Na$^+$ Excretion

Because Na^+ is the most abundant cation in ECF and because Na^+ salts account for over 90% of the osmotically active solute in the plasma and interstitial fluid, the amount of Na^+ in the body is a prime determinant of the ECF volume. Therefore, it is not surprising that multiple regulatory mechanisms have evolved in terrestrial animals to control the excretion of this ion. Through the operation of these regulatory mechanisms, the amount of Na^+ excreted is adjusted to equal the amount ingested over a wide range of dietary intakes, and the individual stays in Na^+ balance. Thus, urinary Na^+ output ranges from less than 1 meq/d on a low-salt diet to 400 meq/d or more when the dietary Na^+ intake is high. In addition, there is a natriuresis when saline is infused intravenously and a decrease in Na^+ excretion when ECF volume is reduced. Variations in Na^+ excretion are effected by changes in the amount filtered (Table 38–10) and the amount reabsorbed in the tubules. The factors affecting the GFR, including tubuloglomerular feedback, are discussed above. Factors affecting Na^+ reabsorption include the circulating level of aldosterone and other adrenocortical hormones, the circulating level of ANP and other natriuretic hormones, the amount of angiotensin II and PGE_2 in the kidneys, and the rate of tubular secretion of H^+ and K^+.

Effects of Adrenocortical Steroids

Adrenal mineralocorticoids such as aldosterone increase tubular reabsorption of Na^+ in association with secretion of K^+ and H^+ and also Na^+ reabsorption with Cl^-

Table 38–9. Quantitative aspects of Na^+ reabsorption in a normal man on a normal Na^+ diet.

GFR = 125 mL/min Plasma HCO_3^- = 27 meq/L Plasma Na^+ = 145 meq/L	
Na^+ filtered per minute	**18,125 μeq**
Reabsorbed with Cl^-	14,585 μeq
Reabsorbed while reabsorbing 3375 μeq of HCO_3^-	3,375 μeq
Reabsorbed in association with formation of titratable acidity and ammonia	50 μeq
Reabsorbed in association with secretion of K^+	50 μeq
Total Na^+ reabsorbed per minute	**18,060 μeq**

Table 38–10. Changes in Na$^+$ excretion that would occur as a result of changes in GFR if there were no concomitant changes in Na$^+$ reabsorption.

GFR (mL/min)	Plasma Na$^+$ (μeq/mL)	Amount Filtered (μeq/min)	Amount Reabsorbed (μeq/min)	Amount Excreted (μeq/min)
125	145	18,125	18,000	125
127	145	18,415	18,000	415
124.1	145	18,000	18,000	0

(see Chapter 20). When these hormones are injected into adrenalectomized animals, there is a latent period of 10–30 minutes before their effects on Na$^+$ reabsorption become manifest, because of the time required for the steroids to alter protein synthesis via their action on DNA. Mineralocorticoids may also have more rapid membrane-mediated effects, but these are not apparent in terms of Na$^+$ excretion in the whole animal. The mineralocorticoids act primarily on the cortical collecting ducts. As noted in Chapter 20, they act on P cells to increase the number of active ENaCs in the apical membranes of these cells (Figure 38–24).

In Liddle's syndrome, mutations in the genes that code for the β subunit and less commonly the γ sub-unit of the ENaCs cause them to become constitutively active in the kidney. This leads to Na$^+$ retention and hypertension.

Other Humoral Effects

Reduction of dietary intake of salt increases aldosterone secretion (see Figure 20–27), producing marked but slowly developing decreases in Na$^+$ excretion. A variety of other humoral factors affect Na$^+$ reabsorption. PGE$_2$ causes a natriuresis, possibly by inhibiting Na$^+$-K$^+$ ATPase and possibly by increasing intracellular Ca^{2+}, which in turn inhibits Na$^+$ transport via ENaCs. Endothelin and IL-1 cause natriuresis, probably by increasing the formation of PGE$_2$. ANP and related molecules increase intracellular cGMP, and this inhibits transport via the ENaCs. Inhibition of Na$^+$-K$^+$ ATPase by the other natriuretic hormone, which appears to be endogenously produced ouabain (see Chapter 24), also increases Na$^+$ excretion. Angiotensin II increases reabsorption of Na$^+$ and HCO$_3^-$ by an action on the proximal tubules. There is an appreciable amount of angiotensin-converting enzyme in the kidneys, and the kidneys convert 20% of the circulating angiotensin I reaching them to angiotensin II. In addition, angiotensin I is generated in the kidneys.

Prolonged exposure to high levels of circulating mineralocorticoids does not cause edema in otherwise normal individuals because eventually the kidneys escape from the effects of the steroids. This **escape phenomenon**, which may be due to increased secretion of ANP, is discussed in Chapter 20 and illustrated in Figure 20–24. It appears to be reduced or absent in nephrosis, cirrhosis, and heart failure, and patients with these diseases continue to retain Na$^+$ and become edematous when exposed to high levels of mineralocorticoids.

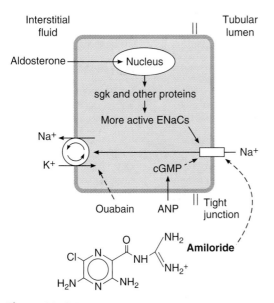

Figure 38–24. Renal P cell. Na$^+$ enters via the ENaCs in the apical membrane and is pumped into the interstitial fluid by Na$^+$-K$^+$ ATPases in the basolateral membrane. Aldosterone activates the genome to produce sgk and other proteins, and the number of active ENaCs is increased.

REGULATION OF K$^+$ EXCRETION

Much of the filtered K$^+$ is removed from the tubular fluid by active reabsorption in the proximal tubules (Table 38–5), and K$^+$ is then secreted into the fluid by the distal tubular cells. The rate of K$^+$ secretion is proportionate to the rate of flow of the tubular fluid through the distal portions of the nephron, because with rapid flow there is less opportunity for the tubular K^+ concentration to rise to a value that stops further secretion. In the absence of complicating factors, the amount secreted is approximately equal to the K^+ intake, and K^+ balance is maintained. In the distal tubules, Na^+ is generally reabsorbed and K^+ is secreted. There is no rigid one-for-one exchange, and much of the movement of K^+ is passive. However, there is electrical coupling in the sense that intracellular migration of Na^+ tends to lower the potential difference across the tubular cell, and this favors movement of K^+ into the tubular lumen. Since Na^+ is also reabsorbed in association with H^+ secretion, there is competition for the Na^+ in the tubular fluid. K^+ excretion is decreased when the amount of Na^+ reaching the distal tubule is small, and it is also decreased when H^+ secretion is increased. When total body K^+ is high, H^+ secretion is inhibited, apparently because of intracellular alkalosis; K^+ secretion and excretion are therefore facilitated. Conversely, the cells are acidic in K^+ depletion, and K^+ secretion declines. Apparently the K^+ secretory mechanism is capable of "adaptation," because the amount of K^+ excreted gradually increases when a constant large dose of a potassium salt is administered for a prolonged period.

Hypokalemia is common and can be severe. In addition to its occurrence when there is excessive loss in the urine, it is occasionally seen in patients with excess loss in diarrheic stools, in patients in whom K^+ is shifted into cells by insulin (see Chapter 19) or β-adrenergic agonists, and in patients with a prolonged low intake of K^+. **Hyperkalemia** is a more dangerous condition because of its effects on the heart (see Chapter 28), but it rarely occurs unless renal function is depressed.

DIURETICS

Although a detailed discussion of diuretic agents is outside the scope of this book, consideration of their mechanisms of action constitutes an informative review of the factors affecting urine volume and electrolyte excretion. These mechanisms are summarized in Table 38–11. Water, alcohol, osmotic diuretics, xanthines, and acidifying salts have limited clinical usefulness, and the vasopressin antagonists are used primarily for research. However, many of the other agents on the list are used extensively in medical practice.

The carbonic anhydrase-inhibiting drugs are only moderately effective as diuretic agents, but because they inhibit acid secretion by decreasing the supply of carbonic acid, they have far-reaching effects. Not only is Na^+ excretion increased because H^+ secretion is de-

Table 38–11. Mechanism of action of various diuretics.

Agent	Mechanism of Action
Water	Inhibits vasopressin secretion.
Ethanol	Inhibits vasopressin secretion.
Antagonists of V_2 vasopressin receptors	Inhibit action of vasopressin on collecting duct.
Large quantities of osmotically active substances such as mannitol and glucose	Produce osmotic diuresis.
Xanthines such as caffeine and theophylline	Decrease tubular reabsorption of Na^+ and increase GFR.
Acidifying salts such as $CaCl_2$ and NH_4Cl	Supply acid load; H^+ is buffered, but an anion is excreted with Na^+ when the ability of the kidneys to replace Na^+ with H^+ is exceeded.
Carbonic anhydrase inhibitors such as acetazolamide (Diamox)	Decrease H^+ secretion, with resultant increase in Na^+ and K^+ excretion.
Metolazone (Zaroxolyn), thiazides such as chlorothiazide (Diuril)	Inhibit the Na^+-Cl^- cotransporter in the early portion of the distal tubule.
Loop diuretics such as furosemide (Lasix), ethacrynic acid (Edecrin), and bumetanide	Inhibit the Na^+-K^+-$2Cl^-$ cotransporter in the medullary thick ascending limb of the loop of Henle.
K^+-retaining natriuretics such as spironolactone (Aldactone), triamterene (Dyrenium), and amiloride (Midamor)	Inhibit Na^+-K^+ "exchange" in the collecting ducts by inhibiting the action of aldosterone (spironolactone) or by inhibiting the ENaCs (amiloride).

creased, but also HCO_3^- reabsorption is depressed; and because H^+ and K^+ compete with each other and with Na^+, the decrease in H^+ secretion facilitates the secretion and excretion of K^+.

Another determinant of the rate of K^+ secretion is the amount of Na^+ delivered to the Na^+-K^+ "exchange" site in the collecting ducts. Thiazides, furosemide, ethacrynic acid, and bumetanide act proximal to this site, and the resultant increase in Na^+ delivery increases K^+ secretion. The K^+ loss is appreciable, and K^+ depletion is one of the common complications of treatment with these agents. Furosemide and the other loop diuretics act by inhibiting the Na^+-K^+-$2Cl^-$ transporter in the thick limb of the loop of Henle, and thiazides act by inhibiting the Na^+-Cl^- cotransporter in the early portion of the distal tubule. On the other hand, the so-called K^+-retaining diuretics act in the collecting duct to inhibit Na^+-K^+ exchange by inhibiting the action of aldosterone or blocking ENaCs.

EFFECTS OF DISORDERED RENAL FUNCTION

A number of abnormalities are common to many different types of renal disease. The secretion of renin by the kidneys and the relation of the kidneys to hypertension are discussed in Chapters 24 and 33. A frequent finding in various forms of renal disease is the presence in the urine of protein, leukocytes, red cells, and **casts,** which are bits of proteinaceous material precipitated in the tubules and washed into the bladder. Other important consequences of renal disease are loss of the ability to concentrate or dilute the urine, uremia, acidosis, and abnormal retention of Na^+.

Proteinuria

In many renal diseases and in one benign condition, the permeability of the glomerular capillaries is increased, and protein is found in the urine in more than the usual trace amounts **(proteinuria).** Most of this protein is albumin, and the defect is commonly called **albuminuria.** The relation of charges on the glomerular membrane to albuminuria is discussed above. The amount of protein in the urine may be very large, and, especially in nephrosis, the urinary protein loss may exceed the rate at which the liver can synthesize plasma proteins. The resulting hypoproteinemia reduces the oncotic pressure, and the plasma volume declines, sometimes to dangerously low levels, while edema fluid accumulates in the tissues.

A benign condition that causes proteinuria is a poorly understood change in renal hemodynamics which in some otherwise normal individuals causes protein to appear in urine formed when they are in the standing position **(orthostatic albuminuria).** Urine formed when these individuals are lying down is protein-free.

Loss of Concentrating & Diluting Ability

In renal disease, the urine becomes less concentrated and urine volume is often increased, producing the symptoms of **polyuria** and **nocturia** (waking up at night to void). The ability to form a dilute urine is often retained, but in advanced renal disease, the osmolality of the urine becomes fixed at about that of plasma, indicating that the diluting and concentrating functions of the kidney have both been lost. The loss is due in part to disruption of the countercurrent mechanism, but a more important cause is a loss of functioning nephrons. When one kidney is removed surgically, the number of functioning nephrons is halved. The number of osmoles excreted is not reduced to this extent, and so the remaining nephrons must each be filtering and excreting more osmotically active substances, producing what is in effect an osmotic diuresis. In osmotic diuresis, the osmolality of the urine approaches that of plasma (see above). The same thing happens when the number of functioning nephrons is reduced by disease. The increased filtration in the remaining nephrons eventually damages them, and thus more nephrons are lost. The damage resulting from increased filtration may be due to progressive fibrosis in the proximal tubule cells, but this is unsettled. However, the eventual result of this positive feedback is loss of so many nephrons that complete renal failure with **oliguria** or even **anuria** results.

Uremia

When the breakdown products of protein metabolism accumulate in the blood, the syndrome known as **uremia** develops. The symptoms of uremia include lethargy, anorexia, nausea and vomiting, mental deterioration and confusion, muscle twitching, convulsions, and coma. The BUN and creatinine levels are high, and the blood levels of these substances are used as an index of the severity of the uremia. It probably is not the accumulation of urea and creatinine per se but rather the accumulation of other toxic substances—possibly organic acids or phenols—that produces the symptoms of uremia.

The toxic substances that cause the symptoms of uremia can be removed by dialyzing the blood of uremic patients against a bath of suitable composition in an artificial kidney **(hemodialysis).** Patients can be kept alive and in reasonable health for many months on dialysis, even when they are completely anuric or have had both kidneys removed. However, the treatment of choice today is certainly transplantation of a kidney from a suitable donor.

Other features of chronic renal failure include anemia, which is caused primarily by failure to produce erythropoietin (see Chapter 24), and secondary hyperparathyroidism due to 1,25-dihydroxycholecalciferol deficiency (see Chapter 21).

Acidosis

Acidosis is common in chronic renal disease because of failure to excrete the acid products of digestion and metabolism (see Chapter 39). In the rare syndrome of **renal tubular acidosis,** there is specific impairment of the ability to make the urine acidic, and other renal functions are usually normal. However, in most cases of chronic renal disease the urine is maximally acidified, and acidosis develops because the total amount of H^+ that can be secreted is reduced because of impaired renal tubular production of NH_4^+.

Abnormal Na⁺ Metabolism

Many patients with renal disease retain excessive amounts of Na^+ and become edematous. There are at least three causes of Na^+ retention in renal disease. In acute glomerulonephritis, a disease that affects primarily the glomeruli, there is a marked decrease in the amount of Na^+ filtered. In the nephrotic syndrome, an increase in aldosterone secretion contributes to the salt retention. The plasma protein level is low in this condition, and so fluid moves from the plasma into the interstitial spaces and the plasma volume falls. The decline in plasma volume triggers the increase in aldosterone secretion via the renin-angiotensin system. A third cause of Na^+ retention and edema in renal disease is **heart failure** (see Chapter 33). Renal disease predisposes to heart failure, partly because of the hypertension it frequently produces.

FILLING OF THE BLADDER

The walls of the ureters contain smooth muscle arranged in spiral, longitudinal, and circular bundles, but distinct layers of muscle are not seen. Regular peristaltic contractions occurring one to five times per minute move the urine from the renal pelvis to the bladder, where it enters in spurts synchronous with each peristaltic wave. The ureters pass obliquely through the bladder wall and, although there are no ureteral sphincters as such, the oblique passage tends to keep the ureters closed except during peristaltic waves, preventing reflux of urine from the bladder.

EMPTYING OF THE BLADDER

Anatomic Considerations

The smooth muscle of the bladder, like that of the ureters, is arranged in spiral, longitudinal, and circular bundles. Contraction of this muscle, which is called the **detrusor muscle,** is mainly responsible for emptying the bladder during urination (micturition). Muscle bundles pass on either side of the urethra, and these

fibers are sometimes called the **internal urethral sphincter,** although they do not encircle the urethra. Farther along the urethra is a sphincter of skeletal muscle, the sphincter of the membranous urethra (**external urethral sphincter**). The bladder epithelium is made up of a superficial layer of flat cells and a deep layer of cuboidal cells. The innervation of the bladder is summarized in Figure 38–25.

Micturition

The physiology of micturition and the physiologic basis of its disorders are subjects about which there is much confusion. Micturition is fundamentally a spinal reflex facilitated and inhibited by higher brain centers and, like defecation, subject to voluntary facilitation and inhibition. Urine enters the bladder without producing much increase in intravesical pressure until the viscus is well filled. In addition, like other types of smooth muscle, the bladder muscle has the property of plasticity; when it is stretched, the tension initially produced is not maintained. The relation between intravesical pressure and volume can be studied by inserting a catheter and emptying the bladder, then recording the pressure while the bladder is filled with 50-mL increments of water or air (**cystometry**). A plot of intravesical pressure against the volume of fluid in the bladder is called

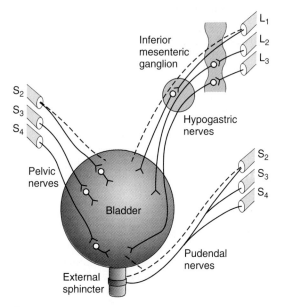

Figure 38–25. Innervation of the bladder. Dashed lines indicate sensory nerves. Parasympathetic innervation is shown at the left, sympathetic at the upper right, and somatic at the lower right.

a **cystometrogram** (Figure 38–26). The curve shows an initial slight rise in pressure when the first increments in volume are produced; a long, nearly flat segment as further increments are produced; and a sudden, sharp rise in pressure as the micturition reflex is triggered. These three components are sometimes called segments Ia, Ib, and II. The first urge to void is felt at a bladder volume of about 150 mL, and a marked sense of fullness at about 400 mL. The flatness of segment Ib is a manifestation of the law of Laplace (see Chapter 30). This law states that the pressure in a spherical viscus is equal to twice the wall tension divided by the radius. In the case of the bladder, the tension increases as the organ fills, but so does the radius. Therefore, the pressure increase is slight until the organ is relatively full.

During micturition, the perineal muscles and external urethral sphincter are relaxed; the detrusor muscle contracts; and urine passes out through the urethra. The bands of smooth muscle on either side of the urethra apparently play no role in micturition, and their main function is believed to be the prevention of reflux of semen into the bladder during ejaculation.

The mechanism by which voluntary urination is initiated remains unsettled. One of the initial events is relaxation of the muscles of the pelvic floor, and this may cause a sufficient downward tug on the detrusor muscle to initiate its contraction. The perineal muscles and external sphincter can be contracted voluntarily, preventing urine from passing down the urethra or interrupting the flow once urination has begun. It is through the learned ability to maintain the external sphincter in a contracted state that adults are able to delay urination until the opportunity to void presents itself. After urination, the female urethra empties by gravity. Urine remaining in the urethra of the male is expelled by several contractions of the bulbocavernosus muscle.

Reflex Control

The bladder smooth muscle has some inherent contractile activity; however, when its nerve supply is intact, stretch receptors in the bladder wall initiate a reflex contraction that has a lower threshold than the inherent contractile response of the muscle. Fibers in the pelvic nerves are the afferent limb of the voiding reflex, and the parasympathetic fibers to the bladder that constitute the efferent limb also travel in these nerves. The reflex is integrated in the sacral portion of the spinal cord. In the adult, the volume of urine in the bladder that normally initiates a reflex contraction is about 300–400 mL. The sympathetic nerves to the bladder play no part in micturition, but they do mediate the contraction of the bladder muscle that prevents semen from entering the bladder during ejaculation (see Chapter 23).

There is no small motor nerve system to the stretch receptors in the bladder wall; but the threshold for the voiding reflex, like the stretch reflexes, is adjusted by the activity of facilitatory and inhibitory centers in the brain stem. There is a facilitatory area in the pontine region and an inhibitory area in the midbrain. After transection of the brain stem just above the pons, the threshold is lowered and less bladder filling is required to trigger it, whereas after transection at the top of the midbrain, the threshold for the reflex is essentially normal. There is another facilitatory area in the posterior hypothalamus. In humans with lesions in the superior frontal gyrus, the desire to urinate is reduced and there is also difficulty in stopping micturition once it has commenced. However, stimulation experiments in animals indicate that other cortical areas also affect the process. The bladder can be made to contract by voluntary facilitation of the spinal voiding reflex when it contains only a few milliliters of urine. Voluntary contraction of the abdominal muscles aids the expulsion of urine by increasing the intra-abdominal pressure, but voiding can be initiated without straining even when the bladder is nearly empty.

Abnormalities of Micturition

There are three major types of bladder dysfunction due to neural lesions: (1) the type due to interruption of the

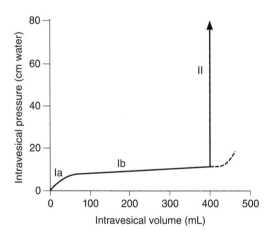

Figure 38–26. Cystometrogram in a normal human. The numerals identify the three components of the curve described in the text. The dashed line indicates the pressure-volume relations that would have been found had micturition not occurred and produced component II. (Modified and reproduced, with permission, from Tanagho EA, McAninch JW: *Smith's General Urology,* 15th ed. McGraw-Hill, 2000.)

afferent nerves from the bladder; (2) the type due to interruption of both afferent and efferent nerves; and (3) the type due to interruption of facilitatory and inhibitory pathways descending from the brain. In all three types the bladder contracts, but the contractions are generally not sufficient to empty the viscus completely, and residual urine is left in the bladder.

Effects of Deafferentation

When the sacral dorsal roots are cut in experimental animals or interrupted by diseases of the dorsal roots such as **tabes dorsalis** in humans, all reflex contractions of the bladder are abolished. The bladder becomes distended, thin-walled, and hypotonic, but there are some contractions because of the intrinsic response of the smooth muscle to stretch.

Effects of Denervation

When the afferent and efferent nerves are both destroyed, as they may be by tumors of the cauda equina or filum terminale, the bladder is flaccid and distended for a while. Gradually, however, the muscle of the "decentralized bladder" becomes active, with many contraction waves that expel dribbles of urine out of the urethra. The bladder becomes shrunken and the blad-

der wall hypertrophied. The reason for the difference between the small, hypertrophic bladder seen in this condition and the distended, hypotonic bladder seen when only the afferent nerves are interrupted is not known. The hyperactive state in the former condition suggests the development of denervation hypersensitization even though the neurons interrupted are preganglionic rather than postganglionic.

Effects of Spinal Cord Transection

During spinal shock, the bladder is flaccid and unresponsive. It becomes overfilled, and urine dribbles through the sphincters **(overflow incontinence).** After spinal shock has passed, the voiding reflex returns, although there is, of course, no voluntary control and no inhibition or facilitation from higher centers when the spinal cord is transected. Some paraplegic patients train themselves to initiate voiding by pinching or stroking their thighs, provoking a mild mass reflex (see Chapter 12). In some instances, the voiding reflex becomes hyperactive. Bladder capacity is reduced, and the wall becomes hypertrophied. This type of bladder is sometimes called the **spastic neurogenic bladder.** The reflex hyperactivity is made worse by, and may be caused by, infection in the bladder wall.

Regulation of Extracellular Fluid Composition & Volume

39

INTRODUCTION

This chapter is a review of the major homeostatic mechanisms that operate, primarily through the kidneys and the lungs, to maintain the **tonicity,** the **volume,** and the **specific ionic composition,** particularly the **H$^+$** concentration, of the ECF. The interstitial portion of this fluid is the fluid environment of the cells, and life depends upon the constancy of this "internal sea" (see Chapter 1).

DEFENSE OF TONICITY

The defense of the tonicity of the ECF is primarily the function of the vasopressin-secreting and thirst mechanisms. The total body osmolality is directly proportionate to the total body sodium plus the total body potassium divided by the total body water, so that changes in the osmolality of the body fluids occur when there is a disproportion between the amount of these electrolytes and the amount of water ingested or lost from the body (see Chapter 1). When the effective osmotic pressure of the plasma rises, vasopressin secretion is increased and the thirst mechanism is stimulated. Water is retained in the body, diluting the hypertonic plasma, and water intake is increased (Figure 39–1). Conversely, when the plasma becomes hypo-

tonic, vasopressin secretion is decreased and "solute-free water" (water in excess of solute) is excreted. In this way, the tonicity of the body fluids is maintained within a narrow normal range. In health, plasma osmolality ranges from 280 to 295 mosm/kg of H_2O, with vasopressin secretion maximally inhibited at 285 mosm/kg and stimulated at higher values (see Figure 14–14). The details of the way the regulatory mechanisms operate and the disorders that result when their function is disrupted are considered in Chapters 14 and 38.

DEFENSE OF VOLUME

The volume of the ECF is determined primarily by the total amount of osmotically active solute in the ECF. The composition of the ECF is discussed in Chapter 1. Since Na$^+$ and Cl$^-$ are by far the most abundant osmotically active solutes in ECF, and since changes in Cl$^-$ are to a great extent secondary to changes in Na$^+$, the amount of Na$^+$ in the ECF is the most important determinant of ECF volume. Therefore, the mechanisms that control Na$^+$ balance are the major mechanisms defending ECF volume. There is, however, a volume control of water excretion as well; a rise in ECF volume inhibits vasopressin secretion, and a decline in ECF volume produces an increase in the secretion of this hormone. Volume stimuli override the osmotic regulation of vasopressin secretion. Angiotensin II stimulates aldosterone and vasopressin secretion. It also causes thirst and constricts blood vessels, which help to maintain blood pressure. Thus, angiotensin II plays a key role in the body's response to hypovolemia (Figure 39–2). In addition, expansion of the ECF volume increases the secretion of ANP and BNP by the heart, and this causes natriuresis and diuresis (see Chapter 24).

In disease states, loss of water from the body (**dehydration**) causes a moderate decrease in ECF volume, because water is lost from both the intracellular and extracellular fluid compartments; but loss of Na$^+$ in the stools (diarrhea), urine (severe acidosis, adrenal insufficiency), or sweat (heat prostration) decreases ECF vol-

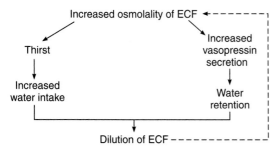

Figure 39–1. Mechanisms for defending ECF tonicity. The dashed arrow indicates inhibition. (Courtesy of J Fitzsimmons.)

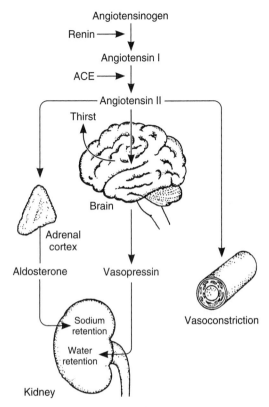

Figure 39–2. Defense of ECF volume by angiotensin II. ACE, angiotensin-converting enzyme.

ume markedly and eventually leads to shock. The immediate compensations in shock operate principally to maintain intravascular volume (see Chapter 33), but they also affect Na^+ balance. In adrenal insufficiency, the decline in ECF volume is due not only to loss of Na^+ in the urine but also to its movement into cells (see Chapter 20).

Because of the key position of Na^+ in volume homeostasis, it is not surprising that more than one mechanism has evolved to control the excretion of this ion. The filtration and reabsorption of Na^+ in the kidneys and the effects of these processes on Na^+ excretion are discussed in Chapter 38. When ECF volume is decreased, blood pressure falls. Glomerular capillary pressure declines, and the GFR therefore falls, reducing the amount of Na^+ filtered. Tubular reabsorption of Na^+ is increased, in part because the secretion of aldosterone is increased. Aldosterone secretion is controlled in part by a feedback system in which the change that initiates increased secretion is a decline in mean intravascular pressure (see Chapters 20 and 24). Other changes in Na^+ excretion occur too

rapidly to be due solely to changes in aldosterone secretion. For example, rising from the supine to the standing position increases aldosterone secretion. However, Na^+ excretion is decreased within a few minutes, and this rapid change in Na^+ excretion occurs in adrenalectomized subjects. It is probably due to hemodynamic changes and possibly to decreased ANP secretion.

DEFENSE OF SPECIFIC IONIC COMPOSITION

Special regulatory mechanisms maintain the levels of certain specific ions in the ECF as well as the levels of glucose and other nonionized substances important in metabolism (see Chapters 17 and 19). The feedback of Ca^{2+} on the parathyroids and the calcitonin-secreting cells to adjust their secretion maintains the ionized calcium level of the ECF (see Chapter 21). The Mg^+ concentration is subject to close regulation, but the mechanisms controlling Mg^+ metabolism are incompletely understood.

The mechanisms controlling Na^+ and K^+ content are part of those determining the volume and tonicity of ECF and are discussed above. The levels of these ions are also dependent upon the H^+ concentration, and pH is one of the major factors affecting the anion composition of ECF.

DEFENSE OF H^+ CONCENTRATION

The mystique that envelopes the subject of acid-base balance makes it necessary to point out that the core of the problem is not "buffer base" or "fixed cation" or the like but simply the maintenance of the H^+ concentration of the ECF. The mechanisms regulating the composition of the ECF are particularly important as far as this specific ion is concerned, because the machinery of the cells is very sensitive to changes in H^+ concentration. Intracellular H^+ concentration, which can be measured by using microelectrodes, pH-sensitive fluorescent dyes, and phosphorus magnetic resonance, is different from extracellular pH and appears to be regulated by a variety of intracellular processes. However, it is sensitive to changes in ECF H^+ concentration.

The pH notation is a useful means of expressing H^+ concentrations in the body, because the H^+ concentrations happen to be low relative to those of other cations. Thus, the normal Na^+ concentration of arterial plasma that has been equilibrated with red blood cells is about 140 meq/L, whereas the H^+ concentration is 0.00004 meq/L (Table 39–1). The pH, the negative logarithm of 0.00004, is therefore 7.4. Of course, a decrease in pH of 1 unit, eg, from 7.0 to 6.0, represents a

Table 39–1. H⁺ concentration and pH of body fluids.

| | H⁺ Concentration | | |
	meq/L	mol/L	pH
Gastric HCl	150	0.15	0.8
Maximal urine acidity	0.03	3×10^{-5}	4.5
Plasma { Extreme acidosis	0.0001	1×10^{-7}	7.0
Normal	0.00004	4×10^{-8}	7.4
Extreme alkalosis	0.00002	2×10^{-8}	7.7
Pancreatic juice	0.00001	1×10^{-8}	8.0

tenfold increase in H⁺ concentration. It is important to remember that the pH of blood is the pH of **true plasma**—plasma that has been in equilibrium with red cells—because the red cells contain hemoglobin, which is quantitatively one of the most important blood buffers (see Chapter 35).

H⁺ Balance

The pH of the arterial plasma is normally 7.40 and that of venous plasma slightly lower. Technically, **acidosis** is present whenever the arterial pH is below 7.40, and **alkalosis** is present whenever it is above 7.40, although variations of up to 0.05 pH unit occur without untoward effects. The H⁺ concentrations in the ECF that are compatible with life cover an approximately 5-fold range, from 0.00002 meq/L (pH 7.70) to 0.0001 meq/L (pH 7.00).

Amino acids are utilized in the liver for gluconeogenesis, leaving as products NH_4^+ and HCO_3^- from their amino and carboxyl groups (Figure 39–3). The NH_4^+ is incorporated into urea and the protons that are formed are buffered intracellularly by HCO_3^-, so little NH_4^+ and HCO_3^- escape into the circulation. However, metabolism of sulfur-containing amino acids produces H_2SO_4, and metabolism of phosphorylated amino acids such as phosphoserine produces H_3PO_4. These strong acids enter the circulation and present a major H⁺ load to the buffers in the ECF. The H⁺ load from amino acid metabolism is normally about 50 meq/d. The CO_2 formed by metabolism in the tissues is in large part hydrated to H_2CO_3 (see Chapter 35), and the total H⁺ load from this source is over 12,500 meq/d. However, most of the CO_2 is excreted in the lungs, and only small quantities of the H⁺ remain to be excreted by the kidneys. Common sources of extra acid loads are strenuous exercise (lactic acid), diabetic ketosis (acetoacetic acid and β-hy-

droxybutyric acid), and ingestion of acidifying salts such as NH_4Cl and $CaCl_2$, which in effect add HCl to the body. Failure of diseased kidneys to excrete normal amounts of acid is also a cause of acidosis. Fruits are the main dietary source of alkali. They contain Na⁺ and K⁺ salts of weak organic acids, and the anions of these salts are metabolized to CO_2, leaving $NaHCO_3$ and $KHCO_3$ in the body. $NaHCO_3$ and other alkalinizing salts are sometimes ingested in large amounts, but a more common cause of alkalosis is loss of acid from the body as a result of vomiting of gastric juice rich in HCl. This is, of course, equivalent to adding alkali to the body.

Buffering

Buffering is defined in Chapter 1 and mentioned in Chapter 35 in the context of CO_2 transport in the body. It is of key importance in maintaining H⁺ homeostasis.

The Henderson-Hasselbalch Equation

The general equation for a buffer system is

$$HA \rightleftarrows H^+ + A^-$$

A⁻ represents any anion and HA the undissociated acid. If an acid stronger than HA is added to a solution containing this system, the equilibrium is shifted to the left. Hydrogen ions are "tied up" in the formation of

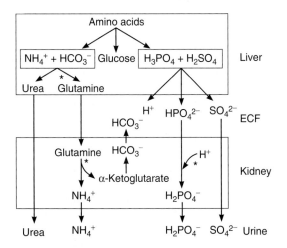

Figure 39–3. Role of the liver and kidneys in the handling of metabolically produced acid loads. Sites where regulation occurs are indicated by asterisks. (Modified and reproduced, with permission, from Knepper MA et al: Ammonium, urea, and systemic pH regulation. Am J Physiol 1987;235:F199.)

more undissociated HA, so the increase in H^+ concentration is much less than it would otherwise be. Conversely, if a base is added to the solution, H^+ and OH^- react to form H_2O; but more HA dissociates, limiting the decrease in H^+ concentration. By the law of mass action, the product of the concentrations of the products in a chemical reaction divided by the product of the concentration of the reactants at equilibrium is a constant:

$$\frac{[H^+][A^-]}{[HA]} = K$$

If this equation is solved for H^+ and put in pH notation (pH is the negative log of $[H^+]$), the resulting equation is that originally derived by Henderson and Hasselbalch to describe the pH changes resulting from addition of H^+ or OH^- to any buffer system (**Henderson-Hasselbalch equation**):

$$pH = pK + \log\frac{[A^-]}{[HA]}$$

It is apparent from these equations that the buffering capacity of a system is greatest when the amount of free anion is equal to the amount of undissociated HA, ie, when $[A^-]/[HA] = 1$, so that $\log [A^-]/[HA] = 0$ and pH = pK. This is why the most effective buffers in the body would be expected to be those with pKs close to the pH in which they operate. The pH of the blood is normally 7.4; that of the cells is probably about 7.2; and that of the urine varies from 4.5 to 8.0.

It should be noted that the equilibrium constant, K, applies only to infinitely dilute solutions in which interionic forces are negligible. In body fluids, it is more appropriate to use the apparent ionization constant, K'.

Buffers in Blood

In the blood, proteins—particularly the **plasma proteins**—are effective buffers because both their free carboxyl and their free amino groups dissociate:

$$RCOOH \rightleftarrows RCOO^- + H^+$$

$$pH = pK'_{RCOOH} + \log\frac{[RCOO^-]}{[RCOOH]}$$

$$RNH_3^+ \rightleftarrows RNH_2 + H^+$$

$$pH = pK'_{RNH_3} + \log\frac{[RNH_2]}{[RNH_3^+]}$$

Another important buffer system is provided by the dissociation of the imidazole groups of the histidine residues in **hemoglobin**:

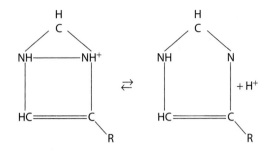

In the pH 7.0–7.7 range, the free carboxyl and amino groups of hemoglobin contribute relatively little to its buffering capacity. However, the hemoglobin molecule contains 38 histidine residues, and on this basis—plus the fact that hemoglobin is present in large amounts—the hemoglobin in blood has six times the buffering capacity of the plasma proteins. In addition, the action of hemoglobin is unique because the imidazole groups of deoxyhemoglobin dissociate less than those of oxyhemoglobin, making Hb a weaker acid and therefore a better buffer than HbO_2. Titration curves for Hb and HbO_2 are shown in Figure 39–4.

The third major buffer system in blood is the **carbonic acid-bicarbonate** system:

$$H_2CO_3 \rightleftarrows H^+ + HCO_3^-$$

The Henderson-Hasselbalch equation for this system is

$$pH = pK + \log\frac{[HCO_3^-]}{[H_2CO_3]}$$

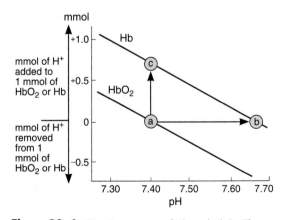

Figure 39–4. Titration curves of Hb and HbO_2. The arrow from a to c indicates the number of millimoles of H^+ that can be added without pH shift. The arrow from a to b indicates the pH shift on deoxygenation.

The pK for this system in an ideal solution is low (about 3), and the amount of H_2CO_3 is small and hard to measure accurately. However, in the body, H_2CO_3 is in equilibrium with CO_2.

$$H_2CO_3 \rightleftarrows CO_2 + H_2O$$

If the pK is changed to pK′ (see above) and $[CO_2]$ is substituted for $[H_2CO_3]$, the pK′ is 6.1.

$$pH = 6.10 + \log \frac{[HCO_3^-]}{[CO2]}$$

The clinically relevant form of this equation is as follows:

$$pH = 6.10 + \log \frac{[HCO_3^-]}{0.0301 \, P_{CO_2}}$$

since the amount of dissolved CO_2 is proportionate to the partial pressure of CO_2 and the solubility coefficient of CO_2 in mmol/L/mm Hg is 0.0301. $[HCO_3^-]$ cannot be measured directly, but pH and P_{CO_2} can be measured with suitable accuracy with pH and P_{CO_2} glass electrodes, and $[HCO_3^-]$ can then be calculated.

The pK′ of this system is still low relative to the pH of the blood, but the system is one of the most effective buffer systems in the body because the amount of dissolved CO_2 is controlled by respiration. In addition, the plasma concentration of HCO_3^- is regulated by the

Table 39–2. Principal buffers in body fluids.

Blood	$H_2CO_3 \rightleftarrows H^+ + HCO_3^-$
	$HProt \rightleftarrows H^+ + Prot^-$
	$HHb \rightleftarrows H^+ + Hb^-$
Interstitial fluid	$H_2CO_3 \rightleftarrows H^+ + HCO_3^-$
Intracellular fluid	$HProt \rightleftarrows H^+ + Prot^-$
	$H_2PO_4^- \rightleftarrows H^+ + HPO_4^{2-}$

kidneys. When H^+ is added to the blood, HCO_3^- declines as more H_2CO_3 is formed. If the extra H_2CO_3 were not converted to CO_2 and H_2O and the CO_2 excreted in the lungs, the H_2CO_3 concentration would rise. When enough H^+ has been added to halve the plasma HCO_3^-, the pH would have dropped from 7.4 to 6.0. However, not only is all the extra H_2CO_3 that is formed removed, but also the H^+ rise stimulates respiration and therefore produces a drop in P_{CO_2}, so that some additional H_2CO_3 is removed. The pH thus falls only to 7.2 or 7.3 (Figure 39–5).

The reaction $CO_2 + H_2O \rightleftarrows H_2CO_3$ proceeds slowly in either direction unless the enzyme **carbonic anhydrase** is present. There is no carbonic anhydrase in plasma, but there is an abundant supply in red blood cells. It is also found in high concentration in gastric acid-secreting cells (see Chapter 26) and in renal tubular cells (see Chapter 38). Carbonic anhydrase is a protein with a molecular weight of 30,000 that contains an atom of zinc in each molecule. It is inhibited by cyanide, azide, and sulfide. The sulfonamides also inhibit this enzyme, and sulfonamide derivatives have been used clinically as diuretics because of their inhibitory effects on carbonic anhydrase in the kidney (see Chapter 38).

The system $H_2PO_4^- \rightleftarrows H^+ + HPO_4^{2-}$ has a pK of 6.80. In the plasma, the phosphate concentration is too low for this system to be a quantitatively important buffer, but it is important intracellularly, and it frequently plays a significant role in the urine (see Chapter 38).

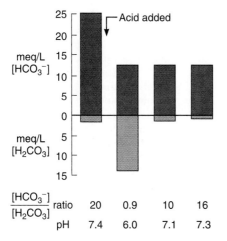

Figure 39–5. Buffering by the $H_2CO_3–HCO_3^-$ system in blood. The bars are drawn as if buffering occurred in separate steps in order to show the effect of the initial reaction, the reduction of H_2CO_3 to its previous value, and its further reduction by the increase in ventilation. In this case, $[H_2CO_3]$ is actually the concentration of dissolved CO_2, so that the meq/L values for it are arbitrary.

Buffering in Vivo

Buffering in vivo is of course not limited to the blood. The principal buffers in the blood, interstitial fluid, and intracellular fluid are listed in Table 39–2. The principal buffers in cerebrospinal fluid and urine are the bicarbonate and phosphate systems. In metabolic acidosis, only 15–20% of the acid load is buffered by the $H_2CO_3–HCO_3^-$ system in the ECF, and most of the remainder is buffered in cells. In metabolic alkalosis, about 30–35% of the OH^- load is buffered in cells, whereas in respiratory acidosis and alkalosis, almost all the buffering is intracellular.

In animal cells, the principal regulators of intracellular pH are HCO_3^- transporters. Those characterized to date include the $Cl^-HCO_3^-$ exchanger **band 3** (see Chapter 35), three Na^+-HCO_3^- cotransporters, and a K^+-HCO_3^- cotransporter.

Summary

When a strong acid is added to the blood, the major buffer reactions are driven to the left. The blood levels of the three "buffer anions"—Hb^- (hemoglobin), $Prot^-$ (protein), and HCO_3^-—consequently drop. The anions of the added acid are filtered into the renal tubules. They are accompanied ("covered") by cations, particularly Na^+, because electrochemical neutrality is maintained. By processes that are discussed in Chapter 38, the tubules replace the Na^+ with H^+ and in so doing reabsorb equimolar amounts of Na^+ and HCO_3^-, thus conserving the cations, eliminating the acid, and restoring the supply of buffer anions to normal. When CO_2 is added to the blood, similar reactions occur, except that since it is H_2CO_3 that is formed, the plasma HCO_3^- rises rather than falls.

Respiratory Acidosis & Alkalosis

A rise in arterial P_{CO_2} due to decreased ventilation causes **respiratory acidosis.** The CO_2 that is retained is in equilibrium with H_2CO_3, which in turn is in equilibrium with HCO_3^-, so that the plasma HCO_3^- rises and a new equilibrium is reached at a lower pH. This can be indicated graphically on a plot of plasma HCO_3^- concentration versus pH (Figure 39–6). Conversely, a decline in P_{CO_2} causes **respiratory alkalosis.**

The initial changes shown in Figure 39–6 are those that occur independently of any compensatory mechanism; ie, they are those of **uncompensated** respiratory acidosis or alkalosis. In either situation, changes are produced in the kidneys, which then tend to **compensate** for the acidosis or alkalosis, adjusting the pH toward normal.

Renal Compensation

HCO_3^- reabsorption in the renal tubules depends not only on the filtered load of HCO_3^-, which is the product of the GFR and the plasma HCO_3^- level, but also on the rate of H^+ secretion by the renal tubular cells, since HCO_3^- is reabsorbed by exchange for H^+. The rate of H^+ secretion—and hence the rate of HCO_3^- reabsorption—is proportionate to the arterial P_{CO_2}, probably because the more CO_2 that is available to form H_2CO_3 in the cells, the more H^+ can be secreted (see Chapter 38). Furthermore, when the P_{CO_2} is high,

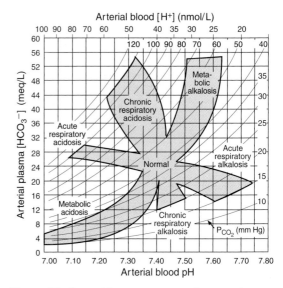

Figure 39–6. Acid-base nomogram showing changes in the P_{CO_2} (curved lines), plasma HCO_3^-, and pH of arterial blood in respiratory and metabolic acidosis. Note the shifts in HCO_3^- and pH as acute respiratory acidosis and alkalosis are compensated, producing their chronic counterparts. (Reproduced, with permission, from Cogan MG, Rector FC Jr: Acid-base disorders. In: *The Kidney*, 4th ed. Brenner BM, Rector FC Jr [editors]. Saunders, 1991.)

the interior of most cells becomes more acidic (see Chapter 35). In respiratory acidosis, renal tubular H^+ secretion is therefore increased, removing H^+ from the body; and even though the plasma HCO_3^- is elevated, HCO_3^- reabsorption is increased, further raising the plasma HCO_3^-. This renal compensation for respiratory acidosis is shown graphically in the shift from acute to chronic respiratory acidosis in Figure 39–6. Cl^- excretion is increased, and plasma Cl^- falls as plasma HCO_3^- is increased. Conversely, in respiratory alkalosis, the low P_{CO_2} hinders renal H^+ secretion, HCO_3^- reabsorption is depressed, and HCO_3^- is excreted, further reducing the already low plasma HCO_3^- and lowering the pH toward normal (Figure 39–6).

Metabolic Acidosis

When acids stronger than HHb and the other buffer acids are added to blood, **metabolic acidosis** is produced; and when the free H^+ level falls as a result of addition of alkali or removal of acid, **metabolic alkalosis** results. If, for example, H_2SO_4 is added, the H^+ is buffered and the Hb^-, $Prot^-$, and HCO_3^- levels in plasma drop. The H_2CO_3 formed is converted to H_2O

and CO_2, and the CO_2 is rapidly excreted via the lungs. This is the situation in **uncompensated** metabolic acidosis (Figure 39–7). Actually, the rise in plasma H^+ stimulates respiration, so that the P_{CO_2}, instead of rising or remaining constant, is reduced. This **respiratory compensation** raises the pH even further. The **renal** compensatory mechanisms then bring about the excretion of the extra H^+ and return the buffer systems to normal.

Renal Compensation

The anions that replace HCO_3^- in the plasma in metabolic acidosis are filtered, each with a cation (principally Na^+), thus maintaining electrical neutrality. The renal tubular cells secrete H^+ into the tubular fluid in exchange for Na^+; and for each H^+ secreted, one Na^+ and one HCO_3^- are added to the blood (see Chapter 38). The limiting urinary pH of 4.5 would be reached rapidly and the total amount of H^+ secreted would be small if there were no buffers in the urine that "tied up" H^+. However, secreted H^+ reacts with HCO_3^- to form CO_2 and H_2O (bicarbonate reabsorption); with HPO_4^{2-} to form $H_2PO_4^-$; and with NH_3 to form NH_4^+. In this way large amounts of H^+ can be secreted, permitting correspondingly large amounts of HCO_3^- to be returned to (in the case of bicarbonate reabsorption) or added to the depleted body stores and large numbers of the cations to be reabsorbed. It is only when the acid load is very large that cations are lost with the anions, producing diuresis and depletion of body cation stores. In chronic acidosis, glutamine synthesis in the liver is increased, using some of the NH_4^+ that usually is converted to urea (Figure 39–3), and the glutamine provides the kidneys with an additional source of NH_4^+ (see Chapter 38). NH_3 secretion increases over a period of days (adaptation of NH_3 secretion; see Chapter 38), further improving the renal compensation for acidosis. In addition, the metabolism of glutamine in the kidneys produces α-ketoglutarate, and this in turn is decarboxylated, producing HCO_3^-, which enters the bloodstream and helps buffer the acid load (Figure 39–3).

The overall reaction in blood when a strong acid such as H_2SO_4 is added is

$$2NaHCO_3 + H_2SO_4 \rightarrow Na_2SO_4 + 2H_2CO_3$$

For each mole of H^+ added, 1 mole of $NaHCO_3$ is lost. The kidney in effect reverses the reaction:

$$Na_2SO_4 + 2H_2CO_3 \rightarrow 2NaHCO_3 + 2H^+ + SO_4^{2-}$$

and the H^+ and SO_4^{2-} are excreted. Of course, H_2SO_4 is not excreted as such, the H^+ appearing in the urine as titratable acidity and NH_4^+.

In metabolic acidosis, the respiratory compensation tends to inhibit the renal response in the sense that the induced drop in P_{CO_2} hinders acid secretion, but it also decreases the filtered load of HCO_3^- and so its net inhibitory effect is not great.

Metabolic Alkalosis

In metabolic alkalosis, the plasma HCO_3^- level and pH rise (Figure 39–7). The respiratory compensation is a decrease in ventilation produced by the decline in H^+ concentration, and this elevates the P_{CO_2}. This brings the pH back toward normal while elevating the plasma HCO_3^- level still further. The magnitude of this compensation is limited by the carotid and aortic chemoreceptor mechanisms, which drive the respiratory center if there is any appreciable fall in the arterial P_{O_2}. In metabolic alkalosis, more renal H^+ secretion is expended in reabsorbing the increased filtered load of HCO_3^-; and if the HCO_3^- level in plasma exceeds 26–28 meq/L, HCO_3^- appears in the urine. The rise in P_{CO_2} inhibits the renal compensation by facilitating acid secretion, but its effect is relatively slight.

Clinical Evaluation of Acid-Base Status

Some examples of acid-base disturbances are shown in Table 39–3.

In evaluating disturbances of acid-base balance, it is important to know the pH and HCO_3^- content of ar-

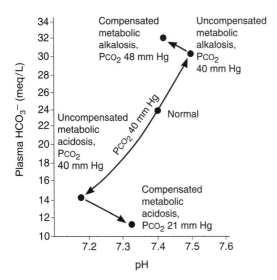

Figure 39–7. Changes in true plasma pH, HCO_3^-, and P_{CO_2} in metabolic acidosis and alkalosis. (This is called a Davenport diagram and is based on Davenport HW: *The ABC of Acid-Base Chemistry*, 6th ed. Univ of Chicago Press, 1974.)

Table 39–3. Plasma pH, HCO_3^-, and PCO_2 values in various typical disturbances of acid-base balance.[1]

Condition	Arterial Plasma			Cause
	pH	HCO_3^- (meq/L)	PCO_2 (mm Hg)	
NORMAL	7.40	24.1	40	
Metabolic acidosis	7.28	18.1	40	NH_4Cl ingestion
	6.96	5.0	23	Diabetic acidosis
Metabolic alkalosis	7.50	30.1	40	$NaHCO_3^-$ ingestion
	7.56	49.8	58	Prolonged vomiting
Respiratory acidosis	7.34	25.0	48	Breathing 7% CO_2
	7.34	33.5	64	Emphysema
Respiratory alkalosis	7.53	22.0	27	Voluntary hyperventilation
	7.48	18.7	26	Three-week residence at 4000-m altitude

[1]In the diabetic acidosis and prolonged vomiting examples, respiratory compensation for primary metabolic acidosis and alkalosis has occurred, and the PCO_2 has shifted from 40 mm Hg. In the emphysema and high-altitude examples, renal compensation for primary respiratory acidosis and alkalosis has occurred and has made the deviations from normal of the plasma HCO_3^- larger than they would otherwise be.

terial plasma. Reliable pH determinations can be made with a pH meter and a glass pH electrode. The HCO_3^- content of plasma cannot be measured directly, but the PCO_2 can be measured with a CO_2 electrode and the HCO_3^- concentration calculated, as noted above. The PCO_2 is 7–8 mm Hg higher and the pH 0.03–0.04 unit lower in venous than arterial plasma because venous blood contains the CO_2 being carried from the tissues to the lungs. Therefore, the calculated HCO_3^- concentration is about 2 mmol/L higher. However, if this is kept in mind, free-flowing venous blood can be substituted for arterial blood in most clinical situations.

A measurement that is of some value in the differential diagnosis of metabolic acidosis is the **anion gap.** This gap, which is something of a misnomer, refers to the difference between the concentration of cations other than Na^+ and the concentration of anions other than Cl^- and HCO_3^- in the plasma. It consists for the most part of proteins in the anionic form, HPO_4^{2-}, SO_4^{2-}, and organic acids, and a normal value is about 12 meq/L. It is increased when the plasma concentration of K^+, Ca^{2+}, or Mg^+ is decreased; when the concentration of or the charge on plasma proteins is increased; or when organic anions such as lactate or foreign anions accumulate in blood. It is decreased when cations are increased or when plasma albumin is decreased. The anion gap is increased in metabolic acidosis due to ketoacidosis, lactic acidosis, and other forms of acidosis in which organic anions are increased.

It is not increased in hyperchloremic acidosis due to ingestion of NH_4Cl or carbonic anhydrase inhibitors.

The Siggaard-Andersen Curve Nomogram

Use of the Siggaard-Andersen curve nomogram (Figure 39–8) to plot the acid-base characteristics of arterial blood is helpful in clinical situations. This nomogram has PCO_2 plotted on a log scale on the vertical axis and pH on the horizontal axis. Thus, any point to the left of a vertical line through pH 7.40 indicates acidosis, and any point to the right indicates alkalosis. The position of the point above or below the horizontal line through a PCO_2 of 40 mm Hg defines the effective degree of hypoventilation or hyperventilation.

If a solution containing $NaHCO_3$ and no buffers were equilibrated with gas mixtures containing various amounts of CO_2, the pH and PCO_2 values at equilibrium would fall along the dashed line on the left in Figure 39–8 or a line parallel to it. If buffers were present, the slope of the line would be greater; and the greater the buffering capacity of the solution, the steeper the line. For normal blood containing 15 g of hemoglobin/dL, the **CO_2 titration line** passes through the 15-g/dL mark on the hemoglobin scale (on the underside of the upper curved scale) and the point where the $PCO_2 = 40$ mm Hg and pH = 7.40 lines intersect, as shown in Figure 39–8. When the hemoglobin content of the blood is low, there is significant loss of buffering

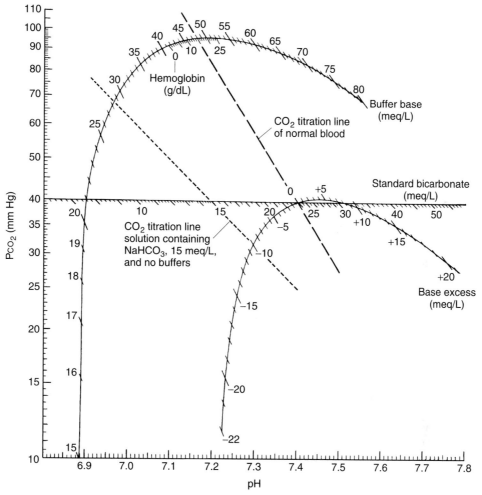

Figure 39–8. Siggaard-Andersen curve nomogram. (Courtesy of O Siggaard-Andersen and Radiometer, Copenhagen, Denmark.)

capacity, and the slope of the CO_2 titration line diminishes. However, blood of course contains buffers in addition to hemoglobin, so that even the line drawn from the zero point on the hemoglobin scale through the normal P_{CO_2}-pH intercept is steeper than the curve for a solution containing no buffers.

For clinical use, arterial blood or arterialized capillary blood is drawn anaerobically and its pH measured. The pHs of the same blood after equilibration with each of two gas mixtures containing different known amounts of CO_2 are also determined. The pH values at the known P_{CO_2} levels are plotted and connected to provide the CO_2 titration line for the blood sample. The pH of the blood sample before equilibration is plotted on this line, and the P_{CO_2} of the sample is read off the vertical scale. The **standard bicarbonate** con-

tent of the sample is indicated by the point at which the CO_2 titration line intersects the bicarbonate scale on the $P_{CO_2} = 40$ mm Hg line. The standard bicarbonate is not the actual bicarbonate concentration of the sample but, rather, what the bicarbonate concentration would be after elimination of any respiratory component. It is a measure of the alkali reserve of the blood, except that it is measured by determining the pH rather than the total CO_2 content of the sample after equilibration. Like the alkali reserve, it is an index of the degree of metabolic acidosis or alkalosis present.

Additional graduations on the upper curved scale of the nomogram (Figure 39–8) are provided for measuring **buffer base** content; the point where the CO_2 calibration line of the arterial blood sample intersects this scale shows the meq/L of buffer base in the sample.

The buffer base is equal to the total number of buffer anions (principally Prot⁻, HCO_3^-, and Hb⁻; see Chapter 35) that can accept hydrogen ions in the blood. The normal value in an individual with 15 g of hemoglobin per deciliter of blood is 48 meq/L.

The point at which the CO_2 calibration line intersects the lower curved scale on the nomogram indicates the **base excess.** This value, which is positive in alkalosis and negative in acidosis, is the amount of acid or base that would restore 1 L of blood to normal acid-base composition at a P_{CO_2} of 40 mm Hg. It should be noted that a base deficiency cannot be completely corrected simply by calculating the difference between the normal standard bicarbonate (24 meq/L) and the actual standard bicarbonate and administering this amount of $NaHCO_3$ per liter of blood; some of the added HCO_3^- is converted to CO_2 and H_2O, and the CO_2 is lost in the lungs. The actual amount that must be added is roughly 1.2 times the standard bicarbonate deficit, but the lower curved scale on the nomogram, which has been developed empirically by analyzing many blood samples, is more accurate.

In treating acid-base disturbances, one must, of course, consider not only the blood but also all the body fluid compartments. The other fluid compartments have markedly different concentrations of buffers. It has been determined empirically that administration of an amount of acid (in alkalosis) or base (in acidosis) equal to 50% of the body weight in kilograms times the blood base excess per liter will correct the acid-base disturbance in the whole body. At least when the abnormality is severe, however, it is unwise to attempt such a large correction in a single step; instead, about half the indicated amount should be given and the arterial blood acid-base values determined again. The amount required for final correction can then be calculated and administered. It is also worth noting that, at least in lactic acidosis, $NaHCO_3$ decreases cardiac output and lowers blood pressure, so it should be used with caution.

REFERENCES FOR SECTION VIII: FORMATION & EXCRETION OF URINE

Adrogué HJ, Madius NE: Management of life-threatening acid-base disorders. N Engl J Med 1998;338:26.

Anderson K-E: Pharmacology of lower urinary tract smooth muscles and penile erectile tissue. Pharmacol Rev 1993;45:253.

Brenner BM, Rector FC Jr (editors): *The Kidney,* 6th ed. 2 vols. Saunders, 1999.

Brezis M, Rosen S: Hypoxia of the renal medulla—Its implications for disease. N Engl J Med 1995;332:647.

Brown D, Stow JL: Protein trafficking and polarity in kidney epithelium: From cell biology to physiology. Physiol Rev 1996;76:245.

Davenport HW: *The ABC of Acid-Base Chemistry,* 6th ed. Univ of Chicago Press, 1974.

DiBona GF, Kopp UC: Neural control of renal function. Physiol Rev 1997;77:75.

Fyfe GF, Quinn A, Canessa CM: Structure and function of the Mec-ENaC family of ion channels. Semin Nephrol 1998;18:138.

Garcia NH, Ramsey CR, Knox FG: Understanding the role of paracellular transport in the proximal tubule. News Physiol Sci 1998;13:38.

Gebow PA: Disorders associated with an altered anion gap. Kidney Int 1985;27:472.

Gennari FJ: Hypokalemia. N Engl J Med 1998;339:451.

Haas M: The Na-K-Cl cotransporters. Am J Physiol 1994;256:C869.

Halperin ML: *Fluid, Electrolyte, and Acid-Base Physiology,* 3rd ed. Saunders, 1998.

Hutch JA: *Anatomy and Physiology of the Bladder, Trigone and Urethra.* Appleton-Century-Crofts, 1972.

Ito S: Role of nitric oxide in glomerular arterioles and macula densa. News Physiol Sci 1994;9:115.

Klahr S, Miller SB: Acute oliguria. N Engl J Med 1998;338:671.

Knepper MA, Packer R, Good DW: Ammonium transport in the kidney. Physiol Rev 1989;69:179.

Madsen KM, Tisher CC: Structural-functional relationships along the distal nephron. Am J Physiol 1986;250:F1.

Mené P, Simonson MS, Dunn MJ: Physiology of the mesangial cell. Physiol Rev 1989;19:1347.

Morel F: Sites of hormone action in the mammalian nephron. Am J Physiol 1981;240:F159.

Morris PJ (editor): *Kidney Transplantation: Principles and Practice,* 5th ed. Saunders, 2001.

Nielsen S et al: Aquaporins in the kidney: from molecules to medicine. Physiol Rev 2002;82:205.

Pastan S, Bailey J: Dialysis therapy. N Engl J Med 1998;338:1428.

Robinson AG, Verbalis JG: Diabetes insipidus. Curr Ther Endocrinol Metab 1997;6:1.

Roos A, Boron WF: Intracellular pH. Physiol Rev 1981;61:296.

Scheinman SJ et al: Genetic disorders of renal electrolyte transport. N Engl J Med 1999;340:1177.

Schild L: The ENaC channel as the primary determinant of two human diseases: Liddle syndrome and pseudohypoaldosteronism. Nephrologie 1996;17:395.

Schmidt-Nielson K: Countercurrent systems in animals. Sci Am (May) 1981;244:118.

Seldin DW, Giebisch G (editors): *The Kidney: Physiology and Pathophysiology,* 2nd ed. 3 vols. Raven Press, 1991.

Spring KR: Epithelial fluid transport—a century of investigation. News Physiol Sci 1999;14:92.

Stella A, Zanchetti A: Functional role of renal afferents. Physiol Rev 1991;71:659.

Suthanthiran M, Strom TB: Renal transplantation. N Engl J Med 1994;331:365.

Wolf G, Neilson EG: Angiotensin II as a renal cytokine. News Physiol Sci 1994;9:40.

Wright FS: Intrarenal regulation of glomerular filtration rate. J Hypertens 1984;2(Suppl 1):105.

Zeidel ML: Hormonal regulation of inner medullary collecting duct sodium transport. Am J Physiol 1993;265:F159.

Self-Study: Objectives, Essay Questions, & Multiple-Choice Questions

CHAPTER 1

This chapter is a review of the cellular, molecular, and general basis of medical physiology. The material in the chapter should help students to—

- Name the different fluid compartments in the body, the size of each, and ways in which their sizes can be measured.
- Define moles, equivalents, and osmoles.
- List and compare the various passive and active forces that produce movement of substances across cell membranes.
- Define osmosis, and give examples of its role in moving fluid from one location in the body to another.
- Describe and give examples of secondary active transport.
- Define and explain the resting membrane potential.
- Know the various organelles in cells and the functions of each.
- Know the chemical nature and physiologic significance of the compounds that make up the cell membrane.
- Understand in general terms the structure of DNA and RNA and the role these nucleotides and other substances in the cell play in the process of protein synthesis.
- Define the processes of exocytosis and endocytosis, and describe the contribution of each to normal cell function.
- Describe the principal ways that the chemical messengers in the extracellular fluid produce changes inside cells, including changes in gene expression.
- Describe the role of G proteins as intracellular signaling molecules.
- List the main theories advanced to explain aging.
- Define homeostasis, and give examples of homeostatic mechanisms.

General Questions

1. What is the role of Na^+-K^+ ATPase in physiology?
2. Compare the compositions of plasma, interstitial fluid, and intracellular fluid. Explain the differences.
3. What happens to resting membrane potential when the extracellular K^+ concentration is increased from 5 to 10 meq/L? What happens when the extracellular Na^+ concentration is increased from 142 to 155 meq/L? Why?
4. Why do red blood cells swell and eventually burst when they are placed in a solution of 0.3% sodium chloride?
5. What are the steps between DNA and the formation of a protein that is secreted by the cell?
6. Describe the structure and function of the proteins found in cell membranes.
7. What is the function of calmodulin? Why is this function important?
8. What is receptor-mediated endocytosis?
9. Discuss aging from the point of view of its cellular and molecular mechanisms that underlie it.

Multiple-Choice Questions

In questions 1–19, select the single best answer.

1. Cell membranes
   (A) contain relatively few protein molecules
   (B) are impermeable to fat-soluble substances
   (C) in some tissues permit the transport of glucose at a greater rate in the presence of insulin
   (D) are freely permeable to electrolytes but not to proteins
   (E) have a stable composition throughout the life of the cell
2. The primary force moving water molecules from the blood plasma to the interstitial fluid is

(A) active transport

(B) cotransport with H^+

(C) facilitated diffusion

(D) cotransport with Na^+

(E) filtration

3. Second messengers

(A) are substances that interact with first messengers outside cells

(B) are substances that bind to first messengers in the cell membrane

(C) are hormones secreted by cells in response to stimulation by another hormone

(D) mediate the intracellular responses to many different hormones and neurotransmitters

(E) are not formed in the brain

4. Which of the following is involved in the regulation of apoptosis?

(A) Glucose-6-phosphatase

(B) Adenylyl cyclase

(C) Nitric oxide synthase

(D) Aquaporin 2

(E) Cytochrome c

5. Proteins that are secreted by cells are

(A) not synthesized on membrane-bound ribosomes

(B) initially synthesized with a signal peptide or leader sequence at their C terminal

(C) found in vesicles and secretory granules

(D) moved across the cell membranes by endocytosis

(E) secreted in a form that is larger than the form present in the endoplasmic reticulum

6. Which of the following cell membrane channels has a single pore surrounded by five protein subunits?

(A) CLC Cl^- channels

(B) Aquaporins

(C) Potassium channels

(D) Glucose transporters

(E) $GABA_A$ channels

7. Deuterium oxide and inulin are injected into a normal 30-year-old man. The volume of distribution of deuterium oxide is found to be 42 L and that of inulin 14 L.

(A) The man's intracellular fluid volume is about 14 L

(B) The man's intracellular fluid volume is about 28 L

(C) The man's plasma volume is about 7 L

(D) The man's interstitial fluid volume is about 9 L

(E) The man's total body water cannot be determined from these data

8. Which of the following diseases is due to a mutation in the mitochondrial genome?

(A) Cystic fibrosis

(B) Leber's hereditary optic neuropathy

(C) Hypothalamic diabetes insipidus

(D) Familial male precocious puberty

(E) Hirschsprung's disease

9. Which of the following receptors does *not* span the cell membrane seven times?

(A) β-Adrenergic receptor

(B) Rhodopsin

(C) 5-HT_{IC} receptor

(D) Mineralocorticoid receptor

(E) LH receptor

10. Which of the following acts directly on the nucleus to produce physiologic effects?

(A) IL-1β

(B) Growth hormone

(C) Ghrelin

(D) Dopamine

(E) Estradiol

11. Which of the following is a lysosomal storage disease?

(A) Cystic fibrosis

(B) Tay-Sachs disease

(C) Duchene muscular dystrophy

(D) Testotoxicosis

(E) Long QT syndrome

In questions 12–15, match the disease in each question with the lettered item that is most closely associated with it. Each lettered item may be selected once, more than once, or not at all.

(A) Abnormal receptor for catecholamines

(B) Abnormal G protein

(C) Antibodies against receptors

(D) Deficiency of receptors for extracellular protein

(E) Abnormal receptor for extracellular protein

12. Many cases of Graves' disease

13. Some cases of pseudohypoparathyroidism

14. Some cases of familial hypercholesterolemia

15. Some cases of acromegaly

In questions 16–20, which refer to Figure 1–A, match the function in each question with the lettered part of the cell that is most closely associated with it. Each lettered item may be selected once, more than once, or not at all.

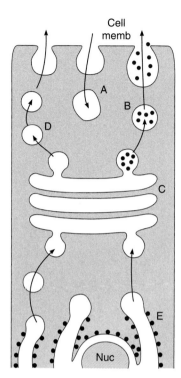

Figure 1–A. Nuc, nucleus; Cell memb, cell membrane.

16. Removal of signal peptides
17. Hydrolysis of prohormones
18. Recycling of cell membrane receptors
19. Glycosylation of proteins
20. Constitutive secretion

*In questions 21–25, indicate whether the first item is greater than (**G**), the same as (**S**), or less than (**L**) the second item.*

21. Hydrogen ion concentration in lysosomes
    G S L
    Hydrogen ion concentration in cytoplasm of cells
22. Contribution of normal concentration of plasma glucose to total plasma osmolality
    G S L
    Contribution of normal concentrations of plasma Na^+ to total plasma osmolality
23. Calculated plasma volume when some of the dye used to measure it is unknowingly injected subcutaneously instead of intravenously
    G S L
    Calculated plasma volume when all of the dye used to measure it is injected intravenously

24. Concentration of Ca^{2+} in intracellular fluid
    G S L
    Concentration of Ca^{2+} in interstitial fluid
25. Size of mitochondrial genome
    G S L
    Size of nuclear genome

CHAPTER 2

This is the first of four chapters concerned with nerves and muscles and the interactions between them. It deals with the morphology of neurons, the electrical and ionic events that underlie their excitation, and their ability to conduct impulses. Glia is also considered. The material in the chapter should help students to—

- Name the various parts of a neuron and the functions of each.
- Distinguish between unmyelinated and myelinated neurons, describe the chemical nature of myelin, and summarize the differences in the ways in which unmyelinated and myelinated neurons conduct impulses.
- Define orthograde axoplasmic transport and retrograde axoplasmic transport and describe the molecular motors involved in each.
- Define excitation and conduction, and describe the changes in ionic movements that underlie electrotonic potentials and action potentials.
- Explain the following characteristics of a nerve impulse: latent period, firing level, spike potential, after-depolarization, and after-hyperpolarization.
- Describe and explain the compound action potential of mixed nerves.
- List the various nerve fiber types found in humans, and comment on their significance in terms of the normal and abnormal function of peripheral nerves.
- Describe the various types of glial cells and describe the function of each type.
- List the established neutrotrophins and the kinds of neurons on which they act. Describe their actions and their mechanism of action.

General Questions

1. Of what benefit to the animal is the evolution of myelinated neurons?
2. Compare the function of dendrites with the function of axons.
3. Describe and explain the effects of tetrodotoxin on the resting membrane potential and the action potential of nerve cells.

4. Which nerve fiber types would you expect to find in a sympathetic nerve that runs from the celiac ganglion to the intestine? Explain your answer.

5. How do the absolute and relative refractory periods correlate in time with the various phases of the action potential? Why is the excitability of the nerve reduced during these periods?

Multiple-Choice Questions

In questions 1–6, select the single best answer.

1. An axon is connected to stimulating and recording electrodes as shown in Figure 2–A. The distance from the anode of the stimulating electrode to the intracellular recording electrode is 6 cm and from the cathode of the stimulating electrode to the intracellular recording electrode is 4.5 cm. When the axon is stimulated, the latent period is 1.5 ms. What is the conduction velocity of the axon?
   (A) 15 m/s
   (B) 30 m/s
   (C) 40 m/s
   (D) 67.5 m/s
   (E) This cannot be determined from the information given

2. What percentage of the human genes are involved in the formation and function of the nervous system?
   (A) 5
   (B) 15
   (C) 25
   (D) 40
   (E) 60

3. Which of the following has the slowest conduction velocity?
   (A) Aα fibers
   (B) Aβ fibers
   (C) Aγ fibers
   (D) B fibers
   (E) C fibers

4. A man falls into a deep sleep with one arm under his head. This arm is paralyzed when he awakens, but it tingles, and pain sensation in it is still intact. The reason for the loss of motor function without loss of pain sensation is that in the nerves to his arm—
   (A) A fibers are more susceptible to hypoxia than B fibers
   (B) A fibers are more sensitive to pressure than C fibers
   (C) C fibers are more senitive to pressure than A fibers
   (D) motor nerves are more affected by sleep than sensory nerves
   (E) sensory nerves are nearer the bone than motor nerves and hence are less affected by pressure

5. Which part of a neuron has the highest concentration of Na⁺ channels per square millimeter of cell membrane?
   (A) Dendrites
   (B) Cell body near dendrites
   (C) Initial segment
   (D) Axonal membrane under myelin
   (E) None of the above

6. Which of the following statements about nerve growth factor is *not* true?
   (A) It is made up of three polypeptide subunits
   (B) It is found in high concentration in the submandibular salivary glands of female mice
   (C) It is necessary for the growth and development of the sympathetic nervous system
   (D) It is picked up by nerves from the organs they innervate
   (E) It is present in the brain

In questions 7–10, indicate whether the first item is greater than (G), the same as (S), or less than (L) the second item.

7. Number of glial cells in brain
   G S L
   Number of neurons in brain

8. Rate of conduction in small-diameter axons
   G S L
   Rate of conduction in large-diameter axons

9. Excitability of neuron when resting membrane potential is increased to −80 mV
   G S L
   Excitability of neuron when resting membrane potential is reduced to −60 mV

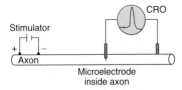

Figure 2–A.

10. Total duration of compound action potential of a mixed nerve 5 mm from stimulating electrode

   G S L

   Total duration of compound action potential of a mixed nerve 50 mm from stimulating electrode

CHAPTER 3

This chapter describes the morphologic and functional characteristics of the three types of muscle found in the body: skeletal, cardiac, and smooth muscle. The material in the chapter should help students to—

- Describe the gross and microscopic anatomy of skeletal muscle, including the cross-striations, the relation of actin to myosin, and the sarcotubular system.
- Describe and explain the interaction between actin and myosin.
- List the sequence of electrical and ionic events leading from an action potential in the motor nerve to contraction of a skeletal muscle, and discuss the significance of each.
- Compare isometric and isotonic contractions.
- Define dystrophin, discuss its normal function, and describe the diseases that occur when it is abnormal or absent.
- Describe the sources of energy for muscle contraction, and explain how energy is transferred to the contractile mechanism.
- Define oxygen debt, and describe its role in muscle function during exercise.
- Describe the differences between fast and slow skeletal muscles.
- Define the term motor unit, and discuss the effects of denervation on skeletal muscle.
- Describe the ionic events underlying the action potential in cardiac muscle.
- Describe the fluctuations in membrane potential seen in pacemaker tissue and the ionic events responsible for the prepotential.
- Compare the electrical and mechanical events in smooth and cardiac muscle with those in skeletal muscle, and compare their responses to acetylcholine and norepinephrine.

General Questions

1. Define total tension, passive tension, active tension, resting length, and equilibrium length. How does the relationship of actin to myosin explain the length-tension curve of skeletal muscle?

2. "Muscle is a machine for converting chemical into mechanical energy." Analyze and discuss this statement.

3. What is the minimum stimulation frequency at which tetanus occurs in a fast muscle with a twitch duration of 7.5 ms, and in a slow muscle with a twitch duration of 100 ms?

4. Discuss the factors that permit gradation of skeletal muscle responses in a living, intact animal.

5. Discuss the length-tension relationship in cardiac muscle in normal and abnormal hearts.

Multiple-Choice Questions

In questions 1–10, select the single best answer.

1. The action potential of skeletal muscle
   (A) has a prolonged plateau phase
   (B) spreads inward to all parts of the muscle via the T tubules
   (C) causes the immediate uptake of Ca^{2+} into the lateral sacs of the sarcoplasmic reticulum
   (D) is longer than the action potential of cardiac muscle
   (E) is not essential for contraction

2. The functions of tropomyosin in skeletal muscle include
   (A) sliding on actin to produce shortening
   (B) releasing Ca^{2+} after initiation of contraction
   (C) binding to myosin during contraction
   (D) acting as a "relaxing protein" at rest by covering up the sites where myosin binds to actin
   (E) generating ATP, which it passes to the contractile mechanism

3. The cross-bridges of the sarcomere in skeletal muscle are made up of
   (A) actin
   (B) myosin
   (C) troponin
   (D) tropomyosin
   (E) myelin

4. The contractile response in skeletal muscle
   (A) starts after the action potential is over
   (B) does not last as long as the action potential
   (C) produces more tension when the muscle contracts isometrically than when the muscle contracts isotonically
   (D) produces more work when the muscle contracts isometrically than when the muscle contracts isotonically
   (E) decreases in magnitude with repeated stimulation

5. Gap junctions

(A) are absent in cardiac muscle

(B) are present but of little functional importance in cardiac muscle

(C) are present and provide the pathway for rapid spread of excitation from one cardiac muscle fiber to another

(D) are absent in smooth muscle

(E) connect the sarcotubular system to individual skeletal muscle cells

In questions 6–10, select

A   if the item is associated with (a) below,

B   if the item is associated with (b) below,

C   if the item is associated with both (a) and (b), and

D   if the item is associated with neither (a) nor (b)

(a) Strength of contraction of a given muscle

(b) Duration of contraction of the muscle

6. Length of muscle at the start of contraction

7. Type of myosin heavy chain isoform in the muscle

8. Degree to which firing in the motor nerve to the muscle is asynchronous

9. Collagen content of muscle

10. Dystrophin content of muscle

In questions 11–13, match the statement in each question with the lettered substance that is most closely associated with it. Each lettered substance may be selected once, more than once, or not at all.

(A) Epinephrine

(B) Inhibin

(C) Testosterone

(D) Thyroid hormones

(E) Cortisol

11. Decreased muscle glycogen content

12. Increased muscle mass

13. Increased speed of contraction

*In questions 14–16, indicate whether the first item is greater than (**G**), the same as (**S**), or less than (**L**) the second item.*

14. Duration of the increase in Ca^{2+} influx during the action potential of cardiac muscle

G S L

Duration of the increase in Ca^{2+} influx during the action potential of skeletal muscle

15. Amount of dystrophin at neuromuscular junction

G S L

Amount of utrophin at neuromuscular junction

16. Excitability of uterine smooth muscle after administration of progesterone

G S L

Excitability of uterine smooth muscle after administration of estrogen

CHAPTER 4

This chapter is concerned with the synaptic junctions between neurons, with the neuromuscular junction, and with the neurotransmitters and receptors involved in transmission at these junctions. The material in the chapter should help students to—

• Describe the main morphologic features of synapses.

• Distinguish between chemical and electrical transmission at synapses.

• Define convergence and divergence in neural networks, and discuss their implications.

• Describe fast and slow excitatory and inhibitory postsynaptic potentials, outline the ionic fluxes that underlie them, and explain how the potentials interact to generate action potentials.

• Define and give examples of direct inhibition, indirect inhibition, presynaptic inhibition, and postsynaptic inhibition.

• Define synaptic plasticity, and discuss its relation to learning.

• List 12 neurotransmitters and the principal sites in the nervous system at which they are released.

• Define the term opioid peptide, list the principal opioid peptides in the body, and name the precursor molecules from which they come.

• Describe the receptors for catecholamines, acetylcholine, 5-HT, and opioids.

• Summarize the steps involved in the biosynthesis, release, action, and removal from the synaptic cleft of the various synaptic transmitters.

• Define and explain habituation, sensitization, long-term potentiation (LTP), and long-term depression (LTD).

• Describe the neuromuscular junction, and explain how action potentials in the motor neuron at the junction lead to contraction of the skeletal muscle.

• Define and explain denervation hypersensitivity.

General Questions

1. Conduction can occur in either direction along an axon but in only one direction at a chemical synaptic junction. Why?

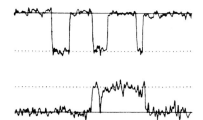

Figure 4–A. Patch clamp records obtained from two different neurons. Downward deflection indicates inward current pulses. Upward deflection indicates outward current pulses.

2. What types of chemicals are known to serve as synaptic transmitters?

3. One of the records in Figure 4–A was obtained with acetylcholine acting as an excitatory transmitter, and the other was obtained with γ-aminobutyric acid (GABA) acting as an inhibitory transmitter. Which is which? Explain your answer.

4. What is the role of reuptake in the metabolism of neurotransmitters? How is it produced? What drugs affect it?

5. "In the human forebrain, the ratio of synapses to neurons has been calculated to be 40,000 to 1." Explain this statement, and discuss its physiologic implications.

6. Discuss the physiology of the glutamate receptors in the brain.

7. Although some neurons secrete neurotransmitters, others secrete hormones, ie, chemical messengers that enter the bloodstream. Name the hormones secreted in this fashion, and identify and locate the neurons involved.

8. The nerve gases that were developed for chemical warfare generally inhibit acetylcholinesterase. Why does inhibition of acetylcholinesterase produce death?

9. Compare the endings of somatic motor nerves in skeletal muscle with the endings of autonomic motor nerves in cardiac and smooth muscle.

Multiple-Choice Questions

In questions 1–19, select the single best answer. In questions 1–6, match the statement in each question with the lettered structure that is most closely associated with it. Each lettered structure may be selected once, more than once, or not at all.

(A) Microtubules

(B) Active zones

(C) Nuclei

(D) Clear vesicles

(E) Granulated vesicles

1. Contain acetylcholine

2. Contain catecholamines

3. Contain steroid receptors

4. Contain glutamate

5. Site of extensive exocytosis

6. Associated with kinesin

In questions 7–10, match the receptor in each question with the one lettered statement that is most closely associated with it. Each lettered statement may be used once, more than once, or not at all.

(A) Made up of multiple subunits

(B) Blocked by Zn^{2+}

(C) Coupled to G_s

(D) Found in high concentration in the nucleus accumbens

(E) Inhibited by strychnine

7. β_1-Adrenergic receptors

8. D_3 dopamine receptor

9. NMDA receptor

10. $GABA_A$ receptor

11. Which of the following is *not* synthesized in postganglionic sympathetic neurons?

(A) L-Dopa

(B) Dopamine

(C) Norepinephrine

(D) Histamine

(E) Acetylcholine

12. Fast inhibitory postsynaptic potentials (IPSPs)

(A) are a consequence of presynaptic inhibition

(B) interact with other fast and slow potentials to move the membrane potential of the postsynaptic neuron toward or away from the firing level

(C) can be produced by an increase in Na^+ conductance

(D) can be produced by an increase in Ca^{2+} conductance

(E) occur in skeletal muscle

13. Initiation of an action potential in skeletal muscle by stimulating its motor nerve

(A) requires spatial facilitation

(B) requires temporal facilitation

(C) is inhibited by a high concentration of Ca^{2+} at the neuromuscular junction

(D) requires the release of norepinephrine

(E) requires the release of acetylcholine

14. Which of the following is a ligand-gated ion channel?

(A) VIP receptor

(B) Norepinephrine receptor

(C) $GABA_A$ receptor

(D) $GABA_B$ receptor

(E) Metabotropic glutamine receptor

15. Which of the following is most likely *not* to be involved in production of LTP?

(A) NO

(B) Ca^{2+}

(C) NMDA receptors

(D) Membrane hyperpolarization

(E) Membrane depolarization

16. Which of the following synaptic transmitters is *not* a peptide, polypeptide, or protein?

(A) Substance P

(B) Met-enkephalin

(C) β-Endorphin

(D) Serotonin

(E) Dynorphin

17. Activation of which of the following receptors would be expected to decrease anxiety?

(A) Nicotinic cholinergic receptors

(B) Glutamate receptors

(C) $GABA_A$ receptors

(D) Glucocorticoid receptors

(E) α_1-Adrenergic receptors

18. Which of the following receptors is coupled to a heterotrimeric G protein?

(A) Glycine receptor

(B) $GABA_B$ receptor

(C) Nicotinic acetylcholine receptor at myoneural junction

(D) $5\text{-}HT_3$ receptor

(E) ANP receptor

19. Which of the following would *not* be expected to enhance noradrenergic transmission?

(A) A drug that increases the entry of arginine into neurons

(B) A drug that enhances tyrosine hydroxylase activity

(C) A drug that enhances dopamine β-hydroxylase activity

(D) A drug that inhibits monoamine oxidase

(E) A drug that inhibits norepinephrine reuptake

In questions 20–21, indicate whether the first item is greater than **(G)**, *the same as* **(S)**, *or less than* **(L)** *the second item.*

20. Number of times a β_2-adrenergic receptor spans the cell membrane

G S L

Number of times an M_1 muscarinic acetylcholine receptor spans the cell membrane

21. Amount of gephyrin in postsynaptic density

G S L

Amount of gephyrin in presynaptic active zones

CHAPTER 5

This chapter describes the way action potentials are produced in sensory nerves, lists the senses, and summarizes the principles and laws that govern the coding of sensory information. The material in the chapter should help students to—

- Define the term sensory receptor.
- Discuss the various classifications of the senses.
- Name the types of sensory receptors found in the skin, and discuss their relation to the four cutaneous senses: touch, cold, warmth, and pain.
- Define generator potential, and describe the relationship between the size of a generator potential in a sensory receptor and the number of impulses generated in its sensory nerve.
- Define adaptation, the doctrine of specific nerve energies, and the law of projection.
- Explain how the intensity, location, and quality of stimuli are coded.

General Questions

1. Discuss sensory units and their recruitment as the intensity of a stimulus increases.

2. What is the difference between tonic and phasic receptors? Which physiologic functions are subserved by each type?

3. What part of a pacinian corpuscle produces its generator potential? Where are action potentials in the sensory nerve produced? What is the function of the multiple lamellas of connective tissue in the capsule that cover the ending of the sensory nerve?

4. Why do people see lightning and hear thunder rather than hear lightning and see thunder?

5. Amputees may feel pain and other sensations in a limb that is no longer there, and this can be a distressing medical problem. Explain the occurrence of pain in the absent limb. How would you treat a patient with such a "phantom limb"?

Multiple-Choice Questions

In questions 1–7, select the single best answer.

1. Pacinian corpuscles are
   (A) a type of temperature receptor
   (B) usually innervated by Aδ nerve fibers
   (C) rapidly adapting touch receptors
   (D) slowly adapting touch receptors
   (E) pain receptors

2. Adaptation to a sensory stimulus produces
   (A) a diminished sensation when other types of sensory stimuli are withdrawn
   (B) a more intense sensation when a given stimulus is applied repeatedly
   (C) a sensation localized to the hand when the nerves of the brachial plexus are stimulated
   (D) a diminished sensation when a given stimulus is applied repeatedly over time
   (E) a decreased firing rate in the sensory nerve from the receptor when one's attention is directed to another matter

3. Which of the following sensations is *not* generated by impulses initiated in naked nerve endings?
   (A) Touch
   (B) Pain
   (C) Cold
   (D) Tickle
   (E) Taste

4. A sensation felt in the right hand *cannot* be produced by stimulation of
   (A) the afferent neurons from the hand as they enter the spinal cord
   (B) the spinothalamic tract neurons on which the afferent neurons from the hand terminate
   (C) the hand area of the thalamus on the left side of the body
   (D) the hand area of the sensory cortex on the left side of the body
   (E) the hand area of the sensory cortex on the right side of the body

5. Which of the following receptors and sense organs are *incorrectly* paired?
   (A) Rods and cones : eye
   (B) Receptors sensitive to sodium : taste buds
   (C) Hair cells : olfactory mucous membranes
   (D) Receptors sensitive to stretch : carotid sinus
   (E) Glomus cells : carotid body

6. Which of the following sensations is most subject to facilitation and inhibition in the central nervous system?
   (A) Touch
   (B) Warmth
   (C) Pain
   (D) Hearing
   (E) Taste

7. Which of the following does *not* contain cation channels that are activated by mechanical distortion, producing depolarization?
   (A) Cones
   (B) Pacinian corpuscles
   (C) Hair cells in cochlea
   (D) Hair cells in semicircular canals
   (E) Hair cells in utricle

In questions 8–9, indicate whether the first item is greater than (G), *the same as* (S), *or less than* (L) *the second item.*

8. Amplitude of generator potential with weak stimulus
   G S L
   Amplitude of generator potential with strong stimulus

9. Amplitude of action potentials in sensory nerve with weak stimulus
   G S L
   Amplitude of action potentials in sensory nerve with strong stimulus

CHAPTER 6

This chapter describes the components that make up the reflex arc, the monosynaptic and polysynaptic spinal reflexes, and the properties and functions of muscle spindles. The material in the chapter should help students to—

- State the Bell-Magendie law, and summarize its physiologic implications.
- Distinguish between and compare monosynaptic and polysynaptic reflexes.
- Give examples of stretch reflexes, including those that are frequently tested clinically.
- Describe the muscle spindles and analyze their function as part of a feedback system that controls muscle length.
- Describe the Golgi tendon organs and analyze their function as part of a feedback system that maintains muscle force.
- Define reciprocal innervation, inverse stretch reflex, clonus, and lengthening reaction.
- Define and explain fractionation and occlusion.

General Questions

1. What is a "final common path"? Discuss its physiologic basis and importance.
2. Define local sign, and explain the mechanism responsible for it.
3. Why does a strong noxious stimulus produce a prolonged withdrawal response?
4. In a normal subject, the central delay for a given reflex response was found to be 0.9 ms. Is it likely that the response is mediated by a monosynaptic or a polysynaptic pathway? Explain your answer.
5. How are reflex responses graded and modified?

Multiple-Choice Questions

In questions 1–4, select the single best answer.

1. The inverse stretch reflex
   (A) has a lower threshold than the stretch reflex
   (B) is a monosynaptic reflex
   (C) is a disynaptic reflex with a single interneuron inserted between the afferent and efferent limbs
   (D) is a polysynaptic reflex with many interneurons inserted between the afferent and efferent limbs
   (E) requires the discharge of central neurons that release acetylcholine

2. When γ motor neuron discharge increases at the same time as α motor neuron discharge to muscle–
   (A) there is prompt inhibition of discharge in spindle Ia afferents
   (B) the contraction of the muscle is prolonged
   (C) the muscle will not contract
   (D) the number of impulses in spindle Ia afferents is smaller than when α discharge alone is increased
   (E) the number of impulses in spindle Ia afferents is greater than when α discharge alone is increased

3. Which of the following is *not* characteristic of a reflex action?
   (A) Modification by impulses from various parts of the central nervous system
   (B) May involve simultaneous contraction of some muscles and relaxation of others
   (C) May involve either somatic or visceral responses but never both simultaneously
   (D) Always involves transmission across at least one synapse
   (E) Frequently occurs without conscious perception

4. Withdrawal reflexes are *not*
   (A) initiated by nociceptive stimuli
   (B) prepotent

   (C) prolonged if the stimulus is strong
   (D) absent several months after transection of the spinal cord
   (E) dependent on local sign for their exact pattern

Questions 5–7 refer to Figure 6–A.

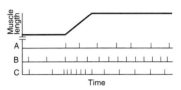

Figure 6–A.

In Figure 6–A, A, B, and C are records of action potentials in afferent nerve fibers from a muscle before and after the muscle was stretched, as shown by the change in muscle length.

5. Which record is from a fiber that innervates a nuclear bag fiber?
6. Which record is from a fiber that innervates a nuclear chain fiber?
7. Which record is from a fiber that innervates a Golgi tendon organ?

In questions 8–10, indicate whether the first item is greater than (G), the same as (S), or less than (L) the second item.

8. Response of a skeletal muscle to stretch after cutting the sensory nerves that supply it
   G S L
   Response of a skeletal muscle to stretch after cutting the motor nerves that supply it
9. During stretch reflex, rate of discharge in motor nerve to protagonist muscles
   G S L
   During stretch reflex, rate of discharge in motor nerve to antagonist muscles
10. Rhythmic contractile response to sustained stretch of a muscle when rate of γ efferent discharge is high
    G S L
    Rhythmic contractile response to sustained stretch of a muscle when rate of γ efferent discharge is low

CHAPTER 7

This chapter describes the sensory mechanisms responsible for touch, proprioception, warmth, cold, pain, itching, and sensations produced by combinations of these fundamental sensations. Pain is considered in detail, including the characteristics of so-called fast pain and slow pain, deep pain, and pain from visceral structures. Inflammation and neuropathic pain are also discussed. The material in the chapter should help students to—

- Outline the neural connections and pathways that mediate sensory input from the skin, deep tissues, and viscera to the cerebral cortex.
- Describe the structure and function of the substantia gelatinosa.
- Describe the areas in which the sensory projection neurons from the thalamus terminate in the sensory cortex.
- Name the types of nerve fibers that mediate warmth and cold in peripheral nerves.
- Name the receptors that mediate pain, and explain the differences between fast and slow pain.
- Compare superficial, deep, and visceral pain.
- Explain hyperalgesia and allodynia.
- Describe and explain referred pain.
- List the main drugs and procedures that have been used for relief of pain. For each, give the rationale for its use and comment on its clinical effectiveness.
- Explain vibratory sensibility, two-point discrimination, and stereognosis.

General Questions

1. Which chemical agents may initiate impulses in pain fibers?
2. Why do tumors that arise in the center of the upper spinal cord cause loss of pain and temperature sensations first in the upper segments of the body and only later in the lower segments? What would you expect to be the corresponding order of loss of proprioception?
3. Describe the locations of the two touch pathways in the spinal cord. What type of touch information does each conduct?
4. Discuss P factor and the role it plays in muscle pain.
5. Why is visceral pain poorly localized? Why is it particularly unpleasant?
6. What are opioid peptides? How do they alter pain sensation?

Multiple-Choice Questions

In questions 1–12, select the single best answer.

1. The distance by which two touch stimuli must be separated to be perceived as two separate stimuli is greatest on

(A) the lips
(B) the palm of the hand
(C) the back of the scapula
(D) the dorsum of the hand
(E) the tips of the fingers

2. A 49-year-old blond woman has diffuse, severe loss of vibration sensitivity. Which of the following tests would give you the most information about the cause of her loss?

(A) X-rays of her spine
(B) Determination of her reflex reaction time
(C) Measurement of the protein content of her cerebrospinal fluid
(D) A biopsy of her gastric mucosa
(E) A complete urinalysis

3. Stimulation of which of the following might be expected to produce itching?

(A) Dorsal root C fibers
(B) B fibers in peripheral nerves
(C) Dorsal columns of the spinal cord
(D) Touch receptors
(E) Motor fibers in the ventral roots of the spinal nerves

4. Visceral pain

(A) shows relatively rapid adaptation
(B) is mediated by B fibers in the dorsal roots of the spinal nerves
(C) can sometimes be relieved by applying an irritant to the skin
(D) most closely resembles "fast pain" produced by noxious stimulation of the skin
(E) can be produced by marked and prolonged stimulation of touch receptors

5. An anterolateral cordotomy is performed that produces relief of pain in the right leg. It is effective because it interrupts the

(A) left dorsal column
(B) left ventral spinothalamic tract
(C) right lateral spinothalamic tract
(D) left lateral spinothalamic tract
(E) right corticospinal tract

6. Which of the following does *not* exert an analgesic effect?

(A) Morphine
(B) Cholinergic antagonists
(C) Adrenergic antagonists
(D) Substance P antagonists
(E) Anandamide

7. A man loses his right hand in a farm accident. Four years later, he has episodes of severe pain in the missing hand (phantom limb pain). A detailed PET scan study of his cerebral cortex might be expected to show
   (A) expansion of the right hand area in his right somatic sensory area I (SI)
   (B) expansion of the right hand area in his left SI
   (C) a metabolically inactive spot where his hand area in his left SI would normally be
   (D) projection of fibers from neighboring sensory areas into the right-hand area of his right SI
   (E) projection of fibers from neighboring sensory areas into the right hand area of his left SI

8. Causalgia is
   (A) associated with loss of hair in the affected area
   (B) abolished by sectioning the sensory nerves in the affected area
   (C) relieved by drugs that block glycine receptors
   (D) relieved by drugs that block α_1 adrenergic receptors
   (E) relieved by drugs that block muscarinic cholinergic receptors

In questions 9–13, select

   A if the item is associated with (a) below,
   B if the item is associated with (b) below,
   C if the item is associated with both (a) and (b), and
   D if the item is associated with neither (a) nor (b)
     (a) Capsaicin receptors
     (b) Aα nerve fibers

9. Cold
10. Warmth
11. Pain
12. Touch
13. Muscle strength

*In questions 14 and 15, indicate whether the first item is greater than (**G**), the same as (**S**), or less than (**L**) the second item.*

14. Rate of conduction in afferent neurons from warmth receptors
   G S L
Rate of conduction in afferent neurons from pressure receptors

15. Effects of parietal lobe lesions on stereognosis
   G S L
Effect of frontal lobe lesions on stereognosis

CHAPTER 8

This chapter reviews the functional anatomy of the visual system, then analyzes in order the processes involved in vision: formation of the visual image on the retina, conversion of light energy to electrical responses in the rods and cones, processing of impulses in the retina, transmission of impulses via the lateral geniculate bodies to the visual cortex, and the events that occur in the visual cortex and other parts of the cortex. The material in the chapter should help students to—

- Describe the various parts of the eye, and list the functions of each.
- Trace the neural pathways that transmit visual information from the rods and cones to the visual cortex.
- Explain how light rays in the environment are brought to a focus on the retina and the role of accommodation in this process.
- Define hyperopia, myopia, astigmatism, presbyopia, and strabismus.
- Describe the electrical responses produced by rods and cones, and explain how these responses are produced.
- Describe the electrical responses seen in bipolar cells, horizontal cells, amacrine cells, and ganglion cells, and comment on the function of each type of cell.
- Describe the responses of cells in the visual cortex and the functional organization of the dorsal and ventral pathways to the parietal cortex.
- Define and explain dark adaptation and visual acuity.
- Describe the neural pathways involved in color vision.
- Name the four types of eye movements and the function of each.

General Questions

1. What type of visual field defect is produced by each of the following lesions and why?
   a. A lesion of one optic nerve
   b. A lesion of one optic tract
   c. A lesion of the optic chiasm
   d. Lesions in various parts of the geniculocalcarine tract
   e. Bilateral lesions destroying the primary visual cortex (area 17)

2. What is the near point of vision? Why does it recede throughout life?

3. How many different kinds of photosensitive pigments are found in the human retina? Discuss their

chemistry, how they produce electrical responses in the retina, and their relation to color vision.

4. What are "on center" and "off center" cells? Define lateral inhibition, and comment on its general physiologic significance.

5. What are orientation columns and ocular dominance columns? How are they mapped experimentally? What are their functions?

6. Discuss parallel processing of visual information in the cerebral cortex.

7. What is binocular vision? Why is it limited to part of the visual field? Discuss its role and the role of monocular visual processes in depth perception.

Multiple-Choice Questions

In questions 1–12, select the single best answer.

1. If the principal focal distance of a lens is 0.75 m, its refractive power is
   (A) 0.25 diopter
   (B) 0.75 diopter
   (C) 1.0 diopter
   (D) 1.33 diopters
   (E) 10.3 diopters

2. Visual accommodation involves
   (A) increased tension on the lens ligaments
   (B) a decrease in the curvature of the lens
   (C) relaxation of the sphincter muscle of the iris
   (D) contraction of the ciliary muscle
   (E) increased intraocular pressure

3. The fovea of the eye
   (A) has the lowest light threshold
   (B) is the region of highest visual acuity
   (C) contains only red and green cones
   (D) contains only rods
   (E) is situated over the head of the optic nerve

4. Which of the following parts of the eye has the greatest concentration of rods?
   (A) Ciliary body
   (B) Iris
   (C) Optic disk
   (D) Fovea
   (E) Parafoveal region

5. The following events that occur in rods in response to light are listed in random sequence:
   1. Activation of transducin
   2. Decreased release of synaptic transmitter
   3. Structural changes in rhodopsin
   4. Closure of Na^+ channels
   5. Decrease in intracellular cGMP

What is the sequence in which they normally occur?
   (A) 2, 1, 3, 5, 4
   (B) 1, 2, 3, 5, 4
   (C) 5, 3, 1, 4, 2
   (D) 3, 1, 5, 4, 2
   (E) 3, 1, 4, 5, 2

6. Vitamin A is a precursor for the synthesis of
   (A) somatostatin
   (B) retinene$_1$
   (C) the pigment of the iris
   (D) scotopsin
   (E) aqueous humor

7. Abnormal color vision is 20 times more common in men than women because most cases are caused by an abnormal
   (A) dominant gene on the Y chromosome
   (B) recessive gene on the Y chromosome
   (C) dominant gene on the X chromosome
   (D) recessive gene on the X chromosome
   (E) recessive gene on chromosome 22

8. Which of the following is *not* involved in color vision?
   (A) Activation of a pathway that signals differences between S cone responses and the sum of L and M cone responses
   (B) Geniculate layers 3–6
   (C) P pathway
   (D) Area V3A of visual cortex
   (E) Area V8 of visual cortex

In questions 9–12, match the abnormality in each question with the lettered disease that is most closely associated with it. Each lettered disease may be selected once, more than once, or not at all.

   (A) Hyperopia
   (B) Myopia
   (C) Heteronymous hemianopia
   (D) Strabismus
   (E) Nyctalopia

9. Tumor of the anterior pituitary gland

10. Vitamin A deficiency

11. Failure of visual images to fall on corresponding points in the retina

12. Long anteroposterior diameter of the eyeball

In questions 13–14, indicate whether the first item is greater than (G), the same as (S), or less than (L) the second item.

13. Number of action potentials in right visual cortex after stimulation of medial half of right retina

    G S L

    Number of action potentials in left visual cortex after stimulation of medial half of right retina

14. Visual acuity in a person with 20/15 vision

    G S L

    Visual acuity in a person with 15/20 vision

CHAPTER 9

This chapter is concerned with the auditory system and the vestibular system. The material in the chapter should help students to—

- Describe the way that movements of molecules in the air are converted into impulses generated in hair cells in the cochlea.
- Trace the path of auditory impulses in the neural pathways from the cochlear hair cells to the auditory cortex, and discuss the function of the auditory cortex.
- Explain how pitch, loudness, and timbre are coded in the auditory pathways.
- Describe the various forms of deafness.
- Explain how the receptors in the semicircular canals detect rotational acceleration and how the receptors in the saccule and utricle detect linear acceleration.
- List the major sensory inputs that provide the information which is synthesized in the brain into the sense of position in space.

General Questions

1. Otosclerosis is a disease in which the foot plate of the stapes becomes rigidly attached to the oval window. Why does this cause deafness?

2. What is the function of the tectorial membrane in the cochlea?

3. Which sound frequencies are audible to humans? At which sound frequency is the auditory threshold the lowest?

4. Define and explain masking.

5. What is the tympanic reflex, and what is its function?

6. Compare ossicular, air, and bone conduction.

7. Compare and contrast processing of visual information and auditory information in the cerebral cortex.

8. Name five genetic mutations that cause congenital deafness. In each case, explain why the genetic abnormality causes loss of hearing. Which are syndromic and which are nonsyndromic?

Multiple-Choice Questions

In questions 1–14, select the single best answer.

1. Sound intensity is measured in
   (A) diopters
   (B) daltons
   (C) torrs
   (D) decibels
   (E) pounds

2. In humans, the primary auditory cortex is located in the
   (A) limbic system
   (B) posterior part of the occipital lobe
   (C) posterior part of the parietal lobe
   (D) postcentral gyrus
   (E) superior part of the temporal lobe

3. Postrotatory nystagmus is caused by continued movement of
   (A) aqueous humor over the ciliary body in the eye
   (B) cerebrospinal fluid over the parts of the brain stem that contain the vestibular nuclei
   (C) endolymph toward the helicotrema
   (D) endolymph in the semicircular canals, with consequent bending of the cupula and stimulation of hair cells
   (E) perilymph over hair cells that have their processes embedded in the tectorial membrane

4. The basilar membrane of the cochlea
   (A) is unaffected by movement of fluid in the scala vestibuli
   (B) covers the oval window and the round window
   (C) vibrates in a pattern determined by the form of the traveling wave in the fluids in the cochlea
   (D) is under tension
   (E) vibrates when the body is subjected to linear acceleration

5. Some diseases damage the hair cells in the ear. When the damage to the outer hair cells is greater than the damage to the inner hair cells,
   (A) the perception of vertical acceleration is disrupted.
   (B) the K^+ concentration in endolymph is decreased.
   (C) the K^+ concentration in perilymph is decreased.
   (D) there is severe hearing loss.
   (E) the affected hair cells fail to shorten when exposed to sound.

6. Which of the following are *incorrectly* paired?

(A) Tympanic membrane : manubrium of malleus

(B) Helicotrema : apex of cochlea

(C) Foot plate of stapes : oval window

(D) Otoliths : semicircular canals

(E) Basement membrane : cochlea

7. The direction of nystagmus is vertical when a subject is rotated

(A) after warm water is put in one ear

(B) with the head tipped backward

(C) after cold water is put in both ears

(D) with the head tipped sideways

(E) after section of one vestibular nerve

8. The human planum temporale is

(A) concerned with auditory-visual coordination

(B) concerned with detection of linear acceleration

(C) concerned with localization of sound

(D) larger in the right cerebral hemisphere than in the left

(E) generally larger in musicians with perfect pitch than in musicians without perfect pitch

9. In the utricle, tip links in hair cells are involved in

(A) formation of perilymph

(B) depolarization of the stria vascularis

(C) movements of the basement membrane

(D) perception of sound

(E) regulation of distortion-activated ion channels

10. A patient enters the hospital for evaluation of deafness. He is found to also have an elevated plasma renin, although his blood pressure is 118/75 mm Hg. Mutation of what single gene would be likely to explain these findings?

(A) The gene for barttin

(B) The gene for Na^+ channel

(C) The gene for renin

(D) The gene for cystic fibrosis transmembrane conductance regulator

(E) The gene for tyrosine hydroxylase

In questions 11–14, which refer to Figure 9–A, select

A   if the item is associated with (a) below,

B   if the item is associated with (b) below,

C   if the item is associated with both (a) and (b), and

D   if the item is associated with neither (a) nor (b)

(a) Sound transmission

(b) Linear acceleration

11. Structure V

12. Structure W

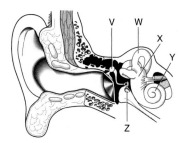

Figure 9–A. Human ear.

13. Structure X

14. Structure Y

15. Structure Z

In questions 16–18, indicate whether the first item is greater than **(G)**, *the same as* **(S)**, *or less than* **(L)** *the second item.*

16. Na^+ concentration in perilymph

G S L

Na^+ concentration in endolymph

17. Maximum sound frequency audible to humans

G S L

Maximum sound frequency audible to dogs

18. Change in membrane potential when stereocilia of a saccular hair cell are pushed toward the kinocilium

G S L

Change in membrane potential when stereocilia of a saccular hair cell are pushed at a right angle to the kinocilium

CHAPTER 10

This chapter considers olfaction and taste from the point of view of signal transduction, pathways to the central nervous system and central representation. In the case of olfaction, the molecular and neural mechanisms that make it possible to discriminate among different odors are considered. The material in the chapter should help students to—

• Describe the olfactory receptors and the way in which impulses are initiated in them.

• Outline the pathways by which impulses generated in the olfactory mucous membrane reach the cerebral cortex.

• Describe and analyze olfactory sensitivity, discrimination, and adaptation.

• Describe the essential features of the taste buds.

• Outline the taste pathways.

- List the substances that produce the primary tastes, and comment on how the signal for each is transduced.

General Questions

1. Compare and contrast smell and taste. How do they interact?
2. Humans can discriminate between many different odors. What are the mechanisms that make this possible?
3. Discuss the relationship between olfaction and sexual behavior.
4. What part do pain fibers in the trigeminal nerves play in olfaction?
5. Name the chemoreceptors in the body.
6. A number of seemingly unrelated substances other than sugars taste sweet. List some of these compounds, and comment on their use in the treatment of various conditions and diseases.
7. Discuss taste as a motivation for aversive learning.

Multiple-Choice Questions

In questions 1–8, select the single best answer.

1. Impulses generated by olfactory receptors in the nasal mucous membrane
   (A) pass through the substantia nigra
   (B) pass through the internal capsule
   (C) are relayed to the olfactory cortex via the hypothalamus
   (D) pass to the mitral cells and from there directly to the olfactory cortex
   (E) pass to the mitral cells and from there to the olfactory cortex via the taste area
2. Impulses generated in the taste buds of the tongue reach the cerebral cortex via the
   (A) thalamus
   (B) internal capsule
   (C) dorsal roots of the first cervical spinal nerves
   (D) trochlear nerve
   (E) hypoglossal nerve
3. Which of the following senses is most strongly associated with aversive conditioning?
   (A) Pressure
   (B) Vision
   (C) Linear acceleration
   (D) Olfaction
   (E) Taste

4. Which of the following does *not* increase the ability to discriminate many different odors?
   (A) Many different receptors
   (B) Pattern of olfactory receptors activated by a given odorant
   (C) Projection of different mitral cell axons to different parts of the brain
   (D) High β-arrestin content in olfactory neurons
   (E) Sniffing
5. Deficiency of which of the following substances causes the combination of diminished sense of smell and hypogonadism?
   (A) Folic acid
   (B) A transcription factor
   (C) An adhesion molecule
   (D) GnRH
   (E) Nerve growth factor
6. Which of the following are *incorrectly* paired?
   (A) α-Gusducin : sweet taste
   (B) α-Gusducin : bitter taste
   (C) Nucleus tractus solitarius : blood pressure
   (D) Insular cortex : smell
   (E) Ebener's glands : taste acuity

In questions 7–10, match the item in each question with the lettered structure that is most closely associated with it. Each lettered structure may be used once, more than once, or not at all.

   (A) Sensory nucleus of trigeminal nerve
   (B) Main olfactory mucous membrane
   (C) Toxic substances
   (D) Vomeronasal organ (or its homolog in humans)
   (E) Monosodium glutamate
7. Pheromones
8. Bitter taste
9. Umami taste
10. Sexual identity

In questions 11 and 12, indicate whether the first item is greater than (G), the same as (S), or less than (L) the second item.

11. Olfactory sensitivity in a 20-year-old woman
    G S L
    Olfactory sensitivity in a 70-year-old woman
12. Number of different G proteins involved in transduction of taste stimuli
    G S L
    Number of different G proteins involved in transduction of olfactory stimuli

CHAPTER 11

This chapter deals with alertness and sleep, the mechanisms that produce these states, and the correlation between them and the electrical activity of the brain. The material in the chapter should help students to—

- Describe the primary types of rhythms that make up the EEG.
- Summarize the behavioral and EEG characteristics of each of the stages of slow-wave sleep and the mechanisms responsible for their production.
- Summarize the electroencephalographic and other characteristics of rapid eye movement (REM) sleep, and describe the mechanisms responsible for its production.
- Describe the pattern of normal nighttime sleep in adults and the variations in this pattern from birth to old age.
- List the main clinical uses of the EEG.

General Questions

1. Define thalamocortical oscillations, reticular activation, and sleep zones. What is the role of each of these plus intracortical activity in the production of the EEG?
2. Define somnambulism and narcolepsy, and discuss their relationship to normal sleep.
3. Compare REM and non-REM sleep.
4. What is the difference between the specific sensory relay nuclei and the nonspecific projection nuclei of the thalamus?

Multiple-Choice Questions

In questions 1–12, select the single best answer.

1. In a healthy, alert adult sitting with the eyes closed, the dominant EEG rhythm observed with electrodes over the occipital lobes is
   (A) delta (0.5–4 Hz)
   (B) theta (4–7 Hz)
   (C) alpha (8–13 Hz)
   (D) beta (18–30 Hz)
   (E) fast, irregular low-voltage activity
2. The EEG record in Figure 11–A is characteristic of
   (A) deep sleep
   (B) a learning response produced by a painful stimulus
   (C) REM sleep
   (D) a psychomotor seizure
   (E) an absence seizure

Figure 11–A.

3. The electrical records in Figure 11–B are characteristic of
   (A) deep sleep
   (B) general-onset seizures
   (C) REM sleep
   (D) psychomotor seizures
   (E) an absence seizure
4. High-frequency stimulation of which of the following does *not* produce the alerting response?
   (A) Sciatic nerve
   (B) Lateral spinothalamic tract
   (C) Midbrain reticular formation
   (D) Medial lemniscus above the midbrain
   (E) Intralaminar nuclei of the thalamus
5. A localized lesion in which of the following structures would be expected to produce prolonged coma?
   (A) Nucleus tractus solitarius
   (B) Locus ceruleus
   (C) Right frontal lobe
   (D) Both frontal lobes
   (E) Periaqueductal region at top of midbrain
6. Regular rhythmic fluctuations in electrical activity are observed in the cerebral cortex and thalamus. In addition, they are seen in the
   (A) mediobasal portion of the hypothalamus
   (B) cerebellar cortex
   (C) midbrain reticular formation
   (D) amygdala
   (E) pons
7. In which of the following is blood flow *not* increased in REM sleep?
   (A) Primary visual cortex
   (B) Anterior cingulate cortex

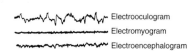

Figure 11–B. Reproduced, with permission, from Kales A et al: Sleep and dreams. Recent research on clinical aspects. Ann Intern Med 1968;68:1078.

(C) Pons

(D) Visual association cortex

(E) Amygdalas

8. Narcolepsy is triggered by abnormalities in

(A) skeletal muscles

(B) medulla oblongata

(C) hypothalamus

(D) olfactory bulb

(E) neocortex

In questions 9–14, select

    A if the item is associated with (a) below,

    B if the item is associated with (b) below,

    C if the item is associated with both (a) and (b), and

    D if the item is associated with neither (a) nor (b)

        (a) Slow-wave sleep

        (b) REM sleep

9. Dreaming

10. Sleep spindles

11. Sleepwalking

12. Bed-wetting

13. Penile erection

14. PGO spikes

In questions 15 and 16, indicate whether the first item is greater than **(G)**, *the same as* **(S)**, *or less than* **(L)** *the second item.*

15. Frequency of episodes of REM sleep after first falling asleep

    G S L

    Frequency of episodes of REM sleep after 6 hours of sleep

16. Percentage of total sleep time spent in REM sleep in neonates

    G S L

    Percentage of total sleep time spent in REM sleep in adults

CHAPTER 12

This chapter is a review of somatic motor function, with emphasis on the contributions of the spinal cord, medulla, midbrain, basal ganglia, cerebral cortex, and cerebellum. The material in the chapter should help students to—

- Describe how skilled movements are planned and carried out.
- Name the posture-regulating parts of the CNS and discuss the role of each.

- Define spinal shock, and describe the initial and long-term changes in spinal reflexes that follow transection of the spinal cord.
- Define decerebrate and decorticate rigidity, and comment on the cause and physiologic significance of each.
- Describe the basal ganglia, and list the pathways that interconnect them, along with the neurotransmitters in each pathway.
- Describe and explain the symptoms of Parkinson's disease and Huntington's disease.
- List the pathways to and from the cerebellum and the connections of each within the cerebellum.
- Discuss the functions of the cerebellum and the neurologic abnormalities produced by diseases of this part of the brain.

General Questions

1. What is meant by the terms upper motor neuron and lower motor neuron? Contrast the effects of lower motor neuron lesions with those of lesions affecting each of the types of upper motor neurons.

2. What is the Babinski sign? What is its physiologic and pathologic significance?

3. What is the mass reflex? Why does it occur after transection of the spinal cord?

4. Define athetosis, ballism, and chorea, and describe the disease processes that produce each of them.

5. List three drugs and two surgical procedures used in the treatment of Parkinson's disease, and explain why each is of value.

6. List five types of neurons found in the cerebellar cortex, and describe the morphology and function of each.

7. What is an intention tremor? Why does it occur in cerebellar disease?

Multiple-Choice Questions

In questions 1–20, select the single best answer.

Questions 1–4 refer to Figure 12–A. Structures identified by letter may be selected once, more than once, or not at all.

1. Which of the labeled structures projects primarily to the brain stem areas concerned with postural control?

2. Which of the labeled structures is concerned with elaboration of somatic sensory perception?

3. Which of the labeled structures is concerned with the primary perception of tactile stimuli?

4. Which of the labeled structures is the supplementary motor area?

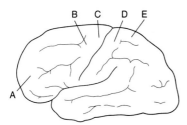

Figure 12–A. Lateral view of the human cerebral cortex.

5. A primary function of the basal ganglia is
   (A) sensory integration
   (B) short-term memory
   (C) planning voluntary movement
   (D) neuroendocrine control
   (E) slow-wave sleep
6. The therapeutic effect of L-dopa in patients with Parkinson's disease eventually wears off because
   (A) antibodies to dopamine receptors develop
   (B) inhibitory pathways grow into the basal ganglia from the frontal lobe
   (C) there is an increase in circulating α-synuclein
   (D) the normal action of NGF is disrupted
   (E) the dopaminergic neurons in the subtstania nigra continue to degenerate
7. Increased neural activity before a skilled voluntary movement is *first* seen in the
   (A) spinal motor neurons
   (B) precentral motor cortex
   (C) midbrain
   (D) cerebellum
   (E) cortical association areas
8. After falling down a flight of stairs, a young woman is found to have partial loss of voluntary movement on the right side of her body and loss of pain and temperature sensation on the left side below the midthoracic region. It is probable that she has a lesion
   (A) transecting the left half of the spinal cord in the lumbar region
   (B) transecting the left half of the spinal cord in the upper thoracic region
   (C) transecting sensory and motor pathways on the right side of the pons
   (D) transecting the right half of the spinal cord in the upper thoracic region
   (E) transecting the dorsal half of the spinal cord in the upper thoracic region

9. Patients with transected spinal cords frequently have a negative nitrogen balance because
   (A) they develop hypercalcemia, and this causes dissolution of the protein in bone
   (B) they are paralyzed below the level of the transection
   (C) they lack the afferent input that normally maintains growth hormone secretion
   (D) they have difficulty voiding, and this causes nitrogen to accumulate in the urine in the bladder
   (E) their ACTH response to stress is reduced
10. Which of the following diseases is *not* caused by overexpression of a trinucleotide repeat?
   (A) Alzheimer's disease
   (B) Fragile X syndrome
   (C) Spinocerebellar ataxia, type 3
   (D) Huntington's disease
   (E) Friedreich's ataxia

In questions 11–20, match the item in each question with the lettered structure that is most closely associated with it. Each lettered structure may be selected once, more than once, or not at all.

   (A) Amygdala
   (B) Cerebellum
   (C) Internal capsule
   (D) Hippocampus
   (E) Striatum

11. Emotional responses
12. Ischemic stroke
13. Parkinson's disease
14. Intention tremor
15. Huntington's disease
16. Alzheimer's disease
17. Spastic hemiparesis
18. Aphasia
19. Sexual behavior
20. Vestibulo-ocular reflex

CHAPTER 13

This chapter is concerned specifically with the motor component of the peripheral autonomic nervous system that innervates the viscera. Visceral afferents are discussed in Chapters 5 and 7, and the central regulation of visceral function in Chapter 14. The material in the chapter should help students to—

• Describe the location of the cell bodies of preganglionic sympathetic and parasympathetic neurons in the central nervous system and identify the nerves by which their axons leave the central nervous system.

- Describe the location of postganglionic sympathetic and parasympathetic neurons and the pathways they take to the visceral structures they innervate.
- Name the neurotransmitters or hormones that are secreted by each of the following:
    a. Preganglionic autonomic neurons
    b. Postganglionic sympathetic neurons innervating intestinal smooth muscle
    c. Postganglionic sympathetic neurons innervating sweat glands
    d. Postganglionic parasympathetic neurons
    e. Small, intensely fluorescent (SIF) cells
    f. Adrenal medullary cells
- Outline the functions of the autonomic nervous system.
- List the ways that drugs act to increase or decrease the activity of the components of the autonomic nervous system.

General Questions

1. Cannon called mass sympathetic discharge the "preparation for flight or fight." How does sympathetic discharge prepare the individual for flight or fight? Does the sympathetic nervous system have any other functions?
2. The cholinergic division of the autonomic nervous system has been called the anabolic nervous system. Discuss the actions of the cholinergic division that justify this label.
3. Where are the cell bodies of the postganglionic sympathetic neurons that supply the head? How do they reach the visceral effectors they innervate?

Multiple-Choice Questions

In questions 1 and 2, select the single best answer.

1. Complete denervation of the small intestine would be expected to
    (A) cause atrophy of the enteric nervous system
    (B) increase the resting rate of intestinal peristalsis
    (C) have little effect on the resting rate of intestinal peristalsis
    (D) abolish the intestinal response to stretch
    (E) cause peristalsis to become chaotic and irregular

2. Which of the following drugs would *not* be expected to increase sympathetic discharge or mimic the effects of increased sympathetic discharge?
    (A) Prazosin
    (B) Neostigmine
    (C) Amphetamine
    (D) Isoproterenol
    (E) Methoxamine

In questions 3–8, one or more of the answers may be correct. Select

- **A** if (1), (2), and (3) are correct;
- **B** if (1) and (3) are correct;
- **C** if (2) and (4) are correct;
- **D** if only (4) is correct; and
- **E** if all are correct

3. Administration of physostigmine, a drug that inhibits acetylcholinesterase, would be expected to
    (1) increase the secretion of gastric juice
    (2) increase the rate of melatonin synthesis and secretion from the pineal gland
    (3) increase the rate of epinephrine secretion from the adrenal medulla
    (4) decrease the secretion of glucagon

4. In low doses, nicotine
    (1) acts directly on intestinal smooth muscle, causing it to contract
    (2) increases the rate of discharge of postganglionic parasympathetic neurons
    (3) decreases the release of acetylcholine in sympathetic ganglia
    (4) increases the rate of discharge of postganglionic sympathetic neurons

5. In the parasympathetic nervous system
    (1) the axons of the preganglionic neurons are in the dorsal roots of the spinal nerves in the sacral region
    (2) the axons of the preganglionic neurons are in the ventral roots of the spinal nerves in the lower thoracic and lumbar regions
    (3) dopamine is the neurotransmitter responsible for transmission from preganglionic neurons to postganglionic neurons
    (4) the axons of postganglionic neurons are usually shorter than those of preganglionic neurons

6. Administration of a drug that blocks conversion of L-dopa to dopamine would be expected to
    (1) increase the diameter of the trachea and bronchi

(2) disrupt the function of the SIF cells in the sympathetic ganglia

(3) decrease peristaltic activity in the small intestine

(4) decrease the amount of norepinephrine in the circulating blood

7. Administration of a drug that blocks β-adrenergic receptors would be expected to

(1) decrease the heart rate

(2) decrease the force of cardiac contraction

(3) decrease the secretion of renin from the kidneys

(4) decrease the secretion of insulin from the B cells in the pancreatic islets

8. Polypeptides found in the autonomic nervous system include

(1) gonadotropin-releasing hormone (GnRH)

(2) neuropeptide Y

(3) gastrin-releasing peptide (GRP)

(4) vasoactive intestinal polypeptide (VIP)

In questions 9 and 10, indicate whether the first item is greater than (**G**), *the same as* (**S**), *or less than* (**L**) *the second item.*

9. Gastric acid secretion after removal of the celiac ganglion

G S L

Gastric acid secretion after vagotomy

10. Number of B fibers in white ramus communicans

G S L

Number of B fibers in gray ramus communicans

CHAPTER 14

This chapter reviews central integration of autonomic and related visceral functions at the level of the spinal cord, medulla oblongata, and hypothalamus. The material in the chapter should help students to—

• Describe the autonomic reflexes integrated at the level of the spinal cord.

• Describe vomiting, including its autonomic and somatic components, and discuss the initiation of the vomiting reflex by afferents from the gastrointestinal tract and the area postrema.

• Describe the anatomic connections between the hypothalamus and the pituitary gland and the functional significance of each connection.

• Discuss body rhythms, including the role of the SCN in their regulation and their genetic control.

• Describe the contribution of the hypothalamus and the role of other factors in the regulation of food intake.

• List the factors that control water intake, and outline the way they exert their effects.

• Describe the synthesis, processing, storage, and secretion of the hormones of the posterior pituitary.

• Discuss the effects of vasopressin, the receptors on which it acts, and how its secretion is regulated.

• Discuss the effects of oxytocin, the receptors on which it acts, and how its secretion is regulated.

• Name the hypophysiotropic hormones, and outline the effects that each has on anterior pituitary function.

• List the mechanisms by which heat is produced in and lost from the body, and comment on the differences in temperature in the hypothalamus, rectum, oral cavity, and skin.

• List the temperature-regulating mechanisms, and describe the way in which they are integrated under hypothalamic control to maintain normal body temperature.

• Discuss the pathophysiology of fever.

General Questions

1. What is behavioral thermoregulation? Give examples, and describe how they operate to maintain body temperature.

2. What functions do the hypophysiotropic hormones have in addition to regulation of anterior pituitary secretion?

3. What hormones act directly on the hypothalamus, and how do they exert their effects?

4. Draw graphs showing the relation of thirst to plasma osmolality, plasma vasopressin to plasma osmolality, and plasma vasopressin to extracellular fluid volume.

5. Draw the curve relating metabolic rate to body temperature from 95 to 106 °F (35–41 °C), and explain the shape of the curve.

6. Some individuals drink large quantities of water for psychologic reasons (psychogenic polydipsia). Others have diabetes insipidus due to damage to the posterior lobe of the pituitary and the hypothalamus. Still others have one of the two forms of nephrogenic diabetes insipidus, in which the kidneys do not respond to vasopressin. All present with polyuria (increased urine output) and polydipsia (increased water intake). How would you differentiate among the four possible causes?

7. What is the effect of bilateral lesions of the paraventricular nuclei on the adrenocortical response to stress?

Multiple-Choice Questions

In questions 1–23, select the single best answer.

1. Thirst is stimulated by
   (A) increases in plasma osmolality and volume
   (B) an increase in plasma osmolality and a decrease in plasma volume
   (C) a decrease in plasma osmolality and an increase in plasma volume
   (D) decreases in plasma osmolality and volume
   (E) injection of vasopressin into the hypothalamus

2. A gain-of-function mutation in which of the following genes would be expected to cause obesity in humans?
   (A) leptin
   (B) leptin receptor
   (C) thyrotropin β subunit
   (D) pro-opiomelanocortin
   (E) melanin-concentrating hormone

3. When an individual is naked in a room in which the air temperature is 21 °C (69.8 °F) and the humidity 80%, the greatest amount of heat is lost from the body by
   (A) elevated metabolism
   (B) respiration
   (C) urination
   (D) vaporization of sweat
   (E) radiation and conduction

4. Secretion of which of the following hormones does *not* increase at night?
   (A) Growth hormone
   (B) ACTH
   (C) Melatonin
   (D) Insulin
   (E) Prolactin

Questions 5–10 refer to Figure 14–A. Each lettered structure may be selected once, more than once, or not at all.

5. Which of the labeled structures has the highest concentration of angiotensin II receptors?

6. Which of the labeled structures has the highest concentration of corticotropin-releasing hormone (CRH)?

7. Which of the labeled structures has the highest concentration of thyrotropin-releasing hormone (TRH)?

8. Which of the labeled structures has the highest concentration of dopamine?

9. Which of the labeled structures has the highest concentration of growth hormone-releasing hormone (GRH)?

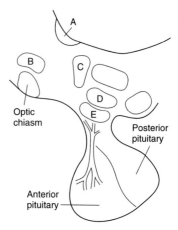

Figure 14–A. Hypothalamus and pituitary.

10. Which of the labeled structures is most involved in the regulation of circadian rhythms?

In questions 11–17, match the functions in each question with the lettered part of the central nervous system that is most closely associated with it. Each lettered part may be selected once, more than once, or not at all.

   (A) Spinal cord
   (B) Medulla oblongata
   (C) Midbrain
   (D) Hypothalamus
   (E) Amygdala

11. Fine control of blood pressure
12. Satiety
13. Primitive sexual reflexes
14. Swallowing
15. Fear
16. Water intake
17. Shivering

In questions 18–23, select

   A if the item is associated with (a) below,
   B if the item is associated with (b) below,
   C if the item is associated with both (a) and (b), and
   D if the item is associated with neither (a) nor (b)
   (a) V_{1A} vasopressin receptors
   (b) V_2 vasopressin receptors

18. Activation of G_s
19. Vasoconstriction
20. Increase in intracellular inositol triphosphate

21. Movement of aquaporin

22. Proteinuria

23. Milk ejection

CHAPTER 15

This chapter is concerned with the limbic system and with emotion, addiction, and sexual and other instinctive and stereotyped behaviors. The material in the chapter should help students to—

- Describe in general terms the structure and function of the limbic system.
- Discuss the brain regions and hormones involved in the regulation of sexual behavior in both sexes.
- Summarize the effects of sex hormones on the brain in fetal and early neonatal life.
- Describe the brain systems that mediate repeated self-stimulation and avoidance of stimulation.
- Discuss the relation of the brain reward system to addiction.
- Outline the anatomy of the serotonergic pathways in the brain, and summarize their known and suspected functions.
- Outline the anatomy of the noradrenergic (norepinephrine-secreting) pathways in the brain, and summarize their known and suspected functions.
- Outline the anatomy of the adrenergic (epinephrine-secreting) pathways in the brain, and summarize their known and suspected functions.
- Outline the anatomy of the dopaminergic pathways in the brain, and summarize their known and suspected functions.
- Outline the anatomy of the histaminergic pathways in the brain, and summarize their known and suspected functions.
- Outline the opioid peptide-secreting pathways in the brain, and summarize their known and suspected functions.

General Questions

1. What cholinergic neuronal system projects from the basal forebrain to the cerebral cortex, and particularly to the medial portions of the temporal lobes? What is the function of this system?

2. Discuss rage and violence from the point of view of the brain areas involved and how they might operate.

3. What are pheromones? What is the evidence that they affect behavior in humans?

4. Some scientists talk about the "female hypothalamus" and the "male hypothalamus." What are the differences between the hypothalami of the two sexes, and how are these differences produced?

5. What pharmacologic agents have been used to treat depression, and why?

6. What drugs cause hallucinations, and what is their mechanism of action?

Multiple-Choice Questions

In questions 1–17, select the single best answer. Questions 1–6 are made up of a statement and an explanation. Select

    A if both the statement and the explanation are true and they are related,

    B if both the statement and explanation are true but they are unrelated,

    C if the statement is true but the explanation is false,

    D if the statement is false but the explanation is true, and

    E if the statement and the explanation are both false

1. L-Dopa is of benefit in the treatment of Parkinson's disease because it increases the dopamine content of the tuberoinfundibular dopaminergic neurons

2. Drugs that decrease serotonin reuptake are of benefit in the treatment of some depressive illnesses because they increase extracellular norepinephrine in the brain

3. Procedures that increase the release of opioid peptides in the central nervous system decrease the effect of painful stimuli because the periaqueductal gray and related areas contain opioid receptors to which the peptides can bind

4. The antipsychotic activity of the major tranquilizers parallels their ability to block serotonin receptors because the serotonergic neural systems in the brain are concerned, among other things, with normal mental function

5. Women with a congenital adrenocortical enzyme defect that causes large amounts of androgens to be secreted have somewhat more masculine behavior than normal women because the androgens act on the brain during fetal development

6. Increased anxiety is produced by $GABA_B$ receptors, so treatment with $GABA_B$ antagonists has an anti-anxiety effect.

In questions 7–13, select

    A if the item is associated with (a) below,

    B if the item is associated with (b) below,

    C if the item is associated with both (a) and (b), and

D if the item is associated with neither (a) nor (b)

(a) Increases extracellular dopamine in the nucleus accumbens

(b) Produces sedation

7. Heroin

8. Alcohol

9. Cocaine

10. Nicotine

11. Amphetamine

12. Epibatidine

13. A benzodiazepine such as diazepam (Valium)

14. Which of the following would be expected to reduce maternal behavior?

(A) Lesions of the occipital cortex

(B) Lesions of the mediobasal hypothalamus

(C) Loss of expression of the gene for fos-B

(D) Ovariectomy

(E) Hysterectomy

15. The cell bodies of the principal histaminergic neurons in the brain are located in the

(A) preoptic area of the hypothalamus

(B) central nucleus of the amygdala

(C) subfornical organ

(D) locus ceruleus

(E) mediobasal hypothalamus

16. Which of the following drugs is a tranquilizer?

(A) LSD

(B) Atropine

(C) Ecstasy (MDMA)

(D) Lithium

(E) Chlorpromazine

17. In which of the following diseases is there reason to believe that the mesocortical system of dopaminergic neurons is hyperactive?

(A) Parkinson's disease

(B) Manic-depressive psychosis

(C) Schizophrenia

(D) Hyperprolactinemia

(E) Migraine

18. Perception of disgust appears to depend on the

(A) frontal cortex

(B) postcentral gyrus on the right side

(C) postcentral gyrus on the left side

(D) insular cortex on the right side

(E) insular cortex on the left side

In questions 19–20, indicate whether the first item is greater than **(G),** *the same as* **(S),** *or less than* **(L)** *the second item.*

19. Area of brain where animals will avoid stimulation through an implanted electrode

G S L

Area of brain where animals will work to receive repeated stimulation through an implanted electrode

20. Amount of norepinephrine in the raphe nuclei

G S L

Amount of serotonin in the raphe nuclei

CHAPTER 16

This chapter is concerned with the "higher functions of the nervous system," and specifically with learning, memory, the specialized natures of the left and right cerebral hemispheres, and the language functions of the human brain. The material in the chapter should help students to—

• Describe the various types of memory.

• Give examples of habituation and sensitization occurring in humans, and discuss the molecular events that may underlie them.

• Define conditioned reflexes, describe their properties, and analyze their physiologic basis.

• List the parts of the brain that appear to be involved in memory in mammals, and summarize the proposed role of each in memory processing and storage.

• Define the terms categorical hemisphere and representational hemisphere, and summarize the difference between the hemispheres and their relationship to handedness.

• Define and explain agnosia, unilateral neglect, dyslexia, and prosopagnosia.

• Summarize the differences between fluent and nonfluent aphasia, and explain each type on the basis of its pathophysiology.

• Describe the abnormalities of brain structure and function found in Alzheimer's disease.

General Questions

1. What methods are used to study learning and memory in humans and experimental animals?

2. Why does a mother often sleep through many different kinds of noise but wake when her baby cries?

3. What is biofeedback? Explain it in physiologic terms.

4. List and explain the abnormalities of brain function produced by sectioning the corpus callosum.

5. List and explain the differences between explicit and implicit memory.

6. Discuss the functions of Wernicke's area, Broca's area, the angular gyrus, and the basal ganglia in the production of speech.

Multiple-Choice Questions

In questions 1–20, select the single best answer.

1. Retrograde amnesia
   (A) is abolished by prefrontal lobotomy
   (B) responds to drugs that block dopamine receptors
   (C) is commonly precipitated by a blow on the head
   (D) is increased by administration of vasopressin
   (E) is due to damage to the brain stem

2. The representational hemisphere
   (A) is the right cerebral hemisphere in most right-handed individuals
   (B) is the left cerebral hemisphere in most left-handed individuals
   (C) includes the part of the brain concerned with language functions
   (D) is the site of lesions in most patients with aphasia
   (E) is morphologically identical to the opposite nonrepresentational hemisphere

3. The optic chiasm and corpus callosum are sectioned in a dog, and with the right eye covered, the animal is trained to bark when it sees a red square. The right eye is then uncovered and the left eye covered. The animal will now
   (A) fail to respond to the red square because the square does not produce impulses that reach the right occipital cortex
   (B) fail to respond to the red square because the animal has bitemporal hemianopia
   (C) fail to respond to the red square if the posterior commissure is also sectioned
   (D) respond to the red square only after retraining
   (E) respond promptly to the red square in spite of the lack of input to the left occipital cortex

4. The effects of bilateral loss of hippocampal function include
   (A) disappearance of remote memories
   (B) loss of working memory
   (C) loss of the ability to encode events of the recent past in long-term memory
   (D) loss of the ability to recall faces and forms but not the ability to recall printed or spoken words

   (E) production of inappropriate emotional responses when recalling events of the recent past

5. Which of the following are *incorrectly* paired?
   (A) Lesion of the parietal lobe of the representational hemisphere : unilateral inattention and neglect
   (B) Loss of cholinergic neurons in the nucleus basalis of Meynert and related areas of the forebrain : loss of recent memory
   (C) Lesions of mamillary bodies: loss of recent memory
   (D) Lesion of the angular gyrus in the categorical hemisphere : nonfluent aphasia
   (E) Lesion of Broca's area in the categorical hemisphere : slow speech

6. The representational hemisphere is better than the categorical hemisphere at
   (A) language functions
   (B) recognition of objects by their form
   (C) understanding printed words
   (D) understanding spoken words
   (E) mathematical calculations

7. Which of the following is *not* characteristic of conditioned reflexes?
   (A) Failure to form conditioned reflexes when the conditioned and unconditioned stimuli are separated by more than 2 minutes
   (B) Disappearance of the conditioned reflex if it is not reinforced from time to time
   (C) Rapid formation of conditioned reflexes when the conditioned stimulus is pleasant for the animal
   (D) Failure to form conditioned reflexes when the conditioned stimulus is unpleasant to the animal
   (E) Failure to form conditioned reflexes when, after the conditioned stimulus, there is a distracting stimulus before the unconditioned stimulus

8. A lesion of Wernicke's area (the posterior end of the superior temporal gyrus) in the categorical hemisphere causes patients to
   (A) lose short-term memory
   (B) speak in a slow, halting voice
   (C) experience déjà vu
   (D) talk rapidly but make little sense
   (E) lose the ability to recognize faces

In questions 9–15, match the function in each question with the lettered structure that is most closely associated with it. Each lettered structure may be selected once, more than once, or not at all.

(A) Corpus callosum
(B) Frontal lobe
(C) Pons
(D) Amygdala
(E) Temporal lobe

9. REM sleep
10. Interhemispheric transfer of learning
11. Emotional memories
12. Audition
13. Experimental neurosis
14. Visual association
15. Recognition of faces
16. Site of working memory

In questions 17–21, select

   A   if the item is associated with (a) below,
   B   if the item is associated with (b) below,
   C   if the item is associated with both (a) and (b),
     and
   D   if the item is associated with neither (a) nor (b)
     (a) Alzheimer's disease
     (b) Huntington's disease

17. Affects basal ganglia
18. Affects explicit memory
19. Is associated with an increased number of CAG trinucleotide repeats
20. Affects β-amyloid protein
21. Can be cured if treatment is started early in the disease

CHAPTER 17

This chapter provides an overview of energy balance and metabolism as a background for consideration of the functions of the endocrine glands. The topics considered are energy balance; the metabolic rate; the genesis and functions of ATP; the metabolism of carbohydrates, proteins, and fats; important fat derivatives, including leukotrienes and prostaglandins; and nutrition. The material in the chapter should help students to—

- Define metabolic rate, calorie, and respiratory quotient.
- State the caloric value per unit weight of carbohydrate, protein, and fat.
- Discuss the biosynthesis and functions of ATP.
- List the principal dietary hexoses, and summarize the main mechanisms that regulate glucose synthesis and breakdown.

- Describe the function of the citric acid cycle in the metabolism of glucose, amino acids, and fatty acids.
- Define the glucostatic function of the liver, and explain how the liver carries out this function.
- Define amino acid, polypeptide, and protein.
- List the main sources of uric acid in the body and the main metabolic products produced from uric acid.
- Describe the metabolic responses to starvation.
- List the major classes of lipids in the body, and describe the principal characteristics of each.
- Define ketone bodies, and describe their formation and metabolism.
- Describe the sources of the free fatty acids in plasma, their metabolic fate, and the principal factors regulating the fatty acid level in plasma.
- Describe the exogenous and endogenous pathways by which lipids are transported in the body, and summarize the processes involved in cholesterol metabolism.
- Outline the major pathways involved in the formation of leukotrienes, lipoxins, thromboxanes, prostacyclin, and prostaglandins, and list the main functions of each.
- List the components in a normal diet that will maintain weight and health.
- Define the term vitamin, name the major vitamins, and summarize the effects of deficiency and excess of each vitamin.

General Questions

1. The oxygen consumption of a fasting animal at rest was found to be 30 mL/kg body weight/min. What is the basal metabolic rate (BMR) (in kcal/kg/ 24 h)? What is the approximate size of this animal?
2. What is the thrifty gene hypothesis? How might it be related to obesity?
3. What is a "directional flow valve" in metabolism? Why are such valves important?
4. What changes in carbohydrate metabolism occur during exercise?
5. Compare the metabolism of fructose with that of glucose. How do they interact?
6. Discuss gout.
7. What are essential amino acids?
8. What is ubiquitination? What are its physiologic roles?
9. Which metabolites of arachidonic acid are important in the body? Why has a major effort been devoted to the development of drugs that modify arachidonic acid metabolism?

10. Discuss the pathophysiology of obesity. What is the significance of the fact that in almost all obese humans, the circulating leptin level is positively correlated with their body mass index, ie, with the degree to which they are obese, even though leptin reduces appetite in normal nonobese individuals?

Multiple-Choice Questions

In questions 1–20, select the single best answer.

1. In the body, metabolism of 10 g of protein would produce approximately
   (A) 1 kcal
   (B) 41 kcal
   (C) 410 kcal
   (D) 4100 kcal
   (E) 41 cal

2. A man with a respiratory quotient (RQ) of 70
   (A) has been eating a high-fat diet
   (B) has been eating a high-protein diet
   (C) has been fasting for 24 hours
   (D) has eaten nothing but carbohydrates for 24 hours
   (E) is dehydrated

3. Which of the following daily intakes (in grams) of carbohydrates (C), protein (P), and fat (F) would be best for a healthy 80-kg male construction worker?
   (A) C: 420; P: 80; F: 125
   (B) C: 420; P: 130; F: 100
   (C) C: 325; P: 80; F: 65
   (D) C: 550; P: 80; F: 65
   (E) C: 340; P: 50; F: 65

4. The major lipoprotein source of the cholesterol used in cells is
   (A) chylomicrons
   (B) intermediate-density lipoproteins (IDL)
   (C) albumin-bound free fatty acids
   (D) LDL
   (E) HDL

5. Which of the following produces the most high-energy phosphate compounds?
   (A) Aerobic metabolism of 1 mol of glucose
   (B) Anaerobic metabolism of 1 mol of glucose
   (C) Metabolism of 1 mol of galactose
   (D) Metabolism of 1 mol of amino acid
   (E) Metabolism of 1 mol of long-chain fatty acid

6. Which of the following produces the most glucose when metabolized in the body?
   (A) 1 mol of serotonin
   (B) 1 mol of alanine
   (C) 1 mol of oleic acid
   (D) 1 mol of thyroxine (T_4)
   (E) 1 mol of acetoacetic acid

7. Which of the following would *not* produce an increase in the plasma level of free fatty acids?
   (A) A drug that increases the level of intracellular cAMP
   (B) A drug that activates β_3-adrenergic receptors
   (C) A drug that inhibits hormone-sensitive lipase
   (D) A drug that decreases the metabolic clearance of glucagon
   (E) A drug that inhibits phosphodiesterase

8. When LDL enters cells by receptor-mediated endocytosis, which of the following does *not* occur?
   (A) Decrease in the formation of cholesterol from mevalonic acid
   (B) Increase in the intracellular concentration of cholesteryl esters
   (C) Increase in the transfer of cholesterol from the cell to HDL
   (D) Decrease in the rate of synthesis of LDL receptors
   (E) Decrease in the cholesterol in endosomes

9. Which of the following does *not* increase the output of glucose from the liver?
   (A) Induction of phosphorylase
   (B) Insulin
   (C) Glucagon
   (D) PGC-1
   (E) Epinephrine

In questions 10–15, select
   A if the item is associated with (a) below,
   B if the item is associated with (b) below,
   C if the item is associated with both (a) and (b), and
   D if the item is associated with neither (a) nor (b)
      (a) Cyclooxygenase
      (b) 5-Lipoxygenase

10. Arachidonic acid
11. Palmitic acid
12. Leukotrienes
13. Thromboxanes
14. Aspirin
15. Lipoxins

In questions 16–20, match each disease or condition in the questions with the lettered substance that is most closely associated with it. Each lettered substance may be selected once, more than once, or not at all.

(A) Folic acid

(B) Vitamin C

(C) Vitamin B$_1$

(D) Vitamin K

(E) Uric acid

16. Primary gout

17. Secondary gout

18. Scurvy

19. Beriberi

20. Spina bifida

CHAPTER 18

This chapter reviews the structure and function of the thyroid gland and its hormones, T$_4$ and T$_3$. The major diseases of the thyroid gland are also considered. The material in the chapter should help students to—

- Describe the gross and microscopic anatomy of the thyroid gland.
- Summarize the main features of iodine metabolism.
- Compare the structure of T$_4$, T$_3$, and RT$_3$.
- Outline the steps involved in the biosynthesis of thyroid hormones, their storage in the colloid, and the transfer of thyroid hormones from the colloid to the bloodstream.
- Name the proteins that bind thyroid hormones in the plasma, describe the relationship between bound and free thyroid hormones, and summarize the mechanisms that regulate thyroid hormone binding.
- Outline the principal pathways by which thyroid hormones are metabolized.
- List the main physiologic actions of thyroid hormones.
- Describe the mechanism of action of thyroid hormones.
- Outline the processes involved in the regulation of secretion of thyroid hormones.
- List the principal drugs that affect thyroid function, and describe the mechanism by which each exerts its effect.
- List the major diseases of the thyroid and their principal symptoms and signs.

General Questions

1. The curves in Figure 18–A show the changes in basal metabolic rate that follow administration of a single dose of T$_3$ or T$_4$ to an individual with previ-

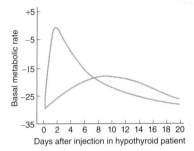

Figure 18–A.

ously untreated hypothyroidism. Which curve is produced by T$_3$ and which by T$_4$? Explain the differences between the two curves.

2. What is the role of TRH in the day-to-day regulation of thyroid function?

3. Discuss the causes and the pathophysiology of goiter. Why do people who eat large amounts of cabbage develop goiters?

4. Why do children with cretinism have short stature and mental deficiency?

5. Patients with thyrotoxicosis often have dramatic remission of many of their signs and symptoms upon administration of β-adrenergic-blocking drugs such as propranolol. Which signs and symptoms are improved, and which are not? Why?

6. Which hormones produce an increase in metabolic rate? How are these effects produced?

7. What are the embryologic origins of the thyroid gland, the parathyroid glands, and the parafollicular cells of the thyroid gland? Why is it important for clinicians to know about the embryologic development of these tissues?

Multiple-Choice Questions

In questions 1–13, select the single best answer.

1. In which of the following conditions is it *least* likely that the TSH response to TRH will be greater than normal?

(A) Hypothyroidism due to tissue resistance to thyroid hormone

(B) Hypothyroidism due to disease destroying the thyroid gland

(C) Hyperthyroidism due to circulating antithyroid antibodies with TSH activity

(D) Hyperthyroidism due to diffuse hyperplasia of thyrotropes of the anterior pituitary

(E) Iodine deficiency

2. Methimazole causes
   (A) a decrease in thyroid size and an increase in plasma T_4
   (B) a decrease in both thyroid size and plasma T_4
   (C) an increase in both thyroid size and plasma T_4
   (D) an increase in thyroid size and a decrease in plasma T_4
   (E) no change in thyroid size and a decrease in plasma T_4

3. A young woman has puffy skin and a hoarse voice. Her plasma TSH concentration is low but increases markedly when she is given TRH. She probably has
   (A) hyperthyroidism due to a thyroid tumor
   (B) hypothyroidism due to a primary abnormality in the thyroid gland
   (C) hypothyroidism due to a primary abnormality in the pituitary gland
   (D) hypothyroidism due to a primary abnormality in the hypothalamus
   (E) hyperthyroidism due to a primary abnormality in the hypothalamus

4. The coupling of monoiodotyrosine and diiodotyrosine and the iodination of thyroglobulin is blocked by
   (A) divalent cations
   (B) monovalent anions such as perchlorate
   (C) TSH
   (D) TRH
   (E) thiourylenes such as methimazole

5. The thyroidal uptake of radioactive iodine is *not* affected by
   (A) calcitonin
   (B) T_3
   (C) TSH
   (D) methimazole
   (E) the iodine content of the diet

6. The metabolic rate is *least* affected by an increase in the plasma level of
   (A) TSH
   (B) TRH
   (C) TBG
   (D) free T_4
   (E) free T_3

7. Which of the following is *not* essential for normal biosynthesis of thyroid hormones?
   (A) Iodine
   (B) Ferritin
   (C) Thyroglobulin
   (D) Protein synthesis
   (E) TSH

8. Which of the following would be *least* affected by injections of TSH?
   (A) Thyroidal uptake of iodine
   (B) Synthesis of thyroglobulin
   (C) Cyclic AMP in thyroid cells
   (D) Cyclic GMP in thyroid cells
   (E) Size of the thyroid

9. Hypothyroidism due to disease of the thyroid gland is associated with increased plasma levels of
   (A) cholesterol
   (B) albumin
   (C) RT_3
   (D) iodide
   (E) TBG

10. Which of the following is *most* likely to bring about improvement in exophthalmos?
    (A) Administration of T_4
    (B) Administration of drugs that inhibit the production of T lymphocytes
    (C) Administration of testosterone
    (D) Hypophysectomy
    (E) Thyroidectomy

11. Which of the following are *incorrectly* paired?
    (A) Type 1 (insulin-dependent) diabetes mellitus: antibodies against B cells
    (B) Myasthenia gravis : antibodies against nicotinic acetylcholine receptors
    (C) Multiple sclerosis : antibodies against myelin
    (D) Hashimoto's thyroiditis : antibodies against TSH
    (E) Graves' disease : antibodies that stimulate TSH receptors

12. Thyroid hormone receptors bind to DNA in which of the following forms?
    (A) A heterodimer with the prolactin receptor
    (B) A heterodimer with the growth hormone receptor
    (C) A heterodimer with the retinoid X receptor
    (D) A heterodimer with the insulin receptor
    (E) A heterodimer with the progesterone receptor

13. Increasing intracellular I^- due to the action of the Na^+/I^- symporter is an example of
    (A) Endocytosis
    (B) Passive diffusion
    (C) Na^+ and K^+ cotransport
    (D) Primary action transport
    (E) Secondary active transport

In questions 14–16, indicate whether the first item is greater than (G), the same as (S), or less than (L) the second item.

14. Food intake in myxedema

G S L

Food intake in thyrotoxicosis

15. Number of α-MHC molecules in the heart in myxedema

G S L

Number of α-MHC molecules in the heart in thyrotoxicosis

16. Amount of endocytosis at the colloid-thyroid cell border following injection of TSH

G S L

Amount of endocytosis at the colloid-thyroid cell border following injection of T$_4$

CHAPTER 19

This chapter is concerned with insulin, the other hormones of the pancreatic islets, and the many additional hormones in the body that affect carbohydrate metabolism. Diabetes mellitus is discussed in detail in this context. The material in the chapter should help students to—

- List the hormones that affect the plasma glucose concentration, and briefly describe the action of each.
- Describe the structure of the pancreatic islets, and name the hormones secreted by each of the cell types in the islets.
- Describe the structure of insulin, and outline the steps involved in its biosynthesis and release into the bloodstream.
- List the consequences of insulin deficiency, and explain how each of these abnormalities is produced.
- Describe insulin receptors, the way they mediate the effects of insulin, and the way they are regulated.
- Describe the effects of hypoglycemia, and summarize the homeostatic mechanisms that combat hypoglycemia.
- Describe the types of glucose transporters found in the body and the function of each.
- List the major factors that affect the secretion of insulin.
- Describe the structure of glucagon and its relation to glicentin.
- List the physiologically significant effects of glucagon.
- List the principal factors that affect the secretion of glucagon.

- Describe the probable physiologic effects of somatostatin in the pancreas.
- Outline the mechanisms by which thyroid hormones, adrenal glucocorticoids, catecholamines, and growth hormone affect carbohydrate metabolism.
- Define insulin-dependent (type 1) and non-insulin-dependent (type 2) diabetes mellitus, and describe the principal features and probable causes of each type.

General Questions

1. A patient has hypoglycemia. He may have an insulin-secreting tumor or functional hypoglycemia, but there is also reason to believe that he may be surreptitiously injecting himself with insulin. How would you differentiate among these possibilities? Which tests would you order, and what would each tell you?

2. Which of the curves in Figure 19–A would you expect to see (1) in a normal individual, (2) in an individual with type 1 diabetes mellitus, and (3) in an individual with type 2 diabetes mellitus? Explain the differences in the curves.

3. Why is ketosis more common and more severe in type 1 than in type 2 diabetes?

4. Mice in which the gene for insulin receptor substrate-1 is knocked out show moderate growth retardation and insulin resistance. However, mice in which the insulin receptor gene is knocked out show marked growth retardation and die soon after birth. What does this tell you about the mechanism of action of insulin?

5. Why is the diabetes seen in patients with somatostatinomas usually mild?

6. Which hormones are produced from the preproglucagon molecule? Where is each produced and what is its function?

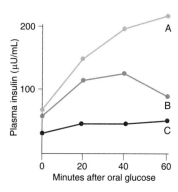

Figure 19–A.

7. Insulin is sometimes described as "the hormone of plenty" and glucagon as "the hormone of starvation." Discuss the appropriateness of these terms in the context of the physiologic effects of each hormone and their interactions.

8. What is proinsulin? What is its relation to preproinsulin and insulin, and what happens to it in the body?

Multiple-Choice Questions

In questions 1–11, select the single best answer.

1. Which of the following are *incorrectly* paired?
   (A) B cells : insulin
   (B) D cells : somatostatin
   (C) A cells : glucagon
   (D) Pancreatic exocrine cells : chymotrypsinogen
   (E) F cells : gastrin

2. Which of the following are *incorrectly* paired?
   (A) Epinephrine : increased glycogenolysis in skeletal muscle
   (B) Insulin : increased protein synthesis
   (C) Glucagon : increased gluconeogenesis
   (D) Progesterone : increased plasma glucose level
   (E) Growth hormone : increased plasma glucose level

3. Which of the following would be *least* likely to be seen 14 days after a rat is injected with a drug that kills all of its pancreatic B cells?
   (A) A rise in the plasma H^+ concentration
   (B) A rise in the plasma glucagon concentration
   (C) A fall in the plasma HCO_3^+ concentration
   (D) A fall in the plasma amino acid concentration
   (E) A rise in plasma osmolality

4. Glucagon is *not* normally found in the
   (A) brain
   (B) pancreas
   (C) gastrointestinal tract
   (D) plasma
   (E) adrenal glands

5. When the plasma glucose concentration falls to low levels, a number of different hormones help combat the hypoglycemia. After intravenous administration of a large dose of insulin, the return of a low blood sugar level to normal is delayed in
   (A) adrenal medullary insufficiency
   (B) glucagon deficiency
   (C) combined adrenal medullary insufficiency and glucagon deficiency

(D) thyrotoxicosis
(E) acromegaly

6. Which of the following is *not* produced by total pancreatectomy?
   (A) Steatorrhea
   (B) Increased plasma levels of free fatty acids
   (C) Decreased plasma P_{CO_2}
   (D) No change or a rise in the plasma insulin level
   (E) No change or a rise in the plasma glucagon level

7. Insulin increases the entry of glucose into
   (A) all tissues
   (B) renal tubular cells
   (C) the mucosa of the small intestine
   (D) most neurons in the cerebral cortex
   (E) skeletal muscle

8. The mechanism by which glucagon produces an increase in the plasma glucose concentration involves
   (A) binding of glucagon to DNA in the nucleus of target cells
   (B) binding of glucagon to receptors in the cytoplasm of target cells
   (C) activation of G_s in target cells
   (D) increased binding of Ca^{2+} in the cytoplasm of target cells
   (E) inhibition of insulin secretion

9. Glucagon increases glycogenolysis in liver cells but ACTH does not because
   (A) cortisol increases the plasma glucose level
   (B) liver cells have an adenylyl cyclase different from that in adrenocortical cells
   (C) ACTH cannot enter the nucleus of liver cells
   (D) the membranes of liver cells contain receptors different from those in adrenocortical cells
   (E) liver cells contain a protein that inhibits the action of ACTH

10. A meal rich in proteins containing the amino acids that stimulate insulin secretion but low in carbohydrate does not cause hypoglycemia because
   (A) the meal causes a compensatory increase in T_4 secretion
   (B) cortisol in the circulation prevents glucose from entering muscle
   (C) glucagon secretion is also stimulated by the meal
   (D) the amino acids in the meal are promptly converted to glucose
   (E) insulin does not bind to insulin receptors if the plasma concentration of amino acids is elevated

11. Glucose increases plasma insulin by a process that involves
    (A) GLUT 1
    (B) GLUT 2
    (C) GLUT 3
    (D) GLUT 4
    (E) SGLT 1

In questions 12–15, match the signs and symptoms in each question with the lettered item that is most appropriate. Each lettered item may be selected once, more than once, or not at all.

(A) Microvascular complication of diabetes
(B) Macrovascular complication of diabetes
(C) Neuropathic complication of diabetes
(D) Uncommon in type 1 diabetes
(E) Caused by chronic hypoglycemia

12. Progressive loss of vision
13. Rising plasma creatinine, ameliorated by angiotensin-converting enzyme inhibitors
14. Impotence and abdominal pain
15. Severe chest pain and nausea

In questions 16–18, indicate whether the first item is greater than (G), the same as (S), or less than (L) the second item.

16. Insulin secretion after glucose is given intravenously
    G S L
    Insulin secretion after the same amount of glucose is given orally
17. Plasma glucose level after surgical pancreatectomy
    G S L
    Plasma glucose level after administration of a toxin such as alloxan that selectively kills B cells
18. Plasma concentration of C peptide in fasting subject
    G S L
    Plasma concentration of C peptide in fed subject

CHAPTER 20

This chapter is concerned with the multiple functions of the adrenal glands: the secretion of catecholamines by the adrenal medulla, glucocorticoids and sex hormones by the zona fasciculata and zona reticularis of the adrenal cortex, and aldosterone by the zona glomerulosa of the adrenal cortex. There is a brief review of the clinical abnormalities produced by excess and deficiency of each of these hormones. The material in the chapter should help students to—

- Name the three catecholamines secreted by the adrenal medulla, and summarize their biosynthesis, metabolism, and function.
- List some of the stimuli that increase adrenal medullary secretion, and describe the way they bring about the increase.
- Differentiate between C_{18}, C_{19}, and C_{21} steroids, and give examples of each.
- Outline the steps involved in steroid biosynthesis in the adrenal cortex.
- Name the plasma proteins that bind adrenocortical steroids, and discuss the role they play in adrenocortical physiology.
- Name the major site of adrenocortical hormone metabolism and the principal metabolites produced from glucocorticoids, adrenal androgens, and aldosterone.
- Describe the mechanisms by which glucocorticoids and aldosterone produce changes in cellular function.
- List and briefly describe the physiologic and pharmacologic effects of glucocorticoids.
- Contrast the physiologic and pathologic effects of adrenal androgens.
- Describe the mechanisms that regulate secretion of glucocorticoids and adrenal sex hormones.
- List the actions of aldosterone.
- Describe the mechanisms that regulate aldosterone secretion.
- Describe the main features of the diseases caused by excess or deficiency of each of the hormones of the adrenal gland.

General Questions

1. Which of the metabolites of norepinephrine and epinephrine would you measure to determine whether a patient with a pheochromocytoma had a norepinephrine- or epinephrine-secreting tumor?
2. Which enzymes are unique to each of the three zones of the adrenal cortex, ie, found only in that particular zone? What are the physiologic consequences of this unique distribution?
3. What are the functions of the human fetal adrenal cortex?
4. What would you expect to be the consequences of 17α-hydroxylase deficiency? Explain your answer.
5. Why do some patients with congenital adrenal hyperplasia and virilization have high blood pressure but others have excessive loss of sodium and hypotension? Explain your answer.

6. Patients with nephrosis sometimes have low plasma 17-hydroxycorticoid levels but do not develop signs and symptoms of adrenal insufficiency. Why?

7. List the possible causes of Cushing's syndrome. How would you treat each of them?

8. What are the advantages and disadvantages of long-term, high-dose treatment with glucocorticoids in diseases such as rheumatoid arthritis and asthma? What problems occur when steroid treatment is stopped suddenly?

9. After hypophysectomy, the responsiveness of the adrenal cortex to ACTH is reduced. Why is the responsiveness reduced, and how would you restore it to normal?

10. What is the current view of the mechanism of action of aldosterone?

Multiple-Choice Questions

In questions 1–18, select the single best answer.
In questions 1–6, which refer to Figure 20–A, select the letter designating the part of the steroid-secreting cell that has a high concentration of the material. Each lettered part may be selected once, more than once, or not at all.

1. Cholesterol ester hydrolase
2. ACTH receptor mRNA

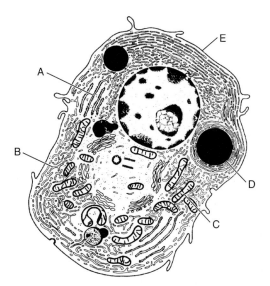

Figure 20–A. Steroid-secreting cell. (Reproduced, with permission, from Fawcett DW, Long JA, Jones AL: The ultrastructure of endocrine glands. Recent Prog Horm Res 1969;25:315.)

3. ACTH receptors
4. 3β-Hydroxysteroid dehydrogenase
5. CYP11B1
6. LDL receptors

7. Which of the following is produced only by *large amounts* of glucocorticoids?
   (A) Normal responsiveness of fat depots to norepinephrine
   (B) Maintenance of normal vascular reactivity
   (C) Increased excretion of a water load
   (D) Inhibition of the inflammatory response
   (E) Inhibition of ACTH secretion

8. Which of the following are *incorrectly* paired?
   (A) Gluconeogenesis : cortisol
   (B) Free fatty acid mobilization : dehydroepiandrosterone
   (C) Muscle glycogenolysis : epinephrine
   (D) Kaliuresis : aldosterone
   (E) Hepatic glycogenesis : insulin

9. Which of the following hormones has the shortest plasma half-life?
   (A) Corticosterone
   (B) Renin
   (C) Dehydroepiandrosterone
   (D) Aldosterone
   (E) Norepinephrine

10. Mole for mole, which of the following has the greatest effect on Na+ excretion?
    (A) Progesterone
    (B) Cortisol
    (C) Vasopressin
    (D) Aldosterone
    (E) Dehydroepiandrosterone

11. Mole for mole, which of the following has the greatest effect on plasma osmolality?
    (A) Progesterone
    (B) Cortisol
    (C) Vasopressin
    (D) Aldosterone
    (E) Dehydroepiandrosterone

12. The secretion of which of the following would be *least* affected by a decrease in extracellular fluid volume?
    (A) CRH
    (B) Arginine vasopressin
    (C) Dehydroepiandrosterone
    (D) Estrogens
    (E) Aldosterone

13. A young man presents with a blood pressure of 175/110 mm Hg. He is found to have a high circulating aldosterone but a low circulating cortisol. Glucocorticoid treatment lowers his circulating aldosterone and lowers his blood pressure to 140/85 mm Hg. He probably has an abnormal

    (A) 17α-hydroxylase

    (B) 21β-hydroxylase

    (C) 3β-hydroxysteroid dehydrogenase

    (D) aldosterone synthase

    (E) cholesterol desmolase

14. A 32-year-old woman presents with a blood pressure of 155/96 mm Hg. In response to questioning, she admits that she loves licorice and eats some at least three times a week. She probably has a low level of

    (A) type 2 11β-hydroxysteroid dehydrogenase activity

    (B) ACTH

    (C) 11β-hydroxylase activity

    (D) glucuronyl transferase

    (E) norepinephrine

15. Which of the following is *not* characteristic of primary hyperaldosteroidism?

    (A) Low plasma rennin activity

    (B) Normal plasma ACTH

    (C) Hypertension

    (D) High plasma Na+

    (E) Low plasma K+

16. In its action in cells, aldosterone

    (A) increases transport of ENaCs from the cytoplasm to the cell membrane

    (B) does not act on the cell membrane

    (C) binds to a receptor in the nucleus

    (D) may activate a heat shock protein

    (E) also binds to glucocorticoid receptors

In questions 17–20, match the statement in each question with the lettered steroid conversion that is most closely associated with it. Each lettered conversion may be selected once, more than once, or not at all.

    (A) Progesterone → corticosterone

    (B) Corticosterone → aldosterone

    (C) 17-Hydroxypregnenolone → dehydroepiandrosterone

    (D) 17-Hydroxyprogesterone → 11-deoxycortisol

    (E) Cholesterol → pregnenolone

17. Primary site of action of ACTH

18. Blocked in congenital 17α-hydroxylase deficiency

19. Produces a 17-ketosteroid

20. Inhibited in congenital 21-hydroxylase deficiency

CHAPTER 21

This chapter reviews the metabolism of calcium and phosphorus, the physiology of bone, and the three hormones that are the primary regulators of calcium metabolism after birth: 1,25-dihydroxycholecalciferol, parathyroid hormone, and calcitonin. In addition, there is a brief review of the effects of other hormones on calcium metabolism. The material in the chapter should help students to—

- Describe the distribution of calcium and phosphorus in the body and the forms in which they exist in plasma.

- Name the factors affecting plasma Ca^{2+} concentration, and discuss the mechanism by which each exerts its effects.

- Name the types of cells found in bone, and describe the function of each.

- Describe the formation of vitamin D in the skin, its subsequent hydroxylation in the liver and kidneys, and the actions of its biologically active metabolites.

- Name the factors affecting the 1α-hydroxylase that forms 1,25-dihydroxycholecalciferol in the kidneys, and describe the action of each.

- Describe the biosynthesis, actions, and metabolism of parathyroid hormone.

- Identify the source of calcitonin, its chemical nature, and its principal actions.

- Summarize the effects of glucocorticoids, growth hormone, and growth factors on Ca^{2+} metabolism.

General Questions

1. How would you identify the cells in which parathyroid hormone and calcitonin are produced?

2. Name the cells in the body on which 1,25-dihydroxycholecalciferol, parathyroid hormone, and calcitonin act, and describe the effects of these hormones on each of the cells.

3. How do bones grow? What is epiphysial closure, and what produces it?

4. What are the signs and symptoms of hyperparathyroidism, and what causes each of them?

5. On the axes in Figure 21–A, plot the relationship of plasma Ca^{2+} concentration to plasma parathyroid hormone concentration.

6. Define osteomalacia, osteogenesis imperfecta, osteopetrosis, and osteoporosis. Discuss the causes and treatment of osteoporosis.

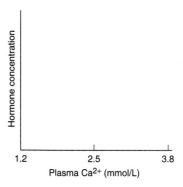

Figure 21–A.

7. There was a sharp increase in the incidence of bone disease with the onset of the Industrial Revolution in the 19th century, but the incidence has since decreased. Which bone disease increased, and why? Why has the incidence declined in the 20th century?

Multiple-Choice Questions

In questions 1–11, select the single best answer.

1. A patient with parathyroid deficiency 10 days after inadvertent damage to the parathyroid glands during thyroid surgery would probably have
   (A) low plasma phosphate and Ca^{2+} levels and tetany
   (B) low plasma phosphate and Ca^{2+} levels and tetanus
   (C) a low plasma Ca^{2+} level, increased muscular excitability, and a characteristic spasm of the muscles of the upper extremity (Trousseau's sign)
   (D) high plasma phosphate and Ca^{2+} levels and bone demineralization
   (E) increased muscular excitability, a high plasma Ca^{2+} level, and bone demineralization

2. A high plasma Ca^{2+} level causes
   (A) bone demineralization
   (B) increased formation of 1,25-dihydroxycholecalciferol
   (C) decreased secretion of calcitonin
   (D) decreased blood coagulability
   (E) increased formation of 24,25-dihydroxycholecalciferol

3. Which of the following is *not* involved in regulating plasma Ca^{2+} levels?
   (A) Kidneys
   (B) Skin
   (C) Liver
   (D) Lungs
   (E) Intestine

4. Which of the following exerts the greatest effect on parathyroid hormone secretion?
   (A) Plasma phosphate concentration
   (B) Calcitonin
   (C) 1,25-Dihydroxycholecalciferol
   (D) Total plasma calcium concentration
   (E) Plasma Ca^{2+} concentration

5. 1,25-Dihydroxycholecalciferol affects intestinal Ca^{2+} absorption through a mechanism that
   (A) includes alterations in the activity of genes
   (B) activates adenylyl cyclase
   (C) decreases cell turnover
   (D) changes gastric acid secretion
   (E) is comparable to the action of polypeptide hormones

6. Which of the following would you expect to find in a patient whose diet has been low in calcium for 2 months?
   (A) Increased formation of 24,25-dihydroxycholecalciferol
   (B) Decreased amounts of calcium-binding protein in intestinal epithelial cells
   (C) Increased parathyroid hormone secretion
   (D) A high plasma calcitonin concentration
   (E) Increased plasma phosphates

7. Which of the following would tend to *increase* the plasma Ca^{2+} level?
   (A) Decreased intestinal absorption of Ca^{2+}
   (B) Elevated plasma protein level
   (C) High plasma pH
   (D) Some types of cancer
   (E) Severe renal disease

8. In which of the following diseases is plasma PTH often elevated?
   (A) Pseudohypoparathyroidism
   (B) Adrenal failure
   (C) Cancer
   (D) Congestive heart failure
   (E) Precocious puberty

9. In osteogenesis imperfecta, which of the following is defective?
   (A) Phosphate deposition in trabecular bone
   (B) Structure of PTHrP
   (C) Osteoblasts
   (D) Osteoclasts
   (E) Bone collagen

10. In osteopetrosis, which of the following is defective?

(A) Phosphate deposition in trabecular bone

(B) Structure of PTHrP

(C) Osteoblasts

(D) Osteoclasts

(E) Bone collagen

11. PTH and PTHrP activate the same receptor, but their effects are different. Which of the following is the most likely explanation of this observation?

(A) For unknown reasons, PTHrP acts mainly on fetal tissues, whereas PTH acts mainly on adult tissues.

(B) Plasma PTH is high when plasma PTHrP is low, and vice versa.

(C) PTH acts on the brain, whereas PTHrP does not.

(D) When PTH binds to the common receptor, it activates adenylyl cyclase, whereas PTHrP activates phospholipase C.

(E) There is an additional receptor in the body that responds to PTH but not to PTHrP.

In questions 12 and 13, indicate whether the first item is greater than (G), the same as (S), or less than (L) the second item.

12. Percentage of ingested Ca^{2+} absorbed from the intestine in a subject fed a high-calcium diet.

G S L

Percentage of ingested Ca^{2+} absorbed from the intestine in a subject fed a low-calcium diet

13. Concentration of PTHrP in milk

G S L

Concentration of PTHrP in plasma collected at the same time

CHAPTER 22

Topics covered in this chapter are the morphology of the pituitary gland, pro-opiomelanocortin and its processing, the intermediate lobe of the pituitary, growth hormone, the somatomedins IGF-I and IGF-II, and the physiology of growth. The material in the chapter should help students to—

- Name the hormones secreted by the pituitary gland in humans, and list the main functions of each.
- Describe the embryonic origin of the lobes of the pituitary gland and their structure in adult humans.
- Describe the mechanisms responsible for changes in skin coloration in species in which such changes take place, and name the hormones involved.

- Describe the pro-opiomelanocortin molecule and its processing in the anterior and intermediate lobes of the pituitary gland.
- Describe the structure of the main form of growth hormone, and list its actions.
- Describe the relation of IGF-I and IGF-II to the actions of growth hormone.
- Name the three hypothalamic hypophysiotropic hormones that regulate growth hormone secretion and the principal stimuli that bring about increases or decreases in growth hormone secretion.
- Describe the growth hormone receptor, the way it functions, and its relation to circulating growth hormone-binding protein.
- List the factors needed for normal growth, and describe the contribution of each during prenatal and postnatal development.
- Describe and explain the main features of hypopituitarism, acromegaly, and Nelson's syndrome.

General Questions

1. Which endocrine disease causes alterations in skin pigmentation in humans? Which hormones are involved?

2. Although the same pro-opiomelanocortin molecule is found in the anterior and intermediate lobes of the pituitary gland and in neurons in the brain, different products are formed from it in each location. How can different products be formed from the same molecule? What processes are involved?

3. How would you decide whether an increase in growth hormone secretion was due to decreased secretion of somatostatin or increased secretion of GRH?

4. Why does destruction of the posterior lobe of the pituitary gland cause marked polyuria whereas destruction of the whole pituitary gland causes only transient polyuria?

5. How is nutrition related to growth?

6. What are the endocrinologic causes of dwarfism, and how does each lead to short stature?

7. How is human growth hormone (hGH) related to human chorionic somatomammotropin (hCS)?

Multiple-Choice Questions

In questions 1–8, select the single best answer.

1. Which of the following hormones exerts the least effect on growth?

(A) Growth hormone

(B) Testosterone

(C) T_4

(D) Insulin

(E) Vasopressin

2. Which of the following are *incorrectly* paired?

(A) Intermediate lobe : corticotropin-like intermediate lobe peptide (CLIP)

(B) Hypothyroidism: low plasma growth hormone concentration

(C) Gigantism : increased growth hormone secretion before puberty

(D) African pygmies : low plasma growth hormone

(E) Acromegaly : high plasma IGF-I concentration

3. Which of the following pituitary hormones is an opioid peptide?

(A) α-Melanocyte-stimulating hormone (α-MSH)

(B) β-MSH

(C) ACTH

(D) Growth hormone

(E) β-Endorphin

4. Which of the following hormones is *not* made up of α- and β-subunits?

(A) TSH

(B) LH

(C) FSH

(D) hCG

(E) Prolactin

5. Which of the following is *not* characteristic of hypopituitarism?

(A) Cachexia

(B) Infertility

(C) Pallor

(D) Low basal metabolic rate

(E) Intolerance to stress

6. Which of the following is *not* synthesized in both endocrine glands and the brain?

(A) Somatostatin

(B) Cortisol

(C) Dopamine

(D) ACTH

(E) ERβ

7. A scientist finds that infusion of growth hormone into the median eminence of the hypothalamus in experimental animals inhibits the secretion of growth hormone, and concludes that this proves that growth hormone feeds back to inhibit GRH secretion. Do you accept this conclusion?

(A) No, because growth hormone does not cross the blood-brain barrier

(B) No, because the infused growth hormone could be stimulating dopamine secretion

(C) No, because substances placed in the median eminence could be transported to the anterior pituitary

(D) Yes, because systemically administered growth hormone inhibits growth hormone secretion

(E) Yes, because growth hormone binds GRH, inactivating it

8. The growth hormone receptor

(A) activates G_s

(B) requires dimerization to exert its effects

(C) must be internalized to exert its effects

(D) resembles the IGF-I receptor

(E) resembles the ACTH receptor

In questions 9–12, match the condition in each question with the lettered abnormality causing dwarfism that is most closely associated with it. Each lettered item may be used once, more than once, or not at all.

(A) Fibroblast growth factor receptor 3 gene defect

(B) Chronic abuse and neglect

(C) Defective growth hormone receptors

(D) Thyroid hormone deficiency

(E) Increased circulating gonadal steroids

9. Laron dwarfism

10. Cretinism

11. Achondroplasia

12. Kaspar Hauser syndrome

13. Precocious puberty

In questions 14 and 15, indicate whether the first item is greater than (G), the same as (S), or less than (L) the second item.

14. Plasma growth hormone level after 30 minutes of sleep

G S L

Plasma growth hormone level after 8 hours of sleep

15. Insulin sensitivity in hypopituitarism

G S L

Insulin sensitivity in acromegaly

CHAPTER 23

This chapter is a review of the physiology of the reproductive system in adult males and females. It also considers pregnancy and lactation, sexual differentiation in the fetus, puberty, and menopause. The material in the chapter should help students to—

- Name the important hormones secreted by the Leydig cells and Sertoli cells of the testes and by the graafian follicles and corpora lutea of the ovaries.
- Outline the role of chromosomes, hormones, and related factors in sex determination and development.
- Summarize the hormonal and other changes that occur at puberty in males and females.
- Outline the hormonal and other changes that occur at menopause.
- List the principal stimuli and drugs that affect prolactin secretion.
- Outline the steps involved in spermatogenesis, from the primitive germ cells to mature, motile spermatozoa.
- Describe the mechanisms that produce erection and ejaculation.
- Know the general structure of testosterone, and describe its biosynthesis, transport, metabolism, and actions.
- Describe the processes involved in regulation of testosterone secretion.
- Describe the changes that occur in the ovaries, uterus, cervix, vagina, and breasts during the menstrual cycle.
- Know the general structures of 17β-estradiol and progesterone, and describe their biosynthesis, transport, metabolism, and actions.
- Describe the roles of the pituitary and the hypothalamus in the regulation of ovarian function, and the role of feedback loops in this process.
- Describe the hormonal changes that accompany pregnancy and parturition.
- Outline the processes involved in development of the breasts, production of milk, milk ejection, and termination of lactation.

General Questions

1. What are the effects on subsequent sexual development when a human female fetus is exposed to excess androgens in utero?
2. What are the effects on subsequent sexual development when normal testes fail to develop in a human male fetus?
3. What is the effect of hypophysectomy on the uterus? Which other operation would produce the same results?
4. What physiologic changes occur in females during sexual intercourse?
5. What is the function of the acrosome?

Day of menstrual cycle

Figure 23–A.

6. Using the axes in Figure 23–A, diagram the changes in plasma LH, FSH, estrogen, progesterone, and inhibin A and B occurring during the human menstrual cycle.
7. What is the testicular feminizing syndrome, and what causes it?
8. What is the evidence that the onset of puberty is under neural control?
9. Using your knowledge of reproductive physiology, list the possible ways that contraception could be produced. How many of these are actually being used by the general public? Comment on the strengths and weaknesses of the five methods that you believe are most commonly used.
10. Discuss the relation of hormones to cancer.

Multiple-Choice Questions

In questions 1–17, select the single best answer. Questions 1–5 refer to Figure 23–B.

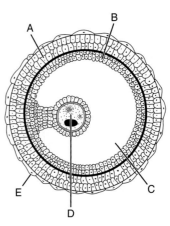

Figure 23–B. Ovarian follicle.

1. Which of the labeled structures has a higher concentration of angiotensin II than plasma does?

2. Which of the labeled structures produces the most androstenedione?

3. Which of the labeled structures produces estradiol that mainly enters the follicular fluid?

4. Which of the labeled structures contains the smallest amount of DNA per cell?

5. In which of the labeled structures is cell division arrested in prophase?

6. If a young woman has high plasma levels of T_3, cortisol, and renin activity but her blood pressure is only slightly elevated and she has no symptoms or signs of thyrotoxicosis or Cushing's syndrome, the most likely explanation is that
   (A) she has been treated with TSH and ACTH
   (B) she has been treated with T_3 and cortisol
   (C) she is in the third trimester of pregnancy
   (D) she has an adrenocortical tumor
   (E) she has been subjected to chronic stress

7. Full development and function of the seminiferous tubules require
   (A) somatostatin
   (B) LH
   (C) oxytocin
   (D) FSH
   (E) androgens and FSH

8. In humans, fertilization usually occurs in the
   (A) vagina
   (B) cervix
   (C) uterine cavity
   (D) uterine tubes
   (E) abdominal cavity

9. In human males, testosterone is produced mainly by the
   (A) Leydig cells
   (B) Sertoli cells
   (C) seminiferous tubules
   (D) epididymis
   (E) vas deferens

10. Home-use kits for determining a woman's fertile period depend on the detection of one hormone in the urine. This hormone is
    (A) FSH
    (B) progesterone
    (C) estradiol
    (D) hCG
    (E) LH

11. Puberty does not normally occur in humans under the age of 8 years, because before that age
    (A) the tissues are unresponsive to gonadal steroids
    (B) the ovaries and testes are unresponsive to gonadotropins
    (C) the pituitary cannot manufacture adequate amounts of gonadotropins
    (D) the brain secretes a substance that inhibits the responsiveness of the gonads to gonadotropins
    (E) the hypothalamus fails to secrete GnRH in a pulsatile fashion

12. Castration cells are found in the
    (A) uterus
    (B) prostate
    (C) placenta
    (D) anterior pituitary
    (E) hypothalamus

13. Decidual cells are found in the
    (A) uterus
    (B) prostate
    (C) placenta
    (D) anterior pituitary
    (E) hypothalamus

14. Which of the following is *not* a steroid?
    (A) 17α-Hydroxyprogesterone
    (B) Estrone
    (C) Relaxin
    (D) Pregnenolone
    (E) Etiocholanolone

15. Which of the following is *not* a male secondary sexual characteristic?
    (A) A beard
    (B) An increased incidence of acne
    (C) A deep voice
    (D) Increased fat in the buttocks
    (E) An enlarged penis

16. Sildenafil (Viagra) causes transient difficulty with color perception because
    (A) It dilates the blood vessels in the retina
    (B) It constricts the blood vessels in the retina
    (C) Related isoforms of guanylyl cyclase are found in the retina and in the penis
    (D) Related isoforms of phosphodiesterase are found in the retina and in the penis
    (E) The retina and the penis arise from the same embryonic structure

17. Which of the following probably triggers the onset of labor?

(A) ACTH in the fetus

(B) ACTH in the mother

(C) Prostaglandins

(D) Oxytocin

(E) Placental renin

In questions 18–22, indicate whether the first item is greater than (G), the same as (S), or less than (L) the second item.

18. Size of the clitoris in a woman with congenital 17α-hydroxylase deficiency

    G S L

    Size of the clitoris in a woman with congenital 21β-hydroxylase deficiency

19. Size of the clitoris in a woman with congenital 11β-hydroxylase deficiency

    G S L

    Size of the clitoris in a woman with congenital 21β-hydroxylase deficiency

20. Plasma LH concentration in a 59-year-old man

    G S L

    Plasma LH concentration in a 59-year-old woman

21. hCG production in the first trimester of pregnancy

    G S L

    hCG production in the third trimester of pregnancy

22. hCS production in the first trimester of pregnancy

    G S L

    hCS production in the third trimester of pregnancy

CHAPTER 24

This chapter is concerned with renin and erythropoietin, two of the three hormones produced by the kidneys; with ANP, BNP, and CNP, the natriuretic peptides produced by the heart, the brain, and other tissues; and with melatonin, the pineal hormone. The material in the chapter should help students to—

- Outline the cascade of reactions that lead to the formation of angiotensin II and its metabolites in the circulation.
- Describe the juxtaglomerular apparatus, and list the factors that regulate its secretion.
- List the functions of angiotensin II and the receptors on which it acts to carry out these functions.
- Describe the site and mechanism of action of erythropoietin, and the feedback regulation of its secretion.

- Describe the structure and functions of ANP, BNP, and CNP and the receptors on which they act.
- Diagram the steps involved in the formation of melatonin from serotonin in the pineal gland, discuss the proposed function of melatonin, and describe the regulation of melatonin secretion.

General Questions

1. Discuss the renin-angiotensin system from the point of view of the similarities and differences between it and the kallikrein system.

2. Why are patients with chronic renal failure anemic? How would you treat their anemia?

3. Using the axes in Figure 24–A, plot the relationship between dietary sodium intake and (1) the plasma renin activity, (2) the plasma concentration of ANP, and (3) the plasma concentration of cortisol.

4. Discuss the relationship of the renin-angiotensin system to clinical hypertension. How would you explain the fact that angiotensin-converting enzyme inhibitors lower blood pressure in hypertensive patients even though their plasma renin activities are normal or low?

5. Define Goldblatt hypertension, Bartter's syndrome, and secondary hyperaldosteronism.

6. One of the two curves in Figure 24–B was obtained without and one with 10^{-6} M losartan in the medium. Which curve is which? Explain the shape of the curves and the horizontal displacement of one from the other. What do the two curves tell you about the type of angiotensin receptor involved in the regulation of aldosterone secretion?

7. Discuss the concept of clearance receptors as it applies to ANP and related polypeptides.

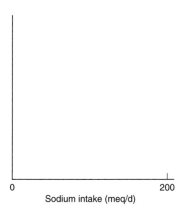

0   200

Sodium intake (meq/d)

Figure 24–A.

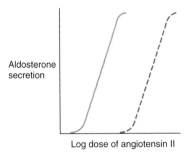

Aldosterone secretion

Log dose of angiotensin II

Figure 24–B. Aldosterone secretion by adrenocortical tissue in vitro.

8. Melatonin secretion is high at night and low during the day. How does the pineal gland "know" that it is day or night?

Multiple-Choice Questions

In questions 1–16, select the single best answer.

1. Renin is secreted by
   (A) cells in the macula densa
   (B) cells in the proximal tubules
   (C) cells in the distal tubules
   (D) juxtaglomerular cells
   (E) cells in the peritubular capillary bed
2. Erythropoietin is secreted by
   (A) cells in the macula densa
   (B) cells in the proximal tubules
   (C) cells in the distal tubules
   (D) juxtaglomerular cells
   (E) cells in the peritubular capillary bed
3. Melatonin secretion would probably *not* be increased by
   (A) stimulation of the superior cervical ganglia
   (B) intravenous infusion of tryptophan
   (C) intravenous infusion of epinephrine
   (D) stimulation of the optic nerve
   (E) induction of pineal hydroxyindole-*O*-methyltransferase
4. When a woman who has been on a low-sodium diet for 8 days is given an intravenous injection of captopril, a drug that inhibits angiotensin-converting enzyme, one would expect
   (A) her blood pressure to rise because her cardiac output would fall
   (B) her blood pressure to rise because her peripheral resistance would fall

(C) her blood pressure to fall because her cardiac output would fall
   (D) her blood pressure to fall because her peripheral resistance would fall
   (E) her plasma renin activity to fall because her circulating angiotensin I level would rise
5. Which of the following would be expected to cause an increase in ANP secretion from the heart?
   (A) Prolactin
   (B) Growth hormone
   (C) Erythropoietin
   (D) Constriction of the ascending aorta
   (E) Constriction of the inferior vena cava
6. Which of the following would *not* be expected to increase renin secretion?
   (A) Administration of a drug that blocks angiotensin-converting enzyme
   (B) Administration of a drug that blocks AT_1 receptors
   (C) Administration of a drug that blocks β-adrenergic receptors
   (D) Constriction of the aorta between the celiac artery and the renal arteries
   (E) Administration of a drug that reduces ECF volume
7. Activation of receptors for ANP increases target cell
   (A) cAMP
   (B) IP_3
   (C) protein kinase A activity
   (D) guanylyl cyclase activity
   (E) GTP
8. Erythropoietin
   (A) contains zinc
   (B) contains iron
   (C) is an important ligand for iron in the tissues
   (D) stimulates renin secretion
   (E) acts on some but not all stem cells in bone marrow
9. Which of the following is *least* likely to contribute to the beneficial effects of angiotensin-converting enzyme inhibitors in the treatment of congestive heart failure?
   (A) Vasodilation
   (B) Decreased cardiac growth
   (C) Decreased cardiac afterload
   (D) Increased plasma renin activity
   (E) Decreased plasma aldosterone

In questions 10–16, match the statement in each question with the lettered hormone that is most closely associated with it. Each lettered hormone may be selected once, more than once, or not at all.

(A) Thyroxine

(B) Erythropoietin

(C) Arginine vasopressin

(D) Melatonin

(E) Aldosterone

10. Increased by injection of ACTH

11. Primarily involved in the regulation of extracellular fluid volume

12. Primarily involved in the regulation of extracellular fluid osmolality

13. A prohormone

14. Primarily involved in the regulation of the metabolic rate

15. An indole

16. Affected by alterations in iron metabolism

CHAPTER 25

This chapter is concerned with digestion and absorption, and Chapter 26 deals with the details of how the gastrointestinal tract and its associated glands function to aid these processes. The material in Chapter 25 should help students to—

- Name the principal digestive enzymes, their precursors, their substrates, and the products of the action of the enzymes.
- Define brush border, unstirred layer, and glycocalyx.
- Outline the processes involved in the conversion of dietary carbohydrates to glucose and other hexoses in the intestine.
- Summarize the processes involved in the absorption of hexoses from the intestine into the bloodstream.
- Describe the conversion of dietary protein into amino acids and small peptides in the intestine.
- Summarize the processes involved in the absorption of amino acids and small peptides from the intestine into the bloodstream.
- Outline the events occurring during digestion of fats.
- Describe the processes by which fatty acids and other lipids are absorbed from the intestine into the bloodstream.
- Summarize the processes that regulate the absorption of water, Na^+, K^+, HCO_3^-, Ca^{2+}, and iron from the gastrointestinal tract.

General Questions

1. Many adults who are not of northern or western European origin have abdominal symptoms after ingestion of milk. What are the symptoms, what causes them, and how would you treat the condition?

2. What is secondary active transport and why is it important in the gastrointestinal tract? Why is it called secondary rather than primary active transport?

3. After a meal, the lumens of the stomach and the intestine contain powerful proteolytic enzymes. Why do these enzymes not digest the pancreas, which produces them, or the wall of the intestine?

4. Intestinal absorption of calcium and iron are both subject to feedback control but by very different mechanisms. Compare and contrast the mechanisms.

5. What are M cells? What is their function?

Multiple-Choice Questions

In questions 1–12, select the single best answer.

1. The pathway from the intestinal lumen to the circulating blood for a short-chain fatty acid (< 10 carbon atoms) is

   (A) intestinal mucosal cell → chylomicrons → lymphatic duct → systemic venous blood

   (B) intestinal mucosal cell → hepatic portal vein blood → systemic venous blood

   (C) space between mucosal cells → lymphatic duct → systemic venous blood

   (D) space between mucosal cells → chylomicrons → lymphatic duct → systemic venous blood

   (E) intestinal mucosal cell → LDL → hepatic portal vein blood → systemic venous blood

2. Maximum absorption of short-chain fatty acids produced by bacteria occurs in the

   (A) stomach

   (B) duodenum

   (C) jejunum

   (D) ileum

   (E) colon

3. Water is absorbed in the jejunum, ileum, and colon and excreted in the feces. Arrange these in order of the amount of water absorbed or excreted from greatest to smallest.

   (A) Colon, jejunum, ileum, feces

   (B) Feces, colon, ileum, jejunum

   (C) Jejunum, ileum, colon, feces

(D) Colon, ileum, jejunum, feces

(E) Feces, jejunum, ileum, colon

4. Drugs and toxins that increase the cAMP content of the intestinal mucosa cause diarrhea because they

(A) increase Na^+-K^+ cotransport in the small intestine

(B) increase K^+ secretion into the colon

(C) inhibit K^+ absorption in the crypts of Lieberkühn

(D) increase Na^+ absorption in the small intestine

(E) increase Cl^- secretion into the intestinal lumen

5. Which of the following are *incorrectly* paired?

(A) Pancreatic α-amylase : starch

(B) Elastase : tissues rich in elastin

(C) Enteropeptidase : fatty acids

(D) Rennin : coagulated milk

(E) Lingual lipase : digestion in the stomach

6. Calcium absorption is increased by

(A) hypercalcemia

(B) oxalates in the diet

(C) iron overload

(D) 1,25-dihydroxycholecalciferol

(E) increased Na^+ absorption

In questions 7–12, select

A   if the item affects (a) below,

B   if the item affects (b) below,

C   if the item affects both (a) and (b), and

D   if the item affects neither (a) nor (b)

   (a) Absorption of glucose

   (b) Absorption of amino acids

7. Plasma K^+

8. Na^+ in the intestinal lumen

9. Plasma insulin

10. Phloridzin

11. Trypsin in the intestinal lumen

12. Ferritin in the intestinal mucosa

CHAPTER 26

This chapter is a detailed consideration of the way in which the gastrointestinal tract and its associated glands, the salivary glands, liver, and pancreas, function in carrying out digestion and absorption of food. The material in the chapter should help students to—

- Describe the structure and function of the enteric nervous system.

- Define the basic electric rhythm (BER) and the migrating motor complex (MMC), and describe the function of each in the regulation of gastrointestinal motility.

- List the principal gastrointestinal hormones, the sites where each is secreted, and the main physiologic function of each of these hormones.

- Summarize the functions of the mouth, the salivary glands, and the esophagus.

- Outline the functional anatomy and histology of the stomach.

- Describe how acid is secreted by cells in the gastric mucosa.

- Describe the mechanisms that regulate the secretion and motility of the stomach.

- Describe the relationship between cyanocobalamin and intrinsic factor.

- List the main components of pancreatic juice, and outline the mechanisms that regulate its secretion.

- Describe the functional anatomy of the liver, and discuss the formation of bile.

- Discuss the function of the gallbladder and the processes that regulate the passage of bile to the intestinal lumen.

- List the types of movement seen in the small intestine and the function of each.

- Describe the events in the colon that lead to defecation, and outline the neural pathways that control this semireflex response.

General Questions

1. What factors regulate the secretion of the exocrine portion of the pancreas?

2. Discuss the abnormalities of gastrin secretion seen in disease states.

3. What abnormalities would you expect to be produced by resection of the terminal portion of the ileum? Why?

4. What is heartburn? Discuss its pathophysiology and ways it might be treated.

5. What causes gallstones? How would you treat them?

6. Several surgical procedures have been recommended for the treatment of severe obesity that fails to respond to other forms of treatment. What are these procedures? How do they cause weight loss? What are their long-term complications?

7. The bacteria in the colon exist in a symbiotic relationship with the host. How does the host benefit from this relationship, and what harmful or potentially harmful effects may occur to the host?

Multiple-Choice Questions

In questions 1–25, select the single best answer. Questions 1–4 refer to Figure 26–A.

1. Which of the labeled structures is the site where maltase is found?
2. Which of the labeled structures is most concerned with integration of peristalsis?
3. Which of the labeled structures has the most rapid cell turnover?
4. Which of the labeled structures contains the cell bodies of chemoreceptors?

Questions 5–8 refer to Table 26–A.

5. Which set of data would be found in a patient with a gallstone obstructing the common bile duct?
6. Which set of data would be found in a patient with intravascular hemolysis?
7. Which set of data would be found in a patient with infectious hepatitis?
8. Which set of data would be found in a patient with a resection of the ileum?
9. Removal of the entire colon would be expected to cause
    (A) death
    (B) megaloblastic anemia
    (C) severe malnutrition
    (D) an increase in the blood level of ammonia in patients with cirrhosis of the liver
    (E) decreased urinary urobilinogen
10. In infants, defecation often follows a meal. The cause of colonic contractions in this situation is
    (A) the gastroileal reflex
    (B) increased circulating levels of CCK
    (C) the gastrocolic reflex
    (D) increased circulating levels of somatostatin
    (E) the enterogastric reflex

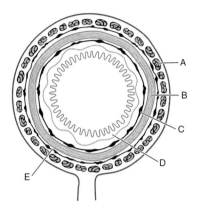

Figure 26–A. Cross section of small intestine. (Reproduced, with permission, from Bell GH, Emslie-Smith D, Paterson CR: *Textbook of Physiology and Biochemistry,* 9th ed. Churchill Livingstone, 1976.)

11. After a meal rich in carbohydrates is ingested, insulin secretion is probably stimulated by
    (A) GLP-1 (7–36) amide
    (B) CCK
    (C) serotonin
    (D) VIP
    (E) gastrin
12. The symptoms of the dumping syndrome (discomfort after meals in patients with intestinal short circuits such as anastomosis of the jejunum to the stomach) are caused in part by
    (A) increased blood pressure
    (B) increased secretion of glucagon
    (C) increased secretion of CCK
    (D) hypoglycemia
    (E) hyperglycemia

Table 26–A.

Pattern	Plasma Bilirubin		Plasma Alkaline Phosphatase	Hematocrit	Plasma Bile Acids
	Direct	Indirect			
A	—	—	—	—	↓
B	↑	↑↑	—	—	—
C	↑↑	↑	↑	—	↓
D	↑	↑↑	—	↓	—
E	↑	↑↑	↑	—	↑

— = no change; ↑ = increase; ↑↑ = marked increase; ↓ = decrease.

13. Which of the following has the highest pH?
    (A) Gastric juice
    (B) Bile in the gallbladder
    (C) Pancreatic juice
    (D) Saliva
    (E) Secretions of the intestinal glands

14. After complete hepatectomy, a rise would be expected in the blood level of
    (A) glucose
    (B) fibrinogen
    (C) 25-hydroxycholecalciferol
    (D) conjugated bilirubin
    (E) estrogens

15. Which of the following would *not* be produced by total pancreatectomy?
    (A) Vitamin E deficiency
    (B) Hyperglycemia
    (C) Metabolic acidosis
    (D) Weight gain
    (E) Decreased absorption of amino acids

Many different abnormalities cause diarrhea. In questions 16–20, match the diarrhea-producing disorder in each question with the lettered abnormality most closely associated with it. Each lettered abnormality may be selected once, more than once, or not at all.

    (A) Increased cAMP in enterocytes
    (B) Increased gastric acid secretion
    (C) Increased bile acids in the colon
    (D) Alteration in intestinal mucin composition
    (E) Abnormal digestion of carbohydrates

16. Ulcerative colitis
17. Operations that bypass the terminal ileum
18. Certain strains of *E coli*
19. Lactase deficiency
20. Zollinger-Ellison syndrome

In questions 21–25, match the listed abnormality with the condition it produces.

    (A) Congenital defect in the distal portion of the myenteric plexus
    (B) Elevated levels of direct-acting plasma bilirubin
    (C) Excess gastric acid secretion
    (D) Heartburn
    (E) Sprue

21. Reflux of gastric contents into the esophagus
22. Allergy to wheat gluten

23. Obstruction of the common bile duct
24. Megacolon
25. Zollinger-Ellison syndrome

CHAPTER 27

This chapter reviews the functions of blood and lymph plus the formed elements of the blood: white blood cells, red blood cells, and platelets. The immune system is also discussed. The material in the chapter should help students to—

- List the various types of cells found in the blood and the precursor cells for each type.
- Describe the functions of neutrophils.
- Describe the functions of monocytes.
- Compare and contrast innate and acquired immunity.
- Describe the structure and function of platelets and the way they discharge their granules.
- Name the common blood types, and describe how blood is typed and cross-matched.
- Describe the blood-clotting and anti-clotting systems and the clinical importance of each system.

General Questions

1. Compare and contrast the composition of blood plasma and lymph.
2. How is the production of white blood cells, platelets, and red blood cells adjusted to meet the varying needs of the individual? Name the humoral factors involved.
3. What are the tissue macrophages? What is their origin, and how do they contribute to body defenses?
4. What are the functions of natural killer cells? How do these cells differ from CD4 and CD8 T lymphocytes?
5. Mutant genes that cause the production of abnormal hemoglobins are common in humans. What determines whether a given mutation in hemoglobin is harmless or harmful?
6. Why is blood clotting abnormal in patients with vitamin K deficiency?

Multiple-Choice Questions

In questions 1–18, select the single best answer. In questions 1–5, match the immunoglobulin in each question with the lettered characteristics that are most closely associated with it. Each lettered characteristic may be selected once, more than once, or not at all.

(A) Complement fixation

(B) Secretory immunity

(C) Release of histamine from basophils and mast cells

(D) Tetramer

(E) Antigen recognition by B cells

1. IgA
2. IgD
3. IgE
4. IgG
5. IgM

Questions 6–9 refer to Figure 27–A.

6. Which of the patterns indicates a mismatch if red cells from a donor are mixed with plasma from a recipient?

7. Which of the patterns would be seen if plasma from an individual with blood type B were mixed with red cells from an individual with blood type O?

8. Which of the patterns would be seen if plasma from an individual with blood type O were mixed with red cells from an individual with blood type B?

9. In which of the patterns would the hemoglobin in the plasma be highest?

10. Which of the following is *not* primarily a function of blood plasma?

(A) Transport of hormones

(B) Maintenance of red cell size

(C) Transport of chylomicrons

(D) Transport of antibodies

(E) Transport of O_2

11. A hematocrit of 41% means that in the sample of blood analyzed

(A) 41% of the hemoglobin is in the plasma

(B) 41% of the total blood volume is made up of blood plasma

(C) 41% of the total blood volume is made up of red and white blood cells and platelets

(D) 41% of the hemoglobin is in red blood cells

(E) 41% of the formed elements in blood are red blood cells

12. In normal human blood

(A) the eosinophil is the most common type of white blood cell

(B) there are more lymphocytes than neutrophils

(C) the iron is mostly in hemoglobin

(D) there are more white cells than red cells

(E) there are more platelets than red cells

13. Lymphocytes

(A) all originate from the bone marrow after birth

(B) are unaffected by hormones

(C) convert to monocytes in response to antigens

(D) interact with eosinophils to produce platelets

(E) are part of the body's defense against cancer

14. Production of the type of cell shown in Figure 27–B is increased by

(A) interleukin-2 (IL-2)

(B) granulocyte colony-stimulating factor (G-CSF)

(C) erythropoietin

(D) interleukin-4 (IL-4)

(E) interleukin-5 (IL-5)

15. The cell shown in Figure 27–B probably does *not* contain

(A) actin

(B) free radicals

(C) myeloperoxidase

(D) cathepsins

(E) a Y chromosome

16. In which of the following diseases is the structure of the hemoglobins that are produced normal but their amount reduced?

(A) Chronic blood loss

(B) Sickle cell anemia

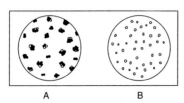

A   B

Figure 27–A. Red blood cells in plasma.

Figure 27–B.

(C) Hemolytic anemia

(D) Thalassemia

(E) Transfusion reactions

17. Which of the following plasma proteins is *not* synthesized primarily in the liver?

(A) Angiotensinogen

(B) IGF-1

(C) Angiotensin II-converting enzyme

(D) α_2-Macroglobulin

(E) Fibrinogen

18. Cells responsible for innate immunity are activated most commonly by

(A) glucocorticoids

(B) pollen

(C) carbohydrate sequences in bacterial cell walls

(D) eosinophils

(E) cytoplasmic protein of bacteria

In questions 19–20, indicate whether the first item is greater than (**G**), *the same as* (**S**), *or less than* (**L**) *the second item.*

19. Number of light chains in a basic immunoglobulin molecule

G S L

Number of heavy chains in a basic immunoglobulin molecule

20. Amount of CD4 glycoprotein on surface of helper T cell

G S L

Amount of CD4 glycoprotein on surface of cytotoxic T cell

CHAPTER 28

This chapter is a review of the conduction system of the heart and the way the impulse that generates a normal heartbeat spreads from the SA node through the AV node, the bundle of His, and the Purkinje system to all parts of the ventricles. The genesis of the ECG, some of its abnormalities, and cardiac arrhythmias are also considered. The material in the chapter should help students to—

- Describe the structure and function of the conduction system of the heart, and compare the action potentials in each part of it with those in cardiac muscle.
- Describe the way the ECG is recorded, the waves of the ECG, and the relationship of the ECG to the electrical axis of the heart.
- Name the common cardiac arrhythmias, and describe the processes that produce them.

- Know how and when to carry out cardiac massage.
- List the principal early and late ECG manifestations of myocardial infarction, and explain the early changes in terms of the underlying ionic events that produce them.
- Describe the ECG changes and the changes in cardiac function produced by alterations in the ionic composition of the body fluids.

General Questions

1. What is sinus arrhythmia? How is it produced? What is its clinical significance?

2. What is an ectopic focus of excitation? What is its pathophysiologic significance?

3. List and explain the effects of slow Ca^{2+} channel blocking drugs on the heart. In which clinical conditions are they of value?

4. Compare and contrast the Wolff-Parkinson-White syndrome and the Lown-Ganong-Levine syndrome. What are the underlying defects that produce these two conditions?

5. What is calcium rigor, and why does it occur?

Multiple-Choice Questions

In questions 1–8, select the single best answer. Questions 1–4 refer to Figure 28–A.

1. Which of the labels identifies the part of the ECG that corresponds to ventricular repolarization?

2. Which of the labels identifies the Q wave?

3. Which of the labels identifies the part of the ECG that corresponds to maximum opening of ventricular Na^+ channels?

4. Which of the labels identifies the part of the ECG that corresponds to maximum opening of ventricular Ca^{2+} channels?

5. Which of the following normally has the most prominent prepotential?

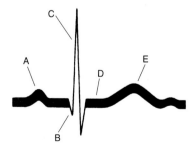

Figure 28–A. ECG.

(A) Sinoatrial node

(B) Atrial muscle cells

(C) Bundle of His

(D) Purkinje fibers

(E) Ventricular muscle cells

6. In second-degree heart block

(A) the ventricular rate is lower than the atrial rate

(B) the ventricular ECG complexes are distorted

(C) there is a high incidence of ventricular tachycardia

(D) stroke volume is decreased

(E) cardiac output is increased

7. Carotid sinus massage sometimes stops supraventricular tachycardia because

(A) it decreases sympathetic discharge to the SA node

(B) it increases vagal discharge to the SA node

(C) it increases vagal discharge to the conducting tissue between the atria and the ventricles

(D) it decreases sympathetic discharge to the conducting tissue between the atria and the ventricles

(E) it increases the refractory period of the ventricular myocardium

8. Currents caused by opening of which of the following channels contribute to the repolarization phase of the action potential of ventricular muscle fibers?

(A) Na^+ channels

(B) Cl^- channels

(C) Ca^{2+} channels

(D) K^+ channels

(E) HCO_3^- channels

In questions 9–11, one or more than one of the answers may be correct. Select

A   if (1), (2), and (3) are correct;

B   if (1) and (3) are correct;

C   if (2) and (4) are correct;

D   if only (4) is correct; and

E   if all are correct

9. In complete heart block

(1) fainting may occur because the atria are unable to pump blood into the ventricles

(2) ventricular fibrillation is common

(3) the atrial rate is lower than the ventricular rate

(4) fainting may occur because of prolonged periods during which the ventricles fail to contract

10. Reentry is a common cause of

(1) paroxysmal atrial tachycardia

(2) paroxysmal nodal tachycardia

(3) atrial fibrillation

(4) sinus arrhythmia

11. In a patient with the long QT syndrome, one might find

(1) a mutation in a cardiac Na^+ channel.

(2) abnormal endolymph in the middle ear.

(3) a loss-of-function mutation in the **HERG** gene.

(4) an abnormal cardiac Na^+-Ca^{2+} antiport.

*In questions 12–15, indicate whether the first item is greater than (**G**), the same as (**S**), or less than (**L**) the second item.*

12. Height of T waves when the plasma K^+ level is 5.5 meq/L

    G S L

    Height of T waves when the plasma K^+ level is 8.5 meq/L.

13. In normal adults, height of the R wave in lead V_1 of the ECG

    G S L

    In normal adults, height of the R wave in lead V_5 of the ECG

14. Interval between an atrial premature beat and the next normal beat

    G S L

    Interval between a ventricular premature beat and the next normal beat

15. Slope of prepotential in SA node after stimulation of vagus

    G S L

    Slope of prepotential in SA node after stimulation of sympathetic nerves to heart

CHAPTER 29

This chapter considers the mechanical events in the cardiac cycle as the heart pumps the blood through the circulation. It also considers heart sounds and murmurs, and measurement and regulation of cardiac output. The material in the chapter should help students to—

• Describe in sequence the events that occur in the heart during the cardiac cycle.

• Outline the changes in the duration of systole and diastole that occur with changes in heart rate, and discuss their physiologic consequences.

- Describe the arterial pulse and jugular venous pulse.
- Describe and explain the first and second heart sounds and the occasionally observed third and fourth heart sounds.
- State the timing of the murmurs produced by aortic stenosis, aortic insufficiency, mitral stenosis, and mitral insufficiency.
- List the factors affecting cardiac output and the effect of each.
- Summarize the factors governing oxygen consumption by the heart.

General Questions

1. Which methods are commonly used to measure cardiac output? What are the advantages and disadvantages of each method?
2. It takes several months for nerves to grow into transplanted hearts. However, before the nerves regrow in patients with transplanted hearts, exercise increases their cardiac output. What is the mechanism involved, and how does it operate?
3. Explain postextrasystolic potentiation.
4. What is the ejection fraction? Which conditions cause it to increase, and which conditions cause it to decrease? Why?
5. Using the axes shown in Figure 29–A, plot the changes in cardiac output as the heart rate increases. Explain the curve that you draw.

Multiple-Choice Questions

In questions 1–15, select the single best answer. Questions 1–5 refer to Figure 29–B. Select the letter in the figure that identifies the following.

1. Stroke volume
2. Point where aortic valve opens

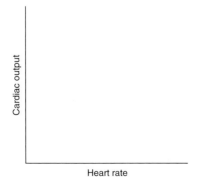

Figure 29–A.

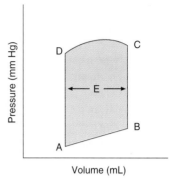

Figure 29–B. Pressure-volume curve of left ventricle.

3. Point where aortic valve closes
4. Point where mitral valve opens
5. Point where mitral valve closes
6. The second heart sound is caused by
   (A) closure of the aortic and pulmonary valves
   (B) vibrations in the ventricular wall during systole
   (C) ventricular filling
   (D) closure of the mitral and tricuspid valves
   (E) retrograde flow in the vena cava
7. The fourth heart sound is caused by
   (A) closure of the aortic and pulmonary valves
   (B) vibrations in the ventricular wall during systole
   (C) ventricular filling
   (D) closure of the mitral and tricuspid valves
   (E) retrograde flow in the vena cava
8. The dicrotic notch on the aortic pressure curve is caused by
   (A) closure of the mitral valve
   (B) closure of the tricuspid valve
   (C) closure of the aortic valve
   (D) closure of the pulmonary valve
   (E) rapid filling of the left ventricle
9. During exercise, a man consumes 1.8 L of oxygen per minute. His arterial O_2 content is 190 mL/L, and the O_2 content of his mixed venous blood is 134 mL/L. His cardiac output is approximately
   (A) 3.2 L/min
   (B) 16 L/min
   (C) 32 L/min
   (D) 54 L/min
   (E) 160 mL/min
10. The work performed by the left ventricle is substantially greater than that performed by the right ventricle, because in the left ventricle

(A) the contraction is slower

(B) the wall is thicker

(C) the stroke volume is greater

(D) the preload is greater

(E) the afterload is greater

11. Starling's law of the heart

(A) does not operate in the failing heart

(B) does not operate during exercise

(C) explains the increase in heart rate produced by exercise

(D) explains the increase in cardiac output that occurs when venous return is increased

(E) explains the increase in cardiac output when the sympathetic nerves supplying the heart are stimulated

In questions 12–15, match the item in each question with the lettered abnormality that is most closely associated with it. Each lettered abnormality may be selected once, more than once, or not at all.

(A) Aortic stenosis

(B) Aortic insufficiency

(C) Mitral stenosis

(D) Patent ductus arteriosus

(E) Pulmonary hypertension

12. Bounding Corrigan pulse

13. Loud, snapping second heart sound

14. Diastolic murmur loudest over the apex of the heart

15. Decreased perfusion of coronary arteries

CHAPTER 30

This chapter considers the blood vessels and lymphatics and the movement of fluids through them. It includes the functional anatomy of the vessels and vascular smooth muscle, the principles and forces that govern pressure and flow in them, and the factors affecting exchange across capillary walls. The material in the chapter should help students to—

- Describe in relative terms the diameter, wall thickness, and total cross-sectional area of the aorta, smaller arteries, arterioles, capillaries, venules, and veins.

- Describe the relationship between flow, pressure, and resistance in the vascular system.

- Define laminar flow and critical closing pressure.

- State the Poiseuille-Hagen formula for flow in blood vessels, and explain on the basis of this formula why the radius of a vessel is such an important determinant of flow.

- Define the law of Laplace, and list three examples of its operation in the body.

- Describe in detail how blood pressure in humans is measured by the auscultatory method and the palpation method.

- Describe the Starling forces that determine the net movement of fluid across the capillary wall. Define flow-limited exchange and diffusion-limited exchange, and describe the variations in capillary permeability and structure seen in different parts of the body.

General Questions

1. What is a Windkessel? Why are the aorta and large arteries called Windkessel vessels? What is the physiologic significance of the Windkessel effect?

2. What is Bernoulli's principle? Discuss its significance in cardiovascular physiology.

3. A young man has a blood pressure of 130/73 mm Hg. What is his pulse pressure? What is his mean arterial pressure? How did you calculate each of these pressures?

4. The pressure in a large artery is 60,000 dynes/cm^2, and the radius of the artery is 0.5 cm. What is the wall tension in the artery? What would the wall tension be if the diameter were 1 cm?

5. What are Korotkoff's sounds, and what produces them?

6. Where in the cardiovascular system is turbulent flow normally found? What factors make flow change from streamline to turbulent?

7. Describe the regulation of Ca^{2+} in vesicles in vascular smooth muscle. Why is this important?

Multiple-Choice Questions

In questions 1–12, select the single best answer. Questions 1–4 refer to Table 30–A.

One hind limb of an anesthetized experimental animal is denervated and attached to instruments so that it is possible to measure the pressure at the arterial and venous ends of the capillaries, the weight of the limb, its blood flow, and the arteriovenous oxygen difference across it. Various substances are then administered. The results obtained are listed in Table 30–A.

1. Which pattern of changes would be observed if the substance produced arteriolar dilation?

2. Which pattern would be observed if the substance produced increased capillary permeability?

3. Which pattern would be observed if the substance produced constriction of the veins?

Table 30–A.

Pattern	Pressure at Arterial End of Capillary	Pressure at Venous End of Capillary	Weight of Limb	Blood Flow in Limb	Arteriovenous Oxygen Difference
A	—	—	↑	—	—
B	—	↑	—	↓	—
C	↓	↓	—	↓	↑
D	↑	↑	↑	↑	↓
E	↑	↑	↑	↓	—

— = no change; ↑ = increase; ↓ = decrease.

4. Which pattern would be observed if the substance produced a decrease in systemic blood pressure?

5. Which of the following has the highest *total* cross-sectional area in the body?
   (A) Arteries
   (B) Arterioles
   (C) Capillaries
   (D) Venules
   (E) Veins

6. Lymph flow from the foot is
   (A) increased when an individual rises from the supine to the standing position
   (B) increased by massaging the foot
   (C) increased when capillary permeability is decreased
   (D) decreased when the valves of the leg veins are incompetent
   (E) decreased by exercise

7. The pressure in a capillary in skeletal muscle is 35 mm Hg at the arteriolar end and 14 mm Hg at the venular end. The interstitial pressure is 0 mm Hg. The colloid osmotic pressure is 25 mm Hg in the capillary and 1 mm Hg in the interstitium. The net force producing fluid movement across the capillary wall at its arteriolar end is
   (A) 3 mm Hg out of the capillary
   (B) 3 mm Hg into the capillary
   (C) 10 mm Hg out of the capillary
   (D) 11 mm Hg out of the capillary
   (E) 11 mm Hg into the capillary

8. Which of the following does *not* occur as blood passes through the systemic capillaries?
   (A) Its hematocrit increases
   (B) Its hemoglobin dissociation curve shifts to the left
   (C) Its protein content increases

   (D) Its pH decreases
   (E) Its red blood cells increase in size

9. The velocity of blood flow
   (A) is higher in the capillaries than the arterioles
   (B) is higher in the veins than in the venules
   (C) is higher in the veins than the arteries
   (D) falls to zero in the descending aorta during diastole
   (E) is reduced in a constricted area of a blood vessel

10. When the radius of the resistance vessels is increased, which of the following is increased?
    (A) Systolic blood pressure
    (B) Diastolic blood pressure
    (C) Viscosity of the blood
    (D) Hematocrit
    (E) Capillary blood flow

11. When the viscosity of the blood is increased, which of the following is increased?
    (A) Mean blood pressure
    (B) Radius of the resistance vessels
    (C) Radius of the capacitance vessels
    (D) Central venous pressure
    (E) Capillary blood flow

12. A pharmacologist discovers a drug that stimulates the production of VEGF receptors. He is excited because his drug might be of value in the treatment of
    (A) coronary artery disease
    (B) cancer
    (C) emphysema
    (D) diabetes insipidus
    (E) dysmenorrhea

13. Why is the dilator response to injected acetylcholine changed to a constrictor response when the endothelium is damaged?

(A) More Na^+ is generated

(B) More bradykinin is generated

(C) The damage lowers the pH of the remaining layers of the artery

(D) The damage augments the production of endothelin by the endothelium

(E) The damage interferes with the production of NO by the endothelium

14. Sildenafil (Viagra) has seen extensive use in aiding the production of erections in men. However, it can also produce transient inability to distinguish between blue and green. This is because it

(A) dilates arteries in the penis and the retina

(B) produces selective constriction of arteries in the penis and the retina

(C) reduces the concentration of cGMP in the penis and the retina

(D) inhibits the production of endothelin in the penis and the retina

(E) inhibits the phosphodiesterases found in the penis and the retina

In questions 15–18, which of the labeled structures in Figure 30–A

15. is the site of the initial visible lesion in atherosclerosis?

16. is reduced in size in the presence of angiotensin II?

17. is responsible for maintaining blood flow during diastole?

18. responds to shear stress via receptors responsive to stretch?

In questions 19–20, indicate whether the first item is greater than (G), the same as (S), or less than (L) the second item.

19. Hydrostatic pressure in venules at the level of the heart

G S L

Hydrostatic pressure in veins at the level of the heart

20. Permeability of capillaries in the lungs

G S L

Permeability of capillaries in the liver

CHAPTER 31

This chapter reviews the local mechanisms, paracrine and endocrine humoral agents, and neural mechanisms that act together to maintain blood pressure and blood flow to the various organs at rest and during a wide variety of situations. The material in the chapter should help students to—

- Define autoregulation, discuss its role in physiology, and summarize the theories that have been advanced to explain its occurrence.

- List the important vasodilator metabolites, and discuss their role in the regulation of tissue perfusion.

- List the principal vasoregulatory factors secreted by endothelial cells, and describe the function of each.

- Name the principal hormones that affect arterial blood pressure, and know the physiologic role of each.

- Outline the neural mechanisms that control arterial blood pressure and heart rate, including the receptors, afferent and efferent pathways, central integrating pathways, and effector mechanisms involved.

- Describe the direct effects of CO_2 and hypoxia on the vasomotor areas in the medulla oblongata.

General Questions

1. Why does increased sympathetic activity cause an increase in right atrial pressure? What are the mechanisms and pathways involved?

2. What is the effect of intravenous injection of the vasodilator nitroprusside on the RR interval of the ECG? How long does it last?

3. Compare the function of the carotid and aortic baroreceptors with the function of the baroreceptors in the atria and great veins.

4. Which hormones affect the heart rate and how do they produce their effects?

5. Which peptides are found in nerves innervating blood vessels? What is the function of each?

6. Name the compounds produced by the endothelium that affect vascular function.

7. What are the effects of baroreceptor denervation?

8. A pharmacologist injects a drug into a dog and finds that it produces a prompt increase in mean

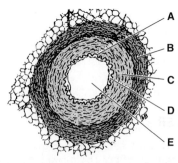

Figure 30–A. Cross section of small artery.

arterial pressure. List the possible mechanisms by which it could be acting to produce this increase, and outline a series of experiments that would permit you to decide the exact mechanism of action of the drug.

Multiple-Choice Questions

In questions 1–10, select the single best answer.

1. Which of the following are *incorrectly* paired?
   (A) Angiotensin II-converting enzyme : kinin metabolism
   (B) Stimulation of the vagus nerve in the neck : bradycardia
   (C) Prostacyclin : vasodilation
   (D) Increased pressure in the carotid sinuses : decreased sympathetic discharge to arterioles
   (E) Increased H_2 in tissues : vasoconstriction in tissues

2. When a pheochromocytoma (tumor of the adrenal medulla) suddenly discharges a large amount of epinephrine into the circulation, the patient's heart rate would be expected to
   (A) increase because the increase in blood pressure stimulates the carotid and aortic baroreceptors
   (B) increase because epinephrine has a direct chronotropic effect on the heart
   (C) increase because of increased tonic parasympathetic discharge to the heart
   (D) decrease because the increase in blood pressure stimulates the carotid and aortic chemoreceptors
   (E) decrease because of increased tonic parasympathetic discharge to the heart

3. Vasopressin secretion is increased by
   (A) increased pressure in the right ventricle
   (B) decreased pressure in the right ventricle
   (C) increased pressure in the right atrium
   (D) decreased pressure in the right atrium
   (E) increased pressure in the aorta

4. Catecholamines acting on α-adrenergic receptors
   (A) increase the contractility of cardiac muscle
   (B) increase the rate of discharge of the sinoatrial node
   (C) increase cardiac output
   (D) constrict coronary arteries by a direct action on these blood vessels
   (E) dilate blood vessels in skeletal muscle

5. Which of the following would be expected to raise blood pressure?

   (A) Prostacyclin
   (B) A drug that inhibits angiotensin II-converting enzyme
   (C) A drug that inhibits NO synthase
   (D) A drug that inhibits V_1 vasopressin receptors
   (E) A drug that inhibits endothelin-converting enzyme

Many of the vasoactive substances produced by endothelial cells are also produced by other cells and have different functions in other parts of the body. In questions 6–10, select

   A   if the item is associated with (a) below,
   B   if the item is associated with (b) below,
   C   if the item is associated with both (a) and (b), and
   D   if the item is associated with neither (a) nor (b)
       (a) Endothelin-1
       (b) NO

6. Brain
7. Intestine
8. Kidneys
9. Lungs
10. Immune cells

In questions 11–16, match the vasoactive compound in each question with the lettered mechanism of action that is most closely associated with it. Each lettered item may be selected once, more than once, or not at all.

   (A) Activates guanylyl cyclase
   (B) Activates protein kinase A
   (C) Activates phospholipase C
   (D) Activates phospholipase A_2
   (E) Activates tyrosine kinase

11. Angiotensin II
12. Endothelin-1
13. Nitric oxide
14. Bradykinin
15. VIP
16. Substance P
17. Which of the vasoactive compounds listed above (11–16) cause(s) vasoconstriction in vivo?
18. Which of the vasoactive compounds (11–16) cause(s) vasodilation in vivo?

CHAPTER 32

This chapter describes special aspects of the circulation of the brain including the blood-brain barrier, the heart, the splanchnic region (including the liver), the skin, and the fetus and placenta. The material in the chapter should help students to—

- Give approximate values for blood flow per unit weight and blood flow per organ at rest in the major organs of the body.
- List the unique gross and microscopic aspects of the circulation of the brain.
- Describe the formation, absorption, and functions of cerebrospinal fluid.
- Outline the characteristics of the blood-brain barrier, and comment on its importance in clinical medicine.
- Describe the circumventricular organs, and list their general functions.
- Summarize the main anatomic features of the coronary circulation.
- List the chemical and neural factors that regulate the coronary circulation, and describe the role of each.
- Outline the unique features of the circulation of the liver and the splanchnic bed, and understand the reservoir function of the splanchnic circulation.
- Describe the triple response produced by firmly stroking the skin, and explain each of its components.
- Describe the operation of the placenta as the "fetal lung."
- Diagram the circulation of the fetus before birth, and list the changes that occur in it at birth.

General Questions

1. What is the axon reflex? What is the evidence for its existence? Which neurotransmitter is probably involved?
2. What is reactive hyperemia? Where does it occur?
3. What is a neurohemal organ? What are the principal neurohemal organs in the body?
4. How can PET and functional MRI be used to study the metabolism and blood flow of the brain and the effects of drugs on it?
5. Discuss the use of radionuclides in the study of the coronary circulation.
6. What is the Monro-Kellie doctrine, and what are its physiologic consequences?
7. What is the hepatic acinus? Why is the acinar organization of the liver important, and what is its pathophysiologic significance?

Multiple-Choice Questions

In questions 1–13, select the single best answer. Questions 1–5 refer to Figure 32–A.

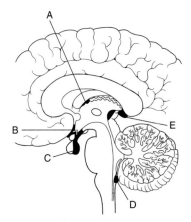

Figure 32–A. Sagittal view of the human brain.

1. Which of the labeled structures is the main site at which hypothalamic hormones leave the brain?
2. Which of the labeled structures is the main site at which vasopressin acts to decrease cardiac output?
3. Which of the labeled structures is the chemoreceptor area that is a trigger zone for vomiting?
4. Which of the labeled structures is the main site where changes in plasma osmolality act to alter vasopressin secretion?
5. Which of the labeled structures is the main site at which angiotensin II acts to increase water intake?
6. Blood in which of the following vessels normally has the lowest P_{O_2}?
   (A) Maternal artery
   (B) Maternal uterine vein
   (C) Maternal femoral vein
   (D) Umbilical artery
   (E) Umbilical vein
7. The pressure differential between the heart and the aorta is least in the
   (A) left ventricle during systole
   (B) left ventricle during diastole
   (C) right ventricle during systole
   (D) right ventricle during diastole
   (E) left atrium during systole
8. Injection of tissue plasminogen activator (t-PA) would probably be most beneficial
   (A) after at least 1 year of uncomplicated recovery following occlusion of a coronary artery
   (B) after at least 2 months of rest and recuperation following occlusion of a coronary artery
   (C) during the second week after occlusion of a coronary artery

(D) during the second day after occlusion of a coronary artery

(E) during the second hour after occlusion of a coronary artery

9. Which of the following organs has the greatest blood flow per 100 g of tissue?

(A) Brain

(B) Heart muscle

(C) Skin

(D) Liver

(E) Kidneys

10. Which of the following organs has the most permeable capillaries?

(A) Brain

(B) Posterior pituitary gland

(C) Liver

(D) Small intestine

(E) Kidneys

11. Which of the following does *not* dilate arterioles in the skin?

(A) Increased body temperature

(B) Epinephrine

(C) Bradykinin

(D) Substance P

(E) Vasopressin

12. A baby boy is brought to the hospital because of convulsions. In the course of a workup, his body temperature and plasma glucose are found to be normal, but his cerebrospinal fluid glucose is 12 mg/dL (normal, 65 mg/dL). A possible explanation of his condition is

(A) constitutive activation of GLUT 3 in neurons

(B) SGLT 1 deficiency in astrocytes

(C) GLUT 5 deficiency in cerebral capillaries

(D) GLUT 1 55K deficiency in cerebral capillaries

(E) GLUT 1 45K deficiency in microglia

13. Recently, there has been great interest in the role of inflammation in the production of atherosclerosis. Evidence for this includes

(A) chronic low-grade fever

(B) increased proliferation of the intima in the medium-sized and small arteries

(C) high cholesterol contents of intestinal bacteria

(D) possible examples of molecular mimicry

(E) premature rupture of atherosclerotic plaques

In questions 14–16, indicate whether the first item is greater than (G), the same as (S), or less than (L) the second item.

14. Amount of water transported into the cerebrospinal fluid

G S L

Amount of water transported out of the cerebrospinal fluid

15. Blood flow in the endocardial portion of the myocardium during systole

G S L

Blood flow in the epicardial portion of the myocardium during systole

16. Amount of blood brought to the liver by the hepatic artery

G S L

Amount of blood brought to the liver by the portal vein

CHAPTER 33

This chapter summarizes the cardiovascular adjustments that occur on assuming the upright position; during space flight; and during exercise. It also describes the pathophysiology of inflammation and of shock, including syncope (fainting). Hypertension, heart failure, and the cardiovascular compensations that occur in these conditions are also considered. The material in the chapter should help students to—

- Outline the compensatory mechanisms that maintain blood pressure on rising from the supine to the standing position.
- Describe the effects of positive and negative gravitational acceleration (*g*) on the body.
- Describe and explain the circulatory changes that occur during exercise.
- Outline the main beneficial effects of athletic training on the cardiovascular system and other systems in the body.
- Define shock, name its major causes, and summarize the main abnormalities that occur during each type of shock. Summarize the short- and long-term compensatory mechanisms that operate in each type.
- List the common causes of fainting.
- List the common causes of high blood pressure in humans, and divide them into those that are now curable and those that are not.
- List the main symptoms of heart failure, and describe how each is produced.

General Questions

1. Compare the composition of blood from exercising skeletal muscle with that of blood from resting muscle, and explain the differences.

2. What are the similarities and differences between hypovolemic shock and low-resistance shock?

3. What is refractory shock? How can it be prevented?

4. What are the physiologic advantages and disadvantages of treating shock by raising the foot of the bed?

5. Why do some people faint when they rise from the supine to the standing position?

6. "Hypertension is a syndrome, not a disease." Discuss this statement and its implications.

7. List five ways that chronic hypertension can be produced in experimental animals. What produces the elevation of blood pressure in each case? What are the similarities and differences between each form of experimental hypertension and human hypertension?

8. Why is vital capacity reduced in heart failure? What is the effect of posture on this reduction?

9. Why are angiotensin II-converting enzyme inhibitors of value in the treatment of congestive heart failure?

10. Mr. Smith, a 55-year-old man who weighs 71 kg, feels tired after relatively mild exertion. On physical examination, he is found to have a moderately intense basal systolic murmur. During cardiac catheterization, the following data were obtained:

Heart rate (beats/min): 64

Ventilation (L/min): 4.98

O_2 consumption (mL/min): 246

Pulmonary arterial O_2 (mL of O_2/L of blood): 137

Pulmonary venous O_2 (mL of O_2/L of blood): 189

Right ventricular pressure (mm Hg): 27/2

Pulmonary artery pressure (mm Hg): 27/10

Left ventricular pressure (mm Hg): 169/0

Aortic pressure (mm Hg): 108/72

What is Mr. Smith's cardiac output? What is his stroke volume? What is the most likely diagnosis?

Multiple-Choice Questions

In questions 1–13, select the single best answer.

Each of the patients in questions 1–5 has a blood pressure of 85/40 mm Hg. *Match the condition in each question with the lettered emergency treatment that is most appropriate. Each lettered treatment may be selected once, more than once, or not at all.*

(A) Injection of dopamine

(B) Infusion of concentrated human albumin

(C) Infusion of isotonic saline solution

(D) Injection of epinephrine

(E) Infusion of whole blood

1. Girl stung by a bee

2. Man bleeding from a stab wound

3. Woman burned over 35% of her body in a kitchen fire

4. Woman who had a myocardial infarction 24 hours previously

5. Man with severe diarrhea

6. Plasma renin activity is most likely to be lower than normal

(A) in congestive heart failure

(B) in hemorrhagic shock

(C) in shock due to infection with gram-negative bacteria

(D) in essential hypertension

(E) during quiet standing

7. From your knowledge of physiology, which of the following would you *least* expect to see during prolonged space flight?

(A) Atrophy of the heart

(B) Loss of bone mass

(C) Nausea and vomiting

(D) Increased circulating lymphocytes

(E) Atrophy of skeletal muscles

8. Which of the following takes longest to return to normal after 1 L of blood is removed from a normal individual?

(A) Plasma aldosterone concentration

(B) Blood pressure

(C) Renin secretion

(D) Plasma volume

(E) Number of red blood cells in peripheral blood

9. Which of the following is *least* likely to cause sustained hypertension?

(A) Chronically increased secretion of the adrenal medulla

(B) Chronically increased secretion of the zona fasciculata and zona reticularis of the adrenal cortex

(C) Chronically increased secretion of the zona glomerulosa of the adrenal cortex

(D) Chronically increased secretion of the posterior pituitary gland

(E) Chronic treatment with oral contraceptives

10. Which of the following is *not* increased during isotonic exercise?

(A) Respiratory rate

(B) Stroke volume

(C) Heart rate

(D) Total peripheral resistance

(E) Systolic blood pressure

11. Which of the following are *incorrectly* paired?

(A) Loss of blood : hypotension

(B) Negative *g* : blackout

(C) Increased cardiac output : exercise

(D) Renal artery constriction : increased blood pressure

(E) Decreased total peripheral resistance : fainting

12. Which of the following would you expect to decrease in a normal individual who stands quietly in the same position for 1 hour?

(A) Hematocrit

(B) Diameter of the thigh

(C) Plasma renin activity

(D) Plasma vasopressin concentration

(E) Central venous pressure

13. Which of the following is *least* likely to cause fainting?

(A) Pressure on the carotid sinus

(B) Paroxysmal atrial fibrillation

(C) Autonomic insufficiency

(D) Strong emotion

(E) Myocardial infarction

In questions 14–15, indicate whether the first item is greater than (**G**), *the same as* (**S**), *or less than* (**L**) *the second item.*

14. Amount of NF-$_\kappa$B in cell nuclei after infusion of epinephrine

G S L

Amount of NF-$_\kappa$B in cell nuclei after infusion of cortisol

15. Total blood flow to brain during exercise

G S L

Total blood flow to brain at rest

CHAPTER 34

This chapter is an analysis of the functions of the respiratory system, including the properties of gases, the way that the lungs and chest operate to produce inspiration and expiration, gas exchange in the lungs, the special features of the pulmonary circulation, lung defense mechanisms, and the metabolic functions of the lungs. The material in the chapter should help students to—

- Define partial pressure, and calculate the partial pressure of each of the important gases in the atmosphere at sea level.

- Draw a graph of the changes in intrapulmonary and intrapleural pressure and lung volume that occur during inspiration and expiration.

- List the passages through which air passes from the exterior to the alveoli, and describe the cells that line each of them.

- List the major muscles involved in respiration, and state the role of each.

- Define tidal volume, inspiratory reserve volume, expiratory reserve volume, and residual volume, and give approximate values for each in a normal adult.

- Define compliance, and give examples of diseases in which it is abnormal.

- Describe the chemical composition and function of surfactant.

- List the factors that determine alveolar ventilation.

- Define diffusion capacity, and compare the diffusion of O_2 with that of CO_2 in the lungs.

- Compare the pulmonary and systemic circulations, listing the main differences between them.

- Describe the metabolic functions of the lungs.

General Questions

1. Discuss the law of Laplace as it relates to pulmonary function.

2. In the standing position, how does blood flow at the apex of each lung compare with blood flow at the base? How does ventilation compare? Explain the differences.

3. What is a hysteresis loop? Give an example in respiratory physiology, and explain its occurrence.

4. What are the functions of epithelial cells and pulmonary macrophages?

5. What is the difference between the anatomic and physiologic dead spaces? What role do they play in pulmonary disease?

6. What factors affect the diffusing capacity of the lungs for O_2?

7. Why is the P_{O_2} of blood in the aorta slightly less than the P_{O_2} of blood in the pulmonary veins?

8. What are the components that contribute to the work of breathing? Discuss the normal and pathologic conditions that cause alterations in each component.

Multiple-Choice Questions

In questions 1–10, select the single best answer.

1. On the summit of Mt. Everest, where the barometric pressure is about 250 mm Hg, the partial pressure of O_2 is about

(A) 0.1 mm Hg
(B) 0.5 mm Hg
(C) 5 mm Hg
(D) 50 mm Hg
(E) 100 mm Hg

2. The approximate amount of gas left in the lungs after maximal forced expiration in a normal woman is
(A) zero
(B) 0.1 L
(C) 1.1 L
(D) 3.1 L
(E) 4.2 L

3. The tidal volume in a normal man at rest is about
(A) 0.5 L
(B) 1.2 L
(C) 2.5 L
(D) 4.9 L
(E) 6.0 L

4. What is the approximate dead space of a normal 70-kg man breathing through a tube that has a radius of 5 mm and a length of 100 cm?
(A) 150 mL
(B) 180 mL
(C) 230 mL
(D) 280 mL
(E) 350 mL

5. Which of the following is responsible for the movement of O_2 from the alveoli into the blood in the pulmonary capillaries?
(A) Active transport
(B) Filtration
(C) Secondary active transport
(D) Facilitated diffusion
(E) Passive diffusion

Questions 6–8 refer to Figure 34–A.

6. Which of the labeled curves would you expect to see in a normal individual?

7. Which of the labeled curves would you expect to see in a patient with severe pulmonary fibrosis?

8. Which of the labeled curves would you expect to see in a patient with advanced emphysema?

9. Which of the following causes relaxation of bronchial smooth muscle?
(A) Leukotrienes
(B) Vasoactive intestinal polypeptide (VIP)
(C) Acetylcholine

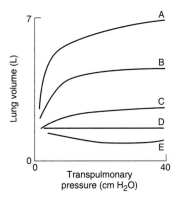

Figure 34–A. Lung volumes at various transpulmonary pressures.

(D) Cool air
(E) Sulfur dioxide

10. Airway resistance
(A) is increased if the lungs are removed and inflated with saline
(B) does not affect the work of breathing
(C) is increased in paraplegic patients
(D) is increased in asthma
(E) makes up 80% of the work of breathing

In questions 11–13, one or more than one of the answers may be correct. Select

A   if (1), (2), and (3) are correct;
B   if (1) and (3) are correct;
C   if (2) and (4) are correct;
D   if only (4) is correct; and
E   if all are correct

11. Surfactant lining the alveoli
(1) helps prevent alveolar collapse
(2) is decreased in hyaline membrane disease
(3) is decreased in the lungs of heavy smokers
(4) is a mixture of proteins and lipids

12. Activation of which of the following receptors causes constriction of pulmonary arteries?
(1) Adrenergic α_2 receptors
(2) Muscarinic M_3 receptors
(3) Histamine H_2 receptors
(4) Endothelin ET_A receptors

13. Which of the following would be expected to cause a reduction in pulmonary ventilation?
(1) Transection of both phrenic nerves
(2) Transection of the spinal cord at the first thoracic level

(3) A large dose of morphine

(4) Pulmonary fibrosis

In questions 14 and 15, indicate whether the first item is greater than (G), the same as (S), or less than (L) the second item.

14. FEV_1 in a normal individual

    G S L

    FEV_1 in a patient with asthma

15. Angiotensin II concentration in blood in pulmonary veins

    G S L

    Angiotensin II concentration in blood in renal veins

CHAPTER 35

This chapter describes the flow of O_2 from the lungs to the tissues and the flow of CO_2 from the tissues to the lungs, with emphasis on the physical and chemical mechanisms that greatly augment the ability of the blood to carry O_2 and CO_2. The material in the chapter should help students to—

- Describe the manner in which O_2 flows "downhill" from the lungs to the tissues and CO_2 flows "downhill" from the tissues to the lungs.
- Describe the reactions of O_2 with hemoglobin and the oxygen-hemoglobin dissociation curve.
- List the important factors affecting the affinity of hemoglobin for O_2 and the physiologic significance of each.
- Describe myoglobin, and outline its physiologic role.
- List the reactions that increase the amount of CO_2 in the blood, and draw the CO_2 dissociation curve for arterial and venous blood.

General Questions

1. Why is the Cl^- concentration inside red blood cells in venous blood greater than that in red blood cells in arterial blood? Describe the mechanisms responsible for this difference.

2. Why does oxyhemoglobin bind less H^+ than reduced hemoglobin?

3. What is the Bohr effect? How is it brought about, and what is its physiologic significance?

4. What is 2,3-diphosphoglycerate (2,3-DPG)? Define P_{50}, and describe how 2,3-DPG affects the P_{50} in blood.

5. What is the role of carbonic anhydrase in red blood cells?

Multiple-Choice Questions

In questions 1–7, select the single best answer.

1. Most of the CO_2 transported in the blood is
    (A) dissolved in plasma
    (B) in carbamino compounds formed from plasma proteins
    (C) in carbamino compounds formed from hemoglobin
    (D) bound to Cl^-
    (E) in HCO_3^-

2. Which of the following has the greatest effect on the ability of blood to transport oxygen?
    (A) Capacity of the blood to dissolve oxygen
    (B) Amount of hemoglobin in the blood
    (C) pH of plasma
    (D) CO_2 content of red blood cells
    (E) Temperature of the blood

3. Which of the following is *not* true of the system

    $$CO_2 + H_2O \overset{1}{\rightleftarrows} H_2CO_3 \overset{2}{\rightleftarrows} H^+ + HCO_3^-?$$

    (A) Reaction 1 is catalyzed by carbonic anhydrase
    (B) Because of reaction 2, the pH of blood declines during breath holding
    (C) Reaction 1 occurs in the kidneys
    (D) Reaction 1 occurs primarily in plasma
    (E) The reactions move to the left when there is excess H^+ in the tissues

In questions 4–7, select

   A if the item is associated with (a) below,
   B if the item is associated with (b) below,
   C if the item is associated with both (a) and (b), and
   D if the item is associated with neither (a) nor (b)
       (a) Involved in O_2 transport in blood
       (b) Involved in CO_2 transport in blood

4. Carbamino compounds

5. Myoglobin

6. Hydrogen ions

7. Plasma proteins

In questions 8 and 9, indicate whether the first item is greater than (G), the same as (S), or less than (L) the second item.

8. Affinity of fetal hemoglobin (hemoglobin F) for O_2

    G S L

    Affinity of adult hemoglobin (hemoglobin A) for O_2

9. Number of osmotically active particles in red blood cells in arterial blood

G S L

Number of osmotically active particles in red blood cells in venous blood

CHAPTER 36

This chapter is a review of the mechanisms that regulate respiration. These include the fundamental rhythm generator in the medulla oblongata; chemical control via the carotid and aortic chemoreceptors and the chemoreceptors located on the ventral surface of the medulla oblongata; and nonchemical control via inputs from other centers and receptors in the lungs, muscles, tendons, and joints. The material in the chapter should help students to—

- Locate the pre-Böttzinger complex, and describe its role in producing spontaneous respiration.
- Identify the location and function of the dorsal and ventral groups of respiratory neurons, the pneumotaxic center, and the apneustic center in the brain stem.
- List the specific respiratory functions of the vagus nerves and the respiratory receptors in the carotid body, the aortic body, and the ventral surface of the medulla oblongata.
- Describe and explain the ventilatory responses to increased CO_2 concentrations in the inspired air.
- Describe and explain the ventilatory responses to decreased O_2 concentrations in the inspired air.
- Describe the effects of each of the main nonchemical factors that influence respiration.

General Questions

1. What are the Hering-Breuer reflexes?
2. What role does the blood-brain barrier play in the regulation of respiration?
3. What is Ondine's curse?
4. What is the breaking point? Discuss the factors that affect it.
5. How does hypoxia generate increased numbers of impulses in the afferent nerves from the carotid chemoreceptors? Discuss the cellular and molecular mechanisms involved.

Multiple-Choice Questions

In questions 1–9, select the single best answer.

1. The main respiratory control neurons
    (A) send out regular bursts of impulses to expiratory muscles during quiet respiration

    (B) are unaffected by stimulation of pain receptors
    (C) are located in the pons
    (D) send out regular bursts of impulses to inspiratory muscles during quiet respiration
    (E) are unaffected by impulses from the cerebral cortex

2. Intravenous lactic acid increases ventilation. The receptors responsible for this effect are located in the
    (A) medulla oblongata
    (B) carotid bodies
    (C) lung parenchyma
    (D) aortic baroreceptors
    (E) trachea and large bronchi

3. Spontaneous respiration ceases after
    (A) transection of the brain stem above the pons
    (B) transection of the brain stem at the caudal end of the medulla
    (C) bilateral vagotomy
    (D) bilateral vagotomy combined with transection of the brain stem at the superior border of the pons
    (E) transection of the spinal cord at the level of the first thoracic segment

4. The following physiologic events that occur in vivo are listed in random order:
    (1) Decreased cerebrospinal fluid pH
    (2) Increased arterial P_{CO_2}
    (3) Increased cerebrospinal fluid P_{CO_2}
    (4) Stimulation of medullary chemoreceptors
    (5) Increased alveolar P_{CO_2}
    What is the usual sequence in which they occur when they affect respiration?
    (A) 1, 2, 3, 4, 5
    (B) 4, 1, 3, 2, 5
    (C) 3, 4, 5, 1, 2
    (D) 5, 2, 3, 1, 4
    (E) 5, 3, 2, 4, 1

5. The following events that occur in the carotid bodies when they are exposed to hypoxia are listed in random order:
    (1) Depolarization of type I glomus cells
    (2) Excitation of afferent nerve endings
    (3) Reduced conductance of hypoxia-sensitive K^+ channels in type I glomus cells
    (4) Ca^{2+} entry into type I glomus cells
    (5) Decreased K^+ efflux
    What is the usual sequence in which they occur upon exposure to hypoxia?

(A) 1, 3, 4, 5, 2

(B) 1, 4, 2, 5, 3

(C) 3, 4, 5, 1, 2

(D) 3, 1, 4, 5, 2

(E) 3, 5, 1, 4, 2

6. Stimulation of the central (proximal) end of a cut vagus nerve would be expected to

(A) increase heart rate

(B) stimulate inspiration

(C) inhibit coughing

(D) raise blood pressure

(E) cause apnea

7. Injection of a drug that stimulates the carotid bodies would be expected to cause

(A) a decrease in the pH of arterial blood

(B) a decrease in the P_{CO_2} of arterial blood

(C) an increase in the HCO_3^- concentration of arterial blood

(D) an increase in urinary Na^+ excretion

(E) an increase in plasma Cl^-

8. Variations in which of the following components of blood or cerebrospinal fluid do *not* affect respiration?

(A) Arterial HCO_3^- concentration

(B) Arterial H^+ concentration

(C) Arterial Na^+ concentration

(D) Cerebrospinal fluid CO_2 concentration

(E) Cerebrospinal fluid H^+ concentration

9. Intractable hiccups sometimes respond to

(A) increasing arterial P_{O_2}

(B) injection of acetylcholine

(C) injection of dopamine antagonists

(D) increasing arterial pH

(E) hyperventilation

In questions 10–11, indicate whether the first item is greater than (G), the same as (S), or less than (L) the second item.

10. Proportion of the alteration in ventilation produced by changes in arterial P_{CO_2} that is due to medullary chemoreceptors

G S L

Proportion of the alteration in ventilation produced by changes in arterial P_{CO_2} that is due to peripheral chemoreceptors

11. Blood flow per gram of tissue to carotid bodies

G S L

Blood flow per gram of tissue to brain

CHAPTER 37

This chapter analyzes the changes in respiration that occur with exercise, during exposure to altitude, and in other forms of hypoxia, including hypoxia produced by various diseases. Hypercapnia and hypocapnia are also considered, along with increased barometric pressure, drowning, and artificial respiration. The material in the chapter should help students to—

- Describe the effects of exercise on ventilation and O_2 exchange in the tissues.
- Define hypoxia, and describe its four principal forms.
- Describe the acute effects of high altitude on respiration, and discuss acclimatization to altitude.
- Define and give examples of ventilation-perfusion imbalance.
- List and explain the effects of carbon monoxide on the body.
- Summarize the abnormalities that occur in emphysema, asthma, pulmonary hypertension, and pulmonary embolism.
- List and explain the adverse effects of excess O_2.
- Describe the effects of hypercapnia and hypocapnia, and give examples of conditions that can cause them.
- Define periodic breathing, and explain its occurrence in various disease states.
- Describe in detail the technique of mouth-to-mouth resuscitation, and explain how it maintains life.

General Questions

1. By means of a diagram, illustrate the relationship between the partial pressure of O_2 in blood and the percent saturation of hemoglobin at rest, and the changes produced in this relationship by exercise.

2. What is the oxygen debt mechanism? How is an oxygen debt measured? What is the value of the mechanism to the individual?

3. Discuss fatigue from the point of view of its causes, prevention, and physiologic significance.

4. What is Cheyne-Stokes respiration? Explain its occurrence.

5. In patients with severe respiratory failure who are hypercapnic and hypoxic, administration of O_2 may stop respiration and even cause death if artificial respiration is not instituted. Why?

6. Diving not only is performed by professionals but, with the development of SCUBA equipment, has become a popular sport. Which medical problems can be produced by diving, and how would you treat them?

7. Discuss the pathophysiology of asthma.

Multiple-Choice Questions

In questions 1–16, select the single best answer. Questions 1–5 refer to Table 37–A.

1. Which set of data would you expect to find in a normal person?
2. Which set of data would you expect to find in a person at rest breathing air at an altitude of 3000 m (10,000 feet; barometric pressure = 520 mm Hg)?
3. Which set of data would you expect to find in a patient with a large left-to-right shunt?
4. Which set of data would you expect to find in a patient with a collapsed lung?
5. Which set of data would you expect to find in a normal subject at rest breathing 100% O_2 in an unpressurized airplane cabin at 10,000 m (32,800 feet; barometric pressure = 180 mm Hg)?

Questions 6 and 7 refer to Table 37–B.

6. Which set of data would you expect to find in a man who is breathing through a tube that greatly increases his dead space?
7. Which set of data would you expect to find in a woman who has flown to an altitude of 4500 m (14,750 feet) in the open cockpit of an airplane?
8. In which of the following conditions is CO_2 retention most likely to occur?
   (A) Climbing a high mountain
   (B) Ventilatory failure
   (C) Carbon monoxide poisoning
   (D) Lung failure
   (E) Hysterical hyperventilation
9. Pulmonary fibrosis would be expected to produce
   (A) histotoxic hypoxia
   (B) stagnant hypoxia
   (C) decreased vital capacity
   (D) cyanosis
   (E) emphysema

Table 37–A. Po_2 (mm Hg).

	Superior Vena Cava	Right Ventricle	Alveolar Gas	Left Ventricle
A	45	40	104	94
B	40	35	105	94
C	40	35	60	55
D	40	35	104	55
E	45	80	104	94

Table 37–B. Arterial blood values.

	Pco_2 (mm Hg)	pH	HCO_3^- (meq/L)
A	60	7.25	29
B	25	7.30	12
C	42	7.55	35
D	25	7.50	20
E	60	7.05	15

10. Which of the following deleterious effects would probably *not* be produced by chronic cigarette smoking?
    (A) Patches of atelectasis
    (B) Myocardial ischemia
    (C) Loss of elastic tissue in the lung
    (D) Increased anatomic dead space
    (E) Increased carbonmonoxyhemoglobin in blood
11. O_2 delivery to the tissues would be reduced to the greatest extent in
    (A) a normal subject breathing 100% O_2 on top of Mt. Everest
    (B) a normal subject running a marathon at sea level
    (C) a patient with carbon monoxide poisoning
    (D) a patient who has ingested cyanide
    (E) a patient with moderately severe metabolic acidosis
12. Which of the following is *not* a manifestation of oxygen toxicity?
    (A) Irritation of the respiratory tract
    (B) Difficulty in seeing because of retrolental fibroplasia
    (C) Convulsions
    (D) Lung cysts in infants
    (E) Rapture of the deep
13. Which of the following drugs would be most useful in treating high-altitude illness?
    (A) Erythropoietin
    (B) Spironolactone
    (C) Erythromycin
    (D) Amiloride
    (E) Acetazolamide

In questions 14–18, match the item in each question with the lettered disease most closely associated with it. Each lettered disease may be selected once, more than once, or not at all.

(A) Asthma
(B) Pulmonary hypertension

(C) Emphysema

(D) Hyaline membrane disease

(E) Acute respiratory distress syndrome

14. Shock

15. Cigarette smoking

16. Prematurity

17. Denuding of airway epithelium

18. Cor pulmonale

CHAPTER 38

This chapter is concerned with renal function, including glomerular filtration and the factors affecting it, tubular reabsorption and secretion and their control, and the function of the ureters and bladder. The endocrine functions of the kidney are also reviewed. The material in the chapter should help students to—

- Describe the morphology of a typical nephron and its blood supply.
- Define autoregulation, and list the major theories advanced to explain autoregulation in the kidneys.
- List the hormones secreted by the kidneys and tell what each one does.
- Outline the functions of the renal nerves.
- Define glomerular filtration rate, describe how it can be measured, and list the major factors affecting it.
- Define and discuss tubuloglomerular feedback and glomerulotubular balance.
- Discuss tubular reabsorption and secretion of glucose and K^+.
- Outline tubular handling of Na^+.
- Summarize tubular handling of Cl^-, HCO_3^-, urea, and uric acid in terms of amounts filtered, secreted, reabsorbed, and excreted in urine.
- Describe how the countercurrent mechanism in the kidney operates to produce a hypertonic or hypotonic urine.
- Outline the processes involved in the secretion of H^+ into the tubules, and discuss the significance of these processes in the regulation of acid-base balance.
- List the major classes of diuretics and how each operates to increase urine flow.
- Describe the voiding reflex and draw a cystometrogram.

General Questions

1. Why, when the Tm of a substance that is secreted by the tubules is reached, does clearance of the substance decrease as its plasma concentration increases?

2. Why is the renal medulla especially sensitive to hypoxic damage?

3. What is the physiologic role of the mesangial cells in the glomeruli, and how do they carry it out?

4. Compare the cellular mechanisms responsible for H^+ secretion in the proximal tubule with those in the distal tubule and those in the gastric mucosa.

5. Discuss the mechanisms responsible for the adaptation of NH_4^+ excretion that develops over a period of days in prolonged acidosis.

6. Why is acidosis a common complication of chronic renal disease? How would you treat it?

7. The following observations were made on a patient:

Plasma HCO_3^-: 20 meq/L

GFR: 125 mL/min

24-hour urine volume: 1500 mL

Urinary HCO_3^-: 25 meq/L

Urinary NH_4^+: 75 meq/L

Urinary titratable acidity: 50 meq/L

(A) Approximately how much HCO_3^- is being reabsorbed per 24 hours?

(B) How much Na^+ is being reabsorbed with the HCO_3^-?

(C) How much H^+ is being secreted by the renal tubules per 24 hours?

Multiple-Choice Questions

In questions 1–17, select the single best answer. In questions 1–5, which refer to Table 38–A, match the disease or condition in each question with the lettered pattern of laboratory findings in Table 38–A that is most closely associated with it. Each lettered pattern may be selected once, more than once, or not at all.

1. Diabetes insipidus

2. Nephrosis

Table 38–A.

	24-Hour Urine Volume (L)	Ketones	Glucose	Protein
A	1.4	+	0	0
B	6.2	2+	4+	0
C	1.6	0	0	4+
D	6.4	0	0	0
E	0.4	0	0	0

3. Fasting

4. Dehydration

5. Diabetes mellitus

6. In the presence of vasopressin, the greatest fraction of filtered water is absorbed in the
   (A) proximal tubule
   (B) loop of Henle
   (C) distal tubule
   (D) cortical collecting duct
   (E) medullary collecting duct

7. In the absence of vasopressin, the greatest fraction of filtered water is absorbed in the
   (A) proximal tubule
   (B) loop of Henle
   (C) distal tubule
   (D) cortical collecting duct
   (E) medullary collecting duct

8. If the clearance of a substance which is freely filtered is less than that of inulin
   (A) there is net reabsorption of the substance in the tubules
   (B) there is net secretion of the substance in the tubules
   (C) the substance is neither secreted nor reabsorbed in the tubules
   (D) the substance becomes bound to protein in the tubules
   (E) the substance is secreted in the proximal tubule to a greater degree than in the distal tubule

9. Glucose reabsorption occurs in the
   (A) proximal tubule
   (B) loop of Henle
   (C) distal tubule
   (D) cortical collecting duct
   (E) medullary collecting duct

10. On which of the following does aldosterone exert its greatest effect?
    (A) Glomerulus
    (B) Proximal tubule
    (C) Thin portion of the loop of Henle
    (D) Thick portion of the loop of Henle
    (E) Cortical collecting duct

11. What is the clearance of a substance when its concentration in the plasma is 10 mg/dL, its concentration in the urine is 100 mg/dL, and urine flow is 2 mL/ min?
    (A) 2 mL/min
    (B) 10 mL/min

(C) 20 mL/min

(D) 200 mL/min

(E) Clearance cannot be determined from the information given

12. As urine flow increases during osmotic diuresis
    (A) the osmolality of urine falls below that of plasma
    (B) the osmolality of urine increases because of the increased amounts of nonreabsorbable solute in the urine
    (C) the osmolality of urine approaches that of plasma because plasma leaks into the tubules
    (D) the osmolality of urine approaches that of plasma because an increasingly large fraction of the excreted urine is isotonic proximal tubular fluid
    (E) the action of vasopressin on the renal tubules is inhibited

Questions 13–15 refer to the data in Table 38–B, which were obtained in a normal woman.

13. The glomerular filtration rate in the experimental period is
    (A) four times that in the control period
    (B) twice that in the control period
    (C) the same as in the control period
    (D) one-half that in the control period
    (E) one-quarter that in the control period

14. The clearance of urea in the experimental period is
    (A) increased, probably because the urine flow is increased
    (B) increased, probably because the urine glucose concentration is increased
    (C) the same as in the control period

Table 38–B.

	Control Period	Experimental Period
Arterial plasma		
Inulin (mg/mL)	0.004	0.004
Glucose (mg/dL)	100 (1 mg/mL)	300 (3 mg/mL)
Urea (mmol/mL)	5	5
Urine		
Inulin (mg/mL)	0.4	0.2
Glucose (mg/mL)	0	5
Urea (mmol/mL)	300	160
Urine flow (mL/min)	1	2

(D) decreased, probably because the urine flow is increased

(E) decreased, probably because the amount of urea filtered is decreased

15. The Tm for glucose in this woman
    (A) is 100 mg/min
    (B) is 130 mg/min
    (C) is 200 mg/min
    (D) is 290 mg/min
    (E) cannot be calculated from the data given

16. To produce a concentrated urine, vasopressin
    (A) introduces the movement of aquaporin from the cytoplasm to the cell membrane of proximal tubule cells
    (B) increases the movement of aquaporin-2 from the cytoplasm to the cell membrane of collecting duct cells
    (C) has no effect on molecular motors in proximal tubule cells
    (D) has no effect on molecular motors in collecting duct cells
    (E) increases Na^+ reabsorption in the thick ascending limb of Henle

17. A gain-of-function mutation in the gene for which of the following proteins is associated with increased Na^+ retention and hypertension with a normal or low plasma aldosterone level (Liddle's syndrome)?
    (A) Aldosterone synthase
    (B) An inward rectifier K^+ channel
    (C) 11β-Hydroxylase
    (D) V_2 receptor
    (E) Epithelial Na^+ channel

Questions 18–22 refer to Figure 38–A. Select the letter or letters identifying the appropriate part of the nephron.

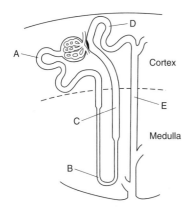

Figure 38–A. Juxtamedullary nephron.

Each lettered item may be selected once, more than once, or not at all. Note that in some instances, the correct answer may be more than one letter.

18. Site(s) at which furosemide acts
19. Site(s) at which thiazides act
20. Site(s) at which tubular fluid osmolality exceeds that of plasma by the greatest amount
21. Site(s) at which Na^+ is actively reabsorbed
22. Site(s) at which K^+ is secreted

CHAPTER 39

This chapter is a review of the homeostatic mechanisms that operate to maintain the osmolality, volume, and ionic composition of the extracellular fluid within normal limits. This includes the concentration of H^+, with consideration of respiratory and metabolic acidosis and alkalosis. The material in the chapter should help students to—

• Describe how the tonicity (osmolality) of the extracellular fluid is maintained by alterations in water intake and vasopressin secretion.

• Describe how the volume of the extracellular fluid is maintained by alterations in renin and aldosterone secretion.

• Name the mechanisms that operate to maintain the constancy of plasma concentrations of glucose and Ca^{2+}.

• Define acidosis and alkalosis, and give (in meq/L and pH) the normal mean and the range of H^+ concentrations in blood that are compatible with health.

• List the principal buffers in blood, interstitial fluid, and intracellular fluid, and, using the Henderson-Hasselbalch equation, describe what is unique about the bicarbonate buffer system.

• Describe the changes in blood chemistry that occur during the development of metabolic acidosis and metabolic alkalosis, and the respiratory and renal compensations for these conditions.

• Describe the changes in blood chemistry that occur during the development of respiratory acidosis and respiratory alkalosis, and the renal compensation for these conditions.

General Questions

1. What are the main sources of the acid loads presented to the body in everyday living? What are some common diseases that cause increased acid loads in the body, and how is the load produced in each?

2. Compare and contrast metabolic acidosis and respiratory acidosis.

3. Describe and explain the alterations in extracellular fluid volume and acid-base balance that occur in patients with chronically elevated plasma aldosterone concentrations due to primary hyperaldosteronism.

4. Describe by means of diagram plotting plasma HCO_3^- concentration against pH (a Davenport diagram) the immediate and more long-term changes that occur in the acid-base balance of a normal individual who hyperventilates for 5 minutes.

Multiple-Choice Questions

In questions 1–10, select the single best answer. Questions 1–4 refer to the numbered points on the nomogram in Figure 39–A. Match the numbered point in each question with the lettered condition that is most closely associated

with it. Each lettered condition may be selected once, more than once, or not at all.

(A) Values seen in a mountain climber after several weeks at high altitude

(B) Values seen in long-standing severe emphysema

(C) Values seen in diabetic coma

(D) Values seen after 5 minutes of hyperventilation

(E) Values seen after prolonged vomiting

1. Point 1

2. Point 2

3. Point 3

4. Point 4

5. Dehydration increases the plasma concentration of all the following hormones *except*

(A) vasopressin

(B) angiotensin II

(C) aldosterone

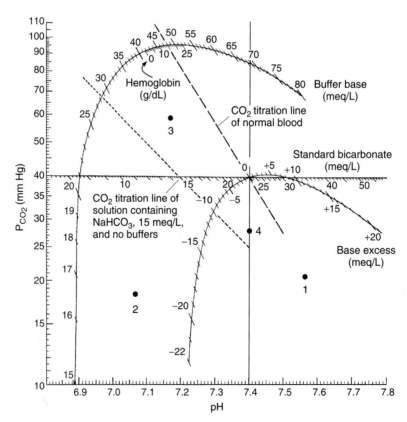

Figure 39–A. Siggaard-Andersen curve nomogram. (Courtesy of O Siggaard-Andersen and Radiometer, Copenhagen, Denmark.)

(D) norepinephrine

(E) atrial natriuretic peptide

6. Which of the following is an important buffer in interstitial fluid?

   (A) Hemoglobin

   (B) Other proteins

   (C) Carbonic acid

   (D) H_2PO_4

   (E) Compounds containing histidine

7. Increasing alveolar ventilation increases the blood pH because

   (A) it activates neural mechanisms that remove acid from the blood

   (B) it makes hemoglobin a stronger acid

   (C) it increases the P_{O_2} of the blood

   (D) it decreases the P_{CO_2} in the alveoli

   (E) the increased muscle work of increased breathing generates more CO_2

8. In uncompensated metabolic alkalosis

   (A) the plasma pH, the plasma HCO_3^- concentration, and the arterial P_{CO_2} are all low

   (B) the plasma pH is high and the plasma HCO_3^- concentration and arterial P_{CO_2} are low

   (C) the plasma pH and the plasma HCO_3^- concentration are low and the arterial P_{CO_2} is normal

   (D) the plasma pH and the plasma plasma HCO_3^- concentration are high and the arterial P_{CO_2} is normal

   (E) the plasma pH is low, the plasma HCO_3^- concentration is high, and the arterial P_{CO_2} is normal

9. In a patient with a plasma pH of 7.10, the $[HCO_3^-]/[H_2CO_3]$ ratio in plasma is

   (A) 20

   (B) 10

   (C) 2

   (D) 1

   (E) 0.1

10. In a patient who has become dehydrated, body water should be replaced by intravenous infusion of

    (A) distilled water

    (B) 0.9% sodium chloride solution

    (C) 5% glucose solution

    (D) hyperoncotic albumin

    (E) 10% glucose solution

Answers to Quantitative & Multiple-Choice Questions

CHAPTER 1

1 (C)	2 (E)	3 (D)	4 (E)	5 (C)
6 (E)	7 (B)	8 (B)	9 (D)	10 (E)
11 (B)	12 (C)	13 (B)	14 (E)	15 (B)
16 (E)	17 (B)	18 (A)	19 (C)	20 (D)
21 (G)	22 (L)	23 (G)	24 (L)	25 (L)

CHAPTER 2

1 (B)	2 (D)	3 (E)	4 (B)	5 (C)
6 (B)	7 (G)	8 (L)	9 (L)	10 (L)

CHAPTER 3

General question 3 Fast muscle, > 133; slow muscle, > 10

1 (B)	2 (D)	3 (B)	4 (C)	5 (C)
6 (C)	7 (C)	8 (C)	9 (A)	10 (C)
11 (A)	12 (C)	13 (D)	14 (G)	15 (L)
16 (L)				

CHAPTER 4

1 (D)	2 (E)	3 (C)	4 (D)	5 (B)
6 (A)	7 (B)	8 (D)	9 (A)	10 (A)
11 (D)	12 (B)	13 (E)	14 (C)	15 (D)
16 (D)	17 (C)	18 (B)	19 (A)	20 (S)
21 (G)				

CHAPTER 5

1 (C)	2 (D)	3 (E)	4 (E)	5 (C)
6 (C)	7 (A)	8 (L)	9 (S)	

CHAPTER 6

1 (C)	2 (E)	3 (C)	4 (D)	5 (C)
6 (B)	7 (A)	8 (S)	9 (G)	10 (G)

CHAPTER 7

1 (C)	2 (D)	3 (A)	4 (C)	5 (D)
6 (B)	7 (E)	8 (D)	9 (D)	10 (A)
11 (A)	12 (B)	13 (B)	14 (L)	15 (G)

CHAPTER 8

1 (D)	2 (D)	3 (B)	4 (E)	5 (D)
6 (B)	7 (D)	8 (D)	9 (C)	10 (E)
11 (D)	12 (B)	13 (B)	14 (L)	15 (G)

CHAPTER 9

1 (D)	2 (E)	3 (D)	4 (C)	5 (E)
6 (D)	7 (D)	8 (E)	9 (E)	10 (A)
11 (A)	12 (B)	13 (B)	14 (C)	15 (A)
16 (G)	17 (L)	18 (G)		

CHAPTER 10

1 (D)	2 (A)	3 (E)	4 (D)	5 (C)
6 (D)	7 (D)	8 (C)	9 (E)	10 (D)
11 (G)	12 (L)			

CHAPTER 11

1 (C)	2 (E)	3 (C)	4 (D)	5 (E)
6 (B)	7 (A)	8 (C)	9 (B)	10 (A)
11 (A)	12 (A)	13 (B)	14 (B)	15 (L)
16 (G)				

CHAPTER 12

1 (B)	2 (E)	3 (D)	4 (B)	5 (C)
6 (E)	7 (E)	8 (D)	9 (B)	10 (A)
11 (A)	12 (C)	13 (E)	14 (B)	15 (E)
16 (D)	17 (C)	18 (B)	19 (A)	20 (B)

CHAPTER 13

1 (C)	2 (A)	3 (A)	4 (C)	5 (C)
6 (C)	7 (E)	8 (E)	9 (G)	10 (G)

CHAPTER 14

1 (B)	2 (E)	3 (E)	4 (D)	5 (A)
6 (C)	7 (C)	8 (E)	9 (E)	10 (B)
11 (B)	12 (D)	13 (A)	14 (B)	15 (E)

16 (D)   17 (D)   18 (B)   19 (A)   20 (A)
21 (B)   22 (D)   23 (D)

CHAPTER 15

1 (B)   2 (C)   3 (A)   4 (B)   5 (A)
6 (E)   7 (C)   8 (C)   9 (C)   10 (A)
11 (A)   12 (D)   13 (B)   14 (C)   15 (E)
16 (E)   17 (C)   18 (E)   19 (L)   20 (L)

CHAPTER 16

1 (C)   2 (A)   3 (E)   4 (C)   5 (D)
6 (B)   7 (D)   8 (D)   9 (C)   10 (A)
11 (D)   12 (E)   13 (B)   14 (E)   15 (E)
16 (B)   17 (B)   18 (A)   19 (B)   20 (A)
21 (D)

CHAPTER 17

General question 1: BMR, 208 kcal/kg/24 h; animal is small.

1 (B)   2 (A)   3 (D)   4 (D)   5 (E)
6 (B)   7 (C)   8 (E)   9 (B)   10 (C)
11 (D)   12 (B)   13 (A)   14 (A)   15 (D)
16 (E)   17 (E)   18 (B)   19 (C)   20 (A)

CHAPTER 18

1 (C)   2 (D)   3 (D)   4 (E)   5 (A)
6 (C)   7 (B)   8 (D)   9 (A)   10 (B)
11 (D)   12 (C)   13 (E)   14 (L)   15 (L)
16 (G)

CHAPTER 19

1 (E)   2 (D)   3 (D)   4 (E)   5 (C)
6 (D)   7 (E)   8 (C)   9 (D)   10 (C)
11 (B)   12 (A)   13 (A)   14 (C)   15 (B)
16 (L)   17 (L)   18 (L)

CHAPTER 20

1 (D)   2 (A)   3 (E)   4 (B)   5 (C)
6 (E)   7 (D)   8 (B)   9 (E)   10 (D)
11 (C)   12 (D)   13 (D)   14 (A)   15 (D)
16 (A)   17 (E)   18 (C)   19 (C)   20 (D)

CHAPTER 21

1 (C)   2 (E)   3 (D)   4 (E)   5 (A)
6 (C)   7 (D)   8 (A)   9 (E)   10 (D)
11 (E)   12 (L)   13 (G)

CHAPTER 22

1 (E)   2 (D)   3 (E)   4 (E)   5 (A)
6 (B)   7 (C)   8 (B)   9 (C)   10 (D)
11 (A)   12 (B)   13 (E)   14 (G)   15 (G)

CHAPTER 23

1 (C)   2 (A)   3 (B)   4 (D)   5 (D)
6 (C)   7 (E)   8 (D)   9 (A)   10 (E)
11 (E)   12 (D)   13 (A)   14 (C)   15 (D)
16 (D)   17 (A)   18 (L)   19 (S)   20 (L)
21 (G)   22 (L)

CHAPTER 24

1 (D)   2 (E)   3 (D)   4 (D)   5 (D)
6 (C)   7 (D)   8 (E)   9 (D)   10 (E)
11 (E)   12 (C)   13 (A)   14 (A)   15 (D)
16 (B)

CHAPTER 25

1 (B)   2 (E)   3 (C)   4 (E)   5 (C)
6 (D)   7 (D)   8 (C)   9 (D)   10 (A)
11 (B)   12 (D)

CHAPTER 26

1 (D)   2 (E)   3 (D)   4 (C)   5 (C)
6 (D)   7 (B)   8 (A)   9 (D)   10 (C)
11 (A)   12 (D)   13 (C)   14 (E)   15 (D)
16 (D)   17 (C)   18 (A)   19 (E)   20 (B)
21 (D)   22 (E)   23 (B)   24 (A)   25 (C)

CHAPTER 27

1 (B)   2 (E)   3 (C)   4 (A)   5 (A)
6 (A)   7 (B)   8 (A)   9 (A)   10 (E)
11 (C)   12 (C)   13 (E)   14 (B)   15 (E)
16 (D)   17 (C)   18 (C)   19 (S)   20 (G)

CHAPTER 28

1 (E)   2 (B)   3 (C)   4 (D)   5 (A)
6 (A)   7 (C)   8 (D)   9 (D)   10 (A)
11 (A)   12 (L)   13 (L)   14 (L)   15 (L)

CHAPTER 29

1 (E)   2 (C)   3 (D)   4 (A)   5 (B)
6 (A)   7 (C)   8 (C)   9 (C)   10 (E)
11 (D)   12 (B)   13 (E)   14 (C)   15 (A)

CHAPTER 30

General question 3: Pulse pressure, 57 mm Hg; mean arterial pressure, 92 mm Hg. General question 4: 30,000 dynes/cm; 60,000 dynes/cm.

1 (D)	2 (A)	3 (E)	4 (C)	5 (C)
6 (B)	7 (D)	8 (B)	9 (B)	10 (E)
11 (A)	12, (A)	13 (E)	14 (E)	15 (D)
16 (E)	17 (A)	18 (D)	19 (G)	20 (L)

CHAPTER 31

1 (E)	2 (B)	3 (D)	4 (D)	5 (C)
6 (C)	7 (B)	8 (C)	9 (B)	10 (B)
11 (C)	12 (D)	13 (A)	14 (C)	15 (B)
16 (C)	17 (11, 12)	18 (13, 14, 15, 16)		

CHAPTER 32

1 (C)	2 (D)	3 (D)	4 (B)	5 (A)
6 (D)	7 (A)	8 (E)	9 (E)	10 (C)
11 (E)	12 (D)	13, (D)	14 (S)	15 (L)
16 (L)				

CHAPTER 33

General question 10: Cardiac output, 4.73 L/min; stroke volume, 73.9 mL. Diagnosis: aortic stenosis.

1 (D)	2 (E)	3 (B)	4 (A)	5 (C)
6 (D)	7 (D)	8 (E)	9 (D)	10 (D)
11 (B)	12 (E)	13 (B)	14 (G)	15 (S)

CHAPTER 34

1 (D)	2 (C)	3 (A)	4 (C)	5 (E)
6 (B)	7 (C)	8 (A)	9 (B)	10 (D)
11 (E)	12 (D)	13 (A)	14 (G)	15 (G)

CHAPTER 35

1 (E)	2 (B)	3 (D)	4 (B)	5 (D)
6 (C)	7 (C)	8 (G)	9 (L)	

CHAPTER 36

1 (D)	2 (B)	3 (B)	4 (D)	5 (E)
6 (E)	7 (B)	8 (C)	9 (C)	10 (G)
11 (G)				

CHAPTER 37

1 (B)	2 (C)	3 (E)	4 (D)	5 (A)
6 (A)	7 (D)	8 (B)	9 (D)	10 (D)
11 (C)	12 (E)	13 (E)	14 (E)	15 (C)
16 (D)	17 (A)	18 (B)		

CHAPTER 38

General question 7: (a) 35625 meq of HCO_3^-; (b) 35625 meq of Na^+; (c) $35625 + 105 + 75 = 37425$ meq of H^+.

1 (D)	2 (C)	3 (A)	4 (E)	5 (B)
6 (A)	7 (A)	8 (A)	9 (A)	10 (E)
11 (C)	12 (D)	13 (C)	14 (A)	15 (D)
16 (B)	17 (E)	18 (C)	19 (D)	20 (B)
21 (A), (C), (D), (E)	22 (D), (E)			

CHAPTER 39

1 (D)	2 (C)	3 (B)	4 (A)	5 (E)
6 (C)	7 (D)	8 (D)	9 (B)	10 (C)

Appendix

GENERAL REFERENCES

Many large, comprehensive textbooks of physiology are available. The following are among the best that have been revised in the last 10 years:

Berne RM, Levy MN (editors): *Physiology,* 3rd ed. Mosby, 1993.

Guyton AC, Hall JE: *Textbook of Medical Physiology,* 10th ed. Saunders, 2000.

Johnson LR (editor): *Essential Medical Physiology,* Raven Press, 1992.

McPhee, Lingappa, and Ganong have recently published the fourth edition of a pathophysiology text designed to introduce students to clinical medicine:

McPhee SJ, Lingappa VR, Ganong WF: *Pathophysiology of Disease: An Introduction to Clinical Medicine,* 4th ed. McGraw-Hill, 2003.

An outstanding cell physiology text is:

Alberts B et al: *Molecular Biology of the Cell,* 4th ed. Garland, 2002.

A standard anatomy reference is:

Bannister CH et al (editors): *Gray's Anatomy: The Anatomical Basis of Medicine and Surgery,* 38th ed. Churchill Livingstone, 1995.

References for the invaluable imaging techniques that have become a key part of modern physiology and medicine include:

Haaga JR, Alfidi RJ (editors): *Computed Tomography of the Whole Body,* 2nd ed. Mosby, 1988.

Von Schultress GR: *Clinical Positron Emission Tomography.* Lippincott Williams & Wilkins, 1999.

Excellent summaries of current research on selected aspects of physiology can be found in the news and views section of *Nature* and the Perspectives that appear in *Science.* Summary articles and various types of valuable reviews appear in the *New England Journal of Medicine.* These include articles that review current topics in physiology and biochemistry with the aim of providing up-to-date information for practicing physicians. There are valuable short reviews in News in Physiological Sciences, published by the International Union of Physiological Sciences and the American Physiological Society. The most pertinent serial review publications are *Physiological Reviews, Pharmacological Reviews, Annual Review of Physiology,* and other volumes of the Annual Reviews series. The *Handbook of Physiology,* published by Oxford University Press, New York, has separate volumes that cover all aspects of physiology. The articles in the *Handbook* are valuable but extremely detailed reviews.

NORMAL VALUES & THE STATISTICAL EVALUATION OF DATA

The approximate ranges of values in normal humans for some commonly measured plasma constituents are summarized in the table on the inside back cover. A worldwide attempt has been under way to convert to a single standard nomenclature by using SI (Système International) units. The system is based on the seven dimensionally independent physical quantities summarized in Table 1. Units derived from the basic units are summarized in Table 2, and the prefixes used to refer to decimal fractions and multiples of these and other units are listed in Table 3. There are a number of complexities involved in the use of these units—for example, the problem of expressing enzyme units—and they have been slow in making their way into the medical literature. In this book, the values in the text are in traditional units, but they are followed in key instances by values in SI units. In addition, values in SI units are listed beside values in more traditional units in the table on the inside back cover of this book.

The accuracy of the methods used for laboratory measurements varies. It is important in evaluating any single measurement to know the possible errors in making the measurement. For chemical determinations on body fluids, these include errors in obtaining the sample and the inherent error of the chemical method.

Table 1. Basic units.

Quantity	Name	Symbol
Length	meter	m
Mass	kilogram	kg
Time	second	s
Electric current	ampere	A
Thermodynamic temperature	kelvin	K
Luminous intensity	candela	cd
Amount of substance	mole	mol

Table 2. Some derived SI units.

Quantity	Unit Name	Unit Symbol
Area	square meter	m^2
Clearance	liter/second	L/s
Concentration		
Mass	kilogram/liter	kg/L
Substance	mole/liter	mol/L
Density	kilogram/liter	kg/L
Electric	volt	V
potential		
Energy	joule	J
Force	newton	N
Frequency	hertz	Hz
Pressure	pascal	Pa
Temperature	degree Celsius	°C
Volume	cubic meter	m^3
	liter	L

However, the values obtained by using even the most accurate methods vary from one normal individual to the next as a result of what is usually called **biologic variation.** This variation occurs because in any system as complex as a living organism or tissue there are many variables that affect the particular measurement. Variables such as age, sex, time of day, time since last meal, etc., can be taken into account. Numerous other variables cannot, and for this reason the values obtained differ from individual to individual.

The magnitude of the normal range for any given physiologic or clinical measurement can be calculated by standard statistical techniques if the measurement has been made on a suitable sample of the normal population (preferably more than 20 individuals). It is important to know not only the average value in this sample but also the extent of the deviation of the individual values from the average.

The average (**arithmetic mean, M**) of the series of values is readily calculated:

$$M = \frac{\Sigma X}{n}$$

where

  Σ = Sum of
  X = Individual values
  n = Number of individual values in the series

The average deviation is the mean of the deviations of each of the values from the arithmetic mean. From a mathematical point of view, a better measure of the deviation is the **geometric mean** of the deviations from

the mean. This is called the **standard deviation of the sample (s):**

$$s = \sqrt{\frac{\Sigma(M - X)^2}{n - 1}}$$

The term n − 1, rather than n, is used for complex mathematical reasons. s should be distinguished from the standard deviation of the mean of the whole population, which is designated σ. However, if the sample is truly representative, s and σ will be comparable.

Another commonly used index of the variation is the **standard error of the mean (SEM):**

$$SEM = \frac{\sigma}{\sqrt{n}}$$

Strictly speaking, the SEM indicates the reliability of the sample mean as representative of the true mean of the general population from which the sample was drawn.

A **frequency distribution** curve can be constructed from the individual values in a population by plotting the frequency with which any particular value occurs in the series against the values. If the group of individuals tested was homogeneous, the frequency distribution curve is usually symmetric (Figure 1), with the highest frequency corresponding to the mean and the width of

Table 3. Standard prefixes.

Prefix[1]	Abbreviation	Magnitude
exa-	E	10^{18}
peta-	P	10^{15}
tera-	T	10^{12}
giga-	G	10^{9}
mega-	M	10^{6}
kilo-	k	10^{3}
hecto-	h	10^{2}
deca-	da	10^{1}
deci-	d	10^{-1}
centi-	c	10^{-2}
milli-	m	10^{-3}
micro-	μ	10^{-6}
nano-	n, mμ	10^{-9}
pico-	p, mμ	10^{-12}
femto-	f	10^{-15}
atto-	a	10^{-18}

[1]These prefixes are applied to SI and other units. For example, a micrometer (μm) is 10^{-6} meter (also called a micron); a picoliter (pL) is 10^{-12} liter, and a kilogram (kg) is 10^{3} grams. Also applied to seconds, units, moles, hertz, volts, farads, ohms, curies, equivalents, osmoles, etc.

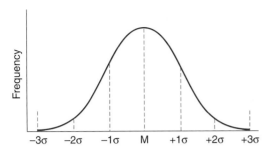

Figure 1. Curve of normal distribution (frequency distribution curve of values from a homogeneous population).

the curve varying with σ (curve of **normal distribution**). Within an ideal curve of normal distribution, the percentage of observations that fall within various ranges is shown in Table 4. The mean and s of a representative sample are approximately the mean and σ of the whole population. It is therefore possible to predict from the mean and s of the sample the probability that any particular value in the general population is normal. For example, if the difference between such a value and the mean is equal to 1.96 s, the chances are 1 out of 20 (5 out of 100) that it is normal. Conversely, of course, the chances are 19 out of 20 that it is abnormal.

Statistical analysis is also useful in evaluating the significance of the difference between two means. In physiologic and clinical research, measurements are often made on a group of animals or patients given a particular treatment. These measurements are compared with similar measurements made on a control group that ideally has been exposed to exactly the same conditions except that the treatment has not been given. If a particular mean value in the treated group is different from the corresponding mean for the control group, the question arises whether the difference is due to the treatment or to chance variation. The probability that the difference represents chance variation can be estimated in many instances by using **Student's t test.** The value t is the ratio of the difference in the means of two series (M_a and M_b) to the uncertainty in these means. The formula used to calculate t is

$$t = \frac{M_a - M_b}{\sqrt{\dfrac{(n_a + n_b)\,[(n_a - 1)s_a^2 + (n_b - 1)s_b^2]}{n_a n_b (n_a + n_b - 2)}}}$$

where n_a and n_b are the number of individual values in series a and b, respectively. When $n_a = n_b$, the equation for t becomes simplified to

$$t = \frac{M_a - M_b}{\sqrt{(SEM_a)^2 + (SEM_b)^2}}$$

The higher the value of t, the less the probability that the difference represents chance variation. This probability also decreases as the number of individuals (n) in each group rises, because the greater the number of measurements, the smaller the error in the measurements. A mathematical expression of the probability (P) for any value of t at different values of n can be found in tables in most texts on statistics. P is a fraction which expresses the probability that the difference between two means was due to chance variation. Thus, for example, if the P value is 0.10, the probability that the difference was due to chance is 10% (1 chance in 10). A P value of < 0.001 means that the chances that the difference was due to random variation are less than 1 in 1000. When the P value is < 0.05, most investigators call the difference "statistically significant"; ie, it is concluded that the difference is due to the operation of some factor other than chance. The use of t tests is appropriate only for comparison of two groups. If they are used for comparison in experiments involving more than two groups, a systematic error is introduced that makes the probability of deciding that there is a significant difference too high. In this situation, **analysis of variance (ANOVA)** is the appropriate statistical test. This and other techniques are discussed in statistics texts.

These elementary methods and many others available for statistical analysis in the research laboratory and the clinic provide a valuable objective means of evaluation. Statistical significance does not arbitrarily mean physiologic significance, and the reverse is sometimes true; but replacement of evaluation by subjective impression with analysis by statistical methods is an important goal in the medical sciences.

Useful books on statistics include:

Dawson B, Trapp RG: *Basic and Clinical Biostatistics,* 3rd ed. McGraw-Hill, 2001.

Rosner B: *Fundamentals of Biostatistics.* Duxbury, 1982.

The SI for the Health Professions: World Health Organization, 1977.

Zar JH: *Biostatistical Analysis.* Prentice-Hall, 1974.

Table 4. Percentage of values in a population which will fall within various ranges within an ideal curve of normal distribution.

Mean ± σ	68.27%
Mean ± 1.96 σ	95.00%
Mean ± 2 σ	95.45%
Mean ± 3 σ	99.73%

ABBREVIATIONS & SYMBOLS COMMONLY USED IN PHYSIOLOGY

Use of abbreviations and acronyms has become so extensive in modern physiology and related sciences that it would be impossible to present a complete list here. Furthermore, most proteins that are isolated today are simply assigned a name made up of letters and numbers that bears no relation to their function. Therefore, the following list is incomplete and in some ways arbitrary. However, it should identify for the reader many of the terms and symbols commonly used in physiology.

[ ]	Concentration of
Δ	Change in (*Example:* ΔV = change in volume.) In steroid nomenclature, Δ followed by a number (eg, Δ^{4-}) indicates the position of a double bond47
s	Standard deviation of a whole population
Ia, Ib, II, III, IV nerve fibers	Types of fibers in sensory nerves (see Chapter 2)
μ	Micro, 10^{-6}; see Table 3, above
a	Atto-, 10^{-18}; see Table 3, above
A (Å)	Angstrom unit(s) (10^{-10} m, 0.1 nm); *also* alanine
A^-	General symbol for anion
$A_1, A_2, A_{1B}, A_{2B}, B, O$	Major blood groups
AASH	Adrenal androgen-stimulating hormone
ABC	ATP-binding cassette
A, B, and C nerve fibers	Types of fibers in peripheral nerves (see Chapter 2)
ABP	Androgen-binding protein
ACE	Angiotensin-converting enzyme
Acetyl-CoA	Acetyl-coenzyme A
Ach	Acetylcholine
ACTH	Adrenocorticotropic hormone
Acyl-CoA	General symbol for an organic compound–coenzyme A ester
ADH	Antidiuretic hormone (vasopressin)
ADP	Adenosine diphosphate
AGEs	Advanced glycosylation end products
AHG	Antihemophilic globulin
Ala	Alanine
ALS	Amyotrophic lateral sclerosis
ALT	Alanine aminotransferase
AM	Adrenomedullin
AME	Apparent mineralocorticoid excess
AMP	Adenosine 5'-monophosphate
AMPA	α-Amino-3-hydroxy-5-methylisoxazole-4-proprinate
ANOVA	Analysis of variance
ANP	Atrial natriuretic peptide
4-AP	4-Aminopyridine
AP-1	Activator protein 1
APC	Activated protein C; *also* antigen-presenting cell
Apo E	Apolipoprotein E
APP	Amyloid precursor protein
APUD cells	Amine precursor uptake and decarboxylation cells that secrete hormones
ARDS	Acute respiratory distress syndrome
ARF	ADP-ribosylation factor
Arg	Arginine
β-ARK	β-Adrenergic kinase
Asn	Asparagine
Asp	Aspartic acid
AST	Aspartate aminotransferase
atm	Atmosphere: 1 atm = 760 torr = mean atmospheric pressure at sea level
ATP	Adenosine triphosphate
A-V difference	Arteriovenous concentration difference of any given substance
AV node	Atrioventricular node
AVP	Arginine vasopressin
aVR, aVF, aVL	Augmented unipolar electrocardiographic leads
AV valves	Atrioventricular valves of heart
ATPD	See Table 34-1
ATPS	See Table 34-1
BDNF	Brain-derived neurotrophic factor
BER	Basic electric rhythm
BGP	Bone Gla protein
BMI	Body mass index
BMP	Bone morphogenic protein
BMR	Basal metabolic rate
BNP	Brain natriuretic peptide
BSP	Sulfobromophthalein
BTPS	See Table 34-1
BUN	Blood urea nitrogen
c	Centi-, 10^{-2}; see Table 3, above
C	Celsius; *also* cysteine
CART	Cocaine- and amphetamine-regulated transcript
C followed by subscript	Clearance; eg, C_{In}, clearance of inulin
C peptide	Connecting peptide
C_{19} steroids	Steroids containing 19 carbon atoms
C_{21} steroids	Steroids containing 21 carbon atoms
cADPR	Cyclic adenosine diphosphate ribose

cal	The calorie (gram calorie)
Cal	1000 calories; kilocalorie
CAM	Cell adhesion molecule
CaMKII	Ca^{2+}/calmodulin-dependent kinase II
cAMP	Cyclic adenosine 3′,5′-monophosphate
CBF	Cerebral blood flow
CBG	Corticosteroid-binding globulin, transcortin
cc	Cubic centimeters
CCK	Cholecystokinin
CCK-PZ	Cholecystokinin-pancreozymin
CDs	Clusters of differentiation
CEH	Cholesteryl ester hydrolase
CFF	Critical fusion frequency
c-fos	One of the immediate early response genes
CFTR	Cystic fibrosis transmembrane conductance regulator
cGMP	Cyclic 3′,5′-guanosine monophosphate
CGP	Chorionic growth hormone-prolactin (same as hCS)
CGRP	Calcitonin gene-related peptide
CH_2O	"Free water clearance"
Ci	Curie
c-jun	One of the immediate early response genes
CLIP	Corticotropin-like intermediate-lobe polypeptide
$CMRO_2$	Cerebral metabolic rate for oxygen
CNG	Cyclic nucleotide-gated
CNP	The third natriuretic peptide
CNS	Central nervous system
CNTF	Ciliary neurotrophic factor
CoA	Coenzyme A
COHb	Carbonmonoxyhemoglobin
Compound B	Corticosterone
Compound E	Cortisone
Compound F	Cortisol
Compound S	11-Deoxycortisol
COMT	Catechol-*O*-methyltransferase
C peptide	Connecting peptide
CPR	Cardiopulmonary resuscitation
cps	Cycles per second, hertz
CR	Conditioned reflex
CREB	cAMP-responsive element binding protein
CRH, CRF	Corticotropin-releasing hormone (factor)
CRO	Cathode-ray oscilloscope
CS	Conditioned stimulus
CSF	Cerebrospinal fluid; also colony-stimulating factor
CT	Computed tomography
C terminal	—COOH end of a peptide or protein
CTP	Cytidine triphosphate
CV	Closing volume
CVLM	Caudal ventrolateral medulla
CVR	Cerebral vascular resistance
cyclic AMP	Cyclic adenosine 3′,5′-monophosphate
Cys	Cysteine
CZI	Crystalline zinc insulin
d	Day
D	Geometric isomer of L form of chemical compound
D	Aspartic acid
5′DI, 5′DII, 5′DIII	Type I, II, and III iodotyrosine deiodinases
Da	Dalton, unit of mass equal to one-twelfth the mass of the carbon-12 atom, or about 1.66×10^{-24} g
DAG	Diacylglycerol
dB	Decibel
DBH	Dopamine β-hydroxylase
DBP	Vitamin D-binding protein
DDAVP	1-Deamino-8-D-arginine vasopressin
DDD	Derivative of DDT that inhibits adrenocortical function
DEA, DHEA, DHA	Dehydroepiandrosterone
DEAS, DHEAS, DHAS	Dehydroepiandrosterone sulfate
DFP	Diisopropyl fluorophosphate
DHT	Dihydrotestosterone
DHTs	Dihydroxy derivatives of eicosatetraenoic acid
DIT	Diiodotyrosine
DMT	*N,N*-Dimethyltryptamine
DNA	Deoxyribonucleic acid
D/N ratio	Ratio of dextrose (glucose) to nitrogen in the urine
D_2O	Deuterium oxide (heavy water)
DOCA	Deoxycorticosterone acetate
DOM	2,5-Dimethoxy-4-methylamphetamine
DOMA	3,4-Dihydroxymandelic acid
Dopa	Dihydroxyphenylalanine, L-dopa
DOPAC	3,4-Dihydroxyphenylacetic acid
DOPEG	3,4-Dihydroxyphenylglycol
DOPET	3,4-Dihydroxyphenylethanol
DPG,2,3-DPG	2,3-Diphosphoglycerate
DPL	Dipalmitoyl lecithin
DPN	Diphosphopyridine nucleotide
DPNH	Reduced diphosphopyridine nucleotide
DPPC	Dipalmitoylphosphatidyl-choline
e	Base for natural logarithms = 2.7182818 . . .
E	Glutamic acid

E followed by subscript number	Esophageal electrocardiographic lead, followed by number of centimeters it is inserted into esophagus
E cells	Expiratory neurons
E_1	Estrone
E_2	Estradiol
EACA	Epsilon-aminocaproic acid
ECF	Extracellular fluid
eCG	Equine chorionic gonadotropin
ECG, EKG	Electrocardiogram
ECL cells	Enterochromaffin-like cells
ECoG	Electrocorticogram
EDRF	Endothelium-derived relaxing factor
EDTA	Ethylenediaminetetraacetic acid
EEG	Electroencephalogram
EET	Epoxyeicosatetraenoic acid
EETs	Epoxyeicosatrienoic acids
EGF	Epidermal growth factor
Egr	Early growth response factor
EJP	Excitatory junction potential
EMG	Electromyogram
ENaC	Epithelial sodium channel
EP	Endogenous pyrogen
EPSP	Excitatory postsynaptic potential
eq	Equivalent
ERα, ERβ	Estrogen receptors α and β
ERG	Electroretinogram
ERKO	Estrogen receptor knockout
ERP	Event-related potential
ERPF	Effective renal plasma flow
ET-1, ET-2, ET-3	Endothelin-1, -2, -3
ETP	Electron transport particle
f	Femto-, 10^{-15}; see Table 3, above
F	Fahrenheit; *also* phenylalanine
FAD	Flavin adenine dinucleotide
FDP	Fibrinogen degradation products
FEV_1	Forced expiratory volume in first second of forced expiration after maximum inspiration
FFA	Unesterified free fatty acid (also called NEFA, UFA)
FGF	Fibroblast growth factor
FGFR	Fibroblast growth factor receptor
FLAP	5-Lipoxygenase-activating protein
FMN	Flavin mononucleotide
fMRI	Functional magnetic resonance imaging
FSH	Follicle-stimulating hormone
ft	Foot or feet
g, gm	Gram or grams
g	Unit of force; 1 *g* equals the force of gravity on the earth's surface
G	Glucose; *also* giga-, 10^9; see Table 3, above; *also* glycine
GABA	Gamma-aminobutyrate
GABA-T	GABA transaminase
GAD	Glutamate decarboxylase
GBG	Gonadal steroid-binding globulin
G-CSF	Granulocyte colony-stimulating factor
GFR	Glomerular filtration rate
GH	Growth hormone
GHS	Growth hormone secretagogue
GIH, GIF	Growth hormone-inhibiting hormone or factor
GIP	Gastric inhibitory peptide
Gla	Gamma-carboxyglutamic acid
GLI	Glicentin
Gln	Glutamine
GLP-1, -2	Glucagon-like polypeptide-1, -2
GLP-1 (7-36) amide	Glucagon-like polypeptide-1 (7-36) amide
Glu	Glutamic acid
GLUT	Glucose transporter
Gly	Glycine
GM-CSF	Granulocyte-macrophage colony-stimulating factor
GnRH	Gonadotropin-releasing hormone; same as LHRH
G6PD	Glucose-6-phosphate dehydrogenase
GRA	Glucocorticoid-remediable aldosteronism
GRH, GRF	Growth hormone-releasing hormone
GRP	Gastrin-releasing polypeptide
GRPP	Glicentin-related polypeptide
GTP	Guanosine triphosphate
h	Hour
H	Histidine
HA	General symbol for an acid
Hb	Deoxygenated hemoglobin
HbA_{1c}	Hemoglobin A_{1c}
HBE	His bundle electrogram
HbO_2	Oxyhemoglobin
HCC, 25-HCC	25-Hydroxycholecalciferol, a metabolite of vitamin D_3
hCG	Human chorionic gonadotropin
hCS	Human chorionic somatomammotropin
Hct	Hematocrit
HDL	High-density lipoprotein
hGH	Human growth hormone
5-HIAA	5-Hydroxyindoleacetic acid
HIFs	Hypoxia-inducible factors
HIOMT	Hydroxyindole-*O*-methyltransferase
His	Histidine
HIV	Human immunodeficiency virus
HLA	Human leukocyte antigen
HMG-CoA reductase	3-Hydroxy-3-methylglutaryl coenzyme A reductase

HO2	Subtype of heme oxygenase
hPL	Human placental lactogen (same as hCS)
HPNS	High-pressure nervous syndrome
HS-CoA	Reduced coenzyme A
HSP	Heat shock protein
H substance	Histamine-like capillary vasodilator
5-HT	Serotonin
hTRα, hTRβ	Human thyroid hormone receptor α, β
HVA	Homovanillic acid
Hyl	Hydroxylysine
Hyp	4-Hydroxyproline
Hz	Hertz, unit of frequency. 1 cycle per second = 1 hertz
I	Isoleucine
ICA cells	Intrinsic cardiac adrenergic cells
ICAM	Intracellular adhesion molecule
I cells	Inspiratory neurons; *also* intercalated cells in the renal tubules
IDDM	Insulin-dependent diabetes mellitus
IDL	Intermediate-density lipoprotein
IFN	Interferon
IGF-I, IGF-II	Insulin-like growth factors I and II
^{123}**I-IMP**	^{123}I-labeled iodoamphetamine
IJP	Inhibitory junction potential
IL	Interleukin
Ile, Ileu	Isoleucine
IML	Intermediolateral gray column
In	Inulin
IP$_3$	Inositol 1,4,5-trisphosphate, inositol triphosphate
IPSP	Inhibitory postsynaptic potential
IRDS	Infant respiratory distress syndrome
ITP	Inosine triphosphate
IU	International unit
IUD	Intrauterine device
J	Joule (SI unit of energy)
JAK	Janus tyrosine kinase
JG cells	Juxtaglomerular cells
k	Kilo-, 10^3; see Table 3, above
K	Lysine; also kilodalton
kcal (Cal)	Kilocalorie (1000 calories)
K$_E$	Exchangeable body potassium
Kf	Glomerular ultrafiltration coefficient
L	Geometric isomer of D form of chemical compound
L	Leucine
LATS	Long-acting thyroid stimulator
LCAT	Lecithin-cholesterol acyltransferase
LCR	Locus control region
LDH	Lactate dehydrogenase
LDL	Low-density lipoprotein
LES	Lower esophageal sphincter
Leu	Leucine
LH	Luteinizing hormone

LIF	Leukemia inhibitory factor
ln	Natural logarithm
log	Logarithm to base 10
LPH	Lipotropin
LHRH	Luteinizing hormone-releasing hormone; same as GnRH
LRP	LDL receptor-related protein
LSD	Lysergic acid diethylamide
LTD	Long-term depression
LTP	Long-term potentiation
LVET	Left ventricular ejection time
Lys	Lysine
m	Meter(s); *also* milli-, 10^{-3}; see Table 3, above
M	Molarity (mol/L); *also* mega-, 10^6; see Table 3, above; *also* methionine
M cells	Microfold cells
M1	Motor cortex
MAO	Monoamine oxidase
MBC	Maximal breathing capacity (same as MVV)
M-CSF	Macrophage colony-stimulating factor
MCH	Melanin-concentrating hormone
MCP-1	Monocyte chemoattractant protein 1
MDMA	3,4-Methylenedioxymethamphetamine
Met	Methionine
MGP	Matrix Gla protein
MHC	Major histocompatibility complex; *also* myosin heavy chain
mho	Unit of conductance; the reciprocal of the ohm
MHPG	3-Methoxy-4-hydroxy-phenylglycol
min	Minute
MIS	Müllerian inhibiting substance; same as MRF, müllerian regression factor
MIT	Monoiodotyrosine
MMC	Migrating motor complex
mol	Mole, gram-molecular weight
MPGF	Major proglucagon fragment
MPP$^-$	1-Methyl-4-phenylpyridinium
MPR	Mannose-6-phosphate receptor
MPTP	1-Methyl-4-phenyl-1,2,5,6-tetrahydropyridine
MRI	Magnetic resonance imaging
mRNA	Messenger RNA
MSH	Melanocyte-stimulating hormone
MT	3-Methoxytyramine
MTOC	Microtubule-organizing center
multi-CSF	Multipotential colony-stimulating factor
MVV	Maximal voluntary ventilation
n	nano-, 10^{-9}; see Table 3, above

N	Normality (of a solution): *also* newton (SI unit of force); *also* asparagine
NAD⁺	Nicotinamide adenine dinucleotide; same as DPN
NADH	Dihydronicotinamide adenine dinucleotide; same as DPNH
NADP⁺	Nicotinamide adenine dinucleotide phosphate; same as TPN
NADPH	Dihydronicotinamide adenine dinucleotide phosphate; same as TPNH
Na$_E$	Exchangeable body sodium
NEAT	Nonexercise activity thermogenesis
NEFA	Unesterified (nonesterified) free fatty acid (same as FFA)
NEP	Neutral endopeptidase
NF-$_\kappa$B	Nuclear factor B
NGF	Nerve growth factor
NIDDM	Non-insulin-dependent diabetes mellitus
NIS	Sodium/iodide symporter
NMDA	*N*-Methyl-D-aspartate
NO	Nitric oxide
NOS	Nitric oxide synthase
NPH insulin	Neutral protamine Hagedorn insulin
NPN	Nonprotein nitrogen
NREM sleep	Nonrapid eye movement (spindle) sleep
NSAID	Nonsteroidal anti-inflammatory drug
NSILA	Nonsuppressible insulin-like activity
NSILP	Nonsuppressible insulin-like protein
N terminal	$-NH_2$ (amino) group end of peptide or protein
NT-3, NT-4/5	Neurotrophin 3, neurotrophin 4/5
NTS	Nucleus of the tractus solitarius
O	Indicates absence of a sex chromosome, eg, XO as opposed to XX or XY
OBP	Odorant-binding protein
25-OHD$_3$	25-Hydroxycholecalciferol
1,25-(OH)$_2$D$_3$	1,25-Dihydroxycholecalciferol
OGF	Ovarian growth factor
osm	Osmole
OVLT	Organum vasculosum of the lamina terminalis
p	Pico-, 10^{-12}; see Table 3, above
P	Proline
P$_0$	Protein zero
P followed by subscript	Plasma concentration, eg, P_{Cr} = plasma creatinine concentration; *also* permeability coefficient, eg, P_{Na^+} = permeability coefficient for Na⁺; *also* pressure (see respiratory symbols, below)
P450	Cytochrome P450

P$_{50}$	Partial pressure of O_2 at which hemoglobin is half-saturated with O_2
Pa	Pascal (SI unit of pressure)
P Ab	Antibodies against thyroid peroxidase
PACAP	Pituitary adenylyl cyclase-activating poly-peptide
PAF	Platelet-activating factor
PAH	*p*-Aminohippuric acid
PAM	Pulmonary alveolar macrophage
PBI	Protein-bound-iodine
P cells	Principal cells in the renal tubules; *also* pacemaker cells of SA and AV nodes
PCR	Polymerase chain reaction
PDECGF	Platelet-derived endothelial cell growth factor
PDGF	Platelet-derived growth factor
PEEP	Positive end-expiratory pressure breathing
PEP	Preejection period
PET	Positron emission tomography
P factor	Hypothetical pain-producing substance produced in ischemic muscle
PGO spikes	Ponto-geniculo-occipital spikes in REM sleep
pH	Negative logarithm of the H⁺ concentration of a solution
Phe	Phenylalanine
PHM-27	Peptide histidyl methionine, produced along with VIP from preproVIP in humans
Pi	Inorganic phosphate
PIH, PIF	Prolactin-inhibiting hormone
PK	Negative logarithm of the equilibrium constant for a chemical reaction
PLC	Phospholipase C
PMN	Polymorphonuclear neutrophilic leukocyte
PMS	Premenstrual syndrome
PNMT	Phenylethanolamine-*N*-methyltransferase
POMC	Pro-opiomelanocortin
PPAR	Peroxisome proliferation-activation receptor
PRA	Plasma renin activity
PRC	Plasma renin concentration
PRH, PRF	Prolactin-releasing hormone
PRL	Prolactin
Pro	Proline
Prot⁻	Protein anion
5-PRPP	5-Phosphoribosyl pyrophosphate
PTA	Plasma thromboplastin antecedent (clotting factor XI)

PTC	Plasma thromboplastin component (clotting factor IX); *also* phenyl-thiocarbamide
PTH	Parathyroid hormone
PTHrP	Parathyroid hormone-related protein
(pyro)Glu	Pyroglutamic acid
PYY	Polypeptide YY
PZI	Protamine zinc insulin
Q	Glutamine
QS$_2$	Total electromechanical systole
R	General symbol for remainder of a chemical formula, eg, an alcohol is R–OH; also gas constant; also respiratory exchange ratio; *also* arginine
RAS	Reticular activating system
rbc	Red blood cell
rCBF	Regional cerebral blood flow
rCMRO$_2$	Regional cerebral metabolic rate for oxygen
RDS	Respiratory distress syndrome
Re	Reynolds number
REF	Renal erythropoietic factor
REM sleep	Rapid eye movement (paradoxical) sleep
Reverse T$_3$	3,3′,5′-Triiodothyronine; isomer of triiodothyronine
RFLP	Restriction fragment length polymorphism
Rh factor	Rhesus group of red cell agglutinogens
RGS	Regulator of G protein signaling
RMV	Respiratory minute volume
RNA	Ribonucleic acid
RPF	Renal plasma flow
RQ	Respiratory quotient
R state	State of heme in hemoglobin that increases O$_2$ binding
R unit	Unit of resistance in cardiovascular system; mm Hg divided by mL/s
RVLM	Rostral ventrolateral medulla
RXR	Retinoid X receptor
s	Second; also standard deviation of a sample
S	Serine
SA	Specific activity
SA node	Sinoatrial node
SCF	Stem cell factor
SCFA	Short-chain fatty acid
SCL	Protein produced by stem cell leukemia gene
SCN	Suprachiasmatic nucleus
SCUBA	Self-contained underwater breathing apparatus
SDA	Specific dynamic action

SEM	Standard error of the mean
Ser	Serine
SERMs	Selective estrogen receptor modulators
SF-1	Steroid factor-1
SFO	Subfornical organ
S$_f$ units	Svedberg units of flotation
sgk	Serum- and glucocorticoid-regulated kinase
SGLT 1	Sodium-dependent glucose transporter 1
SGOT	Serum glutamic-oxaloacetic transaminase
SH	Sulfhydryl
SI units	Units of the Système International
SIADH	Syndrome of inappropriate hypersecretion of antidiuretic hormone (vasopressin)
SIDS	Sudden infant death syndrome
SIF cells	Small, intensely fluorescent cells in sympathetic ganglia
smg	Small GTP-binding protein
SMS	Stiff-man syndrome
SOCC	Store-operated calcium channel
SOD	Superoxide dismutase
SPCA	Proconvertin (clotting factor VII)
SPECT	Single photon emission computed tomography
SPI	One of the shear stress activated transcription factors
sq cm	Square centimeter, cm^2
SRE	Serum response element
SRIF	Somatotropin release-inhibiting factor; same as GIH
sRNA	Soluble or transfer RNA
SRY	Product of sex-determining region of Y chromosome
SS 14	Somatostatin 14
SS 28	Somatostatin 28
SS 28 (1-12)	Polypeptide related to somatostatin that is found in tissues
SSRE	Shear stress response element
STAT	Signal transducer and activator of transcription
STH	Somatotropin, growth hormone
STPD	See Table 34-1
STX	Saxitoxin
Substance P	Polypeptide found in brain and other tissues
T	Absolute temperature; *also* threonine
T$_3$	3,5,3′-Triiodothyronine
T$_4$	Thyroxine
TBG	Thyroxine-binding globulin
TBPA	Thyroxine-binding prealbumin (now called transthyretin)

TBW	Total body water
99mTc-PYP	Technetium 99m stannous pyrophosphate
Tc cells	Cytotoxic T cells
TDF	Testis-determining factor
TEA	Tetraethylammonium
TETRAC	Tetraiodothyroacetic acid
TF/P	Concentration of a substance in renal tubular fluid divided by its concentration in plasma
TG Ab	Antibodies against thyroglobulin
TGF	Transforming growth factor
THC	Δ^9-Tetrahydrocannabinol
Thr	Threonine
TLRs	Toll-like receptors
Tm	Renal tubular maximum
TM	Thrombomodulin
TNF	Tumor necrosis factor
torr	1/760 atm = 1.00000014 mm Hg; unit for various pressures in the body
t-PA	Tissue-type plasminogen activator
TPN	Triphosphopyridine nucleotide
TPNH	Reduced triphosphopyridine nucleotide
TR	Thyroid hormone receptor
TRH, TRF	Thyrotropin-releasing hormone or factor
tRNA	Transfer RNA; same as sRNA
TRP	Transient receptor potential
Trp, Try, Tryp	Tryptophan
TSF	Thrombopoietic-stimulating factor, thrombopoietin
TSH	Thyroid-stimulating hormone
TSH-R [block] Ab	TSH receptor-blocking antibodies
TSH-R [stim] Ab	TSH receptor-stimulating antibodies
TSI	Thyroid-stimulating immunoglobulins
T/S ratio	Thyroid/serum iodide ratio
T state	State of heme in hemoglobin that decreases O_2 binding
TTX	Tetrodotoxin
Tyr	Tyrosine
U	Unit
U followed by subscript	Urine concentration, eg, U_{Cr} = urine creatinine concentration
UCP	Uncoupling protein
UDPG	Uridine diphosphoglucose
UDPGA	Uridine diphosphoglucuronic acid
UFA	Unesterified free fatty acid (same as FFA)
UL	Unstirred layer
u-PA	Urokinase-type plasminogen activator
URF	Uterine-relaxing factor; relaxin
US	Unconditioned stimulus
UTP	Uridine triphosphate
V	Volume; volt; *also* valine
V_1, V_2, etc	Unipolar chest electrocardiographic leads
Val	Valine
VCAM	Vascular cell adhesion molecule
VEGF	Vascular endotheleial growth factor
VF	Unipolar left leg electrocardiographic lead
VIP	Vasoactive intestinal peptide
$\dot{V}$	Volume per unit time
VL	Left arm unipolar electrocardiographic lead
VLDL	Very low density lipoprotein
VLM	Ventrolateral medulla
VMA	Vanillylmandelic acid (3-methoxy-4-hydroxymandelic acid)
$\dot{V}O_{2max}$	Maximal oxygen consumption
VOR	Vestibulo-ocular reflex
VR	Unipolar right arm electrocardiographic lead
W	Tryptophan
wbc	White blood cell
X chromosome	One of the sex chromosomes in humans
X zone	Inner zone of adrenal cortex in some young mammals
Y	Tyrosine
Y chromosome	One of the sex chromosomes in humans

Some Standard Respiratory Symbols (See *Handbook of Physiology,* Section 3: The Respiratory System. American Physiological Society, 1986.)

General Variables
V Gas volume
V̇ Gas volume/unit of time. (Dot over a symbol indicates rate)
P Gas pressure
P̄ Mean gas pressure
f Respiratory frequency (breaths/unit of time)
D Diffusing capacity
F Fractional concentration in dry gas phase
R Respiratory exchange ratio = V_{CO_2}/V_{O_2}
Q Volume of blood
Localization (Subscript letters)
I Inspired gas
E Expired gas

A Alveolar gas
T Tidal gas
D Dead space gas
B Barometric
a Arterial blood
c Capillary blood
v Venous blood

Molecular Species
Indicated by chemical formula printed as subscript
Examples
P_{IO_2} = Pressure of oxygen in inspired air
V_D = Dead space gas volume

Equivalents of Metric, United States, & English Measures (Values rounded off to 2 decimal places.)

Length
1 kilometer = 0.62 mile
1 mile = 5280 feet = 1.61 kilometers
1 meter = 39.37 inches
1 inch = 1/12 foot = 2.54 centimeters

Volume
1 liter = 1.06 US liquid quart
1 US liquid quart = 32 fluid ounces = 1/4 US gallon = 0.95 liter
1 milliliter = 0.03 fluid ounce
1 fluid ounce = 29.57 milliliters
1 US gallon = 0.83 English (Imperial) gallon

Weight
1 kilogram = 2.20 pounds (avoirdupois) = 2.68 pounds (apothecaries')
1 pound (avoirdupois) = 16 ounces = 453.60 grams
1 grain = 65 milligrams

Energy
1 kilogram-meter = 7.25 foot-pounds
1 foot-pound = 0.14 kilogram-meters

Temperature
To convert Celsius degrees into Fahrenheit, multiply by 9/5 and add 32
To convert Fahrenheit degrees into Celsius, subtract 32 and multiply by 5/9

Greek Alphabet

Symbol		Name	Symbol		Name
A	α	alpha	N	ν	nu
B	β	beta	Ξ	ξ	xi
Γ	γ	gamma	O	o	omicron
Δ	δ	delta	Π	π	pi
E	ε	epsilon	P	ρ	rho
Z	ζ	zeta	Σ	σ, ς	sigma
H	η	eta	T	τ	tau
Θ	θ	theta	Y	υ	upsilon
I	ι	iota	Φ	ϕ	phi
K	κ	kappa	X	χ	chi
Λ	λ	lambda	Ψ	ψ	psi
M	μ	mu	Ω	ω	omega

Index

Note: Page numbers in **boldface** type indicate a major discussion. A *t* following a page number indicates tabular material, and an *f* following a page number indicates a figure. Drugs are listed under their generic names. When a drug trade name is listed, the reader is referred to the generic name.

The Neuron